Lecture Notes in Computer Science 13793

More information about this series at https://link.springer.com/bookseries/558

Shweta Agrawal · Dongdai Lin (Eds.)

Advances in Cryptology – ASIACRYPT 2022

28th International Conference on the Theory
and Application of Cryptology and Information Security
Taipei, Taiwan, December 5–9, 2022
Proceedings, Part III

 Springer

Editors
Shweta Agrawal
Indian Institute of Technology Madras
Chennai, India

Dongdai Lin
Chinese Academy of Sciences
Beijing, China

ISSN 0302-9743 ISSN 1611-3349 (electronic)
Lecture Notes in Computer Science
ISBN 978-3-031-22968-8 ISBN 978-3-031-22969-5 (eBook)
https://doi.org/10.1007/978-3-031-22969-5

This Springer imprint is published by the registered company Springer Nature Switzerland AG
The registered company address is: Gewerbestrasse 11, 6330 Cham, Switzerland

Preface

The 28th Annual International Conference on Theory and Application of Cryptology and Information Security (ASIACRYPT 2022) was held in Taiwan during December 5–9, 2022.

The conference covered all technical aspects of cryptology, and was sponsored by the International Association for Cryptologic Research (IACR).

We received a total of 364 submissions from all over the world, and the Program Committee (PC) selected 98 papers for publication in the proceedings of the conference. The two program chairs were supported by a PC consisting of 79 leading experts in aspects of cryptology. Each submission was reviewed by at least three PC members (or their sub-reviewers). The strong conflict of interest rules imposed by IACR ensure that papers are not handled by PC members with a close working relationship with the authors. The two program chairs were not allowed to submit a paper, and PC members were limited to two submissions each. There were approximately 331 external reviewers, whose input was critical to the selection of papers.

The review process was conducted using double-blind peer review. The conference operated a two-round review system with a rebuttal phase. After the reviews and first-round discussions the PC selected 224 submissions to proceed to the second round and the authors were then invited to participate in an interactive rebuttal phase with the reviewers to clarify questions and concerns. The second round involved extensive discussions by the PC members.

Alongside the presentations of the accepted papers, the program of ASIACRYPT 2022 featured two invited talks by Jian Guo and Damien Stehlé. The conference also featured a rump session which contained short presentations on the latest research results of the field.

The four volumes of the conference proceedings contain the revised versions of the 98 papers that were selected. The final revised versions of papers were not reviewed again and the authors are responsible for their contents.

Using a voting-based process that took into account conflicts of interest, the PC selected the three top papers of the conference: "Full Quantum Equivalence of Group Action DLog and CDH, and More" by Hart Montgomery and Mark Zhandry, "Cryptographic Primitives with Hinting Property" by Navid Alamati and Sikhar Patranabis, and "SwiftEC: Shallue–van de Woestijne Indifferentiable Function to Elliptic Curves" by Jorge Chavez-Saab, Francisco Rodriguez-Henriquez, and Mehdi Tibouchi. The authors of all three papers were invited to submit extended versions of their manuscripts to the Journal of Cryptology.

Many people have contributed to the success of ASIACRYPT 2022. We would like to thank the authors for submitting their research results to the conference. We are very grateful to the PC members and external reviewers for contributing their knowledge and expertise, and for the tremendous amount of work that was done with reading papers and contributing to the discussions. We are greatly indebted to Kai-Min Chung and Bo-Yin Yang, the General Chairs, for their efforts and overall organization. We thank

Bart Preneel, Ron Steinfeld, Mehdi Tibouchi, Jian Guo, and Huaxiong Wang for their valuable suggestions and help. We are extremely grateful to Shuaishuai Li for checking all the LaTeX files and for assembling the files for submission to Springer. We also thank the team at Springer for handling the publication of these conference proceedings.

December 2022   Shweta Agrawal
                                                                  Dongdai Lin

Organization

General Chairs

Kai-Min Chung   Academia Sinica, Taiwan
Bo-Yin Yang   Academia Sinica, Taiwan

Program Committee Chairs

Shweta Agrawal   Indian Institute of Technology, Madras, India
Dongdai Lin   Institute of Information Engineering, Chinese
     Academy of Sciences, China

Program Committee

Divesh Aggarwal   National University of Singapore, Singapore
Adi Akavia   University of Haifa, Israel
Martin Albrecht   Royal Holloway, University of London, UK
Ghada Almashaqbeh   University of Connecticut, USA
Benny Applebaum   Tel Aviv University, Israel
Lejla Batina   Radboud University, Netherlands
Carsten Baum   Aarhus University, Denmark
Sonia Belaïd   CryptoExperts, France
Mihir Bellare   University of California, San Diego, USA
Andrej Bogdanov   Chinese University of Hong Kong, China
Christina Boura   Université de Versailles, France
Ran Canetti   Boston University, USA
Jie Chen   East China Normal University, China
Yilei Chen   Tsinghua University, China
Jung Hee Cheon   Seoul National University, South Korea
Ilaria Chillotti   Zama, France
Michele Ciampi   The University of Edinburgh, UK
Craig Costello   Microsoft Research, USA
Itai Dinur   Ben-Gurion University, Israel
Nico Döttling   Helmholtz Center for Information Security
     (CISPA), Germany
Maria Eichlseder   Graz University of Technology, Austria
Saba Eskandarian   University of North Carolina at Chapel Hill, USA
Marc Fischlin   TU Darmstadt, Germany

Pierre-Alain Fouque   Rennes University and Institut Universitaire de
                               France, France
Steven D. Galbraith   University of Auckland, New Zealand
Chaya Ganesh   Indian Institute of Science, India
Juan Garay   Texas A&M University, USA
Sanjam Garg   University of California, Berkeley and NTT
                               Research, USA
Daniel Genkin   Georgia Institute of Technology, USA
Jian Guo   Nanyang Technological University, Singapore
Siyao Guo   New York University Shanghai, China
Mohammad Hajiabadi   University of Waterloo, Canada
Mike Hamburg   Rambus Inc, USA
David Heath   Georgia Institute of Technology, USA
Viet Tung Hoang   Florida State University, USA
Xinyi Huang   Fujian Normal University, China
Takanori Isobe   University of Hyogo, Japan
Tetsu Iwata   Nagoya University, Japan
Khoongming Khoo   DSO National Laboratories, Singapore
Elena Kirshanova   Immanuel Kant Baltic Federal University, Russia
Ilan Komargodski   Hebrew University of Jerusalem and NTT
                               Research, Israel
Gregor Leander   Ruhr-Universität Bochum, Germany
Qipeng Liu   Simons Institute for the Theory of Computing,
                               USA
Tianren Liu   Peking University, China
Shengli Liu   Shanghai Jiao Tong University, China
Zhe Liu   Nanjing University of Aeronautics and
                               Astronautics, China
Hemanta Maji   Purdue University, USA
Giulio Malavolta   Max Planck Institute for Security and Privacy,
                               Germany
Bart Mennink   Radboud University Nijmegen, Netherlands
Tal Moran   Reichman University, Israel
Pratyay Mukherjee   Swirlds/Hedera, USA
Omkant Pandey   State University of New York at Stony Brook,
                               USA
Anat Paskin-Cherniavsky   Ariel University, Israel
Alain Passelègue   Inria and ENS Lyon, France
Svetla Petkova-Nikova   KU Leuven, Belgium and University of Bergen,
                               Norway
Duong Hieu Phan   Télécom Paris, France
Cécile Pierrot   Inria, France
Silas Richelson   UC Riverside, USA

Yu Sasaki	NTT Corporation, Japan
Tobias Schneider	NXP Semiconductors, Austria
Dominique Schröder	Friedrich-Alexander-Universität Erlangen-Nürnberg, Germany
abhi shelat	Northeastern University, USA
Mark Simkin	Ethereum Foundation, USA
Ling Song	Jinan University, Guangzhou, China
Fang Song	Portland State University, USA
Pratik Soni	Carnegie Mellon University, USA
Akshayaram Srinivasan	Tata Institute of Fundamental Research, India
Damien Stehlé	ENS de Lyon, France
Ron Steinfeld	Monash University, Australia
Qiang Tang	University of Sydney, Australia
Yiannis Tselekounis	Carnegie Mellon University, USA
Meiqin Wang	Shandong University, China
Xiaoyun Wang	Tsinghua University, China
David Wu	University of Texas at Austin, USA
Wenling Wu	Institute of Software, Chinese Academy of Sciences, China
Shota Yamada	AIST, Japan
Takashi Yamakawa	NTT Corporation, Japan
Jiang Zhang	State Key Laboratory of Cryptology, China

Additional Reviewers

Behzad Abdolmaleki
Calvin Abou Haidar
Damiano Abram
Bar Alon
Pedro Alves
Ravi Anand
Anurag Anshu
Victor Arribas
Thomas Attema
Christian Badertscher
Anubhab Baksi
Zhenzhen Bao
James Bartusek
Christof Beierle
Ritam Bhaumik
Alexander Bienstock
Olivier Blazy
Alex Block
Maxime Bombar

Charlotte Bonte
Carl Bootland
Katharina Boudgoust
Lennart Braun
Marek Broll
Chris Brzuska
BinBin Cai
Matteo Campanelli
Federico Canale
Avik Chakraborti
Suvradip Chakraborty
John Chan
Rohit Chatterjee
Long Chen
Yu Long Chen
Hongyin Chen
Shan Chen
Shiyao Chen
Rongmao Chen

Nai-Hui Chia
Arka Rai Choudhuri
Jiali Choy
Qiaohan Chu
Hien Chu
Eldon Chung
Sandro Coretti-Drayton
Arjan Cornelissen
Maria Corte-Real Santos
Anamaria Costache
Alain Couvreur
Nan Cui
Benjamin R. Curtis
Jan-Pieter D'Anvers
Joan Daemen
Wangchen Dai
Hannah Davis
Luca De Feo
Gabrielle De Micheli

Thomas Debris-Alazard
Amit Deo
Patrick Derbez
Julien Devevey
Siemen Dhooghe
Benjamin Dowling
Leo Ducas
Yen Ling Ee
Jonathan Eriksen
Daniel Escudero
Muhammed F. Esgin
Thomas Espitau
Andre Esser
Hulya Evkan
Jaiden Fairoze
Joël Felderhoff
Hanwen Feng
Joe Fitzsimons
Antonio Flórez-Gutiérrez
Pouyan Forghani
Cody Freitag
Georg Fuchsbauer
Pierre Galissant
Tommaso Gagliardoni
Daniel Gardham
Pierrick Gaudry
Romain Gay
Chunpeng Ge
Rosario Gennaro
Paul Gerhart
Satrajit Ghosh
Ashrujit Ghoshal
Niv Gilboa
Aarushi Goel
Aron Gohr
Jesse Goodman
Mike Graf
Milos Grujic
Aurore Guillevic
Aldo Gunsing
Chun Guo
Hosein Hadipour
Mathias Hall-Andersen
Shuai Han
Helena Handschuh

Lucjan Hanzlik
Yonglin Hao
Keisuke Hara
Patrick Harasser
Jingnan He
Rachelle Heim-Boissier
Minki Hhan
Shoichi Hirose
Seungwan Hong
Akinori Hosoyamada
James Hsin-Yu Chiang
Zhicong Huang
Senyang Huang
Chloé Hébant
Ilia Iliashenko
Laurent Imbert
Joseph Jaeger
Palak Jain
Ashwin Jha
Mingming Jiang
Zhengzhong Jin
Antoine Joux
Eliran Kachlon
Bhavana Kanukurthi
Alexander Karenin
Shuichi Katsumata
Mojtaba Khalili
Hamidreza Khorasgani
Dongwoo Kim
Duhyeong Kim
Young-Sik Kim
Fuyuki Kitagawa
Kamil Kluczniak
Yashvanth Kondi
Rajendra Kumar
Noboru Kunihiro
Fukang Liu
Russell W. F. Lai
Jason LeGrow
Jooyoung Lee
Hyung Tae Lee
Byeonghak Lee
Charlotte Lefevre
Zeyong Li
Yiming Li

Hanjun Li
Shun Li
Xingjian Li
Xiao Liang
Benoît Libert
Damien Ligier
Chao Lin
Chengjun Lin
Yunhao Ling
Eik List
Jiahui Liu
Feng-Hao Liu
Guozhen Liu
Xiangyu Liu
Meicheng Liu
Alex Lombardi
Patrick Longa
Wen-jie Lu
Yuan Lu
Donghang Lu
You Lyu
Reinhard Lüftenegger
Bernardo Magri
Monosij Maitra
Mary Maller
Lenka Mareková
Mark Marson
Takahiro Matsuda
Alireza Mehrdad
Simon-Philipp Merz
Pierre Meyer
Michael Meyer
Peihan Miao
Tarik Moataz
Hart Montgomery
Tomoyuki Morimae
Fabrice Mouhartem
Tamer Mour
Marta Mularczyk
Michael Naehrig
Marcel Nageler
Yusuke Naito
Mridul Nandi
Patrick Neumann
Ruth Ng

Ky Nguyen
Khoa Nguyen
Ngoc Khanh Nguyen
Jianting Ning
Oded Nir
Ryo Nishimaki
Olga Nissenbaum
Semyon Novoselov
Julian Nowakowski
Tabitha Ogilvie
Eran Omri
Hiroshi Onuki
Jean-Baptiste Orfila
Mahak Pancholi
Omer Paneth
Lorenz Panny
Roberto Parisella
Jeongeun Park
Rutvik Patel
Sikhar Patranabis
Alice Pellet-Mary
Hilder Vitor Lima Pereira
Ludovic Perret
Thomas Peyrin
Phuong Pham
Guru Vamsi Policharla
Sihang Pu
Luowen Qian
Chen Qian
Kexin Qiao
Willy Quach
Rahul Rachuri
Srinivasan Raghuraman
Adrian Ranea
Shahram Rasoolzadeh
Christian Rechberger
Krijn Reijnders
Maxime Remaud
Ling Ren
Mahshid Riahinia
Peter Rindal
Mike Rosulek
Adeline Roux-Langlois
Paul Rösler

Yusuke Sakai
Kosei Sakamoto
Amin Sakzad
Simona Samardjiska
Olga Sanina
Roozbeh Sarenche
Santanu Sarker
Tobias Schmalz
Markus Schofnegger
Jacob Schuldt
Sruthi Sekar
Nicolas Sendrier
Akash Shah
Yaobin Shen
Yixin Shen
Yu Shen
Danping Shi
Rentaro Shiba
Kazumasa Shinagawa
Omri Shmueli
Ferdinand Sibleyras
Janno Siim
Siang Meng Sim
Luisa Siniscalchi
Yongsoo Song
Douglas Stebila
Lukas Stennes
Igors Stepanovs
Christoph Striecks
Ling Sun
Siwei Sun
Bing Sun
Shi-Feng Sun
Akira Takahashi
Abdul Rahman Taleb
Chik How Tan
Adrian Thillard
Sri Aravinda Krishnan
    Thyagarajan
Yan Bo Ti
Elmar Tischhauser
Yosuke Todo
Junichi Tomida
Ni Trieu

Monika Trimoska
Yi Tu
Aleksei Udovenko
Rei Ueno
Mayank Varia
Daniele Venturi
Riad Wahby
Roman Walch
Mingyuan Wang
Haoyang Wang
Luping Wang
Xiao Wang
Yuejun Wang
Yuyu Wang
Weiqiang Wen
Chenkai Weng
Benjamin Wesolowski
Yusai Wu
Yu Xia
Zhiye Xie
Shengmin Xu
Guangwu Xu
Sophia Yakoubov
Hailun Yan
Rupeng Yang
Kang Yang
Qianqian Yang
Shao-Jun Yang
Li Yao
Hui Hui Yap
Kan Yasuda
Weijing You
Thomas Zacharias
Yupeng Zhang
Kai Zhang
Lei Zhang
Yunlei Zhao
Yu Zhou
Chenzhi Zhu
Paul Zimmermann
Lukas Zobernig
matthieu rambaud
Hendrik Waldner
Yafei Zheng

Sponsoring Institutions

- Platinum Sponsor: ZAMA
- Gold Sponsor: BTQ, Hackers in Taiwan, Technology Innovation Institute
- Silver Sponsor: Meta (Facebook), Casper Networks, PQShield, NTT Research, WiSECURE
- Bronze Sponsor: Mitsubishi Electric, Algorand Foundation, LatticeX Foundation, Intel, QSancus, IOG (Input/Output Global), IBM

Contents – Part III

Practical Cryptography

New Algorithms and Analyses for Sum-Preserving Encryption   3
  Sarah Miracle and Scott Yilek

Towards Case-Optimized Hybrid Homomorphic Encryption: Featuring
the Elisabeth Stream Cipher 32
  Orel Cosseron, Clément Hoffmann, Pierrick Méaux,
  and François-Xavier Standaert

Revisiting Related-Key Boomerang Attacks on AES Using
Computer-Aided Tool 68
  Patrick Derbez, Marie Euler, Pierre-Alain Fouque,
  and Phuong Hoa Nguyen

On Secure Ratcheting with Immediate Decryption 89
  Jeroen Pijnenburg and Bertram Poettering

Strongly Anonymous Ratcheted Key Exchange 119
  Benjamin Dowling, Eduard Hauck, Doreen Riepel, and Paul Rösler

Encryption to the Future: A Paradigm for Sending Secret Messages
to Future (Anonymous) Committees 151
  Matteo Campanelli, Bernardo David, Hamidreza Khoshakhlagh,
  Anders Konring, and Jesper Buus Nielsen

Authenticated Encryption with Key Identification 181
  Julia Len, Paul Grubbs, and Thomas Ristenpart

Privacy-Preserving Authenticated Key Exchange in the Standard Model 210
  You Lyu, Shengli Liu, Shuai Han, and Dawu Gu

On the Field-Based Division Property: Applications to MiMC, Feistel
MiMC and GMiMC 241
  Jiamin Cui, Kai Hu, Meiqin Wang, and Puwen Wei

Advanced Encryption

Traceable Receipt-Free Encryption 273
  Henri Devillez, Olivier Pereira, and Thomas Peters

Efficient Searchable Symmetric Encryption for Join Queries 304
  Charanjit Jutla and Sikhar Patranabis

Knowledge Encryption and Its Applications to Simulatable Protocols
with Low Round-Complexity 334
  Yi Deng and Xinxuan Zhang

Compact and Tightly Selective-Opening Secure Public-key Encryption
Schemes 363
  Jiaxin Pan and Runzhi Zeng

Identity-Based Matchmaking Encryption from Standard Assumptions 394
  Jie Chen, Yu Li, Jinming Wen, and Jian Weng

Anonymous Public Key Encryption Under Corruptions 423
  Zhengan Huang, Junzuo Lai, Shuai Han, Lin Lyu, and Jian Weng

Memory-Tight Multi-challenge Security of Public-Key Encryption 454
  Joseph Jaeger and Akshaya Kumar

Zero Knowledge

Short-lived Zero-Knowledge Proofs and Signatures 487
  Arasu Arun, Joseph Bonneau, and Jeremy Clark

Non-interactive Zero-Knowledge Proofs to Multiple Verifiers 517
  Kang Yang and Xiao Wang

Rotatable Zero Knowledge Sets: Post Compromise Secure Auditable
Dictionaries with Application to Key Transparency 547
  *Brian Chen, Yevgeniy Dodis, Esha Ghosh, Eli Goldin,
  Balachandar Kesavan, Antonio Marcedone, and Merry Ember Mou*

Quantum Algorithms

Nostradamus Goes Quantum 583
  Barbara Jiabao Benedikt, Marc Fischlin, and Moritz Huppert

Synthesizing Quantum Circuits of AES with Lower T-depth and Less
Qubits 614
  Zhenyu Huang and Siwei Sun

Exploring SAT for Cryptanalysis: (Quantum) Collision Attacks Against
6-Round SHA-3 645
  Jian Guo, Guozhen Liu, Ling Song, and Yi Tu

Lattice Cryptanalysis

Log-S-unit Lattices Using Explicit Stickelberger Generators to Solve
Approx Ideal-SVP 677
    Olivier Bernard, Andrea Lesavourey, Tuong-Huy Nguyen,
    and Adeline Roux-Langlois

On Module Unique-SVP and NTRU 709
    Joël Felderhoff, Alice Pellet-Mary, and Damien Stehlé

A Non-heuristic Approach to Time-Space Tradeoffs and Optimizations
for BKW 741
    Hanlin Liu and Yu Yu

Improving Bounds on Elliptic Curve Hidden Number Problem for ECDH
Key Exchange 771
    Jun Xu, Santanu Sarkar, Huaxiong Wang, and Lei Hu

Author Index 801

Practical Cryptography

New Algorithms and Analyses for Sum-Preserving Encryption

Sarah Miracle$^{(\boxtimes)}$ and Scott Yilek$^{(\boxtimes)}$

University of St. Thomas, St. Paul, USA
{sarah.miracle,syilek}@stthomas.edu

Abstract. We continue the study of sum-preserving encryption schemes, in which the plaintext and ciphertext are both integer vectors with the same sum. Such encryption schemes were recently constructed and analyzed by Tajik, Gunasekaran, Dutta, Ellia, Bobba, Rosulek, Wright, and Feng (NDSS 2019) in the context of image encryption. Our first main result is to prove a mixing-time bound for the construction given by Tajik et al. using path coupling. We then provide new sum-preserving encryption schemes by describing two practical ways to rank and unrank the values involved in sum-preserving encryption, which can then be combined with the rank-encipher-unrank technique from format-preserving encryption. Finally, we compare the efficiency of the Tajik et al. construction and our new ranking constructions based on performance tests we conducted on prototype implementations.

Keywords: Sum-preserving encryption · Image encryption · Format-preserving encryption

1 Introduction

A sum-preserving encryption scheme, recently studied by Tajik, Gunasekaran, Dutta, Ellis, Bobba, Rosulek, Wright, and Feng [18] in the context of image encryption, is a symmetric encryption scheme with an encryption algorithm that takes as its plaintext input a vector of integers, and outputs as a ciphertext another vector of integers with the same sum as the plaintext vector. For applications, the vector components of both the plaintext and ciphertext will typically be integers from 0 up to d, where d is called the component bound.

Tajik et al. introduced definitions and provided a practical construction of sum-preserving encryption in order to build a separate primitive called thumbnail-preserving encryption [12,20], which is a type of image encryption in which a much smaller version, called a thumbnail, of an encrypted image matches the thumbnail of the unencrypted image. The key idea is that, since sum-preserving encryption turns the plaintext vector of integers into a ciphertext vector with the same *length* and the same *sum*, then the mean will also be preserved. Creating a thumbnail of an image involves replacing a $b \times b$ block of pixels with the average pixel value in the block, so it follows that applying

S. Agrawal and D. Lin (Eds.): ASIACRYPT 2022, LNCS 13793, pp. 3–31, 2022.
https://doi.org/10.1007/978-3-031-22969-5_1

sum-preserving encryption to each block will result in a ciphertext image with the same thumbnail as the original image.

Sum-preserving encryption can be viewed as a special type of format-preserving encryption (FPE). Format-preserving encryption schemes were originally studied by Brightwell and Smith [4] and were eventually formally defined and analyzed by Bellare, Ristenpart, Rogaway, and Stegers [2]. They have since been widely studied, have found numerous applications, and have even been standardized [6,10]. Looking forward, one of the main techniques for constructing FPE schemes, the rank-encipher-unrank construction analyzed in [2], will be an important tool for constructing sum-preserving encryption schemes.

While Tajik et al. focused on the use of sum-preserving encryption in the context of images, the fact that the primitive allows one to encrypt a vector of data while maintaining a common statistical measure like mean opens up the possibility of numerous applications for more general dataset encryption. For example, suppose an instructor would like to encrypt her final exam scores each semester before archiving them (which might be especially important given student privacy laws like FERPA in the United States), yet she would still like to be able to go back and compute exam averages to compare different course sections or compare across semesters. Sum-preserving encryption seems to fit the instructor's requirements perfectly.

At the same time, as we will discuss later when introducing security notions, there is the possibility that the sum or mean themselves leak important information about the plaintexts. To see this, consider the extreme example in which the same instructor gave a particularly easy final exam one semester and every student got a score of 100 out of 100. Encrypting a vector of all 100 s in a sum-preserving way will just result in another vector of all 100 s, so every student's exam score would be revealed by the ciphertext! Nevertheless, while such extreme examples are concerning and are somewhat reminiscent of some of the security issues that arise with other property-preserving encryption schemes (e.g., order-preserving encryption [3]), we emphasize that sum-preserving encryption has already found practical application in the context of image encryption, and can likely be used safely in a number of other application settings, some of which we discuss later in the paper.

1.1 Previous Construction

In order to build thumbnail encryption schemes, Tajik et al. set their sights on building a sum-preserving encryption scheme on vectors with values 0–255. Suppose $(m_1, \ldots, m_n)$ with sum S is such a vector. With the obvious connection to format-preserving encryption, they first explore using rank-encipher-unrank, one of the common techniques for building FPE schemes, for sum-preserving encryption. With rank-encipher-unrank, plaintexts are first ranked, meaning mapped to the set of integers $\{0, \ldots, N - 1\}$, where N is the number of possible plaintexts. Then a cipher with integer domain is applied to get another integer in this same range, before finally an unrank algorithm maps that integer back into the original plaintext domain. Tajik et al. observe that it should be technically possible to apply rank-encipher-unrank to sum-preserving encryption by first representing the plaintext vector using the stars-and-bars representation from combinatorics to get a vector $1^{m_1}01^{m_2}0\ldots01^{m_n}$, where each component of the plaintext vector is given in unary with 0 s as separators. Such binary strings, with S total 1 s (since S is the sum we wish to preserve), are a regular language, and thus known techniques for ranking deterministic finite automata (DFAs) could be applied [2]. Unfortunately, Tajik et al. point out such a method would have very high time complexity and would be impractical for applications.

Tajik et al. then show that, while ranking vectors of length n with a particular sum in general seems hard, it is actually possible to rank vectors of length 2 with a particular sum simply and efficiently. Given this ability to rank (and thus rank-encipher-unrank) vectors of length 2, they go on to give a method to encrypt longer vectors. The idea is to proceed in rounds and, in each round, match up adjacent vector components and apply rank-encipher-unrank to each pair. Since each pair is enciphered in a sum-preserving way, the entire vector will also maintain its sum. Then the entire vector is shuffled before the next round, effectively randomizing the points that will be matched up in the next round. Because each round matches up points and then applies rank-encipher-unrank to each matched pair, we refer to this algorithm as the matching-based construction for the rest of the paper.

Tajik et al. observe their algorithm can be modeled as a Markov chain and the number of necessary rounds to achieve security is then tied to the mixing time of this chain. They discuss how the mixing time would relate to the eigenvalues of the Markov chain transition matrix but, because such matrices are so large in practice, were unable to explicitly compute these values. They go on to give some heuristic arguments for what secure round choices might be, and ultimately test performance with 1000, 3000, and 5000 rounds, but their paper does not provide a mixing time proof for the matching-based construction.

1.2   Our Results

We continue the study of sum-preserving encryption and produce two main results. First, we provide the first mixing-time proof of the matching-based construction of Tajik et al., using a path-coupling technique due to Dyer and Greenhill [7]. Second, we show it actually is possible (and practical) to use rank-encipher-unrank to build sum-preserving encryption for vectors of length n by giving new algorithms for directly ranking and unranking such vectors. Further, we create prototype implementations of both the matching-based construction of Tajik et al. and our new ranking constructions, and show that our ranking constructions have significant performance benefits in a number of applications, including the thumbnail encryption application that was the original motivation for sum-preserving encryption. We now discuss each of these contributions in more detail.

MIXING-TIME PROOF OF THE TAJIK ET AL. CONSTRUCTION. Our first contribution is to formally analyze the matching-based construction given by

Tajik et al. [18] and give a bound on the mixing time. We begin by framing their algorithm as a Markov chain $\mathcal{M}_S$ with state space the set of all vectors of length n with component bound d and sum S. Next we prove that the mixing time $\tau_{\mathcal{M}_S}$ of their Markov chain $\mathcal{M}_S$ satisfies $\tau_{\mathcal{M}_S}(\epsilon) \leq n \ln(\min(dn, 2S)\epsilon^{-1})$.

Our proof uses a path coupling technique due to Dyer and Greenhill [7]. To apply path coupling, we carefully select a custom distance metric and design a coupling. Path coupling allows us to only consider a subset of pairs of configurations namely those that differ on only two points and show that using our coupling the expected distance between two configurations will decrease after a single step of $\mathcal{M}_S$. If the shuffling selected by $\mathcal{M}_S$ pairs the two points that differ together the distance decreases to zero. However if these points are not paired together the situation is much more complex and requires a detailed coupling and careful analysis. Much of the complexity comes because the two chains will often have a different number of possible valid next configurations (since the pairs of points have different sums) and thus each individual configuration is selected with a different probability in each chain. The complete proof is given in Sect. 3.

New Algorithms for Ranking/Unranking Sum-Preserving Vectors. Second, we give algorithms for directly ranking and unranking sum-preserving vectors based on two different total orders. The first is the standard lexicographical order. Stein [17] previously described an algorithm for unranking such vectors using lexicographical order, for use in random sampling. Their algorithm relies on pre-computing a table C_d where position (n, S) stores the number of vectors of length n with sum S and component bound d. We improve on this algorithm by using a cumulative sum table where each position (i, j) stores $\sum_{k=i}^{n} \mathsf{C}_d(k, j)$. Additionally we give a dynamic programming based justification for the computations required to fill the table (Sect. 4.3). Finally, in Sect. 4.1 we give a ranking algorithm which also uses the cumulative sum table.

In order to handle applications with larger parameters d and n, we develop a second set of rank and unrank algorithms based on a new total ordering we call recursive block order. While recursive block order uses ideas that are reminiscent to those used in orderings of monomials, specifically block order (see e.g., [8]) and graded order (see e.g. [5]), these are combined differently and applied recursively unlike in monomial orderings. At a high-level, in recursive block order configurations are first ordered based on the sum of the first $n/2$ points. Configurations with smaller "first-half" sums have lower rank. Configurations with the same "first-half" sum are then ordered recursively according to the first $n/2$ points or, if these are identical, the last $n/2$ points. We formally define this order, give both rank and unrank algorithms, and go over an example in Sects. 4.2 and 4.5.

One of the main advantages of recursive block order is that it requires fewer rows of the C table to be computed and stored. Specifically it only requires at most $2 \log n$ rows (and only $\log n$ if n is a power of 2). However, the dynamic programming approach to filling the C table does not allow us to take advantage of this and requires all n rows be computed. To address this, we present an alternative way to compute only the values needed from the C table using well-known formulas derived from generating functions (see e.g. [1,15]).

PERFORMANCE COMPARISON. We created prototype implementations of both the Tajik et al. matching-based construction and our two ranking-base constructions and ran a number of performance tests for a variety of applications with a wide range of parameter choices for the vector length n and component bound d. In short, we found that for a number of applications, including the thumbnail encryption application that originally motivated sum-preserving encryption, our new ranking-based constructions are significantly more efficient, even when the matching-based construction is used with round numbers well below the bounds we prove in Sect. 3. The matching-based construction appears to be the superior choice when n is small but d is large (e.g., $n = 30$ and $d = 100000$). When n is very large, in the thousands or higher, then neither the matching-based construction nor our ranking constructions perform well, and new approaches are likely needed. We discuss these and other details of our performance analysis in Sect. 5.

2   Background on Sum-Preserving Encryption

In this section we define what a sum-preserving encryption scheme is, discuss security goals, describe some example applications, and give details on previous constructions. We note that much of this section closely follows the work of Tajik et al. [18].

2.1   Syntax

We now give a formal definition of a sum-preserving encryption scheme. Let $n \geq 1$ be an integer called the *vector length*, and $d > 0$ be an integer called the *component bound*. A d-bounded vector of length n is a vector of integers $(x_1, \ldots, x_n)$ with $0 \leq x_i \leq d$. We denote by $(\mathbb{Z}_{d+1})^n$ the set of all d-bounded vectors of length n. A sum-preserving encryption scheme for $(\mathbb{Z}_{d+1})^n$ is a pair of algorithms (Enc, Dec) with the following properties.

- The (deterministic) encryption algorithm Enc : $\mathcal{K} \times \{0,1\}^* \times (\mathbb{Z}_{d+1})^n \to (\mathbb{Z}_{d+1})^n$ takes as input a key $K \in \mathcal{K}$, a nonce $T \in \{0,1\}^*$, and message $M \in (\mathbb{Z}_{d+1})^n$, and outputs a ciphertext $C \in (\mathbb{Z}_{d+1})^n$ which is also a d-bounded vector of length n. Importantly, the encryption algorithm must be *sum-preserving*, which means for all keys $K \in \mathcal{K}$, all nonces $T \in \{0,1\}^*$, and all messages $M \in (\mathbb{Z}_{d+1})^n$, it is true that $\sum M = \sum \mathsf{Enc}(K, T, M)$, meaning the sum of the message vector components $\sum M = \sum_{i=1}^n m_i$ must be equal to the sum of the ciphertext components $\sum \mathsf{Enc}(K, T, M) = \sum_{i=1}^n c_i$, where $\mathsf{Enc}(K, T, M) = (c_1, \ldots, c_n)$.
- The decryption algorithm Dec : $\mathcal{K} \times \{0,1\}^* \times (\mathbb{Z}_{d+1})^n \to (\mathbb{Z}_{d+1})^n$ takes as input a key $K \in \mathcal{K}$, a nonce $T \in \{0,1\}^*$, and ciphertext $C \in (\mathbb{Z}_{d+1})^n$ and outputs a message vector $M \in (\mathbb{Z}_{d+1})^n$.

For correctness we require that for all $K \in \mathcal{K}$, all $T \in \{0,1\}^*$, and all $M \in (\mathbb{Z}_{d+1})^n$, it must be the case that $\mathsf{Dec}(K, T, \mathsf{Enc}(K, T, M)) = M$.

Connection to bounded integer compositions. We note that d-bounded vectors of length n with the sum of vector components equal to an integer S are also called d-bounded (or restricted) n-part compositions of S or d-bounded n-compositions of S. There has been much previous work studying restricted compositions. Much of the previous work has surrounded the problems of enumerating all compositions and counting the number of such compositions (see e.g., [1, 13–15, 19]). We will use the terminology bounded n-composition throughout the paper.

2.2   Examples

To help better understand sum-preserving encryption, we now discuss some example applications where it has either already been used or could potentially be used. We will also revisit these same examples in Sect. 5 when we evaluate our prototype implementations.

Example 1: Thumbnail-Preserving Image Encryption. Tajik et al. previously used sum-preserving encryption to build a type of image encryption called Thumbnail-preserving encryption, in which the thumbnail of an encrypted image exactly matches the thumbnail of the unencrypted image. We can view image data as a matrix of pixel values 0-255 with dimensions $h \times w \times 3$, where each of the three $h \times w$ matrices represents an RGB channel. Tajik et al. showed one can do thumbnail-preserving encryption by taking $m \times m$ blocks and encrypting them in a sum-preserving way. Then, when forming a thumbnail by replacing each $m \times m$ block with a single pixel that is its average, the sum-preserving property of encryption ensures the encrypted blocks have the same average as the unencrypted image blocks. Thus, to use this construction we need sum-preserving encryption for vectors of length $n = m \cdot m$, the total number of pixels in a block, and component bound $d = 255$, the maximum value of a pixel.

Example 2: Exam Scores. Suppose an instructor has a class with 300 students, and vector of final exam scores which are each integers 0–100. The instructor may want to encrypt this vector of scores in a way that the exam average can be calculated from the encrypted vector alone. In this case we can use sum-preserving encryption with vector length $n = 300$ and component bound $d = 100$.

Example 3: Employee Salaries. A small company with 30 employees wants to encrypt a vector of employee salaries, which are integers between 30,000 and 100,000, in a way such that the average salary can be computed from the encrypted salary vector. In this case we potentially have a few options for using sum-preserving encryption. We could set $n = 30$, and $d = 100000$; in this case the encrypted salaries will range from 0 to 100000, so we could get encrypted salaries below the lowest actual salary of 30,000, and the top salary of 100,000 will also potentially be revealed by the vector, since no value will be above that.

Due to this latter issue, the encryptor might instead choose to use a larger choice of d. But, as we will discuss later in the implementation section, choosing a larger d in this example could have significant performance consequences.

Example 4: Rating Dataset. Suppose a website has 5,000 user ratings, each integers that are 0 to 4 stars. The website might want to encrypt this dataset in a way that allows the average rating to still be computed. Sum-preserving encryption with $n = 5000$ and $d = 4$ could be used.

We note that for some example applications, there may be possible solutions other than using sum-preserving encryption. For example, one could encrypt the data using a standard symmetric encryption scheme, and then simply append the sum or mean of the original data to the ciphertext. However, sum-preserving encryption has the benefit of also maintaining the format of the original message, so it may be the superior solution when it is necessary, or even just more convenient, for ciphertexts to still be d-bounded vectors.

2.3   Security Notions

Like previous work on format-preserving encryption and sum-preserving encryption, we aim to build sum-preserving encryption schemes that are indistinguishable from randomly chosen sum-preserving permutations on the same domains. Formally, let $\mathsf{Enc} : \mathcal{K} \times \{0,1\}^* \times (\mathbb{Z}_{d+1})^n \to (\mathbb{Z}_{d+1})^n$ be a sum-preserving encryption algorithm on d-bounded vectors of length n and let $\mathsf{Dec} : \mathcal{K} \times \{0,1\}^* \times (\mathbb{Z}_{d+1})^n \to (\mathbb{Z}_{d+1})^n$ be the corresponding decryption algorithm. To define PRP security, we say the prp-advantage of an adversary A is

$$\mathbf{Adv}_{\mathsf{Enc}}^{\mathrm{prp}}(A) = \Pr\left[A^{\mathsf{Enc}_K(\cdot,\cdot)} \Rightarrow 1 \right] - \Pr\left[A^{\pi(\cdot,\cdot)} \Rightarrow 1 \right].$$

The adversary is given access to either an encryption oracle that takes as input a nonce and a message vector, or a randomly chosen family of permutations $\pi : \{0,1\}^* \times (\mathbb{Z}_{d+1})^n \to (\mathbb{Z}_{d+1})^n$. We can also target a stronger security notion, strong PRP security, if we additionally give the adversary either a decryption oracle or an inverse permutation family π^{-1}. The previous work of Tajik et al. on sum-preserving encryption further restricted the definition above to only consider adversaries that never repeat a nonce input to their oracles. They called such adversaries *nonce respecting* (NR), and argued that this security notion is meaningful for applications. Following Tajik et al., we will primarily focus on security against NR adversaries, but will also discuss how to achieve the stronger notions and the corresponding effects on performance.

Discussion. It is important to note there are inherent limitations in the security we can achieve with sum-preserving encryption, even when achieving the security notions just described. To see this, consider the exam score application described above. If the exam is particularly easy and all 300 students achieve a score of

100/100, then the resulting message vector with 100 repeated 300 times will encrypt to exactly the same vector and end up revealing every student's score. In other applications, this may not be as problematic. For example, in the thumbnail encryption example, a large block of white pixels will all have the maximum pixel value of 255 and thus encrypt to another block of all white pixels. Yet if our goal is to use such image encryption to hide details of the image like facial features, then a large block of pixels staying the same color does not seem so damaging.

2.4   Previous Constructions

After defining sum-preserving encryption, Tajik et al. observe that one option for constructing such schemes would be to use the rank-encipher-unrank construction of [2], which is one of the common ways of building format-preserving encryption schemes. We say that a set $\mathcal{X}$ has an efficient ranking if there is an efficient algorithm rank : $\mathcal{X} \rightarrow \{0, \ldots, |\mathcal{X}| - 1\}$ mapping elements of the set to integers, and then an efficient inverse function unrank mapping integers back into the set $\mathcal{X}$. The idea behind rank-encipher-unrank is, given a point $x \in \mathcal{X}$ to encipher, one can first rank x to get an integer n_x in the range $\{0, \ldots, |\mathcal{X}| - 1\}$. Then one applies a cipher that works on that integer domain to get another integer n_y in the same domain. Finally, applying unrank to n_y yields another point $y \in \mathcal{X}$, which acts as the ciphertext.

Algorithms for ranking based on the DFA representation of a language are known [2], so Tajik et al. observe it is technically possible to use this paradigm for constructing sum-preserving encryption. The key idea is to use the stars-and-bars representation of the message space: the individual elements of the vector to encrypt are represented in unary and the symbol 0 can be used as a separator between these unary sequences, so the vector $(3, 1, 2)$ would be 11101011. It is easy to come up with a regular expression for such binary strings, and from there a DFA to be used in ranking. Unfortunately, Tajik et al. argue this would be far too computationally expensive to be useful.

Tajik et al. then observe that, while ranking vectors of length n with a particular sum in general would be too computationally expensive, it is actually straightforward to rank vectors of length 2. Let $(a, b) \in \{0, \ldots, d\}^2$ have sum $S = a + b$. Then rank$((a, b)) = a$ if $S \leq d$ and $d + 1 - a$ otherwise, and unrank$(t) = (t, S - t)$ if $S \leq d$ and $(d + 1 - r, S - d - 1 + r)$ otherwise.

Given a way to rank and unrank vectors of size 2, Tajik et al. then give an efficient construction for encrypting longer vectors while preserving their sum. Their algorithm proceeds in rounds. In each round, match up all of the points in the vector with their neighbor and, for each pair of neighbors, apply rank-encipher-unrank using the ranking and unranking formulas for length-2 vectors just described. As a result, each pair of neighboring points in the vector will be replaced by a new pair with the same sum, and the overall sum of the vector will be maintained. The points of the vector are then randomly shuffled between rounds (e.g., with the Knuth shuffle) so that the next round finds new neighboring points matched up.

3 Analyzing the Matching-based Approach of Tajik et al.

In this section we formally analyze the matching-based construction given by Tajik et al. [18] and described in Sect. 2.4. We begin by framing the algorithm formally as a Markov chain $\mathcal{M}_S$ (Sect. 3.1) and then use Markov chain analysis techniques to bound the mixing time (Sects. 3.2 and 3.3). Finally in Sect. 3.4 we apply our bound to each example application given in Sect. 2.2.

We begin by formally defining the mixing time. The time a Markov chain $\mathcal{M}$ takes to converge to its stationary distribution μ is measured in terms of the distance between μ and $\mathcal{P}^t$, the distribution at time t. Let $\mathcal{P}^t(x, y)$ be the t-step transition probability and Ω be the state space. The *mixing time* of $\mathcal{M}$ is

$$\tau_{\mathcal{M}}(\epsilon) = \min\{t : ||\mathcal{P}^{t'} - \mu|| \leq \epsilon, \forall t' \geq t\},$$

where $||\mathcal{P}^t - \mu|| = \max_{x \in \Omega} \frac{1}{2} \sum_{y \in \Omega} |\mathcal{P}^t(x, y) - \mu(y)|$ is the *total variation distance* at time t.

3.1 The Sum-Preserving Markov Chain $\mathcal{M}_S$

Let the state space $\Omega_{S,d,n} = \Omega$ be the set of all d-bounded n-compositions of S. Specifically, given a configuration x, each point $x(i)$ for $1 \leq i \leq n$ satisfies $0 \leq x(i) \leq d$ and the sum of the points satisfies $\sum_{i=1}^{n} x(i) = S$. We analyze the following Markov chain which is equivalent to the construction given by Tajik et al. [18].

The Sum-Preserving Shuffle Markov chain $\mathcal{M}_S$

Starting at any valid configuration $x_0 \in \Omega$, iterate the following:

- At time t, choose a random shuffling R on all points uniformly at random (u.a.r.).
- Pair adjacent points in R to create a perfect matching M.
- Independently, for each matched pair of points $(p_i, p_j) \in M$ select values for $x_{t+1}(p_i)$ and $x_{t+1}(p_j)$ u.a.r. from all valid choices that preserve the sum. Namely, all choices that satisfy $x_t(p_i) + x_t(p_j) = x_{t+1}(p_i) + x_{t+1}(p_j)$, $x_{t+1}(i) \leq d$, and $x_{t+1}(j) \leq d$.

Next we show that this Markov chain is irreducible (i.e. the state space Ω is connected) and thus has a unique stationary distribution (see e.g., [11]) and that the stationary distribution is uniform.

Lemma 1. *The Markov chain $\mathcal{M}_S$ is irreducible and has the uniform distribution on $\Omega_{S,d,n}$ as its unique stationary distribution.*

Proof. We will prove that $\mathcal{M}_S$ is irreducible by defining a distance metric ϕ and showing that there is always a move of $\mathcal{M}_S$ that decreases the distance between any two configurations. By repeatedly decreasing the distance it is thus possible to create a path between any two valid configurations. Define the distance ϕ between two configurations x and y as follows.

$$\phi(x,y) = \sum_{i=1}^{n} |x(i) - y(i)|. \tag{1}$$

We claim there is always a valid transition of $\mathcal{M}_S$ that will decrease the distance between x and y. Select a point p^+ which is larger in x than in y (i.e. $x(p^+) > y(p^+)$) and a point p^- that is smaller in x than in y. This is always possible since $x \neq y$ and the sum of the points in x is the same as the sum of the points in y. Next, decrease p^+ by 1, increase p^- by 1, and leave other points the same in x. This creates a valid configuration x' such that $\phi(x',y) < \phi(x,y)$ and (x, x') is a valid transition of $\mathcal{M}_S$. To see that this is a valid transition in $\mathcal{M}_S$ select a shuffling where these points are adjacent, it is clear there is a valid selection for each pair of matched points that gives the desired transition.

Next, we will show that for all $x, y \in \Omega, P(x,y) = P(y,x)$ and thus by detailed balance the uniform distribution must be the stationary distribution (see e.g., [11]). Consider any $x, y \in \Omega$. Let $\mathcal{T}_{x,y}$ be the set of all shufflings that allow a transition from x to y (i.e. starting from x if one of these shufflings is selected there is a way to select the values in the second step of the chain to match y.) Note that if $P(x,y) = 0$ then $\mathcal{T}_{x,y} = \emptyset$. It is clear that $\mathcal{T}_{y,x} = \mathcal{T}_{x,y}$ and for each shuffling $t \in \mathcal{T}_{x,y}$ each matched pair has the same sum in x as in y implying that both chains have the same valid choices in the last step of $\mathcal{M}_S$. Thus the probability of moving from x to y if t is selected is the same in both configurations and therefore $P(x,y) = P(y,x)$.   □

3.2   Bounding the Mixing Time of $\mathcal{M}_S$

In order to bound the mixing time of $\mathcal{M}_S$ we will use the path coupling method due to Dyer and Greenhill [7] which is an extension of the well-known coupling method (see e.g., [9]). A coupling of Markov chains with transition matrix P is a stochastic process $(X_t, Y_t)_{t=0}^{\infty}$ on $\Omega \times \Omega$ such that X_t and Y_t are both Markov chains with stationary distribution P and if $X_t = Y_t$, then $X_{t+1} = Y_{t+1}$. In other words when viewed in isolation each of the chains X and Y simulate the original chain and once they agree, they will always agree. Informally, the coupling time is the time until the two chains agree. A common technique is to select an appropriate coupling and then use the coupling time to bound the mixing of time of the Markov chain. This is often done by defining a distance metric and showing that in expectation the distance between any two arbitrary pairs of configurations is decreasing. The path coupling technique, which common in the Markov chain community, only requires considering pairs of states that are close according to the selected distance metric. In our case we will again use the Manhattan distance metric and will only need to consider pairs of states that differ on exactly 2 points.

More formally, we will use the following path coupling theorem due to Dyer and Greenhill [7].

Theorem 1. *Let ϕ be an integer valued metric defined on $\Omega \times \Omega$ which takes values in $\{0, ..., B\}$. Let U be a subset of $\Omega \times \Omega$ such that for all $(x_t, y_t) \in \Omega \times \Omega$ there exists a path $x_t = z_0, z_1, ..., z_r = y_t$ between x_t and y_t such that $(z_i, z_{i+1}) \in U$ for $0 \leq i < r$ and $\sum_{i=0}^{r-1} \phi(z_i, z_{i+1}) = \phi(x_t, y_t)$. Let $\mathcal{M}$ be a Markov chain on Ω with transition matrix P. Consider any random function $f : \Omega \rightarrow \Omega$ such that $\Pr[f(x) = y] = P(x, y)$ for all $x, y \in \Omega$, and define a coupling of the Markov chain by $(x_t, y_t) \rightarrow (x_{t+1}, y_{t+1}) = (f(x_t), f(y_t))$. If there exists $\beta < 1$ such that $E[\phi(x_{t+1}, y_{t+1})] \leq \beta \phi(x_t, y_t)$, for all $(x_t, y_t) \in U$, then the mixing time of $\mathcal{M}$ satisfies*

$$\tau(\epsilon) \leq \frac{\ln(B\epsilon^{-1})}{1 - \beta}.$$

To apply the theorem, we will present a coupling and show that the expected distance between any pair of configurations that differ by exactly two points decreases by at least β after each step of the Markov chain (for an appropriately choosen β). Note that here B is the maximum distance between two configurations using the selected distance metric.

Next, we prove the following theorem giving an upper bound on the mixing time of $\mathcal{M}_S$. Note that the mixing time (defined at the beginning of Sect. 3) bounds the number of steps of $\mathcal{M}_S$ needed and does not include the time to implement each step. We believe our bound can be improved.

Theorem 2. *Let n be the vector length, d the component bound, and S be the fixed sum. Given these definition, the mixing time $\tau_{\mathcal{M}_S}(\epsilon)$ of the sum-preserving Markov chain $\mathcal{M}_S$ on state space $\Omega_{n,d,S}$ satisfies*

$$\tau_{\mathcal{M}_S}(\epsilon) \leq n \ln(\min(dn, 2S)\epsilon^{-1}).$$

Proof. In order to apply Theorem 1 we begin by formally defining ϕ, U, and then bound B and β. As before, we define the distance $\phi(x, y)$ as in Eq. 1 as the L_1 norm (the Manhattan distance). Let U be the set of configurations x and y which differ on exactly 2 points. Next, we will show that for all $(x, y) \in \Omega \times \Omega$ there exists a path $x = z_0, z_1, ..., z_r = y$ between x and y such that $(z_i, z_{i+1}) \in U$ for $0 \leq i < r$ and $\sum_{i=0}^{r-1} \phi(z_i, z_{i+1}) = \phi(x, y)$.

Consider the path between any true arbitrary configurations given above in the proof that $\mathcal{M}_S$ is irreducible (Lemma 1). We claim this path satisfies the conditions. Given a configuration z_i to determine the next step in the path z_{i+1} two points p^+ and p^- are chosen where $z_i(p^+) > y(p^+)$ and $z_i(p^-) < y(p^-)$. A shuffling is selected where these points are paired together and these are the only two points that are modified. Namely, $z_{i+1}(p^+) = z_i(p^+) - 1$, $z_{i+1}(p^-) = z_i(p^-) - 1$, and for all other points p, $z_i(p) = z_{i+1}(p)$. It is thus clear that $(z_i, z_{i+1}) \in U$. It is easily seen that for $0 \leq i < r$, $\phi(z_i, z_{i+1}) = 2$ and $\phi(z_{i-1}, y) = \phi(z_i, y) - 2$ and thus the second distance requirement is satisfied.

To bound the maximum distance B recall that S is the sum over all points $S = \sum_{i=1}^{n} x(i)$. Since each point can contribute at most d to ϕ and there are n points, B satisfies

$$B \leq \min(dn, 2S).$$

In our coupling we will use the same shuffling in both x and y. If the two points that differ get paired together in the shuffling then the valid choices for all paired points are identical in both x and y, our coupling will choose the same configurations in both chains, and the distance will decrease to 0. This happens with probability $1/(n-1) > 1/n$. Otherwise, both of the two points that differ between x and y will get paired with points that are the same in both x and y. For all other pairs that do not include these points, our coupling will make the same choice in both x and y. We will prove in Lemma 2 that no matter what values the 2 points that differ have, there is a way to couple them so that the distance never increases. Thus we have

$$E[\phi(x_{t+1}, y_{t+1})] \leq \left(\frac{1}{n}\right) \cdot 0 + \left(\frac{n-1}{n}\right) \cdot \phi(x_t, y_t) = \left(\frac{n-1}{n}\right) \cdot \phi(x_t, y_t).$$

Letting $\beta = \frac{n-1}{n}$, $B \leq \min(dn, 2S)$, and applying Theorem 1 gives

$$\tau_{\mathcal{M}_S}(\epsilon) \leq n \ln(\min(dn, 2S)\epsilon^{-1}).$$

$\square$

3.3   Proof of Lemma 2 Coupling Two Points

It remains to show we can construct a coupling of any two points with different values where the expected change in distance is zero. We will prove the following lemma.

Lemma 2. *Given two arbitrary points i and j with $x_t(i) = y_t(i)$ and $x_t(j) \neq y_t(j)$, there exists a coupling such that*

$$|x_t(j) - y_t(j)| \geq |x_{t+1}(i) - y_{t+1}(i)| + |x_{t+1}(j) - y_{t+1}(j)|.$$

Proof. Recall that for two points i and j with $x_t(i) + x_t(j) = S_x$, the values $x_{t+1}(i), x_{t+1}(j)$ are chosen uniformly from all possible choices with $x_{t+1}(i) + x_{t+1}(j) = S_x$, $x_{t+1}(i) \leq d$, and $x_{t+1}(j) \leq d$. For example, if $S_x = 4$ and $d = 3$ the options are $(1, 3), (2, 2), (3, 1)$ and each is selected with probability $1/3$. Since $x_t(j) \neq y_t(j)$, the two chains will often have a different number of possible choices (see Fig. 1) and thus each individual configuration is selected with a different probability in each chain.

We begin by creating an ordering of the possible choices for $(x_{t+1}(i), x_{t+1}(j))$ (and similarly for $(y_{t+1}(i), y_{t+1}(j))$) and then show how we will carefully pair the choices to ensure that the distance never increases. This is especially difficult because when the two chains have a different number of possible choices, one configuration in x will needed to be paired with multiple configurations in y (or vice versa). We can view the coupling as creating a weighted bipartite graph (as shown in Fig. 1) where one partition is the possible choices in x and the other is the choices in y. A valid coupling is a set of edges between the partitions where

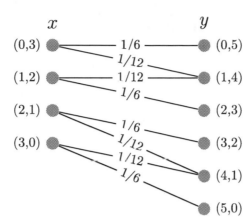

Fig. 1. The weighted bipartite graph visualizing the coupling for $S_x = 3$, $S_y = 5$, and $d = 5$.

each edges is assigned a probability such that the sum of the edges adjacent to each configuration is equal to the probability of that configuration.

We will order the choices by increasing value of $x_{t+1}(i)$ (or $y_{t+1}(i)$) starting from the lowest possible value (i.e. lexicographical order). For example, if $x_t(i) + x_t(j) = 3$ there are 4 choices ordered $(0,3), (1,2), (2,1), (3,0)$ and each has probability $1/4$. With configurations in this order, we will always couple a configuration in x to the lowest possible configuration in y while maintaining the correct probabilities. More specifically, let $S_x = x_t(i) + x_t(j)$ and $S_y = y_t(i) + y_t(j)$. We will begin with the case that $x_t(i) + x_t(j) <= d$, $y_t(i) + y_t(j) <= d$, and $S_x < S_y = \delta + S_x$. We will handle the general case later. In this case, x has $S_x + 1$ possible configurations ordered $(0, S_x), (1, S_x - 1), \ldots, (S_x, 0)$. Similarly, y has $S_y + 1$ possible configurations ordered $(0, S_y), (1, S_y - 1), \ldots, (S_y, 0)$. We start by pairing (adding an edge between) the lowest configurations (i.e. $(0, S_x)$ and $(0, S_y)$) with probability $1/(S_y + 1)$. At this point, configuration $(0, S_x)$ still has probability $1/(S_x + 1) - 1/(S_y + 1)$ remaining. We will then pair it to the next lowest configuration in y (i.e. $(1, S_y - 1)$) with the remaining probability or $1/(S_y + 1)$ whichever is smaller. We will continue pairing $(0, S_x + 1)$ with the lowest configuration in y that has unused probability (i.e. it's adjacent edges do not add to $1/(S_y + 1)$) until the edges assigned add to the correct probability $1/(S_x + 1)$. We then iterate through the remaining configurations in x (in the ordering above) using the same algorithm. Specifically, pairing each with the lowest configuration(s) in y that have remaining probability. For example, the case $x_t(i) = y_t(i) = 2$, $x_t(j) = 1$, and $y_t(j) = 4$ is shown in Fig. 1.

It remains to show that using the coupling described above the distance will never increase. Specifically we will show for each coupled pair (i.e. configurations connected by an edge), the distance between those configurations is at most $\delta = |x_t(j) - y_t(j)|$. Let $(i, S_x - i)$ for $0 \leq i \leq S_x$ be any valid configuration for x. We want to show that given our coupling the edges leaving this configuration only go to configurations in y at distance at most δ. Specifically we need to show that all edges are to configurations in the range $\{(i, S_y - i), \ldots, (i + \delta, S_y - i - \delta)\}$

(using the ordering defined above). Let $\mathbb{P}_x(i,j)$ be the probability of all x points in the range $\{(0,S_x),\ldots,(i,j)\}$ (and similarly define $\mathbb{P}_y(i,j)$). Recall that our coupling will match a point in x to the lowest point(s) in y that have remaining probability. Thus to show that $(i, S_x - i)$ only gets mapped to points in the appropriate range (i.e. the points in y at distance δ) we need to show that the point in y just before that range $(i-1, S_y - i - 1)$ will have no remaining probability and that (i, S_x-i) will not get matched to points after $(i+\delta, S_y-i-\delta)$. Specifically it suffices to prove the following.

Proposition 1.

1. *For $0 < i \leq S_x$, $\mathbb{P}_x(i-1, S_x - i - 1) \geq \mathbb{P}_y(i-1, S_y - i - 1)$.*
2. *For $0 \leq i < S_x$, $\mathbb{P}_x(i, S_x - i) \leq \mathbb{P}_y(i+\delta, S_y - i - \delta)$.*

Proof. Since there are $S_x + 1$ configurations for x and $S_y + 1 = S_x + \delta + 1$ configurations for y we have the following

$$\mathbb{P}_x(i,j) = \frac{i+1}{S_x + 1}, \quad \mathbb{P}_y(i,j) = \frac{i+1}{S_x + \delta + 1}.$$

Given these definitions, it is straightforward to prove the proposition using basic algebra.   □

It remains to consider the more general case where either $x_t(i) + x_t(j) > d$ or $y_t(i) + y_t(j) > d$ (or both). We will use the same coupling described above where a configuration x is paired with the lowest configuration(s) in y with remaining probability. Similar to what we did in the first case, for each valid configuration for x we will map the configuration to two configurations in y. We will prove that these two configurations and all configurations between them are at distance at most δ. Then we will show that the x configuration will never get coupled to a y configuration outside of this range.

   Let n_x be the number of valid x configurations and n_y be the number of valid y configurations. Without loss of generality, we will assume that $n_x \leq n_y$. Let (x_1, x_2) be the lowest configuration for x and (y_1, y_2) be the lowest configuration for y. As before, if $S_x \leq d$ then $x_1 = 0, x_2 = S_x$ and there are $x_2 - x_1 + 1 = S_x + 1$ configurations. However if $S_x > d$ then $x_1 = d - n_x, x_2 = d$ and there are $x_2 - x_1 + 1 = 2d - S_x + 1$ configurations $(S_x - d, d), (d - S_x + 1, d - 1), \ldots, (d, S_x - d)$. We begin by proving that these initial configurations are at distance $\delta = |S_x - S_y|$.

Lemma 3. *Let $\delta = |S_x - S_y|$ be the initial distance between x and y. Assuming $n_x \leq n_y$ and the possible configurations are ordered as described above we have.*

1. *The lowest configurations (x_1, x_2) and (y_1, y_2) as defined above satisfy.*

$$\phi((x_1, x_2), (y_1, y_2)) \leq \delta.$$

2. *The highest configurations (x_2, x_1) and (y_2, y_1) as defined above satisfy.*

$$\phi((x_2, x_1), (y_2, y_1)) \leq \delta.$$

3. For $0 < c \le x_2 - x_1$, $\phi((x_1 + c, x_2 - c), (y_1 + c, y_2 - c)) \le \delta$.
4. For $0 < c \le x_2 - x_1$, $\phi((x_2 - c, x_1 + c), (y_2 - c, y_1 + c)) \le \delta$.

Proof. Here we will consider four cases based on how S_x and S_y compare to d. If both $S_x, S_y \le d$ then $(x_1, x_2) = (0, S_x)$ and $(y_1, y_2) = (0, S_y)$. Here we have

$$|x_1 - y_1| + |x_2 - y_2| = |0 - 0| + |S_x - S_y| = \delta.$$

Next, we will consider the case where exactly one of S_x, S_y is greater than d. If $S_x > d$ and $S_y \le d$. In this case $(x_1, x_2) = (d - S_x, d)$ and $(y_1, y_2) = (0, S_y)$. Here we have

$$|x_1 - y_1| + |x_2 - y_2| = |S_x - d| + |d - S_y| = S_x - S_y = \delta.$$

If instead $S_y > d$ and $S_x \le d$ the argument is identical. Finally if both $S_x, S_y > d$ we have $(x_1, x_2) = (d - S_x, d)$ and $(y_1, y_2) = (d - S_y, d)$. Here we have

$$|x_1 - y_1| + |x_2 - y_2| = |(S_x - d) - (S_y - d)| + |d - d| = |S_x - S_y| = \delta.$$

It immediately follows that the final configurations (x_2, x_1) and (y_2, y_1) are also at distance at most δ. Similarly it can easily be shown by the definition of the L_1 distance metric that third and fourth statements are true. □

To begin, if $n_x = n_y$ then we will match each point $(x_1 + c, x_2 - c)$ with exactly one point $(y_1 + c, y_2 - c)$ with weight $1/n_x - 1/n_y$. By Lemma 3 these points are distance at most δ and we are done.

Next, assume $n_x < n_y$ and consider any general configuration for x, $(x_1 + c, x_2 - c)$. We will show that this configuration will only be matched with configurations in between (and including) $(y_1 + c, y_2 - c)$ and $(y_2 - (x_2 - x_1 - c), y_1 + (x_2 - x_1 - c))$. By Lemma 3 parts 3 and 4, these configurations are both at distance at most δ from $(x_1 + c, x_2 - c)$. It also immediate follows that any points between these are at distance at most d from $(x_1 + c, x_2 - c)$.

Next we show that $(x_1 + c, x_2 - c)$ will only be coupled to points in y between and including $(y_1 + c, y_2 - c)$ and $(y_2 - (x_2 - x_1 - c), y_1 + (x_2 - x_1 - c))$. Again as in the first case to do this we need to show that $(x_1 + c, x_2 - c)$ will never be matched with anything below $(y_1 + c, y_2 - c)$ or anything above $(y_2 - (x_2 - x_1 - c), y_1 + (x_2 - x_1 - c))$. Specifically we prove the following.

1. For $0 < c \le x_2 - x_1$, $\mathbb{P}_x(x_1 + c - 1, x_2 - c + 1) > \mathbb{P}_y(y_1 + c - 1, y_2 - c + 1)$.
2. For $0 \le c < x_2 - x_1$, $\mathbb{P}_x(x_1 + c, x_2 - c) \le \mathbb{P}_y(y_2 - (x_2 - x_1 - c), y_1 + (x_2 - x_1 - c))$.

Recall that $\mathbb{P}_x(x_1 + c - 1, x_2 - c + 1)$ is the probability of all configurations for x up to and including $(x_1 + c - 1, x_2 - c + 1)$. This includes c configurations each with probability $1/n_x$ and thus $\mathbb{P}_x(x_1 + c - 1, x_2 - c + 1) = c/n_x$. Similarly $\mathbb{P}_y(y_1 + c - 1, y_2 - c + 1) = c/n_y$. Thus our first statement follows directly from the fact that $n_x < n_y$.

Using the fact that $n_x = x_2 - x_1 + 1$ and $n_y = y_2 - y_1 + 1$ we can prove the second statement as follows.

$$c \leq x_2 - x_1$$
$$c + 1 \leq x_2 - x_1 + 1$$
$$c + 1 \leq n_x$$
$$(n_y - n_x)(c + 1) \leq (n_y - n_x)n_x$$
$$n_y(c + 1) \leq (n_y - n_x)n_x + n_x(c + 1)$$
$$(c + 1)/n_x \leq (n_y - n_x + c + 1)/n_y$$
$$(c + 1)/n_x \leq (y_2 - y_1 - x_2 + x_1 + c + 1)/n_y$$
$$\mathbb{P}_x(x_1 + c, x_2 - c) \leq \mathbb{P}_y(y_2 - (x_2 - x_1 - c), y_1 + (x_2 - x_1 - c)).$$

$\square$

3.4   Applying the Mixing Bound to Examples

Finally, we apply our upper bound on the mixing time (Theorem 2) to each of the example applications given in Sect. 2.2. The results are given below in Table 1. While our theorem gives the first formal proof bounding the mixing time of $\mathcal{M}_S$ that we are aware of, we expect that it is not a tight bound and further improvements are possible. Thus, the number of rounds given in the table while provably sufficient are likely more than needed.

Note that in our bound we have the term $\min(dn, 2S)$. Since we have not specified a specific sum in any of the examples we used dn for the bounds in the table. If the desired sum S satisfies $S < dn/2$ then our theorem could be used to obtain a smaller upper bound.

Table 1. Our upper bound for the rounds needed in the matching-based algorithm of Tajik et al.

Application	$\epsilon = 10^{-10}$	$\epsilon = 2^{-80}$
10×10 image block ($n = 100, d = 255$)	3,318	6,560
16×16 image block ($n = 256, d = 255$)	8,733	17,034
32×32 image block ($n = 1024, d = 255$)	36,351	69,555
Exam scores ($n = 300, d = 100$)	10,001	19,729
Salaries ($n = 30, d = 100000$)	1,139	2,111
Ratings ($n = 5000, d = 4$)	164,647	326,777

4 Approaches Based on Ranking

In this section we describe algorithms for ranking and unranking d-bounded n-compositions of S based on two different total orderings. The first is the standard lexicographical order (Sect. 4.1) and the second is a new ordering we call recursive block ordering (Sect. 4.2). Both orderings rely heavily on pre-computed information which we describe in Sect. 4.3. Finally in Sects. 4.4 and 4.5 we describe the unrank algorithms for both orderings.

4.1 Lexicographical Ranking

In this section, we will use the lexicographical ordering on d-bounded n-compositions of S to generate a ranking. We will use the notation $<_L$ and $>_L$ to refer to lexicographical order. Specifically let $x, y \in \Omega_{S,d,n}$ be two arbitrary configurations such that $x \neq y$ and let i be the smallest integer such that $x(i) \neq y(i)$. If $x(i) < y(i)$ then $x <_L y$ otherwise $y >_L x$. For an example, see Fig. 2.

Configuration	Rank	Configuration	Rank
(0,3,3)	0	(2,3,1)	5
(1,2,3)	1	(3,0,3)	6
(1,3,2)	2	(3,1,2)	7
(2,1,3)	3	(3,2,1)	8
(2,2,2)	4	(3,3,0)	9

Fig. 2. The lexicographical ranking of configurations in $\Omega_{S,d,n} = \Omega_{6,3,3}$.

We begin by describing our ranking algorithm using the running example $S = 6, d = 3$, and $n = 3$ shown in Fig. 2. To rank a configuration x according to our ordering we start by determining how many configurations start with a number strictly less than $x(1)$. For example if $x = (2, 3, 1)$ there are three such configurations: $(0, 3, 3), (1, 2, 3)$, and $(1, 3, 2)$ (see Fig. 2). Next we determine the position of $(2, 3, 1)$ among configurations in $\Omega_{6,3,3}$ that start with 2. Since these configurations all start with 2 the remaining sum must add to $6 - 2 = 4$ and there are $3 - 1 = 2$ remaining points. Thus this is equivalent to determining the rank of $(3, 1)$ in $\Omega_{6-2,3,3-1} = \Omega_{4,3,2}$ which is two. We then add the numbers together to get 5, the rank of $(2, 3, 1)$. More generally, let $C_d(n, S)$ be the number of n-compositions with sum S and component bound d and $\mathsf{l\text{-}rank}_{n,S}(x)$ be the lexicographical rank of x in $\Omega_{S,d,n}$. Given these definitions, we can define the rank recursively as follows.

```
1: procedure L-RANK((x₁, . . . , xₙ), S, CUM)
2:     rankSoFar ← 0
3:     startN ← n
4:     for i ← 1 to startN do
5:         if xᵢ ≠ 0 then
6:             rankSoFar ← rankSoFar + CUM[n − 1, S − xᵢ + 1]
7:             if S < CUM.length then
8:                 rankSoFar ← rankSoFar − CUM[n − 1, S + 1]
9:             end if
10:        end if
11:        S ← S − xᵢ
12:        n ← n - 1
13:    end for
14:    return rankSoFar
15: end procedure
```

Fig. 3. Lexicographic ranking algorithm with the cumulative sum table CUM.

$$\mathsf{l\text{-}rank}_{n,S}(x_1, \ldots, x_n) =$$
$$\begin{cases} 0 & \text{if } n = 1 \\ \sum_{0 \le i < x_1} C_d(n-1, S-i) + \mathsf{l\text{-}rank}_{n-1, S-x_1}(x_2, \ldots, x_n) & \text{if } n > 1 \end{cases}$$

It is straightforward to implement an algorithm for ranking given the definition above and a pre-computed C table storing the needed values.

In order to improve the efficiency of our ranking algorithm we will actually store the cumulative sum so position (i, j) in our table will store $\sum_{k=i}^{n} C_d(k, j)$. Using this cumulative sum table CUM, we give a more efficient ranking algorithm in Fig. 3. Note that the efficiency gain comes from not having to compute a sum in each iteration of the for loop.

4.2 Recursive Block Ranking

Next, we give an alternative ranking algorithm based on a new total ordering we call *recursive block order*. While recursive block order uses ideas that are reminiscent to those used in orderings of monomials, specifically block order (see e.g., [8]) and graded order (see e.g. [5]), these are combined differently and applied recursively unlike in monomial orderings. We will use the notation $<_B$ and $>_B$ to refer to recursive block order. Throughout this section we will assume n is a power of two. We can fairly easily generalize our ordering and algorithms to work when n is not a power of two and briefly describe the needed alterations at the end of the section. Let $x, y \in \Omega_{S,d,n}$ be two arbitrary configurations such that $x \neq y$. To compare x and y using recursive block order, we let x_L be the first $n/2$ points in x, and x_R be the remaining points. Similarly define y_L as the first $n/2$ points in y and y_L to be the remaining points. We begin by considering the sum S_{x_L} of the points in x_L and similarly S_{y_L} (the sum of the points in y_L).

If $S_{x_L} < S_{y_L}$ then $x <_B y$. Similarly, if $S_{x_L} > S_{y_L}$ then $x >_B y$. If the sums are equal then we will apply our ordering recursively. Specifically, if $x_L \neq y_L$ then $x <_B y$ if $x_L <_B y_L$ and $x >_B y$ if $x_L >_B y_L$. Finally if x_L is identical to y_L then $x <_B y$ if $x_R <_B y_R$ and $x >_B y$ if $x_R >_B y_R$. As a base case, if n is 1 then there is only one configuration with rank 0. See Fig. 4 for an example of recursive block order. We summarize these conditions in the table below.

$$S_{x_L} < S_{y_L} \qquad\qquad\qquad\qquad\qquad \implies x <_B y \qquad (2)$$
$$S_{x_L} > S_{y_L} \qquad\qquad\qquad\qquad\qquad \implies x >_B y \qquad (3)$$
$$S_{x_L} = S_{y_L} \text{ and } x_L <_B y_L \qquad\qquad \implies x <_B y \qquad (4)$$
$$S_{x_L} = S_{y_L} \text{ and } x_L >_B y_L \qquad\qquad \implies x >_B y \qquad (5)$$
$$S_{x_L} = S_{y_L} \text{ and } x_L = y_L \text{ and } x_R <_B x_L \implies x <_B y \qquad (6)$$
$$S_{x_L} = S_{y_L} \text{ and } x_L = y_L \text{ and } x_R >_B x_L \implies x >_B y \qquad (7)$$

Configuration	Rank	Configuration	Rank
(0,2,3,3)	0	(2,1,2,3)	7
(1,1,3,3)	1	(2,1,3,2)	8
(2,0,3,3)	2	(3,0,2,3)	9
(0,3,2,3)	3	(3,0,3,2)	10
(0,3,3,2)	4	(1,3,1,3)	11
(1,2,2,3)	5	(1,3,2,2)	12
(1,2,3,2)	6	(1,3,3,1)	13

Fig. 4. The recursive block ranking of some configurations in $\Omega_{S,d,n} = \Omega_{8,3,4}$.

Based on the above definition of recursive block order we will give a recursive ranking algorithm. Again our algorithm will rely on pre-computed values stored in the C table where $C_d(n, S)$ is the number of n-compositions with sum S and component bound d. Let b-rank$_n(x)$ be the recursive block rank of x in $\Omega_{S_x,d,n}$. In order to compute the rank of a configuration x we need to determine the number of configurations y such that $x >_B y$. Using the above definition of recursive block order such configurations fit into three cases given by Eqs. 3, 5, and 7 in the above table. Our algorithm will compute the number of configuration for each case and add the three to determine the rank of x. Consider first the case given by Eq. 3. Here, we need to compute the number of configurations y that satisfy $S_{x_L} > S_{y_L}$. For each possible smaller sum S_{y_L} (i.e. any sum less than S_{x_L}) we compute the number of configurations with this left sum. We do this by using the C table to determine the number of choices for y_L and multiplying by the number of choices for y_R. This is exactly $\sum_{s=0}^{S_{x_L}-1} C(n/2, s) \cdot C(n/2, S - s)$. The second case is given by Eq. 5, configurations with the same left sum but $y_L <_B x_R$. The number of such configurations y_L is given by b-rank$_{n/2}(x_L)$.

```
 1: procedure B-RANK((x₁, ..., xₙ), S, C)
 2:     if n == 1 then
 3:         return 0
 4:     end if
 5:     X_L ← (x₁, ..., x_{n/2})
 6:     X_R ← (x_{n/2+1}, ..., xₙ)
 7:     leftSum ← SUM(x₁, ..., x_{n/2})
 8:     rightSum ← SUM(x_{n/2+1}, ..., xₙ)
 9:     case3 ← 0
10:     for s ← 0 to leftSum -1 do
11:         case3 ← case3 + C[n/2, s] · C[n/2, S − s]
12:     end for
13:     case5 ← C[n/2, rightSum]·B-RANK(X_L, leftSum, C)
14:     case7 ← B-RANK(X_R, rightSum, C)
15:     return case3 + case5 + case7
16: end procedure
```

Fig. 5. Recursive block ranking algorithm.

For each of these configurations there are $C(n/2, S_{x_R})$ choices for y_R. Thus the total number of such configurations is b-rank$_{n/2}(x_L) \cdot C(n/2, S_{x_R})$. Finally the last case corresponding to Eq. 7 is configurations with $y_L = x_L$ and $y_R <_B x_R$. There are b-rank$_{n/2}(x_R)$ such configurations. Combining these gives the following recurrence for b-rank$_n(x)$:

$$\sum_{s=0}^{S_{x_L}-1} C(n/2, s)\cdot C(n/2, S-s)+\text{b-rank}_{n/2}(x_L)\cdot C(n/2, S_{x_R})+\text{b-rank}_{n/2}(x_R) \quad (8)$$

It is straightforward to design a recursive algorithm based on the above recurrence. Note that as a base case, if $n = 1$ then b-rank$_1(x) = 0$ since there is only one configuration. Based on this equation we give a recursive algorithm for determining the recursive block rank in Fig. 5. A key advantage of this algorithm is that it does not requirement the entire C table. As seen from Eq. 8 in order to compute the rank of a configuration with n points we look at row $n/2$ of the C table and make two recursive calls both with $n/2$ points. Thus we only need to pre-compute the C table for rows that are powers of 2 resulting in only $\log n$ rows. Note that if n is not a power of 2 it is straightforward to generalize the rank algorithm and Theorem 4 below shows that we only need to at most double the number of rows that need to be pre-computed.

Lemma 4. *The recursive calls in each level of the recursion tree for the recursive block ranking algorithm will either all be the same size or have two unique sizes that differ by one.*

Proof. We will prove this using induction. As a base case the first call has a single size. We will assume inductively that the conditions are satisfied at a level i and show that they will continue to be satisfied at the next level. Recall that at each step of the recursion the size will be split in half. We will consider 2 different cases. If there is only one size x at level i then if x is even there will continue to be one size $x/2$ at the next level. If x is odd then there will be exactly two sizes that differ by one $(x-1)/2$ and $(x+1)/2$ at the next level. The second case we will consider is if there are two sizes x and $x+1$ that differ by one at level i. If x is even then at the next step the calls will all have sizes $x/2$ or $x/2+1$. If x is odd then at the next step the calls will all have sizes $(x-1)/2$ or $(x+1)/2$. In all cases the conditions of the theorem are satisfied.

4.3   Pre-computing the C Table

To support the rank and unrank algorithms for lexicographical and recursive block orderings we will need access to additional information stored in a table. We consider two approaches for pre-computing the information needed based on previous work in the area of counting restricted compositions. The first is based on dynamic programming techniques. While this algorithm is faster per table entry, it requires all values in the table to be computed. The second method is based on generating functions and while slower, does not rely on previously computed entries. This allows us to only compute the tables entries needed and is thus useful for the recursive block ranking algorithm. Again let $C_d(n, S)$ be the number of n-compositions with sum S and maximum value d.

Filling the Table with Dynamic Programming. First we describe a dynamic programming based approach to filling the C table. We begin by developing and justifying a recurrence for $C_d(n, S)$. Note that this same recurrence is given previously by Abramson [1] although with a different more combinatorial explanation. For each n-composition with sum $S > 0, n > 1$ the first point is either 0 or some number greater than 0. The number of such compositions that start with 0 is equal to $C_d(n-1, S)$. For those that start with a number greater than 0, consider the quantity $C_d(n, S-1)$. For each of the compositions counted by $C_d(n, S-1)$ we can add one to the first point and obtain a n-composition with sum S and first point greater than 0. However, if $S > d$ some of these compositions will have the first point set to $d+1$ which is not a valid composition. Thus, in this case, $C_d(n, S-1)$ is over counting the number of n-compositions that start with a point greater than 0. If a n-composition starts with $d+1$, the remaining points are a $(n-1)$-compositions of the remaining sum $S - (d+1)$. Thus the excess number of compositions is exactly $C_d(n-1, S-(d+1))$. Combining these ideas gives the following recurrence for the C table.

$$C_d(n, S) = \begin{cases} 1 & \text{if } n = 1, S \leq d \\ 0 & \text{if } n = 1, S > d \\ 1 & \text{if } n = 0 \\ C_d(n-1, S) + C_d(n, S-1) & \text{if } n > 0, S \leq d \\ C_d(n-1, S) + C_d(n, S-1) - C_d(n-1, S-d-1) & \text{otherwise} \end{cases}$$

It is straightforward to see how using a nested loop and the above recurrence we can use fill the C table with dimensions $n \times (n * d)$ in time $\Theta(n^2 d)$. The value $n * d$ is chosen since this is the maximum possible sum. Note that Stein [17] uses a very similar approach to fill the table but does not use the dynamic programming framework or provide a justification for the recurrence. As mentioned above, in order to improve the efficiency of our lexicographical ranking and unranking algorithms we will actually store the cumulative sum so position (i, j) in the CUM table will store $\sum_{k=i}^{n} C_d(k, j)$.

Filling the Table with Generating Functions. As an alternative approach to fill in the table, we can use formulas derived from a generating function view of the problem. This approach to counting restricted compositions is well established (see e.g. [1], [15]) and we briefly provide the details here for completeness. We will need the following well-known polynomial expansions:

$$\frac{1 - x^{n+1}}{1 - x} = 1 + x + x^2 + \ldots + x^n \tag{9}$$

$$(1 + x)^n = 1 + \binom{n}{1} x + \binom{n}{2} x^2 + \ldots + \binom{n}{n} x^n \tag{10}$$

$$\frac{1}{(1 - x)^n} = 1 + \binom{1 + n - 1}{1} x + \binom{2 + n - 1}{2} x^2 + \ldots \tag{11}$$

If $f(x)$ is a polynomial, we will use the notation $[x^k] f(x)$ to mean the coefficient of x^k in $f(x)$. So for example, $[x^3](1 + x)^{15}$ would mean the coefficient of x^3 in the expansion of $(1 + x)^{15}$, which from the identities above we can see is $\binom{15}{3}$.

We can use generating functions to compute the value $C_d(n, S)$ by noting that $C_d(n, S)$ will actually be the coefficient of the x^S term in the polynomial

$$(1 + x + x^2 + \ldots + x^d)^n .$$

Given this, we can then use the above identities to derive a formula for $C_d(n, S)$.

Lemma 5. *Consider vectors of length n with component bound d that sum to S. Then the number of such vectors is given by*

$$C_d(n, S) = [x^S](1 + x + x^2 + \ldots + x^d)^n = \sum_{k=0}^{n} (-1)^k \binom{n}{k} \binom{n + S - (d+1)k - 1}{n - 1}$$

Proof. From Eq. (9) we can see that

$$[x^S](1 + x + x^2 + \ldots + x^d)^n = [x^S]\left(\frac{1 - x^{d+1}}{1 - x}\right)^n$$

$$= [x^S](1 - x^{d+1})^n \cdot \frac{1}{(1 - x)^n}$$

We can apply Equations (10) and (11) to the two parts of this last polynomial and see that

$$(1 - x^{d+1})^n = 1 - \binom{n}{1}x^{d+1} + \ldots + (-1)^k \binom{n}{k}x^{k(d+1)} + \ldots + (-1)^n \binom{n}{n}x^{n(d+1)}$$

and

$$\frac{1}{(1 - x)^n} = 1 + \binom{1 + n - 1}{1}x + \binom{2 + n - 1}{2}x^2 + \ldots$$

Given these two equations, we are interested in ways to get x^S. We need to account for possibly each term in the first equation combining with a term in the second equation. Specifically, if we have $x^{k(d+1)}$ from the first equation, then to get x^S we need the term $x^{S-k(d+1)}$ from the second. Summing up over all such possibilities and using the fact that $\binom{n}{k} = \binom{n}{n-k}$ we get

$$C_d(n, S) = \sum_{k=0}^{n}(-1)^k \binom{n}{k}\binom{S - k(d + 1) + n - 1}{S - k(d + 1)}$$

$$= \sum_{k=0}^{n}(-1)^k \binom{n}{k}\binom{n + S - k(d + 1) - 1}{n - 1}$$

$$\square$$

In the next sections we describe the unrank algorithms for both lexicographical order and recursive block order. Both of these algorithms rely on the same pre-computed information as the associated rank algorithms.

4.4 Lexicographical Unrank

Next we describe the unranking algorithm for lexicographical order l-unrank. Note this is similar to the algorithm given by Stein [17] with the exception of using the cumulative sum table. We begin by describing our unrank algorithm again using the running example $S = 6, d = 3$, and $n = 3$ shown in Fig. 2. To unrank an integer r according to lexicographical order we start by determining x_1. We can do this again using our C table. For example we know that $x_1 = 0$ if the rank r is less than the number of configurations that start with 0 or $r < C_d(n-1, S)$. Similarly $x_1 = 1$ if $C_d(n-1, S) <= r < C_d(n-1, S-1)+C_d(n-1, S)$. In our example, x_1 is 1 if $1 \leq r < 3$. More generally,

$$x_1 = \max j : \sum_{i=S-j}^{S} C_d(n - 1, i) < r.$$

```
 1: procedure L-UNRANK(rank, S, n, CUM)
 2:     (x₁, …, xₙ) ← (0, …, 0)
 3:     startN ← n
 4:     for i ← 1 to startN -1 do
 5:         if S < CUM.length then
 6:             offset ← CUM[n − 1, S + 1]
 7:         else
 8:             offset ← 0
 9:         end if
10:         while CUM[n − 1, S − xᵢ]−offset ≤ rank do
11:             xᵢ ← xᵢ + 1
12:         end while
13:         n ← n − 1
14:     end for
15:     x_startN ← S
16:     return x
17: end procedure
```

Fig. 6. Lexicographic unranking algorithm with the cumulative sum table CUM.

We can use this idea to define the unrank algorithm recursively. At each step i we will first determine x_i using the formula above and then recursively call the unrank algorithm to determine the remaining points l-unrank$_{n-1, S-x_i}$ $(r - \sum_{i=S-j+1}^{S} C_d(n − 1, i))$. Based on this idea we give an l-unrank algorithm in Fig. 6 that again uses the cumulative sum table. Note that since we are using the cumulative sum table, the algorithm could be improved by replacing the while loop with a variation on binary search. More specifically, look at the value in the first position, the second, the fourth and so forth until a value greater than rank is found and then perform binary search.

4.5 Recursive Block Unrank

Finally we describe the unranking algorithm for the recursive block ordering b-unrank. At a high-level our algorithm begins by determining the sum of the left $n/2$ points. Next we determine the rank of x_L (i.e. b-rank$_{n/2}(x_L)$) and the rank of x_R. Finally we apply our unrank algorithm recursively to each of these ranks to determine x_L and x_R. Throughout this section we will use the running example $S = 8, d = 3$, and $n = 4$ shown in Fig. 4.

We begin by showing how to determine the left sum S_{x_L}. In our example we know that $S_{x_L} = 2$ if the rank r is less than the total number of configurations with left sum 0,1, or 2. Since there are no configurations with left sum 0 or 1 the left sum is 2 if $r < C_d(n/2, 2) \cdot C_d(n/2, S − 2) = 3 \cdot 1 = 3$. More generally,

$$S_{x_L} = \min s : r < \sum_{i=0}^{s} C_d(n/2, i) \cdot C_d(n/2, S − i).$$

```
 1: procedure B-UNRANK(rank, S, n, C)
 2:     if n== 1 then
 3:         return (rank)
 4:     end if
 5:     leftPointsSmallerSum ← 0
 6:     leftSum ← 0
 7:     while leftPointsSmallerSum <= rank do        ▷ Determine the left sum
 8:         leftPointsSmallerSum += C[n/2, leftSum] · C[n/2, n − leftSum]
 9:         leftSum ← leftSum + 1
10:     end while                    ▷ The loop went one past the correct sum
11:     leftPointsSmallerSum -= C[n/2, leftSum] · C[n/2, n − leftSum]
12:     leftSum ← leftSum - 1
13:     rightPoints ← C[n/2, S − leftSum]
14:     leftRank ← (rank - leftPointsSmallerSum) / rightPoints
15:     rightRank ← (rank - leftPointsSmallerSum) % rightPoints
16:     left ← B-UNRANK(leftRank, leftSum, n/2, C)
17:     right ← B-UNRANK(rightRank, S - leftSum, n/2, C)
18:     return (left, right)
19: end procedure
```

Fig. 7. Recursive block unranking algorithm.

Next we will determine the rank of x_L. Note that each possible configuration of the left n points will occur in the ranking one time for each possible right configuration or $C_d(n/2, S − S_{x_L})$ times. For example, in our running example when $S_{x_L} = 3$ then there are 2 possible configurations for x_R namely $(2, 3)$ and $(3, 2)$ so each possible left configuration shows up in the ordering consecutively 2 times. Thus to determine the rank of the left configuration we first subtract the number of configurations with smaller sum from the rank and then divide by the number of right configurations $C_d(n/2, S − S_{x_L})$. To determine the right rank we again subtract the number of configurations with smaller sum and then determine the remainder when divided by $C_d(n/2, S − S_{x_L})$. Finally once we have the left and right ranks we can apply our b-unrank algorithm recursively to each rank. We give our b-unrank algorithm in Fig. 7.

5   Implementation and Performance Comparison

To compare the matching-based algorithm of Tajik et al. to the ranking algorithms we proposed in the previous section, we created prototype implementations of both and ran a number of performance tests. All implementations were done in Python 3.9.7 and performance tests were conducted on a machine with an Intel Core i5-8265U CPU @ 1.60 GHz (1 socket, 4 cores per socket, and 2 threads per core) and 8 GB of RAM, running 64-bit Ubuntu Linux 18.04. Before looking at performance results, we discuss more details of our implementations.

5.1   Implementation Details

Matching-Based Algorithm. We implemented the matching-based algorithm of Tajik et al. as we described it earlier. In each round, adjacent points are matched and rank-encipher-unrank is applied to each pair. We shuffled the points between rounds using the Knuth shuffle, which was suggested by Tajik et al. Like the previous work, we are considering NR security, so we implemented each rank-encipher-unrank using addition mod N, where N depends on the points being paired up and their sum. For our performance tests, we considered 50, 500, and 1000 rounds.

Ranking-Based Algorithms. We implemented both the lexicographic and recursive block ranking algorithms from the previous section to use with rank-encipher-encrypt. For lexicographic, we implemented the rank and unrank algorithms that use the cumulative sum table (Fig. 3 and Fig. 6). We also considered NR security, so the encryption portion was implemented using addition mod N, where N is the maximum rank given a vector length, component bound, and sum.

For the lexicographic ranking algorithm, we used a 2d numpy array with datatype object for the $C_d(n, S)$ table so that we could store unlimited precision Python integers in the arrays. This is necessary since the numbers in the table get very large; for the 16×16 image block example, the maximum value in the table is over 2000 bits.

For the recursive block ranking algorithm, since we only need a much smaller number of rows, we stored the needed rows of the $C_d(n, S)$ table in a Python dictionary indexed by n. We generated each entry using the generating function formula in Lemma 5. We also used the Python decorator `functools.cache` to memoize the choose function and speed up any repeated n choose k calculations that take place while generating the table.

5.2   Performance Tests and Results

To compare the performance of the matching-based and ranking-based solutions, we considered the applications mentioned earlier in Sect. 2.2 which have various parameter choices for n and d. For each application, we generated five vectors with n uniformly random elements chosen from 0 to d and measured the average time to encrypt using the iPython `%time` built-in "magic" command.

For the matching-based algorithm, we tested three choices of rounds, 50, 500, and 1000. The results are shown in Table 2. For the ranking-based solutions, we additionally measured the time to generate the C_d tables and also the eventual sizes of those tables. We determined the size by applying the Python `sys.getsizeof` method to each integer in the table, plus the result of applying `sys.getsizeof` to the table data structure itself. The results are shown in Table 3. It is interesting to note that the lexicographic ranking algorithm failed on the 32×32 block image encryption application and on the ratings application, due to the table size getting too large for the system to handle. The recursive block ranking, on the other hand, was successful on those two applications.

Table 2. Performance results for our implementation of the matching-based algorithm of Tajik et al.

Application	50 rounds	500 rounds	1000 rounds
10×10 image block ($n = 100, d = 255$)	0.06 s	0.39 s	0.77 s
16×16 image block ($n = 256, d = 255$)	0.11 s	0.98 s	1.98 s
32×32 image block ($n = 1024, d = 255$)	0.41 s	3.91 s	7.81 s
Exam scores ($n = 300, d = 100$)	0.12 s	1.14 s	2.26 s
Salaries ($n = 30, d = 100000$)	0.02 s	0.13 s	0.25 s
Ratings ($n = 5000, d = 4$)	1.84 s	18.5 s	37.8 s

Table 3. Performance results for our implementation of the ranking algorithms from the previous section. Encryption (Enc.) time is the time to do rank-encipher-unrank.

Application	Lexicographic			Recursive Block		
	Table size	Table time	Enc. time	Table size	Table time	Enc. time
10×10	162 MB	1.81 s	0.009 s	9 MB	0.87 s	0.06 s
16×16	1885 MB	13.1 s	0.013	32 MB	5.8 s	.18 s
32×32	Fail	–	–	271 MB	316 s	3.50 s
Exams	988 MB	7.17 s	0.011 s	15 MB	3.81 s	.096 s
Salaries	4756 MB	79 s	0.34 s	672 MB	40.5 s	4.2 s
Ratings	Fail	–	–	25 MB	271 s	0.40 s

5.3   Discussion

Looking at the results in Tables 2 and 3, we can come to a few conclusions.

For Thumbnail Encryption, Ranking is Likely the Better Choice. We can see from the tables that for the 10×10 and 16×16 image block encryption applications, the lexicographic ranking encryption time is faster than the matching-based solution with only 50 rounds. Even at the 32×32 block size, the recursive block ranking has faster encryption than 500 rounds of the matching-based solution. Since a large image contains thousands of such blocks, using the faster ranking-based solutions could be especially beneficial and well-worth the time and memory cost of generating and storing the table.

For Small n but Large d, the Matching-Based Algorithm is a Good Choice. From the salaries application $n = 30, d = 100000$ we can see that the matching-based algorithm performs well and, even at 1000 rounds, encrypts in $1/4$ of a second. The ranking solutions, on the other hand, are starting to hit their upper limits as far as table size. Specifically, the lexicographic ranking table took nearly 80 s to generate and ended up at 4756 MB in size. Even with this large table, encryption time for lexicographic ranking was still slightly slower than 1000 rounds of the matching-based algorithm. We also tried increasing n and d slightly to 50 and 250,000 and, as expected, the table size ended up being too large for our testing machine.

Ranking Solutions Should Scale Better if More Security is Desired. We have done the performance tests with NR (nonce-respecting) security in mind, which is consistent with previous work. However, if we instead want to target something like strong PRP security, then the ranking solutions should still be reasonably performant, while the matching solution will likely significantly slow down. The reason for this is that with the matching solution encryptions happen $n/2$ times (once for each pair of points) each round, and each of these, which are just additions mod N with NR security, would likely need to be replaced with a stronger cipher. With the ranking solutions, there is just a single cipher call that would need to be swapped for something stronger. One challenge with this, however, is that the numbers are very large with the ranking solutions (e.g., 2000 bit numbers in the 16×16 image example), so the inner cipher in rank-encipher-unrank would need to support such large sizes. A strong, variable-length cipher such as those in [16] may be a good choice. A closer look at these issues and whether NR security or strong PRP security is the right target for sum-preserving encryption applications would be interesting future work.

For larger n, New Constructions are Likely Necessary. We can see from the ratings application that as n gets large both the matching solution and the ranking solutions start to struggle. The matching solution with 1000 rounds takes over 30 s to encrypt the vector with 5000 ratings; if we instead wanted to encrypt a vector with 1 million ratings, we might estimate it to take 10 min or longer! With the ranking solutions, our lexicographic table generation already failed on the ratings application with $n = 5000$. The recursive block ranking, at first glance, appears to be an acceptable solution, but the table generation time appears to drastically increase as n grows. Already at $n = 5000$ the table took almost 5 min to generate, and in additional tests with $n = 10000$ the table was not finished after 30 min. Scaling this up to something like 1 million ratings is just not practical. A better solution for very large n is likely to be a combination of the matching and ranking solutions. The matching algorithm is already based on the idea of applying rank-encipher-encrypt to smaller subvectors (of size 2), so a natural extension is to apply our ranking algorithms to increase the size of the subvectors enciphered each round. It is not immediately clear to us whether our proof from Sect. 3 could be adapted to this situation, so we leave proving a mixing time bound on this combined algorithm to future work.

Acknowledgements. We thank the anonymous Asiacrypt 2022 reviewers for providing detailed comments and suggestions for improving the presentation of our results.

References

1. Abramson, M.: Restricted combinations and compositions. Fibonacci Quart. **14**, 439–452 (1976)
2. Bellare, M., Ristenpart, T., Rogaway, P., Stegers, T.: Format-preserving encryption. In: Jacobson, M.J., Rijmen, V., Safavi-Naini, R. (eds.) SAC 2009. LNCS, vol. 5867, pp. 295–312. Springer, Heidelberg (2009). https://doi.org/10.1007/978-3-642-05445-7_19

3. Boldyreva, A., Chenette, N., Lee, Y., O'Neill, A.: Order-preserving symmetric encryption. In: Joux, A. (ed.) EUROCRYPT 2009. LNCS, vol. 5479, pp. 224–241. Springer, Heidelberg (2009). https://doi.org/10.1007/978-3-642-01001-9_13

4. Brightwell, M., Smith, H.: Using datatype-preserving encryption to enhance data warehouse security. In: National Information Systems Security Conference (NISSC) (1997)

5. Cox, D.A., Little, J., O'Shea, D.: Ideals, Varieties, and Algorithms: An Introduction to Computational Algebraic Geometry and Commutative Algebra, 3rd edn. Springer (2010)

6. Dworkin, M.: Recommendation for block cipher modes of operation: Methods for format preserving-encryption. NIST Special Publication 800–38G. https://doi.org/10.6028/NIST.SP.800-38G(2016)

7. Dyer, M., Greenhill, C.: A more rapidly mixing Markov chain for graph colorings. Random Struct. Algorithms **13**, 285–317 (1998)

8. El Kahoui, M., Rakrak, S.: Structure of Grobner bases with respect to block orders. Math. Comput. 76, 2181–2187 (2007). https://doi.org/10.1090/S0025-5718-07-01972-2

9. Hoang, V.T., Rogaway, P.: On generalized Feistel networks. In: Rabin, T. (ed.) CRYPTO 2010. LNCS, vol. 6223, pp. 613–630. Springer, Heidelberg (2010). https://doi.org/10.1007/978-3-642-14623-7_33

10. Institute, A.N.S.: Financial services - symmetric key cryptography for the financial services industry - format-preserving encryption. ANSI X9.124 Standard (2020). https://webstore.ansi.org/standards/ascx9/ansix91242020

11. Levin, D.A., Peres, Y., Wilmer, E.L.: Markov Chains and Mixing Times. American Mathematical Society (2006)

12. Marohn, B., Wright, C.V., Feng, W., Rosulek, M., Bobba, R.B.: Approximate thumbnail preserving encryption. In: Multimedia Privacy and Security - MPS@CCS 2017, pp. 33–43. ACM (2017)

13. Opdyke, J.: A unified approach to algorithms generating unrestricted and restricted integer compositions and integer partitions. J. Math. Modelling Algorithms **9**, 53–97 (2010). https://doi.org/10.1007/s10852-009-9116-2

14. Page, D.R.: Generalized algorithm for restricted weak composition generation. J. Math. Modelling Algorithms Oper. Res. **12**(4), 345–372 (2012). https://doi.org/10.1007/s10852-012-9194-4

15. Riordan, J.: An Introduction to Combinatorial Analysis. Dover Publications, Inc. (1980). https://doi.org/10.1515/9781400854332

16. Shrimpton, T., Terashima, R.S.: A modular framework for building variable-input-length tweakable ciphers. In: Sako, K., Sarkar, P. (eds.) ASIACRYPT 2013. LNCS, vol. 8269, pp. 405–423. Springer, Heidelberg (2013). https://doi.org/10.1007/978-3-642-42033-7_21

17. Stein, T.: Uniform random samples for second-order restricted k-compositions. January 2020. http://essay.utwente.nl/80718/

18. Tajik, K., et al.: Balancing image privacy and usability with thumbnail-preserving encryption. In: NDSS 2019. The Internet Society, February 2019

19. Walsh, T.: Loop-free sequencing of bounded integer compositions. JCMCC. J. Combinat. Math. Combinat. Comput. **33**, 323–345 (2000)

20. Wright, C.V., Feng, W., Liu, F.: Thumbnail-preserving encryption for JPEG. In: ACM Workshop on Information Hiding and Multimedia Security - IH&MMSec 2015, pp. 141–146. ACM (2015)

Towards Case-Optimized Hybrid Homomorphic Encryption

Featuring the **Elisabeth** Stream Cipher

Orel Cosseron[1,2,3(✉)] , Clément Hoffmann[4] , Pierrick Méaux[5] ,
and François-Xavier Standaert[4]

[1] INRIA Lyon, Lyon, France
[2] ENS de Lyon, LIP, Lyon, France
[3] LTCI, Telecom Paris, Institut Polytechnique de Paris, Paris, France
orel.cosseron@ens-lyon.fr
[4] UCLouvain, ICTEAM/ELEN/Crypto Group, Louvain-la-Neuve, Belgium
[5] Luxembourg University, SnT, Luxembourg, Luxembourg

Abstract. Hybrid Homomorphic Encryption (HHE) reduces the amount of computation client-side and bandwidth usage in a Fully Homomorphic Encryption (FHE) framework. HHE requires the usage of specific symmetric schemes that can be evaluated homomorphically efficiently. In this paper, we introduce the paradigm of Group Filter Permutator (GFP) as a generalization of the Improved Filter Permutator paradigm introduced by Méaux *et al.*. From this paradigm, we specify Elisabeth , a family of stream cipher and give an instance: Elisabeth-4 . After asserting the security of this scheme, we provide a Rust implementation of it and ensure its performance is comparable to state-of-the-art HHE. The true strength of Elisabeth lies in the available operations server-side: while the best HHE applications were limited to a few multiplications server-side, we used data sent through Elisabeth-4 to homomorphically evaluate a neural network inference. Finally, we discuss the improvement and loss between the HHE and the FHE framework and give ideas to build more efficient schemes from the Elisabeth family.

1 Introduction

State-of-the-art. Hybrid Homomorphic Encryption (HHE) is a powerful solution to limit the performance overheads and ciphertext expansion that a direct application of Fully Homomorphic Encryption (FHE) on private data causes. Rather than directly encrypting private data homomorphically, its high-level idea is to encrypt a symmetric key with the (expensive) FHE scheme and send it to the server. Next, a client can send private data encrypted with the symmetric cipher to the server, which will homomorphically carry out symmetric decryption before performing the private computations [36].

O. Cosseron—Part of this work was done while the first author was working in Zama, Paris, France.

S. Agrawal and D. Lin (Eds.): ASIACRYPT 2022, LNCS 13793, pp. 32–67, 2022.
https://doi.org/10.1007/978-3-031-22969-5_2

While HHE can in principle be instantiated with any symmetric cipher, it was rapidly observed that standard ciphers like the AES may not be the best for this purpose [16,23]. Researchers therefore started to investigate the improved performances that specialized ciphers (e.g. with limited multiplicative complexity and/or depth) can lead to. Popular examples of such ciphers include LowMC [2,3], Kreyvium [5], FLIP & FiLIP [34,35] or RASTA & DASTA [19,29].

As a first step, these ciphers have been optimized towards making their homomorphic decryption as efficient and low-noise as possible, independent of the applications. Recent benchmarks like [20] exhibit that parallel ciphers (e.g. LowMC, RASTA, or DASTA) can reach slightly better throughputs than serial ciphers (e.g. Kreyvium, FLIP or FiLIP) at the cost of higher latency. Yet, while all these ciphers lead to practical performances for their homomorphic decryption, it remains that they are specified in $\mathbb{F}_2$ which, despite being convenient for symmetric cryptography, is not ideal when considering the global optimization goal of performing signal processing on encrypted data. For example, standard machine learning (e.g. classification) algorithms operate on reals or integers, which can then be quantized in $\mathbb{Z}_q$ with preferably large q's. Hence, applying HHE with ciphers specified in $\mathbb{F}_2$ requires expensive conversions from $\mathbb{F}_2$ to $\mathbb{Z}_q$.

Contributions. Building on this state-of-the-art, the general goal of this paper is to show that the optimization of HHE can benefit from considering the full toolchain going from the symmetric cipher to the data processing to perform homomorphically. For this purpose, we first aim to design a symmetric encryption scheme that not only allows efficient (stand-alone) homomorphic decryption but is also well suited for practically-relevant applications such as classification with machine learning algorithms. We next aim to put forward the tradeoff between the HHE performances and the constraints that it implies for machine learning classification.

We first generalize the filter permutator design approach used by the FLIP and FiLIP stream ciphers to arbitrary groups and instantiate this new construction by specifying the Elisabeth stream cipher family. Besides the goal of enabling homomorphic computations in $\mathbb{Z}_q$ without expensive conversions, this new design approach is driven by the recent work of Chillotti et al. [11], which generalizes the TFHE scheme from [10] in order to enable efficient Programmable BootStrapping (PBS). PBS is pushing for ciphers that combine (negacyclic) Look Up Tables (LUTs) and additions in $\mathbb{Z}_q$ with q a power of 2. We select $q = 2^4$ for our first instance of Elisabeth, hence denoted Elisabeth-4, by analyzing the current cost of a PBS as a function of the LUT sizes. Overall, Elisabeth-4 only requires two levels of PBS and a few additions in $\mathbb{Z}_q$, therefore enabling a very efficient homomorphic decryption. We study the security of the group filter permutator approach and our proposed instantiations. We also show that the (still stand-alone) performances of Elisabeth-4 compare well with (and sometimes improve) the aforementioned state-of-the-art ciphers optimized for FHE.

Next, we discuss the application of HHE to a classification based on a simple neural network. As a first step in this direction, we use the Fashion-MNIST dataset for this purpose [40]. It contains 28×28 grayscale images of 70,000 fashion products from 10 categories. This case study allows us to benchmark a full protocol, including the (one-time) key exchange, the homomorphic (symmetric) decryption, the homomorphic classification, and the result decryption. It confirms the interest in the HHE approach compared to the direct application of FHE and the interest of Elisabeth-4 over binary ciphers (in particular FLIP and FiLIP which share a similar design approach). It also emphasizes the tradeoff between the accuracy of the machine learning classification and the HHE performances.

We conclude the paper with a critical discussion of its limitations, which allows us to identify important directions for further research. In particular, we put forward (i) how improving TFHE to enable efficient PBS for larger LUTs could improve the accuracy of machine learning classifications with larger instances of Elisabeth, and (ii) how public preprocessing and other optimizations could improve the efficiency of homomorphic classification.

Related Works. The observation that non-binary ciphers could be beneficial to HHE has also been made in other recent works. In [27], the authors describe MASTA a variant of RASTA adapted to $\mathbb{Z}_q$ where $q = 2^{16} + 1$. In [13], the authors describe the cipher HERA which is defined over $\mathbb{Z}_q$ with $q > 2^{16}$. In [28], the authors describe the RUBATO cipher, adapted to $\mathbb{Z}_q$ where $q \approx 2^{25}$. In [20], the authors describe the family of ciphers PASTA which is defined over $\mathbb{F}_p$ with p a 16-bit prime. They also discuss how it can be combined with the homomorphic computation of a 5×5 matrix-vector multiplication. Their main difference with our proposal is the need for a larger modulus, which is not compatible with the efficient PBS of the generalized TFHE scheme we target. As a result, they are not ideal for application to our machine learning classification problem with this scheme.

We insist that we make no claim regarding the possibility to adapt HERA, PASTA, or other FHE-friendly ciphers to our case study. In particular, some specific choices of parameters such as the selection of $q = 2^4$, are motivated by current technological constraints (which also motivates the group generalization of filter permutators). Instances of Elisabeth with larger q's (discussed in conclusion) or HERA-like & PASTA-like ciphers with smaller moduli will admittedly look more and more similar, and it is an interesting long-term problem to determine which cipher to use for which range of applications. By contrast, our claim is that taking into account the data processing necessary to perform homomorphic operations when designing an FHE-friendly cipher can lead to improved global performances, as the comparison between Elisabeth-4 and FiLIP showcases. In this respect, we believe focusing on machine learning algorithms is a relevant direction since they typically raise important privacy concerns [37]. Yet, since the comparative performances of HHE schemes may be application-dependent, our results also suggest the identification of relevant case studies for benchmarking as an interesting topic of discussion for the FHE research community.

2   Background

General notations. We use $[n]$ to denote $\{1,\ldots,n\}$ and more generally $[a,b]$ for the set of integers c such that $a \leq c \leq b$. We will use the notation $\mathbb{G}$ to denote a group with operation $+$, 0 for its neutral element, and $-$ to denote the inverse. For a vector $a \in \mathbb{G}^n$ we denote $\mathsf{w_H}(a)$ its Hamming weight: $\mathsf{w_H}(a) = |\{a_i \neq 0, i \in [n]\}|$. For two vectors $a, b \in \mathbb{G}^n$ we denote $\mathsf{d_H}(a)$ their Hamming distance: $\mathsf{d_H}(a,b) = \mathsf{w_H}(a-b)$. log refers to the base-2 logarithm.

2.1   TFHE

By definition, a fully homomorphic encryption scheme can evaluate any arbitrary operation on its ciphertexts. While this is technically true, depending on the FHE scheme, some operations will be way easier to compute than others. The efficiency of an HHE scheme therefore deeply relies on how the symmetric cipher can use the efficient operations of the FHE scheme.

The Elisabeth stream cipher family has been conceived to take advantage of the efficient operations of the FHE scheme TFHE [10] and implemented using the operations available in the Concrete library [9].

In this section, we recall the core concepts of TFHE needed to understand our contributions, for a detailed explanation of TFHE operations we refer the reader to [9], from which we took most of the notations.

Encryption and Decryption. The encryption scheme at the core of TFHE[1] is based on Learning With Error (LWE). A variant relying on the General Learning With Error (GLWE) assumption introduced in [4] is used to perform some homomorphic operations. The generality of GLWE allows encompassing constructions relying on LWE [38], Ring LWE (RLWE) [32] or trade-offs between both under the same structure [31].

Let $\mathbb{B}$ be the set $\{0, 1\}$, $q = 2^p$, $\Delta = 2^\delta$ and $N = 2^d$ be three powers of two. We note $\mathbb{B}_N[X] := \mathbb{B}[X]/\langle X^N + 1\rangle$ and $\mathbb{Z}_{q,N}[X] := \mathbb{Z}_q[X]/\langle X^N + 1\rangle$.

Definition 1 (GLWE ciphertexts [4]). *An encryption $\boldsymbol{c}$ of a message $\mu \in \mathbb{Z}_{q/\Delta,N}[X]$ under secret key $\mathfrak{s} := (\mathfrak{s}_1(X),\ldots,\mathfrak{s}_k(X)) \in \mathbb{B}_N[X]^k$ is given by $\boldsymbol{c} \in (\mathbb{Z}_{q,N}[X])^{k+1}$ such that:*

$$\boldsymbol{c} := \left(a_1(X),\ldots,a_k(X), b(X) := \sum_{i=1}^{k} a_i(X) \cdot \mathfrak{s}_i(X) + \Delta \cdot \mu(X) + \varepsilon(X)\right)$$

where the coefficients of the a_i are uniformly sampled over $\mathbb{Z}_q$ and the ones of ε are sampled from a small discretized Gaussian noise.

---

[1] i.e. Regev's encryption scheme [38].

We call $\boldsymbol{a} := (a_1, \ldots, a_n)$ the **mask** and b the **body** of the ciphertext. Given a ciphertext $\boldsymbol{c}$ and the secret key $\boldsymbol{s}$, one decrypts $\boldsymbol{c}$ by computing $\mu^*(X) := b(X) - \sum_{i=1}^{k} a_i(X) \cdot s_i(X) = \Delta \cdot \mu(X) + \varepsilon(X)$.

By choosing Δ such that $\Delta > \|\varepsilon\|_\infty$, one can recover μ from μ^* by dividing each coefficient by Δ (which is equivalent to shifting each coefficient δ bits to the right). This means that the value of δ fixes the noise budget, that is the maximal number of bits one allows the coefficients of ε to be written on. So, for a given value of q, a bigger noise budget means a smaller message size.

In the following of the paper, we refer to LWE for a GLWE ciphertext with $N = 1$ and to RLWE for a GLWE ciphertext with $k = 1$. The LWE ciphertexts are the core of the TFHE scheme, whereas the other GLWE ciphertexts only play a role in the product and bootstrapping.

Addition and Multiplication. GLWE ciphertexts encrypted under the same secret key can be added coefficient-wise, the result of which gives an encryption of the sum of the plaintexts in $\mathbb{Z}_{q,N}[X]$. By iterating, it is then possible to multiply a ciphertext by a plain scalar. Note that, in both cases, the noises increase, resulting in linear growth of the noise inside the ciphertext.

To perform the multiplication of a GLWE ciphertext by an encrypted (polynomial) constant while managing the noise growth, TFHE uses another kind of ciphertext. These ciphertexts are the generalization of the GSW scheme [24] corresponding to GLWE, hence dubbed General GSW ciphertexts. Then, an external product between GGSW and GLWE ciphertexts allows the multiplication between encrypted plaintexts.

Definition 2 (GGSW Ciphertexts). *Let $B = 2^b \in \mathbb{N}$ and $\ell \in \mathbb{N}$. A GGSW ciphertext $\mathbf{C}$ of a message $m \in \mathbb{Z}_{q,N}[X]$ under secret key $\boldsymbol{s} := (s_1, \ldots, s_k) \in \mathbb{B}_N[X]^k$ is defined as:*

$$\mathbf{C} = \left(\mathrm{GLWE}_{\boldsymbol{s}}\left(-s_i \tfrac{q}{B^j} m\right)\right)_{(i,j) \in [k+1] \times [\ell]} \in \left(\mathbb{Z}_{q,N}[X]\right)^{\ell(k+1) \times (k+1)},$$

with $s_{k+1} = -1$.

We call B the basis and ℓ the number of levels of the ciphertext $\mathbf{C}$. A GSW (resp. RGSW) ciphertext is a GGSW ciphertext made from LWE (resp. RLWE).

External Product. The external product between GGSW and GLWE ciphertexts gives a GLWE ciphertext of the product of the input plaintexts. The external product of a GGSW ciphertext $\mathbf{C}$ and a GLWE ciphertext $\boldsymbol{c}$ is defined as $\mathbf{C} \boxdot \boldsymbol{c} = \mathrm{Decomp}(\boldsymbol{c}) \cdot \mathbf{C}$, where Decomp is a transformation that flattens a vector of $k + 1$ polynomials into a vector of $\ell(k + 1)$ polynomials with coefficients in $[-B/2; B/2]$.

Programmable Boostrapping. Most of the homomorphic operations make the noise grow. In order to achieve fully homomorphic encryption, it is necessary to have an operation to decrease the noise. This is what bootstrapping is used for, by resetting the noise to a level independent of the noise level in the input ciphertext. In addition to noise control, TFHE's bootstrapping on LWE ciphertexts, also noted PBS for Programmable BootStrapping, allows the evaluation of a negacyclic look-up table L. Instead of giving a ciphertext of μ, it directly gives a bootstrapped ciphertext of $L[\bar{\mu}]$, where $\bar{\mu}$ is a noisy rescaling of μ on $[0, 2N - 1]$ and N is determined by the GLWE scheme used. More information on PBS can be found in Appendix A.

Definition 3 (Negacyclic Look-Up Table (NLUT)). *A negacyclic look-up table over a group $\mathbb{G}$ is a look-up table L of length $2N$ that verifies the following property:*

$$\forall i \in [0, N - 1], L[i + N] = -L[i] \in \mathbb{G},$$

where $=$ and $-$ denote equivalence and inverse defined on $\mathbb{G}$.

Keyswitch. The ciphertext output by the bootstrapping might be encrypted under a different, possibly larger, key than the input ciphertext. The keyswitching operation allows to convert a ciphertext encrypted under s_{in} into a ciphertext encrypted under s_{out} thanks to a keyswitching key $\mathbf{K}$. $\mathbf{K}$ is a GGSW encryption of 1, up to two details:

1. The last ℓ rows of the matrix are cropped;
2. The ciphertexts composing each line are encrypted under s_{out} rather than s_{in}.

The keyswitching of $c = (a, b)$ is then computed as $(0, b) - \text{Decomp}(a) \cdot \mathbf{K}$. It allows switching back to the scheme with the first key.

2.2   Boolean Functions and Cryptographic Criteria

In this part, we introduce Boolean functions and relevant cryptographic criteria (taken from [6]) that we will use to study the security of the main stream-cipher construction we propose in this article. Then we generalize the classic cryptographic criteria on Boolean functions to functions from $\mathbb{G}^n$ to $\mathbb{G}$ where $\mathbb{G}$ is a group.

Definition 4 ((Vectorial) Boolean function). *An (n, m) vectorial Boolean function F is a function from $\mathbb{F}_2^n$ to $\mathbb{F}_2^m$. When $m = 1$ the $(n, 1)$ vectorial Boolean function is simply referred to as a Boolean function and we denote the space of n-variable Boolean function as $\mathcal{B}_n$. We call coordinate functions of F the m Boolean functions f_i associating for each $x \in \mathbb{F}_2^n$ the i-th binary output of $F(x)$. We call component functions of F the $2^m - 1$ non-trivial linear combinations of the coordinate functions of F.*

Definition 5 (Algebraic Normal Form (ANF) and degree). *We call Algebraic Normal Form of a Boolean function f its n-variable polynomial representation over* $\mathbb{F}_2$ *(i.e. belonging to* $\mathbb{F}_2[x_1,\ldots,x_n]/(x_1^2 + x_1,\ldots,x_n^2 + x_n)$*):*

$$f(x) = \sum_{I \subseteq [n]} a_I \left(\prod_{i \in I} x_i \right) = \sum_{I \subseteq [n]} a_I x^I,$$

where $a_I \in \mathbb{F}_2$. *The algebraic degree of f equals the global degree of its ANF:* $\deg(f) = \max_{\{I \mid a_I = 1\}} |I|$ *(with the convention that* $\deg(0) = -\infty$*).*

Definition 6 (Algebraic Immunity). *The algebraic immunity of a Boolean function* $f \in \mathcal{B}_n$, *denoted as* $\mathsf{AI}(f)$, *is defined as:*

$$\mathsf{AI}(f) = \min_{g \neq 0}\{\deg(g) \mid fg = 0 \text{ or } (f \oplus 1)g = 0\},$$

where $\deg(g)$ *is the algebraic degree of g. The function g is called an annihilator of f (or* $f \oplus 1$*).*

Definition 7 (Balancedness and Resiliency). *A Boolean function* $f \in \mathcal{B}_n$ *is said to be balanced if its output is uniformly distributed, that is:* $|\{x \mid f(x) = 0\}| = |\{x \mid f(x) = 1\}|$.

A Boolean function f is called be m-resilient if any of its restrictions obtained by fixing at most m of its coordinates is balanced. We denote by $\mathsf{res}(f)$ *the maximum resiliency (also called resiliency order) m of f and set* $\mathsf{res}(f) = -1$ *if f is unbalanced.*

Definition 8 (Nonlinearity). *The nonlinearity* NL *of an n-variable Boolean function f, where n is a positive integer, is the minimum Hamming distance between f and all the affine functions in* $\mathcal{B}_n$:

$$\mathsf{NL}(f) = \min_{g,\,\deg(g) \leq 1} \{d_H(f,g)\},$$

with $d_H(f,g) = \#\{x \in \mathbb{F}_2^n \mid f(x) \neq g(x)\}$ *the Hamming distance between f and g; and with* $g(x) = a \cdot x + \varepsilon$, $a \in \mathbb{F}_2^n, \varepsilon \in \mathbb{F}_2$ *(where* $\cdot$ *is some inner product in* $\mathbb{F}_2^n$; *any choice of an inner product will give the same definition).*

Additionally, we denote NL^d *the order-d nonlinearity of f, the minimum Hamming distance between f, and all the functions of degree at most d.*

Definition 9 (Direct Sum). *Let f be a Boolean function of n variables and g a Boolean function of m variables, f and g depending on distinct variables, the direct sum h of f and g is defined by:*

$$h(x,y) = \mathsf{DS}(f,g) = f(x) \oplus g(y), \quad \text{where } x \in \mathbb{F}_2^n \text{ and } y \in \mathbb{F}_2^m.$$

Additionally we denote $\mathsf{DS}^t(f)$ *when is the direct sum of t times the function f.*

Extending to Groups. The notions defined above on Boolean functions can easily be extended to functions from $\mathbb{G}^n$ to $\mathbb{G}$, we denote these extended notions by the subscript $\mathbb{G}$.

Definition 10 (Cryptographic criteria over $\mathbb{G}$). *For a function f from $\mathbb{G}^n$ to $\mathbb{G}$ we denote:*

- $\deg_{\mathbb{G}}(f)$ *the degree over* $\mathbb{G}$. *It corresponds to the minimum degree over the polynomial representations of f in the polynomial ring $(\mathbb{G}, \cdot)[x_1, \ldots, x_n]$ when such representations exist.*
- $\mathsf{res}_{\mathbb{G}}(f)$ *the resiliency order over* $\mathbb{G}$. *f is balanced if and only if:*

$$\forall a \in \mathbb{G} : |\{x \mid f(x) = a\}| = |\mathbb{G}|^{n-1}.$$

  f is m-resilient if all the sub-functions obtained by fixing up to m variables are balanced.
- $\mathsf{NL}_{\mathbb{G}}(f)$ *the nonlinearity over* $\mathbb{G}$. *The nonlinearity is taken as the minimum Hamming distance between f and the affine functions: $a_0 + \sum_{i=1}^{n} a_i x_i$ where the a_i describe $\mathbb{G}^{n+1}$. Additionally, $\mathsf{NL}_{\mathbb{G}}^d(f)$ denotes the order-d nonlinearity over $\mathbb{G}$.*

We also denote $\mathsf{DS}_{\mathbb{G}}(f, g)$ the direct sum $f(x) + g(y)$ where $x \in \mathbb{G}^n$ and $y \in \mathbb{G}^m$ and f and g are two n-variable and m-variable functions defined on distinct variables.

3   Group Filter Permutator and Elisabeth-4

The improved filter permutator (IFP) paradigm [34] led to binary stream ciphers efficient for HHE [30,34,35] where the homomorphic evaluation (of the stream cipher decryption) is reduced to the evaluation of a unique filtering function. Its connection with Goldreich's pseudorandom generator [25], the different security analyses [15,34,35] and cryptanalysis studies [8,21], built confidence in the soundness of this design. They make it a natural starting point to design HHE with a non-binary symmetric scheme.

Section 3.1 introduces the group filter permutator which generalizes the IFP paradigm over $\mathbb{F}_2$ to any group. We then move towards the instantiation of group filter permutators. For this purpose, we observe that concrete choices of parameters are driven by technological constraints (e.g., operations that are efficient in TFHE (C)). Section 3.2 discusses these choices and explains how future advances in the Concreteimplementation could benefit the stream cipher. The discussion leads to the design of the stream cipher family Elisabeth and an instantiation in Sect. 3.3 (full algorithm are detailed in Appendix B.2). Finally, the performances of the transciphering alone are provided in Sect. 3.4.

3.1 Design and Stream Cipher Families

The group filter permutator is defined by a group $\mathbb{G}$ with operation noted $+$, a forward secure PRNG, a key size N, a subset size n, and a filtering function f from $\mathbb{G}^n$ to $\mathbb{G}$. To encrypt m elements of $\mathbb{G}$ under a secret key $K \in \mathbb{G}^N$, the public parameters of the PRNG are chosen and then the following process is executed for each key-stream s_i (for $i \in [m]$):

- The PRNG is updated, its output determines a subset, a permutation, and a length-n vector of $\mathbb{G}$.
- the subset S_i is chosen, as a subset of n elements over N,
- the permutation P_i (a re-ordering) from n to n elements is chosen,
- the vector, called whitening and denoted by w_i, from $\mathbb{G}^n$ is chosen,
- the key-stream element s_i is computed as $s_i = f(P_i(S_i(K)) + w_i)$, where $+$ denotes the element-wise addition of $\mathbb{G}$.

The GFP, depicted in Fig. 1, is a generalization of the improved filter permutator [34] where $\mathbb{G} = \mathbb{F}_2$. The XOR is replaced by the addition of $\mathbb{G}$ and the Boolean function by a function from $\mathbb{G}^n$ to $\mathbb{G}$.

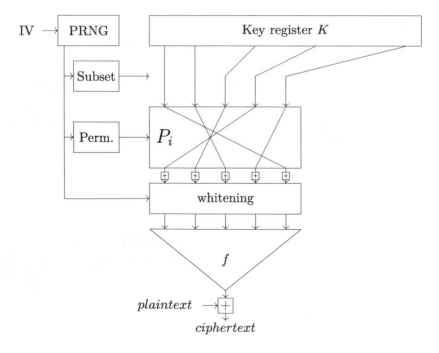

Fig. 1. The group filter permutator design

3.2   Look-Up Tables Specifications

In order to be efficiently evaluated homomorphically, Elisabeth filters should rely on the most native TFHE homomorphic functions, which are modular additions and NLUTs (see Definition 3) evaluation. The use of NLUTs leads to some non-standard constraints for the design of Elisabeth. In particular, they are entirely determined by the first half of their values and cannot be bijective. Indeed, 0 can only have an even number of antecedents: given a 2^n size NLUT L, for each $i \in [0, 2^{n-1} - 1]$ such that $L[i] = 0$, we also have $L[i + 2^{n-1}] = -L[i] = 0$. As will be shown next, those peculiarities do not raise fundamental problems for the security of the cipher and it is possible to reach the necessary cryptographic criteria to ensure 128 bits of security with a cipher based on NLUTs and modular additions.

On the choice of the NLUTs size. The choice of the size of the NLUTs is driven by several conflicting arguments: on the one hand, smaller NLUTs increase the amount of NLUTs that needs to be evaluated to reach a given level of security, which therefore slows down the evaluation of the circuit. On the other hand, bigger NLUTs lead to heavier computation: indeed, the NLUTs are homomorphically computed through a PBS, which is essentially a sequence of polynomial multiplications. The bigger the NLUT, the bigger the polynomials involved in the PBS. Since the cost of a polynomial multiplication is of $\mathcal{O}(N \log N)$, where N is the polynomial degree, there is a clear interest in taking smaller NLUTs. That being said, Elisabeth's final goal is to ensure the evaluation of neural networks over the output of the transciphering. This goal is more easily achieved if the output ciphertext contains bigger chunks of data. To find out where the trade-off between smaller polynomials and bigger chunks of data lies, we designed Table 1 which gives the minimal size of polynomial needed to evaluate a NLUT of given size for different LWE dimensions in TFHE. These values are hard minimums computed before any specific choice of design for the encryption scheme, i.e. if we wanted to evaluate a NLUT on a ciphertext that contains just enough noise to ensure 128 bits of security. Every operation in the scheme performed before a PBS could increase the noise in the ciphertext up to a point where bigger polynomials are needed for the same NLUT size. Considering all the previous arguments, we found that NLUTs of 4 bits would be the best trade-off as it relaxes as much as possible the constraint on polynomial sizes while keeping enough bits of message in a single ciphertext. We chose not to use bigger messages for several reasons: first, this guarantees that once the specifics of our design have been set, the polynomial degree should remain small. Second, this ensures that Elisabeth-4 will require a smaller noise budget to be evaluated than any computation that could be performed afterward. By doing so, one ensures that transciphering will not be the noisiest part of the evaluation server-side, and thus can be computed with exactly the same parameters as the following circuit. In a nutshell, in the vast majority of usecases, adding Elisabeth-4 before the evaluation of a circuit will not require new parameters to be computed. With the current limitations of TFHE, using a bigger length for the NLUTs could not only multiply the transciphering time by more than two but also slow down the whole evaluation server-side. Nevertheless, the generality of the design enables it to adapt easily to powers of two of the same order once future improvement of TFHE regarding big(ger) integer computation would make it worthwhile to use bigger NLUTs.

Table 1. Polynomial degrees required to evaluate NLUTs containing bigger elements, in function of the LWE dimension. The LWE dimension is the number of elements in the mask of a ciphertext. The NLUT length is both the number of elements in the NLUT and the size of the group $\mathbb{G}$ over which the NLUT is defined.

LWE dimension NLUT length	512	630	650	688	710	750	800	830	1024	2048	4096
16	512	512	512	512	512	512	512	512	512	1024	1024
32	1024	1024	1024	1024	1024	1024	1024	1024	1024	2048	2048
64	2048	2048	2048	2048	2048	2048	2048	2048	2048	4096	4096
128	4096	4096	4096	4096	4096	4096	4096	4096	4096	8192	8192
256	8192	8192	8192	8192	8192	8192	8192	8192	8192	16384	16384

3.3   Elisabeth-4

In order to benefit from Concrete [9] efficient operations, the design of Elisabeth is a GFP defined over $\mathbb{G} = \mathbb{Z}_{2^\ell}$ where the filtering function f is composed by additions over $\mathbb{G} = \mathbb{Z}_{2^\ell}$ and NLUTs that could be evaluated in parallel. A simple way to follow these guidelines and allow a simpler security analysis is to make the filter f rely on the direct sum construction: f is the addition (over $\mathbb{G}$) of functions acting on independent variables. To be even simpler, each sub-function of f can be chosen to be the same function, which gives a highly parallelizable design. Accordingly, splitting n in t parts of m (different) variables, Elisabeth's filters have the following shape:

$$f(x_1, \ldots, x_n) = \mathsf{DS}_{\mathbb{G}}^t(g(x_1, \ldots, x_m)).$$

The general idea of Elisabeth 's filters design is to use the fewest levels possible of NLUT, in order to parallelize computation efficiently.

At the moment, as explained in Sect. 3.2, Concrete's implementation leads to fix $\mathbb{G} = \mathbb{Z}_{16}$. Only one level of NLUTs of this size would not guarantee a secure scheme for 128-bit security[2]. Hence, for this NLUT size we propose a filter with two levels of NLUTs, following the 5-to-1 construction depicted in Fig. 2. The

---

[2] Based on the analysis in Sect. 4, an adversary could retrieve the key by solving an algebraic system of degree at most 4 over $\mathbb{F}_2$.

high-level idea of this construction is to mix 4 variables to get strong enough algebraic properties and to use the fifth one to balance the output.

It gives the following instantiation of Elisabeth-4: $\mathbb{G} = \mathbb{Z}_{16}$, $N = 256$, $n = 60$, $t = 12$, $m = 5$, and g is the 5-to-1 function. As will be shown in Sect. 4, a random choice of NLUTs has a high probability to lead to secure instances of Elisabeth. Therefore, and to mitigate the possibility of trapdoor the design by targeting particular relations over $\mathbb{G}$, we selected the 8 NLUTs required for g by hashing the sentence "Welcome to Elisabeth, heir of FiLIP!", using the following process[3]:

- The sentence is encoded in UTF-8 and the 256 bits hash is computed using SHA-2. Those 256 bits are then split into 8 blocks of 32 bits. Each NLUT S_i is determined using the i-th block.
- The 32 bits are themselves split in 8 blocks of 4 bits, which directly gives the 8-th first coefficients of the size-16 NLUT, which determines entirely the NLUT.

All the NLUTs obtained with this process are explicitly given in Appendix B.1. We analyze this particular choice of NLUTs in Sect. 4.

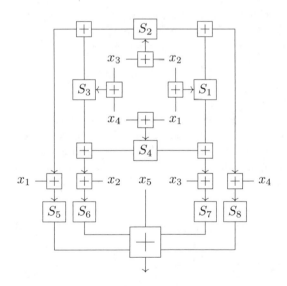

Fig. 2. Elisabeth-4 's 5 to 1 inner function.

[3] A python implementation of this protocol can be found here: https://github. com/princess-elisabeth/sboxes_generation.

3.4 Implementation and Performances

Implementation. We implemented Elisabeth-4 over TFHE's library Concrete [9][4]. The symmetric key used by Elisabeth-4 is encrypted under an LWE key s_{in} of length n, and the PBSs are performed under an RLWE key of length k and a polynomial degree N. Since a side effect of the PBS is to switch the secret key under which a message is encrypted from s_{in} to s_{out}, a KeySwitch is needed after each PBS[5]. This KeySwitch can either happen right after the PBS, or after several operations. Since a KeySwitch is a rather noisy operation, the preferred approach was to perform it as close to the next PBS as possible. KeySwitches from s_{out} to s_{in} were thus computed after the sum of the output of 2 NLUTs and before an input ciphertext was added to this sum. The reason why this input ciphertext is summed after rather than before the KeySwitch is that it is encrypted under the secret key s_{in}. For the same reason, a KeySwitch is needed at the final step of the inner function, where we sum the fresh ciphertext of x_5 with the rest of the inner function. This KeySwitch can either be performed on x_5, from s_{in} to s_{out} with another KeySwitching key, or on every other part of the final sum, with the same KeySwitching key than earlier. The first approach allows faster computation, since this KeySwitch can be performed in parallel to the rest of the evaluation of the inner function, but increases the amount of data sent to the server before transciphering and has for side-effect that the output ciphertext and the symmetric key are not encrypted under the same LWE key. See Fig. 3 for an illustration of the KeySwitches integration inside Elisabeth-4's circuit.

Parameters. Depending on the positioning of the KeySwitches, two sets of homomorphic parameters have been chosen for Elisabeth-4, specified in Table 2. These parameters have been selected to ensure 128 bits of security and 4 bits of message integrity while optimizing evaluation time and key sizes. The integrity of the message has been controlled using the noise formulas given in [12] and the security is estimated based on the LWE security estimator [1].

Table 2. Set of recommended FHE parameters for Elisabeth-4.

TFHE Parameters											
Mode	n	k	N	$\log(\sigma_{LWE})$	$\log(\sigma_{GLWE})$	PBS		KeySwitch		inverse KeySwitch	
						$\log(B)$	ℓ	$\log(B)$	ℓ	$\log(B)$	ℓ
two KS	784	3	512	−18.6658	−38.4997	19	1	6	2	19	1
single KS	863			−20.7494				7		-	-

---

[4] https://www.github.com/princess-elisabeth/Elisabeth.
[5] Or, more precisely, between two PBS.

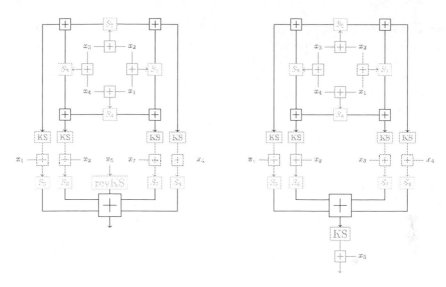

(a) Two KeySwitch keys.   (b) Single KeySwitch key.

Fig. 3. Elisabeth-4's homomorphic circuit. A gray edge represents operations under secret key s_{in} while black edges represent operations under s_{out}. Yellow boxes are operations that take inputs encrypted under s_{in} and output ciphertexts encrypted under s_{out} and orange boxes represent the opposite. (Color figure online)

Benchmark. Since Elisabeth-4 has been designed for server usage, it has been tested on a 64-core computer equipped with AMD Ryzen Threadripper 3990X 64-Core Processor[6]. Results are in Table 3.

Table 3. Comparison of running time of different versions of Elisabeth-4.

Timings		
Cipher	Time per ciphertext (ms)	Time per bit (ms)
Elisabeth-4 (two KS)	91.143	22.786
Elisabeth-4 (single KS)	103.810	25.953
Elisabeth-4 (two KS, monothreaded)	1485.0	371.25
Elisabeth-4 (single KS, monothreaded)	1648.6	412.15

As Elisabeth-4 has been designed for TFHE, it will mainly be compared to FiLIP, which has had its own specific implementation for TFHE [30]. In order to make our comparisons more relevant, FiLIP has been reimplemented over Concrete[7], with the same parameters as described in [30]. For completeness, we

---

[6] Note that this particular design is initially optimized for 48 cores.

[7] https://github.com/princess-elisabeth/FiLIP.

also provide a rough comparison of Elisabeth-4 with other state of the art encryption schemes for transciphering, namely LowMC [2], Kreyvium [5], RASTA [19], FiLIP [34], DASTA [29], MASTA [27] and PASTA [20], all of which have been benchmarked with the tools provided by [20], and HERA [14], on the same machine used for benchmarking Elisabeth-4. Table 4 collects the results of these benchmarks.

Note that since these symmetric schemes have been designed with different FHE cryptosystems in mind, comparisons between their performances must be considered with care. For example, some homomorphic schemes are very efficient at packing, that is storing many cleartexts inside one ciphertext, but depending on the usecase following the transciphering, unpacking these cleartexts can be pretty costly. From a high-level perspective, these results confirm that ciphers that are more serial are also more interesting for latency, whereas the ones more parallel are better suited for throughput. But they are not indicative of any real-world performance in a specific usecase, which motivates the investigations of Sect. 5.

Table 4. Comparison of running time of different 128-bit security HHE, using either TFHE (TFHE (C)or Concrete), BFV (HELib) or CKKS (CKKS). The parameters used are the ones recommended by their original papers for 128 bits of security.

Cipher	Homomorphic library	Time per ciphertext (s)	Time per bit (ms)
LowMC	TFHE (C)	4283.678	16733
Kreyvium	TFHE (C)	208.255	208255
RASTA 6	TFHE (C)	2424.503	6907
FiLIP 144	Concrete	0.134	134
FiLIP 1216	Concrete	0.586	586
FiLIP 1280	Concrete	0.627	627
DASTA 6	TFHE (C)	2387.674	6802
Elisabeth-4 (two KS)	Concrete	0.091	22.75
Elisabeth-4 (single KS)	Concrete	0.104	26
LowMC	HELib	853.302	3333.21
Kreyvium	HELib	8.222	8222
RASTA 6	HELib	163.131	464.76
DASTA 6	HELib	156.935	447.11
MASTA 5	HELib	22.096	20.31
PASTA 4	HELib	9.827	18.06
HERA	CKKS	14.747	0.01

Note also that the seemingly less efficient performances of some cryptosystems used in combination with TFHE (C)are explained by a trade-off on the homomorphically encrypted key size: in order to make the ciphertexts used smaller, most of their operations are performed through gate bootstrapping, which significantly slow down the computation. FiLIP, which does not rely on this approach, runs much faster but needs way bigger keys. Table 5 gives a comparison of these sizes with Elisabeth-4, when transciphering towards TFHE (C). In the case of Elisabeth-4, we also included in this value the size of the KeySwitch and PBS keys needed to evaluate the transciphering algorithm. Data that would have been sent even without transciphering are not taken into account. Also note that rather than sending or storing a full GLWE or GGSW ciphertext, it is always possible to send only the body part of the ciphertext along with the seed used to generate its mask. When several ordered ciphertexts are sent, a single seed can be provided for all of them. It is supposed in Table 5 that such compressions are used whenever possible.

Table 5. Comparison of public key sizes of FiLIP and Elisabeth-4 when used in combination with TFHE.

Cipher	Key sizes (kB)
FiLIP 144	1610613
FiLIP 1216	402653
FiLIP 1280	1610613
Elisabeth-4 (two KS)	20
Elisabeth-4 (single KS)	8

Tables 4 and 5 show that Elisabeth-4 gives a reduction of the data overhead compared to FiLIP by a factor 20000 to 200000, for a speedup of a factor 5 to 27.5 per bit. Moreover, Elisabeth-4 outputs are 4-bits integer, which FiLIP could only achieve through a costly series of PBS. Indeed, to reconstruct a multi-bit message, every bit but the most significant one transciphered by FiLIP must be shifted to the right as needed before being summed together. Each shift being performed through a PBS, a 4-bits message would thus cost an additional 3 PBS to be transciphered. The capacity to output multibit messages will prove useful in actual usecases.

4 Security Analysis

4.1 Security of the Group Filter Permutator

The group filter permutator generalizes the IFP paradigm over $\mathbb{F}_2$ [34] to any group. Accordingly, for the security analysis, we consider how the attacks known on IFP apply in the general case, and we discuss new attacks arising from the group. The security of IFP is discussed in [34], it is built on top of former analyses on filter permutator [35] (FP) and the connection with Goldreich's pseudorandom generator [25]. The main differences between the FP paradigm and the IFP paradigm are the addition of a subset selection before the permutation and a whitening. These modifications allow to mitigate the impact of guess and determine attacks on FP [21] and restricted inputs cryptanalysis [8]. The security analyses on IFP consider that no additional weakness arises from the PRNG which is chosen forward secure to avoid malleability. The subsets, permutations, and whitening are derived with no bias, hence the pseudorandom system obtained is not chosen in an adversarial way and the attacks applying on IFP are the ones targeting the filtering functions. These attacks are adaptations from the one applying on the filtered register, regrouped in [34] as algebraic-like attacks, correlation-like attacks, and guess and determine strategies.

The complexity of the different attacks on IFP is derived from parameters of the filtering function f. In the IFP case, f is a Boolean function and the significant cryptographic criteria of Boolean functions have been studied over the last decades for filtered registers. We argue that if the GFP is defined over a field, such as $\mathbb{F}_q$, the natural generalizations of such Boolean cryptographic criteria over $\mathbb{F}_q$ would give a strong basis for the security analysis. Since GFP is defined over $\mathbb{G}$ we analyze the security based on the properties over $\mathbb{G}$ and over a field allowing a polynomial representation of f. At a high level since the "$+$" operation is linear in $\mathbb{G}$ the first view angle aims at verifying the filter breaks the linearity enough. The goal of the second perspective (over $\mathbb{G}'$) is to prevent attacks arising from a cryptographic weakness of the function f in another representation easy to obtain for an adversary. We consider the attacks in the single-key setting, in the known plaintext-ciphertext pairs model, focusing on key-recovery attacks. We make the standard assumption that an adversary is limited by a data complexity of $2^{\frac{\lambda}{2}} = 2^{64}$.

For Elisabeth-4 since $\mathbb{G} = \mathbb{Z}_{16}$ we study the properties of the filter f as a function from $\mathbb{G}^n$ to $\mathbb{G}$. Fixing the representation of each element of $\mathbb{G}$ as an element of $\mathbb{G}' = \mathbb{F}_2^4$ based on its binary decomposition fixes a vectorial Boolean function F from $\mathbb{F}_2^{4n}$ to $\mathbb{F}_2^4$. We study F by analyzing the properties of its 15 non-null component Boolean functions, the 15 functions from $\mathbb{F}_2^{4n}$ to $\mathbb{F}_2$.

Algebraic Attacks. The general principle of algebraic attacks is to consider an algebraic system of equations derived for the key-stream: an attacker knowing plaintext-ciphertext pairs can rewrite key-stream outputs as multivariate polynomials in the secret key elements, and solve the system to recover the key. Various techniques can be used to solve such an algebraic system, as simple as linearization (considering each monomial as a new variable) followed by Gaussian elimination and as complex as approaches using Grobner bases (such as [22]). For the security estimation, we will use the approaches using linearization, as they allow to have a good estimation of the attack complexity from the algebraic degree of the system and the number of variables, whereas the complexity of more complex attacks is often difficult to assess or interpolate from toy examples. For a system of equations of degree d with N unknowns the attack has complexity $\mathcal{O}(D)^\omega$, where D is the number of monomials of degree lower or equal to d, and ω is the exponent from the Gaussian elimination (fixed to $\log(7)$ for our estimations).

Analyzing the security of Elisabeth-4 over $\mathbb{G} = \mathbb{Z}_{16}$, an adversary could target a system of equations in the polynomial ring $K[x_1, \ldots, x_N] = (\mathbb{G}, \cdot)[x_1, \ldots, x_N]$ where $\cdot$ denotes the usual product in $\mathbb{G}$. In this case where $\mathbb{G} = \mathbb{Z}_{16}$, K is a ring but not a field, hence not all the functions from K^n to K have a polynomial representation. When f does not correspond to a polynomial over $K[x_1, \ldots, x_N]$

we consider no algebraic attack applies on the representation over $\mathbb{G}$. Considering $K[x]$, since for all $x \in \mathbb{G}$ it holds $x^8 = x^4$, any polynomial in the univariate representation is equivalent to a polynomial of degree at most 7, hence there are at most 16^8 polynomial evaluations. In comparison, there are 16^{16} evaluations from $\mathbb{G}$ to $\mathbb{G}$. Accordingly, we assume that with high probability the function f used for the filter does not admit a polynomial representation over $K[x_1, \ldots, x_n]$, which avoids this attack, and experimentally we verify that the function does not correspond to a low degree polynomial.

Over $\mathbb{F}_2$ more specific attacks have been exhibited. The algebraic attack of Courtois-Meier [18] on filtered LFSR applies to the GFP: instead of considering a system of equations given by a filtering function f (in our case one of the non-null component functions of F) the attacker finds low-degree functions g and h such that $fg = h$ and derive a system of degree $\mathsf{AI}(f) = \max\left(\deg(g), \deg(h)\right)$ which has the same solutions as the initial system. Unlike the algebraic degree that can be up to n for an n-variable Boolean function, it has been shown in [18] that the algebraic immunity (AI) is at most $\lceil (n+1)/2 \rceil$. Targeting this algebraic system of lower degree the corresponding algebraic-attack has a complexity of $\mathcal{O}(D)^\omega$ where $D = \sum_{i=1}^{\mathsf{AI}(f)} \binom{N}{i}$. The fast algebraic attack [17] is a variant of the previous attack where the adversary considers functions g and h such that $fg = h$ as before with g and h still of low degree but with h of degree potentially higher than $\mathsf{AI}(f)$. This attack can be more efficient than the previous one when the relations between the consecutive key-stream equations allow eliminating the monomials of degree between $\deg(g)$ and $\deg(h)$ in the system. In the case of filtered LFSR, the linear relation given by the LFSR allows to eliminate the monomials of degree larger than $\deg(g)$ using the Berlekamp-Massey algorithm, hence it gives a better attack than the standard algebraic one. The permutations used in the GFP do not provide such linear relations for consecutive key-stream equations, hence the elimination step would be more costly in this context. On a filtered LFSR the security estimation is $\mathcal{O}(D \log^2(D) + ED \log(D) + E^\omega)$ where $D = \sum_{i=1}^{\deg(h)} \binom{N}{i}$ and $E = \sum_{i=1}^{\deg(g)} \binom{N}{i}$. This estimation will be used as an indicator rather than a sharp limit, considering that the complexity of the best attack of the algebraic kind would lie between this (too low) bound and the (too high) one given by the algebraic attack of Courtois-Meier.

Correlation Attacks. Two kinds of correlation attacks are considered on IFP, the first one uses the bias of the output (or a sub-part of the output) from uniform to mount a distinguishing attack, the second one consists in approximating the key-stream equations by low-degree equations and in retrieving the key by solving a noisy linear system such as a Learning Parity with Noise (LPN) instance. Two main criteria permit to estimate the impact of such attacks. The first one is the number of key elements to fix before having a non-uniform output, often referred to as resiliency order for Boolean functions. Generalizing the concept to any group $\mathbb{G}$, a resiliency order of k means that the output of a function from $\mathbb{G}^n$ to $\mathbb{G}$ remains uniformly distributed in $\mathbb{G}$ even fixing up to k variables, and there exists an affectation of $k+1$ variables such that the output is not uniform. The second criterion is the quality of the approximation (equivalently the importance of the bias), it measures the Hamming distance between f and a low-degree function. In particular, the distance to affine functions, the nonlinearity, often leads to the best attacks in this context. For l the best affine approximation of f, the nonlinearity $\mathsf{NL}_\mathbb{G}$ corresponds to $\mathsf{d}_\mathsf{H}(f, l)$ and an adversary can consider the system of equation given by f as a noisy linear system given by l with noise parameter $\mathsf{NL}_\mathbb{G}/(|\mathbb{G}|^n)$. Since affine non-constant functions are well distributed (each element appears exactly $|\mathbb{G}|^{n-1}$ times), if the best affine approximation is not a constant function then the nonlinearity gives a bound on the bias from uniform. Consequently, both for the perspective from $\mathbb{G}$ and from $\mathbb{G}'$ we will ensure a non-trivial resiliency order to prevent these attacks, and we will consider the following rationals for the attack based on low-degree approximation. For an approximation of degree d over $\mathbb{G}$:

- D is the number of monomials of degree up to d,
- m is a positive integer lower than or equal to the number of equations available to the adversary,
- $\mathsf{NL}_\mathbb{G}^d$ is the Hamming distance between f and its best degree d approximation,
- $\mathsf{P}_{D,m}$ is the probability of having at least D equations over m with no error (Computed from a binomial law with error parameter $\mathsf{NL}_\mathbb{G}^d/(|\mathbb{G}|^n)$),

We will ensure that $\binom{m}{D}\mathsf{P}_{D,m}^{-1}D^\omega > 2^\lambda$, since this complexity bound matches the one of an attack consisting in solving a linearized system with enough noiseless equations.

In particular for Elisabeth-4, over $\mathbb{Z}_{16}$ we will ensure a nontrivial resiliency and a sufficient distance to degree-1 and degree-2 functions. Over $\mathbb{F}_2$ we will ensure a nontrivial resiliency and a sufficient nonlinearity. We assume that considering higher degree approximations will lead to less efficient attacks since the gain in approximating the function is lost by the increasing number of monomials.

Guess and Determine Attacks. The cryptanalysis of [21] on an instance of FP showed the strength of guess and determine attacks on this design when too simple filters are used. The principle of the attack is to guess some elements of the key and attack the resulting system with one of the previously described attacks. Guess and determine attacks perform better than the former ones when the cost of guessing elements (try the attack for the $|\mathbb{G}|^\ell$ possibilities) is compensated by the simplicity of the remaining system. As for IFP, in the GFP the subset selection and whitening decrease the interest of guess and determine attacks since only a sub-part of the fixed elements of the key area in the selected subset for each equation, and the whitening randomizes which sub-function is obtained. In [35] the complexity of the guess and determine attacks is estimated based on the cryptographic parameter of the weakest function that can be obtained by fixing ℓ variables. In [34] an involved algorithm is used to compute the complexity of the attack targeting sub-functions with parameters lower than a threshold and leverage it by the probability of obtaining such sub-functions by fixing ℓ

bits. Since this algorithm is practical only for bit-fixing stable families of functions (see [7]), we will use the simpler approach of [35]. For the different attacks listed above, the attack complexity is derived from the value of a filter's parameter, hence we will consider the attacks obtained with weaker values taking into account the minimal number of guesses needed to reach this value. Hence, the complexity estimation is $|\mathbb{G}|^\ell C(k)$ where $C(k)$ is the complexity for a parameter with value k and ℓ the minimal number of guesses to obtain a sub-function with this parameter value.

4.2   Security of Elisabeth-4

As described in Sect. 3.3 the design of Elisabeth relies on the use of direct sums, repeating the same basis function in a few variables. In this part we show how to derive the security estimations for Elisabeth-4 from the parameters of its basis function, analyzing it over $\mathbb{Z}_{16}$ and over $\mathbb{F}_2$.

Analysis over $\mathbb{Z}_{16}$. Over $\mathbb{Z}_{16}$ the filtering function can be written as:

$$f(x_1, \ldots, x_n) = \mathsf{DS}_\mathbb{G}(\mathsf{DS}_\mathbb{G}^{12}(h), \mathsf{DS}_\mathbb{G}^{12}(x_i)),$$

where h is the function from $\mathbb{G}^4$ to $\mathbb{G}$ given by the combinations of the 8 NLUTs given in Sect. 3.3 before adding x_5. We experimentally compute the resiliency, $\mathsf{NL}_\mathbb{G}$ and $\mathsf{NL}_\mathbb{G}^2$ of h and the sub-functions obtained from h by fixing 1 and 2 inputs, and list the results in Table 6. Then, the following proposition (Proposition 1) enables to bound the parameters of a direct sum over $\mathbb{G}$ and it is used to bound the parameters of the 60-variable filter f and its sub-functions obtained by guessing a limited number of inputs. Finally, we summarize the bounds on f parameters and the associated attack complexity in Table 7 following the analysis of Sect. 4.1. Additionally, we provide a primary analysis on the polynomials and functions over $\mathbb{G}$ which studies how selecting random NLUTs benefits the security over $\mathbb{G}$.

Proposition 1 (G-direct sum properties). *Let $h = \mathsf{DS}(f, g)$ be the direct sum of f and g with n and m $\mathbb{G}$-variables respectively. Then $\mathsf{DS}_\mathbb{G}(f, g)$ has the following properties:*

1. *Resiliency:* $\mathsf{res}_\mathbb{G}(h) \geq \mathsf{res}_\mathbb{G}(f) + \mathsf{res}_\mathbb{G}(g) + 1$.
2. *Nonlinearity:* $\mathsf{NL}_\mathbb{G}(h) \geq \max(|\mathbb{G}|^n \mathsf{NL}_\mathbb{G}(g), |\mathbb{G}|^m \mathsf{NL}_\mathbb{G}(f))$.
3. *Order-2 nonlinearity:* $\mathsf{NL}_\mathbb{G}^2(h) \geq \max(|\mathbb{G}|^n \mathsf{NL}_\mathbb{G}^2(g), |\mathbb{G}|^m \mathsf{NL}_\mathbb{G}^2(f))$.

Proof. We begin with the result on $\mathsf{res}_\mathbb{G}(h)$. For all $t \in \mathbb{N} \cup -1$ such that $t \leq \mathsf{res}_\mathbb{G}(f) + \mathsf{res}_\mathbb{G}(g) + 1$, all choice of t variables corresponds to t_1 variables of f and t_2 of g such that $t = t_1 + t_2$. If $t_1 \leq \mathsf{res}_\mathbb{G}(f)$ the corresponding $(n - t_1)$-variable sub-function f' of f is balanced hence the direct sum $f' + g'$ is also balanced. If $t_1 > \mathsf{res}_\mathbb{G}(f)$ then $t_2 \leq \mathsf{res}_\mathbb{G}(g)$, therefore the corresponding $(m - t_2)$-variable sub-function g' of g is balanced hence the direct sum $f' + g'$ is also balanced. It allows us to conclude $\mathsf{res}_\mathbb{G}(h) \geq t$.

For the nonlinearity, an $n + m$ affine function can be written as:

$$\ell_{a,b} = a_0 + \sum_{i=1}^{n} a_i x_i + \sum_{j=1}^{m} b_j y_j,$$

where the x_i (resp. y_j) denote the variables from the f part (resp. g). Then:

$$d_H(h, \ell_{a,b}) = \sum_{v \in \mathbb{G}^m} d_H(f + g(v), \ell_a + b \cdot v) \geq |\mathbb{G}|^m \mathsf{NL}_\mathbb{G}(f).$$

Since f and g play similar roles, the same arguments lead to $d_H(h, \ell_{a,b}) \geq |\mathbb{G}|^n \mathsf{NL}_\mathbb{G}(g)$, giving the final result.

Finally, the result for the order-2 nonlinearity can be directly adapted from the proof for the nonlinearity. When the m variables of g are fixed, on one side h equals f plus a constant, and on the other side any $n + m$ degree-at-most-2 function equals a degree-at-most-2 function depending only in the variables of f. Hence, the order-2 nonlinearity can be written as the sum of $|\mathbb{G}|^m$ distances (between f and a degree-at-most-2 function), each one greater than or equal to $\mathsf{NL}_\mathbb{G}^2(f)$. Noting that f and g play a similar role allows us to conclude.

Table 6. Minimum $\mathbb{G}$-cryptographic parameters for h after fixing up to ℓ inputs. For 0 and 1 guess $\mathsf{NL}_\mathbb{G}(h)$ and $\mathsf{NL}_\mathbb{G}^2(h)$ are bounded using the worst value obtained with 2 guesses.

Number of fixed variables ℓ	0	1	2
Resiliency $\mathsf{res}_\mathbb{G}$	-1	-1	-1
Nonlinearity $\mathsf{NL}_\mathbb{G}$	≥ 55296	≥ 3456	216
Order-2 nonlinearity $\mathsf{NL}_\mathbb{G}^2$	≥ 53248	≥ 3328	208

Table 7. Elisabeth-4, minimal parameter bounds up to ℓ fixed inputs and complexity estimations for a key size $N = 256$ (1024 bits). The parameter bounds come from applying Proposition 1 with the values from Table 6 and the complexity estimations come from Sect. 4.1. We write ">> 128" when the complexity estimation overreaches several hundreds.

Number of guesses ℓ	0	[1, 12]	[13, 23]	[24, 35]
Resiliency	11	$11 - \ell$	-1	-1
$\mathsf{NL}_\mathbb{G}/|\mathbb{G}|^{n-\ell}$	≥ 0.84	≥ 0.84	≥ 0.84	0.84
Correlation attack complexity (bits)	>> 128	>> 128	>> 128	>> 128
$\mathsf{NL}_\mathbb{G}^2/|\mathbb{G}|^{n-\ell}$	≥ 0.81	≥ 0.81	≥ 0.81	0.81
Order-2 Correlation attack complexity (bits)	>> 128	>> 128	>> 128	>> 128

The results from Table 7 show that the attacks considered on $\mathbb{G}$, correlation attacks using approximations of degree 1 or 2, are impracticable with the chosen filter.

Polynomials and Functions from $\mathbb{G}^4$ to $\mathbb{G}$

As explained in Sect. 4.1, not all functions from $\mathbb{G}^N$ to $\mathbb{G}$ can be mapped to polynomials in the polynomial ring $K[x_1, \ldots, x_N] = (\mathbb{G}, \cdot)[x_1, \ldots, x_N]$. With the same arguments, not all functions from $\mathbb{G}^4$ to $\mathbb{G}$ admit a polynomial representation in $K[x_1, \ldots, x_4]$, and we show that in particular h does not correspond to a polynomial using a simple necessary condition to be a polynomial. Moreover, we show that most of the functions from this space are far in Hamming distance from the polynomials. More precisely, the probability of a random function to agree with a polynomial on at least half of the inputs is lower than 2^{-128}.

Proposition 2. *Let f a function from $\mathbb{G}^4$ to $\mathbb{G}$. If $f(8,8,8,8) - f(0,0,0,0) \notin \{0,8\}$ then f is not a polynomial of $(\mathbb{G}, \cdot)[x_1, \ldots, x_4]$.*

Proof. All monomial of $(\mathbb{G}, \cdot)[x_1, \ldots, x_4]$ have the form $x_1^{i_1} x_2^{i_2} x_3^{i_3} x_4^{i_4}$ where i_1 to i_4 are positive integers. In $(0,0,0,0)$ all non-trivial monomial gives 0 and in $(8,8,8,8)$ if one exponent is greater than 1 or more than one exponent is not null a monomial gives 0, the remaining non trivial monomials give 8. Let p be a polynomial of $(\mathbb{G}, \cdot)[x_1, \ldots, x_4]$:

$$p(x_1, x_2, x_3, x_4) = \sum_{i_1, i_2, i_3, i_4 \in \mathbb{N}} a_{i_1, i_2, i_3, i_4} x_1^{i_1} x_2^{i_2} x_3^{i_3} x_4^{i_4},$$

where $a_{i_1, i_2, i_3, i_4} \in \mathbb{G}$. Then, $p(0,0,0,0) = a_{0,0,0,0}$ and $p(8,8,8,8) = a_{0,0,0,0} + 8(a_{1,0,0,0} + a_{0,1,0,0} + a_{0,0,1,0} + a_{0,0,0,1})$ hence $p(8,8,8,8) - p(0,0,0,0) \in \{0,8\}$, allowing to conclude.

Proposition 3. *Let $\mathcal{G}$ be the space of functions from $\mathbb{G}^4$ to $\mathbb{G}$, the probability that a function f taken uniformly at random in $\mathcal{G}$ agrees with a polynomial of $(\mathbb{G}, \cdot)[x_1, x_2, x_3, x_4]$ on at least half of the inputs is lower than 2^{-128}.*

Proof. For this proof we use the formalization of error correcting codes. Each function in $\mathcal{G}$ can be uniquely represented by its vector of outputs in $\mathbb{G}^{16^4}$, and the polynomials of $(\mathbb{G}, \cdot)[x_1, x_2, x_3, x_4]$ form a linear code of the vector-space $(\mathbb{G}, \cdot)^{16^4}$ with parameter $[n, k, d]$. First, we determine an upper bound on the number of polynomials with different evaluations (which corresponds to an upper bound on the code dimension k). Then we determine an upper bound on the number of elements of $\mathbb{G}^{16^4}$ at Hamming distance lower than $n/2 + 1$. Finally, we compare the last bound with the number of functions in $\mathcal{G}$ to conclude.

For $x \in \mathbb{G}$ the value of x^4 is 0 if x is even and 1 if x is odd, since x^8 has the same property, all polynomial of $(\mathbb{G}, \cdot)[x]$ has a representative of degree at most 7. Thereafter, for the polynomial ring $(\mathbb{G}, \cdot)[x_1, x_2, x_3, x_4]$ all monomial $x_1^{i_1} x_2^{i_2} x_3^{i_3} x_4^{i_4}$ with exponents $i_1, i_2, i_3, i_4 \in \mathbb{N}$ has a representative with $i_1, i_2, i_3, i_4 \in [0, 7]$. Since it gives at most $8^4 = 2^{12}$ monomials (that is $k \leq 2^{12}$), there are at most $16^{2^{12}} = 2^{2^{14}}$ different polynomial representations.

Then, we use the standard packing/covering bound to determine the maximum number of elements covered by the union of Hamming balls of radius $n/2$ centered around the code elements (here the polynomials). Each ball contains

$B = \sum_{i=0}^{n/2}(|\mathbb{G}| - 1)^i \binom{n}{i}$ elements, giving a total number of $2^{2^{14}}B$ elements: the upper bound reached if no ball intersect.

Finally, since there are $(16)^{16^4} = 2^{2^{18}}$ elements in $\mathcal{G}$ we can compute the proportion p of functions coinciding on a least half of there inputs with a polynomial:

$$p \leq \frac{2^{2^{14}}B}{2^{2^{18}}} = \frac{\sum_{i=0}^{2^{15}} 15^i \binom{2^{16}}{i}}{2^{15 \cdot 2^{14}}} \approx 2^{-2^{15.67}} < 2^{-128}.$$

Proposition 3 shows that most of the functions from $\mathbb{G}^4$ to $\mathbb{G}$ cannot be well approximated by a polynomial. The same idea extends to functions from $\mathbb{G}^n$ to $\mathbb{G}$, therefore it makes the attack consisting in solving a noisy polynomial system over $\mathbb{G}$ very unrealistic when the filter is a random function. Since h is not taken at random from $\mathcal{G}$, and neither is f, we cannot directly use this argument. This is why we additionally provide bounds for the nonlinearity and distance from degree-2 polynomials.

Analysis over $\mathbb{F}_2$. To perform the analysis over $\mathbb{F}_2$ we introduce Beth-4, a variation of Elisabeth-4. The variation consists in replacing some additions over $\mathbb{Z}_{16}$ by coordinate-wise XOR, it gives a scheme which security is easier to analyze and conjectured at most as secure as Elisabeth-4.

Formally, Beth-4 is a stream cipher following the GFP paradigm with $\mathbb{G}' = \mathbb{F}_2^4$ where the operation is the addition modulo 2 coordinate-wise. The addition of the whitening and the addition of the key-stream with the plaintext are considered in $\mathbb{G}'$, only the additions inside the 5-to-1 function remain in $\mathbb{G}$. More precisely, the filtering function F from $(\mathbb{F}_2^4)^n$ to $\mathbb{F}_2^4$ is computed as $\mathsf{DS}_{\mathbb{G}'}(H(x_{20i-19}, \cdots, x_{20i}), i \in [4n/5])$ where H is determined by the truth table obtained by considering the canonical embedding of $\mathbb{G}$ over $\mathbb{G}'$ on the inputs and outputs of the ($\mathbb{G}^5$ to $\mathbb{G}$) function g defined in Sect. 3.3.

Since the design of Beth-4 uses direct sums and 20-variable Boolean functions its security can be analyzed easily following the approach on IFP [34]. Seen over $\mathbb{G}' = \mathbb{F}_2^4$, the addition in $\mathbb{G}$ corresponds to the addition (in $\mathbb{F}_2^4$) more nonlinear combinations of the coordinate functions due to caries. Therefore, the algebraic system given by Elisabeth-4 is more complex over $\mathbb{F}_2$ than the one of Beth-4, and we assume that Elisabeth-4 is at least as secure as Beth-4 for the different attacks we consider. Accordingly, in the following part, we determine the parameters of the 20-variable component functions of the basis function of Beth-4. Then, we bound the significant cryptographic parameters of the $4n$-variable component Boolean functions. Finally, we give the complexity estimations for the security of Beth-4, we consider these values as lower bounds for the complexity estimations on the security of Elisabeth-4.

Over $\mathbb{F}_2^4$ the filtering function can be written as $F(x_1, \ldots, x_{4n}) = \mathsf{DS}_{\mathbb{G}'}^{12}(H)$, where H is the function from $\mathbb{F}_2^{20}$ to $\mathbb{F}_2^4$ given by the NLUTs and addition in $\mathbb{G}$ illustrated in Sect. 3.3. We experimentally compute the parameters of H and give them in Table 8. Then, the following lemmas allow to determine the cryptographic parameters of the function F, with or without guessed bits. The structure of direct sum of F component functions is preserved even guessing bits, hence Lemma 1 on the cryptographic properties of direct sums enables to bound the parameters (degree, nonlinearity, resiliency) of the obtained functions from the parameters of H. Lemma 2 gives a lower bound on the AI of a direct sum based on the degree of its sub-parts, it allows us to bound the AI of the obtained functions from the degree of the components of H (with or without guessing bits). Finally, the bounds on F's parameters are given in Table 9, together with the complexity estimations following the analysis of Sect. 4.1.

Table 8. Minimum Boolean cryptographic parameters over H's components after fixing up to ℓ binary inputs.

ℓ	0	1	2	3
deg	12	10	9	8
res	0	-1	-1	-1
NL	509440	249216	120768	57920

Lemma 1 (Boolean direct sum properties (e.g. [35] Lemma 3)). *Let $h = \mathsf{DS}(f, g)$ be the direct sum of f and g n and m-variable Boolean functions respectively. Then $\mathsf{DS}(f, g)$ has the following cryptographic properties:*

1. *Degree:* $\deg(h) = \max(\deg(f), \deg(g))$.
2. *Algebraic immunity:* $\max(\mathsf{AI}(f), \mathsf{AI}(g)) \leq \mathsf{AI}(h) \leq \mathsf{AI}(f) + \mathsf{AI}(g)$.
3. *Resiliency:* $\mathsf{res}(h) = \mathsf{res}(f) + \mathsf{res}(g) + 1$.
4. *Nonlinearity:* $\mathsf{NL}(h) = 2^m \mathsf{NL}(f) + 2^n \mathsf{NL}(g) - 2\mathsf{NL}(f)\mathsf{NL}(g)$.

Lemma 2 ([33] Lemma 6). *Let $t \in \mathbb{N}^*$, and $f_1, \ldots, f_t$ be t Boolean functions, if for $r \in [t]$ there exists r different indexes $i_1, \cdots, i_r$ of $[t]$ such that $\forall j \in [r], \deg(f_{i_j}) \geq j$ then $\mathsf{AI}(\mathsf{DS}(f_1, \ldots, f_t)) \geq r$.*

Table 9. Beth-4, minimal parameter bounds up to ℓ fixed binary inputs and complexity estimations (in bits) for key sizes $N = 256$ (1024 bits) and $N = 1024$ (4096 bits). The complexity estimations come from Sect. 4.1, when ℓ lies in an interval the complexity given is the minimal one for this interval of guesses. AA refers to the algebraic attack and FAA to the fast algebraic attack. The complexity of the fast algebraic attack is computed taking $\deg(g) = 1$ and $\deg(h) = \mathsf{AI}(f) + 1$.

Number of guesses ℓ	0	$[1, 10]$	11	$[12, 23]$	$[24, 35]$
Resiliency	11	$11 - \ell$	0	-1	-1
NL/$2^{4n-\ell}$	0.49	≥ 0.48	≥ 0.48	≥ 0.47	≥ 0.46
Correlation attack complexity (bits)	$>> 128$	$>> 128$	$>> 128$	$>> 128$	$>> 128$
Algebraic immunity	12	12	11	10	9
AA complexity (bits) $4N = 1024$	229	230	222	205	198
AA complexity (bits) $4N = 4096$	288	289	276	254	242
FAA complexity (bits) $4N = 1024$	110	111	114	108	113
FAA complexity (bits) $4N = 4096$	136	137	139	131	134

The results on the security of Beth-4 in Table 9 show that the correlation attacks are impracticable with the chosen filter, which goes in the same direction as the analysis over $\mathbb{G}$. Regarding the attacks on the algebraic kind, the estimations for the fast algebraic attacks (FAA) are slightly below the threshold of 128 bits of security when we consider a key size corresponding to 1024 bits. The last row of the table shows that taking a bigger key (equivalent to 4096 bits) is sufficient to reach the threshold of 128 bits for this attack, and for the cipher instance since the FAA gives by far the lower attack complexity. Since as explained in Sect. 4.1 the complexity of FAA is underestimated on GFP, and the level of security of Elisabeth-4 is assumed higher than the one of Beth-4, we consider that the key size of 1024 bits is an adequate choice for a security level of 128 bits, and an interesting starting point for further cryptanalyses.

5 Case Study

Homomorphic encryption is useful to prevent privacy issues and transciphering will typically be used on personal data in a real-world scenario. Nowadays, algorithms handling personal data, and already running server-side, are mainly neural networks. Therefore, we next present a neural network evaluation with HHE. The homomorphic evaluation of neural networks (i.e., the inference part) using TFHE has already been studied in [11]. To the best of our knowledge, our following investigations are the first to evaluate the combination of such a complex FHE computation with a transciphering, which will allow us to discuss the interest of HHE in more general terms.

We decided to evaluate a neural network over the standard dataset Fashion MNIST (FMNIST)[8]. The neural network performs classification on 28×28 pixels grayscale clothes pictures. This dataset has been created to provide a more challenging problem than MNIST, which is a handwritten text recognition dataset. We note that this dataset itself does not pose privacy concerns and an even more relevant usecase would be to classify medical data streams (e.g. monitored by a smartwatch). Yet, it is a standard and well-documented one and serves as an interesting first step for benchmarking. The challenges raised by more complex data sets are discussed in the conclusions.

5.1 Network Design Under FHE Constraints

To make the homomorphic evaluation of the network as fast as possible as well as taking into account the fact that Elisabeth-4 outputs 4-bit messages, the model has to be designed with some constraints in mind.

TFHE for Machine Learning. As presented in Sect. 2.1, TFHE is efficient for performing modular additions, multiplications with public scalars, and evaluation of homomorphic NLUTs. While most functions are not negacyclic, it is

---

[8] https://github.com/zalandoresearch/fashion-mnist.

actually possible to evaluate them through a PBS thanks to a small trick: any arbitrary LUT of length N can be transformed into a negacyclic one of length $2N$ by appending its negation to it. When evaluating this lookup table, the user is only interested in the value written in the first half of the table, that is the slots 0 to $N-1 = 2^{d-1} - 1$. These are the slots for which the most significant bit is set to 0, so to ensure that the value retrieved from the table is in the first half, one must guarantee that the MSB of the message is 0. This is done by adding a bit of padding in the MSB of the plaintext. Since by design Elisabeth-4 does not require this bit of padding, it has been decided to encrypt it along with the message, leaving 3 effective bits for the actual message. Note that a recent work from Chillotti *et al.* [12] suggests that bits of padding inside a PBS can be avoided and could become obsolete in the future. The ability to compute arbitrary functions, especially non-linear ones, through a PBS gives the possibility to efficiently evaluate the activation function of a neural network. Every other operation in between two activation functions is linear and can be efficiently computed with TFHE. The reader can also refer to [11] for more details on the homomorphic evaluation of neural networks.

Neural Network. In order to classify FMNIST pictures, the Convolutional Neural Network (CNN) [26] relies on a convolutional layer to extract features, then on linear layers in order to perform the classification. Since the homomorphic inference is performed on 3-bit data (instead of originally 8 bits), the CNN has been trained on unencrypted quantized data using Weight Decay [26] to prevent overfitting. The reader will find below a picture of a grayscale picture from the FMNIST dataset, which is degraded from 8 bits to 3 bits of quantization in the right part.

(a) Original data (8 bits shades)   (b) Quantized data (3 bits shades)

Fig. 4. 784-pixels Fashion MNIST pictures

We use a sigmoid function on our data at the entrance of the neural network. This (non-standard) operation is performed for 2 reasons:

– The sigmoid bounds the norm of the data we manipulate. Having smaller-norm data reduce the noise growth during the homomorphic operation.

– We are impelled to perform a PBS between Elisabeth-4 and the CNN, for the reasons developed in the previous Subsect. 5.1 (mainly in order to add a bit of padding). While doing it, we can evaluate the sigmoid without any additional cost.

In order to perform homomorphic inference, the CNN expects a bootstrapped LWE ciphertext encrypted under the input key s_{in}. This implies that Elisabeth is used with the single KeySwitch parameters in order to save a KeySwitch at the exit of Elisabeth-4's evaluation. The scalar weights inside the linear layers are given in clear so that these layers are evaluated efficiently with TFHE. Still, for noise management reasons, these weights ought to remain small (inside a magnitude of 20 for the FMNIST implementation). The CNN does not output normalized probability nor use a softmax function, as those operations would be costly when performed homomorphically.

As detailed in the next subsection, not using these tools does not have a big impact on the resulting accuracy, which remains high when we evaluate the adapted-for-homomorphic CNN with unencrypted data.

5.2   Performance

In order to reduce the amount of data sent to the server, we chose to use a single PBS key and a single KeySwitch key for the whole circuit, rather than different keys for the transciphering and the homomorphic inference. The parameters chosen with these constraints are given in Table 10. In clear, the trained network gave 84.37% accuracy over 10,000 randomized inputs. When tested over the same inputs, encrypted this time, the network kept a pretty similar accuracy of 84.18%. The loss of accuracy in an homomorphized network is studied in [11].

One homomorphic inference took 427.23 seconds on average over 100 samples, versus 5.74 seconds per homomorphic inference without transciphering. This means that the transciphering took 537.62 ms per 4-bit message (with $28^2 = 784$ messages). This increase in transciphering time is explained by the constraints put by the neural network on the PBS and KeySwitch keys it shares with Elisabeth-4. This is expected since Elisabeth-4 was designed so that its parameters would be less restrictive than most usecases, in order to make sure that transciphering would neither require its own set of public keys nor slow down the rest of the computation. Note that we chose here to minimize bandwidth consumption, but computation time could be further improved by sending Elisabeth's optimal PBS and KeySwitch keys along with the one used by the neural network. That way, Elisabeth could run in optimal time, reducing the transciphering time to $28 \times 28 \times 0.091143 = 71.46$ seconds. The initial data overload on the bandwidth would thus be higher and longer to compensate, but the inference time would be greatly reduced. More details on this are given in Fig. 6. Also note that the transciphering algorithm[9] can be performed offline even before the client starts sending their inputs, thus saving time during the online phase.

---

[9] Up to the point of summing with the ciphertext.

Evaluating 1000 inferences of this model with Elisabeth-4 requires sending 400 MB over the network, against 5, 337 GB if LWE ciphertexts were sent directly without compression, and 6, 272 MB if only LWE bodies and their corresponding seed were sent. This represents a saving of at least 93.62% of bandwidth. Figure 5 compares the bandwidth saved by Elisabeth-4 and FiLIP over seeded LWE ciphertexts. It both confirms the interest of the HHE framework in general, and the significant gains that Elisabeth-4 provides over FiLIP for our usecase.

Table 10. Set of FHE parameters used for Elisabeth-4 combined with a neural network.

TFHE Parameters								
n	k	N	$\log(\sigma_{\mathrm{LWE}})$	$\log(\sigma_{\mathrm{GLWE}})$	PBS		KeySwitch	
					$\log(B)$	ℓ	$\log(B)$	ℓ
754	1	2048	-17.8745	-52.0036	7	6	2	8

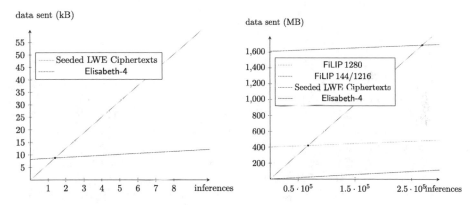

Fig. 5. Data sent per inference with different methods of compression. Elisabeth-4 has a better usage of bandwidth if more than two inferences are performed. On the other side, FiLIP needs 67366 or 269424 inferences (depending on the version) to become cheaper in bandwidth usage than seeded ciphertexts.

Our implementation is available at:

https://github.com/princess-elisabeth/elisabeth_usecase.

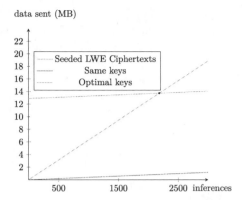

Fig. 6. Using Elisabeth-4 with its optimal parameters implies an overhead on bandwidth usage since another PBS and KeySwitch have to be sent to the server. That being the case, this overhead is compensated compared to seed-compression in only 2185 inferences. Note also that these inferences would take roughly 47 hours to compute, whereas less than 400 inferences could be computed in the same amount of time using the same keys for both Elisabeth-4 and the neural network.

6   Conclusion

Our results show that HHE can highly benefit from the optimization of a symmetric encryption scheme depending on the operations to be performed homomorphically. We introduced the family of stream ciphers Elisabeth for this purpose and instantiated it with 4-bit components, motivated by current constraints of the TFHE Concretelibrary. While the stand-alone homomorphic evaluation of Elisabeth-4 only brings similar performances to state-of-the-art competitors, its combination with the homomorphic classification of grayscale images from the Fashion MNIST dataset leads to significant improvements. In particular, the transciphering does not impact the accuracy of the inference (compared to the direct homomorphic processing of fresh ciphertexts) and it enables large gains in terms of bandwidth.

These results lead to several challenging open problems. First, they are based on an admittedly simple usecase, where the quantization of the images in 3 bits does not significantly affect its accuracy. It is expected that more challenging case studies (e.g. medical data, where fine-grain details can be essential for the classification) may require either more granular data (which would be calling for larger NLUTs) or more complex models (which may change the cost balance between the transciphering and the homomorphic inference). Considering larger instances of Elisabeth is an interesting direction regarding the first concern. Optimizing machine learning for homomorphic computations and leveraging transformations that would reduce the load of homomorphic computations to perform are interesting directions regarding the second concern. For example, the application of a Principal Component Analysis (PCA) [39], could be used to reduce the number of dimensions of the Fashion MNIST images from 28×28

down to 10, however with a penalty on the accuracy of approximately 10%. PCA therefore enables to choose a compromise between the amount of data sent and the accuracy. Other types of features selection would certainly deserve further investigation.

Another possible improvement client-side would be the encryption time. Even though Elisabeth-4 encryption algorithm is reasonably fast, the generation of its permutations and whitening values could be optimized and would represent an interesting development.

Finally, studying how to apply other FHE-friendly symmetric encryption and FHE schemes to practical-relevant case studies will be important, in order to better assess which combination works best in which context. Besides, and independent of the interest of the HHE framework, improving the loss of accuracy that homomorphic implementations may imply, as observed with TFHE and the Concretelibrary for example, is a critical optimization goal to further incentivize the application of FHE in contexts where privacy requires it. We believe having a first (and more) benchmark(s) is an essential seed to drive research towards achieving these important goals.

Acknowledgments. François-Xavier Standaert is a senior research associate of the Belgian Fund for Scientific Research (F.R.S.-FNRS). Pierrick Méaux was supported by the ERC Advanced grant CLOUDMAP (num. 787390). This work has been funded in part by the European Union through the ERC consolidator grant SWORD (num. 724725), and by the PEPR Cyber France 2030 programme (ANR-22-PECY-0003). We thank Arthur Mcyre for his help with the neural network, Samuel Tap for the parameters and Pascal Paillier, Damien Stehlé and Alain Passelègue for interesting discussions.

Supplementary material

A   Details on the Programmable Bootstrapping

In this part, we detail how works the PBS to apply an NLUT and bootstrap a LWE ciphertext at the same time. First, we recall the definition of a cryptographic multiplexer, and then we explain how the computation of a length-$2N$ NLUT is incorporated inside the bootstrapping.

Definition 11 (CMux). *A Cryptographic Multiplexer (CMux in short) is an operator that, given two GLWE ciphertexts $\mathbf{c}_0$ and $\mathbf{c}_1$ of μ_0 and μ_1 respectively and a GGSW encryption $\mathbf{B}$ of a bit b, outputs a reencryption of μ_b. This can be done in a single external product by computing $\mathbf{B} \boxdot (\mathbf{c}_1 - \mathbf{c}_0) + \mathbf{c}_0$.*

Given $\mathscr{L}(X) = \sum_{i=0}^{N} l_i X^i \in \mathbb{Z}_{q,N}[X]$, one can see that the constant term of $X^{-(i+N)}\mathscr{L}(X) = -X^{-i}\mathscr{L}(X)$ is $-l_i$, meaning that a negacyclic look-up table of length $2N$ can be represented as a polynomial of $\mathbb{Z}_{q,N}[X]$: accessing the i-th value of the look-up table is equivalent to multiplying the polynomial by X^{-i} then keeping the constant term.

Let $\boldsymbol{c}_L$ be a GLWE encryption of such a polynomial[10] and $\boldsymbol{c}$ be a LWE encryption of a message μ. The goal of bootstrapping is to secretly select a value of L based on the value of μ. Since there are $2N$ slots in L and since μ can take q different values, one first needs to rescale μ from $[0, q]$ to $[0, 2N]$, an operation known as Modulus Switching. This is simply done by multiplying each coefficient of $\boldsymbol{c}$ by $2N/q$, then by rounding to the closest integer. Let call $(\bar{a}_1, \ldots, \bar{a}_n, \bar{b})$ the scaled coefficients obtained, and $\bar{\mu}$ the value of μ rescaled. One now has to select the $\bar{\mu}$-th slot of the NLUT. This can be done approximately by multiplying the polynomial $\mathscr{L}$ by $X^{-\bar{b} + \sum \bar{a}_i s_i} = X^{-\bar{\mu} - \bar{\varepsilon}}$, where the $\bar{a}_i$ and $\bar{b}$ are publicly known. Thus, computing $\mathsf{ACC} \leftarrow \boldsymbol{c}_L \cdot X^{-\bar{b}}$ can be done immediatly. Multiplying by $X^{\sum \bar{a}_i s_i}$ is done iteratively thanks to a series of CMuxes: by using GGSW encryption $\mathbf{S}_i$ of the bits of the LWE secret key s_1 to s_n, one computes $\mathsf{ACC} \leftarrow \mathsf{CMux}(\mathbf{S}_i, X^{\bar{a}_i}\,\mathsf{ACC}, \mathsf{ACC}) = \mathsf{ACC} \cdot X^{\bar{a}_i s_i}$. This yields an encryption of a polynomial which constant coefficient is $L[\bar{\mu}^*]$. Since it is not possible to directly compute the rounding of $\bar{\mu}^* = \bar{\mu} + \bar{\varepsilon}$ homomorphically to recover $L[\bar{\mu}]$, the only alternative is to introduce redundancy in L, so that $L[\bar{\mu}^*] = L[\bar{\mu}]$ for small enough values of $\bar{\varepsilon}$. The actual number of values that the lookup table can hold thus depends on both the degree of the polynomial and the maximal size $\bar{\varepsilon}$ can take: the bigger the polynomial the more redundancy can be introduced, the smaller ε and the lesser redundancy is needed. Now, given a GLWE encryption $(\boldsymbol{a}, \boldsymbol{b})$ of $\sum \mu_i X^i$ under the secret key $\boldsymbol{s} = (\sum s_{1,i} X^i, \ldots, \sum s_{k,i} X^i)$ with $\boldsymbol{a} = (\sum a_{1,i} X^i, \ldots, \sum a_{k,i} X^i)$ and $b = \sum b_i X^i$, one can build an LWE encryption of μ_0 under the secret key $(s_{1,0}, \ldots, s_{1,N-1}, \ldots, s_{k,0}, \ldots, s_{k,N-1})$ as:

$$(a_{1,0}, -a_{1,N-1}, \ldots, -a_{1,1}, \ldots, a_{k,0}, -a_{k,N-1}, \ldots, -a_{k,1}, b_0)$$

This operation is called sample extraction and does not increase the noise in the ciphertext. A complete bootstrap cycle then consists of these three operations: modulus switching, blind rotation of the negacyclic look-up table, and sample extraction. The output noise of the bootstrapped ciphertext is independent of the input ciphertext and depends only on the number of CMuxes that has been performed, which in turn depends on the length of the LWE key. The noise caused by each CMux depends on the degree N of the polynomials as well as on the basis and number of levels of the GGSW used.

B   Elisabeth-4 Specifications

This appendix describes the details of Elisabeth-4 's implementation.

B.1   NLUTs Table

We specify the Negacyclic Look-Up Tables used for Elisabeth-4 implementation. Remember that the second half of each NLUT's value is entirely determined by the first.

---

[10] Note that it is always possible to build a trivial encryption of $\mathscr{L}$ by appending it to a vector of zeros.

	0	1	2	3	4	5	6	7	8	9	10	11	12	13	14	15
S_1	3	2	6	12	10	0	1	11	13	14	10	4	6	0	15	5
S_2	4	11	4	4	4	15	9	12	12	5	12	12	12	1	7	4
S_3	11	10	12	2	2	11	13	14	5	6	4	14	14	5	3	2
S_4	5	9	13	2	11	10	12	5	11	7	3	14	5	6	4	11
S_5	3	0	11	8	13	14	13	11	13	0	5	8	3	2	3	5
S_6	8	13	12	12	3	15	12	7	8	3	4	4	13	1	4	9
S_7	4	2	9	13	10	12	10	7	12	14	7	3	6	4	6	9
S_8	10	2	5	5	3	13	15	1	6	14	11	11	13	3	1	15

B.2   Elisabeth-4 Algorithms

In this specification, notations from the article are used. For the reader's comfort, let us remind here a few of them:

- $\mathbb{G} = \mathbb{Z}_{16}$.
- S_i denotes the NLUTs.
- Addition ($+$) between two vectors of $\mathbb{G}^k$ denotes the adddition coefficient by coefficient and substraction ($-$) denotes the inverse of the addition.

We define the Elisabeth-4 encryption scheme as its key generation, encryption and decryption algorithm. Both encryption and decryption use the keystream algorithm.

Algorithm 1: Elisabeth-4 .KeyGen()

---

output: Secret key $sk \in \mathbb{G}^{256}$
begin

  $sk \xleftarrow{\$} \mathbb{G}^{256}$
  return sk

Algorithm 2: Elisabeth-4 .KeyStream(sk)

---

input   : Secret key $sk \in \mathbb{G}^{256}$
          Integer $k \in \mathbb{N}$
output: Stream $stream \in G^k$
begin

   for i in range(k) **do**

      `// aborted Knuth shuffling`

      for j in range(60) **do**

         $r = $ `random_int`$(j, 256)$

                  `// Use a forward-secure PRG as described in [34]`

         $w = $ `random_int`$(0, 16)$ `// Random element of` $\mathbb{G}$

         `swap`$(s[j], s[r])$

         $s[j] = (s[j] + w)\%16$ `// Whitening`

      $Acc = 0$

      for b in range(12) **do**

         for j in range(5) **do**

            $x_j = s_{5b+j}$

         for j in range(4) **do**

            $y_j = S_j \left[x_j + x_{(j+1)\%4} \right]$

         $r = 0$

         for j in range(4) **do**

            $r\ += S_{4+j} \left[x_j + y_{(j+1)\%4} + y_{(j+2)\%4} \right]$

         $r\ += x_4$

         $Acc\ += r$

      $stream_i = Acc$

   return $stream$

Algorithm 3: Elisabeth-4 .Enc(m, sk)

---

input   : Message $m \in \mathbb{G}^k$
          Secrek key $sk \in \mathbb{G}^{256}$
output: Ciphertext $c \in \mathbb{G}^k$
begin

   $s = $ Elisabeth-4 .KeyStream(sk)

   $c = m + s$

   return c

Algorithm 4: Elisabeth-4 .Dec(c, sk)

---

input   : Ciphertext $c \in \mathbb{G}^k$
          Secrek key $sk \in \mathbb{G}^{256}$
output: Message $m \in \mathbb{G}^k$
begin

   $s = $ Elisabeth-4 .KeyStream(sk)

   $m = c - s$

   return m

References

1. Albrecht, M.R., Player, R., Scott, S.: On the concrete hardness of learning with errors. J. Math. Cryptol. **9**(3), 169–203 (2015)
2. Albrecht, M.R., Rechberger, C., Schneider, T., Tiessen, T., Zohner, M.: Ciphers for MPC and FHE. In: Oswald, E., Fischlin, M. (eds.) EUROCRYPT 2015. LNCS, vol. 9056, pp. 430–454. Springer, Heidelberg (2015). https://doi.org/10.1007/978-3-662-46800-5_17
3. Albrecht, M.R., Rechberger, C., Schneider, T., Tiessen, T., Zohner, M.: Ciphers for MPC and FHE. IACR Cryptology ePrint Archive **2016**, 687 (2016)
4. Brakerski, Z., Gentry, C., Vaikuntanathan, V.: (Leveled) fully homomorphic encryption without bootstrapping. In: Goldwasser, S. (ed.) ITCS 2012, pp. 309–325. ACM, Jan. 2012
5. Canteaut, A., Carpov, S., Fontaine, C., Lepoint, T., Naya-Plasencia, M., Paillier, P., Sirdey, R.: Stream ciphers: a practical solution for efficient homomorphic-ciphertext compression. J. Cryptology **31**(3), 885–916 (2018). https://doi.org/10.1007/s00145-017-9273-9
6. Carlet, C.: Boolean Functions for Cryptography and Coding Theory. Cambridge University Press (2021)
7. Carlet, C., Méaux, P.: A complete study of two classes of boolean functions: Direct sums of monomials and threshold functions. IEEE Trans. Inf. Theory **68**(5), 3404–3425 (2022)
8. Carlet, C., Méaux, P., Rotella, Y.: Boolean functions with restricted input and their robustness; application to the FLIP cipher. IACR Trans. Symmetric Cryptol. **3**, 2017 (2017)
9. Chillotti, I., Joye, M., Ligier, D., Orfila, J.-B., Tap, S.: Concrete: concrete operates on ciphertexts rapidly by extending tfhe. In: 8th Workshop on Encrypted Computing and Applied Homomorphic Cryptography (WAHC 2020) (2020)
10. Chillotti, I., Gama, N., Georgieva, M., Izabachène, M.: TFHE: fast fully homomorphic encryption over the torus. J. Cryptol. **33**(1), 34–91 (2020)
11. Chillotti, I., Joye, M., Paillier, P.: Programmable bootstrapping enables efficient homomorphic inference of deep neural networks. In: Dolev, S., Margalit, O., Pinkas, B., Schwarzmann, A. (eds.) CSCML 2021. LNCS, vol. 12716, pp. 1–19. Springer, Cham (2021). https://doi.org/10.1007/978-3-030-78086-9_1
12. Chillotti, I., Ligier, D., Orfila, J.-B., Tap, S.: Improved programmable bootstrapping with larger precision and efficient arithmetic circuits for TFHE. In: Tibouchi, M., Wang, H. (eds.) ASIACRYPT 2021. LNCS, vol. 13092, pp. 670–699. Springer, Cham (2021). https://doi.org/10.1007/978-3-030-92078-4_23
13. Cho, J., Ha, J., Kim, S., Lee, B., Lee, J., Lee, J., Moon, D., Yoon, H.: Transciphering framework for approximate homomorphic encryption (full version). IACR Cryptol. ePrint Arch., p. 1335 (2020)
14. Cho, J., Ha, J., Kim, S., Lee, B., Lee, J., Lee, J., Moon, D., Yoon, H.: Transciphering framework for approximate homomorphic encryption. In: Tibouchi, M., Wang, H. (eds.) ASIACRYPT 2021. LNCS, vol. 13092, pp. 640–669. Springer, Cham (2021). https://doi.org/10.1007/978-3-030-92078-4_22
15. Cogliati, B., Tanguy, T.: Multi-user security bound for filter permutators in the random oracle model. Designs, Codes and Cryptography, September 2018
16. Coron, J.-S., Lepoint, T., Tibouchi, M.: Scale-invariant fully homomorphic encryption over the integers. In: Krawczyk, H. (ed.) PKC 2014. LNCS, vol. 8383, pp. 311–328. Springer, Heidelberg (2014). https://doi.org/10.1007/978-3-642-54631-0_18

17. Courtois, N.: Fast algebraic attacks on stream ciphers with linear feedback. In: Boneh, D. (ed.) CRYPTO 2003, pp. 176–194 (2003)
18. Courtois, N.T., Meier, W.: Algebraic attacks on stream ciphers with linear feedback. In: Biham, E. (ed.) EUROCRYPT 2003. LNCS, vol. 2656, pp. 345–359. Springer, Heidelberg (2003). https://doi.org/10.1007/3-540-39200-9_21
19. Dobraunig, C., Eichlseder, M., Grassi, L., Lallemand, V., Leander, G., List, E., Mendel, F., Rechberger, C.: Rasta: a cipher with low ANDdepth and few ANDs per bit. In: Shacham, H., Boldyreva, A. (eds.) CRYPTO 2018. LNCS, vol. 10991, pp. 662–692. Springer, Cham (2018). https://doi.org/10.1007/978-3-319-96884-1_22
20. Dobraunig, C., Grassi, L., Helminger, L., Rechberger, C., Schofnegger, M., Walch, R.: Pasta: a case for hybrid homomorphic encryption. IACR Cryptol. ePrint Arch., p. 731 (2021)
21. Duval, S., Lallemand, V., Rotella, Y.: Cryptanalysis of the FLIP family of stream ciphers. In: Robshaw, M., Katz, J. (eds.) CRYPTO 2016. LNCS, vol. 9814, pp. 457–475. Springer, Heidelberg (2016). https://doi.org/10.1007/978-3-662-53018-4_17
22. Faugère, J.-C.: A new efficient algorithm for computing groebner bases. J. Pure Appl. Algebra **139**, 61–88 (1999)
23. Gentry, C., Halevi, S., Smart, N.P.: Homomorphic evaluation of the AES circuit. In: Safavi-Naini, R., Canetti, R. (eds.) CRYPTO 2012. LNCS, vol. 7417, pp. 850–867. Springer, Heidelberg (2012). https://doi.org/10.1007/978-3-642-32009-5_49
24. Gentry, C., Sahai, A., Waters, B.: Homomorphic encryption from learning with errors: conceptually-simpler, asymptotically-faster, attribute-based. In: Canetti, R., Garay, J.A. (eds.) CRYPTO 2013. LNCS, vol. 8042, pp. 75–92. Springer, Heidelberg (2013). https://doi.org/10.1007/978-3-642-40041-4_5
25. Goldreich, O.: Candidate one-way functions based on expander graphs. Electronic Colloquium on Computational Complexity (ECCC), 7(90) (2000)
26. Goodfellow, I.J., Bengio, Y., Courville, A.C.: Deep Learning. MIT Press, Adaptive computation and machine learning (2016)
27. Ha, J., Kim, S., Choi, W., Lee, J., Moon, D., Yoon, H., Cho, J.: Masta: an he-friendly cipher using modular arithmetic. IEEE Access **8**, 194741–194751 (2020)
28. Ha, J., Kim, S., Lee, B., Lee, J., Son, M.: Noisy ciphers for approximate homomorphic encryption. In: Dunkelman, O., Dziembowski, S. (eds.) EUROCRYPT 2022. LNCS, vol. 13275. Springer, Cham. https://doi.org/10.1007/978-3-031-06944-4_20
29. Hebborn, P., Leander, G.: Dasta - alternative linear layer for rasta. IACR Trans. Symmetric Cryptol. **2020**(3), 46–86 (2020)
30. Hoffmann, C., Méaux, P., Ricosset, T.: Transciphering, Using FiLIP and TFHE for an efficient delegation of computation. In: Bhargavan, K., Oswald, E., Prabhakaran, M. (eds.) INDOCRYPT 2020. LNCS, vol. 12578, pp. 39–61. Springer, Cham (2020). https://doi.org/10.1007/978-3-030-65277-7_3
31. Langlois, A., Stehlé, D.: Worst-case to average-case reductions for module lattices. Des. Codes Crypt. **75**(3), 565–599 (2015)
32. Lyubashevsky, V., Peikert, C., Regev, O.: On ideal lattices and learning with errors over rings. In: Gilbert, H. (ed.) EUROCRYPT 2010. LNCS, vol. 6110, pp. 1–23. Springer, Heidelberg (2010). https://doi.org/10.1007/978-3-642-13190-5_1
33. Méaux, P.: On the algebraic immunity of direct sum constructions. Discret. Appl. Math. **320**, 223–234 (2022)
34. Méaux, P., Carlet, C., Journault, A., Standaert, F.-X.: Improved filter permutators for efficient FHE: better instances and implementations. In: Hao, F., Ruj, S., Sen Gupta, S. (eds.) INDOCRYPT 2019. LNCS, vol. 11898, pp. 68–91. Springer, Cham (2019). https://doi.org/10.1007/978-3-030-35423-7_4

35. Méaux, P., Journault, A., Standaert, F.-X., Carlet, C.: Towards stream ciphers for efficient FHE with low-noise ciphertexts. In: Fischlin, M., Coron, J.-S. (eds.) EUROCRYPT 2016. LNCS, vol. 9665, pp. 311–343. Springer, Heidelberg (2016). https://doi.org/10.1007/978-3-662-49890-3_13
36. Naehrig, M., Lauter, K.E., Vaikuntanathan, V.: Can homomorphic encryption be practical? In: CCSW, pp. 113–124. ACM (2011)
37. Papernot, N., McDaniel, P.D., Sinha, A., Wellman, M.P.: Sok: security and privacy in machine learning. In: EuroS&P, pp. 399–414. IEEE (2018)
38. Regev, O.: On lattices, learning with errors, random linear codes, and cryptography. In: Gabow, H.N., Fagin, R. (eds.) 37th ACM STOC, pp. 84–93. ACM Press, May 2005
39. Shalev-Shwartz, S., Ben-David, S.: Understanding Machine Learning - From Theory to Algorithms. Cambridge University Press (2014)
40. Xiao, H., Rasul, K., Vollgraf, R.: Fashion-mnist: a novel image dataset for benchmarking machine learning algorithms. CoRR, abs/1708.07747 (2017)

Revisiting Related-Key Boomerang Attacks on AES Using Computer-Aided Tool

Patrick Derbez[1]([✉]), Marie Euler[1,2][iD], Pierre-Alain Fouque[1][iD], and Phuong Hoa Nguyen[1]

[1] Univ Rennes, CNRS, IRISA, Rennes, France
{patrick.derbez,marie.euler,pierre-alain.fouque,
phuong-hoa.nguyen}@irisa.fr
[2] Direction Générale de l'Armement, Rennes, France

Abstract. In recent years, several MILP models were introduced to search automatically for boomerang distinguishers and boomerang attacks on block ciphers. However, they can only be used when the key schedule is linear. Here, a new model is introduced to deal with nonlinear key schedules as it is the case for AES. This model is more complex and actually it is too slow for exhaustive search. However, when some hints are added to the solver, it found the current best related-key boomerang attack on AES-192 with 2^{124} time, 2^{124} data, and $2^{79.8}$ memory complexities, which is better than the one presented by Biryukov and Khovratovich at ASIACRYPT 2009 with complexities $2^{176}/2^{123}/2^{152}$ respectively. This represents a huge improvement for the time and memory complexity, illustrating the power of MILP in cryptanalysis.

Keywords: AES · MILP · Boomerang attacks

1 Introduction

Boomerang attacks have been first discovered by Wagner in [Wag99] to attack block ciphers such as Khufu, COCONUT98, FEAL, and CAST at the end of the nineties. This technique is a differential-like attack where one can merge two high probability differential trails to attack more rounds and circumvent security proofs against differential attacks. In particular, Wagner applied this technique on COCONUT98 which has been proven resistant against differential attacks using the decorrelation technique introduced by Vaudenay [Vau03]. Many techniques can be used to prove the resistance of a cipher against differential attacks, for instance the AES highlighted the wide trail strategy, but it seems more difficult to prevent boomerang or any of its variants. Consequently, recent researches on cryptanalysis focus on this powerful technique.

Patrick Derbez, Pierre-Alain Fouque and Phuong Hoa Nguyen were partially supported by the French Agence Nationale de la Recherche through the DeCrypt project under Contract ANR-18-CE39-0007.

© International Association for Cryptologic Research 2022
S. Agrawal and D. Lin (Eds.): ASIACRYPT 2022, LNCS 13793, pp. 68–88, 2022.
https://doi.org/10.1007/978-3-031-22969-5_3

Assuming two differentials $\alpha \to \beta$ and $\gamma \to \delta$ (with probability p and q) on two halves of the algorithm, and considering them as independent, one can build a boomerang distinguisher with probability $p^2 q^2$. The basic idea consists in encrypting pairs of messages with a fixed difference α, assuming that a high proportion of them will follow the differential trail, and then apply another difference δ on both ciphertexts. After decrypting these new ciphertexts, one could expect to have a difference α between the plaintexts with a significant probability. Such distinguishers can be extended using standard methods to attacks recovering the secret key.

Since Wagner's paper, a series of variants of boomerang attacks (amplified boomerang, rectangle and sandwich) has drawn much attention on this technique. However, in 2011, Murphy *et al.* provided evidence on AES and DES that the independence argument was not always valid [Mur11]: some boomerang distinguishers have probability 0 *i.e.*, the boomerang never comes back. Moreover, Kircanski showed in [Kir15] using an SAT solver that some previous rectangle and boomerang attacks were based on incompatible characteristics.

Following such issues, the Boomerang Connectivity Table (BCT) has been introduced in [CHP+18] as a new tool to study the connection between the two trails. It can be seen as a precomputation of all the possible boomerangs at the S-box level. BCT solved the problem of incompatibility in boomerang distinguishers pointed out by Murphy. Since then, many improvements and further research into the BCT technique have enriched boomerang attacks.

Boomerang attacks have been applied successfully on AES [Bir04] and they are the most efficient attacks on AES-192 and AES-256 [BKN09,BK09]. At first glance, it is surprising that for these two versions, related-key boomerang attacks allow to cover all the rounds of the cipher and do not break a reduced-round version. However, it is well-known that for these versions, the diffusion in the key schedule is slow. These papers rely on the fact that there are very short local collisions if we can control the key schedule and the state of AES in a related-key attack. Such model is useful since AES was considered to construct a hash function at the end of the 2000s [GKM+09]. In [BKN09], Biryukov, Khovratovich and Nikolic present an attack on AES-256 using 2^{35} keys and 2^{96} for each key in time and data. The same year, in [BK09], Biryukov and Khovratovich improve the related key attack using only 4 keys on AES-192 with $2^{176}/2^{123}/2^{152}$ time/data/memory complexities and using 4 keys on AES-256 with $2^{99.5}/2^{99.5}/2^{77}$ time/data/memory complexities. In [BDK+10], Biryukov *et al.* show new attacks on 10-round AES-256 up to time complexity 2^{45}, data 2^{44} and memory 2^{33}, illustrating the fact that these attacks can be very effective in practice on round-reduced versions, which have been used most recently. The latter attack has been further extended by Biryukov and Khovratovich to 13 rounds with complexity 2^{76} in time, data and memory, but still in the related subkey model [BK10]. More recently, Dunkelman et al. in [DKRS20] have introduced a new technique, called retracing boomerang. By creating some correlations, they have been able to improve the probability of the distinguisher to $p^2 q$. They apply it on round-reduced AES and show that on 5-round AES we can recover the secret key with very low data complexity ($2^{16.5}$).

Some automatic tools have been developed for AES-128 in [BDF11] to look for low data complexity attacks on round-reduced AES. Then, more advanced differential attacks have been investigated in [DF13, DFJ13] leading to the best attack on 7-round AES-128 with time/data/memory complexity around 2^{100} and 8 and 9 rounds for AES-192 and AES-256 respectively. Li et al. have extended the former attack on AES-192 up to 9 rounds in [LJW14]. More recently, Mixed-integer Linear Programming (MILP) have been considered to automatically find boomerang distinguishers and attacks. For instance, Liu and Sasaki in [LS19], Song et al. in [SQH19] and Delaune et al. in [DDV20] have looked for the best distinguishers on the SKINNY blockcipher. In particular, Song's work shows that many boomerang distinguishers from [LGS17] against SKINNY and AES have a higher probability than expected. Finally, in [QDW+21], Qin et al. extend MILP capabilities to look for related-key rectangle attacks on SKINNY and ForkSkinny. This is a much harder problem and it has been done previously with ad-hoc tools in [BDF11, DF16] for Meet-in-the-middle (MITM) attacks on AES and impossible differential attacks on AES, mCRYPTON, SIMON, IDEA, KTANTAN, PRINCE and ZORRO. One of the most difficult task is to estimate the complexity of the attacks.

The current best time complexity for an attack on all rounds of AES-192 is $2^{189.7}$ for biclique attacks, and 2^{176} for related-key attacks. Table 1 shows some existing attacks against AES-192.

Table 1. Summary of existing attacks against AES-192. Note that biclique attacks against AES-192 are only accelerated exhaustive searches, with a complete loop on the key space.

Key size	Rounds	Time	Data	Memory	Type	Reference
192 bits	8/12	2^{172}	2^{107}	2^{96}	MITM	[DFJ13]
	9/12	$2^{182.5}$	2^{117}	$2^{165.5}$		[LJW14]
	10/12	2^{183}	2^{124}	N/A	Related-key Rectangle	[KHP07]
		2^{156}	2^{156}	2^{65}	Related-key Differential	[GLMS18]
	12/12	2^{176}	2^{123}	2^{152}	Related-key Boomerang	[BK09]
		$2^{190.16}$	2^{80}	2^{8}	Biclique	[BKR11]
		$2^{190.83}$	2	2^{60}		[BCGS14]
		$2^{189.76}$	2^{48}	2^{60}		[TW15]
		2^{124}	2^{124}	$2^{79.8}$	Related-key Boomerang	Sect. 3

Our model and source codes will be publicly available at https://gitlab.inria.fr/pderbez/asia-2022-aes

Our Contributions

Looking for attacks instead of distinguishers is a harder problem. It is worth mentioning that for instance our attack on AES-192 is built from a distinguisher that has a lower probability than Biryukov's attack, but that is easier to propagate through the rest of the cipher. To this end, we have to tweak MILP models that look only for boomerang distinguishers and that work only on *linear key schedules*. When there are differences in a nonlinear key, they may not be predictable and the differences intervening in the trails cannot merely be described as free or controlled as in [DDV20]. This makes the MILP model more complex as it is discussed in Sect. 4.2. Finally, as mentioned before, one important feature is the computation of the probability (Sect. 4.3) and objective function to evaluate the cost of the attacks which is actually a rough approximation (Sect. 4.4).

Then, we propose the best related-key boomerang attack on AES-192 known so far and we recover the one on AES-256 by Biryukov and Khovratovich, showing that our tool is working.

Organization of the Paper

We will begin by giving an overview of AES and related key boomerang attacks in Sect. 2. Then we will describe our new attack on AES-192 in Sect. 3. Next, in Sect. 4, we will recall the MILP model introduced in [DDV20] to search for boomerang distinguishers and explain how we adapted it to find our attack. Finally, Sect. 5 concludes the paper.

2   AES and Boomerang Attacks

2.1   Description of AES

The Advanced Encryption Standard [DR02] is a Substitution-Permutation Network (SPN) that can be instantiated using three different key sizes: 128, 192, and 256. The 128-bit plaintext initializes the internal state viewed as a 4×4 matrix of bytes as values in the finite field $\mathbb{F}_{256}$, which is defined using the irreducible polynomial $x^8 + x^4 + x^3 + x + 1$ over $\mathbb{F}_2$. Depending on the version of the AES, N_r rounds are applied to that state: $N_r = 10$ for AES-128, $N_r = 12$ for AES-192 and $N_r = 14$ for AES-256. Each of the N_r AES rounds (Fig. 1) applies four operations to the state matrix (except in the last round where the **MixColumns** operation is missing):

- **AddRoundKey** (AK) adds a 128-bit subkey to the state.
- **SubBytes** (SB) applies the same 8-bit to 8-bit invertible S-Box S 16 times in parallel on each byte of the state.
- **ShiftRows** (SR) shifts the i-th row left by i positions.
- **MixColumns** (MC) replaces each of the four columns C of the state by $M \times C$ where M is a constant 4×4 maximum distance separable matrix over $\mathbb{F}_{256}$.

After the N_r-th round has been applied, a final subkey is added to the internal state to produce the ciphertext. The key expansion algorithms to produce the $N_r + 1$ subkeys are described in Fig. 2 for each keysize. We refer to the original publication [DR02] for further details.

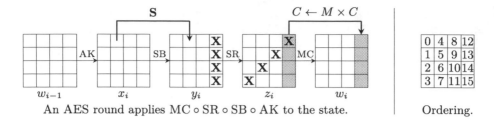

An AES round applies MC ∘ SR ∘ SB ∘ AK to the state.   Ordering.

Fig. 1. Description of one AES round and the ordering of bytes in an internal state.

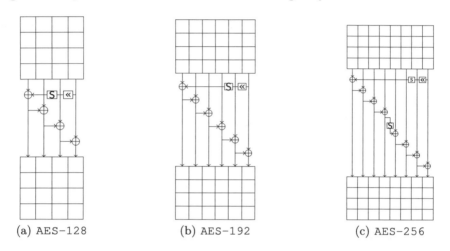

(a) AES-128   (b) AES-192   (c) AES-256

Fig. 2. Key schedules of the variants of the AES: AES-128, AES-192 and AES-256.

Notations. In this paper, we count the AES rounds from 0 and we refer to a particular byte of an internal state x by $x[i]$, as depicted in Fig. 1. Moreover, in the ith round, we denote the internal state after **AddRoundKey** by x_i, after **SubBytes** by y_i, after **ShiftRows** by z_i and after **MixColumns** by w_i. To refer to the difference in a state x, we use the notation Δx. The first added subkey is the master key k_{-1}, and the one added after round i is denoted k_i.

2.2 Probability of Boomerang Distinguishers

In a boomerang distinguisher, a cipher E is regarded as the composition of two sub-ciphers E_0 and E_1 so that $E = E_1 \circ E_0$. Suppose there exist both a differential $\gamma \to \theta$ for E_0 and a differential $\lambda \to \delta$ for E_1 with probabilities p and q respectively.

If we assume the two differentials are independent then we obtain a boomerang distinguisher of probability:

$$\mathbb{P}\left(E^{-1}(E(P) \oplus \delta) \oplus E^{-1}(E(P \oplus \gamma) \oplus \delta) = \gamma\right) = p^2 q^2.$$

But in practice the independence assumption does not always hold, especially at the junction of both the lower and upper differentials, and many counterexamples have already been found [WKD07, Mur11, Kir15]. However, since the work of Delaune *et al.* [DDV20], we know how to precisely compute the probability of a boomerang characteristic for any SPN, assuming round independence. The idea is to compute the probability of transitions for each S-box independently, using five different tables covering all the possible cases.

Definition 1. *Given a n-bit S-box S and four differences* $\gamma, \theta, \lambda, \delta \in \mathbb{F}_2^n$, *the DDT, BCT [CHP+18], U(pper)/L(ower)BCT [WP19] and EBCT [DDV20] are defined as*

$$DDT(\gamma, \theta) = \#\{x \in \mathbb{F}_2^n \mid S(x) \oplus S(x \oplus \gamma) = \theta\}$$

$$BCT(\gamma, \delta) = \#\{x \in \mathbb{F}_2^n \mid S^{-1}(S(x) \oplus \delta) \oplus S^{-1}(S(x \oplus \gamma) \oplus \delta) = \gamma\}$$

$$UBCT(\gamma, \theta, \delta) = \#\left\{x \in \mathbb{F}_2^n \;\middle|\; \begin{array}{l} S(x) \oplus S(x \oplus \gamma) = \theta \\ S^{-1}(S(x) \oplus \delta) \oplus S^{-1}(S(x \oplus \gamma) \oplus \delta) = \gamma \end{array}\right\}$$

$$LBCT(\gamma, \lambda, \delta) = \#\left\{x \in \mathbb{F}_2^n \;\middle|\; \begin{array}{l} S(x) \oplus S(x \oplus \lambda) = \delta \\ S^{-1}(S(x) \oplus \delta) \oplus S^{-1}(S(x \oplus \gamma) \oplus \delta) = \gamma \end{array}\right\}$$

$$EBCT(\gamma, \theta, \lambda, \delta) = \#\left\{x \in \mathbb{F}_2^n \;\middle|\; \begin{array}{l} S(x) \oplus S(x \oplus \gamma) = \theta \\ S(x) \oplus S(x \oplus \lambda) = \delta \\ S^{-1}(S(x) \oplus \delta) \oplus S^{-1}(S(x \oplus \gamma) \oplus \delta) = \gamma \end{array}\right\}$$

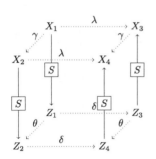

Fig. 3. Boomerang through one S-box

Figure 3 helps to understand the notations used in these definitions. Note that all those tables are particular cases of the Extended BCT (EBCT) in which some differences are free. Intuitively, the UBCT (resp. LBCT) corresponds to the junction between the upper (resp. lower) trail and the boomerang switch, while the EBCT deals with the middle rounds of the switch. From them, one can determine the associated probabilities by dividing by 2^n. Then the probability of a boomerang distinguisher is merely the product of the probabilities of the transitions through all the S-boxes of the characteristic. For a specific S-box, the table to be used is determined by the set of its inputs/outputs which are set to a fixed value in the upper and the lower trail. This technique does not need middle rounds to be defined, so it handles smoothly the switch between the lower and the upper differentials of a boomerang.

2.3   Boomerang Attacks on AES

In [BK09], Biryukov and Khovratovich described the first attacks against the full versions of both AES-192 and AES-256 working for all keys. These attacks are related-key boomerang attacks relying on the low diffusion in the AES key schedules.

Related Keys. In the related-key model, the attacker is allowed to ask for the encryption and/or the decryption of messages under different unknown keys, related by chosen properties. More precisely, the attacker should be able to provide an algorithm $\mathcal{A}$ which takes as input a master key k_A and outputs the related keys.

For boomerang attacks, the algorithm $\mathcal{A}$ typically specifies the differences between some key bits. Because the key schedules of AES each involves non-linear S-boxes, we need to take care of the values of the differences. In particular, specifying non-zero differences both at the input and output of an S-box makes the algorithm $\mathcal{A}$ unable to generate outputs for all keys, leading to a weak-key attack. This is because there is no non-trivial differential transition through the AES S-box that holds with probability one. Regarding related-key boomerang attacks on AES, this means that the differences in the key bytes going through S-boxes in the middle of the distinguisher will most likely be null. This can be observed in both attacks as depicted in Fig. 4 and Fig. 5: the last column of almost each subkeys (as well as the fourth column for AES-256) is fully inactive for at least one of the trails.

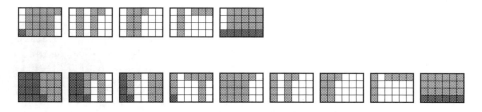

Fig. 4. Difference in the subkeys for the attack against AES-192 in [BK09]. First line is Δk, second line is ∇k. No difference are depicted in white bytes, known differences in green bytes and unknown differences in blue bytes. (Color figre online)

Boomerang Attack Against AES-256. It is based on a boomerang distinguisher of probability 2^{-96} covering all rounds but the first one. The distinguisher is extended by one round at the beginning as depicted in Fig. 6. The four keys k_A, k_B, k_C and k_D are generated such that they have the differences as specified in Fig. 5.

Then, the attack procedure is quite straightforward:

1. Ask for the encryption through both k_A and k_B of a structure of $2^{9 \times 8}$ plaintexts such that bytes 0, 1, 2, 3, 5, 9, 10, 13 and 15 take all the possible values while the remaining ones are constant.

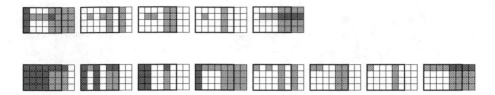

Fig. 5. Difference in the subkeys for the attack against AES-256 in [BK09]. First line is Δk, second line is ∇k. No difference are depicted in white bytes, known differences in green bytes and unknown differences in blue bytes.(Color figure online)

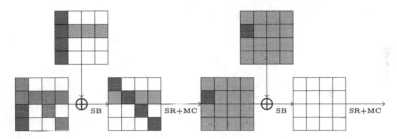

Fig. 6. Cancellations of difference in the first round of the attack against AES-256 in [BK09]. No difference in white bytes, known differences in green bytes and unknown differences in blue and gray bytes. Differences in blue bytes fully depend on the keys and known differences.(Color figure online)

2. For each plaintext p_A, look at the corresponding ciphertext c_A, apply the right difference to compute c_C and ask for its decryption under key k_C. Store the resulting plaintext p_C into a hash table indexed by the 7 constant bytes as well as $p_A[2] \oplus p_C[2]$ and $p_A[3] \oplus p_C[3]$. Indeed, while unknown, the difference in bytes 2 and 3 of the plaintexts should be equal for both pairs to satisfy the distinguisher.
3. Repeat the previous step from plaintext p_B and key k_D.
4. Look for collisions in the hash table to obtain $2^{72+72-56-16} = 2^{72}$ possible quartets.
5. For each quartet, regarding the 2 pairs (p_A, p_B) and (p_C, p_D), we know the difference at the input and at the output of $9 \times 2 = 18$ S-boxes. Each of them has on average one solution for the corresponding values and thus each quartet leads on average to one value for $18 \times 2 = 36$ bytes of key (9 for each key). Each time a value is reached, we increase a counter.
6. Because the probability of the distinguisher is 2^{-96} and the probability that one pair of the structure passes the first round is 2^{-72}, we need on average $2^{96+72-2\times72} = 2^{24}$ structures to get one right quartet. Thus we repeat the previous steps until a counter reaches the value 3 which should hardly correspond to a wrong value. This requires around $2^{25.5}$ structures.
7. The remaining key bytes are gradually recovered as detailed in [BK09].

This attack thus requires to encrypt and decrypt $2^{25.5+72} = 2^{97.5}$ messages for each of the four related keys and to process the same number of quartets leading to an overall complexity of $4 \times 2^{97.5} = 2^{99.5}$. Note that in the original attack the authors propose to perform more filtering on the quartets before increasing the counter, reducing the memory complexity to $2^{77.5}$.

Boomerang Attack Against AES-192. The attack against this AES version is similar to the attack against AES-256. But the key schedule of AES-192 has a better diffusion and, in particular the differences in the ciphertexts are no longer fully known. As a result, Biryukov and Khovratovich describe a procedure to recover the keys with a data complexity of 2^{123}, a time complexity of 2^{176} and a memory complexity of 2^{152}. We refer to [BK09] for further details.

3   New Attack on AES-192

In this section we describe our new attack against full AES-192 which is actually very similar to the ones of Biryukov and Khovratovich. We found a slightly better boomerang distinguisher that can be much easily turned into a key-recovery attack.

3.1   Related Keys

Generating keys k_B, k_C and k_D from an original key k_A actually relies on a boomerang distinguisher with probability 1 on the key schedule algorithm. First we note that AES key schedules are such that all subkeys can be constructed from one of them. Thus, starting from k_A we apply the chosen difference on the second subkey and compute the corresponding key k_B. Then for both k_A and k_B we apply the chosen difference on the eighth subkey and compute k_C and k_D respectively. Because the differences form a boomerang of probability 1, we can ensure that the difference between keys k_C and k_D on the second subkey is equal to the difference between keys k_A and k_B.

In our new attack, we use related keys with differences as depicted in Fig. 7. Actual values of the differences are given in Table 2.

3.2   The Attack

Our attack is based on the boomerang trail depicted in Fig. 8. The boomerang distinguisher covers all rounds but the first and the last one and has probability $2^{-2(4 \times 6)} \times 2^{-2(5 \times 6)} = 2^{-108}$. Internal state differences are given in Appendix A. It is worth mentioning here that the boomerang distinguisher used in the original attack had probability 2^{-110}, highlighting that a small change in the distinguisher might lead to a much better attack.

The attack procedure is as follows:

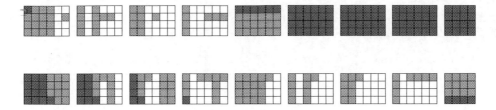

Fig. 7. Key schedule for this attack. The subkeys for the upper trail are represented above the ones of the lower trail

1. We first observe in Fig. 7 that the differences in the last row of the ciphertext all come from the unknown difference at the output of the S-box of $k_{11}[12]$ (the active byte on the last column of the round-key before the last round-key). Thus differences in $8+3 = 11$ linear combinations of bytes are fixed. Then, because the number of bytes for which the difference is known is bigger for the plaintexts than for the ciphertexts, it is better to start the attack from the ciphertexts. Thus we first ask for the decryption under k_A of a structure of $2^{5 \times 8} = 2^{40}$ ciphertexts such that bytes 1, 2, 5, 6, 9, 10, 13 and 14 are constant as well as $c[3] \oplus c[7]$, $c[3] \oplus c[11]$ and $c[3] \oplus c[15]$, and while bytes 0, 3, 4, 8 and 12 take all the possible values.
2. We ask for the decryption of a similar structure under k_C, taking care that the constant values match the required difference in the 11 linear combination of bytes given above.
3. In the trail, 2 plaintext bytes have an unknown difference. Luckily, for both of them we know the expected difference after application of the S-box and thus there are only 2^{14} possible differences for the plaintexts. Hence, for each ciphertext c_A and each of the 2^{14} possible differences, we look at the corresponding plaintext p_A, apply the difference to compute p_B and ask for its encryption under key k_B. We finally store the resulting ciphertext c_B into a hash table.
4. We repeat the previous step from ciphertext c_C and key k_D.
5. We now look for collisions in the hash table on the 11 linear combinations of bytes and should obtain on average $2^{2(40+14)-11 \times 8} = 2^{20}$ possible quartets.
6. For each quartet, regarding the 4 pairs (p_A, p_B), (p_C, p_D), (c_A, c_C) and (c_B, c_D), we know the difference at the input and at the output of $7 \times 2 = 14$ S-boxes (2 from (p_A, p_B), 2 from (p_C, p_D), 5 from (c_A, c_C) and 5 from (c_B, c_D)). In particular, this applies a $14 - 4 = 10$-bit filter (4 Sboxes are already used at Step 3) on the quartets and thus only $2^{20-10} = 2^{10}$ of them pass this test. Each of them leads on average to 2^{14} values for 28 bytes of key (7 for each key). Each time a value is reached, we increase a counter.
7. Because the probability of the distinguisher is 2^{-108} and the probability that one pair of the structure passes the last round is 2^{-40}, we need on average $2^{108+40-2 \times 40} = 2^{68}$ structures to get one right quartet. Thus we repeat the previous steps until a counter reaches the value 2. This requires around 2^{69} structures.

8. The remaining key bytes are gradually recovered using a right quartet and available data.

Data Complexity. In this attack we decrypt $2^{69+40} = 2^{109}$ messages under the keys k_A and k_C respectively but we encrypt $2^{69+40+14} = 2^{123}$ messages under the keys k_B and k_D.

The Counter. As described in the procedure, we have to update a counter around $2^{10+14+69} = 2^{93}$ times. This can be reduced to 2^{79} by storing sequences of 14 ordered pairs of 2 bytes instead of sequences of 28 bytes. Indeed, given the input and output differences of an S-box, the symmetric of any solution for the actual values is a solution as well. Furthermore, this barely affects the success of the procedure since more than half of the sequences of 28 bytes are actually sequences of 14 ordered pairs of 2 bytes.

Noisy Quartets. We generate through the procedure 2^{79} sequences of 14 ordered pairs of 2 bytes. Theoretically, there are more than $2^{8 \times 28 - 1} = 2^{223}$ such sequences. Hence, we expect on average 2^{-49} noisy quartets increasing the same counter.

Memory Complexity. The hash table used during Step 5 contains $2 \times 2^{40+14} = 2^{55}$ messages. We also need to store 2^{79} sequences of 28 bytes for the counter. Thus the memory complexity is $2^{79.8}$ 128-bit blocks.

Time Complexity. Filling the hash table 2^{69} times requires to process $2 \times 2^{109} + 2 \times 2^{123}$ messages and the counter is updated 2^{79} times. Regarding the missing key bytes, we used the tool developed by Bouillaguet *et al.* [BDF11] which found a procedure to enumerate all their possible values using only the constraints on the round keys in 2^{104} operations. The idea is once we know the value of the 4 keys on one byte, we know the differences on this byte in both trails. In particular we obtain the differences in some of the blue bytes leading to the knowledge of new actual values and so on. Then the 2^{104} solutions can be tested against available data.

All in all, the data complexity is $2 \times 2^{109} + 2 \times 2^{123} \approx 2^{124}$, the memory complexity is $2^{79.8}$ and the time complexity is $2 \times 2^{109} + 2 \times 2^{123} + 2^{79} \approx 2^{124}$.

4   New MILP Model

Our new attack against AES-192 was found using a new MILP model dedicated to AES. In this section we thus describe this new model and discuss its limitation.

4.1   Previous Works

In 2020, both Delaune *et al.* [DDV20] and Hadipour *et al.* [HBS21] independently proposed new MILP models to search for boomerang distinguishers and applied them to the block ciphers SKINNY. Recently, at EUROCRYPT'22, Dong *et al.* [DQSW21] improved those models by adding some new constraints

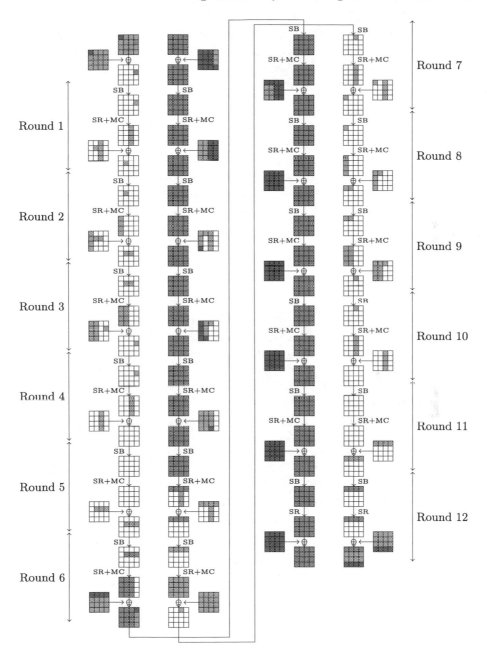

Fig. 8. Boomerang attack against `AES-192`. We recall that white stands for no difference, blue for a set difference, green for a known difference and gray for a free variable. (Color figure online)

and a new objective function to directly search for the best rectangle attacks. Their attacks were applied to two more rounds than the previous attack on SKINNY and found better attacks on other block ciphers such as ForkSkinny, Deoxys, GIFT, Serpent.

All those models highly rely on the linearity of the key schedule and the simplicity of the internal linear layer of each target. Thus they are not well-suited to study AES. When the key schedule is nonlinear, the differences in it may be unpredictable. Therefore, the differences intervening in the trails cannot merely be described as free or controlled as in [DDV20]: some differences take a known specific value, some take an unknown (coming from the key) specific value and some are free (they can take any value uniformly). Following the previous model of [DDV20], we thus introduce a new model to search for boomerang attacks.

4.2   New Variables and Constraints

For each step and for each trail, each of the 16 differences of the state or the round key has to be described by three values answering the three following questions: is it null? is it known? is it set to a specific value? Thus for any byte a of an internal state or a round key we define three binary variables a^z, a^k, a^s containing the Boolean answers to those questions. Because we directly want to search for attacks and not only distinguishers we also add an extra binary variable a^d indicating whether the byte a belongs to the distinguisher or to the key-recovery phase.

There are several straightforward constraints involving those variables. The first and most important one is $a^z \leq a^k \leq a^s$ which states that if the difference is zero then it is known and if it is known it is set to a specific value. Furthermore, if a is a key variable then its difference cannot be free and thus $a^s = 1$. We also impose that each variable a belongs to the distinguisher in either the upper trail or the lower one which is translated into the constraint:

$$a^{d,lo} + a^{d,up} \geq 1.$$

Let us now describe precisely the constraints for each inner component of AES.

SubBytes. Let $b = S(a)$ where S is the AES S-box. The first simple constraint is $a^z = b^z$ since if the input or output difference is null then both the input and output differences are null.

ShiftRows. This operation being a permutation of the bytes it does not affect the variables.

MixColumns. Let $(b_1, b_2, b_3, b_4) = MC(a_1, a_2, a_3, a_4)$. Because the matrix used in this operation is Maximum Distance Separable (MDS) we can simplify the constraints between the variables into

$$a_1^u + \ldots + b_4^u \in \{0, 1, 2, 3, 8\}, \text{ for } u \in \{z, k, s, d\}.$$

This can be easily translated into MILP constraints by adding an extra binary variable e and enforcing:

$$8 - a_1^u - \ldots - b_4^u \geq 5e$$
$$8 - a_1^u - \ldots - b_4^u \leq 8e$$

AddRoundKey. In this operation, variables are related by an equation of the form $a \oplus b \oplus c = 0$. This corresponds to the 3 inequalities:

$$a^u - b^u - c^u \geq -1$$
$$b^u - c^u - a^u \geq -1$$
$$c^u - a^u - b^u \geq -1$$

which ensure $a^u + b^u + c^u \neq 2$ for $u \in \{z, k, s, d\}$.

We now need some constraints about the variables that belong to the distinguisher and the other ones. First we force that all key variables belong to the distinguisher. For the state variables, belonging to the distinguisher is a property propagated with probability 1. According to the notation introduced in Sect. 2, this means that for any r and i we have the following constraints:

propagation through **SubBytes**: $\begin{cases} x_r[i]^{d,up} \leq y_r[i]^{d,up} \\ x_r[i]^{d,lo} \geq y_r[i]^{d,lo} \end{cases}$

propagation through **MixColumns**: $\begin{cases} 4z_r[i]^{d,up} \leq \sum_{j=0}^{3} w_r[4\lfloor i/4 \rfloor + j]^{d,up} \\ 4w_r[i]^{d,lo} \leq \sum_{j=0}^{3} z_r[4\lfloor i/4 \rfloor + j]^{d,lo} \end{cases}$

In order to simplify the computation of the probability of the inner distinguisher, and more generally to simplify the whole attack, we add several extra constraints, mainly to ensure that transitions through the linear layers happen with probability 1. Here, we use the property that the constraint $a + b + c \geq 1$ only removes the solution $a = b = c = 0$ and its variants (e.g. $a + b + 1 - c \geq 1$ only removes $a = b = 1 - c = 0$). The new constraints are:

- Do not control difference outside the distinguisher:

$$\begin{cases} x_r[i]^{d,up} + 1 - x_r[i]^{s,up} + y_r[i]^{z,up} \geq 1 \\ y_r[i]^{d,lo} + 1 - y_r[i]^{s,lo} + x_r[i]^{z,lo} \geq 1 \\ w_r[i]^{d,up} + w_r[i]^{s,up} + 1 - z_r[4\lfloor i/4 \rfloor + j]^{s,up} \geq 1 & \text{for } j \in \{0,1,2,3\} \\ z_r[i]^{d,lo} + z_r[i]^{s,lo} + 1 - w_r[4\lfloor i/4 \rfloor + j]^{s,lo} \geq 1 & \text{for } j \in \{0,1,2,3\} \end{cases}$$

- Transitions through the linear layers happen with probability 1:

$$\begin{cases} 1 - z_r[i]^{d,up} + z_r[i]^{s,up} + 1 - w_r[4\lfloor i/4 \rfloor + j]^{s,up} \geq 1 & \text{for } j \in \{0,1,2,3\} \\ 1 - w_r[i]^{d,lo} + w_r[i]^{s,lo} + 1 - z_r[4\lfloor i/4 \rfloor + j]^{s,lo} \geq 1 & \text{for } j \in \{0,1,2,3\} \end{cases}$$

- Do not take back control inside the distinguisher:

$$\begin{cases} 1 - x_r[i]^{d,up} + 1 - y_r[i]^{s,up} + x_r[i]^{s,up} \geq 1 \\ 1 - y_r[i]^{d,lo} + 1 - x_r[i]^{s,lo} + y_r[i]^{s,lo} \geq 1 \\ x_r[i]^{s,up} + x_r[i]^{s,lo} \geq x_r[i]^{d,up} + y_r[i]^{d,lo} - 1 \\ y_r[i]^{s,up} + y_r[i]^{s,lo} \geq x_r[i]^{d,up} + y_r[i]^{d,lo} - 1 \end{cases}$$

Finally, as explained in Sect. 2.3, we want to ensure that all transitions through the keyschedule happen with probability 1. In particular, if a is a key variable and $b = S(a)$, we need to ensure that if the difference in both a and b are known then it is zero:

$$\begin{cases} 2a^{k,up} + b^{k,up} + b^{k,lo} \le 2a^{z,up} + 2a^{z,lo} + 2 \\ 2a^{k,lo} + b^{k,up} + b^{k,lo} \le 2a^{z,up} + 2a^{z,lo} + 2 \end{cases}$$

Note that the above inequalities involve both trails because there is no non-trivial transition through the BCT occurring with probability one.

4.3 Computing Probabilities

The probability of the inner distinguisher is computed as the product of the probability of each individual S-box transition. However, since some differences can be set but unknown, we have to extend the definitions of the BCT, UBCT, LBCT, EBCT and DDT tables. More precisely, given $b = S(a)$, we need to compute the probability of the transition for each value of $a^{z,up}$, $a^{k,up}$, $a^{s,up}$, $b^{z,up}$, $b^{k,up}$, $b^{s,up}$, $a^{z,lo}$, $a^{k,lo}$, $a^{s,lo}$, $b^{z,lo}$, $b^{k,lo}$ and $b^{s,lo}$. In practice, only 59 configurations are possible and for each of them we have to compute the associated probability. The novelty here is that some of the differences cannot be chosen to maximize the probability. For instance let consider the transition $\Delta_{in} \longrightarrow \Delta_{out}$ through the AES S-box. It is well known that we can choose $(\Delta_{in}, \Delta_{out})$ so that the probability of this transition is 2^{-6}. But now let assume that Δ_{in} is set to an unknown non-zero value and we have to choose Δ_{out}. Whatever the choice we make for it, in 126 cases the transition holds with probability 2^{-7}, in 1 case it holds with probability 2^{-6} and in 128 cases in holds with probability 0. Translated to the distinguisher, we would be able to compute the probability that the probability of the distinguisher is not zero and, in that case, its average probability. Unfortunately, performing such precise computation for all configurations was out of reach. Instead, we only computed the average probability and would say that the transition $\Delta_{in} \longrightarrow \Delta_{out}$ holds with probability 2^{-8} in the studied case.

Overall we found 11 different possible probabilities: 2^0, $2^{-5.4}$, 2^{-6}, 2^{-8}, 2^{-12}, $2^{-13.4}$, 2^{-14}, 2^{-16}, 2^{-20}, $2^{-21.4}$ and 2^{-24}. Using classical techniques to lower the number of inequalities (mainly using the Quine-McCluskey algorithm), we were able to include the computation of the probability into our MILP model by using 5 extra binary variables and 33 inequalities per S-box.

Because the distinguisher should allow to actually distinguish the block cipher from a random permutation, we added a constraint to ensure that its probability is higher than 2^{-127}.

4.4 Objective Function

Precisely evaluating the complexity of a boomerang attack is highly non-trivial and thus we chose to explore another direction. In our opinion, what matters the most for the complexity of the whole attack is on the one hand the probability of the distinguisher (p_{dist}, the $-\log_2$ of the probability) and on the other hand the number of bytes in which the differences are known in both the plaintexts and the ciphertexts. Furthermore, variables set to a specific values are more interesting than free ones since they may depend on the same unknown differences. This is actually the case in our new attack against AES-192. Thus, we set as objective the following expression:

$$2 \times \left(\sum_{i=0}^{15} p[i]^{k,up} + c[i]^{k,lo} \right) + 6 \times \left(\sum_{i=0}^{15} p[i]^{s,up} + c[i]^{s,lo} \right) - p_{dist},$$

and we asked the MILP solver Gurobi[1] to maximize it. Note that we can choose other coefficients than $(2, 6)$ as long as they both are positive and sum to 8. It mainly depends on how confident we are that unknown but set differences will be related to each other.

4.5 Callback

The problem with our model is that we cannot exhaust all the possible relations between the variables. For instance, whenever 5 variables of the same column of z_r and x_{r+1} are known, a linear combination of the round key bytes is known as well. We used a callback to overcome this issue. When the MILP solver found a solution, the callback checks whether it is a valid solution, and otherwise removes it via lazy constraints. We refer interested readers to [DL22] for more information regarding *lossy modelization*.

Assuming we have an equation of the form $\alpha_1 a_1 \oplus \ldots \oplus \alpha_n a_n = \beta$ where the α_i's and β are constant, we need to add to the MILP model the constraints:

$$a_1^z + \ldots + a_n^z \neq n - 1$$
$$a_1^k + \ldots + a_n^k \neq n - 1$$
$$a_1^s + \ldots + a_n^s \neq n - 1$$
$$a_1^d + \ldots + a_n^d \neq n - 1$$

Because of the main constraints of the model, it is quite unlikely that the two last constraints are violated. However, it happens regularly for the two first ones. Checking if one such constraint is violated is actually pretty simple. We first perform a Gauss-Jordan elimination on the system of equations describing AES, echelonizing on the variables a for which $a^u = 0$ in the solution. Then we go through those equations and for each of them we check whether it satisfies the $a_1^u + \ldots + a_n^u \neq n - 1$. If one equation does not, and say for instance that $a_1^u = 0$, we add the constraint $a_2^u + \ldots + a_n^u \leq n - 2 + a_1^u$ to the model.

During the callback, we check as well whether generating the keys can be done using a boomerang of probability 1. Given the system of equations, we first echelonize on state variables a for which $a^{k,up} = a^{k,lo} = 0$. Then we recursively

---

[1] https://www.gurobi.com/.

echelonize on the key variables a for which $a^{k,up} = a^{k,lo} = 0$ and appearing *linearly* in the remaining equations (i.e. only a or $S(a)$ appears). At the end of the process, all the remaining variables should be known. Otherwise a lazy constraint is added to the model.

4.6 Limitations

Actually our model was too slow to exhaust boomerang attacks on AES. We identified two main problems:

1. When solving the relaxed problem in which all variables are not restricted to integers, Gurobi does set $a^z = a^k = a^s$ whenever it is possible. For an integer solution this would be either $(0, 0, 0)$ which corresponds to a free variable or $(1, 1, 1)$ which corresponds to a zero difference. Unfortunately, this scales badly with our constraints related to the probability of the distinguisher as this leads to a probability equals to 1. Thus the bound in Gurobi is moving very slowly.
2. Looking at the solutions for which the callback has to add a lazy constraint, we noticed that in most cases Gurobi sets a column of x_{r+1} and 3 bytes of the same column of z_r with a null difference while the corresponding column on k_r was fully set to non-zero difference (known or unknown). The problem is that on itself this configuration is possible but, in practice, it rarely passes the callback constraint regarding the keys generation process.

Thus we had to add some additional constraints to the model. More precisely, we ran the model by setting the number of active S-boxes in most of the relevant states (3 on the upper trail and 3 on the lower one). As a result, we obtained the attack against AES-192 described in Sect. 3. We also recovered the attack of Biryukov *et al.* against AES-256.

Note that the model is very sensitive to those extra constraints. In practice, when setting the right number of active S-boxes for 6 well-chosen states, Gurobi takes less than an hour to output the optimal pattern. But for instance if we only set the number of active S-boxes to be at most 3 (for the same 6 states), then Gurobi was still far from the optimal pattern after few days. Thus we believe it is worth improving the modelization of the problem to ensure the boomerang attack we found against AES-192 is truly optimal.

5   Conclusion

In this paper we described a new related-(sub)keys attack against full AES-192. Its complexity is 2^{52} times lower than the original attack of Biryukov and Khovratovich published at ASIACRYPT'09 while relying on a slightly better distinguisher. This highlights once again that directly searching the attack is very important as distinguishers with similar probabilities might lead to key-recovery attacks with very different complexities. Contrary to AES-256, AES-192 has a faster diffusion which makes the search of such attacks harder and is a good testbed for our tool.

We also described the MILP model which helped us to find this attack. We believe this model can still be improved a lot and opens an interesting research direction regarding automatic search of boomerang attacks with nonlinear key schedule.

A   New Distinguisher on AES-192

Table 2. Key schedule difference in the AES-192 trail

ΔK^0	?	21	21	21	00	00	ΔK^1	21	00	21	00	00	00	ΔK^2	21	21	00	00	00	00
	3e	3e	3e	3f	00	01		3e	00	3e	01	01	00		3e	3e	00	01	00	00
	1f	1f	1f	1f	00	00		1f	00	1f	00	00	00		1f	1f	00	00	00	00
	1f	1f	1f	1f	00	00		1f	00	1f	00	00	00		1f	1f	00	00	00	00
ΔK^3	21	00	00	00	00	00	ΔK^4	?	?	?	?	?	?	ΔK^5	?	?	?	?	?	?
	3e	00	00	01	01	01		3e	3e	3e	3f	3e	3f		?	?	?	?	?	?
	1f	00	00	00	00	00		1f	1f	1f	1f	1f	1f		?	?	?	?	?	?
	1f	00	00	00	00	00		1f	1f	1f	1f	1f	1f		?	?	?	?	?	?
ΔK^6	?	?	?	?	?	?	ΔK^7	?	?	?	?	?	?	ΔK^8	?	?	?	?	?	?
	?	?	?	?	?	?		?	?	?	?	?	?		?	?	?	?	?	?
	?	?	?	?	?	?		?	?	?	?	?	?		?	?	?	?	?	?
	?	?	?	?	?	?		?	?	?	?	?	?		?	?	?	?	?	?
∇K^0	?	?	?	f8	33	f8	∇K^1	?	?	33	cb	f8	00	∇K^2	?	f8	cb	00	f8	f8
	?	?	?	7c	7c	7c		?	?	7c	00	7c	00		?	7c	00	00	7c	7c
	?	?	?	7c	7c	7c		?	?	7c	00	7c	00		?	7c	00	00	7c	7c
	?	?	?	?	84	84		?	?	?	00	84	00		?	?	00	00	84	84
∇K^3	f8	00	cb	cb	33	cb	∇K^4	f8	f8	33	f8	cb	00	∇K^5	f8	00	33	cb	00	00
	7c	00	00	00	7c	00		7c	7c	7c	7c	00	00		7c	00	7c	00	00	00
	7c	00	00	00	7c	00		7c	7c	7c	7c	00	00		7c	00	7c	00	00	00
	?	00	00	00	84	00		84	84	84	84	00	00		84	00	84	00	00	00
∇K^6	f8	f8	cb	00	00	00	∇K^7	f8	00	cb	cb	cb	cb	∇K^8	f8	f8	33	f8	33	f8
	7c	7c	00	00	00	00		7c	00	00	00	00	00		7c	7c	7c	7c	7c	7c
	7c	7c	00	00	00	00		7c	00	00	00	00	00		7c	7c	7c	7c	7c	7c
	84	84	00	00	00	00		84	00	00	00	00	00		?	?	?	?	?	?

Table 3. Internal state difference in the AES-192 trail

ΔP	?	00	00	00	Δy^1	00	00	00	00	Δy^2	00	00	00	00	Δy^3	00	00	00	00
	00	00	00	?		00	00	00	1f		00	1f	00	00		00	1f	1f	00
	00	00	00	00		00	00	00	00		00	00	00	00		00	00	00	00
	00	00	00	00		00	00	00	00		00	00	00	00		00	00	00	00
Δy^4	00	00	00	00	Δy^5	00	00	00	00	Δy^6	00	00	00	00	Δy^7	?	?	?	?
	00	00	00	1f		00	00	00	00		00	?	?	?		?	?	?	?
	00	00	00	00		00	00	00	00		00	00	00	00		?	?	?	?
	00	00	00	00		00	00	00	00		00	00	00	00		?	?	?	?
∇y^6	7c	7c	7c	7c	∇y^7	00	00	7c	00	∇y^8	7c	00	00	00	∇y^9	7c	7c	00	00
	00	00	00	00		00	00	00	00		00	00	00	00		00	00	00	00
	00	00	00	00		00	00	00	00		00	00	00	00		00	00	00	00
	00	00	00	00		00	00	00	00		00	00	00	00		00	00	00	00
∇y^{10}	00	00	7c	00	∇y^{11}	00	00	00	00	∇y^{12}	?	?	?	?	ΔC	?	?	?	?
	00	00	00	00		00	00	00	00		00	00	00	00		7c	7c	7c	7c
	00	00	00	00		00	00	00	00		00	00	00	00		7c	7c	7c	7c
	00	00	00	00		00	00	00	00		00	00	00	00		?	?	?	?

References

[BCGS14] Bogdanov, A., Chang, D., Ghosh, M., Sanadhya, S.K.: Bicliques with minimal data and time complexity for AES. In: Lee, J., Kim, J. (eds.) ICISC 2014. LNCS, vol. 8949, pp. 160–174. Springer, Cham (2015). https://doi.org/10.1007/978-3-319-15943-0_10

[BDF11] Bouillaguet, C., Derbez, P., Fouque, P.-A.: Automatic search of attacks on round-reduced AES and applications. In: Rogaway, P. (ed.) CRYPTO 2011. LNCS, vol. 6841, pp. 169–187. Springer, Heidelberg (2011). https://doi.org/10.1007/978-3-642-22792-9_10

[BDK+10] Biryukov, A., Dunkelman, O., Keller, N., Khovratovich, D., Shamir, A.: Key recovery attacks of practical complexity on AES-256 variants with up to 10 rounds. In: Gilbert, H. (ed.) EUROCRYPT 2010. LNCS, vol. 6110, pp. 299–319. Springer, Heidelberg (2010). https://doi.org/10.1007/978-3-642-13190-5_15

[Bir04] Biryukov, A.: The boomerang attack on 5 and 6-round reduced AES. In: Dobbertin, H., Rijmen, V., Sowa, A. (eds.) AES 2004. LNCS, vol. 3373, pp. 11–15. Springer, Heidelberg (2005). https://doi.org/10.1007/11506447_2

[BK09] Biryukov, A., Khovratovich, D.: Related-key cryptanalysis of the full AES-192 and AES-256. In: Matsui, M. (ed.) ASIACRYPT 2009. LNCS, vol. 5912, pp. 1–18. Springer, Heidelberg (2009). https://doi.org/10.1007/978-3-642-10366-7_1

[BK10] Biryukov, A., Khovratovich, D.: Feasible attack on the 13-round AES-256. IACR Cryptol. ePrint Arch., page 257 (2010)

[BKN09] Biryukov, A., Khovratovich, D., Nikolić, I.: Distinguisher and related-key attack on the full AES-256. In: Halevi, S. (ed.) CRYPTO 2009. LNCS, vol. 5677, pp. 231–249. Springer, Heidelberg (2009). https://doi.org/10.1007/978-3-642-03356-8_14

[BKR11] Bogdanov, A., Khovratovich, D., Rechberger, C.: Biclique cryptanalysis of the full AES. In: Lee, D.H., Wang, X. (eds.) ASIACRYPT 2011. LNCS, vol. 7073, pp. 344–371. Springer, Heidelberg (2011). https://doi.org/10.1007/978-3-642-25385-0_19

[CHP+18] Cid, C., Huang, T., Peyrin, T., Sasaki, Yu., Song, L.: Boomerang connectivity table: a new cryptanalysis tool. In: Nielsen, J.B., Rijmen, V. (eds.) EUROCRYPT 2018. LNCS, vol. 10821, pp. 683–714. Springer, Cham (2018). https://doi.org/10.1007/978-3-319-78375-8_22

[DDV20] Delaune, S., Derbez, P., Vavrille, M.: Catching the fastest boomerangs application to SKINNY. IACR Trans. Symmetric Cryptol. **2020**(4), 104–129 (2020)

[DF13] Derbez, P., Fouque, P.-A.: Exhausting Demirci-Selçuk meet-in-the-middle attacks against reduced-round AES. In: Moriai, S. (ed.) FSE 2013. LNCS, vol. 8424, pp. 541–560. Springer, Heidelberg (2014). https://doi.org/10.1007/978-3-662-43933-3_28

[DF16] Derbez, P., Fouque, P.-A.: Automatic search of meet-in-the-middle and impossible differential attacks. In: Robshaw, M., Katz, J. (eds.) CRYPTO 2016. LNCS, vol. 9815, pp. 157–184. Springer, Heidelberg (2016). https://doi.org/10.1007/978-3-662-53008-5_6

[DFJ13] Derbez, P., Fouque, P.-A., Jean, J.: [Improved key recovery attacks on reduced-round , in the single-key setting]. In: Johansson, T., Nguyen, P.Q. (eds.) EUROCRYPT 2013. LNCS, vol. 7881, pp. 371–387. Springer, Heidelberg (2013). https://doi.org/10.1007/978-3-642-38348-9_23

[DKRS20] Dunkelman, O., Keller, N., Ronen, E., Shamir, A.: The retracing boomerang attack. In: Canteaut, A., Ishai, Y. (eds.) EUROCRYPT 2020. LNCS, vol. 12105, pp. 280–309. Springer, Cham (2020). https://doi.org/10.1007/978-3-030-45721-1_11

[DL22] Derbez, P., Lambin, B.: Fast MILP models for division property. IACR Trans. Symmetric Cryptol. **2022**(2), 289–321 (2022)

[DQSW21] Dong, X., Qin, L., Sun, S.,Wang. X.: Key guessing strategies for linear key-schedule algorithms in rectangle attacks. IACR Cryptol. ePrint Arch., p. 856 (2021)

[DR02] Daemen, J., Rijmen, V.: The Design of Rijndael: AES - The Advanced Encryption Standard. Information Security and Cryptography, Springer Berlin, Heidelberg (2002). https://doi.org/10.1007/978-3-662-04722-4

[GKM+09] Gauravaram, P., et al.: Grøstl - a SHA-3 candidate. In: Handschuh, H., Lucks, S., Preneel, B., Rogaway, P. (eds.) Symmetric Cryptography, 11.01. - 16.01.2009, vol. 09031 of Dagstuhl Seminar Proceedings. Schloss Dagstuhl - Leibniz-Zentrum für Informatik, Germany (2009)

[GLMS18] Gérault, D., Lafourcade, P., Minier, M., Solnon, C.: Revisiting AES related-key differential attacks with constraint programming. Inf. Process. Lett. **139**, 24–29 (2018)

[HBS21] Hadipour, H., Bagheri, N., Song, L.: Improved rectangle attacks on SKINNY and CRAFT. IACR Trans. Symmetric Cryptol. **2021**(2), 140–198 (2021)

[KHP07] Kim, J., Hong, S., Preneel, B.: Related-Key rectangle attacks on reduced AES-192 and AES-256. In: Biryukov, A. (ed.) FSE 2007. LNCS, vol. 4593, pp. 225–241. Springer, Heidelberg (2007). https://doi.org/10.1007/978-3-540-74619-5_15

[Kir15] Kircanski. A.: Analysis of boomerang differential trails via a sat-based constraint solver URSA. In: Malkin, T., Kolesnikov, V., Lewko, A.B., Polychronakis, M. (eds.) Applied Cryptography and Network Security - 13th International Conference, ACNS 2015, New York, NY, USA, 2–5 June 2015, Revised Selected Papers, volume 9092 of LNCS, pp. 331–349. Springer, Cham (2015). https://doi.org/10.1007/978-3-319-28166-7

[LGS17] Liu, G., Ghosh, M., Song, L.: Security analysis of SKINNY under related-tweakey settings (long paper). IACR Trans. Symmetric Cryptol. **2017**(3), 37–72 (2017)

[LJW14] Li, L., Jia, K., Wang, X.: Improved single-key attacks on 9-round AES-192/256. In: Cid, C., Rechberger, C. (eds.) FSE 2014. LNCS, vol. 8540, pp. 127–146. Springer, Heidelberg (2015). https://doi.org/10.1007/978-3-662-46706-0_7

[LS19] Liu, Y., Sasaki, Yu.: Related-key boomerang attacks on GIFT with automated trail search including BCT effect. In: Jang-Jaccard, J., Guo, F. (eds.) ACISP 2019. LNCS, vol. 11547, pp. 555–572. Springer, Cham (2019). https://doi.org/10.1007/978-3-030-21548-4_30

[Mur11] Murphy, S.: The return of the cryptographic boomerang. IEEE Trans. Inf. Theory **57**(4), 2517–2521 (2011)

[QDW+21] Qin, L., Dong, X., Wang, X., Jia, K., Liu, Y.: Automated search oriented to key recovery on ciphers with linear key schedule applications to boomerangs in SKINNY and ForkSkinny. IACR Trans. Symmetric Cryptol. **2021**(2), 249–291 (2021)

[SQH19] Song, L., Qin, X., Hu, L.: Boomerang connectivity table revisited. Application to SKINNY and AES. IACR Trans. Symmetric Cryptol. **2019**(1):118–141 (2019)

[TW15] Tao, B., Wu, H.: Improving the biclique cryptanalysis of AES. In: Foo, E., Stebila, D. (eds.) Information Security and Privacy - 20th Australasian Conference, ACISP 2015, Brisbane, QLD, Australia, June 29 - July 1, 2015, Proceedings, volume 9144, LNCS, pp. 39–56. Springer, Cham (2015). https://doi.org/10.1007/978-3-319-19962-7

[Vau03] Vaudenay, S.: Decorrelation: A theory for block cipher security. J. Cryptol. **16**(4), 249–286 (2003)

[Wag99] Wagner, D.: The boomerang attack. In: Knudsen, L. (ed.) FSE 1999. LNCS, vol. 1636, pp. 156–170. Springer, Heidelberg (1999). https://doi.org/10.1007/3-540-48519-8_12

[WKD07] Wang, G., Keller, N., Dunkelman, O.: The delicate issues of addition with respect to xor differences. In: Adams, C., Miri, A., Wiener, M. (eds.) SAC 2007. LNCS, vol. 4876, pp. 212–231. Springer, Heidelberg (2007). https://doi.org/10.1007/978-3-540-77360-3_14

[WP19] Wang, H., Peyrin, T.: Boomerang switch in multiple rounds. Application to AES variants and Deoxys. IACR Trans. Symmetric Cryptol. **2019**(1):142–169 (2019)

On Secure Ratcheting with Immediate Decryption

Jeroen Pijnenburg[1] and Bertram Poettering[2]([✉])[iD]

[1] Royal Holloway, University of London, Egham Hill, Egham, Surrey, UK
[2] IBM Research Europe – Zurich, Säumerstr 4, 8803 Rüschlikon, Switzerland
poe@zurich.ibm.com

Abstract. Ratcheting protocols let parties securely exchange messages in environments in which state exposure attacks are anticipated. While, unavoidably, some promises on confidentiality and authenticity cannot be upheld once the adversary obtains a copy of a party's state, ratcheting protocols aim at confining the impact of state exposures as much as possible. In particular, such protocols provide *forward security* (after state exposure, past messages remain secure) and *post-compromise security* (after state exposure, participants auto-heal and regain security).

Ratcheting protocols serve as core components in most modern instant messaging apps, with billions of users per day. Most instances, including Signal, guarantee *immediate decryption* (ID): Receivers recover and deliver the messages wrapped in ciphertexts immediately when they become available, even if ciphertexts arrive out-of-order and preceding ciphertexts are still missing. This ensures the continuation of sessions in unreliable communication networks, ultimately contributing to a satisfactory user experience. While initial academic treatments consider ratcheting protocols without ID, Alwen *et al.* (EC'19) propose the first ID-aware security model, together with a provably secure construction. Unfortunately, as we note, in their protocol a receiver state exposure allows for the decryption of all prior *undelivered* ciphertexts. As a consequence, from an adversary's point of view, intentionally preventing the delivery of a fraction of the ciphertexts of a conversation, and corrupting the receiver (days) later, allows for correctly decrypting all suppressed ciphertexts. The same attack works against Signal.

We argue that the level of (forward-)security realized by the protocol of Alwen *et al.*, and mandated by their security model, is considerably lower than both intuitively expected and technically possible. The main contributions of our work are thus a careful revisit of the security notions for ratcheted communication in the ID setting, together with a provably secure proof-of-concept construction. One novel component of our model is that it reflects the progression of *physical time*. This allows for formally requiring that (undelivered) ciphertexts automatically *expire* after a configurable amount of time.

Please find the full version at https://ia.cr/2022/995.

S. Agrawal and D. Lin (Eds.): ASIACRYPT 2022, LNCS 13793, pp. 89–118, 2022.
https://doi.org/10.1007/978-3-031-22969-5_4

1   Introduction

We consider a communication model between two parties, Alice and Bob, as it occurs in real-world instant messaging (e.g., in smartphone-based apps like Signal). A key principle in this context is that the parties are only very loosely synchronized. For instance, a "ping-pong" alteration of the sender role is not assumed but parties can send concurrently, i.e., whenever they want to. Further, specifically in phone-based instant messaging, a generally unpredictable network delay has to be tolerated: While some messages are received split seconds after they are sent, it may happen that other messages are delivered only with a considerable delay.[1] We refer to this type of communication (with no enforced structure and arbitrary network delays) as *asynchronous*. We say that asynchronous communication has *in-order* delivery if messages always arrive at the receiver in the order they were sent (what Alice sends first is received by Bob before what she sends later); otherwise, if in-order delivery cannot be guaranteed by the network, we say that the communication has *out-of-order* delivery.

The central cryptographic goals in instant messaging are that the confidentiality and integrity of messages are maintained. As communication sessions are routinely long-lived (e.g., go on for months), and as mobile phones are so easily lost, stolen, confiscated, etc., the resilience of solutions against *state exposure attacks* has been accepted as pivotal. In such an attack, the adversary obtains a full copy of the attacked user's program state.[2] We say that a protocol provides *forward security* if after a state exposure the already exchanged messages remain secure (in particular confidential), and we say that it provides *post-compromise security* if after a state exposure the attacked participant heals automatically and regains full security.

Past research efforts succeeded with proposing various security models and constructions for the (in-order) asynchronous communication setting with state exposures [10,14,17–19,24,26]. The rule of thumb "the stronger the model the more costly the solution" applies also to the ratcheting domain, and the indicated works can be seen as positioned at different points in the security-vs-cost trade-off space. For instance, the security models of [17,24] are the strongest (for excluding no attacks beyond the trivial ones) but seem to necessitate HIBE-like building blocks [6], while [10,14,18] work with a relaxed healing requirement (either parties do not recover completely or recovery is delayed) that can be satisfied with DH-inspired constructions.

While the works discussed above exclusively consider communication with in-order delivery, popular instant-messaging solutions like Signal are specifically designed to tolerate out-of-order delivery [22, Sec. 2.6] in order to best deal with

---

[1] E.g., delays of hours can occur if a phone is switched off over night or during a long-distance flight.

[2] Program states could leak because of malware executed on the user's phone, by analyzing backup images of a phone's memory that are stored insufficiently encrypted in the cloud, by analyzing memory residues on swap drives, etc. Less technical conditions include that users are legally or illegally coerced to reveal their states.

the needs of users who want to effectively communicate despite temporary network outages, radio dead spots, etc. Given this means that the protocols cannot rely on ciphertexts arriving in the order they were sent, let alone that they arrive at all, the *immediate decryption* (**ID**) property of such protocols demands that independently of the order in which ciphertexts are received, and independently of the ciphertexts that might still be missing, any ciphertext shall be decryptable for immediate display in the moment it arrives.[3] The ID property received first academic attention in an article by Alwen, Coretti, and Dodis (**ACD**) [2]. As the authors point out, while virtually all practical secure messaging solutions do support ID, most rigorous treatments do not. The work of ACD aims at closing this gap. We revisit and refine their results.

The main focus of ACD is on the Double Ratchet (**DR**) primitive which is one of the core components of the Signal protocol [13,22]. DR was specifically developed to allow for simultaneously achieving forward and post-compromise security in ID-supporting instant messaging. ACD contribute a formal security model for this primitive and detail how instant messaging can be constructed from it. This approach, taken by itself, does not guarantee that their solution is secure also in an intuitive sense: As everywhere else in cryptography, if a model turns out to be weak in practical cases, so may be the protocols implementing it. Indeed, we identified an attack that should not be successful against a secure ID-supporting instant messaging protocol, yet if applied against the ACD protocol (or Signal) it leads to the full decryption of arbitrarily selected ciphertexts.

Our attack is surprisingly simple: Assume Alice encrypts, possibly spread over a timespan of months, a sequence of messages $m_1, \ldots, m_L$ and sends the resulting ciphertexts $c_1, \ldots, c_L$ to Bob. An adversary that is interested in learning the target message m_1, arranges that all ciphertexts with exception of c_1 arrive at their destination. By the ID property, Bob decrypts the ciphertexts $c_2, \ldots, c_L$ delivered to him and recovers the messages $m_2, \ldots, m_L$. Further, expecting that the missing c_1 is eventually delivered, he consciously remains in the position to eventually decrypt c_1. But if Bob can decrypt c_1, the adversary, after obtaining Bob's key material via a state exposure, can decrypt c_1 as well, revealing the target message m_1. Note that the attack is not restricted to targeting specifically the first ciphertext; it would similarly work against any other ciphertext, or against a selection of ciphertexts, and the adversary would in all cases fully recover the target messages from just one state exposure. That is, for an adversary who wants to learn specific messages of a conversation secured with Signal or the protocol of ACD, it suffices to suppress the delivery of the corresponding ciphertexts and arrange for a state exposure at some later time. This obviously contradicts the spirit of FS.

Main Conceptual Contributions. Our attack seems to indicate that the immediate decryption (ID) and forward security (FS) goals, by their very nature, are mutually exclusive, meaning that one can have the one or the other, but not both. Our interpretation is less black and white and involves refining both the ID and the FS notions. We argue that, while out-of-order delivery and ID features

---

[3] In the user interface, placeholders could indicate messages that are still missing.

are indeed necessary to deal with unreliable networks, it also makes sense to put a cap on the acceptable amount of transmission delay. For concreteness, let threshold δ specify a maximum delay that messages traveling on the network may experience (including when transmissions are less reliable). Then ciphertexts that are sent at a time t_1 and arrive at a time t_2 should be deemed useful and decryptable only if $\Delta(t_1, t_2) \leq \delta$, while they should be considered expired and thus disposable if $\Delta(t_1, t_2) > \delta$. Once a threshold δ on the delay is fixed, the ID notion can be weakened to demand the correct decryption of ciphertexts only if the latter are at most δ old, and the FS notion can be weakened to protect past messages under state exposure only if they are older than δ (or already have been decrypted). As we show, once the two notions have been weakened in this sense, they fit together without contradicting each other. That is, this article promotes the idea of integrating a notion of progressing physical time into the ID and FS definitions so that their seemingly inherent rivalry is resolved and one can have both properties at the same time.

Our models and constructions see δ as a configurable parameter. The value to pick depends on the needs of the participants. For instance, if Alice and Bob are political activists operating under an oppressive regime, choosing $\delta < 10$ mins might be useful; more relaxed users might want to choose $\delta = 1$ week. Note that for $\delta = \infty$ our definitions 'degrade' to the no-expiration setting of ACD.

Main Technical Contributions. We start with a compact description of our three main technical contributions. We expand on the topics subsequently.

In a nutshell, the contributions of this article are: (1) We introduce the concept of evolving physical time to formal treatments of secure messaging. This allows us to express requirements on the automatic expiration of ciphertexts after a definable amount of time. (2) We propose new security models for secure messaging with immediate decryption (ID). Our approach is to have the security definitions disregard the unavoidable trivial attacks but nothing else; this renders our models particularly strong. By incorporating the progressing of physical time into our notions, our FS and ID definitions are not in conflict with each other. (3) We contribute a proof-of-concept protocol that provably satisfies our security notions. Efficiency-wise our protocol might be less convincing than the ACD protocol and Signal, but it is definitely more secure.

(1) MODELING PHYSICAL TIME. Among the many possible approaches to formalizing evolving physical time, the likely most simple option is sufficient for our purposes. In our treatments we assume that participants have access to a local clock device that notifies them periodically through events referred to as *ticks* about the elapse of a configurable amount of time.[4] The clocks of all participants are expected to be configured to the same ticking frequency (e.g., one tick every one minute), but otherwise our synchronization demands are very moderate: The only aspect relevant for us is that when Alice sends a ciphertext at a time t_1 (according to her clock) and Bob receives the ciphertext at a time t_2 (accord-

---

[4] Modern computing environments provide such a service right away. For instance, in Linux, via the setitimer system call or the alarm standard library function.

ing to his clock), then we expect that the difference $\Delta(t_1, t_2)$ be meaningful to declare ciphertexts fresh or expired. More precisely, we deem ciphertexts with $\Delta(t_1, t_2) \leq \delta$, for a configurable threshold δ, fresh and thus acceptable, while we consider all other ciphertexts expired and thus discardable. Note here that threshold δ specifies both a maximum on the tolerated network delay and on a possibly emerging clock drift between the sender's and the receiver's clock. The right choice of threshold δ is an implementation detail which controls the robustness-security trade-off.[5] See above for a discussion on how to choose δ.

(2) SECURITY MODELS. We develop security models for secure messaging with out-of-order delivery and immediate decryption (ID). We claim two main improvements over prior definitions: (a) We incorporate physical time into all correctness and security notions. For instance, when formulating the correctness requirements, we do not demand the correct decryption of expired ciphertexts, and our confidentiality definitions deem state exposure based message recovery attacks successful if the targeted ciphertext is expired. (b) We formalize the maximum level of attainable security (under state exposures). Recall that ACD was designed for analyzing Double Ratchet based constructions which were proven to achieve only limited security already in the in-order delivery setting [17,24].[6] In contrast, our models are designed to exclude the unavoidable 'trivial' attacks but nothing else, thus guaranteeing the best-possible security. (In the full version we review examples of such trivial attacks. We also list attacks that are included in our model but excluded by the ACD model.)

(3) OUR CONSTRUCTION. We propose a proof-of-concept construction that provably satisfies our security definitions. Its cryptographic core is formed by two specialized types of key encapsulation mechanism (KEM): a KeKEM and a KuKEM. In a nutshell, our KeKEM (key-evolving KEM) primitive is a type of KEM where public and secret keys can be linearly updated 'to the next epoch', almost like in forward-secure PKE. In contrast, our KuKEM (key-updatable KEM) primitive allows for updating keys based on provided auxiliary input strings. In both cases, key updates provide forward secrecy, i.e., 'the updates cannot be undone'.[7] Together with additional more standard building blocks like (stateful) signatures,

---

[5] One might wonder about the resilience of computer clocks against desynchronization attacks where the adversary aims at desynchronizing participants. We note that instant messaging apps are typically run on mobile devices that have access to multiple independent clock sources (e.g., a local clock, NTP, GSM, and GNSS) that can be compared and relied upon when consistent. Only the strongest adversaries can arrange for a common deviation of all these clock sources simultaneously and even in this case our solutions degrade gracefully: If all clocks stop, the security of our solution doesn't degrade below the security defined by ACD.

[6] In a nutshell, DR provides optimal security only if used for ping-pong structured communication [17,24]. In contrast, the constructions of [17,24] provide security for *any* (in-order) communication pattern, though require stronger primitives than DR.

[7] We note that similar KEM variants have been proposed and used in prior work on instant messaging [6,17,24], so in this article we claim novelty for neither the concepts nor the constructions.

we finally obtain a secure instant messaging protocol. In addition to the cryptographic core, a considerable share of our protocol specification is concerned with data management: the KeKEM and KuKEM primitives require that senders and receivers perform their updates in a strictly synchronized fashion; if ciphertexts arrive out of order, careful bookkeeping is required to let the receiver update in the right order and at the right time.

When compared to the constructions of ACD and Signal, our construction is admittedly less efficient, primarily because (a) we employ the KuKEM and KeKEM primitives that seem to require a considerable computational overhead, and (b) the ciphertexts of our protocol are larger. Concerning (a), we note that prior work like [17,24] that achieves strongest possible security for the much less involved in-order instant messaging case uses the same primitives, and that results [6] indicate that their use is actually unavoidable. We conclude from this that the computational overhead that the primitives bring with them seems to represent the due price to pay for the extra security. A similar statement can be made concerning (b): If an instant messaging conversation is such that the sender role strictly alternates between Alice and Bob, then the ciphertext overhead of our protocol, when compared to Signal, is just a couple of bytes per message. If the sender role does not strictly alternate, the ciphertext size grows linearly in the number of messages that the sender still has to confirm to have arrived. Recalling that the non-alternating case is precisely the one where Signal fails to provide best-possible security, the ciphertext overhead seems to be fair given the extra security that is achieved.

Related Work

We start with providing a more detailed comparison of our results with those of the prior work mentioned above. We first remark that our results generalize the findings of [17,24]: If in our models the physical time is 'frozen', messages are always delivered, and messages are delivered in-order, they express exactly the same security guarantees as [17,24]. It is clear that as soon as time starts ticking our model is stronger: We allow state exposures once ciphertexts 'expire', while this concept does not exist in [17,24]. For out-of-order delivery the picture is more complicated: Note that when messages are delivered in-order, optimal security demands that user states immediately 'cryptographically diverge' when receiving an unauthentic ciphertext, but for out-of-order delivery the situation becomes more nuanced. Consider the scenario where Alice sends a message and is then state-exposed. Using the obtained state information, the adversary could now trivially and perfectly impersonate Alice towards Bob for the second message. That is, if Bob receives the second ciphertext first, there is no (cryptographic) way for him to tell whether it is authentic or not, i.e., to distinguish whether Alice sent or the adversary injected it. If the ciphertext was indeed sent by Alice, correctness would require that Bob remains able to decrypt the first ciphertext. Thus, the latter also has to hold if the ciphertext is unauthentic. Hence, in contrast to the setting with in-order delivery, in the out-of-order setting there are inherent limits to how much the states of Alice and Bob can 'cryptographically diverge' once unauthentic ciphertexts are processed.

Multiple weaker security definitions for secure messaging have been proposed [2,10,14,18]. We provide a brief overview about what makes their security notions suboptimal. In [10,14] the adversary is forbidden to impersonate a user when a secure key is being established. Hence, in this case the authors do not require recovery from a state exposure (which enables an impersonation attack). In [2,18] the construction can take longer than strictly necessary to recover from state exposures. This is encoded in the security games by artificially labeling certain win conditions as trivial. See [9] for an extensive treatment of the limitations of the ACD model. Moreover, in both works the user states are not required to immediately 'cryptographically diverge' for future ciphertexts when accepting an unauthentic ciphertext. We note that an important difference between our KuKEM and Healable key-updating Public key encryption (HkuPke) introduced in [18] is that HkuPke key updates are based on secret update information, while our KuKEM is updated with adversarially controlled associated data.

The security definitions of [2,17,18] assume a slightly different understanding of what it means to expose a participant. Our understanding is that exposures reveal the current protocol state of a participant to the adversary, while their approach is rather that exposures reveal the randomness used for the next sending operation. The two views seem ultimately incomparable, and likely one can find arguments for both sides. One argument that supports our approach is that modern computing environments have RNGs that *constantly* refresh their state based on unpredictable events (e.g., the RDRAND instruction of Intel CPUs or the urandom device in Linux) so that if one of the situations listed in Footnote 2 leads to a state exposure then it still can be assumed that the randomness used for the next sending operation is indeed safe. A third view considers state exposures to leak a party's state except for signing keys [1], which seems unrealistic (to us).

See [12] for a treatment of secure messaging in the UC setting.

Our work is not the first to consider a notion of physical time in a cryptographic treatment. See [25] for modeling approaches using linear counters, or [11,20,21] for encrypting data 'to the future'.

Recent work in the group messaging setting [4] similarly designs their protocol in a modular way and captures security in game based definitions. A main component, *continuous group key agreement* (CGKA) was first defined in [3] and the analysis of [5] shows, even in the passive case, no known CGKA protocol achieves optimal security without using HIBE.

Organization. This article considers the security and constructions of what we refer to as bidirectional out-of-order messaging protocols, abbreviated BOOM. In Sect. 3 we define the security model. In Sect. 4 we introduce non-interactive components that we employ in our construction. This includes the mentioned KuKEM and KeKEM primitives. In Sect. 5 we finally present our construction.

2   Notation

We write T or 1, and F or 0, for the Boolean constants True and False, respectively. For $t_1, t_2 \in \mathbb{N}$ we let $\Delta(t_1, t_2) := t_2 - t_1$ if $t_1 \leq t_2$ and $\Delta(t_1, t_2) := 0$ if $t_1 > t_2$. For $a, b \in \mathbb{N}$, $a \leq b$, we let $[a \mathinner{..} b] := \{a, \ldots, b\}$ and $[b] := [0 \mathinner{..} b]$ and $[\![a \mathinner{..} b]\!] := [a \mathinner{..} (b-1)]$ and $[\![b]\!] := [0 \mathinner{..} (b-1)]$. We further write $[\![\infty]\!]$ for the set of natural numbers $\mathbb{N} = \{0, \ldots\}$. Note that $[\![0]\!]$ represents the empty set.

We specify scheme algorithms and security games in pseudocode. In such code we write $var \leftarrow exp$ for evaluating expression exp and assigning the result to variable var. If var is a set variable and exp evaluates to a set, we write $var \xleftarrow{\cup} exp$ shorthand for $var \leftarrow var \cup exp$ and $var \xleftarrow{\cap} exp$ shorthand for $var \leftarrow var \cap exp$. A vector variable can be appended to another vector variable with the concatenation operator $\shortparallel$, and we write $var \xleftarrow{\shortparallel} exp$ shorthand for $var \leftarrow var \shortparallel exp$. We do *not* overload the $\shortparallel$ operator to also indicate string concatenation, i.e., the objects $\mathsf{a} \shortparallel \mathsf{b}$ and ab are not the same. We use $[\,]$ notation for associative arrays (i.e., the 'dictionary' data structure): Once the instruction $A[\cdot] \leftarrow exp$ initialized all items of array A to the default value exp, individual items can be accessed as per $A[idx]$, e.g., updated and extracted via $A[idx] \leftarrow exp$ and $var \leftarrow A[idx]$, respectively, for any expression idx.

Unless explicitly noted, any scheme algorithm may be randomized. We use $\langle\,\rangle$ notation for stateful algorithms: If alg is a (stateful) algorithm, we write $y \leftarrow alg\langle st \rangle(x)$ shorthand for $(st, y) \leftarrow alg(st, x)$ to denote an invocation with input x and output y that updates its state st. (Depending on the algorithm, x and/or y may be missing.) Importantly, and in contrast to most prior works, we assume that *any* algorithm of a cryptographic scheme may fail or abort, even if this is not explicitly specified in the syntax definition. This approach is inspired by how modern programming languages deal with error conditions via *exceptions*: Any code can at any time 'throw an exception' which leads to an abort of the current code and is passed on to the calling instance. In particular, if in our game definitions a scheme algorithm aborts, the corresponding game oracle immediately aborts as well (and returns to the adversary).

Security games are parameterized by an adversary, and consist of a main game body plus zero or more oracle specifications. The adversary is allowed to call any of the specified oracles. The execution of the game starts with the main game body and terminates when a 'Stop with exp' instruction is reached, where the value of expression exp is taken as the outcome of the game. If the outcome of a game G is Boolean, we write $\Pr[\mathrm{G}(\mathcal{A})]$ for the probability (over the random coins of G and $\mathcal{A}$) that an execution of G with adversary $\mathcal{A}$ results in the outcome T or 1. We define shorthand notation for specific combinations of game-ending instructions: While in computational games we write 'Win' for 'Stop with T', in distinguishing games we write 'Win' for 'Stop with b' (where b is the challenge bit). In any case we write 'Lose' for 'Stop with F'. Further, for a Boolean condition C, we write 'Require C' for 'If $\neg C$: Lose', 'Penalize C' for 'If C: Lose', 'Reward C' for 'If C: Win', and 'Promise C' for 'If $\neg C$: Win'.

3   Syntax and Security of BOOM

We formalize Bidirectional Out-of-Order Messaging (BOOM) protocols. The scheme API assumes the four algorithms *init, send, recv, tick* and a timestamp decoding function *ts*. The *init, send, recv* algorithms are akin to prior work and implement instance initialization, message sending, and message receiving, respectively.[8] The *tick* algorithm enables a user's instance to track the progression of physical time: It is assumed to be periodically invoked by the computing platform (e.g., once every second), and has no visible effect beyond updating the instance's internal state. This allows us to model physical time with an integer counter that indicates the number of occurred *tick* invocations of the corresponding participant. Independently of physical time, a notion of logical time is induced by the sequence in which messages are processed by a sender: We track logical time with an integer counter that indicates the number of occurred *send* invocations of the corresponding participant. The logical time associated with a sending operation is also referred to as the operation's sending index. Whenever a ciphertext is produced, we assume a production timestamp is attached to it. Formally, we demand that, given a ciphertext, the timestamp decoding function *ts* recovers the physical time and logical time of the sender at the point when it created the ciphertext by invoking the *send* algorithm. The timestamp notion will prove crucial to formulate conditions related to ciphertext expiration.

We proceed with defining the syntax, the semantics (execution environment and correctness), and the security notions associated with BOOM protocols.

SYNTAX. A (two-party) BOOM scheme for an associated-data space $\mathcal{AD}$ and a message space $\mathcal{M}$ consists of a state space $\mathcal{ST}$, a ciphertext space $\mathcal{C}$, algorithms *init, send, recv, tick*, and a timestamp decoding function *ts*. Algorithm *init* generates initial states $st_A, st_B \in \mathcal{ST}$ for the participants. Algorithm *send* takes a state $st \in \mathcal{ST}$, an associated-data string $ad \in \mathcal{AD}$, and a message $m \in \mathcal{M}$, and outputs an (updated) state $st' \in \mathcal{ST}$ and a ciphertext $c \in \mathcal{C}$. Algorithm *recv* takes a state $st \in \mathcal{ST}$, an associated-data string $ad \in \mathcal{AD}$, and a ciphertext $c \in \mathcal{C}$, and outputs an (updated) state $st' \in \mathcal{ST}$, an acknowledgment set $A \subseteq \mathbb{N}$, and a message $m \in \mathcal{M}$. (The understanding of output A is that when c was generated by the peer, then for all $i \in A$ the peer had received the ciphertext with sending index i.) Algorithm *tick* takes a state $st \in \mathcal{ST}$ and outputs an (updated) state $st' \in \mathcal{ST}$. Function *ts* takes a ciphertext $c \in \mathcal{C}$ and recovers a logical timestamp (sending index) $lt \in \mathbb{N}$ and a physical timestamp $pt \in \mathbb{N}$. If $\mathcal{P}(\mathbb{N})$ denotes the powerset of set $\mathbb{N}$, the BOOM API is thus as follows:

$$init \to \mathcal{ST} \times \mathcal{ST} \qquad tick\langle\mathcal{ST}\rangle \qquad ts\colon \mathcal{C} \to \mathbb{N} \times \mathbb{N}$$

$$\mathcal{AD} \times \mathcal{M} \to send\langle\mathcal{ST}\rangle \to \mathcal{C} \qquad \mathcal{AD} \times \mathcal{C} \to recv\langle\mathcal{ST}\rangle \to \mathcal{P}(\mathbb{N}) \times \mathcal{M}$$

---

[8] More precisely, our *recv* algorithm has a dedicated output for reporting to the invoking user which of the priorly sent own messages have been received by the peer; this output does not exist in prior work.

Semantics. We give game based definitions of correctness and security. Recall that the form of secure messaging that we consider supports the out-of-order processing of ciphertexts. This property, of course, has to be reflected in all games, rendering them more complex than those of prior works that deal with easier settings. To manage this complexity, we carefully developed our games such that they share, among each other, as many code lines and game variables as possible. In particular, the games can be seen as derived by individualizing a common basic game body in order to express specific aspects of functionality or security. This individualization is done by inserting an appropriate small set of additional code lines.[9] (For instance, the game defining authenticity adds lines of code that identify and flag forgery events.) In the following we explain first the BASIC game and then its refinements FUNC, AUTH, and CONF.[10]

Game BASIC. We first take a quick glance over the BASIC game of Fig. 1, deferring the discussion of details to the upcoming paragraphs. The game body [G00–G18] initializes some variables [G00–G13], invokes the *init* algorithm to initialize states for two users A and B [G17], and invokes the adversary [G18]. The adversary has access to four oracles, each of which takes an input $u \in \{A, B\}$ to specify the targeted user. The Tick oracle gives access to the *tick* algorithm [T00], the Send oracle gives access to the *send* algorithm [S00,S07], and the Recv oracle, besides internally recovering the logical and physical sending timestamps of an incoming ciphertext [R00], gives access to the *recv* algorithm [R01,R29]. Finally, the Expose oracle reveals the current protocol state of a user to the adversary [E06]. The game variables and remaining code lines are related to monitoring the actions of the adversary, allowing for identifying specific game states and tracking the transitions between them. In particular we identified the user-specific states *in-sync* and *authoritative*, the ciphertext properties *sync-preserving*, *sync-damaging*, *certifying*, and *vouching*, and the transitions *losing sync*, *poisoning*, and *healing*, as relevant in the BOOM setting. We explain these concepts one by one.

We say that protocol actors are synchronized if their views on the communication is consistent. A little more precisely, a participant Alice is in-sync with her peer Bob if all ciphertexts that Alice received are identical with ciphertexts that Bob priorly sent. The complete definition, formalized as part of the BASIC game as discussed below, further requires that the employed associated-data inputs are matching, and that the processing of ciphertexts of an out-of-sync peer also renders the receiver out-of-sync. If Alice is in-sync with Bob, we refer to ciphertexts that Alice can receive without losing sync as sync-preserving; the ciphertexts that would render her out-of-sync are referred to as sync-damaging.[11]

---

[9] Removing or modifying existing lines will not be necessary. That said, restricting the options to only add new lines might lead to also introducing a small number of redundancies that could allow for simplifications.

[10] The BASIC game itself is not used to model any kind of functionality or security. It merely describes the execution environment.

[11] The in-sync notion first surfaced in [7] in the context of unidirectional channels. It was extended in [23] to handle bidirectional communication and associated-data

Game BASIC($\mathcal{A}$)

G00 For $u \in \{A, B\}$:
G01 $lt_u \leftarrow 0$
G02 $pt_u \leftarrow 0$
G03 $is_u \leftarrow T$
G04 $SC_u \leftarrow \emptyset$
G05 $CERT_u \leftarrow \emptyset$
G06 $VF_u[\cdot] \leftarrow [\![\infty]\!]$
G07 $AU_u \leftarrow [\![\infty]\!]$
G13 $poisoned_u \leftarrow F$
G17 $(st_A, st_B) \leftarrow init$
G18 Invoke $\mathcal{A}$

Oracle Tick(u)

T00 $tick\langle st_u \rangle$
T01 $pt_u \leftarrow pt_u + 1$

Oracle Send(u, ad, m)

S00 $c \leftarrow send\langle st_u \rangle(ad, m)$
S02 If is_u:
S03 $SC_u \overset{\cup}{\leftarrow} \{(ad, c)\}$
S06 $lt_u \leftarrow lt_u + 1$
S07 Return c

Oracle Expose(u)

E01 If is_u:
E02 $VF_u[\![lt_u]\!] \overset{\cap}{\leftarrow} [\![lt_u]\!]$
E03 $AU_u \overset{\cap}{\leftarrow} [\![lt_u]\!]$
E06 Return st_u

Oracle Recv(u, ad, c)

R00 $(lt, pt) \leftarrow ts(c)$
R01 $(A, m) \leftarrow recv\langle st_u \rangle(ad, c)$
R06 If $(ad, c) \in SC_{\bar{u}}$:
R07   If is_u:
R08 $CERT_u \overset{\cup}{\leftarrow} [lt]$
R09 $AU_{\bar{u}} \overset{\cup}{\leftarrow} VF_{\bar{u}}[lt]$
R17 If $(ad, c) \notin SC_{\bar{u}}$:
R18   If is_u:
R19   If $lt \notin AU_{\bar{u}}$:
R20 $poisoned_u \leftarrow T$
R23 $is_u \leftarrow F$
R29 Return (A, m)

Fig. 1. Game BASIC. We refer the reader to Footnote 14 for the interpretation of $VF_u[\![lt_u]\!]$ in line [E02]. We write $\bar{u}$ for the element such that $\{u, \bar{u}\} = \{A, B\}$.

As we consider communication algorithms that are stateful, any ciphertext created by a participant may depend on, and may implicitly reflect, the full prior communication history of that participant. That is, if from a sequence of sent ciphertexts only a subset of ciphertexts arrive, then from what *did* arrive the receiver should be able to extract information linked to what was sent before but is still missing. In particular, any ciphertext that is received in-sync should allow for identifying which earlier-sent though later-delivered ciphertexts are authentic. We correspondingly say that in-sync received ciphertexts certify the ciphertexts sent *earlier* by the same sender.

Ciphertexts can also make promises about the future: Every received ciphertext may carry (cryptographic) information that is used to authenticate later ciphertexts (of the same sender, up to their next exposure). Here we say that ciphertexts (cryptographically) vouch for the ones sent *later* by the same participant.

We finally discuss attack classes that are enabled by exposing the states of users: Once a participant's state becomes known by exposure, it is trivial to impersonate the user, simply by invoking the scheme algorithms with the captured state. We refer to states of a participant as authoritative if their actions can *not* be trivially emulated by the adversary in this way. If an impersonation happens right after an exposure, as the adversary can perfectly and permanently emulate all actions of the impersonated party, in addition to all authenticity and confidentiality guarantees being lost, there is also no option to recover into a safe state. We refer to the transition into such a setting, more precisely to the action of exploiting the state exposure of one participant by delivering an impersonating ciphertext to the other participant, as poisoning the latter. A second option of

---

strings. Our definitions are based on [23], but adapted to tolerate the out-of-order delivery of ciphertexts.

the adversary after exposing a state is to remain passive (in particular, not to poison the partner). In this case the ░healing░ property of ratcheting-based secure messaging protocols shall automatically fully restore safe operations.

Coming back to the BASIC game of Fig. 1, we describe how the above concepts are reflected in the game variables and code lines. We start with the game body [G00–G18]. If $u \in \{A, B\}$ refers to one of the two participants, integer lt_u ('logical time') reflects the logical time of u; integer pt_u ('physical time') reflects the physical time of u; Boolean flag is_u ('in-sync') indicates whether u is in-sync with their peer $\bar{u}$; set SC_u ('sent ciphertexts') records the associated-data–ciphertext pairs sent by u; set $CERT_u$ ('certified') indicates which of the peer $\bar{u}$'s sending indices have been certified by receiving an in-sync ciphertext from them; for each sending index i, set $VF_u[i]$ ('vouches for') indicates for which sending indices of u the ciphertext with index i can vouch for; set AU_u indicates for which sending indices participant u is authoritative; flag $poisoned_u$ indicates whether u was poisoned.

We next explain how these variables are updated throughout the game. The cases of lt_u [G01,S06] and pt_u [G02,T01] are clear. Flag is_u is initialized to T [G03], and cleared [R23] in the moment that u receives a ciphertext that the peer $\bar{u}$ either didn't send, or did send but after becoming out-of-sync [R17] (in conjunction with [S02,S03], see next sentence).[12] Set SC_u is initialized empty [G04] and populated [S03] for each sending operation in which u is in-sync [S02].[13] Set $CERT_u$ is initialized empty [G05] and, when a sync-preserving ciphertext is received [R06,R06insync], populated with all indices prior to, and including, the current one [R08]. All entries of array-of-sets VF_u are initialized to 'all-indices' [G06], expressing that, by default, each sending index cryptographically vouches for its entire future (and past). This changes when u's state is exposed, as impersonating u then becomes trivial; the game reflects this by updating all VF_u entries related to the time preceding the exposure so that the corresponding ciphertexts do not vouch for ciphertexts that are created after the exposure [E02].[14] Set AU_u is initialized to 'all-indices' [G07], and indices are removed from it by exposing u's state, and added back to it by letting u heal; more precisely, while exposing u's state removes all indices starting with the current one (marking the entire future as non-authoritative) [E03], receiving a sync-preserving ciphertext from peer $\bar{u}$ [R06,R06insync] adds the vouched-for entries back [R09] (re-establishing authoritativeness up to the next exposure). Finally, flag $poisoned_u$ is initialized clear [G13], and set [R20] when a sync-damaging ciphertext is received (i.e., one that was not sent by peer $\bar{u}$ [R17] and is the first one making u lose sync [R18]) that was trivially injected after an

---

[12] The mechanism of considering participants out-of-sync once they process (unmodified) ciphertexts from out-of-sync peers is taken from [23], see Footnote 11.

[13] Note that the sending index of any ciphertext is uniquely recoverable (with function ts), implying that each execution of [S03] adds a new element to the set (collisions cannot occur).

[14] Line [E02] should be read as 'For all $0 \le i < lt_u$: $VF_u[i] \leftarrow VF_u[i] \cap [\![lt_u]\!]$' and expresses that all entries of $VF_u[\cdot]$ that correspond with prior sending indices are trimmed so that they cover no indices that succeed the current one (including).

exposure of peer $\bar{u}$'s state (technically: was crafted for a non-authoritative index [R19]).

This completes the description of the BASIC game. We refine it in the following to obtain three more games, but the basic working mechanisms of the oracles and variables remain the same.

GAME FUNC. We specify the expected functionality (a.k.a. correctness) of a BOOM protocol by formulating requirements on how it shall react to receiving valid and invalid ciphertexts. Concretely, in Fig. 2 we specify the corresponding FUNC game as an extension of the BASIC game from Fig. 1. In the figure, the code lines marked with neither ○ nor ● are taken verbatim from the BASIC game, and the lines marked with ○ are the ones to be added to obtain the FUNC game. (Ignore the lines marked with ● for now.) The FUNC game tests for a total of seven conditions, letting the adversary 'win' if any one of them is not fulfilled. Five of the conditions are checked for all operations (in-sync *and* out-of-sync): The conditions are (1) that the *ts* decoding function correctly indicates the logical and physical creation time of ciphertexts [S01]; (2) that no sending index is received twice (single delivery of ciphertexts) [G10,R02,R25] (set RI_u records 'received indices'); (3) that expired ciphertexts are not delivered (the reported sender's physical time *pt* is compared with the receiver's physical time pt_u, tolerating a lag of up to δ time units) [R03]; (4) that physical timestamps increase as logical timestamps do [G11,R04,R26] (set RT_u records 'received timestamps');[15] and (5) that the reported acknowledgment set A never shrinks and never lists never-sent indices [G12,R05,R27] (set RA_u records 'received acknowledgments'). Two additional conditions are checked for certified ciphertexts (this includes all in-sync ciphertexts, as they certify themselves [R06,R07,R08]): The conditions are (6) that the *recv* algorithm accurately reports the acknowledgment set A [R13] (recall that set RI_u holds the received indices [G10,R25], allowing to associate this set with each (in-sync) sending operation [G08,S04], so that set $\mathrm{SR}_{\bar{u}}[i]$ [S04,R13] indicates the indices that participant $\bar{u}$ received from u before $\bar{u}$ used sending index i in their sending operation); and (7) that encrypted messages are correctly recovered via decryption [G09,S05,R14] (array SM_u records 'sent messages'). We say that a BOOM protocol is functional if the advantage $\mathbf{Adv}^{\mathrm{func}}(\mathcal{A}) := \Pr[\mathrm{FUNC}(\mathcal{A})]$ is negligibly small for all realistic adversaries $\mathcal{A}$.

GAME AUTH. Our authenticity notion focuses on the protection of the integrity of ciphertexts (INT-CTXT). In Fig. 2 we specify the corresponding AUTH game as an extension of the BASIC game from Fig. 1. In the figure, the code lines marked with neither ○ nor ● are taken verbatim from the BASIC game, and the lines marked with ● are the ones to be added to obtain the AUTH game. (This time, ignore the lines marked with ○.) A BOOM scheme provides AUTH security if any adversarial manipulation (or injection) of ciphertexts is detected and rejected. Taking into account that associated-data strings need to be protected in the same vein, as a first approximation the notion could be formalized by

---

[15] A relation $R \subseteq \mathbb{N} \times \mathbb{N}$ is monotone [R04] if for all $(x,y),(x',y') \in R$ we have $x \leq x' \Rightarrow y \leq y'$.

Game FUNC($\mathcal{A}$)   //with ∘
Game AUTH($\mathcal{A}$)   //with •
G00 For $u \in \{A, B\}$:
G01 $lt_u \leftarrow 0$
G02 $pt_u \leftarrow 0$
G03 $is_u \leftarrow T$
G04 $SC_u \leftarrow \emptyset$
G05 $CERT_u \leftarrow \emptyset$
G06 $VF_u[\cdot] \leftarrow [\![\infty]\!]$
G07 $AU_u \leftarrow [\![\infty]\!]$
∘ G08 $SR_u[\cdot] \leftarrow \bot$
∘ G09 $SM_u[\cdot] \leftarrow \bot$
∘ G10 $RI_u \leftarrow \emptyset$
∘ G11 $RT_u \leftarrow \emptyset$
∘ G12 $RA_u \leftarrow \emptyset$
G17 $(st_A, st_B) \leftarrow init$
G18   Invoke $\mathcal{A}$
G19   Lose

Oracle Tick(u)
T00 $tick\langle st_u \rangle$
T01 $pt_u \leftarrow pt_u + 1$

Oracle Send(u, ad, m)
S00 $c \leftarrow send\langle st_u \rangle(ad, m)$
∘ S01 Promise $ts(c) = (lt_u, pt_u)$
S02 If is_u:
S03 $SC_u \overset{\cup}{\leftarrow} \{(ad, c)\}$
∘ S04 $SR_u[lt_u] \leftarrow RI_u$
∘ S05 $SM_u[lt_u] \leftarrow m$
S06 $lt_u \leftarrow lt_u + 1$
S07 Return c

Oracle Recv(u, ad, c)
R00 $(lt, pt) \leftarrow ts(c)$
R01 $(A, m) \leftarrow recv\langle st_u \rangle(ad, c)$
∘ R02 Promise $lt \notin RI_u$
∘ R03 Promise $\Delta(pt, pt_u) \leq \delta$
∘ R04 Promise $RT_u \cup \{(lt, pt)\}$ monotone
∘ R05 Promise $RA_u \subseteq A \subseteq [\![lt_u]\!]$
R06 If $(ad, c) \in SC_{\bar{u}}$:
R07   If is_u:
R08 $CERT_u \overset{\cup}{\leftarrow} [lt]$
R09 $AU_{\bar{u}} \overset{\cup}{\leftarrow} VF_{\bar{u}}[lt]$
∘ R12   If $lt \in CERT_u$:
∘ R13   Promise $A = RA_u \cup SR_{\bar{u}}[lt]$
∘ R14   Promise $m = SM_{\bar{u}}[lt]$
R17 If $(ad, c) \notin SC_{\bar{u}}$:
• R18   If is_u:
• R21   Reward $lt \in AU_{\bar{u}}$
• R22   Reward $lt \in CERT_u$
R23 $is_u \leftarrow F$
∘ R25 $RI_u \overset{\cup}{\leftarrow} \{lt\}$
∘ R26 $RT_u \overset{\cup}{\leftarrow} \{(lt, pt)\}$
∘ R27 $RA_u \leftarrow A$
R29 Return (A, m)

Oracle Expose(u)
E01 If is_u:
E02 $VF_u[\![lt_u]\!] \overset{\cap}{\leftarrow} [\![lt_u]\!]$
E03 $AU_u \overset{\cap}{\leftarrow} [\![lt_u]\!]$
E06 Return st_u

Fig. 2. Games FUNC and AUTH. The FUNC game includes the lines marked with ∘ but not the ones marked with •. The AUTH game includes the lines marked with • but not the ones marked with ∘.

adding the instruction 'Reward $is_u \wedge (ad, c) \notin SC_{\bar{u}}$' to the Recv oracle.[16] Note however that delivering a forged ciphertext to a participant u is trivial if the state of their peer $\bar{u}$ is exposed, and thus a small refinement is due. Recalling that set $AU_{\bar{u}}$ lists the sending indices for which participant $\bar{u}$ is authoritative, i.e., their actions not trivially emulatable, we reward the adversary only if the forgery is made for an index contained in this set [R17,R18,R21]. Recall further that in-sync delivered ciphertexts certify prior ciphertexts by the same sender, even if the latter ciphertexts are delivered out-of-sync. In the game we thus reward the adversary also if it forges on a certified index [R17,R22]. We say that a BOOM protocol provides authenticity if the advantage $\mathbf{Adv}^{\mathrm{auth}}(\mathcal{A}) := \Pr[\mathrm{AUTH}(\mathcal{A})]$

---

[16] The instruction should be read as 'Reward the adversary if it makes an in-sync participant accept an associated-data–ciphertext pair for which at least one of associated-data and ciphertext is not authentic'.

is negligibly small for all realistic adversaries $\mathcal{A}$. We refer the reader to the full version for a formalization of the trivial attack excluded by the AUTH game, and an overview of similar but non-trivial attacks that are allowed.

GAMES $\text{CONF}^0, \text{CONF}^1$. Our confidentiality notion is formulated in the style of left-or-right indistinguishability under active attacks (IND-CCA). In Fig. 3 we specify corresponding CONF^0 and CONF^1 games. The games are derived from the BASIC game by adding the lines marked with $\bullet$ plus two new oracles: The left-or-right Chal oracle [C00–C08], which behaves similar to the Send oracle but processes one of two possible input messages [C03] depending on bit b that encodes which game CONF^b is played, and the Decide oracle [D00] that lets the adversary control the return value of the game. (A successful adversary manages to correlate this return value with bit b.)

Three new game variables keep track of the actions of the adversary: Variable lx_u ('last exposure', [G14,E04]) indicates the index of the last exposure of user u. Set CH_u ('challenge') represents the set of sending indices for which a challenge query has been posed for u that peer $\bar{u}$ still should be able to validly decrypt. Indices are added to this set in the Chal oracle [G15,C06], and they are removed from it as a reaction to three events. (1) The corresponding ciphertext becomes invalid because the receiver already processed a ciphertext (the same or a different one) with the same index [R28] (see the corresponding guarantee in the FUNC game [G10,R02,R25]). (2) It becomes invalid because it expired based on physical time: To capture the latter condition we denote with

$$\text{ITC}(u) := \{lt : \exists ad, c, pt \text{ s.t. } (ad, c) \in \text{SC}_{\bar{u}} \wedge ts(c) - (lt, pt) \wedge \Delta(pt, pt_u) \leq \delta\}$$

('in-time ciphertexts') for participant u the set of sending indices of ciphertexts produced by peer $\bar{u}$ for which the difference between generation time pt and the physical time pt_u of the receiver is less than δ. With the progression of physical time the game removes those indices from set $\text{CH}_{\bar{u}}$ that are not an element of $\text{ITC}(u)$ [T02]. (See the corresponding guarantee in the FUNC game [R03].) (3) Receiving an out-of-sync ciphertext renders u's state incompatible to decrypt *future* challenge queries. Hence all future indices are removed from the challenge set [R24]. Observe this corresponds with [C06]: indices are only added to CH_u for an in-sync peer. Finally, flag xp_u ('exposed') indicates whether the state of u has to be considered known to the adversary after a state exposure. This flag is initially cleared [G16], set when u's state is exposed [E05], and reset if u heals by letting peer $\bar{u}$ receive an in-sync ciphertext created after the last exposure [R10,R11].

We next explain how the new variables help identifying four different trivial attack conditions. The first two conditions consider cases where posing a Chal query needs to be prevented because the receiver state is known due to impersonation or exposure: (1) if participant $\bar{u}$'s state was exposed and $\bar{u}$ is impersonated to u, i.e., u is poisoned, all future encryptions by u for $\bar{u}$ are trivially decryptable, simply because the adversary can emulate *all* actions of $\bar{u}$ [C01]; (2) encryptions by an in-sync sender u for a state-exposed receiver $\bar{u}$ are trivially decryptable (recall that flag $xp_{\bar{u}}$ traces the latter condition) [C02]. The next condition considers cases where posing an Expose query needs to be prevented because an already made Chal query would become trivial to break: (3) if participant $\bar{u}$ generated (challenge) ciphertext c for u, and the latter should still be able to validly decrypt c, then exposing u makes c trivially decryptable [E00]. The last condition is unrelated to exposures: (4) if participant u in-sync decrypts a ciphertext, by correctness the resulting message is identical to the encrypted message, and thus has to be suppressed by the Recv oracle by overwriting it [R15,R16]. (Note how line R16 corresponds with line R14 of FUNC.) This concludes the description of games CONF^b. We say that a BOOM protocol provides confidentiality if the advantage $\mathbf{Adv}^{\text{conf}}(\mathcal{A}) := |\Pr[\text{CONF}^1(\mathcal{A})] - \Pr[\text{CONF}^0(\mathcal{A})]|$ is negligibly small for all realistic adversaries $\mathcal{A}$. We refer the reader to the full version for a formalization of the trivial attacks excluded by the CONF game, and similar but non-trivial attacks that are allowed.

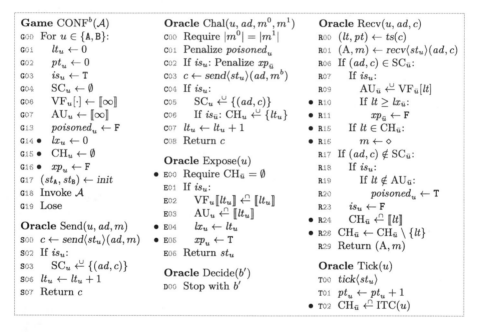

Fig. 3. Games $\text{CONF}^0, \text{CONF}^1$. See text for the definition of function ITC [T02].

4 Non-interactive Primitives

In Sect. 3 we defined the syntax and security of BOOM protocols and we will provide a secure construction in Sect. 5. The current section is dedicated to presenting a set of cryptographic building blocks, in the spirit of public key encryption (PKE) and signature schemes (SS), that will play crucial roles in our construction. Recall that a defining property of a BOOM protocol is that it provides maximum resilience against (continued) state exposure attacks, preventing all but trivial attacks. If a construction would rely on regular PKE or SS schemes as building blocks, the secret keys of the latter would leak on state exposure, which in most cases would inevitably clear the way for an attack on confidentiality or authenticity. We hence employ stateful variants of PKE and SS that process their internal keying material after each use to an updated 'refreshed' version that limits the options of a state-exposing adversary to harm only future operations. Some of the building blocks proposed here additionally fold an *associated data* input into their state, and the assumption is that sender and receiver (i.e., signer and verifier, or encryptor and decryptor) update their states with consistent such inputs.[17]

While the specifics of our building blocks might be different from those of prior work, it can be generally considered well-understood how to construct such primitives. For instance, a forward-secure SS [8], which is a primitive close to one of ours, can be built by coupling each signing operation with the generation of a fresh signature key pair, the public component of which is signed and thus authenticated along with the message; after the signing operation is complete, the original signing key is disposed of and replaced by the freshly generated one. Adding the support of auxiliary associated-data strings into such a scheme is trivial (just authenticate the string along with the message) and is less a cryptographic challenge than an exercise of maintaining the right data structures in the sender/receiver state. Similarly, forward-secure PKE [11], which is a primitive close to one of ours as well, is routinely built from hierarchical identity-based encryption (HIBE) by associating key validity epochs with the nodes of a binary tree. Variants of forward-secure PKE that support key updates that depend on auxiliary associated-data strings have been proposed in prior work as well [17,24], using design approaches that can be seen as minor variations of the original tree-based idea from [11].

For our BOOM construction in Sect. 5 we require three independent forward-secure public key primitives which we refer to as *updatable signature scheme*, *key-updatable KEM*, and *key-evolving KEM*, respectively. We specify their syntax and explain the expected behavior below. We formalize the details and propose concrete constructions in the full version. We note that our security definitions and constructions can be seen as following immediately from the syntax and expected functionality: While the security definitions give the adversary the option to expose the state of any participant any number of times, and formalize the best-possible security that is feasible under such a regime (i.e., maximum resilience against state exposure attacks), the constructions, which all follow the approaches of [8,11,17,24] discussed above, are engineered to re-generate fresh key material whenever an opportunity for this arises.

---

[17] Unlike regular signature schemes where for each signer there can be many independent verifiers, and unlike regular public key encryption where for each decryptor there can be many encryptors, for the primitives we consider in the current section a strict one-to-one correspondence between sender and receiver is assumed.

4.1   Updatable Signature Schemes (USS)

Like a regular signature scheme, a USS has algorithms $gen, sign, vfy$, where $sign$ creates a signature on a given message and vfy verifies that a given signature is valid for a given message. The particularity of USS is that signing and verification keys can be updated, and that signatures only verify correctly if these updates are performed consistently. More precisely, signing and verification keys are replaced by signing and verification *states*, and update algorithms $updss, updvs$ (for 'update signing state' and 'update verification state', respectively) can update these states to a new version, taking also an associated-data input into account. Multiple such update operations can be performed in succession, on both sides. Signatures of the signer are recognized as valid by the verifier only if the updates of both parties are in-sync, i.e., are performed with the same sequence of update strings. Our security model provides the means to the adversary to expose the state of the parties between any two update operations, and requires unforgeability with maximum resilience to such exposures.

Formally, a key-updatable signature scheme for a message space $\mathcal{M}$ and an associated-data space $\mathcal{AD}$ consists of a signing state space $\mathcal{SS}$, a verification state space $\mathcal{VS}$, a signature space Σ, and algorithms $gen, sign, vfy, updss, updvs$ with APIs

$$gen \to \mathcal{SS} \times \mathcal{VS} \qquad \mathcal{M} \to sign\langle \mathcal{SS} \rangle \to \Sigma \qquad \mathcal{VS} \times \mathcal{M} \times \Sigma \to vfy$$

$$\mathcal{AD} \to updss\langle \mathcal{SS} \rangle \qquad \mathcal{AD} \to updvs\langle \mathcal{VS} \rangle.$$

Note that the vfy algorithm doesn't have an explicit output. The assumption behind this is that the algorithm signals acceptance by terminating normally, while it signals rejection by aborting. (See Sect. 2 on the option of any algorithm to abort.) We expect of a correct USS that for all $(ss, vs) \in [gen]$, if ss and vs are updated by invoking $updss\langle ss \rangle (\cdot)$ and $updvs\langle vs \rangle (\cdot)$ with the same sequence $ad_1, \ldots, ad_l \in \mathcal{AD}$ of associated data, then for all $m \in \mathcal{M}$ and $\sigma \in [sign\langle ss \rangle (m)]$ we have that $vfy(vs, m, \sigma)$ accepts. See the full version for examples of the expected functionality a formalization of correctness and security, and a construction.

4.2   Key-Updatable KEM (KuKEM)

A key-updatable key encapsulation mechanism is a stateful KEM variant with algorithms gen, enc, dec and update properties like for USS: both the encapsulator and the decapsulator can update their public/secret state material with algorithms $updps, updss$ (for 'update public state' and 'update secret state', respectively) that also take an associated-data input into account. The decapsulator, if updated in-sync with the encapsulator, can successfully decapsulate ciphertexts. Our security model formalizes IND-CCA-like security in a model supporting exposing the state of both parties, with the explicit requirement that state exposures neither harm the confidentiality of keys encapsulated for past epochs, nor the confidentiality of keys encapsulated with diverged states.

Formally, a key-updatable key encapsulation mechanism for a key space $\mathcal{K}$ and an associated-data space $\mathcal{AD}$, consists of a secret state space $\mathcal{SS}$, a public state space $\mathcal{PS}$, a ciphertext space $\mathcal{C}$, KEM algorithms gen, enc, dec and state update algorithms $updps, updss$ with APIs

$$gen \to \mathcal{SS} \times \mathcal{PS} \qquad \mathcal{PS} \to enc \to \mathcal{K} \times \mathcal{C} \qquad \mathcal{SS} \times \mathcal{C} \to dec \to \mathcal{K}$$

$$\mathcal{AD} \to updps\langle \mathcal{PS} \rangle \qquad \mathcal{AD} \to updss\langle \mathcal{SS} \rangle.$$

We expect of a correct KuKEM that for all $(ss, ps) \in [gen]$, if ss and ps are updated by invoking $updps\langle ps \rangle(\cdot)$ and $updss\langle ss \rangle(\cdot)$ with the same sequence $ad_1, \ldots, ad_l \in \mathcal{AD}$ of associated data, then for all $(k, c) \in [enc(ps)]$ and $k' \in [dec(ss, c)]$ we have that $k = k'$. See the full version for a formalization of correctness and security, and a construction.

4.3   Key-Evolving KEM (KeKEM)

A key-evolving key encapsulation mechanism consists of algorithms gen, enc, dec like a regular KEM, but, as above, public and secret keys are replaced by public and secret states, respectively, that can be updated. More precisely, the encapsulator's and decapsulator's states can be updated 'to the next epoch' by invoking the *evolveps* (for 'evolve public state') algorithm and the *evolvess* (for 'evolve secret state') algorithm, respectively. Note, however, that if a secret state is updated, the decryptability of ciphertexts generated for older epochs is not automatically lost; rather, ciphertexts associated to multiple epochs remain decryptable until epochs are explicitly declared redundant by invoking the *expire* algorithm.[18] Our security model formalizes IND-CCA-like security in a model supporting exposing the state of both parties, with the explicit requirement that state exposures do not harm the confidentiality of keys encapsulated for expired epochs. Note that our formalization of KeKEMs does not support updating states with respect to an associated-data input.

Formally, a key-evolving key encapsulation mechanism for a key space $\mathcal{K}$ consists of a secret state space $\mathcal{SS}$, a public state space $\mathcal{PS}$, a ciphertext space $\mathcal{C}$, KEM algorithms gen, enc, dec and state update algorithms *evolveps, evolvess, expire* with APIs

$$\mathbb{N} \to gen \to \mathcal{SS} \times \mathcal{PS} \qquad \mathcal{PS} \to enc \to \mathcal{K} \times \mathcal{C} \qquad \mathcal{SS} \times \mathbb{N} \times \mathcal{C} \to dec \to \mathcal{K}$$

$$evolveps\langle \mathcal{PS} \rangle \qquad evolvess\langle \mathcal{SS} \rangle \qquad expire\langle \mathcal{SS} \rangle.$$

In the KeKEM setting it makes sense to number the epochs. Note that the *dec* algorithm expects, besides the secret state and the ciphertext, an explicit indication of the epoch number for which the ciphertext was created. For simplicity, one would like to provide an absolute time to the *dec* algorithm, e.g. Unix time, rather than the time offset relative to the generation time. For this

---

[18] The *expire* algorithm expires always to oldest currently supported epoch. That is, active epochs of KeKEMs always span a continuous interval.

reason, the *gen* algorithm takes in an epoch number which can be used to specify the generation time and thus the first epoch need not necessarily start at zero. Then the state can internally compute the relative offset on decapsulation. As the full definition is quite involved and thus deferred to the full version, we illustrate the functionality of a correct KeKEM using an example: If we invoke $(ss, ps) \leftarrow gen(5)$ to generate a state pair (and associating the number 5 with the first state), then invoking 2-times $evolveps\langle ps \rangle$ followed by $(k, c) \leftarrow enc(ps)$, and then 4-times $evolvess\langle ss \rangle$, then invoking $dec(ss, 7, c)$ will return k until $expire\langle ss \rangle$ has been invoked for the third time (expiring epochs 5, 6, and finally 7). See the full version for a formalization of correctness and security, and a construction.

5   Interactive Primitives and BOOM

This section exposes our Bidirectional Out-of-Order Messaging (BOOM) protocol, in three steps. In Sect. 5.1 we first present a BOOM-signature scheme, which uses the USS introduced in Sect. 4.1 as building block. This scheme will be used by our final BOOM construction in a black box manner by calling its *sign* and *vfy* procedures on each message to add an authenticity layer. Next, we present a BOOM-KEM scheme in Sect. 5.2. Our final BOOM construction will query the BOOM-KEM in a black box manner by calling its *enc* and *dec* procedures to obtain encryption keys for each message. The BOOM-KEM uses the KuKEM and KeKEM building blocks introduced in Sect. 4.2 and Sect. 4.3, to ensure the BOOM scheme can achieve confidentiality with its keys. The BOOM construction will additionally invoke its *upd* procedure to reflect the passing of time and the *expire* procedure to indicate we no longer wish to be able to obtain 'old' decryption keys.

Despite the strong building blocks defined in Sect. 4, our BOOM protocols are complex and involved. These difficulties stem from the data structures required to manage out-of-order delivery of ciphertexts. These data structures obscure the cryptographically novel core of our construction and render it difficult to interpret. Therefore, we have separated the authenticity tool and the confidentiality tool and present them in their own right. Note that this modularization implies certain data structures will be duplicated across each tool, but an implementation could consolidate them.

5.1   BOOM-Signature Scheme

In Sect. 5.3 we will use a specialized signature scheme to achieve authenticity for our BOOM construction. In this section we describe the inner workings of this cryptographic tool.

SYNTAX. A BOOM-signature scheme for a message space $\mathcal{M}$ consists of a state space $\mathcal{ST}$, a signature space Σ, algorithms *init*, *sign*, *vfy*, and a (logical) timestamp decoder *ts* as follows:

$$init \rightarrow \mathcal{ST} \times \mathcal{ST} \quad \mathcal{M} \rightarrow sign\langle \mathcal{ST} \rangle \rightarrow \Sigma \quad \mathcal{M} \times \Sigma \rightarrow vfy\langle \mathcal{ST} \rangle \quad ts \colon \Sigma \rightarrow \mathbb{N}.$$

CONSTRUCTION. We provide a construction for a BOOM-signature scheme in Fig. 4. The construction consists of four procedures: *init*, *sign*, *vfy* and *ts*. The *init* procedure initializes the states for two users A and B. The *sign* procedure is stateful and will output a signature σ for any message m, updating its state in the process. The *vfy* procedure is also stateful and will verify any pair $(m, \sigma) \in \mathcal{M} \times \Sigma$. If σ is a correct signature on m, the state will update and *vfy* will return control to the caller. If the signature does not correctly verify, the *vfy* procedure will abort. The *ts* function returns the logical time (measured in signer invocations).

On a very high level, *sign* generates a fresh USS key pair every iteration to recover from (potential) state exposures and signs the hash of its sent transcript, while *vfy* updates its state with the messages that have been received, so states will diverge if the adversary injects a message, while managing out of order delivery. We will now describe the variables and code lines in more detail.

Proc *init*
100 For $u \in \{A, B\}$:
101      $lt_u \leftarrow 0$; $lt_u^* \leftarrow 0$
102      $S_u[\cdot] \leftarrow \bot$; $V_u[\cdot] \leftarrow \bot$
103      $(ss_u, vs_u^*) \leftarrow$ USS.gen
104      $P_u \leftarrow \emptyset$
105      $AS_u[\cdot] \leftarrow \bot$
106      $AS_u[lt_u] \leftarrow H()$
107      $av_u \leftarrow H()$
108      $st_u := (\ldots)$
109 Return (st_A, st_B)

Proc *sign*$\langle st_u \rangle(m)$
s00 $(ss, vs) \leftarrow$ USS.gen
s01 $h \leftarrow H(m \;\|\; lt_u \;\|\; P_u \;\|\; vs \;\|\; S_u[\![lt_u]\!])$
s02 $\sigma \leftarrow$ USS.sign$\langle ss_u \rangle(h)$
s03 $AS_u[lt_u + 1] \leftarrow H(AS_u[lt_u] \;\|\; h \;\|\; \sigma)$
s04 $S_u[lt_u] \leftarrow (h, P_u, vs, \sigma)$
s05 $\sigma \overset{\cdot}{\leftarrow} lt_u \;\|\; P_u \;\|\; vs \;\|\; S_u[\![lt_u]\!]$
s06 $lt_u \leftarrow lt_u + 1$
s07 $(ss_u, P_u) \leftarrow (ss, \emptyset)$
s08 Return σ

Proc *ts*(σ)
t00 Parse $\sigma \;\|\; lt \;\|\; P \;\|\; vs \;\|\; S[\![lt]\!] \leftarrow \sigma$
t01 Return lt

Proc *vfy*$\langle st_u \rangle(m, \sigma)$
v00 Parse $\sigma \;\|\; lt \;\|\; P \;\|\; vs \;\|\; S[\![lt]\!] \leftarrow \sigma$
v01 $h \leftarrow H(m \;\|\; lt \;\|\; P \;\|\; vs \;\|\; S[\![lt]\!])$
v02 While $lt_u^* \leq lt$:
v03   If $lt_u^* = lt$:
v04 $(P', vs') \leftarrow (P, vs)$
v05 $(h', \sigma') \leftarrow (h, \sigma)$
v06   Else:
v07 $(h', P', vs', \sigma') \leftarrow S[\![lt_u^*]\!]$
v08 $vs^* \leftarrow vs_u^*$
v09   For $i \in P'$:
v10   USS.updvs$\langle vs^* \rangle(AS_u[i])$
v11   Require USS.vfy$\langle vs^*, h', \sigma' \rangle$
v12 $vs_u^* \leftarrow vs'$
v13 $av_u \leftarrow H(av_u \;\|\; h' \;\|\; \sigma')$
v14 $V_u[lt_u^*] \leftarrow (h', \sigma')$
v15   USS.updss$\langle ss_u \rangle(av_u)$
v16 $lt_u^* \leftarrow lt_u^* + 1$
v17 $P_u \overset{\cup}{\leftarrow} \{lt_u^*\}$
v18 If $lt_u^* > lt$:
v19   Require $V_u[lt] \neq \diamond$
v20   Require $V_u[lt] = (h, \sigma)$
v21 $V_u[lt] \leftarrow \diamond$

Fig. 4. BOOM-signature construction. We use an updatable signature scheme (USS) as building block. Function H is assumed to be a collision-resistant hash function. The *vfy* procedure aborts if parsing fails.

For each user $u \in \{A, B\}$ we initialize the signing index lt_u and the verifying index lt_u^* [i01], and the arrays S_u and V_u, which will store information about signed and verified messages, respectively [i02]. We generate pairs of USS signing and verification keys [i03] and initialize the set P_u of messages processed by the current signing key to be empty [i04]. We initialize the accumulated signed transcript AS_u [i05], set the first index [i06] and initialize the accumulated verified transcript av_u [i07]. Finally, we store everything in the users' states [i08].

The *sign* procedure first generates a new USS key pair [s00]. Next, it computes the hash of the message m, the signing index lt_u, the set of processed messages P_u, the verification state, and the array S_u [s01]. It signs the hash with its old key [s02]. It accumulates the new hash and signature in AS_u [s03] and stores the hash, processed set, verification key and signature in S_u [s04]. The signing index, processed set, verification key and array S_u are appended to the signature [s05]. It increments the signing index lt_u [s06] and stores the new signing key along with an empty processed set [s07], before returning the signature [s08].

The *vfy* procedure parses the additional information embedded in the signature [v00] and recomputes the hash [v01]. If the verifying index lt_u^* is less or equal than lt, the verifier will iteratively check signatures until it catches up [v02–v17]. To be concrete, if $lt_u^* = lt$ it will use the current value for the hash and signature [v05] or if $lt_u^* < lt$ it will obtain these values from $S[lt_u^*]$ [v07]. It will update a copy of its verification key for all indices the signer has processed since generating its signing key [v08–v10] and verify the signature [v11]. Note it uses the transcript for its signed messages to update its verification key, which should match the transcript for the verified messages the signer has used to update its signing key. If signature verification passes, it will replace the verification key [v12]. Note that if USS.*vfy* failed the verification key remains unchanged, as if [v09–v10] were never executed. Next, it accumulates the hash and signature in its verified transcript av_u [v13], stores the hash and signature in $V_u[lt_u^*]$ for later comparison [v14] and it will update its signing state with av_u [v15]. It increments the index lt_u^* [v16] and add lt_u^* to P_u to indicate it has processed this message into its signing key [v17]. If the verifying index lt_u^* was strictly greater than lt, the verifier will check if an entry exists for this index [v19] and compare whether it is equal to the value of the hash and signature [v20]. At last, the verifier will remove the entry in V_u for index lt to prevent double delivery [v21].

Note for simplicity we omit code lines to 'clean up' variables that are no longer needed. These lines are not required for security, but would help for efficiency. For example, if a party learns its peer has processed signature i, it will no longer have to include the first i entries of S_u in its next signature.

5.2   BOOM-KEM Scheme

In Sect. 5.3 we will use a specialized KEM to achieve confidentiality for our BOOM construction. In this section we describe the inner workings of this cryptographic tool.

SYNTAX. A BOOM-KEM scheme for a key space $\mathcal{K}$ consists of a state space $\mathcal{ST}$, a ciphertext space $\mathcal{C}$, and algorithms *init*, *upd*, *expire*, *enc*, *dec* and the timestamp decoder *ts* that recovers the logical and physical time.

$$init \to \mathcal{ST} \times \mathcal{ST} \qquad upd\langle \mathcal{ST} \rangle \qquad expire\langle \mathcal{ST} \rangle \qquad ts \colon \mathcal{C} \to \mathbb{N} \times \mathbb{N}$$

$$\mathcal{AD} \rightarrow enc\langle\mathcal{ST}\rangle \rightarrow \mathcal{K} \times \mathcal{C} \qquad \mathcal{AD} \times \mathcal{C} \rightarrow dec\langle\mathcal{ST}\rangle \rightarrow \mathcal{K}.$$

CONSTRUCTION. Internally our BOOM-KEM construction will invoke the KuKEM primitive introduced in Sect. 4.2, the KeKEM primitive introduced in Sect. 4.3, and a secure KEM combiner K such that if at least one of the input keys is indistinguishable from a uniformly random string of equal length, then so is the output key. In this article we will consider K a random oracle. An implementation could use the CCA secure combiner presented in [16].

We noted both our KuKEM and KeKEM building block can be built generically from hierarchical identity-based encryption (HIBE, [15]). This strong component, while inefficient, should come as no surprise as it has already been proposed by [17] and [24] in the much simpler setting where every message is always delivered, and always in order. Moreover, recent work [6] shows that if an exposure additionally reveals the random coins used for the next *send* operation, the use of KuKEM is required to achieve confidentiality. They hypothesize the same implication holds without revealing the random coins and provide a strong intuition, but a formal proof remains an open problem.

We remark that both our KuKEM and KeKEM can be built from a single HIBE instance if one immediately delegates the master secret key to a 'KuKEM identity' and to a 'KeKEM identity'. We avoid doing so for two reasons. First of all, these primitives correspond to two perpendicular security goals. It is conceptually easier to grasp if we do not intertwine them. Secondly, KeKEM can be built from a forward-secure KEM, which is a simpler primitive than the HIBE-KEM used for KuKEM. Thus it may also be more efficient to separate them.

We provide a construction for a BOOM-KEM in Fig. 5. The construction consists of six procedures: *init*, *enc*, *dec*, *expire*, *upd* and *ts*. A correct decryption procedure *dec* is determined by the encryption procedure: it mirrors the operations in *enc*. As deriving the *dec* procedure is a rather vacuous technical exercise we have omitted it from Fig. 5 to focus on the more interesting cryptographic procedures instead. We have also omitted the *ts* procedure which simply parses the timestamps embedded in each ciphertext. A full reconstruction of all BOOM-KEM procedures is provided in the full version. The construction is quite technical but the general idea is to generate a new KuKEM and a new KeKEM instance with every *enc* invocation for post-compromise security. We update the KeKEM for forward secrecy in physical time, and the KuKEM for forward secrecy in logical time. The *enc* procedure will output a key dependent on the output of the KuKEM encapsulation procedure, the KeKEM encapsulation procedure and the associated data input.

We remark the physical time updates must be a separate primitive as simply updating the KuKEM would render the users out-of-sync. For example, consider the scenario where Alice sends a message, updating her KuKEM. Now physical time advances and both Alice and Bob would update their KuKEM. Finally, Bob receives Alice's message and updates his KuKEM. Clearly the updates have occurred in a different order, hence correctness would fail.

We note our security notion implies ciphertexts must contain information about prior ciphertexts. To see that ciphertexts cannot be independent, consider

---

Proc *init*

i00 For $u \in \{\mathtt{A}, \mathtt{B}\}$:

i01 $lt_u \leftarrow 0;\ lt_u^* \leftarrow 0$

i02 $ft_u \leftarrow 0;\ pt_u \leftarrow 0$

i03 $\mathrm{AS}_u[\cdot] \leftarrow \bot$

i04 $\mathrm{AR}_u[\cdot] \leftarrow \bot$

i05 $\mathrm{AS}_u[lt_u] \leftarrow H()$

i06 $\mathrm{AR}_u[lt_u^*] \leftarrow H()$

i07 $E_u[\cdot] \leftarrow \bot$

i08 $U_u[\cdot] \leftarrow \bot$

i09 $(E_u[lt_u], \varepsilon_u^*) \leftarrow \mathrm{ke.gen}(pt_u)$

i10 $(U_u[lt_u], v_u^*) \leftarrow \mathrm{ku.gen}$

i11 $\mathrm{KC}_u[\cdot] \leftarrow \bot$

i12 $\mathrm{DK}_u[\cdot] \leftarrow \bot$

i13 $st_u := (\dots)$

i14 Return $(st_\mathtt{A}, st_\mathtt{B})$

Proc *expire*$\langle st_u \rangle$

x00 $ft_u \leftarrow ft_u + 1$

x01 For $i \in [lt_u]$:

x02 $\mathrm{ke.expire}\langle E_u[i] \rangle$

Proc *enc*$\langle st_u \rangle(ad)$

e00 $(k_0, c_0) \leftarrow \mathrm{ke.enc}(\varepsilon_u^*)$

e01 $(k_1, c_1) \leftarrow \mathrm{ku.enc}(v_u^*)$

e02 $\mathrm{KC}_u[lt_u] \leftarrow (lt_u^*, c_1)$

e03 $(E_u[lt_u], \varepsilon) \leftarrow \mathrm{ke.gen}(pt_u)$

e04 $(U_u[lt_u], v) \leftarrow \mathrm{ku.gen}$

e05 $\mathrm{ku.updss}\langle U_u[lt_u] \rangle(\mathrm{AR}_u[lt_u^*])$

e06 $c \leftarrow lt_u \parallel pt_u \parallel \varepsilon \parallel v$

e07 $c \overset{\shortparallel}{\leftarrow} c_0 \parallel \mathrm{KC}_u[*] \parallel \mathrm{AS}_u[*]$

e08 $adc \leftarrow ad \parallel c$

e09 $k \leftarrow K(k_0, k_1; adc)$

e10 $lt_u \leftarrow lt_u + 1$

e11 $\mathrm{AS}_u[lt_u] \leftarrow H(\mathrm{AS}_u[lt_u - 1] \parallel adc)$

e12 $\mathrm{ku.updps}\langle v_u^* \rangle(\mathrm{AS}[lt_u])$

e13 Return (k, c)

Proc *upd*$\langle st_u \rangle$

u00 $pt_u \leftarrow pt_u + 1$

u01 $\mathrm{ke.evolveps}\langle \varepsilon_u^* \rangle$

u02 For $i \in [lt]_u$:

u03 $\mathrm{ke.evolvess}\langle E_u[i] \rangle$

Fig. 5. BOOM-KEM construction. Building blocks are a KeKEM, whose algorithms are prefixed with 'ke.', a KuKEM, whose algorithms are prefixed with 'ku.' and a KEM combiner K.

an adversary that exposes Alice and creates two ciphertexts. The adversary will deliver the second ciphertext to Bob, rendering Bob out-of-sync. Now the adversary can challenge Alice, making her send her first ciphertext, and since Bob is out-of-sync, expose Bob. If Bob were able to decrypt any ciphertext with logical index 1, the adversary could now decrypt Alice's challenge ciphertext and win the confidentiality game. Hence, the second ciphertext must 'pin' the first.

We achieve this with the KEM/DEM encryption paradigm. The *enc* procedure will embed past KuKEM ciphertexts in the current ciphertext. When receiving a ciphertext, the *dec* procedure will decapsulate all embedded KuKEM ciphertexts, store the DEM keys and destroy its capability to decapsulate again. Reconsidering our example above, Bob is now only able to decrypt the first ciphertext if it was encrypted with the same DEM key he obtained from the second ciphertext, and Bob has no capability to decapsulate another KuKEM ciphertext. The probability that Alice and the adversary had generated the same KuKEM ciphertext for the first ciphertext is negligible.

We now discuss the procedures in more detail, starting with *init*. For each user the *init* procedure initializes a sending index lt_u, a receiving index lt_u^*, the first physical time that is still recoverable ft_u and the current physical time pt_u [i01–i02]. It initializes the array AS for the accumulated sent transcript and AR for the accumulated received transcript [i03–i06]. The accumulated transcripts will be used to update the KuKEM states, ensuring the user states diverge when users go out-of-sync. Because ciphertexts may be delivered out-of-order, or not at all, each user will be maintaining several instances of each primitive, ready to decapsulate ciphertexts for any of them. However, it will always encapsulate to the latest one. Hence we initialize storage for multiple secret states, but only one public state, and we store the first KeKEM and KuKEM instance [i07–i10]. Finally, we initialize the array KC to store KuKEM ciphertexts [i11] and the array DK to store DEM keys [i12], as described in the general construction overview.

The *enc* procedure encapsulates keys for both the KeKEM and the KuKEM [e00–c01], and stores the KuKEM ciphertext in KC, along with its receiver index lt_u^*, indicating which public states were used for encapsulation [e02]. Next, it generates a new instance for both the KeKEM and the KuKEM [e03–e04]. It will immediately update the secret state for the KuKEM with the received transcript [e05], as the adversary is allowed unrestricted expose queries if we are out-of-sync. The *enc* procedure combines the KEM ciphertexts into one ciphertext, adds the freshly generated public states, and includes the indices and the sending transcript such that the receiver can correctly update its state [e06–e07]. Subsequently, it uses the KEM-combiner K to produce a key, using the associated data and ciphertext as context [e09]. Finally, it increments the sending index lt_u [e10], accumulates the associated data and ciphertext into its transcript [e11] and updates its public KuKEM state with it [e12].

The *upd* procedure is quite straightforward: it simply updates the public state and evolves the secret states for all its KeKEM instances as physical time advances. Similarly, the *expire* procedure will update all secret states.

Note that for simplicity we have omitted code lines to 'clean up' variables that are no longer needed. These lines are not required for security, but would help for efficiency. For example when a user has either received or expired all messages encapsulated for its i-th KeKEM and KuKEM instance, it can drop instance i, as later keys will always be encapsulated to later instances. As another example we remark that, after receiving an acknowledgment from the other user they have received message i, a user would no longer have to embed all their KuKEM ciphertexts for indices less than or equal to i in their current ciphertext.

5.3   BOOM Construction

We first introduce a functional protocol and discuss it in detail before delving into the full BOOM construction that achieves authenticity and confidentiality. The functional protocol consists of all the unmarked code lines in Fig. 6. The protocol has four procedures: the initialization procedure *init*, which initializes the users' initial states; the sending procedure *send*, which takes a state, associated data and a message, updates the state and outputs a ciphertext; the receiving procedure *recv*, which takes a state, associated data and a ciphertext, updates the state and outputs a message; and the time progression algorithm *tick*, which updates the state.

For each user u, the *init* procedure initializes the logical time lt_u and lt_u^* [i03], the physical time pt_u [i04], the set of received indices RI_u [i04], the set of

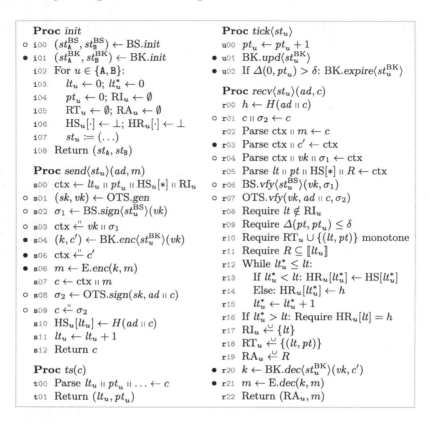

Proc *init*
- i00 $(st_A^{BS}, st_B^{BS}) \leftarrow BS.init$
- i01 $(st_A^{BK}, st_B^{BK}) \leftarrow BK.init$
- i02 For $u \in \{A, B\}$:
- i03 $lt_u \leftarrow 0;\ lt_u^* \leftarrow 0$
- i04 $pt_u \leftarrow 0;\ RI_u \leftarrow \emptyset$
- i05 $RT_u \leftarrow \emptyset;\ RA_u \leftarrow \emptyset$
- i06 $HS_u[\cdot] \leftarrow \perp;\ HR_u[\cdot] \leftarrow \perp$
- i07 $st_u := (\dots)$
- i08 Return (st_A, st_B)

Proc *send*$\langle st_u \rangle (ad, m)$
- s00 $ctx \leftarrow lt_u \parallel pt_u \parallel HS_u[*] \parallel RI_u$
- s01 $(sk, vk) \leftarrow OTS.gen$
- s02 $\sigma_1 \leftarrow BS.sign\langle st_u^{BS} \rangle (vk)$
- s03 $ctx \stackrel{\shortparallel}{\leftarrow} vk \parallel \sigma_1$
- s04 $(k, c') \leftarrow BK.enc\langle st_u^{BK} \rangle (vk)$
- s05 $ctx \stackrel{\shortparallel}{\leftarrow} c'$
- s06 $m \leftarrow E.enc(k, m)$
- s07 $c \leftarrow ctx \parallel m$
- s08 $\sigma_2 \leftarrow OTS.sign(sk, ad \parallel c)$
- s09 $c \stackrel{\shortparallel}{\leftarrow} \sigma_2$
- s10 $HS_u[lt_u] \leftarrow H(ad \parallel c)$
- s11 $lt_u \leftarrow lt_u + 1$
- s12 Return c

Proc *ts*(c)
- t00 Parse $lt_u \parallel pt_u \parallel \dots \leftarrow c$
- t01 Return (lt_u, pt_u)

Proc *tick*$\langle st_u \rangle$
- u00 $pt_u \leftarrow pt_u + 1$
- u01 $BK.upd\langle st_u^{BK} \rangle$
- u02 If $\Delta(0, pt_u) > \delta$: $BK.expire\langle st_u^{BK} \rangle$

Proc *recv*$\langle st_u \rangle (ad, c)$
- r00 $h \leftarrow H(ad \parallel c)$
- r01 $c \parallel \sigma_2 \leftarrow c$
- r02 Parse $ctx \parallel m \leftarrow c$
- r03 Parse $ctx \parallel c' \leftarrow ctx$
- r04 Parse $ctx \parallel vk \parallel \sigma_1 \leftarrow ctx$
- r05 Parse $lt \parallel pt \parallel HS[*] \parallel R \leftarrow ctx$
- r06 $BS.vfy\langle st_u^{BS} \rangle (vk, \sigma_1)$
- r07 $OTS.vfy(vk, ad \parallel c, \sigma_2)$
- r08 Require $lt \notin RI_u$
- r09 Require $\Delta(pt, pt_u) \leq \delta$
- r10 Require $RT_u \cup \{(lt, pt)\}$ monotone
- r11 Require $R \subseteq [\![lt_u]\!]$
- r12 While $lt_u^* \leq lt$:
- r13   If $lt_u^* < lt$: $HR_u[lt_u^*] \leftarrow HS[lt_u^*]$
- r14   Else: $HR_u[lt_u^*] \leftarrow h$
- r15 $lt_u^* \leftarrow lt_u^* + 1$
- r16 If $lt_u^* > lt$: Require $HR_u[lt] = h$
- r17 $RI_u \stackrel{\cup}{\leftarrow} \{lt\}$
- r18 $RT_u \stackrel{\cup}{\leftarrow} \{(lt, pt)\}$
- r19 $RA_u \stackrel{\cup}{\leftarrow} R$
- r20 $k \leftarrow BK.dec\langle st_u^{BK} \rangle (vk, c')$
- r21 $m \leftarrow E.dec(k, m)$
- r22 Return (RA_u, m)

Fig. 6. The functional construction consists of the unmarked lines. The authentic construction adds the lines marked with ∘. The BOOM construction consists of all lines. BS is the BOOM-signature scheme construction in Fig. 4, BK is the BOOM-KEM construction in Fig. 5, OTS is a (one-time) signature scheme, and E is a symmetric encryption scheme.

received timestamps RT_u [i05], the set of received acknowledgments RA_u [i05], and the arrays of hashed sent ciphertexts HS_u and hashed received ciphertexts HR_u [i06]. The *tick* procedure increments the user's physical time pt_u [u00].

The *send* procedure takes associated data ad and message m as input. It creates context ctx which includes the user's current time (lt_u, pt_u), the hashes of previously sent ciphertexts $HS_u[*]$ and the set of received indices RI_u [s00]. The context ctx together with the message m will form the ciphertext c [s07]. Finally, it stores the hash $H(ad \parallel c)$ of the associated data and the ciphertext [s10], increments the logical time lt_u [s11] and returns the ciphertext [s12].

The *recv* procedure first hashes the ciphertext [r00] and subsequently parses it to obtain the message m [r02] and the context variables $lt, pt, HS[*]$ and R [r05]. Now, recall that 'Require C' is short for 'If $\neg C$: Abort'. Thus the *recv* procedure performs four sanity checks to guarantee functionality. (1) A ciphertext has not yet been received for this logical time lt [r08]. (2) The ciphertext is fresh, that is the Δ difference between its physical creation time pt and the user's time pt_u is 'small' [r09]. (3) Time is monotonic: a message that is newer in logical time must be newer in physical time [r10]. (4) Only sent messages can be acknowledged [r11]: Bob cannot acknowledge having received a message that Alice never sent. Next, $recv$ handles the out-of-order delivery. While u's receiving index lt_u^* is smaller or equal than lt, it will iteratively update its array HR_u with the received hashes that it obtains from $\mathrm{HS}[*]$ or the current ciphertext itself [r12–r15]. If u's receiving index is greater than lt, it will require the hash h of the current ciphertext is equal to the stored value for that index $\mathrm{HR}_u[lt]$ [r16]. Finally, it will update its set of received indices RI_u [r17], its set of received timestamps RT_u [r18], its set of received acknowledgments RA_u [r19], and return (RA_u, m) [r22].

We extend the functional protocol to an authentication protocol by including the lines marked with ∘. The *init* procedure now initializes a BOOM-signature scheme BS [i00]. The *send* procedure generates a fresh one-time signature key pair (sk, vk) [s01], and calls BS.*sign* to obtain a signature σ_1 on the verification key vk [s02]. We add vk and σ_1 to the context ctx [s03]. We use the signing key sk to sign the associated data ad and ciphertext c [s022], and append the signature σ_2 to c [s09]. The *recv* procedure will parse the newly added signatures and verification key [r01,r04]. It will first verify the signature on the verification key vk by calling BS.*vfy* [r06]. Then it uses vk to verify the signature on the associated data and ciphertext [r062].

It may appear peculiar not to sign the ciphertext directly with the BOOM-signature. However, this design decision is made to simplify the confidentiality construction. If we sign the ciphertext directly, the adversary could expose the user to obtain its signing key and generate a new signature for the ciphertext. Indeed, this would not break authenticity as the forgery is trivial. Nonetheless, if the adversary submits the ciphertext to the Recv oracle with a different signature, the oracle will decrypt and return the (challenge) message. Now, because the one-time signature key pair is generated during the *send* procedure, it cannot be exposed. Thus, if the adversary succeeds in creating a valid but different signature, this would break the strong unforgeability property.

This brings us to the lines marked with •. Including these lines provides confidentiality, resulting in our BOOM protocol. The *init* procedure now also initializes a BOOM-KEM BK [i01]. The *send* procedure provides the BOOM-KEM with the verification key as context when requesting (k, c') [s04], appends c' to the context [s05], and uses k to encrypt the message [s06].

The adversary could have exposed the sender's state and created a (trivial) forgery by generating its own one-time signature pair. The Recv oracle would accept the ciphertext and attempt to decrypt it. Therefore, it is critical for confidentiality that the key derivation is dependent on the verification key [s04]. The *recv* procedure parses the newly added c' [r03] and inputs it, along with vk, to BK.*dec* to retrieve k [r20]. Subsequently, *recv* uses k to decrypt m [r20rypt].

The *tick* procedure now calls BK.*upd* [u01] because its state must advance over time, even when no messages are exchanged, to achieve forward secrecy in physical time. Once time has advanced δ times it will start calling BK.*expire* [u02] to indicate we no longer desire to be able to decrypt 'old' messages. Neither of these procedures require the physical time as input because they advance linearly over time, with the expire procedure lagging behind the update procedure. This completes the description of our BOOM protocol in Fig. 6.

Our construction provides authenticity and confidentiality. The proofs are in the full version.

Theorem 1. *Let π be the BOOM construction in Fig. 6, let* AUTH *be the authenticity game in Fig. 2 that calls π's procedures in its oracles, let H be a perfectly collision resistant hash function, and let $\mathcal{A}$ be an adversary that makes at most q_s Send queries. Then there exists an adversary $\mathcal{A}'$ of comparable efficiency such that*

$$\mathbf{Adv}_\pi^{\mathrm{AUTH}}(\mathcal{A}) \leq q_s \cdot \left(\mathbf{Adv}_{\mathrm{OTS}}^{\mathrm{SUF}}(\mathcal{A}') + \mathbf{Adv}_{\mathrm{USS}}^{\mathrm{AUTH}}(\mathcal{A}') \right).$$

Theorem 2. *Let π be the BOOM construction in Fig. 6, let* CONF *be the confidentiality game in Fig. 3 that calls π's procedures in its oracles and let $\mathcal{A}$ be an adversary that makes at most q_c Chal queries and ϵ the probability that the adversary successfully computes a pre-image of the random oracle. Then there exists an adversary $\mathcal{A}'$ of comparable efficiency such that $\mathbf{Adv}_\pi^{\mathrm{CONF}}(\mathcal{A}) \leq$*

$$2q_c \left(\mathbf{Adv}_{\mathrm{keKEM}}^{\mathrm{CONF}}(\mathcal{A}') + \mathbf{Adv}_{\mathrm{kuKEM}}^{\mathrm{CONF}}(\mathcal{A}') + \mathbf{Adv}_{\mathrm{E}}^{\mathrm{CONF}}(\mathcal{A}') \right) + \mathbf{Adv}_\pi^{\mathrm{AUTH}}(\mathcal{A}') + \epsilon.$$

6   Conclusion

After ACD [2] observed that research on secure messaging protocols routinely only considers settings with a guaranteed in-order delivery of messages, while most real-world protocols like Signal are actually designed for out-of-order delivery, we reassess the model and construction of ACD and argue that the intuitive notion of forward secrecy is not provided. We identify that the reason for this is the lack of modeling of physical time, which is required to express that ciphertexts may time out and expire. We hence develop new security models for the out-of-order delivery setting with immediate decryption. Our model incorporates the concept of physical clocks and implements a maximally strong corruption model. We finally design a proof-of-concept protocol that provably satisfies it.

References

1. Alwen, J., et al.: CoCoA: concurrent continuous group key agreement. In: Dunkelman, O., Dziembowski, S. (eds.) EUROCRYPT 2022. LNCS, vol. 13276, pp. 815–844. Springer, Cham (2022). https://doi.org/10.1007/978-3-031-07085-3_28
2. Alwen, J., Coretti, S., Dodis, Y.: The double ratchet: security notions, proofs, and modularization for the signal protocol. In: Ishai, Y., Rijmen, V. (eds.) EUROCRYPT 2019, Part I. LNCS, vol. 11476, pp. 129–158. Springer, Cham (2019). https://doi.org/10.1007/978-3-030-17653-2_5

3. Alwen, J., Coretti, S., Dodis, Y., Tselekounis, Y.: Security analysis and improvements for the IETF MLS standard for group messaging. In: Micciancio, D., Ristenpart, T. (eds.) CRYPTO 2020, Part I. LNCS, vol. 12170, pp. 248–277. Springer, Cham (2020). https://doi.org/10.1007/978-3-030-56784-2_9

4. Alwen, J., Coretti, S., Dodis, Y., Tselekounis, Y.: Modular design of secure group messaging protocols and the security of MLS. In: Vigna, G., Shi, E. (eds.) ACM CCS 2021, pp. 1463–1483. ACM Press (2021). https://doi.org/10.1145/3460120.3484820

5. Alwen, J., Coretti, S., Jost, D., Mularczyk, M.: Continuous group key agreement with active security. In: Pass, R., Pietrzak, K. (eds.) TCC 2020, Part II. LNCS, vol. 12551, pp. 261–290. Springer, Cham (2020). https://doi.org/10.1007/978-3-030-64378-2_10

6. Balli, F., Rösler, P., Vaudenay, S.: Determining the core primitive for optimally secure ratcheting. In: Moriai, S., Wang, H. (eds.) ASIACRYPT 2020, Part III. LNCS, vol. 12493, pp. 621–650. Springer, Cham (2020). https://doi.org/10.1007/978-3-030-64840-4_21

7. Bellare, M., Kohno, T., Namprempre, C.: Authenticated encryption in SSH: provably fixing the SSH binary packet protocol. In: Atluri, V. (ed.) ACM CCS 2002. pp. 1 11. ACM Press (2002). https://doi.org/10.1145/586110.586112

8. Bellare, M., Miner, S.K.: A forward-secure digital signature scheme. In: Wiener, M. (ed.) CRYPTO 1999. LNCS, vol. 1666, pp. 431–448. Springer, Heidelberg (1999). https://doi.org/10.1007/3-540-48405-1_28

9. Bienstock, A., Fairoze, J., Garg, S., Mukherjee, P., Raghuraman, S.: A more complete analysis of the Signal Double Ratchet algorithm. In: Dodis, Y., Shrimpton, T. (eds.) CRYPTO 2022. LNCS, vol. 13507, pp. 784–813. Springer, Cham (2022). https://doi.org/10.1007/978-3-031-15802-5_27

10. Caforio, A., Durak, F.B., Vaudenay, S.: On-demand ratcheting with security awareness. Cryptology ePrint Archive, Report 2019/965 (2019). https://eprint.iacr.org/2019/965

11. Canetti, R., Halevi, S., Katz, J.: A forward-secure public-key encryption scheme. In: Biham, E. (ed.) EUROCRYPT 2003. LNCS, vol. 2656, pp. 255–271. Springer, Heidelberg (2003). https://doi.org/10.1007/3-540-39200-9_16

12. Canetti, R., Jain, P., Swanberg, M., Varia, M.: Universally composable end-to-end secure messaging. In: Dodis, Y., Shrimpton, T. (eds.) Lecture Notes in Computer Science. LNCS, vol. 13508, pp. 3–33. Springer, Cham (2022). https://doi.org/10.1007/978-3-031-15979-4_1

13. Cohn-Gordon, K., Cremers, C., Dowling, B., Garratt, L., Stebila, D.: A formal security analysis of the signal messaging protocol. In: 2017 IEEE European Symposium on Security and Privacy (EuroS&P), pp. 451–466 (2017)

14. Durak, F.B., Vaudenay, S.: Bidirectional asynchronous ratcheted key agreement with linear complexity. In: Attrapadung, N., Yagi, T. (eds.) IWSEC 2019. LNCS, vol. 11689, pp. 343–362. Springer, Cham (2019). https://doi.org/10.1007/978-3-030-26834-3_20

15. Gentry, C., Silverberg, A.: Hierarchical ID-based cryptography. In: Zheng, Y. (ed.) ASIACRYPT 2002. LNCS, vol. 2501, pp. 548–566. Springer, Heidelberg (2002). https://doi.org/10.1007/3-540-36178-2_34

16. Giacon, F., Heuer, F., Poettering, B.: KEM combiners. In: Abdalla, M., Dahab, R. (eds.) PKC 2018, Part I. LNCS, vol. 10769, pp. 190–218. Springer, Cham (2018). https://doi.org/10.1007/978-3-319-76578-5_7

17. Jaeger, J., Stepanovs, I.: Optimal channel security against fine-grained state compromise: the safety of messaging. In: Shacham, H., Boldyreva, A. (eds.) CRYPTO 2018, Part I. LNCS, vol. 10991, pp. 33–62. Springer, Cham (2018). https://doi.org/10.1007/978-3-319-96884-1_2
18. Jost, D., Maurer, U., Mularczyk, M.: Efficient ratcheting: almost-optimal guarantees for secure messaging. In: Ishai, Y., Rijmen, V. (eds.) EUROCRYPT 2019, Part I. LNCS, vol. 11476, pp. 159–188. Springer, Cham (2019). https://doi.org/10.1007/978-3-030-17653-2_6
19. Jost, D., Maurer, U., Mularczyk, M.: A unified and composable take on ratcheting. In: Hofheinz, D., Rosen, A. (eds.) TCC 2019, Part II. LNCS, vol. 11892, pp. 180–210. Springer, Cham (2019). https://doi.org/10.1007/978-3-030-36033-7_7
20. Li, C., Palanisamy, B.: Timed-release of self-emerging data using distributed hash tables. In: 2017 IEEE 37th International Conference on Distributed Computing Systems (ICDCS), pp. 2344–2351 (2017)
21. Liu, J., Jager, T., Kakvi, S.A., Warinschi, B.: How to build time-lock encryption. Des. Codes Crypt. **86**(11), 2549–2586 (2018). https://doi.org/10.1007/s10623-018-0461-x
22. Marlinspike, M., Perrin, T.: The Double Ratchet Algorithm (2016). https://signal.org/docs/specifications/doubleratchet/doubleratchet.pdf
23. Marson, G.A., Poettering, B.: Security notions for bidirectional channels. IACR Trans. Symm. Cryptol. **2017**(1), 405–426 (2017). https://doi.org/10.13154/tosc.v2017.i1.405-426
24. Poettering, B., Rösler, P.: Towards bidirectional ratcheted key exchange. In: Shacham, H., Boldyreva, A. (eds.) CRYPTO 2018, Part I. LNCS, vol. 10991, pp. 3–32. Springer, Cham (2018). https://doi.org/10.1007/978-3-319-96884-1_1
25. Schwenk, J.: Modelling time for authenticated key exchange protocols. In: Kutyłowski, M., Vaidya, J. (eds.) ESORICS 2014, Part II. LNCS, vol. 8713, pp. 277–294. Springer, Cham (2014). https://doi.org/10.1007/978-3-319-11212-1_16
26. Yan, H., Vaudenay, S.: Symmetric asynchronous ratcheted communication with associated data. In: Aoki, K., Kanaoka, A. (eds.) IWSEC 2020. LNCS, vol. 12231, pp. 184–204. Springer, Cham (2020). https://doi.org/10.1007/978-3-030-58208-1_11

Strongly Anonymous Ratcheted Key Exchange

Benjamin Dowling[1], Eduard Hauck[2]([✉]) [ID], Doreen Riepel[2] [ID], and Paul Rösler[3] [ID]

[1] University of Sheffield, Sheffield, UK
b.dowling@sheffield.ac.uk
[2] Ruhr-Universität Bochum, Bochum, Germany
{eduard.hauck,doreen.riepel}@rub.de
[3] New York University, New York, USA
paul.roesler@cs.nyu.edu

Abstract. Anonymity is an (abstract) security goal that is especially important to threatened user groups. Therefore, widely deployed communication protocols implement various measures to hide different types of information (i.e., metadata) about their users. Before actually defining anonymity, we consider an attack vector about which targeted user groups can feel concerned: continuous, temporary exposure of their secrets. Examples for this attack vector include intentionally planted viruses on victims' devices, as well as physical access when their users are detained.

Inspired by *Signal's Double-Ratchet Algorithm*, *Ratcheted* (or *Continuous*) *Key Exchange* (RKE) is a novel class of protocols that increase *confidentiality* and *authenticity* guarantees against temporary exposure of user secrets. For this, an RKE regularly renews user secrets such that the damage due to past and future exposures is minimized; this is called *Post-Compromise Security* and *Forward-Secrecy*, respectively.

With this work, we are the first to leverage the strength of RKE for achieving strong *anonymity* guarantees under temporary exposure of user secrets. We extend existing definitions for RKE to capture attacks that interrelate ciphertexts, seen on the network, with secrets, exposed from users' devices. Although, at first glance, strong authenticity (and confidentiality) conflicts with strong anonymity, our anonymity definition is as strong as possible without diminishing other goals.

We build strongly anonymity-, authenticity-, and confidentiality-preserving RKE and, along the way, develop new tools with applicability beyond our specific use-case: *Updatable and Randomizable Signatures* as well as *Updatable and Randomizable Public Key Encryption*. For both new primitives, we build efficient constructions.

Keywords: RKE · CKE · Ratcheted key exchange · Continuous key exchange · Anonymity · Secure messaging · State exposure · Post-compromise security

The full version of this article is available in the IACR eprint archive as article 2022/1187, at https://eprint.iacr.org/2022/1187.

S. Agrawal and D. Lin (Eds.): ASIACRYPT 2022, LNCS 13793, pp. 119–150, 2022.
https://doi.org/10.1007/978-3-031-22969-5_5

1   Introduction

ANONYMITY. Traditionally, anonymity means that participants of a session cannot be *identified*. As we will argue below, this notion of anonymity is very narrow. Furthermore, in the context of this work, it is not immediately clear what the identity of a session participant actually is. The reason for this is that we consider a modular protocol stack that consists of a *Session Initialization Protocol* (SIP; e.g., an authenticated key exchange) and an independent, subsequent *Session Protocol* (SP; e.g., a symmetric channel or a ratcheted key exchange). According to this modular composition paradigm, only the SIP actually deals with users and their identities, and groups them into session participants who execute the subsequent SP. While the SP may assign different roles to its session participants, the SP is (usually) agnostic about their identities. Thus, it cannot reveal identities by definition. Nevertheless, the context of an SP session and the role of its participant therein may suffice to identify the underlying identity.

SESSION PROTOCOLS. In this work, we focus on anonymity for SPs. Roughly, we call an SP *anonymity-preserving* if its execution reveals nothing about its context, including the session participants, the protocol session itself, the status of a session, etc. We note that real-world deployment of an anonymity-preserving SP requires more than that—e.g., an anonymous SIP, a delivery protocol that transmits anonymous traffic across the Internet, or a mechanism that ensures a large enough set of potential protocol users. While these external components are outside the scope of our work, we mind the broader execution environment of SPs to direct our definitions.

EXPOSURE OF SECRETS. Intuitively, anonymity complements standard security goals, such as confidentiality and authenticity, by requiring that *publicly observable* context data (or *metadata*) remains hidden. More specifically, anonymity means that ciphertexts on the network cannot be interrelated. In this work, we augment this perspective by considering adversaries against anonymity who can expose information that is *secretly stored* by the targeted users. Consequently, our notion of anonymity requires that it is hard to interrelate these exposed user secrets with publicly visible data.

Temporary exposure of user secrets is a realistic threat, especially against cryptographic protocols with long-lasting sessions. The most prominent example for this type of long-term protocols is secure messaging where sessions almost never terminate and, hence, can last for several years. Therefore, anticipating the exposure of participants' locally stored secrets during the lifetime of a session is advisable.

RATCHETED KEY EXCHANGE. Inspired by Signal's Double-Ratchet Algorithm [32], *Ratcheted Key Exchange* (RKE) is an SP primitive that provides security in the presence of adversaries who can expose session participants' local secrets. The core idea of RKE is that the participants continuously establish new symmetric session keys. Following the modular composition paradigm, these keys can be used by another subsequent SP, for instance, to encrypt payload data symmetrically. While establishing session keys, the participants update and renew all their local secrets to recover from potential past exposures (Post-Compromise Security; PCS), and delete old secrets before a potential future exposure occurs (Forward-Secrecy; FS). So far, RKE was only used for preserving *secrecy* and *authenticity* of session keys under the exposure of secrets. In order to also achieve strong *anonymity* under exposure of secrets, we are the first to take advantage of RKE.

Examining RKE constructions, one may doubt that this secrecy- and authenticity-preserving primitive can be extended to also realize strong anonymity: On the one hand, authenticity and anonymity generally tend to be incompatible security goals. On the other hand, for continuously performing updates, participants locally store structured information that is often encoded in sent and received ciphertexts, or has traceable relations to the secrets stored by other session participants. Avoiding this structure (and hiding all relations between sender secrets, ciphertexts, and receiver secrets) is highly non-trivial.

We start with extending RKE syntactically to account for an environment in which preserving anonymity is crucial. Then, we specify a security definition that captures strong anonymity under exposure of secrets. This new definition is compatible with strong secrecy and authenticity notions of RKE.

Flavors of RKE. To reduce complexity and maintain clarity, we consider *unidirectional* RKE [5,8,34], which is a simple, natural notion of RKE that restricts communication between two session participants, Alice and Bob, to flow only from the former to the latter. We leave it an open, highly non-trivial[1] problem for future work to extend our results to more complex *bidirectional* RKE (e.g., [25,33,34]), RKE with *immediate decryption* (e.g., [3]), RKE in *static* groups (e.g., [14]) and *dynamic* groups (e.g., [4,9,37]), resilient to *concurrent* operations (e.g., [2,10]), etc. In the full version of this paper [17], we take a look at the "unidirectional core" of each *two-party* RKE construction from the literature and present successful attacks against anonymity for all of them. We refrain from also presenting (non-trivial) attacks against constructions from the *group* setting without having a suitable anonymity definition that formally separates *trivial* attacks from *non-trivial* ones.[2]

FURTHER RELATED WORK. The literature of anonymity-preserving cryptography ranges from key-private public key encryption (e.g., [7,23,27]) to anonymous

---

[1] Immediate extension and generalization of our results seems unlikely, given the remarkable gap of complexity between non-anonymous unidirectional RKE and more advanced non-anonymous types of RKE.

[2] Note that all CGKA (or "group RKE") constructions reveal structural information like the group size via (publicly) sent ciphertexts. (Moreover, these constructions let users store information about other members in the local user states, and most constructions rely on an active server that participates in the protocol execution.) However, without a formal, satisfiable anonymity definition, it is unclear which information can theoretically be hidden, even by an ideal CGKA construction.

signatures (e.g., [21,41]) to privacy-preserving key exchange (e.g., [24,38,42]) to anonymous onion encryption (e.g., [15,35]) and many other primitives. In principle, our definitions are in line with these notions insofar that we require indistinguishability of "everything that the adversary sees" for a real RKE execution (i.e., ciphertexts and exposed user secrets) from independently sampled equivalents. While some previous works furthermore cover non-cryptographic properties such as anonymous delivery mechanisms (see, e.g., [15]), our work abstracts these external components. To the best of our knowledge, anonymity under (temporary and continuous) exposure of user secrets has not been formally studied before.

Nevertheless, anonymity, privacy, and deniability is generally considered relevant in the domain of secure messaging. For example, the Signal messenger implements the Sealed Sender mechanism [39] to hide the identities of senders. Yet, this mechanism is stateless and uses static long-term secrets, which means that it is insecure under the exposure of receiver secrets. Besides this, several attacks against the deployment of Sealed Sender [30,40] undermine its anonymity guarantees. The Sealed Sender mechanism is related to instances of the Noise protocol framework [18,31] that also claims to reach various notions of anonymity. Yet, the established symmetric session key in a Noise protocol session is static, which means that its exposure breaks anonymity, too. Finally, there is an ongoing discussion about privacy and deniability in the MLS standardization initiative [6] that is yet to be concluded.[3] Related to this, Emura et al. [20] informally propose changes to an early version of MLS by Cohn-Gordon et al. [14] in order to hide the identities of group members. As mentioned above, this is a rather weak form of anonymity. Finally, we note that none of our definitions requires deniability and none of our constructions reaches deniability.

CONTRIBUTIONS. Our main contributions are defining *anonymity* for *Ratcheted Key Exchange* (RKE) and designing a construction that provably satisfies this definition. However, we do not naïvely adopt and extend prior notions of RKE, but we take a fresh look at this primitive, keeping in mind the overall execution environment in which anonymity is important.

Along the way, we develop two new tools that we use to build our final RKE construction. The first tool, *Updatable and Randomizable Public Key Encryption* (urPKE), realizes anonymous PKE with randomizable encryption keys and updatable key pairs. We believe this has applications beyond our work, for example, to Updatable PKE [4,16,26]. The second tool, *Updatable and Randomizable Signatures* (urSIG), simultaneously provides strong anonymity and authenticity guarantees. Roughly, it achieves strong unforgeability of signatures if the signing key is uncorrupted. Furthermore, the signer can derive multiple signing keys that work for the same verification key. However, it should be hard to derive the verification key from a signing key and, beyond that, hard to distinguish whether two signing keys correspond to the same verification key. Surprisingly,

---

[3] See the discussion thread initiated here: https://mailarchive.ietf.org/arch/msg/mls/-1VF95d8od0lF__AFj2WMvk5SQXE/.

both urPKE and urSIG can be built efficiently from cryptographic standard components.

Due to the page limit, we focus on *anonymity* of RKE and its building blocks in the main body of this paper. All novel definitions, constructions, and proofs regarding other security goals such as authenticity and secrecy (which are valuable contributions), are summarized in the subsequent technical overview (Sect. 1.1). The full details of these summarized results can be found in the full version [17].

1.1 Technical Overview

UNIDIRECTIONAL RATCHETED KEY EXCHANGE. Definitions and constructions of *Ratcheted Key Exchange* (RKE) in the literature are highly complex. Since we are the first to consider *anonymity* for this primitive, we want to focus on the core challenges that arise due to the interplay of strong anonymity, confidentiality, and authenticity. Furthermore, we present novel, insightful solutions for these challenges. Thus, for didactic reasons, we condense the question of how to define and construct anonymous RKE by considering the simplest variant of this primitive—so called *Unidirectional* RKE (URKE) [5,8,34]. As we will see, definitions and constructions of anonymous RKE become complex even for this simple unidirectional variant.

An RKE session between two users begins with the initialization that produces a secret state for each user RKE.init $\rightarrow_\$$ (stS, stR). (In practice, this abstract initialization can be instantiated by using an authenticated key exchange protocol.) The users then continuously use their secret states to asynchronously send ciphertexts to their partners. These ciphertexts establish fresh symmetric keys (for the use in subsequent, higher layer SPs) and refresh the secrets in both users' states. While a fully *bidirectional* RKE scheme allows both users to establish new symmetric keys, a *unidirectional* RKE scheme assigns different roles to the two users: only one user (Alice) sends ciphertexts to establish new keys RKE.snd(stS, ad) $\rightarrow_\$$ (stS, c, k) and the other user (Bob) receives these ciphertexts to compute these (same) established keys RKE.rcv(stR, c, ad) $\rightarrow_\$$ (stR, k). Either way, secrets in both users' states are continuously renewed by these operations.

STANDARD SECURITY GOALS. Secrecy and authenticity of established symmetric keys for URKE have been studied in prior work [5,8,34]. These works extend standard secrecy and authenticity notions by allowing the adversary to expose the secret states of Alice and Bob before *and* after each of their send and receive operations, respectively.

Key Secrecy. For *secrecy* of URKE [34], we require that all symmetric keys established by Alice are indistinguishable from random keys unless Bob's corresponding secret state was exposed earlier. More precisely, the symmetric key established by Alice's i_k-th ciphertext must be secure, unless Bob's secret state was exposed already after successfully processing the first i_x ciphertexts from Alice, where $i_x < i_k$. By correctness, Bob's (exposed) state after processing Alice's first i_x ciphertexts can always be used to successfully process the subsequent $i_k - i_x$ ciphertexts from Alice and then compute the i_k-th symmetric key. This notion captures *post-compromise security* (PCS) and *forward-secrecy* (FS) on Alice's side, since all her established symmetric keys must remain secure independent of whether her secret state is ever exposed. It also captures a strong notion of FS on Bob's side, since exposures of his state must not impact the secrecy of a key established with ciphertext i_k under two conditions: (1) the exposures occurred after Bob received ciphertext i'_x, and $i_k \leq i'_x$, or (2) Bob falsely accepted an earlier ciphertext $i_f, i_f < i_k$ that was not sent by Alice and Bob was exposed subsequently at point i'_x, and $i_f \leq i'_x$. This requires that Bob's state becomes incompatible with Alice's state immediately after accepting a forged ciphertext.

Authenticity. *Authenticity* for URKE [19] requires that Bob must not falsely accept a ciphertext i_f, unless Alice's matching secret state was exposed. More precisely, after successfully accepting $i_f - 1$ ciphertexts from Alice, Bob must reject the i_f-th ciphertext if it was not sent by Alice, unless Alice's secret state was exposed after sending the i_x-th ciphertext, where $i_x = i_f - 1$. We call such a successful trivial ciphertext forgery a *trivial impersonation*.

Robustness and Recover Security. We consider two additional properties for URKE: *robustness* and *recover security*. The former requires that Bob will not change his state when rejecting a ciphertext. Thus, Bob can uphold his communication with Alice even if he sometimes receives (and rejects) false ciphertexts that did not result in a trivial impersonation. When considering (receiver) anonymity, robustness is a valuable feature as it allows Bob to perform "trial decryptions" to check if a ciphertext was meant for him or not. Furthermore, consider a setting in which Bob is the receiver of many independent URKE sessions. Due to (sender) anonymity, he may not know the sender of a ciphertext, so he can "trial decrypt" the ciphertext with all of his receiver states until one of them accepts. We conclude that robustness is a crucial property for anonymous RKE. Recover security [19] requires that, whenever Bob falsely accepts a trivial impersonation ciphertext, he will never again accept a ciphertext sent by Alice. This ensures that an adversary who conducted a successful trivial impersonation cannot hide this attack by letting Alice and Bob resume their communication.

For comprehensibility, we make the simplifying assumption that Alice always samples "good" randomness for her send operations. While "bad" randomness can be a realistic threat in some scenarios, we note that URKE under bad randomness—beyond causing more complex definitions and constructions—must rely on strong and inefficient HIBE-like building blocks as Balli et al. [5] prove. We leave it an open problem to extend our results to stronger threat models.

KNOWN CONSTRUCTIONS. RKE constructions only achieving the above properties can be built from standard public key encryption (PKE) and one-time signatures (OTS) [19,25,34]. The idea is that Alice (1) generates fresh PKE key pair (ek_i, dk_i) and OTS key pair (vk_i, sk_i) with every send operation i. She then

(2) encrypts the new decryption key dk_i with the prior encryption key ek_{i-1}, and she (3) signs the resulting PKE ciphertext as well as the new verification key vk_i with the prior signing key sk_{i-1}. The composed URKE ciphertext consists of PKE ciphertext, new verification key, and signature. Alice deletes all prior values as well as the new decryption key dk_i and sends the composed URKE ciphertext to Bob, who verifies the signature, decrypts the PKE ciphertext, and stores (dk_i, vk_i). An additional hash-chain over the entire sent (resp. received) transcript maintains consistency between Alice and Bob, and additional encrypted key material sent from Alice to Bob establishes the symmetric session keys.

Shortcomings. To understand why the above construction does not provide anonymity, note that standard (one-time) signatures can reveal the corresponding verification key. Thus, it can be easy to link two subsequent URKE ciphertexts by testing whether the signature contained in one ciphertext verifies under the verification key contained in the other. (More detailed attacks against anonymity of existing two-party RKE constructions are in the full version [17].) To overcome this limitation, one could simply encrypt the verification key along with the transmitted decryption key. This prevents adversaries who only see ciphertexts transmitted on the network from linking these ciphertexts and, thereby, attributing them to the same URKE session. As we will argue next, this weak level of anonymity is inadequate for settings in which ratcheted key exchange is deployed.

DEFINING (STRONG) ANONYMITY. The main goal of ratcheted key exchange is to continuously establish symmetric keys that remain secure even if the involved users' secret states are temporarily exposed earlier (PCS) and/or later (FS). Hence, if temporary state exposure is considered a realistic threat against secrecy of keys, it is also a realistic threat against anonymity. Consequently, we allow an adversary against anonymity to expose both Alice's and Bob's states.

Ciphertext Anonymity. In a first attempt to define anonymity, we follow the standard concept from the literature: We require that ciphertexts sent from Alice to Bob cannot be distinguished from ciphertexts sent in an independent URKE session from Clara to David, even if the adversary can expose Alice's and Bob's secret states. In this preliminary notion that we call *ciphertext anonymity*, adversaries can perform a trivial exposure that we have to forbid in order to obtain a sound definition. Forbidding this attack, ciphertext anonymity requires that Alice's i_c-th ciphertext must be indistinguishable from a ciphertext sent in an independent URKE session, unless Bob's secret state was exposed already after successfully processing the first i_x ciphertexts from Alice, where $i_x < i_c$. Note that by authenticity, Bob's (exposed) state after processing Alice's first i_x ciphertexts can always be used to verify whether the subsequent $i_c - i_x$ ciphertexts were sent by Alice or by an independent user. This notion captures *post-compromise anonymity* (PCA) and *forward-anonymity* (FA) on Alice's side, since all her ciphertexts must remain anonymous independent of whether her secret state is ever exposed. It also captures a strong notion of FA on Bob's side, since exposures of his state must remain harmless for the anonymity of a ciphertext i_c under two conditions: (1) the exposures were conducted after Bob received ciphertext i'_x, and $i_c \leq i'_x$, or (2) Alice was trivially impersonated towards Bob with an earlier ciphertext i_f, and $i_f < i_c$ and Bob was exposed after ciphertext i'_x, and $i_f \leq i'_x$.

Full Anonymity. Our above description of ciphertext anonymity is not fully formal and the attentive reader may have identified a gap. Consider an adversary who exposes Alice's state twice, once before seeing a ciphertext on the network and once afterwards. By only checking if Alice's state changed between these exposures, the adversary can determine if the ciphertext was sent by Alice. (Note that by authenticity, Alice's state must change with every send operation whereas the state does not change as long as Alice remains inactive.)

To mitigate the threat that Alice's exposed URKE states reveal whether she sent something, we extend the syntax of URKE by adding algorithm RKE.rr(stS) $\rightarrow_{\$}$ stS that (re-)randomizes her state on demand. Executing this algorithm between two exposures, Alice's state can be changed independent of whether she sent a ciphertext. Thus, she can hide if she was the sender of a ciphertext that the adversary observed.

Before specifying a corresponding (stronger) notion of anonymity, we present another threat against anonymity. Consider an adversary who can observe all URKE ciphertexts sent from Alice's device. At some point, this adversary exposes all secrets Alice stores on her device. If Alice has only one stored URKE state, the adversary knows that all observed URKE ciphertexts were sent with this state in the same single session. Since Alice may want to hide how many URKE sessions are running on her device, and how many URKE ciphertexts are sent in each of these sessions, she may want to set up "dummy" URKE states. This scenario motivates that we require for anonymity that Alice's and Bob's secret states must be indistinguishable from independent secret sender and receiver states, respectively—beyond requiring that ciphertexts between Alice and Bob must be indistinguishable from ciphertexts sent in an independent session.

In summary, we require that all secret states that an adversary exposes and all ciphertexts that an adversary observes on a network must be indistinguishable from independent secret states and ciphertexts, respectively, unless correctness, secrecy, and authenticity impose conditions that inevitably allow for distinguishing them. This notion of anonymity is extremely strong and its precise pseudo-code definition is rather complex. However, the basic concept is relatively simple.

Security Experiment. An adversary $\mathcal{A}$ against anonymity plays a game in which it has adaptive access to the following oracles: Snd, RR, Rcv, Expose$_S$, Expose$_R$. Internally, these oracles execute Alice's RKE.snd algorithm, outputting the resulting ciphertext, Alice's RKE.rr algorithm, Bob's RKE.rcv algorithm, and expose Alice's and Bob's current secret states stS and stR, respectively. Access to these oracles is standard in the literature on RKE (except for oracle RR for the additional RKE.rr algorithm). In addition, the adversary can adaptively query oracles that depend on a challenge bit b that is randomly sampled at the beginning of the game:

- ChallSnd equals oracle Snd iff $b = 0$; otherwise, it temporarily initializes a new, independent URKE session with algorithm RKE.init, uses the temporary sender to send a ciphertext with algorithm RKE.snd, and outputs this ciphertext (the temporary URKE session is discarded immediately afterwards); oracle Rcv silently ignores ciphertexts created by ChallSnd under $b = 1$
- ChallExpose$_S$ equals oracle Expose$_S$ iff $b = 0$; otherwise, it initializes a new, independent session with algorithm RKE.init (as above) and outputs the resulting secret sender state
- ChallExpose$_R$ equals oracle Expose$_R$ iff $b = 0$; otherwise, it behaves as oracle ChallExpose$_S$ under $b = 1$, except that it outputs the resulting temporary secret receiver state

The adversary wins the game if it determines challenge bit b without performing a trivial attack that inevitably reveals this challenge bit.

Identifying Trivial Attacks. To complete the above anonymity definition, all attacks that trivially reveal the challenge bit have to be identified, detected, and forbidden. Our aim is to detect these attacks as precisely as possible such that the restrictions limit the adversary as little as possible (leading to a strong definition of anonymity). Interestingly, one class of trivial attacks is particularly hard to detect in a precise way for the anonymity game: trivial impersonations. To give a simple, clarifying example for this, we consider the following adversarial schedule of oracle queries: (1) ChallExpose$_S$ $\rightarrow$ stS$_b$, (2) Rcv(c'), where c' is crafted by the adversary[4], (3) Expose$_R$ $\rightarrow$ stR.

We begin with the case $b = 1$, which means that the adversary plays in the random world. In this world, exposed state stS$_b$ = stS$_1$ is a random sender state that corresponds to a hidden temporary receiver state independent of Bob's actual receiver state stR at step (1). Thus, by authenticity, Bob should not accept any adversarially crafted ciphertext c' in this case. Put differently, impersonating Alice towards Bob is non-trivial for this adversarial behavior in the random world. This means that Bob will reject c' with high probability and the exposed receiver state of Bob in step (3) remains stR, which is independent of the sender state stS$_1$ exposed in step (1).

In contrast, if $b = 0$, which means that the adversary plays in the real world, exposed sender state stS$_b$ = stS$_0$ corresponds to the real receiver state of Bob stR at step (1). Hence, stS$_0$ can be used to craft a valid ciphertext forgery c' that trivially impersonates Alice towards Bob. If the adversary, indeed, performs such a trivial impersonation by executing RKE.snd(stS$_0$) $\rightarrow_\$$ (stS$'$, c', k') and querying Rcv(c'), Bob will compute RKE.rcv(stR, c') $\rightarrow$ (stR$'$, k').[4] The state of Bob stR$'$ that is exposed in final step (3) corresponds to the state stS$'$ that the adversary computed (in their head) during the impersonation. By authenticity, a pair of

---

[4] For simplicity, we ignore the associated data input ad here.

corresponding states $(\mathrm{stS'}, \mathrm{stR'})$ can always be identified as such by sending with the sender state and receiving the result with the receiver state.

Our full anonymity game must, consequently, forbid the final exposure in step (3) because otherwise the adversary can determine the challenge bit from the exposed state.

The presented trivial attack serves as the simplest example for multiple, more complicated trivial impersonations that our game must detect, which we describe in Sect. 4.2.

MAIN COMPONENTS OF CONSTRUCTION. At a first glance, our new URKE construction that fulfills the above anonymity notion follows the design principle of prior non-anonymous URKE constructions described earlier. That means intuitively, in every send operation, Alice (1) generates new PKE and OTS key pairs, (2) encrypts fresh secrets to Bob with which he can compute his matching new PKE decryption key (and the symmetric session key), and she (3) signs the resulting PKE ciphertext. Yet, the exact details of our construction are far more sophisticated. We proceed with presenting the most important anonymity requirements and the corresponding solutions implemented in our construction.

Hiding the Signature. Without presenting the full details of our anonymity definition yet, we note that it imposes the following intuitive requirements: (1) adversaries are allowed to see all (challenge) ciphertexts between sender and receiver; (2) seen (challenge) ciphertexts must remain anonymous even if Alice's state was ever exposed by the adversary before; (3) the authenticity notion presented above imposes the use of asymmetric authentication methods (i.e., signatures) from Alice to Bob. Thus, Alice must have a signing key stored in her state (due to (3)) that is potentially known by the adversary (due to (2)) and, simultaneously, her ciphertexts must be authenticated by corresponding signatures in an anonymous way (due to (1)+(2)+(3)). To ensure that the adversary cannot link matching signing keys (from Alice's exposed states) and signatures (in the sent ciphertexts), our construction encrypts signatures. This encryption of signatures is implemented deterministically with a symmetric key that is encrypted in the PKE ciphertext. Thus, the signature remains confidential while the signed ciphertext is determined before the signature is created, which maintains authenticity and anonymity.

Randomizing Signing Keys Anonymously. The second property required by our anonymity notion focuses on Alice's sender states before and after executing the RKE.rr algorithm. The two sender states of Alice, exposed before and after executing the RKE.rr algorithm, respectively, must be indistinguishable from two freshly generated, independent sender states. That means, an adversary must not learn whether the signing keys, stored in both states of Alice, produce signatures that are valid under the same verification key.[5] For this, we introduce the new notion of *Updatable and Randomizable Signatures* (urSIG) below.

---

[5] Note that RKE.rr only randomizes Alice's state without any interaction with Bob.

Randomizing Encryption Keys Anonymously. Much like the relationship between two signing keys must be hidden by state randomizations, two PKE encryption keys, stored in Alice's exposed states, should not be easily linked. Namely, (a) encryption keys must look random, (b) there must be an routine that re-randomizes them, and (c) it cannot be determined which ciphertexts were created by them. For this, we introduce the new notion of *Updatable and Randomizable Public Key Encryption* (urPKE) below.

UPDATABLE AND RANDOMIZABLE PUBLIC KEY ENCRYPTION. We start with a high level overview of urPKE. As mentioned above, urPKE encryption keys must look random, be re-randomizable, and look independent of the ciphertexts that they produce. Our construction is based on ElGamal encryption. The encryption key consists of ek $\leftarrow (g^r, g^{xr})$, where r and x are random exponents and $x = $ dk is the decryption key. For re-randomizing the encryption key, we apply the same random exponent r' to both of its components $(\text{ek}_0^{r'}, \text{ek}_1^{r'})$. Encryption of message m takes a random exponent s to create ciphertext $c \leftarrow (\text{ek}_0^s, \text{H}(\text{ek}_0^s, \text{ek}_1^s) \oplus m)$. Decryption follows immediately via $m \leftarrow \text{H}(c_0, c_0^{\text{dk}}) \oplus c_1$.

This idea has applications beyond our specific use-case. For example, we point out how our construction can be extended to realize anonymous Updatable PKE [4,16,26] that is broadly used in the literature of RKE and secure messaging.

UPDATABLE AND RANDOMIZABLE SIGNATURES. The security requirements for our new signature primitive urSIG are more challenging. Concretely, an urSIG scheme must provide the following properties: (a) verification keys must look random, (b) deriving the matching verification key from a signing key must be hard, and, beyond this, (c) determining whether two signing keys can produce signatures valid under the same (unknown) verification key must be hard. While ostensibly related to *Designated Verifier Signatures*, urSIG is a novel, incomparable primitive.

Construction Idea. Although the above requirements appear contradictory, we provide a simple construction. The idea is based on Lamport signatures [28]. Intuitively, we start generating the signing key by sampling $2 \cdot \ell$ pre-images $\text{sk}'_{i,b}, (i, b) \in [\ell] \times \{0, 1\}$. To derive the matching verification key, we apply a one-way function on each pre-image $\text{vk}'_{i,b} \leftarrow \text{f}(\text{sk}'_{i,b})$. Finally, we generate a PKE key pair (ek, dk) that allows ciphertext re-randomization. The final verification key consists of the decryption key dk and all images $\text{vk}'_{i,b}$. The final signing key consists of the encrypted pre-images $\text{sk}_{i,b} \leftarrow \text{rPKE.enc}(\text{ek}, \text{sk}'_{i,b})$. To re-randomize Alice's verification key, she re-randomizes each component ciphertext $\text{sk}_{i,b}$. The signature of message $m = (m_1, \ldots, m_\ell)$ consists of the respective signing key components $\sigma \leftarrow (\text{sk}_{1,m_1}, \ldots, \text{sk}_{\ell,m_\ell})$. To verify the signature, Bob decrypts each component and applies the one-way function for comparison with his verification-key component.

For strong unforgeability, we use a technique similar to the CHK transform [13, 29] by employing a strongly unforgeable OTS that signs the actual message. The scheme above then signs the verification key of the strongly unforgeable OTS.

Shrinking Signatures. A drawback of this basic urSIG scheme is that it has large verification keys and large signatures. To mitigate the latter, we instantiate the above construction with a bilinear map $e\colon \mathbb{G}_1 \times \mathbb{G}_2 \to \mathbb{G}_T$, where $\mathbb{G}_1$ is the ciphertext space of the PKE scheme and $\mathbb{G}_2$ and $\mathbb{G}_T$ are chosen such that they are of sufficient size. This allows for aggregation of signing key components $(\text{sk}_{1,m_1}, \dots, \text{sk}_{\ell,m_\ell})$ to obtain a compact signature σ; this aggregation is inspired by BLS signatures [11,12]. The full details of this construction are in Sect. 6.

2 Preliminaries

We write $h \xleftarrow{\$} \mathcal{S}$ to denote that the variable h is uniformly sampled from finite set $\mathcal{S}$. For integers $N, M \in \mathbb{N}$, we define $[N, M] := \{N, N + 1, \dots, M\}$ (which is the empty set for $M < N$) and $[N] := [0, N - 1]$. We use bold notation $\boldsymbol{v}$ to denote vectors. We define $\xleftarrow{\cup} \top$ as the operation which appends $\top$ to the data structure it was called upon. If the data structure is a set, then $\top$ is added to the set. If the data structure is a vector then $\top$ is appended to the end.

We write $\mathcal{A}^{\mathcal{B}}$ to denote that algorithm $\mathcal{A}$ has oracle access to algorithm $\mathcal{B}$ during its execution. To make the randomness ω of an algorithm $\mathcal{A}$ on input x explicit, we write $\mathcal{A}(x; \omega)$. Note that in this notation, $\mathcal{A}$ is deterministic. For a randomised algorithm $\mathcal{A}$, we use the notation $y \in \mathcal{A}(x)$ to denote that y is a possible output of $\mathcal{A}$ on input x.

Basic cryptographic assumptions and definitions used in our proofs are given in the full version [17].

3 Ratcheted Key Exchange

Throughout this paper, we consider unidirectional communication, as defined in several flavors in previous works [5,8,34]. Thus, messages flow from a fixed sender to a fixed receiver; there is no communication from the receiver to the sender. We now define the syntax and properties of unidirectional ratcheted key exchange.

Syntax. A unidirectional ratcheted key exchange scheme RKE consists of four algorithms RKE.init, RKE.snd, RKE.rcv and RKE.rr, where the algorithms are defined as follows.

- $(\text{stS}, \text{stR}) \xleftarrow{\$} \text{RKE.init}$ returns a sender and receiver state.
- $(\text{stS}, c, k) \xleftarrow{\$} \text{RKE.snd}(\text{stS}, \text{ad})$ on input a sender state stS and associated data ad, outputs an updated sender state stS, a ciphertext c, and a key k.
- $(\text{stR}, k) \leftarrow \text{RKE.rcv}(\text{stR}, c, \text{ad})$ on input a receiver state stR, a ciphertext c and associated data ad, outputs an updated receiver state stR and a key k or a failure symbol $\bot$.
- $\text{stS} \xleftarrow{\$} \text{RKE.rr}(\text{stS})$ on input a sender state stS, outputs an randomized sender state stS.

The encapsulation space $\mathcal{C}$ and the key space $\mathcal{K}$ are defined via the support of the RKE.snd algorithm. Let $\mathcal{AD} := \{0,1\}^*$ be the space of associated data.

State Randomization. All algorithms except RKE.rr are standard in the literature of RKE. This new randomization algorithm is designed for settings in which the sender wants strong anonymity. Assume Alice has at least one running RKE session in which she sends periodically. To obfuscate both the number of running RKE sessions and the number of real ciphertexts sent in each, Alice can generate "dummy" RKE sender states. Whenever Alice executes RKE.snd with one of her states, she can re-randomize all remaining states via RKE.rr. Looking ahead, our definition of anonymity requires that all sender states are indistinguishable from a freshly generated sender state, ensuring that it is hard to identify the state that was just used for sending.[6]

Basic Consistency Requirements. In the full version [17], we formally specify three basic consistency notions for RKE: *Robustness*, *Correctness*, and *Recover Security*. Robustness requires that whenever algorithm $(\text{stR}', k) \leftarrow$ RKE.rcv(stR, c, ad) rejects a ciphertext c and associated data ad (and outputs $k = \perp$), the output receiver state stR' must be unchanged (i.e., $\text{stR} = \text{stR}'$), which is crucial for ensuring strong anonymity. Correctness requires that, as long as Bob only accepts ciphertexts sent by Alice (i.e., accepts no forged messages from the attacker), keys output by Bob match those output by Alice. Finally, recover security ensures that it is hard to perform a trivial impersonation of Alice towards Bob without being detected eventually. More concretely, whenever Bob computes a key that does not match the corresponding key computed by Alice, Bob must never accept another ciphertext from Alice.

3.1   Secrecy and Authenticity

We provide compact notions of key-indistinguishability and authenticity for RKE in the full version [17]. In both games, the adversary can control the protocol execution via oracles Snd, RR, Rcv that internally run the respective algorithms. Furthermore, the adversary can expose the sender state and receiver state via oracles Expose_S and Expose_R, respectively.

Secrecy. In game KIND, which models secrecy of session keys, the adversary can additionally query ChallSnd. This oracle internally executes algorithm RKE.snd and, depending on random challenge bit b, either outputs the computed key k (if $b = 0$) or a uniformly random key k' (if $b = 1$). To correctly guess the challenge bit b, the adversary can query all oracles with two limitations. These limitations

---

[6] A corresponding randomization algorithm for the receiver state is meaningless in the *unidirectional* RKE setting since, as soon as Bob's state is exposed, he cannot hope for any security guarantees after that.

depend on whether the receiver accepted a ciphertext (via Rcv) that was not sent by the sender (via Snd resp. ChallSnd). If the receiver never accepted a malicious ciphertext, we say the receiver is *in sync*. As long as the receiver is *in sync*, querying Expose$_R$ is only permitted if all ciphertexts output by ChallSnd were given to Rcv in the same order. Otherwise, exposing the receiver would reveal challenges still in transit. For the same reason, querying ChallSnd is forbidden if the receiver was exposed while *in sync*.

Authenticity. In game AUTH, the adversary wins when the receiver accepts a ciphertext (via Rcv) that was not sent by the sender (via Snd resp. ChallSnd). The only restriction is that Expose$_S$ must not have been queried after the last ciphertext, accepted by the receiver *in sync* (in Rcv), was sent (via Snd resp. ChallSnd). This condition rules out trivial impersonations.

4   Anonymous Ratcheted Key Exchange

In anonymous ratcheted key exchange, any interaction of a fixed RKE instance, consisting of a fixed sender and receiver, should be indistinguishable from an interaction of a fresh RKE instance which is sampled uniformly at random. This includes not only the indistinguishability of ciphertexts and keys, but also the internal states. We capture these core requirements for our anonymity security experiment in so-called utopian games below.

As opposed to KIND and AUTH, there are far more trivial attacks that need to be considered. We elaborate on how we model security such that we can identify and prevent trivial attacks, and give a detailed security notion for anonymity in this section. Following the approach of Rogaway and Zhang [36], we give the core of our definition (which we call *utopian games*), ignoring trivial attacks for now.

Utopian Games. The definition of our utopian games $U - ANON^b$ is given in Fig. 1. Our definitions are "real-or-random"-style and games are parameterized by a bit b, where $U - ANON^0$ denotes the real world execution, and in $U - ANON^1$ all outputs of challenge oracles are random. At the beginning of the game, $U - ANON^b_{RKE}$ samples the initial sender and receiver states and provides several oracles to the adversary. As usual for RKE security, the adversary can control the message flow and obtain internal states via oracles Snd, Rcv, RR, Expose$_S$ and Expose$_R$.

The remaining oracles provide the adversary with some challenge depending on b. We define three different challenge oracles:

- ChallSnd models indistiguishability of ciphertexts and keys. It should be hard to distinguish if the ciphertexts and keys are produced by running RKE.snd on the real sender state ($U - ANON^0$) or a random sender state ($U - ANON^1$).
- ChallExpose$_S$ models indistinguishability of sender states. In $U - ANON^0$ this oracle outputs the real sender state, whereas in $U - ANON^1$ it outputs a random sender state. At this point, we store the corresponding receiver

Game U-ANON$^b_{RKE}(\mathcal{A})$	Oracle RR	Oracle ChallExpose$_S$
00 $(stS, stR) \xleftarrow{\$} $ RKE.init	08 $stS \xleftarrow{\$} $ RKE.rr(stS)	17 If $b = 0$:
01 $ceStR \leftarrow \bot$	09 Return	18 Return stS
02 $b' \xleftarrow{\$} \mathcal{A}$		19 $(stS', ceStR) \xleftarrow{\$} $ RKE.init
03 Stop with b'	**Oracle ChallSnd(ad)**	20 Return stS'
	10 If $b = 0$:	
Oracle Snd(ad)	11 $(stS, c, k) \xleftarrow{\$} $ RKE.snd(stS, ad)	**Oracle Expose$_R$**
04 $(stS, c, k) \xleftarrow{\$} $ RKE.snd(stS, ad)	12 If $b = 1$:	21 Return stR
05 Return (c, k)	13 $(stS', _) \xleftarrow{\$} $ RKE.init	
	14 $(_, c, k) \xleftarrow{\$} $ RKE.snd(stS', ad)	**Oracle ChallExpose$_R$**
Oracle Rcv$(c, $ad)	15 Return (c, k)	22 If $b = 0$:
06 $(stR, k) \leftarrow $ RKE.rcv(stR, c, ad)		23 Return stR
07 Return $[\![k \neq \bot]\!]$:	**Oracle Expose$_S$**	24 $(_, stR') \xleftarrow{\$} $ RKE.init
	16 Return stS	25 Return stR'

Fig. 1. Utopian games $\mathsf{U} - \mathsf{ANON}^b$ for anonymity, where $b \in \{0, 1\}$ and RKE is a ratcheted key exchange scheme.

state in an additional variable ceStR which we require later to define trivial attacks.

- ChallExpose$_R$ models indistinguishability of receiver states and is defined as in ChallExpose$_S$, only it instead outputs the real receiver state ($\mathsf{U} - \mathsf{ANON}^0$) or a random receiver state ($\mathsf{U} - \mathsf{ANON}^1$).

4.1   Anonymity Definition

In this section, we show how to extend the utopian games to a full anonymity security notion for RKE (cf. Fig. 2). Since identifying trivial attacks is quite involved and needs a lot of additional book-keeping, the subsequent text aims to give an in-depth description of our game-based definition on a syntactical level. It provides the framework to prevent trivial attacks and should help the reader to understand how all the tracing logic works. Apart from that, the security game ANON^b_{RKE} basically builds upon the logic of the corresponding utopian game $\mathsf{U} - \mathsf{ANON}^b$. A more high-level perspective and, in particular, descriptions of the actual trivial attacks are given in the subsequent Sect. 4.2.

For comprehensibility, we assume that an RKE scheme, analyzed with our anonymity notion, offers recover security, correctness, as well as authenticity. It is notable that an adversary breaking authenticity also trivially breaks anonymity (cf. the full version [17]).

Execution Model. Depending on the bit b, game ANON^b_{RKE} either simulates the real world as captured in utopian game $\mathsf{U} - \mathsf{ANON}^0_{RKE}$ or the random world as captured in utopian game $\mathsf{U} - \mathsf{ANON}^1_{RKE}$ (cf. Fig. 1). In the following, we will write $\mathsf{U} - \mathsf{ANON}_0$ and $\mathsf{U} - \mathsf{ANON}_1$ for brevity. Hence, ANON^b runs the utopian game $\mathsf{U} - \mathsf{ANON}_b$ as a subroutine and we allow access to all oracles. For example, we denote oracle access by $\mathsf{U} - \mathsf{ANON}_b.\mathsf{Snd}(\mathrm{ad})$, which will run a send query in $\mathsf{U} - \mathsf{ANON}_b$ on input ad. We also allow access to internal variables. For example, we write $\mathsf{U} - \mathsf{ANON}_b.stR$ to access the current receiver state in $\mathsf{U} - \mathsf{ANON}_b$.

To ensure that the game ANON^b can identify trivial attacks, we also need to observe what would have happened if we had run the same sequence of queries

Game ANON$_{RKE}^b$($\mathcal{A}$)

00 U-ANON$_b$ ← U-ANON$_{RKE}^b$
01 For $d \in \{0,1\}$:
02   $(s_d, r_d) \leftarrow (0,0)$
03   $imp_d \leftarrow$ fal
04 $(stR, cstR, cad) \leftarrow ([\cdot],[\cdot],[\cdot])$
05 $stR[0] \leftarrow$ U-ANON$_b$.stR
06 $(c, cc, rcvd) \leftarrow (\emptyset, \emptyset, \emptyset)$
07 $(xS, cxS) \leftarrow (\emptyset, [\cdot])$
08 $(xS, cxS, xR, cxR) \leftarrow$ (fal, fal, fal, fal)
09 $b' \xleftarrow{\$} \mathcal{A}$
10 Stop with b'

Oracle RR

11 U-ANON$_b$.RR
12 · $(xS, cxS) \leftarrow$ (fal, fal)
13 Return

Oracle Expose$_S$

14 ▷ If $cxS =$ tru: Require $(s_0, \_) \notin cxS$
15 ▷ If $xS =$ tru $\wedge (s_0, s_1) \notin xS$:
16 ▷   Require $(\_, s_1) \notin xS$
17 ◇ If $imp_0 = imp_1 =$ fal:
18 ◇   Require $cxR =$ fal
19 $stS \leftarrow$ U-ANON$_b$.Expose$_S$
20 $xS \xleftarrow{\cup} \{(s_0, s_1)\}$
21 · $xS \leftarrow$ tru
22 Return stS

Oracle Expose$_R$

23 i Require unique = tru
24 ◁ Require $cxR =$ fal
25 ◇ Require $imp_0 = imp_1$
26 If $imp_0 = imp_1 =$ fal:
27 ⊕   For all $\hat{s} \in cc$ require $\hat{s} \le r_0$
28 ◇   Require $(r_0, \_) \notin cxS$
29 $stR \leftarrow$ U-ANON$_b$.Expose$_R$
30 · $xR \leftarrow$ tru
31 Return stR

Oracle ChallExpose$_S$

32 ▷ If $xS =$ tru $\vee cxS =$ tru:
33 ▷   Require $(s_0, \_) \notin cxS \wedge (s_0, \_) \notin xS$
34 ◇ If $imp_0 = imp_1 =$ fal:
35 ◇   Require $xR = cxR =$ fal
36 $stS_b \leftarrow$ U-ANON$_b$.ChallExpose$_S$
37 i If $b = 0$: $cstR$.append($stR[s_0]$)
38 i If $b = 1$: $cstR$.append(U-ANON$_1$.ceStR)
39 cxS.append((s_0, s_1))
40 · $cxS \leftarrow$ tru
41 Return stS_b

Oracle Snd(ad)

42 ⊕ If $imp_0 = imp_1 =$ fal: Require $cxR =$ fal
43 $(c, k) \xleftarrow{\$}$ U-ANON$_b$.Snd(ad)
44 cad.append(c, ad)
45 $c \xleftarrow{\cup} \{s_0\}$
46 $s_0 \xleftarrow{+} 1$, $s_1 \xleftarrow{+} 1$
47 i $(stR[s_b], \_) \leftarrow$ RKE.rcv($stR[s_b - 1], c$, ad)
48 Return (c, k)

Oracle ChallSnd(ad)

49 ⊕ If $imp_0 =$ fal: Require $xR = cxR =$ fal
50 $(c_b, k_b) \xleftarrow{\$}$ U-ANON$_b$.ChallSnd(ad)
51 cad.append(c_b, ad)
52 $cc \xleftarrow{\cup} \{s_0\}$
53 $s_0 \xleftarrow{+} 1$
54 i If $b = 0$: $(stR[s_0], \_) \leftarrow$ RKE.rcv($stR[s_0 - 1], c_0$, ad)
55 Return (c_b, k_b)

Oracle Rcv(c, ad)

56 $succ_b \leftarrow$ U-ANON$_b$.Rcv(c, ad)
57 If $\exists \hat{r} \ge \min(r_0, r_1)$ s.t. $(c, ad) = cad[\hat{r}]$
58   If $b = 0$:
59     $r_1' \leftarrow \min(c \setminus rcvd)$
60     $succ_1 \leftarrow \neg imp_1 \wedge [\![r_1' = \hat{r}]\!]$
61   If $succ_1$: $rcvd \xleftarrow{\cup} \{\hat{r}\}$
62   If $b = 1$: $succ_0 \leftarrow \neg imp_0 \wedge [\![r_0 = \hat{r}]\!]$
63   If $succ_{1-b}$: $r_{1-b} \xleftarrow{+} 1$
64 i Else: // check for impersonations
65 i   Let $\mathcal{S} \subseteq xS$ s.t. $(\_, r_1) \in xS$
66 i   If $|\mathcal{S}| > 1 \wedge (r_0, \_) \in \mathcal{S}$: unique ← fal
67 i   For $(\hat{r}_0, \hat{r}_1) \in \mathcal{S}$
68 i   If RKE.rcv($stR[\hat{r}_b], c$, ad) $\neq (\_, \perp)$:
69 i $imp_0 \leftarrow imp_0 \vee [\![r_0 = \hat{r}_0]\!]$
70 i $imp_1 \leftarrow$ tru
71 i   If imp_{1-b}: $r_{1-b} \xleftarrow{+} 1$
72 i $\mathcal{I} \leftarrow \{i \mid cxS[i] = (\hat{r}_0, \hat{r}_1)$ s.t. $\hat{r}_b = r_b\}$
73 i   For $i \in \mathcal{I}$
74 i   If RKE.rcv($cstR[i], c$, ad) $\neq (\_, \perp)$:
75 i $imp_0 \leftarrow imp_0 \vee [\![r_0 = \hat{r}_0]\!]$, where $\hat{r}_0 = cxS[i][0]$
76 i   If imp_{1-b}: $r_{1-b} \xleftarrow{+} 1$
77   If $succ_b$: $r_b \xleftarrow{+} 1$
78 Return

Oracle ChallExpose$_R$

79 ◁ Require $xR = cxR =$ fal
80 ◇ Require $(r_0, \_) \notin xS \wedge (r_0, \_) \notin cxS$
81 ◇ Require $imp_0 =$ fal
82 ⊕ If $imp_1 =$ fal: Require $s_0 = r_0$
83 $stR_b \leftarrow$ U-ANON$_b$.ChallExpose$_R$
84 · $cxR \leftarrow$ tru
85 Return stR_b

Fig. 2. Full anonymity games ANON^b for $b \in \{0,1\}$, where lines in dashed boxes disallow trivial attacks. We further distinguish between different trivial attacks (cf. Sect. 4.2): Lines marked with ⊕ are due to correctness relations, those marked with ▷, ◁ are due to state equality relations on sender resp. receiver side, those marked with ◇ are due to matching state relations, and i indicates an impersonation requirement.

in the other utopian game $1 - b$. We will explain this in more detail in Sect. 4.2. First, we introduce additional book-keeping variables and describe our oracles.

Send Queries. Oracles Snd and ChallSnd take as input a string ad which it forwards to utopian game $U - ANON_b$ to compute a ciphertext and key (c, k). All tuples (c, ad) are stored in a list $\boldsymbol{cad}$. Additionally, we have counters (s_0, s_1) to keep track of the number of ciphertexts sent in game $U - ANON_b$ and the number of ciphertexts that would have been sent in $U - ANON_{1-b}$. On a Snd query, we increment both counters. Since Snd results in updated sender states, we already store the corresponding updated receiver state in a list $\boldsymbol{stR}$ by running the RKE.rcv algorithm locally (line 47). Note that the first entry of $\boldsymbol{stR}$ at position 0 is set to the initial receiver state $U - ANON_b$.stR when the game is initialized (line 05). We additionally store the current counter value s_0 in a set $\boldsymbol{c}$.

On a ChallSnd query, we only increment s_0 because the real sender state is not used in $U - ANON_1$. Thus, we also only need to store the corresponding receiver state in case $b = 0$ (line 54). The value of the counter s_0 is additionally stored in the challenge set $\boldsymbol{cc}$.

Exposures and Randomizations. Oracles Expose_S and Expose_R forward queries to the utopian game and output the real sender state stS (resp. receiver state stR). Additionally, the current sender counters (s_0, s_1) are added to a set $\boldsymbol{xS}$. We use boolean flags xS resp. xR to indicate that the sender resp. receiver was exposed.

Challenge exposures are handled similarly, however now we use a list $\boldsymbol{cxS}$ to store tuples (s_0, s_1) of a query to ChallExpose_S. Thus, we have another list $\boldsymbol{cstR}$ to additionally store the corresponding receiver state of the exposed sender state. When $b = 0$, we simply copy the state stored in $\boldsymbol{stR}$ and for $b = 1$, we store the receiver state $U - ANON_1$.ceStR (belonging to the randomly chosen sender state stS_1). We use boolean flags cxS resp. cxR to register a challenge sender resp. receiver exposure.

A randomization query via RR will reset the sender flags to fal, thus modeling post-compromise anonymity on the sender's side. Note that we do not need to track the time of a receiver exposure. Once exposed, all subsequent updated states can be computed locally by the adversary.

Before describing Rcv behaviour, we want to highlight the importance of impersonations. We use boolean flags imp_0, imp_1 to indicate an impersonation in $U - ANON_0$ or $U - ANON_1$. Both are initialized to fal and will be set to tru if a sequence of queries leads to an impersonation in the corresponding utopian game. Note that sequences of queries may lead to impersonations in both, none or one utopian game(s).[7] Thus, we need track whether an impersonation would

---

[7] An impersonation may occur in one of the games when sender and receiver states are not updated *simultaneously*. The sequence of oracle calls ChallSnd, Expose_S with a subsequent impersonation attempt issued to Rcv will only impersonate $U - ANON_1$, since in $U - ANON_0$ the challenge ciphertext needs to be received first.

have happened. While it is easy to check the impersonation state of the simulated game $U-ANON_b$, i.e., the value of imp_b, it is more involved to determine imp_{1-b}. We will explain how this can be done below.

Receive Queries. Oracle Rcv advances receiver states. Since the adversary only sees ciphertexts of $U - ANON_b$, we first forward the adversary's query (c, ad) to $U-ANON_b$. Similarly to the counters (s_0, s_1), we use counters (r_0, r_1) to track the number of successfully received ciphertexts in games $U-ANON_0$ and $U-ANON_1$. For $U-ANON_{1-b}$, we can determine these numbers from the sequence of queries. We introduce another book-keeping set **rcvd**, which stores the counter values of send queries stored in **c** that have been successfully received in $U - ANON_1$, allowing us to keep track of which tuples stored in **cad** have been processed by $U - ANON_1$. Now, independent of whether this ciphertext has been received successfully, we proceed in three steps.

CHECK FOR IN-ORDER-RECEIVE (LINES 57–63). If the adversary intends to receive a ciphertext output by Snd or ChallSnd (which we check by comparing the query to the list **cad**) we need to decide if this query would have been accepted in $U - ANON_{1-b}$. Let $\hat{r}$ be the index in **cad** such that the tuple stored in $\textbf{cad}[\hat{r}]$ matches the adversary's query. If $b = 0$, we need to decide whether this query would lead to a successful receive in $U - ANON_1$. At this point, we only care about ciphertexts from Snd since challenge ciphertexts in $U - ANON_1$ are produced by a random state. We denote the index of the next ciphertext in **cad** that belongs to a send query by r_1'. Note that we can compute r_1' using sets **c** and **rcvd**. We say that $U - ANON_1$ accepts this ciphertext if $\hat{r} = r_1'$ and we will add $\hat{r}$ to **rcvd**. If $b = 1$, it is easy to decide whether a ciphertext would have been accepted in $U - ANON_0$, since we only need to compare $\hat{r}$ with r_0. Since any ciphertext stored in **cad** should not be accepted after an impersonation, the statements in lines 60, 62 will always evaluate to false.

CHECK FOR IMPERSONATIONS AFTER $Expose_S$ (LINES 65–71). We know that an exposed sender state can lead to an impersonation, depending on when exposure occurred and which ciphertexts have been received. Since we require authenticity, an impersonation can *only* occur after an exposed sender state. Thus, in $U - ANON_1$ an impersonation will only be successful if the counter value r_1 is in the set **xS**. We add all the relevant tuples to a set $\mathcal{S}$. Ignore line 66 for now. We iterate over all entries $(\hat{r}_0, \hat{r}_1) \in \mathcal{S}$ and use $\textbf{stR}[\hat{r}_b]$ to check if the ciphertext decrypts under that state. If so, this may be an impersonation, which we will decide next. Since we always have $\hat{r}_1 = r_1$, a successful decryption implies an impersonation in $U - ANON_1$, so we set imp_1 to tru. If $\hat{r}_0 = r_0$, then we had an impersonation in $U - ANON_0$ as well. By RECOV security, once a sender is impersonated, the receiver will no longer accept their ciphertexts. Thus once $imp_0 \leftarrow$ tru, imp_0 will always be tru independent of the counter comparison, which is captured by the "or" statement in line 69. The result of this check will be the same in both games $ANON^0$ and $ANON^1$, unless the case in line 66 happens. For an example of a sequence of queries triggering this case, we refer to the full version [17]. Note that if there exist multiple entries such that $(\hat{r}_0, \hat{r}_1)$ in $\mathcal{S}$, but $r_0 \neq \hat{r}_0$ for all, then imp_0 will always be set to the same value.

CHECK FOR IMPERSONATIONS AFTER ChallExpose_S (LINES 72–76) Impersonation can also occur using the sender state output by ChallExpose_S. Similarly to the previous step, we first identify relevant entries in the list $\boldsymbol{cxS}$. In particular, we look for all entries $(\hat{r}_0, \hat{r}_1)$, where $\hat{r}_b = r_b$. Since $\boldsymbol{cxS}$ is a list and we stored the corresponding receiver states at the same position in list $\boldsymbol{cstR}$, we need to find the position of the tuples $(\hat{r}_0, \hat{r}_1)$ and store these indices in a set $\mathcal{I}$. This structure is needed, since entries in $\boldsymbol{cxS}$ are not necessarily unique.[8] Now we proceed as in the previous step. An impersonation in $\mathsf{U} - \mathsf{ANON}_0$ has occurred if the counter $\hat{r}_0$ in $\boldsymbol{cxS}$ equals the current counter r_0. Note that in $\mathsf{U} - \mathsf{ANON}_1$, there will not be an impersonation since the real receiver state should accept a ciphertext output by a random sender state. Again, the outcome is the same for both games ANON^b. For $b = 0$, this can be observed by the fact that $\mathcal{I}$ maps to indices where $\hat{r}_0 = r_0$ and thus $\boldsymbol{cstR}[i] = \boldsymbol{cstR}[j]$ for all $i, j \in \mathcal{I}$ and the check only depends on the successful decryption using the current state. For $b = 1$, since all entries in $\boldsymbol{cstR}$ contain different receiver states, there will be at most one state that decrypts the ciphertext. Thus, $\hat{r}_0$ is uniquely defined and imp_0 is only set to tru if $\hat{r}_0 = r_0$ (or if it has already been tru before).

We will increase the counter r_{1-b} if the impersonation was successful. At the very end, we will also increase counter r_b if the query was accepted in the first place. This concludes the description of Rcv.

4.2   Identifying Trivial Attacks

If we ignore trivial attacks, the adversary easily distinguishes ANON^0 from ANON^1, since relations between outputs differ between games. We group these relations into four categories: ability to decrypt, state equality, matching states, and impersonations. In our pseudocode, we indicate restrictions on the adversary with a symbol corresponding to a relation group. We briefly explain the relations below, and we provide justification for all requirements in the full version [17].

Ability to Decrypt (Marked with $\oplus$). Our correctness definition captures that a ciphertext computed with the sender state can always be decrypted with the corresponding receiver state. Due to this, lines marked with ($\oplus$) trace sequences of oracle queries that allow an adversary to determine if a given ciphertext decrypts successfully under an exposed receiver state in one game but not the other, revealing the bit b.

Equality of States (Marked with $\triangleright$, $\triangleleft$). For both sender ($\triangleright$) and receiver ($\triangleleft$) exposures, our anonymity game allows the *direct* exposure of a real state and *challenge* exposures which will output either a real or random state. Depending on the sequence of queries, the output of two *subsequent* calls to Expose_S or

---

[8] Imagine a sequence of queries ChallExpose_S, RR, ChallExpose_S. In this case, the sender counters s_0, s_1 do not change. Also the receiver states appended to $\boldsymbol{cstR}_0$ are the same, but the (random) receiver states appended to $\boldsymbol{cstR}_1$ are different, which is crucial for identifying impersonations.

ChallExpose$_S$ may inevitably be the same in ANON^0 but not in ANON^1, which we detect with the marked code lines to prevent that this inconsistency trivially reveals bit b.

Matching States (Marked with ⋄). We also consider sequences of queries that may expose one party and challenge-expose the other. It is easy to see that the adversary can test whether two such states are linked (which leaks bit b) by creating a ciphertext with the exposed sender state and trial-decrypt with the receiver state.

Impersonations (Marked with i *).* As argued earlier, it is crucial to determine whether a sequence of queries leads to an impersonation in any of the games ANON^0 and ANON^1. Only then, we can detect whether the relations above lead to a trivial attack. However, sometimes it is not possible to uniquely determine the impersonation status in game ANON^{1-b}. Whenever this is the case, we need to disallow receiver exposures since the receiver's state leaks whether the impersonation attempt was successful.

Finally, we formalise the advantage of an adversary against RKE anonymity.

Definition 1. *Consider the games* ANON^b *for* $b \in \{0,1\}$ *in Fig. 2. We define the advantage of an adversary* $\mathcal{A}$ *against anonymity of a ratcheted key exchange scheme* RKE *as*

$$\mathsf{Adv}^{\mathsf{ANON}}_{\mathcal{A},\mathsf{RKE}} := \left| \Pr[\mathsf{ANON}^0_{\mathsf{RKE}}(\mathcal{A}) \Rightarrow 1] - \Pr[\mathsf{ANON}^1_{\mathsf{RKE}}(\mathcal{A}) \Rightarrow 1] \right|.$$

5   Updatable and Randomizable PKE

We construct two types of PKE with related properties: a randomizable PKE scheme (rPKE) and an updatable and randomizable PKE scheme (urPKE). An rPKE scheme is used in the updatable and randomizable signature scheme (cf. Sect. 6.2) and urPKE is a direct building block in the overall construction of ratcheted key exchange (cf. Sect. 7).

5.1   Randomizable PKE

In the following, we define the syntax and properties of an rPKE scheme.

Syntax. A randomizable public-key encryption scheme rPKE consists of four algorithms rPKE.gen, rPKE.enc, rPKE.dec, rPKE.rr, which are defined as follows:

- $(\mathrm{ek}, \mathrm{dk}) \xleftarrow{\$} \mathsf{rPKE.gen}$ outputs an encryption key and a decryption key.
- $c \xleftarrow{\$} \mathsf{rPKE.enc}(\mathrm{ek}, m)$ takes an ek, message m and returns an encryption c.
- $m \leftarrow \mathsf{rPKE.dec}(\mathrm{dk}, c)$ takes dk, c and outputs the decrypted message m.
- $(\mathrm{ek}, c) \xleftarrow{\$} \mathsf{rPKE.rr}(\mathrm{ek}, c)$ returns randomized ek and c.

Compared to a standard public-key encryption scheme, the additional feature lies in the rPKE.rr algorithm that allows to (re-)randomize encryption keys and ciphertexts while preserving correctness. More formally, we require that for all (ek, dk) $\in$ rPKE.gen, $m \in \mathcal{M}$, for random $c \xleftarrow{\$}$ rPKE.enc(ek, m) and for an arbitrary number of randomizations (ek, c) $\xleftarrow{\$}$ rPKE.rr(ek, c), we have that rPKE.dec(dk, c) = m.

We want to use an rPKE scheme as building block of the signature scheme in Sect. 6. For this, we will need some additional properties that we define below.

Homomorphic Property. An rPKE scheme is called *homomorphic* if for an arbitrary but fixed public key (ek, __) $\in$ rPKE.gen, there exists a group homomorphism rPKE.enc: $(\mathcal{M}, \otimes) \times (\mathcal{R}, \oplus) \mapsto (\mathcal{C}, \otimes)$, where $\mathcal{M}, \mathcal{R}, \mathcal{C}$ are message space, randomness space and ciphertext space of the rPKE and $\oplus, \otimes$ are the corresponding group operations. More explicitly,

$$\text{rPKE.enc}(\text{ek}, m_1; r_1) \otimes \text{rPKE.enc}(\text{ek}, m_2; r_2) = \text{rPKE.enc}(\text{ek}, m_1 \otimes m_2; r_1 \oplus r_2) \ ,$$

where $r_1, r_2 \in \mathcal{R}$ and $\otimes$ is taken component-wise.

Further, we want randomizations to be (computationally) indistinguishable, which we capture in the following definition.

Definition 2 (IND $-$ R). *Let* rPKE *be a randomizable public key encryption scheme. We require that a pair of encryption key and ciphertext that has been randomized via* rPKE.rr *is indistinguishable from a freshly generated encryption key and ciphertext. More formally, we define the advantage of a distinguisher $\mathcal{D}$ for arbitrary $2\ell \in \mathbb{Z}$, $(m_0, \ldots, m_{2\ell}) \in \mathcal{M}^{2\ell}$ as*

$$\text{Adv}_{\mathcal{D},\text{rPKE}}^{\text{IND}-\text{R}} := \big| \Pr[\mathcal{D}(\text{ek}, c_0, \ldots, c_\ell, \text{ek}', c_0', \ldots, c_\ell') \Rightarrow 1]$$
$$- \Pr[\mathcal{D}(\text{ek}, c_0, \ldots, c_\ell, \hat{\text{ek}}, \hat{c}_0, \ldots, \hat{c}_\ell) \Rightarrow 1] \big| \ ,$$

where (ek, __) $\xleftarrow{\$}$ rPKE.gen, $c_i \xleftarrow{\$}$ rPKE.enc(ek, m_i), (ek', $c_0', \ldots, c_\ell'$) $\leftarrow$ rPKE.rr (ek, $c_0, \ldots, c_\ell$), $(\hat{\text{ek}}, \_) \xleftarrow{\$}$ rPKE.gen, $\hat{c}_0, \ldots, \hat{c}_\ell \xleftarrow{\$}$ rPKE.enc(ek, $m_{\ell+1}, \ldots, m_{2\ell}$).

CONSTRUCTION. In Fig. 3, we construct an rPKE scheme based on the ElGamal KEM and PKE scheme. Thus, we denote the corresponding scheme by rPKE$_{\text{EG}}$. An encryption key basically consists of an ElGamal encapsulation and KEM key. The encryption and randomization algorithms then use the homomorphic property of ElGamal.

Lemma 1. *Scheme* rPKE$_{\text{EG}}$ *is homomorphic. Furthermore, it satisfies indistinguishability of randomizations under the* DDH *assumption. In particular, for any adversary $\mathcal{A}$, there exists an adversary $\mathcal{B}$ against* DDH *such that*

$$\text{Adv}_{\mathcal{A},\text{rPKE}_{\text{EG}}}^{\text{IND}-\text{R}} \leq \text{Adv}_{\mathcal{B},\mathbb{G}}^{\text{DDH}} \ .$$

Proc rPKE.gen	**Proc** rPKE.enc(ek, m)	**Proc** rPKE.dec(dk, c)	**Proc** rPKE.rr(ek, $c_0, \ldots, c_\ell$)
00 $x, r \overset{\$}{\leftarrow} \mathbb{Z}_p$	04 Parse ek as (ek_0, ek_1)	09 Parse c as (c_0, c_1)	12 Parse ek as (ek_0, ek_1)
01 dk $\leftarrow x$	05 $s \overset{\$}{\leftarrow} \mathbb{Z}_p$	10 $m \leftarrow c_1 \cdot c_0^{-dk}$	13 $r' \overset{\$}{\leftarrow} \mathbb{Z}_p$
02 ek $\leftarrow (g^r, g^{xr})$	06 $c_0 \leftarrow ek_0^s$	11 Return m	14 ek' $\leftarrow (ek_0^{r'}, ek_1^{r'})$
03 Return (ek, dk)	07 $c_1 \leftarrow ek_1^s \cdot m$		15 For $i \in [\ell]$:
	08 Return (c_0, c_1)		16 Parse c_i as (c_i^0, c_i^1)
			17 $s_i' \overset{\$}{\leftarrow} \mathbb{Z}_p$
			18 $c_i' \leftarrow (c_i^0 \cdot ek_0^{s_i'}, c_i^1 \cdot ek_1^{s_i'})$
			19 Return $(ek', c_0', \ldots, c_\ell')$

Fig. 3. Randomizable PKE scheme rPKE$_{EG}$.

5.2 Updatable and Randomizable PKE

In this section, we introduce the primitive of an updatable and randomizable PKE, which will be used in our construction of ratcheted key exchange. The syntax is similar to that of rPKE, but it extends it with the ability to update the key pair. We briefly sketch the differences below.

SYNTAX. An updatable and randomizable public-key encryption scheme urPKE consists of six algorithms urPKE.gen, urPKE.enc, urPKE.dec, urPKE.rr, urPKE.nextDk and urPKE.nextEk, where the first three algorithms are defined as for rPKE and the remaining ones follow the syntax:

- ek $\overset{\$}{\leftarrow}$ urPKE.rr(ek) outputs a randomized encryption key ek.
- dk $\leftarrow$ urPKE.nextDk(dk, r) updates the decryption key with randomness r.
- ek $\leftarrow$ urPKE.nextEk(ek, r) updates the encryption key with randomness r.

Note that the main difference to rPKE is that the randomization algorithm urPKE.rr randomizes only the encryption key.

We now require the following additional properties.

Instance Independence. We say a urPKE scheme is *instance-independent* if for uniformly chosen randomness r and any key pair (ek, dk) in the support of urPKE.gen, the two distributions (urPKE.nextEk(ek, r), urPKE.nextDk(dk, r)) and (ek', dk') $\overset{\$}{\leftarrow}$ urPKE.gen are the same.

Indistinguishability of Randomizations. Similar to rPKE, we require for $\mathsf{IND} - \mathsf{R}$ (formally defined in the full version [17]) security that an encryption key that has been randomized is indistinguishable from a freshly generated encryption key. In particular, the two distributions (ek, ek$_1$) and (ek, ek$_2$), where (ek, _) $\overset{\$}{\leftarrow}$ urPKE.gen, ek$_1 \leftarrow$ urPKE.rr(ek), (ek$_2$, _) $\overset{\$}{\leftarrow}$ urPKE.gen should be (computationally) indistinguishable under chosen ciphertext attacks.

Ciphertext Anonymity. For *ciphertext anonymity* of urPKE we require that ciphertexts generated by a particular (and possibly exposed) encryption key are indistinguishable from ciphertexts generated by a freshly chosen encryption key under chosen ciphertext attacks. We provide a more fine-grained game-based definition in the full version [17].

CONSTRUCTION. We construct an updatable and randomizable PKE scheme based on hashed ElGamal, which was first proven to be IND − C secure in [1]. The construction is also similar to the secretly key-updatable encryption scheme of [26], thus we will only sketch it here. We give the full scheme in the full version [17], including the proofs of the properties mentioned above.

Algorithms urPKE.gen, urPKE.enc, urPKE.dec follow the ideas from rPKE, only that they hash the ElGamal KEM key used for encryption. Since the ciphertext does not need to be randomized, urPKE.rr can still be performed in the same way as the randomization of the encryption key in rPKE.rr. Algorithms urPKE.nextDk and urPKE.nextEk asynchronously update the decryption and encryption key by exponentiation with some uniformly chosen randomness.

6   Updatable and Randomizable One-Time Signatures

In this section we introduce our new signature primitive, namely updatable and randomizable one-time signatures. The property of updatability refers to asynchronous updates of the signing and verification keys. Randomizability refers to the randomization of signing keys. These will be crucial to provide anonymity guarantees of our ratcheted key exchange scheme.

Challenges. The main technical difficulty in designing the signature scheme lies in maintaining unforgeability while achieving randomizability of signing keys. More specifically, randomization must be implemented in a way such that both the original signing key and one of its randomized versions produce signatures that are *unforgeable* (if neither of both signing keys is corrupted); furthermore, signatures from both signing keys must verify under the same single verification key. Simultaneously, seeing the original and the randomized signing key should be indistinguishable from seeing two independently sampled signing keys. (Note that, by unforgeability, two independent signing keys will *not* produce signatures valid under the same verification key.)

We conjecture that updatability of a signature scheme is easy for most algebraic signature schemes. Unforgeability usually reduces to hardness of inverting some one-way function mapping from signing keys to verification keys. So it must be hard to invert verification keys to get valid signing keys. Our randomization requirements, intuitively, demand this for the opposite direction, too: obtaining verification keys from signing keys must be hard. Strictly speaking, we require an even stronger property: Without having the verification key, signing keys and their signatures look random, independent of whether they correspond to the same verification key. This might seem contradictory or, at least, very strong.

OUTLINE. As a warm-up, we start with a definition and construction of updatable one-time signatures in Sect. 6.1. Then, we will extend the construction to updatable and randomizable one-time signatures in Sect. 6.2. To achieve randomizability, we use the ElGamal-based rPKE scheme introduced in Sect. 5.

6.1   Warm-Up: Updatable Signatures

Syntax. An updatable signature scheme uSIG consists of five algorithms uSIG.gen, uSIG.sig, uSIG.vfy, uSIG.nextSk, uSIG.nextVk. Let $\mathcal{M}$ be the message space and $\mathcal{R}$ be the randomness space. Then the algorithms are defined as follows:

- $(\text{vk}, \text{sk}) \xleftarrow{\$} \text{uSIG.gen}$ generates a verification key vk and signing key sk.
- $\sigma \xleftarrow{\$} \text{uSIG.sig}(\text{sk}, m)$ takes sk and a message m and returns a signature σ.
- $\{0, 1\} \leftarrow \text{uSIG.vfy}(\text{vk}, m, \sigma)$ takes vk, m and σ and returns a bit indicating whether σ is a valid signature for m.
- $\text{sk} \leftarrow \text{uSIG.nextSk}(\text{sk}, r)$ asynchronously updates sk with randomness r.
- $\text{vk} \leftarrow \text{uSIG.nextVk}(\text{vk}, r)$ asynchronously updates vk with randomness r.

Correctness. Apart from the standard correctness requirement, we require that updates yield valid verification and signing keys. More formally, we require the following:

(1) $\forall (\text{sk}, \text{vk}) \in \text{uSIG.gen}, m \in \mathcal{M}$:

$$\Pr[\text{uSIG.vfy}(\text{vk}, \sigma, m) = 1 \mid \sigma \xleftarrow{\$} \text{uSIG}(\text{sk}, m)] = 1$$

(2) $\forall (\text{sk}, \text{vk}) \in \text{uSIG.gen}, r \in \mathcal{R}$:

$$(\text{uSIG.nextSk}(\text{sk}, r), \text{uSIG.nextVk}(\text{vk}, r)) \in \text{uSIG.gen}$$

Intuition Updatability. At the core of our construction lies a slight variation of Lamport one time signature scheme, where signing keys are group elements. To shrink the size of signatures and to mitigate the lack of updateability we instantiate the hash function with a hash function fulfilling one-wayness and the homomorphic property. By one-wayness the unforgeability property of Lamport signature scheme is unchanged and by the homomorphic property we can i) optimize the signature length to a single element in the target group ii) update signing and verification key.

To achieve this we use pairings. Let $\mathcal{G}$ be a pairing group with bilinear map $e \colon \mathbb{G}_1 \times \mathbb{G}_2 \to \mathbb{G}_T$. By the XDH assumption, DDH is hard in group $\mathbb{G}_1$ and CDH is hard in groups $\mathbb{G}_1$ and $\mathbb{G}_2$. For fixed $g_2 \in \mathbb{G}_2$, we then set $\mathsf{H}(h) := e(h, g_2)$. Clearly the homomorphic property of H follows from bilinearity of the pairing,

$$e(m_1, g_2) \cdot e(m_2, g_2) = e(m_1 \cdot m_2, g_2) .$$

By the FAPI-2 Assumption [22], H is a one way function.

CONSTRUCTION. Our construction of an updatable one-time signature scheme is given in Fig. 4. It follows the idea of the one-time Lamport signature scheme, where we replace the hash function of the original scheme with a Type-II pairing. Thus, let $\mathcal{G}$ be a pairing group and $\mathsf{H} : \{0, 1\}^* \to \{0, 1\}^\ell$ a hash function. Secret keys consist of 2ℓ group elements in $\mathbb{G}_1$ and verification keys consist of 2ℓ group elements in $\mathbb{G}_T$. For the signature generation, we borrow the approach of aggregated BLS signatures [11,12]. Additionally following the "Hash-and-Sign" approach, we first hash the message using H and then interpret the hash value bit-wise. For the ith bit we choose the ith element of the signing key depending on the bit value. The signature σ will then be the product of ℓ group elements. Verification uses the pairing to compute $e(\sigma, g_2)$ and compares the result to the product of the respective ℓ target group elements.

The idea for updating the signing and verification key is that we can multiply each group element of the signing key $\text{sk}_{i,b}$ with another group element $R_{i,b}$. Verification keys can be updated by multiplying the respective target group element with $e(R_{i,b}, g_2)$.

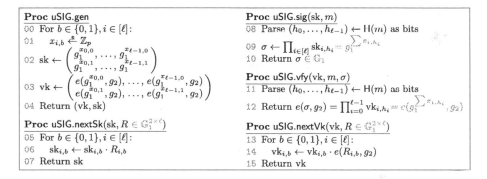

Fig. 4. Updatable one-time signature scheme uSIG for a pairing group $\mathcal{G} = (p, \mathbb{G}_1, \mathbb{G}_2, \mathbb{G}_T, e, g_1, g_2)$, where $H: \{0,1\}^* \mapsto \{0,1\}^\ell$ is a hash function.

In the full version [17] we prove one-time existential unforgeability of the scheme.

6.2 Extension to Updatable and Randomizable Signatures

Syntax. An updatable and randomizable signature scheme urSIG shares the syntax of an updatable signature scheme, i.e., the algorithms urSIG.gen, urSIG.sig, urSIG.vfy, urSIG.nextSk, urSIG.nextVk are defined analogously. Additionally, there is a sixth algorithm urSIG.rr, which is defined as follows

– sk $\xleftarrow{\$}$ urSIG.rr(sk) randomizes the signing key sk.

Correctness. We extend correctness requirements (1), (2) from the previous section by the following: We require that for all (vk, sk) $\in$ urSIG.gen, $m \in \mathcal{M}$, an arbitrary number of randomizations resulting in an randomized signing key sk $\xleftarrow{\$}$ urSIG.rr(sk), a signature $\sigma \xleftarrow{\$}$ urSIG.sig(sk, m) still verifies correctly.

Below we define a similar security property as for randomizable PKE schemes, which will be needed in the anonymity proof of our ratcheted key exchange scheme.

In the full version [17] define additional security properties that are needed for authenticity and recover security.

Definition 3 (Indistinguishability of Randomizations). *Let* urSIG *be a an updatable and randomizable signature scheme. We require that a signing key that has been randomized using* urSIG.rr *is indistinguishable from a freshly generated signing key. More formally, we define the advantage of a distinguisher* $\mathcal{D}$ *as*

$$\mathsf{Adv}_{\mathcal{D},\mathsf{urSIG}}^{\mathsf{IND-R}} := \left| \Pr[\mathcal{D}(\mathrm{sk}, \mathrm{sk}_0) \Rightarrow 1] - \Pr[\mathcal{D}(\mathrm{sk}, \mathrm{sk}_1) \Rightarrow 1] \right|,$$

where the probability is taken over $(\mathrm{sk}, \mathrm{vk}) \xleftarrow{\$} \mathsf{urSIG.gen}$, $\mathrm{sk}_0 \leftarrow \mathsf{urSIG.rr}(\mathrm{sk})$ *and* $(\mathrm{sk}_1, \_) \xleftarrow{\$} \mathsf{urSIG.gen}$ *and the internal randomness of* $\mathcal{D}$.

OUR CONSTRUCTION. In Fig. 5 we extend the updatable signature scheme in Fig. 4 by the randomizable PKE in Fig. 3 to get an updatable and randomizable one-time signature scheme.

Recall that signing keys in our updatable one-time signature scheme are group elements. In order to achieve signing key randomization, the idea is to encrypt those signing keys with ElGamal. However, this means that the ElGamal encryption key must be part of the overall signing key and thus in turn be randomized as well. Therefore, we do not use plain ElGamal encryption, but our randomizable PKE encryption scheme $\mathsf{rPKE_{EG}}$.

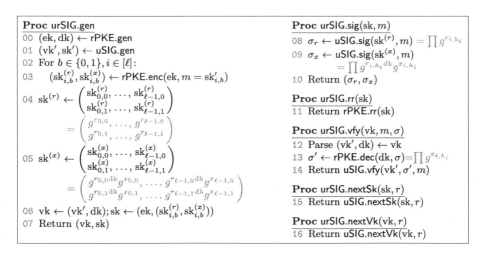

Fig. 5. Our updatable and randomizable one-time signature scheme urSIG[rPKE, uSIG].

Finally, to achieve strong unforgeability we use the CHK transformation [13, 29] using a strongly unforgeable signature.

7   Construction of Anonymous RKE

Our construction of anonymous unidirectional RKE in Fig. 6 elegantly arises from the two primitives presented in the last sections, urPKE and urSIG. Beyond this, we use a hash function (modeled as a random oracle) and a pseudorandom generator (PRG).

Proc RKE.init	**Proc RKE.rr(stS)**
00 $(ek, dk) \xleftarrow{\$} urPKE.gen$	16 $(ek, sk) \leftarrow stS$
01 $(vk, sk) \xleftarrow{\$} urSIG.gen$	17 $ek \xleftarrow{\$} urPKE.rr(ek)$
02 $stS \leftarrow (ek, sk)$	18 $sk \xleftarrow{\$} urSIG.rr(sk)$
03 $stR \leftarrow (dk, vk)$	19 $stS \leftarrow (ek, sk)$
04 Return (stS, stR)	20 Return stS

Proc RKE.snd(stS, ad)	**Proc RKE.rcv(stR, c, ad)**
05 $(ek, sk) \leftarrow stS$	21 $k \leftarrow \bot$
06 $k_H, k_S \xleftarrow{\$} \mathcal{K}$	22 $(dk, vk) \leftarrow stR$
07 $(vk_{next}, sk_{next}) \leftarrow urSIG.gen$	23 $(c_{urPKE}, \sigma') \leftarrow c$
08 $c_{urPKE} \xleftarrow{\$} urPKE.enc(ek, (k_H, k_S, vk_{next}))$	24 $(k_H, k_S, vk_{next}) \leftarrow urPKE.dec(dk, c_{urPKE})$
09 $\sigma \xleftarrow{\$} urSIG.sig(sk, (c_{urPKE}, ad))$	25 Require $(k_H, k_S, vk_{next}) \neq \bot$
10 $c \leftarrow (c_{urPKE}, \sigma \oplus PRG(k_S))$	26 If $urSIG.vfy(vk, (c_{urPKE}, ad)), \sigma' \oplus PRG(k_S))$
11 $(k, r_{urPKE}, r_{urSIG}) \leftarrow H(k_H, c, ad)$	27 $(k, r_{urPKE}, r_{urSIG}) \leftarrow H(k_H, c, ad)$
12 $sk \leftarrow urSIG.nextSk(sk_{next}, r_{urSIG})$	28 $vk \leftarrow urSIG.nextVk(vk_{next}, r_{urSIG})$
13 $ek \leftarrow urPKE.nextEk(ek, r_{urPKE})$	29 $dk \leftarrow urPKE.nextDk(dk, r_{urPKE})$
14 $stS \leftarrow (ek, sk)$	30 $stR \leftarrow (dk, vk)$
15 Return (stS, k, c)	31 Return (stR, k)

Fig. 6. Construction of our RKE scheme $RKE[urPKE, urSIG, H, PRG]$.

Construction. On initialization, a urPKE key pair and a urSIG key pair is generated, both of which are split between Alice's and Bob's state. Randomization of Alice's state works componentwise. When sending, Alice (1) generates a fresh signature key pair, (2) encrypts the new verification key as well as random symmetric keys, and (3) signs the resulting ciphertext with her prior signing key. (4) The signature is encrypted with one of the encrypted symmetric keys. Using the random oracle on input of the other symmetric key, the composed ciphertext, and the associated data string, Alice (5) derives the final session key as well as two pseudorandom strings which update her two state components (encryption key and signing key). Bob performs the corresponding decryption, verification, hash evaluation, and key updates when receiving.

Consistency and Authenticity. By the *correctness* properties of urPKE and urSIG, this URKE construction is correct, too. The construction provides *robustness* since Bob either accepts with an actual session key (if decryption and verification succeed) or his state remains unchanged. We formally prove *recover security* of this construction in the full version [17]. On an intuitive level, each fresh signing key is "entangled" with the ciphertext that transmits it via the key update in line 12. This means that Bob will only accept signatures from a signing key if he received the corresponding verification key with the originally transmitted ciphertext. Based on unforgeability of the urSIG scheme and collision resistance of the random oracle, this mechanism maintains recover security. *Authenticity* similarly follows from the signature scheme's unforgeability, which we prove in the full version [17].

Secrecy. In the presence of a passive adversary, the secrecy of session keys follows directly from the confidentiality of the urPKE scheme. In case of a trivial impersonation—which, by authenticity, is the only successful way to let Bob accept a forged ciphertext—, we need the consistency guarantees of the urSIG scheme and the hash function to prove that Bob's state immediately diverges incompatibly from Alice's state. We prove this informal claim in the full version [17].

Anonymity. Below we establish our main theorem, namely anonymity of our RKE construction. Additionally, we provide theorems and proofs for robustness, recover security, authenticity and key indistinguishability in the full version [17].

Theorem 1 (Anonymity of $\mathsf{RKE}[\mathsf{urPKE}, \mathsf{urSIG}, \mathsf{H}, \mathsf{PRG}]$). *Let* $\mathsf{H} : \{0,1\}^* \to \{0,1\}^\lambda$ *be a random oracle. Let* urPKE *be an updatable and randomizable PKE scheme. Let* urSIG *be an updatable and randomizable one-time signature scheme. Let* PRG *be a pseudorandom generator. We show that* $\mathsf{RKE}[\mathsf{urPKE}, \mathsf{urSIG}, \mathsf{H}, \mathsf{PRG}]$ *is secure with respect to* ANON, *such that*

$$\mathsf{Adv}_{\mathsf{RKE}}^{\mathsf{ANON}} \leq (q_S + q_{CS}) \cdot \mathsf{Adv}_{\mathsf{urPKE}}^{\mathsf{ANON}} + q_{CS} \cdot \mathsf{Adv}_{\mathsf{PRG}}$$
$$+ (q_{CE} + q_{CS}) \cdot (\mathsf{Adv}_{\mathsf{urSIG}}^{\mathsf{IND-R}} + \mathsf{Adv}_{\mathsf{urPKE_{EG}}}^{\mathsf{IND-R}}) + \frac{1}{2^\lambda}.$$

where q_S, q_{CS}, *and* q_{CE} *are the number of queries to oracles* Snd, ChallSnd, *and* $\mathsf{ChallExpose}_S$, *respectively.*

We provide a proof sketch below and defer the full proof to the full version [17].

Proof (Sketch). Conceptually, the proof consists of three steps. First we show on the sender side that after calls to oracles Snd and ChallSnd, the sender states are statistically independent from prior ones. Similarly, after successful calls to oracle Rcv, the receiver state is statistically independent from prior ones. The forward *anonymity* and post-compromise *anonymity* guarantees follow from this state independence. We prove this independence via $(q_S + q_{CS})$ applications of the instance independence of urPKE.

In the second step, we replace all outputs of challenge oracles in the real world with independently sampled values. We get this for free for oracle $\mathsf{ChallExpose}_R$, since, by definition of our trivial attack detection and instance independence, the adversary may call oracle $\mathsf{ChallExpose}_R$ only on receiver states which are statistically independent from any other oracle output. To replace the output of oracle ChallSnd with random, we employ two hybrid arguments. In the first hybrid argument, we show that the adversary cannot distinguish whether we replaced challenge ciphertexts c_{urPKE} with random ciphertexts, implying a loss factor of $(q_S + q_{CS}) \cdot \mathsf{Adv}_{\mathsf{urPKE}}^{\mathsf{ANON}}$. In the second hybrid argument, we replace all outputs of the PRG in oracle ChallSnd with random, implying a loss factor of $q_{CS} \cdot \mathsf{Adv}_{\mathsf{PRG}}$. To replace the outputs of oracle $\mathsf{ChallExpose}_S$ with uniform random values, we again give two hybrid arguments. Here we loose a total factor of $q_{CE} \cdot (\mathsf{Adv}^{\mathsf{IND-R}}_{\mathsf{urSIG}} + \mathsf{Adv}^{\mathsf{IND-R}}_{\mathsf{urPKE_{EG}}})$. Finally, in the third step of the proof, we show that the adversary cannot distinguish how often the sender state was advanced. Recall that oracle ChallSnd is the only oracle which updates the sender state depending on bit b. In order for the adversary to see a difference in updated sender states, the adversary must expose the sender prior to and after a call to oracle ChallSnd. By definition of the trivial attacks, the adversary must call oracle RR before exposing the sender a second time. Using a hybrid argument, we replace the sender state after a call to RR by uniform random values in both worlds. Thus the adversary learns with both sender state exposures two independent distributions of sender states, which implies a total loss factor of $q_{CS} \cdot (\mathsf{Adv}^{\mathsf{IND-R}}_{\mathsf{urSIG}} + \mathsf{Adv}^{\mathsf{IND-R}}_{\mathsf{urPKE_{EG}}})$.

Acknowledgments. We thank Kenny Paterson, Eike Kiltz, and Joël Alwen for recurring very inspiring discussions during our work on this article. Special thanks goes to Kenny for hosting Paul as a visitor at ETH Zürich, which led to launching this research project.

Doreen Riepel was funded by the Deutsche Forschungsgemeinschaft (DFG, German Research Foundation) under Germany's Excellence Strategy - EXC 2092 CASA - 390781972.

References

1. Abdalla, M., Bellare, M., Rogaway, P.: The oracle Diffie-Hellman assumptions and an analysis of DHIES. In: Naccache, D. (ed.) CT-RSA 2001. LNCS, vol. 2020, pp. 143–158. Springer, Heidelberg (2001). https://doi.org/10.1007/3-540-45353-9_12

2. Alwen, J., et al.: CoCoA: concurrent continuous group key agreement. In: Dunkelman, O., Dziembowski, S. (eds.) EUROCRYPT 2022. LNCS, vol. 13276, pp. 815–844. Springer, Cham (2022). https://doi.org/10.1007/978-3-031-07085-3_28

3. Alwen, J., Coretti, S., Dodis, Y.: The double ratchet: security notions, proofs, and modularization for the signal protocol. In: Ishai, Y., Rijmen, V. (eds.) EUROCRYPT 2019. LNCS, vol. 11476, pp. 129–158. Springer, Cham (2019). https://doi.org/10.1007/978-3-030-17653-2_5

4. Alwen, J., Coretti, S., Dodis, Y., Tselekounis, Y.: Security analysis and improvements for the IETF MLS standard for group messaging. In: Micciancio, D., Ristenpart, T. (eds.) CRYPTO 2020. LNCS, vol. 12170, pp. 248–277. Springer, Cham (2020). https://doi.org/10.1007/978-3-030-56784-2_9

5. Balli, F., Rösler, P., Vaudenay, S.: Determining the core primitive for optimally secure ratcheting. In: Moriai, S., Wang, H. (eds.) ASIACRYPT 2020. LNCS, vol. 12493, pp. 621–650. Springer, Cham (2020). https://doi.org/10.1007/978-3-030-64840-4_21

6. Barnes, R., Beurdouche, B., Robert, R., Millican, J., Omara, E., Cohn-Gordon, K.: The messaging layer security (MLS) protocol. Internet-Draft draft-ietf-mls-protocol-14, IETF (2020)

7. Bellare, M., Boldyreva, A., Desai, A., Pointcheval, D.: Key-privacy in public-key encryption. In: Boyd, C. (ed.) ASIACRYPT 2001. LNCS, vol. 2248, pp. 566–582. Springer, Heidelberg (2001). https://doi.org/10.1007/3-540-45682-1_33

8. Bellare, M., Singh, A.C., Jaeger, J., Nyayapati, M., Stepanovs, I.: Ratcheted encryption and key exchange: the security of messaging. In: Katz, J., Shacham, H. (eds.) CRYPTO 2017. LNCS, vol. 10403, pp. 619–650. Springer, Cham (2017). https://doi.org/10.1007/978-3-319-63697-9_21

9. Bienstock, A., Dodis, Y., Garg, S., Grogan, G., Hajiabadi, M., Rösler, P.: On the worst-case inefficiency of CGKA. In: Kiltz, E., Vaikuntanathan, V. (eds.) TCC 2022. LNCS, vol. 13748, pp. 213–243. Springer, Cham (2022). https://doi.org/10.1007/978-3-031-22365-5_8

10. Bienstock, A., Dodis, Y., Rösler, P.: On the price of concurrency in group ratcheting protocols. In: Pass, R., Pietrzak, K. (eds.) TCC 2020. LNCS, vol. 12551, pp. 198–228. Springer, Cham (2020). https://doi.org/10.1007/978-3-030-64378-2_8

11. Boneh, D., Gentry, C., Lynn, B., Shacham, H.: Aggregate and verifiably encrypted signatures from bilinear maps. In: Biham, E. (ed.) EUROCRYPT 2003. LNCS, vol. 2656, pp. 416–432. Springer, Heidelberg (2003). https://doi.org/10.1007/3-540-39200-9_26

12. Boneh, D., Lynn, B., Shacham, H.: Short signatures from the Weil pairing. In: Boyd, C. (ed.) ASIACRYPT 2001. LNCS, vol. 2248, pp. 514–532. Springer, Heidelberg (2001). https://doi.org/10.1007/3-540-45682-1_30

13. Canetti, R., Halevi, S., Katz, J.: Chosen-ciphertext security from identity-based encryption. In: Cachin, C., Camenisch, J.L. (eds.) EUROCRYPT 2004. LNCS, vol. 3027, pp. 207–222. Springer, Heidelberg (2004). https://doi.org/10.1007/978-3-540-24676-3_13

14. Cohn-Gordon, K., Cremers, C., Garratt, L., Millican, J., Milner, K.: On ends-to-ends encryption: asynchronous group messaging with strong security guarantees. In: ACM CCS 2018 (2018)

15. Degabriele, J.P., Stam, M.: Untagging tor: a formal treatment of onion encryption. In: Nielsen, J.B., Rijmen, V. (eds.) EUROCRYPT 2018. LNCS, vol. 10822, pp. 259–293. Springer, Cham (2018). https://doi.org/10.1007/978-3-319-78372-7_9

16. Dodis, Y., Karthikeyan, H., Wichs, D.: Updatable public key encryption in the standard model. In: Nissim, K., Waters, B. (eds.) TCC 2021. LNCS, vol. 13044, pp. 254–285. Springer, Cham (2021). https://doi.org/10.1007/978-3-030-90456-2_9

17. Dowling, B., Hauck, E., Riepel, D., Rösler, P.: Strongly anonymous ratcheted key exchange. Cryptology ePrint Archive, Paper 2022/1187. https://eprint.iacr.org/2022/1187

18. Dowling, B., Rösler, P., Schwenk, J.: Flexible authenticated and confidential channel establishment (fACCE): analyzing the noise protocol framework. In: Kiayias, A., Kohlweiss, M., Wallden, P., Zikas, V. (eds.) PKC 2020. LNCS, vol. 12110, pp. 341–373. Springer, Cham (2020). https://doi.org/10.1007/978-3-030-45374-9_12

19. Durak, F.B., Vaudenay, S.: Bidirectional asynchronous ratcheted key agreement with linear complexity. In: Attrapadung, N., Yagi, T. (eds.) IWSEC 2019. LNCS, vol. 11689, pp. 343–362. Springer, Cham (2019). https://doi.org/10.1007/978-3-030-26834-3_20

20. Emura, K., Kajita, K., Nojima, R., Ogawa, K., Ohtake, G.: Membership privacy for asynchronous group messaging. Cryptology ePrint Archive, Report 2022/046. https://eprint.iacr.org/2022/046

21. Fischlin, M.: Anonymous signatures made easy. In: Okamoto, T., Wang, X. (eds.) PKC 2007. LNCS, vol. 4450, pp. 31–42. Springer, Heidelberg (2007). https://doi.org/10.1007/978-3-540-71677-8_3

22. Galbraith, S.D., Hess, F., Vercauteren, F.: Aspects of pairing inversion. Cryptology ePrint Archive, Report 2007/256. https://eprint.iacr.org/2007/256

23. Grubbs, P., Maram, V., Paterson, K.G.: Anonymous, robust post-quantum public key encryption. In: Dunkelman, O., Dziembowski, S. (eds.) EUROCRYPT 2022. LNCS, vol. 13277, pp. 402–432. Springer, Cham (2022). https://doi.org/10.1007/978-3-031-07082-2_15
24. Ishibashi, R., Yoneyama, K.: Post-quantum anonymous one-sided authenticated key exchange without random oracles. In: Hanaoka, G., Shikata, J., Watanabe, Y. (eds.) PKC 2022. LNCS, vol. 13178, pp. 35–65. Springer, Cham (2022). https://doi.org/10.1007/978-3-030-97131-1_2
25. Jaeger, J., Stepanovs, I.: Optimal channel security against fine-grained state compromise: the safety of messaging. In: Shacham, H., Boldyreva, A. (eds.) CRYPTO 2018. LNCS, vol. 10991, pp. 33–62. Springer, Cham (2018). https://doi.org/10.1007/978-3-319-96884-1_2
26. Jost, D., Maurer, U., Mularczyk, M.: Efficient ratcheting: almost-optimal guarantees for secure messaging. In: Ishai, Y., Rijmen, V. (eds.) EUROCRYPT 2019. LNCS, vol. 11476, pp. 159–188. Springer, Cham (2019). https://doi.org/10.1007/978-3-030-17653-2_6
27. Kohlweiss, M., Maurer, U., Onete, C., Tackmann, B., Venturi, D.: Anonymity-preserving public-key encryption: a constructive approach. In: De Cristofaro, E., Wright, M. (eds.) PETS 2013. LNCS, vol. 7981, pp. 19–39. Springer, Heidelberg (2013). https://doi.org/10.1007/978-3-642-39077-7_2
28. Lamport, L.: Constructing digital signatures from a one-way function. Technical report SRI-CSL-98, SRI International Computer Science Laboratory (1979)
29. MacKenzie, P., Reiter, M.K., Yang, K.: Alternatives to non-malleability: definitions, constructions, and applications. In: Naor, M. (ed.) TCC 2004. LNCS, vol. 2951, pp. 171–190. Springer, Heidelberg (2004). https://doi.org/10.1007/978-3-540-24638-1_10
30. Martiny, I., Kaptchuk, G., Aviv, A.J., Roche, D.S., Wustrow, E.: Improving signal's sealed sender. In: NDSS 2021 (2021)
31. Perrin, T.: The noise protocol framework. http://noiseprotocol.org/noise.html, revision 34
32. Perrin, T., Marlinspike, M.: The double ratchet algorithm. https://whispersystems.org/docs/specifications/doubleratchet/doubleratchet.pdf
33. Poettering, B., Rösler, P.: Asynchronous ratcheted key exchange. Cryptology ePrint Archive, Report 2018/296. https://eprint.iacr.org/2018/296
34. Poettering, B., Rösler, P.: Towards bidirectional ratcheted key exchange. In: Shacham, H., Boldyreva, A. (eds.) CRYPTO 2018. LNCS, vol. 10991, pp. 3–32. Springer, Cham (2018). https://doi.org/10.1007/978-3-319-96884-1_1
35. Rogaway, P., Zhang, Y.: Onion-ae: Foundations of nested encryption. Proc. Priv. Enhancing Technol. (2018)
36. Rogaway, P., Zhang, Y.: Simplifying game-based definitions. In: Shacham, H., Boldyreva, A. (eds.) CRYPTO 2018. LNCS, vol. 10992, pp. 3–32. Springer, Cham (2018). https://doi.org/10.1007/978-3-319-96881-0_1
37. Rösler, P., Mainka, C., Schwenk, J.: More is less: on the end-to-end security of group chats in signal, WhatsApp, and Threema. In: IEEE EuroS&P 2018 (2018)
38. Schäge, S., Schwenk, J., Lauer, S.: Privacy-preserving authenticated key exchange and the case of IKEv2. In: Kiayias, A., Kohlweiss, M., Wallden, P., Zikas, V. (eds.) PKC 2020. LNCS, vol. 12111, pp. 567–596. Springer, Cham (2020). https://doi.org/10.1007/978-3-030-45388-6_20
39. Signal: Sealed sender. https://signal.org/blog/sealed-sender/, blog post
40. Tyagi, N., Len, J., Miers, I., Ristenpart, T.: Orca: blocklisting in sender-anonymous messaging. In: USENIX Security 2022 (2022)

41. Yang, G., Wong, D.S., Deng, X., Wang, H.: Anonymous signature schemes. In: Yung, M., Dodis, Y., Kiayias, A., Malkin, T. (eds.) PKC 2006. LNCS, vol. 3958, pp. 347–363. Springer, Heidelberg (2006). https://doi.org/10.1007/11745853_23
42. Zhao, Y.: Identity-concealed authenticated encryption and key exchange. In: ACM CCS 2016 (2016)

Encryption to the Future

A Paradigm for Sending Secret Messages to Future (Anonymous) Committees

Matteo Campanelli[1]([✉]), Bernardo David[3], Hamidreza Khoshakhlagh[2], Anders Konring[3], and Jesper Buus Nielsen[2]

[1] Protocol Labs, San Francisco, USA
matteo@protocol.ai
[2] Aarhus University, Aarhus, Denmark
{hamidreza,jbn}@cs.au.dk
[3] IT University of Copenhagen, Copenhagen, Denmark
{beda,konr}@itu.dk

Abstract. A number of recent works have constructed cryptographic protocols with flavors of adaptive security by having a randomly-chosen anonymous committee run at each round. Since most of these protocols are stateful, transferring secret states from past committees to future, but still unknown, committees is a crucial challenge. Previous works have tackled this problem with approaches tailor-made for their specific setting, which mostly rely on using a blockchain to orchestrate auxiliary committees that aid in the state hand-over process. In this work, we look at this challenge as an important problem on its own and initiate the study of Encryption to the Future (EtF) as a cryptographic primitive. First, we define a notion of an EtF scheme where time is determined with respect to an underlying blockchain and a lottery selects parties to receive a secret message at some point in the future. While this notion seems overly restrictive, we establish two important facts: 1. if used to encrypt towards parties selected in the "far future", EtF implies witness encryption for NP over a blockchain; 2. if used to encrypt only towards parties selected in the "near future", EtF is not only sufficient for transferring state among committees as required by previous works, but also captures previous tailor-made solutions. To corroborate these results, we provide a novel construction of EtF based on witness encryption over commitments (cWE), which we instantiate from a number of standard assumptions via a construction based on generic cryptographic primitives. Finally, we show how to use "near future" EtF to obtain "far future" EtF with a protocol based on an auxiliary committee whose communication complexity is *independent* of the length of plaintext messages being sent to the future.

M. Campanelli—Work done in part while the author was affiliated to Aarhus University.

S. Agrawal and D. Lin (Eds.): ASIACRYPT 2022, LNCS 13793, pp. 151–180, 2022.
https://doi.org/10.1007/978-3-031-22969-5_6

1   Introduction

Most cryptographic protocols assume that parties' identities are publicly known. This is a natural requirement, since standard secure channels are identified by a sender and a receiver. However, this status quo also makes it easy for adaptive (or proactive) adversaries to readily identify which parties are executing a protocol and decide on an optimal corruption strategy. In more practical terms, a party with a known identity (*e.g.* IP address) is at risk of being attacked.

A recent line of work [3, 15, 16] has investigated means for avoiding adaptive (or proactive) corruptions by having different randomly chosen committees of anonymous parties execute each round of a protocol. The rationale is that parties whose identities are unknown cannot be purposefully corrupted. Hence, having each round of a protocol executed by a fresh anonymous committee makes the protocol resilient to such powerful adversaries. However, this raises a new issue:

> *How can past committees efficiently transfer secret states to future yet-to-be-assigned anonymous committees?*

1.1   Motivation: Role Assignment

The task of sending secret messages to a committee member that will be elected in the future can be abstracted as *role assignment*, a notion first introduced in [3] and further developed in [15]. This task consists of sending a message to an abstract role R at a given point in the future. A role is just a bit-string describing an abstract role, such as R = "party number j in round sl of the protocol Γ". Behind the scenes, there is a mechanism that samples the identity of a random party P_i and associates this machine to the role R. Such a mechanism allows anyone to send a message m to R and have m arrive at P_i chosen at some point in the future to act as R. A crucial point is: no one should know the identity of P_i even though P_i learns that it is chosen to act as R.

The approaches proposed in [3, 15, 16] for realizing role assignment all use an underlying Proof-of-Stake (PoS) blockchain (*e.g.* [9]). On a blockchain, a concrete way to implement role assignment is to sample a fresh key pair $(\mathsf{sk_R}, \mathsf{pk_R})$ for a public key encryption scheme, post $(\mathsf{R}, \mathsf{pk_R})$ on the blockchain and somehow send $\mathsf{sk_R}$ to a random P_i without leaking the identity of this party to anyone. Once $(\mathsf{R}, \mathsf{pk_R})$ is known, every party has a target-anonymous channel to P_i and is able to encrypt under $\mathsf{pk_R}$ and post the ciphertext on the blockchain. Notice that using time-lock puzzles (or similar notions) is not sufficient for achieving this notion, since only the party or parties elected for a role should receive a secret message encrypted for that role, while time-lock puzzles allow any party to recover the message if they invest enough computing time.

A shortcoming of the approaches of [3, 15, 16] is that, besides an underlying blockchain, they require an auxiliary committee to aid in generating $(\mathsf{sk_R}, \mathsf{pk_R})$ and selecting P_i. In the case of [3], the auxiliary committee performs cheap operations but can adversarially influence the probability distribution with which P_i

is chosen. In the case of [15, 16], the auxiliary committee cannot bias this probability distribution but must perform very expensive operations (using Mix-Nets or FHE; see also Sect. 1.3). Moreover, these approaches have another caveat: they can only be used to select P_i to act as R according to a probability distribution *already known* at the time the auxiliary committee outputs (R, pk_R). Hence, they only allow sending messages to future committees that have been *recently* elected. Later we explicitly consider this weaker setting—where we want to communicate with a "near-future" committee (i.e., whose distribution is known)—and dub it "Encryption to the Current Winner[1]" (ECW).

In this paper we further investigate solutions to the role-assignment problem[2]. Taking a step back from specific solutions to this problem, we strive to obtain non-interactive solutions to *encrypting* to a future role with IND-CPA security *without* the aid of an auxiliary committee. We improve on solutions relying on interaction with an auxiliary committee and shed light on the hardness of achieving a fully non-interactive solution. We also discuss how to extend our approach to IND-CCA2 security and how to allow winners of a role to authenticate themselves when sending a message, achieving both goals using standard assumptions.

1.2   Our Contributions

We look at the issue of sending messages to future roles as a problem on its own and introduce the Encryption to the Future (EtF) primitive as a central tool to solve it. Apart from defining this primitive and showing constructions based on previous works, we propose constructions based on new insights and investigate limits of EtF in different scenarios. Our general constructions for EtF work by lifting a weaker primitive, namely encryption for the aforementioned "near-future" setting, or ECW. Before providing further details, we summarize our contributions as follows:

---

[1] The word "winner" here refers to the party who is selected to perform a role according to the underlying lottery of the PoS blockchain (see remainder of introduction).

[2] The family of protocols we consider actually has two role-related aspects to solve. The first—and the focus of this paper—is the aforementioned *role assignment (RA)* *which deals with the sending of messages to parties selected to perform future roles of a protocol while hiding the identities of such parties*. The other aspect is *role execution* (RX) which focuses on the execution of the specific protocol that runs on top of the RA mechanism, i.e., what messages are sent to which roles and what specification the protocol implements. In [15] the so-called *You Only Speak Once* (YOSO) model is introduced for studying RX. In the YOSO model the protocol execution is between abstract roles which can each speak only once. Later these can then be mapped to physical machines using an RA mechanism. The work of [15] shows that given RA in a synchronous model, any well-formed ideal functionality can be implemented in the YOSO model with security against malicious adaptive corruption of a minority of machines. Concretely, [15] gives an ideal functionality for RA and shows that a YOSO protocol for abstract roles can be compiled into the RA-hybrid model to give a protocol secure against adaptive attacks.

- A definition for the notion of Encryption to the Future (EtF) in terms of an underlying blockchain and an associated lottery scheme that selects parties in the future to receive messages for a role. We study the strength of EtF as a primitive and prove that a non-interactive EtF scheme allowing for encryption towards parties selected at arbitrary points in the future implies a flavor of witness encryption for NP over a blockchain (referred to as BWE).
- A novel construction of Encryption to the Current Winner (ECW), *i.e.* EtF where the receiver of a message is determined by the *current* state of the blockchain, which can be instantiated *without auxiliary committees* from standard assumptions via a construction based on generic primitives.
- A transformation from ECW to EtF through an auxiliary committee holding a *small* state, i.e., with communication complexity *independent* of plaintext size $|m|$ (in contrast to [3,15,16] where a committee's state grows with $|m|$).
- An application of ECW as a central primitive for realizing role assignment in protocols that require it (*e.g.* [3,15,16]).

Our EtF notion arguably provides a useful abstraction for the problem of transferring secret states to secret committees. Our ECW construction is the first primitive to realize role assignment without the need for an auxiliary committee. Moreover, building on new insights from our EtF notion and constructions, we show the first protocol for obtaining role assignment with no constraints on when parties are chosen to act as the role. While our protocol uses auxiliary committees, it improves on previous work by only requiring a communication complexity *independent* of the plaintext length. We elaborate on our results, discussing the intuition behind the notion of EtF, its constructions and its fundamental limits. We also invite the reader to use Fig. 1 as reference for the discussion below.

Encryption to the Future (EtF)—Sect. 3. As in previous works [3,15,16], an EtF scheme is defined with respect to an underlying PoS blockchain. We naturally use core features of the PoS setting to define what "future" means. The vast majority of PoS blockchains (*e.g.* [9]) associates a *slot number* to each block and uses a lottery for selecting parties to generate blocks according to a stake distribution (*i.e.* the probability a party is selected is proportional to the stake the party controls). Thus, in EtF, we let a message be encrypted towards a party that is selected by the underlying blockchain's lottery scheme at a given future slot. We can generalize this and let the lottery select parties for multiple roles associated to each slot (so that committees consisting of multiple parties can be elected at a single point in time). We note that the goal of defining EtF with respect to an underlying blockchain is to construct it without having to assume very strong primitives such as (extractable) witness encryption for NP[3]. Moreover, it is necessary to provide a non-interactive EtF scheme with a means to publicly verify whether a given party has won the lottery to perform a certain role. Since this lottery predicate's output must hold for all parties, we need

---

[3] While one *might* define EtF in more general settings, namely without a blockchain, it is unclear how to obtain *interesting* instantiations, that is from standard primitives.

a consensus mechanism that allows for all parties to agree on lottery parameters/outputs while allowing for third parties to verify this result. An important point of our EtF definition is that it does not impose any constraints on the underlying blockchain's lottery scheme (*e.g.* it is not required to be anonymous) or on the slot when a party is supposed to be chosen to receive a message sent to a given role (*i.e.* party selection for a given role may happen w.r.t. a *future* stake distribution).

Relation to "Blockchain Witness Encryption" (BWE)—Sect. 8. In order to study how hard it is to realize EtF, we show that EtF implies a version of witness encryption [14] over a blockchain (similar to that of [18] but without relying on committees). The crux of the proof: if we can encrypt a message towards a role assigned to a party only at an arbitrary point in the future, then we can easily construct a witness encryption scheme exploiting EtF and a smart contract on the EtF's underlying blockchain. We also prove the opposite direction (BWE implies EtF), showing that the notions are similar from a feasibility standpoint. This shows another crucial point: to implement non-interactive EtF, we would plausibly need strong assumptions (e.g., full-blown WE). This follows by observing that existing constructions of WE over blockchains (e.g., [18]) are interactive in the sense that they rely on a committee that holds all encrypted messages in secret shared form and periodically re-share them. On the other hand, in the interactive setting, we show a construction of EtF with improved communication complexity that is *independent* from the size (or amount) of EtF encrypted messages: the committee only needs to hold an IBE master secret key (secret shared) and compute secret keys for specific identities. We note that the goal of constructing BWE from EtF is not to provide a concrete instantiation based on existing blockchains but rather to provide evidence that EtF is hard to construct from standard assumptions. The underlying blockchain protocol and lottery we use are standard Proof-of-Stake based blockchains with a VRF-based lottery and smart contracts. The only non-realistic assumption we make is that the stake is distributed in arbitrarily (i.e. it is all locked inside one smart contract) which is an assumption on how the blockchain is operated rather than on how it is constructed or why it is secure.

Encryption to the Current Winner (ECW)—Sect. 3. By the previous result we know that, unless we turn to strong assumptions, we may not construct a fully non-interactive EtF (i.e., without auxiliary committees); therefore, we look for efficient ways to construct EtF under standard assumptions while minimizing interaction. As a first step towards such a construction, we define the notion of Encryption to a Current Winner (ECW), which is a restricted version of EtF where messages can only be encrypted towards parties selected for a role whose lottery parameters are available for the *current* slot, the one in which we encrypt (this is as in previous constructions [3,15,16]). Unrestricted EtF, on the other hand, allows for encrypting a message toward lottery winners that will be determined at any arbitrary point in the future, including parties who only join the protocol execution far in the future (after the ciphertext has been generated).

Constructing ECW (non-interactively)—Sect. 5. We show that it is possible to construct a fully non-interactive ECW scheme from standard assumptions. Our construction relies on a milder flavor of witness encryption, which we call Witness Encryption over Commitments (cWE) and define it in Sect. 4. This primitive is significantly more restricted than full-fledged WE (see also discussion in Remark 2), but still powerful enough: we show in Sect. 5.1 that ECW can be constructed in a black-box manner from cWE, which in turn can be constructed from oblivious transfer and garbled circuits [7]. This construction improves over the previous results [3,15,16] since it does not rely on auxiliary committees.

Instantiating YOSO MPC using ECW —Sect. 6. The notion of ECW is more restricted than EtF, but it can still be useful in applications. We show how to use it as a building block for the YOSO MPC protocol of [15]. Here, each of the rounds in an MPC protocol is executed by a different committee. This same committee will simultaneously transfer its secret state to the next (near-future) committee, which in turn remains anonymous until it transfers its own secret state to the next committee, and so on. This setting clearly matches what ECW offers as a primitive, but it also introduces a few more requirements: 1. ECW ciphertexts must be non-malleable, *i.e.* we need an IND-CCA secure ECW scheme; 2. Only one party is selected for each role; 3. A party is selected for a role at random with probability proportional to its relative stake on the underlying PoS blockchain; 4. Parties selected for roles remain anonymous until they choose to reveal themselves; 5. A party selected for a role must be able to authenticate messages on behalf of the role, *i.e.* publicly proving that it was selected for a certain role and that it is the author of a message. We show that all of these properties can be obtained departing from an IND-CPA secure ECW scheme instantiated over a natural PoS blockchain (*e.g.* [9]). First, we observe that VRF-based lottery schemes implemented in many PoS blockchains are sufficient to achieve properties 1, 2 and 3. We then observe that natural block authentication mechanisms used in such PoS blockchains can be used to obtain property 4. Finally, we show that standard techniques can be used to obtain an IND-CCA secure ECW scheme from an IND-CPA secure ECW scheme.

Constructing EtF from ECW (interactively)—Sect. 7. Since we argued the implausibility of constructing EtF non-interactively from standard assumptions, we study how to transform an ECW scheme into an unrestricted EtF scheme when given access to an auxiliary committee but with "low communication" (and still from standard assumptions). We explain what we mean by "low communication" by an example of its opposite: in previous works [3,15,16] successive committees were required to store and reshare secret shares of every message to be sent to a party selected in the future. That is, their communication complexity grows both with the number and the amount and length of the encrypted messages. In contrast, our solution has communication complexity independent of the plaintext length. How our transformation from ECW to EtF works: we associate each role in the future to a unique identity of an Identity Based Encryption scheme (IBE); to encrypt a message towards a role we apply the encryption of the IBE scheme. When, at any point in the future, a party for that role is selected, a committee generates and delivers the corresponding secret key for that role/identity. To realize the latter step, we apply YOSO MPC instantiated from ECW as shown in Sect. 6. In contrast to previous schemes, our auxiliary committee only needs to hold shares of the IBE's master secret key and so it performs communication/computation dependent on the security parameter but not on the length/amount of messages encrypted to the future.

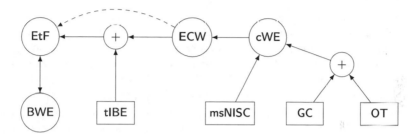

Fig. 1. Dependency diagram for primitives in this work. Legend: primitives wrapped in circles are introduced in this work; $A \to B$: *"We can construct B from A"*; $A \dashrightarrow B$: *"A is a special case of B"*.

1.3   Previous Works

We compare previous works related to our notions of EtF and ECW (encryption to future and current winner, respectively) in Fig. 2.

Type	Scheme	Communication	Committee?	Interaction?
	CaBKaS [3]	$O(1)$	yes	yes
	RPIR [16]	$O(1)$	yes	yes
ECW	cWE (MS-NISC) (Sec. 4.2)	$O(N)$	no	no*
	cWE (GC+OT) (Sec. 4.2)	$O(N)$	no	no*
	IBE (Sec. 7)	$O(1)$	yes	yes
EtF	WEB [18]	$O(M)$	yes	yes
	Full-fledged WE	$O(1)$	no	no

Fig. 2. The column "Committee?" indicates whether a committee is required. The column "Communication" refers to the communication complexity in terms of the number of all parties N, and the number of plaintexts (called deposited secrets in [18]) M of a given fixed length. We denote by an asterisk non-interactive solutions that require sending a first reusable message during the initial step.

Encryption to the Current Winner (ECW). We recall that ECW is an easier setting than EtF: both the stake distribution and the randomness extracted from the blockchain are static and known at the time of encryption. This means that all of the parameters except the secret key of the lottery winner are available to the encryption algorithm. We now survey works that solved this problem and compare them to our solutions:

- "Can a Blockchain Keep a Secret?" (CaBKaS) [3]. The work of [3] addresses the setting where a dynamically changing committee (over a public blockchain) maintains a secret. The main challenge in order for the committee to *securely* reshare its secret can be summarized as: how to select a small committee from a large population of parties so that everyone can send secure messages to the committee members without knowing who they are? The solution of [3] is to select the "secret-holding" committee by having another committee, a "nominating committee", that nominates members of the former (while the members of the nominating committee are self-nominated).
  One can see the nominating committee as a tool providing the ECW functionality. A major caveat in such a solution, however, is that to guarantee an honest majority in the committees, [3] can only tolerate up to 1/4 as the fraction of corrupted parties. This is because corrupted nominators can always select corrupted parties, whereas honest nominators may select corrupted parties by chance. We can improve this through our non-interactive ECW: we can remove the nominating committee and just let the current committee ECW-encrypt their secret shares to the roles of the next committee.
- "Random-Index PIR" (RPIR) [16]. The recent work of [16] defines a new flavour of Private Information Retrieval (PIR) called Random-index PIR (or RPIR) that allows each committee to perform the nomination task by themselves. While RPIR improves on [3] (not requiring a nominating committee and tolerating up to 1/2 of corrupted parties), its constructions are inefficient, either based on Mix-Nets or Fully Homomorphic Encryption (FHE). The construction based on Mix-Nets uses k shufflers, where k is the security parameter, and has an impractical communication complexity of $O(nk^2)$, where n is the number of public keys that each shuffler broadcasts. The FHE-based construction gives a total communication complexity of $O(k^3)$ where $O(k)$ is the length of an FHE decryption share.

WE over Commitments (cWE). Benhamouda and Lin [4] defined a type of witness encryption, called "Witness Encryption for NIZK of Commitments". In their setting, parties first commit to their private inputs once and for all. Later, an encryptor can produce a ciphertext so that any party with a committed input that satisfies the relation (specified at encryption time) can decrypt. More accurately, who can decrypt is any party *with a NIZK showing that the committed input satisfies the relation*. The authors construct this primitive based on standard assumptions in asymmetric bilinear groups.

In our work, we generalize the encryption notion in [4], formalize it as cWE and finally use it to construct ECW. While the original construction of [4] fits the definition of cWE, we observe it is an overkill for our application. Specifically our setting does not require NIZKs to be involved in encryption/decryption. We instead give more efficient instantiations based on two-party Multi-Sender Non-Interactive Secure Computation (MS-NISC) protocols and Oblivious Transfer plus Garbled Circuits.

Encryption to the Future (EtF). The general notion of EtF is significantly harder to realize than ECW (as we show in Sect. 8). Below we discuss natural ideas to obtain EtF. They can be seen as illustrating two extremes where our approach (Sect. 7) lies in the middle.

- Non-Interactive—Using Witness Encryption [14]: One trivial approach to realize EtF is to use full-fledged general Witness Encryption [14] (WE) for the arithmetic relation $\mathcal{R}$ being the lottery predicate such that the party who holds a winning secret key sk can decrypt the ciphertext. However, constructing a general witness encryption scheme [14] which we can instantiate reliably is still an open problem. Existing constructions rely on very strong assumptions such as multilinear maps, indistinguishability obfuscation or other complexity theoretical conjectures [2]. The challenges in applying this straightforward solution are not surprising given our result showing that EtF implies a flavor of WE.
- Interactive—Multiple Committees and Continuous Executions of ECW: A simple way to achieve an interactive version of EtF is to first encrypt secret shares of a message towards members of a committee that then re-share their secrets towards members of a future anonymous committee via an invocation of ECW (in our instantiations or those in [3] and [16]). This is essentially the solution proposed in CaBKaS [3] where committees interact in order to carry a secret (on the blockchain) into the future. Notice that, for a fixed security parameter and corruption ratio, the communication complexity of the protocol executed by the committee in this solution depends on the plaintext message length. On the other hand, for a fixed security parameter and corruption ratio, the communication complexity of our committee-based transformation from ECW to EtF is *constant*.

Other Works. Using blockchains in order to construct non-interactive primitives with game-based security has been previously considered in [17]. Other approaches for transferring secret state to future committees have been proposed in [18], although anonymity is not a concern in this setting. On the other hand, using anonymity to overcome adaptive corruption has been proposed in [12], although this work considers anonymous channels among a fixed set of parties.

2 Preliminaries

Notation. For any positive integer n, $[n]$ denotes the set $\{1, \ldots, n\}$. We use λ to denote the security parameter. We write $a \xleftarrow{\$} S$ to denote that a is sampled according to distribution S, or uniformly randomly if S is a set. We write $A(x; r)$ to denote the output of algorithm A given an input x and a random tape r.

2.1 Proof-of-Stake (PoS) Blockchains

In this work we rely on PoS-based blockchain protocols. In such a protocol, each participant is associated with some stake in the system. A process called leader election encapsulates a lottery mechanism that ensures (of all eligible parties) each party succeeds in generating the next block with probability proportional to its stake in the system. In order to formally argue about executions of such protocols, we depart from the framework presented in [17] which, in turn, builds on the analysis done in [13] and [21]. We invite the reader to re-visit the abstraction used in [17]. We present a summary of the framework in the full version [7] and discuss below the main properties we will use in the remainder of this paper. Moreover, we note that in [17] it is proven that there exist PoS blockchain protocols with the properties described below, e.g. Ouroboros Praos [9].

Blockchain Structure. A genesis block $B_0 = \{(\mathsf{Sig.pk}_1, \mathsf{aux}_1, \mathsf{stake}_1), \ldots,$ $(\mathsf{Sig.pk}_n, \mathsf{aux}_n, \mathsf{stake}_n), \mathsf{aux}\}$ associates each party P_i to a signature scheme public key $\mathsf{Sig.pk}_i$, an amount of stake stake_i and auxiliary information aux_i (i.e. any other relevant information required by the blockchain protocol, such as verifiable random function public keys). A blockchain $\mathbf{B}$ relative to a genesis block B_0 is a sequence of blocks $B_1, \ldots, B_n$ associated with a strictly increasing sequence of slots $\mathsf{sl}_1, \ldots, \mathsf{sl}_m$ such that $B_i = (\mathsf{sl}_j, H(B_{i-1}), \mathsf{d}, \mathsf{aux})$. Here, sl_j indicates the time slot that B_i occupies, $H(B_{i-1})$ is a collision resistant hash of the previous block, d is data and aux is auxiliary information required by the blockchain protocol (e.g. a proof that the block is valid for slot sl_j). We denote by $\mathbf{B}^{\lceil \ell}$ the chain (sequence of blocks) $\mathbf{B}$ where the last ℓ blocks have been removed and if $\ell \geq |\mathbf{B}|$ then $\mathbf{B}^{\lceil \ell} = \epsilon$. Also, if $\mathbf{B}_1$ is a prefix of $\mathbf{B}_2$ we write $\mathbf{B}_1 \preceq \mathbf{B}_2$. Each party participating in the protocol has public identity P_i and most messages will be a transaction of the following form: $m = (P_i, P_j, \mathsf{q}, \mathsf{aux})$ where P_i transfers q coins to P_j along with some optional, auxiliary information aux.

Blockchain Setup and Key Knowledge. As in [9], we assume that the genesis block is generated by an initialization functionality $\mathcal{F}_{\mathsf{INIT}}$ that registers all parties' keys. Moreover, we assume that primitives specified in separate functionalities in [9] as incorporated into $\mathcal{F}_{\mathsf{INIT}}$. $\mathcal{F}_{\mathsf{INIT}}$ is executed by the environment $\mathcal{Z}$ as defined below and is parameterized by a stake distribution associating each party P_i to an initial stake stake_i. Upon being activated by P_i for the first time, $\mathcal{F}_{\mathsf{INIT}}$ generates a signature key pair $\mathsf{Sig.sk}_i, \mathsf{Sig.pk}_i$, auxiliary information aux_i and a lottery witness $\mathsf{sk}_{L,i}$, which will be defined as part of the lottery predicate in Sect. 2.1, sending $(\mathsf{Sig.sk}_i, \mathsf{Sig.pk}_i, \mathsf{aux}_i, \mathsf{sk}_{L,i}, \mathsf{stake}_i)$ to P_i as response. After all parties have activated $\mathcal{F}_{\mathsf{INIT}}$, it responds to requests for a genesis block by providing $B_0 = \{(\mathsf{Sig.pk}_1, \mathsf{aux}_1, \mathsf{stake}_1), \ldots, (\mathsf{Sig.pk}_n, \mathsf{aux}_n, \mathsf{stake}_n), \mathsf{aux}\}$, where aux is generated according to the underlying blockchain consensus protocol.

Since $\mathcal{F}_{\mathsf{INIT}}$ generates keys for all parties, we capture the fact that even corrupted parties have registered public keys and auxiliary information such that they know the corresponding secret keys. Moreover, when our EtF constructions are used as part of more complex protocols, a simulator executing the EtF

and its underlying blockchain with the adversary will be able to predict which ciphertexts can be decrypted by the adversary by simulating $\mathcal{F}_{\mathsf{INIT}}$ and learning these keys. This fact will be important when arguing the security of protocols that use our notion of EtF.

Evolving Blockchains. In order to define an EtF scheme, some concept of future needs to be established. In particular we want to make sure that the initial chain $\mathbf{B}$ has "correctly" evolved into the final chain $\tilde{\mathbf{B}}$. Otherwise, the adversary can easily simulate a blockchain where it wins a future lottery and finds itself with the ability to decrypt. Fortunately, the *Distinguishable Forking* property provides just that (see full version [7] and [17] for more details). A sufficiently long chain in an honest execution can be distinguished from a fork generated by the adversary by looking at the combined amount of stake proven in such a sequence of blocks. We encapsulate this property in a predicate called evolved$(\cdot, \cdot)$. First, let $\Gamma^V = (\mathsf{UpdateState}^V, \mathsf{GetRecords}, \mathsf{Broadcast})$ be a blockchain protocol with validity predicate V and where the $(\alpha, \beta, \ell_1, \ell_2)$-*distinguishable forking* property holds. And let $\mathbf{B} \leftarrow \mathsf{GetRecords}(1^\lambda, \mathsf{st})$ and $\tilde{\mathbf{B}} \leftarrow \mathsf{GetRecords}(1^\lambda, \tilde{\mathsf{st}})$.

Definition 1 (Evolved Predicate). *An evolved predicate is a polynomial time function* evolved *that takes as input blockchains* $\mathbf{B}$ *and* $\tilde{\mathbf{B}}$

$$\mathsf{evolved}(\mathbf{B}, \tilde{\mathbf{B}}) \in \{0, 1\}$$

It outputs 1 iff $\mathbf{B} = \tilde{\mathbf{B}}$ *or the following holds (i)* $V(\mathbf{B}) = V(\tilde{\mathbf{B}}) = 1$; *(ii)* $\mathbf{B}$ *and* $\tilde{\mathbf{B}}$ *are consistent i.e.* $\mathbf{B}^{\lceil \kappa} \preceq \tilde{\mathbf{B}}$ *where* κ *is the common prefix parameter; (iii) Let* $\ell' = |\tilde{\mathbf{B}}| - |\mathbf{B}|$ *then it holds that* $\ell' \geq \ell_1 + \ell_2$ *and* $\mathsf{u} - \mathsf{stakefrac}(\tilde{\mathbf{B}}, \ell' - \ell_1) > \beta$.

Blockchain Lotteries. Earlier we mentioned the concept of leader election in PoS-based blockchain protocols. In this kind of lottery any party can win the right to become a slot leader with a probability proportional to its relative stake in the system. Usually, the lottery winner wins the right to propose a new block for the chain, introduce new randomness to the system or become a part of a committee that carries out some computation. In our encryption scheme we take advantage of this inherent lottery mechanism.

Independent Lotteries. In some applications it is useful to conduct multiple independent lotteries for the same slot sl. Therefore we associate each slot with a set of roles $\mathsf{R}_1, \ldots, \mathsf{R}_n$. Depending on the lottery mechanism, each pair $(\mathsf{sl}, \mathsf{R}_i)$ may yield zero, one or multiple winners. Often, a party can locally compute if it, in fact, is the lottery winner for a given role and the evaluation procedure may equip the party with a proof for others to verify. The below definition details what it means for a party to win a lottery.

Definition 2 (Lottery Predicate). *A lottery predicate is a polynomial time function* lottery *that takes as input a blockchain* $\mathbf{B}$, *a slot* sl, *a role* R *and a lottery witness* $\mathsf{sk}_{L,i}$ *and outputs 1 if and only if the party owning* $\mathsf{sk}_{L,i}$ *won the*

lottery for the role R *in slot* sl *with respect to the blockchain* **B**.
Formally, we write

$$\mathsf{lottery}(\mathbf{B}, \mathsf{sl}, \mathsf{R}, \mathsf{sk}_{L,i}) \in \{0,1\}$$

It is natural to establish the set of lottery winning keys $\mathcal{W}_{\mathbf{B},\mathsf{sl},\mathsf{R}}$ for parameters $(\mathbf{B}, \mathsf{sl}, \mathsf{R})$. This is the set of eligible keys satisfying the lottery predicate.

2.2   Commitment Schemes

We recall the syntax for a commitment scheme $\mathsf{C} = (\mathsf{Setup}, \mathsf{Commit})$ below:

- $\mathsf{Setup}(1^\lambda) \to \mathsf{ck}$ outputs a commitment key. The commitment key ck defines a message space $\mathcal{S}_m$ and a randomizer space $\mathcal{S}_r$.
- $\mathsf{Commit}(\mathsf{ck}, \mathsf{s}; \rho) \to \mathsf{cm}$ outputs a commitment given as input a message $\mathsf{s} \in \mathcal{S}_m$ and randomness $\rho \in \mathcal{S}_r$.

We require a commitment scheme to satisfy the standard properties of *binding* and *hiding*. It is binding if no efficient adversary can come up with two pairs $(\mathsf{s}, \rho), (\mathsf{s}', \rho')$ such that $\mathsf{s} \neq \mathsf{s}'$ and $\mathsf{Commit}(\mathsf{ck}, \mathsf{s}; \rho) = \mathsf{Commit}(\mathsf{ck}, \mathsf{s}'; \rho')$ for $\mathsf{ck} \leftarrow \mathsf{Setup}(1^\lambda)$. The scheme is hiding if for any two $\mathsf{s}, \mathsf{s}' \in \mathcal{S}_m$, no efficient adversary can distinguish between a commitment of s and one of s'.

Extractability. In our construction of ECW from cWE (Sect. 5.1), we require our commitments to satisfy an additional property which allows to *extract* message and randomness of a commitment. In particular we assume that our setup outputs both a commitment key and a trapdoor td and that there exists an algorithm Ext such that $\mathsf{Ext}(\mathsf{td}, \mathsf{cm})$ outputs (s, ρ) such that $\mathsf{cm} = \mathsf{Commit}(\mathsf{ck}, \mathsf{s}; \rho)$. We remark we can generically obtain this property by attaching to the commitment a NIZK argument of knowledge that shows knowledge of opening, i.e., for the relation $\mathcal{R}^{\mathrm{opn}}(\mathsf{cm}_i; (\mathsf{s}, \rho)) \iff \mathsf{cm}_i = \mathsf{Commit}(\mathsf{ck}, \mathsf{s}; \rho)$.

2.3   (Threshold) Identity Based Encryption

In an IBE scheme, users can encrypt simply with respect to an *identity* (rather than a public key). Given a master secret key, an IBE can generate secret keys that allows to open to specific identities. In our construction of EtF (Sect. 7.1) we rely on a *threshold variant of IBE* (TIBE) where no single party in the system holds the master secret key. Instead, parties in a committee hold a partial master secret key msk_i. Like other threshold protocols, threshold IBE can be generically obtained by "lifting" an IBE through a secret sharing with homomorphic properties (see for example [20]).

Threshold IBE. A TIBE system consists of the following algorithms.

$\Pi_{\mathsf{TIBE}}.\mathsf{Setup}(1^\lambda, n, k) \to (\mathsf{sp}, \mathsf{vk}, \vec{\mathsf{msk}})$: It outputs some public system parameters sp (including mpk), verification key vk, and vector of master secret key shares $\vec{\mathsf{msk}} = (\mathsf{msk}_1, \ldots, \mathsf{msk}_n)$ for n with threshold k. We assume that all algorithms takes sp as input implicitly.

$\Pi_{\mathsf{TIBE}}.\mathsf{ShareKG}(i, \mathsf{msk}_i, \mathsf{ID}) \rightarrow \theta = (i, \hat{\theta})$: It outputs a private key share $\theta = (i, \hat{\theta})$ for ID given a share of the master secret key.

$\Pi_{\mathsf{TIBE}}.\mathsf{ShareVerify}(\mathsf{vk}, \mathsf{ID}, \theta) \rightarrow 0/1$: It takes as input the verification key vk, an identity ID, and a share of master secret key θ, and outputs 0 or 1.

$\Pi_{\mathsf{TIBE}}.\mathsf{Combine}(\mathsf{vk}, \mathsf{ID}, \vec{\theta}) \rightarrow \mathsf{sk}_{\mathsf{ID}}$: It combines the shares $\vec{\theta} = (\theta_1, \ldots, \theta_k)$ to produce a private key $\mathsf{sk}_{\mathsf{ID}}$ or $\bot$.

$\Pi_{\mathsf{TIBE}}.\mathsf{Enc}(\mathsf{ID}, m) \rightarrow \mathsf{ct}$: It encrypts message m for identity ID and outputs a ciphertext ct.

$\Pi_{\mathsf{TIBE}}.\mathsf{Dec}(\mathsf{ID}, \mathsf{sk}_{\mathsf{ID}}, \mathsf{ct}) \rightarrow m$: It decrypts the ciphertext ct given a private key $\mathsf{sk}_{\mathsf{ID}}$ for identity ID.

Correctness. A TIBE scheme Π_{TIBE} should satisfy two correctness properties:

1. For any identity ID, if $\theta = \Pi_{\mathsf{TIBE}}.\mathsf{ShareKG}(i, \mathsf{msk}_i, \mathsf{ID})$ for $\mathsf{msk}_i \in \vec{\mathsf{msk}}$, then $\Pi_{\mathsf{TIBE}}.\mathsf{ShareVerify}(\mathsf{vk}, \mathsf{ID}, \theta) = 1$.

2. For any ID, if $\vec{\theta} = \{\theta_1, \ldots, \theta_k\}$ where $\theta_i = \Pi_{\mathsf{TIBE}}.\mathsf{ShareKG}(i, \mathsf{msk}_i, \mathsf{ID})$, and $\mathsf{sk}_{\mathsf{ID}} = \Pi_{\mathsf{TIBE}}.\mathsf{Combine}(\mathsf{vk}, \mathsf{ID}, \vec{\theta})$, then for any $m \in \mathcal{M}$ and $\mathsf{ct} = \Pi_{\mathsf{TIBE}}.\mathsf{Enc}(\mathsf{ID}, m)$ we have $\Pi_{\mathsf{TIBE}}.\mathsf{Dec}(\mathsf{ID}, \mathsf{sk}_{\mathsf{ID}}, \mathsf{ct}) = m$.

Structural Property: TIBE as IBE + Secret Sharing. We model threshold IBE in a modular manner from IBE and assume it to have a certain structural property: that it can be described as an IBE "lifted" through a homomorphic secret-sharing [5,6,20]. TIBE constructions can often be described as such. We assume this structural property to present our proofs for EtF modularly, but we remark our results do not depend on it and they hold for an arbitrary TIBE. For lack of space we refer the reader to the full version for details.

Assume a secure IBE (the non-threshold variant of TIBE). We can transform it into a threshold IBE using homomorphic secret sharing algorithms (Share, EvalShare, Combine). A homomorphic secret sharing scheme is a secret sharing scheme with an extra property: given a shared secret, it allows to compute a share of a function of the secret on it. The correctness of the homomorphic scheme requires that running $y_i \leftarrow \mathsf{EvalShare}(\mathsf{msk}_i, f)$ on msk_i output of Share and then running Combine on (a large enough set of) the y_i-s produces the same output as $f(\mathsf{msk})$. We also require that Combine can reconstruct msk from a large enough set of the msk_i-s. For security we assume we can simulate the shares not available to the adversaries (if the adversary holds at most $T = k$ shares). For the resulting TIBE's security we assume that, for an adversary holding at most T shares, we can simulate: master secret key shares not held by the adversary (*msk shares simulation*) and shares of the id-specific keys (*key-generation simulation*) for the same shares. We finally assume we can verify that each of the id-specific key shares are authenticated (*robustness*) and that shares of the master secret key can be reshared (*proactive resharing*).

3   Modelling EtF

In this section, we present a model for encryption to the future winner of a lottery. In order to argue about a notion of future, we use the blocks of an underlying blockchain ledger and their relative positions in the chain to specify points in time. Intuitively, our notion allows for creating ciphertexts that can only be decrypted by a party that is selected to perform a certain role R at a future slot sl according to a lottery scheme associated with a blockchain protocol. The winner of the lottery at a point in the future with respect to a blockchain state $\tilde{\mathbf{B}}$ is determined by the lottery predicate defined in Sect. 2.1, *i.e.* the winner is the holder of a lottery secret key sk such that $\mathsf{lottery}(\tilde{\mathbf{B}}, \mathsf{sl}, \mathsf{R}, \mathsf{sk}) = 1$. However, notice that the winner might only be determined by a blockchain state produced in the future as a result of the blockchain protocol execution. This makes it necessary for the ciphertext to encode an initial state $\mathbf{B}$ of the blockchain that allows for verifying that a future state $\tilde{\mathbf{B}}$ (presented at the time of decryption) has indeed been produced as a result of correct protocol execution. This requirement is captured by the evolving blockchain predicate defined in Sect. 2.1, *i.e.* $\mathsf{evolved}(\mathbf{B}, \tilde{\mathbf{B}}) = 1$ iff $\tilde{\mathbf{B}}$ is obtained as a future state of executing the blockchain protocol departing from $\mathbf{B}$.

Definition 3 (Encryption to the Future). *A pair of PPT algorithms $\mathcal{E} = (\mathsf{Enc}, \mathsf{Dec})$ in the context of a blockchain Γ^V is an EtF-scheme with evolved predicate* evolved *and a lottery predicate* lottery. *The algorithms work as follows.*

Encryption. $\mathsf{ct} \leftarrow \mathsf{Enc}(\mathbf{B}, \mathsf{sl}, \mathsf{R}, m)$ *takes as input an initial blockchain $\mathbf{B}$, a slot* sl, *a role R and a message m. It outputs a ciphertext* ct *- an encryption to the future.*

Decryption. $m/\bot \leftarrow \mathsf{Dec}(\tilde{\mathbf{B}}, \mathsf{ct}, \mathsf{sk})$ *takes as input a blockchain state $\tilde{\mathbf{B}}$, a ciphertext* ct *and a secret key sk and outputs the original message m or $\bot$.*

An EtF must satisfy the following properties:

Correctness. An EtF-scheme is said to be correct if for honest parties i and j, there exists a negligible function μ such that for all $\mathsf{sk}, \mathsf{sl}, \mathsf{R}, m$:

$$\left| \Pr \left[\begin{array}{l} \mathsf{view} \leftarrow \mathsf{EXEC}^\Gamma(\mathcal{A}, \mathcal{Z}, 1^\lambda) \\ \mathbf{B} = \mathsf{GetRecords}(\mathsf{view}_i) \\ \tilde{\mathbf{B}} = \mathsf{GetRecords}(\mathsf{view}_j) \\ \mathsf{ct} \leftarrow \mathsf{Enc}(\mathbf{B}, \mathsf{sl}, \mathsf{R}, m) \\ \mathsf{evolved}(\mathbf{B}, \tilde{\mathbf{B}}) = 1 \end{array} : \begin{array}{l} \mathsf{lottery}(\tilde{\mathbf{B}}, \mathsf{sl}, \mathsf{R}, \mathsf{sk}) = 0 \\ \vee \, \mathsf{Dec}(\tilde{\mathbf{B}}, \mathsf{ct}, \mathsf{sk}) = m \end{array} \right] - 1 \right| \leq \mu(\lambda)$$

Security. We establish a game between a challenger $\mathcal{C}$ and an adversary $\mathcal{A}$. In Sect. 2.1 we describe how $\mathcal{A}$ and $\mathcal{Z}$ execute a blockchain protocol. In addition, we now let the adversary interact with the challenger in a game $\mathsf{Game}_{\Gamma, \mathcal{A}, \mathcal{Z}, \mathcal{E}}^{\mathsf{IND\text{-}CPA}}$ described in Algorithm 1. The game can be summarized as follows:

1. $\mathcal{A}$ executes the blockchain protocol Γ together with $\mathcal{Z}$ and at some round r chooses a blockchain $\mathbf{B}$, a role R for the slot sl and two messages m_0 and m_1 and sends it all to $\mathcal{C}$.
2. $\mathcal{C}$ chooses a random bit b and encrypts the message m_b with the parameters it received and sends ct to $\mathcal{A}$.
3. $\mathcal{A}$ continues to execute the blockchain until some round $\tilde{r}$ where the blockchain $\tilde{\mathbf{B}}$ is obtained and $\mathcal{A}$ outputs a bit b'.

If the adversary is a lottery winner for the challenge role R in slot sl, the game outputs a random bit. If the adversary is not a lottery winner for the challenge role R in slot sl, the game outputs $b \oplus b'$. The reason for outputting a random guess in the game when the challenge role is corrupted is as follows. Normally the output of the IND-CPA game is $b \oplus b'$ and we require it to be 1 with probability $1/2$. This models that the guess b' is independent of b. This, of course, cannot be the case when the challenge role is corrupted. We therefore output a random guess in these cases. After this, any bias of the output away from $1/2$ still comes from b' being dependent on b.

Algorithm 1. $\text{Game}^{\text{IND-CPA}}_{\Gamma, \mathcal{A}, \mathcal{Z}, \mathcal{E}}$

$\text{view}^r \leftarrow \text{EXEC}^\Gamma_r(\mathcal{A}, \mathcal{Z}, 1^\lambda)$ ▷ $\mathcal{A}$ executes Γ with $\mathcal{Z}$ until round r

$(\mathbf{B}, \text{sl}, \text{R}, m_0, m_1) \leftarrow \mathcal{A}(\text{view}^r_\mathcal{A})$ ▷ $\mathcal{A}$ outputs challenge parameters

$b \xleftarrow{\$} \{0, 1\}$

$\text{ct} \leftarrow \text{Enc}(\mathbf{B}, \text{sl}, \text{R}, m_b)$

$\text{st} \leftarrow \mathcal{A}(\text{view}^r_\mathcal{A}, \text{ct})$ ▷ $\mathcal{A}$ receives challenge ct

$\text{view}^{\tilde{r}} \leftarrow \text{EXEC}^\Gamma_{(\text{view}^r, \tilde{r})}(\mathcal{A}, \mathcal{Z}, 1^\lambda)$ ▷ Execute from view^r until round $\tilde{r}$

$(\tilde{\mathbf{B}}, b') \leftarrow \mathcal{A}(\text{view}^{\tilde{r}}_\mathcal{A}, \text{st})$

if $\text{evolved}(\mathbf{B}, \tilde{\mathbf{B}}) = 1$ **then**   ▷ $\tilde{\mathbf{B}}$ is a valid evolution of $\mathbf{B}$

  if $sk^\mathcal{A}_{L,j} \notin \mathcal{W}_{\tilde{\mathbf{B}}, \text{sl}, \text{R}}$ **then**   ▷ $\mathcal{A}$ does not win role R

   return $b \oplus b'$

  end if

end if

return $\hat{b} \xleftarrow{\$} \{0, 1\}$

Definition 4 (IND-CPA Secure EtF). *An EtF-scheme $\mathcal{E} = (\text{Enc}, \text{Dec})$ in the context of a blockchain protocol Γ executed by PPT machines $\mathcal{A}$ and $\mathcal{Z}$ is said to be IND-CPA secure if, for any $\mathcal{A}$ and $\mathcal{Z}$, there exists a negligible function μ such that for $\lambda \in \mathbb{N}$:*

$$\left| 2 \cdot \Pr\left[\text{Game}^{\text{IND-CPA}}_{\Gamma, \mathcal{A}, \mathcal{Z}, \mathcal{E}} = 1\right] - 1 \right| \leq \mu(\lambda)$$

Remark 1 (On the requirement of Proof-of-Stake for EtF). The EtF notion requires the guarantee that an honest chain should be verifiable without interaction with the network (i.e. verified by the EtF ciphertext). While this is possible for Proof-of-Stake (PoS) blockchains, in a Proof-of-Work (PoW) blockchain the adversary can always simulate a chain where it generates all blocks. In general we require a blockchain in order to model time (via block height) for EtF.

3.1   ECW as a Special Case of EtF

In this section we focus on a special class of EtF. We call schemes in this class ECW schemes. ECW is particularly interesting since the underlying lottery is always conducted with respect to the current blockchain state. This has the following consequences

1. $\mathbf{B} = \tilde{\mathbf{B}}$ means that evolved$(\mathbf{B}, \tilde{\mathbf{B}}) = 1$ is trivially true.
2. The winner of role R in slot sl is already defined in $\mathbf{B}$.

It is easy to see that this kind of EtF scheme is simpler to realize since there is no need for checking if the blockchain has "correctly" evolved. Furthermore, all lottery parameters like stake distribution and randomness extracted from the blockchain are static. Thus, an adversary has no way to move stake between accounts in order to increase its chance of winning the lottery.

   Note that, when using an ECW scheme, the lottery winner is already decided at encryption time. In other words, there is no delay and the moment a ciphertext is produced the receiver is chosen.

4   Witness Encryption over Commitments (cWE)

Here, we describe witness encryption over commitments that is a relaxed notion of witness encryption. In witness encryption parties encrypt to a public input for some NP statement. In cWE we have two phases: first parties provide a (honestly generated) commitment cm of their private input s. Later, anybody can encrypt to a public input for an NP statement which *also* guarantees correct opening of the commitment. Importantly, in applications, the first message in our model can be reused for many different invocations.

Remark 2 (Comparing cWE and WE). We observe that cWE is weaker than standard WE because of its deterministic flavor. In standard WE we encrypt without having any "pointer" to an alleged witness, but in cWE it requires the witness to be implicitly *known* at encryption time through the commitment (to which it is bound). That is why—as for the weak flavors of witness encryption in [4]—we believe it would be misleading to just talk about WE. This is true in particular since we show cWE can be constructed from standard assumptions such as oblivious transfer and garbled circuits (see full version [7]), whereas constructions of WE from standard assumptions are still an open problem or require strong primitives like indistinguishability obfuscation. Finally we stress a difference with the trivial "interactive" WE proposed in [14] (Sect. 1.3): cWE is still non-interactive after producing a once-and-for-all reusable commitment.

4.1   Definition

The type of relations we consider are of the following form: a statement $\mathsf{x} = (\mathsf{cm}, C, y)$ and a witness $\mathsf{w} = (\mathsf{s}, \rho)$ are in the relation (i.e., $(\mathsf{x}, \mathsf{w}) \in \mathcal{R}$) iff "cm commits to some secret value s using randomness ρ, and $C(\mathsf{s}) = y$". Here, C is a circuit in some circuit class $\mathcal{C}$ and y is the expected output of the function. Formally, we define witness encryption over commitments as follows:

Definition 5 (Witness encryption over commitments). *Let* C = (Setup, Commit) *be a non-interactive commitment scheme. A cWE-scheme for witness encryption over commitments with circuit class $\mathcal{C}$ and commitment scheme* C *consists of a pair of algorithms* Π_{cWE} = (Enc, Dec):

Encryption phase. ct ← Enc(ck, x, m) *on input a commitment key* ck, *a statement* x = (cm, C, y) *such that* $C \in \mathcal{C}$, *and a message* $m \in \{0, 1\}^*$, *generates a ciphertext* ct.

Decryption phase. $m/\bot$ ← Dec(ck, ct, w) *on input a commitment key* ck, *a ciphertext* ct, *and a witness* w, *returns a message* m *or* $\bot$.

A cWE should satisfy *correctness* and *semantic security* as defined below.

(Perfect) Correctness. An honest prover with a statement x = (cm, C, y) and witness w = (s, ρ) such that cm = Commit(ck, s; ρ) and $C(\mathsf{s}) = y$ can always decrypt with overwhelming probability. More precisely, a cWE with circuit class $\mathcal{C}$ and commitment scheme C has perfect correctness if for all $\lambda \in \mathbb{N}$, $C \in \mathcal{C}$, ck $\in$ Range(C.Setup), s $\in \mathcal{S}_m$, randomness $\rho \in \mathcal{S}_r$, commitment cm ← C.Commit(ck, s; ρ), and bit message $m \in \{0, 1\}^*$, it holds that

$$\Pr\left[\mathsf{ct} \leftarrow \mathsf{Enc}(\mathsf{ck}, (\mathsf{cm}, C, C(\mathsf{s})), m); m' \leftarrow \mathsf{Dec}(\mathsf{ck}, \mathsf{ct}, (\mathsf{s}, \rho)) \ : \ m = m'\right] = 1$$

(Weak) Semantic Security. Intuitively, encrypting with respect to a false statement (with honest commitment) produces indistinguishable ciphertexts. Formally, there exists a negligible function μ such that for all $\lambda \in \mathbb{N}$, all auxiliary strings aux and all PPT adversaries $\mathcal{A}$:

$$\left| 2 \cdot \Pr\left[\begin{array}{c} \mathsf{ck} \leftarrow \mathsf{C.Setup}(1^\lambda) \\ (\mathsf{st}, \mathsf{s}, \rho, C, y, m_0, m_1) \leftarrow \mathcal{A}(\mathsf{ck}, \mathsf{aux}) \\ \mathsf{cm} \leftarrow \mathsf{C.Commit}(\mathsf{ck}, \mathsf{s}; \rho); b \xleftarrow{\$} \{0, 1\} \ : \ \mathcal{A}(\mathsf{st}, \mathsf{ct}) = b \\ \mathsf{ct} \leftarrow \mathsf{Enc}(\mathsf{ck}, (\mathsf{cm}, C, y), m_b) \\ \mathsf{ct} := \bot \text{ if } C(\mathsf{s}) = y, \ C \notin \mathcal{C} \text{ or } |m_0| \neq |m_1| \end{array} \right] - 1 \right| \leq \mu(\lambda)$$

Later, to show the construction of ECW from cWE, we need a stronger notion of semantic security where the adversary additionally gets to see ciphertexts of the challenge message under true statements with unknown to $\mathcal{A}$ witnesses. We formalize this property in the full version [7] and show that weak semantic security together with hiding of the commitment imply strong semantic security.

4.2 Constructions of cWE

From Multi-Sender 2P-NISC [1]. A cWE scheme can be constructed from protocols for Multi-Sender (reusable) Non-Interactive Secure Computation (MS-NISC) [1]. In such protocols, there is a receiver R with input x who first broadcasts an encoding of its input, and then later every sender S_i with input y_i can send a single message to R that conveys only $f(x, y_i)$. This is done while preserving privacy of inputs and correctness of output.

In the full version [7] we provide a detailed explanation of how to construct cWE using MS-NISC as in [1]. We here state the main points of the construction.

Let f be the function that on input $y = (\mathbf{x}, k)$ and $x = \mathbf{w}$ outputs k if and only if $(\mathbf{x}, \mathbf{w}) \in \mathcal{R}$. This will be the underlying function for the MS-NISC protocol. We then obtain a cWE scheme over the relation $\mathcal{R}$ in the following way:

1. First, the receiver commits to its witness $\mathbf{w}$ by providing an encoding of it as its first message in the MS-NISC protocol.
2. Secondly, to encrypt m under statement $\mathbf{x}$, a sender samples a key k of size $|m|$ and provides an encoding of $(\mathbf{x}, k)$ as the second message in the MS-NISC protocol and sends the ciphertext $\mathsf{ct} = m \oplus k$ to the receiver.
3. Finally, the receiver obtains the key as the output of $f(x = \mathbf{w}, y = (\mathbf{x}, k)) = k$ iff $\mathbf{w}$ is a valid witness for the statement $\mathbf{x}$ encoded in the second message. And it decrypts the ciphertext $m = \mathsf{ct} \oplus k$.

We observe that the above construction actually yields a stronger notion of cWE where the statement $\mathbf{x}$ is private which is not a requirement in our setting. This asymmetry between sender and receiver privacy was also observed by others [19] and it opens the door for efficient constructions using oblivious transfer (OT) and privacy-free garbled circuits as described in [23]. More details on the more efficient construction of cWE using OT and garbled circuits are provided inthe full version [7].

5   Construction of ECW

Here we show a novel construction of ECW from cWE. We then show alternative constructions through instantiations from previous work.

5.1   ECW from cWE

In this section we realize the notion of ECW from cWE. We define our scheme with respect to a set of parties $\mathcal{P} = \{P_1, \ldots, P_n\}$ executing a blockchain protocol Γ as described in Sect. 2.1, $i.e.$ each party P_i has access to the blockchain ledger and is associated to a tuple $(\mathsf{Sig.pk}_i, \mathsf{aux}_i, \mathsf{st}_i)$ registered in the genesis block for which it has corresponding secret keys $(\mathsf{Sig.sk}_i, \mathsf{sk}_{L,i})$. Our construction uses as a main building block a witness encryption scheme over commitments $\Pi_{\mathsf{cWE}} = (\mathsf{Enc}_{\mathsf{cWE}}, \mathsf{Dec}_{\mathsf{cWE}})$; we assume the commitments to be extractable. The class of circuits $\mathcal{C}$ of Π_{cWE} includes the lottery predicate $\mathsf{lottery}(\mathbf{B}, \mathsf{sl}, \mathsf{R}, \mathsf{sk}_{L,i})$. We let each party publish an initial commitment of its witness. This way we can do without any interaction for encryption/decryption through a one-time setup where parties publish the commitments over which all following encryptions are done. We construct our ECW scheme Π_{ECW} as follows:

System Parameters: We assume that a commitment key $\mathsf{Setup}(1^\lambda) \to \mathsf{ck}$ is contained in the genesis block B_0 of the underlying blockchain.
Setup Phase: All parties $P_i \in \mathcal{P}$ proceed as follows:

1. Compute a commitment $\mathsf{cm}_i \leftarrow \mathsf{Commit}(\mathsf{ck}, \mathsf{sk}_{L,i}; \rho_i)$ to $\mathsf{sk}_{L,i}$ using randomness ρ_i. We abuse the notation and define P_i's secret key as $\mathsf{sk}_{L,i} \| \rho_i$.
2. Compute a signature $\sigma_i \leftarrow \mathsf{Sig}_{\mathsf{Sig}.\mathsf{sk}_i}(\mathsf{cm}_i)$.
3. Publish $(\mathsf{cm}_i, \sigma_i)$ on the blockchain by executing $\mathsf{Broadcast}(1^\lambda, (\mathsf{cm}_i, \sigma_i))$.

Encryption $\mathsf{Enc}(\mathbf{B}, \mathsf{sl}, \mathsf{R}, m)$: Construct a circuit C that encodes the predicate $\mathsf{lottery}(\mathbf{B}, \mathsf{sl}, \mathsf{R}, \mathsf{sk}_{L,i})$, where $\mathbf{B}$, sl and R are hardcoded and $\mathsf{sk}_{L,i}$ is the witness. Let $\mathcal{P}_{Setup}$ be the set of parties with non-zero relative stake and a valid setup message $(\mathsf{cm}_i, \sigma_i)$ published in the common prefix $\mathbf{B}^{\lceil\kappa}$ (if P_i has published more than one valid $(\mathsf{cm}_i, \sigma_i)$, only the latest one is considered). For every $P_i \in \mathcal{P}_{Setup}$, compute $\mathsf{ct}_i \leftarrow \mathsf{Enc}_{\mathsf{cWE}}(\mathsf{ck}, \mathsf{x}_i = (\mathsf{cm}_i, C, 1), m)$. Output $\mathsf{ct} = \big(\mathbf{B}, \mathsf{sl}, \mathsf{R}, \{\mathsf{ct}_i\}_{P_i \in \mathcal{P}_{Setup}}\big)$.

Decryption $\mathsf{Dec}(\mathbf{B}, \mathsf{ct}, \mathsf{sk})$: Given $\mathsf{sk} := \mathsf{sk}_{L,i} \| \rho_i$ such that $\mathsf{cm}_i = \mathsf{Commit}(\mathsf{ck}, \mathsf{sk}_{L,i}; \rho_i)$ and $\mathsf{lottery}(\mathbf{B}, \mathsf{sl}, \mathsf{R}, \mathsf{sk}_{L,i}) = 1$ for parameters $\mathbf{B}, \mathsf{sl}, \mathsf{R}$ from ct, output $m \leftarrow \mathsf{Dec}_{\mathsf{cWE}}(\mathsf{ck}, \mathsf{ct}_i, (\mathsf{sk}_{L,i}, \rho_i))$. Otherwise, output $\bot$.

Theorem 1. *Let* $\mathsf{C} = (\mathsf{Setup}, \mathsf{Commit})$ *be a non-interactive extractable commitment scheme and* $\Pi_{\mathsf{cWE}} = (\mathsf{Enc}_{\mathsf{cWE}}, \mathsf{Dec}_{\mathsf{cWE}})$ *be a strong semantically secure cWE over* C *for a circuit class* $\mathcal{C}$ *encoding the lottery predicate* $\mathsf{lottery}(\mathbf{B}, \mathsf{sl}, \mathsf{R}, \mathsf{sk}_{L,i})$ *as defined in Sect. 4. Let* Γ *be a blockchain protocol as defined in Sect. 2.1.* Π_{ECW} *is an* IND-CPA*-secure ECW scheme as per Definition 4.*

The proof is provided in the full version [7].

5.2   Other Instantiations

ECW from Target Anonymous Channels [3,16]. As mentioned before, another approach to construct ECW can be based on a recent line of work that aims to design secure-MPC protocols where parties should remain anonymous until they speak [3,15,16]. The baseline of these results is to establish a communication channel to random parties, while preserving their anonymity. It is quite clear that such anonymous channels can be used to realize our definition of ECW for the underlying lottery predicate that defines to whom the anonymous channel is established. Namely, to encrypt m to a role R at a slot sl with respect to a blockchain state $\mathbf{B}$, create a target anonymous channel to $(\mathsf{R}, \mathsf{sl})$ over $\mathbf{B}$ by using the above approaches and send m via this channel. Depending on the lottery predicate that specifies which random party the channel is created for, a recipient with the secret key who wins this lottery can retrieve m. To include some concrete examples, the work of Benhamouda et al. [3] proposed the idea of using a "nomination" process, where a nominating committee chooses a number of random parties $\mathcal{P}$, look up their public keys, and publish a re-randomization of their key. This allows everyone to send messages to $\mathcal{P}$ while keeping their anonymity. The work of [3] answered this question differently by delegating the nomination task to the previous committees without requiring a nominating committee. That is, the previous committee runs a secure-MPC protocol to choose a random subset of public keys, and broadcasts the rerandomization of the keys. To have a MPC protocol that scales well with the total number of parties, they define a new flavour of private information retrieval (PIR) called random-index PIR (or RPIR) and show how each committee—playing the role of the RPIR client—can select the next committee with the complexity only proportional to the size of the committee. There are two constructions of RPIR proposed in [16], one based on Mix-Nets and the other based on FHE. Since the purpose of the constructions described is to establish a target-anonymous channel to a random party, one can consider them as examples of a stronger notion of ECW with anonymity and a specific lottery predicate that selects *a single* random party from the entire population as the winner.

ECW from [10]. Derler and Slamanig [10] (DS) constructed a variant of WE for a restricted class of algebraic languages. In particular, a user can conduct a Groth-Sahai (GS) proof for the satisfiability of some pairing-product equations (PPEs). Such a proof contains commitments to the witness using randomness only known by this user. The proof can be used by anyone to encrypt a message resulting in a ciphertext which can only be decrypted by knowing this randomness. More formally, they consider a type of WE associated with a proof system $\Pi = (\mathsf{Setup}, \mathsf{Prove}, \mathsf{Verify})$ consisting of two rounds. In the first round, a recipient computes and broadcasts $\pi \leftarrow \mathsf{Prove}(\mathsf{crs}, \mathsf{x}, \mathsf{w})$. Later, a user can verify the proof and encrypt a message m under (x, π) if $\mathsf{Verify}(\mathsf{crs}, \mathsf{x}, \pi) = 1$. We note that the proof π does not betray the user conducting the proof and therefore it can use an anonymous broadcast channel to communicate the proof to the encrypting party in order to obtain anonymous ECW. Moreover, although GS proofs may look to support only a restricted class of statements based on PPEs, they are expressive enough to cover all the statements arising in pairing-based cryptography. This indicates the applicability of this construction for any VRF-based lottery where the VRF is algebraic and encodable as a set of PPEs. Further details are provided inthe full version [7]. This interactive ECW just described yields an improvement in communication complexity at the cost of having an extra round of interaction.

From Signatures of Knowledge. Besides the above instantiations, we point out a (potentially more inefficient) abstract construction from zero-knowledge signatures of knowledge (SoK) [8] (roughly, a non-malleable non-interactive zero-knowledge proof). This is similar in spirit to the previous instantiation and can be seen as a generalization. Assume each party has a (potentially ephemeral) public key. At the time the lottery winner has been decided, the winners can post a SoK showing knowledge of the secret key corresponding to their pk *and* that their key is a winner of the lottery. To encrypt, one would first verify the SoK and then encrypt with respect to the corresponding public key.

6   YOSO Multiparty Computation from ECW

In this section we show how ECW can be used as the crucial ingredient in setting up a YOSO MPC. So far we have only focused on IND-CPA secure ECW, which falls short of role assignment in the sense of [15]. In general role assignment requires the following properties which are not provided by ECW (or EtF):

1. Multiple parties must be able to send messages to the same role (in most applications this requires IND-CCA).
2. Parties must authenticate messages on behalf of a role they executed in the past (authentication from the past)
3. A party assigned to a given role must stay covert until the role is executed.

We will define a number of properties needed for EtF to realize applications such as role assignment. We start by looking at CCA security for an EtF scheme. We then introduce the notion of Authentication from the Past (AfP) and definition of unforgeability and privacy guarantees. Finally, we introduce the notion of YOSO-friendly blockchains that have inbuilt lotteries with properties that are needed to conduct YOSO MPC and corresponding EtF and AfP schemes.

6.1   IND-CCA EtF

In this section we define what it means for an EtF to be IND-CCA secure. This security property is useful in many applications where more encryptions are done towards the same slot and role. As in the definition of IND-CPA, we establish a game between a challenger C and an adversary A. We introduce a decryption oracle, $\mathcal{O}_{\mathsf{EtF}}$, which on input ct returns the decryption of ciphertext. Furthermore, the $\mathcal{O}_{\mathsf{EtF}}$ maintains a list of ciphertext queries $\mathcal{Q}_{\mathsf{EtF}}$. Algorithm 2 shows the details of the game.

Algorithm 2. $\mathsf{Game}_{\Gamma,\mathcal{A},\mathcal{Z},\mathsf{E}}^{\mathsf{IND\text{-}CCA2}}$

$\mathsf{view}^r \leftarrow \mathsf{EXEC}_r^\Gamma(\mathcal{A}^{\mathcal{O}_{\mathsf{EtF}}}, \mathcal{Z}, 1^\lambda)$ $\triangleright$ $\mathcal{A}$ executes Γ with $\mathcal{Z}$ until round r

$(\mathbf{B}, \mathsf{sl}, \mathsf{R}, m_0, m_1) \leftarrow \mathcal{A}^{\mathcal{O}_{\mathsf{EtF}}}(\mathsf{view}_\mathcal{A}^r)$ $\triangleright$ $\mathcal{A}$ outputs challenge parameters

$b \xleftarrow{\$} \{0,1\}$

$\mathsf{ct} \leftarrow \mathsf{Enc}(\mathbf{B}, \mathsf{sl}, \mathsf{R}, m_b)$

$\mathsf{st} \leftarrow \mathcal{A}^{\mathcal{O}_{\mathsf{EtF}}}(\mathsf{view}_\mathcal{A}^r, \mathsf{ct})$ $\triangleright$ $\mathcal{A}$ receives challenge ct

$\mathsf{view}^{\tilde{r}} \leftarrow \mathsf{EXEC}_{(\mathsf{view}^r, \tilde{r})}^\Gamma(\mathcal{A}^{\mathcal{O}_{\mathsf{EtF}}}, \mathcal{Z}, 1^\lambda)$ $\triangleright$ Execute from view^r until round $\tilde{r}$

$(\tilde{\mathbf{B}}, b') \leftarrow \mathcal{A}^{\mathcal{O}_{\mathsf{EtF}}}(\mathsf{view}_\mathcal{A}^{\tilde{r}}, \mathsf{st})$

if $\mathsf{evolved}(\mathbf{B}, \tilde{\mathbf{B}}) = 1$ **then** $\triangleright$ $\tilde{\mathbf{B}}$ is a valid evolution of $\mathbf{B}$

   if $\mathsf{sk}_{L,j}^\mathcal{A} \notin \mathcal{W}_{\tilde{\mathbf{B}}, \mathsf{R}, \mathsf{sl}} \wedge \mathsf{ct} \notin \mathcal{Q}_{\mathsf{EtF}}$ **then** $\triangleright$ $\mathcal{A}$ does not win role R

    return $b \oplus b'$

   end if

end if

return $g \xleftarrow{\$} \{0,1\}$

Definition 6 (IND-CCA2 Secure EtF). *Formally, an EtF-scheme $\mathcal{E}$ is said to be IND-CCA2 secure in the context of a blockchain protocol Γ executed by PPT machines $\mathcal{A}$ and $\mathcal{Z}$ if there exists a negligible function μ such that for $\lambda \in \mathbb{N}$:*

$$\left| 2 \cdot \Pr\left[\mathsf{Game}_{\Gamma,\mathcal{A},\mathcal{Z},\mathcal{E}}^{\mathsf{IND\text{-}CCA2}} = 1\right] - 1 \right| \leq \mu(\lambda)$$

To add IND-CCA2 security to an IND-CPA secure EtF scheme (as defined in Definition 4) we can use standard transformations such as [11, 22]. In the transformation based on [22] we could add to the setup of the blockchain a CRS for a simulation-sound extractable NIZK. When encrypting m to a role R the sender will send along a proof of knowledge of the plaintext m. We get the challenge ciphertext from the IND-CPA game and use the ZK property to simulate the NIZK proof. We can use the extraction trapdoor of the proof system to simulate the CCA decryption oracles by simulation soundness. When the IND-CCA2 adversary makes a guess, we make the same guess. The details of the construction and proof follow using standard techniques and are omitted. On the other hand, the popular transformation of [11] allows for simulating CCA decryption oracles by observing the adversary's queries to a random oracle, which should not be an issue since an EtF scheme is likely already running on top of a blockchain which is secure in the random oracle model. We leave the construction of concretely efficient IND-CCA2 EtF as future work.

6.2   Authentication from the Past (AfP)

When the winner of a role R_1 sends a message m to a future role R_2 then it is typically also needed that R_2 can be sure that the message m came from a party P which, indeed, won the role R_1. Most PoS blockchains deployed in practice have a lottery where a certificate can be released proving that P won the role R_1. In order to formalize this concept, we introduce an AfP scheme with a corresponding EUF-CMA game representing the authentication property.

Definition 7 (Authentication from the Past). *A pair of PPT algorithms* $\mathcal{U} = (\mathsf{Sign}, \mathsf{Verify})$ *is a scheme for authenticating messages as a winner of a lottery in the past in the context of blockchain* Γ *with lottery predicate* lottery.

Authenticate. $\sigma \leftarrow \mathsf{AfP.Sign}(\mathbf{B}, \mathsf{sl}, \mathsf{R}, \mathsf{sk}, m)$ *takes as input a blockchain* $\mathbf{B}$*, a slot* sl*, a role* R*, a secret key* sk*, and a message* m*. It outputs a signature* σ *that authenticates the message* m*.*

Verify. $\{0, 1\} \leftarrow \mathsf{AfP.Verify}(\tilde{\mathbf{B}}, \mathsf{sl}, \mathsf{R}, \sigma, m)$ *uses the blockchain* $\tilde{\mathbf{B}}$ *to ensure that* σ *is a signature on* m *produced by the secret key winning the lottery for slot* sl *and role* R*.*

Furthermore, an AfP-scheme has the following properties:

Correctness. *An AfP-scheme is said to be correct if for honest parties* i *and* j*, there exists a negligible function* μ *such that for all* sk, sl, R, m*:*

$$\left| Pr \left[\begin{array}{l} \mathsf{view} \leftarrow EXEC^\Gamma(\mathcal{A}, \mathcal{Z}, 1^\lambda) \\ \mathbf{B} = \mathsf{GetRecords}(\mathsf{view}_i) \\ \tilde{\mathbf{B}} = \mathsf{GetRecords}(\mathsf{view}_j) \\ \sigma \leftarrow \mathsf{AfP.Sign}(\mathbf{B}, \mathsf{sl}, \mathsf{R}, \mathsf{sk}, m) \end{array} : \begin{array}{l} \mathsf{lottery}(\mathbf{B}, \mathsf{sl}, \mathsf{R}, \mathsf{sk}) = 0 \\ \vee\ \mathsf{lottery}(\tilde{\mathbf{B}}, \mathsf{sl}, \mathsf{R}, \mathsf{sk}) = 0 \\ \vee\ \mathsf{AfP.Verify}(\tilde{\mathbf{B}}, \mathsf{sl}, \mathsf{R}, \sigma, m) = 1 \end{array} \right] - 1 \right| \leq \mu(\lambda)$$

In other words, an AfP on a message from an honest party with a view of the blockchain $\mathbf{B}$ *can attest to the fact that the sender won the role* R *in slot* sl. *If another party, with blockchain* $\tilde{\mathbf{B}}$ *agrees, then the verification algorithm will output 1.*

Security. *We here describe the game detailed in Algorithm 3 representing the security of an AfP scheme. The algorithm represents a standard EUF-CMA game where the adversary has access to a signing oracle* $\mathcal{O}_{\mathsf{AfP}}$ *which it can query with a slot* sl, *a role* R *and a message* m_i *and obtain AfP signatures* $\sigma_i = \mathsf{AfP.Sign}(\mathbf{B}, \mathsf{sl}, \mathsf{R}, \mathsf{sk}_j, m_i)$ *where* $\mathsf{sk}_j \in \mathcal{W}_{\mathbf{B},\mathsf{sl},\mathsf{R}}$ *i.e.* $\mathsf{lottery}(\mathbf{B}, \mathsf{sl}, \mathsf{R}, \mathsf{sk}_j) = 1$. *The oracle maintains the list of queries* $\mathcal{Q}_{\mathsf{AfP}}$.

Formally, an AfP-scheme $\mathcal{U}$ *is said to be EUF-CMA secure in the context of a blockchain protocol* Γ *executed by PPT machines* $\mathcal{A}$ *and* $\mathcal{Z}$ *if there exists a negligible function* μ *such that for* $\lambda \in \mathbb{N}$:

$$\Pr\left[\mathsf{Game}^{\mathsf{EUF-CMA}}_{\Gamma,\mathcal{A},\mathcal{Z},\mathcal{U}} = 1\right] \le \mu(\lambda)$$

Algorithm 3. $\mathsf{Game}^{\mathsf{EUF\text{-}CMA}}_{\Gamma,\mathcal{A},\mathcal{Z},\mathcal{U}}$

$\mathsf{view} \leftarrow \mathsf{EXEC}^{\Gamma}(\mathcal{A}, \mathcal{Z}, 1^{\lambda})$ ▷ $\mathcal{A}$ executes Γ with $\mathcal{Z}$

$(\mathbf{B}, \mathsf{sl}, \mathsf{R}, m', \sigma') \leftarrow \mathcal{A}^{\mathcal{O}_{\mathsf{AfP}}}(\mathsf{view}_{\mathcal{A}})$

if $(m' \in \mathcal{Q}_{\mathsf{AfP}}) \vee (\mathsf{sk}^{\mathcal{A}}_{L,j} \in \mathcal{W}_{\mathbf{B},\mathsf{sl},\mathsf{R}})$ **then**   ▷ $\mathcal{A}^{\mathcal{O}_{\mathsf{AfP}}}$ won or queried illegal m'

    return 0

end if

$\mathsf{view}^{\tilde{r}} \leftarrow \mathsf{EXEC}^{\Gamma}_{(\mathsf{view}^r, \tilde{r})}(\mathcal{A}, \mathcal{Z}, 1^{\lambda})$ ▷ Execute from view^r until round $\tilde{r}$

$\tilde{\mathbf{B}} \leftarrow \mathsf{GetRecords}(\mathsf{view}^{\tilde{r}}_i)$

if $\mathsf{evolved}(\mathbf{B}, \tilde{\mathbf{B}}) = 1$ **then**

    if $\mathsf{AfP.Verify}(\mathbf{B}, \mathsf{sl}, \mathsf{R}, \sigma', m') = 1$ **then**   ▷ $\mathcal{A}$ successfully forged an AfP

        return 1

    end if

end if

return 0

General AfP. In general we can add authentication to a message as follows. Recall that P_i wins R if $\mathsf{lottery}(\mathbf{B}, \mathsf{sl}, \mathsf{R}, \mathsf{sk}_{L,i}) = 1$. Here, $\mathcal{R}(\mathsf{x} = (\mathbf{B}, \mathsf{sl}, \mathsf{R}), \mathbf{w}) = \mathsf{lottery}(\mathsf{x}, \mathbf{w})$ is an NP relation where all parties know x but only the winner knows a witness $\mathbf{w}$ such that $\mathcal{R}(\mathsf{x}, \mathbf{w}) = 1$. We can therefore use a signature of knowledge (SoK) [8] to sign m under the knowledge of $\mathsf{sk}_{L,i}$ such that $\mathsf{lottery}(\mathbf{B}, \mathsf{sl}, \mathsf{R}, \mathsf{sk}_{L,i}) = 1$. This will attest that the message m was sent by a winner of the lottery for R. In [7], we show more efficient construction of AfP by exploring the structure of PoS-based blockchains with VRF lotteries.

6.3 AfP Privacy

Just EUF-CMA security is not sufficient for an AfP mechanism to be YOSO friendly. It must also preserve the privacy guarantees of the lottery predicate, guaranteeing that the adversary does not gain any undue advantage in predicting when a party is selected to perform a role after it uses AfP to authenticate a message. To appreciate this fact, we consider the case where instead of creating a signature of knowledge of $\mathsf{sk}_{L,i}$ on message m we simply use a regular EUF-CMA secure signature scheme to sign the message concatenated with $\mathsf{sk}_{L,i}$, revealing the signature public key, the resulting signature and $\mathsf{sk}_{L,i}$ itself as a means of authentication. By definition, this will still constitute an existentially unforgeable AfP but will also reveal whether the party who owns $\mathsf{sk}_{L,i}$ is the winner when future lotteries are conducted. The specific privacy property we seek is that an adversary, observing AfP tags from honest parties, cannot use this information to enhance its chances in predicting the winners of lotteries for roles for which an AfP tag has not been published. On the other hand, the identity of a party who won the lottery for a given role is not kept private when it publishes an AfP tag on behalf of this role, which is not an issue in a YOSO-setting since corruption after-the-fact is futile. Specifically, we allow an AfP tag to be linked to the identity of the party who generated it. Note, that this kind of privacy is different from notions like k-anonymity since the success of the adversary in guessing lottery winners with high accuracy depends on the stake distribution. The stake distribution is public in most PoS-settings and, thus, a privacy definition must take into account this inherent leakage.

Definition 8 (AfP Privacy). *An AfP scheme $\mathcal{U}$ with corresponding lottery predicate* lottery *is private if a PPT adversary $\mathcal{A}$ is unable to distinguish between the scenarios defined in Algorithm 4 and Algorithm 5 with more than negligible probability in the security parameter.*

Scenario 0 ($b = 0$). *In this scenario (Algorithm 4), $\mathcal{A}$ is first running the blockchain Γ together with the environment $\mathcal{Z}$. At round r, $\mathcal{A}$ is allowed to interact with the oracle $\mathcal{O}_{\mathsf{AfP}}$ (see Definition 7). The adversary then continues the execution until round $\tilde{r}$ where it outputs a bit b'.*

Scenario 1 ($b = 1$). *This scenario (Algorithm 5) is identical to scenario 0 but instead of interacting with $\mathcal{O}_{\mathsf{AfP}}$, the adversary interacts with a simulator* Sim.

Algorithm 4. $b = 0$	**Algorithm 5.** $b = 1$
$\mathsf{view}^r \leftarrow \mathsf{EXEC}_r^{\Gamma}(\mathcal{A}, \mathcal{Z}, 1^\lambda)$	$\mathsf{view}^r \leftarrow \mathsf{EXEC}_r^{\Gamma}(\mathcal{A}, \mathcal{Z}, 1^\lambda)$
$\mathcal{A}^{\mathcal{O}_{\mathsf{AfP}}}(\mathsf{view}_{\mathcal{A}}^r)$	$\mathcal{A}^{\mathsf{Sim}}(\mathsf{view}_{\mathcal{A}}^r)$
$\mathsf{view}^{\tilde{r}} \leftarrow \mathsf{EXEC}_{(\mathsf{view}^r, \tilde{r})}^{\Gamma}(\mathcal{A}, \mathcal{Z}, 1^\lambda)$	$\mathsf{view}^{\tilde{r}} \leftarrow \mathsf{EXEC}_{(\mathsf{view}^r, \tilde{r})}^{\Gamma}(\mathcal{A}, \mathcal{Z}, 1^\lambda)$
return $b' \leftarrow \mathcal{A}^{\mathcal{O}_{\mathsf{AfP}}}(\mathsf{view}_{\mathcal{A}}^{\tilde{r}})$	**return** $b' \leftarrow \mathcal{A}^{\mathsf{Sim}}(\mathsf{view}_{\mathcal{A}}^{\tilde{r}})$

We let $\mathsf{Game}_{\Gamma,\mathcal{A},\mathcal{Z},\mathcal{U}}^{\mathsf{AfP\text{-}PRIV}}$ *denote the game where a coin-flip decides whether the adversary is executed in scenario 0 or scenario 1. We say that the adversary wins the game (i.e. $\mathsf{Game}_{\Gamma,\mathcal{A},\mathcal{Z},\mathcal{U}}^{\mathsf{AfP-PRIV}} = 1$) iff $b' = b$. Finally, an AfP scheme $\mathcal{U}$ is called private in the context of the blockchain Γ executed together with environment $\mathcal{Z}$ if the following holds for a negligible function μ.*

$$\left| 2 \cdot \Pr\left[\mathrm{Game}_{\Gamma,\mathcal{A},\mathcal{Z},\mathcal{U}}^{\mathsf{AfP\text{-}PRIV}} = 1 \right] - 1 \right| \leq \mu(\lambda)$$

6.4   Round and Committee Based YOSO Protocols

Having IND-CCA2 ECW and an EUF-CMA secure and Private AfP, we can establish a round-based YOSO model, where there is a number of rounds $r = 1, 2, \ldots$ and where for each round there are n roles $\mathsf{R}_{r,i}$. We call the role $\mathsf{R}_{r,i}$ "party i in round r". We fix a round length L and associate role $\mathsf{R}_{r,i}$ to slot $\mathsf{sl} = L \cdot r$. This L has to be long enough that in each round the parties executing the roles can decrypt ciphertexts sent to them, execute the steps of the role, compute encryptions to the roles in the next round and post these to the blockchain in time for these to be available to the committee of round $r+1$ before slot $(r+1) \cdot L$. Picking such an L depends crucially on the underlying blockchain and network, and we will here simply assume that it can be done for the blockchain at hand.

Using this setup, the roles $\mathsf{R}_{r,i}$ of round r can use ECW and AfP with the aforementioned properties to send secret authenticated messages to the roles $\mathsf{R}_{r+1,i}$ in round $r + 1$. They find their ciphertexts on the blockchain before slot $r \cdot L$, decrypt using ECW, compute their outgoing messages, encrypt using ECW, authenticate using AfP, and post the ciphertexts and AfP tags on the blockchain.

Honest Majority. In round based YOSO MPC it is critical that we can assume some fraction of honesty in each committee $\mathsf{R}_{r,1}, \ldots, \mathsf{R}_{r,n}$. We discuss here assumptions needed on the lottery for this to hold and how to guarantee it. Assume an adversary that can corrupt parties identified by sk and a lottery assigning parties to roles $\mathsf{R}_{r,i}$. We map the corruption status of parties to roles as follows:

1. If a role $\mathsf{R}_{r,i}$ is won by a corrupted party or by several parties, call the role MALICIOUS. Even if $\mathsf{R}_{r,i}$ is won by two honest parties, they will both execute the role and send outgoing messages, which might violate security.
2. If a role $\mathsf{R}_{r,i}$ is won by exactly one honest party, call it HONEST.
3. If a role $\mathsf{R}_{r,i}$ is not won by any party, call it CRASHED. These roles will not be executed and are therefore equivalent to a crashed party.

Note that because we assume corrupted parties know their lottery witness $\mathsf{sk}_{L,i}$ in our model, we can, in poly-time, extract those witnesses and compute the corruption status of roles. This will be crucial in our reductions. Imagine that a role could be won by an honest party but also by a corrupted party which stays completely silent but decrypts messages sent to the role. If we are not aware of the corrupted party winning the role, we might send a simulated ciphertext to the apparently honest role. The corrupted party also having won the role would be able to detect this. Since any role won by an honest party could also be corrupted by a silent malicious party, simulation would become impossible.

In order to realize YOSO MPC, we will need committees where a majority of the roles are honest according to the description above. We capture this requirement in the definition below and argue how it can be achieved in the full version [7].

Definition 9 (Honest Committee Friendly). *We call a blockchain* Γ *honest committee friendly if there exist* n *and* H *and* T *such that* $H > T$ *s.t. we can define a sequence of roles* $\mathsf{R}_{r,i}$ *for* $r = 1, \ldots, \mathsf{poly}(\lambda)$ *and* $i = 1, \ldots, n$ *for a slot* sl_r *and that for all* r *it holds that except with negligible probability there are at least* H *honest roles in* $\mathsf{R}_{r,1}, \ldots, \mathsf{R}_{r,n}$ *and at most* T *malicious roles. Furthermore, if an honest party executing* $\mathsf{R}_{r,1}, \ldots, \mathsf{R}_{r,n}$ *sends a message at* sl_r, *it is guaranteed to appear on the blockchain before slot* sl_{r+1}.

We are now ready to capture the above discussion using a definition.

Definition 10 (YOSO Friendly Blockchain). *Let* Γ *be a blockchain with a lottery predicate* $\mathsf{lottery}(\mathbf{B}, \mathsf{sl}_r, \mathsf{R}_{r,i}, \mathsf{sk}_{L,i})$ *and let* $\mathcal{E} = (\mathsf{Enc}, \mathsf{Dec})$ *and* $\mathcal{U} = (\mathsf{Sign}, \mathsf{Verify})$ *be an EtF and AfP for* $\mathsf{lottery}(\mathbf{B}, \mathsf{sl}_r, \mathsf{R}_{r,i}, \mathsf{sk}_{L,i})$, *respectively. We call* $(\Gamma, \mathcal{E}, \mathcal{U})$ *YOSO MPC friendly if the following holds:*

1. $\mathcal{E}$ *is an IND-CCA2 secure EtF (Definition 6).*
2. $\mathcal{U}$ *is a secure and private AfP (Definition 7 and Sect. 8).*
3. Γ *is honest committee friendly (Definition 9).*

We will later assume a YOSO friendly blockchain, and we argued above that the existence of a YOSO friendly blockchain is a plausible assumption without having given formal proofs of this. It is interesting future work to prove that a concrete blockchain is a YOSO friendly blockchain in a given communication model. We omit this as our focus is on constructing flavours of EtF.

7   Construction of EtF from ECW and Threshold-IBE

The key intuition about our construction is as follows: we use IBE to encrypt messages to an arbitrary future $(\mathsf{R}, \mathsf{sl}_{\mathsf{fut}})$ pair. When the winners of the role in slot $\mathsf{sl}_{\mathsf{fut}}$ are assigned, we let them obtain an ID-specific key for $(\mathsf{R}, \mathsf{sl}_{\mathsf{fut}})$ from the IBE key-generation algorithm using ECW as a channel. Notice that this key-generation happens in the *present* while the encryption could have happened at any earlier time. We generate the key for $(\mathsf{R}, \mathsf{sl}_{\mathsf{fut}})$ in a threshold manner by assuming that, throughout the blockchain execution, a set of committee members each holds a share of the master secret key msk_i.

7.1   Construction

We now describe our construction. We assume an encryption to the current winner $\Pi_{\mathsf{ECW}} = (\mathsf{Enc}_{\mathsf{ECW}}, \mathsf{Dec}_{\mathsf{ECW}})$ and a threshold IBE scheme Π_{TIBE}. In the setup stage we assume a dealer acting honestly by which we can assign master secret key shares of the TIBE.

Parameters: We assume that the genesis block B_0 of the underlying blockchain contains all the parameters for Π_{ECW}.

Setup Phase: Parties run the setup stage for the Π_{ECW}. The dealer produces $(\mathsf{mpk}, \vec{\mathsf{msk}} = (\mathsf{msk}_1, \ldots, \mathsf{msk}_n))$ from TIBE setup with threshold k. Then it chooses n random parties and gives a distinct msk_i to each. All learn mpk.

Blockchain Execution: The blockchain execution we assume is as in Sect. 3. We additionally require that party i holding a master secret key share msk_i broadcasts $\mathsf{ct}_{(\mathsf{sl},\mathsf{R})}^{\mathsf{sk},i} \leftarrow \mathsf{Enc}_{\mathsf{ECW}}(\mathbf{B}, \mathsf{sl}, \mathsf{R}, \mathsf{sk}_{(\mathsf{sl},\mathsf{R})}^i)$, whenever the winner of role R in slot sl is defined in the blockchain $\mathbf{B}$, where $\mathsf{sk}_{(\mathsf{sl},\mathsf{R})}^i \leftarrow \Pi_{\mathsf{TIBE}}.\mathsf{IDKeygen}(\mathsf{msk}_i, (\mathsf{sl}, \mathsf{R}))$.

Encryption $\mathsf{Enc}(\mathbf{B}, \mathsf{sl}, \mathsf{R}, m)$: Each party generates $\mathsf{ct}_i \leftarrow \Pi_{\mathsf{TIBE}}.\mathsf{Enc}(\mathsf{mpk}, \mathsf{ID} = (\mathsf{sl}, \mathsf{R}), m)$. Output $\mathsf{ct} = (\mathbf{B}, \mathsf{sl}, \mathsf{R}, \{\mathsf{ct}_i\}_{P_i})$.

Decryption $\mathsf{Dec}(\mathbf{B}, \mathsf{ct}, \mathsf{sk})$: Party i outputs $\perp$ if it does not have $\mathsf{sk}_{L,i}$ such that $\mathsf{lottery}(\mathbf{B}, \mathsf{sl}, \mathsf{R}, \mathsf{sk}_{L,i}) = 1$ for parameters $\mathbf{B}, \mathsf{sl}, \mathsf{R}$ from ct. Otherwise, it retrieves enough (above threshold) valid ciphertexts $\mathsf{ct}_{(\mathsf{sl},\mathsf{R})}^{\mathsf{sk},j}$ from the current state of the blockchain and decrypts each through Π_{ECW} obtaining $\mathsf{sk}_{(\mathsf{sl},\mathsf{R})}^j$. It then computes $\mathsf{sk}_{(\mathsf{sl},\mathsf{R})} \leftarrow \Pi_{\mathsf{TIBE}}.\mathsf{Combine}(\mathsf{mpk}, (\mathsf{sk}_{(\mathsf{sl},\mathsf{R})}^j)_j)$. It finally outputs $m \leftarrow \Pi_{\mathsf{TIBE}}.\mathsf{Dec}(\mathsf{sk}_{(\mathsf{sl},\mathsf{R})}, \mathsf{ct})$.

Resharing. We can ensure that the master secret key is proactively reshared by modifying each party so that msk_i-s are reshared and reconstructed in the evolution of the blockchain.

Correctness. Correctness of the construction follows from the correctness of the underlying IBE and the fact that a winning role will be able to decrypt the id-specific key by the correctness of the ECW scheme.

In the following we assume some of the extensions discussed in Sect. 6.

Theorem 2 (informal). *Let Γ^V be a YOSO MPC friendly blockchain, Π_{TIBE} be a robust secure threshold IBE as in Sect. 2.3 with threshold $n/2$, and Π_{ECW} a secure IND-CCA2 ECW. The construction in Sect. 7.1 is a secure EtF.*

We refer the reader to the full version for a proof of security [7].

8   Blockchain WE Versus EtF

In this section we show that an account-based PoS blockchain with sufficiently expressive smart contracts and an EtF scheme for this blockchain implies a notion of witness encryption on blockchains, and *vice versa*. The construction of EtF from BWE is completely straightforward and natural: encrypt to the witness which is the secret key winning the lottery. The construction of BWE from EtF is also straightforward but slightly contrived: it requires that we can restrict the lottery such that only some accounts can win a given role and that the decryptor has access to a constant fraction of the stake on the blockchain and are willing to bind them for the decryption operation. The reason why we still prove the result is that it establishes a connection at the feasibility level. For sufficiently expressive blockchains the techniques allowing to construct EtF and BWE are the same. To get EtF from simpler techniques than those we need for BWE we need to do it in the context of very simple blockchains. In addition, the techniques allowing to get EtF without getting BWE should be such that they prevent the blockchain from having an expressive smart contract layer added. This seems like a very small loophole, so we believe that the result shows that there is essentially no assumptions or techniques which allow to construct EtF which do not also allow to construct BWE. Since BWE superficially looks stronger than EtF the equivalence helps better justify the strong assumptions for constructing EtF.

Definition 11 (Blockchain Witness Encryption). *Consider PPT algorithms* (Gen, Enc, Dec) *in the context of a blockchain* Γ^V *is an BWE-scheme with evolved predicate* evolved *and a lottery predicate* lottery *working as follows:*

Setup. (pv, td) $\leftarrow$ Gen() *generates a public value* pv *and an extraction trapdoor* td. *Initially* pv *is put on* **B**.

Encryption. ct $\leftarrow$ Enc(**B**, W, m) *takes as input a blockchain* **B**, *including the public value, a PPT function* W, *the witness recogniser, and a message* m. *It outputs a ciphertext* ct, *a blockchain witness encryption.*

Decryption. $m/\bot \leftarrow$ Dec($\tilde{\mathbf{B}}$, ct, w) *in input a blockchain state* $\tilde{\mathbf{B}}$, *including the a public value* pv, *a ciphertext* ct *a witness* w, *it outputs a message* m *or* $\bot$.

Correctness. An BWE-scheme is correct if for honest parties i and j, PPT function W, and witness w such that $W(\mathbf{w}) = 1$ the following holds with overwhelming probability: if party i runs ct $\leftarrow$ Enc(**B**, W, m) and party j starts running Dec($\tilde{\mathbf{B}}$, ct, w) in $\tilde{\mathbf{B}}$ evolved from **B**, then eventually Dec($\tilde{\mathbf{B}}$, ct, w) outputs m.

Security. We establish a game between a challenger $\mathcal{C}$ and an adversary $\mathcal{A}$. In Sect. 2.1 we described how $\mathcal{A}$ and $\mathcal{Z}$ execute a blockchain protocol. In addition, we now let the adversary interact with the challenger in a game $\mathrm{Game}^{\mathrm{IND-CPA}}_{\Gamma, \mathcal{A}, \mathcal{Z}, \mathcal{E}}$ which can be summarized as follows.

1. (pv, td) $\leftarrow$ Gen() and put pv on the blockchain.
2. $\mathcal{A}$ executes the blockchain protocol Γ together with $\mathcal{Z}$ and at some round r chooses a blockchain **B**, a function W and two messages m_0 and m_1 and sends it all to $\mathcal{C}$.
3. $\mathcal{C}$ chooses a random bit b and encrypts the message m_b with the parameters it received and sends ct to $\mathcal{A}$.
4. $\mathcal{A}$ continues to execute the blockchain until some round $\tilde{r}$ where the blockchain $\tilde{\mathbf{B}}$ is obtained and the $\mathcal{A}$ outputs a bit b'.

The adversary wins the game if it succeeds in guessing b with probability notably greater than one half without $W(\mathsf{Extract}(\mathsf{td}, \tilde{\mathbf{B}}, \mathsf{ct}, W)) = 1$.

EtF from BWE. We first show the trivial direction of getting EtF from BWE. Let $\Pi_{\mathsf{BWE}} = (\mathsf{Gen}_{\mathsf{BWE}}, \mathsf{Enc}_{\mathsf{BWE}}, \mathsf{Dec}_{\mathsf{BWE}})$ be an BWE scheme. Recall that one wins the lottery if lottery(**B**, sl, R, sk) = 1. We construct a EtF scheme. To encrypt, let W be the function $W(\mathbf{w}) =$ lottery(**B**, sl, R, w) and output $\mathsf{Enc}_{\mathsf{BWE}}(\mathbf{B}, W, m)$. If winning the lottery for (sl, R) then let w be the secret key winning the lottery and output Dec($\tilde{\mathbf{B}}$, ct, w). The proof is straightforward.

BWE from EtF. We describe a proof of this direction in the full version [7].

Acknowledgements. Bernardo David is supported by the Concordium Foundation and by the Independent Research Fund Denmark (IRFD) grants number 9040-00399B (TrA^{2}C), 9131-00075B (PUMA) and 0165-00079B. Hamidreza Khoshakhlagh has been funded by the Concordium Foundation under Concordium Blockchain Research Center, Aarhus. Anders Konring is supported by the IRFD grant number 9040-00399B (TrA^{2}C). Jesper Buus Nielsen is partially funded by the Concordium Foundation; The Danish Independent Research Council under Grant-ID DFF-8021-00366B (BETHE); The Carlsberg Foundation under the Semper Ardens Research Project CF18-112 (BCM).

References

1. Afshar, A., Mohassel, P., Pinkas, B., Riva, B.: Non-interactive secure computation based on cut-and-choose. In: Nguyen, P.Q., Oswald, E. (eds.) EUROCRYPT 2014. LNCS, vol. 8441, pp. 387–404. Springer, Heidelberg (2014). https://doi.org/10.1007/978-3-642-55220-5_22

2. Barta, O., Ishai, Y., Ostrovsky, R., Wu, D.J.: On succinct arguments and witness encryption from groups. In: Micciancio, D., Ristenpart, T. (eds.) CRYPTO 2020, Part I. LNCS, vol. 12170, pp. 776–806. Springer, Cham (2020). https://doi.org/10.1007/978-3-030-56784-2_26

3. Benhamouda, F., et al.: Can a public blockchain keep a secret? In: Pass, R., Pietrzak, K. (eds.) TCC 2020, Part I. LNCS, vol. 12550, pp. 260–290. Springer, Cham (2020). https://doi.org/10.1007/978-3-030-64375-1_10

4. Benhamouda, F., Lin, H.: Mr NISC: multiparty reusable non-interactive secure computation. In: Pass, R., Pietrzak, K. (eds.) TCC 2020, Part II. LNCS, vol. 12551, pp. 349–378. Springer, Cham (2020). https://doi.org/10.1007/978-3-030-64378-2_13

5. Boneh, D., Boyen, X., Halevi, S.: Chosen ciphertext secure public key threshold encryption without random oracles. In: Pointcheval, D. (ed.) CT-RSA 2006. LNCS, vol. 3860, pp. 226–243. Springer, Heidelberg (2006). https://doi.org/10.1007/11605805_15

6. Boyle, E., Gilboa, N., Ishai, Y., Lin, H., Tessaro, S.: Foundations of homomorphic secret sharing. In: Karlin, A.R. (ed.) ITCS 2018. LIPIcs, vol. 94, pp. 21:1–21:21 (2018)

7. Campanelli, M., David, B., Khoshakhlagh, H., Konring, A., Nielsen, J.B.: Encryption to the future: a paradigm for sending secret messages to future (anonymous) committees. Cryptology ePrint Archive, Paper 2021/1423 (2021). https://eprint.iacr.org/2021/1423

8. Chase, M., Lysyanskaya, A.: On signatures of knowledge. In: Dwork, C. (ed.) CRYPTO 2006. LNCS, vol. 4117, pp. 78–96. Springer, Heidelberg (2006). https://doi.org/10.1007/11818175_5

9. David, B., Gaži, P., Kiayias, A., Russell, A.: Ouroboros praos: an adaptively-secure, semi-synchronous proof-of-stake blockchain. In: Nielsen, J.B., Rijmen, V. (eds.) EUROCRYPT 2018, Part II. LNCS, vol. 10821, pp. 66–98. Springer, Cham (2018). https://doi.org/10.1007/978-3-319-78375-8_3

10. Derler, D., Slamanig, D.: Practical witness encryption for algebraic languages and how to reply an unknown whistleblower. Cryptology ePrint Archive, Report 2015/1073 (2015). https://eprint.iacr.org/2015/1073

11. Fujisaki, E., Okamoto, T.: Secure integration of asymmetric and symmetric encryption schemes. In: Wiener, M. (ed.) CRYPTO 1999. LNCS, vol. 1666, pp. 537–554. Springer, Heidelberg (1999). https://doi.org/10.1007/3-540-48405-1_34

12. Garay, J.A., Gelles, R., Johnson, D.S., Kiayias, A., Yung, M.: A little honesty goes a long way - the two-tier model for secure multiparty computation. In: Dodis, Y., Nielsen, J.B. (eds.) TCC 2015, Part I. LNCS, vol. 9014, pp. 134–158. Springer, Heidelberg (2015). https://doi.org/10.1007/978-3-662-46494-6_7

13. Garay, J., Kiayias, A., Leonardos, N.: The bitcoin backbone protocol: analysis and applications. In: Oswald, E., Fischlin, M. (eds.) EUROCRYPT 2015, Part II. LNCS, vol. 9057, pp. 281–310. Springer, Heidelberg (2015). https://doi.org/10.1007/978-3-662-46803-6_10

14. Garg, S., Gentry, C., Sahai, A., Waters, B.: Witness encryption and its applications. In: Boneh, D., Roughgarden, T., Feigenbaum, J. (eds.) 45th ACM STOC, pp. 467–476. ACM Press (2013)

15. Gentry, C., et al.: YOSO: you only speak once - secure MPC with stateless ephemeral roles. In: Malkin, T., Peikert, C. (eds.) CRYPTO 2021, Part II. LNCS, vol. 12826, pp. 64–93. Springer, Cham (2021). https://doi.org/10.1007/978-3-030-84245-1_3

16. Gentry, C., Halevi, S., Magri, B., Nielsen, J.B., Yakoubov, S.: Random-index PIR and applications. In: Nissim, K., Waters, B. (eds.) TCC 2021. LNCS, vol. 13044, pp. 32–61. Springer, Cham (2021). https://doi.org/10.1007/978-3-030-90456-2_2

17. Goyal, R., Goyal, V.: Overcoming cryptographic impossibility results using blockchains. In: Kalai, Y., Reyzin, L. (eds.) TCC 2017, Part I. LNCS, vol. 10677, pp. 529–561. Springer, Cham (2017). https://doi.org/10.1007/978-3-319-70500-2_18

18. Goyal, V., Kothapalli, A., Masserova, E., Parno, B., Song, Y.: Storing and retrieving secrets on a blockchain. Cryptology ePrint Archive, Report 2020/504 (2020). https://eprint.iacr.org/2020/504

19. Jawurek, M., Kerschbaum, F., Orlandi, C.: Zero-knowledge using garbled circuits: how to prove non-algebraic statements efficiently. In: Sadeghi, A.-R., Gligor, V.D., Yung, M. (eds.) ACM CCS 2013, pp. 955–966. ACM Press (2013)

20. Nielsen, J.B.: On protocol security in the cryptographic model. Citeseer (2003)

21. Pass, R., Seeman, L., Shelat, A.: Analysis of the blockchain protocol in asynchronous networks. In: Coron, J.-S., Nielsen, J.B. (eds.) EUROCRYPT 2017, Part II. LNCS, vol. 10211, pp. 643–673. Springer, Cham (2017). https://doi.org/10.1007/978-3-319-56614-6_22

22. Sahai, A.: Non-malleable non-interactive zero knowledge and adaptive chosen-ciphertext security. In: 40th FOCS, pp. 543–553. IEEE Computer Society Press (1999)

23. Zahur, S., Rosulek, M., Evans, D.: Two halves make a whole - reducing data transfer in garbled circuits using half gates. In: Oswald, E., Fischlin, M. (eds.) EUROCRYPT 2015, Part II. LNCS, vol. 9057, pp. 220–250. Springer, Heidelberg (2015). https://doi.org/10.1007/978-3-662-46803-6_8

Authenticated Encryption with Key Identification

Julia Len[1]([✉]), Paul Grubbs[2], and Thomas Ristenpart[1]

[1] Cornell Tech, New York, USA
jl3836@cornell.edu
[2] University of Michigan, Ann Arbor, USA

Abstract. Authenticated encryption with associated data (AEAD) forms the core of much of symmetric cryptography, yet the standard techniques for modeling AEAD assume recipients have no ambiguity about what secret key to use for decryption. This is divorced from what occurs in practice, such as in key management services, where a message recipient can store numerous keys and must identify the correct key before decrypting. To date there has been no formal investigation of their security properties or efficacy, and the ad hoc solutions for identifying the intended key deployed in practice can be inefficient and, in some cases, vulnerable to practical attacks.

We provide the first formalization of nonce-based AEAD that supports key identification (AEAD-KI). Decryption now takes in a vector of secret keys and a ciphertext and must both identify the correct secret key and decrypt the ciphertext. We provide new formal security definitions, including new key robustness definitions and indistinguishability security notions. Finally, we show several different approaches for AEAD-KI and prove their security.

Keywords: Key identification · Authenticated encryption · Key commitment · Key robustness

1 Introduction

Authenticated encryption with associated data (AEAD) is ubiquitously used in practice. Standard formalizations of AEAD schemes model a "single-key" setting where a single sender sends an encrypted message to a single receiver. Though simple to analyze, this model is increasingly divorced from practice in a number of important aspects.

A setting which has received little or no attention in the cryptographic literature is one where a message recipient will store numerous keys and must identify the correct key to use before decryption can proceed. This practice can be seen in cryptographic libraries such as Google's Tink API [31], key management services (KMS) such as for Amazon Web Services (AWS) [5], and multi-user Shadowsocks [29], among others. Notably, AEAD schemes and their security models do not formally address the issue of key identification, producing a gap when translating from cryptographic theory to practice.

© International Association for Cryptologic Research 2022
S. Agrawal and D. Lin (Eds.): ASIACRYPT 2022, LNCS 13793, pp. 181–209, 2022.
https://doi.org/10.1007/978-3-031-22969-5_7

Two approaches for key identification are often used in practice. The first is trial decryption, where one attempts to decrypt the ciphertext under each of the keys held by the recipient. However, this is slow, even for a small number of keys. As such, a second approach is often used: the sender attaches a previously agreed-upon key identifier to each message it sends. For instance, this approach is used by Tink, which derives 5-byte strings as the identifier. The recipient can then efficiently use the identifier to look up the corresponding key. However, this approach does not work in settings where keys must remain anonymous, such as in anonymous messaging protocols. It is also unclear what security properties (for both approaches) are being achieved in the case where there is potential for adversarial modification of key identifiers and/or adversarial choice of some of the recipient keys. Adversarially chosen keys can arise in settings where the sender chooses a secret key to share with the recipient, which have resulted in several recent attacks [2, 14, 18, 22].

One example is multi-user Shadowsocks [29]. Shadowsocks is an anonymity proxy that works by having a client encrypt their traffic under a shared password with a server, which decrypts using AEAD. The multi-user mode allows multiple passwords to be specified for a single server, which means that incoming packets must be trial decrypted under every possible user password. Len et al. [22] describe an attack on this scheme where an attacker can insert a malicious password into this set of passwords, then mount a partitioning oracle attack that enables the attacker to learn some target user's password. Fundamentally, the vulnerability is that Shadowsocks's AEAD has no efficient and secure way to identify the appropriate key.

Our Contributions. We initiate the formal study of AEAD that supports key identification. The starting point is nonce-based AEAD [27], which we extend to include in the formal syntax and semantics of encryption schemes the key identification task: decryption takes in a vector of secret keys as well as a nonce, associated data, and a ciphertext, and must both identify the correct secret key and decrypt the ciphertext. This change, while conceptually simple, immediately introduces a number of complexities. It forces scheme designers to specify how the right key is identified, requires changes to notions of correctness, suggests that we must give new security definitions that speak to issues such as adversaries forcing the wrong key to be identified, and more.

We formalize a new cryptographic primitive called AEAD-KI, or AEAD with key identification. Like AEAD, the primitive is composed of a triple of algorithms for key generation, encryption, and decryption. Key generation takes in what we call a key label so that AEAD-KI keys are composed of the traditional secret key as well as the key label. This label acts as optional public metadata for the key and models techniques in practice, e.g., URLs that suggest where to locate the key or other kinds of static identifiers. Encryption takes in a key, nonce, associated data, and message. Ciphertexts can opt to include a special component, called a key tag. Decryption for AEAD-KI, in turn, accepts a vector of keys, instead of one key as for AEAD. We use a vector instead of a set to preserve information about the order of keys, which could affect the decryption outcome. Given the ciphertext (including the key tag), decryption returns both the key that correctly decrypts the ciphertext as well as the resulting message. If decryption determines that no key correctly decrypts the ciphertext, it simply returns an error symbol $\perp$.

We next consider which security definitions best capture the AEAD-KI setting. Our first goal is to extend the standard AEAD security notions of confidentiality and ciphertext integrity to AEAD-KI. A good starting point is transforming the traditional all-in-one real-versus-random indistinguishability security notion for AEAD due to Rogaway and Shrimpton [28] to the AEAD KI setting. Specifically, we can allow the attacker to interact with multiple encryption key instances, reminiscent of the multi-user setting for encryption [6]. However, this definition only allows for honest keys, which unfortunately does not capture attacks in which a malicious key is somehow inserted into a recipient's key vector. Indeed, attacks have already been shown in practice where malicious keys are given to a recipient to prevent decryption under honest keys [2,14].

We therefore opt for a notion of security for which adversaries can insert malicious keys into key vectors used during decryption queries. This renders more complex how the security game should handle decryption oracles in order to distinguish between honest and malicious keys. To handle these subtleties, we introduce KI-nAE, a new security notion that uses a simulation-based approach for the all-in-one definition. This definition captures a wide class of interference attacks in which a malicious user somehow inserts a malicious key into a recipient's key vector in an attempt to interfere with honest keys.

A security property intrinsic to the key identification setting is key robustness, a security goal first investigated in the context of public-key encryption [1] and later investigated for authenticated encryption [16] (see also [2,8,14,18,22]). Interestingly, robustness here functions as a form of correctness, as ensuring the correct key decrypts a given ciphertext in the presence of many (potentially adversarial) keys can only be guaranteed by robustness. We thus extend the AEAD robustness notion called full robustness (FROB) [16] to the AEAD-KI setting, which is straightforward. Interestingly, this extension however proves insufficient to rule out some attacks. In particular, when decryption is given the correct key within a key set, it should not fail to decrypt; such failures could leak information about the (honest) keys composing a key vector. One way to handle this is with an extended key robustness notion, but an observation due to Mihir Bellare is that one can instead extend correctness to rule out such decryption failures. See the body and the full version of this work for more details.

Approaches to AEAD-KI. We then turn to analyzing security of existing key identification schemes as well as suggesting new ones. A summary of our analyses appears in Fig. 1. We divide key identification into several categories. The first approach utilizes the key label of the key as the key tag itself. Decryption can then find all keys whose label matches the key tag of the ciphertext and perform trial decryptions. (Note the second category, trial decryption, is a special case where all labels are the empty string.) This reflects how, in practice, labels are sometimes not unique and instead used to label a set of keys. Using key

Approach	Description	AEAD FROB?	Key anon.?	Section
Key labels	Key gen. labels each key, sent as part of ciphertext; brute-force decrypt with all keys matching key label in ciphertext	Yes	No	§4
Trial decryption	Special case of key labels where all labels are empty (ε)	Yes	Yes	§4
Static key hint	Ciphertext includes deterministic non-CR hash of key	Yes	No	§5
Static key commitment	Ciphertext includes deterministic CR hash of key	No	No	§5
Dynamic key hint	Ciphertext includes PRF of key & nonce	Yes	Yes	§6
Dynamic key commitment	Ciphertext includes CR-PRF of key & nonce	No	Yes	§6

Fig. 1. Summary of various approaches to AEAD-KI that we consider. For each approach, we also list whether it must use a FROB AEAD scheme and whether it provides key anonymity.

labels obviates achieving key anonymity, since a ciphertext produced by a certain key will always be flagged by the key's label. Nevertheless, brute-force trial decryption, a special case of the key label approach, can achieve anonymity, although with the trade-off of increased computational costs. Since key labels are not unique, they do not provide key commitment to AEAD schemes that are not key committing. Nevertheless, we see this insecure construction arises in practice, such as in the Tink library. Our analysis shows that if one instead uses the key label approach with a key-committing AEAD scheme, then the composition is key committing.

Next we turn to what we call "static" approaches, those where key tags are deterministically computed from the secret key. These are similar to "key check values", legacy schemes for ensuring integrity of the key [17, 26, 30]. Since the key tag for a key never changes, this approach also does not allow for key anonymity. We further divide static key identifiers into two classes: static key hints and static key commitments. Key hints are key tags computed from the key in a non-key-committing way, typically using a non-collision resistant hash of the key. This means that AEAD-KI schemes using key hints, like key labels, will need to trial decrypt on all keys matching the key hint and use a FROB AEAD scheme to achieve key robustness. The benefit of key hints is that they can often be short and efficiently computed. In contrast, static key commitments, the second class of static key identifiers, do commit to the key. These are typically a collision-resistant hash of the key. While static key commitments might be less efficient to compute than key hints, they can be used to build secure AEAD-KI from non-FROB AEAD schemes.

Finally, both key hints and key identifiers can be made "dynamic" to provide key anonymous counterparts of static schemes. This approach uses a nonce when computing the key tag so that the key tag for a key is unique for each encryption call. Like static identifiers, dynamic identifiers can be both key hints and key commitments, with security achieved when combining with any AEAD or FROB AEAD, respectfully.

Further Related Work. Farshim, Orlandi, and Rosie [16] first proposed the set of key robustness notions for symmetric primitives. Their strongest definition, full robustness, represents the goal of key commitment for AEAD schemes. Grubbs et al. [18] suggest a notion of compactly committing AEAD, which is useful in abuse moderation settings. Dodis et al. [14] describe an attack against Facebook Messenger's abuse moderation tooling that relies on AES-GCM, which is not key committing. In this attack, a malicious sender can force an honest recipient to use a malicious key to decrypt a message to abusive content that cannot be reported. Len et al. [22] and Albertini et al. [2] showed that other commonly used AEAD schemes, such as AES-GCM-SIV, ChaCha20-Poly1305, and OCB3, are not key committing and they describe other practical attack scenarios that exploit non-key committing AEAD schemes.

Albertini et al. also propose two approaches to adding key commitment to AEAD schemes. One of these approaches is to compute a collision-resistant PRF of the key, both deterministically and using a nonce for key anonymity. Our static and dynamic identifiers parallel these schemes, but for the AEAD-KI setting. Bellare and Hoang [8] propose a spectrum of new definitions for commitment that capture not just committing to the key but also committing to the nonce, associated data, and plaintext message. They also propose new key-committing AEAD schemes based on AES-GCM and AES-GCM-SIV as well as a construction that transforms a legacy AEAD scheme into one that is key-committing using what they call a *committing PRF*. This construction is similar to that proposed by Albertini et al. (and indeed they note that Albertini et al.'s construction can be viewed as a specific instantiation of their scheme). Their work considers only the single-key setting, but their schemes can be used as the necessary FROB AEAD schemes in our AEAD-KI constructions.

Degabriele et al. [13] propose nonce-set AEAD, which is similar to our AEAD-KI formalism but instead for nonce sets. Their formalism considers decryption accepting a set of nonces and then returning the correct nonce along with the plaintext message.

Chan and Rogaway [12] formalize anonymous authenticated encryption, which requires that ciphertexts maintain strong privacy when considering nonces and associated data. They mention the need for robustness, although they only consider robustness for honestly generated keys, which is implied by the typical AEAD security notion as shown in [16]. Finally, Jaeger and Tyagi [21] consider multi-user simulation-based security definitions for various standard symmetric definitions where keys can be adaptively compromised.

2   Preliminaries

We follow the notational conventions used in [18]. We fix some alphabet Σ, e.g. $\Sigma = \{0,1\}$. For any $x \in \Sigma^*$, let $|x|$ denote its length. We write $x \leftarrow_\$ X$ to denote uniformly sampling from a finite set X. We write $X\|Y$ to denote concatenation of two strings. For a string X of n bits, we will write $X[i, \ldots, j]$ for $1 \leq i < j \leq n$ to mean the substring of X beginning at index i and ending at index j. For notational simplicity, we assume that one can unambiguously parse $Z = X \| Y$ into its two parts, even for strings of varying length. For strings $X, Y \in \{0,1\}^*$ we write $X \oplus Y$ to denote taking the XOR of $X[1, \ldots, \min(|X|, |Y|)] \oplus Y[1, \ldots, \min(|X|, |Y|)]$. For some table of values where y_i is stored at key x_i, denoted as $\mathsf{T}[x_i] \leftarrow y_i$, for a set of keys $X = \{x_i : 1 \leq i \leq \kappa\}$ for some integer κ, $\mathsf{T}[X]$ denotes the set $\{\mathsf{T}[x_i] : x_i \in X\}$. We denote a vector of elements as $[\cdot]$. We denote the value stored at index i in vector $\mathbb{K}$ as $\mathbb{K}[i]$. For vector $\mathbb{K}$, we denote $x \in \mathbb{K}$ as x is an element of $\mathbb{K}$ and $|\mathbb{K}|$ as the number of elements in $\mathbb{K}$. We also denote $\mathbb{K}.\mathsf{add}(x)$ to mean adding x to the end of vector $\mathbb{K}$. We denote $[n]_\ell$ as the ℓ-bit representation of the integer n.

We use code-based games [10] to formalize security notions. Variables' types should be clear from context and are modeled as random variables in the probability distribution defined by the random coins used in execution. $\Pr[\mathrm{G} \Rightarrow y]$ denotes (over the random coins of G) that the game G outputs the value y. For a scheme S, we will sometimes use "a G_S adversary" to describe an adversary in the game G instantiated with the scheme S. For an adversary $\mathcal{A}$, $\mathrm{G}_S^{\mathcal{A}}$ denotes the game G instantiated with the scheme S and specific adversary $\mathcal{A}$. $\Pr[\mathrm{G}_S^{\mathcal{A}} \Rightarrow \mathsf{out}]$ denotes the probability that game G instantiated with scheme S and adversary $\mathcal{A}$ outputs out.

Authenticated Encryption. An authenticated encryption with associated data (AEAD) scheme $\mathrm{AEAD} = (\mathsf{Kg}, \mathsf{Enc}, \mathsf{Dec})$ consists of a triple of algorithms. Associated to any scheme AEAD is a key space $\mathcal{K} \subseteq \Sigma^*$, nonce space $\mathcal{N} \subseteq \Sigma^*$, header space $\mathcal{A} \subseteq \Sigma^*$, message space $\mathcal{M} \subseteq \Sigma^*$, and ciphertext space $\mathcal{C} \subseteq \Sigma^*$. The randomized key generation algorithm Kg outputs a secret key $K \in \mathcal{K}$. Encryption Enc is deterministic and takes as input a 4-tuple $(K, N, AD, M) \in (\Sigma^*)^4$ and outputs ciphertext C or a distinguished error symbol $\bot$. We require that $\mathsf{Enc}(K, N, AD, M) \neq \bot$ if $(K, N, AD, M) \in \mathcal{K} \times \mathcal{N} \times \mathcal{A} \times \mathcal{M}$. Decryption Dec is deterministic and takes as input a tuple $(K, N, AD, C) \in (\Sigma^*)^4$ and outputs value M or $\bot$. An AEAD scheme is correct if for any $(K, N, AD, M) \in \mathcal{K} \times \mathcal{N} \times \mathcal{A} \times \mathcal{M}$ it holds that $\mathsf{Dec}(K, N, AD, \mathsf{Enc}(K, N, AD, M)) = M$.

The security notion we consider for AEAD schemes is nonce-based real-or-random security under chosen-ciphertext attack [28]. We generalize this to a multi-user setting [11] where an adversary can interact with multiple instances of the AEAD scheme, which we call $\mathrm{MU\text{-}nAE}_{\mathrm{AEAD}}$. The game pseudocode is presented in Fig. 2. We restrict our attention to nonce-respecting adversaries, meaning they never query the same nonce twice to ENC for the same key identifier id, and they only query a key identifier $\mathsf{id} < i$ to ENC and DEC. The $\mathrm{MU\text{-}nAE}_{\mathrm{AEAD}}$ advantage of an adversary $\mathcal{A}$ is defined as

$$\mathbf{Adv}_{\mathrm{AEAD}}^{\mathrm{mu\text{-}nae}}(\mathcal{A}) = |\Pr[\mathrm{MU\text{-}nAE1}_{\mathrm{AEAD}}^{\mathcal{A}} \Rightarrow 1] - \Pr[\mathrm{MU\text{-}nAE0}_{\mathrm{AEAD}}^{\mathcal{A}} \Rightarrow 1]|.$$

Full Robustness. We use the full robustness notion for AEAD schemes from Farshim et al. [16], but adapted to the nonce-based setting. Albertini et al. [2]

$$
\begin{array}{|ll|}
\hline
\end{array}
$$

MU-nAE1$_{AEAD}^{\mathcal{A}}$:	MU-nAE0$_{AEAD}^{\mathcal{A}}$:
$i \leftarrow 0$	$i \leftarrow 0$
$b' \leftarrow\!\!\$ \; \mathcal{A}^{\text{GenKey,Enc,Dec}}$	$b' \leftarrow\!\!\$ \; \mathcal{A}^{\text{GenKey,Enc,Dec}}$
Return b'	Return b'
GenKey():	GenKey():
$K \leftarrow\!\!\$ \; \text{AEAD.Kg}()$	$i \leftarrow i + 1$
$T[i] \leftarrow K \; ; \; i \leftarrow i + 1$	
	Enc(id, N, AD, M):
Enc(id, N, AD, M):	$C \leftarrow\!\!\$ \; \{0,1\}^{\text{clen}(|M|)}$
$C \leftarrow \text{AEAD.Enc}(\mathsf{T}[\text{id}], N, AD, M)$	Return C
$\mathsf{c}[N, AD, C] \leftarrow \text{id}$	
Return C	Dec(id, N, AD, C):
	Return $\perp$
Dec(id, N, AD, C):	
If $\mathsf{c}[N, AD, C] = \text{id}$ then return $\perp$	
$M \leftarrow \text{AEAD.Dec}(\mathsf{T}[\text{id}], N, AD, C)$	
Return M	

Fig. 2. Games MU-nAE1 and MU-nAE0 are used for MU-nAE$_{AEAD}$, or multi-user real-or-random security, for scheme AEAD $= (\text{Kg}, \text{Enc}, \text{Dec})$.

also provide a FROB notion for nonce-based AEAD, with slightly different syntax — our formulation is equivalent to theirs. Roughly, FROB security tasks an adversary with providing two keys and a ciphertext such that both keys successfully decrypt the ciphertext. We define the game in Fig. 3.

The FROB$_{AEAD}$ advantage of an adversary $\mathcal{A}$ is defined as

$$
\mathbf{Adv}_{AEAD}^{\text{frob}}(\mathcal{A}) = \Pr\left[\, \text{FROB}_{AEAD}^{\mathcal{A}} \Rightarrow 1 \,\right].
$$

Pseudo-random Functions. We use a multi-user variant of the traditional pseudo-random function (PRF) definition (q.v., [7]) for two functions, where the adversary is given access to oracles for both functions. We define the games in Fig. 3. The game gives the adversary an additional GenKey oracle that allows it to generate multiple keys. The MU-PRF$_F$ advantage of an adversary $\mathcal{A}$ is defined as

$$
\mathbf{Adv}_{F_0,F_1}^{\text{mu-prf}}(\mathcal{A}) = |\Pr\left[\, \text{REAL}_{F_0,F_1}^{\mathcal{A}} \Rightarrow 1 \,\right] - \Pr\left[\, \text{IDEAL}^{\mathcal{A}} \Rightarrow 1 \,\right]|.
$$

Collision Resistance. The collision resistance (CR) game for function $F = \{0,1\}^\kappa \times \{0,1\}^\ell \to \{0,1\}^n$ measures the ability of an adversary to find two key-value pairs such that the evaluation of F on these inputs evaluates to the same output. More formally, the CR$_F$ advantage of an adversary $\mathcal{A}$ is defined as

$$
\mathbf{Adv}_F^{\text{cr}}(\mathcal{A}) = \Pr\left[\, K_0, x_0, K_1, x_1 \leftarrow \mathcal{A} : (K_0, x_0) \neq (K_1, x_1) \wedge F(K_0, x_0) = F(K_1, x_1) \,\right].
$$

Pre-image Resistance. The pre-image resistance game for function $F = \{0,1\}^\kappa \times \{0,1\}^\ell \to \{0,1\}^n$ measures the ability of an adversary to find the

$\text{FROB}^{\mathcal{A}}_{\text{AEAD}}:$	$\text{REAL}^{\mathcal{A}}_{\mathsf{F}_0,\mathsf{F}_1}:$	$\text{IDEAL}^{\mathcal{A}}:$
$K_0, K_1, N, AD, C \leftarrow \mathcal{A}$	$i \leftarrow 0$	$i \leftarrow 0;\ \mathsf{T}_0, \mathsf{T}_1 \leftarrow [\ ]$
If $K_0 = K_1$:	$b \leftarrow\!\!\$\ \mathcal{A}^{\text{GenKey},\text{Func0},\text{Func1}}$	$b' \leftarrow\!\!\$\ \mathcal{A}^{\text{GenKey},\text{Func0},\text{Func1}}$
$\quad$ Return 0	Return b'	Return b'
$M_0 \leftarrow \mathsf{Dec}(K_0, N, AD, C)$		
$M_1 \leftarrow \mathsf{Dec}(K_1, N, AD, C)$	$\underline{\text{GenKey}():}$	$\underline{\text{GenKey}():}$
Return $M_0 \neq \perp \wedge M_1 \neq \perp$	$K \leftarrow\!\!\$\ \{0,1\}^{\kappa}$	$i \leftarrow i + 1$
	$\mathsf{T}[i] \leftarrow K\ ;\ i \leftarrow i + 1$	
		$\underline{\text{Func0}(\mathsf{id}, x):}$
	$\underline{\text{Func0}(\mathsf{id}, x):}$	If $\mathsf{T}_0[\mathsf{id}, x] = \perp$: $\mathsf{T}_0[\mathsf{id}, x] \leftarrow\!\!\$\ \{0,1\}^n$
	Return $\mathsf{F}_0(\mathsf{T}[\mathsf{id}], x)$	Return $\mathsf{T}_0[\mathsf{id}, x]$
	$\underline{\text{Func1}(\mathsf{id}, x):}$	$\underline{\text{Func1}(\mathsf{id}, x):}$
	Return $\mathsf{F}_1(\mathsf{T}[\mathsf{id}], x)$	If $\mathsf{T}_1[\mathsf{id}, x] = \perp$: $\mathsf{T}_1[\mathsf{id}, x] \leftarrow\!\!\$\ \{0,1\}^n$
		Return $\mathsf{T}_1[\mathsf{id}, x]$

Fig. 3. (Left) Game $\text{FROB}_{\text{AEAD}}$ is the full robustness security notion for scheme $\text{AEAD} = (\mathsf{Kg}, \mathsf{Enc}, \mathsf{Dec})$. **(Center/Right)** Games REAL and IDEAL are used for MU-PRF$_{\mathsf{F}_0,\mathsf{F}_1}$, or multi-user PRF security for functions $\mathsf{F}_0 = \{0,1\}^{\kappa} \times \{0,1\}^{\ell_0} \to \{0,1\}^{n_0}$ and $\mathsf{F}_1 = \{0,1\}^{\kappa} \times \{0,1\}^{\ell_0} \to \{0,1\}^{n_0}$.

pre-image of a random range point for F. More formally, the PRE$_{\mathsf{F}}$ advantage of an adversary $\mathcal{A}$ is defined as

$$\mathbf{Adv}^{\text{pre}}_{\mathsf{F}}(\mathcal{A}) = \Pr\left[\,y \leftarrow\!\!\$\ \{0,1\}^n; K, x \leftarrow \mathcal{A}(y) : \mathsf{F}(K, x) = y\,\right].$$

3   Defining AEAD with Key Identification

We start by formalizing the notion of AEAD with key identification (AEAD-KI). AEAD-KI extends AEAD schemes to the setting where a recipient stores multiple keys and must therefore choose which key to use for decryption. At a high level, our formalization extends prior ones on AEAD in the following ways:

- We add a notion of key labels, which are potentially public, application-defined strings associated to secret keys. For notational simplicity, we will redefine a key to be a label, secret key pair.
- Decryption takes as input a vector of keys, instead of a single key. Decryption must determine both which key to use, and the corresponding plaintext. We model the keys used by decryption as a vector, instead of a set, to preserve information about the order.
- Ciphertexts may include a key identification tag, to assist decryption. We will explore a variety of ways to construct key identification tags, each with different security and performance profiles.

These changes to our conceptualization of syntax and semantics of AEAD necessarily require revisiting security as well. Later in this section we propose a new simulator-based all-in-one security definition that captures confidentiality and integrity for AEAD-KI schemes. Furthermore, we will see that the shift to

AEAD-KI introduces a number of subtleties related to ciphertexts potentially being decryptable under more than one key. We will therefore provide new key robustness (also called key commitment) notions for AEAD-KI.

Syntax and Semantics. An AEAD-KI scheme is a triple of algorithms combined with a key space $\mathcal{K} \subseteq \Sigma^*$, key label space $\mathcal{L} \subseteq \Sigma^*$, nonce space $\mathcal{N} \subseteq \Sigma^*$, associated data space $\mathcal{A} \subseteq \Sigma^*$, message space $\mathcal{M} \subseteq \Sigma^*$, ciphertext space $\mathcal{C} \subseteq \Sigma^*$, and key tag space $\mathcal{T} \subseteq \Sigma^*$. We will often leave the spaces implicit and clear from context. Thus we write that an AEAD-KI scheme AEKI = (Kg, Enc, Dec) consists of the following algorithms:

- $K \leftarrow_{\$} \mathsf{AEKI.Kg(kid)}$
  The randomized key generation algorithm takes as input the key label kid to use for the generated secret key. The key label operates as metadata for the secret key. Key generation outputs a key $K \in \mathcal{L} \times \mathcal{K}$, which is a pair composed of the key label and the secret encryption/decryption key. While the encryption key must be kept secret, the key label can be public.
- $(T_k, C) \leftarrow \mathsf{AEKI.Enc}(K, N, AD, M)$
  The nonce-based deterministic encryption algorithm takes as input tuple $(K, N, AD, M) \in (\Sigma^*)^4$ and outputs pair $(T_k, C) \in \mathcal{T} \times \mathcal{C}$ or a distinguished error symbol $\perp$. Notice that encryption returns, in addition to the encrypted plaintext, a bit string T_k, which can be empty. We will refer to this as the key tag, as we discuss more below. Both the key tag and the encrypted plaintext form the ciphertext. We require that if $(K, N, AD, M) \in \mathcal{K} \times \mathcal{N} \times \mathcal{A} \times \mathcal{M}$ then $\mathsf{Enc}(K, N, AD, M) \neq \perp$.
- $(K, M) \leftarrow \mathsf{AEKI.Dec}(\mathbb{K}, N, AD, T_k, C)$
  The decryption algorithm's input is $(\mathbb{K}, N, AD, T_k, C) \in \mathcal{K}^* \times (\Sigma^*)^4$ and its output is $(K, M) \in \mathcal{K} \times \mathcal{M}$ or $\perp$. Decryption is deterministic. Notice that instead of a single key, decryption takes as input a vector of keys $\mathbb{K}$, and we denote vectors of length one or greater by $\mathcal{K}^*$. Furthermore, in addition to the plaintext, decryption returns the corresponding key that produced the plaintext. If no key can decrypt, then the error symbol $\perp$ is returned.

Correctness. When extending AEAD to allow for multiple keys, a meaningful definition of correctness becomes more complex. We expect that when encrypting with a key K to produce ciphertext C, and then decrypting C with any vector $\mathbb{K}$ that includes K, the original plaintext should be recovered. However, this cannot for practical schemes be guaranteed absolutely, since there may exist another key that successfully decrypts the ciphertext. Indeed, the correct outcome that we expect of a scheme that allows decryption to accept multiple keys now becomes more like a security property of key robustness, where we have computational guarantees that decryption succeeds for a single key. We therefore provide here a simpler, absolute correctness definition and later focus on capturing the behavior we want from AEAD-KI through key robustness, which we cover below.

Definition 1. *An AEAD-KI scheme is correct if the following hold:*

(1) *For any* (K, N, AD, M) *it holds that* $\Pr[\,(K', M') = (K, M)\,] = 1$ *where* $(K', M') \leftarrow \mathsf{Dec}([K], N, AD, \mathsf{Enc}(K, N, AD, M))$ *and the probability is over the coins used by encryption;*

(2) *For any* $(\mathbb{K}, N, AD, T_k, C)$ *and* $(K, M) \leftarrow \mathsf{Dec}(\mathbb{K}, N, AD, T_k, C)$ *it must be that either* $(K, M) = \bot$ *or* $K \in \mathbb{K}$; *and*

(3) *For any* $\mathbb{K}, \mathbb{K}'$ *and any* (N, AD, T_k, C), *let* $(K, M) \leftarrow \mathsf{Dec}(\mathbb{K}, N, AD, T_k, C)$ *and* $(K', M') \leftarrow \mathsf{Dec}(\mathbb{K}', N, AD, T_k, C)$. *If* $(K, M) \neq \bot$ *and* $K \in \mathbb{K}'$, *then* $(K', M') \neq \bot$.

The first condition lifts traditional perfect correctness for AEAD to the syntax of AEAD-KI for decryption with a single key. The second condition additionally asks that decryption only ever output a key that was in the key vector. The third correctness condition roughly requires that if Dec outputs some key K for a key set $\mathbb{K}$, any other $\mathbb{K}'$ containing K must decrypt to non-$\bot$. (Note we do not require decryption with $\mathbb{K}'$ outputs K; this property is guaranteed by our key robustness security notion, which we discuss next.) For all schemes we consider, correctness is easily established via inspection of their decryption algorithm; we therefore will omit explicit analysis.

3.1 Key Robustness

As mentioned, in the AEAD-KI setting, key robustness is partly about correctness: we expect that the key used to encrypt a plaintext should be the only one to correctly decrypt the resulting ciphertext. However, when decryption allows for multiple keys—some of which may be adversarially-chosen—this property cannot be satisfied without some form of key robustness. Briefly, key robustness guarantees that only a single key can be used to decrypt a given ciphertext.

Farshim et al. [15] first defined several key robustness notions for AEAD schemes. Their strongest notion is called full robustness (FROB), and requires an adversary to discover two keys that each successfully decrypt an adversarially chosen nonce, associated data, and ciphertext (see Sect. 2). Bellare and Hoang [8] recently introduced even stronger notions, such as their CMT3, which require commitment to not just the key but also nonces and associated data. For simplicity we stick with adapting the FROB notion. Analogous adaptations can be made to lift CMT3 to the key identification setting, but some schemes would require modification to meet them (e.g., including nonce and associated data in key check value computations).

We define KI-FROB security for an AEAD-KI scheme via the game shown in Fig. 4. It requires an adversary to find two key vectors and a nonce, associated data, and ciphertext. Decryption is run with each key vector, and the adversary wins should both decryptions succeed and the returned secret key, message pairs are distinct. Note here that we are focused on the secret key, not key including key label, thereby explicitly excluding as a win having distinct key labels. This is to allow schemes that use multiple key labels for the same key. The KI-FROB$_{\mathsf{AEKI}}$ advantage of an adversary $\mathcal{A}$ is defined as

$$\mathbf{Adv}_{\mathsf{AEKI}}^{\mathrm{ki\text{-}frob}}(\mathcal{A}) = \Pr\left[\,\mathrm{KI\text{-}FROB}_{\mathsf{AEKI}}^{\mathcal{A}} \Rightarrow 1\,\right].$$

$$
\boxed{
\begin{array}{l}
\text{KI-FROB}_{\text{AEKI}}^{\mathcal{A}}: \\[4pt]
\mathbb{K}_0, \mathbb{K}_1, N, AD, T_k, C \leftarrow^\$ \mathcal{A} \\
(K_0, M_0) \leftarrow \mathsf{Dec}(\mathbb{K}_0, N, AD, T_k, C) \\
(K_1, M_1) \leftarrow \mathsf{Dec}(\mathbb{K}_1, N, AD, T_k, C) \\
(\mathsf{kid}_0, K_0^*) \leftarrow K_0 \,;\, (\mathsf{kid}_1, K_1^*) \leftarrow K_1 \\
\text{If } (K_0, M_0) \neq \bot \wedge (K_1, M_1) \neq \bot \wedge K_0^* \neq K_1^*: \\
\quad \text{Return } 1 \\
\text{Return } 0
\end{array}
}
$$

Fig. 4. Game KI-FROB$_{\text{AEKI}}$ represents full robustness for AEAD-KI scheme $\mathsf{AEKI} =$ $(\mathsf{Kg}, \mathsf{Enc}, \mathsf{Dec})$.

Finally, note an important property of our KI-FROB security notion: any scheme meeting it is in some sense agnostic to the *ordering* of keys in the key vector input to decryption. If two different orderings of the same key vector caused different keys to be output, these two key vectors would give a KI-FROB win. (Note that correctness condition (3) above implies that two different orderings of the same key vector must both either output $\bot$ or both non-$\bot$.) Looking ahead, it also means our KI-nAE definition need not account for distinguishing attacks caused by the order of keys in the key vectors; they are ruled out for any KI-FROB scheme.

3.2 All-in-one Confidentiality and Integrity

Just as for an AEAD scheme, we expect an AEAD-KI scheme to maintain confidentiality and integrity. Towards formalizing what this means, one starting point is existing indistinguishability style security definitions for AEAD (e.g., [28]). However, an important modeling question is how to handle key vectors during decryption. This suggests we should start instead with a multi-user style AEAD security notion [11] which allows the adversary to request generation of many keys and obtain encryption of plaintexts under keys of their choice.

To model key anonymity, we may additionally require that adversaries not be able to distinguish between encryptions under different honest keys. To model chosen-ciphertext attacks and, in particular, ciphertext integrity, we face additional choices about how much control to give adversaries over key vectors during decryption. One option would be to allow adversaries to choose the key vector but only allow honestly generated keys to be added to the vector. Unfortunately, this would not capture attacks in which a malicious key is somehow inserted into a recipient's key vector, a scenario that arises in practice. For instance, Albertini et al. [2] describe a vulnerability arising from such a scenario in the context of key rotation within key management services.

We therefore opt for a stronger notion of security for which adversaries can insert malicious keys into key vectors. This renders more complex how the security game should handle decryption oracles, because we need to somehow demarcate between decryption queries that correspond to ciphertext forgeries and ones

$\underline{\text{KI-nAE1}_{\text{AEKI}}^{\mathcal{A}}:}$

$j \leftarrow 0$
$b \leftarrow\!\!\$\ \mathcal{A}^{\text{GenHonestKey},\text{Enc},\text{Dec}}$
Return b

$\underline{\text{GenHonestKey}(\text{kid}):}$
$K \leftarrow\!\!\$\ \text{AEKI.Kg}(\text{kid})$
$T[j] \leftarrow K \ ; \ j \leftarrow j + 1$

$\underline{\text{Enc}(\text{id}, N, AD, M):}$
$(T_k, C) \leftarrow \text{AEKI.Enc}(T[\text{id}], N, AD, M)$
Return (T_k, C)

$\underline{\text{Dec}(\mathbb{K}, N, AD, T_k, C):}$
$\mathbb{K}^* \leftarrow [\,]$
For $(\text{honest}, \text{data}) \in \mathbb{K}$:
  If $\text{honest} = \text{true}$: $\mathbb{K}^*.\text{add}(T[\text{data}])$
  Else: $\mathbb{K}^*.\text{add}(\text{data})$
$(K, M) \leftarrow \text{AEKI.Dec}(\mathbb{K}^*, N, AD, T_k, C)$
Return M

$\underline{\text{KI-nAE0}_{\text{AEKI},S,L_{\text{Enc}}}^{\mathcal{A}}:}$

$\sigma_s \leftarrow\!\!\$\ \text{S.Init}()$
$b \leftarrow\!\!\$\ \mathcal{A}^{\text{GenHonestKey},\text{Enc},\text{Dec}}$
Return b

$\underline{\text{GenHonestKey}(\text{kid}):}$
$\sigma_s \leftarrow\!\!\$\ \text{S.Kg}(\text{kid}, \sigma_s)$

$\underline{\text{Enc}(\text{id}, N, AD, M):}$
$\mathcal{L} \leftarrow\!\!\$\ L_{\text{Enc}}(\text{id}, M)$
$(T_k, C, \sigma_s) \leftarrow\!\!\$\ \text{S.Enc}(N, AD, \mathcal{L}, \sigma_s)$
$\text{C}[\text{id}, N, AD, T_k, C] \leftarrow M$
Return (T_k, C)

$\underline{\text{Dec}(\mathbb{K}, N, AD, T_k, C):}$
For $(\text{honest}, \text{data}) \in \mathbb{K}$:
  If $\text{honest} = \text{true} \wedge \text{C}[\text{data}, N, AD, T_k, C] \neq \perp$:
    Return $\text{C}[\text{data}, N, AD, T_k, C]$
If $\exists(\text{honest}, \text{data}) \in \mathbb{K}$ s.t. $\text{honest} = \text{false}$:
  $(M, \sigma_s) \leftarrow\!\!\$\ \text{S.Dec}(\mathbb{K}, N, AD, T_k, C, \sigma_s)$
  Return M
Return $\perp$

Fig. 5. Game KI-nAE$_{\text{AEKI}}$ is the all-in-one security notion for AEAD-KI schemes.

that should not: an adversary can always generate ciphertexts that decrypt under an adversary-chosen key.

To handle these subtleties, we use a simulation-based approach and an all-in-one confidentiality and ciphertext integrity notion. The pseudocode games for the resulting KI-nAE$_{\text{AEKI}}$ security notion are shown in Fig. 5. KI-nAE$_{\text{AEKI}}$ is parameterized by the simulator, a stateful tuple of algorithms $S = (\text{Init}, \text{Kg}, \text{Enc}, \text{Dec})$. The adversary is given access to an honest key generation oracle GenHonestKey, an encryption oracle Enc, and a decryption oracle Dec. Dec accepts as input a key vector $\mathbb{K}$ as well as ciphertext tuple (N, AD, T_k, C). The key vector $\mathbb{K}$ is composed of tuples $(\text{honest}, \text{data})$, where $\text{honest} = \text{true}$ indicates that the key is honestly generated and data is the game-generated key identifier for the key. If $\text{honest} = \text{false}$, then the key is malicious and data is itself the key.

The game KI-nAE1 models interactions with the real scheme, and therefore calls the relevant AEKI algorithm to answer oracle queries. The ideal game KI-nAE0 instead uses simulator S to generate the oracle outputs for the adversary. Encryption provides leakage to the simulator. The encryption leakage algorithm $L_{\text{Enc}}(\text{id}, M)$ takes as input the game-generated key identifier and the plaintext message and outputs the encryption leakage. We specify two concrete leakage functions, $L_{\text{Enc}}^{\text{id}}$ and $L_{\text{Enc}}^{\text{anon}}$. The non-key anonymous algorithm $L_{\text{Enc}}^{\text{id}}$ returns as leakage both the game-generated key identifier and the size of the plaintext, while $L_{\text{Enc}}^{\text{anon}}$ only returns the size of the plaintext. In the latter case, this of course means that the simulator will have no knowledge of which honest key the adversary chose for encryption.

The decryption oracle in KI-nAE0 works in two parts. In the first part, the table of previous ENC outputs is scanned for each honest key. If the queried ciphertext is in the table, the message is returned. Otherwise, if there are malicious keys in $\mathbb{K}$, the simulator is given its state, the ciphertext tuple, and $\mathbb{K}$.

It may not be immediately obvious why this DEC is the "right" one. The main advantage in defining DEC as we have is that for honest keys this decryption oracle ensures our definition implies a variant of ciphertext integrity for AEAD-KI: no matter the simulator's behavior, crafting a new valid ciphertext for an honest key automatically gives a distinguisher between real and ideal games.

We have two versions of KI-nAE, one key anonymous and one not. For a simulator S, the more general (non-anonymous) KI-nAE$_{AEKI}$ advantage of an adversary $\mathcal{A}$ with respect to S is defined as

$$\mathbf{Adv}^{\text{ki-nae}}_{\text{AEKI},\text{S}}(\mathcal{A}) = \left| \Pr\left[\text{KI-nAE1}^{\mathcal{A}}_{\text{AEKI}} \Rightarrow 1 \right] - \Pr\left[\text{KI-nAE0}^{\mathcal{A}}_{\text{AEKI},\text{S},\text{L}^{\text{id}}_{\text{Enc}}} \Rightarrow 1 \right] \right|.$$

Meanwhile, the key anonymous KI-nAE-KA$_{AEKI}$ advantage of an adversary $\mathcal{A}$ with respect to S is defined as

$$\mathbf{Adv}^{\text{ki-nae-anon}}_{\text{AEKI},\text{S}}(\mathcal{A}) = \left| \Pr\left[\text{KI-nAE1}^{\mathcal{A}}_{\text{AEKI}} \Rightarrow 1 \right] - \Pr\left[\text{KI-nAE0}^{\mathcal{A}}_{\text{AEKI},\text{S},\text{L}^{\text{anon}}_{\text{Enc}}} \Rightarrow 1 \right] \right|.$$

We further note that while we chose to base our definitions on those for nonce-based AEAD from [11,28], we can also adapt these definitions to other settings, such as those for randomized symmetric encryption and the nonce-hiding framework for the AE2 definitions proposed by Bellare et al. [9].

Malicious Keys and Ciphertexts. As discussed before, we opt to capture the adversary providing both malicious keys and malicious ciphertexts since this models real-world settings. However, we must explicitly give the simulator all malicious keys because the adversary knows these keys as well and could otherwise trivially distinguish. This makes it difficult for KI-nAE to capture key robustness notions like KI-FROB. We therefore opt for separate key robustness notions, mirroring the AEAD setting. We believe it is possible to give an all-in-one definition that also implies KI-FROB; we leave this difficult, but interesting, modelling question to future work.

4   Key Labels

One simple technique for key identification is assigning to each key a static label and then prepending the key's label to every ciphertext it produces. In practice, labels typically are randomly-generated strings, URLs indicating where to fetch the key, or even user identifiers. This approach is widely used by both key management services (KMS), such as the Amazon Web Services (AWS) KMS [5], Microsoft Azure Key Vault [24], and Oracle Key Vault [25]; as well as cryptography libraries, such as Google's Tink [31]. Relatedly, the popular cryptographic library Libsodium [23] also recommends using a key label as a way to add robustness to AEAD schemes.

```
KL.Kg(kid):                          KL.Dec(𝕂, N, AD, T_k, C):

K ←$ 𝒦                               For (kid, K) ∈ 𝕂:
Return (kid, K)                        If kid = T_k:
                                          M ← AEAD.Dec(K, N, AD‖kid, C)
KL.Enc(K, N, AD, M):                      If M ≠ ⊥: Return (K, M)
(kid, K*) ← K                        Return ⊥
C ← AEAD.Enc(K*, N, AD‖kid, M)
Return (kid, C)
```

Fig. 6. A typical key label scheme KL which is parameterized by an AEAD scheme.

While key labels appear efficient and straightforward, they have not been formally analyzed. For instance, key labels are often used with non-FROB AEAD schemes, such as AES-GCM and ChaCha20-Poly1305. We will see in our analysis that key labels do not automatically produce a KI-FROB AEAD-KI scheme.

We generalize the key labels construction as KL[AEAD] and provide the pseudocode in Fig. 6. The scheme is parameterized by an AEAD scheme, used for encryption and decryption. To simplify notation, we also refer to the scheme as KL when the specific AEAD scheme used can be arbitrary or is obvious from context. Key generation allows the caller to specify the key label kid, which is then stored as part of the key. We model the label as input to generalize label creation out-of-band and to enable adversarial inputs. Meanwhile, encryption simply uses kid as the key tag. Decryption iterates through 𝕂 to find the first key with an identifier matching the key tag that successfully decrypts the ciphertext.

Analyzing Robustness. Utilizing a key label at first seems like a trivial and practical way to add robustness to any AEAD-KI scheme. A ciphertext with some key identifier can only be decrypted by the corresponding key. However, this method fails when multiple keys can have the same label and a non-FROB AEAD scheme is chosen. Notice, for instance, that KL does not enforce uniqueness of labels. An adversary could choose two AEAD keys so that they have the same label, compute an AEAD key multi-collision ciphertext [22] for these keys, and attach the label to the ciphertext as the key tag. Decryption will then proceed successfully for both keys because their labels match the key tag of the ciphertext.

In the following theorem, we show that an FROB AEAD is both sufficient and necessary for KL to be KI-FROB.

Theorem 1. *Let $\mathcal{A}$ be a KI-FROB adversary for scheme KL[AEAD]. Then we give FROB adversary $\mathcal{B}$ for AEAD such that*

$$\mathbf{Adv}^{\text{ki-frob}}_{\text{KL[AEAD]}}(\mathcal{A}) \leq \mathbf{Adv}^{\text{frob}}_{\text{AEAD}}(\mathcal{B}).$$

Furthermore, let $\mathcal{C}$ be an FROB adversary for AEAD. Then we give a KI-FROB adversary $\mathcal{D}$ for KL[AEAD] such that

$$\mathbf{Adv}^{\text{frob}}_{\text{AEAD}}(\mathcal{C}) \leq \mathbf{Adv}^{\text{ki-frob}}_{\text{KL[AEAD]}}(\mathcal{D}).$$

B runs in time that of A and D runs in time that of C.

Proof. We first construct FROB adversary B against AEAD as follows. B runs A, which returns $\mathbb{K}_0, \mathbb{K}_1, N, AD, T_k, C$. Let $\mathsf{Dec}(\mathbb{K}_0, N, AD, T_k, C)$ and $\mathsf{Dec}(\mathbb{K}_1, N, AD, T_k, C)$ return (K_0, M_0) and (K_1, M_1), respectively, where $K_0 = (\mathsf{kid}_0, K_0^*)$ and $K_1 = (\mathsf{kid}_1, K_1^*)$. Note that A can only win the FROB game if $(K_0, M_0), (K_1, M_1) \neq \perp$ and $K_0^* \neq K_1^*$, so let this be the case. B can return K_0^*, K_1^*, N, AD, C, where again both keys successfully decrypt N, AD, C. Thus, B wins game FROB for AEAD when A wins game KI-FROB for $\mathsf{KL}[\mathsf{AEAD}]$.

We now construct KI-FROB adversary D against KL as follows. D runs C, which returns K_0, K_1, N, AD, C. Then D creates KL key vectors $\mathbb{K}_0 \leftarrow [(0^n, K_0)]$ and $\mathbb{K}_1 \leftarrow [(0^n, K_1)]$ and finally returns $\mathbb{K}_0, \mathbb{K}_1, N, AD, 0^n, C$. Since both K_0, K_1 have the same identifier, which matches the given key tag, the scheme KL will try both keys when decrypting and successfully decrypt. Thus, we have that D wins game KI-FROB for $\mathsf{KL}[\mathsf{AEAD}]$ when C wins game FROB for AEAD. $\square$

Analyzing KI-nAE. We now show that $\mathsf{KL}[\mathsf{AEAD}]$ is KI-nAE secure for leakage algorithm $\mathsf{L}_{\mathsf{Enc}}^{\mathsf{id}}$ when AEAD is both FROB and MU-nAE secure. Notably, encryption cannot be key anonymous: the label is static across calls to ENC and the simulator can only simulate this by knowing which key was queried for encryption. We provide the theorem statement and proof sketch below. The full proof is provided in the full version of this work.

Theorem 2. *Using $\mathsf{L}_{\mathsf{Enc}}^{\mathsf{id}}$, let A be a KI-nAE adversary making at most q queries to its oracles and querying at most m malicious keys. Then we give KI-nAE simulator S and adversaries B, C such that*

$$\mathbf{Adv}_{\mathsf{KL}[\mathsf{AEAD}],\mathsf{S}}^{\mathsf{ki\text{-}nae}}(A) \leq \mathbf{Adv}_{\mathsf{AEAD}}^{\mathsf{frob}}(B) + \mathbf{Adv}_{\mathsf{AEAD}}^{\mathsf{mu\text{-}nae}}(C) + mq/|\mathcal{K}|.$$

Adversaries B, C run in time that of A with an $\mathcal{O}(q)$ overhead and simulator S runs in time $\mathcal{O}(mq)$.

Proof Sketch: The KI-nAE simulator can simulate ENC queries by keeping track of the label for each key and returning it as the key tag along with a random string of the correct length for the encrypted plaintext. For DEC queries, the simulator can iterate through the list of key data given in the key vector $\mathbb{K}$ and check for malicious keys, for which it is directly given the secret key and can decrypt itself for any that have a key label matching the key tag. If there are no malicious keys or none that correctly decrypt, then the simulator returns $\perp$.

We bound the advantage of A with a sequence of game hops. We first transition to a game in which DEC keeps iterating through $\mathbb{K}$ if A provides a malicious key in $\mathbb{K}$ that was honestly generated by a call to GENHONESTKEY. We bound the ability of A to distinguish between these games by $mq/|\mathcal{K}|$, since there are at most m malicious keys and at most q honest keys and A can only at best guess one of the honest keys. We next transition to a game in which DEC skips any malicious key that can decrypt a ciphertext output from a call to ENC. We bound the ability of A to distinguish between these games by the FROB security of AEAD. Finally, we transition to a game in which ENC generates a random string as the encrypted plaintext and DEC skips honest keys in the key vector if they were not used to produce the queried ciphertext from a call to ENC. We bound the ability of the adversary to distinguish between these games by the MU-nAE security of AEAD. Since this last game guarantees that no malicious key can decrypt an honestly generated ciphertext and that no honest key can decrypt a malicious ciphertext, iterating through $\mathbb{K}$ in order in this game is identical to iterating through the honest keys first and then the malicious keys, proving our claim.   □

Using Unique Random Identifiers. Thus far, we have relied on an FROB AEAD scheme, due to the fact that KL allows duplicate key identifiers, which follows practice (e.g., the Tink library). One might instead suggest somehow enforcing uniqueness of key labels at the "application layer", and using a non-FROB AEAD. We argue below that this approach either fails to meet natural security goals or greatly increases the complexity of the application layer; thus, FROB AEADs are superior as the "base" AEAD for an KL-like construction.

To study this question, we first need to express application-level enforcement of unique key labels in our AEKI formalism. A simple way to do this is to have Dec check the identifiers for all keys in $\mathbb{K}$ and output $\bot$ if two different keys have the same identifier. This approach does not meet condition (3) of our correctness notion above. Any key vector with repeated labels will cause decryption to fail, even if it contains the correct key.

More subtly, this approach also fails to provide KI-FROB for the AEKI scheme. To see why, note that because decryption is stateless, uniqueness cannot be checked across different invocations. Thus, if the underlying AEAD is not FROB, an adversary can choose two keys, produce a key multi-collision ciphertext, assign both keys the same label that matches the ciphertext's key tag, and then put the keys in separate key vectors. Decryption will succeed for both key vectors, even though their keys have non-unique labels.

Preventing this attack and providing KI-FROB for the AEKI scheme requires stateful decryption: namely, the application must track all key identifier-secret key pairs seen across *all* decryption operations. This seems difficult to implement correctly and efficiently, and is certain to increase the complexity of the application. Thus, we believe it is better to use FROB AEAD to cryptographically guarantee KI-FROB security of AEKI.

Analyzing Trial Decryption. A special case of the key labels scheme is brute-force trial decryption, which we refer to as TD[AEAD]. This scheme simply assigns the empty string ε as the key label for all keys, meaning there are no key identifiers, and decryption must always trial decrypt for all keys in the key vector. The Tink library, for instance, allows keys to also have "raw" labels, which indicate they have no identifier.

Notably, the multi-user Shadowsocks protocol we describe in the introduction falls into this category. The attack described by Len et al. [22] is made possible by the fact that the Shadowsocks protocol uses a non-FROB AEAD scheme. The benefit of our KI-FROB definition is that it demonstrates that such a scheme is insecure when not using a FROB AEAD scheme.

While trial decryption still requires the use of an FROB AEAD scheme, it is key anonymous. In the below theorem, we show that trial decryption meets our stronger key anonymous encryption leakage model. We provide the full proof in the full version of this work.

Theorem 3. *Using* $\mathsf{L}^{\mathsf{anon}}_{\mathsf{Enc}}$, *let* $\mathcal{A}$ *be a KI-nAE-KA adversary making at most* q *queries to its oracles and querying at most* m *malicious keys. Then we give* KI nΛE-KA *simulator* S *and adversaries* $\mathcal{B}, \mathcal{C}$ *such that*

$$\mathbf{Adv}^{\mathsf{ki\text{-}nae\text{-}anon}}_{\mathsf{TD[AEAD]},\mathsf{S}}(\mathcal{A}) \leq \mathbf{Adv}^{\mathsf{frob}}_{\mathsf{AEAD}}(\mathcal{B}) + \mathbf{Adv}^{\mathsf{mu\text{-}nae}}_{\mathsf{AEAD}}(\mathcal{C}) + mq/|\mathcal{K}|.$$

Adversaries $\mathcal{B}, \mathcal{C}$ *run in time that of* $\mathcal{A}$ *with an* $\mathcal{O}(q)$ *overhead and simulator* S *runs in time* $\mathcal{O}(mq)$.

Proof Sketch: The KI-nAE simulator can simulate ENC queries by returning a random string of the correct length for the encrypted plaintext. For DEC queries, the simulator can iterate through the list of key data given in the key vector $\mathbb{K}$ and check for malicious keys, for which it is directly given the secret key and can decrypt itself. If there are no malicious keys or none that correctly decrypt, then the simulator returns $\perp$.

We bound the advantage of $\mathcal{A}$ with a sequence of game hops. We first transition to a game in which DEC keeps iterating through $\mathbb{K}$ if $\mathcal{A}$ provides a malicious key in $\mathbb{K}$ that was honestly generated by a call to GENHONESTKEY. We bound the ability of $\mathcal{A}$ to distinguish between these games by $mq/|\mathcal{K}|$, since there are at most m malicious keys and at most q honest keys and $\mathcal{A}$ can only at best guess one of the honest keys. We next transition to a game in which DEC skips any malicious key that can decrypt a ciphertext output from a call to ENC. We bound the ability of $\mathcal{A}$ to distinguish between these games by the FROB security of AEAD. Finally, we transition to a game in which ENC generates a random string as the encrypted plaintext and DEC skips honest keys in the key vector if they were not used to produce the queried ciphertext from a call to ENC. We bound the ability of the adversary to distinguish between these games by the MU-nAE security of AEAD. Since this last game guarantees that no malicious key can decrypt an honestly generated ciphertext and that no honest key can decrypt a malicious ciphertext, iterating through $\mathbb{K}$ in order in this game is identical to iterating through the honest keys first and then the malicious keys, proving our claim. □

While key labels are a simple and commonly used approach in practice for identifying keys, here we have shown that a formal analysis surfaces subtleties. In particular, key labels do not provide key anonymity, unless all keys have the same label as in the less efficient trial decryption-based scheme. Moreover, a key label approach must rely on the underlying AEAD scheme being FROB. In the next section, we explore a different tactic called static key identifiers, which computes the identifier from the key itself.

```
KCV.Kg(kid):                          KCV.Dec(𝕂, N, AD, T_k, C):

K ←$ 𝒦                                For K ∈ 𝕂:
Return (kid, K)                          (kid*, K*) ← K
                                         kcv* ← F_kcv(K*) ;  K_e ← KDF(K*)
KCV.Enc(K, N, AD, M):                    If kid*‖kcv* = T_k:
(kid, K*) ← K                                M ← AEAD.Dec(K_e, N, AD‖T_k, C)
kcv ← F_kcv(K*) ;  K_e ← KDF(K*)             If M ≠ ⊥: Return (K, M)
T_k ← kid‖kcv                         Return ⊥
C ← AEAD.Enc(K_e, N, AD‖T_k, M)
Return (T_k, C)
```

Fig. 7. A key check value scheme KCV parameterized by encryption scheme AEAD = (Kg, Enc, Dec) with associated key space $\mathcal{K} = \{0,1\}^k$, key check value function F_{kcv} : $\{0,1\}^\kappa \to \{0,1\}^n$, and encryption key derivation function KDF : $\{0,1\}^\kappa \to \{0,1\}^k$.

5   Static Key Identifiers

In this section we describe a class of AEAD-KI that uses what we call *static key identifiers*. This technique computes a static identifier from the key that along with the key label is used as the key tag. In practice, this static identifier is often referred to as a *key check value*, also known as a key checksum value. We formalize this approach as the AEAD-KI scheme KCV[AEAD, F_{kcv}, KDF], shown in Fig. 7. The scheme has key space $\{0,1\}^\kappa$ and is parameterized by AEAD scheme AEAD = (Kg, Enc, Dec) with associated key space $\mathcal{K} = \{0,1\}^k$, key check value function $\mathsf{F}_{kcv} : \{0,1\}^\kappa \to \{0,1\}^n$, and encryption key derivation function KDF : $\{0,1\}^\kappa \to \{0,1\}^k$. For simplicity, we will sometimes refer to the scheme as KCV when the parameters are obvious.

Key generation generates a secret key and attaches the input label kid as part of the AEAD-KI key. For static identifier schemes, the key tag is composed of both the key label and the key check value. This models what happens in practice, as schemes may use the key label as a way to locate a key or set of keys and then use the key check value as a commitment or integrity check of the key. For instance, AWS uses the Amazon Resource Name (ARN) as a URL for looking up keys, while an extra commitment string verifies this is the correct key [4].

Encryption derives the key check value using the function F_{kcv} and separately derives the AEAD secret key using the function KDF. Encryption adds the key tag to the authentication scope of AEAD by appending it to the authenticated data. Decryption iterates through the key vector to compute each key's identifier using F_{kcv} and find the first key with one matching T_k. If AEAD decryption with this key succeeds, then the corresponding decrypted plaintext is returned.

Key check values have been widely used in practice [2,4,17,26,30] to derive a value from the key, typically using a hash function or block cipher, that can then be used to confirm the integrity of or identify the key during decryption. These static key identifiers can be used in two ways: as static key hints or as static key commitments. Static key hints use a non-CR function to derive an

Application	Key Check Value	KH or KC?
AWS Encryption SDK [4]	$\mathsf{SHA256}(K\,\|\,\mathtt{0x436f6d6d69740102})$	KC
GlobalPlatform [18]	$\mathsf{msb}_{24}(\mathsf{AES}_K(([1]_8)^{16}))$	KH
Telegram [31]	$\mathsf{lsb}_{64}(\mathsf{SHA1}(K))$	KH
PKCS#11 [27]	$\mathsf{msb}_{24}(\mathsf{AES}_K(0^{128}))$	KH

Fig. 8. Key check value functions used in practice. We list whether each function is a key hint (KH) or key commitment (KC).

identifier from the key, meaning that key hints are not unique to a single key. Because they are not used to commit to a key, they can be short and efficiently computed, while still enabling AEAD-KI schemes to narrow down the scope of keys to check during decryption. For instance, one common technique is taking 24 bits from the AES evaluation of the key over some fixed string. However, key hints must be used with an FROB AEAD scheme in practice to guarantee robust key identification.

Conversely, static key commitments do commit to the key and therefore do not need to be used with an FROB AEAD scheme. In practice, this means key commitments must employ a CR key check value function, which can be more computationally intensive and require longer key tags. For instance, Albertini et al. [2] suggest a variant of this method as their "generic solution" for adding key commitment to AEAD schemes. Their Type I and II schemes in particular feature a static identifier by computing two SHA256 hashes over the key and using these values for the key identifier and AEAD encryption key. This scheme has been adopted as the default method of key identification by AWS [4]. We summarize some sample schemes used for both key hints and key commitments in Fig. 8. Later in this section we will show detailed security results for both types of static key identifiers.

Using a Key Derivation Function. Whenever a fixed value is computed from the key as a key tag and then composed with an AEAD scheme that uses the same key, there is the potential that the AEAD scheme uses the same value for its internal computation. Since the key tag is sent in the clear, this could lead to confidentiality or integrity vulnerabilities. Indeed, Iwata and Wang [20] have shown that this does happen in practice. They describe forgery attacks for several variants of CBC-MAC proposed by ISO/IEC 9797-1:2011 [19] when used with the key check value suggested by ANSI X9.24-1:2009 [3].

To simplify the analysis of composing $\mathsf{F}_{\mathsf{kcv}}$ with an AEAD scheme, we also use key derivation function KDF to derive an independent AEAD encryption key. In this section, we will show that KDF will need to be a CR PRF. For similar reasons, Albertini et al. also use a CR PRF to derive a separate AEAD key. This of course results in extra overhead and could be unnecessary if the key identifier is never used in the internal computation of AEAD. One could analyze specific key check value functions and AEAD schemes that can be safely used together without the need for a separate key derivation function; we leave this as an open problem.

5.1   Static Key Hint

Static key hints are a sub-class of static key identifiers which compute the key check value using a non-collision-resistant PRF. This means they can be short and efficiently computed, e.g. using a universal hash or by truncating the output of a hash function or block cipher. While key hints cannot be used to commit to a key, they can be used to narrow down the search space of the given key vector during decryption. However, in order to ensure that key robustness is achieved, these key hints must rely on using an FROB AEAD scheme, as we show below.

Analyzing Robustness. Here we show that static key hints rely on using an FROB AEAD scheme as well as collision-resistant function KDF to achieve KI-FROB.

Theorem 4. *Let $\mathcal{A}$ be an KI-FROB adversary for scheme $\mathsf{KCV}[\mathsf{AEAD},$ $\mathsf{F_{kcv}}, \mathsf{KDF}]$. Then we give adversaries $\mathcal{B}, \mathcal{C}$ such that*

$$\mathbf{Adv}^{\mathrm{ki\text{-}frob}}_{\mathsf{KCV}[\mathsf{AEAD},\mathsf{F_{kcv}},\mathsf{KDF}]}(\mathcal{A}) \leq \mathbf{Adv}^{cr}_{\mathsf{KDF}}(\mathcal{B}) + \mathbf{Adv}^{\mathrm{frob}}_{\mathrm{AEAD}}(\mathcal{C}).$$

Adversaries $\mathcal{B}, \mathcal{C}$ run in time that of $\mathcal{A}$.

Proof. We prove the theorem using a sequence of game hops. Let game G_0 be the game FROB with the call to the decryption algorithm Dec replaced by the pseudocode for $\mathsf{KCV}[\mathsf{AEAD}, \mathsf{F_{kcv}}, \mathsf{KDF}].\mathsf{Dec}()$. Next we transition to game G_1, which is identical to G_0 except that when KDF is called to derive the encryption key from the keys in $\mathbb{K}_1$, it checks if there was some other different key K^* prior to this call that output to the same encryption key derived from the keys in $\mathbb{K}_0$. If this happens, then G_1 will return 0.

We can upper bound the difference in advantage of $\mathcal{A}$ in G_0 and G_1 by the probability that KDF finds a collision. We then provide the CR adversary $\mathcal{B}$ such that its advantage in the CR game for KDF upper bounds this probability. $\mathcal{B}$ runs $\mathcal{A}$, which returns $\mathbb{K}_0, \mathbb{K}_1, N, AD, T_k, C$. $\mathcal{B}$ then checks if there is some key $K_0 \in \mathbb{K}_0$ such that $(\mathsf{kid}_0, K_0^*) \leftarrow K_0$ and $K_e \leftarrow \mathsf{KDF}(K_0^*)$ and some key $K_1 \in \mathbb{K}_1$ such that $(\mathsf{kid}_1, K_1^*) \leftarrow K_1$ and $K_e \leftarrow \mathsf{KDF}(K_1^*)$, and returns K_0^*, K_1^*. Notice that whenever KDF finds the collision in G_1, then $\mathcal{B}$ wins the CR game for KDF.

Finally, we upper bound the advantage of $\mathcal{A}$ in game G_1 by the advantage of the following FROB adversary $\mathcal{C}$ against AEAD. $\mathcal{C}$ runs $\mathcal{A}$, which returns $\mathbb{K}_0, \mathbb{K}_1, N, AD, T_k, C$. Let $\mathsf{Dec}(\mathbb{K}_0, N, AD, T_k, C)$ and $\mathsf{Dec}(\mathbb{K}_1, N, AD, T_k, C)$ return (K_0, M_0) and (K_1, M_1), respectively. Also let $(\mathsf{kid}_0, K_0^*) \leftarrow K_0$ and $(\mathsf{kid}_1, K_1^*) \leftarrow K_1$. Note that $\mathcal{A}$ can only win game G_1 if $(K_0, M_0) \neq \perp$ and $(K_1, M_1) \neq \perp$ and $K_0^* \neq K_1^*$, so let this be the case. $\mathcal{C}$ can return K_0^*, K_1^*, N, AD, C, where both keys successfully decrypt N, AD, C. Thus, $\mathcal{C}$ wins FROB for AEAD when $\mathcal{A}$ wins G_1 for KCV. $\square$

Analyzing KI-nAE. We now show that $\mathsf{KCV}[\mathsf{AEAD}, \mathsf{F}_{\mathsf{kcv}}, \mathsf{KDF}]$ is KI-nAE secure for leakage algorithm $\mathsf{L}_{\mathsf{Enc}}^{\mathsf{id}}$ when AEAD is both FROB and MU-nAE secure, $\mathsf{F}_{\mathsf{kcv}}$ and KDF are multi-user PRFs, and KDF is pre-image resistant. Notably, this means that encryption is not key anonymous. We provide the theorem statement and proof sketch below. We show the full proof in the full version of this work.

Theorem 5. *Using* $\mathsf{L}_{\mathsf{Enc}}^{\mathsf{id}}$, *let* $\mathcal{A}$ *be a* KI-nAE *adversary making at most* q *queries to its oracles, of which* q_k *are to* GENHONESTKEY, *and querying at most* m *malicious keys. Then we give KI-nAE simulator* S *and adversaries* $\mathcal{B}, \mathcal{C}, \mathcal{D}, \mathcal{E}$ *such that*

$$\mathbf{Adv}_{\mathsf{KCV}[\mathsf{AEAD}, \mathsf{F}_{\mathsf{kcv}}, \mathsf{KDF}], \mathsf{S}}^{\mathrm{ki\text{-}nae}}(\mathcal{A}) \leq \mathbf{Adv}_{\mathsf{F}_{\mathsf{kcv}}, \mathsf{KDF}}^{\mathrm{mu\text{-}prf}}(\mathcal{B}) + q_k \cdot \mathbf{Adv}_{\mathsf{KDF}}^{\mathrm{pre}}(\mathcal{C}) + \mathbf{Adv}_{\mathsf{AEAD}}^{\mathrm{frob}}(\mathcal{D})$$

$$+ \mathbf{Adv}_{\mathsf{AEAD}}^{\mathrm{mu\text{-}nae}}(\mathcal{E}) + \frac{q_k^2}{2^{\kappa+1}}.$$

$\mathcal{B}, \mathcal{C}, \mathcal{D}, \mathcal{E}$ *run in time that of* $\mathcal{A}$ *with* $\mathcal{O}(q)$ *overhead and* S *runs in time* $\mathcal{O}(mq)$.

Proof Sketch: The KI-nAE simulator can simulate ENC queries by keeping track of the label for each key and generating a random n-bit string as the key check value kcv. It can then return the label appended with the key check value as the key tag along with a random string of the correct length for the encrypted plaintext. For DEC queries, the simulator can iterate through the list of key data given in the key vector $\mathbb{K}$ and check for malicious keys, for which it is directly given the secret key and can decrypt itself for any that have a matching key tag. If there are no malicious keys or none that correctly decrypt, then the simulator returns $\perp$.

We bound the advantage of $\mathcal{A}$ with a sequence of game hops. We first transition to a game in which calls to $\mathsf{F}_{\mathsf{kcv}}$ and KDF for honest keys are replaced with calls to random functions. We bound the ability of $\mathcal{A}$ to distinguish these games by the MU-PRF security of $\mathsf{F}_{\mathsf{kcv}}$ and KDF. We next transition to a game in which malicious keys in $\mathbb{K}$ queried to DEC are skipped if for KDF they are the pre-image of some honestly generated AEAD encryption key K_e computed in GENHONESTKEY. We bound the ability of $\mathcal{A}$ to distinguish these games by the pre-image resistance security of KDF, multiplied by a factor of q_k. Then we transition to a game in which malicious keys in $\mathbb{K}$ queried to DEC are skipped if they can decrypt some honestly generated ciphertext output by GENHONESTKEY. We bound the ability of $\mathcal{A}$ to distinguish between these games by the FROB security of AEAD. We then transition to a game in which we eliminate collisions when key K is chosen at random from the key space $\mathcal{K}$, for which we use the birthday bound $q_k^2/2^{\kappa+1}$ to bound. Finally, we transition to a game in which ENC generates a random string as the encrypted plaintext and DEC skips honest keys in the key vector if they were not used to produce the queried ciphertext from a call to ENC. We bound the distinguishing advantage by the MU-nAE security of AEAD. Since this last game guarantees that no malicious key can decrypt an honestly generated ciphertext and that no honest key can decrypt a malicious ciphertext, iterating through $\mathbb{K}$ in order in this game is identical to iterating through the honest keys first and then the malicious keys, proving our claim. $\square$

5.2   Static Key Commitment

Static key commitments are the second subclass of static key identifiers. They compute the key check value using using a collision-resistant PRF. While this means they must be longer and less efficient than key hints, they can commit to the key, and thus can be used with non-FROB AEADs, as we now prove.

Theorem 6. *Let $\mathcal{A}$ be a* KI-FROB *adversary for* $\mathsf{KCV}[\mathsf{AEAD}, \mathsf{F_{kcv}}, \mathsf{KDF}]$. *Then we give CR adversary $\mathcal{B}$, running in time that of $\mathcal{A}$, for $\mathsf{F_{kcv}}$ such that*

$$\mathbf{Adv}_{\mathsf{KCV}[\mathsf{AEAD}, \mathsf{F_{kcv}}, \mathsf{KDF}]}^{\mathrm{ki\text{-}frob}}(\mathcal{A}) \leq \mathbf{Adv}_{\mathsf{F_{kcv}}}^{cr}(\mathcal{B}).$$

Proof. We construct adversary $\mathcal{B}$ as follows. It runs $\mathcal{A}$, which returns $\mathbb{K}_0, \mathbb{K}_1, N$, AD, T_k, C. Let the values returned by decryption of the ciphertext for each key vector be $(K_0, M_0), (K_1, M_1)$. Also let $(\mathsf{kid}_0, K_0^*) \leftarrow K_0$ and $(\mathsf{kid}_1, K_1^*) \leftarrow K_1$. We know that $(K_0, M_0), (K_1, M_1) \neq \perp$ and $K_0^* \neq K_1^*$ for $\mathcal{A}$ to win. This also means that $\mathsf{F_{kcv}}(K_0^*) = \mathsf{F_{kcv}}(K_1^*)$. $\mathcal{B}$ can then return K_0, K_1 as a collision for $\mathsf{F_{kcv}}$.   □

Analyzing KI-nAE. While Albertini et al. do not explicitly prove security for this scheme, they claim that it meets their real-or-random AE security definition. However, any encryption scheme that attaches a fixed string to a ciphertext trivially cannot meet this definition. The benefit of our KI-nAE definition is that it captures security for non-key anonymous schemes. Indeed, our result shows that if $\mathsf{F_{kcv}}$ is a pre-image resistant multi-user PRF, KDF is a multi-user PRF, and AEAD is MU-nAE-secure, then KCV is KI-nAE-secure, for leakage $\mathsf{L_{Enc}^{id}}$. We provide the theorem statement and proof sketch here; the full proof is in the full version of this work.

Theorem 7. *Using $\mathsf{L_{Enc}^{id}}$, let $\mathcal{A}$ be a* KI-nAE *adversary making at most q queries to its oracles, of which q_k are to* GenHonestKey, *and querying at most m malicious keys. Then we give adversaries $\mathcal{B}, \mathcal{C}, \mathcal{D}$ such that*

$$\mathbf{Adv}_{\mathsf{KCV}[\mathsf{AEAD}, \mathsf{F_{kcv}}, \mathsf{KDF}], \mathsf{S}}^{\mathrm{ki\text{-}nae}}(\mathcal{A}) \leq \mathbf{Adv}_{\mathsf{F_{kcv}}, \mathsf{KDF}}^{\mathrm{mu\text{-}prf}}(\mathcal{B}) + q_k \cdot \mathbf{Adv}_{\mathsf{F_{kcv}}}^{\mathrm{pre}}(\mathcal{C})$$

$$+ \mathbf{Adv}_{\mathrm{AEAD}}^{\mathrm{mu\text{-}nae}}(\mathcal{D}) + \frac{q_k^2}{2^{\kappa+1}}.$$

$\mathcal{B}, \mathcal{C}, \mathcal{D}$ *run in time that of $\mathcal{A}$ with a $\mathcal{O}(q)$ overhead and S runs in time $\mathcal{O}(mq)$.*

Proof Sketch: The proof uses the same KI-nAE simulator as that for Theorem 5. We again bound the advantage of $\mathcal{A}$ with a sequence of game hops. We first transition to a game in which calls to $\mathsf{F_{kcv}}$ and KDF for honest keys are replaced with calls to random functions. We bound the ability of $\mathcal{A}$ to distinguish these games by the MU-PRF security of $\mathsf{F_{kcv}}$ and KDF. We next transition to a game in which malicious keys in $\mathbb{K}$ queried to Dec are skipped if for $\mathsf{F_{kcv}}$ they are the pre-image of some honestly generated key check value kcv computed in GenHonestKey. We bound the ability of $\mathcal{A}$ to distinguish these games by the pre-image resistance security of $\mathsf{F_{kcv}}$, multiplied by a factor of q_k. We then transition to a game

nKCV.Kg(kid):	nKCV.Dec($\mathbb{K}, N, AD, T_k, C$):
$K \leftarrow\!\!\$\ \mathcal{K}$	$(N_0, N_1) \leftarrow N$
Return (ε, K)	For $K \in \mathbb{K}$:
	$\quad (\varepsilon, K^*) \leftarrow K$
nKCV.Enc(K, N, AD, M):	$\quad \text{kcv}^* \leftarrow F_{\text{kcv}}(K^*, N_0)\,;\ K_e \leftarrow \text{KDF}(K^*)$
$(\varepsilon, K^*) \leftarrow K\,;\ (N_0, N_1) \leftarrow N$	$\quad$ If $\text{kcv}^* = T_k$:
$\text{kcv} \leftarrow F_{\text{kcv}}(K^*, N_0)\,;\ K_e \leftarrow \text{KDF}(K^*)$	$\quad\quad M \leftarrow \text{AEAD.Dec}(K_e, N_1, AD \| T_k, C)$
$T_k \leftarrow \text{kcv}$	$\quad\quad$ If $M \neq \bot$: Return (K, M)
$C \leftarrow \text{AEAD.Enc}(K_e, N_1, AD \| T_k, M)$	Return $\bot$
Return (T_k, C)	

Fig. 9. A nonce-based key check value scheme nKCV parameterized by AEAD $=$ (Kg, Enc, Dec) with key space $\mathcal{K} = \{0,1\}^k$; key check value function $F_{\text{kcv}} : \{0,1\}^\kappa \times \{0,1\}^r \to \{0,1\}^n$, and encryption key derivation function KDF $: \{0,1\}^\kappa \to \{0,1\}^k$.

in which we eliminate collisions when key K is chosen at random from the key space $\mathcal{K}$, for which we use the birthday bound $q_k^2/2^{\kappa+1}$ to bound. Finally, we transition to a game in which ENC generates a random string as the encrypted plaintext and DEC skips honest keys in the key vector if they were not used to produce the queried ciphertext from a call to ENC. We bound the distinguishing advantage by the MU-nAE security of AEAD. Since this last game guarantees that no malicious key can decrypt an honestly generated ciphertext and that no honest key can decrypt a malicious ciphertext, iterating through $\mathbb{K}$ in order in this game is identical to iterating through the honest keys first and then the malicious keys, proving our claim. □

While static key identifiers are versatile in that they can be used either as key hints or key commitments, they unfortunately do not provide key anonymity. Next, we will see how dynamic identifiers enable anonymous key identification.

6   Dynamic Key Identifiers

In this section we describe the key anonymous counterpart to static key identifiers, which we call *dynamic key identifiers*. A dynamic identifier is computed from the secret key during encryption using part of the input nonce. Unlike the static identifier approach, this scheme cannot use key labels as part of the key tag because key labels are fixed for a key and would therefore break key anonymity. We formalize dynamic key identifiers as an AEAD-KI scheme with nKCV[AEAD, F_{kcv}, KDF], shown in Fig. 9. The scheme has key space $\{0,1\}^\kappa$ and is parameterized by encryption scheme AEAD $=$ (Kg, Enc, Dec) with associated key space $\mathcal{K} = \{0,1\}^k$, key check value function $F_{\text{kcv}} : \{0,1\}^\kappa \times \{0,1\}^r \to \{0,1\}^n$, and encryption key derivation function KDF $: \{0,1\}^\kappa \to \{0,1\}^k$. For simplicity, we may refer to the scheme as nKCV when the parameters are obvious.

Encryption now takes in a nonce for which one part is used to derive the key check value and the other is used in the AEAD computation. These nonces do not need to be distinct. Encryption derives the key check value and AEAD key from the secret key using the functions F_{kcv} and KDF, respectively. The key check value kcv is computed on part of the nonce so that it changes for each encryption call. Meanwhile, the AEAD key is computed on just the secret key, meaning the AEAD key is fixed for each nKCV secret key. Unlike for KCV, here the key check value by itself forms the key tag. Encryption adds the key tag to the authentication scope of AEAD by appending it to the authenticated data. Decryption iterates through the key vector to compute each key's identifier using F_{kcv} and find the first key with one matching T_k.

6.1 Dynamic Key Hint

Dynamic key hints are a subclass of dynamic key identifiers which compute the key check value using using a non-collision-resistant PRF. Similar to static key hints, they can be short and more efficiently computed than key commitments. They are useful for narrowing down the search space of the given key vector during decryption. However, in order to ensure that key robustness is achieved, these key hints must rely on using an FROB AEAD scheme, as we show below.

Analyzing Robustness. Here we show that dynamic key hints rely on using a CR function KDF and an FROB AEAD scheme to achieve KI-FROB.

Theorem 8. *Let $\mathcal{A}$ be a KI-FROB adversary for scheme nKCV[AEAD, F_{kcv}, KDF]. Then we give adversaries $\mathcal{B}, \mathcal{C}$ running in time that of $\mathcal{A}$ such that*

$$\mathbf{Adv}^{ki\text{-}frob}_{nKCV[AEAD,F_{kcv},KDF]}(\mathcal{A}) \leq \mathbf{Adv}^{cr}_{KDF}(\mathcal{B}) + \mathbf{Adv}^{frob}_{AEAD}(\mathcal{C}).$$

Proof. We prove the theorem using a sequence of game hops. Let game G_0 be the game KI-FROB with the call to the decryption algorithm Dec replaced by the pseudocode for nKCV[AEAD, F_{kcv}, KDF].Dec(). Next we transition to game G_1, which is identical to G_0 except that when KDF is called to derive the encryption key from the keys in $\mathbb{K}_1$, it checks if there was some other different key K^* prior to this call that output to the same encryption key derived from the keys in $\mathbb{K}_0$. If this happens, then G_1 will return 0.

We can upper bound the difference in advantage of $\mathcal{A}$ in G_0 and G_1 by the probability that KDF finds a collision. We then provide the CR adversary $\mathcal{B}$ such that its advantage in the CR game for KDF upper bounds this probability. $\mathcal{B}$ runs $\mathcal{A}$, which returns $\mathbb{K}_0, \mathbb{K}_1, N, AD, T_k, C$. $\mathcal{B}$ then checks if there is some key $K_0 \in \mathbb{K}_0$ such that $(kid_0, K_0^*) \leftarrow K_0$ and $K_e \leftarrow KDF(K_0^*)$ and some key $K_1 \in \mathbb{K}_1$ such that $(kid_1, K_1^*) \leftarrow K_1$ and $K_e \leftarrow KDF(K_1^*)$, and returns K_0^*, K_1^*. Notice that whenever KDF finds the collision in G_1, then $\mathcal{B}$ wins the CR game for KDF.

Finally, we upper bound the advantage of $\mathcal{A}$ in game G_1 by the advantage of the following FROB adversary $\mathcal{C}$ against AEAD. $\mathcal{C}$ runs $\mathcal{A}$, which returns $\mathbb{K}_0, \mathbb{K}_1, N, AD, T_k, C$. Let $Dec(\mathbb{K}_0, N, AD, T_k, C)$ and $Dec(\mathbb{K}_1, N, AD, T_k, C)$ return (K_0, M_0) and (K_1, M_1), respectively. Also let $(\varepsilon, K_0^*) \leftarrow K_0$, $(\varepsilon, K_1^*) \leftarrow K_1$, and $N_0 \| N_1 \leftarrow N$. Note that $\mathcal{A}$ can only win game G_1 if $(K_0, M_0) \neq \bot$ and $(K_1, M_1) \neq \bot$ and $K_0^* \neq K_1^*$, so let this be the case. $\mathcal{C}$ can return K_0^*, K_1^*, N_1, AD, C, where again both keys decrypt N_1, AD, C. Thus, $\mathcal{C}$ wins FROB for AEAD when $\mathcal{A}$ wins G_1 for nKCV. □

Analyzing KI-nAE. We now show that nKCV[AEAD, $\mathsf{F_{kcv}}$, KDF] is KI-nAE-KA secure for leakage algorithm $\mathsf{L_{Enc}^{anon}}$ when AEAD is both FROB and MU-nAE secure, KDF is a pre-image resistant multi-user PRF, and $\mathsf{F_{kcv}}$ is a multi-user PRF. Notably, this means that encryption is key anonymous. We assume that the adversary $\mathcal{A}$ is a nonce-respecting adversary that never queries the same N_0 or N_1 to ENC. We provide the theorem statement and proof sketch below. The proof is provided in the full version of this work.

Theorem 9. *Using* $\mathsf{L_{Enc}^{anon}}$, *let* $\mathcal{A}$ *be a* KI-nAE *adversary making at most q queries to its oracles, of which q_k are to* GENHONESTKEY, *and querying at most m malicious keys. Then we give adversaries* $\mathcal{B}, \mathcal{C}, \mathcal{D}, \mathcal{E}$ *such that*

$$\mathbf{Adv}_{\mathsf{nKCV[AEAD,F_{kcv},KDF]},S}^{\mathsf{ki\text{-}nae\text{-}anon}}(\mathcal{A}) \leq \mathbf{Adv}_{\mathsf{F_{kcv}},\mathsf{KDF}}^{\mathsf{mu\text{-}prf}}(\mathcal{B}) + q_k \cdot \mathbf{Adv}_{\mathsf{KDF}}^{\mathsf{pre}}(\mathcal{C}) + \mathbf{Adv}_{\mathsf{AEAD}}^{\mathsf{frob}}(\mathcal{D})$$

$$+ \mathbf{Adv}_{\mathsf{AEAD}}^{\mathsf{mu\text{-}nae}}(\mathcal{E}) + \frac{q_k^2}{2^{\kappa+1}}.$$

$\mathcal{B}, \mathcal{C}, \mathcal{D}, \mathcal{E}$ *run in time that of* $\mathcal{A}$ *with a* $\mathcal{O}(q)$ *overhead and* S *runs in time* $\mathcal{O}(mq)$.

Proof Sketch: The KI-nAE simulator can simulate ENC queries by generating a random n-bit string as the key check value kcv. It can then return the key check value as the key tag along with a random string of the correct length for the encrypted plaintext. For DEC queries, the simulator can iterate through the list of key data given in the key vector $\mathbb{K}$ and check for malicious keys, for which it is directly given the secret key and can decrypt itself for any that have a matching key tag. If there are no malicious keys or none that correctly decrypt, then the simulator returns $\bot$.

We bound the advantage of $\mathcal{A}$ with a sequence of game hops. We first transition to a game in which calls to $\mathsf{F_{kcv}}$ and KDF for honest keys are replaced with calls to random functions. We bound the ability of $\mathcal{A}$ to distinguish these games by the MU-PRF security of $\mathsf{F_{kcv}}$ and KDF. We next transition to a game in which malicious keys in $\mathbb{K}$ queried to DEC are skipped if for KDF they are the pre-image of some honestly generated AEAD encryption key K_e computed in GENHONESTKEY. We bound the ability of $\mathcal{A}$ to distinguish these games by the pre-image resistance security of KDF, multiplied by a factor of q_k. Then we transition to a game in which malicious keys in $\mathbb{K}$ queried to DEC are skipped if they can decrypt some honestly generated ciphertext output by GENHONESTKEY. We bound the ability of $\mathcal{A}$ to distinguish between these games by the FROB security of AEAD. We then transition to a game in which we eliminate collisions when key K is chosen at random from the key space $\mathcal{K}$, for which we use the birthday bound $q_k^2/2^{\kappa+1}$ to bound. Finally, we transition to a game in which ENC generates a random string as the encrypted plaintext and DEC skips honest keys in the key vector if they were not used to produce the queried ciphertext from a call to ENC. We bound the distinguishing advantage by the MU-nAE security of AEAD. Since this last game guarantees that no malicious key can decrypt an honestly generated ciphertext and that no honest key can decrypt a malicious ciphertext, iterating through $\mathbb{K}$ in order in this game is identical to iterating through the honest keys first and then the malicious keys, proving our claim.   □

6.2   Dynamic Key Commitment

Dynamic key commitments are the second subclass of dynamic key identifiers; they instead compute the key check value using using a collision-resistant PRF. While this means they must be longer and less efficient than their key hint counterpart, they can be used to commit to the key. This also means that they can be used with non-FROB AEAD schemes. Dynamic key commitments parallel the Type III generic construction from Albertini et al. [2]. They are also similar to the UtC transform proposed by Bellare and Hoang [8], although this scheme is only considered in the traditional single-key setting. The UtC transform uses a committing PRF that takes as input a key and nonce and outputs pair (P, L) such that P is a string that commits to the key. L is then used as the encryption key. We note that a committing PRF can be used in place of KDF and F_{kcv}, although our formalism allows for analyzing the security requirements for deriving the key tag separately from deriving the key.

Furthermore, the NonceWrap scheme proposed by Chan and Rogaway [12] can be considered a type of dynamic key commitment scheme. Their scheme encrypts the ciphertext as $C = \mathrm{AES}(K_1, N \| 0^3 2) \| \mathrm{AES\text{-}GCM}(K_2, N, AD, M)$, where the first string is a 128-bit "header". During decryption, the correct key is found from a set of possible keys by re-computing the header and verifying the 32-bit all-zeros string remains intact. Interestingly, this scheme may be considered a key commitment scheme when the nonce must be specified along with the ciphertext, as in our formalization of KI-FROB. However, if, as the setting in this work intends, the nonce does not need to be specified, then this scheme does not meet KI-FROB and a key-committing AEAD should be used instead.

Analyzing Robustness. Here we show that dynamic key commitments rely only on the collision-resistance of the function F_{kcv} to achieve KI-FROB. In particular, this means that AEAD does not in fact have to be FROB.

Theorem 10. *Let $\mathcal{A}$ be a KI-FROB adversary for $\mathrm{KCV}[\mathrm{AEAD}, F_{kcv}, \mathrm{KDF}]$. Then we give CR adversary $\mathcal{B}$, running in time that of $\mathcal{A}$, for F_{kcv} such that*

$$\mathbf{Adv}^{\mathrm{ki\text{-}frob}}_{\mathrm{nKCV}[\mathrm{AEAD}, F_{kcv}, \mathrm{KDF}]}(\mathcal{A}) \leq \mathbf{Adv}^{cr}_{F_{kcv}}(\mathcal{B}).$$

$\mathcal{B}$ runs in time that of $\mathcal{A}$.

Proof. We construct CR adversary $\mathcal{B}$ against F_{kcv} as follows. $\mathcal{B}$ runs $\mathcal{A}$, which returns $\mathbb{K}_0, \mathbb{K}_1, N, AD, T_k, C$. Let $\mathrm{Dec}(\mathbb{K}_0, N, AD, T_k, C)$ and $\mathrm{Dec}(\mathbb{K}_1, N, AD, T_k, C)$ return (K_0, M_0) and (K_1, M_1), respectively. Also let $(\varepsilon, K_0^*) \leftarrow K_0$, $(\varepsilon, K_1^*) \leftarrow K_1$, and $N_0 \| N_1 \leftarrow N$. We know that $(K_0, M_0) \neq \bot$ and $(K_1, M_1) \neq \bot$ and $K_0^* \neq K_1^*$ for $\mathcal{A}$ to win. $\mathcal{B}$ can then return $(K_0^*, N_0), (K_1^*, N_0)$

as a collision for F_{kcv} since $F_{kcv}(K_0^*, N_0) = F_{kcv}(K_1^*, N_0) = T_k$. We therefore have that $\mathcal{B}$ wins game CR for F_{kcv} when $\mathcal{A}$ wins game KI-FROB for nKCV.

Analyzing KI-nAE. We now show that $nKCV[AEAD, F_{kcv}, KDF]$ is KI-nAE-KA secure for leakage algorithm L_{Enc}^{anon} when AEAD MU-nAE secure, KDF is a multi-user PRF, and F_{kcv} is a CR multi-user PRF. Again, this means that encryption is key anonymous. We assume that the adversary $\mathcal{A}$ is a nonce-respecting adversary that never queries the same N_0 or N_1 across queries to ENC. We provide the theorem statement below; the full proof is provided in the full version of this work.

Theorem 11. *Using* L_{Enc}^{anon}, *let* $\mathcal{A}$ *be a KI-nAE adversary making at most* q *queries to its oracles, of which* q_k *are to* GENHONESTKEY *and* q_e *are to* ENC, *and querying at most* m *malicious keys. Then we give adversaries* $\mathcal{B}, \mathcal{C}, \mathcal{D}$ *such that*

$$\mathbf{Adv}_{nKCV[AEAD, F_{kcv}, KDF], S}^{ki\text{-}nae\text{-}anon}(\mathcal{A}) \leq \mathbf{Adv}_{F_{kcv}, KDF}^{mu\text{-}prf}(\mathcal{B}) + q_e \cdot \mathbf{Adv}_{F_{kcv}}^{pre}(\mathcal{C})$$

$$+ \mathbf{Adv}_{AEAD}^{mu\text{-}nae}(\mathcal{D}) + \frac{q_k^2}{2^{\kappa+1}}.$$

$\mathcal{B}, \mathcal{C}, \mathcal{D}$ *run in time that of* $\mathcal{A}$ *with a* $\mathcal{O}(q)$ *overhead and* S *runs in time* $\mathcal{O}(mq)$.

Proof Sketch: The proof uses the same KI-nAE simulator as that for Theorem 9. We again bound the advantage of $\mathcal{A}$ with a sequence of game hops. We first transition to a game in which calls to F_{kcv} and KDF for honest keys are replaced with calls to random functions. We bound the ability of $\mathcal{A}$ to distinguish these games by the MU-PRF security of F_{kcv} and KDF. We next transition to a game in which malicious keys in $\mathbb{K}$ queried to DEC are skipped if for F_{kcv} they are the pre-image of some honestly generated key check value kcv computed in ENC. We bound the ability of $\mathcal{A}$ to distinguish these games by the pre-image resistance security of F_{kcv}, multiplied by a factor of q_e. We then transition to a game in which we eliminate collisions when key K is chosen at random from the key space $\mathcal{K}$, for which we use the birthday bound $q_k^2/2^{\kappa+1}$ to bound. Finally, we transition to a game in which ENC generates a random string as the encrypted plaintext and DEC skips honest keys in the key vector if they were not used to produce the queried ciphertext from a call to ENC. We bound the distinguishing advantage by the MU-nAE security of AEAD. Since this last game guarantees that no malicious key can decrypt an honestly generated ciphertext and that no honest key can decrypt a malicious ciphertext, iterating through $\mathbb{K}$ in order in this game is identical to iterating through the honest keys first and then the malicious keys, proving our claim. □

Acknowledgments. The authors thank Mihir Bellare for suggesting an improved correctness notion and various improvements in security definitions, as well as other helpful feedback on an early draft of the paper. The authors also thank Ian Miers and Nirvan Tyagi for their help in the early stages of this project. Finally, the authors are grateful to the anonymous reviewers of Asiacrypt 2022 for their feedback and suggestions. This work was supported in part by NSF grant CNS-2120651 and the NSF Graduate Research Fellowship under Grant No. DGE-2139899.

References

1. Abdalla, M., Bellare, M., Neven, G.: Robust encryption. In: Micciancio, D. (ed.) TCC 2010. LNCS, vol. 5978, pp. 480–497. Springer, Heidelberg (2010). https://doi.org/10.1007/978-3-642-11799-2_28
2. Albertini, A., Duong, T., Gueron, S., Kölbl, S., Luykx, A., Schmieg, S.: How to abuse and fix authenticated encryption without key commitment. In: USENIX Security (2022)
3. ANSI: Retail financial services symmetric key management Part 1: Using symmetric techniques. Standard, ANSI X9.24-1:2009 (2009)
4. Improved client-side encryption: explicit KeyIds and key commitment (2020). https://aws.amazon.com/blogs/security/improved-client-side-encryption-explicit-keyids-and-key-commitment/
5. Amazon Web Services Key Management Service. https://aws.amazon.com/kms/
6. Bellare, M., Boldyreva, A., Micali, S.: Public-key encryption in a multi-user setting: security proofs and improvements. In: Preneel, B. (ed.) EUROCRYPT 2000. LNCS, vol. 1807, pp. 259–274. Springer, Heidelberg (2000). https://doi.org/10.1007/3-540-45539-6_18
7. Bellare, M., Canetti, R., Krawczyk, H.: Pseudorandom functions revisited: the cascade construction and its concrete security. In: Proceedings of 37th Conference on Foundations of Computer Science, pp. 514–523. IEEE (1996)
8. Bellare, M., Hoang, V.T.: Efficient schemes for committing authenticated encryption. In: Dunkelman, O., Dziembowski, S. (eds.) EUROCRYPT 2022. LNCS, vol. 13276, pp. 845–875. Springer, Cham (2022)
9. Bellare, M., Ng, R., Tackmann, B.: Nonces are noticed: AEAD revisited. In: Boldyreva, A., Micciancio, D. (eds.) CRYPTO 2019. LNCS, vol. 11692, pp. 235–265. Springer, Cham (2019). https://doi.org/10.1007/978-3-030-26948-7_9
10. Bellare, M., Rogaway, P.: The security of triple encryption and a framework for code-based game-playing proofs. In: Vaudenay, S. (ed.) EUROCRYPT 2006. LNCS, vol. 4004, pp. 409–426. Springer, Heidelberg (2006). https://doi.org/10.1007/11761679_25
11. Bellare, M., Tackmann, B.: The multi-user security of authenticated encryption: AES-GCM in TLS 1.3. In: Robshaw, M., Katz, J. (eds.) CRYPTO 2016. LNCS, vol. 9814, pp. 247–276. Springer, Heidelberg (2016). https://doi.org/10.1007/978-3-662-53018-4_10
12. Chan, J., Rogaway, P.: Anonymous AE. In: Galbraith, S.D., Moriai, S. (eds.) ASIACRYPT 2019. LNCS, vol. 11922, pp. 183–208. Springer, Cham (2019). https://doi.org/10.1007/978-3-030-34621-8_7
13. Degabriele, J.P., Karadžić, V., Melloni, A., Münch, J.P., Stam, M.: Rugged pseudorandom permutations and their applications (2022). https://rwc.iacr.org/2022/program.php. Real World Crypto
14. Dodis, Y., Grubbs, P., Ristenpart, T., Woodage, J.: Fast message franking: from invisible salamanders to encryptment. In: Shacham, H., Boldyreva, A. (eds.) CRYPTO 2018. LNCS, vol. 10991, pp. 155–186. Springer, Cham (2018). https://doi.org/10.1007/978-3-319-96884-1_6
15. Farshim, P., Libert, B., Paterson, K.G., Quaglia, E.A.: Robust encryption, revisited. In: Kurosawa, K., Hanaoka, G. (eds.) PKC 2013. LNCS, vol. 7778, pp. 352–368. Springer, Heidelberg (2013). https://doi.org/10.1007/978-3-642-36362-7_22
16. Farshim, P., Orlandi, C., Rosie, R.: Security of symmetric primitives under incorrect usage of keys. IACR Trans. Symmetric Cryptology (2017)

17. GlobalPlatform Technology Card Specification Version 2.3.1. Standard, GlobalPlatform (2018). https://globalplatform.org/wp-content/uploads/2018/05/GPC_CardSpecification_v2.3.1_PublicRelease_CC.pdf
18. Grubbs, P., Lu, J., Ristenpart, T.: Message franking via committing authenticated encryption. In: Katz, J., Shacham, H. (eds.) CRYPTO 2017. LNCS, vol. 10403, pp. 66–97. Springer, Cham (2017). https://doi.org/10.1007/978-3-319-63697-9_3
19. ISO/IEC: Information technology - security techniques - message authentication codes (MACs) - part 1: Mechanisms using a block cipher. Standard, ISO/IEC 9797–1:2011 (2011)
20. Iwata, T., Wang, L.: Impact of ANSI X9.24-1:2009 key check value on ISO/IEC 9797-1:2011 MACs. In: Cid, C., Rechberger, C. (eds.) FSE 2014. LNCS, vol. 8540, pp. 303–322. Springer, Heidelberg (2015). https://doi.org/10.1007/978-3-662-46706-0_16
21. Jaeger, J., Tyagi, N.: Handling adaptive compromise for practical encryption schemes. In: Micciancio, D., Ristenpart, T. (eds.) CRYPTO 2020. LNCS, vol. 12170, pp. 3–32. Springer, Cham (2020). https://doi.org/10.1007/978-3-030-56784-2_1
22. Len, J., Grubbs, P., Ristenpart, T.: Partitioning oracle attacks. In: USENIX Security (2021)
23. libsodium AEAD. https://doc.libsodium.org/secret-key_cryptography/aead
24. Microsoft Key Vault. https://azure.microsoft.com/en-us/services/key-vault/#product-overview
25. Oracle Key Vault. https://www.oracle.com/security/database-security/key-vault/
26. PKCS #11 cryptographic token interface base specification version 2.40. Standard, OASIS (2015). http://docs.oasis-open.org/pkcs11/pkcs11-base/v2.40/os/pkcs11-base-v2.40-os.pdf
27. Rogaway, P.: Nonce-based symmetric encryption. In: Roy, B., Meier, W. (eds.) FSE 2004. LNCS, vol. 3017, pp. 348–358. Springer, Heidelberg (2004). https://doi.org/10.1007/978-3-540-25937-4_22
28. Rogaway, P., Shrimpton, T.: A provable-security treatment of the key-wrap problem. In: Vaudenay, S. (ed.) EUROCRYPT 2006. LNCS, vol. 4004, pp. 373–390. Springer, Heidelberg (2006). https://doi.org/10.1007/11761679_23
29. Shadowsocks (2020). https://shadowsocks.org/en/index.html
30. Telegram mobile protocol. https://core.telegram.org/mtproto/description
31. Google Tink library. https://developers.google.com/tink

Privacy-Preserving Authenticated Key Exchange in the Standard Model

You Lyu[1,2], Shengli Liu[1,2,4](✉), Shuai Han[2,3], and Dawu Gu[1]

[1] Department of Computer Science and Engineering, Shanghai Jiao Tong University, Shanghai 200240, China
{vergil,slliu,dwgu}@sjtu.edu.cn
[2] State Key Laboratory of Cryptology, P.O. Box 5159, Beijing 100878, China
dalen17@sjtu.edu.cn
[3] School of Cyber Science and Engineering, Shanghai Jiao Tong University, Shanghai 200240, China
[4] Westone Cryptologic Research Center, Beijing 100070, China

Abstract. Privacy-Preserving Authenticated Key Exchange (PPAKE) provides protection both for the session keys and the identity information of the involved parties. In this paper, we introduce the concept of robustness into PPAKE. Robustness enables each user to confirm whether itself is the target recipient of the first round message in the protocol. With the help of robustness, a PPAKE protocol can successfully avoid the heavy redundant communications and computations caused by the ambiguity of communicants in the existing PPAKE, especially in broadcast channels.

We propose a generic construction of robust PPAKE from key encapsulation mechanism (KEM), digital signature (SIG), message authentication code (MAC), pseudo-random generator (PRG) and symmetric encryption (SE). By instantiating KEM, MAC, PRG from the DDH assumption and SIG from the CDH assumption, we obtain a specific robust PPAKE scheme in the standard model, which enjoys forward security for session keys, explicit authentication and forward privacy for user identities. Thanks to the robustness of our PPAKE, the number of broadcast messages per run and the computational complexity per user are constant, and in particular, independent of the number of users in the system.

Keywords: Authenticated key exchange · Privacy · Robustness

1 Introduction

Authenticated Key Exchange (AKE) enables two parties to authenticate each other and compute a shared session key. It has been widely deployed over Internet, like IPsec IKE (Internet Key Exchange), TLS, Tor, Google's QUIC protocol, etc. Generally, AKE focuses on the protection of session keys between two parties against adversaries implementing both passive and active attacks. As a well-studied topic, a variety of AKE schemes have been proposed, but little

© International Association for Cryptologic Research 2022
S. Agrawal and D. Lin (Eds.): ASIACRYPT 2022, LNCS 13793, pp. 210–240, 2022.
https://doi.org/10.1007/978-3-031-22969-5_8

attention was paid to privacy of user identities in AKE. The research on Privacy-Preserving AKE (PPAKE) was ignited by the chasing of privacy protection. For instance, SKEME [14], TLS 1.3 [3], Tor [8] and private airdrop [12] all take user privacy as one of important design principles. Recently two proposals for PPAKE arise [19,20], aiming to provide protection for user identity besides their session keys. Next we overview the recent two works, namely SSL-PPAKE [20] and RSW-PPAKE [19].

SSL-PPAKE. In [20], Schäge, Schwenk, and Lauer (SSL) isolated a generic PPAKE construction from TLS 1.3, QUIC, IPsec IKE, SSH and certain patterns of NOISE to achieve user identity protection. We name it SSL-PPAKE.

SSL-PPAKE [20] has 4 rounds. In the first two rounds, P_i and P_j run a basic Diffie-Hellman (DH) handshake to obtain a shared DH key $K = g^{xy}$. In the last two rounds, P_i and P_j use the shared DH key $K = g^{xy}$ to protect protocol messages that contain identity-related data such as identities, public keys or digital signatures. As pointed out in [20], due to the lack of authenticity in the first two rounds, the SSL-PPAKE suffers a weakness on preserving the privacy of initiator's identity. More precisely, let us consider a broadcast channel with μ users as an example. First we identify three facts about SSL-PPAKE.

Fact 1. In the 1st round, to protect the identity of its intended target recipient P_j, initiator P_i has to broadcast g^x in the system. As a result, every user is able to receive g^x.

Fact 2. In the 2nd round, every user P_{j_k} has to respond to P_i by broadcasting $g^{y_{j_k}}$, here $j_k \in [\mu] \backslash \{i\}$, since P_{j_k} is uncertain about the intended recipient.

Fact 3. In the 3rd round, P_i receives all the messages $\{g^{y_{j_k}}\}_{j_k \in [\mu] \backslash \{i\}}$, but it is not able to identify the right message sent from the intended party P_j and has to computes all DH keys $\{K_{i,j_k} = g^{xy_{j_k}}\}_{j_k \in [\mu] \backslash \{i\}}$. Consequently, P_i has to encrypt the message in the third round with each K_{i,j_k} individually to obtain $\mu - 1$ ciphertext $C_{j_k} = \mathsf{SE.Enc}(K_{i,j_k}, i|pk_i|auth_i)$ and broadcast the $\mu - 1$ ciphertexts to all users. Here $\mathsf{SE.Enc}$ denotes a symmetric encryption algorithm, and $auth_i$ denotes the authentication part of the protocol.

Now let us see how an adversary reveals the identity of the initiator. After receiving g^x from P_i, the adversary can simply select $\tilde{y}$ and send $g^{\tilde{y}}$ to P_i. According to the facts, P_i will broadcast $\tilde{C} = \mathsf{SE.Enc}(\tilde{K} = g^{x\tilde{y}}, i|pk_i|auth_i)$ in the 3rd round. Then the adversary can compute $\tilde{K} = (g^x)^{\tilde{y}}$ and easily decrypt $\tilde{C}$ with $\tilde{K}$ to obtain the identity information $i|pk_i$.

RSW-PPAKE. To deal with the active attacks on the SSL-PPAKE scheme, Ramacher, Slamanig and Weninger (RSW) [19] proposed three solutions in the Random Oracle model.[1] The first one has 3 rounds and assumes pre-shared key between every pair of users. It resorts to the pre-shared key to accomplish authentication. The third one converts an AKE to a PPAKE by encrypting every

---

[1] No security proofs are provided for the three schemes in [19] and its full-version is still not available.

message of AKE with communication peer's public key. However, it does not achieve forward privacy for user identities. If any user's secret key is corrupted, the adversary can break forward privacy by decrypting the ciphertexts in the previous runs to reveal the used identities. The second solution has 4 rounds and does not possess forward privacy when the responder is corrupted. Here we recall the second scheme and show the weakness on its forward privacy.

- In the first two rounds, similar to SSL-PPAKE, a Diffie-Hellman handshake is implemented to share key $K = g^{xy}$ between P_i and P_j. Meanwhile, P_i has to handshake with every P_{j_k} and share $K_{i,j_k} = g^{xy_{j_k}}$ with P_{j_k}, $j_k \in [\mu] \backslash \{i\}$.
- In the 3rd round, P_i uses P_j's public key pk_j to encrypt a random string r and obtains $C = \mathsf{PKE.Enc}(pk_j, r)$, where $\mathsf{PKE.Enc}$ denotes a public-key encryption algorithm. Then it uses K to encrypt C to obtain a $c_0 = \mathsf{SE.Enc}(K, C)$. P_i signs $i|j|c_0|g^x|g^y$ to get the signature σ_i and encrypts its certificate cert_i and σ_i with a derived key $K' = H(K, r, g^x, g^y)$, resulting in $c_1 = \mathsf{SE.Enc}(K', \mathsf{cert}_i|\sigma_i)$. In the real scenario, P_i cannot identify the right K from $\{K_{i,j_k}\}_{j_k \in [\mu] \backslash \{i\}}$, thus has to use each K_{i,j_k} to obtain (c_{0,j_k}, c_{1,j_k}). Finally, P_i broadcasts $\{(c_{0,j_k}, c_{1,j_k})\}_{j_k \in [\mu] \backslash \{i\}}$ to all users.
- In the 4th round, each user j_k decrypts every pair in $\{(c_{0,j_k}, c_{1,j_k})\}_{j_k \in [\mu] \backslash \{i\}}$ with its Diffie-Hellman key $K_{i,j_k} = g^{xy_{j_k}}$, trying to recover $\mathsf{cert}_i|\sigma_i$. Only the right responder P_j can certify the validity of $\mathsf{cert}_i|\sigma_i$ and recover r. After that, P_j knows its partner is P_i. Then P_j broadcasts the hash value $h := H(r, i|j|g^x|g^y|c_0|c_1)$ to P_i.
- Finally, P_i checks if $h = H(r, i|j|g^x|g^y|c_0|c_1)$ holds (to authenticate P_j).

The attack is similar to that on SSL-PPAKE but here on forward privacy of RSW-PPAKE. After receiving g^x from P_i, the adversary $\mathcal{A}$ can simply select $\tilde{y}$ and send $g^{\tilde{y}}$ to P_i. Then $\mathcal{A}$ also shares a key $\tilde{K} = g^{x\tilde{y}}$ with P_i. In the second phase, there must exist $(\tilde{c}_0, \tilde{c}_1) \in \{(c_{0,j_k}, c_{1,j_k})\}_{j_k \in [\mu] \backslash \{i\}}$ such that $(\tilde{c}_0, \tilde{c}_1)$ is computed with $\tilde{K}$. So $\mathcal{A}$ can always recover $C = \mathsf{SE.Dec}(\tilde{K}, \tilde{c}_0)$. Later $\mathcal{A}$ corrupts P_j and obtains sk_j. Then $\mathcal{A}$ decrypts C with sk_j to recover $r = \mathsf{PKE.Dec}(sk_j, C)$. Finally $\mathcal{A}$ can identify P_i, P_j by finding $i, j, c_{0,j}, c_{1,j}$ such that $h = H(r, i|j|g^x|g^y|c_{0,j}|c_{1,j})$.

Our Approach to PPAKE. From the above analysis, we know that the SSL-PPAKE provides no protection for the initiator's identity, and the RSW-PPAKE loses forward privacy for identities of both the initiator and the responder when the responder is corrupted.

The reason for the attacks lies in the facts that each user replies the initiator and the initiator cannot identify the message sent from the intended peer in the 2nd round. Thus the initiator has to reply messages to each individual user in the third round. This leaks too much information, of which the adversary can take advantage to break privacy of PPAKE, as shown before.

At the same time, these facts also lead to another drawback: the communication band of the protocol is as large as $O(\mu)$ and each user's computational complexity is as high as $O(\mu)$, since each user has to compute or deal with $\mu - 1$

messages in the 3rd round. Here μ is the number of users in the system. See Fig. 1.

In this paper, we study how to avoid the above attacking problems and improve efficiency of PPAKE. Our idea in a nutshell is to make PPAKE robust.

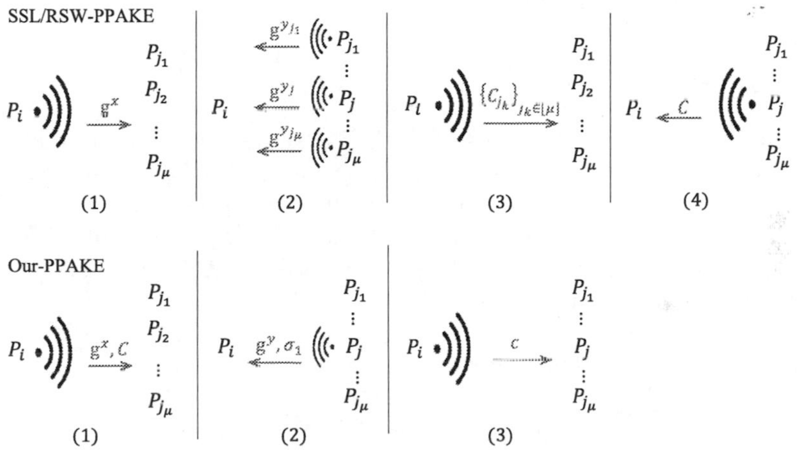

Fig. 1. The upper part is the information flows of rounds (1) (2) (3) (4) in SSL-PPAKE and RSW-PPAKE [19,20]. The lower part is the information flows of rounds (1) (2) (3) in our robust PPAKE. Here the parties communicate over a broadcast channel.

Robustness of PPAKE. We introduce the concept of robustness. It requires that only one party P_j is able to ascertain that the message in the 1st round is for him/her, hence correctly reply a message in the 2nd round.

Our robust PPAKE makes use of a key encapsulation mechanism KEM, a signature scheme SIG, a message authentication code MAC, a pseudo-random generator PRG and a symmetric encryption SE. The public/secret key pair (pk, sk) of KEM and the verification/signing key (vk, ssk) of SIG serve as the long-term key of a user. Our PPAKE has 3 rounds and is shown below. See Fig. 1.

Round 1 $(P_i \Rightarrow P_j)$: P_i broadcasts g^x and a ciphertext C to P_j, where $(C, N) \leftarrow$ KEM.Encap(pk_j) with N the key encapsulated in C.

Round 2 $(P_i \Leftarrow P_j)$: P_j decrypts C with its secret key sk_j to recover N, then it uses N as the MAC key to compute a MAC tag $\sigma_1 = \mathsf{MAC}(N, g^x|C)$. P_j broadcasts (g^y, σ_1). We require that when decrypting C, only P_j succeeds and all other parties will get a special failure symbol $\perp$, which is guaranteed by the robustness of KEM (see more details later). Consequently, only P_j responds in this round, and all other parties (except P_i and P_j) will terminate the protocol in time.

Round 3 $(P_i \Rightarrow P_j)$: P_i checks the validity of σ_1 and computes the Diffie-Hellman key $K = g^{xy}$. Furthermore, it derives a session key k and a symmetric key k' from K via $(k, k') \leftarrow \mathsf{PRG}(K)$. It signs the message $g^x|C|\sigma_1|g^y$ to get the signature σ_2. Then it uses k' to encrypt its identity i and σ_2 to obtain $c \leftarrow \mathsf{SE.Enc}(k', i|\sigma_2)$. P_i broadcasts c.

Similarly, P_j can obtain (k, k') from K and decrypt c to get $i|\sigma_2$. By checking the validity of σ_2 with P_i's verification key vk_i, P_j ascertains its partner's identity i and accepts k as the session key.

We refer to Fig. 5 in Sect. 4 for the details of our PPAKE construction. Below is a high-level analysis of our PPAKE.

- _Robustness._ For the robustness of PPAKE, we require that the underlying KEM is robust in such a sense: if C is generated with pk_j, then decrypting C with any other secret key sk_{j_k} will result in a decryption failure.
- _Explicit mutual authentication._ The authenticity of P_j is guaranteed by KEM and MAC, and the authenticity of P_i is guaranteed by SIG. Hence our PPAKE has explicit mutual authentication.
- _Forward security for session keys._ After excluding active attacks by authenticity, $K = g^{xy}$ is pseudo-random by the DDH assumption. Hence, the session key k, as output of PRG, is pseudo-random as well. Thanks to the ephemeral randomness of x and y, session keys have forward security.
- _Privacy for user identities._ The privacy for user identities relies on KEM and SE. We require that C does not leak information about pk_j computationally, and this is formalized by IK-CCA security. As a function output of C, σ_1 does not leak any information either. Meanwhile, g^x and g^y are randomly chosen and independent of i and j. Moreover, ciphertext c protects i and P_i's signature σ_2. Therefore, identity information i, j is well-protected.
- _Forward privacy for user identities._ The forward privacy holds if the initiator P_i is corrupted by $\mathcal{A}$, since the knowledge of the signing key ssk_i does not help $\mathcal{A}$ to learn user's identity in previous runs of PPAKE (recall that the user privacy is guaranteed by KEM and SE). On the other hand, if the responder P_j is corrupted by $\mathcal{A}$, because of the robustness, the knowledge of sk_j can help $\mathcal{A}$ to identify j as long as decrypting C in the previous runs of PPAKE does not result in decryption failure. This suggests that the disclosure of responder's identity j is unavoidable due to the robustness of our PPAKE in the case of responder corruption. However, the initiator's identity i is still well-protected. Therefore, our PPAKE achieves semi-forward privacy when the responder P_j is corrupted and full forward privacy when the initiator P_i is corrupted.
- _Constant communication and computational complexity._ Thanks to the robustness of our PPAKE, the number of broadcast messages per run and the computational complexity per user are constant in our PPAKE, while those in the SSL-PPAKE and RSW-PPAKE schemes are linear to the number μ of users.

Our Contribution. We summarize our contribution in this paper. We introduce the concept of robustness into PPAKE, and present a formalized security model for robust PPAKE. In the security model, we consider adversary's passive attacks, active attacks, corruptions of users' long-term keys, and revealing of session keys. Based on the security model, we define user authenticity, forward security for session keys, and forward privacy for user identities.

We propose a generic construction of 3-round robust PPAKE from KEM, SIG, MAC, PRG and SE. By instantiating KEM, MAC, PRG from the DDH assumption and SIG from the CDH assumption (together with a one-time pad SE), we obtain a specific PPAKE scheme in the standard model.

- Our PPAKE scheme enjoys explicit mutual authentication, forward security for session keys and forward privacy for user identities, and resists those attacks on SSL-PPAKE and RSW-PPAKE.
- Our PPAKE scheme is efficient in the sense that both the communication complexity of the protocol and the computational complexity per user is independent of the number of users, thanks to its robustness.

The comparison of our scheme with other PPAKE schemes is shown in Table 1.

Table 1. Comparison among the PPAKE schemes, where μ refers to the number of users. **Comm** denotes the communication complexity of the protocols in terms of the number of group elements. **Comp** denotes the computational complexity per user, where $O(\mu)$ means that **Comp** is linear to μ and $O(1)$ means that **Comp** is independent of μ. "**#**" denotes the number of rounds in the protocol. **Forward Security** is for session keys, where "weak" prevents adversary from modifying the messages sent by the two parties. **Privacy** denotes the privacy of user identity in case of no user corruption. **Forward Privacy** denotes the forward privacy of user identity. **CrpI** denotes forward privacy when initiator is corrupted. **CrpR** denotes forward privacy when responder is corrupted. **I** (**R**) checks whether the privacy of initiator's (responder's) identity is preserved. **Mutual Auth** denotes whether the PPAKE scheme achieves mutual authentication. **Std** denotes whether the security of PPAKE is proved in the standard model.

PPAKE schemes	Comm	Comp	#	Forward security	Privacy		Forward privacy				Mutual Auth	Std
							CrpI		CrpR			
					I	R	I	R	I	R		
IY [13]	6	$O(1)$	2	Weak	✓	×	✓	×	✓	×	×	✓
SKEME [14]	16	$O(1)$	3	✓	✓	✓	×	×	×	×	✓	×
SSL [20]	5μ	$O(\mu)$	4	✓	×	✓	×	✓	×	✓	✓	✓
RSW [19]	$7\mu-5$	$O(\mu)$	4	✓	✓	✓	✓	✓	×	×	✓	×
Ours	12	$O(1)$	3	✓	✓	✓	✓	✓	✓	×	✓	✓

On Modeling (Forward) Privacy in PPAKE. Our PPAKE works not only for broadcast channel, but also for any public channel, as long as the identifiers like IP or MAC addresses leak no identity information (as considered in [20] and [19]). In these channels, after receiving a message from an initiator, every user may give a response when not aware whether itself is the target recipient.

Some of previous works [1,2,15,21] consider the settings of pre-shared symmetric long-term keys (or passwords) among each pair of users. In this setting, it is easy to achieve authentication, but the assumption is too strong. Most recent work [13] considered a special client-server setting, where client has no long-term key. In this case, the client can be perfectly anonymous but authentication for client is lost.

Our security model, like the security models of SSL-PPAKE [20] and RSW-PPAKE [19], considers that many parties communicate over a public channel. However, We consider a more comprehensive scenario than [19,20].

Recall that [19,20] consider the scenario in which the sender and responder in PPAKE are agent servers, and behind each server sits many users. The adversary implements passive and active attacks over the channel between the sender (agent server) and receiver (agent server) but has no access to the channel between the agent server and the end users. The privacy for user identity in [19,20] essentially said that the adversary cannot tell which user the agent server is delegating during the communications. In our paper, we are considering intact end-to-end user communications rather than limited communications between agent servers. For the sake of privacy protection, messages must not contain user identity explicitly, hence have to be broadcasted to all end users. Each end user may respond the message even if she/he is not the target recipient. Consequently, the initiator may have to deal with a pile of messages from different recipients. Covering end-to-end user communications must consider adversary accessing the channel connecting the end users. Hence, our security model allows adversary's eavesdropping, message insertion/modification/deletion over the broadcast channel which connects end-users. Moreover, as pointed out in [20], their security model only guarantees the privacy of user identities in accepted sessions. Our model also protects user privacy for incomplete sessions and failed sessions.

We stress that our model protects the forward privacy of user identities as much as possible while achieving robustness. To achieve robustness, the first message must be tied to the responder's long term secret key. Once the responder is corrupted, the adversary can identify whether the responder has received messages (but may still do not know the identity of the initiator). Hence, the forward privacy for responder when itself is corrupted is mutually exclusive with the robustness of PPAKE. Consequently, the best forward privacy for robust PPAKE to achieve is semi-forward privacy when the responder is corrupted and full forward privacy when the initiator is corrupted. As shown in Table 1, our PPAKE scheme achieves the best forward privacy as a robust PPAKE, and provides 3 out of 4 kinds of forward privacy, which is the most compared with other PPAKE schemes.

2   Preliminary

Let $\emptyset$ denote an empty string. If x is defined by y or the value of y is assigned to x, we write $x := y$. For $\mu \in \mathbb{N}$, define $[\mu] := \{1, 2, \ldots, \mu\}$. Denote by $x \leftarrow_\$ \mathcal{X}$

the procedure of sampling x from set $\mathcal{X}$ uniformly at random. Let $|\mathcal{X}|$ denote the number of elements in $\mathcal{X}$. All our algorithms are probabilistic unless states otherwise. We use $y \leftarrow \mathcal{A}(x)$ to define the random variable y obtained by executing algorithm $\mathcal{A}$ on input x. We use $y \in \mathcal{A}(x)$ to indicate that y lies in the support of $\mathcal{A}(x)$. We also use $y \leftarrow \mathcal{A}(x; r)$ to make explicit the random coins r used in the probabilistic computation. If X and Y have identical distribution, we simply denote it by $X \equiv Y$.

In the full version [18], we review the definition of digital signature and its security notion of strongly existential unforgeability (sEUF-CMA), the definition of message authentication code (MAC) and its security notion of strongly existential unforgeability (sEUF-CMA), the definition of pseudo-random generator (PRG) and its pseudo-randomness, and the definition of ciphertext diversity and semantic security of symmetric encryption (SE).

2.1 Key Encapsulation Mechanism

Definition 1 (KEM). *A key encapsulation mechanism (KEM) scheme* KEM = (KEM.Setup, KEM.Gen, Encap, Decap) *consists of four algorithms:*

- KEM.Setup : *The setup algorithm outputs public parameters* $\mathsf{pp}_{\mathsf{KEM}}$, *which determines an encapsulation key space* $\mathcal{K}$, *a public key space* $\mathcal{PK}$, *a secret key space* $\mathcal{SK}$, *and a ciphertext space* $\mathcal{CT}$.
- KEM.Gen : *Taking* $\mathsf{pp}_{\mathsf{KEM}}$ *as input, the key generation algorithm outputs a pair of public key and secret key* $(pk, sk) \in \mathcal{PK} \times \mathcal{SK}$.
- Encap(pk) : *Taking pk as input, the encapsulation algorithm outputs a pair of ciphertext* $C \in \mathcal{CT}$ *and encapsulated key* $K \in \mathcal{K}$.
- Decap(sk, C) : *Taking as input sk and C, the deterministic decapsulation algorithm outputs* $K \in \mathcal{K} \cup \{\bot\}$.

The correctness of KEM *requires that for all* $\mathsf{pp}_{\mathsf{KEM}} \in$ KEM.Setup, $(pk, sk) \in$ KEM.Gen($\mathsf{pp}_{\mathsf{KEM}}$), *and* $(C, K) \in$ Encap(pk), *it holds that* Decap(sk, C) = K.

We recall the IND-CPA and IND-CCA security of KEM.

Definition 2 (IND-CPA/IND-CCA Security for KEM). *For a key encapsulation mechanism* KEM, *the advantage functions of an adversary* $\mathcal{A}$ *are defined by* $\mathsf{Adv}^{\mathsf{CPA}}_{\mathsf{KEM}}(\mathcal{A}) := \left| \Pr\left[\mathsf{Exp}^{\mathsf{CPA}\text{-}0}_{\mathsf{KEM},\mathcal{A}} \Rightarrow 1\right] - \Pr\left[\mathsf{Exp}^{\mathsf{CPA}\text{-}1}_{\mathsf{KEM},\mathcal{A}} \Rightarrow 1\right] \right|$ *and* $\mathsf{Adv}^{\mathsf{CCA}}_{\mathsf{KEM}}(\mathcal{A}) :=$ $\left| \Pr\left[\mathsf{Exp}^{\mathsf{CCA}\text{-}0}_{\mathsf{KEM},\mathcal{A}} \Rightarrow 1\right] - \Pr\left[\mathsf{Exp}^{\mathsf{CCA}\text{-}1}_{\mathsf{KEM},\mathcal{A}} \Rightarrow 1\right] \right|$, *where the experiments* $\mathsf{Exp}^{\mathsf{CPA}\text{-}b}_{\mathsf{KEM},\mathcal{A}}$ *and* $\mathsf{Exp}^{\mathsf{CCA}\text{-}b}_{\mathsf{KEM},\mathcal{A}}$ *for* $b \in \{0, 1\}$ *are defined in Fig. 2. The IND-CPA/IND-CCA security for* KEM *requires* $\mathsf{Adv}^{\mathsf{CPA}}_{\mathsf{KEM}}(\mathcal{A})/\mathsf{Adv}^{\mathsf{CCA}}_{\mathsf{KEM}}(\mathcal{A}) = \mathsf{negl}(\lambda)$ *for all PPT* $\mathcal{A}$.

We recall the security notion indistinguishability of keys under chosen-ciphertext attack (IK-CCA Security) formalized by Bellare *et al.* in [5].

Definition 3 (IK-CCA Security for KEM). *For a key encapsulation mechanism* KEM, *the advantage function of an adversary* $\mathcal{A}$ *is defined*

$$
\begin{array}{|l|l|}
\hline
\mathsf{Exp}_{\mathsf{KEM},\mathcal{A}}^{\mathsf{CPA\text{-}b}}, \ \mathsf{Exp}_{\mathsf{KEM},\mathcal{A}}^{\mathsf{CCA\text{-}b}}: & \mathcal{O}_{\mathrm{DEC}}(C): \\
\hline
\mathsf{pp}_{\mathsf{KEM}} \leftarrow \mathsf{KEM.Setup}; \ (pk, sk) \leftarrow \mathsf{KEM.Gen}(\mathsf{pp}_{\mathsf{KEM}}) & \\
(C^*, K_0^*) \leftarrow \mathsf{Encap}(pk); \quad K_1^* \leftarrow \mathcal{K} & \text{If } C = C^*: \text{Return } \perp \\
& K \leftarrow \mathsf{Decap}(sk, C) \\
b' \leftarrow \mathcal{A}^{\mathcal{O}_{\mathrm{DEC}}(\cdot)}(pk, C^*, K_b^*) & \text{Return } K \\
\text{Return } b' & \\
\hline
\end{array}
$$

Fig. 2. The IND-CPA security experiment $\mathsf{Exp}_{\mathsf{KEM},\mathcal{A}}^{\mathsf{CPA\text{-}b}}$ and the IND-CCA security experiment $\mathsf{Exp}_{\mathsf{KEM},\mathcal{A}}^{\mathsf{CCA\text{-}b}}$ of KEM, where in the latter the adversary can query the decapsulation oracle $\mathcal{O}_{\mathrm{DEC}}(\cdot)$.

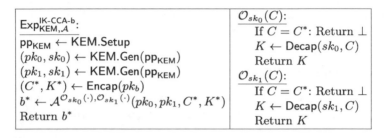

Fig. 3. The IK-CCA security experiment $\mathsf{Exp}_{\mathsf{KEM},\mathcal{A}}^{\mathsf{IK\text{-}CCA\text{-}b}}$.

with $\mathsf{Adv}_{\mathsf{KEM}}^{\mathsf{IK\text{-}CCA}}(\mathcal{A}) := \left| \Pr\left[\mathsf{Exp}_{\mathsf{KEM},\mathcal{A}}^{\mathsf{IK\text{-}CCA\text{-}0}} \Rightarrow 1 \right] - \Pr\left[\mathsf{Exp}_{\mathsf{KEM},\mathcal{A}}^{\mathsf{IK\text{-}CCA\text{-}1}} \Rightarrow 1 \right] \right|$, where the experiment $\mathsf{Exp}_{\mathsf{KEM},\mathcal{A}}^{\mathsf{IK\text{-}CCA\text{-}b}}$ for $b \in \{0,1\}$ is defined in Fig. 3. The IK-CCA security for KEM requires that $\mathsf{Adv}_{\mathsf{KEM}}^{\mathsf{IK\text{-}CCA}}(\mathcal{A}) = \mathsf{negl}(\lambda)$ for all PPT $\mathcal{A}$.

Next we introduce the robustness and encapsulated key uniformity of KEM.

Definition 4 (Robustness of KEM). *A key encapsulation mechanism* KEM *has robustness if for all* $\mathsf{pp}_{\mathsf{KEM}} \in \mathsf{KEM.Setup}(1^\lambda)$, *it holds that*

$$
\Pr\left[\begin{array}{l} (pk_1, sk_1) \leftarrow \mathsf{KEM.Gen}(\mathsf{pp}_{\mathsf{KEM}}); \\ (pk_2, sk_2) \leftarrow \mathsf{KEM.Gen}(\mathsf{pp}_{\mathsf{KEM}}); C_1 \leftarrow \mathsf{Encap}(pk_1) \end{array} : \mathsf{Decap}(sk_2, C_1) \neq \perp \right] = \mathsf{negl}(\lambda).
$$

Definition 5 (Encapsulated Key Uniformity of KEM). *A key encapsulation mechanism* KEM *has encapsulated key uniformity if for all* $\mathsf{pp}_{\mathsf{KEM}} \in \mathsf{KEM.Setup}(1^\lambda)$, *it holds that*

- $\forall r \in \mathcal{R}$, *it holds that*

$$
\{K | r' \leftarrow_\$ \mathcal{R}', (pk, sk) \leftarrow \mathsf{KEM.Gen}(\mathsf{pp}_{\mathsf{KEM}}; r'), (C, K) \leftarrow \mathsf{Encap}(pk; r)\} \equiv \{K | K \leftarrow_\$ \mathcal{K}\},
$$

- $\forall (pk, sk) \in \mathsf{KEM.Gen}(\mathsf{pp}_{\mathsf{KEM}})$, *it holds that*

$$
\{K | r \leftarrow_\$ \mathcal{R}, (C, K) \leftarrow \mathsf{Encap}(pk; r)\} \equiv \{K | K \leftarrow_\$ \mathcal{K}\},
$$

where $\mathcal{R}, \mathcal{R}'$ *are the randomness spaces involved in* Encap *and* Gen *respectively.*

Definition 6 (γ-PK-Diversity of KEM). *A key encapsulation mechanism* KEM *has γ-pk-diversity if for all* $pp_{KEM} \in Setup(1^\lambda)$, *it holds that*

$$\Pr\left[\begin{array}{l} r \leftarrow_\$ \mathcal{R}; (pk, sk) \leftarrow_\$ KEM.Gen(pp_{KEM}; r); \\ r' \leftarrow_\$ \mathcal{R}; (pk', sk') \leftarrow_\$ KEM.Gen(pp_{KEM}; r') \end{array} : pk = pk'\right] = 2^{-\gamma},$$

where $\mathcal{R}$ is the randomness space involved in KEM.Gen *algorithm.*

3 Privacy-Preserving Authenticated Key Exchange

3.1 Definition of Privacy-Preserving Authenticated Key Exchange

Definition 7 (PPAKE). *A privacy-preserving authenticated key exchange scheme* PPAKE = (PPAKE.Setup, PPAKE.Gen, PPAKE.Protocol) *consists of two probabilistic algorithms and an interactive protocol.*

- PPAKE.Setup(1^λ): *The setup algorithm takes as input the security parameter* 1^λ, *and outputs the public parameter* pp_{PPAKE}.
- PPAKE.Gen(pp_{PPAKE}, i): *The generation algorithm takes as input* pp_{PPAKE} *and a party identity i, and outputs a key pair (pk_i, sk_i).*
- PPAKE.Protocol($P_i(res_i) \rightleftharpoons P_j(res_j)$): *The protocol involves two parties P_i and P_j, who have access to their own resources,* $res_i := (sk_i, pp_{PPAKE}, \{pk_u\}_{u \in [\mu]})$ *and* $res_j := (sk_j, pp_{PPAKE}, \{pk_u\}_{u \in [\mu]})$, *respectively. Here μ is the total number of users. After execution, P_i outputs a flag* $\Psi_i \in \{\emptyset, accept, reject\}$, *and a session key k_i (k_i might be empty string $\emptyset$), and P_j outputs (Ψ_j, k_j) similarly.*

Correctness of PPAKE. *For all* $pp_{PPAKE} \in$ PPAKE.Setup(1^λ), *for any distinct and honest parties P_i and P_j with $(pk_i, sk_i) \leftarrow$* PPAKE.Gen(pp_{PPAKE}, i) *and* $(pk_j, sk_j) \leftarrow$ PPAKE.Gen(pp_{PPAKE}, j), *after the execution of* PPAKE.Protocol($P_i(res_i) \rightleftharpoons P_j(res_j)$), *it holds that $\Psi_i = \Psi_j =$* **accept** *and* $k_i = k_j \neq \emptyset$.

Definition 8 (Robustness of PPAKE). *A* PPAKE *scheme is robust if for any party P_i who initializes the protocol, then with overwhelming probability, only P_i's intended peer P_j is able to determine the validity of the first message sent by P_i when following the protocol specifications.*

Remark 1. The correctness and robustness of PPAKE implies the following: in the scenario of honest setting (i.e., all users are honest in the system), if P_i broadcasts the first message and its intended peer is P_j, then only P_j is able to ascertain that the message is for him/her and hence responds to this message.

3.2   Security Model and Security Definitions for PPAKE

We will adapt the security model formalized by [4,11,16], which in turn followed the model proposed by Bellare and Rogaway [6]. We also include replay attacks [17]. In addition, we extend the security model so that the (forward) privacy for user identity is taken into account.

Our security notions for PPAKE include user authenticity, forward security for session key, and forward-privacy for user identity. These are characterized by three security experiments named $\mathsf{Exp}_{\mathsf{PPAKE},\mu,\ell,\mathcal{A}}^{\mathsf{AUTH}}$, $\mathsf{Exp}_{\mathsf{PPAKE},\mu,\ell,\mathcal{A}}^{\mathsf{IND}}$ and $\mathsf{Exp}_{\mathsf{PPAKE},\mu,\ell,\mathcal{A}}^{\mathsf{Privacy}}$. In those experiments, we will formalize oracles for adversary $\mathcal{A}$. The passive and active attacks by adversary $\mathcal{A}$ is formalize by its querying to oracles and obtaining answers from oracles. Note that the adversary can copy, delay, erase, replay, and interpolate the messages transmitted over the public channels, obtains some session keys from the PPAKE protocol instances, corrupt some users by obtaining their long-term secret keys, etc.

3.2.1   Oracles

Firstly, we define oracles and their static variables to formalize the behaviour of users and the attacks by the adversary. Suppose there are at most μ users $P_1, P_2, \ldots, P_\mu$, and each user will involve at most ℓ instances. P_i is formalized by a series of oracles, $\pi_i^1, \pi_i^2, \ldots, \pi_i^\ell$.

Oracle π_i^s. Oracle π_i^s will take a message as input and output a new message, simulating user P_i's execution of s-th PPAKE protocol instance. Each oracle π_i^s has access to P_i's resource $\mathsf{res}_i := (sk_i, \mathsf{pp}_{\mathsf{PPAKE}}, \mathsf{PKList} := \{pk_u\}_{u \in [\mu]})$. π_i^s also has its own variables $\mathsf{var}_i^s := (st_i^s, \mathsf{Pid}_i^s, k_i^s, \Psi_i^s)$.
   - st_i^s : State information that has to be stored for π_i^s's next round in the execution of the protocol.
   - Pid_i^s : The intended communication peer's identity.
   - $k_i^s \in \mathcal{K}$: The session key computed by π_i^s. Here $\mathcal{K}$ is the session key space. We assume that $\emptyset \in \mathcal{K}$.
   - $\Psi_i^s \in \{\emptyset, \mathbf{accept}, \mathbf{reject}\}$: Ψ_i^s indicates whether π_i^s has completed the protocol execution and accepted k_i^s.

At the beginning, $(st_i^s, \mathsf{Pid}_i^s, k_i^s, \Psi_i^s)$ are initialized to $(\emptyset, \emptyset, \emptyset, \emptyset)$. We declare that $k_i^s \neq \emptyset$ if and only if $\Psi_i^s = \mathbf{accept}$.

Next, we formalize the oracles that dealing with $\mathcal{A}$'s queries as follows.

Oracle $\mathsf{Send}(i, s, j, \mathsf{MsgList})$. For the query $(i, s, j, \mathsf{MsgList})$, it means that $\mathcal{A}$ invokes π_i^s with $\mathsf{MsgList}$, making π_i^s to play the role of initiator with j being the intended communication peer. Oracle π_i^s will deal with each message in $\mathsf{MsgList}$ to generate new messages $\mathsf{MsgList}'$ according to the protocol specification and update its own variables $\mathsf{var}_i^s = (st_i^s, \mathsf{Pid}_i^s, k_i^s, \Psi_i^s)$. The output messages $\mathsf{MsgList}'$ is returned to $\mathcal{A}$. If $\mathsf{MsgList} = \emptyset$, $\mathcal{A}$ asks oracle π_i^s to send the first round message to j (via broadcast channel).

If $\mathsf{Send}(i, s, j, \mathsf{MsgList})$ is the τ-th query asked by $\mathcal{A}$ and π_i^s changes Ψ_i^s to $\mathbf{accept}$ after that, then we say that π_i^s is τ-accepted.

Oracle Respond(OList, MsgList). For the query (OList, MsgList), it means that $\mathcal{A}$ chooses an oracle set OList $= \{\pi_j^t\}$ to respond messages in MsgList. For $\forall \pi_j^t \in$ OList, π_j^t executes the PPAKE protocol with messages in MsgList as a potential recipient, and its variables $\mathrm{var}_j^t = (st_j^t, \mathrm{Pid}_j^t, k_j^t, \Psi_j^t)$ are updated accordingly. Those responding messages generated by OList constitute message set MsgList$'$. The output message set MsgList$'$ is returned to $\mathcal{A}$.

Oracle Corrupt(i). Upon $\mathcal{A}$'s query i, the oracle reveals to $\mathcal{A}$ the long-term secret key sk_i of party P_i. After this corruption, $\pi_i^1, \ldots, \pi_i^\ell$ will stop answering any query from $\mathcal{A}$. If Corrupt(i) is the τ-th query asked by $\mathcal{A}$, we say that P_i is τ-corrupted. If $\mathcal{A}$ has never asked Corrupt(i), we say that P_i is ∞-corrupted.

Oracle RegisterCorrupt(i, pk). $\mathcal{A}$'s query (i, pk) suggests that $\mathcal{A}$ registers a new party $P_i(i > \mu)$. The oracle distributes $(i, pk_i := pk)$ to all users. In this case, we say that P_i is 0-corrupted.

Oracle SessionKeyReveal(i, s). The query (i, s) means that $\mathcal{A}$ asks the oracle to reveal π_i^s's session key. If $\Psi_i^s \neq$ **accept**, the oracle returns $\bot$. Otherwise, the oracle returns the session key k_i^s of π_i^s. If SessionKeyReveal(i, s) is the τ-th query asked by $\mathcal{A}$, we say that π_i^s is τ-revealed. If $\mathcal{A}$ has never asked SessionKeyReveal(i, s), we say that π_i^s is ∞-revealed.

Oracle TestKey(i, s). The query (i, s) means that $\mathcal{A}$ chooses the session key of π_i^s for challenge (test). If $\Psi_i^s \neq$ **accept**, the oracle returns $\bot$. Otherwise, the oracle sets $k_0 = k_i^s$, samples $k_1 \leftarrow_\$ \mathcal{K}$. The oracle returns k_b to $\mathcal{A}$, where b is the random bit chosen by the challenger.

Oracle TestPrivacy(i_0, j_0, i_1, j_1). $\mathcal{A}$'s query is the privacy challenge and it consists of two pairs of identities (i_0, j_0) and (i_1, j_1). The oracle builds μ new oracles $\{\pi_u^0\}_{u \in [\mu]}$. Let $\pi_{i_b}^0$ initialize the PPAKE protocol with $\pi_{j_b}^0$ being the intended peer. After the initialization by $\pi_{i_b}^0$, the adversary is allowed to interfere the protocol execution. The transcript of the protocol execution is returned to $\mathcal{A}$, where b is the random bit chosen by the challenger.

3.2.2   Security Experiments of PPAKE

Now we are ready to describe the PPAKE experiments serving for authentication, forward security for session key, and forward privacy for user identity.

Recall that μ is the number of users and ℓ is maximum number of protocol executions per user. The security experiment $\mathsf{Exp}_{\mathsf{PPAKE}, \mu, \ell, \mathcal{A}}^{\mathsf{X}}$, where $\mathsf{X} \in \{\mathsf{AUTH}, \mathsf{IND}, \mathsf{Privacy}\}$, is played between challenger $\mathcal{C}$ and adversary $\mathcal{A}$.

1. $\mathcal{C}$ runs PPAKE.Setup to get PPAKE public parameter $\mathsf{pp}_{\mathsf{PPAKE}}$.
2. For each party P_i, $\mathcal{C}$ runs PPAKE.Gen($\mathsf{pp}_{\mathsf{PPAKE}}, i$) to get the long-term key pair (pk_i, sk_i). Next it chooses a random bit $b \leftarrow_\$ \{0, 1\}$ and provides $\mathcal{A}$ with the public parameter $\mathsf{pp}_{\mathsf{PPAKE}}$ and the list of public keys PKList $:= \{pk_i\}_{i \in [\mu]}$.
3. $\mathcal{A}$ has access to oracles Send, Respond, Corrupt, RegisterCorrupt, SessionKeyReveal, TestKey, TestPrivacy by issuing queries in an adaptive way. Note that $\mathcal{A}$ can issue only one query either to TestKey or to TestPrivacy, but not both. The oracles will reply the corresponding answers to $\mathcal{A}$ as long as the queries lead no trivial attacks.

$$\underline{\mathsf{Exp}_{\mathsf{PPAKE},\mu,\ell,\mathcal{A}}}$$
$\mathsf{pp}_{\mathsf{PPAKE}} \leftarrow \mathsf{PPAKE.Setup}$
For $i \in [\mu]$:
  $(pk_i, sk_i) \leftarrow \mathsf{PPAKE.Gen}(\mathsf{pp}_{\mathsf{PPAKE}}, i)$
  $crp_i = \mathbf{false}$ //Corruption variable
$\mathsf{PKList} := \{pk_i\}_{i\in[\mu]}; b \leftarrow_\$ \{0,1\}$
For $(i,s) \in [\mu] \times [\ell]$:
  $\mathsf{var}_i^s := (\mathsf{Pid}_i^s, k_i^s, \Psi_i^s, st_i^s) = (\emptyset, \emptyset, \emptyset, \emptyset)$
  $\mathsf{Aflag}_i^s := \mathbf{false}$ //Whether Pid_i^s is corrupted when π_i^s accepts
  $T_i^s := \mathbf{false}; kRev_i^s = \mathbf{false}$ // Test Key Reveal variables
  $T_{key} := \mathbf{false}; T_{id} := \mathbf{false}$ //TestKey, TestPrivacy Oracle variables
  $\mathsf{TUsers} := \emptyset$ //Record users queried in TestID Oracle

$\mathcal{A}^{\mathcal{O}_{\mathsf{PPAKE}}(\cdot)}(\mathsf{pp}_{\mathsf{PPAKE}}, \mathsf{PKList})$ //$\mathcal{O}_{\mathsf{PPAKE}}$ = Send,Respond,Corrupt,RegisterCorrupt
$\mathsf{Win}_{\mathsf{Auth}} := \mathbf{false}$ SessionKeyReveal,TestKey or TestPrivacy
$\mathsf{Win}_{\mathsf{Auth}} := \mathbf{true}$, If $\exists(i,s) \in [\mu] \times [\ell]$ s.t.
(1) $\Psi_i^s = \mathbf{accept}$
(2) $\mathsf{Aflag}_i^s = \mathbf{false}$
(3) (3.1) $\vee$ (3.2) $\vee$ (3.3). Let $j := \mathsf{Pid}_i^s$
  (3.1) $\nexists t \in [\ell]$ s.t. $\mathsf{Partner}(\pi_i^s \leftarrow \pi_j^t)$
  (3.2) $\exists t \in [\ell], (j',t') \in [\mu] \times [\ell]$ with $(j,t) \neq (j',t')$ s.t.
      $\mathsf{Partner}(\pi_i^s \leftarrow \pi_j^t) \cap \mathsf{Partner}(\pi_i^s \leftarrow \pi_{j'}^{t'})$
  (3.3) $\exists t \in [\ell], (i',s') \in [\mu] \times [\ell]$ with $(i,s) \neq (i',s')$ s.t.
      $\mathsf{Partner}(\pi_i^s \leftarrow \pi_j^t) \cap \mathsf{Partner}(\pi_{i'}^{s'} \leftarrow \pi_j^t)$
Return $\mathsf{Win}_{\mathsf{Auth}}$

If $T_{key} = \mathbf{true} \wedge T_{id} = \mathbf{true}$: Return$(\bot)$
// Query on TestKey and TestPrivacy are mutually exclusive

$b^* \leftarrow \mathcal{A}^{\mathcal{O}_{\mathsf{PPAKE}}(\cdot)}(\mathsf{pp}_{\mathsf{PPAKE}}, \mathsf{PKList})$ //$\mathcal{O}_{\mathsf{PPAKE}}$ = Send,Respond,Corrupt,TestKey
$\mathsf{Win}_{\mathsf{Ind}} := \mathbf{false}$ SessionKeyReveal RegisterCorrupt
If $b^* = b \wedge T_{key} = \mathbf{true}$:
  $\mathsf{Win}_{\mathsf{Ind}} := \mathbf{true}$
Return $\mathsf{Win}_{\mathsf{Ind}}$

$b^* \leftarrow \mathcal{A}^{\mathcal{O}_{\mathsf{PPAKE}}(\cdot)}(\mathsf{pp}_{\mathsf{PPAKE}}, \mathsf{PKList})$ //$\mathcal{O}_{\mathsf{PPAKE}}$ = Send,Respond,Corrupt,TestPrivacy
$\mathsf{Win}_{\mathsf{Privacy}} := \mathbf{false}$ SessionKeyReveal RegisterCorrupt
$\mathsf{Win}_{\mathsf{Privacy}} := \mathbf{true}$, If
(1) $b^* = b \ \wedge \ T_{id} := \mathbf{true}$:
(2) Let $\mathsf{TUsers} := (i_0, j_0, i_1, j_1)$,
$(crp_{j_0} = \mathbf{false} \wedge crp_{j_1} = \mathbf{false}) \vee j_0 = j_1$ //avoid **TA5**
Return $\mathsf{Win}_{\mathsf{Privacy}}$

$\underline{\pi(i, s, j, \mathsf{MsgList}):}$
If $\Psi_i^s = \mathbf{reject} \vee \Psi_i^s = \mathbf{accept}$: Return $\bot$
$\mathsf{MsgList}' := \emptyset$
If $\mathsf{MsgList} = \emptyset$:
  π_i^s generates the first message msg' for user P_j
  update $(st_i^s, \mathsf{Pid}_i^s, \Psi_i^s, k_i^s)$
  Return $\{msg'\}$
For each $msg \in \mathsf{MsgList}$:
  If π_i^s accepts msg:
    π_i^s generates the next message msg' of PPAKE
    $\mathsf{MsgList}' := \mathsf{MsgList}' \cup \{msg'\}$
    update $(st_i^s, \mathsf{Pid}_i^s, \Psi_i^s, k_i^s)$
Return $\mathsf{MsgList}'$

$\underline{\mathsf{Tran}(i,j):}$ //Return the transcript
Build μ new oracles $\pi_t^0, t \in [\mu]$
$\mathsf{MsgList} := \emptyset; \mathsf{Transcript} := \emptyset; \mathsf{TfirstMsg} := \emptyset$
While $(\Psi_i^0 = \emptyset \wedge \Psi_j^0 = \emptyset)$ do:
  If $\mathsf{MsgList} = \emptyset$:   //The adversary can not insert messages in the first round
    $msg' \leftarrow \pi(i, 0, j, \emptyset)$
    $\mathsf{MsgList}' := \{msg'\}; \mathsf{TfirstMsg} := msg'$
  If $\mathsf{MsgList} \neq \emptyset$:   //The adversary can insert messages in the non-first round
    $\mathsf{MsgList}' := \emptyset;$
    For $msg \in \mathsf{MsgList}$:
      $msg' \leftarrow \pi(i, 0, j, msg)$
      $\mathsf{MsgList}' := \mathsf{MsgList}' \cup \{msg'\}$
    $\mathsf{InsertList} \leftarrow \mathcal{A}(\mathsf{MsgList}, \mathsf{MsgList}')$
    $\mathsf{MsgList}' := \mathsf{MsgList}' \cup \mathsf{InsertList}$
    $\mathsf{Transcript} := \mathsf{Transcript} \cup \mathsf{MsgList}'$
    $\mathsf{MsgList} := \mathsf{MsgList}'; \mathsf{MsgList}' := \emptyset$
    For each $j' \in [\mu]$ and each $msg \in \mathsf{MsgList}$
      $msg' \leftarrow \pi(j', 0, \emptyset, msg)$
      $\mathsf{MsgList}' := \mathsf{MsgList}' \cup \{msg'\}$
  If $\neg(\Psi_i^0 = \emptyset \wedge \Psi_j^0 = \emptyset)$:Return $\mathsf{Transcript}$

$\mathsf{InsertList} \leftarrow \mathcal{A}(\mathsf{MsgList}, \mathsf{MsgList}')$
$\mathsf{MsgList}' := \mathsf{MsgList}' \cup \mathsf{InsertList}$
$\mathsf{Transcript} := \mathsf{Transcript} \cup \mathsf{MsgList}'$
$\mathsf{MsgList} := \mathsf{MsgList}'$
Return $\mathsf{Transcript}$

$\underline{\mathsf{Partner}(\pi_i^s \leftarrow \pi_j^t):}$ //Checking whether $\mathsf{Partner}(\pi_i^s \leftarrow \pi_j^t)$
If $\Psi_i^s \neq \mathbf{accept}$: Return 0;
If $\Psi_i^s \neq j$: Return 0;
check wheter the outputs of π_i^s are the inputs of π_j^t
upon the acceptance of π_i^s, and vice verse.
If the transcirpts are consistent: Return 1;
Return 0;

$\underline{\mathcal{O}_{\mathsf{PPAKE}}(\mathsf{query}):}$
If query $= \mathsf{RegisterCorrupt}(u, pk_u)$:
  If $u \in [\mu]$: Return $\bot$
  $\mathsf{PKList} := \mathsf{PKList} \cup \{pk_u\}$
  $crp_u := \mathbf{true}$
  Return PKList

If query $= \mathsf{Send}(i, s, j, \mathsf{MsgList})$:
  If $i \notin [\mu] \vee s \notin [\ell] \vee j \notin [\mu]$: Return $\bot$
  If $\mathsf{Pid}_i^s = \emptyset$: $\mathsf{Pid}_i^s := j$
  If $\mathsf{Pid}_i^s \neq j$: Return $\bot$
  $\mathsf{MsgList}' := \emptyset$
  If $\mathsf{MsgList} = \emptyset$:
    $msg' \leftarrow \pi(i, s, j, msg)$
    Return $\mathsf{MsgList}' = \{msg'\}$
  For $msg \in \mathsf{MsgList}$
    $msg' \leftarrow \pi(i, s, j, msg)$
    $\mathsf{MsgList}' := \mathsf{MsgList}' \cup \{msg'\}$
  Return $\mathsf{MsgList}'$

If query $= \mathsf{Respond}(\mathsf{OList}, \mathsf{MsgList})$:
  If $T_{id} = \mathbf{true} \wedge ((j_0,*) \in \mathsf{OList} \vee (j_1,*) \in \mathsf{OList})$
    $\wedge \mathsf{TfirstMsg} \cap \mathsf{MsgList} \neq \emptyset$:
      Return $\bot$ //avoid **TA6**
  If $\exists(j,t) \in \mathsf{OList} \wedge (j,t) \notin [\mu] \times [\ell]$: Return $\bot$
  $\mathsf{MsgList}' := \emptyset$
  If $crp_j = \mathbf{false}$:
    For each $(j,t) \in \mathsf{OList}$, and each $msg \in \mathsf{MsgList}$:
      $msg' \leftarrow \pi(j, t', \emptyset, msg)$
      $\mathsf{MsgList}' := \mathsf{MsgList}' \cup \{msg'\}$
  Return $\mathsf{MsgList}'$

If query $= \mathsf{Corrupt}(i)$:
  If $i \notin [\mu]$: Return $\bot$
  $crp_i := \mathbf{true}$
  Return sk_i

If query $= \mathsf{SessionKeyReveal}(i, s)$:
  If $i \notin [\mu] \vee s \notin [\ell]$: Return $\bot$
  If $\Psi_i^s \neq \mathbf{accept}$: Return $\bot$
  If $T_i^s = \mathbf{true}$: Return $\bot$ //avoid **TA2**
  Let $j := \mathsf{Pid}_i^s$
  If $\exists t \in [\ell]$ s.t. $\mathsf{Partner}(\pi_i^s \leftrightarrow \pi_j^t)$:
    If $T_j^t = \mathbf{true}$: Return $\bot$ //avoid **TA3**
  $kRev_i^s := \mathbf{true};$
  Return k_i^s

If query $= \mathsf{TestKey}(i, s)$:
//This oracle can be only queried once
  $T_{key} := \mathbf{true}$
  If $\Psi_i^s \neq \mathbf{accept}$:
    Return $\bot$
  If $\mathsf{Aflag}_i^s = \mathbf{true} \vee kRev_i^s = \mathbf{true}$
    Return $\bot$ //avoid **TA1, TA2**
  $T_i^s := \mathbf{true}; k_0 := k_i^s; k_1 \leftarrow_\$ \mathcal{K};$
  Return k_b

If query $= \mathsf{TestPrivacy}(i_0, j_0, i_1, j_1)$:
//This oracle can be only queried once
  $T_{id} := \mathbf{true}$
  If $crp_{i_0} \vee crp_{j_0} \vee crp_{i_1} \vee crp_{j_1}$:
    Return $\bot$ //avoid **TA4**
  $\mathsf{TUsers} = (i_0, j_0, i_1, j_1)$
  If $b = 0$: Return $\mathsf{Tran}(i_0, j_0)$
  Else: Return $\mathsf{Tran}(i_1, j_1)$

Fig. 4. The security experiments $\mathsf{Exp}^{\mathsf{AUTH}}_{\mathsf{PPAKE},\mu,\ell,\mathcal{A}}$(with plain text and ⬛text⬛), $\mathsf{Exp}^{\mathsf{IND}}_{\mathsf{PPAKE},\mu,\ell,\mathcal{A}}$(with plain text and ▨text▨), $\mathsf{Exp}^{\mathsf{Privacy}}_{\mathsf{PPAKE},\mu,\ell,\mathcal{A}}$(with plain text and text). The list of trivial attacks is given in Table 2.

4. At the end of the experiment, $\mathcal{A}$ terminates with an output b^*.
5. If $b^* = b$, the experiment returns 1; otherwise the experiment returns 0.

$\mathsf{Exp}^{\mathsf{IND}}_{\mathsf{PPAKE},\mu,\ell,\mathcal{A}}$: If $\mathcal{A}$ ever queried oracle TestKey (only once), then $\mathsf{Exp}^{\mathsf{X}}_{\mathsf{PPAKE},\mu,\ell,\mathcal{A}}$ $= \mathsf{Exp}^{\mathsf{IND}}_{\mathsf{PPAKE},\mu,\ell,\mathcal{A}}$, which is the experiment for forward security of session key. Through TestKey, adversary $\mathcal{A}$ obtains a real session key k_i^s of target oracle π_i^s or a random key. The forward security of session key requires that it is hard for any PPT $\mathcal{A}$ to distinguish the two cases.

$\mathsf{Exp}^{\mathsf{Privacy}}_{\mathsf{PPAKE},\mu,\ell,\mathcal{A}}$: If $\mathcal{A}$ ever queried oracle TestPrivacy (only once), then $\mathsf{Exp}^{\mathsf{X}}_{\mathsf{PPAKE},\mu,\ell,\mathcal{A}} = \mathsf{Exp}^{\mathsf{Privacy}}_{\mathsf{PPAKE},\mu,\ell,\mathcal{A}}$, which is the experiment for forward privacy of user identity. Through TestPrivacy, $\mathcal{A}$ obtains a protocol transcript, which is either the interaction of $\pi_{i_0}^0$ and $\pi_{j_0}^0$ or the interaction of $\pi_{i_1}^0$ and $\pi_{j_1}^0$. The forward privacy requires that it is hard for any PPT $\mathcal{A}$ to distinguish the two cases.

$\mathsf{Exp}^{\mathsf{AUTH}}_{\mathsf{PPAKE},\mu,\ell,\mathcal{A}}$: If $\mathcal{C}$ checks whether event $\mathsf{Win}_{\mathsf{Auth}}$ happens ($\mathsf{Win}_{\mathsf{Auth}}$ is defined in Definition 10) at the end of the experiment (either $\mathsf{Exp}^{\mathsf{IND}}_{\mathsf{PPAKE},\mu,\ell,\mathcal{A}}$ or $\mathsf{Exp}^{\mathsf{Privacy}}_{\mathsf{PPAKE},\mu,\ell,\mathcal{A}}$), this experiment is also regarded as $\mathsf{Exp}^{\mathsf{AUTH}}_{\mathsf{PPAKE},\mu,\ell,\mathcal{A}}$, which is the experiment for authenticity. Roughly speaking, the authenticity of PPAKE requires that if an oracle π_i^s accepts a session key, then there must exist a unique oracle π_j^t such that the two oracles have essentially established partnership. Meanwhile, the authenticity makes sure that replay attacks are prevented in the sense that no oracle can make two distinct oracles accepts.

Details of the three experiments are given in Fig. 4.

To precisely describe the security notions for PPAKE, we have to forbid some trivial attacks by $\mathcal{A}$. To clearly describe trivial attacks, we first define partner.

Definition 9 (Partner). *We say that an oracle π_i^s is partnered to π_j^t, denoted as* $\mathsf{Partner}(\pi_i^s \leftarrow \pi_j^t)$*, if the following requirements hold:*

- π_i^s *accepts with* $\Psi_i^s = \boldsymbol{accept}$ *and* $\mathsf{Pid}_i^s = j$.
- *Upon the time π_i^s accepts, the transcript of π_i^s is consistent with that of π_j^t, i.e., the outputs of π_i^s are the inputs of π_j^t, and vice verse.*

We write $\mathsf{Partner}(\pi_i^s \leftrightarrow \pi_j^t)$ *if* $\mathsf{Partner}(\pi_i^s \leftarrow \pi_j^t)$ *and* $\mathsf{Partner}(\pi_j^t \leftarrow \pi_i^s)$.

We will keep track of the following variables for each party P_i and oracle π_i^s:

- crp_i: whether user i is corrupted.
- Aflag_i^s: whether the intended partner is corrupted when π_i^s accepts.
- $kRev_i^s$: whether the session key k_i^s was revealed.
- T_i^s : whether π_i^s was tested.
- T_{id} : whether oracle TestPrivacy is queried.
- T_{key} : whether oracle TestKey is queried.

For forward security for session key, we identify three trivial attacks.

TA1 Suppose that when user i (formalize by π_i^s) accepts a session key k_i^s, its partner j (formalize by π_j^t) has already been corrupted by $\mathcal{A}$, then it is quite possible that $\mathcal{A}$ impersonated j to obtain the shared session key k_i^s. In this case k_i^s cannot be tested by $\mathsf{TestKey}(i, s)$, otherwise, it will be a trivial attack.

TA2 If a session key k_i^s is accepted by user i (formalized by π_i^s) and is also revealed to $\mathcal{A}$, then k_i^s cannot be tested, otherwise, it will be a trivial attack.

TA3 If two users (formalize by oracles π_i^s and π_j^t) are partnered with each other and session key k_i^s of π_i^s is revealed to $\mathcal{A}$, then session key k_j^t of π_j^t cannot be tested due to $k_i^s = k_j^t$. Otherwise, it will be a trivial attack.

For the forward privacy for user identity, we identify three trivial attacks.

TA4 If user i is corrupted, then the adversary is able to impersonate the user in a PPAKE protocol after the corruption. After the protocol execution, the adversary will know the identity of its communicant peer. Hence, this is a trivial attack on privacy of PPAKE when testing i with $\mathsf{TestPrivacy}$.

TA5 The robustness of a PPAKE makes sure that only one target recipient j is able to use its secret key sk_j to correctly respond the first round message. If the secret key sk_j of the target recipient is corrupted by $\mathcal{A}$, no privacy on j is guaranteed. This is a trivial attack on forward privacy of robust PPAKE.

TA6 If the adversary can observe the response of each user after the user receives the first message, then the identity of the responding user is clear to the adversary. Hence, this is also a trivial attack on the privacy of robust PPAKE. This trivial attack can be extended to any core part of the first message. To exclude this trivial attack, if the adversary sees the first round message, it is not allowed to feed a message containing the core part of the first round message to other users and observe their responses.

In Table 2, we list the above trivial attacks **TA1**--**TA3** in $\mathsf{Exp}_{\mathsf{PPAKE},\mu,\ell,\mathcal{A}}^{\mathsf{IND}}$ and trivial attacks **TA4**--**TA6** in $\mathsf{Exp}_{\mathsf{PPAKE},\mu,\ell,\mathcal{A}}^{\mathsf{Privacy}}$.

3.2.3 Security Notions for PPAKE

Definition 10 (Authentication of PPAKE). *Let* $\mathsf{Win}_{\mathsf{Auth}}$ *denote the event that* $\mathcal{A}$ *breaks authentication in the security experiment* $\mathsf{Exp}_{\mathsf{PPAKE},\mu,\ell,\mathcal{A}}^{\mathsf{AUTH}}$ *(see Fig. 4).* $\mathsf{Win}_{\mathsf{Auth}}$ *happens iff* $\exists (i, s) \in [\mu] \times [\ell]$, *s.t.*

(1) π_i^s *is* τ-*accepted.*
(2) P_j *is* $\hat{\tau}$-*corrupted with* $j := \mathsf{Pid}_i^s$ *and* $\hat{\tau} > \tau$.
(3) Either (3.1) or (3.2) or (3.3) happens. Let $j := \mathsf{Pid}_i^s$.
(3.1) There is no oracle π_j^t *that* π_i^s *is partnered to.*
(3.2) There exist two distinct oracles π_j^t *and* $\pi_{j'}^{t'}$, *to which* π_i^s *is partnered.*
(3.3) There exist two oracles $\pi_{i'}^{s'}$ *and* π_j^t *with* $(i', s') \neq (i, s)$, *such that both* π_i^s *and* $\pi_{i'}^{s'}$ *are partnered to* π_j^t.

The advantage of an adversary $\mathcal{A}$ *in* $\mathsf{Exp}_{\mathsf{PPAKE},\mu,\ell,\mathcal{A}}^{\mathsf{AUTH}}$ *is defined as*

$$\mathsf{Adv}_{\mathsf{PPAKE},\mu,\ell,\mathcal{A}}^{\mathsf{AUTH}} := \Pr\left[\mathsf{Exp}_{\mathsf{PPAKE},\mu,\ell,\mathcal{A}}^{\mathsf{AUTH}} \Rightarrow 1\right] = \Pr_{\exists (i,s)}\left[(1) \wedge (2) \wedge ((3.1) \vee (3.2) \vee (3.3))\right].$$

Table 2. Trivial attacks **TA1**––**TA3** for security experiment $\mathsf{Exp}^{\mathsf{IND}}_{\mathsf{PPAKE},\mu,\ell,\mathcal{A}}$. **TA4**–**TA6** for security experiment $\mathsf{Exp}^{\mathsf{Privacy}}_{\mathsf{PPAKE},\mu,\ell,\mathcal{A}}$. Note that $\mathsf{Aflag}^s_i = \mathbf{false}$ is implicitly contained in **TA2, TA3** because of **TA1**.

Types	Trivial attacks	Explanation
TA1	$T^s_i = \mathbf{true} \wedge \mathsf{Aflag}^s_i = \mathbf{true}$	π^s_i is tested but π^s_i's partner is corrupted when π^s_i accepts session key k^s_i
TA2	$T^s_i = \mathbf{true} \wedge kRev^s_i = \mathbf{true}$	π^s_i is tested and its session key k^s_i is revealed
TA3	$T^s_i = \mathbf{true} \wedge \mathsf{Partner}(\pi^s_i \leftrightarrow \pi^t_j) \wedge kRev^t_j = \mathbf{true}$	π^s_i is tested, π^s_i and π^t_j are partnered to each other, and π^t_j's session key k^t_j is revealed
TA4	$T_{id} = \mathbf{true} \wedge (crp_{i_0} = \mathbf{true} \vee crp_{j_0} = \mathbf{true} \vee crp_{i_1} = \mathbf{true} \vee crp_{j_1} = \mathbf{true})$	When $\mathsf{TestPrivacy}(i_0, j_0, i_1, j_1)$ is queried, one of i_0, j_0, i_1, j_1 has been corrupted
TA5	$T_{id} = \mathbf{true} \wedge b^* = b \wedge (crp_{j_0} = \mathbf{true} \vee crp_{j_1} = \mathbf{true}) \wedge j_0 \neq j_1$	$\mathsf{TestPrivacy}(i_0, j_0, i_1, j_1)$ has been queried, and either j_0 or j_1 has been corrupted when checking $b^* = b$
TA6	$T_{id} = \mathbf{true} \wedge \mathcal{A}$ queried Respond (OList, MsgList) s.t. $((j_0, *) \in \mathsf{OList} \vee (j_1, *) \in \mathsf{OList}) \wedge \mathsf{TfirstMsg} \cap \mathsf{MsgList} \neq \emptyset$	$\mathsf{TestPrivacy}(i_0, j_0, i_1, j_1)$ is queried, TfirstMsg is the first message in transcript, $\mathcal{A}$ sees the output $\pi^t_{j_0}(\mathsf{MsgList})$ or $\pi^t_{j_1}(\mathsf{MsgList})$ for some $t \in [\ell]$ via querying Respond with messages MsgList containing essential information of TfirstMsg

Remark 2. Given $(1) \wedge (2)$, (3.1) indicates a successful impersonation of P_i, (3.2) suggests one instance of P_i has multiple partners, and (3.3) corresponds to a successful replay attack. Definition 10 captures mutual explicit authentication since π^s_i is either an initiator or a responder.

Definition 11 (Forward Security for Session Key of PPAKE). *In experiment* $\mathsf{Exp}^{\mathsf{IND}}_{\mathsf{PPAKE},\mu,\ell,\mathcal{A}}$ *(see Fig. 4), Let* b^* *be* $\mathcal{A}$*'s output. Then* $\mathsf{Exp}^{\mathsf{IND}}_{\mathsf{PPAKE},\mu,\ell,\mathcal{A}} \Rightarrow 1$ *iff* $b^* = b$*. The advantage of* $\mathcal{A}$ *in* $\mathsf{Exp}^{\mathsf{IND}}_{\mathsf{PPAKE},\mu,\ell,\mathcal{A}}$ *is defined as*

$$\mathsf{Adv}^{\mathsf{IND}}_{\mathsf{PPAKE},\mu,\ell,\mathcal{A}} := \left| \Pr\left[\mathsf{Exp}^{\mathsf{IND}}_{\mathsf{PPAKE},\mu,\ell,\mathcal{A}} \Rightarrow 1 \right] - 1/2 \right|.$$

Forward security for session key asks $\mathsf{Adv}^{\mathsf{IND}}_{\mathsf{PPAKE},\mu,\ell,\mathcal{A}} \leq \mathsf{negl}(\lambda)$ *for all PPT* $\mathcal{A}$.

Definition 12 (Forward Privacy for User Identity of PPAKE). *Suppose that* $\mathcal{A}$ *queries* $\mathsf{TestPrivacy}(i_0, j_0, i_1, j_1)$ *and* b^* *is* $\mathcal{A}$*'s output in* $\mathsf{Exp}^{\mathsf{Privacy}}_{\mathsf{PPAKE},\mu,\ell,\mathcal{A}}$ *(see Fig. 4). Define event* $\mathsf{Win}_{\mathsf{Privacy}}$ *as* $b^* = b$ *and neither* j_0 *nor* j_1 *are corrupted unless* $j_0 = j_1$ *(i.e.* $(crp_{j_0} = \mathbf{false} \wedge crp_{j_1} = \mathbf{false}) \vee j_0 \neq j_1$*). Then* $\mathsf{Exp}^{\mathsf{Privacy}}_{\mathsf{PPAKE},\mu,\ell,\mathcal{A}} \Rightarrow 1$ *iff* $\mathsf{Win}_{\mathsf{Privacy}}$ *happens. Forward privacy for user identity requires that for all PPT* $\mathcal{A}$*, its advantage function* $\mathsf{Adv}^{\mathsf{Privacy}}_{\mathsf{PPAKE},\mu,\ell,\mathcal{A}}$ *satisfies*

$$\mathsf{Adv}^{\mathsf{Privacy}}_{\mathsf{PPAKE},\mu,\ell,\mathcal{A}} := \left| \Pr\left[\mathsf{Exp}^{\mathsf{Privacy}}_{\mathsf{PPAKE},\mu,\ell,\mathcal{A}} \Rightarrow 1 \right] - 1/2 \right| \leq \mathsf{negl}(\lambda).$$

Remark 3 (Difference with security models in [19, 20]*).* In the security models in [19, 20], the initiator only deals with one responding message with accept or reject and does not take into account other users' responses. This feature excludes the application of their PPAKE schemes in broadcast channels or similar scenarios. In our security model, the initiator receives and processes all messages from other users. This is especially important in the scenario where every user may give a response when not aware whether itself is the target recipient. More precisely, in our security model, the adversarial behaviors are reflected by the formalization that $\mathcal{A}$ designates *a list of messages* for π_i^s to deal with by Send or Respond queries. In comparison, the security models in [19, 20] only consider the case that π_i^s deals with a single message and after that π_i^s will stop responding to other messages (from other users).

Remark 4 (The best forward privacy for robust PPAKE). The best forward privacy for a robust PPAKE scheme is full forward privacy for initiator and semi-forward privacy for responder. The reason is as follows. If the responder P_j is corrupted, the robustness of PPAKE enables the adversary to use the responder's secret key to test the first round messages in previous sessions so as to determine whether P_j is the intended recipient. Therefore, this is the optimal forward privacy for robust PPAKE to achieve: full forward privacy for initiator (no matter initiator or responder is corrupted) and forward privacy for responder when initiator is corrupted.

4   Generic Construction of PPAKE and Its Security Proof

We propose a generic construction of PPAKE = (PPAKE.Setup, PPAKE.Gen, PPAKE.Protocol) with session key space $\mathcal{K}_1$ from the following building blocks.

- A signature scheme SIG = (SIG.Setup, SIG.Sign, SIG.Ver).
- A key encapsulation mechanism scheme KEM = (KEM.Setup, Encap, Decap) with encapsulation key space $\mathcal{K}$.
- A one-time key encapsulation mechanism scheme otKEM = (otKEM.Setup, otEncap, otDecap) with the encapsulation key space $\mathcal{K}'$.
- A message authentication code scheme MAC = (MAC.Tag, MAC.Ver) with key space $\mathcal{K}$.
- A symmetric encryption scheme SE = (SEnc, SDec) with key space $\mathcal{K}_2$.
- A pseudo-random generator PRG : $\mathcal{K}' \rightarrow \mathcal{K}_1 \times \mathcal{K}_2$.

Our generic construction is given in Fig. 5.

PPAKE.Setup: The setup algorithm generates the public parameter $pp_{PPAKE} :=$ ($pp_{SIG}, pp_{KEM}, pp_{otKEM}$) by running SIG.Setup, KEM.Setup and otKEM.Setup.

PPAKE.Gen: The key generation algorithm takes as input pp_{PPAKE} and a user identity i, and generates a key pair (vk_i, ssk_i) for SIG and a key pair (pk_i, sk_i) for KEM. The public key of user i is (pk_i, vk_i) and the secret key is (ssk_i, sk_i).

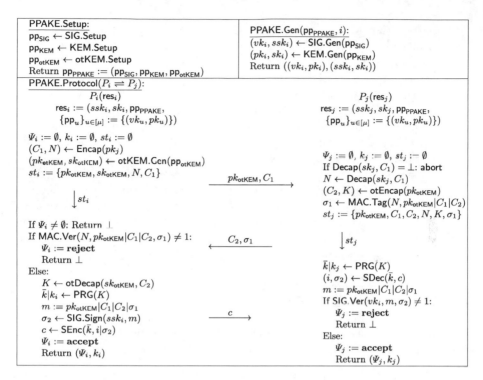

Fig. 5. Generic construction of PPAKE

PPAKE.Protocol($P_i \rightleftharpoons P_j$): The protocol between two parties P_i and P_j is as follows. Each party has access to their own resources $\text{res}_i = (ssk_i, sk_i, \text{pp}_{\text{PPAKE}}, \{\text{pp}_u\}_{u\in[\mu]})$ and $\text{res}_j = (ssk_j, sk_j, \text{pp}_{\text{PPAKE}}, \{\text{pp}_u\}_{u\in[\mu]})$ which contain the corresponding secret key, the public parameter and a list PKList consisting of the public keys of all users. Each party initializes its local variables Ψ_i, k_i and st_i with the empty string. The protocol consists of three rounds of communications.

The First Round: When party P_i initiates a session with party P_j in PPAKE, P_i computes $(C_1, N) \leftarrow \text{Encap}(pk_j)$ and generates an ephemeral key pair $(pk_{\text{otKEM}}, sk_{\text{otKEM}}) \leftarrow \text{otKEM.Gen}(\text{pp}_{\text{otKEM}})$. It then sends (pk_{otKEM}, C_1) to P_j and stores $(pk_{\text{otKEM}}, sk_{\text{otKEM}}, N, C_1)$ as its state st_i.

The Second Round: After receiving message (pk_{otKEM}, C_1), P_j computes $N \leftarrow \text{Decap}(sk_j, C_1)$. If $N = \bot$, then P_j aborts, indicating that it is not the intended recipient of this message. Otherwise, P_j invokes $(C_2, K) \leftarrow \text{otEncap}(pk_{\text{otKEM}})$. It uses N as the MAC key to compute a tag $\sigma_1 \leftarrow \text{MAC}(N, pk_{\text{otKEM}}|C_1|C_2)$. Then it sends (C_2, σ_1) to P_i and stores $(pk_{\text{otKEM}}, C_1, C_2, \sigma_1, N, K)$ as its state st_j.

The Third Round: After receiving message (C_2, σ_1), P_i retrieves its state $st_i = (pk_{\text{otKEM}}, sk_{\text{otKEM}}, N, C_1)$. It verifies the validity of σ_1 by checking whether $\mathsf{MAC.Tag}(N, pk_{\mathsf{otKEM}}|C_1|C_2, \sigma_1) = 1$ with the help of N. If invalid, it rejects this message. Otherwise, it continues the protocol by computing $K \leftarrow \mathsf{Decap}(sk_{\mathsf{otKEM}}, C_2)$. It then generates $\bar{k}|k_i \leftarrow \mathsf{PRG}(K)$, where $\bar{k}$ is used as the secret key for SE and k_i as its session key. P_i uses its signing key ssk_i to sign $pk_{\mathsf{otKEM}}|C_1|C_2|\sigma_1$ and obtain the signature $\sigma_2 \leftarrow \mathsf{SIG.Sign}(ssk_i, pk_{\mathsf{otKEM}}|C_1|C_2|\sigma_1)$. Then it encrypts the identity i and the signature σ_2 with $\bar{k}$ and obtains $c \leftarrow \mathsf{SEnc}(\bar{k}, i|\sigma_2)$. It broadcasts the ciphertext c, and sets $\Psi_i = \mathbf{accept}$ and outputs (Ψ_i, k_i), indicating its acceptance of k_i as its session key.

After receiving c, P_j retrieves its state $st_j = (pk_{\mathsf{otKEM}}, C_1, C_2, \sigma_1, N, K)$ and generates $(\bar{k}, k_j) \leftarrow \mathsf{PRG}(K)$. It then uses $\bar{k}$ to decrypt the ciphertext c and obtains $(i, \sigma_2) \leftarrow \mathsf{SDec}(\bar{k}, c)$. Next it checks that the validity of (i, σ_2) by checking $\mathsf{SIG.Ver}(vk_i, pk_{\mathsf{otKEM}}|C_1|C_2|\sigma_1, \sigma_2) = 1$. P_j rejects in case of invalid. Otherwise, it sets $\Psi_j = \mathbf{accept}$ and outputs (Ψ_j, k_j), indicating its acceptance of k_j as its session key with P_i.

Correctness. Correctness of PPAKE follows directly from the correctness of $\mathsf{SIG}, \mathsf{KEM}, \mathsf{otKEM}, \mathsf{MAC}$ and SE.

Robustness. Robustness of PPAKE follows directly from the robustness of KEM, which guarantees that only P_j has $\mathsf{Decap}(sk_j, C_1) \neq \bot$.

Theorem 1. *For the* PPAKE *construction in Fig. 5, suppose that the underlying* SIG *is sEUF-CMA secure,* MAC *is sEUF-CMA secure,* KEM *is IND-CCA secure and IK-CCA secure,* otKEM *is IND-CPA secure and has the properties of key uniformity and public key diversity, and* PRG *is a pseudo-random generator, and* SE *is semantic secure and has the property of ciphertext diversity, then the* PPAKE *construction has explicit mutual authenticity, forward security and forward privacy.*

Before the proof, we will first define two sets Sent_i^s and Recv_i^s for oracle π_i^s. Set Sent_i^s will store outgoing messages of the oracle and Recv_i^s will store valid incoming messages, respectively. We stress that valid messages in Recv_i^s are those incoming messages that pass the verification of MAC or SIG.

We know that $\mathsf{Partner}(\pi_i^s \leftarrow \pi_j^t)$ holds if the following conditions are satisfied.

- $\mathsf{Pid}_i^s = j$ and $\Psi_i^s = \mathbf{accept}$.
- If π_i^s is the initiator, i.e., π_i^s has sent the first message, then $\mathsf{Sent}_i^s = \mathsf{Recv}_j^t = \{(pk_{\mathsf{otKEM}}, C_1)\}$ and $\mathsf{Recv}_i^s = \mathsf{Sent}_j^t = \{(C_2, \sigma_1)\}$.
- If π_i^s is the responder, i.e., π_i^s has received the first message, then $\mathsf{Sent}_i^s = \mathsf{Recv}_j^t = \{(C_2, \sigma_1)\}$, and $\mathsf{Recv}_i^s = \mathsf{Sent}_j^t = \{(pk_{\mathsf{otKEM}}, C_1), c\}$.

Besides, we define a set S recording all the pairs (i, s) such that $\mathsf{Win}_{\mathsf{Auth}} = \mathbf{true}$.

Proof of Explicit Mutual Authenticity. To prove authenticity for PPAKE, we now describe a sequence of games $\mathsf{G}_0 \text{--} \mathsf{G}_3$ and show that the advantage of $\mathcal{A}$ in adjacent games. The full codes of $\mathsf{G}_0 \text{--} \mathsf{G}_3$ are also given in Fig. 6. Define Win_i as the event of $\mathsf{Win}_{\mathsf{Auth}} = \mathbf{true}$ in $\mathsf{G}_i \wedge (i^*, s^*) \in S$, where $(i^*, s^*) \leftarrow_\$ [\mu] \times [\ell]$.

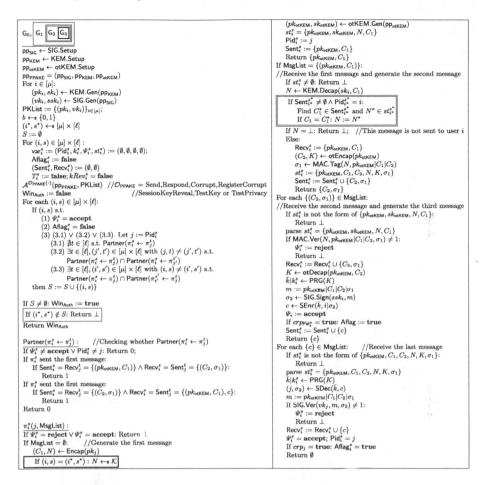

Fig. 6. Games G_0––G_3 for authenticity of PPAKE. Queries to $\mathcal{O}_{\mathsf{PPAKE}} \in$ {Send, Respond, Corrupt, RegisterCorrupt, SessionKeyReveal, TestPrivacy, TestKey} are defined as in the original game in Fig. 4 and omitted here.

Game G_0: G_0 is the original experiment $\mathsf{Exp}^{\mathsf{AUTH}}_{\mathsf{PPAKE},\mu,\ell,\mathcal{A}}$. In addition, challenger $\mathcal{C}$ uses Sent^s_i and Recv^s_i recording valid incoming valid messages and outgoing messages for π^s_i. This is only a conceptual change. So, $\Pr\left[(i^*, s^*) \in S \mid \mathsf{Win}_{\mathsf{Auth}} = \mathbf{true}\right] = \Pr\left[\mathsf{Win}_0\right]/\Pr\left[\mathsf{Win}_{\mathsf{Auth}} = \mathbf{true}\right] \geq \frac{1}{\mu\ell}$. Then

$$\Pr\left[\mathsf{Win}_{\mathsf{Auth}} = \mathbf{true}\right] \leq \mu\ell \cdot \Pr\left[\mathsf{Win}_0\right]. \tag{1}$$

Game G_1: In G_1, challenger $\mathcal{C}$ first chooses $(i^*, s^*) \leftarrow_{\$} [\mu] \times [\ell]$. At the end of G_1, if $(i^*, s^*) \notin S$, G_1 aborts by returning $\bot$. Then for the specific pair (i^*, s^*),

$$\Pr\left[\mathsf{Win}_1\right] = \Pr\left[\mathsf{Win}_0\right] = \Pr_{(i^*, s^*)}\left[(1) \wedge (2) \wedge (3)\right]. \tag{2}$$

Game G_2: In G_2, if $\pi_{i^*}^{s^*}$ is a responder, G_2 is the same as G_1. If $\pi_{i^*}^{s^*}$ is an initiator and $\mathsf{Pid}_{i^*}^{s^*} = j^*$, $\mathsf{Sent}_{i^*}^{s^*} \neq \emptyset$, $\mathcal{C}$ changes the behavior of $\pi_{j^*}^t$ for $t \in [\ell]$.

Note $\mathsf{Sent}_{i^*}^{s^*} \neq \emptyset$ implies that $\exists (pk_{\mathsf{otKEM}}^*, C_1^*) \in \mathsf{Sent}_{i^*}^{s^*}$, where $(pk_{\mathsf{otKEM}}^*, sk_{\mathsf{otKEM}}^*) \leftarrow \mathsf{otKEM.Gen}(\mathsf{pp}_{\mathsf{otKEM}})$ and $(C_1^*, N^*) \leftarrow \mathsf{Encap}(pk_{j^*})$. Meanwhile, $\pi_{i^*}^{s^*}$ also has state $st_{i^*}^{s^*} = \{pk_{\mathsf{otKEM}}^*, sk_{\mathsf{otKEM}}^*, N^*, C_1^*\}$. Then for $\forall t \in [\ell]$, if $(pk_{\mathsf{otKEM}}^*, C_1) \in \mathsf{Recv}_{j^*}^t$, oracle $\pi_{j^*}^t(pk_{\mathsf{otKEM}}^*, C_1)$ will compute N' by $N \leftarrow \mathsf{Decap}(sk_{j^*}, C_1)$ in G_1. But in G_2, $\pi_{j^*}^t(pk_{\mathsf{otKEM}}^*, C_1)$ computes N' in the following way.

- $C_1 = C_1^*$: $\pi_{j^*}^t$ borrows N^* from $st_{i^*}^{s^*}$ and sets $N := N^*$.
- $C_1 \neq C_1^*$: $\pi_{j^*}^t$ computes $N \leftarrow \mathsf{Decap}(sk_{j^*}, C_1)$ (as in G_1).

Due to the correctness of KEM, we have

$$\Pr[\mathsf{Win}_2] = \Pr[\mathsf{Win}_1]. \tag{3}$$

Game G_3: In G_3, if $\pi_{i^*}^{s^*}$ is a responder, G_2 is the same as G_1. If $\pi_{i^*}^{s^*}$ is an initiator, then the encapsulation key N^* is randomly chosen with $N^* \leftarrow_{\$} \mathcal{K}$, instead of $N^* \leftarrow \mathsf{Encap}(pk_{j^*})$ as in G_2.

Lemma 1. $|\Pr[\mathsf{Win}_2] - \Pr[\mathsf{Win}_3]| \leq \mu \cdot \mathsf{Adv}_{\mathsf{KEM}}^{\mathsf{CCA}}(\mathcal{B}_{\mathsf{KEM}})$.

The formal proof of Lemma 1 is given in the full version [18]. Here we sketch the proof. We construct adversary $\mathcal{B}_{\mathsf{KEM}}$ against IND-CCA security of KEM scheme. $\mathcal{B}_{\mathsf{KEM}}$ will simulates G_2/G_3 for $\mathcal{A}$. $\mathcal{B}_{\mathsf{KEM}}$ gets its challenge (C^*, K^*) w.r.t. pk^*, it sets $pk_{j^*} := pk^*$ with $j^* \leftarrow_{\$} [\mu]$, and embeds C^* into $\pi_{i^*}^{s^*}$'s output message $(pk_{\mathsf{otKEM}}^*, C_1^* := C^*)$ and embeds K^* into its state $st_{i^*}^{s^*} := (pk_{\mathsf{otKEM}}^*, sk_{\mathsf{otKEM}}^*, N^* = K^*, C_1^* = C^*)$. $\mathcal{B}_{\mathsf{KEM}}$ also asks its own DECAP oracle $\mathcal{O}_{\mathsf{Decap}}$ to simulate decapsulation of $C_1 \neq C^*$ for oracle $\pi_{j^*}^t(pk_{\mathsf{otKEM}}^*, C_1)$. Finally, $\mathcal{B}_{\mathsf{KEM}}$ outputs 1 iff Win occurs and $j^* = \mathsf{Pid}_{i^*}^{s^*}$. If K^* is an encapsulated key for C^*, $\mathcal{B}_{\mathsf{KEM}}$ simulates G_2; if K^* is random, $\mathcal{B}_{\mathsf{KEM}}$ simulates G_3. Since $j^* = \mathsf{Pid}_{i^*}^{s^*}$ with probability $1/\mu$, we have $|\Pr[\mathsf{Win}_2] - \Pr[\mathsf{Win}_3]| \leq \mu \cdot \mathsf{Adv}_{\mathsf{KEM}}^{\mathsf{CCA}}(\mathcal{B}_{\mathsf{KEM}})$.

Next, we analyze $(1), (2), (3.1), (3.2), (3.3)$ in G_3 so as to determine $\Pr[\mathsf{Win}_{\mathsf{Auth}}]$.

We define the event $\mathsf{NoPartner}(i, s)$ as $(1) \wedge (2) \wedge (3.1)$ happens for (i, s). Equivalently, π_i^s accepts, the intended partner $j := \mathsf{Pid}_i^s$ is uncorrupted when π_i^s accepts, and there does not exist $t \in [\ell]$ such that $\mathsf{Partner}(\pi_i^s \leftarrow \pi_j^t)$.

Lemma 2. In G_3, we have $\Pr_{(i^*, s^*)}[(1) \wedge (2) \wedge (3.1)]$

$$= \Pr[\mathsf{NoPartner}(i^*, s^*)] \leq \mathsf{Adv}_{\mathsf{MAC}}^{\mathsf{sEUF\text{-}CMA}}(\mathcal{B}_{\mathsf{MAC}}) + \mu \cdot \mathsf{Adv}_{\mathsf{SIG}}^{\mathsf{sEUF\text{-}CMA}}(\mathcal{B}_{\mathsf{SIG}}).$$

This proof of Lemma 2 relies on the sEUF-CMA security of SIG and MAC.

We consider the probability of event $\mathsf{NoPartner}(i^*, s^*)$ in two cases: $\pi_{i^*}^{s^*}$ is an initiator and $\pi_{i^*}^{s^*}$ is a responder. In the first case, $\pi_{i^*}^{s^*}$ must have received a message (C_2^*, σ_1^*) such that σ_1^* is a valid MAC tag for some non-consistent message $pk^*_{\mathsf{otKEM}}|C^*_1|C^*_2$, yielding a fresh and valid forgery for MAC. In the second case, $\pi^{s^*}_{i^*}$ must have received non-consistent messages $(pk^*_{\mathsf{otKEM}}, C^*_1)$ and c^* whose decryption results in (j^*, σ^*_2), and σ^*_2 must be a valid signature for message $pk^*_{\mathsf{otKEM}}|C^*_1|C^*_2|\sigma^*_1$. Due to the ciphertext diversity of SE, $c \neq c^*$ implies that $(j^*, \sigma^*_2) \neq (j', \sigma^*_2)$. If $\mathsf{NoPartner}(i^*, s^*)$ happens, then $(pk^*_{\mathsf{otKEM}}|C^*_1|C^*_2|\sigma^*_1, \sigma^*_2)$ must be a fresh and valid message-signature pair, yielding a successful forgery for SIG. The formal proof is given in the full version [18].

Furthermore, considering the random selection of (i^*, s^*), in G_3 we have

$$\Pr_{\exists(i,s)}[(1) \wedge (2) \wedge (3.1)] \leq \mu\ell \cdot (\mathsf{Adv}^{\mathsf{sEUF\text{-}CMA}}_{\mathsf{MAC}}(\mathcal{B}_{\mathsf{MAC}}) + \mu \cdot \mathsf{Adv}^{\mathsf{sEUF\text{-}CMA}}_{\mathsf{SIG}}(\mathcal{B}_{\mathsf{SIG}})). \quad (4)$$

By Lemma 1 and Eq. (1) (2) (3) and (4), we have the following corollary.

Corollary 1. *In* $\mathsf{Exp}^{\mathsf{AUTH}}_{\mathsf{PPAKE},\mu,\ell,\mathcal{A}}$, *it holds that* $\Pr_{\exists(i,s)}[(1) \wedge (2) \wedge (3.1)]$

$$\leq (\mu\ell) \cdot \left(\mu \cdot \mathsf{Adv}^{\mathsf{CCA}}_{\mathsf{KEM}}(\mathcal{B}_{\mathsf{KEM}}) + \mathsf{Adv}^{\mathsf{sEUF\text{-}CMA}}_{\mathsf{MAC}}(\mathcal{B}_{\mathsf{MAC}}) + \mu \cdot \mathsf{Adv}^{\mathsf{sEUF\text{-}CMA}}_{\mathsf{SIG}}(\mathcal{B}_{\mathsf{SIG}})\right).$$

Lemma 3. *In* G_3, *we have*

$$\Pr_{(i^*,s^*)}[(1) \wedge (2) \wedge (3.2)] \leq (\mu\ell)^2 \cdot \left(\mathsf{Adv}^{\mathsf{ps}}_{\mathsf{PRG}}(\mathcal{B}_{\mathsf{PRG}}) + \frac{1}{|\mathcal{K}_2|}\right).$$

If $(1) \wedge (2) \wedge (3.2)$ happens for (i^*, s^*) in G_3, then $\pi^{s^*}_{i^*}$ will accept with session key $k^{s^*}_{i^*}$ and there exist two oracles π^t_j and $\pi^{t'}_{j'}$ subject to $\mathsf{Partner}(\pi^{s^*}_{i^*} \leftarrow \pi^t_j)$ and $\mathsf{Partner}(\pi^{s^*}_{i^*} \leftarrow \pi^{t'}_{j'})$. Then $\pi^{s^*}_{i^*}$ must share the same session key with both π^t_j and $\pi^{t'}_{j'}$, which happens with negligible probability, due to the independent randomness in $\pi^{s^*}_{i^*}$, π^t_j and $\pi^{t'}_{j'}$, the key uniformity of otKEM, and the pseudorandomness of PRG. The formal proof is shown in the full version [18].

By Lemma 3 and Eq. (1) (2) (3), we have the following corollary.

Corollary 2. *In* $\mathsf{Exp}^{\mathsf{AUTH}}_{\mathsf{PPAKE},\mu,\ell,\mathcal{A}}$, *we have*

$$\Pr_{\exists(i,s)}[(1) \wedge (2) \wedge (3.2)] \leq (\mu\ell)^3 \cdot \left(\mathsf{Adv}^{\mathsf{ps}}_{\mathsf{PRG}}(\mathcal{B}_{\mathsf{PRG}}) + \frac{1}{|\mathcal{K}_2|}\right) + (\mu^2\ell) \cdot \mathsf{Adv}^{\mathsf{CCA}}_{\mathsf{KEM}}(\mathcal{B}_{\mathsf{KEM}}).$$

Lemma 4. *In* $\mathsf{Exp}^{\mathsf{AUTH}}_{\mathsf{PPAKE},\mu,\ell,\mathcal{A}}$, *we have*

$$\Pr_{\exists(i,s)}[(1) \wedge (2) \wedge (3.3)] \leq \Pr_{\exists(i,s)}[(1) \wedge (2) \wedge (3.2)] + (\mu\ell)^2 \cdot 2^{-\gamma}.$$

Proof. If $\exists(i^*, s^*)$ satisfies $(1) \wedge (2) \wedge (3.3)$, then $\Psi^{s^*}_{i^*} = \mathbf{accept}$, $\mathsf{Aflag}^{s^*}_{i^*} = \mathbf{false}$, $\mathsf{Partner}(\pi^{s^*}_{i^*} \leftarrow \pi^t_j)$ and $\mathsf{Partner}(\pi^{s^*}_{i'} \leftarrow \pi^t_j)$. We consider the following two cases.

- **Initiator** $\pi^{s^*}_{i^*}$. According to the definition, we know that $\mathsf{Partner}(\pi^{s^*}_{i^*} \leftarrow \pi^t_j)$ and $\mathsf{Partner}(\pi^{s'}_{i'} \leftarrow \pi^t_j)$ implies $(pk^*_{\mathsf{otKEM}}, C^*_1) \in \mathsf{Sent}^{s^*}_{i^*} = \mathsf{Recv}^t_j$, $(pk'_{\mathsf{otKEM}}, C'_1) \in \mathsf{Sent}^{s'}_{i'} = \mathsf{Recv}^t_j$, $(C^*_2, \sigma^*_1) \in \mathsf{Recv}^{s^*}_{i^*} = \mathsf{Sent}^t_j$,

$(C_2', \sigma_1') \in \mathsf{Recv}_{i'}^{s'} = \mathsf{Sent}_j^t$. Then it holds that $(pk_{\mathsf{otKEM}}^*, C_1^*, C_2^*) = (pk_{\mathsf{otKEM}}', C_1', C_2')$. According to the γ -pk-diversity of otKEM, we know that $\Pr\left[pk_{\mathsf{otKEM}}' = pk_{\mathsf{otKEM}}\right] = 2^{-\gamma}$. Therefore, $(1) \wedge (2) \wedge (3.3)$ happens for (i^*, s^*) and (i', s') with probability at most $2^{-\gamma}$. As there are at most $(\mu\ell)^2$ choices of (i^*, s^*) and (i', s'), we can upper bound the probability of event $(1) \wedge (2) \wedge (3.3)$ by $(\mu\ell)^2 \cdot 2^{-\gamma}$ in this case.

- **Responder** $\pi_{i^*}^{s^*}$. In this case, $\mathsf{Partner}(\pi_{i^*}^{s^*} \leftarrow \pi_j^t)$ implies $\mathsf{Partner}(\pi_j^t \leftarrow \pi_{i^*}^{s^*})$ and $\mathsf{Partner}(\pi_{i'}^{s'} \leftarrow \pi_j^t)$ implies $\mathsf{Partner}(\pi_j^t \leftarrow \pi_{i'}^{s'})$. This further implies that $(1) \wedge (2) \wedge (3.2)$ happens for (j, t). Therefore, we can upper bound the probability of event $(1) \wedge (2) \wedge (3.3)$ by $(1) \wedge (2) \wedge (3.2)$ in this case.

Combining the above two cases yields Lemma 4. $\qquad\qquad\qquad\square$

Finally, the authenticity of PPAKE follows from Corollary 1, 2 and Lemma 4 and

$$\Pr\left[\mathsf{Win}_{\mathsf{Auth}}\right] \leq 3\mu^2\ell \cdot \mathsf{Adv}_{\mathsf{KEM}}^{\mathsf{CCA}}(\mathcal{B}_{\mathsf{KEM}}) + \mu\ell \cdot \mathsf{Adv}_{\mathsf{MAC}}^{\mathsf{sEUF\text{-}CMA}}(\mathcal{B}_{\mathsf{MAC}}) + (\mu\ell)^2 \cdot 2^{-\gamma}$$
$$+ 2(\mu\ell)^3 \cdot \left(\mathsf{Adv}_{\mathsf{PRG}}^{\mathsf{ps}}(\mathcal{B}_{\mathsf{PRG}}) + \frac{1}{|\mathcal{K}_2|}\right) + \mu^2\ell \cdot \mathsf{Adv}_{\mathsf{SIG}}^{\mathsf{sEUF\text{-}CMA}}(\mathcal{B}_{\mathsf{SIG}}).$$
$$\tag{5}$$

Proof of forward security for session key. We now consider another sequence of games $\mathbf{G}_0\text{--}\mathbf{G}_5$ and analyze $\mathcal{A}$'s advantages in these games. Let Win_i denote the event that $\mathbf{G}_i$ outputs 1, i.e. $\mathcal{A}$'s output bit satisfies $b^* = b$ in $\mathbf{G}_i$. Let $adv_i := |\Pr\left[\mathsf{Win}_i\right] - 1/2|$. Then $|adv_i - adv_{i+1}| \leq |\Pr\left[\mathsf{Win}_i\right] - \Pr\left[\mathsf{Win}_{i+1}\right]|$. The full codes of $\mathbf{G}_0\text{--}\mathbf{G}_4$ are presented in Fig. 7.

Game $\mathbf{G}_0$: $\mathbf{G}_0$ is the original experiment $\mathsf{Exp}_{\mathsf{PPAKE},\mu,\ell,\mathcal{A}}^{\mathsf{IND}}$. We add the sets Sent_i^s and Recv_i^s which is only a conceptual change. So,

$$\mathsf{Adv}_{\mathsf{PPAKE},\mu,\ell,\mathcal{A}}^{\mathsf{IND}} := |\Pr\left[\mathsf{Win}_0\right] - 1/2| = adv_0. \tag{6}$$

Game $\mathbf{G}_1$: Challenger $\mathcal{C}$ will check whether event $\mathsf{Win}_{\mathsf{Auth}}$ occurs in $\mathbf{G}_1$. If $\mathsf{Win}_{\mathsf{Auth}}$ occurs, $\mathcal{C}$ will abort the game by returning 0. Otherwise, $\mathbf{G}_1$ is the same as $\mathbf{G}_0$. Then $|\Pr\left[\mathsf{Win}_0\right] - \Pr\left[\mathsf{Win}_1\right]| \leq \Pr\left[\mathsf{Win}_{\mathsf{Auth}}\right]$. By (5), we have

$$|adv_0 - adv_1| \leq 3\mu^2\ell \cdot \mathsf{Adv}_{\mathsf{KEM}}^{\mathsf{CCA}}(\mathcal{B}_{\mathsf{KEM}}) + \mu\ell \cdot \mathsf{Adv}_{\mathsf{MAC}}^{\mathsf{sEUF\text{-}CMA}}(\mathcal{B}_{\mathsf{MAC}}) + (\mu\ell)^2 \cdot 2^{-\gamma}$$
$$+2(\mu\ell)^3 \cdot \left(\mathsf{Adv}_{\mathsf{PRG}}^{\mathsf{ps}}(\mathcal{B}_{\mathsf{PRG}}) + \frac{1}{|\mathcal{K}_2|}\right) + \mu^2\ell \cdot \mathsf{Adv}_{\mathsf{SIG}}^{\mathsf{sEUF\text{-}CMA}}(\mathcal{B}_{\mathsf{SIG}}) \tag{7}$$

Game $\mathbf{G}_2$: In $\mathbf{G}_2$, if event Hit does not occur, $\mathcal{C}$ will return a random bit $\theta \leftarrow_\$ \{0, 1\}$. Otherwise, $\mathbf{G}_2$ is the same as $\mathbf{G}_1$. Event Hit is defined as follows. Randomly choose $(i^*, s^*, j^*, t^*) \leftarrow_\$ ([\mu] \times [\ell])^2$. If $\mathcal{A}$ queried $\mathsf{TestKey}(i, s)$ and $\mathsf{TestKey}(i, s)$ did not reply $\perp$, then π_i^s must accept and $\mathsf{Aflag}_i^s = \mathbf{false}$. Accordingly, π_i^s must have a unique partner π_j^t such that $\mathsf{Partner}(\pi_i^s \leftarrow \pi_j^t)$. So $\mathsf{TestKey}(i, s)$ uniquely determines a tuple (i, s, j, t). Event Hit occurs if and only if $(i^*, s^*, j^*, t^*) = (i', s', j', t')$. Obviously, $\Pr\left[\mathsf{Hit}\right] = 1/(\mu\ell)^2$. We have $\Pr\left[\mathsf{Win}_2\right] = \Pr\left[\mathsf{Hit}\right] \cdot \Pr\left[\mathsf{Win}_1\right] + \Pr\left[\overline{\mathsf{Hit}}\right] \cdot \frac{1}{2} = \Pr\left[\mathsf{Hit}\right] \cdot (\frac{1}{2} \pm adv_1) + \Pr\left[\overline{\mathsf{Hit}}\right] \cdot \frac{1}{2} = \frac{1}{2} \pm \frac{1}{(\mu\ell)^2} \cdot adv_1$. Hence,

$$adv_1 = (\mu\ell)^2 \cdot adv_2. \tag{8}$$

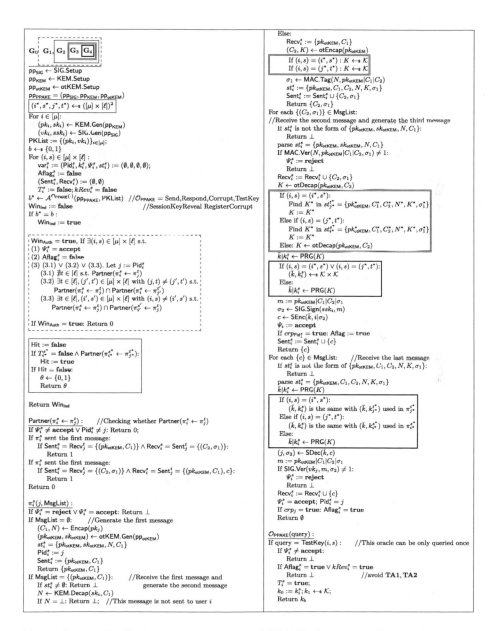

Fig. 7. Games $\mathbf{G_0}$-$\mathbf{G_4}$ for forward security of PPAKE. Queries to $\mathcal{O}_{\mathsf{PPAKE}}$ where query $\in \{\mathsf{Send}, \mathsf{Respond}, \mathsf{Corrupt}, \mathsf{RegisterCorrupt}, \mathsf{SessionKeyReveal}\}$ are defined as in the original game in Fig. 4.

Game G_3: In G_3, the encapsulation key K shared $\pi_{i^*}^{s^*}$ and $\pi_{j^*}^{t^*}$ is generated by $K \leftarrow_\$ \mathcal{K}$. Recall that in G_2, $\pi_{i^*}^{s^*}$ computes K with $(C, K) \leftarrow \mathsf{otEncap}(pk_{\mathsf{otKEM}})$ while $\pi_{j^*}^{t^*}$ computes K with $K \leftarrow \mathsf{otDecap}(sk_{\mathsf{otKEM}}, C)$.

Lemma 5. $|adv_2 - adv_3| \leq |\Pr[\mathsf{Win}_2] - \Pr[\mathsf{Win}_3]| \leq \mathsf{Adv}_{\mathsf{otKEM}}^{\mathsf{CPA}}(\mathcal{B}_{\mathsf{otKEM}})$.

Recall that in G_2, if $\pi_{i^*}^{s^*}$ accepts session key $k_{i^*}^{s^*}$ and $\mathsf{Aflag}_{i^*}^{s^*} = \mathbf{false}$, then there must exist $\pi_{j^*}^{t^*}$ such that $\mathsf{Partner}(\pi_{i^*}^{s^*} \leftarrow \pi_{j^*}^{t^*})$. To prove this lemma, we construct an adversary $\mathcal{B}_{\mathsf{otKEM}}$ against the CPA security of otKEM. Given the challenge (C^*, K^*) w.r.t pk^*, $\mathcal{B}_{\mathsf{otKEM}}$ embeds C^* as C_2^* and pk^* as pk_{otKEM}^* in the transcript between $\pi_i^{s^*}$ and $\pi_j^{t^*}$ and sets K^* in the state $st_{i^*}^{s}$ or $st_{j^*}^{t}$. Finally, $\mathcal{A}$ outputs a guessing bit b^*. If $b^* = b$, $\mathcal{B}_{\mathsf{otKEM}}$ outputs 1; otherwise, $\mathcal{B}_{\mathsf{otKEM}}$ outputs 0.

If K^* is the encapsulated key for C^*, then $\mathcal{B}_{\mathsf{otKEM}}$ perfectly simulates G_2 for $\mathcal{A}$; it K^* is random, then $\mathcal{B}_{\mathsf{otKEM}}$ perfectly simulates G_3 for $\mathcal{A}$. Then, we have $|adv_2 - adv_3| \leq |\Pr[\mathsf{Win}_2] - \Pr[\mathsf{Win}_3]| \leq \mathsf{Adv}_{\mathsf{otKEM}}^{\mathsf{CPA}}(\mathcal{B}_{\mathsf{otKEM}})$.

The detailed proof is shown in the full version [18].

Game G_4: In G_4, the symmetric key and session key of $\pi_{i^*}^{s^*}$ and $\pi_{j^*}^{t^*}$ are uniformly sampled by $(\bar{k}, k_{i^*}^{s^*} = k_{j^*}^{t^*}) \leftarrow_\$ \mathcal{K}_1 \times \mathcal{K}_2$. Recall that in G_3, they are generated by $\bar{k} | k_{i^*}^{s^*} \leftarrow \mathsf{PRG}(K)$. Due to the pseudo-randomness of PRG, we have

$$|adv_3 - adv_4| \leq |\Pr[\mathsf{Win}_3] - \Pr[\mathsf{Win}_4]| \leq \mathsf{Adv}_{\mathsf{PRG}}^{\mathsf{ps}}(\mathcal{B}_{\mathsf{PRG}}). \tag{9}$$

Now that the session key of $\pi_{i^*}^{s^*}$ is randomly chosen with $k_{i^*}^{s^*} \leftarrow_\$ \mathcal{K}$, we have

$$adv_4 = |\Pr[\mathsf{Win}_4] - 1/2| = 0. \tag{10}$$

Finally, the forward security of PPAKE follows from Lemma 5 and Eq. (6)–(10).

Proof of Forward Privacy for User Identity. To this end, we now consider another sequence of games G_0'-G_7'. Let Win_i denote the event that $\mathsf{Win}_{\mathsf{Privacy}} = \mathbf{true}$ in G_i'. Let $adv_i := |\Pr[\mathsf{Win}_i] - 1/2|$. Then $|adv_i - adv_{i+1}| := |\Pr[\mathsf{Win}_i] - \Pr[\mathsf{Win}_{i+1}]|$. The full codes of G_0'-G_7' are presented in Fig. 8.

Game G_0': G_0' is the original experiment $\mathsf{Exp}_{\mathsf{PPAKE}, \mu, \ell, \mathcal{A}}^{\mathsf{Privacy}}$. We also add the sets Sent_i^s and Recv_i^s which is only a conceptual change. So,

$$\mathsf{Adv}_{\mathsf{PPAKE}, \mu, \ell, \mathcal{A}}^{\mathsf{Privacy}} := |\Pr[\mathsf{Win}_{\mathsf{Privacy}}] - 1/2| = adv_0 \tag{11}$$

Game G_1': At the end of G_1', challenger $\mathcal{C}$ will check whether event $\mathsf{Win}_{\mathsf{Auth}}$ occurs. If $\mathsf{Win}_{\mathsf{Auth}}$ occurs, $\mathcal{C}$ will abort the game by returning 0. Otherwise, G_1' is the same as G_0'. Due to the difference lemma and (5), we have

$$|adv_0 - adv_1| \leq 3\mu^2\ell \cdot \mathsf{Adv}_{\mathsf{KEM}}^{\mathsf{CCA}}(\mathcal{B}_{\mathsf{KEM}}) + \mu\ell \cdot \mathsf{Adv}_{\mathsf{MAC}}^{\mathsf{sEUF\text{-}CMA}}(\mathcal{B}_{\mathsf{MAC}}) + (\mu\ell)^2 2^{-\gamma}$$
$$+ \mu^2\ell \cdot \mathsf{Adv}_{\mathsf{SIG}}^{\mathsf{sEUF\text{-}CMA}}(\mathcal{B}_{\mathsf{SIG}}) + 2(\mu\ell)^3 \cdot \left(\mathsf{Adv}_{\mathsf{PRG}}^{\mathsf{ps}}(\mathcal{B}_{\mathsf{PRG}}) + \tfrac{1}{|\mathcal{K}_2|}\right) \tag{12}$$

Game G_2': In G_2', upon $\mathcal{A}$'s query to oracle $\mathsf{Tran}(i, j)$, π_i^0 and π_j^0 will not respond to any message in InsertList sent by $\mathcal{A}$. Note that each oracle responds to only one valid message. If this valid message is not sent by $\mathcal{A}$, then $\mathbf{G}'_2$ is the same as $\mathbf{G}'_1$. If this valid message is sent by $\mathcal{A}$ (the message can only be inserted in the second round or third round of our protocol), then this will lead to occurrence of event $\mathsf{NoPartner}(i, 0)$, which is impossible. Hence, $\mathbf{G}'_2$ is identical to $\mathbf{G}'_1$, and

$$adv_1 = adv_2. \tag{13}$$

Now we define an event named Hit. When $\mathcal{A}$ queries $\mathsf{TestPrivacy}(i, j, i', j')$, a unique tuple (i, j, i', j') is determined. Even Hit happens iff $(i_0^*, j_0^*, i_1^*, j_1^*) = (i, j, i', j')$, where $(i_0^*, j_0^*, i_1^*, j_1^*) \leftarrow_\$ [\mu]^4$ is sample at the beginning the game. Note that $(i_0^*, j_0^*, i_1^*, j_1^*)$ follows a uniform distribution, so we have $\Pr[\mathsf{Hit}] = \frac{1}{\mu^4}$.

Game $\mathbf{G}'_3$: At the end of $\mathbf{G}'_3$, if event Hit does not occur, $\mathcal{C}$ will return a random bit $\theta \leftarrow_\$ \{0, 1\}$ instead of detecting event Win. Otherwise, $\mathbf{G}'_3$ is the same as $\mathbf{G}'_2$. We have $\Pr[\mathsf{Win}_3] = \Pr[\mathsf{Hit}] \cdot \Pr[\mathsf{Win}_2] + \Pr[\overline{\mathsf{Hit}}] \cdot \frac{1}{2} = \Pr[\mathsf{Hit}] \cdot (\frac{1}{2} \pm adv_2) + \Pr[\overline{\mathsf{Hit}}] \cdot \frac{1}{2} = \frac{1}{2} \pm \frac{1}{\mu^4} \cdot adv_2$. As a result,

$$adv_2 = \mu^4 \cdot adv_3. \tag{14}$$

Game $\mathbf{G}'_4$: In $\mathbf{G}'_4$, the encapsulation key K shared by $\pi_{i_b^*}^0$ and $\pi_{j_b^*}^0$ is generated by $K \leftarrow_\$ \mathcal{K}$, instead of $(C, K) \leftarrow \mathsf{otEncap}(pk)$ and $K \leftarrow \mathsf{otDecap}(C)$ as in $\mathbf{G}'_3$. Similar to the proof of Lemma 5, we have

$$|adv_3 - adv_4| \leq \mathsf{Adv}_{\mathsf{KEM}}^{\mathsf{CPA}}(\mathcal{B}_{\mathsf{otKEM}}). \tag{15}$$

Game $\mathbf{G}'_5$: In $\mathbf{G}'_5$, the symmetric key and session key of $\pi_{i_b^*}^0$ and $\pi_{j_b^*}^0$ are generated by $(\bar{k}, k_{i_b^*}^0) = (\bar{k}, k_{j_b^*}^0) \leftarrow_\$ \mathcal{K}_1 \times \mathcal{K}_2$ instead of $\mathsf{PRG}(K)$ as in $\mathbf{G}'_4$. Hence,

$$|adv_4 - adv_5| \leq \mathsf{Adv}_{\mathsf{PRG}}^{\mathsf{ps}}(\mathcal{B}_{\mathsf{PRG}}). \tag{16}$$

Game $\mathbf{G}'_6$: In $\mathbf{G}'_6$, If $j_0 = j_1$, then $\mathbf{G}'_6$ is the same as $\mathbf{G}'_5$. Otherwise, $\pi_{i_b^*}^0$ generates C_1^* by $(C_1^*, N) \leftarrow \mathsf{Encap}(pk_{j_0^*})$, instead of $(C_1^*, N) \leftarrow \mathsf{Encap}(pk_{j_b^*})$ as in $\mathbf{G}'_5$. By IK-CCA security of KEM, we know that (C_1^*, N) w.r.t $pk_{j_0^*}$ is indistinguishable to that w.r.t $pk_{j_1^*}$. So we have Lemma 6 with proof shown in the full version [18].

Lemma 6. $|adv_5 - adv_6| \leq |\Pr[\mathsf{Win}_5] - \Pr[\mathsf{Win}_6]| \leq \mathsf{Adv}_{\mathsf{KEM}}^{\mathsf{IK\text{-}CCA}}(\mathcal{B}_{\mathsf{KEM}})$.

Game $\mathbf{G}'_7$: $\mathbf{G}'_7$ is almost the same as $\mathbf{G}'_6$, except for the answer generation of oracle $\mathsf{TestPrivacy}(i, j, i', j')$ (which is $\mathsf{TestPrivacy}(i_0^*, j_0^*, i_1^*, j_1^*)$). In $\mathbf{G}'_7$, c^* is an encryption of (i_1^*, σ_2^*) where σ_2^* is computed using the signing key $ssk_{i_1^*}$. However, in $\mathbf{G}'_6$, c^* is an encryption of (i_b^*, σ_2^*) with σ_2^* a signature generated by the signing key $ssk_{i_b^*}$. The semantic security of SE makes sure that this change is indistinguishable, as shown in Lemma 7.

Lemma 7. $|adv_6 - adv_7| \leq |\Pr[\mathsf{Win}_6] - \Pr[\mathsf{Win}_7]| \leq \mathsf{Adv}_{\mathsf{SE}}^{\mathsf{Sem}}(\mathcal{B}_{\mathsf{SE}})$.

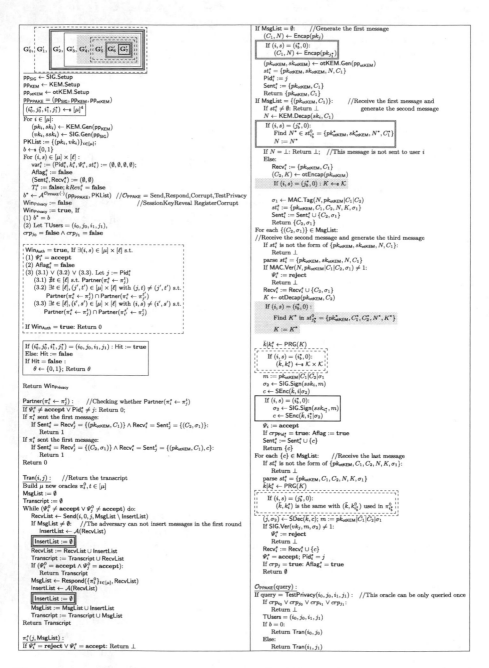

Fig. 8. Games $\mathbf{G}_0'--\mathbf{G}_7'$ for forward privacy of PPAKE. Queries to $\mathcal{O}_{\mathsf{PPAKE}}$ where query $\in \{\mathsf{Send}, \mathsf{Respond}, \mathsf{Corrupt}, \mathsf{RegisterCorrupt}, \mathsf{SessionKeyReveal}\}$ are defined as in the original game in Fig. 4.

The formal proof is given in the full version [18].

Finally, in $\mathbf{G}_7'$, all the messages in $\mathsf{Transcript} = \{(pk^*_{\mathsf{otKEM}}, C_1^*), (C_2^*, \sigma_1^*), c^*\}$ are independent of b, so we have

$$adv_7 = |\Pr[\mathsf{Win}_7] - 1/2| = 0. \tag{17}$$

Finally, the forward privacy of PPAKE follows from Lemma 6, 7 and (11)–(17).

5   Instantiations of PPAKE

In this section, we present concrete instantiations for the building blocks of our PPAKE including KEM, otKEM, SIG, MAC, PRG and SE. This yields a specific PPAKE scheme based on the DDH assumption over a cyclic group $\mathbb{G}$ and the CDH assumption over a bilinear group in the standard model. The details of the instantiations are shown in the full version [18].

KEM. We employ the Cramer-Shoup KEM (CS-KEM) scheme over a cyclic group $\mathbb{G}$ of order q. It is well known that CS-KEM is IND-CCA secure. Its public parameter is $(\mathbb{G}, q, g_1, g_2)$. Now we show its robustness. Given a ciphertext $C = (u_1, u_2, v) \in \mathbb{G}^3$ under public key $pk = (c = g_1^{x_1} g_2^{x_2}, d = g_1^{y_1} g_2^{y_2}, h = g_1^{z_1} g_2^{z_2}) \in \mathbb{G}^3$, we know that $u_1 = g^r, u_2 = g^r$ and $v = c^r d^{\alpha r} = u_1^{x_1 + \alpha y_1} u_2^{x_2 + \alpha y_2}$, where α is the hash value of (u_1, u_2). When decrypting C with another independent and random secret key $(x_1', x_2', y_1', y_2', z_1', z_2')$, we have that $\Pr\left[v = u_1^{x_1' + \alpha y_1'} u_2^{x_2' + \alpha y_2'}\right]$ with probability $2/q$. Therefore, C will be rejected except with probability $2/q$.

otKEM. We employ the Elgamal-KEM scheme over a cyclic group $\mathbb{G}$ of order q. It is well known that Elgamal-KEM is IND-CPA secure. The public key is given by $pk = g^x \in \mathbb{G}$ and the ciphertext is $C = g^y \in \mathbb{G}$ and the encapsulated key is $K = g^{xy}$. The encapsulated key $K = g^{xy}$ is uniformly distributed, when either the secret key $sk = x$ or the randomness y used in otKEM.Encap is independently and randomly chosen over $\mathbb{Z}_q$. Hence, ElGamal-KEM has encapsulated key uniformity. Meanwhile, when $x, x' \leftarrow_s \mathbb{Z}_q$, two public keys $pk = g^x = g^{x'} = pk'$ collide, i.e., $pk = g^x = g^{x'} = pk'$ with probability $1/q$. Hence it has $\log q$-pk-diversity.

SIG. We employ the BSW signature scheme [7] over a bilinear group with bilinear map $e : \mathbb{G}' \times \mathbb{G}' \to \mathbb{G}_1$. Its sEUF-CMA security is based on the CDH assumption over $\mathbb{G}'$. Its signature space is $\Sigma = \mathbb{G}'^2 \times \mathbb{Z}_q$.

MAC. We use the MAC scheme [9] over a cyclic group $\mathbb{G}$ of order q. Its sEUF-CMA security is based on the DDH assumption over $\mathbb{G}$. The MAC key is $(\omega, x, x') \in \mathbb{Z}_q^3$ and the tag for message m is given by $\sigma = (u, v_1, v_2) \in \mathbb{G}^3$, where u is uniformly chosen, $v_1 = u^\omega$ and $v_2 = u^{x\ell + x'}$ with ℓ the hash value of (u, v_1, m). Its tag space is $\mathbb{G}^3$.

PRG. We use the PRG scheme [10], where PRG : $\mathbb{Z}_q \to \mathbb{Z}_q^5$. The PRG scheme is based on the DDH assumption over a cyclic group of order q.

SE. We can use one time pad over $\mathbb{Z}_q$ as our SE scheme, which has information-theoretical semantic security. The secret key space, the plain text space and the cipher text space is $\mathcal{K} = \mathcal{M} = \mathcal{C} = \mathbb{Z}_q$ with q a prime.

Assembling the above schemes according to our generic construction, we have a specific PPAKE scheme, with communication complexity $(\mathbb{G}+3\mathbb{G})+(\mathbb{G}+3\mathbb{G})+(2\mathbb{G}'+2\mathbb{Z}_q) = 8\mathbb{G} + 2\mathbb{G}' + 2\mathbb{Z}_q$. The security of the PPAKE scheme is based on the DDH assumption over $\mathbb{G}$ and the CDH assumption over the bilinear group $\mathbb{G}'$. The detail of the scheme is shown in the full version [18].

Acknowledgements. We would like to thank the anonymous reviewers for their helpful comments. Shengli Liu and You Lyu were partially supported by National Natural Science Foundation of China (NSFC No. 61925207), Guangdong Major Project of Basic and Applied Basic Research (2019B030302008), and the National Key R&D Program of China under Grant 2022YFB2701500. Shuai Han was partially supported by National Natural Science Foundation of China (Grant No. 62002223), Shanghai Sailing Program (20YF1421100), and Young Elite Scientists Sponsorship Program by China Association for Science and Technology (YESS20200185).

References

1. Abdalla, M., Izabachène, M., Pointcheval, D.: Anonymous and transparent gateway-based password-authenticated key exchange. In: Franklin, M.K., Hui, L.C.K., Wong, D.S. (eds.) CANS 2008. LNCS, vol. 5339, pp. 133–148. Springer, Heidelberg (2008). https://doi.org/10.1007/978-3-540-89641-8_10
2. Alwen, J., Hirt, M., Maurer, U., Patra, A., Raykov, P.: Anonymous authentication with shared secrets. In: Aranha, D.F., Menezes, A. (eds.) LATINCRYPT 2014. LNCS, vol. 8895, pp. 219–236. Springer, Cham (2015). https://doi.org/10.1007/978-3-319-16295-9_12
3. Arfaoui, G., Bultel, X., Fouque, P., Nedelcu, A., Onete, C.: The privacy of the TLS 1.3 protocol. Proc. Priv. Enhancing Technol. **2019**(4), 190–210 (2019). https://doi.org/10.2478/popets-2019-0065
4. Bader, C., Hofheinz, D., Jager, T., Kiltz, E., Li, Y.: Tightly-secure authenticated key exchange. In: Dodis, Y., Nielsen, J.B. (eds.) TCC 2015, Part I. LNCS, vol. 9014, pp. 629–658. Springer, Heidelberg (2015). https://doi.org/10.1007/978-3-662-46494-6_26
5. Bellare, M., Boldyreva, A., Desai, A., Pointcheval, D.: Key-privacy in public-key encryption. In: Boyd, C. (ed.) ASIACRYPT 2001. LNCS, vol. 2248, pp. 566–582. Springer, Heidelberg (2001). https://doi.org/10.1007/3-540-45682-1_33
6. Bellare, M., Rogaway, P.: Entity authentication and key distribution. In: Stinson, D.R. (ed.) CRYPTO 1993. LNCS, vol. 773, pp. 232–249. Springer, Heidelberg (1994). https://doi.org/10.1007/3-540-48329-2_21
7. Boneh, D., Shen, E., Waters, B.: Strongly unforgeable signatures based on computational Diffie-Hellman. In: Yung, M., Dodis, Y., Kiayias, A., Malkin, T. (eds.) PKC 2006. LNCS, vol. 3958, pp. 229–240. Springer, Heidelberg (2006). https://doi.org/10.1007/11745853_15
8. Dingledine, R., Mathewson, N., Syverson, P.F.: Tor: The second-generation onion router. In: Blaze, M. (ed.) Proceedings of the 13th USENIX Security Symposium, 9–13 August 2004, San Diego, CA, USA, pp. 303–320. USENIX (2004). http://www.usenix.org/publications/library/proceedings/sec04/tech/dingledine.html

9. Dodis, Y., Kiltz, E., Pietrzak, K., Wichs, D.: Message authentication, revisited. In: Pointcheval, D., Johansson, T. (eds.) EUROCRYPT 2012. LNCS, vol. 7237, pp. 355–374. Springer, Heidelberg (2012). https://doi.org/10.1007/978-3-642-29011-4_22

10. Farashahi, R.R., Schoenmakers, B., Sidorenko, A.: Efficient pseudorandom generators based on the DDH assumption. In: Okamoto, T., Wang, X. (eds.) PKC 2007. LNCS, vol. 4450, pp. 426–441. Springer, Heidelberg (2007). https://doi.org/10.1007/978-3-540-71677-8_28

11. Gjøsteen, K., Jager, T.: Practical and tightly-secure digital signatures and authenticated key exchange. In: Shacham, H., Boldyreva, A. (eds.) CRYPTO 2018, Part II. LNCS, vol. 10992, pp. 95–125. Springer, Cham (2018). https://doi.org/10.1007/978-3-319-96881-0_4

12. Heinrich, A., Hollick, M., Schneider, T., Stute, M., Weinert, C.: PrivateDrop: practical privacy-preserving authentication for apple airdrop. In: Bailey, M., Greenstadt, R. (eds.) 30th USENIX Security Symposium, USENIX Security 2021(August), pp. 11–13, pp. 3577–3594. USENIX Association (2021), https://www.usenix.org/conference/usenixsecurity21/presentation/heinrich

13. Ishibashi, R., Yoneyama, K.: Post-quantum anonymous one-sided authenticated key exchange without random oracles. In: Hanaoka, G., Shikata, J., Watanabe, Y. (eds.) PKC 2022, Part II. LNCS, vol. 13178, pp. 35–65. Springer, Cham (2022). https://doi.org/10.1007/978-3-030-97131-1_2

14. Krawczyk, H.: SKEME: a versatile secure key exchange mechanism for internet. In: Ellis, J.T., Neuman, B.C., Balenson, D.M. (eds.) 1996 Symposium on Network and Distributed System Security, (S)NDSS 1996, San Diego, CA, USA, 22–23 February 1996, pp. 114–127. IEEE Computer Society (1996). https://doi.org/10.1109/NDSS.1996.492418

15. Lee, M.-F., Smart, N.P., Warinschi, B., Watson, G.J.: Anonymity guarantees of the UMTS/LTE authentication and connection protocol. Int. J. Inf. Secur. 13(6), 513–527 (2014). https://doi.org/10.1007/s10207-014-0231-3

16. Li, Y., Schäge, S.: No-match attacks and robust partnering definitions: defining trivial attacks for security protocols is not trivial. In: Thuraisingham, B., Evans, D., Malkin, T., Xu, D. (eds.) Proceedings of the 2017 ACM SIGSAC Conference on Computer and Communications Security, CCS 2017, Dallas, TX, USA, 30 October–03 November 2017, pp. 1343–1360. ACM (2017). https://doi.org/10.1145/3133956.3134006

17. Liu, X., Liu, S., Gu, D., Weng, J.: Two-pass authenticated key exchange with explicit authentication and tight security. In: Moriai, S., Wang, H. (eds.) ASIACRYPT 2020, Part II. LNCS, vol. 12492, pp. 785–814. Springer, Cham (2020). https://doi.org/10.1007/978-3-030-64834-3_27

18. Lyu, Y., Liu, S., Han, S., Gu, D.: Privacy-preserving authenticated key exchange in the standard model. Cryptology ePrint Archive, Paper 2022/1217 (2022). https://eprint.iacr.org/2022/1217

19. Ramacher, S., Slamanig, D., Weninger, A.: Privacy-preserving authenticated key exchange: stronger privacy and generic constructions. In: Bertino, E., Shulman, H., Waidner, M. (eds.) ESORICS 2021, Part II. LNCS, vol. 12973, pp. 676–696. Springer, Cham (2021). https://doi.org/10.1007/978-3-030-88428-4_33

20. Schäge, S., Schwenk, J., Lauer, S.: Privacy-preserving authenticated key exchange and the case of IKEv2. In: Kiayias, A., Kohlweiss, M., Wallden, P., Zikas, V. (eds.) PKC 2020, Part II. LNCS, vol. 12111, pp. 567–596. Springer, Cham (2020). https://doi.org/10.1007/978-3-030-45388-6_20
21. Yang, X., Jiang, H., Hou, M., Zheng, Z., Xu, Q., Choo, K.-K.R.: A provably-secure two-factor authenticated key exchange protocol with stronger anonymity. In: Au, M.H., et al. (eds.) NSS 2018. LNCS, vol. 11058, pp. 111–124. Springer, Cham (2018). https://doi.org/10.1007/978-3-030-02744-5_8

On the Field-Based Division Property: Applications to MiMC, Feistel MiMC and GMiMC

Jiamin Cui[1,3], Kai Hu[2], Meiqin Wang[1,3,4(✉)], and Puwen Wei[1,3]

[1] School of Cyber Science and Technology, Shandong University, Qingdao, Shandong, China
cuijiamin@mail.sdu.edu.cn, {mqwang,pwei}@sdu.edu.cn
[2] School of Physical and Mathematical Sciences, Nanyang Technological University, Singapore, Singapore
kai.hu@ntu.edu.sg
[3] Key Laboratory of Cryptologic Technology and Information Security, Ministry of Education, Shandong University, Qingdao, Shandong, China
[4] Quan Cheng Shandong Laboratory, Jinan, China

Abstract. Recent practical applications using advanced cryptographic protocols such as multi-party computations (MPC) and zero-knowledge proofs (ZKP) have prompted a range of novel symmetric primitives described over large finite fields, characterized as arithmetization-oriented (AO) ciphers. Such designs, aiming to minimize the number of multiplications over fields, have a high risk of being vulnerable to algebraic attacks, especially to the higher-order differential attack. Thus, it is significant to carefully evaluate the growth of their algebraic degree. However, the degree estimation for AO ciphers has been a challenge for cryptanalysts due to the lack of general and accurate methods.

In this paper, we extend the division property, a state-of-the-art framework for finding the upper bound of the algebraic degree over binary fields, to the scope of $\mathbb{F}_{2^n}$. It is a generic method to detect the algebraic degree for AO ciphers, even applicable to Feistel ciphers which have no better bounds than the trivial exponential one. In this general division property, our idea is to evaluate whether the polynomial representation of a block cipher contains some specific monomials. With a deep investigation of the arithmetical feature, we introduce the propagation rules of monomials for field-based operations, which can be efficiently modeled using the bit-vector theory of SMT. Then the new searching tool for degree estimation can be constructed due to the relationship between the algebraic degree and the exponents of monomials.

We apply our new framework to some important AO ciphers, including Feistel MiMC, GMiMC, and MiMC. For Feistel MiMC, we show that the algebraic degree grows significantly slower than the native exponential bound. For the first time, we present a secret-key higher-order differential distinguisher for up to 124 rounds, much better than the 83-round distinguisher for Feistel MiMC permutation proposed at CRYPTO 2020. We also exhibit a full-round zero-sum distinguisher with a data complexity of 2^{251}. Our method can be further extended for the general Feistel

© International Association for Cryptologic Research 2022
S. Agrawal and D. Lin (Eds.): ASIACRYPT 2022, LNCS 13793, pp. 241–270, 2022.
https://doi.org/10.1007/978-3-031-22969-5_9

structure with more branches and exhibit higher-order differential distinguishers against the practical instance of GMiMC for up to 50 rounds. For MiMC in SP-networks, our results correspond to the exact algebraic degree proved by Bouvier et al. We also point out that the number of rounds in MiMC's specification is not sufficient to guarantee the security against the higher-order differential attack for MiMC-like schemes with different exponents. The investigation of different exponents provides some guidance on the cipher design.

Keywords: Degree evaluation · Division property · Finite field · MiMC · Feistel network

1 Introduction

The recent progress of advanced cryptographic protocols such as multi-party computations (MPC) and zero-knowledge proofs (ZKP) has motivated new insights into the design paradigm. These innovative primitives, characterized as *arithmetization-oriented* (AO) *ciphers*, focus more on the arithmetic metrics. In the case of MPC-friendly constructions, the goal is to minimize the number of multiplications in large finite fields. Examples include MiMC [3] and its generalizations Feistel MiMC and GMiMC [2,3], HADESMiMC [24], VISION and RESCUE [4] and CIMINION [21].

AO cipher designs are quite different from the traditional ones. Instead of symmetric primitives whose non-linear layers are usually composed of relatively small S-boxes (typically 4 or 8 bits), AO ciphers tend to use the non-linear function with an explicit and compact algebraic representation over large finite fields (e.g., power maps like $x \mapsto x^d$ for some odd integer d). Statistical attacks such as differential [9] and linear cryptanalysis [31], which are two of the most powerful classical cryptanalytic tools, appear not to threaten the security of these new primitives. Consequently, algebraic attacks, especially the higher-order differential attack [29], usually determine their overall security level. As a concrete example, Eichlseder et al. proposed a new upper bound on the algebraic degree for low-degree key-alternating ciphers over $\mathbb{F}_{2^n}$ [22], based on which they successfully mounted a key-recovery attack on full-round MiMC. Fairly speaking, the algebraic degree is the most crucial security property of AO ciphers. It is of great importance to devise new tools for their degree estimations.

Related Work. Different methods and tools for degree evaluation have always been an important topic in the literature. Trivially, the algebraic degree of the composition of two functions F and G is bounded by $\deg(F \circ G) \leq \deg(F) \cdot \deg(G)$. However, if iterated, the resulting exponential bound fails to show the real growth of the algebraic degree for many cryptographic primitives, especially after a high number of rounds. The first improvement of the trivial bound was proposed by Canteaut and Videau at EUROCRYPT 2002 [15]. Later, Boura et al. focused on the iterated SPN schemes over $\mathbb{F}_{2^n}^t$ and presented a tighter upper bound [12]. Subsequently, more improved upper bounds for SPN schemes were proposed through comprehensive consideration of the underlying building blocks. By further studying the influence of the algebraic degree of F^{-1}, Boura and Canteaut [10] proposed a tighter bound than [12]. Recently at FSE 2022 [18], the influence of the linear layer on the algebraic degree was also noticed and the current best bounds for SPN schemes with large and low-degree S-boxes over $\mathbb{F}_{2^n}^t$ were presented. Moreover, for Even-Mansour schemes, a special case of SPN schemes, Eichlseder et al. pointed out that the algebraic degree grows linearly with the number of rounds [22] for ciphers with low-degree round functions. As an application, they managed to give a higher-order differential distinguisher on almost full MiMC. Very recently, by carefully tracing the evolution of the exponents, Bouvier et al. presented a tighter bound for ciphers based on iterated power functions [14], leading to the exact algebraic degree estimation for MiMC. However, there is no improved bound for Feistel schemes except the trivial bound. Consequently, although the general method is more universal, if we are not able to exploit the information of the components in a more fine-grained way, the resulting algebraic degree will not be accurate enough.

Besides the above-mentioned methods, another approach for degree estimation is based on division property, a state-of-the-art framework for finding integral property proposed by Todo at EUROCRYPT 2015 [34]. It is currently the optimal way to estimate the algebraic degree in terms of accuracy as pointed out in [17]. The division property was initially word-oriented and then extended to bit level [35], referred to as the bit-based division property and three-subset bit-based division property [35]. Subsequently, there was a lot of research focusing on this topic to explain the imperfect nature inherent or extend the application scope with the help of automatic approaches [11,13,20,26,30,33,37,38]. At EUROCRYPT 2020, Hao et al. proposed the three-subset bit-based division property without unknown subset (3SDPwoU) [25] and achieves perfect accuracy. The monomial prediction proposed by Hu et al. [27] is another language of division property from a complete polynomial viewpoint. It allows us to precisely determine whether or not a specific monomial appears in the ANF. Besides, they also provide a framework to detect the integral properties more precisely than but with similar efficiency as the two-subset division property for block ciphers. Throughout this paper, we use the division porperty and monomial prediction to denote the same technology without making strict distinctions. Despite of their powerfulness, the division property/monomial prediction requires the ANF of local components, which is too complicated to be calculated or stored in practice for large finite fields. Even if we know the ANF, the existing tools cannot handle the modeling for S-boxes with a size larger than 32 bits [36] in practical time to the best of our knowledge. Overall, the bit-based division property fails to be useful for AO ciphers. However, AO ciphers can be directly regarded as multivariate polynomials over public variables (e.g., plaintext variables) and secret variables (e.g., key variables) in $\mathbb{F}_{2^n}$. This inspires us to focus on the algebraic essentials of division property and thus take benefit from the concise polynomial representations over fields.

1.1 Our Contribution

In this paper, we extend the division property, a state-of-the-art method for finding integral properties over binary fields, to the scope of the binary extension field $\mathbb{F}_{2^n}$, called *general monomial prediction* (GMP). It is a generic method to evaluate the algebraic degree for ciphers over fields, in the way of studying whether or not the polynomial representation of a block cipher described over $\mathbb{F}_{2^n}$ contains some specific monomials by decomposing the cipher into a sequence of simpler functions and tracing the monomial propagations. We then propose the propagation rules of the monomials based on the arithmetical features and model them with the aid of the bit-vector theory of Satisfiability Modulo Theories (SMT). Finally, by tracing the evolution of exponents for the monomials, we construct an SMT-based searching tool for the degree estimation of ciphers over $\mathbb{F}_{2^n}$. We apply our algorithm to some important arithmetization-oriented block ciphers, including MiMC, Feistel MiMC, GMiMC, and their variants. The full version of the paper can be found in [1]. All the source codes are available at https://github.com/iljido/GeneralMonomialPrediction.

- For Feistel MiMC, we show in particular that its algebraic degree grows obviously slower than the originally believed one. More precisely, after an initial linear growth, the algebraic degree grows rather slow for a long period, along with several large plateaus until reaching the maximal degree. While the previous work only handles the permutation Feistel MiMC, using our results, for the first time we present a secret-key higher-order differential distinguisher covering a total of 124 rounds. It is 41 rounds more than the previous best distinguisher of permutation Feistel MiMC. We also establish a known-key zero-sum distinguisher for the full-round Feistel MiMC over $\mathbb{F}_{2^n}$ with a data complexity of 2^{251}. Our method can be extended to more branches and we successfully find the currently longest secret-key higher-order differential distinguisher for practical instance of block cipher GMiMC reaching up to 50 rounds, 10 rounds longer than the previous best distinguisher.
- We also investigate the algebraic degree of MiMC-like schemes with generic exponents d. For exponents of the form $d = 2^l - 1$, we extend the higher-order differential distinguishers by one or two more rounds for different instances compared to the currently best results in [22]. For exponents of the form $d = 2^l + 1$, we find distinguishers with lower data complexities for $d = 5, 9, 17$. Our results for MiMC with $d = 3$ are consistent with the exact algebraic degree proved in [14]. Based on our results, we point out that the formula for the number of rounds used in MiMC specification [3] is not sufficient to guarantee security against the higher-order differential attack. This investigation of different exponents provides some guidance on the design.
- Moreover, we present a comprehensive analysis of the degree growth of MiMC-like schemes in (unbalanced) Feistel networks and prove a theoretical upper bound that improves the trivial exponential bound.

All the results are summarized in Table 1, Table 2 and Table 3. Our experiments are implemented in the AMD EPYC 7302 CPU @ 3.0 GHz with 8 threads.

Table 1. Higher-order differential distinguishers for $\mathsf{FeistelMiMC}_3(129, r)$.

Security	#Rounds	Target		Attack		Time	Source
		Permutation	Block Cipher	#Rounds	Cost		
129	164	✓	–	82	$2^{127\dagger}$	–	[8]
		✓	✓	**82**	$\mathbf{2^{127}}$	< 1 min	**Sect.** 5.1
258	166	✓	–	83	2^{129}	–	[8]
		✓	✓	**83**	$\mathbf{2^{129}}$	< 1 min	**Sect.** 5.1
		✓	✓	**124**	$\mathbf{2^{257}}$	< 5 min	**Sect.** 5.1

$\dagger$ This complexity is calculated using the formula in [8] with subgroup of size 2^{127}.

Table 2. Zero-sum distinguishers for $\mathsf{FeistelMiMC}_3(129, r)$.

Security	#Rounds	Attack		Source
		#Rounds	Cost	
129	164	162	$2^{127\dagger}$	[8]
		163	$\mathbf{2^{127}}$	**Sect.** 5.1
258	166	164	2^{129}	[8]
		165	$\mathbf{2^{129}}$	**Sect.** 5.1
		166	$\mathbf{2^{251}}$	**Sect.** 5.1

$\dagger$ This complexity is calculated using the formula in [8] with subgroup of size 2^{127}.

1.2   Outline

The rest of this paper is organized as follows. In Sect. 2, we revisit some background knowledge about polynomial representations, the monomial prediction, and SMT solvers. In Sect. 3, we propose the principle of *general monomial prediction* and present the new searching model for degree estimation. For a better insight into the degree estimation, we prove a theoretical upper bound on the algebraic degree for ciphers in (unbalanced) Feistel-networks with low-degree round functions in Sect. 4. Section 5 shows the applications to MiMC, Feistel MiMC, and GMiMC. We conclude the paper in Sect. 6.

2   Preliminaries

2.1   Notations

Let $\mathbb{F}_2^n$ denote the n-dimensional vector space over the finite field $\mathbb{F}_2$. $\mathbb{F}_{2^n}^t$ denotes the t-fold Cartesian product of the binary extension field $\mathbb{F}_{2^n}$. For any n-bit vector $\boldsymbol{u} = (u[0], \cdots, u[n-1]) \in \mathbb{F}_2^n$, the Hamming weight of $\boldsymbol{u}$ is $wt(\boldsymbol{u}) = \sum_{i=0}^{n-1} u[i]$. For any $a \in \mathbb{F}_{2^n}$, we have $a = \sum_{i=0}^{n-1} a[i] \cdot 2^i$ for $a[i] \in \{0, 1\}$ and $wt(a) = \sum_{i=0}^{n-1} a[i]$. For any $a, a' \in \mathbb{F}_{2^n}$, we define $a \preceq a'$ if $a[i] \leq a'[i]$ for all

Table 3. Distinguishers for different instances of $\mathsf{MiMC}_d(129, r)$.

	d/l	$n \cdot \log_d(2)$	Attack		Source
			#Rounds	Cost	
$d = 2^l - 1$	7/3	46	45	2^{127}	[22]
			45	$\mathbf{2^{124}}$	**Sect.** 5.2
			46	2^{127}	**Sect.** 5.2
	15/4	34	32	2^{126}	[22]
			32	$\mathbf{2^{125}}$	**Sect.** 5.2
			33	$\mathbf{2^{125}}$	**Sect.** 5.2
	31/5	27	25	2^{124}	[22]
			25	$\mathbf{2^{121}}$	**Sect.** 5.2
			27	2^{128}	**Sect.** 5.2
$d = 2^l + 1$	3/1	82	80	2^{128}	[22]
			81	2^{127}	[14]
			81	2^{127}	**Sect.** 5.2
	5/2	56	54	2^{125}	[14]
			54	$\mathbf{2^{124}}$	**Sect.** 5.2
			55	2^{128}	[22]
			55	2^{127}	[14]
			55	2^{127}	**Sect.** 5.2
	9/3	41	40	2^{127}	[22]
			40	2^{125}	[14]
			40	$\mathbf{2^{124}}$	**Sect.** 5.2
			41	2^{128}	[14]
			41	2^{127}	**Sect.** 5.2
	17/4	32	31	2^{127}	[22]
			32	2^{128}	[14]
			32	2^{127}	**Sect.** 5.2

i, $a \succeq a'$ if $a[i] \geq a'[i]$ for all i. We use $\oplus$ as addition over $\mathbb{F}_2$ or $\mathbb{F}_{2^n}$. 0^n or 1^n represents the all-zeros or all-ones vector of length n, respectively.

Polynomial Representations. Let $F: \mathbb{F}_{2^n}^t \rightarrow \mathbb{F}_{2^n}$ be a function over $\mathbb{F}_{2^n}[x_0, x_1, \cdots, x_{t-1}]/\left\langle x_0^{2^n} - x_0, x_1^{2^n} - x_1, \cdots, x_{t-1}^{2^n} - x_{t-1} \right\rangle$. F can be uniquely expressed by a polynomial over $\mathbb{F}_{2^n}$ with t variables $x_0, x_1, \cdots, x_{t-1} \in \mathbb{F}_{2^n}$, as

$$F(x_0, \cdots, x_{t-1}) = \sum_{\boldsymbol{v} = (v_0, \cdots, v_{t-1}) \in \{0, 1, \cdots, 2^n - 1\}^t} \varphi(\boldsymbol{v}) \cdot \pi_{\boldsymbol{v}}(\boldsymbol{x}) \qquad (1)$$

where the coefficient $\varphi(\boldsymbol{v}) \in \mathbb{F}_{2^n}$. We call the degree of a single variable in F as *univariate degree* and the degree of F as a multivariate polynomial as *multivariate degree.* The maximum univariate degree is $2^n - 1$. When $t = 1$, the maximum univariate degree is $2^n - 2$ if F is invertible since the maximal algebraic degree of invertible functions over $\mathbb{F}_{2^n}$ is $n - 1$.

In Eq. (1), $\pi_v(\boldsymbol{x}) = \prod_{i=0}^{t-1} x_i^{v_i} = x_0^{v_0} \cdot \ldots \cdot x_{t-1}^{v_{t-1}}$ is called a monomial over $\mathbb{F}_{2^n}$. If the coefficient of $\pi_v(\boldsymbol{x})$ in F is a constant $c \neq 0$, we say $\pi_v(\boldsymbol{x})$ is contained by F, denoted by $\pi_v(\boldsymbol{x}) \to F$. Otherwise, if the coefficient of $\pi_v(\boldsymbol{x})$ in F is 0, $\pi_v(\boldsymbol{x})$ is not contained by F, denoted by $\pi_v(\boldsymbol{x}) \nrightarrow F$.

As is well-known, the function F can as well be represented at bit level with $N = n \cdot t$ variables. The i-th output element is defined by the coordinate function

$$F_i(y_0, \cdots, y_{N-1}) = \sum_{\boldsymbol{u} = (u_0, \cdots, u_{N-1}) \in \{0,1\}^N} \rho_i(\boldsymbol{u}) \cdot \pi_u(\boldsymbol{y}). \tag{2}$$

The coefficient $\rho_i(\boldsymbol{u}) \in \mathbb{F}_2$ can be computed by the *Möbius transform.* $\pi_u(\boldsymbol{y}) = \prod_{i=0}^{N-1} y_i^{u_i} = y_0^{u_0} \cdot \ldots \cdot y_{N-1}^{u_{N-1}}$ is called a monomial. If the coefficient of $\pi_u(\boldsymbol{y})$ in F_i is 1, we say $\pi_u(\boldsymbol{y})$ is contained by F_i, denoted by $\pi_u(\boldsymbol{y}) \to F_i$. Otherwise, $\pi_u(\boldsymbol{y})$ is not contained by F_i, denoted by $\pi_u(\boldsymbol{y}) \nrightarrow F_i$.

This representation is also referred to as *algebraic normal form* (ANF) of Boolean functions. Essentially, we can see that the polynomial representation and ANF of F are equivalent when $n = 1$.

Definition 1 (ANF and Algebraic Degree). *Let $f \colon \mathbb{F}_2^n \to \mathbb{F}_2$ be a Boolean function. Its algebraic normal form (ANF) is given as*

$$f(\boldsymbol{x}) = f(x[0], x[1], \cdots, x[n-1]) = \bigoplus_{\boldsymbol{u} \in \mathbb{F}_2^n} \rho(\boldsymbol{u}) \cdot \boldsymbol{x}^{\boldsymbol{u}} \tag{3}$$

where the coefficient $\rho(\boldsymbol{u}) \in \mathbb{F}_2$ and $\boldsymbol{x}^{\boldsymbol{u}} = \prod_{i=0}^{n-1} x[i]^{u[i]}$. Then the algebraic degree of f is defined as

$$\delta(f) = \max\{wt(\boldsymbol{u}) \mid \boldsymbol{u} \in \mathbb{F}_{2^n}, \rho(\boldsymbol{u}) \neq 0\}.$$

If $\boldsymbol{f} \colon \mathbb{F}_2^n \to \mathbb{F}_2^m$ is a vectorial Boolean function, then the algebraic degree is defined as the maximal algebraic degree of its coordinate functions f_i, i.e., $\delta(\boldsymbol{f}) = \max\{\delta(f_i) \mid 0 \leq i < m\}$.

The link between the algebraic degree and the univariate degree of a vectorial Boolean function is well-known.

Proposition 1 ([16]). *For any univariate function $F \colon \mathbb{F}_{2^n} \to \mathbb{F}_{2^n}$ as*

$$F(x) = \sum_{v \in \{0,1,\cdots,2^n-1\}} \varphi(v) \cdot x^v,$$

the algebraic degree of F as a vectorial Boolean function is the maximum Hamming weight of the exponents for the non-vanishing monomials, i.e.,

$$\delta(F) = \max_{0 \leq v \leq 2^n - 1} \{wt(v) \mid \varphi(v) \neq 0\}.$$

Corollary 1. *For $x_0, x_1, \cdots, x_{t-1} \in \mathbb{F}_{2^n}$, the algebraic degree of a monomial $\pi_u(\boldsymbol{x}) = x_0^{u_0} \cdot \ldots \cdot x_{t-1}^{u_{t-1}}$ is given by $\sum_{i=0}^{t-1} wt(u_i)$.*

2.2   Monomial Prediction

In this paper, we mainly take the framework of the monomial prediction to simplify the exposition. The monomial prediction, proposed by Hu et al. in [27], is another language of division property from a pure algebraic perspective. By counting the so-called monomial trails, the monomial prediction can determine if a monomial of the plaintext or IV appears in the polynomial of the output of the cipher, proved to be equivalent to the three-subset bit-based division property without unknown subsets [25].

Let $f \colon \mathbb{F}_2^{n_0} \to \mathbb{F}_2^{n_r}$ be a composite vectorial Boolean function of a sequence of smaller functions $f^{(i)} \colon \mathbb{F}_2^{n_i} \to \mathbb{F}_2^{n_{i+1}}$, $0 \le i \le r - 1$, as

$$f = f^{(r-1)} \circ f^{(r-2)} \circ \cdots \circ f^{(0)}$$

where $x^{(i+1)} = f^{(i)}(x^{(i)})$. Considering the function $f^{(i)}$, if the ANF of $f^{(i)}$ is available, we can find the monomial $\pi_{u^{(i+1)}}(x^{(i+1)})$ that contains the monomial $\pi_{u^{(i)}}(x^{(i)})$ for any $u^{(i)}$ easily, denoted by $\pi_{u^{(i)}}(x^{(i)}) \to \pi_{u^{(i+1)}}(x^{(i+1)})$. If we can find an r-round transition connecting $\pi_{u^{(0)}}(x^{(0)})$ and $\pi_{u^{(r)}}(x^{(r)})$ as

$$\pi_{u^{(0)}}(x^{(0)}) \to \pi_{u^{(1)}}(x^{(1)}) \to \cdots \to \pi_{u^{(r)}}(x^{(r)}),$$

then the r-round transition is denoted by $\pi_{u^{(0)}}(x^{(0)}) \rightsquigarrow \pi_{u^{(r)}}(x^{(r)})$, called a *monomial trail*. The set of all monomial trails from $\pi_{u^{(0)}}(x^{(0)})$ to $\pi_{u^{(r)}}(x^{(r)})$ are denoted by $\pi_{u^{(0)}}(x^{(0)}) \bowtie \pi_{u^{(r)}}(x^{(r)})$. The size of the monomial trails determines whether $\pi_{u^{(0)}}(x^{(0)}) \to \pi_{u^{(r)}}(x^{(r)})$. If there is no trail from $\pi_{u^{(0)}}(x^{(0)})$ to $\pi_{u^{(r)}}(x^{(r)})$, we say $\pi_{u^{(0)}}(x^{(0)}) \not\rightsquigarrow \pi_{u^{(r)}}(x^{(r)})$ and hence $\pi_{u^{(0)}}(x^{(0)}) \not\to \pi_{u^{(r)}}(x^{(r)})$.

2.3   SMT Solvers

A recent approach to construct automatic tools is to formulate the searching problems into some mathematical problems and delegate the solving task to the powerful off-the-shelf solvers. The Satisfiability Modulo Theories (SMT) [6] is a problem of determining whether logical formulas in the first-order logic is satisfiable. It is a generalization of the Boolean Satisfiability Problem (SAT) [19]. SMT formulas provide much richer modeling language than SAT formulas such as bit-vectors, which give more flexibility in the interpretation of mathematical problems.

A bit-vector variable is a string of Boolean variables that can represent either a bit-vector or an integer. The set of the basic bit-vector operations is a combination of arithmetic operations and bit-wise operations. We list the operations used in the following sections in Table 4.

There are many public available solvers to solve SMT problems. We construct our model using the CVC [7] input language and take STP [23] and Cryptominisat5 [32] as our solvers in the paper. For more details about STP and CVC, readers are encouraged to refer to http://stp.github.io/.

Table 4. Basic bit-vector operations.

$x \wedge y$	bit-wise AND of x and y	$x + y$	addition of x and y
$x \vee y$	bit-wise OR of x and y	$x \times y$	multiplition of x and y
$x \oplus y$	bit-wise XOR of x and y	$x = y$	x is equal to y
$x \parallel y$	concatenation of x and y	$x \neq y$	x is not equal to y
$x \ll i$	x left shift by i bits	$x \leq y$	x is less than or equal to y
$x \gg i$	x right shift by i bits	$x \geq y$	x is greater than or equal to y

3   General Monomial Prediction

Let $y = F(x)$ be a function from $\mathbb{F}_{2^n}^t$ to $\mathbb{F}_{2^n}^s$. We focus on the exponents of x so that the algebraic degree can be estimated based on its relationship with the Hamming weight of exponents. In Sect. 3.1, we will introduce how to trace the transition of the exponents by generalizing the monomial prediction from $\mathbb{F}_2$ to the finite field $\mathbb{F}_{2^n}$, referred to as *general monomial prediction* (GMP). Since any function F can be represented as a sequence of basic operations such as XOR, AND, COPY, m-COPY, and POWER, we give the propagation rules for these basic functions by investigating the arithmetical features in Sect. 3.2 and provide their SMT models in Sect. 3.3. Finally in Sect. 3.4, by setting the initial constraints and stopping rules appropriately, the problem of degree estimation for ciphers over fields can be converted into an SMT problem and solved efficiently.

3.1   Definition of General Monomial Prediction

Let $y = F(x)$ be a function from $\mathbb{F}_{2^n}^t$ to $\mathbb{F}_{2^n}^s$, where $x = (x_0, \cdots, x_{t-1})$ and $y = (y_0, \cdots, y_{s-1})$. By *general monomial prediction* we mean the problem of whether a particular monomial y^v is contained by x^u, denoted by $x^u \rightarrow y^v$. Notice that we make no distinction between the secret variables and public variables here and they are all treated as symbolic variables. While it is a trivial problem if the polynomial representation of F is available, F is usually too complicated to be computed or stored in practice for most symmetric primitives and we are limited to knowing the local components of F.

Let $F \colon \mathbb{F}_{2^n}^{t_0} \rightarrow \mathbb{F}_{2^n}^{t_r}$ be a composite function over $\mathbb{F}_{2^n}$ consisting of a sequence of smaller functions $F^{(i)} \colon \mathbb{F}_{2^n}^{t_i} \rightarrow \mathbb{F}_{2^n}^{t_{i+1}}$, $0 \leq i \leq r - 1$, as

$$F = F^{(r-1)} \circ F^{(r-2)} \circ \cdots \circ F^{(0)}.$$

We assume that $x^{(i)}$ and $x^{(i+1)}$ are the input and output variables of $F^{(i)}$, where $x^{(i)} = (x_0^{(i)}, \cdots, x_{t_i-1}^{(i)})$. Each $x_j^{(i)}$ is a variable over $\mathbb{F}_{2^n}$. For a given pair of $(u^{(i)}, u^{(i+1)})$, if the polynomial representation of $F^{(i)}$ is available, one can determine whether $\pi_{u^{(i)}}(x^{(i)}) \rightarrow \pi_{u^{(i+1)}}(x^{(i+1)})$. We emphasize that $\pi_{u^{(i)}}(x^{(i)}) \rightarrow$

$\pi_{\boldsymbol{u}^{(i+1)}}(\boldsymbol{x}^{(i+1)})$ if and only if the coefficient of $\pi_{\boldsymbol{u}^{(i)}}(\boldsymbol{x}^{(i)})$ in $\pi_{\boldsymbol{u}^{(i+1)}}(\boldsymbol{x}^{(i+1)})$ is a constant $c \neq 0$. If there exists a trail such that

$$\pi_{\boldsymbol{u}^{(0)}}(\boldsymbol{x}^{(0)}) \to \cdots \pi_{\boldsymbol{u}^{(i)}}(\boldsymbol{x}^{(i)}) \to \cdots \to \pi_{\boldsymbol{u}^{(r)}}(\boldsymbol{x}^{(r)}),$$

there exists a trail connecting $\pi_{\boldsymbol{u}^{(0)}}(\boldsymbol{x}^{(0)})$ and $\pi_{\boldsymbol{u}^{(r+1)}}(\boldsymbol{x}^{(r+1)})$, which naturally leads to the definition of *general monomial trail*.

Definition 2 (General Monomial Trail). *Let $F^{(i)}$ be a sequence of polynomials over $\mathbb{F}_{2^n}$ for $0 \leq i < r$, while $\boldsymbol{x}^{(i+1)} = F^{(i)}(\boldsymbol{x}^{(i)})$. We call a sequence of monomials $(\pi_{\boldsymbol{u}^{(0)}}(\boldsymbol{x}^{(0)}), \pi_{\boldsymbol{u}^{(1)}}(\boldsymbol{x}^{(1)}), \cdots, \pi_{\boldsymbol{u}^{(r)}}(\boldsymbol{x}^{(r)}))$ an r-round general monomial trail connecting $\pi_{\boldsymbol{u}^{(0)}}(\boldsymbol{x}^{(0)})$ and $\pi_{\boldsymbol{u}^{(r)}}(\boldsymbol{x}^{(r)})$ with respect to the composite function $F = F^{(r-1)} \circ F^{(r-2)} \circ \cdots \circ F^{(0)}$ if*

$$\pi_{\boldsymbol{u}^{(0)}}(\boldsymbol{x}^{(0)}) \to \pi_{\boldsymbol{u}^{(1)}}(\boldsymbol{x}^{(1)}) \to \cdots \to \pi_{\boldsymbol{u}^{(r)}}(\boldsymbol{x}^{(r)}).$$

If there is at least one general monomial trail connecting $\pi_{\boldsymbol{u}^{(0)}}(\boldsymbol{x}^{(0)})$ and $\pi_{\boldsymbol{u}^{(r)}}(\boldsymbol{x}^{(r)})$, we write $\pi_{\boldsymbol{u}^{(0)}}(\boldsymbol{x}^{(0)}) \rightsquigarrow \pi_{\boldsymbol{u}^{(r)}}(\boldsymbol{x}^{(r)})$. Otherwise, $\pi_{\boldsymbol{u}^{(0)}}(\boldsymbol{x}^{(0)}) \not\rightsquigarrow \pi_{\boldsymbol{u}^{(r)}}(\boldsymbol{x}^{(r)})$. When $n = 1$, general monomial trail is equivalent to monomial trail.

Proposition 2. $\pi_{\boldsymbol{u}^{(0)}}(\boldsymbol{x}^{(0)}) \rightsquigarrow \pi_{\boldsymbol{u}^{(r)}}(\boldsymbol{x}^{(r)})$ if $\pi_{\boldsymbol{u}^{(0)}}(\boldsymbol{x}^{(0)}) \to \pi_{\boldsymbol{u}^{(r)}}(\boldsymbol{x}^{(r)})$, and thus $\pi_{\boldsymbol{u}^{(0)}}(\boldsymbol{x}^{(0)}) \not\rightsquigarrow \pi_{\boldsymbol{u}^{(r)}}(\boldsymbol{x}^{(r)})$ implies $\pi_{\boldsymbol{u}^{(0)}}(\boldsymbol{x}^{(0)}) \not\to \pi_{\boldsymbol{u}^{(r)}}(\boldsymbol{x}^{(r)})$.

Proof. We proceed by induction on r. Assuming that this proposition holds for $r < s$, we now prove that it also holds for $r = s$. When $r = s$, we expand $\pi_{\boldsymbol{u}^{(s)}}(\boldsymbol{x}^{(s)})$ on $\pi_{\boldsymbol{u}^{(s-1)}}(\boldsymbol{x}^{(s-1)})$ as

$$\pi_{\boldsymbol{u}^{(s)}}(\boldsymbol{x}^{(s)}) = \bigoplus_{\pi_{\boldsymbol{u}^{(s-1)}}(\boldsymbol{x}^{(s-1)}) \to \pi_{\boldsymbol{u}^{(s)}}(\boldsymbol{x}^{(s)})} \varphi(\boldsymbol{u}^{(s-1)}) \cdot \pi_{\boldsymbol{u}^{(s-1)}}(\boldsymbol{x}^{(s-1)}), \varphi(\boldsymbol{u}^{(s-1)}) \neq 0.$$

Since $\pi_{\boldsymbol{u}^{(0)}}(\boldsymbol{x}^{(0)}) \to \pi_{\boldsymbol{u}^{(s)}}(\boldsymbol{x}^{(s)})$, there is at least one $\pi_{\boldsymbol{u}^{(s-1)}}(\boldsymbol{x}^{(s-1)})$ contained by $\pi_{\boldsymbol{u}^{(s)}}(\boldsymbol{x}^{(s)})$ satisfying $\pi_{\boldsymbol{u}^{(0)}}(\boldsymbol{x}^{(0)}) \to \pi_{\boldsymbol{u}^{(s-1)}}(\boldsymbol{x}^{(s-1)})$. According to the assumption that $\pi_{\boldsymbol{u}^{(0)}}(\boldsymbol{x}^{(0)}) \rightsquigarrow \pi_{\boldsymbol{u}^{(s-1)}}(\boldsymbol{x}^{(s-1)})$, we have $\pi_{\boldsymbol{u}^{(0)}}(\boldsymbol{x}^{(0)}) \rightsquigarrow \pi_{\boldsymbol{u}^{(s)}}(\boldsymbol{x}^{(s)})$. $\square$

Example 1. Let $x_0, x_1, y, z \in \mathbb{F}_{2^3}$ with the irreducible polynomial $f(x) = x^3 + x + 1$. $z = 2y^3$, $y = x_0^3 \oplus 2x_0 \oplus x_1^2$.

Considering the monomial x_0^5, we can compute all the monomials of y as

$$y^0 \equiv 1,$$
$$y^1 \equiv x_0^3 \oplus 2x_0 \oplus x_1^2,$$
$$y^2 \equiv x_0^6 \oplus 4x_0^2 \oplus x_1^4,$$
$$y^3 \equiv 2x_0^7 \oplus x_0^6 x_1^2 \oplus \underline{4x_0^5} \oplus x_0^3 x_1^4 \oplus 3x_0^3 \oplus 4x_0^2 x_1^2 \oplus x_0^2 \oplus 2x_0 x_1^4 \oplus x_1^6,$$
$$y^4 \equiv \underline{x_0^5} \oplus 6x_0^4 \oplus x_1,$$
$$y^5 \equiv 6x_0^7 \oplus 2x_0^6 \oplus x_0^5 x_1^2 \oplus \underline{7x_0^5} \oplus 6x_0^4 x_1^2 \oplus x_0^3 x_1 \oplus 2x_0 x_1 \oplus x_0 \oplus x_1^3,$$
$$y^6 \equiv 4x_0^7 \oplus x_0^6 x_1 \oplus 6x_0^6 \oplus x_0^5 x_1^4 \oplus 6x_0^4 x_1^4 \oplus x_0^4 \oplus 6x_0^3 \oplus 4x_0^2 x_1 \oplus x_1^5$$

$$y^7 \equiv 6x_0^7 x_1^4 \oplus 4x_0^7 x_1^2 \oplus 2x_0^7 x_1 \oplus 2x_0^6 x_1^4 \oplus x_0^6 x_1^3 \oplus 6x_0^6 x_1^2 \oplus 6x_0^6 \oplus x_0^5 x_1^6 \oplus 7x_0^5 x_1^4,$$

$$\oplus 4x_0^5 x_1 \oplus \underline{2x_0^5} \oplus 6x_0^4 x_1^6 \oplus x_0^4 x_1^2 \oplus 7x_0^4 \oplus x_0^3 x_1^5 \oplus 6x_0^3 x_1^2 \oplus 3x_0^3 x_1 \oplus 4x_0^3$$

$$\oplus 4x_0^2 x_1^3 \oplus x_0^2 x_1 \oplus 6x_0^2 \oplus 2x_0 x_1^5 \oplus x_0 x_1^4 \oplus 3x_0 \oplus x_1^7.$$

Similarly, we can compute all the monomial of z

$$z^0 \equiv y^0, \ z^1 \equiv \underline{2y^3}, \ z^2 \equiv 4y^6, z^3 \equiv 4y^3 \equiv 3y^2,$$

$$z^4 \equiv 6y^{12} \equiv \underline{6y^5}, z^5 \equiv 7y^{15} \equiv 7y, z^6 \equiv 5y^{18} \equiv \underline{5y^4}, z^7 \equiv y^{21} \equiv \underline{y^7}.$$

There are four monomial trails connecting x_0^5 and monomials of z:

$$x_0^5 \to y^3 \to z^1, \ x_0^5 \to y^4 \to z^6, \ x_0^5 \to y^5 \to z^4, \ x_0^5 \to y^7 \to z^7.$$

Comparison with Word-Based Division Property. At a first glance, the general monomial prediction is similar to the word-based division property as both of them are described at the *word* level. However, we emphasize that they are completely different, especially in the way of extracting information. While the word-based division property can only exploit the information of the degree, our general monomial prediction can essentially utilize the internal structure of the ciphers in a more fine-grained way. Actually, it is more like the bit-based division property since *word* is the minimum unit of polynomials over $\mathbb{F}_{2^n}$.

Comparison with Bit-Based Division Property. From the example above we can see that the obvious difference between the general monomial prediction and bit-based division property is the range of the variables. Given a specific monomial m, there are two possible cases for the coefficient c in the ANF of a block cipher: $c = 1$ or $c = 0$, i.e., the ANF contains exactly m or the ANF does not contain m. However, since the coefficient c for ciphers over fields ranges over the 2^n elements of $\mathbb{F}_{2^n}$, the existence of a monomial m represents multiple states. As long as $c \neq 0$, the monomial m does appear in the polynomial representations. Recall that two-subset bit-based division property can essentially allow us to derive one of two possible results: the ANF of a block cipher does not contain any multiple of the monomial m, or we do not know any thing about the monomial. Given a specific monomial m for ciphers over fields, we can derive one of the following two results for the general monomial prediction according to Proposition 2:

- The monomial m with a corresponding coefficient $c \neq 0$ does not appear in the polynomial representation if there is no general monomial trail from m to the polynomial representation of the block cipher,
- We do not know anything about the monomial.

Essentially, we believe that the concept of the general monomial prediction is more common with the conventional bit-based division property. Moreover, we emphasize that due to the field-based structure for ciphers described over $\mathbb{F}_{2^n}$, the word-based/bit-based division property fails to be useful in this case.

3.2 Propagation Rules of Basic Field-Based Operations

Considering a sequence of monomials

$$(\pi_{\boldsymbol{u}^{(0)}}(\boldsymbol{x}^{(0)}), \pi_{\boldsymbol{u}^{(1)}}(\boldsymbol{x}^{(1)}), \cdots \pi_{\boldsymbol{u}^{(r)}}(\boldsymbol{x}^{(r)})),$$

where $\boldsymbol{x}^{(i)}$ and $\boldsymbol{x}^{(i+1)}$ are the input and output of $\boldsymbol{F}^{(i)}$. Each pair $(\boldsymbol{u}^{(i)}, \boldsymbol{u}^{(i+1)})$ is a valid monomial trail through $\boldsymbol{F}^{(i)}$ if and only if $\pi_{\boldsymbol{u}^{(i)}}(\boldsymbol{x}^{(i)}) \rightarrow \pi_{\boldsymbol{u}^{(i+1)}}(\boldsymbol{x}^{(i+1)})$ (Notice that $\boldsymbol{x}^{(i)}$ and $\boldsymbol{x}^{(i+1)}$ are only symbolic variables.). However, each $\boldsymbol{u}^{(i)}$ is defined in $\mathbb{F}_{2^n}$ where the size of n is typically larger than 32. So we are not able to depict the possible propagations by simple observation or exhaustive search. As any arithmetization-oriented cipher can be represented as a sequence of basic operations such as XOR, AND, COPY, m-COPY, and POWER, we carefully investigate the arithmetical feature of operations and prove the propagation rules. Our propagation rules put no restrictions on the irreducible polynomial since we do not care about the exact value of the coefficients.

Rule 1 (Field-based XOR). *Let F be a function compressed by an XOR over $\mathbb{F}_{2^n}$, where the input $\boldsymbol{x} = (x_0, x_1, \cdots, x_{n-1})$ and the output $\boldsymbol{y}$ is calculated as $\boldsymbol{y} = (x_0 \oplus x_1, x_2, \cdots, x_{n-1})$. Considering a monomial of $\boldsymbol{x}$ as $\boldsymbol{x}^{\boldsymbol{u}}$, the monomial $\boldsymbol{y}^{\boldsymbol{v}}$ contains $\boldsymbol{x}^{\boldsymbol{u}}$ iff*

$$\boldsymbol{v} = (v, u_2, \cdots, u_{n-1}),$$

where $v = u_0 + u_1$, $v \succeq u_0$.

Proof. We have

$$(x_0 \oplus x_1)^v \equiv \bigoplus_{0 \leq u_0 \leq v} p_v(u_0) \cdot (x_0^{u_0} x_1^{v-u_0}),$$

where $p_v(u_0) = 1$ if $\binom{v}{u_0}$ is odd and $p_v(u_0) = 0$ if $\binom{v}{u_0}$ is even. $\binom{v}{u_0}$ is the binomial coefficient. Clearly, $\binom{v}{u_0}$ is odd if and only if $u_0 \preceq v$ according to the Lucas's theorem. Therefore, if $x_0^{u_0} \cdot x_1^{u_1} \rightarrow (x_0 \oplus x_1)^v$, there must be $p_v(u_0) = 1$ and we have $u_0 + u_1 = v, u_0 \preceq v$. Conversely, if $u_0 + u_1 = v, u_0 \preceq v$, we have $p_v(u_0) = 1$ and $x_0^{u_0} \cdot x_1^{u_1} \rightarrow (x_0 \oplus x_1)^v$. □

Rule 2 (Field-based AND). *Let F be a function compressed by an AND over $\mathbb{F}_{2^n}$, where the input $\boldsymbol{x} = (x_0, x_1, \cdots, x_{n-1})$ and the output $\boldsymbol{y}$ is calculated as $\boldsymbol{y} = (x_0 x_1, x_2, \cdots, x_{n-1})$. Considering a monomial of $\boldsymbol{x}$ as $\boldsymbol{x}^{\boldsymbol{u}}$, the monomial $\boldsymbol{y}^{\boldsymbol{v}}$ contains $\boldsymbol{x}^{\boldsymbol{u}}$ iff*

$$\boldsymbol{v} = (u_0, u_2, \cdots, u_{n-1}),$$

where $(u_0, u_1) = (i, i)$, for $0 \leq i \leq 2^n - 1$.

Proof. Since

$$(x_0 x_1)^v = (x_0 x_1)^{u_0} = x_0^{u_0} x_1^{u_0} = x_0^{u_0} x_1^{u_1},$$

we have $x_0^{u_0} x_1^{u_1} \rightarrow (x_0 x_1)^v$ if $v = u_0 = u_1 = i$ for $0 \leq i \leq 2^n - 1$. Conversely if $v = u_0 = u_1 = i$ for $0 \leq i \leq 2^n - 1$, there must be $x_0^{u_0} x_1^{u_1} = (x_0 x_1)^i = (x_0 x_1)^v$ and $x_0^{u_0} x_1^{u_1} \rightarrow (x_0 x_1)^v$. □

Rule 3 (Field-based COPY). *Let F be a COPY function over $\mathbb{F}_{2^n}$, where the input $\boldsymbol{x} = (x_0, x_1, \cdots, x_{n-1})$ and the output $\boldsymbol{y}$ is calculated as $\boldsymbol{y} = (x_0, x_0, x_1, \cdots, x_{n-1})$. Considering a monomial of $\boldsymbol{x}$ as $\boldsymbol{x}^{\boldsymbol{u}}$, the monomial $\boldsymbol{y}^{\boldsymbol{v}}$ contains $\boldsymbol{x}^{\boldsymbol{u}}$ iff*

$$\boldsymbol{v} = (v_0, v_1, u_1, u_2, \cdots, u_{n-1}),$$

where

$$(v_0, v_1) = \begin{cases} (0,0), & \text{if } u_0 = 0; \\ (i, u_0 - i), (j, u_0 + 2^n - 1 - j), & \text{else.} \end{cases}$$

for $0 \leq i \leq u_0, u_0 \leq j \leq 2^n - 1$.

Proof. For $u_0 = 0$, if $x_0^{u_0} \rightarrow x_0^{v_0 + v_1}$, there must be $v_0 + v_1 = 0$, which implies $(v_0, v_1) - (0,0)$. Conversely if $u_0 = 0$ and $(v_0, v_1) = (0,0)$, we have $x_0^{u_0} = x_0^{v_0+v_1} = x_0^0$ and $x_0^{u_0} \rightarrow x_0^{v_0+v_1}$.

Let us now consider $u_0 \neq 0$. When $v_0 + v_1 \leq 2^n - 1$, if $x_0^{u_0} \rightarrow x_0^{v_0+v_1}$ we have $u_0 = v_0 + v_1$ and it holds that $(v_0, v_1) = (i, u_0 - i)$ for $0 \leq i \leq u_0$. Conversely if $(v_0, v_1) = (i, u_0 - i)$ for $0 \leq i \leq u_0$, we have $v_0 + v_1 = u_0$ and $x_0^{u_0} \rightarrow x_0^{v_0+v_1}$. When $v_0 + v_1 > 2^n - 1$, we have $v_0 + v_1 = t + 2^n - 1$ and $x_0^{v_0+v_1} \equiv x^t$, $0 < t \leq 2^n - 1$. If $x_0^{u_0} \rightarrow x_0^{v_0+v_1}$, we have $x_0^{u_0} \rightarrow x_0^t$ and $u_0 = t$. Therefore it holds that $(v_0, v_1) = (j, u_0 + (2^n - 1) - j)$ for $u_0 \leq j \leq 2^n - 1$. Conversely if $(v_0, v_1) = (j, u_0 + (2^n - 1) - j)$ for $u_0 \leq j \leq 2^n - 1$, we have $v_0 + v_1 = u_0 + 2^n - 1$ and $x_0^{u_0} \equiv x_0^{u_0+2^n-1} \equiv x_0^{v_0+v_1}$. Then $x_0^{u_0} \rightarrow x_0^{v_0+v_1}$. $\qquad\square$

Rule 4 (Field-based POWER). *Let F be a POWER function over $\mathbb{F}_{2^n}$, where the input $\boldsymbol{x} = (x_0, x_1, \cdots, x_{n-1})$ and the output $\boldsymbol{y}$ is calculated as $\boldsymbol{y} = (x_0^d, x_1, \cdots, x_{n-1})$, for $\gcd(d, 2^n - 1) = 1$. Considering a monomial of $\boldsymbol{x}$ as $\boldsymbol{x}^{\boldsymbol{u}}$, the monomial $\boldsymbol{y}^{\boldsymbol{v}}$ contains $\boldsymbol{x}^{\boldsymbol{u}}$ iff*

$$\boldsymbol{u} = (v, u_1, \cdots, u_{n-1}),$$

where

$$v = \begin{cases} u_0, & \text{if } u_0 = 0 \text{ or } 2^n - 1; \\ (d^{-1})u_0 \bmod (2^n - 1), & \text{else.} \end{cases}$$

Proof. For $u_0 = 0$, if $(x_0)^0 \rightarrow (x_0)^{dv}$ then v must be 0. Conversely if $v = 0$, we have $(x_0)^{u_0} \rightarrow (x_0)^{dv}$. For $u_0 = 2^n - 1$, if $(x_0)^{2^n-1} \rightarrow (x_0)^{dv}$, then v must be $2^n - 1$. Conversely if $v = 2^n - 1$, we have $(x_0)^{d \times (2^n-1)} \equiv x_0^{2^n-1}$ and $x_0^{2^n-1} \rightarrow x_0^{2^n-1}$.

If $u_0 \neq 0$ and $u_0 \neq 2^n - 1$, if $x_0^{u_0} \rightarrow (x_0^d)^v$ we have $u_0 = dv \bmod (2^n - 1)$. Conversely if $u_0 = dv \bmod (2^n - 1)$, we have $(x_0^d)^v \equiv x_0^{dv \bmod (2^n-1)} \equiv x_0^{u_0}$. Therefore $(x_0)^{u_0} \rightarrow (x_0)^{dv}$. Then we have $v = (d^{-1})u_0 \bmod (2^n - 1)$. $\qquad\square$

Rule 5 (Field-based m-COPY). *Let F be a m-COPY function over $\mathbb{F}_{2^n}$, where the input $\boldsymbol{x} = (x_0, x_1, \cdots, x_{n-1})$ and the output $\boldsymbol{y}$ is calculated as $\boldsymbol{y} =$*

$(\underbrace{x_0, \cdots, x_0}_{m}, x_1, \cdots, x_{n-1})$. *Considering a monomial of $\boldsymbol{x}$ as $\boldsymbol{x}^u$, the monomial $\boldsymbol{y}^v$ contains $\boldsymbol{x}^u$ iff*

$$\boldsymbol{v} = (v_0, v_1, \cdots, v_{m-1}, u_1, u_2, \cdots, u_{n-1}),$$

where

$$(v_0, \cdots, v_{m-1}) = \begin{cases} (0, 0, \cdots, 0), & \text{if } u_0 = 0; \\ (i_0^s, \cdots, i_{m-2}^s, i_{m-1}^s) \text{ for } 0 \le s \le m-1, & \text{else.} \end{cases}$$

Here, $i_{m-1}^s = u_0 + (s-1)(2^n - 1) - \sum_{j=0}^{m-2} i_j^s$, $0 \le i_j^s \le 2^n - 1$ for $0 \le j < m$.

Proof. For $u_0 = 0$, if $x_0^0 \to x_0^{v_0 + v_1 + \cdots + v_{m-1}}$ there must be $(v_0, v_1, \cdots, v_{m-1}) = (0, 0, \cdots, 0)$. Conversly, if $(v_0, v_1, \cdots, v_{m-1}) = (0, 0, \cdots, 0)$ we have $(x_0)^{u_0} \to (x_0)^{v_0 + v_1 + \cdots + v_{m-1}}$.

Let us consider $u_0 \ne 0$. We have $0 < v_0 + v_1 + \cdots + v_{m-1} \le m \cdot (2^n - 1)$. When $s(2^n - 1) < v_0 + v_1 + \cdots + v_{m-1} \le (s+1)(2^n - 1)$, $0 \le s \le m-1$, we have $v_0 + \cdots + v_{m-1} = t + (s-1)(2^n - 1)$, $0 < t \le 2^n - 1$.

If $x_0^{u_0} \to x_0^{v_0 + v_1 + \cdots + v_{m-1}}$, we have $x_0^{u_0} \to x_0^t$ and thus $u_0 = t$. Therefor it holds that $(v_0, v_1, \cdots, v_{m-1}) = (i_0^s, i_1^s, \cdots, i_{m-2}^s, u_0 + (s-1)(2^n - 1) - \sum_{j=0}^{m-2} i_j^s)$ for $0 \le i_j^s \le 2^n - 1$.

Conversely if $(v_0, v_1, \cdots, v_{m-1}) = (i_0^s, i_1^s, \cdots, i_{m-2}^s, u_0 + (s-1)(2^n - 1) - \sum_{j=0}^{m-2} i_j^s)$ for $0 \le i_j^s \le 2^n - 1$, we have $v_0 + v_1 + \cdots + v_{m-1} = u_0 + (s-1)(2^n - 1)$ and $x_0^{u_0} \equiv x_0^{u_0 + (s-1)(2^n - 1)} \equiv x_0^{v_0 + v_1 + \cdots + v_{m-1}}$. Then $x^{u_0} \to x_0^{v_0 + v_1 + \cdots + v_{m-1}}$. □

Example 2. Let $x_0, x_1, y, z \in \mathbb{F}_{2^3}$ with the irreducible polynomial $f(x) = x^3 + x + 1$. $y = (x_0 \oplus 3x_1)^3$. Compute (u_0, u_1) when $x_0^{u_0} \cdot x_1^{u_1} \rightsquigarrow y^v, v = 2$.

Consider $z = x_0 \oplus 3x_1$, then $y = z^3$. Then we need to compute all the monomial trails $x_0^{u_0} x_1^{u_1} \rightsquigarrow z^w \rightsquigarrow y^v$. According the Rule 4, we have $w = 3v \mod (7) = 6 \mod (7)$, $w = 6$. As $w = u_0 + u_1$ and $u_0 \preceq w$, we have $u_0 = 0, 2, 4, 6$. Then we deduce that $(u_0, u_1, w, v) = (6, 0, 6, 2), (4, 2, 6, 2), (2, 4, 6, 2), (0, 6, 6, 2)$ by the propagation rules. It is verified by

$$y^2 = x_0^6 \oplus 5x_0^4 \cdot x_1^2 \oplus 7x_0^2 \cdot x_1^4 \oplus 6x_1^6 \to (u_0, u_1, v) = (6, 0, 2), (4, 2, 2), (2, 4, 2), (0, 6, 2).$$

3.3   Bit-Vector Models for Field-Based Operations

In this subsection, we take advantage of the bit-theory of SMT and translate the propagations into a system of equations involving both arithmetic operations and bit-based operations. The solutions to the constraints are all the possible monomial trails through the basic operations. Moreover, we avoid arithmetic multiplications and arithmetic modular to obtain efficient bit-vector constraints. The models for XOR, AND, COPY, m-COPY, and POWER are introduced as follows.

Model 1 (Field-based XOR). *Let* $(u_0, u_1) \xrightarrow{XOR} (v)$ *denote the monomial trails through the field-based XOR function, where two n-bit words are compressed to one n-bit word using an XOR operation. Then, it can be depicted using the following constraints:*

$$\begin{cases} u_0 + u_1 = v, \\ v \;\wedge\; u_0 = u_0, \\ u_0, u_1, v \text{ are } n\text{-bit variables.} \end{cases}$$

The constraint $v \;\wedge\; u_0 = u_0$ excludes the invalid trails for $v \not\succeq u_0$.

Model 2 (Field-based AND). *Let* $(u_0, u_1) \xrightarrow{AND} (v)$ *denote the monomial trails through the field-based AND function, where two n-bit words are compressed to one n-bit word using an AND operation. Then, it can be depicted using the following constraints:*

$$\begin{cases} u_0 = v, \\ u_1 = v, \\ u_0, u_1, v \text{ are } n\text{-bit variables.} \end{cases}$$

Model 3 (Field-based COPY). *Let* $(u) \xrightarrow{COPY} (v_0, v_1)$ *denote the monomial trails through the field-based COPY function, where one n-bit word is copied to two n-bit words using a COPY operation. Then, it can be depicted using the following constraints:*

$$\begin{cases} v_0 + v_1 + t = t \;||\; u, \\ u \;||\; t \neq 0^n \;||\; 1^1, \\ u, v_0, v_1 \text{ are } n\text{-bit variables}, \\ t \text{ is a 1-bit variable.} \end{cases}$$

Proof. For $u = 0$, the only valid trail is $v_0 + v_1 = 0$ since $t \neq 1$. For $u \neq 0$, we have $v_0 + v_1 = u$ when $t = 0$ and $v_0 + v_1 + 1 = 1 \;||\; u = u + 2^n$ when $t = 1$. $\square$

Model 4 (Field-based m-COPY). *Let* $(u) \xrightarrow{m-COPY} (v_0, v_1, \cdots, v_{m-1})$ *denote the monomial trails through the field-based m-COPY function, where one n-bit word is copied to m n-bit words using an m-COPY operation. Then, it can be depicted using the following constraints:*

$$\begin{cases} v_0 + v_1 \cdots + v_{m-1} + t = t \;||\; u, \\ u \;||\; t \neq 0^n \;||\; q, \; 0 < q \leq m - 1 \\ t \leq m - 1, \\ u, v_0, v_1 \text{ are } n\text{-bit variables}, \\ t \text{ is a } s\text{-bit variable}, s = \lfloor \log_2(m-1) \rfloor + 1. \end{cases}$$

The constraints

$$u \parallel t \neq 0^n \parallel q, \; 0 < q \leq m - 1$$

is implemented in STP solver with an IF-THEN-ELSE branch statement as follows

ASSERT (IF $u = 0^n$ THEN $t = 0^s$ ELSE $t \geq 0^s$);

Model 5 (Field-based POWER). *Let* $(u) \xrightarrow{POWER} (v)$ *denote the monomial trails through the field-based POWER function, where one n-bit word is transmitted to another n-bit word using a POWER operation,* $\gcd(d, 2^n - 1) = 1$, *Then, it can be depicted using the following constraints:*

$$\begin{cases} d \times v + t = t \parallel u, \\ t \leq d - 1, \\ u, v \text{ are } n\text{-bit variables}, \\ t \text{ is an } s\text{-bit variable}, s = \lfloor log_2(d-1) \rfloor + 1. \end{cases}$$

Moreover, when $d = 2^l + 1$ *or* $d = 2^l - 1$, *we can avoid multiplications and give more efficient constraints as:*

$$\begin{cases} (v \ll l) \pm v + t = t \parallel u, \\ t \leq d - 1, \\ u, v \text{ are } n\text{-bit variables}, \\ t \text{ is an } s\text{-bit variable}, s = \lfloor log_2(d-1) \rfloor + 1. \end{cases}$$

Proof. When $u = 0$, we have $d \times v = t \times (2^n - 1)$. Since $\gcd(d, 2^n - 1) = 1$, we have $\gcd(d, t \times (2^n - 1)) \leq t < d$, then d is not divisible by $t \times (2^n - 1)$ if $t \neq 0$. Then we have $v = 0$.

When $u = 2^n - 1$, $d \times v = (1 + t) \times (2^n - 1)$. If $t = d - 1$, $v = 2^n - 1$. If $t \neq d - 1$, there are no solutions since d is not divisible by $(t + 1) \times (2^n - 1)$ for $0 < t + 1 < d$.

When $u \neq 0$ and $u \neq 2^n - 1$, we have $d \times v = u + t \times (2^n - 1)$ and thus $u = dv \mod (2^n - 1)$. □

3.4   Detecting the Upper Bound of the Algebraic Degree

In this subsection, we describe how to detect the upper bound of the algebraic degree for block ciphers considering round keys. All the round keys $\boldsymbol{k}^{(i)}$ are regarded as independent input variables defined over $\mathbb{F}_{2^n}$ for $0 \leq i < r$. Suppose the input of the statement is defined over $\mathbb{F}_{2^n}^t$, that is, the length of the word size is n and the number of words is t. For $0 \leq i < r$, let $\pi_{\boldsymbol{u}^{(i)}}(\boldsymbol{x}^{(i)})$ denote the input monomials of the i-th round function, respectively. Then $\pi_{\boldsymbol{u}^{(r)}}(\boldsymbol{x}^{(r)})$ denotes a monomial of ciphertext we are interested in and is usually set as a unit vector to study a certain word of the ciphertext in practice. $\pi_{\boldsymbol{v}^{(i)}}(\boldsymbol{k}^{(i)})$ denotes the monomial of the i-th round key. We use equations to add constraints for variables $\boldsymbol{u}^{(i)}$, $\boldsymbol{u}^{(i+1)}$, and $\boldsymbol{v}^{(i)}$ according to the function between them. The monomial trails through the public function described in the models are introduced in Sect. 3.3. Notice that the monomials $\pi_{\boldsymbol{v}^{(i)}}(\boldsymbol{k}^{(i)})$ are treated equivalently as $\pi_{\boldsymbol{u}^{(i)}}(\boldsymbol{x}^{(i)})$ when we add constraints.

Initial Constraints. According to Proposition 2, if we want to determine whether a specific monomial $\pi_{\tilde{\boldsymbol{u}}^{(r)}}(\boldsymbol{x}^{(r)})$ does not contain any (key-related) monomial $\pi_{\boldsymbol{v}^{(0)},\cdots,\boldsymbol{v}^{(r)}}(\boldsymbol{k}^{(0)},\cdots,\boldsymbol{k}^{(r)})\cdot\pi_{\tilde{\boldsymbol{u}}^{(0)}}(\boldsymbol{x}^{(0)})$, we only need to check whether there exist some trails from the monomial $\pi_{\boldsymbol{v}^{(0)},\cdots,\boldsymbol{v}^{(r)}}(\boldsymbol{k}^{(0)},\cdots,\boldsymbol{k}^{(r)})\cdot\pi_{\tilde{\boldsymbol{u}}^{(0)}}(\boldsymbol{x}^{(0)})$ to $\pi_{\tilde{\boldsymbol{u}}^{(r)}}(\boldsymbol{x}^{(r)})$. Given an initial vector $I_u = (\tilde{u}_0^{(0)},\cdots,\tilde{u}_{t-1}^{(0)})$, where $\tilde{u}_i^{(0)} \in \mathbb{F}_{2^n}$, we use

$$u_i^{(0)} = \tilde{u}_i^{(0)} \text{ for } 0 \leq i < t$$

to add the initial constraints on $\boldsymbol{u}^{(0)}$ and search for the general monomial trails. Notice that we do not add any constraints on $(\boldsymbol{v}^{(0)},\cdots,\boldsymbol{v}^{(r)})$ since they are all free variables over $\mathbb{F}_{2^n}$.

However, in the higher-order differential attacks [29], we are interested in the algebraic degree of F. If the algebraic degree of F is $\delta(F)$, then we have $\bigoplus_{v\in\mathcal{V}\oplus c} F(v) = 0$ if the dimension of the affine vector space $\mathcal{V} \oplus c$ is strictly greater than $\delta(F)$. We then use Corollary 1 that the algebraic degree of monomial $x_0^{u_0} \cdot ... \cdot x_{t-1}^{u_{t-1}}$ is given by $\sum_{i=0}^{t-1} wt(u_i)$. Therefore, if we want to determine the algebraic degree of a certain monomial $\pi_{\boldsymbol{u}^{(r)}}(\boldsymbol{x}^{(r)})$, we only need to check whether $\pi_{\boldsymbol{u}^{(r)}}(\boldsymbol{x}^{(r)})$ contains any term in the set $\mathbb{S}_l$ for $d \leq l \leq \Lambda$, where

$$\mathbb{S}_l = \{\pi_{\boldsymbol{v}^{(0)},\cdots,\boldsymbol{v}^{(r)}}(\boldsymbol{k}^{(0)},\cdots,\boldsymbol{k}^{(r)})\cdot\pi_{\boldsymbol{u}^{(0)}}(\boldsymbol{x}^{(0)}) \mid \sum_{i=0}^{t-1} wt(u_i^{(0)}) = l\} \qquad (4)$$

and Δ denotes the maximum algebraic degree. According to Proposition 2, if the monomial $\pi_{\boldsymbol{u}^{(r)}}(\boldsymbol{x}^{(r)})$ contains no monomials in $\mathbb{S}_l$ for $d \leq l \leq \Delta$, the algebraic degree $\delta(F)$ is strictly less than d and the upper bound of the algebraic degree is $d - 1$.

Stopping Rules. If we consider the algebraic degree of the i'th ciphertext word, then we use

$$\begin{cases} u_i^{(r)} = 1, & \text{if } i = i', \\ u_i^{(r)} = 0, & \text{if } i \neq i'. \end{cases}$$

to add the stopping rules on $\boldsymbol{u}^{(r)}$.

Detecting the Upper Bound of the Algebraic Degree. Let us denote the stopping constraints as $\Gamma = (\underbrace{0,\cdots,0}_{i'},1,\underbrace{0,\cdots,0}_{t-i'-1})$. If we want to determine whether the upper bound of the algebraic degree for a certain monomial $\pi_{\boldsymbol{u}^{(r)}}(\boldsymbol{x}^{(r)})$ is $d-1$, we only need to check whether $\pi_{\boldsymbol{u}^{(r)}}(\boldsymbol{x}^{(r)})$ contains any term

---

Algorithm 1: $\delta = \mathsf{SearchDegree}(\mathcal{M}_r, \Delta, \Gamma)$

Input: The r-round SMT model $\mathcal{M}_r$, the maximum algebraic degree Δ, the stopping constraints Γ

Output: The algebraic degree δ

1 $\mathcal{M} \leftarrow \mathcal{M}_r$;
2 $\delta = 0$;
3 **for** $i = 0; i < t; i \leftarrow i + 1$ **do**
4 $\quad\lfloor\ \mathcal{M}.con \leftarrow u_i^{(r)} = \Gamma[i]$;
5 **for** $i = \Delta; i \geq 0; i \leftarrow i - 1$ **do**
6 $\quad\mid\quad \mathcal{M}.con \leftarrow \sum_j wt(u_j^{(0)}) = i$;
7 $\quad\mid\quad$ solve the r-round SMT model $\mathcal{M}$;
8 $\quad\mid\quad$ **if** *the problem is satisfiable* **then**
9 $\quad\mid\quad\quad\mid\quad \delta = i$;
10 $\quad\mid\quad\quad\lfloor$ **break;**

11 **return** δ;

---

in the set $\mathbb{S}_l$. $\mathbb{S}_l$ is defined in Equation (4). For $l \leq \Delta$, if there is no general monomial trail from any monomial contained by $\mathbb{S}_l$ to $\pi_{\boldsymbol{u}}^{(r)}(\boldsymbol{x}^{(r)})$ for $d \leq l \leq \Delta$, the upper bound of the algebraic degree is $d - 1$. Therefore, we use constraint

$$\sum_i wt(u_i^{(0)}) = l \tag{5}$$

from $l = \Delta$ in a decreasing order to add the initial constraint on $\boldsymbol{u}^{(0)}$. Δ denotes the theoretical upper bound of the algebraic degree with a maximum value of $n \cdot t - 1$ for permutations.

The framework of the whole algorithm is illustrated in Algorithm 1. When the SMT solver finds the solution for the first satisfiable problem, an assignment of the variables that makes the problem satisfiable is obtained. Then the searching process finishes and we have found the upper bound of the algebraic degree δ.

4   Theoretical Upper Bound for MiMC-Like Constructions in (Unbalanced) Feistel Network

In this section, for a better insight into the behavior of the degree growth, we firstly investigate the algebraic degree of MiMC-like constructions in the (unbalanced) Feistel network. Based on the upper bound given in [22] valid for the MiMC-like construction in Even-Mansour schemes, we propose a new linear upper bound in Sect. 4.1. Besides, we show that higher-order differential distinguisher can be established using the special structure of the function in Sect. 4.2.

4.1 Theoretical Upper Bound on the Algebraic Degree

Considering $E_k \colon \mathbb{F}_{2^n}^t \to \mathbb{F}_{2^n}^t$ as an (unbalanced) Feistel-network cipher, $t \geq 2$. The j-th (expanding) round function is defined as

$$(x_0^{(j)}, x_1^{(j)}, \cdots, x_{t-1}^{(j)}) \leftarrow (x_1^{(j-1)} \oplus F(x_0^{(j-1)}), \cdots, x_{t-1}^{(j-1)} \oplus F(x_0^{(j-1)}), x_0^{(j-1)}), \quad (6)$$

where $F(x) := (x \oplus k^{(j-1)})^d$. $k = (k^{(0)}, \cdots, k^{(r-1)})$ denotes a sequence of round keys.

We firstly focus on the univariate degree of E_k. The maximum Hamming weight of the exponent for a single variable x_i in $x_j^{(r)}$ is represented by $\delta_{x_i}(x_j^{(r)})$.

Example 3. Let $y = x_0^2 x_1^7 \oplus x_0^5 x_1^3 \oplus x_0^8 x_1^5$, then we have

$$\delta_{x_0}(y) = 2, \delta_{x_1}(y) = 3.$$

Recalling the upper bound given in [22] valid for the MiMC-like construction in Even-Mansour schemes, we apply this idea to the (unbalanced) Feistel network and prove the following Lemma 1.

Lemma 1. *Considering $E_k \colon \mathbb{F}_{2^n}^t \to \mathbb{F}_{2^n}^t$ as an (unbalanced) Feistel-network cipher represented as in Equation (6). For $0 \leq i, j < t$, $r \geq 1$, we have*

$$\delta_{x_i}(x_j^{(r)}) \leq \begin{cases} \min\{\lfloor \log_2(d^{r-(i+\theta(j))} + 1) \rfloor, n\}, & \text{if } r > i + \theta(j), \\ 1, & \text{else if } r = i - j + t \cdot \theta(j), \\ 0, & \text{else.} \end{cases}$$

where

$$\theta(j) = \begin{cases} 0, \text{ if } 0 \leq j < t - 1, \\ 1, \text{ if } j = t - 1. \end{cases}$$

Proof. Notice that the degree of x_i grows differently in the different branches of the (unbalanced) Feistel network. When $j < t - 1$, we have that the maximum exponents of x_i in $x_j^{(i)}$ is d. Then the exponents of x_i in r-th round are upper bounded by d^{r-i} if $r > i$. This means that the upper bound of $\delta_{x_i}(x_j^{(r)})$ is the maximum integer l that satisfies $2^l - 1 \leq d^{r-i}$. Since $\delta_{x_i}(x_j^{(r)}) \leq n$ we have $\delta_{x_i}(x_j^{(r)}) \leq \min\{\lfloor \log_2(d^{r-i} + 1) \rfloor, n\}$. For the case of $r \leq i$, by simple observation the maximum exponents of x_i in $x_j^{(i-j)}$ is 1 when $i > j$ and $x_j^{(r)}$ does not contain the variable x_i otherwise. Hence we have $\delta_{x_i}(x_j^{(r)}) = 1$ if $r = i - j$ and $\delta_{x_i}(x_j^{(r)}) = 0$ otherwise.

When $j = t - 1$, due to the structure of the (unbalanced) Feistel network we have $\delta_{x_i}(x_{t-1}^{(r)}) = \delta_{x_i}(x_0^{(r-1)})$. Then we can derive $\delta_{x_i}(x_j^{(r)})$ in the same way before. Since the maximum exponents of x_i in $x_j^{(i+1)}$ is d, we have that the exponents in r-th round are upper bounded by $d^{r-(i+1)}$ and $\delta_{x_i}(x_j^{(r)}) \leq \min\{\lfloor \log_2(d^{r-(i+1)} + 1) \rfloor, n\}$ for $r > i+1$. For the case of $r \leq i$, the maximum exponents of x_i in $x_j^{(i-j+1)}$ is 1 and we have $\delta_{x_i}(x_j^{(r)}) = 1$ if $r = i - j + 1$. □

Using Corollary 1, we propose the upper bound on the algebraic degree of MiMC-like constructions in (unbalanced) Feistel network as follows.

Proposition 3. *Considering $E_k \colon \mathbb{F}_{2^n}^t \to \mathbb{F}_{2^n}^t$ as a (unbalanced) Feistel-network cipher represented as in Equation (6). Let $\mathcal{R}_j = \log_d(2^n - 1) + (t - 1 + \theta(j))$, The algebraic degree of E_k satisfies that*

$$\delta(x_j^{(r)}) \leq \begin{cases} \sum_{i=0}^{t-1} \delta_{x_i}(x_j^{(r)}), & \text{if } r < \mathcal{R}_j, \\ t \cdot n - 1, & \text{else.} \end{cases}$$

The proof can be found in the full version of the paper [1].

Discussion of Proposition 3. By the proof of Proposition 3, we can see that the growth of the algebraic degree is almost linear. When $r \geq \mathcal{R}_j$, $\delta(x_i^{(r)})$ is always $t \cdot n - 1$. However, we point out that it is not always the case. Taking the polynomial $(x_0 \oplus k_0)^{27}(x_1 \oplus k_1)^{18}$ as a simple example, the algebraic degree is $wt(27) + wt(18) = 6$ for the appearance of monomial $x_0^{27} x_1^{18}$. However, when $n = 3$, actually we have

$$(x_0 \oplus k_0)^{27}(x_1 \oplus k_1)^{18} \equiv (x_0 \oplus k_0)^6 (x_1 \oplus k_1)^4$$

and the algebraic degree is $wt(6) + wt(4) = 4$. The deviation is caused by the offset of exponents when the degree is over $2^n - 1$, which is difficult to give a condition to guarantee when a particular monomial will change. Moreover, the relatively simple algebraic structure also leads to sparser terms and this decreases the practical degree.

4.2   Constructing Higher-Order Differential Distinguishers by Considering Different Numbers of Branches

As is well known, if the algebraic degree of F is strictly smaller than $d - 1$, then for any subspace $V \subseteq \mathbb{F}_2^N$ with dimension $s \geq d$, we have $\bigoplus_{x \in V \oplus c} F(x) = 0$. Unfortunately, the opposite does not hold in general. Even if the summing over all the $x \in V \oplus c$ of dimension $s \leq d$ always results in a zero-sum, we cannot make sure if the algebraic degree is d since it can be caused by some special structure of the function. However, this provides us with a new approach for detecting the higher-order differential distinguisher. For multivariate polynomial $F_k \colon \mathbb{F}_{2^n}^t \to \mathbb{F}_{2^n}$ defined as

$$F_k(x_0, \cdots, x_{t-1}) = \sum_{\boldsymbol{u} = (u_0, \cdots, u_{t-1}) \in \{0, 1, \cdots, 2^n - 1\}^t} \varphi_k(\boldsymbol{u}) \cdot \prod_{i=0}^{t-1} x_i^{u_i}, \boldsymbol{k} \in \mathbb{F}_{2^n}^r.$$

We limit ourselves to consider the sum of the Hamming weight for the exponents of different branches, denoted by $\sum_{i \in \mathbb{X}} \delta_{x_i}(x_j^{(r)})$ for some certain j. $\mathbb{X}$ represents the set of the branches we are interested in. Then by tracing the upper bound of $\sum_{i \in \mathbb{X}} \delta_{x_i}(x_j^{(r)})$ precisely, we can establish higher-order differential distinguishers. The following theorem is a corollary of Proposition 1 in [8].

Corollary 2. *Let* $F_{\boldsymbol{k}} : \mathbb{F}_{2^n}^t \rightarrow \mathbb{F}_{2^n}$ *be multivariate polynomial defined as*

$$F_{\boldsymbol{k}}(x_0, \cdots, x_{t-1}) = \sum_{\boldsymbol{u}=(u_0,\cdots,u_{t-1}) \in \{0,1,\cdots,2^n-1\}^t} \varphi_{\boldsymbol{k}}(\boldsymbol{u}) \cdot \prod_{i=0}^{t-1} x_i^{u_i}, \boldsymbol{k} \in \mathbb{F}_{2^n}^r.$$

If there exist m *variables* $x_{j_0}, x_{j_1}, \cdots, x_{j_{m-1}}$ *satisfies that for each non-vanishing monomial in* $F_{\boldsymbol{k}}$ *there is* $\bigoplus_{w=0}^{m-1} hw(u_{j_w}) \leq s-1$, *we have* $\bigoplus_{v \in \mathcal{V} \oplus c} F_{\boldsymbol{k}}(v) = 0.$ $\mathcal{V} = \{(l_0, l_1, \cdots, l_{t-1}) \mid (l_{j_0}, l_{j_1}, \cdots, l_{j_{m-1}}) \in V\}$ *for any affine subspace* $V \subseteq \mathbb{F}_{2^{m \times n}}$ *of dimension at least* s.

Proof. Each non-vanishing monomial of $F_{\boldsymbol{k}}$ can be written in the form of $\varphi_{\boldsymbol{k}}(\boldsymbol{u}) \cdot x_0^{u_0} x_1^{u_1} \cdot \cdots \cdot x_{t-1}^{u_{t-1}}$ with $\varphi_{\boldsymbol{k}}(\boldsymbol{u}) \neq 0$. Since the dimension of V is at least s, then we have

$$\sum_{(x_{j_0}, \cdots, x_{j_{m-1}}) \in V} x_{j_0}^{u_{j_0}} \cdot \ldots \cdot x_{j_{m-1}}^{u_{j_{m-1}}} = 0$$

and for each monomial of $F_{\boldsymbol{k}}$ we have

$$\sum_{(x_0, \cdots, x_{t-1}) \in \mathcal{V} \oplus c} \varphi_{\boldsymbol{k}}(\boldsymbol{u}) \cdot \prod_{i=0}^{t-1} x_i^{u_i} = 0.$$

Consequently, $\bigoplus_{v \in \mathcal{V} \oplus c} F_{\boldsymbol{k}}(v) = 0$.

5 Applications to Feistel MiMC, MiMC and GMiMC

We apply our algorithm to some competitive arithmetization-oriented block ciphers, including MiMC and its generalization Feistel MiMC and GMiMC. All of them use $x \mapsto x^3$ as their round function, but are based on different design strategies. The original MiMC introduced by Albrecht et al. [3] is a family of block ciphers dedicated to applications that support operations in large finite fields posing largest performance bottleneck. Due to its outstanding performance in applications such as MPC, SNARKs and STARKs, it quickly became the optimal choice for many use cases. In the same specification, a variant of MiMC was proposed by inserting the original design into the Feistel structure, named Feistel MiMC or MiMC-$2n/n$. This first application of Feistel networks in AO ciphers brings more flexibility of being able to rely on a larger field size. In that spirit, Albrecht et al. proposed GMiMC [2], a family of block ciphers based on different types of Feistel networks which can operate on different numbers of branches.

5.1 Application to Feistel MiMC

In this subsection, we focus on Feistel MiMC, an r-round block cipher in Feistel network with n-bit block size and the same key size operating on $\mathbb{F}_{2^n}$. The i-th round function $F^{(i)}$ is depicted in Fig. 1 and defined as

$$(x_0^{(i)}, x_1^{(i)}) \leftarrow (x_1^{(i-1)} \oplus (x_0^{(i-1)} \oplus k^{(i-1)})^d, x_0^{(i-1)})$$

$k = (k^{(0)}, \ldots, k^{(r-1)})$ denotes a sequence of r round subkeys. The round constants are omitted for simplicity since they can be regarded as part of the round keys and do not affect the upper bound of the algebraic degree. We denote Feistel MiMC specified by exponent d and block size n as $\mathsf{FeistelMiMC}_d(n, r)$. When $d = 3$, the number of rounds to achieve n-bit security is $r_n = 2n \cdot \log_3(2) + 1$ and the number of rounds to achieve $2n$-bit security is $r_N = \lceil 2n \cdot \log_3(2) \rceil + 3$. As far as we know, the best higher-order differential distinguisher in the literature is the 83-round one proposed in [8] for the permutation Feistel MiMC.

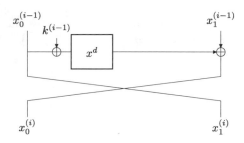

Fig. 1. The round function $F^{(i)}$ of $\mathsf{FeistelMiMC}_d(n, r)$.

Detect the algebraic degree of $\mathsf{FeistelMiMC}_d$. Let $\pi_{u^{(0)}}(x^{(0)})$ denote the monomials of the input statements of $\mathsf{FeistelMiMC}_d$. $\pi_{u^{(i)}}(x^{(i)})$ and $\pi_{v^{(i)}}(k^{(i)})$ denote the output statements of the i-th round function $F^{(i)}$ and the monomial of the i-th round key, respectively. We introduce auxiliary variables and the whole SMT model $\mathcal{M}_r$ is described as

$$
\mathcal{M}_r \leftarrow
\begin{cases}
(u_0^{(i)}) \xrightarrow{COPY} (u_1^{(i+1)}, m^{(i)}) & for\ 0 \leq i < r, \\
(v^{(i)}, m^{(i)}) \xrightarrow{XOR} (p_i) & for\ 0 \leq i < r, \\
(p^{(i)}) \xrightarrow{POWER} (q^{(i)}) & for\ 0 \leq i < r, \\
(q^{(i)}, u_1^{(i)}) \xrightarrow{XOR} (u_0^{(i+1)}) & for\ 0 \leq i < r,
\end{cases}
$$

By setting the initial constraints and stopping rules, Algorithm 1 is implemented to search the algebraic degree of $\mathsf{FeistelMiMC}_d(n, r)$. We denote the algebraic degree of the left branch and the right branch by $\delta(x_0^{(r)})$ and $\delta(x_1^{(r)})$, respectively. Without loss of generality, we only search for $\delta(x_0^{(r)})$ due to the structure of the Feistel network, i.e., $\delta(x_1^{(r)}) = \delta(x_0^{(r-1)})$.

Comparison of Our Results with Theoretical Bounds. We have practically verified our results on small-scale instances of $\mathsf{FeistelMiMC}_3(n, r)$ with block size $n = 13$ and found that our detected bounds correspond to the practical results. A concrete comparison of different degree bounds for $\delta(x_0)$ is given in Fig. 2 for $\mathsf{FeistelMiMC}_3(129, r)$. The trivial upper bound is defined as 2^r. Meanwhile, we also depict the trivial lower bound to explicitly understand the security margin. Indeed, since the monomials $x_0^{3^r}$ and $x_1^{3^{r-1}}$ always appears in $x_0^{(r)}$ independently from the choice of round constants or round keys, we can define the trivial lower bound as $\max\{wt(3^r), wt(3^{r-1})\}$.

We notice that both our detected bound and our theoretical bound present a linear growth in the initial stage. A more substantial difference appears when the algebraic degree is reaching the maximal, namely when $r > \mathcal{R}_j$. While the theoretical bound predicts that $\delta(x_0^{(r)})$ reaches the maximal degree directly after the linear growth, the detected bounds show that there is still a long stage of slow growth before achieving the maximum algebraic degree. During some consecutive rounds, the algebraic degree even remains constant, called a *plateau* in [14]. Some plateaus cover a few rounds, e.g., $\delta(x_0^{(r)})$ stays constant at 254 for only 2 rounds. The largest plateau appears when the algebraic degree is $2n - 2$. It remains constant for an especially long time, covering a total of 27 rounds. Our results indicate that the stage of slow growth significantly influences the growth of the algebraic degree. Therefore, more rounds than previously predicted may be necessary to guarantee security against high-order differential distinguishers. For $\mathsf{FeistelMiMC}_3(129, r)$, our detected bound can produce the distinguisher for 124 rounds, which extends the theoretical distinguisher for 41 rounds.

We would like to mention that our Algorithm 1 can also be applied to search for the univariate degree $\delta_{x_i}(x_j^{(r)})$ by slightly modifying the initial constraints as

$$wt(u_i^{(0)}) = l$$

for $i \in \{0, 1\}$, respectively. Then if $wt(u_i^{(0)}) \le s - 1$, we can always construct a higher-order differential distinguisher with a data complexity of 2^s for branch i. As an example, for $\mathsf{FeistelMiMC}_3(129, r)$, we can exhibit a distinguisher with data complexity $2^{127}/2^{129}$ for 82/83 rounds, both resulting in a zero-sum in the output of the right branch.

Comparison of Our Bounds with Different Exponents. We also applied our algorithm to different exponents d and observe the influence of exponents on the degree growth. A simple observation of Fig. 3 is that the linear rate of the initial linear growth goes up with d and the number of rounds for the slow growth (i.e., from the round $\mathcal{R}_0$ until reaching the maximal algebraic degree) is reduced. The number of rounds for the slow growth is 42 rounds for $d = 3$ whereas it is 30 rounds for $d = 5$ despite the Hamming weight of d is the same. For $d = 31$, the number of rounds for the slow growth is only 15, with the largest plateau covering 10 rounds.

Known-Key Zero-Sum Distinguisher for the Full $\mathsf{FeistelMiMC}_3(129, r)$. The known-key distinguishers for block ciphers were introduced by Knudsen and Rijmen at ASIACRYPT 2007 [28] and have been a major research direction in

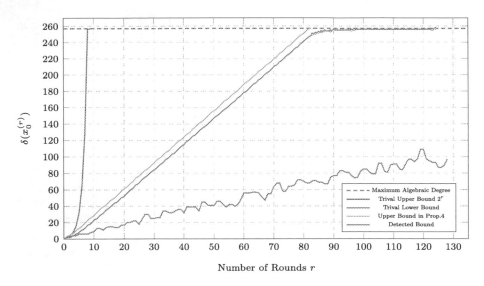

Fig. 2. Comparison of different degree bounds $\delta(x_0^{(r)})$ for FeistelMiMC$_3$(129, r).

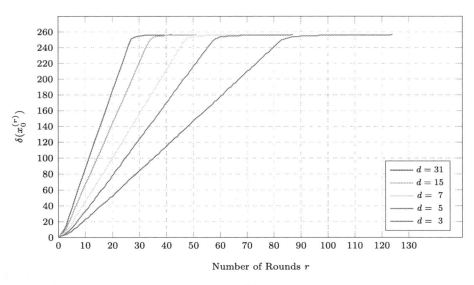

Fig. 3. Comparison of our degree bounds $\delta(x_0^{(r)})$ for FeistelMiMC$_d$(129, r) with different exponents d.

cryptanalysis since then. There is no secret material involved in the computation and the attacker aims to find a structural property for the cipher which an ideal cipher would not have. It is also related to the distinguishers for permutation since the analysis is often done in the known-key model. A well-known powerful distinguisher for the known-key setting is the so-called zero-sum distinguisher [5]. The idea is based on the inside-out approach, where the attacker starts from the middle rounds and chooses a set of internal states so that the sum of the inputs and outputs are all zero when computing backwards and forwards.

Let us consider $\mathsf{FeistelMiMC}_3(129, r)$. By choosing a subspace of the right input branch of dimension 127, the distinguisher can be extended forwards for 82 rounds, with the output of the right branch achieving zero-sum property. The inverse of $\mathsf{FeistelMiMC}_3(129, r)$ still follows the Feistel network. When computing backwards, the active branch is now the left one. Then the distinguisher can be extended backwards for 81 rounds, resulting in a zero-sum property in the right output branch. This eventually leads to a distinguisher with complexity 2^{127} for a total of $82 + 81 = 163$ rounds. Besides, by saturating the branch x_1, we can further derive a zero-sum distinguisher of $83 + 82 = 165$ rounds. Moreover, since the upper bound of the algebraic degree $\delta(x_0^{(r)})$ is 250 for 83 rounds, we can establish a zero-sum distinguisher covering a total of $83 \times 2 = 166$ rounds by choosing a subspace of dimension $D = 251$. With the largest non-trivial vector space $\mathbb{F}_2^{257}$, we can deduce the longest zero-sum distinguisher covering a total of $124 \times 2 = 248$ rounds, much more than $r_N = 166$ rounds for $2n$-bit security.

5.2   Application to MiMC

In this subsection, we consider the algebraic degree of different variants of MiMC and investigate some possible choices for the generic exponents d. MiMC [3] is an r-round key-alternating block cipher with an n-bit block size and the same key size. Each round consists of three steps: a key addition with the master key k, a round constant addition of $c_i \in \mathbb{F}_{2^n}$, and the application of the non-linear function $R_d := x^d$ over $\mathbb{F}_{2^n}$ with $(d, 2^n - 1) = 1$. After r round iterations, an additional k is added at last. To simplify the representation, we equivalently regard $k \oplus c_i$ as the round key k_i and the instance $\mathsf{MiMC}_d(n, r)$ is defined by

$$\mathsf{MiMC}_d(n, r) := R_d(\cdots R_d(R_d(x \oplus k_0) \oplus k_1) \cdots) \oplus k_r \tag{7}$$

where $R_d(x) := x^d$.

Comparison of Different Choices for Exponents d. Referring to the analysis proposed in MiMC, the best choice of the exponents seems to be of the form $d = 2^l - 1$ for integer l. We then apply our searching algorithm for $\mathsf{MiMC}_d(129, r)$ with $d \in \{3, 7, 15, 31\}$, respectively. Appendix A in [1] compares the different upper bounds of the algebraic degree. $\delta_{\mathrm{MP}}^{(d,r)}$ denotes the algebraic degree found by our algorithm, while $\delta_{[\mathrm{EGL}+20]}^{(d,r)}$ denotes the theoretical upper bounds in [22] given as $\delta_{[\mathrm{EGL}+20]}^{(d,r)} = \lfloor \log_2(d^r + 1) \rfloor$. We also verified our bounds on small-scale instances and the observed degree is denoted by $\delta^{(d,r)}$.

We observe that for $\mathsf{MiMC}_d(129, r)$ with $d \in \{3, 7, 15, 31\}$, the higher-order differential distinguisher can be established for up to $81, 46, 33, 27$ rounds, respectively. However, according to the formula for the number of rounds used in MiMC specification [3], the total rounds are $82, 46, 34, 27$ rounds, respectively. The distinguishers even can cover the full-round $\mathsf{MiMC_d}$ for $d = 7$ and 31 while the security margin is only 1 round for $d = 3$ and 15. Therefore, it invalidates the security claims of the designers and we expect that more rounds than previously predicted in MiMC-like schemes are necessary to guarantee the security against the higher-order differential distinguisher.

We also investigate the algebraic degree of $\mathsf{MiMC}_d(n, r)$ with $d = 2^l + 1$. Besides the theoretical bound $\delta^{(d,r)}_{[\text{EGL}+20]}$, the work of [14] proposed another theoretical bound for $d = 2^l + 1$, represented by $\delta^{(d,r)}_{[\text{BCP22}]}$. $\delta^{(d,r)}_{[\text{BCP22}]}{}^1$ is given as

$$\delta^{(d,r)}_{[\text{BCP22}]} = \begin{cases} 2 \times \lceil k_r/2 - 1 \rceil, k_r = \lfloor r\log_2(d) \rfloor, & \text{for } l = 1, \\ \lfloor r\log_2(d) \rfloor - l + 1, & \text{for } l > 1. \end{cases}$$

When $d = 3$, $\delta^{(3,r)}_{[\text{BCP22}]}$ is exact for up to more than 16000 rounds of MiMC.

Appendix A in [1] compare the different upper bounds of the algebraic degree for $\mathsf{MiMC}_d(129, r)$ for exponents $d = 2^l + 1$. When $d = 3$, our detected bound seems to coincide with $\delta^{(3,r)}_{[\text{BCP22}]}$, the exact algebraic degree. However, with the increase of d, the theoretical bounds $\delta^{(d,r)}_{[\text{EGL}+20]}$ and $\delta^{(d,r)}_{[\text{BCP22}]}$ do not match the observed bound $\delta^{(d,r)}$ as well as $d = 3$, even if the weight of d is the same. Instead, the upper bound found by our algorithm seems to coincide to the observed degree well. Overall, our bounds provide a more precise evaluation of the algebraic degree. This leads to the full-round or almost full-round higher-order differential distinguishers for different instances of $\mathsf{MiMC_d}$. All the results are summarized in Table 3.

5.3   Application to GMiMC

In this subsection, we focus on $\mathsf{GMiMC_{erf}}$, which achieves the best performance among all the variants of GMiMC and has been chosen in the StarkWare challenges. We denote $\mathsf{GMiMC_{erf}}$ specified by branch number t and block size n as $\mathsf{GMiMC}(n, t)$ for simplicity. It is an r-round block cipher in unbalanced Feistel network with an expanding round function, defined as

$$(x_0^{(i)}, x_1^{(i)}, \cdots, x_{t-1}^{(i)}) \leftarrow (x_1^{(i-1)} \oplus F(x_0^{(i-1)}), \cdots, x_{t-1}^{(i-1)} \oplus F(x_0^{(i-1)}), x_0^{(i-1)}),$$

where $x_j^{(i)}$ denotes the input of the j-th branch for round i. F represents the cubic mapping over finite field as

$$F(x) := (x \oplus k^{(i-1)})^3.$$

$k = (k^{(0)}, \cdots, k^{(r-1)})$ is a sequence of round keys and we omit the round constants for simplicity. The overall round function of $\mathsf{GMiMC}(n, t)$ is illustrated in Fig. 4.

---

1   [14] also gives an improved bound when $d \neq 3$. However, the cost for computing the Hamming weight is exponential in r, which means that the bound is infeasible to be determined computationally.

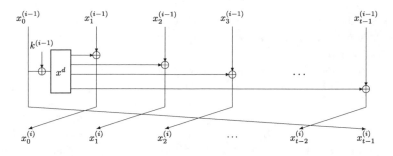

Fig. 4. The round function of $\mathsf{GMiMC}(n,t)$.

Higher-Order Differential Distinguisher for $\mathsf{GMiMC}_{\mathsf{erf}}$. With Model 4, we can apply Algorithm 1 to search for the higher-order differential distinguisher by slightly modifying the initial constraints as

$$\sum_{i \in \mathbb{X}} wt(u_i^{(0)}) = l.$$

$\mathbb{X}$ denotes the set of branches we focus on, For a concrete example, we search for the degree growth of $\mathsf{GMiMC}(33,8)$.

By saturating three branches (x_5, x_6, x_7) of $\mathsf{GMiMC}(33,8)$, our algorithm finds zero-sum in all the output variables after 29 rounds. Due to the structure of the unbalanced Feistel structure, we can extend the distinguishers for $t-1$ more rounds according to Proposition 3 in [8], as a total of 36 rounds. Moreover, we can modify the initial constraint as

$$\sum_{i=0}^{t-1} wt(u_i^{(0)}) = t \cdot n - 1$$

and search for the longest higher-order differential distinguisher. If the model is infeasible, then the corresponding algebraic degree is strictly less than $n \cdot t - 1$ and we can always construct the distinguisher with the largest non-trivial vector space. The longest distinguisher we can find covers a total of 43 rounds with all the output branches achieving zero-sum property, which can be naturally extended to 50 rounds in the same way as before. It is 10 rounds longer than the distinguisher for permutation $\mathsf{GMiMC}(33,8)$ found in [8].

6   Conclusion

While the traditional block ciphers defined over $\mathbb{F}_2$ possess a far-developed analysis toolbox, there is a lack of cryptanalytic methods for the novel arithmetization-oriented ciphers due to the quite different design constraints. In this paper, we introduce a novel extension of the division property, called *general monomial prediction*. It is a generic technique to detect the algebraic degree for ciphers over

$\mathbb{F}_{2^n}$ by evaluating whether the polynomial representation of a block cipher contains some specific monomials. Through tracing the transition of the exponents, we develop a searching tool for the degree estimation of ciphers based on the relationship between the exponents of monomials and the algebraic degree. We apply our algorithm to some competitive arithmetization-oriented block ciphers including MiMC, Feistel MiMC, and GMiMC. As a result, we successfully find the currently best degree bounds and get much longer distinguishers than previous results for several instances. Overall, our methods provide a better estimation for the algebraic degree in case of ciphers over the finite field $\mathbb{F}_{2^n}$ and furthermore, help to establish a more accurate number of rounds necessary to guarantee the security level.

Acknowledgements. We thank the anonymous reviewers for their valuable comments. This work is supported by the National Natural Science Foundation of China (Grant No. 62032014), the National Key Research and Development Program of China (Grant No. 2018YFA0704702), the Major Basic Research Project of Natural Science Foundation of Shandong Province, China (Grant No. ZR202010220025). Kai Hu is supported by the "ANR-NRF project SELECT". Puwen Wei is supported by the Shandong Provincial Natural Science Foundation (No. ZR2020MF053).

References

1. https://eprint.iacr.org/2022/1210.pdf
2. Albrecht, M.R., et al.: Feistel structures for MPC, and more. In: Sako, K., Schneider, S., Ryan, P.Y.A. (eds.) ESORICS 2019. LNCS, vol. 11736, pp. 151–171. Springer, Cham (2019). https://doi.org/10.1007/978-3-030-29962-0_8
3. Albrecht, M., Grassi, L., Rechberger, C., Roy, A., Tiessen, T.: MiMC: efficient encryption and cryptographic hashing with minimal multiplicative complexity. In: Cheon, J.H., Takagi, T. (eds.) ASIACRYPT 2016. LNCS, vol. 10031, pp. 191–219. Springer, Heidelberg (2016). https://doi.org/10.1007/978-3-662-53887-6_7
4. Aly, A., Ashur, T., Ben-Sasson, E., Dhooghe, S., Szepieniec, A.: Design of symmetric-key primitives for advanced cryptographic protocols. IACR Trans. Symmetric Cryptol. **2020**(3), 1–45 (2020)
5. Aumasson, J.P., Meier, W.: Zero-sum distinguishers for reduced Keccak-f and for the core functions of Luffa and Hamsi. Rump Session Cryptographic Hardware Embed. Syst.-CHES **2009**, 67 (2009)
6. Barrett, C., Sebastiani, R., Seshia, S.A., Tinelli, C.: Satisfiability modulo theories. In: Handbook of Satisfiability, volume 185 of Frontiers in Artificial Intelligence and Applications, pp. 825–885. IOS Press (2009)
7. Barrett, C., Tinelli, C.: CVC3. In: Damm, W., Hermanns, H. (eds.) CAV 2007. LNCS, vol. 4590, pp. 298–302. Springer, Heidelberg (2007). https://doi.org/10.1007/978-3-540-73368-3_34
8. Beyne, T., et al.: Out of oddity – new cryptanalytic techniques against symmetric primitives optimized for integrity proof systems. In: Micciancio, D., Ristenpart, T. (eds.) CRYPTO 2020. LNCS, vol. 12172, pp. 299–328. Springer, Cham (2020). https://doi.org/10.1007/978-3-030-56877-1_11
9. Biham, E., Shamir, A.: Differential cryptanalysis of DES-like cryptosystems. In: Menezes, A.J., Vanstone, S.A. (eds.) CRYPTO 1990. LNCS, vol. 537, pp. 2–21. Springer, Heidelberg (1991). https://doi.org/10.1007/3-540-38424-3_1

10. Boura, C., Canteaut, A.: On the influence of the algebraic degree of f^{-1} on the algebraic degree of G ∘ F. IEEE Trans. Inf. Theory **59**(1), 691–702 (2013)
11. Boura, C., Canteaut, A.: Another view of the division property. In: Robshaw, M., Katz, J. (eds.) CRYPTO 2016. LNCS, vol. 9814, pp. 654–682. Springer, Heidelberg (2016). https://doi.org/10.1007/978-3-662-53018-4_24
12. Boura, C., Canteaut, A., De Cannière, C.: Higher-order differential properties of KECCAK and *Luffa*. In: Joux, A. (ed.) FSE 2011. LNCS, vol. 6733, pp. 252–269. Springer, Heidelberg (2011). https://doi.org/10.1007/978-3-642-21702-9_15
13. Boura, C., Coggia, D.: Efficient MILP modelings for sboxes and linear layers of SPN ciphers. IACR Trans. Symmetric Cryptol. **2020**(3), 327–361 (2020)
14. Bouvier, C., Canteaut, A., Perrin, L.: On the algebraic degree of iterated power functions. Cryptology ePrint Archive, Report 2022/366 (2022) https://ia.cr/2022/366
15. Canteaut, A., Videau, M.: Degree of composition of highly nonlinear functions and applications to higher order differential cryptanalysis. In: Knudsen, L.R. (ed.) EUROCRYPT 2002. LNCS, vol. 2332, pp. 518–533. Springer, Heidelberg (2002). https://doi.org/10.1007/3-540-46035-7_34
16. Carlet, C., Charpin, P., Zinoviev, V.A.: Codes, bent functions and permutations suitable for des-like cryptosystems. Des. Codes Cryptogr. **15**(2), 125–156 (1998)
17. Chen, S., Xiang, Z., Zeng, X., Zhang, S.: On the relationships between different methods for degree evaluation. IACR Trans. Symmetric Cryptol. **2021**(1), 411–442 (2021)
18. Cid, C., Grassi, L., Gunsing, A., Lüftenegger, R., Rechberger, C., Schofnegger, M.: Influence of the linear layer on the algebraic degree in sp-networks. IACR Trans. Symmetric Cryptol. **2022**(1), 110–137 (2022)
19. Cook, S.A.: The complexity of theorem-proving procedures. In: Proceedings of the 3rd Annual ACM Symposium on Theory of Computing, vol. 1971, pp. 151–158. ACM (1971)
20. Derbez, P., Fouque, P.: Increasing precision of division property. IACR Trans. Symmetric Cryptol. **2020**(4), 173–194 (2020)
21. Dobraunig, C., Grassi, L., Guinet, A., Kuijsters, D.: CIMINION: symmetric encryption based on Toffoli-Gates over large finite fields. In: Canteaut, A., Standaert, F.-X. (eds.) EUROCRYPT 2021. LNCS, vol. 12697, pp. 3–34. Springer, Cham (2021). https://doi.org/10.1007/978-3-030-77886-6_1
22. Eichlseder, M., et al.: An algebraic attack on ciphers with low-degree round functions: application to full MiMC. In: Moriai, S., Wang, H. (eds.) ASIACRYPT 2020. LNCS, vol. 12491, pp. 477–506. Springer, Cham (2020). https://doi.org/10.1007/978-3-030-64837-4_16
23. Ganesh, V., Dill, D.L.: A decision procedure for bit-vectors and arrays. In: Damm, W., Hermanns, H. (eds.) CAV 2007. LNCS, vol. 4590, pp. 519–531. Springer, Heidelberg (2007). https://doi.org/10.1007/978-3-540-73368-3_52
24. Grassi, L., Lüftenegger, R., Rechberger, C., Rotaru, D., Schofnegger, M.: On a generalization of substitution-permutation networks: the hades design strategy. In: Canteaut, A., Ishai, Y. (eds.) EUROCRYPT 2020. LNCS, vol. 12106, pp. 674–704. Springer, Cham (2020). https://doi.org/10.1007/978-3-030-45724-2_23
25. Hao, Y., Leander, G., Meier, W., Todo, Y., Wang, Q.: Modeling for three-subset division property without unknown subset. In: Canteaut, A., Ishai, Y. (eds.) EUROCRYPT 2020. LNCS, vol. 12105, pp. 466–495. Springer, Cham (2020). https://doi.org/10.1007/978-3-030-45721-1_17

26. Hebborn, P., Lambin, B., Leander, G., Todo, Y.: Lower bounds on the degree of block ciphers. In: Moriai, S., Wang, H. (eds.) ASIACRYPT 2020. LNCS, vol. 12491, pp. 537–566. Springer, Cham (2020). https://doi.org/10.1007/978-3-030-64837-4_18

27. Hu, K., Sun, S., Wang, M., Wang, Q.: An algebraic formulation of the division property: revisiting degree evaluations, cube attacks, and key-independent sums. In: Moriai, S., Wang, H. (eds.) ASIACRYPT 2020. LNCS, vol. 12491, pp. 446–476. Springer, Cham (2020). https://doi.org/10.1007/978-3-030-64837-4_15

28. Knudsen, L.R., Rijmen, V.: Known-key distinguishers for some block ciphers. In: Kurosawa, K. (ed.) ASIACRYPT 2007. LNCS, vol. 4833, pp. 315–324. Springer, Heidelberg (2007). https://doi.org/10.1007/978-3-540-76900-2_19

29. Lai, X.: Higher order derivatives and differential cryptanalysis. In: Communications and Cryptography, pp. 227–233. Springer, Boston (1994)

30. Lambin, B., Derbez, P., Fouque, P.: Linearly equivalent s-boxes and the division property. Des. Codes Cryptogr. **88**(10), 2207–2231 (2020). https://doi.org/10.1007/s10623-020-00773-4

31. Matsui, M.: Linear cryptanalysis method for DES cipher. In: Helleseth, T. (ed.) EUROCRYPT 1993. LNCS, vol. 765, pp. 386–397. Springer, Heidelberg (1994). https://doi.org/10.1007/3-540-48285-7_33

32. Soos, M., Nohl, K., Castelluccia, C.: Extending SAT solvers to cryptographic problems. In: Kullmann, O. (ed.) SAT 2009. LNCS, vol. 5584, pp. 244–257. Springer, Heidelberg (2009). https://doi.org/10.1007/978-3-642-02777-2_24

33. Sun, L., Wang, W., Wang, M.: Automatic search of bit-based division property for ARX ciphers and word-based division property. In: Takagi, T., Peyrin, T. (eds.) ASIACRYPT 2017. LNCS, vol. 10624, pp. 128–157. Springer, Cham (2017). https://doi.org/10.1007/978-3-319-70694-8_5

34. Todo, Y.: Structural evaluation by generalized integral property. In: Oswald, E., Fischlin, M. (eds.) EUROCRYPT 2015. LNCS, vol. 9056, pp. 287–314. Springer, Heidelberg (2015). https://doi.org/10.1007/978-3-662-46800-5_12

35. Todo, Y., Morii, M.: Bit-based division property and application to SIMON family. In: Peyrin, T. (ed.) FSE 2016. LNCS, vol. 9783, pp. 357–377. Springer, Heidelberg (2016). https://doi.org/10.1007/978-3-662-52993-5_18

36. Udovenko, A.: Convexity of division property transitions: theory, algorithms and compact models. In: Tibouchi, M., Wang, H. (eds.) ASIACRYPT 2021. LNCS, vol. 13090, pp. 332–361. Springer, Cham (2021). https://doi.org/10.1007/978-3-030-92062-3_12

37. Wang, S., Hu, B., Guan, J., Zhang, K., Shi, T.: MILP-aided method of searching division property using three subsets and applications. In: Galbraith, S.D., Moriai, S. (eds.) ASIACRYPT 2019. LNCS, vol. 11923, pp. 398–427. Springer, Cham (2019). https://doi.org/10.1007/978-3-030-34618-8_14

38. Xiang, Z., Zhang, W., Bao, Z., Lin, D.: Applying MILP method to searching integral distinguishers based on division property for 6 lightweight block ciphers. In: Cheon, J.H., Takagi, T. (eds.) ASIACRYPT 2016. LNCS, vol. 10031, pp. 648–678. Springer, Heidelberg (2016). https://doi.org/10.1007/978-3-662-53887-6_24

Advanced Encryption

Traceable Receipt-Free Encryption

Henri Devillez$^{(\boxtimes)}$, Olivier Pereira, and Thomas Peters

UCLouvain – ICTEAM – Crypto Group, 1348 Louvain-la-Neuve, Belgium
`henri.devillez@uclouvain.be`

Abstract. CCA-like game-based security definitions capture confidentiality by asking an adversary to distinguish between honestly computed encryptions of chosen *plaintexts*. In the context of voting systems, such guarantees have been shown to be sufficient to prove ballot privacy (Asiacrypt'12). In this paper, we observe that they fall short when one seeks to obtain receipt-freeness, that is, when corrupted voters who submit chosen *ciphertexts* encrypting their vote must be prevented from proving how they voted to a third party.

Since no known encryption security notion can lead to a receipt-free ballot submission process, we address this challenge by proposing a novel publicly verifiable encryption primitive coined Traceable Receipt-free Encryption (TREnc) and a new notion of traceable CCA security filling the definitional gap underlined above.

We propose two TREnc instances, one generic achieving stronger guarantees for the purpose of relating it to existing building blocks, and a dedicated one based on SXDH. Both support the encryption of group elements in the standard model, while previously proposed encryption schemes aiming at offering receipt-freeness only support a polynomial-size message space, or security in the generic group model.

Eventually, we demonstrate how a TREnc can be used to build receipt-free protocols, by following a standard blueprint.

Keywords: New primitive · Public-key encryption · Receipt-freeness

1 Introduction

A protocol offers receipt-freeness when players are unable to demonstrate to a third party which input they provided during a protocol execution. The need for receipt-freeness is most acute in order to prevent vote selling in the context of elections [7], which is our motivating application.

Receipt-free voting. In voting protocols, the random coins used by the voters can often be used as a receipt. For instance, in the famous protocol by Cramer et al. [20], of which a variant is used by the IACR in its own elections, a voter encrypts his vote with the election public key, and the resulting ciphertext is posted on a public bulletin board in order to support the verifiability of the election. If the voter decides to reveal to a third party the randomness used in the encryption process, that party can re-encrypt the claimed vote intent with

© International Association for Cryptologic Research 2022
S. Agrawal and D. Lin (Eds.): ASIACRYPT 2022, LNCS 13793, pp. 273–303, 2022.
https://doi.org/10.1007/978-3-031-22969-5_10

the randomness provided by the voter and verify that the resulting ciphertext appears on the bulletin board: the randomness used for encryption is, in effect, a receipt for the vote.

Since the seminal work of Benaloh [7], numerous protocols explored mechanisms that would guarantee that the random coins used by a voter are insufficient to explain his ballot as it is posted on the bulletin board for the needs of verification. In a first line of works [7,25,38], every possible voting choice is encrypted, the resulting ciphertexts are rerandomized and shuffled by the election authorities and made available to the voter. Furthermore, the permutations applied during the shuffle are also transmitted to the voter using secure channels. The voter then picks the ciphertext encoding his choice, and submits it for display on the bulletin board. Such a protocol guarantees that the voter ignores the randomness used to encrypt his ballot, and the protocol is designed in such a way that the voter is unable to prove which permutation he received, typically using designated-verifier zero-knowledge proofs. Such protocols are however quite demanding in terms of resources, as they require to encrypt a number of ciphertext proportional to the number of voting options, and a communication bandwidth to the voters that is proportional to the number of authorities. The more recent protocol of Kiayias et al. [28] faces similar challenges in terms of complexity, and also only considers a weaker form of receipt-freeness that focuses on voters preparing their ballot honestly.

More recently, Blazy et al. [10] proposed a simpler voting flow supporting receipt-freeness based on signatures on randomizable ciphertexts (SRC): the voters encrypt their vote and sign the resulting ciphertext, which is then transmitted to a re-encryption authority that re-randomizes the ciphertext, adapts the signature accordingly and posts the result on the bulletin board. The voter remains able to verify that a vote with a valid signature is posted on the board on his behalf, but is unable to explain the vote content thanks to the re-randomization step. Furthermore the SRC guarantees that the content of the encrypted ballot cannot be modified during the re-encryption process. This approach was further refined by Chaidos et al. [15], who also propose a simple game-based definition of receipt-freeness, which we adopt here, and more efficient SRCs keep being proposed [5,15].

This approach makes the ballot submission process asymptotically optimal for the voter, in the sense of Cramer et al. [20]: the protocol complexity for the voter becomes logarithmic in the number of voting options and independent of the number of election authorities, contrary to a dependency that is at least linear in both these factors when the approaches of [7,25,28,38] are used.

Receipt-free ballot submission. These works, by offering a simple ballot submission process in one pass, raise the natural question of identifying a public key encryption primitive that would support a receipt-free ballot submission process. Such a primitive would support a modular analysis of voting protocols that would be built around it, including various tallying approaches (based on mix-nets and homomorphic tallying for instance), and approaches to individual verifiability (based on the so-called Benaloh challenge [6] or on code voting for example [17]).

This question has been answered in the context of private (rather than receipt-free) ballot submission: it is well-known that a CCA-secure encryption scheme can be used to obtain a private ballot submission, a requirement that can be relaxed to NM-CPA security when the tally takes place in a single decryption round [9,22,41].

These works highlight the importance of some form of non-malleability in a submission process. From a practical point of view, non-malleability is needed in order to be able to detect (and prevent) non-independent ballot submissions (e.g., ballot copies) that would violate the privacy of the vote. From a technical point of view, security proofs require the availability of a decryption oracle used to extract the votes submitted by the adversary.

CCA security is however problematic in the context of receipt-free ballot submission, since we need to be able to re-randomize encrypted votes, so that the voter cannot explain the vote content anymore. The exploration of CCA-like security notions that would support some form of controlled malleability has been a fertile research area, which resulted in the definition of the notions of replayable-CCA (RCCA) security [14], homomorphic-CCA (HCCA) security [36], and controlled-malleable CCA (CM-CCA) security [16] for instance. As far as we know, all these works rely on the same CCA blueprint, in which an adversary submits one or more messages to a challenger, who answers either with an honest encryption of the messages or with something else, and the adversary must decide what he received with the help of a decryption oracle that accepts to decrypt any ciphertext that is not "recognizably" related to the challenge ciphertexts. The same holds in any other encryption primitives with CCA-like security with enhanced decryption capabilities. While they give more flexible ways to decrypt ciphertexts (based on identities, attributes and so on [11,26,37,39]), the challenge ciphertext is computed when the adversary sends a chosen *message*. Eventually, and following an observation which dates back at least to [25], deniable encryption [12] also only focusses on *honestly* computed ciphertexts that can then be explained for any other plaintexts.

This blueprint is however inadequate when turning to encryption schemes that would support the design of protocols that support receipt-freeness: in such a setting, we need to consider an adversary who sends to the challenger chosen *ciphertexts*, that may not be computed as a random encryption of a plaintext vote: they could have been *maliciously* computed.

Our contributions

1) TCCA security. In this work, we investigate for the first time the implication of defining the notion of *traceable* CCA security (TCCA), a CCA-like security notion in which adversarially-chosen ciphertexts are submitted in the challenge phase. The challenge ciphertext is produced by *randomizing* one ciphertext or another, and we recognize derivatives of the challenge ciphertext thanks to a non-malleable public *trace* which is present in any ciphertext. To avoid trivial attacks, both ciphertexts given in the challenge phase must trace to each other, i.e., they must have the same trace.

This makes it possible for voters to submit a ciphertext of their choice, which will then be re-randomized by an authority, and can still be tracked by the voters using the trace.

For honestly produced ciphertexts, our security notion also implies traditional confidentiality properties, so that ballot privacy remains guaranteed should the re-randomizing authority be corrupted. So, non-malleability really serves two purposes here: *(1)* it guarantees that the re-randomizing authority cannot produce a ciphertext that would be related to an honestly produced one and have a different trace (which would violate ballot privacy), and *(2)* it guarantees that the re-randomizing authority cannot produce a ciphertext that would have the same trace as a given one but would decrypt to a different plaintext.

2) TREnc. We introduce Traceable Receipt-free Encryption (TREnc) as a new primitive with the following features:

- *Traceability.* Honestly generated ciphertexts are traceable in the sense that it is infeasible to modify the encrypted message;
- *Randomizability.* Valid ciphertexts are fully re-randomizable, up to the trace;
- *TCCA security.* Given a pair of ciphertexts that trace to each other, it is unfeasible to guess which one is randomized, even with access to a decryption oracle which decrypts any ciphertexts that do not trace to the challenge ciphertext, except before the challenge phase.

We also provide:

1. A generic TREnc that can be instantiated from existing building blocks that offer security in the standard model, and whose CRS is public-coin;
2. A pairing-based TREnc under the SXDH assumption in the standard model, where the public key only contains 13 first-source group elements and 6 s-source group elements, and the ciphertext contains 13 first-source group elements and 5 s-source group elements.

Both approaches improve on the state of the art: the previous SRC-based solutions either require costly bit-by-bit encryption [10,15], or only offer security in the generic group model [5].

3) A TREnc based voting scheme. Eventually, we show how to turn a TREnc into a simple voting scheme in a generic way, following the *Enc2Vote* blueprint previously used to turn a CCA-secure encryption scheme into a private voting scheme [9].

We demonstrate that the resulting voting scheme satisfies a notion of receipt-freeness that is equivalent in spirit to the one of Chaidos et al. [15], but fixes a small technical issue in that definition that makes their security game trivial to win (making it impossible to build a protocol that is receipt-free according to their definition).

Other related works. We focus on offering receipt-freeness in the context of voting, which is the context in which receipt-freeness was introduced [7], and which remains the main application context in which receipt-freeness is desired. Voting can however be seen as a special type of secure function evaluation protocol, in which specific tallying functions are evaluated and, as such, the notion of receipt-freeness, and the related notion of coercion-resistance have also been defined in the general multi-party computation setting [4,13,35,40]. We keep our focus on the voting context in order to clarify various design choices that are most meaningful in the voting setting compared to the general MPC setting: our primitive is targetted for a ballot submission process in which voters submit their ballot in one pass and do not communicate with each other, contrary to most MPC protocols, and we design mechanisms in which the ballot submission process can be fast, even on devices with limited computational power, while the verification of an election may require a longer period of time and use a dedicated computing infrastructure. Despite our focus on voting, it may be the case that TREnc mechanisms find applications in other contexts.

2   Traceable Receipt-Free Encryption

We propose a new public key encryption primitive and associated security notions that would support the receipt-free submission of votes in a protocol. As a first task, we identify the fundamental ingredients that are needed for our new encryption primitive.

An encryption scheme. We expect voters to submit their vote in an encrypted form, in order to guarantee the privacy of the votes.

Receipt-free encryption. Voters willing to sell their vote may choose to submit an arbitrary encrypted vote, which may be in the range of honestly produced ciphertexts but sampled according to a different distribution, or even just a sequence of bits that would not be within the range of the encryption mechanism. By deviating from the normal encryption process, the voter hopes to obtain a receipt that could be used to demonstrate his vote intent to a third party.

If the encrypted vote that is tallied is produced by the voter only, then the voter will always have a receipt: the random coins used to encrypt the ballot. In order to avoid this, we rely on the existence of a semi-trusted authority: that authority will be trusted to prevent a dishonest voter from obtaining a receipt for his vote, but will not be trusted for the correctness of the election result, and will not be trusted for the privacy of votes encrypted by honest voters.

Concretely, in order to achieve receipt-freeness, this semi-trusted authority tests the validity of a voter submitted ciphertext (without the need of any secret key) and re-randomizes every valid ciphertext before posting it on a public bulletin board.

Traceable Receipt-Free Encryption. In order to make it possible for a voter to check that his ballot has not been unduly modified by this semi-trusted re-randomizing authority, it must be possible to extract a *trace* from any valid ciphertext. A honest re-randomization process would keep the trace is unchanged, hence making ciphertexts *traceable*, while no corrupted authority should be able to modify a ciphertext in such a way that it would decrypt to a different vote while keeping the trace unchanged.

Furthermore, we need to make sure that this trace cannot serve as a receipt for the vote. In order to make sure that it is the case, we split the encryption process in two steps, that guarantee that any trace can be associated to any possible vote intent. Concretely, an encryption starts with the generation of a secret *link key*, which is then used, together with the encryption public key, to encrypt any possible vote. This guarantees that, even if a voter leaks the link key associated to his ballot as a receipt, the ballot could still encrypt any vote.[1]

2.1   Syntax

We now have the ingredients that we need to define a Traceable Receipt-Free Encryption scheme, or TREnc.

Definition 1 (Traceable Receipt-Free Encryption). *A* Traceable Receipt-Free Encryption *scheme (TREnc) is a public key encryption scheme* (Gen, Enc, Dec) *that is augmented with a 5-tuple of algorithms* (LGen, LEnc, Trace, Rand, Ver):

- LGen(pk; r): *The link generation algorithm takes as input a public encryption key* pk *in the range of* Gen *and randomness* r, *and outputs a link key* lk.
- LEnc(pk, lk, m; r): *The linked encryption algorithm takes as input a pair of public/link keys* (pk, lk), *a message* m *and randomness* r *and outputs a ciphertext.*
- Trace(pk, c) : *The tracing algorithm takes as input a public key* pk, *a ciphertext* c *and outputs a trace* t. *We call* t *the trace of* c.
- Rand(pk, c; r): *The randomization algorithm takes as input a public key* pk, *a ciphertext* c *and randomness* r *and outputs another ciphertext.*
- Ver(pk, c): *The verification algorithm takes as input a public key* pk, *a ciphertext* c *and outputs* 1 *if the ciphertext is valid,* 0 *otherwise.*

In many cases, we will omit the randomness r *from our notations. It is then assumed that it is selected uniformly at random.*

We require several correctness properties from the additional algorithms of a TREnc. The first requires that encrypting a message m by picking a link key lk

---

[1] Of course, this also means that, if a corrupted re-randomizing authority obtains a voter's secret link key (e.g., by corrupting the voter's voting client), then it might be able to produce a ciphertext that encrypts a different vote intent but would still trace to the original voter trace. Just as other attacks related to corrupted voting clients, such attacks can be prevented by traditional continuous ballot testing procedures [6], in which a voter would have the option to ask an authority to spoil a ballot posted on the bulletin board, which would then be verifiability decrypted for verification, and later replaced by a fresh new ballot produced by the voter, using a fresh link key.

using LGen and computing LEnc(pk, lk, m) produces a ciphertext that is identically distributed to a fresh encryption of m using Enc. The second requires that the Trace of a ciphertext does not depend on the message that is encrypted. The third requires that randomizing a ciphertext does not change the corresponding plaintext neither the corresponding trace. The last requires that every honestly computed ciphertext passes the verification algorithm.

Definition 2 (TREnc correctness). *We require that a TREnc scheme satisfies the following correctness requirements.*

Encryption compatibility. *For every* pk *in the range of* Gen *and message* m, *the distributions of* Enc(pk, m) *and* LEnc(pk, LGen(pk), m) *are identical.*

Link traceability. *For every* pk *in the range of* Gen, *every* lk *in the range of* LGen(pk), *the encryptions of every pair of messages* (m_0, m_1) *trace to the same trace, that is, it always holds that* Trace(pk, LEnc(pk, lk, m_0)) = Trace(pk, LEnc(pk, lk, m_1)).

Publicly Traceable Randomization. *For every* pk *in the range of* Gen, *every message* m *and every* c *in the range of* Enc(pk, m), *we have that* Dec(sk, c) = Dec(sk, Rand(pk, c)) *and* Trace(pk, c) $-$ Trace(pk, Rand(pk, c)).

Honest verifiability. *For every* pk *in the range of* Gen *and every messages* m, *it holds that* Ver(pk, Enc(pk, m)) = 1

2.2 Security Definitions

Verifiability We require several security properties from a TREnc. Our first property is fairly standard: a TREnc is verifiable if the Ver algorithm guarantees that a ciphertext is within the range of Enc. In other words, the ciphertext can be explained by some message m, some link key lk, and some coins, even if they are not easily computable.

Definition 3 (Verifiability). *A TREnc is* verifiable *if for every* PPT *adversary, the following probability is negligible in* λ:

$$\Pr[\mathsf{Ver}(\mathsf{pk}, c) = 1 \ and \ c \notin \mathsf{Enc}(\mathsf{pk}, \cdot) | (\mathsf{pk}, \mathsf{sk}) \leftarrow \mathsf{Gen}(1^\lambda); c \leftarrow \mathcal{A}(\mathsf{pk}, \mathsf{sk})].$$

TCCA *security.* We now turn to our central security definition, *security against traceable chosen ciphertexts attacks*, or TCCA security, which differs from all existing CCA-like notions by letting the adversary submit pairs of ciphertexts instead of pairs of messages, reflecting that we need security in front of adversarially chosen ciphertexts. In the TCCA security game (Fig. 1), the adversary receives the public key and has access to a decryption oracle, as usual. It then submits a pair of ciphertexts that must be valid and have identical traces. One of the ciphertexts is randomized and returned to the adversary, who must decide which one it is. After receiving this challenge ciphertext, the adversary can still query the decryption oracle, but only on ciphertexts that have a trace different of his challenge ciphertext. So, the challenger must faithfully decrypt pre-challenge ciphertexts that have the same trace as the challenge ciphertext. Looking ahead, this decryption capability offers an easy but necessary means allowing simulating the result of an election when proving receipt-freeness.

TCCA security guarantees that, if a voter submits a ciphertext that is randomized before it is posted on a public bulletin board, then the resulting ciphertext becomes indistinguishable from any other ciphertext that would have the same trace, and we know from the link traceability that the encryption of any vote could have that trace. This essentially guarantees the absence of a vote receipt.

Definition 4 (TCCA). *A TREnc is* TCCA *secure if for every* PPT *adversary* $\mathcal{A} = (\mathcal{A}_1, \mathcal{A}_2)$ *the experiment* $\mathsf{Exp}_{\mathcal{A}}^{\mathrm{tcca}}(\lambda)$ *defined in Fig. 1 (left) returns* 1 *with a probability negligibly close in* λ *to* $\frac{1}{2}$.

$\mathsf{Exp}_{\mathcal{A}}^{\mathrm{tcca}}(\lambda)$	$\mathsf{Exp}_{\mathcal{A}}^{\mathrm{trace}}(\lambda)$
$(\mathsf{pk}, \mathsf{sk}) \leftarrow\!\!\$\ \mathsf{Gen}(1^{\lambda})$	$(\mathsf{pk}, \mathsf{sk}) \leftarrow\!\!\$\ \mathsf{Gen}(1^{\lambda})$
$(c_0, c_1, \mathsf{st}) \leftarrow\!\!\$\ \mathcal{A}_1^{\mathsf{Dec}(\cdot)}(\mathsf{pk})$	$(m, \mathsf{st}) \leftarrow\!\!\$\ \mathcal{A}_1(\mathsf{pk}, \mathsf{sk})$
$b \leftarrow\!\!\$\ \{0, 1\}$	$c \leftarrow\!\!\$\ \mathsf{Enc}(\mathsf{pk}, m)$
if $\mathsf{Trace}(\mathsf{pk}, c_0) \neq \mathsf{Trace}(\mathsf{pk}, c_1)$ **or**	$c^{\star} \leftarrow\!\!\$\ \mathcal{A}_2(c, \mathsf{st})$
$\mathsf{Ver}(\mathsf{pk}, c_0) = 0$ **or** $\mathsf{Ver}(\mathsf{pk}, c_1) = 0$ **then return** b	**if** $\mathsf{Trace}(\mathsf{pk}, c) = \mathsf{Trace}(\mathsf{pk}, c^{\star})$ **and**
$c^{\star} \leftarrow\!\!\$\ \mathsf{Rand}(\mathsf{pk}, c_b)$	$\mathsf{Ver}(\mathsf{pk}, c^{\star}) = 1$ **and** $\mathsf{Dec}(\mathsf{sk}, c^{\star}) \neq m$
$b' \leftarrow\!\!\$\ \mathcal{A}_2^{\mathcal{O}\mathsf{Dec}^{\star}(\cdot)}(c^{\star}, \mathsf{st})$	**then return** 1
return $b' = b$	**else return** 0

Fig. 1. TCCA and trace experiments. In the TCCA experiment, $\mathcal{A}_2$ has access to a decryption oracle $\mathcal{O}\mathsf{Dec}^{\star}(\cdot)$ which, on input c, returns $\mathsf{Dec}(c)$ if $\mathsf{Trace}(\mathsf{pk}, c) \neq \mathsf{Trace}(\mathsf{pk}, c^{\star})$ and test otherwise.

It is naturally possible to write a multi-challenge version of the $\mathsf{Exp}_{\mathcal{A}}^{\mathrm{TCCA}}(\lambda)$ experiment, which we call $q - \mathrm{TCCA}$, in which the adversary can submit q pairs of ciphertexts. This leads to an equivalent definition, as demonstrated in the full version [21]. We also stress that in the challenge query the adversary may know the random coins underlying c_0 and c_1 and may have drawn them from a specific secret distribution. The randomization leading to the challenge ciphertext $c^{\star}$ should thus erase any subliminal information binding $c^{\star}$ to the message in c_b. This definition introduces some technical difficulty when it comes to proving the TCCA security as it becomes harder to program the public key to ease the transition toward a game where we are able to inject an independent message in the plaintext in an undetectable way. Indeed, we have no clue at the setup time about the distribution of (c_0, c_1) and their common trace while the emulation of $\mathsf{Rand}(\mathsf{pk}, c_b)$ must preserve it without even knowing the underlying link keys.

TCCA security is reminiscent of the notion of publicly detectable replayable-CCA (pd-RCCA) security proposed by Canetti et al. [14]. The $\mathrm{pd} - \mathrm{RCCA}$ security game is essentially the same as the CCA game, except for two main differences: a publicly computable equivalence relation is defined on ciphertexts and, after the challenge ciphertext has been received, the challenger will refuse to decrypt any ciphertext that is equivalent to this challenge ciphertext. Furthermore, ciphertexts that are in the same equivalence class must decrypt to the same message (for completeness, the full definition is available in the full version [21]). The pd-RCCA security game looks appealing in the context of voting, because it captures this idea of having the possibility to re-randomize ciphertexts while also keeping a trace that could be detected through the equivalence relation. And, indeed, RCCA-secure encryption has been used in previous proposals of receipt-free voting schemes [15].

There are three central differences, though, which motivate the introduction of the TCCA security game.

- The challenge ciphertexts of the pd − RCCA security game are always honestly computed and, as such, pd − RCCA security does not offer any guarantee in the face of maliciously produced ciphertexts, as it would be the case when a voter tries to obtain a receipt for his vote.
- Contrary to pd − RCCA security, it can be observed that TCCA security says nothing about the hiding property of the Enc algorithm, since the adversary must distinguish based on outputs of Rand. An extreme case could define Enc as the identity function, Trace as mapping to a single constant trace, and Rand actually performing the encryption work, and this could still offer a TCCA secure scheme. The confidentiality requirements on Enc will be handled through the traceability and strong randomization properties below.
- There is no requirement for TCCA security that trace equivalent ciphertexts decrypt to the same message: a single link key can be used to encrypt any message, and all the resulting ciphertexts would have the same trace (by the link traceability correctness property). We recall that this non-binding feature is essential for receipt-free voting.

As such, TCCA security is not comparable to pd-RCCA security. It is shown in the full version of the paper [21] that, under (different) additional conditions, implications can be proven in both directions for a natural variant of pd-RCCA security adapted to TREnc schemes.

Traceability and Strong Randomization. While TCCA security relates to a model in which the voting client may be corrupted but the re-randomization server is honest, we now focus on two central properties that are important when the voting client is honest and the re-randomization server might be corrupted.

The traceability property guarantees to the sender of a honestly encrypted message that no efficient adversary would be able to produce another ciphertext that traces to the same trace and would decrypt to a different message, even if the adversary knows the secret decryption key. So, even if a TREnc offers some form of ciphertext malleability, its traceability implies the non-malleability of the plaintexts. This is an important feature for the verifiability of a voting system: as long as the link key used to encrypt a vote remains secret, and the voter submits a single ciphertext encrypted with that link key, the voter is guaranteed that any ciphertext that would trace to his original ciphertext encrypts his original vote.

(But, of course, using the link key, it remains possible to produce ciphertexts with the same trace that would decrypt to any vote.)

Definition 5 (Traceability). *A TREnc is* traceable *if for every* PPT *adversary* $\mathcal{A} = (\mathcal{A}_1, \mathcal{A}_2)$, *the experiment* $\mathsf{Exp}_{\mathcal{A}}^{\text{trace}}(\lambda)$ *defined in Fig. 1 (right) returns* 1 *with a probability negligible in* λ.

The second property, strong randomization, requires that the output of the Rand algorithm applied to any valid ciphertext is distributed just as a random encryption of the same message with the same link key.

Definition 6 (Strong Randomization). *A TREnc is* strongly randomizable *if for every* $c \in$ LEnc(pk, lk, m) *with* pk *in the range of* Gen *and* lk *in the range of* LGen(pk), *the following computational indistinguishability relation holds:*

$$\mathsf{Rand}(\mathsf{pk}, c) \approx_c \mathsf{LEnc}(\mathsf{pk}, \mathsf{lk}, m)$$

Requiring strong randomization together with TCCA security guarantees that Enc actually hides messages. CPA security comes easily: when the CPA adversary sends (m_0, m_1) to the TCCA adversary, the TCCA adversary can encrypt the 2 messages using a single random link key and send them to the TCCA challenger, which will return a randomization of one of them. Strong randomization guarantees that this is distributed exactly like an encryption of one of the two messages, and we can send the result to the CPA adversary, who will then offer the answer expected for the TCCA game. We show a stronger implication to RCCA security in the full version of the paper [21].

3   Towards a Generic TREnc

We are now interested in exploring how a TREnc could be designed from existing tools. The core TREnc security feature comes from the TCCA security game, in which the adversary submits a pair of ciphertexts with identical traces and receives a re-randomization of one of them. If we want relate this game to a more standard RCCA-style security definition in which the adversary submits a pair of plaintext and receives an encryption of one of them, we need to be able to translate a re-randomization query on two ciphertexts into an encryption query on the two corresponding plaintexts. But there is an additional constraint that needs to be satisfied: the ciphertext resulting from the encryption query needs to have the same trace as the original ciphertexts. In other words, we need to be able to decrypt the challenge ciphertexts from the TCCA game, but also to extract the link key that they contain. We capture this last idea in an augmented version of a TREnc, which we call *extractable TREnc*.

3.1   Extractable TREncs

Essentially, an extractable TREnc makes it possible to produce encryption keys together with a trapdoor using a TrapGen algorithm. Using that trapdoor, it becomes possible to extract, from any ciphertext, a link key that makes it possible to produce new ciphertexts with the same trace as the original one. This in turn implies the possibility to break the traceability of the scheme.

Definition 7 (Extractable TREnc). *An extractable TREnc is a TREnc with two additional algorithms* TrapGen *and* LExtr*:*

- TrapGen(1^λ)*: The trapdoor generation algorithm takes as input the security parameter and outputs a tuple of public/secret/trapdoor keys* (pk, sk, tk)*. We require the distribution of the* (pk, sk) *pairs produced by* TrapGen(1^λ) *to be identical to the one of the outputs of* Gen(1^λ)*.*
- LExtr(tk, c)*: The link extraction algorithm takes as input the trapdoor key and a ciphertext and returns a link key* lk *such that, if c is in the range of* Enc(pk, ·) *with* pk *in the range of* Gen*, then c is in the range of* LEnc(pk, lk, ·)*.*

It is fairly natural to require that ciphertexts can only be consistent with one single link key, hence guaranteeing a unique link key extraction.

Definition 8 (Unique Extraction). *An extractable TREnc has* unique extraction *if, for every* (pk, sk, tk) *in the range of* TrapGen *and* lk *in the range of* LGen(pk)*, we have that:*

- LExtr(c, tk) = lk *whenever* $c \in$ LEnc(pk, lk, ·)*;*
- LExtr(c_0, tk) = LExtr(c_1, tk) *whenever we have* Trace(pk, c_0) = Trace(pk, c_1) *and* $c_0, c_1 \in$ Enc(pk, ·)*.*

3.2   A TREnc Flavored Variant of pd-RCCA Security

Based on an extractable TREnc, we now propose an RCCA-like security definition, pd*-RCCA-security, which shares much of the spirit of the pd − RCCA notion of Canetti et al. [14], but is rather tailored as a useful intermediary notion for achieving TCCA security: we will show that any pd*-RCCA-secure extractable TREnc is also TCCA secure. Eventually, we will show how to achieve pd*-RCCA-security from existing tools.

Definition 9 (pd*-RCCA). *An extractable TREnc is* pd*-RCCA*-secure if for any* PPT *adversary* $\mathcal{A} = (\mathcal{A}_1, \mathcal{A}_2)$*, the experiment* $\mathrm{Exp}_{\mathcal{A}}^{\mathrm{pd}^*\text{-RCCA}}(\lambda)$ *in Fig. 2 returns 1 with a probability negligibly close on λ to $\frac{1}{2}$.*

Just as in the pd − RCCA security definition, our adversary receives a public key, then can make decryption queries, make a challenge query on a pair of plaintexts, receive an encryption c^* of one of them, and then make more decryption queries, provided that they are not about ciphertexts that are equivalent to c^*. Here the notion of equivalent ciphertext is defined by ciphertexts with identical traces, which does not imply that they decrypt to the same plaintext, contrary to the compatibility requirement of pd − RCCA security. The extra features of pd*-RCCA security, which come naturally in the context of an extractable TREnc, are that:

$$\mathsf{Exp}_{\mathcal{A}}^{\mathrm{pd}^{*}\text{-}\mathrm{rcca}}(\lambda)$$

$(\mathsf{pk}, \mathsf{sk}, \mathsf{tk}) \leftarrow\!\!\$\ \mathsf{TrapGen}(1^{\lambda})$

$(m_0, m_1, \mathsf{lk}, \mathsf{st}) \leftarrow\!\!\$\ \mathcal{A}_1^{\mathsf{Dec}(\mathsf{sk}, \cdot), \mathsf{LExtr}(\mathsf{tk}, \cdot)}(\mathsf{pk})$

$b \leftarrow\!\!\$\ \{0, 1\}$

if $\mathsf{lk} = \bot$ **then** $c^* \leftarrow\!\!\$\ \mathsf{Enc}(\mathsf{pk}, m_b)$

else $c^* \leftarrow\!\!\$\ \mathsf{LEnc}(\mathsf{pk}, \mathsf{lk}, m_b)$

$b' \leftarrow\!\!\$\ \mathcal{A}_2^{\mathcal{O}\mathsf{Dec}^*(\mathsf{sk}, \cdot), \mathsf{LExtr}(\mathsf{tk}, \cdot)}(c^*, \mathsf{st})$

return $b = b'$

Fig. 2. pd*-RCCA experiment. Here, $\mathcal{O}\mathsf{Dec}^*(\mathsf{sk}, c)$ is a decryption oracle that returns test if $\mathsf{Trace}(\mathsf{pk}, c) = \mathsf{Trace}(\mathsf{pk}, c^*)$ and $\mathsf{Dec}(\mathsf{sk}, c)$ otherwise.

- On top of having access to a decryption oracle, the adversary has access to a LExtr oracle giving him the ability to extract the link key from any ciphertext.
- During his challenge query, the adversary can provide a link key on top of its two plaintexts: the challenge ciphertext will then be computed using that link key.

As announced, a pd*-RCCA-secure and strongly randomizable extractable TREnc is also TCCA-secure.

Theorem 1. *If a TREnc scheme T is extractable, strongly randomizable, and pd*-RCCA-secure, then it is TCCA secure. More precisely, if the advantages of any PPT adversary at strong randomization and pd*-RCCA experiment are respectively bounded by ε_{SR} and ε, then for any PPT adversary $\mathcal{A}$, we have $Pr[\mathsf{Exp}_{\mathcal{A}, T}^{\mathrm{tcca}}(\lambda) = 1] \leq \frac{1}{2} + \varepsilon_{SR} + \varepsilon$.*

Proof. (See the full version for details.) The decryption queries from the TCCA adversary are forwarded to the pd*-RCCA challenger. When the TCCA adversary makes his challenge query on (c_0, c_1), the reduction obtains the corresponding link key and plaintexts by querying the pd*-RCCA challenger, and sends them as pd*-RCCA challenge. The resulting ciphertext is correctly distributed thanks to strong randomizability, and has the correct trace thanks to the extractability. The winning probability of the TCCA adversary is then negligibly close to the winning probability of the resulting pd*-RCCA adversary. □

3.3 Building a pd*-RCCA-Secure Extractable TREnc

We are now ready to build a TREnc. As a first natural building block, we use a signature on randomizable ciphertexts (SRC), as introduced by Blazy et al. [10]. In an SRC, any signed ciphertext can be publicly re-randomized, and the signature can be publicly adapted so that it remains valid for the new ciphertext.

We can easily obtain the structure of a TREnc from an SRC by defining the LGen function as setting lk as a fresh signing key for the SRC, and the LEnc function as encrypting the plaintext using a randomizable encryption scheme, then signing that ciphertext using lk. The trace of a ciphertext would then be the signature verification key.

This offers a promising skeleton, but it is not sufficient to obtain pd*-RCCA security: as it is, the adversary could simply remove the signature from the challenge ciphertext, sign that ciphertext with a fresh key in order to obtain a different trace, and ask for the decryption of the result, which would be granted.

A natural solution to this problem is to link the trace to the ciphertext using tag-based encryption [29] mechanism. In a tag-based encryption scheme, the encryption and decryption functions take an arbitrary tag as an extra input, and the decryption of a ciphertext with an incorrect tag will fail. We rely on the standard notion of weak-CCA security for tag based encryption [34], which is the CCA security game excepted that the challenge ciphertext is produced using an *adversarially chosen* tag, and that no decryption query can be made using that tag (only) *after* the challenge phase. This security game nicely fits our pd*-RCCA security game, in which the trace derived from the link key submitted by the adversary can be used as a tag, and guarantees that no ciphertext can be modified in such a way that it successfully decrypts with a tag that is different of the original one. We note that we must be able to decrypt pre-challenge queries that already contains the "challenge tag" of the adversary, which prevents us from only relying on (weak) selective-tag security.

But we still need to be able to extract the link key from a tag-based ciphertext. This can be done fairly easily, by augmenting our encryption process with the requirement to encrypt the link key using a randomizable CPA-secure encryption scheme, and to add a randomizable ZK proof that the encrypted link key is indeed the one that is used as tag for the tag-based encryption. Extraction would then simply proceed by decrypting that CPA ciphertext. (In particular, it does not rely on any extraction property of the ZK proof system: we just need its soundness).

To summarize, we build an extractable TREnc from the following ingredients:

- A randomizable weakly CCA secure tag-based encryption scheme (TBGen, TBEnc, TBDec).
- An SRC compatible with the tag-based encryption scheme, which includes a signature scheme (SGen, Sign, SVer).
- A randomizable CPA secure public key encryption scheme (EGen, EEnc, EDec).
- A randomizable NIZK proof system (Prove, VerifyProof) that, on input (c_{tbe}, c_{extr}) and associated public keys, demonstrates that c_{extr} is an encryption with EEnc of the signing key whose corresponding verification key has been used as a tag in order to compute c_{tbe} using TBEnc.

And the blueprint of our TREnc is as follows:

- TrapGen uses TBGen to produces a key pair (tpk, tsk), and EGen to produce a key pair (e_{extr}, d_{extr}). It returns pk = (tpk, e_{extr}), sk = tsk and tk = d_{extr}.
- LGen sets lk as a signing key obtained from Sign. We assume that the corresponding signature verification key vk can be derived from lk.
- LEnc encrypts the message m as follows:
  - $c_{tbe} = \text{TBEnc(tpk, lk, }m)$;
  - $\sigma = \text{Sign(lk, }c_{tbe})$;

- $c_{extr} = \mathsf{EEnc}(\mathsf{e_{extr}}, \mathsf{lk})$;
- $\pi = \mathsf{Prove}(c_{tbe}, c_{extr}; \mathsf{lk})$

The ciphertext is made of these 4 elements, together with vk.

- Dec returns $\mathsf{TBDec}(\mathsf{tsk}, \mathsf{vk}, c_{tbe})$
- Trace returns vk.
- Rand re-randomizes c_{tbe}, adapts σ accordingly, re-randomizes c_{extr}, and re-randomizes and adapts π.
- Ver accepts a TREnc ciphertext if the two ciphertexts that it contains are valid, if the signature is valid, and if the ZK proof verifies.
- LExtr returns $\mathsf{EDec}(\mathsf{tsk}, c_{extr})$.

A complete description of this generic TREnc, together with proofs of its security, is available in the full version of the paper [21], where we rely on the standard security notions of all the above ingredients. Exploring whether some of these notions can be relaxed is an interesting scope for further research. Finally, we mention that pairing-based realizations exist for all these ingredients and that it would be appealing to understand how to construct a secure post-quantum TREnc. The main obstacle we see relates to the controlled-malleability feature of our new primitive (i.e., any ciphertext must support an unbounded number of randomization) which makes it less straightforward to realize in general, and for instance based on lattices.

Remark 1. This section showed that the notion of extractable TREnc offers a convenient companion for a TREnc: it is possible to build an extractable TREnc from relatively common, yet strong, building blocks, and the proof of TCCA security of this TREnc comes relatively easily because we can design a pd − RCCA-like security notion for extractable TREnc that implies our new TCCA security notion. The resulting construction is however expected to be fairly expensive since, in the standard model, all known instantiations of the building blocks relies on a bit-by-bit decomposition of the message or the secret singing key of which the ciphertext must contain a (malleable) ZK proof of. Nevertheless, providing this extractability feature is an artifice for the construction that is not necessary for the security of the TREnc, but as far as we know there is no obvious generic construction leading to a TREnc without extractability. In the next section we turn to the construction of an ad-hoc efficient instance of a TREnc based on a standard computational assumption that also avoid the costly bit-by-bit decomposition.

4   Pairing-Based Construction Under SXDH

This section provides a secure TREnc in the standard model, only relying on the SXDH assumption and on a CRS. Contrary to our previous construction, this one is not extractable – extractability was just a convenience but does not offer any security benefit. This allows us to get a more efficient solution, here, in asymmetric bilinear groups. Moreover, our construction enjoys a short public-key and short ciphertexts as they only contain a constant number of group elements to encrypt a full group element, contrary to previous proposals that required to process the message bit by bit [10, 15].

We first introduce the cryptographic assumptions on which we will rely, as well as the main existing building block that we will use: linearly homomorphic structure-preserving signatures.

4.1 Computational Setting

We rely on an efficient Setup algorithm that generates common public parameters pp. Given a security parameter λ, $\mathsf{Setup}(1^\lambda)$ generates a bilinear group pp $= (\mathbb{G}, \hat{\mathbb{G}}, \mathbb{G}_T, p, e, g, \hat{g}, \hat{h})$ of prime order $p > 2^{\mathrm{poly}(\lambda)}$ for some polynomial poly, where $g \leftarrow\!\!\$\, \mathbb{G}$ and $\hat{g}, \hat{h} \leftarrow\!\!\$\, \hat{\mathbb{G}}$ are random generators and $e : \mathbb{G} \times \hat{\mathbb{G}} \rightarrow \mathbb{G}_T$ is a bilinear map. In this setting, we rely on the SXDH assumption, which states that the DDH problem must be hard in both $\mathbb{G}$ and $\hat{\mathbb{G}}$. Following the Groth-Sahai standard notation, we also define the linear map $\iota : \mathbb{G} \rightarrow \mathbb{G}^2$ with $\iota : Z \mapsto (1, Z)$.

4.2 Linearly Homomorphic Structure-Preserving Signatures

A central tool for our efficient TREnc construction is linearly homomorphic structure-preserving signatures. The structure preserving [2] [1] property makes it possible to sign messages that are group elements (and not just bits as in schemes based on the Waters signature), while the additional linearly homomorphic feature, introduced by Libert et al. [32], will be used to make the signatures randomizable while guaranteeing the non-malleability of the plaintext.

Keygen(pp, n): given the public parameter pp and the (polynomial) space dimension $n \in \mathbb{N}$, choose $\chi_i, \gamma_i \leftarrow\!\!\$\, \mathbb{Z}_p$ and compute $\hat{g}_i = \hat{g}^{\chi_i} \hat{h}^{\gamma_i}$, for $i = 1$ to n. The private key is $\mathsf{sk} = \{(\chi_i, \gamma_i)\}_{i=1}^n$ and the public key is $\mathsf{pk} = \{\hat{g}_i\}_{i=1}^n \in \mathbb{G}^n$.

Sign(sk, $(M_1, \ldots, M_n)$): to sign a vector $(M_1, \ldots, M_n) \in \mathbb{G}^n$ using $\mathsf{sk} = \{(\chi_i, \gamma_i)\}_{i=1}^n$, output $\sigma = (Z, R) = \left(\prod_{i=1}^n M_i^{\chi_i}, \prod_{i=1}^n M_i^{\gamma_i} \right)$.

SignDerive(pk, $\{(\omega_i, \sigma^{(i)})\}_{i=1}^\ell$): given pk as well as ℓ tuples $(\omega_i, \sigma^{(i)})$, parse $\sigma^{(i)}$ as $\sigma^{(i)} = (Z_i, R_i)$ for $i = 1$ to ℓ. Return the triple $\sigma = (Z, R) \in \mathbb{G}$, where $Z = \prod_{i=1}^\ell Z_i^{\omega_i}$, $R = \prod_{i=1}^\ell R_i^{\omega_i}$.

Verify(pk, $\sigma, (M_1, \ldots, M_n)$): given $\sigma = (Z, R) \in \mathbb{G}^2$ and $(M_1, \ldots, M_n)$, return 1 if and only if $(M_1, \ldots, M_n) \neq (1_{\mathbb{G}}, \ldots, 1_{\mathbb{G}})$ and (Z, R) satisfies

$$e(Z, \hat{g}) \cdot e(R, \hat{h}) = \prod_{i=1}^n e(M_i, \hat{g}_i). \tag{1}$$

4.3 Intuition of Our Construction

To encrypt a message $m \in \mathbb{G}$, we combine a CPA encryption $\boldsymbol{c} = (c_0, c_1, c_2)$ of the form $c_0 = m \cdot f^\theta$, $c_1 = g^\theta$, $c_2 = h^\theta$ and a randomizable publicly verifiable proof that $\log_g c_1 = \log_h c_2$, à la Cramer-Shoup. For that purpose, we can rely on the idea to include a one-time LHSP signature on top of $\boldsymbol{c}$ as first suggested in [32]. That means that the public key contains an LHSP signature Σ on (g, h) so that we can derive a signature on $(g, h)^\theta$ if indeed (c_1, c_2) lies in $\text{span}\langle(g, h)\rangle$ by computing $\pi = \Sigma^\theta$. Such a proof is quasi-adaptive [27] as the CRS depends on the language of which we have to prove membership. Here, the public key includes a CRS that contains a signature on the basis of the linear subspace $\text{span}\langle(g, h)\rangle$ of $\mathbb{G}^2$. Given $\boldsymbol{c}$ and the LHSP signature π one can easily randomize the ciphertext as follows: compute $\boldsymbol{c}' = \boldsymbol{c} \cdot (f, g, h)^{\theta'}$, and adapt the proof $\pi' = \pi \cdot \Sigma^{\theta'}$. While this solution is perfectly randomizable and the signing key allows to perfectly simulate the proof, it only provides a CCA1 security. Still, this technique has been enhanced to provide tag-based simulation-sound proof system which is reminiscent to building CCA-like secure encryption. The underlying technique is to generate a one-time key pair $(\mathsf{opk}, \mathsf{osk})$ of some one-time signature scheme that will be discussed in the next paragraph, and to define the tag as $\tau = H(\mathsf{opk})$, for some collision-resistant hash function H,[2] before computing π that (c_1, c_2) lies in $\text{span}\langle(g, h)\rangle$ based on τ. The ciphertext is then completed by signing $(\boldsymbol{c}, \pi)$ with osk, resulting in the ciphertext $(\boldsymbol{c}, \pi, \sigma, \mathsf{opk})$. A natural solution would be to borrow the first solution due to [33] but it only provides selective-tag simulation soundness. Since we will be using opk as the trace of our TREnc construction, the TCCA security implies that our underlying tag-based encryption must achieve tag-based weak CCA security, and selective-tag security is not enough. Indeed, the tag $\tau^* = H(\mathsf{opk}^*)$ involved in the challenge ciphertext may be chosen by the adversary at any time. Furthermore, we must be able to answer *any* pre-challenge decryption queries, so even those that already used τ^*. That means that we cannot program the public key to embed τ^* that will help us to incorporate an SXDH instance in the computation of the challenge ciphertext. Fortunately, by including a signature Σ_u on $(g, h, 1, 1)$ and another signature Σ_v on $(1, 1, g, h)$ in the public key, given a tag τ, the computation of $\pi = (\Sigma_u^\tau \Sigma_v)^\theta$ due to [30] is an LHSP signature on $(c_1^\tau, c_2^\tau, c_1, c_2)$ which gives us the expected security and still enjoys a perfect randomizability, but for the given tag τ (and trace opk), only, which is still what we were looking for.

Now, we come back on the signature σ of $(\boldsymbol{c}, \pi)$. Usually, $(\mathsf{opk}, \mathsf{osk})$ is a key pair of a strongly unforgeable signature scheme providing non-malleability of ciphertext. However, we want to keep the malleability of the ciphertext as we want to be able to fully randomize it up to opk that will serve as our trace, but we also want to retain the non-malleability of the encrypted message m to satisfy traceability. Here again, in the standard model under SXDH, LHSP signature scheme comes in handy. If our fresh key pair $(\mathsf{opk}, \mathsf{osk})$ is generated from a one-time LHSP signature scheme, we can fix the message and preserve the randomizability of $(\boldsymbol{c}, \pi)$ by computing one-time LHSP signatures σ_1 on (g, c_0, c_1, c_2) and σ_2 on $(1, f, g, h)$. Like this, when we randomize $(\boldsymbol{c}, \pi, \sigma_1, \sigma_2)$ as $\boldsymbol{c}' = \boldsymbol{c} \cdot (f, g, h)^{\theta'}$, $\pi' = \pi \cdot (\Sigma_u^\tau \Sigma_v)^{\theta'}$ we can also adapt the signature $\sigma_1' = \sigma_1 \cdot \sigma_2^{\theta'}$ on $(g, c_0', c_1', c_2') = (g, c_0, c_1, c_2) \cdot (1, f, g, h)^{\theta'}$, and simply keep σ_2. While the correctness follows by inspection, we have several comments to make that are less obvious. First, the reason why we are no more able to modify m is

---

[2] H must not only be second-preimage resistance as in [29] since the adversary can choose opk^* adaptively.

due to the presence of the constant g that must be the first component of the signed vector associated to σ_1 and any adaptation σ_1'. Modifying m requires computing a one-time LHSP signature on a vector necessarily outside the span generated by (g, c_0, c_1, c_2) and $(1, f, g, h)$. Second, the signature σ_2 is unchanged during the randomization. Still, it is a signature on a fixed vector and the one-time LHSP signing algorithm is deterministic. Moreover, if we have two distinct signatures on a single vector we can solve SXDH. That means that any other adversarially generated ciphertext for opk (as the adversary might know osk) will have to share σ_2 and our randomizability holds. Third, while the tag-based simulation-sound QA-NIZK proof π can be simulated if we embed a random triple (F, G, H) into (c_0, c_1, c_2) we also have to produce a valid looking adaptation of σ_1 while we do not know osk^*. To avoid extracting osk from a costly bit-by-bit proof of knowledge in the standard model since osk consists of random scalars,[3] we would like to add $(1, F, G, H)$ in the public key and requires the ciphertext to further compute a signature σ_3 on it with osk. However, if we reveal $(1, F, G, H)$, computing $(g, c_0^*, c_1^*, c_2^*) = (g, c_0, c_1, c_2) \cdot (1, f, g, h)^{\theta^*} \cdot (1, F, G, H)^{\rho*}$ allows deriving a valid $\sigma_1^* = \sigma_1 \cdot \sigma_2^{\theta^*} \cdot \sigma_3^{\rho^*}$ but $\boldsymbol{c}^* = (c_0^*, c_1^*, c_2^*)$ will not be random even if (F, G, H) is random. Fortunately, it is actually sufficient for the traceability to use $(\mathsf{opk}, \mathsf{osk})$ to sign the shorter vectors (g, c_0, c_1), $(1, f, g)$ and $(1, F, G)$, and keep H away from the adversary's view to have a statistically random $\boldsymbol{c}^*$ in the reduction. When $(1, F, G)$ is not in $\mathrm{span}\langle(1, f, g)\rangle$, the proof π simply prevents the adversary from randomizing ciphertexts with $(1, F, G)$ without losing validity.

For technical reason, we hide σ_1 in the ciphertext and make a randomizable NIWI Groth-Sahai proof to show the randomizability and the TCCA security of the scheme. While we can adapt the σ_1 component when we randomize one of the two ciphertexts given by the adversary in the challenge phase (or in the randomization experiment), and that trace to each other, since the adversary might know osk it might infer more information about how we adapt this signature into σ_1^* if we left it in the clear.

4.4 Description

$\mathsf{Gen}(1^\lambda)$: Choose bilinear groups $(\mathbb{G}, \hat{\mathbb{G}}, \mathbb{G}_T)$ of prime order $p > 2^{\mathrm{poly}(\lambda)}$ together with $g, h \leftarrow_\$ \mathbb{G}$ and $\hat{g}, \hat{h} \leftarrow_\$ \hat{\mathbb{G}}$.
1. Pick random $\alpha, \beta \leftarrow_\$ \mathbb{Z}_p$ and set $f = g^\alpha h^\beta$.
2. Pick $\delta \leftarrow_\$ \mathbb{Z}_p$ and compute $(F, G, H) = (f, g, h)^\delta$.
3. Generate a Groth-Sahai CRS $\mathsf{crs}_w = (\vec{w}_1, \vec{w}_2) \in \mathbb{G}^4$ to commit to groups elements of $\mathbb{G}$, where $\vec{w}_1 = (w_{11}, w_{12})$ and $\vec{w}_w = (w_{21}, w_{22})$ are generated in the perfect NIWI mode, i.e., $\mathsf{crs}_w \leftarrow_\$ \mathbb{G}^4$.

---

[3] There is no fully structure-preserving signature schemes under SXDH and none with full randomizability (except in the generic group model [24]), which might still not be enough to be combined with a ciphertext as an SRC). And, we are not aware of any fully structure-preserving LHSP signature scheme, where the secret keys only contain source group elements.

4. Define the vector $\mathbf{v} = (g, h)$ and generate 2 key pairs $(\mathsf{sk}_u, \mathsf{pk}_u)$ and $(\mathsf{sk}_u, \mathsf{pk}_u)$ for the one-time linearly homomorphic signature of Sect. 4.2 in order to sign vectors of dimension $n = 2$, given the common public parameters $\hat{g}, \hat{h}$. Let $\mathsf{pk}_u = \{\hat{u}_1, \hat{u}_2\}$ and $\mathsf{pk}_v = \{\hat{v}_1, \hat{v}_2\}$. Using sk_u (resp. sk_v), generate a one-time LHSP signature $\Sigma_u = (Z_u, R_u)$ (resp. $\Sigma_v = (Z_v, R_v)$) on $\mathbf{v}$. In other words, for $\mathsf{pk}_{\mathsf{lhsp}}^{\mathsf{qazk}} = \{\hat{u}_1, \hat{u}_2, \hat{v}_1, \hat{v}_2\}$, Σ_u, Σ_v are one-time LHSP signatures on the rows of the matrix

$$\mathbf{P} = \begin{pmatrix} g & h & 1 & 1 \\ 1 & 1 & g & h \end{pmatrix}.$$

The private key consists of $\mathsf{SK} = (\alpha, \beta)$ and the public key $\mathsf{PK} \in \mathbb{G}^{13} \times \hat{\mathbb{G}}^6$ is

$$\mathsf{PK} = \left(f, \ g, \ h, \ F, \ G, \ \mathsf{crs}_w, \ \Sigma_u, \ \Sigma_v, \ \mathsf{pk}_{\mathsf{lhsp}}^{\mathsf{qazk}}, \ \hat{g}, \ \hat{h} \right).$$

Enc(PK, m): to encrypt a message $m \in \mathbb{G}$, first run $\mathsf{LGen}(\mathsf{PK})$: Generate a key pair $(\mathsf{osk}, \mathsf{opk})$ for the one-time linearly homomorphic signature of Sect. 4.2 from the public generators $\hat{g}, \hat{h}$ in order to sign vectors of dimension 3. Let $\mathsf{lk} = \mathsf{osk} = \{(\eta_i, \zeta_i)\}_{i=1}^3$ be the private key, of which the corresponding public key is $\mathsf{opk} = \{\hat{f}_i\}_{i=1}^3$. Then, conduct the following steps of $\mathsf{LEnc}(\mathsf{PK}, \mathsf{lk}, m)$:
1. Pick $\theta \leftarrow_\$ \mathbb{Z}_p$ and compute the CPA encryption $\mathbf{c} = (c_0, c_1, c_2)$, where $c_0 = mf^\theta$, $c_1 = g^\theta$ and $c_2 = h^\theta$, and keep the random coin θ.
   Next steps 2–3 are dedicated to the tracing part.
2. To allow tracing, authenticate the row space of the matrix $\mathbf{T} = (T_{i,j})_{1 \le i,j \le 3}$

$$\mathbf{T} = \begin{pmatrix} g & c_0 & c_1 \\ 1 & f & g \\ 1 & F & G \end{pmatrix} \tag{2}$$

by using $\mathsf{lk} = \mathsf{osk}$. Namely, sign each row $\vec{T}_i = (T_{i,1}, T_{i,2}, T_{i,3})$ of $\mathbf{T}$ resulting in $\boldsymbol{\sigma} = (\sigma_i)_{i=1}^3 \in \mathbb{G}^6$, where $\sigma_i = (Z_i, R_i) \in \mathbb{G}^2$.
3. To allow strong randomizability, commit to σ_1 using the Groth-Sahai CRS crs_w by computing $C_Z = \iota(Z_1)\vec{w}_1^{z_1}\vec{w}_2^{z_2}$ and $C_R = \iota(R_1)\vec{w}_1^{r_1}\vec{w}_2^{r_2}$. To ensure that σ_1 is a valid one-time LHSP signature on (g, c_0, c_1) compute the proof $\hat{\pi}_{\mathsf{sig}} = (\hat{P}_1, \hat{P}_2) \in \hat{\mathbb{G}}^2$ such that $\hat{P}_1 = \hat{g}^{z_1}\hat{h}^{r_1}$ and $\hat{P}_2 = \hat{g}^{z_2}\hat{h}^{r_2}$.
   Next step 4 shows the validity of $\mathbf{c}$ associated to the tag $\tau = H(\mathsf{opk})$.
4. Given θ and $\tau = H(\mathsf{opk})$, compute a randomizable simulation-sound proof that $(c_1, c_2) \in \mathrm{span}\langle(g, h)\rangle$. Namely, derive the LHSP signature $\pi = (\Sigma_u^\tau \Sigma_v)^\theta =: (Z_\pi, R_\pi)$ on the vector $(c_1^\tau, c_2^\tau, c_1, c_2) = ((g, h, 1, 1)^\tau (1, 1, g, h))^\theta$.

Output the ciphertext

$$\mathsf{CT} = \left(\mathbf{c}, C_Z, C_R, \sigma_2, \sigma_3, \pi, \hat{\pi}_{\mathsf{sig}}, \mathsf{opk} = \{\hat{f}_i\}_{i=1}^3 \right) \in \mathbb{G}^{13} \times \hat{\mathbb{G}}^5$$

Trace(PK, CT): Parse PK and CT as above, and output opk in the obvious way.

Rand(PK, CT): If PK and CT do not parse as the outputs of Gen and Enc, abort. Otherwise, conduct the following steps:

1. Parse the CPA encryption part $\mathbf{c} = (c_0, c_1, c_2)$, pick $\theta' \leftarrow\!\!\$\; \mathbb{Z}_p$ and compute $\mathbf{c}' = \mathbf{c} \cdot (f, g, h)^{\theta'}$, so that $c'_0 = c_0 f^{\theta'}$, $c'_1 = c_1 g^{\theta'}$ and $c'_2 = c_2 h^{\theta'}$.

2. Implicitly adapt the committed signature σ_1 of the tracing part. First, compute $\tilde{\sigma}_1 = (\tilde{Z}_1, \tilde{R}_1) = (Z_2^{\theta'}, R_2^{\theta'}) = \sigma_2^{\theta'}$, which consists of a one-time LHSP signature on $(1, f, g)^{\theta'}$ for opk. Second, adapt the commitments $C'_Z = C_Z \cdot \iota(\tilde{Z}_1)\vec{w}_1^{z'_1}\vec{w}_2^{z'_2}$ and $C'_R = C_R \cdot \iota(\tilde{R}_1)\vec{w}_1^{r'_1}\vec{w}_2^{r'_2}$, for some random scalars $z'_1, z'_2, r'_1, r'_2 \leftarrow\!\!\$\; \mathbb{Z}_p$, which should commit to the valid one-time LHSP signature $\sigma'_1 = \sigma_1 \sigma_2^{\theta'}$ on (g, c'_0, c'_1) for opk. Third, adapt the proof $\hat{\pi}_{\mathsf{sig}} = (\hat{P}_1, \hat{P}_2)$ as $\hat{\pi}'_{\mathsf{sig}} = (\hat{P}'_1, \hat{P}'_2)$, where $\hat{P}'_1 = \hat{P}_1 \cdot \hat{g}^{z'_1}\hat{h}^{r'_1}$ and $\hat{P}'_2 = \hat{P}_2 \cdot \hat{g}^{z'_2}\hat{h}^{r'_2}$.

3. Adapt the proof of the validity of the CPA ciphertext. Namely, computes $\pi' = \pi \cdot (\Sigma_u^\tau \Sigma_v)^{\theta'} = (Z_\pi (Z_u^\tau Z_v)^{\theta'}, R_\pi (R_u^\tau R_v)^{\theta'})$, where $\tau = H(\mathsf{opk})$.

Output the re-randomized ciphertext

$$\mathsf{CT} = \left(\mathbf{c}', C'_Z, C'_R, \sigma_2, \sigma_3, \pi', \hat{\pi}'_{\mathsf{sig}}, \mathsf{opk}\right).$$

Ver(PK, CT): First, abort and output 0 if PK or CT does not parse properly. Second, verify the validity of the signatures σ_2 and σ_3 on the 2 last rows $\{\vec{T}_i\}_{i=2}^3$ of the matrix $\mathbf{T}$, and output 0 if it does not hold. Third, verify that:

1. The committed signature of the tracing part is valid, i.e., $\sigma_1 = (Z_1, R_1)$ is a valid one-time LHSP signature on the vector (g, c_1, c_2). To hold, the commitments C_Z, C_R and the proof $\hat{\pi}_{\mathsf{sig}} = (\hat{P}_1, \hat{P}_2)$ must satisfy

$$E(C_Z, \hat{g}) \cdot e(C_R, \hat{h}) = E(\iota(g), \hat{f}_1) \cdot E(\iota(c_0), \hat{f}_2) \cdot E(\iota(c_1), \hat{f}_3)$$
$$\cdot E(\vec{w}_1, \hat{P}_1) \cdot E(\vec{w}_2, \hat{P}_2)); \qquad (3)$$

2. The proof that the CPA ciphertext is valid, i.e., $\pi = (Z_\pi, R_\pi)$ is a valid one-time LHSP signature on the vector $(c_1^\tau, c_2^\tau, c_1, c_2)$, which must satisfy

$$e(Z_\pi, \hat{g}) \cdot e(R_\pi, \hat{h}) = e(c_1, \hat{u}_1^\tau \hat{v}_1) \cdot e(c_2, \hat{u}_2^\tau \hat{v}_2), \qquad (4)$$

where $\tau = H(\mathsf{opk})$.

If at least one of theses checks fails, output 0, otherwise, output 1.

Dec(SK, PK, CT): If $\mathsf{Ver}(\mathsf{PK}, \mathsf{CT}) = 0$, output $\perp$. Otherwise, given $\mathsf{SK} = (\alpha, \beta)$ and $\mathbf{c} = (c_0, c_1, c_2)$ included in CT, compute and output $m = c_0 \cdot c_1^{-\alpha} \cdot c_2^{-\beta}$.

4.5   Security

The security of our pairing-based TREnc relies solely on the SXDH assumption. We first show the verifiability of this TREnc as it eases the analysis of the traceability and the randomizability properties. The verifiability essentially relies on the unforgeability of LHSP signatures since it also implies the (simulation-) soundness of the (quasi-adaptive) proof of (subspace) membership. We refer to the full version of the paper [21] for all the proofs of the theorems.

Theorem 2. *The above TREnc is verifiable under the SXDH assumption. More precisely, for any adversary $\mathcal{A}$, we have $\Pr[\mathsf{Exp}_{\mathcal{A}}^{\mathrm{ver}}(\lambda) = 1] \leq \varepsilon_{\mathsf{sxdh}} + 1/p$.*

Theorem 3. *The above TREnc Π is traceable under the SXDH assumption. More precisely, for any adversary $\mathcal{A}$, we have $\Pr[\mathsf{Exp}_{\mathcal{A},\Pi}^{\mathrm{trace}}(\lambda) = 1] \leq 2 \cdot \varepsilon_{\mathsf{sxdh}} + 2/p$.*

Strong randomizability essentially relies on the verifiability, which shows that computationally-bounded adversary only produces (except with negligible probability) valid ciphertexts that are honest (but, possibly with biased randomness), and the perfect randomization of honest ciphertexts.

Theorem 4. *The above TREnc is strongly randomizable under the SXDH assumption. More precisely, for any adversary $\mathcal{A} = (\mathcal{A}_1, \mathcal{A}_2)$, where $\mathcal{A}_2$ is possibly unbounded, we have $\Pr[\mathsf{Exp}_{\mathcal{A}}^{\mathrm{rand}}(\lambda) = 1] \leq \varepsilon_{\mathsf{sxdh}} + 2/p$.*

Theorem 5. *The above TREnc is TCCA-secure under the SXDH assumption and the collision resistance of the hash function. More precisely, we have $\Pr[\mathsf{Exp}_{\mathcal{A},\Pi}^{\mathrm{tcca}}(\lambda) = 1] \leq \frac{1}{2} + \varepsilon_{\mathsf{cr}} + 2 \cdot \varepsilon_{\mathsf{sxdh}} + \Omega(2^{-\lambda})$.*

4.6   Efficiency

This TREnc instance is reasonably efficient. In particular, in order to encrypt a message, which is typically the bottleneck in voting applications because it must run more or less transparently on low-end voter devices, we can encrypt one group element using 29 exponentiations in $\mathbb{G}$ and 10 exponentiations in $\hat{\mathbb{G}}$. This group element would make it possible to encode up to a few hundred bits in practice, depending on the chosen security parameter.

In contrast, the SRC aiming at similar applications and used in the BeleniosRF election system [15] requires 33 exponentiations in $\mathbb{G}$ and 22 exponentiations in $\hat{\mathbb{G}}$ for the (signed) encryption of only 1 bit. In general, their construction requires $11k + 22$ exponentiations in $\mathbb{G}$ and $10k + 12$ exponentiations in $\hat{\mathbb{G}}$ in order to encrypt k bits. These estimates are based on the reference code of the SRC, since the paper does not entirely specify the algorithms (especially how commitments and proofs are computed).

5   Voting Scheme Based on Traceable Receipt-Free Encryption

Traceable Receipt-Free Encryption schemes are particularly well suited for the design of voting systems offering receipt freeness, that is, systems in which voters cannot demonstrate how they voted to a third party.

We are now formalizing the notion of voting system (Sect. 5.1) and receipt-freeness (Sect. 5.2), using a definition closely related to the one of Chaidos et al. [15], while fixing two technical issues that it contains, then show how to build a receipt-free voting scheme from a TREnc (Sect. 5.3).

5.1   Definitions and Notations

We define voting protocols in a way that largely follows the SOK from Bernhard et al. [8] and BeleniosRF [15]. In our voting protocols, we consider the following parties:

- The *voters* are participating in the election and are willing to cast a ballot representing their vote intent.
- The *election administrator* is organizing the election and is responsible for coordinating the protocol execution.
- The *ballot box manager* is gathering the ballots of the voters on a bulletin board BB and provides a public view PBB of those ballots, for verifiability.
- The *trustees* are responsible for correctly tallying the ballot box and providing a proof of the correctness of the tally. We consider a k-threshold tallying system, that is k honest trustees are required to compute the tally of the election.

These parties are standard entities in the voting literature. In some cases, we will also refer to the ballot box manager as the rerandomizing server, in order to make its receipt-freeness related role more visible. We also define a family of deterministic results functions ρ_m which given m votes, returns the result of the election for these votes. The following definition encompasses the procedures used in a voting system.

Definition 10 (Voting System). *A Voting System is a tuple of probabilistic polynomial-time algorithms* (SetupElection, Vote, ProcessBallot, TraceBallot, Valid, Append, Publish, VerifyVote, Tally, VerifyResult) *associated to a result function* $\rho_m : \mathbb{V}^m \cup \{\bot\} \rightarrow \mathbb{R}$ *where* $\mathbb{V}$ *is the set of valid votes and* $\mathbb{R}$ *is the result space such that:*

- SetupElection(1^λ): *on input security parameter* 1^λ, *generate the public and secret keys* (pk, sk) *of the election.*
- Vote(id, v): *when receiving a voter* id *and a vote* v, *outputs a ballot* b *and auxiliary data* aux. *It will also be possible to call* Vote(id, v, aux) *in order to obtain a ballot (without auxiliary data this time) for vote* v *using* aux. *This auxiliary data will be useful to define security and enables the creation of ballots that share the same* aux.
- ProcessBallot(b): *on input ballot* b, *outputs an updated ballot* b'. *In our case,* b' *would be a rerandomization of* b.
- TraceBallot(b): *on input ballot* b, *outputs a tag* t. *The tag is the information that a voter can use to trace his ballot, using* VerifyVote.
- Valid(BB, b): *on input ballot box* BB *and ballot* b, *outputs* 0 *or* 1. *The algorithm outputs* 1 *if and only if the ballot is valid.*
- Append(BB, b): *on input ballot box* BB *and ballot* b, *appends* ProcessBallot(b) *to* BB *if* Valid(BB, b) = 1.
- Publish(BB): *on input ballot box* BB, *outputs the public view* PBB *of* BB, *which is the one that is used to verify the election. Depending on the context, it may be used to remove some voter credentials for instance.*

– VerifyVote(PBB, t): *on input public ballot box* PBB *and tag* t, *outputs* 0 *or* 1. *This algorithm is used by voters to check if their ballot has been processed and recorded properly.*
– Tally(BB, sk): *on input ballot box* BB *and private key of the election* sk, *outputs the tally* r *and a proof* Π *that the tally is correct w.r.t. the result function* ρ_m.
– VerifyResult(PBB, r, Π): *on input public ballot box* PBB, *result of the tally* r *and proof of the tally* Π, *outputs* 0 *or* 1. *The algorithm outputs* 1 *if and only if* Π *is a valid proof that* r *is the election result, computed w.r.t.* ρ_m, *corresponding to the ballots on* PBB.

For all of these algorithms except SetupElection, *the public key of the election* pk *is an implicit argument.*

These algorithms are used as follows in a typical election, the election authorities first generate the election public and secret keys with SetupElection. Then, using the public key of the election, each voter can prepare a ballot bwith the Vote algorithm and send it to the ballot box manager. The voter also keeps TraceBallot(b) in order to be able to trace its ballot on the election public bulletin board. Each time the ballot box manager receives a ballot, it checks if it is valid with the Valid algorithm. If this is the case, it runs the ProcessBallot algorithm on it and appends the resulting ballot to the ballot box using Append.

The ballot box manager also applies Publish on the ballot box in order to obtain the content that is made available on a public bulletin board PBB. Voters can check that their ballot has been correctly recorded on PBBusing VerifyVote.

Eventually, the trustees run the Tally algorithm on the ballot box in order to compute the election result and a proof of correctness of this result. Anyone can use these, together with the content of PBB, in order to verify the election result using VerifyResult.

This definition differs from [8,15] in two important ways. First, we introduce the TraceBallot algorithm. Such a procedure is implicit other voting system descriptions, often because voters simply check the presence on PBBof their ballot, in which case TraceBallotwould simply be the identity function. In our case, TraceBallotmust extract the signature verification key that is generated at voting time by Vote, making this algorithm non-trivial.

The correctness guarantees of the various algorithms listed above are as usual and follow the intuitions given above. We only formalize the correctness guarantee of TraceBallot, which is novel.

Definition 11 (Tracing correctness). *For every* v, BB, (b, aux) ← Vote(id, v) *and* t ← TraceBallot(b), *after* Append(BB, b) *we have that* VerifyVote(Publish(BB), t) = 1 *with overwhelming probability.*

As a second difference, we omit the voter registration procedure, of which we make no use here: it is used in some protocols in order to obtain some forms of delegated verifiability where an extra authority is partially trusted to handle voter credentials, but this is not our focus. To make things concrete here, one can imagine that voter authentication is handled with a process similar to the one used in Helios [3], where the ballot box manager distributes credentials (e.g., passwords) to the voters and publishes the voter names next to their ballot on PBB. Voters who did not vote can then verify that there is no ballot recorded for them, and auditors can sample voters and contact them to perform similar verification steps. (We make no claim regarding the effectiveness of this process in practice – it is just here for context.)

5.2   Receipt-Freeness

We adopt here a definition of receipt-freeness that is similar to the one of Chaidos et al [15]. Various other definitions exist, but they are either too informal for our purpose (e.g., [25]), or focus on the stronger notion of coercion resistance, in which voters need to adopt a specific counter-strategy depending on instructions of the coercer (e.g., [4,31,35,40]).

This definition requires that voters should not be able to pick a ballot, possibly from a distribution that deviates from the honest one, in such a way that no third party, by looking at the election public bulletin board, and knowing exactly how the voter's announced ballot was built, is able to decide whether that ballot was submitted by the voter rather than another ballot that could encode a vote for a different candidate. This definition also considers that the channels between the voters and the ballot box manager is private and, indeed, without the assumption that such private channels are available, achieving receipt-freeness in a verifiable election is impossible [18].

Definition 12 (Receipt-Freeness). *A voting system $\mathcal{V}$ has receipt-freeness if there exists algorithms* SimSetupElection *and* SimProof *such that no* PPT *adversary $\mathcal{A}$ can distinguish between games* $\mathsf{Exp}_{\mathcal{A},\mathcal{V}}^{\mathsf{srf},0}(\lambda)$ *and* $\mathsf{Exp}_{\mathcal{A},\mathcal{V}}^{\mathsf{srf},1}(\lambda)$ *defined by the oracles in Fig. 3, that is for any efficient algorithm $\mathcal{A}$:*

$$\left| Pr\left[\mathsf{Exp}_{\mathcal{A},\mathcal{V}}^{\mathsf{srf},0}(\lambda) = 1 \right] - Pr\left[\mathsf{Exp}_{\mathcal{A},\mathcal{V}}^{\mathsf{srf},1}(\lambda) = 1 \right] \right|$$

is negligible in λ.

In this game, parameterized by a bit β, the adversary has access to the following oracles:

- $\mathcal{O}$init: initializes the voting system. It generates the public and privates keys of election and returns the public key to the adversary. When $\beta = 1$, a simulated setup may be performed, depending on the computational model, which will offer some trapdoor information that may be needed to produced a simulated tally correctness proof for instance. Eventually, two empty ballot boxes are created: the real ballot box BB_0 that will be tallied and the fake ballot box BB_1. Both boxes will be populated during the game, but the adversary will only see $\mathsf{Publish}(\mathsf{BB}_\beta)$.
- $\mathcal{O}$receiptLR($\mathsf{b}_0, \mathsf{b}_1$): Lets the adversary cast ballots b_0 in the real ballot box BB_0 and b_1 in the fake ballot box BB_1, as long as both ballots are valid and have the same trace. This oracle is the central one for receipt-freeness.

$\mathcal{O}\mathsf{init}(\lambda)$

if $\beta = 0$ **then** $(\mathsf{pk}, \mathsf{sk}) \leftarrow \mathsf{SetupElection}(1^\lambda)$
else $(\mathsf{pk}, \mathsf{sk}, \tau) \leftarrow \mathsf{SimSetupElection}(1^\lambda)$
$\mathsf{BB}_0 \leftarrow \perp; \mathsf{BB}_1 \leftarrow \perp$
return pk

$\mathcal{O}\mathsf{receiptLR}(b_0, b_1)$

if $\mathsf{TraceBallot}(b_0) \neq \mathsf{TraceBallot}(b_1)$
or $\mathsf{Valid}(\mathsf{BB}_0, b_0) = 0$ **or** $\mathsf{Valid}(\mathsf{BB}_1, b_1) = 0$
then return $\perp$
else $\mathsf{Append}(\mathsf{BB}_0, b_0); \mathsf{Append}(\mathsf{BB}_1, b_1)$

$\mathcal{O}\mathsf{board}()$

return $\mathsf{Publish}(\mathsf{BB}_\beta)$

$\mathcal{O}\mathsf{tally}()$

$(r, \Pi) \leftarrow \mathsf{Tally}(\mathsf{BB}_0, \mathsf{sk})$
if $\beta = 1$ **then** $\Pi \leftarrow \mathsf{SimProof}(\mathsf{BB}_1, r)$
return (r, Π)

Fig. 3. Oracles used in the $\mathsf{Exp}^{\mathsf{srf}, \beta}_{\mathcal{A}, \mathcal{V}}(\lambda)$ experiment. The adversary first calls $\mathcal{O}\mathsf{init}$ and then can call Oboard and $\mathsf{OreceiptLR}$ as much as it wants. Finally, the adversary calls $\mathcal{O}\mathsf{tally}$, receives the result of the election and must return its guess, which is the output of the experiment.

- $\mathcal{O}\mathsf{board}$: Returns $\mathsf{Publish}(\mathsf{BB}_\beta)$, which represents its view of the public bulletin board.
- $\mathcal{O}\mathsf{tally}$: Returns the result of the election based on the ballots on BB_0, as well as a proof of correctness of the tally. If $\beta = 1$, this proof is simulated w.r.t. the content derived from BB_1.

Several observations can be made about this game. First, and as expected, it can be seen that the ballot box manager is considered to behave honestly. A dishonest ballot box manager could simply replace $\mathsf{ProcessBallot}$ and $\mathsf{Publish}$ with the identity function, which would make the game trivial to win, independently of any cryptographic operation. The Tally operation is also performed honestly: dishonest talliers could decrypt all the ballots individually, which would again make the game trivial to win. In practice, this assumption can be mitigated by using a distributed decryption process, which is always possible using MPC but can typically be done more efficiently.

Second, this game prompts for the introduction of an extra correctness requirement on the definition of Vote and $\mathsf{TraceBallot}$, in order to make sure that ballots that encode different votes and have the same tag can be computed.

Definition 13 (Ballot traceability for receipt freeness). *For every public key* pk *in the range of* $\mathsf{SetupElection}$, *the ballots produced for every pair of voting choices* (v_0, v_1) *with the same auxiliary data trace to the same tag. That is, for* $b_0, \mathsf{aux} \leftarrow_\$ \mathsf{Vote}(\mathsf{id}, v_0)$, $b_1 \leftarrow_\$ \mathsf{Vote}(\mathsf{id}, v_1, \mathsf{aux})$, *we have* $\mathsf{TraceBallot}(b_0) = \mathsf{TraceBallot}(b_1)$.

Without this extra constraint, we could imagine a $\mathsf{TraceBallot}$ algorithm which returns a tag depending on the vote inside the ballot. For example if $\mathsf{TraceBallot}(b) = b$ as we discussed earlier, then $\mathcal{O}\mathsf{receipt}(b_0, b_1)$ does nothing except if the two ballots are identical and the adversary can never win the receipt-freeness game. It is thus natural to require that a tag returned by $\mathsf{TraceBallot}$ can be reached with any possible voting choice.

The related constraint that TraceBallot(b_0) = TraceBallot(b_1) is actually missing from Chaidos' definition [15], and makes their game trivial to win in most natural case, including with their own protocol: an adversary could simply submit two ballots that have different traces (or are signed with different keys in the wording of their paper), and immediately identify which bulletin board he sees.

Compared to Chaidos' definition, we also removed the $\mathcal{O}$corrupt oracle, as we simply assume that all the voters are under adversarial control. We also omitted their $\mathcal{O}$cast and $\mathcal{O}$voteLR oracles, because $\mathcal{O}$receiptl R subsumes them.

5.3 Voting Scheme

We now explain how a generic voting system can be built from a TREnc. The protocol in itself is of little interest: it essentially follows previous proposals [10, 15]. Its central interest is that defining it from a TREnc makes the proof of its receipt-freeness almost immediate, and independent of any specific TREnc instance.

A detailed pseudocode description is proposed in Fig. 4. The protocol executes as follow. The election authorities set-up the election in the following way. They create an empty bulletin board and create the pair of public and private keys (pk, sk) of the election by running the Gen algorithm of the TREnc scheme with the desired security parameter. pk is distributed to every party taking part to the election and sk is given to the tallier. Note that sk can be generated in a distributed way, so that decryption requires the contribution of multiple trustees – our TREnc constructions are compatible with standard threshold key generation protocols for discrete log based cryptosystems, which can be used as usual since the decryption of a ciphertext is independent from the link key and the Trace algorithm [19,23]. We consider a unique tallier in the following.

When a user wants to cast a vote v, they first generate a link key lk by running the LGen(pk) algorithm, then encrypts the vote with the LEnc(pk, lk, v) algorithm in order to obtain a ballot b, while aux is defined as lk. The voter then sends the encrypted ballot to the ballot box manager of the voting system. It is utterly important that the user erases the link key as soon as possible, as the integrity of their vote may rely on the secrecy of this key. The voter will however store TraceBallot(b) = Trace(b) in order to verify that a ballot with the correct trace eventually appears on the public bulletin board.

When the ballot box manager receives a new ballot b, he verifies the validity of the ballot by checking that Ver(b) succeeds and that no ballot with the same trace was recorded before. Invalid ballots are dropped and valid ones are going through Append(BB, b), which runs ProcessBallot(b) = Rand(pk, b) and appends the result to BB.

The user can verify that their vote is on the bulletin board by checking with the TraceBallot algorithm if any entry in the public bulletin board has the same trace as the one they recorded when they produced their ballot. The traceability property of the TREnc then guarantees that nobody (including the rerandomizing server and the election authorities who hold the decryption key) could have forged another valid ciphertext of another vote linked to this ballot with non-negligible probability.

Once every voter has cast a vote, the tallier can gather the ballots on the bulletin board and compute the result of the election r, as well as a proof of correctness Π. The exact details of this process will depend on the ballot format and the result function ρ_m that the voting protocol requires. One standard way of performing this operation would be to process all the published ballots through a verifiable mixnet: our TREnc ciphertexts are compatible with various standard options that operate on votes encrypted as vectors of group elements, including the Verificatum mixnet [42].

SetupElection (λ)

$(\mathsf{pk}, \mathsf{sk}) \leftarrow \mathsf{Gen}(1^\lambda)$
return pk

TraceBallot(b)

return Trace(b)

ProcessBallot(b)

return Rand(pk, b)

Vote(id, v[, aux])

if aux is specified **then** lk $\leftarrow$ aux
else lk $\leftarrow\$ LGen(pk)$
b $\leftarrow$ LEnc(pk, lk, v)
if aux is not specified **then return** b
else return b, lk

Valid(BB, b)

if Ver(b)$\wedge$
$\forall$b$'$ $\in$ BB, TraceBallot(b$'$) $\neq$ TraceBallot(b)
then return 1 **else return** 0

VerifyVote(PBB, t)

if $\exists b \in$ PBB : Valid(b) $\wedge$ t == TraceBallot(b)
then return 1 **else return** 0

Fig. 4. Instantiation of our voting scheme from a generic TREnc scheme. Publish is simply the identity function. Tally and VerifyResult are instantiated via standard techniques, depending on the result function (homomorphic tallying, verifiable mixnet, . . .).

5.4   Security of the Voting Scheme

This voting scheme has receipt freeness, as claimed in the following theorem.

Theorem 6. *If the TREnc used in the voting scheme is TCCA secure and verifiable and if the proof system used to prove the correctness of the tally is zero-knowledge, then the voting scheme has receipt freeness. More precisely, if the advantage of any adversary at distinguishing a simulator from an honest prover of the proof system is bounded by ε_{ZK} and if the advantage of any adversary at the TCCA experiment is bounded by a negligible function ε_{TCCA}, then every adversary at the receipt freeness game making q_r $\mathcal{O}$receiptLR requests has an advantage bounded by $\varepsilon_{ZK} + q_r \varepsilon_{TCCA}$.*

Proof. The proof uses two different games, where the first one is the receipt-freeness game. In each of those games, we pick a random bit β corresponding to either the real ballot box ($\beta = 0$) or the fake ballot box ($\beta = 1$). The adversary is expected to guess in which case it is and to output a bit β'. We note S_i the event that $\beta = \beta'$ in the i-th game. We show that S_0, the probability of an adversary to win the receipt freeness game, is negligibly close in λ to $\frac{1}{2}$.

Game$_0(\lambda)$: We define Game$_0(\lambda)$ as the original receipt freeness experiment and $\mathcal{A}$ as a PPT adversary for the game. We set SimSetup as the simulation trapdoor for the proof systems used in the tally and SimProof(BB, r) as an algorithm simulating fake proofs of the decryption of the ciphertexts in BB into the plaintexts listed in r. By definition, $\mathcal{A}$ wins the game with probability $Pr[S_0]$.

Game$_1(\lambda)$: If $\beta = 0$, we generate the keys of the election with SimSetup instead of the honest Setup algorithm. We also give a simulated proof of the tally as in the case $\beta = 1$.
Game$_0(\lambda) \to$ Game$_1(\lambda)$: Since the proof system of the tally is zero-knowledge, $|Pr[S_0] - Pr[S_1]| \leq \varepsilon_{ZK}$.

In the second game, the only difference between $\beta - 0$ and $\beta = 1$ are now in the $\mathcal{O}$receiptLR oracle. We can reduce the q $-$ TCCA experiment (the q challenge variant of the TCCA game, which is proven to be equivalent to TCCA security up to a factor q in the full version) to Game$_1(\lambda)$ in the following way. We build an adversary against the q $-$ TCCA challenger by instantiating the voting system and simulating an efficient adversary for Game$_1(\lambda)$. Each time we are asked to append ballots to the bulletin board from a $\mathcal{O}$receiptLR oracle call, we ask the challenger to decrypt them. Then, we give both ballots as a challenge to the challenger and receive a randomized ballot that we append to our bulletin board. There are q_r such requests. After all the requests, we can compute the result of the election as we have the plaintext of every ballot in BB$_0$. Moreover, we can use SimProof to simulate a proof that our bulletin board has been correctly tallied. Hence, this adversary wins the q $-$ TCCA game with the same probability as the simulated adversary wins the second game and $Pr[S_1] \leq q_r \varepsilon_{TCCA}$. We conclude that the probability that the adversary wins the receipt freeness experiment, $Pr[S_0]$, is bounded by $\varepsilon_{ZK} + q_r \varepsilon_{TCCA}$.

It is immediate that our voting scheme also satisfies ballot traceability (Def. 13), thanks to the link traceability of the TREnc (Def. 2).

This demonstrates the receipt-freeness of our protocol against an adversary who sees the public bulletin board, and assuming a honest ballot box manager. Our protocol also offers privacy against a malicious ballot box manager, as demonstrated in the full version. As can be expected, proving privacy against such an adversary requires taking advantage of the strong-randomization property of the TREnc, which was not necessary for receipt-freeness.

It is also important to note that the notions of receipt-freeness and ballot privacy only make sense when applied to voting protocols that satisfy some extra correctness requirements (see Bernhard et al. [8] for instance) – a pathological

Valid process that would just drop all but one ballot could result in this ballot been tallied alone, which could satisfy the definition or receipt-freeness but would obviously be problematic from a privacy point of view. The notions of strong consistence, correctness, and validity, defined in the full version of the paper, address these questions.

We do not detail the verifiability of our voting system, which would require to introduce a substantial machinery. We outline how this could work here:

- Individual verifiability requires that a voter who successfully completes the VerifyVote verification steps can be convinced that his vote is properly recorded. If the voter's voting client is honest, this follows from the traceability property of the TREnc and the single use of lk, which guarantee that any ballot with the same trace as the ballot submitted by the voter would decrypt to the right vote. Detecting a malicious voting client that may encrypt a vote different of the one chosen by the voter is more tricky. One option would be to consider a variation on the Benaloh challenge, in which voters would have the option to decide to spoil a ballot that has been posted on the public bulletin board, and either ask for its decryption, or for the randomness used both during the Vote process and during ProcessBallot. Any newly created ballot would need to be generated using a fresh lk.
- Eligibility verifiability could proceed by adding the voter's name next to each ballot on the public bulletin board, and let auditors check whether these are legitimate voters. Weaker but more convenient options include relying on a trusted authentication server and/or on a PKI.
- Universal verifiability, which guarantees that the tally is computed correctly, would result from the tallying process, e.g., from a verifiable mix-net.

Acknowledgments. This work has been funded in part by the European Union (EU) and the Walloon Region through the FEDER project USERMedia (convention number 501907-379156). Henri Devillez is a research fellow of the Belgian FRIA. Thomas Peters is research associate of the Belgian Fund for Scientific Research (F.R.S.-FNRS). This work has been funded in parts by the F.R.S.–FNRS SeVote project.

References

1. Abe, M., Fuchsbauer, G., Groth, J., Haralambiev, K., Ohkubo, M.: Structure-preserving signatures and commitments to group elements. In: Rabin, T. (ed.) CRYPTO 2010. LNCS, vol. 6223, pp. 209–236. Springer, Heidelberg (2010). https://doi.org/10.1007/978-3-642-14623-7_12
2. Abe, M., Haralambiev, K., Ohkubo, M.: Signing on elements in bilinear groups for modular protocol design. Cryptology ePrint Archive, Report 2010/133 (2010). https://ia.cr/2010/133
3. Adida, B., de Marneffe, O., Pereira, O., Quisquater, J.: Electing a university president using open-audit voting: analysis of real-world use of Helios. In: 2009 Electronic Voting Technology Workshop/Workshop on Trustworthy Elections, EVT/WOTE '09. USENIX Association (2009)

4. Alwen, J., Ostrovsky, R., Zhou, H.-S., Zikas, V.: Incoercible multi-party computation and universally composable receipt-free voting. In: Gennaro, R., Robshaw, M. (eds.) CRYPTO 2015. LNCS, vol. 9216, pp. 763–780. Springer, Heidelberg (2015). https://doi.org/10.1007/978-3-662-48000-7_37

5. Bauer, B., Fuchsbauer, G.: Efficient signatures on randomizable ciphertexts. In: Galdi, C., Kolesnikov, V. (eds.) SCN 2020. LNCS, vol. 12238, pp. 359–381. Springer, Cham (2020). https://doi.org/10.1007/978-3-030-57990-6_18

6. Benaloh, J.: Simple verifiable elections. In: 2006 USENIX/ACCURATE Electronic Voting Technology Workshop, EVT'06. USENIX Association (2006)

7. Benaloh, J.C., Tuinstra, D.: Receipt-free secret-ballot elections (extended abstract). In: Proceedings of the Twenty-Sixth Annual ACM Symposium on Theory of Computing, pp. 544–553. ACM (1994)

8. Bernhard, D., Cortier, V., Galindo, D., Pereira, O., Warinschi, B.: SOK: a comprehensive analysis of game-based ballot privacy definitions. In: 2015 IEEE Symposium on Security and Privacy, pp. 499–516 (2015)

9. Bernhard, D., Pereira, O., Warinschi, B.: How not to prove yourself: pitfalls of the fiat-shamir heuristic and applications to Helios. In: Wang, X., Sako, K. (eds.) ASIACRYPT 2012. LNCS, vol. 7658, pp. 626–643. Springer, Heidelberg (2012). https://doi.org/10.1007/978-3-642-34961-4_38

10. Blazy, O., Fuchsbauer, G., Pointcheval, D., Vergnaud, D.: Signatures on randomizable ciphertexts. In: Catalano, D., Fazio, N., Gennaro, R., Nicolosi, A. (eds.) PKC 2011. LNCS, vol. 6571, pp. 403–422. Springer, Heidelberg (2011). https://doi.org/10.1007/978-3-642-19379-8_25

11. Boneh, D., Sahai, A., Waters, B.: Functional encryption: definitions and challenges. In: Ishai, Y. (ed.) TCC 2011. LNCS, vol. 6597, pp. 253–273. Springer, Heidelberg (2011). https://doi.org/10.1007/978-3-642-19571-6_16

12. Canetti, R., Dwork, C., Naor, M., Ostrovsky, R.: Deniable encryption. In: Kaliski, B.S. (ed.) CRYPTO 1997. LNCS, vol. 1294, pp. 90–104. Springer, Heidelberg (1997). https://doi.org/10.1007/BFb0052229

13. Canetti, R., Gennaro, R.: Incoercible multiparty computation (extended abstract). In: 37th Annual Symposium on Foundations of Computer Science, FOCS '96, pp. 504–513. IEEE Computer Society (1996)

14. Canetti, R., Krawczyk, H., Nielsen, J.B.: Relaxing chosen-ciphertext security. In: Boneh, D. (ed.) CRYPTO 2003. LNCS, vol. 2729, pp. 565–582. Springer, Heidelberg (2003). https://doi.org/10.1007/978-3-540-45146-4_33

15. Chaidos, P., Cortier, V., Fuchsbauer, G., Galindo, D.: BeleniosRF: A noninteractive receipt-free electronic voting scheme. In: Proceedings of the 2016 ACM SIGSAC Conference on Computer and Communications Security (CCS), pp. 1614–1625 (2016)

16. Chase, M., Kohlweiss, M., Lysyanskaya, A., Meiklejohn, S.: Malleable proof systems and applications. In: Pointcheval, D., Johansson, T. (eds.) EUROCRYPT 2012. LNCS, vol. 7237, pp. 281–300. Springer, Heidelberg (2012). https://doi.org/10.1007/978-3-642-29011-4_18

17. Chaum, D.: Surevote: technical overview. In: Proceedings of the Workshop on Trustworthy Elections (WOTE 2001) (2001)

18. Chevallier-Mames, B., Fouque, P.-A., Pointcheval, D., Stern, J., Traoré, J.: On some incompatible properties of voting schemes. In: Chaum, D., et al. (eds.) Towards Trustworthy Elections. LNCS, vol. 6000, pp. 191–199. Springer, Heidelberg (2010). https://doi.org/10.1007/978-3-642-12980-3_11

19. Cortier, V., Galindo, D., Glondu, S., Izabachène, M.: Distributed elgamal à la pedersen: application to helios. In: Proceedings of the 12th annual ACM Workshop on Privacy in the Electronic Society, WPES 2013, pp. 131–142. ACM (2013)

20. Cramer, R., Gennaro, R., Schoenmakers, B.: A secure and optimally efficient multi-authority election scheme. In: Fumy, W. (ed.) EUROCRYPT 1997. LNCS, vol. 1233, pp. 103–118. Springer, Heidelberg (1997). https://doi.org/10.1007/3-540-69053-0_9

21. Devillez, H., Pereira, O., Peters, T.: Traceable receipt-free encryption. Cryptology ePrint Archive (2022)

22. Gennaro, R.: Achieving independence efficiently and securely. In: Proceedings of the Fourteenth Annual ACM Symposium on Principles of Distributed Computing, pp. 130–136. ACM (1995)

23. Gennaro, R., Jarecki, S., Krawczyk, H., Rabin, T.: Secure distributed key generation for discrete-log based cryptosystems. J. Cryptol. 20(1), 51–83 (2007)

24. Groth, J.: Efficient fully structure-preserving signatures for large messages. In: Iwata, T., Cheon, J.H. (eds.) ASIACRYPT 2015. LNCS, vol. 9452, pp. 239–259. Springer, Heidelberg (2015). https://doi.org/10.1007/978-3-662-48797-6_11

25. Hirt, M., Sako, K.: Efficient receipt-free voting based on homomorphic encryption. In: Preneel, B. (ed.) EUROCRYPT 2000. LNCS, vol. 1807, pp. 539–556. Springer, Heidelberg (2000). https://doi.org/10.1007/3-540-45539-6_38

26. Horwitz, J., Lynn, B.: Toward hierarchical identity-based encryption. In: Knudsen, L.R. (ed.) EUROCRYPT 2002. LNCS, vol. 2332, pp. 466–481. Springer, Heidelberg (2002). https://doi.org/10.1007/3-540-46035-7_31

27. Jutla, C.S., Roy, A.: Shorter quasi-adaptive NIZK proofs for linear subspaces. In: Sako, K., Sarkar, P. (eds.) ASIACRYPT 2013. LNCS, vol. 8269, pp. 1–20. Springer, Heidelberg (2013). https://doi.org/10.1007/978-3-642-42033-7_1

28. Kiayias, A., Zacharias, T., Zhang, B.: End-to-end verifiable elections in the standard model. In: Oswald, E., Fischlin, M. (eds.) EUROCRYPT 2015. LNCS, vol. 9057, pp. 468–498. Springer, Heidelberg (2015). https://doi.org/10.1007/978-3-662-46803-6_16

29. Kiltz, E.: Chosen-ciphertext security from tag-based encryption. In: Halevi, S., Rabin, T. (eds.) TCC 2006. LNCS, vol. 3876, pp. 581–600. Springer, Heidelberg (2006). https://doi.org/10.1007/11681878_30

30. Kiltz, E., Wee, H.: Quasi-adaptive NIZK for linear subspaces revisited. Cryptology ePrint Archive, Report 2015/216 (2015). https://ia.cr/2015/216

31. Küsters, R., Truderung, T., Vogt, A.: A game-based definition of coercion-resistance and its applications. In: Proceedings of the 23rd IEEE Computer Security Foundations Symposium, CSF 2010, pp. 122–136. IEEE Computer Society (2010)

32. Libert, B., Peters, T., Joye, M., Yung, M.: Linearly homomorphic structure-preserving signatures and their applications. In: Canetti, R., Garay, J.A. (eds.) CRYPTO 2013. LNCS, vol. 8043, pp. 289–307. Springer, Heidelberg (2013). https://doi.org/10.1007/978-3-642-40084-1_17

33. Libert, B., Peters, T., Joye, M., Yung, M.: Non-malleability from malleability: simulation-sound quasi-adaptive NIZK proofs and CCA2-secure encryption from homomorphic signatures. In: Nguyen, P.Q., Oswald, E. (eds.) EUROCRYPT 2014. LNCS, vol. 8441, pp. 514–532. Springer, Heidelberg (2014). https://doi.org/10.1007/978-3-642-55220-5_29

34. MacKenzie, P., Reiter, M.K., Yang, K.: alternatives to non-malleability: definitions, constructions, and applications. In: Naor, M. (ed.) TCC 2004. LNCS, vol. 2951, pp. 171–190. Springer, Heidelberg (2004). https://doi.org/10.1007/978-3-540-24638-1_10

35. Moran, T., Naor, M.: Receipt-free universally-verifiable voting with everlasting privacy. In: Dwork, C. (ed.) CRYPTO 2006. LNCS, vol. 4117, pp. 373–392. Springer, Heidelberg (2006). https://doi.org/10.1007/11818175_22

36. Prabhakaran, M., Rosulek, M.: Homomorphic encryption with CCA security. In: Aceto, L., Damgård, I., Goldberg, L.A., Halldórsson, M.M., Ingólfsdóttir, A., Walukiewicz, I. (eds.) ICALP 2008. LNCS, vol. 5126, pp. 667–678. Springer, Heidelberg (2008). https://doi.org/10.1007/978-3-540-70583-3_54

37. Sahai, A., Waters, B.: Fuzzy identity-based encryption. In: Cramer, R. (ed.) EURO-CRYPT 2005. LNCS, vol. 3494, pp. 457–473. Springer, Heidelberg (2005). https://doi.org/10.1007/11426639_27

38. Sako, K., Kilian, J.: Receipt-free mix-type voting scheme. In: Guillou, L.C., Quisquater, J.-J. (eds.) EUROCRYPT 1995. LNCS, vol. 921, pp. 393–403. Springer, Heidelberg (1995). https://doi.org/10.1007/3-540-49264-X_32

39. Shamir, A.: Identity-based cryptosystems and signature schemes. In: Blakley, G.R., Chaum, D. (eds.) CRYPTO 1984. LNCS, vol. 196, pp. 47–53. Springer, Heidelberg (1985). https://doi.org/10.1007/3-540-39568-7_5

40. Unruh, D., Müller-Quade, J.: Universally composable incoercibility. In: Rabin, T. (ed.) CRYPTO 2010. LNCS, vol. 6223, pp. 411–428. Springer, Heidelberg (2010). https://doi.org/10.1007/978-3-642-14623-7_22

41. Wikström, D.: Simplified submission of inputs to protocols. In: Ostrovsky, R., De Prisco, R., Visconti, I. (eds.) SCN 2008. LNCS, vol. 5229, pp. 293–308. Springer, Heidelberg (2008). https://doi.org/10.1007/978-3-540-85855-3_20

42. Wikström, D.: A commitment-consistent proof of a shuffle. In: Boyd, C., González Nieto, J. (eds.) ACISP 2009. LNCS, vol. 5594, pp. 407–421. Springer, Heidelberg (2009). https://doi.org/10.1007/978-3-642-02620-1_28

Efficient Searchable Symmetric Encryption for Join Queries

Charanjit Jutla[1] and Sikhar Patranabis[2(✉)]

[1] IBM Research, New York, USA
csjutla@us.ibm.com
[2] IBM Research, Bangalore, India
sikhar.patranabis@ibm.com

Abstract. The Oblivious Cross-Tags (OXT) protocol due to Cash et al. (CRYPTO'13) is a highly scalable searchable symmetric encryption (SSE) scheme that allows fast processing of conjunctive and more general Boolean queries over encrypted relational databases. A long-standing open question has been to extend OXT to also support queries over joins of tables without pre-computing the joins. In this paper, we solve this open question without compromising on the nice properties of OXT with respect to both security and efficiency. We propose Join Cross-Tags (JXT) - a purely symmetric-key solution that supports efficient conjunctive queries over (equi-) joins of encrypted tables without any pre-computation at setup. The JXT scheme is fully compatible with OXT, and can be used in conjunction with OXT to support a wide class of SQL queries directly over encrypted relational databases. JXT incurs a storage cost (over OXT) of a factor equal to the number of potential join-attributes in a table, which is usually compensated by the fact that JXT is a fully symmetric-key solution (as opposed to OXT which relies on discrete-log hard groups). We prove the (adaptive) simulation-based security of JXT with respect to a rigorously defined leakage profile.

1 Introduction

The advent of cloud computing allows individuals and organizations to outsource the storage and processing of large volumes of data to third party servers. However, clients typically do not trust service providers to respect the confidentiality of their data [CZH+13]. This lack of trust is often further reinforced by threats from malicious insiders and external attackers. One solution is to upload data in an encrypted form, with the client keeping the secret key.

Consider a client that offloads an encrypted relational database of (potentially sensitive) credit-card transactions to an untrusted server. At a later point of time, the client might want to issue a query of the form *retrieve all transactions for a particular merchantID for a given time*. Ideally, we want the client to be able to perform this task without revealing any sensitive information

S. Patranabis—Most of the work was done while the author was affiliated with ETH Zürich, Switzerland and Visa Research USA.

S. Agrawal and D. Lin (Eds.): ASIACRYPT 2022, LNCS 13793, pp. 304–333, 2022.
https://doi.org/10.1007/978-3-031-22969-5_11

to the server, such as the actual transactions, the merchantID underlying the given query, etc. Moreover, one could consider even more complicated Boolean queries over additional *attribute-value pairs*. Unfortunately, techniques such as fully homomorphic encryption [Gen09], that potentially allow achieving such an "ideal" notion of privacy, are unsuitable for practical deployment due to large performance overheads.

Searchable Symmetric Encryption. Searchable symmetric encryption (SSE) [SWP00, Goh03, CGKO06, CK10, PRZB11, CJJ+13, CJJ+14, KM18, CNR21] is the study of provisioning symmetric-key encryption schemes with search capabilities. The most general notion of SSE with optimal security guarantees can be achieved using the work of Ostrovsky and Goldreich on Oblivious RAMs [GO96]. More precisely, using these techniques, one can evaluate a functionally rich class of queries on encrypted data without leaking *any* information to the server. However, such an ideal notion of privacy comes at the cost of significant computational or communication overhead.

A large number of existing SSE schemes prefer to trade-off security for practical efficiency by allowing the server to learn "some" information during query execution. The information learnt by the server is referred to as *leakage*. Some examples of leakage include the database size, the query pattern (which queries have the same attribute-value pair) and the access pattern. Practical implementations of such schemes can be made surprisingly efficient and scalable using specially designed data structures. This line of works on efficient SSE schemes that trade-off leakage for efficiency was initiated by Curtmola et al. [CGKO06], who introduced and formalized the simulation-based framework for proving the security of SSE schemes with respect to a given leakage function. Subsequently, Chase and Kamara [CK10] introduced the concept of "structured encryption" - a generalization of SSE to structured databases, along with the corresponding security definitions.

For any SSE scheme to be truly practical, it should at least support conjunctive queries, i.e., given a set of attribute-value pairs $(w_1, \ldots, w_n)$, it should be able to find and return the set of records that match *all* of these attribute-value pairs. The example query above, namely, *"retrieve all transactions for a particular merchantID for a given time"* is an instance of a conjunctive query. There exist dedicated SSE schemes that can support conjunctive, disjunctive and general Boolean queries over attribute-value pairs in relational databases [CJJ+13, CJJ+14, KM17, LPS+18, PM21].

A very important class of queries that any relational database should support are queries over joins of tables. We illustrate the concept of joins by extending the above example. Consider the scenario where the credit-card processor has two tables: (Table A) a transactions table and (Table B) a merchants information table. Instead of the earlier query *"retrieve all transactions for a particular merchantID for a given time"* in Table A, the new query may be *"retrieve all records for a given time* HHMM *in a given city* CC*"* in the *join* of Table A and Table B, where the join is over the attribute merchantID. More formally, the result of such a query is

$$\{((\langle A; r1\rangle, \langle B; r2\rangle)) \mid \exists \, \text{merchantID MID} :$$
$$(\text{recordID} = \langle A; r1\rangle, \text{time} = \text{HHMM}, \text{merchantID} = \text{MID}) \in \text{Table A}$$
$$\textbf{and}$$
$$(\text{recordID} = \langle B; r2\rangle, \text{city} = \text{CC}, \text{merchantID} = \text{MID}) \in \text{Table B}\}$$

Unfortunately, the above schemes are unable to handle such queries on joins of tables without prohibitive pre-computation of joins. This inability to efficiently and flexibly support queries over joins of tables is indeed a major impediment to actual deployment of these schemes. Only a handful of recent works [KM18,CNR21] address search queries over joins of tables; we will review their techniques subsequently.

Oblivious Cross-Tags (OXT). The work of Cash et al. [CJJ+13] showed for the first time how to design an SSE scheme for conjunctive (and more general Boolean) queries, for which (i) the encrypted database has memory requirement that is linear in the size of the database, (ii) searches require a single round of communication (query followed by response), and (iii) the leakage to the server is low. Their scheme, called Oblivious Cross-Tags (OXT), relies on specially *structured* pseudorandom functions (PRFs), such as those that can be enabled using hard discrete-log groups.

Since our work is closely related to OXT, we give a brief overview. In its simplest embodiment, the SSE scheme OXT precomputes an encrypted version of a database (using a secret symmetric key) and stores it at a server that is presumed to be *honest-but-curious*. A client with access to this symmetric key, breaks a (2-) conjunctive query $b = b_1 \wedge b_2$ into two search tokens for the server. The first search token yields all entries for the first conjunct b_1 and the second search token is used to search for exactly the conjunct b_2 using a "*cross-tag helper token*" stored as part of the entries for b_1. The cross-tag helper is independent of the second attribute and hence only one cross-tag helper per record-attribute pair is stored. Since there is one data element anyway for each record-attribute pair, this at most doubles the total space requirement. For example, consider the conjunctive query above: (time HHMM and merchantID MID). The client computes two PRF values (using its secret key): one for (time; HHMM), say p1, and another for (merchantID; MID), say p2. It sends to the server a key k1 derived from p1, and a token $= h^{p2/p1}$ (in a DDH-hard group with generator h). The attributes (time, merchantID) are also revealed to the server.

The server uses k1 to search for an encrypted set (stored in the encrypted database) corresponding to (time; HHMM) as well as uses k1 to decrypt it. Next, for each record, in this decrypted set D, the server can also find a "*cross-tag helper token*" $z = p1 * rind$ (where, rind stands for randomized-record-index). The search token $h^{p2/p1}$ raised to the power z yields a *cross-tag* $h^{p2*rind}$, which is then checked in a lookup-table called XSet. This lookup-table XSet stored with the server has every valid member of the form $h^{p2*rind}$, and hence this check allows the server to confirm the record in D to satisfy the conjunct. Note, the size of this set XSet is exactly the number of records times the number of attributes (as in each record, for each attribute there is exactly one value such as MID). This is exactly the size of the database. That this lookup-table reveals no information, a priori, is proved under the DDH assumption[1]; hence the name *oblivious cross-tag* (OXT).

Note that, both the client and the server have to perform exponentiations (in the DDH-hard group) during this search protocol. Moreover, the number of these exponentiations can be large, as there will be one such exponentiation per entry in the decrypted set D. Similarly, during the setup stage, i.e. when the database is encrypted and the XSet is computed, an exponentiation is required for every attribute-value pair in the database. Hence, the setup maybe computationally intensive for large databases.

1.1   The New JXT Protocol

Our first contribution is to show that if the number of attributes in a table is small, say m, then the encrypted database with a size blowup by a factor m, can achieve the same security as OXT without the use of DDH, and more precisely using only symmetric-key primitives such as PRFs and symmetric-key encryption in the standard model. As a result, the search computation becomes considerably faster as there is no exponentiation (by either the client or the server). Further, the setup becomes much faster, as the XSet computation requires no exponentiations.

Next, as a main contribution of this work, we show that the above modification to OXT also allows us to search over (equi-) joins of tables without any pre-computation of joins[2]. We refer to this new protocol as Join Cross-Tags (abbreviated as JXT). Moreover, since joins are usually performed over a limited set of attributes (e.g. primary keys or high-entropy attributes[3]), the size blowup to the encrypted database is small; more precisely, a T fold blowup, where T is the number of attributes in a table over which joins are allowed. Recall the example join query *"retrieve all records for a given time* HHMM *in a given city* CC*"* in the join of Table A and Table B, where the join is over the attribute merchantID. The JXT protocol can support this query (over encrypted databases) without any pre-computation of joins of the two tables. The only requirement is that both encrypted tables must be configured to support join over the attribute merchantID[4]. As mentioned previously, if merchantID is amongst the few attributes (say, T many) that a table supports for join, the space requirement for that encrypted

---

[1] The actual protocol is slightly more complicated to be fully secure and provably secure under DDH, but the above description gives the main gist of OXT.

[2] Throughout this paper, when we refer to joins, we mean equi joins.

[3] By high entropy attribute we mean the information-theoretic entropy of the column corresponding to the attribute. For example, the attribute *gender* has low entropy, whereas the attribute *name* can have high entropy in a table.

[4] By configuration we mean the (pre-) computation of the encrypted table. We remind the reader that this pre-computation does not involve join pre-computation, as each table is encrypted independently.

table only increases T-fold. Some other tables may not even have the attribute merchantID, but these may have other small number of attributes over which join is allowed.

We provide a more detailed overview of the ideas behind the JXT protocol subsequently in Sect. 1.2.

Comparison with Pre-computation of Joins. It is worth contrasting this approach to one where joins are pre-computed (for instance, in [KM18,CNR21] as discussed under related work in Sect. 1.4 later), and this is best exhibited by considering the above example. The transactions table A is likely to be tall and skinny, i.e. have many records and few attributes. On the other hand, the merchant information table B is likely to be short and wide. However, their join will have at least as many records as table A and at least as many attributes as table B, i.e. tall and wide. This can cause a considerable blowup in storage requirement. Since JXT does not pre-compute joins, it avoids such blowups, as well as other blowups caused by the possibilities of pair-wise joins of many tables. We present a more detailed comparison of JXT with the above pre-processing based approaches to Sect. 1.4.

Modular Setup and Flexible Updates. The JXT approach also allows for a modular setup stage, as well as flexible table additions and updates. Some tables are updated much faster than others, and hence can be re-setup on their own without the need to re-setup tables that are more or less constant[5]. This allows flexibly adding new tables and updating existing tables in an independent manner without having to re-perform setup across all tables in the encrypted database. This is not supported by any of the existing approaches where joins are pre-computed [KM18,CNR21], and constitutes one of the most appealing aspects of our JXT construction.

Storage and Search Overheads. We provide a high-level summary of the overheads incurred by JXT in terms of storage and search processing. Suppose that a table has a total of n attributes, with $T \leq n$ amongst these being "join attributes"; i.e. attributes over which the table can be joined with other tables in the database. Also, assume that the table has a total of m records (equivalently, rows). Then, in JXT, the corresponding encrypted table incurs a storage overhead of $O(mnT)$, which is $O(T)$-fold blowup to the storage required for the plaintext table. Also, given a 2-conjunctive query over the join of two tables that involves an attribute-value pair w_1 from the first table and an attribute-value pair w_2 from the second table, the computational overhead at the server is $O(\ell_1 \ell_2)$, where ℓ_1 and ℓ_2 are the numbers of records matching the attribute-value pairs w_1 and w_2 in the first and second table, respectively.

---

[5] We remark here that the transactions database is encrypted for post-transactional audit, fraud detection, money-laundering detection, machine learning etc. The real-time transactions database is usually updated and used without encryption, as it runs in a secure domain. It is later encrypted on a periodic basis for above additional functionalities.

Compatibility with OXT. An important feature of JXT is that it is fully compatible with OXT. For example, consider the two tables A and B above and suppose table A has few attributes (say e.g. four) and table B has many attributes (say, e.g. twenty). Also, suppose that some of these attributes are the attributes over which joins can be performed. Then, the OXT protocol can be used to support Boolean queries within each table (spanning many attributes), as well as Boolean queries across tables using the JXT part for the join. So for example, the query maybe a 4-conjunct *"retrieve all records for a given time and a given amount in a given city and a given merchant category"* in the join of Table A and Table B, where the join is over the attribute merchantID.

Further, there is a "multi-client" extension of OXT where the client does not own the secret key; instead, an authority owns the secret key and the client computes its PRF based search tokens using an oblivious-PRF (OPRF) protocol with the authority [JJK+13]. JXT is also fully compatible with this multi-client extension of OXT. In fact, JXT can be easily extended to the scenario where different databases are owned by different entities operating under a single authority, and a client can perform a search query over join of tables owned by different entities; this requires that the entities setup their respective encrypted databases using "oblivious" help from the authority.

1.2   The Main Idea of Our JXT Protocol

We now present a brief overview of the main ideas behind our new JXT protocol.

Breaking a Join Query into Sub-queries. To begin with, we show that a 2-conjunctive (join) query $q = q_1 \wedge q_2$ over the join of a pair of tables (say Tables A and B), with the join being over a third attribute, e.g. merchantID, can be broken into normal sub-queries, i.e. non-join queries, that are either over table A or table B. These sub-queries might be simple or conjunctive, but over a single table. We illustrate this with our running example from above, i.e. the query *"retrieve all records for a given time* HHMM *in a given city* CC*"* in the *join* of Table A and Table B, where the join is over the attribute merchantID. In this case, the *first sub-query* $\mathfrak{a}$ can be viewed as the simple keyword search for the attribute-value pair (time = HHMM) in Table A. Now, suppose that for all records matching this first sub-query $\mathfrak{a}$ in Table A, we create a set of the corresponding values of merchantID, say of the form {MID}. We can now define a *second set of sub-queries* $\mathfrak{B}$ in Table B, with one sub-query for each MID in the aforementioned set. Each such sub-query $\mathfrak{b}$ (in the set $\mathfrak{B}$) is of the form *"retrieve all records for a given city* CC *for a given merchantID* MID*"*. Note that $\mathfrak{b}$ is a 2-conjunctive query itself.

Handling Sub-queries. It is easy to see that each sub-query $\mathfrak{b}$ in $\mathfrak{B}$ can be executed securely using the original OXT protocol, if we could somehow use the results of the first sub-query $\mathfrak{a}$ to derive the tokens needed by OXT server to execute $\mathfrak{b}$. In other words, we wish to design a protocol such that executing $\mathfrak{a}$ in Table A generates a set of tokens that an OXT server can use in the same way as it would use tokens issued directly by an OXT client for query $\mathfrak{b}$ in Table B. The challenging part is to implement the search in Table A to mimic this client for OXT query $\mathfrak{b}$ in Table B, without introducing additional rounds of interaction between the client and the server. Achieving this constitutes the technical core of our new JXT protocol.

As explained earlier in the Introduction, in the original OXT protocol, a client with access to the symmetric key, breaks a 2-conjunctive query $\mathfrak{b} = \mathfrak{b}_1 \wedge \mathfrak{b}_2$ into two search tokens for the OXT server. Similarly, to enable the JXT protocol, we would need two search tokens for each $\mathfrak{b}$ in $\mathfrak{B}$, as it is a 2-conjunctive query. The first search token should yield all entries for the first conjunct $\mathfrak{b}_1$ ((city = CC) for our example), and the second search token is used to search for the second conjunct $\mathfrak{b}_2$ ((merchantID = MID), in our example) using a *"cross-tag helper token"* stored (in the OXT server) as part of the entries for the first conjunct. The OXT client computes two PRF values (using its secret key): one for $\mathfrak{b}_1$ (city; CC), say $\mathsf{p1}$, and another for $\mathfrak{b}_2$ (merchantID; MID), say $\mathsf{p2}$. It sends to the server a key $\mathsf{k1}$ derived from $\mathsf{p1}$, and a token $= \mathsf{h}^{\mathsf{p2}/\mathsf{p1}}$ (in a DDH-hard group with generator h). In the JXT protocol, we require that executing $\mathfrak{a}$ in Table A precisely generates this token $\mathsf{h}^{\mathsf{p2}/\mathsf{p1}}$.

Implementing Search in Table A. The first challenge we face is to implement the search in Table A in a manner that mimics the client for the second set of OXT queries in Table B. We achieve this as follows. At a high level, our idea is to amend the encrypted table A to store the PRF value $\mathsf{p2}$ for the merchantID MID in each record in the set keyed by (time; HHMM), such that the search token in query $\mathfrak{a}$ (corresponding to (time = HHMM)) can be used to retrieve these $\mathsf{p2}$ values. However, note that doing this naïvely has two disadvantages. First of all, it would reveal the occurrence of the same $\mathsf{p2}$ value across many different queries. More crucially, this potentially causes a quadratic blowup in storage, since we would need to store the $\mathsf{p2}$ value for each attribute-value pair in the record, when storing it as a set of records keyed by say, (time, HHMM).

We tackle this as follows. If we restrict the join attributes to be a limited set, say of size T, then the blowup is only T-fold. This is a reasonable assumption in practice, since the join attribute is typically either the primary key (i.e. takes a unique value for each record) or a high-entropy attribute, and there are likely to be only a limited number of candidate join attributes per table. In order to hide the occurrences of join attributes across different queries, we embed nonces or counters in the pseudorandom values. As in OXT, this allows us to avoid cross-query leakage.

Implementing Search in Table B. Since we are willing to allow a T-fold blowup in the encrypted database, we now show that with a 2*T-fold blowup, we can actually get rid of the complicated discrete-log based approach of OXT for handling conjunctive queries, at least for the OXT part that we are emulating inside JXT. Recall in OXT, for each first conjunct $\mathfrak{b}_1$ we stored the "cross-tag helper token" $\mathsf{z} = \mathsf{p}_1 * \mathsf{rind}$. Instead, we now store T different helper tokens $\mathsf{z}_t = \mathsf{p}_1 * \mathsf{rind}_t$ (or simply $\mathsf{p}_1 + \mathsf{rind}_t$), where rind_t is a different pseudorandom value for each $t \in [\mathrm{T}]$. This way, the search token for $\mathfrak{b}_2$ need not be $\mathsf{h}^{\mathsf{p2}/\mathsf{p1}}$

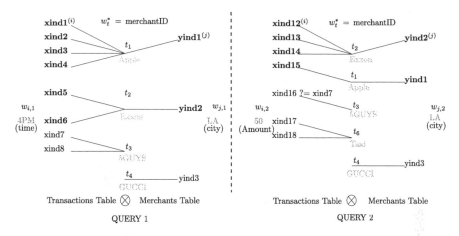

Fig. 1. We illustrate here how the leakage across join queries depends on the ordering of the attribute-value pairs and/or the join ordering. Query 1 is "SELECT * FROM (Transactions JOIN Merchants ON merchantID) WHERE time = 4PM AND city = LA." Query 2 is "SELECT * FROM (Transactions JOIN Merchants ON merchantID) WHERE amount = 50 AND city = LA." The *conditional intersection pattern leakage* reveals that yind1 is same in the two queries. Even though xind16 maybe same as xind7, this information is not leaked. The *join-distribution pattern leakage* also reveals that in the Transactions table, there are additional sets of records, each record in the set sharing the same merchantID (i.e., (redacted) 5GUYS from Query 1 has two records and (redacted) Taxi from Query 2 has two records as well). However, if the order in which the tables are joined is reversed, i.e., the query is over "Merchants JOIN Transactions", then the *join-distribution leakage* is null. This is because merchantID is the *primary key* in the Merchants table. For more details, see Sect. 5.1.

anymore, but just $p2 - p1$. This, when added to the particular cross-tag helper token would yield $rind_t$, which can then be checked for membership[6] in $XSet_t$. Security is maintained because for different t, $rind_t$ is random and independent. The search token for b_2 is now just $p2 - p1$, which is readily obtained from the search for b_1 in table A as described above. Of course, the actual protocol is slightly more complicated as we embed nonces in the PRF values to attain full security. The detailed protocol for JXT appears in Sect. 3 and Figs. 2 and 3.

1.3   The Leakage Profile of JXT

An astute reader may wonder about the leakage of JXT and how it compares to the leakage profile of the OXT protocol. The leakage profile of OXT (i.e. leakage to the server) is known to be technically abstruse, but at the same time a careful analysis also shows that in practice the leakage of OXT is benign given that much

---

[6] Note that we have T different XSets, but their total size is same as the single XSet of OXT.

of this leakage can also be obtained a priori by auxiliary means. We remark that the OXT leakage profile is abstruse mainly because the OXT protocol achieves high scalability while supporting Boolean search queries. Further, the rigorous definition of the leakage profile allows for a simulation-based security proof of OXT. The leakage profile for JXT also follows the same model, with some additional leakage over OXT leakage, which is to be expected because the queries are across tables and express an existential quantifier over the join attribute. Nevertheless, we describe below that in practice the leakage is still benign. In this Introduction section, this is best illustrated using an example as in Fig. 1. A rigorous definition of the leakage profile is given in Sect. 5.1.

The leakage of JXT can be split into six main categories: (a) database size, (b) result pattern of the queries, (c) equality pattern across the queries, (d) size pattern of the queries, (e) conditional intersection pattern across queries, and (f) join-attribute distribution pattern of the queries. While the first five are more or less similar to OXT leakage (but, see Sect. 5.1 for subtle differences), the last one is obviously new to JXT.

We illustrate the concept of join-attribute distribution pattern leakage using the running example. For the query over the join of Tables A and B with respect to the attribute merchantID, it reveals the frequency distribution of values taken by merchantID across records matching the attribute-value pair in the "first slot", i.e. records in Table A matching the time HHMM.

The extent of this leakage depends on the "entropy" of the join attribute merchantID in the first table i.e. Table A. In this particular example, merchantID is the primary key in Table A, and takes a unique value for each record. Hence, in this case, the join-attribute distribution pattern leakage is essentially query-invariant (as each possible value occurs with frequency exactly 1), and hence, benign. In other examples, the join attribute may not be a primary key in the first table. However, if it is still a "high-entropy" attribute in the first table, it is likely to take a unique value for each record (or each value with frequency close to 1), and hence, the leakage can be minimal.

Now, just as in OXT, the client has the choice to order the conjuncts in a query, as well as the order in which the tables are joined. The way OXT is designed is that the first conjunct usually leaks the most information (as the server gets to decrypt information related to the first conjunct). Thus, usually, the attribute that has lesser entropy is not made the first conjunct in a query, as the size pattern leakage has the potential to un-blind the attribute-value pair for a low-entropy attribute (such as gender). A similar design principle is followed in JXT, and the order of the tables being joined can make a difference to the leakage, as illustrated in Fig. 1. In particular, the table in which the join attribute is the primary-key (or high-entropy attribute) should be made the first table in a join query. If such an ordering is always possible, then the additional join-attribute distribution pattern leakage can be null (in case of primary-keys) or minimal (in case of high-entropy attributes).

Security Proofs. We formally prove the (adaptive) simulation-based security of JXT with respect to the above leakage profile (formally defined in Sect. 5.1).

Our proofs follow the same simulation-based framework that was originally proposed by Curtmola et al. in [CGKO06] (and is widely adopted in the SSE literature [CK10, CJJ+13]). Our proofs employ purely symmetric-key primitives such as PRFs and symmetric-key encryption in the standard model.

1.4 Related Work

The SPX Scheme. In [KM18], Kamara and Moataz showed how to encrypt a relational database in such a way that it can efficiently support a large class of SQL queries, including join queries. However, their proposed protocol (called SPX) crucially relies on explicitly pre-computing joins over all attributes that share a common domain. In our context, SPX essentially pre-computes joins of tables over all attributes configured for joins. On the other hand, our proposed JXT protocol avoids all such pre-computation of joins (and the associated storage overheads as discussed earlier).

We note, however, that there are scenarios where SPX provides better query complexity than JXT. Consider a join query over tables Tab_1 and Tab_2, where each individual table has m_1 and m_2 matching records, respectively, but the number of matching records in the join of Tab_1 and Tab_2 is *empty*. This is an extreme case where SPX outperforms JXT since the computational and communication complexity incurred by JXT is $O(m_1 \cdot m_2)$, while that incurred by SPX is $O(1)$.

We also note that the SPX protocol leaks less information during join queries as compared to JXT. This is another consequence of the pre-processing of joins at setup in SPX. In particular, SPX does not incur two kinds of leakage that JXT does: the conditional intersection pattern leakage and the join attribute distribution pattern leakage. We view these types of leakage as tradeoffs for efficiency and flexibility (w.r.t. table updates) that JXT achieves by avoiding pre-computation of joins at setup.

The CNR Scheme. In a more recent work [CNR21], Cash et al. introduced the interesting concept of *partially pre-computed joins*, which potentially has a lower result pattern leakage than is usually expected (we refer to their construction as CNR henceforth). In CNR, the server only learns the projection of the actual result set onto the two tables, and the client has to do extra work to extract the exact set of records matching the join query. However, the storage requirement for their scheme is at least as much as would be required in a scheme that pre-computes joins of tables at setup. Finally, it is not immediately obvious if their scheme is compatible with OXT, which is the state-of-the-art for conjunctive (and more general Boolean) queries. As in SPX, there are scenarios where CNR provides better query complexity than JXT by virtue of the partial pre-computation of joins at setup. Additionally, CNR also does not incur the conditional intersection pattern leakage and the join attribute distribution pattern leakage.

Property-Preserving Encryption. Finally, there exist solutions based on property-preserving encryption (PPE) that allow handling a large class of SQL queries over encrypted relational databases. However, these schemes are vulnerable to *leakage-abuse attacks* [IKK12, NKW15, CGPR15, ZKP16, BKM20]. For example, PPE-based schemes such as CryptDB [PRZB11] typically use deterministic encryption and its variants to support conjunctive (and other Boolean queries), as well as join queries. These techniques typically leak *frequency information* about the underlying plaintext data, which can be potentially exploited in certain settings to completely break query privacy [NKW15]. Our proposed JXT protocol, on the other hand, does not use any PPE-like techniques, and only incurs benign leakage (similar to those in OXT) that are resistant to leakage-abuse attacks (see [BKM20] for an overview of why leakage incurred by schemes such as OXT are not exploitable via leakage-abuse attacks in practice).

1.5   Open Questions and Future Research Directions

Our work gives rise to many interesting open questions and directions of future research. We summarize some of them below.

Joins over Arbitrary and Low-Entropy Attributes. While it is true that JXT does not support joins over arbitrary attributes (in particular, the attributes over which the encrypted database was not configured to support joins), in practice, it is indeed the case that the designer of the tables knows in advance which attributes are likely candidates for joins. We leave it as an open problem to analyze the leakage of the JXT protocol when a join is performed over an attribute which has "low-entropy" in *both* the tables.

Joins over Three or More Tables. We also leave it as an interesting direction of future research to extend JXT to support queries over joins of *three or more* tables (without pre-computation). We do believe that our techniques presented in this paper can be extended to support joins over three (or more) tables. However, a detailed discussion of such an extension is beyond the scope of this paper, as formalizing the implications for its leakage is likely to be non-trivial and could involve some unexpected issues. In addition, a detailed security proof for such an extension would require careful analysis.

Concretely, we expect the key non-triviality to arise in the search protocol, where the client needs to send to the server a significantly more complicated combination of join tokens to enable searching over joins of three (or more) tables, while leaking as little information as possible beyond the leakage for the two-join case. For instance, a naive extension from two-joins to three-joins might leak whether a particular record is in the join of two tables, but not in the join of all three tables. We would ideally want to avoid such "sub-query leakage", which could otherwise lead to attacks.

JXT for Dynamic Databases. We also leave it open to extend JXT to support dynamic addition/deletion of records directly to the encrypted database (e.g., in the spirit of [PM21], which extends OXT to the dynamic setting). Another open problem is to extend JXT to achieve lower result pattern leakage, as in the scheme due to Cash et al. [CNR21] discussed above.

Implementation. We acknowledge that implementing and testing JXT over massive relational databases with TBs of data is important from a performance analysis point of view. However, given the potentially significant implementation-level challenges involved, we leave this as an interesting and challenging follow-up project (similar to the follow-up to OXT by Cash et al. [CJJ+14]).

2   Definitions and Tools

Notation. We write $[n]$ for the set $\{1, \ldots, n\}$. For a vector $\mathbf{v}$ we write $|\mathbf{v}|$ for the dimension (length) of $\mathbf{v}$ and for $i \in [|\mathbf{v}|]$ we write $\mathbf{v}[i]$ for the i-th component of $\mathbf{v}$. All algorithms (including adversaries) are assumed to be probabilistic polynomial-time (PPT) unless otherwise specified. If A is an algorithm, then $y \leftarrow A(x)$ means that the y is the output of A when run on input x. If A is randomized then y is a random variable. We refer to $\lambda \in \mathbb{N}$ as the security parameter, and denote by $\text{poly}(\lambda)$ and $\text{neg}(\lambda)$ any generic (unspecified) polynomial function and negligible function in λ, respectively.[7]

2.1   Relational Databases and Join Queries

Syntax. A relational database $\mathsf{DB} = \{\mathsf{Tab}_i\}_{i \in [N]}$ is represented as a collection of tables. Each table Tab_i is in turn composed of records over a set of attribute-value pairs W_i. For simplicity, we represent Tab_i as a list of tuples of the form $\{(\mathsf{ind}_{i,\ell}, \mathsf{w}_{i,\ell})\}_{\ell \in [L]}$, where each record-identifier $\mathsf{ind}_{i,\ell}$ is a bit-string in $\{0,1\}^\lambda$ and each attribute-value-pair $\mathsf{w}_{i,\ell} \in \mathsf{W}_i$ is an (arbitrary-length) bit-string in $\{0,1\}^*$. For the sake of search it is sufficient to represent a record as its associated attribute-value pair set W_i.

Identifiers. An identifier $\mathsf{ind}_{i,\ell}$ is a value that can be revealed to the server storing the database (for instance, a permutation of the original record indices). It can be used by the server to efficiently retrieve the corresponding (encrypted) record and send it to the client. We assume throughout the paper that any identifier ind corresponding to a record in a table Tab_i is appended with the table number i. In other words, two distinct tables Tab_i and Tab_j cannot have a record identifier ind in common.

Join Attributes. We assume that for each table T_i, the set of all attributes $\{\mathsf{attr}_{i,t}^*\}_{t \in [T]}$ that it shares with other tables in the database DB is fixed at setup and has size upper-bounded by some polynomial function of the security parameter. We refer to such attributes as "join attributes". Looking ahead, these join attributes are used to perform join queries across tables.

---

[7] Note that a function $f : \mathbb{N} \to \mathbb{N}$ is said to be negligible in λ if for every positive polynomial p, $f(\lambda) < 1/p(\lambda)$ when λ is sufficiently large.

Inverted Index. For an attribute-value pair $w \in W_i$, we define $\mathsf{DB}_{\mathsf{Tab}_i}(w)$ as the set of identifiers of records that contain an entry matching w. In other words, $\mathsf{DB}_{\mathsf{Tab}_i}(w)$ is a set of the form:

$$\mathsf{DB}_{\mathsf{Tab}_i}(w) = \{(\mathsf{ind} \mid (\mathsf{ind}, w) \in \mathsf{Tab}_i\}.$$

We refer to the collection of sets $\{\mathsf{DB}_{\mathsf{Tab}_i}(w_\ell)\}_{w_\ell \in W_i}$ as the "inverted index" for the table Tab_i.

Inverted Join Index. For an attribute-value pair $w \in W_i$, we additionally define $\mathsf{DB}_{\mathsf{Tab}_i}^{\mathsf{Join}}(w)$ as the set of identifiers of records that contain an entry matching w, along with the attribute-value pairs corresponding to the join attributes for the same record. In other words, $\mathsf{DB}_{\mathsf{Tab}_i}^{\mathsf{Join}}(w)$ is a set of the form:

$$\mathsf{DB}_{\mathsf{Tab}_i}^{\mathsf{Join}}(w) = \{(\mathsf{ind}, \{w_t^*\}_{t \in [T]}) \mid$$
$$(\mathsf{ind}, w) \in \mathsf{Tab}_i \wedge \forall t \in [T](\mathsf{ind}, w_t^*) \in \mathsf{Tab}_i\}.$$

We refer to the collection of sets $\{\mathsf{DB}_{\mathsf{Tab}_i}^{\mathsf{Join}}(w_\ell)\}_{w_\ell \in W_i}$ as the "inverted join index" for the table Tab_i.

Join Query. A *join query* over a pair of tables Tab_{i_1} and Tab_{i_2} with corresponding attribute-value pair sets W_1 and W_2, respectively, is specified by a tuple

$$q = (i_1, i_2, w_1, w_2, \mathsf{attr}^*),$$

where $w_1 \in W_1$, $w_2 \in W_2$, and attr^* is a special "join attribute" that defines the join relation between the tables Tab_{i_1} and Tab_{i_2} for the query q.

We write $\mathsf{DB}(q)$ to be the set of tuples of the form $(\mathsf{ind}_1, \mathsf{ind}_2)$ that "satisfy" the query q, where ind_1 and ind_2 are identifiers corresponding to records in the tables Tab_{i_1} and Tab_{i_2}, respectively. Formally, this means that for each $(\mathsf{ind}_1, \mathsf{ind}_2) \in \mathsf{DB}(q)$, the following conditions hold simultaneously:

$$((\mathsf{ind}_1, w_1) \in \mathsf{Tab}_{i_1}) \wedge ((\mathsf{ind}_2, w_2) \in \mathsf{Tab}_{i_2}),$$

$$\exists \gamma \text{ s.t. } ((\mathsf{ind}_1, \langle \mathsf{attr}^*, \gamma \rangle) \in \mathsf{Tab}_{i_1}) \wedge ((\mathsf{ind}_2, \langle \mathsf{attr}^*, \gamma \rangle) \in \mathsf{Tab}_{i_2})$$

2.2   SSE Syntax and Security Model

In this section, we formally define searchable symmetric encryption (SSE). Before presenting the formal definition, we present certain assumptions we make in the rest of the paper.

– In the rest of the paper, we assume that any plaintext record is identified by its index ind while, the corresponding encrypted version of the record is identified by a "randomized index" rind (computed as a pseudorandom mapping applied on the original index ind).

- We also assume that the output from the SSE protocol for a given search query are the indices (or identifiers) ind corresponding to the records that satisfy the query. A client program can then use these indices to retrieve the encrypted records and decrypt them. We adopt this formulation because it allows us to decouple the storage of payloads (which can be done in a variety of ways, with varying types of leakage) from the storage of metadata, which is the focus of our protocol (e.g., a client may retrieve the encrypted records from the same server running the query or from a different server, or may only retrieve records not previously cached, etc.)

We note here that a similar formulation is used by almost all existing works on SSE, and more generally structured encryption [CGKO06, CK10, CJJ+13, CJJ+14].

Formal Definition of SSE. A *searchable symmetric encryption (SSE) scheme* Π consists of an algorithm EDBSetup and a protocol Search between the client and server, all fitting the following syntax:

- EDBSetup takes as input a database DB, and outputs a secret key K along with an encrypted database EDB.
- The Search protocol is between a *client* and *server*, where the client takes as input the secret key K and a query q and the server takes as input EDB. At the end, the client outputs a set of identifiers, and the server has no output.

Correctness. We say that an SSE scheme is *correct* if for all inputs DB and queries q, if $(K, \text{EDB}) \xleftarrow{\$} \text{EDBSetup}(\text{DB})$, after running Search with client input (K, q) and server input EDB, the client outputs the set of indices $\text{DB}(q)$.

Adaptive Security of SSE. We recall the semantic security definitions of SSE from [CGKO06, CK10]. The definition is parameterized by a *leakage function* $\mathcal{L}$, which describes what an adversary (the server) is allowed to learn about the database and queries. Formally, security says that the server's view during an adaptive attack (where the server selects the database and queries) can be simulated given only the output of $\mathcal{L}$.

Definition 1. *Let* $\Pi = (\text{EDBSetup}, \text{Search})$ *be an SSE scheme and let* $\mathcal{L}$ *be a stateful algorithm. For algorithms* $\mathcal{A}$ *(denoting the adversary) and* S *(denoting a simulator), we define the experiments (algorithms)* $\mathbf{Real}_{\mathcal{A}}^{\Pi}(\lambda)$ *and* $\mathbf{Ideal}_{\mathcal{A},S}^{\Pi}(\lambda)$ *as follows:*

$\mathbf{Real}_{\mathcal{A}}^{\Pi}(\lambda)$: $\mathcal{A}(1^{\lambda})$ *chooses* DB. *The experiment then runs*

$$(K, \text{EDB}) \leftarrow \text{EDBSetup}(\text{DB}),$$

and gives EDB *to* $\mathcal{A}$. *Then* $\mathcal{A}$ *repeatedly chooses a query* q. *To respond, the game runs the* Search *protocol with client input* (K, q) *and server input* EDB *and gives the transcript and client output to* $\mathcal{A}$. *Eventually* $\mathcal{A}$ *returns a bit that the game uses as its own output.*

Ideal$_{\mathcal{A},S}^{\Pi}(\lambda)$: *The game initializes a counter* cnt $= 0$ *and an empty list* **q**. $\mathcal{A}(1^\lambda)$ *chooses* DB. *The experiment runs* EDB $\leftarrow S(\mathcal{L}(\text{DB}))$ *and gives* EDB *to* $\mathcal{A}$. *Then* $\mathcal{A}$ *repeatedly chooses a query* q. *To respond, the game records this as* **q**$[i]$, *increments* i, *and gives to* $\mathcal{A}$ *the output of* $S(\mathcal{L}(\text{DB}, \mathbf{q}))$. *(Note that here,* **q** *consists of all previous queries in addition to the latest query issued by* $\mathcal{A}$.) *Eventually* $\mathcal{A}$ *returns a bit that the game uses as its own output.*

We say that Π *is* $\mathcal{L}$-*semantically-secure against adaptive attacks if for all adversaries* $\mathcal{A}$ *there exists an algorithm* S *such that*

$$\mid \Pr[\mathbf{Real}_{\mathcal{A}}^{\Pi}(\lambda) = 1] - \Pr[\mathbf{Ideal}_{\mathcal{A},S}^{\Pi}(\lambda) = 1] \mid \leq \text{neg}(\lambda).$$

We note that in the security analysis of our SSE schemes we include the client's output, the set of indices $\text{DB}(\psi(\bar{w}))$, in the adversary's view in the real game, to model the fact that these ind's will be used for retrieval of encrypted record payloads.

Selective Security of SSE. We also consider a weaker version of *selective* security for SSE that is identical to the adaptive security definition except that: (a) in the real world experiment, the adversary $\mathcal{A}$ does not get to choose its queries adaptively, but is required to specify all such queries non-adaptively at the beginning of the protocol along with the plaintext database DB, and receives EDB and the transcript and client output corresponding to each of its queries together at the end of the experiment. Also, (b) in the ideal world experiment, the adversary $\mathcal{A}$ directly receives as output the final response of a non-adaptive simulator S, computed as $S(\mathcal{L}(\text{DB}, \{\mathbf{q}[N]\}_{i \in [Q]}))$, where Q is the total number of queries issued by the adversary $\mathcal{A}$ non-adaptively.

2.3   TSets

We recall the definition of syntax and security for a *tuple set*, or TSet. Intuitively, a TSet allows one to associate a list of fixed-sized data tuples with each attribute-value pair in the database, and later issue related tokens to retrieve these lists. We will use it in our SSE protocols for join queries as an "expanded inverted join index".

TSet Syntax. Formally, a TSet implementation $\Sigma = (\text{TSetSetup}, \text{TSetGetTag}, \text{TSetRetrieve})$ will consist of three algorithms with the following syntax:

- TSetSetup takes as input $\mathbf{T} = (\mathbf{T}_1, \ldots, \mathbf{T}_N)$, where each $\mathbf{T}_i$ for $i \in [N]$ is an array of lists of equal-length bit strings indexed by the elements of W_i, and outputs (TSet, K_T).
- TSetGetTag takes as input the key K_T and a tuple (i, w) and outputs stag_i.
- TSetRetrieve takes as input TSet and stag_i, and returns a list of strings.

TSet Correctness. We say that Σ is *correct* if for all $\{\mathsf{W}_i\}_{i \in [N]}$, all $\mathbf{T} = (\mathbf{T}_1, \ldots, \mathbf{T}_N)$, and any $\mathsf{w} \in \mathsf{W}_i$, we have

$$\text{TSetRetrieve}(\text{TSet}, \text{stag}) = \mathbf{T}_i[\mathsf{w}],$$

when $(\text{TSet}, K_T) \leftarrow \text{TSetSetup}(\mathbf{T})$ and $\text{stag} \leftarrow \text{TSetGetTag}(K_T, (i, \mathsf{w}))$.

Intuitively, $\mathbf{T}$ holds lists of tuples associated with attribute-value pairs and correctness guarantees that the TSetRetrieve algorithm returns the data associated with the given attribute-value pair.

TSet Security. The security goal of a TSet implementation is to hide as much as possible about the tuples in $\mathbf{T} = (\mathbf{T}_1, \dots, \mathbf{T}_N)$ and the attribute-value pairs these tuples are associated to, except for the vectors $\mathbf{T}_i[\mathsf{w}_1], \mathbf{T}_i[\mathsf{w}_2], \dots$ of tuples revealed by the client's queried attribute-value pairs $\mathsf{w}_1, \mathsf{w}_2, \dots$. (For the purpose of TSet implementation we equate client's query with a single attribute-value pair).

The formal definition of security is similar to that of keyword-search based SSE for single-keyword queries. Since the list of tuples associated to searched attribute-value pairs can be viewed as information provided to the server, this information is also provided to the simulator in the security definition below.

We parameterize the TSet security definition with a leakage function $\mathcal{L}_T$ that describes what else the adversary is allowed to learn by looking at the TSet and stag values. For most implementations this leakage will reveal something about the structure of $\mathbf{T}$, and consequently also the structure of DB.

Definition 2. *Let* $\Sigma = ($TSetSetup, TSetGetTag, TSetRetrieve$)$ *be a* TSet *implementation, and let* $\mathcal{A}, S$ *be an adversary and a simulator, and let* $\mathcal{L}_T$ *be a stateful algorithm. We define two games,* $\mathbf{Real}_{\mathcal{A}}^{\Sigma}$ *and* $\mathbf{Ideal}_{\mathcal{A}}^{\Sigma}$ *as follows.*

$\mathbf{Real}_{\mathcal{A}}^{\Sigma}(\lambda)$: $\mathcal{A}(1^{\lambda})$ *outputs* $\{\mathsf{W}_i\}_{i \in [N]}, \mathbf{T} = (\mathbf{T}_1, \dots, \mathbf{T}_N)$ *with the above syntax. The game computes*

$$(\mathsf{TSet}, K_T) \leftarrow \mathsf{TSetSetup}(\mathbf{T}),$$

and gives TSet *to* $\mathcal{A}$. *Then* $\mathcal{A}$ *repeatedly issues queries* $q \in \mathsf{W}$, *and for each* q *the game gives* stag $\leftarrow$ TSetGetTag(K, q) *to* $\mathcal{A}$. *Eventually,* $\mathcal{A}$ *outputs a bit, which the game also uses as its output.*

$\mathbf{Ideal}_{\mathcal{A}, S}^{\Sigma}(\lambda)$: *The game initializes a counter* $i = 0$ *and an empty list* $\mathbf{q}$. $\mathcal{A}(1^{\lambda})$ *outputs* $\{\mathsf{W}_i\}_{i \in [N]}, \mathbf{T} = (\mathbf{T}_1, \dots, \mathbf{T}_N)$ *as above. The game runs* TSet $\leftarrow$ $S(\mathcal{L}_T(\mathbf{T}))$ *and gives* TSet *to* $\mathcal{A}$. *Then* $\mathcal{A}$ *repeatedly issues queries* $q \in \mathsf{W}$, *and for each* q *the game stores* q *in* $\mathbf{q}[i]$, *increments* i, *and gives to* $\mathcal{A}$ *the output of* $S(\mathcal{L}_T(\mathbf{T}, \mathbf{q}), \mathbf{T}[q])$. *Eventually,* $\mathcal{A}$ *outputs a bit, which the game also uses as its output.*

We say that Σ *is a* $\mathcal{L}_T$-*adaptively-secure* TSet *implementation if for all adversaries* $\mathcal{A}$ *there exists an algorithm* S *such that*

$$| \Pr[\mathbf{Real}_{\mathcal{A}}^{\Sigma}(\lambda) = 1] - \Pr[\mathbf{Ideal}_{\mathcal{A}, S}^{\Sigma}(\lambda) = 1] | \leq \mathrm{neg}(\lambda).$$

3   Join Cross-Tags (JXT): SSE for Joins

In this section, we formally describe our new JXT protocol for searching over joins of tables in encrypted relational databases. The JXT protocol consists of two protocols:

- The EDBSetup protocol is a randomized algorithm executed locally at the client. This protocol takes as input the plaintext database and generates the encrypted database EDB, which is to be outsourced to the (untrusted) server. The encrypted database EDB consists of two data structures - the TSet and the XSet. The EDBSetup protocol also generates a secret key K, which is stored locally at the client and is used subsequently to generate query tokens.
- The Search protocol is used to execute queries over joins of encrypted tables in the encrypted database EDB. At a high level, it is a two-party protocol executed jointly by the client and the server, where the client's input is the query to be executed and the server's input is the encrypted database EDB. It consists of a single round of communication (i.e. a query message from the client to the server, followed by a response message from the server to the client). At the end of the protocol, the client is expected to learn the set of record indices (across the two tables) matching the join query.

We now expand in more details on how each of the aforementioned protocols function. In what follows, we assume that: (a) $F : \{0,1\}^\lambda \times \{0,1\}^* \to \{0,1\}^\lambda$ is a family of pseudorandom functions, (b) $\mathsf{SKE} = (\mathsf{Gen}, \mathsf{Enc}, \mathsf{Dec})$ is an IND-CPA secure symmetric-key encryption algorithm with λ-bit keys, and (c) $\Sigma = (\mathsf{TSetSetup}, \mathsf{TSetGetTag}, \mathsf{TSetRetrieve})$ is a secure TSet as defined in Sect. 2.

3.1 The EDBSetup Algorithm of JXT

We now describe the EDBSetup algorithm of JXT. A summary of how the algorithm works appears in Fig. 2.

We note that in JXT, each table Tab is processed independently; so we focus on the processing for a single table. Given a table Tab, let W denote the set of attribute-value pairs across this table Tab. Also, let $\{\mathsf{attr}_t^*\}_{t \in [T]}$ denote the set of T special attributes over which we allow the table Tab to be joined with other tables in the database. We begin by describing how the XSet component of the encrypted database is generated for a given table.

Generating the XSet Table-Wise. For each record with identifier ind in the table Tab, let $\{\mathsf{w}_t^*\}_{t \in [T]}$ denote the set of attribute-value pairs for this record with identifier ind corresponding to the T special "join attributes". For each $t \in [T]$, the EDBSetup algorithm computes the values

$$\mathsf{xind}_t = F(K_I, t, \mathsf{ind}), \quad \mathsf{xw}_t = F(K_W, \mathsf{w}_t^*),$$

where $K_I, K_W \in \{0,1\}^\lambda$ are uniformly sampled keys for the PRF family F. Additionally, the EDBSetup algorithm computes the "cross-tag"

$$\mathsf{xtag}_t = \mathsf{xw}_t + \mathsf{xind}_t,$$

where xw_t and xind_t are as described above. The XSet corresponding to the table Tab is then populated with all such xtag values.

EDBSetup(DB)

1. Sample uniformly random keys $K_I, K_W, K_Z, K_{Z'}, K_{enc}$ for the PRF F and parse the database as $DB = \{Tab_i, W_i\}_{i \in [N]}$.

2. For each table Tab_i, proceed as follows:

    (a) For each pair of record index and join attribute-values of the form $(ind, \{w_t^*\}_{t \in [T]})$ in Tab_i, build the set $XSet[i]$ as follows:

        i. Set the following (for each $t \in [T]$):

        $$xind_t = F(K_I, t, ind) \quad, \quad xw_t = F(K_W, w_t^*) \quad, \quad xtag_t = xw_t + xind_t.$$

        ii. Add the entries $xtag_t$ (one for each $t \in [T]$) to $XSet[i]$.

    (b) For each $w \in W_i$, build the tuple list $\mathbf{T}_i[w]$ as follows:

        i. Set $z_0 = F(K_Z, w \| 0) \quad, \quad z_0' = F(K_{Z'}, w \| 0)$.

        ii. For all $(ind, \{w_t^*\}_{t \in [T]}) \in DB^{Join}_{Tab_i}(w)$ in random order, initialize a counter $cnt = 1$, and proceed as follows:

            A. Set $z_{cnt} = F(K_Z, w \| cnt) \quad, \quad z_{cnt}' = F(K_{Z'}, w \| cnt)$.

            B. For each $t \in [T]$:
                - Set $xind_t = F(K_I, t, ind)$, and $xw_t = F(K_W, w_t^*)$.
                - Set $y_t = xind_t - (z_0 + z_{cnt})$, and $y_t' = xw_t - (z_0' + z_{cnt}')$.

            C. Set $K_{enc,w} = F(K_{enc}, w)$ and compute $ct \leftarrow Enc(K_{enc,w}, ind)$.

            D. Append $(ct, \{y_t, y_t'\}_{t \in [T]})$ to $\mathbf{t}$.

            E. Set $cnt \leftarrow cnt + 1$.

        iii. Set $\mathbf{T}_i[w] = \mathbf{t}$.

3. $(TSet, K_T) \leftarrow TSetSetup(\mathbf{T}_1 \| \ldots \| \mathbf{T}_N)$.

4. Output the key $K = (K_I, K_W, K_Z, K_{Z'}, K_{enc}, K_T)$ and $EDB = (TSet, XSet)$.

Fig. 2. The setup algorithm of Join Cross-Tags (JXT.EDBSetup). We assume that each record index ind in a table Tab_i is appended with the table number i.

Remark. Looking ahead, the XSet is used primarily for membership-testing, hence we can implement this using a Bloom filter to save storage (this is essentially similar to what is done for the XSet in the OXT protocol).

Generating the TSet Table-Wise. We now describe how to generate the TSet component for the table Tab. For each attribute-value pair w in the set W for the table Tab, the EDBSetup algorithm does the following:

– It generates a pair of "padding elements" of the form

$$z_0 = F(K_Z, w \| 0), \quad z_0' = F(K_{Z'}, w \| 0),$$

where K_Z and $K_{Z'}$ are again uniformly sampled keys for the PRF family F.

– Suppose that the attribute-value pair w occurs in a record with identifier ind, and let $\{w_t^*\}_{t \in [T]}$ denote the set of attribute-value pairs for this record with identifier ind corresponding to the T special "join attributes". To each such record, the EDBSetup algorithm assigns a *unique* counter value $cnt \geq 1$ and computes the following additional "padding elements":

$$z_{cnt} = F(K_Z, w \| cnt), \quad z_{cnt}' = F(K_{Z'}, w \| cnt).$$

– In addition, for each $t \in [T]$, the EDBSetup algorithm computes

$$\mathsf{xind}_t = F(K_I, t, \mathsf{ind}), \quad \mathsf{xw}_t = F(K_W, \mathsf{w}_t^*),$$

$$y_t = \mathsf{xind}_t - (z_0 + z_{\mathsf{cnt}}), \quad y_t' = \mathsf{xw}_t - (z_0' + z_{\mathsf{cnt}}').$$

Remark. Note that xind_t and xw_t are generated identically as in the computation of the TSet. In fact, while we duplicate the generation of these elements for ease of understanding, in a real execution of the algorithm, these values can be generated exactly once and re-used for the generation of the XSet and the TSet.

– Finally, the EDBSetup algorithm computes the randomized index ct for the index ind as

$$K_{\mathsf{enc},\mathsf{w}} = F(K_{\mathsf{enc}}, \mathsf{w}), \quad \mathsf{ct} = \mathsf{Enc}(K_{\mathsf{enc},\mathsf{w}}, \mathsf{ind}),$$

where K_{enc} is again a uniformly sampled key for the PRF family F.

Overall, the TSet entry corresponding to the attribute-value pair w consists of an entry corresponding to each record ind containing w, where each such entry is a tuple of the form $(\mathsf{ct}, \{y_t, y_t'\}_{t \in [T]})$, generated as described above. The actual TSet is then generated using the secure T-Set implementation Σ.

3.2 The Search Protocol of JXT

We now describe how the Search protocol works on a join query of the form $q = \left(i, j, \mathsf{w}^{(1)}, \mathsf{w}^{(2)}, \mathsf{attr}_{i,j}^*\right)$, which essentially denotes a query over the join of the tables Tab_i and Tab_j, where the join is computed with respect to the special attribute $\mathsf{attr}_{i,j}^*$, which is a designated "join attribute" for both tables Tab_i and Tab_j. A concise summary of how the protocol works appears in Fig. 3.

At a high level, the search protocol can be divided into three parts:

– [**Round-1 (Client→Server)**]: The client generates a "query message" and sends it across to the server.
– [**Round-2 (Server→Client)**]: The server generates a "response message" and sends it across to the client.
– [**Local Computation (Client)**]: The client performs some local computation to retrieve the final set of record identifiers matching the query.

We describe how each of these parts work.

[**Round-1:**] **Query Message (Client→Server).** The client sends to the server the table indices i and j, along with the join attribute $\mathsf{attr}_{i,j}^*$ over which the query is to be executed. The client also sends to the server the stag values $\mathsf{stag}^{(1)}$ and

---

Search protocol

1. The client has input the key K and a join query $q = (i, j, \mathsf{w}^{(1)}, \mathsf{w}^{(2)}, \mathsf{attr}_{i,j}^*)$, and proceeds as follows:

   − Send to the server $(i, j, \mathsf{attr}_{i,j}^*)$. Locally compute and store

   $$K_{\mathrm{enc},\mathsf{w}^{(1)}} = F(K_{\mathrm{enc}}, \mathsf{w}^{(1)}), \quad K_{\mathrm{enc},\mathsf{w}^{(2)}} = F(K_{\mathrm{enc}}, \mathsf{w}^{(2)}).$$

   − Send to the server $(\mathsf{stag}^{(1)}, \mathsf{stag}^{(2)})$, where

   $$\mathsf{stag}^{(1)} \leftarrow \mathsf{TSetGetTag}(K_T, (i, \mathsf{w}^{(1)})), \quad \mathsf{stag}^{(2)} \leftarrow \mathsf{TSetGetTag}(K_T, (j, \mathsf{w}^{(2)})).$$

   − For $\mathsf{cnt}^{(1)} = 1, 2 \ldots$ and until server sends $\mathsf{stop}^{(1)}$, send to the server

   $$\mathsf{xjointoken}^{(1)}[\mathsf{cnt}^{(1)}] = F(K_{Z'}, \mathsf{w}^{(1)} \| \mathsf{cnt}^{(1)}) + F(K_Z, \mathsf{w}^{(2)} \| 0).$$

   − For $\mathsf{cnt}^{(2)} = 1, 2 \ldots$ and until server sends $\mathsf{stop}^{(2)}$, send to the server

   $$\mathsf{xjointoken}^{(2)}[\mathsf{cnt}^{(2)}] = F(K_{Z'}, \mathsf{w}^{(1)} \| 0) + F(K_Z, \mathsf{w}^{(2)} \| \mathsf{cnt}^{(2)}).$$

2. The server has input $(\mathsf{TSet}, \mathsf{XSet})$ and responds to the messages from the client as follows.

   (a) It sets:

   $$\mathsf{t}^{(1)} \leftarrow \mathsf{TSetRetrieve}(\mathsf{TSet}, (i, \mathsf{stag}^{(1)})), \quad \mathsf{t}^{(2)} \leftarrow \mathsf{TSetRetrieve}(\mathsf{TSet}, (j, \mathsf{stag}^{(2)})).$$

   (b) For $\mathsf{cnt}^{(1)} = 1, \ldots, |\mathsf{t}^{(1)}|$, the server does the following:

       i. Retrieve $(\mathsf{ct}^{(1)}, \{y_t^{(1)}, y'^{(1)}_t\}_{t \in [T]})$ from the $\mathsf{cnt}^{(1)}$-th tuple in $\mathsf{t}^{(1)}$. Let $y'^{(1)}_{t^*}$ be the entry from among $\{y'^{(1)}_t\}_{t \in [T]}$ corresponding to the attribute $\mathsf{attr}_{i,j}^*$.

       ii. Set $\mathsf{xtoken}_{t^*}^{(1)} = \mathsf{xjointoken}^{(1)}[\mathsf{cnt}^{(1)}] + y'^{(1)}_{t^*}$.

   (c) When last tuple in $\mathsf{t}^{(1)}$ is reached, send $\mathsf{stop}^{(1)}$ to the client.

   (d) for $\mathsf{cnt}^{(2)} = 1, \ldots, |\mathsf{t}^{(2)}|$:

       i. Retrieve $(\mathsf{ct}^{(2)}, \{y_t^{(2)}, y'^{(2)}_t\}_{t \in [T]})$ from the $\mathsf{cnt}^{(2)}$-th tuple in $\mathsf{t}^{(2)}$. Let $y_{t^*}^{(2)}$ be the entry from among $\{y_t^{(2)}\}_{t \in [T]}$ corresponding to the attribute $\mathsf{attr}_{i,j}^*$.

       ii. Set $\mathsf{xtoken}_{t^*}^{(2)} = \mathsf{xjointoken}^{(2)}[\mathsf{cnt}^{(2)}] + y_{t^*}^{(2)}$.

   (e) Send $\mathsf{stop}^{(2)}$ to the client.

   (f) For $\mathsf{cnt}^{(1)} = 1, \ldots, |\mathsf{t}^{(1)}|$ and $\mathsf{cnt}^{(2)} = 1, \ldots, |\mathsf{t}^{(2)}|$:
   If $(\mathsf{xtoken}_{t^*}^{(1)} + \mathsf{xtoken}_{t^*}^{(2)}) \in \mathsf{XSet}[j]$, then send $(\mathsf{ct}^{(1)}, \mathsf{ct}^{(2)})$ to the client.

3. For each $(\mathsf{ct}^{(1)}, \mathsf{ct}^{(2)})$ received from the server, the client recovers and outputs:

$$\mathsf{ind}^{(1)} = \mathsf{Dec}(K_{\mathrm{enc},\mathsf{w}^{(1)}}, \mathsf{ct}^{(1)}), \quad \mathsf{ind}^{(2)} = \mathsf{Dec}(K_{\mathrm{enc},\mathsf{w}^{(2)}}, \mathsf{ct}^{(2)}),$$

Fig. 3. The search protocol of Join Cross-Tags (JXT.Search).

$\mathsf{stag}^{(2)}$, which allow the server to recover the TSet entries corresponding to the attribute value pairs $\mathsf{w}^{(1)}$ and $\mathsf{w}^{(2)}$, respectively. In addition, corresponding to each TSet entry for the attribute value pairs $\mathsf{w}^{(1)}$ and $\mathsf{w}^{(2)}$, the client sends across to the server a sequence of terms of the form

$$\mathsf{xjointoken}^{(1)}[1], \mathsf{xjointoken}^{(1)}[2], \ldots$$

$$\mathsf{xjointoken}^{(2)}[1], \mathsf{xjointoken}^{(1)}[2], \ldots$$

until the server sends the signals $\mathsf{stop}^{(1)}$ and $\mathsf{stop}^{(2)}$, respectively, indicating that there are no more TSet entries to process for either attribute-value pair. These terms are generated as follows: for a given counter value $\mathsf{cnt}^{(1)} \in \{1, 2, \ldots\}$, the term $\mathsf{xjointoken}^{(1)}[\mathsf{cnt}^{(1)}]$ is generated as:

$$\mathsf{xjointoken}^{(1)}[\mathsf{cnt}^{(1)}] = F(K_{Z'}, \mathsf{w}^{(1)} \| \mathsf{cnt}^{(1)}) + F(K_Z, \mathsf{w}^{(2)} \| 0).$$

Similarly, for a given counter value $\mathsf{cnt}^{(2)} \in \{1, 2, \ldots\}$, the term $\mathsf{xjointoken}^{(2)}[\mathsf{cnt}^{(2)}]$ is generated as:

$$\mathsf{xjointoken}^{(2)}[\mathsf{cnt}^{(2)}] = F(K_{Z'}, \mathsf{w}^{(1)} \| 0) + F(K_Z, \mathsf{w}^{(2)} \| \mathsf{cnt}^{(2)}).$$

[Round-2:] Response Message (Server → Client). The server uses the stag values sent across by the client to recover the TSet entries corresponding to the attribute-value pairs $\mathsf{w}^{(1)}$ and $\mathsf{w}^{(2)}$. More specifically:

- The server uses $\mathsf{stag}^{(1)}$ (received from the client as part of the first round message) to recover the TSet entries corresponding to the attribute-value pair $\mathsf{w}^{(1)}$ from the T-Set corresponding to table Tab_i. Suppose that each such entry is a tuple of the form

$$(\mathsf{ct}^{(1)}, \{y_t^{(1)}, {y'}_t^{(1)}\}_{t \in [T]}).$$

Also, let ${y'}_{t^*}^{(1)}$ be the entry from among $\{{y'}_t^{(1)}\}_{t \in [T]}$ corresponding to the attribute $\mathsf{attr}_{i,j}^*$ over which the query is being executed. The server computes

$$\mathsf{xtoken}_{t^*}^{(1)} = \mathsf{xjointoken}^{(1)}[\mathsf{cnt}^{(1)}] + {y'}_{t^*}^{(1)}.$$

- Similarly, the server uses $\mathsf{stag}^{(2)}$ (also received from the client as part of the first round message) to recover the TSet entries corresponding to the attribute-value pair $\mathsf{w}^{(2)}$ from the T-Set corresponding to table Tab_j. Suppose that each such entry is a tuple of the form

$$(\mathsf{ct}^{(2)}, \{y_t^{(2)}, {y'}_t^{(2)}\}_{t \in [T]}).$$

Again, let $y_{t^*}^{(2)}$ be the entry from among $\{y_t^{(2)}\}_{t \in [T]}$ corresponding to the attribute $\mathsf{attr}_{i,j}^*$ over which the query is being executed. The server computes

$$\mathsf{xtoken}_{t^*}^{(2)} = \mathsf{xjointoken}^{(2)}[\mathsf{cnt}^{(2)}] + y_{t^*}^{(2)}.$$

Now, for each such pair of TSet entries (where the first entry corresponds to the attribute-value pair $w^{(1)}$ from the T-Set for table Tab_i, and the second corresponds to the attribute-value pair $w^{(2)}$ from the T-Set for table Tab_j), the server computes a candidate xtag value of the form

$$\text{xtag}^{(1,2)} = \text{xtoken}_{t^*}^{(1)} + \text{xtoken}_{t^*}^{(2)},$$

and checks the membership of $\text{xtag}^{(1,2)}$ in the XSet corresponding to Tab_j.

- If the membership-test returns **true**, then the server infers that the pair of records match constitute a matching record in the join of the two tables; hence it sends back the corresponding randomized identifiers $(\text{ct}^{(1)}, \text{ct}^{(2)})$ to the client.
- If the membership test returns **false**, then the server discards the corresponding randomized identifiers.

Local Computation (Client): Finally, the client decrypts the set of randomized record identifiers (i.e., the tuples of the form $(\text{ct}^{(1)}, \text{ct}^{(2)})$) sent across by the client, and decrypts them to retrieve the set of plaintext record identifiers corresponding to the records matching the query q.

Realizing TSet and XSet. We note here that an implementation of JXT can use the same selectively/adaptively secure implementations of TSet (built from purely symmetric-key cryptographic primitives) as used by OXT [CJJ+13]. We also note here that one could equivalently use an encrypted multi-map (EMM) [CK10, KM17, KM19] instead of a TSet in JXT. Also note that during the search protocol, the server uses the XSet purely for membership-testing. This allows implementing the XSet using a Bloom filter, as in OXT. These observations provide evidence for the overall compatibility of JXT with OXT.

Correctness. We now formally argue that the JXT protocol is correct. More concretely, we state and prove the following theorem:

Theorem 1. *Assuming that* SKE *satisfies decryption correctness and Σ is a correct* TSet *implementation, the* JXT *protocol satisfies correctness.*

Proof. Consider a query of the form $q = (i, j, w^{(1)}, w^{(2)}, \text{attr}_{i,j}^*)$, and suppose that there exists an index-pair $(\text{ind}_1, \text{ind}_2) \in DB(q)$. We now argue that the client recovers $(\text{ind}_1, \text{ind}_2)$ as an outcome of the Search protocol. To see this, observe the following. Since the aforementioned conditions hold true, the server must retrieve the following TSet entries corresponding to $w^{(1)}$ and $w^{(2)}$ (this follows from the correctness of the TSet implementation σ):

$$(\text{ct}^{(1)}, \{y_t^{(1)}, y_t'^{(1)}\}_{t \in [T]}), \quad (\text{ct}^{(2)}, \{y_t^{(2)}, y_t'^{(2)}\}_{t \in [T]}),$$

where $\text{ct}^{(1)} = \text{Enc}(K_{\text{enc,w}}, \text{ind}_1)$ and $\text{ct}^{(2)} = \text{Enc}(K_{\text{enc,w}}, \text{ind}_2)$, and letting $y_{t^*}'^{(1)}$ and $y_{t^*}^{(2)}$ be the respective entries corresponding to the attribute $\text{attr}_{i,j}^*$,

$$y_{t^*}'^{(1)} = F(K^{(2)}, \langle \text{attr}_{i,j}^*, \gamma \rangle) - (F(K_{Z'}, w^{(1)} \| 0) + F(K_{Z'}, w^{(1)} \| \text{cnt}^{(1)})),$$

$$y_{t^*}^{(2)} = F(K_I, t^*\mathsf{ind}_2) - (F(K_Z, \mathsf{w}^{(2)} \parallel 0) + F(K_Z, \mathsf{w}^{(2)} \parallel \mathsf{cnt}^{(2)})),$$

for some appropriate counter values $\mathsf{cnt}^{(1)}$ and $\mathsf{cnt}^{(2)}$. In addition, the client sends across to the server the values $\mathsf{xjointoken}^{(1)}[\mathsf{cnt}^{(1)}]$ and $\mathsf{xjointoken}^{(2)}[\mathsf{cnt}^{(2)}]$ where

$$\mathsf{xjointoken}^{(1)}[\mathsf{cnt}^{(1)}] = F(K_{Z'}, \mathsf{w}^{(1)} \parallel \mathsf{cnt}^{(1)}) + F(K_Z, \mathsf{w}^{(2)} \parallel 0),$$

$$\mathsf{xjointoken}^{(2)}[\mathsf{cnt}^{(2)}] = F(K_{Z'}, \mathsf{w}^{(1)} \parallel 0) + F(K_Z, \mathsf{w}^{(2)} \parallel \mathsf{cnt}^{(2)}).$$

Consequently, as per the Search protocol, the server computes

$$\mathsf{xtoken}_{t^*}^{(1)} = F(K^{(2)}, \langle \mathsf{attr}_{i,j}^*, \gamma \rangle) - F(K_{Z'}, \mathsf{w}^{(1)} \parallel 0) + F(K_Z, \mathsf{w}^{(2)} \parallel 0),$$

and

$$\mathsf{xtoken}_{t^*}^{(2)} = F(K_I, t^*\mathsf{ind}_2) - F(K_Z, \mathsf{w}^{(2)} \parallel 0) + F(K_{Z'}, \mathsf{w}^{(1)} \parallel 0).$$

Next, the server computes the candidate xtag as

$$\mathsf{xtag}^{(1,2)} = \mathsf{xtoken}_{t^*}^{(1)} + \mathsf{xtoken}_{t^*}^{(2)} = F(K_I, t^*\mathsf{ind}_2) + F(K^{(2)}, \langle \mathsf{attr}_{i,j}^*, \gamma \rangle).$$

Note that this is nothing but the xtag corresponding to the index-attribute value pair $(\mathsf{ind}_2, \mathsf{w}^* = \langle \mathsf{attr}_{i,j}^*, \gamma \rangle)$. Finally, assuming that the symmetric-key encryption scheme SKE satisfies correctness of decryption, the client recovers the index-pair $(\mathsf{ind}_1, \mathsf{ind}_2)$. A similar argument can be used to show that the client does not recover any index-pair $(\mathsf{ind}_1', \mathsf{ind}_2') \notin \mathsf{DB}(q)$. This completes the proof of correctness for the JXT protocol.

On Bloom Filter and False Positives. We point out that using a Bloom filter to realize the XSet data structure potentially introduces false positives. The rate of such false positives can be programmed by setting the parameters of the Bloom filter, thus yielding a tradeoff between storage and false positive rate. As an alternative to Bloom filter, one could use any non-lossy data structure that allows checking for set-membership. This would prevent false positives, albeit at the cost of some extra storage at the server end.

4   Complexity Analysis of JXT

In this section, we analyze the asymptotic complexity of JXT.

Storage Overhead. We first discuss the storage overhead for each table. Recall that in JXT, the TSet and XSet for the encrypted database are built table-wise. Hence, the total storage overhead for JXT is essentially the sum of the overheads for each individual table. Suppose that a table Tab has a total of n attributes, with $T \leq n$ amongst these being "join attributes"; i.e. attributes over which the table can be joined with other tables in the database. Also, assume that Tab has a total of m records (equivalently, rows). We enumerate below the number of entries in the TSet and XSet corresponding to Tab.

Recall that for each attribute-value pair w in the set W for the table Tab, the TSet stores as many entries as the number of records containing the attribute-value pair w, where each such entry is a tuple of the form: $(ct, \{y_t, y'_t\}_{t \in [T]})$. In other words, each entry is a collection of $(2T + 1)$ objects. Hence, the total number of TSet entries for Tab is $\sum_{w \in W}(2T + 1)|DB_{Tab}(w)|$. But note that $\sum_{w \in W}|DB_{Tab}(w)| = m \cdot n$, where m and n are the total number of records and attributes in the table Tab, respectively. Hence, $|TSet(Tab)| = m \cdot n \cdot (2T + 1)$. In other words, the TSet incurs an $O(T)$-fold overhead over the storage required for the plaintext table Tab.

Next, recall that the XSet for the table Tab has T entries corresponding to each record index ind. More concretely, for each record with identifier ind in the table Tab, let $\{w^*_t\}_{t \in [T]}$ denote the set of attribute-value pairs for this record with identifier ind corresponding to the T special "join attributes". Then, for each $t \in [T]$, the EDBSetup algorithm stores a unique $xtag_t$ entry corresponding to the pair (ind, w^*_t). Thus, we have $|XSet(Tab)| = m \cdot T$.

We note, however, that the XSet is implemented using a Bloom filter; consequently, the storage overhead for the XSet is significantly lower. As in OXT, we expect the overhead for the XSet in JXT to be low enough for the server to be able to store it in the RAM. The TSet will typically be stored on the disk.

Computational and Communication Overheads. We now present an asymptotic analysis for the computational and communication overheads when executing a search query over the joins of two tables Tab_1 and Tab_2. Suppose that the query involves two attribute-value pairs w_1 and w_2, and is to be executed over the join of Tab_1 and Tab_2 w.r.t. the attribute $attr^*$.

Computational Overhead (Client). The client computes the stag values corresponding to w_1 and w_2 using $O(1)$ invocations of the stag-generation algorithm for the TSet (the exact overhead depends on the implementation of the TSet; however, for efficient implementations such as the one for OXT [CJJ+13], this is a constant overhead). The client also computes xjointoken$^{(1)}$ and xjointoken$^{(2)}$ values; the number of such computations is $|DB_{Tab_1}(w_1)| + |DB_{Tab_2}(w_2)|$. Hence, the comp. overhead is $O(|DB_{Tab_1}(w_1)| + |DB_{Tab_2}(w_2)|)$.

Computational Overhead (Server). The server's computation can be broadly divided into two categories: (a) TSet lookups (using the stag values sent across by the client), and (b) xtag computations and membership-checks. The total number of TSet lookups performed by the server corresponding to w_1 and w_2 is again $DB_{Tab_1}(w_1)| + |DB_{Tab_2}(w_2)$. The number of xtag computations is larger; in particular, the server computes (and checks membership of) a candidate xtag entry corresponding to each pair $(xjointoken^{(1)}[cnt^{(1)}], xjointoken^{(2)}[cnt^{(2)}])$. Hence, the computational overhead at the server is $O(|DB_{Tab_1}(w_1)| \cdot |DB_{Tab_2}(w_2)|)$. Note that this computational overhead is unavoidable since in the worst case, we have $|DB(q)| = |DB_{Tab_1}(w_1)| \cdot |DB_{Tab_2}(w_2)|$, and the server must perform at least as much computation as is required to compute and send across to the client the final result set pertaining to the join query.

Communication Overhead. The message from the client to the server consists of $O(\mathsf{DB}_{\mathsf{Tab}_1}(\mathsf{w}_1)| + \mathsf{DB}_{\mathsf{Tab}_2}(\mathsf{w}_2)|)$ terms, while the message from the server to the client consists of $|\mathsf{DB}(q)|$ terms. Hence, the overall communication complexity is $O(|\mathsf{DB}_{\mathsf{Tab}_1}(\mathsf{w}_1)| + |\mathsf{DB}_{\mathsf{Tab}_2}(\mathsf{w}_2)| + |\mathsf{DB}(q)|)$.

Bloom Filter Configuration. We note here that the configuration of the Bloom filter used is expected to influence the performance of an actual implementation of JXT. We propose using Bloom filter configurations similar to those used in implementations of OXT reported in prior work [CJJ+13, CJJ+14].

5   Leakage Profile and Security of JXT

In this section, we formally describe the leakage profile of our JXT protocol(i.e. leakage to the server) for join queries, and prove its security with respect to this leakage profile.

5.1   The Leakage Profile of JXT

We represent a sequence of Q join queries by $\mathbf{q} = (\mathbf{i}_1, \mathbf{i}_2, \mathbf{s}_1, \mathbf{s}_2, \mathbf{attr}^*)$, where an individual join query is represented (as per the definition of join queries introduced in Sect. 2) as a five-tuple $\mathbf{q}[\ell] = (\mathbf{i}_1[\ell], \mathbf{i}_2[\ell], \mathbf{s}_1[\ell], \mathbf{s}_2[\ell], \mathbf{attr}^*[\ell])$. The leakage profile of JXT for such a sequence of join queries is a tuple of the form $\mathcal{L} = (\mathbf{n}, \mathsf{RP}, \mathsf{EP}_1, \mathsf{EP}_2, \mathsf{SP}_1, \mathsf{SP}_2, \mathsf{JD}, \mathsf{IP})$ where:

- $\mathbf{n}$ is an N-sized list, where for each $i \in [N]$, $\mathbf{n}[i]$ represents the total number of occurrences of all attribute-value pairs in W_i across records in table Tab_i.
- RP is the result pattern leakage, i.e. the set of records matching each query. Formally, we represent RP as a Q-sized list, where for each $\ell \in [Q]$, we have $\mathsf{RP}[\ell] = \mathsf{DB}(\mathbf{q}[\ell])$. Here, $\mathsf{DB}(q)$ for $q = \mathbf{q}[\ell]$ is as defined in Sect. 2.
- EP_1 is the equality pattern over $\mathbf{s}_1$ indicating which queries have equal attribute-value pairs in the *first* coordinate. Formally, we represent EP_1 as a $Q \times Q$ table with entries in $\{0, 1\}$, where $\mathsf{EP}_1[\ell, \ell'] = 1$ if $\mathbf{s}_1[\ell] = \mathbf{s}_1[\ell']$, and 0 otherwise.
- Similarly, EP_2 is the equality pattern over $\mathbf{s}_2$ indicating which queries have equal attribute-value pairs in the *second* coordinate. Formally, we represent EP_2 as a $Q \times Q$ table with entries in $\{0, 1\}$, where $\mathsf{EP}_2[\ell, \ell'] = 1$ if $\mathbf{s}_2[\ell] = \mathbf{s}_2[\ell']$, and 0 otherwise.
- SP_1 is the size pattern over $\mathbf{s}_1$, i.e. the number of records matching the *first* attribute-value pair in each join query. Formally, we represent SP_1 as a Q-sized list, where for each $\ell \in [Q]$, we have $\mathsf{SP}_1[\ell] = |\mathsf{DB}_{\mathsf{Tab}_{\mathbf{i}_1[\ell]}}(\mathbf{s}_1[\ell])|$.
- Similarly, SP_2 is the size pattern over $\mathbf{s}_2$, i.e. the number of records matching the *second* attribute-value pair in each join query. Formally, we represent SP_2 as a Q-sized list, where for each $\ell \in [Q]$, we have $\mathsf{SP}_2[\ell] = |\mathsf{DB}_{\mathsf{Tab}_{\mathbf{i}_2[\ell]}}(\mathbf{s}_2[\ell])|$.

- JD is the join attribute distribution pattern over s_1, which is represented as a collection of Q multi-sets. The ℓth entry in this collection, i.e., $\mathsf{JD}[\ell]$ is a multi-set of (global) randomized encodings of the join attribute values corresponding to the join attribute $\mathbf{attr}^*[\ell]$ in the records matching the attribute-value pair $s_1[\ell]$ in the table $\mathsf{Tab}_{i_1[\ell]}$. More formally, for each $\ell \in [Q]$, we have the multi-set[8]

$$\mathsf{JD}[\ell] = \{\mathsf{encode}(\mathsf{val}^*) : (\mathsf{ind}, s_1[\ell]) \in \mathsf{Tab}_{i_1[\ell]} \text{ and } (\mathsf{ind}, \langle \mathbf{attr}^*[\ell], \mathsf{val}^* \rangle) \in \mathsf{Tab}_{i_1[\ell]}\}.$$

- IP is the conditional intersection pattern, which is a $Q \times Q$ table with entries defined as explained next. For each $\ell, \ell' \in [Q]$, $\mathsf{IP}[\ell, \ell']$ is an empty set if one of the following conditions holds:

  - $(\mathbf{i}_1[\ell], \mathbf{i}_2[\ell], \mathbf{attr}^*[\ell]) \neq (\mathbf{i}_1[\ell'], \mathbf{i}_2[\ell'], \mathbf{attr}^*[\ell'])$.
  - $\mathsf{JD}[\ell] \cap \mathsf{JD}[\ell']$ is empty.

  Otherwise, $\mathsf{IP}[\ell, \ell']$ is non-empty, and is defined as the intersection of all record identifiers matching the keywords $s_2[\ell]$ and $s_2[\ell']$ in the table $\mathsf{Tab}_{i_2[\ell]}$. More formally, we have $\mathsf{IP}[\ell, \ell'] = \mathsf{DB}_{\mathsf{Tab}_{i_2[\ell]}}(s_2[\ell]) \cap \mathsf{DB}_{\mathsf{Tab}_{i_2[\ell]}}(s_2[\ell'])$.

5.2 Discussion on Leakage Components and Comparison with OXT

In this section, we present a discussion on the various components in the leakage profile of JXT, and also compare the same with OXT. Note that one fundamental difference between JXT and OXT is that while JXT supports queries over joins of multiple tables, OXT only supports "unilateral queries", where each such query is defined over a single table. This difference manifests in subtle distinctions between the leakage profiles for JXT and OXT, as described below.

The n-Leakage. Suppose that a table Tab_i has a total of n_i attributes (equivalently, columns) and a total of m_i records (equivalently, rows). We note here that $\mathbf{n}[i]$ is nothing but $n_i \cdot m_i$, i.e., the total number of entries in the table. We note that this information (or an upper bound thereof) is leaked by almost all existing SSE schemes in the literature with efficient search capabilities [CGKO06, CJJ+13, CJJ+14, LPS+18], including OXT.

Result Pattern. The RP leakage allows the server to learn the set of identifiers corresponding to records in the result set for the query. Such a leakage is considered benign and is incurred by nearly all existing SSE schemes (notably [CGKO06, CK10, CJJ+13, CJJ+14]), including OXT. However, one subtle difference with OXT is that in JXT, the RP leakage spans across multiple tables, while in OXT, the RP leakage is confined to a single table. This, of course, is a direct consequence of the fact that JXT handles queries over the join of multiple tables, which OXT does not support.

Remark. We note here that our analysis of the result pattern leakage of JXT is rather conservative; an astute reader may observe that during the Search

---

[8] Note that a multi-set additionally reveals the frequency of each entry.

protocol in JXT, the server does not learn the actual plaintext identifiers for the records in the result set; it only learns the randomized/encrypted versions of these identifiers, which are then locally decrypted at the client.

Equality and Size Patterns. The EP leakage reveals repetitions of attribute-value pairs across join queries (including the "coordinate" of the join query where the repetition occurs), while the SP leakage reveals the individual frequency of each attribute-value pair in a given join query. The EP leakage can be mitigated by having more than one TSet entry per attribute-value pair, and the client using stag values that point to different entries for the same attribute-value pair across multiple queries, while the SP leakage can be potentially mitigated by artificially "padding" the number of TSet entries corresponding to each attribute-value pair; this would leak an upper bound rather than the exact frequency.

The EP and SP leakage can be viewed as consequences of our strategy of avoiding join pre-computations in the setup phase (and the corresponding blowup in storage overheads); since JXT processes each table individually rather than pre-computing their joins, processing a query over the join of two tables inevitably requires some independent searches over the individual TSet entries for each table. In particular, the EP and SP leakage of JXT are conceptually similar to the EP and SP leakage in OXT, albeit with the difference that in OXT, a unilateral search query over two conjuncts (referred to in OXT as the s-term and the x-term) incurs these leakage for only one of the terms (the s-term). We view the additional leakage in JXT as a necessary trade-off for the additional capability of handling join queries (or more concretely, existential quantifiers over the join attributes) with comparable efficiency.

Join Attribute Distribution Pattern. The JD leakage is new to JXT and is a direct consequence of the fact that it handles queries over joins of tables. For a given query over the join of two tables with respect to a common attribute (say attr*), it reveals the frequency distribution of values taken by attr* across records matching the attribute-value pair in the "first slot". The extent of this leakage depends on the "entropy" of the join attribute in the first table. For example, consider the case where the join-attribute is a primary key or a "high-entropy" key in the first table. In this case, it is likely to take a unique value for each record, and hence the JD leakage is essentially query-invariant (as each possible value occurs with frequency close to 1), and hence, benign. Thus the JD leakage can be mitigated by planning join queries (i.e., by ordering the attribute-value pairs in the "first" and "second" slots) such that join attribute is a primary/"high-entropy" key in the first table.

Conditional Intersection Pattern. The IP leakage of JXT is quite subtle; for a pair of queries over the join of the *same* tables over the *same* common attribute attr*, it reveals the intersection of records matching the attribute-value pairs in the "second slot" provided that the attribute-value pairs in the "first slot" have matching records with identical ⟨attr*, val*⟩ entries. This leakage is conditioned on the fact that the attribute-value pairs in the "first slot" have such matching records; if such matching records do not exist, then this leakage is empty.

We note that the IP leakage is essentially guaranteed to be empty in either of the following scenarios: (a) the join-attribute is a primary key or a "high-entropy" key in the first table, in which case, it is likely to take a unique value for each record, or (b) the attribute-value pairs in the "second slot" share the same attribute but different values, in which case, they cannot match with the same record. In particular, similar to the JD leakage, the IP leakage can also be mitigated by planning join queries such that join attribute is a primary/"high-entropy" key in the first table. This bears similarities with the IP leakage of OXT, where the leakage can be minimized by re-arranging the conjuncts in each unilateral query such that the s-term has low frequency.

To summarize, the overall leakage profile of JXT bears many similarities with the leakage profile of OXT, and can be made benign in practice by simple query-planning strategies that do not compromise on practical search performance.

5.3 The Security Theorems for JXT

In this section, we state the theorems for the selective and adaptive security of JXT.

Selective Security of JXT. Let $\mathcal{L}$ be the leakage profile of JXT as described in Sect. 5.1. We state the following theorem.

Theorem 2. *Assuming that F is a secure PRF family, SKE is an IND-CPA secure symmetric-key encryption scheme, and Σ is a n-selectively secure TSet implementation, the JXT protocol is $\mathcal{L}$-semantically simulation-secure against selective attacks.*

Proof Overview. We defer a detailed proof of this theorem to the full version of our paper [JP21]. The proof of selective security proceeds via a sequence of games between a simulator S and an adversary $\mathcal{A}$, where the first game is identical to the "ideal-world" game played between the simulator and the adversary $\mathcal{A}$ as described in Definition 1, while the final game is identical to the "real-world" game played between the simulator S and the adversary $\mathcal{A}$. We establish formally that the view of the adversary $\mathcal{A}$ in each pair of consecutive games is computationally indistinguishable.

The crux of our selective security proof is that the simulator for JXT can initialize the XSet to uniformly random elements, and program the outputs of the PRFs accordingly while making sure that the search tokens corresponding to a given join query are generated in a consistent manner. The programming is done given the leakage of JXT corresponding to the various search queries. Additionally, the simulator for JXT can directly invoke the simulator for the selectively secure TSet to simulate the TSet entries at setup and the corresponding stag values during searches. We refer to the full version of our paper [JP21] for the detailed description of the simulator, as well as for descriptions of the hybrids that allows us to prove the indistinguishability of this simulation from a real execution of JXT.

Adaptive Security of JXT. Again, let $\mathcal{L}$ be the leakage profile of JXT as described in Sect. 5.1.

Theorem 3. *Assuming that F is a secure PRF family, SKE is an IND-CPA secure symmetric-key encryption scheme, and Σ is a n-adaptively secure TSet implementation, the JXT protocol is $\mathcal{L}$-semantically simulation-secure against adaptive attacks.*

Proof Overview. We again defer a detailed proof of this theorem to the full version of our paper [JP21]. In our adaptive proof of security, we assume an instantiation of adaptively secure TSet in the standard model. While the original construction of TSet [CJJ+13] requires random oracles for adaptive security, the authors of [CJJ+13] also discuss an alternative instantiation of adaptively secure TSets in the standard model without incurring additional rounds of communication. The idea is to send the actual addresses in the TSet (i.e., the outputs of PRF evaluation) directly to the server instead of sending the PRF key (or the stag), and allowing the server to compute PRF outputs on its own. In the context of our JXT protocol, using the standard model instantiation of TSet increases the communication overhead for the TSet component, but not the (asymptotic) communication overhead for the overall search protocol.

Acknowledgments. We thank the anonymous reviewers of IACR ASIACRYPT 2022 for their helpful comments and suggestions.

References

[BKM20] Blackstone, L., Kamara, S., Moataz, T.: Revisiting leakage abuse attacks. In: NDSS 2020 (2020)

[CGKO06] Curtmola, R., Garay, J.A., Kamara, S., Ostrovsky, R.: Searchable symmetric encryption: improved definitions and efficient constructions. In: ACM CCS 2006, pp. 79–88 (2006)

[CGPR15] Cash, D., Grubbs, P., Perry, J., Ristenpart, T.: Leakage-abuse attacks against searchable encryption. In: ACM CCS 2015, pp. 668–679 (2015)

[CJJ+13] Cash, D., Jarecki, S., Jutla, C., Krawczyk, H., Roşu, M.-C., Steiner, M.: Highly-scalable searchable symmetric encryption with support for Boolean queries. In: Canetti, R., Garay, J.A. (eds.) CRYPTO 2013. LNCS, vol. 8042, pp. 353–373. Springer, Heidelberg (2013). https://doi.org/10.1007/978-3-642-40041-4_20

[CJJ+14] Cash, D., et al.: Dynamic searchable encryption in very-large databases: data structures and implementation. In: NDSS 2014 (2014)

[CK10]   Chase, M., Kamara, S.: Structured encryption and controlled disclosure. In: Abe, M. (ed.) ASIACRYPT 2010. LNCS, vol. 6477, pp. 577–594. Springer, Heidelberg (2010). https://doi.org/10.1007/978-3-642-17373-8_33

[CNR21]   Cash, D., Ng, R., Rivkin, A.: Improved structured encryption for SQL databases via hybrid indexing. In: Sako, K., Tippenhauer, N.O. (eds.) ACNS 2021. LNCS, vol. 12727, pp. 480–510. Springer, Cham (2021). https://doi.org/10.1007/978-3-030-78375-4_19

[CZH+13] Chu, C.-K., Zhu, W.T., Han, J., Liu, J.K., Xu, J., Zhou, J.: Security concerns in popular cloud storage services. IEEE Perv. Comput. **12**(4), 50–57 (2013)

[Gen09]   Gentry, C.: Fully homomorphic encryption using ideal lattices. In: ACM STOC 2009, pp. 169–178 (2009)

[GO96]   Goldreich, O., Ostrovsky, R.: Software protection and simulation on oblivious rams. J. ACM **43**(3), 431–473 (1996)

[Goh03]   Goh, E.-J.: Secure indexes. IACR Cryptology ePrint Archive 2003/216 (2003)

[IKK12]   Islam, M.S., Kuzu, M., Kantarcioglu, M.: Access pattern disclosure on searchable encryption: ramification, attack and mitigation. In: NDSS 2012 (2012)

[JJK+13] Jarecki, S., Jutla, C.S., Krawczyk, H., Rosu, M.-C., Steiner, M.: Outsourced symmetric private information retrieval. In: ACM CCS 2013, pp. 875–888 (2013)

[JP21]   Jutla, C.S., Patranabis, S.: Efficient searchable symmetric encryption for join queries (full version). IACR Cryptology ePrint Archive, p. 1471 (2021)

[KM17]   Kamara, S., Moataz, T.: Boolean searchable symmetric encryption with worst-case sub-linear complexity. In: Coron, J.-S., Nielsen, J.B. (eds.) EUROCRYPT 2017. LNCS, vol. 10212, pp. 94–124. Springer, Cham (2017). https://doi.org/10.1007/978-3-319-56617-7_4

[KM18]   Kamara, S., Moataz, T.: SQL on structurally-encrypted databases. In: Peyrin, T., Galbraith, S. (eds.) ASIACRYPT 2018. LNCS, vol. 11272, pp. 149–180. Springer, Cham (2018). https://doi.org/10.1007/978-3-030-03326-2_6

[KM19]   Kamara, S., Moataz, T.: Computationally volume-hiding structured encryption. In: Ishai, Y., Rijmen, V. (eds.) EUROCRYPT 2019. LNCS, vol. 11477, pp. 183–213. Springer, Cham (2019). https://doi.org/10.1007/978-3-030-17656-3_7

[LPS+18] Lai, S., et al.: Result pattern hiding searchable encryption for conjunctive queries. In: ACM CCS 2018, pp. 745–762 (2018)

[NKW15]   Naveed, M., Kamara, S., Wright, C.V.: Inference attacks on property-preserving encrypted databases. In: ACM CCS 2015, pp. 644–655 (2015)

[PM21]   Patranabis, S., Mukhopadhyay, D.: Forward and backward private conjunctive searchable symmetric encryption. In: NDSS 2021 (2021)

[PRZB11] Popa, R.A., Redfield, C.M.S., Zeldovich, N., Balakrishnan, H.: CryptDB: protecting confidentiality with encrypted query processing. In: ACM SOSP 2011, pp. 85–100 (2011)

[SWP00]   Song, D.X., Wagner, D.A., Perrig, A.: Practical techniques for searches on encrypted data. In: IEEE S&P 2000, pp. 44–55 (2000)

[ZKP16]   Zhang, Y., Katz, J., Papamanthou, C.: All your queries are belong to us: the power of file-injection attacks on searchable encryption. In: USENIX Security Symposium 2016, pp. 707–720 (2016)

Knowledge Encryption and Its Applications to Simulatable Protocols with Low Round-Complexity

Yi Deng[1,2(✉)] and Xinxuan Zhang[1,2]

[1] State Key Laboratory of Information Security, Institute of Information Engineering, Chinese Academy of Sciences, Beijing, China
{deng,zhangxinxuan}@iie.ac.cn
[2] School of Cyber Security, University of Chinese Academy of Sciences, Beijing, China

Abstract. We introduce a new notion of public key encryption, *knowledge encryption*, for which its ciphertexts can be reduced to the public-key, i.e., any algorithm that can break the ciphertext indistinguishability can be used to extract the (partial) secret key. We show that knowledge encryption can be built solely on any two-round oblivious transfer with game-based security, which are known based on various standard (polynomial-hardness) assumptions, such as the DDH, the Quadratic(N^{th}) Residuosity or the LWE assumption.

We use knowledge encryption to construct the *first three-round* (weakly) simulatable oblivious transfer. This protocol satisfies (fully) simulatable security for the receiver, and weakly simulatable security $((T, \epsilon)$-simulatability) for the sender in the following sense: for any polynomial T and any inverse polynomial ϵ, there exists an efficient simulator such that the distinguishing gap of any distinguisher of size less than T is at most ϵ.

Equipped with these tools, we construct a variety of fundamental cryptographic protocols with low round-complexity, *assuming only the existence of two-round oblivious transfer with game-based security*. These protocols include three-round delayed-input weak zero knowledge argument, three-round weakly secure two-party computation, three-round concurrent weak zero knowledge in the BPK model, and a *two-round* commitment with weak security under selective opening attack. These results improve upon the assumptions required by the previous constructions. Furthermore, all our protocols enjoy the above (T, ϵ)-simulatability (stronger than the distinguisher-dependent simulatability), and are quasi-polynomial time simulatable under the same (polynomial hardness) assumption.

1 Introduction

We study the problem of constructing generic public-key encryption with a natural property that the public key can be reduced to its ciphertexts, i.e., any algorithm that breaks the ciphertext indistinguishability can be used to extract

S. Agrawal and D. Lin (Eds.): ASIACRYPT 2022, LNCS 13793, pp. 334–362, 2022.
https://doi.org/10.1007/978-3-031-22969-5_12

the (partial) secret key. We call such a public-key encryption scheme *knowledge encryption*. Although we often have the impression of public key encryption that only the one holding the secret key can decrypt/distinguish a ciphertext, almost none of known constructions *provably* achieves this property. Instead, they only guarantee that, if an algorithm can break the ciphertext indistinguishability, then we can use it to find a solution to a random instance of certain hard problem (rather than finding the corresponding secret key). The only exception we aware of is the public-key encryption based on Rabin's trapdoor permutations, for which one can establish the equivalence between breaking the ciphertext indistinguishability and finding a secret key.

Essentially, the decryption of a knowledge encryption scheme can be viewed as a *proof of knowledge* of the (partial) secret key. From this prospective, the concepts of conditional disclosure of secret (CDS) [1, 4, 23] and witness encryption (WE) [20] in the literature are close to our knowledge encryption. Specifically, a public key of a CDS (WE) scheme is generated from a publicly known instance x (for WE, x serves as the pubic key) of an NP language L, and guarantees that if $x \notin L$, then the receiver obtains nothing about the encrypted message.

But the decryption of CDS/WE schemes provides *only* a *sound* proof that the corresponding public key is valid (i.e., $x \in L$), rather than *proof of knowledge* (or, *extractability*) of the witness of $x \in L$. Goldwasser et al. [29] put forward the notion of *extractable* witness encryption, which, similar in spirit to our knowledge encryption, requires that any algorithm that breaks the ciphertext indistinguishability can be used to extract the witness for the instance x. However, their scheme requires rather strong (unfalsifiable) knowledge assumptions.

Motivation. Our study is motivated by the recent works [9,16,34] on cryptographic protocols with low round-complexity beyond the known black-box barriers. At a very high level, the idea of behind these constructions is to design a protocol in such a way that any distinguisher with relatively large distinguishing advantage (inverse polynomial) ϵ can be used to extract certain secret of the adversary, which can be used for a successful simulation (except with probability ϵ). Thus, for a given distinguisher, the simulator now can first exploit the power of it to extract some secret information from the adversary and then simulate in a straightforward manner. This distinguisher-dependent simulation technique was introduced by Jain et al. in [34] and used to achieve delayed-input weak zero knowledge argument and weakly secure two-party computation for certain functionalities in three round, which bypass the well-known lower bounds on the round-complexity [27] and are round-optimal under polynomially hard falsifiable assumptions while black-box reduction/simulation are used to prove the soundness/security for receiver [38]. Bitansky et al. [9] introduced an ingenious homomorphic trapdoor simulation paradigm and presented a three-round weak zero knowledge argument, without requiring "delayed-input" or the simulator to work in distributional setting. Latter, the distinguisher-dependent simulation was also used to achieve oblivious transfer (OT) in three round with distinguisher-dependent simulatable security for the sender [31].

Deng [16] introduced an individual simulation technique and exploited a variant of Rabin encryption (the only known "knowledge encryption") to realize the above-mentioned design idea. The work of [16] proposed a two-round commitment satisfying (T, ϵ)-simulatable security under selective opening attack and a three-round concurrent (T, ϵ)-zero knowledge argument in the bare public-key model (both bypassing the black-box lowerbounds [3,44,45]), where the (T, ϵ)-simulatability is defined as follows: For any polynomial T and any inverse polynomial ϵ, there exists a simulator such that the distinguishing gap of any distinguisher of size less than T is at most ϵ. Note that the (T, ϵ)-simulatability is stronger[1] than the distinguisher-dependent simulatability since it depends only on *the size of the distinguisher* (*not* on the distinguisher per se).

All above protocols require specific number-theoretic assumptions. This state of the art leaves the several intriguing questions:

Can we construct oblivious transfer in three-round that achieves simulatable security for both sides? Can we base the above protocols on more general assumptions?

1.1   Our Contribution

We introduce the notion of knowledge encryption. Like CDS, a knowledge encryption scheme is associated with an NP language L, and the public/secret key pair (pk, sk) is generated from an instance $x \in L$ and its witness w. We let the public key (secret key) contain the instance x (witness w, respectively). We require the following properties from a knowledge encryption scheme:

1 Indistinguishability: ciphertext indistinguishability holds for any $(x, w) \in R_L$;
2 Witness extractability: for any algorithm that can break the ciphertext indistinguishability can be used to extract the witness w (part of the secret key). This holds even when the public key is maliciously generated.
3 Public key simulation: for any $(x, w) \in R_L$, there is a simulator that, taking only x as input, can output a public key that is indistinguishable from the honestly generated one.

We show that knowledge encryption can be built solely on any two-round OT with game-based security, which are known based on various standard (polynomial-hardness) assumptions, such as the DDH [40], the Quadratic(Nth) Residuosity [33] or the LWE assumption [10].

Equipped with knowledge encryption, we obtain the following results assuming *only the existence of two-round OT with game-based security (against polynomial-time adversaries)*:

- **The *first* three-round (T, ϵ)-simulatable OT** with fully simulatable security for the receiver and (T, ϵ)-simulatable security for the sender.

---

[1] Note that the result of [14] that distinguisher-dependent simulatability can be upgraded to (T, ϵ)-simulatability holds only for *zero knowledge* protocols.

Achieving polynomially simulatable security (of any kind) for *both parties* of OT in three rounds has been an elusive. Previous work on three-round OT achieves either *one-sided* (distinguisher-dependent) simulatability for the sender [31], or *game-based* security for both parties [13].

- **A variety of protocols achieving (T, ϵ)-simulatable security**, including three-round delayed-input (T, ϵ)-zero knowledge argument, three-round (T, ϵ)-secure two-party computation for independent-input functionalities, three-round concurrent (T, ϵ)-zero knowledge in the BPK model and *two-round* commitment with (T, ϵ)-security under selective opening attack.

  Prior works on these protocols either require an additional assumption–the existence of dense encryption, or are only known based on the Factoring assumption [16]. The three-round protocol of secure two-party computation in [4] is built on a rather strong assumptions of the existence of succinct randomized encodings scheme, which are only known based on indistinguishable obfuscation. Furthermore, as mentioned before, the (T, ϵ)-simulatability we achieve is stronger than the notion of distinguisher-dependent simulatability achieved by the work of [34].

  Our result on weak zero knowledge is incomparable to the work of [9]: The protocol in [9] requires both LWE and Factoring (or standard Bilinear-Group) assumptions, but the common input need not to be delayed to the last round.

- **Quasi-polynomial time simulatable under polynomial hardness assumption**: All above protocols are quasi-polynomial time simulatable under the same (polynomial hardness) assumption.

  Previous results achieving quasi-polynomial time simulatable security (e.g., see [42] and [35]) usually require quasipolynomial/exponential hardness assumption.

1.2   Technique Overview

Knowledge Encryption. Before describing our construction, we briefly recall the idea behind a CDS scheme for an NP relation R_L. Given input $(x, w) \in R_L$ of length $\lambda + \ell$, the receiver uses the algorithm OT_1 to encode w bit-by-bit, and publishes his public key $(x, \mathsf{OT}_1(w_1), \mathsf{OT}_1(w_2) \cdots, \mathsf{OT}_1(w_\ell))$; to encrypt a bit $m \in \{0, 1\}$, the sender first garbles the following circuit C: on input (x, w, m), C checks if $(x, w) \in R_L$, if so, outputs m; otherwise outputs $\bot$. After obtaining a garbled circuit $\hat{C}$ and the associated labels $\{\mathsf{lab}_{i,b}\}_{i \in [\lambda + \ell + 1], b \in \{0,1\}}$, the sender sends the ciphertext $c := (\hat{C}, \{\mathsf{lab}_{i,x_i}^x\}_{i \in [\lambda]}, \{\mathsf{OT}_2(\mathsf{lab}_{i,0}^w, \mathsf{lab}_{i,1}^w)\}_{i \in [\ell]}, \mathsf{lab}_m^m)$ to the receiver, which retrieves the labels $\{\mathsf{lab}_{i,w_i}^w\}_{i \in [\ell]}$ and then decrypts c using the evaluating algorithm of the garbling scheme.

To achieve the *witness extractability* property, our key idea is to embed a simple decoding mechanism in the above circuit C, which enables us to reduce the instance x to random ciphertexts. Specifically, we let C to take an extra input y of length ℓ and define it as follows: on input $((x, w, y, m)$, if $(x, w) \in R_L$ and $y = 0^\ell$, output m; if $(x, w) \in R_L$ and the Hamming weight of $\|y\|_1 \geq 1$, output

$\Sigma_{i=1}^{\ell} y_i w_i \bmod 2$; if $(x, w) \notin R_L$, output $\perp$. With this modification, when encrypting a bit m, the honest sender always chooses $y = 0^{\ell}$, garbles the above circuit C and then sets the ciphertext to be $c := (\hat{C}, \{\mathsf{lab}_{i,x_i}^x\}_{i \in [\lambda]}, \{\mathsf{OT}_2(\mathsf{lab}_{i,0}^w, \mathsf{lab}_{i,1}^w)\}_{i \in [\lambda]}, \{\mathsf{lab}_{i,0}^y\}_{i \in [\ell]}, \mathsf{lab}_m^m)$.

It is not hard to see that this modification does not affect the *indistinguishability* of the scheme. On the other hand, the *witness extractability* property follows from the following observations. Note first that, for every $i \in [\ell]$, one can always choose a bad y which has 1 on the i-th coordinate and zero on all others, and compute a ciphertext with such a y. Due to the security of the underlying garbling scheme, no polynomial size circuit can distinguish these bad ciphertexts from the honestly-generated ones. Thus, for any polynomial size circuit that decrypts honestly-generated ciphertexts correctly with high probability, when given a bad ciphertext as input, it would output $\Sigma_{i=1}^{\ell} y_i w_i \bmod 2 = w_i$ correctly with almost the same probability. One can apply this reasoning to ciphertext distinguishers and prove the witness extractability property.

An alternative construction from CDS and random-self-reducible encryptions is presented in the full version of this paper [17].

Nearly Optimal (T, ϵ)-Extractor for Knowledge Encryption. Applying the result of [16], we will have a nearly optimal (T, ϵ)-extractor for any (possibly malicious) key generation algorithm of knowledge encryption in the following sense: for any polynomial T and any inverse polynomial ϵ, the extractor outperforms any circuits of size T in extracting the witness for x in the public key except for probability ϵ.

Looking ahead, the (T, ϵ)-simulatability of all our protocols relies on this nearly optimal extractor. When receiving the public key(s) of knowledge encryption from an adversary, the corresponding simulator will run this extractor to extract the witness for x, and if it succeeds, then the simulation can be done; if it fails, then the optimality of the extractor guarantees that no other circuits (distinguishers) of size T can extract the witness either (except for small probability ϵ), and thus the simulator can encrypt a dummy message in its last round, which cannot be told apart from an real execution by any distinguishers of size T except for probability ϵ (by the witness extractability of knowledge encryption.)

Three-Round OT with (T, ϵ)-Simulatability for Both Parties. A natural idea here is to have the receiver generate a pair of public keys $\mathsf{pk}_0, \mathsf{pk}_1$ of knowledge encryption from two NP instances x_0 and x_1, for one of which it knows a valid witness so that it can receive one message encrypted by the sender. However, there are two challenges that arise from this approach:

1 We need to make sure that the receiver knows a witness for *only one* of these two instances (to achieve the sender security), while at the same time one needs to know both witnesses for x_0 and x_1 to extract the two messages from the sender in the proof of receiver security.

2 There is no way for the receiver to tell honest ciphertexts from "bad" ones.

One may think of the following solution to the first challenge: the sender generates some hard instance y (and prove to the receiver that it knows a witness for y in three rounds), and then the receiver proves that it knows either a witness for y or only one of x_0 and x_1 is in the language L (for some suitable language) in a two-round WI protocol. However, among other issues, there is no known two-round WI protocol based on two-round OT.

To this end, we have the sender generate two images y_0 and y_1 of a one-way function f and prove to the receiver that it knows one pre-image of y_0 or y_1 via a three-round WI protocol[2]. Given the pair (y_0, y_1) and input b, the receiver prepares two instances x_0 and x_1 in the following way: it runs the HVZK simulator of the Σ-protocol to obtain an acceptable proof (a, b, z) of knowledge of one preimage of y_0 or y_1, and sets $x_b = (y_0, y_1, a, b)$ and $x_{1-b} = (y_0, y_1, a, 1 - b)$, where $x_i = (y_0, y_1, a, i)$ is said to be a YES instance if and only if there exists a z such that (a, i, z) is acceptable. The receiver now generates pk_b honestly using the valid witness z for $x_b = (y_0, y_1, a, b)$, and runs the key simulator of knowledge encryption to obtain the other public key pk_{1-b}. In the third round, the sender encrypt its two message under the two public keys respectively and send the two ciphertexts to the receiver.

Notice that the receiver does not know a witness for the instance x_{1-b} on the public key pk_{1-b}, since otherwise it would be able to compute a preimage of y_0 or y_1 generated by the sender at random (which is infeasible due to the fact that the WI proof actually hides the two preimages of y_0 or y_1.) This observation, together with the existence of nearly optimal extractor (as mentioned above) that outperforms any other circuits of a-priori bounded size for extracting a witness of x_0 or x_1, one can prove the (T, ϵ)-simulatable security for the sender.

Our proof of the (fully) simulatable security for the receiver departs from the traditional proof strategy that is usually done by extracting the sender's two messages from a WI proof of knowledge. Our simulator extracts the sender's two messages by decryption. Using rewinding strategy[3] the simulator extracts a preimage of y_0 and y_1, then generates two Yes instance x_0 and x_1 and two valid public keys. When receiving the two ciphertexts from the sender, it can decrypt to obtain both messages[4] and send them to the functionality. Note that, although these ciphertexts from the sender may be generated maliciously (as mentioned in the above second challenge) and adaptively (depending on the receiver's public keys), we can still prove the simulatable security for the receiver since the public keys of the receiver in the real model execution and the ones in the ideal model execution are indistinguishable.

---

[2] Note that the three-round WI and the Σ-protocol used in our construction can be based on non-interactive commitment. As noted in [12], combing the recent work of [39] with the work [24], one can build non-interactive commitment from two-round (perfectly correct) OT with game-based security. Thus, two-round OT with game-based security as we define is sufficient for constructing all primitives used in our protocol.

[3] Here we actually need Goldriech-Kahan technique to bound the running time of the extractor, see the detailed proof in the full version of this paper [17].

[4] If the simulator fails to decrypt a ciphertext, it sets the corresponding "plaintext" to be $\perp$.

(T, ϵ)-**zero knowledge and** (T, ϵ)-**secure two-party computation**. At a high level, our construction of (T, ϵ)-zero knowledge protocol follows the paradigm of [2,36]. The prover and the verifier execute a three-round OT as constructed above (denoted by $(\mathsf{OT}_1, \mathsf{OT}_2, \mathsf{OT}_3)$ the three OT step algorithms respectively), where the verifier plays the role of the receiver and chooses a random bit $\beta \leftarrow \{0,1\}$ as the receiver's input in the second round. In the last round of OT, the prover prepares two acceptable Σ-proofs $(\alpha, 0, \gamma_0), (\alpha, 1, \gamma_1)$ for the statement $x \in L$, and sends x and $(\alpha, \mathsf{OT}_3(\gamma_0, \gamma_1))$ to the verifier. Finally, the verifier recovers γ_β from OT and checks whether $(\alpha, \beta, \gamma_\beta)$ is an acceptable proof. In order to reduce the soundness error, we have the prover and the verifier run this protocol λ times in parallel. The (T, ϵ)-zero knowledge of the protocol essentially follows from the (T, ϵ)-simulatable security for sender of the underlying OT and the fact that the nearly optimal extractor guaranteed by Lemma 2 works well for (possibly malicious) parallelized key generator of knowledge encryption.

One can also prove a sort of soundness of the above protocol due to the simulatable security for receiver of the underlying OT. However, we do not know how to show it satisfies *adaptive* soundness/argument of knowledge, which is naturally required in settings where the prover can choose statements to be proven adaptively. Inspired by [34], we use additional knowledge encryption schemes to achieve *adaptive* argument of knowledge. In addition to executing the above protocol, the prover generates two public keys of knowledge encryption and proves to the verifier that one of them is generated honestly in a three-round WI protocol. In the last round, it encrypts each of γ_0 and γ_1 twice under the two public keys, and sends these encryptions along with the third OT messages (which now encode both (γ_0, γ_1) and the *randomnesses* used in these encryptions). We observe that these additional encryptions does not harm zero knowledge property of the above protocol since the WI proof for the sender's two public keys actually hides both secret keys. On the other hand, it does help us achieve *adaptive* argument of knowledge: One can extract a secret key by rewinding the prover and decrypt those encryptions in the original transcript obtained before rewinding, which will reveal a witness for the statement in that transcript.

Equipped with the above three-round OT and weak zero knowledge argument, we follow the GMW paradigm [28] to give a three-round protocol for (T, ϵ)-secure two-party computation for independent-input functionalities. We stress that the (T, ϵ)-simulatable security against malicious receiver of our two-party computation protocol only holds for *independent-input functionalities*, since for the proof of (T, ϵ)-simulatability against malicious receiver to go through, we need to make sure that one can freely sample the sender's input x even when the malicious receiver's input y is fixed. This is roughly also the reason that we achieve (T, ϵ)-zero knowledge only for *delayed-input* argument.

Our protocols of commitment with weak security under selective opening attack and concurrent weak zero knowledge argument (in the BPK model) simply follows by replacing the corresponding encryption scheme in the constructions of [16] with our knowledge encryption (and revising their protocol accordingly so that the simulation can go through with a witness for the instance on the public key of knowledge encryption). Furthermore, when using our construction of (T, ϵ)-zero knowledge argument of knowledge in the extractable commitment of [34], we obtain a three-round extractable commitment from two-round OT with game-based security.

1.3   More Related Work

Related Work on Simulatable Oblivious Transfer. The work of [13,19, 41] achieved fully-simulatable black-box construction of OT in four-round from certified/full domain trapdoor permutations or strongly uniform key agreement protocol, which are also round optimal for black-box constructions [37]. In the common reference string model, fully-simulatable secure (even UC-secure) OT can be achieved in two rounds from various assumptions [18,43], such as DDH, LWE, CDH or LPN assumptions.

Related Work on Two/Multi-party Computation. Katz and Ostrovsky [37] showed that four-round is necessary for black-box two-party computation for general functionalities where only one party receives the output. The construction of four-round black-box two-party computation was constructed in [15,41]. Garg et. al [21] study two-party computations with simultaneous message transmission and give a four-round construction for general functionalities where both parties receive the output. Four-round secure multi-party computation can be constructed from various assumptions [5,32]. Recently, Choudhuri et. al [12] constructed a four-round construction only from four-round fully-simulatable OT. In the CRS model, Benhamouda and Lin [6] and Garg and Srinivasan [22] presented the two-round constructions from two-round semi-malicious OT protocol and NIZK or two-round fully-simulatable OT respectively.

2   Preliminaries

Throughout this paper, we let λ denote the security parameter. Given a positive integer m, a and b, we denote by $[m]$ the set $\{1, 2, \cdots, m\}$, and by $[a, b]$ the set $\{a, a+1, \cdots, b\}$. We often write a string x as a concatenation of its bits, $x = x_1 \| x_2 \| \cdots \| x_n$, where x_i is the i-th bit of x. For a given y, we denote by $\|y\|_1$ the Hamming weight of y. We use the standard abbreviation PPT to denote probabilistic polynomial time. We will use the terms (non-uniform) PPT algorithm and polynomial-size circuits interchangeably. When writing a polynomial-size circuit C, we mean a polynomial-size family of circuits $C = \{C_\lambda\}_{\lambda \in \mathbb{N}}$. For two random ensembles $\mathcal{X} := \{\mathcal{X}_\lambda\}_{\lambda \in \mathbb{N}}$ and $\mathcal{Y} := \{\mathcal{Y}_\lambda\}_{\lambda \in \mathbb{N}}$, we write $\mathcal{X} \stackrel{c}{\approx} \mathcal{Y}$ to mean $\mathcal{X} := \{\mathcal{X}_\lambda\}_{\lambda \in \mathbb{N}}$ and $\mathcal{Y} := \{\mathcal{Y}_\lambda\}_{\lambda \in \mathbb{N}}$ are indistinguishable against all polynomial-size circuits.

Due to space limitations, most of standard definitions (e.g., commitment schemes, Σ-protocol, game-based secure OT, garbled circuits etc.) are deferred to the full version of this paper [17].

2.1 Interactive Argument

Let L be an NP language and R_L be its associated relation. For a given $x \in L$, we use $R_L(x)$ to denote the set of valid witnesses to x. An interactive argument (P,V) for L is a pair of PPT algorithms (called the prover and the verifier), in which the prover P wants to convince the verifier V of a statement $x \in L$. For a given $(x,w) \in R_L$, we denote by $\mathsf{Out}_V(P(w),V)(x)$ the output of V at the end of an execution of (P,V), and by $\mathsf{View}_V^{P(w)}(x)$ the view of V in an interaction.

Definition 1 (Argument). *A protocol (P,V) for an NP language L is an argument if the following two conditions hold:*

- **Completeness**: *For any $x \in L$ and $w \in R_L(x)$, $\mathsf{Out}_V(P(w),V)(x) = 1$.*
- **Computational soundness**: *For any polynomial-size prover P^*, there exists a negligible function $negl(\cdot)$ such that for any $x \notin L$ of length λ,*

$$\Pr[\mathsf{Out}_V(P^*,V)(x) = 1] < negl(\lambda).$$

Additionally, an interactive argument system is called public-coin if at every verifier step, the verifier sends only truly random messages.

Delayed-Input and Adaptive Computational Soundness. We call an argument is *delayed-input* if the statement x is sent to verifier only in the last round. Note that delayed-input argument system would enable a cheating prover to choose a false statement adaptively (depending on the interaction history) to fool the verifier. We consider such an adaptive cheating prover and define adaptive computational soundness in a natural way: A delayed-input argument is called *adaptive computational sound* if its computational soundness condition holds even against adaptive cheating prover.

Argument of Knowledge and Adaptive Argument of Knowledge. The adaptive argument of knowledge property is defined in similar way to the argument of knowledge, except that here we need to deal with the issue that the statement may be chosen adaptively. We follow the definition in [7,8] to define three-round adaptive argument of knowledge.

Definition 2. *A three-round delayed-input argument system with message (a_1, a_2, a_3) for NP language L is called an adaptive argument of knowledge if there exists an oracle extractor E and a polynomial poly such that for any PPT malicious prover P^*, any noticeable function ϵ and any security parameter $\lambda \in \mathbb{N}$:*

$$if \ \ \Pr\left[V(x,(a_1,a_2,a_3)) = 1 \ \middle| \ \begin{array}{c} a_1 \leftarrow P^* \\ a_2 \leftarrow V(\lambda, a_1) \\ x, a_3 \leftarrow P^*(a_1, a_2) \end{array}\right] \geq \epsilon(\lambda),$$

$$then \ \ \Pr\left[\begin{array}{c} V(x,(a_1,a_2,a_3)) = 1 \wedge \\ E^{P^*}(x,(a_1,a_2,a_3)) \notin R_L(x) \end{array} \ \middle| \ \begin{array}{c} a_1 \leftarrow P^* \\ a_2 \leftarrow V(\lambda, a_1) \\ x, a_3 \leftarrow P^*(a_1, a_2) \end{array}\right] \leq negl(\lambda),$$

where E runs in expected time bounded by $poly(\lambda)/\epsilon$.

An argument system is zero knowledge [30] if the view of the (even malicious) verifier in an interaction can be efficiently reconstructed. We consider a weak version of zero-knowledge as defined in [14,16], (T, ϵ)-zero-knowledge, which relaxes the definition of zero-knowledge and requires that, for any polynomial T and inverse polynomial ϵ, there exists an efficient simulator such that the distinguishing gap of any T-size distinguisher is at most ϵ.

Definition 3 $((T, \epsilon)$-Zero-Knowledge). *An argument (P, V) is (T, ϵ)-zero-knowledge if for any polynomial-size malicious verifier V^*, any polynomial T and any inverse polynomial ϵ, there exists a polynomial-size simulator $S = \{S_\lambda\}_{\lambda \in \mathbb{N}}$ such that for any T-size distinguisher $D = \{D_\lambda\}_{\lambda \in \mathbb{N}}$, and any statement $x \in L \cap \{0,1\}^\lambda$, $w \in R_L(x)$:*

$$\left| \Pr\left[D_\lambda(\mathsf{View}_{V^*}^{P(w)}(x)) = 1 \right] - \Pr\left[D_\lambda(S_\lambda(x)) = 1 \right] \right| < \epsilon(\lambda).$$

2.2 Oblivious Transfer

A 1-out-of-2 oblivious transfer protocol (OT) (S, R) is a two-party protocol between a sender S and a receiver R. The sender S has input of two strings (m_0, m_1) and the receiver R has input a bit b. At the end of the protocol, the receiver R learns m_b (and nothing beyond that), whereas the sender S learns nothing about b. We denote the output of receiver $\mathsf{Out}_R(S(m_0, m_1), R(b))(1^\lambda)$.

There are two notable security definitions in the literature, the game-based security [1,40] and the simulation-based security [25]. In this paper our goal is to achieve simulation-based security, which is defined as follows..

Message Space. We let the message space $\mathcal{M}$ to include the special symbol $\perp$, i.e., $\mathcal{M} := \{0,1\}^n \cup \perp$. Jumping ahead, in the proof of receiver's security of our construction, the simulator may extract (by decryption) two messages like $(m, \perp)$ or $(\perp, \perp)$ from a corrupted sender. In this case, the simulator will not abort, instead, it views $\perp$ as a message and send these two messages to the functionality.

Simulation-Based Security. We follow the standard real/ideal paradigm and define the simulation-based security of OT. Roughly, to prove security in the real/ideal paradigm, one first defines an ideal functionality $\mathcal{F}$ executed by a trusted party, then constructs a simulator Sim that interacts with $\mathcal{F}$ and the adversary, and then shows that the output of Sim is indistinguishable from the real execution.

The ideal functionality of OT is provided in Fig. 1.

We denote by $\mathsf{REAL}_{\Pi, R^*(\tau)}(1^\lambda, m_0, m_1, b)$(resp., $\mathsf{REAL}_{\Pi, S^*(\tau)}(1^\lambda, m_0, m_1, b)$) the distribution of the output of the malicious receiver (resp., the malicious sender and the honest receiver) during a real execution of the protocol Π (with m_0, m_1 as inputs of the sender, b as choice bit of the receiver), and by $\mathsf{IDEAL}_{\mathcal{F}_{OT}, \mathsf{Sim}^{R^*(\tau)}}(1^\lambda, m_0, m_1, b)$ (resp., $\mathsf{IDEAL}_{\mathcal{F}_{OT}, \mathsf{Sim}^{S^*(\tau)}}(1^\lambda, m_0, m_1, b)$) the distribution of the output of the malicious receiver (resp., the malicious sender and the honest receiver) during a ideal execution where τ is the auxiliary input.

Functionality $\mathcal{F}_{OT}$

Security parameter: λ

$\mathcal{F}_{OT}$ interacts with a sender S and a receiver R.

- Upon receiving (send, m_0, m_1) from S, where $m_0, m_1 \in \mathcal{M}$, record m_0, m_1 and then send send to R.
- Upon receiving (receive, b) from R, send m_b to R and receive to S and halt.

Fig. 1. The oblivious transfer functionality $\mathcal{F}_{OT}$

Definition 4 (Oblivious Transfer with Simulation-based Security). *A protocol $\Pi = (S, R)$ securely computing $\mathcal{F}_{OT}$ if it satisfies the following properties:*

- **Simulatable Security for Receiver***: For any polynomial-size malicious sender S^*, there exists a polynomial-size simulator* Sim *such that for any auxiliary input $\tau \in \{0,1\}^*$, any $m_0, m_1 \in \{0,1\}^n, b \in \{0,1\}$,*

$$\{\mathsf{REAL}_{\Pi, S^*(\tau)}(1^\lambda, m_0, m_1, b)\} \stackrel{c}{\approx} \{\mathsf{IDEAL}_{\mathcal{F}_{OT}, \mathsf{Sim}^{S^*(\tau)}}(1^\lambda, m_0, m_1, b)\}.$$

- **Simulatable Security for Sender***: For any polynomial-size malicious receiver R^*, there exists a polynomial-size simulator* Sim *such that for any auxiliary input $\tau \in \{0,1\}^*$, any $m_0, m_1 \in \{0,1\}^n, b \in \{0,1\}$,*

$$\{\mathsf{REAL}_{\Pi, R^*(\tau)}(1^\lambda, m_0, m_1, b)\} \stackrel{c}{\approx} \{\mathsf{IDEAL}_{\mathcal{F}_{OT}, \mathsf{Sim}^{R^*(\tau)}}(1^\lambda, m_0, m_1, b)\}.$$

In this paper, we follow the definition of weak simulatability in [14,16] and give a definition of simulatable (T, ϵ)-security for sender of an OT protocol (S, R).

Definition 5 ((T, ϵ)-Simulatable Security for Sender). *For any polynomial-size malicious receiver R^*, any polynomial T, any inverse polynomial ϵ, any auxiliary input distribution $\mathcal{Z}$ and $\tau \leftarrow \mathcal{Z}$, there exists a polynomial-size simulator* Sim *such that for any T-size distinguisher $D = \{D_\lambda\}_{\lambda \in \mathbb{N}}$, any $m_0, m_1 \in \{0,1\}^n, b \in \{0,1\}$:*

$$\big| \Pr[D_\lambda(\mathsf{REAL}_{\Pi, R^*(\tau)}(1^\lambda, m_0, m_1, b))] = 1 \\ - \Pr[D_\lambda(\mathsf{IDEAL}_{\mathcal{F}_{OT}, \mathsf{Sim}(\tau)}(1^\lambda, m_0, m_1, b))] = 1 \big| \leq \epsilon(\lambda). \tag{1}$$

Remark 1. Notice that traditional security definitions (such as the definition of sender's security above) require that the *black-box* simulator can deal with *any* auxiliary input τ, while, in our definition of (T, ϵ)-sender's security, we weaken this requirement by switching the order of the qualifiers and require only that for any auxiliary input τ drawn from a (known) distribution, there is a desired *individual* simulator. We make this change for the reason that, in the proof of (T, ϵ)-simulatability for the sender of our OT protocol, the simulator will apply the nearly-optimal extractor (similar to the one in [16]) for extracting some secret keys from the malicious receiver, and such an extractor is really sensitive and works well only when *all input distributions* (including the auxiliary input distribution) of the malicious receiver are well defined.

Still, as we will see, this weaker notion also has wide applications in protocol composition. We can plug a protocol Π_i satisfying this weaker security into a global protocol Π composed from a series of subprotocols $\Pi_1, \Pi_2, ..., \Pi_n$, and achieve (T, ϵ)-simulation security of Π, *as long as* all these subprotocols are simulatable and specified in advance[5]. One can view all messages from subprotocols $\Pi_{j \neq i}$ as auxiliary input drawn from the distributions over the transcripts of these subprotocols, which are well defined when we simulate the subprotocol Π_i in the proof of (T, ϵ)-simulatability of Π.

2.3   Secure Two-Party Computation

In this subsection we present the definition of secure two-party computation, independent-input functionalities and the (T, ϵ)-security. Parts of the definition of secure two-party computation are taken verbatim from [4]. In this paper, we only consider the case where only one party (a.k.a receiver R) learns the output. The other party is referred to as the sender S. Sender S has input x and receiver R has input y. For a given deterministic functionality F, they execute a protocol to jointly compute $F(x, y)$, and R obtains $F(x, y)$ at the end of execution. As observed in [37], a two-party computation protocol which only one party learns the output can be easily transformed into the one where both parties receive the output by computing a modified functionality that outputs signed values.

We follow the real/ideal paradigm to define the simulation-based security of two-party computation. The ideal model execution proceeds as follows:

Ideal Model Execution. Ideal model execution is defined as follows.

- *Input*: Each party obtains an input, denoted u ($u = x$ for S and $u = y$ for R).
- *Send inputs to trusted party*: The parties now send their inputs to the trusted party. The honest party always sends u to the trusted party. A malicious party may, however, can send a different input to the trusted party.
- *Aborting Adversaries*: An adversarial party can then send a message $\bot$ to the trusted party to abort the execution. Upon receiving this, the trusted party terminates the ideal world execution. Otherwise, the following steps are executed.
- *Trusted party answers receiver R*: Suppose the trusted party receives inputs (x', y') from S and R respectively. It sends the output $\text{out} = F(x', y')$ to receiver.
- *Outputs*: If the receiver R is honest, then it outputs out. The adversarial party (S or R) outputs its entire view.

---

[5] One exceptional case is the UC composition [11], where Π may be composed with arbitrarily unknown protocols.

We denote the adversary participating in the above protocol to be $\mathcal{B}$ and the auxiliary input to $\mathcal{B}$ is denoted by τ. We define $\mathsf{IDEAL}_{\mathcal{F}_{2pc},\mathcal{B}}$ to be the joint distribution over the outputs of the adversary and the honest party from above ideal execution.

Real Model Execution. We next consider the real model in which a real two-party protocol is executed (and there exists no trusted third party). In this case, a malicious party may follow an arbitrary feasible strategy. In particular, the malicious party may abort the execution at any time (and when this happens prematurely, the other party is left with no output).

Let Π be a two-party protocol for computing F. Note that in the two-party case at most one of S, R is controlled by an adversary. We denote the adversarial party to be $\mathcal{A}$ and the auxiliary input to $\mathcal{A}$ is denoted by τ. We define $\mathsf{REAL}_{\Pi,\mathcal{A}}$ to be the joint distribution over the outputs of the adversary and the honest party from the real execution.

Definition 6 (Security). *Let F and Π be described above. We say that Π securely computes F if for every polynomial-size malicious adversary $\mathcal{A}$ in the real world, there exists a polynomial-size adversary $\mathcal{B}$ for the ideal model, such that for any auxiliary input $\tau \in \{0,1\}^*$.*

$$\{\mathsf{REAL}_{\Pi,A(\tau)}(1^\lambda, x, y)\} \overset{c}{\approx} \{\mathsf{IDEAL}_{\mathcal{F}_{2pc},B(\tau)}(1^\lambda, x, y)\}.$$

In this paper, we only consider independent-input functionalities, as defined [34].

Definition 7 (Independent-Input Functionalities). *An independent-input functionality is defined as a functionality between two parties, Alice and Bob. Let $(\mathcal{Q},\mathcal{R},\mathcal{U})$ denote the joint distribution over inputs of both parties, where Alice's input is sampled efficiently from $\mathcal{Q}$ and Bob's input is sampled efficiently from distribution $\mathcal{R}$, and $\mathcal{U}$ denotes their common public input. Then, a functionality F over $(\mathcal{X} = (\mathcal{Q},\mathcal{U}) \times \mathcal{Y} = (\mathcal{R},\mathcal{U}))$ is independent-input for Alice if $\mathcal{Q}$ is independent of $(\mathcal{R},\mathcal{U})$.*

Similar to (T, ϵ)-zero knowledge, we define (T, ϵ)-security for a protocol of two-party computation as follows.

Definition 8 ((T,ϵ)-Security). *Let F and Π be described above. We say Π computes F with (T, ϵ)-security if for any polynomial-size malicious adversary $\mathcal{A}$ in the real model, any polynomial T, any inverse polynomial ϵ, and any auxiliary input distribution $\mathcal{Z}$, there exists a polynomial-size adversary $\mathcal{B}$ in the ideal model, such that for any T-size distinguisher $D := \{D_\lambda\}_{\lambda\in\mathbb{N}}$,*

$$\big| \Pr[D_\lambda(\mathsf{REAL}_{\Pi,A(\tau)}(1^\lambda, x, y))] = 1$$
$$- \Pr[D_\lambda(\mathsf{IDEAL}_{\mathcal{F}_{2pc},B(\tau)}(1^\lambda, x, y))] = 1\big| \leq \epsilon(\lambda).$$

where the probabilities is over the coin of joining parties and $\tau \leftarrow \mathcal{Z}$.

3   Knowledge Encryption and the Nearly Optimal Extractor for Key Generation

We now introduce a new concept of encryption– *knowledge encryption*. Roughly, a knowledge encryption is a public-key encryption scheme for which ciphertexts can be reduced to the public-key, i.e., any algorithm with large (ciphertexts) distinguishing advantage can be used to extract the (partial) secret key. Like CDS/WE schemes, a public-key of a knowledge encryption scheme is generated from a (publicly known) instance x of an NP language L, but it provides stronger security guarantee in that the decryption of knowledge encryption actually constitutes a *proof of knowledge* of the corresponding (partial) secret key: While CDS/WE schemes guarantee that the receiver obtains nothing about the encrypted message when $x \notin L$, knowledge encryption ensures that any receiver that can decrypt ciphertexts must *know* a valid witness of x (and hence $x \in L$). The semantic security of knowledge encryption is required to hold when $(x, w) \in R_L$ and the public key is honestly generated. This is in contrast to that of CDS/WE schemes, which only consider semantic security for false statements.

Definition 9 (Knowledge Encryption). *A knowledge encryption scheme with respect to an NP relation R_L is a triple of PPT algorithms* (KE.Gen, KE.Enc, KE.Dec):

- KE.Gen($1^\lambda, x, w$) : *On input the security parameter $\lambda \in \mathbb{N}$ and statement $x \in L \cap \{0,1\}^\lambda, w \in R_L(x)$,* Gen *outputs a key pair* (pk, sk), *where the public key is of the form* pk $= (\mathsf{k}, x)$.
- KE.Enc(pk, m) : *On input the public key* pk *and a message $m \in \{0,1\}$,* KE.Enc *outputs a ciphertext c.*
- KE.Dec(sk, c) : *On input the secret key* sk *and ciphertext c,* KE.Dec *outputs a message m (if c is undecryptable, we set m to be " $\perp$ ").*

We require the following properties from above scheme:

- **Completeness**: *For any $\lambda \in \mathbb{N}$, $m \in \{0,1\}$ and $x \in L \cap \{0,1\}^\lambda, w \in R_L(x)$:*

$$\Pr\left[\mathsf{KE.Dec}(\mathsf{sk}, c) = m \,\middle|\, \begin{array}{c} (\mathsf{pk}, \mathsf{sk}) \leftarrow \mathsf{KE.Gen}(1^\lambda, x, w) \\ c \leftarrow \mathsf{KE.Enc}(\mathsf{pk}, m) \end{array}\right] = 1.$$

- **Indistinguishability**: *For any polynomial-size distinguisher $D = \{D_\lambda\}_{\lambda \in \mathbb{N}}$, there exists a negligible function negl such that for any security parameter $\lambda \in \mathbb{N}$ and $x \in L \cap \{0,1\}^\lambda, w \in R_L(x)$:*

$$\Pr\left[D_\lambda(\mathsf{pk}, c) = m \,\middle|\, \begin{array}{c} (\mathsf{pk}, \mathsf{sk}) \leftarrow \mathsf{KE.Gen}(1^\lambda, x, w) \\ m \leftarrow \{0,1\}; \ c \leftarrow \mathsf{KE.Enc}(\mathsf{pk}, m) \end{array}\right] < \frac{1}{2} + negl(\lambda).$$

- **Witness Extractability**: *There exists a PPT extractor E satisfying that, for any public key* pk$^* = (\mathsf{k}^*, x)$, *polynomial-size distinguisher $D = \{D_\lambda\}_{\lambda \in \mathbb{N}}$ and inverse polynomial ϵ, if*

$$|\Pr[D_\lambda(\mathsf{KE.Enc}(\mathsf{pk}^*, 0)) = 1] - \Pr[D_\lambda(\mathsf{KE.Enc}(\mathsf{pk}^*, 1)) = 1]| \geq \epsilon,$$

then

$$\Pr[E^{D_\lambda}(\mathsf{pk}^*, 1^{1/\epsilon}) = w \wedge (x, w) \in R_L] \geq 1 - negl(\lambda),$$

where E runs in time polynomial in ϵ^{-1} and λ.

- **Public Key Simulation**: *There exists a PPT simulator* KE.KeySim *such that for any (x, w) where $x \in L \cap \{0,1\}^\lambda, w \in R_L(x)$:*

$$\{\mathsf{KE.Gen}(1^\lambda, x, w)\} \overset{c}{\approx} \{\mathsf{KE.KeySim}(1^\lambda, x)\}.$$

Remark 2. One can also define the security properties of knowledge encryption over a randomly chosen (according to certain distribution) instance x. We choose our definition because it gives great flexibility in applications, especially in the applications where several parties *jointly* compute the instance x for some public key of knowledge encryption, like our construction of three-round OT. However, we note that the distributional version of our definition may admit more instantiations, for example, the public-key encryption based on Rabin's one-way permutation is also a distributional knowledge encryption scheme.

In the rest of this section, we first present how to construct knowledge encryption from two-round OT, and then we will apply techniques of [16] and prove that, for any key generator of knowledge encryption, there exists a nearly optimal extractor for the witness of x such that when it fails, no circuit of a-priori bounded size can distinguish ciphertexts except with small probability.

3.1   Knowledge Encryption from Two-Round OT

In this section, we give a construction of knowledge encryption from two-round OT. At a high level, this construction follows the two-party-function-evaluation approach used in CDS scheme, and relies on the following two ingredients:

- A two-round OT $(\mathsf{OT}_1, \mathsf{OT}_2)$ with game-based security, and,
- A garbling circuit scheme $\mathsf{GC} = (\mathsf{Garble}, \mathsf{Eval})$.

Note that the garbling circuit scheme can be based on any one-way function, which is already implied by the existence of two-round OT with game-based security.

The main idea behind our construction is to modify the circuit C to be garbled in a CDS scheme and embed a simple decoding mechanism in C, which enables us to reduce the instance x to random ciphertexts. Specifically, we let C take an extra input y of length ℓ and define it as follows:

$$C(x, w, y, m) = \begin{cases} m & \text{if } (x, w) \in R_L \text{ and } y = 0^\ell, \\ \Sigma_{i=1}^{\ell} y_i w_i \bmod 2 & \text{if } (x, w) \in R_L \text{ and } \|y\|_1 \geq 1^6, \\ \perp & \text{if } (x, w) \notin R_L. \end{cases} \quad (2)$$

---

[6] In the following proofs, we only consider the case that $\|y\|_1 = 1$. In this case, C will output a coordinate of w, and the extractor will extract the witness bit-by-bit.

The formal description of knowledge encryption for $R_L{}^7$ from two-round OT is shown in Fig. 2.

Knowledge Encryption from Two-Round OT

KE.Gen($1^\lambda, x, w$): Parse $w = w_1 \| w_2 \| \cdots \| w_\ell$, and choose random coins $\{r_i\}_{i\in[\ell]}$, then run the 2-round OT scheme in parallel to generate $\mathsf{k} := (\mathsf{OT}_1(1^\lambda, w_1; r_1), \cdots, \mathsf{OT}_1(1^\lambda, w_\ell; r_\ell))$. Output the public-key $\mathsf{pk} = (\mathsf{k}, x)$ and the secret key $\mathsf{sk} = (w, r_1, \cdots, r_\ell)$.

KE.Enc(pk, m): Set $y = 0^\ell$, and run the GC scheme to generate a garbled circuit $\hat{C}$ with labels $\{\mathsf{lab}^x_{i,b}\}_{i\in[\lambda], b\in\{0,1\}}$, $\{\mathsf{lab}^w_{i,b}\}_{i\in[\ell], b\in\{0,1\}}, \{\mathsf{lab}^y_{i,b}\}_{i\in[\ell], b\in\{0,1\}}, \{\mathsf{lab}^m_b\}_{b\in\{0,1\}}$ for circuit C defined in (2). Output ciphertext
$$c := (\hat{C}, \{\mathsf{lab}^x_{i,x_i}\}_{i\in[\lambda]}, \{\mathsf{OT}_2(\mathsf{lab}^w_{i,0}, \mathsf{lab}^w_{i,1})\}_{i\in[\ell]}, \{\mathsf{lab}^y_{i,0}\}_{i\in[\ell]}, \mathsf{lab}^m_m)$$

KE.Dec(sk, c): Use sk to retrieve $\{\mathsf{lab}^w_{i,w_i}\}_{i\in[\ell]}$ from $\{\mathsf{OT}_2(\mathsf{lab}^w_{i,0}, \mathsf{lab}^w_{i,1})\}_{i\in[\ell]}$, and compute $m \leftarrow \mathsf{Eval}(\hat{C}, \{\mathsf{lab}^x_{i,x_i}\}_{i\in[\lambda]}, \{\mathsf{lab}^w_{i,w_i}\}_{i\in[\ell]}, \{\mathsf{lab}^y_{i,0}\}_{i\in[\ell]}, \mathsf{lab}^m_m)$.

Fig. 2. The construction of knowledge encryption from two-round OT

Theorem 1. *Assuming the existence of two-round OT protocol with computational game-based security, there exists a knowledge encryption scheme.*

Proof. We prove that the construction presented in Fig. 2 is a knowledge encryption scheme. Since the two-round OT with game-based security implies the existence of garbling scheme, our construction can be based solely on the two-round OT with game-based security. Note first that it is easy to verify the completeness property.

Indistinguishability. For a given pair $(x, w) \in R_L$, denote by $\mathcal{D}_m$ the distribution $\{\mathsf{pk}, c | \mathsf{pk} \leftarrow \mathsf{KE.Gen}(1^\lambda, x, w), c \leftarrow \mathsf{KE.Enc}(\mathsf{pk}, m)\}$ for $m = \{0, 1\}$. We prove $\mathcal{D}_0 \overset{c}{\approx} \mathcal{D}_1$ by a standard hybrid argument. Consider the following distributions.

$\mathcal{D}_{1,m}$: the same as $\mathcal{D}_m$ except that the public key is generated by using $(x, w^*) \notin R_L$, i.e., $\mathsf{pk} \leftarrow \mathsf{KE.Gen}(1^\lambda, x, w^*)$ (w.o.l.g.,we assume that such a w^* exists, see footnote 7.)

$\mathcal{D}_{2,m}$: the same as $\mathcal{D}_{1,m}$ except that it computes $\{\mathsf{OT}_2(\mathsf{lab}^{w^*}_{i,w^*_i}, \mathsf{lab}^{w^*}_{i,w^*_i})\}_{i\in[\ell]}$ in the key generation, rather than $\{\mathsf{OT}_2(\mathsf{lab}^{w^*}_{i,0}, \mathsf{lab}^{w^*}_{i,1})\}_{i\in[\ell]}$.

$\mathcal{D}_{3,m}$: the same as $\mathcal{D}_{2,m}$ except that it generates the labels and garbled circuit using the simulator of GC, i.e., $(\hat{C}, \{\mathsf{lab}_{i,b_i}\}) \leftarrow \mathsf{Sim}(1^\lambda, \phi(C), \bot)$.

---

7   For ease of presentation, we assume that for every $x \in L \cap \{0,1\}^\lambda$ there is a string $w^* \in \{0,1\}^\ell$ such that $(x, w^*) \notin R_L$. For any NP relation R_L that does not satisfy this condition, one can easily extend it to a new relation:
$$R'_L := (x, w') \in \{0,1\}^\lambda \times \{0,1\}^{\ell+1} : w' = w\|1 \text{ and } (x, w) \in R_L,$$
for which $w\|0$ is not a valid witness (for any instance x).

Note that the only difference between $\mathcal{D}_m$ and $\mathcal{D}_{1,m}$ is the first OT messages on those positions i where $w_i \neq w_i^*$. Due to the receiver's security of the underlying two-round OT, one can prove that $\mathcal{D}_m \overset{c}{\approx} \mathcal{D}_{1,m}$ by a standard hybrid argument. From the sender's security of the underlying two-round OT, it follows $\mathcal{D}_{1,m} \overset{c}{\approx} \mathcal{D}_{2,m}$. Furthermore, we have $\mathcal{D}_{2,m} \overset{c}{\approx} \mathcal{D}_{3,m}$, since for $(x, w^*) \notin R_L$, the circuit garbled in the distribution $\mathcal{D}_{2,m}$ on input (x, w^*, y, m) always outputs $\bot$. Observing that both $\mathcal{D}_{3,0}$ and $\mathcal{D}_{3,1}$ are generated by the simulator of the garbling scheme and are independent of the message m, one can see that $\mathcal{D}_{3,0} \equiv \mathcal{D}_{3,1}$. This concludes the proof of indistinguishability of our knowledge encryption scheme.

Public Key Simulation. One can easily construct a simulator for simulating the public key: On input x, the simulator chooses $\{r_i\}_{i \in [\ell]}$ at random and outputs $\mathsf{pk} = (\{\mathsf{OT}_1(1^\lambda, 0; r_i)\}_{i \in [\ell]}, x)$. This simulated public key is indistinguishable from the honestly-generated one due simply to the receiver's security of the underlying two-round OT.

Witness Extractability: Here our basic goal is to build an efficient extractor such that for any $\mathsf{pk}^* = (\mathsf{k}^*, x)$ and any distinguisher D[8] with high distinguishing advantage, the extractor, with oracle access to D, can extract a witness for x except for negligible probability.

Fix an arbitrary public key $\mathsf{pk}^* = ((\mathsf{k}^* = (\mathsf{ot}_{1,1}^*, \cdots, \mathsf{ot}_{1,\ell}^*)), x)$. We use the sender's security property (which is against unbounded receiver) of the two-round OT to define $w^* \in \{0,1\}^\ell$ as follows: For each $i \in [\ell]$, if for any (δ_0, δ_1), $\mathsf{OT}_{2,i}(\delta_0, \delta_1)$ is indistinguishable from $\mathsf{OT}_{2,i}(\delta_0, \delta_0)$ against any polynomial-size adversary, $w_i^* = 0$, otherwise $w_i^* = 1$.

Suppose that D is a polynomial-size distinguisher and ϵ is an inverse polynomial such that

$$|\Pr[D(\mathsf{KE.Enc}(\mathsf{pk}^*, 0)) = 1] - \Pr[D(\mathsf{KE.Enc}(\mathsf{pk}^*, 1)) = 1]| \geq \epsilon(\lambda), \qquad (3)$$

we construct a desirable oracle machine E^D to complete the proof of the witness extractability property.

We first argue that the definition of w^*, together with the inequality (3), implies $(x, w^*) \in R_L$. Suppose otherwise $(x, w^*) \notin R_L$. Let $\{\mathcal{D}_{j,m}\}_{j \in [3], m \in \{0,1\}}$ be as above. For every $j \in [3]$ and $m \in \{0,1\}$, Denote by $\mathcal{D}_{j,m}|\mathsf{pk}^*$ the distribution conditioned on pk^*. Then, for each $m \in \{0,1\}$, we have $\mathsf{KE.Enc}(\mathsf{pk}^*, m) \equiv \mathcal{D}_{1,m}|\mathsf{pk}^*$ and $\mathcal{D}_{1,m}|\mathsf{pk}^* \overset{c}{\approx} \mathcal{D}_{2,m}|\mathsf{pk}^*$ (by definition of w^*). Furthermore, applying the same reasoning as in the proof of the indistinguishability property, we also have $\mathcal{D}_{2,m}|\mathsf{pk}^* \overset{c}{\approx} \mathcal{D}_{3,m}|\mathsf{pk}^*$ (for each $m \in \{0,1\}$) and $\mathcal{D}_{3,0}|\mathsf{pk}^* \equiv \mathcal{D}_{3,1}|\mathsf{pk}^*$. Putting together, we conclude that $\mathsf{KE.Enc}(\mathsf{pk}^*, 0)$ and $\mathsf{KE.Enc}(\mathsf{pk}^*, 1)$ are indistinguishable, which contradicts the inequality (3).

We now turn to the construction of the oracle machine E^D assuming the distinguisher D satisfies the inequality (3). Our main idea is to run D on *fake* ciphertexts by manipulating the input y and use its distinguishing advantage to compute the witness w^* bit-by-bit.

---

[8] D might know of the random coins used to sample pk^*.

Denote by $\vec{y}(j)$ the string with the j-th coordinate being 1 and all others being 0. Observe that, by the definition of circuit C, when choosing $\vec{y}(j)$ to compute a ciphertext, it will be decrypted to w_j^*. We formally define such an encryption algorithm $\mathsf{KE.Enc}'(\mathsf{pk}^*, 0)$ as follows: $\mathsf{KE.Enc}'(\mathsf{pk}^*, 0)$ acts exactly the same as $\mathsf{KE.Enc}(\mathsf{pk}^*, 0)$ except that it chooses $y' = \vec{y}(j) = y_1' \| y_2' \| \cdots \| y_\ell'$ (as a result, the i-th label with respect to y generated by $\mathsf{KE.Enc}'(\mathsf{pk}^*, 0)$ is $\mathsf{lab}_{j,1}^y$, rather than $\mathsf{lab}_{j,0}^y$). A ciphertext generated by $\mathsf{KE.Enc}'(\mathsf{pk}^*, 0)$ can be viewed as a ciphertext of w_j^*, and furthermore, the distribution $\mathsf{KE.Enc}'(\mathsf{pk}^*, 0)$ is actually indistinguishable from $\mathsf{KE.Enc}(\mathsf{pk}^*, w_j^*)$. To see this, consider the following distribution $\mathcal{D}_S$: run the simulator Sim for garbling scheme and obtain $(\hat{C}, \{\mathsf{lab}_{i,x_i}^x\}_{i\in[\lambda]}, \{\mathsf{lab}_{w_i^*}^w\}_{i\in[\ell]}, \{\mathsf{lab}_{y_i'}^y\}_{i\in[\ell]}, \mathsf{lab}_m^m) \leftarrow \mathsf{Sim}(1^\lambda, \phi(C), w_j^*)$, and output ciphertext $c = (\hat{C}, \{\mathsf{lab}_{i,x_i}^x\}_{i\in[\lambda]}, \{\mathsf{OT}_2(\mathsf{lab}_{i,w_i^*}^w, \mathsf{lab}_{i,w_i^*}^w)\}_{i\in[\ell]}, \{\mathsf{lab}_{i,y_i'}^y\}_{i\in[\ell]}, \mathsf{lab}_m^m)$.

Note that $w_j^* = C(x, w^*, y' = \vec{y}(j), 0) = C(x, w^*, y = 0^\ell, w_j^*)$, and for this reason, the above ciphertext simulator can be viewed as a simulator for both $\mathsf{KE.Enc}'(\mathsf{pk}^*, 0)$, which garbles C on input $(x, y' = \vec{y}(j), 0)$, and $\mathsf{KE.Enc}(\mathsf{pk}^*, w_j^*)$, which garbles C on input $(x, y = 0^\ell, w_j^*)$. Similarly to the proof of the indistinguishability property, due to the sender's security of the two-round OT and the security of the garbling scheme, one can prove that both $\mathsf{KE.Enc}'(\mathsf{pk}^*, 0)$ and $\mathsf{Enc}(\mathsf{pk}^*, w_j^*)$ are indistinguishable from $\mathcal{D}_S$. Thus,

$$\mathsf{KE.Enc}'(\mathsf{pk}^*, 0)) \stackrel{c}{\approx} \mathsf{KE.Enc}(\mathsf{pk}^*, w_j^*)). \tag{4}$$

This means the distinguisher D can tell apart $\mathsf{KE.Enc}'(\mathsf{pk}^*, 0))$ from $\mathsf{KE.Enc}(\mathsf{pk}^*, 1 - w_j^*))$, which gives rise to the following oracle extraction machine E^D.

$E^D(\mathsf{pk}^*, 1^{1/\epsilon})$

1. For each $j \in [\lambda]$:
   (a) Run D on input $\mathsf{KE.Enc}(\mathsf{pk}^*, 0)$ $\lambda\epsilon^{-2}$ times with fresh randomness (for both D and $\mathsf{KE.Enc}$) each time. Denote by $d_{0,k}$ the output of $D(\mathsf{KE.Enc}(\mathsf{pk}^*, 0))$ in the k-th repetition. Compute $d_0 = \lambda^{-1}\epsilon^2 \Sigma_{k\in[p]} d_{0,k}$.
   (b) Run D on input $\mathsf{KE.Enc}(\mathsf{pk}^*, 1)$ $\lambda\epsilon^{-2}$ times with fresh randomness (for both D and $\mathsf{KE.Enc}$) each time. Denote by $d_{1,k}$ the output of $D(\mathsf{KE.Enc}(\mathsf{pk}^*, 1))$ in the k-th repetition. Compute $d_1 = \lambda^{-1}\epsilon^2 \Sigma_{k\in[p]} d_{1,k}$.
   (c) Run D on input $\mathsf{KE.Enc}'(\mathsf{pk}^*, 0)$ $\lambda\epsilon^{-2}$ times with fresh randomness (for both D and $\mathsf{KE.Enc}$) each time. Denote by $\hat{d}_k$ the output of $D(\mathsf{KE.Enc}'(\mathsf{pk}^*, 0))$ in the k-th repetition. Compute $\hat{d} = \lambda^{-1}\epsilon^2 \Sigma_{k\in[p]} d_{0,k}$.
   (d) If $|d_0 - \hat{d}| > |d_1 - \hat{d}|$, then set $\hat{w}_j = 1$, if else, set $\hat{w}_j = 0$.
2. Output $\hat{w} = \hat{w}_1 \| \hat{w}_2 \| \cdots \| \hat{w}_\ell$.

We denote by u_0 the probability $\Pr[D(\mathsf{KE.Enc}(\mathsf{pk}^*, 0)) = 1]$, by u_1 the probability $\Pr[D(\mathsf{KE.Enc}(\mathsf{pk}^*, 1)) = 1]$ and by $\hat{u}$ the probability $\Pr[D(\mathsf{KE.Enc}'(\mathsf{pk}^*, 0)) = 1]$. By Chernoff bound, we have

$$\Pr[|d_0 - u_0| \geq \delta u_0] \leq 2e^{-\delta^2 u_0 p/3}.$$

Set $\delta u_0 = \epsilon/8$. Due to that $u_0 \leq 1$, we have that $\delta \geq \epsilon/8$. Therefore,

$$\Pr[|d_0 - u_0| \geq \epsilon/8] \leq 2e^{-\lambda/2^6 \cdot 3}. \tag{5}$$

Similarly,

$$\Pr[|d_1 - u_1| \geq \epsilon/8] \leq 2e^{-\lambda/2^6 \cdot 3}, \text{ and} \tag{6}$$

$$\Pr[|\hat{d} - \hat{u}| \geq \epsilon/8] \leq 2e^{-\lambda/2^6 \cdot 3}. \tag{7}$$

From the (in)equalities (3) and (4), we also have $|u_0 - u_1| \geq \epsilon$ and $|\hat{u} - u_{w_j^*}| \leq negl$. Putting together with the inequalities (5),(6),(7), it follows

$$\Pr[|d_{1-w_j^*} - \hat{d}| > |d_{w_j^*} - \hat{d}|] \geq 1 - negl,$$

which implies that,

$$\Pr[w_j^* \neq \hat{w}_j | \hat{w} \leftarrow E^D(\mathsf{pk}^*, 1^{1/\epsilon})] \leq negl(\lambda).$$

Note also that $(x, w^*) \in R_L$, we have

$$\Pr[\hat{w} \leftarrow E^D(\mathsf{pk}^*, 1^{1/\epsilon}) \wedge (x, \hat{w}) \in R_L] \geq 1 - negl(\lambda),$$

as desired.   □

An alternative construction based on RSR encryption and CDS scheme appears in the full version of this paper [17].

3.2   Nearly-Optimal Extractor for Knowledge Encryption

Following [16], we show the existence of the nearly optimal (T, ϵ)-extractor for any (malicious) key generation algorithm of knowledge encryption, which essentially states that, for any ciphertext distinguisher of size T, the probability that the extractor fails to extract a valid witness for the instance x on the public key whereas the ciphertext distinguisher succeeds is less than ϵ. For any (malicious) key generator that generates multiple public keys simultaneously, this property holds for each one of them, even if the distinguisher takes the output of the nearly optimal extractor as input.

For a given polynomial t, denote by $\overline{x}_{[t]}$ the set of t strings $\{x_k\}_{k \in [t]}$. We first recall the lemma on the existence of nearly-optimal (T, ϵ)-extractor for any hard distributions in [16].

Lemma 1 (Nearly-Optimal (T, ϵ)-Extractor for t-Instance Sampler [16]). *Let L be an NP language and poly be the size of the circuits for deciding the NP-language R_L. Let* Samp *be an arbitrarily t-instance sampling algorithm over L with input distribution ensemble $\mathcal{R} := \{\mathcal{R}_\lambda\}_{\lambda \in \mathbb{N}}$. Let $F := \{F_\lambda\}_{\lambda \in \mathbb{N}}$ be a probabilistic (not necessarily efficient-computable) machine.*

1. *For every polynomial T, ϵ^{-1}, there exists a probabilistic circuit family* $\mathsf{Ext} :=$ $\{\mathsf{Ext}_\lambda\}_{\lambda \in \mathbb{N}}$ *of size* $O(\frac{t}{\epsilon}(T+\mathrm{poly}))$ *such that for every* $j \in [t]$, *every probabilistic circuit family* $C := \{C_\lambda\}_{\lambda \in \mathbb{N}}$ *of size* T *and every security parameter* $\lambda \in \mathbb{N}$,

$$\Pr\left[\begin{array}{c} (x_j, w_j^*) \in R_L \wedge \\ (x_j, w_j') \notin R_L \end{array} \middle| \begin{array}{l} r \leftarrow \mathcal{R}; \overline{x}_{[t]} \leftarrow \mathsf{Samp}(1^\lambda, r); \\ \overline{w}_{[t]}' \leftarrow \mathsf{Ext}(\overline{x}_{[t]}, r, F(r)); \\ w_j^* \leftarrow C(\overline{x}_{[t]}, r, F(r), \overline{w}_{[t]}); \end{array}\right] < \epsilon(\lambda).$$

2. *There exists a probabilistic circuit family* $\mathsf{Ext} := \{\mathsf{Ext}_\lambda\}_{\lambda \in \mathbb{N}}$ *of quasi-polynomial size such that for every probabilistic circuit family* $C := \{C_\lambda\}_{\lambda \in \mathbb{N}}$ *of polynomial size, the above probability is negligible.*

The original version of this lemma in [16] considers only a deterministic function F, however, it is easy to verify that the same proof also yields the above lemma with respect to a probabilistic (possibly unbounded) function F.

We consider an arbitrary key generator $\mathsf{KE.Gen}^*$ that outputs t public keys simultaneously. We write its input as r (including possibly its random coins, NP instances and the corresponding witnesses), and assume that r are drawn from certain distribution ensemble $\mathcal{R} := \{\mathcal{R}_\lambda\}_{\lambda \in \mathbb{N}}$.

The following lemma can be viewed as a knowledge encryption version of Lemma 4 in [16] (which holds only with respect to the Rabin's encryption based on factoring). For the sake of completeness, we provide its proof in the full version [17].

Lemma 2. *Let t be a polynomial. Let $\mathsf{KE.Gen}^*$ be any t-public-key generator of knowledge encryption with respect to an NP language L, whose output is of the form $\overline{\mathsf{pk}}_{[t]}^* = \{(\mathsf{k}_k^*, x_k)\}_{k \in [t]}$, and let the input distribution ensemble be $\mathcal{R} := \{\mathcal{R}_\lambda\}_{\lambda \in \mathbb{N}}$. Let $F := \{F_\lambda\}_{\lambda \in \mathbb{N}}$ be a probabilistic (not necessarily efficient-computable) machine.*

1. *For every polynomial T and every inverse polynomial ϵ, there exists a probabilistic circuit family* $\mathsf{Ext} := \{\mathsf{Ext}_\lambda\}_{\lambda \in \mathbb{N}}$ *of polynomial size such that for every* $j \in [t]$, *every probabilistic distinguisher* $D := \{D_\lambda\}_{\lambda \in \mathbb{N}}$ *of size* T *and any security parameter* $\lambda \in \mathbb{N}$,

$$\left| \Pr\left[\begin{array}{c} D(\overline{\mathsf{pk}}_{[t]}^*, c, r, F(r), \overline{w}_{[t]}') = 1 \wedge \\ (x_j, w_j') \notin R_L \end{array} \middle| \begin{array}{l} r \leftarrow \mathcal{R}; \overline{\mathsf{pk}}_{[t]}^* \leftarrow \mathsf{KE.Gen}^*(1^\lambda, r) \\ \overline{w}_{[t]}' \leftarrow \mathsf{Ext}(\overline{\mathsf{pk}}_{[t]}^*, r, F(r)); \\ c \leftarrow \mathsf{KE.Enc}(\mathsf{pk}_j^*, 0); \end{array}\right] - \right.$$
$$\left. \Pr\left[\begin{array}{c} D(\overline{\mathsf{pk}}_{[t]}^*, c, r, F(r), \overline{w}_{[t]}') = 1 \wedge \\ (x_j, w_j') \notin R_L \end{array} \middle| \begin{array}{l} r \leftarrow \mathcal{R}; \overline{\mathsf{pk}}_{[t]}^* \leftarrow \mathsf{KE.Gen}^*(1^\lambda, r) \\ \overline{w}_{[t]}' \leftarrow \mathsf{Ext}(\overline{\mathsf{pk}}_{[t]}^*, r, F(r)); \\ c \leftarrow \mathsf{KE.Enc}(\mathsf{pk}_j^*, 1); \end{array}\right] \right| < \epsilon(\lambda).$$

2. *There exists a probabilistic circuit family* $\mathsf{Ext} := \{\mathsf{Ext}_\lambda\}_{\lambda \in \mathbb{N}}$ *of quasi-polynomial size such that for every probabilistic distinguisher* $D := \{D_\lambda\}_{\lambda \in \mathbb{N}}$ *of polynomial size, the above holds with respect to a negligible function* ϵ.

Remark 3. The proof strategy of [16] for this kind of lemma only works if the algorithms Ext and D take the same input (except that D is also given the output of Ext as input). However, in the security reduction, D usually sees a complete session transcript, but the simulator has only a partial transcript when it applies Ext to extract some secrets from the adversary. This is the reason why we have both Ext and D take an extra input $F(r)$, which represents some messages in a session generated after the point that the simulator did extraction. Although $F(r)$ may not be efficiently computable from the input of Ext, but in our cases, the simulator is able to compute it efficiently with the randomness used in generating certain transcript prefix.

4   Three-Round Simulatable Oblivious Transfer

In this section, we show how to use the knowledge encryption scheme to construct a three-round OT scheme with simulatable security for the receiver and (T, ϵ)-simulatable security for the sender.

Our protocol proceeds as follows. The sender generates two images y_0 and y_1 of a one-way function f and prove to the receiver that it knows one pre-image of y_0 or y_1 via a three-round WI protocol. Given the pair (y_0, y_1) and input b, the receiver prepares two instances x_0 and x_1 in the following way: it runs the HVZK simulator of the Σ-protocol to obtain an acceptable proof (a, b, z) of knowledge of one preimage of y_0 or y_1, and sets $x_b = (y_0, y_1, a, b)$ and $x_{1-b} = (y_0, y_1, a, 1 - b)$, where $x_i = (y_0, y_1, a, i)$ is said to be a YES instance if and only if there exists a z such that (a, i, z) is acceptable. The receiver now generates pk_b honestly using the valid witness z for $x_b = (y_0, y_1, a, b)$, and runs the key simulator of knowledge encryption to obtain the other public key pk_{1-b}. In the third round, the sender encrypts its two message under the two public keys respectively and sends the two ciphertexts to the receiver.

We give a formal description of our construction in Fig. 3, which is based on the following ingredients:

- A one-way function f.
- A three-round public-coin witness indistinguishable argument $(\mathsf{WI}_1, \mathsf{WI}_2, \mathsf{WI}_3)$ with special soundness and negligible soundness error for language L_f.
- A Σ-protocol (a, e, z) with 1-bit challenge for language L_f.
- A knowledge encryption scheme $(\mathsf{KE.Gen}, \mathsf{KE.Enc}, \mathsf{KE.Dec})$ for language L_Σ.

where L_f, L_Σ are defined as follows:

$$L_f := \{(y_0, y_1) | \exists x \text{ s.t. } f(x) = y_0 \vee f(x) = y_1\}$$
$$L_\Sigma := \{(y_0, y_1, a, e) | \exists z \text{ s.t. } (a, e, z) \text{ is an acceptable proof for } (y_0, y_1) \in L\}$$

Three-round Oblivious Transfer Protocol

Sender Input: Security parameter 1^λ and messages $m_0, m_1 \in \{0,1\}^n$.

Receiver Input: Security parameter 1^λ and bit $b \in \{0,1\}$.

- **Sender Message**: Sample $\delta_0, \delta_1 \leftarrow \{0,1\}^\lambda$ at random, compute $y_0 = f(\delta_0)$, $y_1 = f(\delta_1)$ and generate WI_1 as the first message of WI for $(y_0, y_1) \in L_f$. Send $(y_0, y_1, \mathsf{WI}_1)$.

- **Receiver Message**: Generate the second WI message WI_2. Use the HVZK simulator of the Σ-protocol to generate an acceptable Σ-proof (a, b, z) for $(y_0, y_1) \in L_f$ (where b is the receiver's input). Generate $(\mathsf{pk}_b, \mathsf{sk}_b) \leftarrow \mathsf{KE.Gen}(1^\lambda, (y_0, y_1, a, b), z)$ (where $((y_0, y_1, a, b), z) \in R_{L_\Sigma}$) and $\mathsf{pk}_{1-b} \leftarrow \mathsf{KE.KeySim}(1^\lambda, (y_0, y_1, a, 1-b))$. Send $(\mathsf{WI}_2, \mathsf{pk}_0, \mathsf{pk}_1)$.

- **Sender Message**: Write $\mathsf{pk}_i = (\mathsf{k}_i, x_i = ((y_0, y_1, a, i)))$ for $i \in \{0,1\}$, and check if both x_i share the same (y_0, y_1, a). If not, abort; Otherwise, generate the third WI message WI_3 using a random witness and encrypt messages m_i under public key pk_i in bitwise manner: $c_0 \leftarrow \mathsf{KE.Enc}(\mathsf{pk}_0, m_0)$, $c_1 \leftarrow \mathsf{KE.Enc}(\mathsf{pk}_1, m_1)$. Send $(\mathsf{WI}_3, c_0, c_1)$.

- **Receiver's Output**: Check if $(\mathsf{WI}_1, \mathsf{WI}_2, \mathsf{WI}_3)$ is acceptable. If not, output $\bot$; otherwise, output $m_b \leftarrow \mathsf{KE.Dec}(\mathsf{sk}_b, c_b)$ (if c_b is not decryptable, set m_b to be $\bot$)).

Fig. 3. Three-round oblivious transfer protocol

Note that non-interactive commitment can be built from two-round (perfectly correct) OT with game-based security (see footnote 2). Thus, two-round OT with game-based security as we define is sufficient for constructing all primitives used in our protocol.

Theorem 2. *Assuming the existence of two-round OT with game-based security (against polynomial-time adversaries), there exists a three-round OT protocol with fully simulatable security for the receiver and (T, ϵ)-simulatable security for the sender. Furthermore, the same protocol also achieves quasi-polynomial simulatable security for the sender under the same assumption.*

Due to space limitation, we defer the detailed proof to the full version of this paper [17]. Here we only provide a sketch of proof.

proof sketch. The simulatable security for the receiver can be proven using rewinding simulation strategy (once a preimage is obtained by rewinding, the simulator can generate two valid public keys and decrypt both ciphertexts[9] from the sender), but one must be careful in the analysis of the running time of the rewinding simulator, which actually requires the Goldreich-Kahan technique [26] to make sure that the simulator will run in expected polynomial time.

---

[9] Like the honest receiver, the simulator sets the "plaintext" of an undecryptable ciphertext to be $\bot$.

The (T, ϵ)-simulatable security for the sender can be proven by constructing the following simulator. The simulator generates the first message by following the honest sender strategy. Upon receiving two public keys $\mathsf{pk}_0 = (\mathsf{k}_0, x_0), \mathsf{pk}_1 = (\mathsf{k}_1, x_1)$ of knowledge encryption from the malicious receiver, it applies the nearly optimal extractor for the receiver and tries to extract one witness of x_i. For the case that the simulator extracts two witnesses, it aborts the simulation; For the case that the simulator extracts at most one valid witness, it sets $b' = 0$ if a valid z_0 is extracted s.t. $(x_0 = (y_0, y_1, a, 0), z_0) \in R_{L_\Sigma}$ and sets $b' = 1$ if else. Then it sends b' to $\mathcal{F}_{OT}$ and encrypts the message $m_{b'}$ received from $\mathcal{F}_{OT}$ under both public keys $\mathsf{pk}_{b'}$ and $\mathsf{pk}_{1-b'}$. For the first case, we prove that it happens only with negligible probability. For the second case, we will use the (near) optimality of the extractor to prove that the simulation and the real execution are indistinguishable against distinguishers of certain size except for small probability.

When replacing (T, ϵ)-extractor with a quasi-polynomial extractor (guaranteed by Lemma 2) in the simulation of the receiver's view, the second part of Theorem 2 follows.

5   Three-Round Weak Zero-Knowledge Argument of Knowledge

In this section, we construct a delayed-input (T, ϵ)-zero-knowledge argument satisfying adaptive argument of knowledge, which is based on the following ingredients:

- A 3-round OT $(\mathsf{OT}_1, \mathsf{OT}_2, \mathsf{OT}_3)$ presented in Fig. 3.
- A one-way function f.
- A knowledge encryption scheme $(\mathsf{KE.Gen}, \mathsf{KE.Enc}, \mathsf{KE.Dec})$ for language L'_f.
- A 3-round public-coin WI protocol $(\mathsf{WI}_1, \mathsf{WI}_2, \mathsf{WI}_3)$ with special-soundness property for language L_{pk}.
- A Σ-protocol (α, β, γ) with 1-bit challenge space for an NP language L.

where L'_f, L_{pk} are defined as follows:

$$L'_f : \{y | \exists \delta \ s.t. \ f(\delta) = y\}$$

$$L_{pk} : \{\mathsf{pk}_0, \mathsf{pk}_1 | \exists b, \mathsf{sk}_b, r_{\mathsf{KE}}, (y_b, \delta_b) \in L'_f \ s.t. \ (\mathsf{pk}_b, \mathsf{sk}_b) = \mathsf{KE.Gen}(1^\lambda, y_b, \delta_b; r_{\mathsf{KE}})\}$$

We formally present our construction in Fig. 4. Due to space limitation, we give only the statement of our result in this section. The proof can be found in the full version [17].

Theorem 3. *Assuming the existence of two-round OT protocol with game-based security (against polynomial-time adversaries), there exists a three-round delayed-input (T, ϵ)-zero-knowledge adaptive argument of knowledge. Furthermore, the same protocol also satisfies witness hiding and quasi-polynomial simulatable zero knowledge under the same assumption.*

Delayed-input (T, ϵ)-Zero-knowledge Argument of Knowledge

Prover Input: $(x, w) \in R_L$.

- **Prover Message:** Run OT_1 λ times in parallel and obtain $\{\mathsf{ot}_i\}_{i \in [\lambda]}$. Sample $\delta_0, \delta_1 \leftarrow \{0, 1\}^\lambda$ and compute $y_0 = f(\delta_0), y_1 = f(\delta_1)$. Generate two knowledge encryption public keys $(\mathsf{pk}_0, \mathsf{sk}_0) \leftarrow \mathsf{KE.Gen}(1^\lambda, y_0, \delta_0)$, $(\mathsf{pk}_1, \mathsf{sk}_1) \leftarrow \mathsf{KE.Gen}(1^\lambda, y_1, \delta_1)$ and the first message WI_1 of WI for statement $(\mathsf{pk}_0, \mathsf{pk}_1) \in L_{pk}$. Send $(\{\mathsf{ot}_{1,i}\}_{i \in [\lambda]}, \mathsf{pk}_0, \mathsf{pk}_1, \mathsf{WI}_1)$.

- **Verifier Message:** For each $i \in [\lambda]$, sample $\beta_i \leftarrow \{0, 1\}$ and compute $\mathsf{ot}_{2,i} \leftarrow \mathsf{OT}_{2,i}(\beta_i)$ independently. Generate the second message WI_2 of WI. Send $(\{\mathsf{ot}_{2,i}\}_{i \in [\lambda]}, \mathsf{WI}_2)$.

- **Prover Message:** For each $i \in [\lambda]$, generate two Σ-proofs with the same first message (i.e. $(\alpha_i, 0, \gamma_{i,0}), (\alpha_i, 1, \gamma_{i,1})$). For $b = 0, 1$, encrypt $\gamma_{i,b}$ using both of $\mathsf{pk}_0, \mathsf{pk}_1$ separately to obtain $C_{i,b}$, i.e. $C_{i,b} = (\mathsf{KE.Enc}(\mathsf{pk}_0, \gamma_{i,b}), \mathsf{KE.Enc}(\mathsf{pk}_1, \gamma_{i,b}))$. Let $\gamma'_{i,b}$ be the message consisting of $\gamma_{i,b}$ and the randomness used in computing $C_{i,b}$. Compute $\mathsf{ot}_{3,i} \leftarrow \mathsf{OT}_{3,i}(\gamma'_{i,0}, \gamma'_{i,1})$. Generate the third message WI_3 of WI. Send $(x, \{\alpha_i, C_{i,0}, C_{i,1}, \mathsf{ot}_{3,i}\}_{i \in [\lambda]}, \mathsf{WI}_3)$.

- **Verifier's Output:** Recover γ'_{i,β_i} from OT, output 1 if for all $i \in [\lambda]$, $(\alpha_i, \beta_i, \gamma_{i,\beta_i})$ and $\mathsf{WI}_1, \mathsf{Wi}_2, \mathsf{WI}_3$ are acceptable proofs and C_{i,β_i} is indeed the encryptions of γ_{i,β_i} (using the randomness contained in γ'_{i,β_i}).

Fig. 4. Three-round argument system for NP

6   Two-Party Secure Computation

Equipped with the three-round OT and zero knowledge argument constructed in previous sections, we now follow the GMW paradigm [28] to give a three-round protocol for weakly secure two-party computation for independent-input functionalities. We use the following ingredients in our construction:

- A 3-round OT $(\mathsf{OT}_1, \mathsf{OT}_2, \mathsf{OT}_3)$ (presented in Fig. 3).
- A 3-round delayed-input weak zero knowledge argument $(\mathsf{ZK}_1, \mathsf{ZK}_2, \mathsf{ZK}_3)$ (presented in Fig. 4) for language L_{2pc}.
- A garbling circuit scheme $\mathsf{GC} = (\mathsf{Garble}, \mathsf{Eval})$,

where L_{2pc} is defined as follows: $(\hat{C}, \{\mathsf{lab}^x_{i,x_i}\}_{i \in [n]}, \{\mathsf{ot}_{1,i}, \mathsf{ot}_{2,i}, \mathsf{ot}_{3,i}\}_{i \in [n]}) \in L_{2pc}$ if and only if there exists a random tape for the honest sender (on input $\mathsf{ot}_{2,i}$) to generate messages $(\hat{C}, \{\mathsf{lab}^x_{i,x_i}\}_{i \in [n]}, \{c_{i,b} = \mathsf{KE.Enc}(\mathsf{pk}^1_{i,b}, \mathsf{lab}^y_{i,b})\}_{i \in [n], b \in \{0,1\}})(c_{i,b}$ is the ciphertexts in $\mathsf{ot}_{3,i}$ under the public key $\mathsf{pk}^1_{i,b}$ contained in $\mathsf{ot}_{2,i})$.

We assume that the independent-input functionality C maps (x, y) of length $2n$ to a string of length n. The protocol is formally presented in Fig. 5.

Theorem 4. *Assuming the existence of two-round OT protocol with game-based security (against polynomial-time adversaries), there exists a three-round two-party computation protocol for independent-input functionalities that achieves*

3-round Two-party Weak Secure Computation

Sender Input: $x \in \{0,1\}^n$
Receiver Input: $y \in \{0,1\}^n$

- **Sender Message:** Run OT_1 λ times in parallel and obtain $\{\mathsf{ot}_i\}_{i \in [\lambda]}$. Generate the first message ZK_1.
  Send $(\{\mathsf{ot}_{1,i}\}_{i \in [n]}, \mathsf{ZK}_1)$.

- **Receiver Message:** Generate the second message ZK_2. For each $i \in [n]$, compute $\mathsf{ot}_{2,i} \leftarrow \{\mathsf{OT}_{2,i}(y_i)\}_{i \in [n]}$ independently where y_i is the i-th bit of y.
  Send $(\{\mathsf{ot}_{2,i}\}_{i \in [n]}, \mathsf{ZK}_2)$.

- **Sender Message:** Use GC to generate the garbled circuit $\hat{C}$ along with labels $\{\mathsf{lab}_{i,b}^x\}_{i \in [n], b \in \{0,1\}}$, $\{\mathsf{lab}_{i,b}^y\}_{i \in [n], b \in \{0,1\}}$ for functionality C. Compute $\mathsf{ot}_{3,i} \leftarrow \mathsf{OT}_{3,i}(\mathsf{lab}_{i,0}^y, \mathsf{lab}_{i,1}^y)$. Compute ZK_3 for
  $(\hat{C}, \{\mathsf{lab}_{i,x_i}^x\}_{i \in [n]}, \{\mathsf{ot}_{1,i}, \mathsf{ot}_{2,i}, \mathsf{ot}_{3,i}\}_{i \in [n]}) \in L_{2pc}$.
  Send $(\hat{C}, \{\mathsf{lab}_{i,x_i}^x\}_{i \in [n]}, \{\mathsf{ot}_{3,i}\}_{i \in [n]}, \mathsf{ZK}_3)$.

- **Receiver's Output:** Recover lab_{i,y_i}^y from OT, and check if $(\mathsf{ZK}_1, \mathsf{ZK}_2, \mathsf{ZK}_3)$ is acceptable. If not, output $\bot$; otherwise, output
  $\hat{C}(\{\mathsf{lab}_{i,x_i}^x\}_{i \in [n]}, \{\mathsf{lab}_{i,y_i}^y\}_{i \in [n]})$.

Fig. 5. 3-round two-party weak secure computation

(T, ϵ)-*security against malicious receiver and standard security against malicious sender. Furthermore, the same protocol also achieves quasi-polynomial simulatable security against malicious receiver under the same assumption.*

We provide the proof of Theorem 4 in the full version of this paper [17].

7   More Applications

In this section we present direct applications of our results in previous sections to various protocols, including extractable commitment, selective opening secure commitment and concurrent zero knowledge argument in the BPK model. Compared with existing protocols, all our new constructions only rely on two-round OT with game-based security. Since one can prove the security of these new constructions using essentially the same security proof strategies in [16,34], we will not repeat these proofs here.

The work [34] provides a transformation of non-interactive commitment into a three-round extractable commitment via three-round weak zero knowledge argument of knowledge. When using our construction of (T, ϵ)-zero knowledge argument of knowledge in their transformation, we have the following result.

Theorem 5. *Assuming the existence of two-round OT with game-based security (against polynomial-time adversaries), there exists a three-round extractable commitment scheme.*

The commitment with (T, ϵ)-security under selective opening attack and concurrent (T, ϵ)-zero knowledge argument (in the BPK model) in [16] are constructed from Rabin encryption scheme (based on hardness of Factoring). We can also replace the Rabin encryption scheme with our knowledge encryption (and revise their protocol accordingly so that the simulation can go through with a witness for the instance on the public key of knowledge encryption), and obtain the following result.

Theorem 6. *Assuming the existence of two-round OT with game-based security (against polynomial-time adversaries), there exist:*

1. *Two-round commitment scheme with (T, ϵ)-security under selective opening attacks.*
2. *Three-round concurrent (T, ϵ)-zero knowledge argument with concurrent soundness in the BPK model, which also satisfies concurrent witness hiding in the same model.*
3. *All above protocols satisfy (fully) quasi-polynomial simulatable security.*

Acknowledgments. We would like to thank the anonymous reviewers for their valuable suggestions. We are supported by the National Natural Science Foundation of China (Grant No. 61932019 and No. 61772522), the Key Research Program of Frontier Sciences, CAS (Grant No. QYZDB-SSW-SYS035) and Beijing Natural Science Foundation (Grant No. M22003).

References

1. Aiello, B., Ishai, Y., Reingold, O.: Priced oblivious transfer: how to sell digital goods. In: Pfitzmann, B. (ed.) EUROCRYPT 2001. LNCS, vol. 2045, pp. 119–135. Springer, Heidelberg (2001). https://doi.org/10.1007/3-540-44987-6_8
2. Aiello, W., Bhatt, S., Ostrovsky, R., Rajagopalan, S.R.: Fast verification of any remote procedure call: short witness-indistinguishable one-round proofs for NP. In: Montanari, U., Rolim, J.D.P., Welzl, E. (eds.) ICALP 2000. LNCS, vol. 1853, pp. 463–474. Springer, Heidelberg (2000). https://doi.org/10.1007/3-540-45022-X_39
3. Alwen, J., Persiano, G., Visconti, I.: Impossibility and feasibility results for zero knowledge with public keys. In: Shoup, V. (ed.) CRYPTO 2005. LNCS, vol. 3621, pp. 135–151. Springer, Heidelberg (2005). https://doi.org/10.1007/11535218_9
4. Ananth, P., Jain, A.: On secure two-party computation in three rounds. In: Kalai, Y., Reyzin, L. (eds.) TCC 2017. LNCS, vol. 10677, pp. 612–644. Springer, Cham (2017). https://doi.org/10.1007/978-3-319-70500-2_21
5. Badrinarayanan, S., Goyal, V., Jain, A., Kalai, Y.T., Khurana, D., Sahai, A.: Promise zero knowledge and its applications to round optimal MPC. In: Shacham, H., Boldyreva, A. (eds.) CRYPTO 2018. LNCS, vol. 10992, pp. 459–487. Springer, Cham (2018). https://doi.org/10.1007/978-3-319-96881-0_16
6. Benhamouda, F., Lin, H.: k-round multiparty computation from k-round oblivious transfer via garbled interactive circuits. In: Nielsen, J.B., Rijmen, V. (eds.) EUROCRYPT 2018. LNCS, vol. 10821, pp. 500–532. Springer, Cham (2018). https://doi.org/10.1007/978-3-319-78375-8_17

7. Bitansky, N., Brakerski, Z., Kalai, Y., Paneth, O., Vaikuntanathan, V.: 3-message zero knowledge against human ignorance. In: Hirt, M., Smith, A. (eds.) TCC 2016. LNCS, vol. 9985, pp. 57–83. Springer, Heidelberg (2016). https://doi.org/10.1007/978-3-662-53641-4_3

8. Bitansky, N., Canetti, R., Paneth, O., Rosen, A.: On the existence of extractable one-way functions. In: Proceedings of the 45th Annual ACM Symposium on the Theory of Computing - STOC'14, pp. 505–514. ACM Press (2014). https://doi.org/10.1145/2591796.2591859

9. Bitansky, N., Khurana, D., Paneth, O.: Weak zero-knowledge beyond the black-box barrier. In: Proceedings of the 51st Annual ACM SIGACT Symposium on Theory of Computing - STOC'19, pp. 1091–1102. ACM press (2019). https://doi.org/10.1145/3313276.3316382

10. Brakerski, Z., Döttling, N.: Two-message statistically sender-private OT from LWE. In: Beimel, A., Dziembowski, S. (eds.) TCC 2018. LNCS, vol. 11240, pp. 370–390. Springer, Cham (2018). https://doi.org/10.1007/978-3-030-03810-6_14

11. Canetti, R.: Universally composable security: A new paradigm for cryptographic protocols. In: Proceedings of the 42nd Annual Symposium on Foundations of Computer Science - FOCS'01, pp. 136–145. IEEE Computer Society (2001). https://doi.org/10.1109/SFCS.2001.959888

12. Rai Choudhuri, A., Ciampi, M., Goyal, V., Jain, A., Ostrovsky, R.: Round optimal secure multiparty computation from minimal assumptions. In: Pass, R., Pietrzak, K. (eds.) TCC 2020. LNCS, vol. 12551, pp. 291–319. Springer, Cham (2020). https://doi.org/10.1007/978-3-030-64378-2_11

13. Choudhuri, A.R., Ciampi, M., Goyal, V., Jain, A., Ostrovsky, R.: Oblivious transfer from trapdoor permutations in minimal rounds. In: Nissim, K., Waters, B. (eds.) TCC 2021. LNCS, vol. 13043, pp. 518–549. Springer, Cham (2021). https://doi.org/10.1007/978-3-030-90453-1_18

14. Chung, K.-M., Lui, E., Pass, R.: From weak to strong zero-knowledge and applications. In: Dodis, Y., Nielsen, J.B. (eds.) TCC 2015. LNCS, vol. 9014, pp. 66–92. Springer, Heidelberg (2015). https://doi.org/10.1007/978-3-662-46494-6_4

15. Ciampi, M., Ostrovsky, R., Siniscalchi, L., Visconti, I.: Round-optimal secure two-party computation from trapdoor permutations. In: Kalai, Y., Reyzin, L. (eds.) TCC 2017. LNCS, vol. 10677, pp. 678–710. Springer, Cham (2017). https://doi.org/10.1007/978-3-319-70500-2_23

16. Deng, Y.: Individual simulations. In: Moriai, S., Wang, H. (eds.) ASIACRYPT 2020. LNCS, vol. 12493, pp. 805–836. Springer, Cham (2020). https://doi.org/10.1007/978-3-030-64840-4_27

17. Deng, Y., Zhang, X.: Knowledge encryption and its applications to simulatable protocols with low round-complexity. Cryptology ePrint Archive, Paper 2022/1193 (2022). https://eprint.iacr.org/2022/1193

18. Döttling, N., Garg, S., Hajiabadi, M., Masny, D., Wichs, D.: Two-round oblivious transfer from CDH or LPN. In: Canteaut, A., Ishai, Y. (eds.) EUROCRYPT 2020. LNCS, vol. 12106, pp. 768–797. Springer, Cham (2020). https://doi.org/10.1007/978-3-030-45724-2_26

19. Friolo, D., Masny, D., Venturi, D.: A black-box construction of fully-simulatable, round-optimal oblivious transfer from strongly uniform key agreement. In: Hofheinz, D., Rosen, A. (eds.) TCC 2019. LNCS, vol. 11891, pp. 111–130. Springer, Cham (2019). https://doi.org/10.1007/978-3-030-36030-6_5

20. Garg, S., Gentry, C., Sahai, A., Waters, B.: Witness encryption and its applications. In: Proceedings of the 45th Annual ACM Symposium on Theory of Comput-

ing - STOC'13, p. 467–476. ACM press (2013). https://doi.org/10.1145/2488608.2488667

21. Garg, S., Mukherjee, P., Pandey, O., Polychroniadou, A.: The exact round complexity of secure computation. In: Fischlin, M., Coron, J.-S. (eds.) EUROCRYPT 2016. LNCS, vol. 9666, pp. 448–476. Springer, Heidelberg (2016). https://doi.org/10.1007/978-3-662-49896-5_16

22. Garg, S., Srinivasan, A.: Two-round multiparty secure computation from minimal assumptions. In: Nielsen, J.B., Rijmen, V. (eds.) EUROCRYPT 2018. LNCS, vol. 10821, pp. 468–499. Springer, Cham (2018). https://doi.org/10.1007/978-3-319-78375-8_16

23. Gertner, Y., Ishai, Y., Kushilevitz, E., Malkin, T.: Protecting data privacy in private information retrieval schemes. In: Proceedings of the 30th Annual ACM Symposium on Theory of Computing - STOC'98, p. 151–160. ACM press (1998). https://doi.org/10.1145/276698.276723

24. Gertner, Y., Kannan, S., Malkin, T., Reingold, O., Viswanathan, M.: The relationship between public key encryption and oblivious transfer. In: Proceedings of the 41th Annual IEEE Symposium on Foundations of Computer Science - FOCS'00, pp. 325–335. IEEE Computer Society (2000). https://doi.org/10.1109/SFCS.2000.892121

25. Goldreich, O.: Foundations of Cryptography, vol. Basic Applications. Cambridge University Press (2004). https://doi.org/10.1017/CBO9780511721656

26. Goldreich, O., Kahan, A.: How to construct constant-round zero-knowledge proof systems for NP. J. Cryptol. **9**(3), 167–189 (1996). https://doi.org/10.1007/BF00208001

27. Goldreich, O., Krawczyk, H.: On the composition of zero-knowledge proof systems. SIAM J. Comput. **25**(1), 169–192 (1996). https://doi.org/10.1137/S0097539791220688

28. Goldreich, O., Micali, S., Wigderson, A.: How to play any mental game. In: Proceedings of the 19th Annual ACM Symposium on Theory of Computing - STOC'87, pp. 218–229. ACM press (1987). https://doi.org/10.1145/28395.28420

29. Goldwasser, S., Kalai, Y.T., Popa, R.A., Vaikuntanathan, V., Zeldovich, N.: How to run turing machines on encrypted data. In: Canetti, R., Garay, J.A. (eds.) CRYPTO 2013. LNCS, vol. 8043, pp. 536–553. Springer, Heidelberg (2013). https://doi.org/10.1007/978-3-642-40084-1_30

30. Goldwasser, S., Micali, S., Rackoff, C.: The knowledge complexity of interactive proof systems. SIAM J. Comput. **18**(1), 186–208 (1989). https://doi.org/10.1137/0218012

31. Goyal, V., Jain, A., Jin, Z., Malavolta, G.: Statistical zaps and new oblivious transfer protocols. In: Canteaut, A., Ishai, Y. (eds.) EUROCRYPT 2020. LNCS, vol. 12107, pp. 668–699. Springer, Cham (2020). https://doi.org/10.1007/978-3-030-45727-3_23

32. Halevi, S., Hazay, C., Polychroniadou, A., Venkitasubramaniam, M.: Round-optimal secure multi-party computation. In: Shacham, H., Boldyreva, A. (eds.) CRYPTO 2018. LNCS, vol. 10992, pp. 488–520. Springer, Cham (2018). https://doi.org/10.1007/978-3-319-96881-0_17

33. Halevi, S., Kalai, Y.T.: Smooth projective hashing and two-message oblivious transfer. J. Cryptol. **25**(1), 158–193 (2010). https://doi.org/10.1007/s00145-010-9092-8

34. Jain, A., Kalai, Y.T., Khurana, D., Rothblum, R.: Distinguisher-dependent simulation in two rounds and its applications. In: Katz, J., Shacham, H. (eds.) CRYPTO

2017. LNCS, vol. 10402, pp. 158–189. Springer, Cham (2017). https://doi.org/10.1007/978-3-319-63715-0_6

35. Kalai, Y.T., Khurana, D., Sahai, A.: Statistical witness indistinguishability (and more) in two messages. In: Nielsen, J.B., Rijmen, V. (eds.) EUROCRYPT 2018. LNCS, vol. 10822, pp. 34–65. Springer, Cham (2018). https://doi.org/10.1007/978-3-319-78372-7_2

36. Kalai, Y.T., Raz, R.: Probabilistically checkable arguments. In: Halevi, S. (ed.) CRYPTO 2009. LNCS, vol. 5677, pp. 143–159. Springer, Heidelberg (2009). https://doi.org/10.1007/978-3-642-03356-8_9

37. Katz, J., Ostrovsky, R.: Round-optimal secure two-party computation. In: Franklin, M. (ed.) CRYPTO 2004. LNCS, vol. 3152, pp. 335–354. Springer, Heidelberg (2004). https://doi.org/10.1007/978-3-540-28628-8_21

38. Kiyoshima, S.: Black-box impossibilities of obtaining 2-round weak ZK and strong WI from polynomial hardness. In: Nissim, K., Waters, B. (eds.) TCC 2021. LNCS, vol. 13042, pp. 369–400. Springer, Cham (2021). https://doi.org/10.1007/978-3-030-90459-3_13

39. Lombardi, A., Schaeffer, L.: A note on key agreement and non-interactive commitments. Cryptology ePrint Archive, Paper 2019/279 (2019). https://eprint.iacr.org/2019/279

40. Naor, M., Pinkas, B.: Efficient oblivious transfer protocols. In: Proceedings of the 12th Annual Symposium on Discrete Algorithms - SODA'01, pp. 448–457. Society for Industrial and Applied Mathematics (2001)

41. Ostrovsky, R., Richelson, S., Scafuro, A.: Round-optimal black-box two-party computation. In: Gennaro, R., Robshaw, M. (eds.) CRYPTO 2015. LNCS, vol. 9216, pp. 339–358. Springer, Heidelberg (2015). https://doi.org/10.1007/978-3-662-48000-7_17

42. Pass, R.: Simulation in quasi-polynomial time, and its application to protocol composition. In: Biham, E. (ed.) EUROCRYPT 2003. LNCS, vol. 2656, pp. 160–176. Springer, Heidelberg (2003). https://doi.org/10.1007/3-540-39200-9_10

43. Peikert, C., Vaikuntanathan, V., Waters, B.: A framework for efficient and composable oblivious transfer. In: Wagner, D. (ed.) CRYPTO 2008. LNCS, vol. 5157, pp. 554–571. Springer, Heidelberg (2008). https://doi.org/10.1007/978-3-540-85174-5_31

44. Xiao, D.: (Nearly) round-optimal black-box constructions of commitments secure against selective opening attacks. In: Ishai, Y. (ed.) TCC 2011. LNCS, vol. 6597, pp. 541–558. Springer, Heidelberg (2011). https://doi.org/10.1007/978-3-642-19571-6_33

45. Xiao, D.: Errata to *(Nearly) round-optimal black-box constructions of commitments secure against selective opening attacks*. In: Sahai, A. (ed.) TCC 2013. LNCS, vol. 7785, pp. 721–722. Springer, Heidelberg (2013). https://doi.org/10.1007/978-3-642-36594-2_40

Compact and Tightly Selective-Opening Secure Public-key Encryption Schemes

Jiaxin Pan[(✉)] and Runzhi Zeng

Department of Mathematical Sciences, NTNU – Norwegian University of Science and Technology, Trondheim, Norway
{jiaxin.pan,runzhi.zeng}@ntnu.no

Abstract. We propose four public-key encryption schemes with tight simulation-based selective-opening security against chosen-ciphertext attacks (SIM-SO-CCA) in the random oracle model. Our schemes only consist of small constant amounts of group elements in the ciphertext, ignoring smaller contributions from symmetric-key encryption, namely, they have compact ciphertexts. Furthermore, three of our schemes have compact public keys as well.

Known (almost) tightly SIM-SO-CCA secure PKE schemes are due to the work of Lyu et al. (PKC 2018) and Libert et al. (Crypto 2017). They have either linear-size ciphertexts or linear-size public keys. Moreover, they only achieve almost tightness, namely, with security loss depending on the security parameters.

Different to them, our schemes are the *first* ones achieving both tight SIM-SO-CCA security and compactness. Our schemes can be divided into two families:

Direct Constructions. Our first three schemes are constructed directly based on the Strong Diffie-Hellman (StDH), Computational DH (CDH), and Decisional DH assumptions. Both their ciphertexts and public keys are compact. Their security loss is a small constant. Interestingly, our CDH-based construction is the first scheme achieving all these advantages based on a weak, search assumption.

A Generic Construction. Our last scheme is the well-known Fujisaki-Okamoto transformation. We show that it can turn a lossy encryption scheme into a tightly SIM-SO-CCA secure PKE. This transformation preserves both tightness and compactness of the underlying lossy encryption, which is in contrast to the non-tight proof of Heuer et al. (PKC 2015).

Keywords: Selective-opening security · Public-key encryption · Tight security · Random oracle model

1 Introduction

Selective-opening (SO) security is a stronger security notion for encryption schemes. It considers encryption security in the multi-challenge setting. More precisely, an adversary is given multiple challenge ciphertexts and it is allowed

© International Association for Cryptologic Research 2022
S. Agrawal and D. Lin (Eds.): ASIACRYPT 2022, LNCS 13793, pp. 363–393, 2022.
https://doi.org/10.1007/978-3-031-22969-5_13

to corrupt some of them to get the corresponding randomness. SO security guarantees that even with this additional capability an adversary still cannot learn any information about the remaining 'unopened' messages.

The motivation of constructing SO secure encryption is that removing cryptographic information is hard and expensive in practice and adversaries can hack into a user's computer and reveal the randomness used in generating a ciphertext. In some scenario, it is even a requirement to reveal the randomness to publicly verify a user's computation.

DEFINITIONS OF SELECTIVE-OPENING SECURITY. There are two types of definitions for SO security, the indistinguishability-based (IND-based) ones (weak-IND-SO and full-IND-SO) [3,8] and the simulation-based (SIM-based) one (SIM-SO) [3]. They are not polynomial-time equivalent to each other. For SIM-SO security, it requires that for every SO adversary its output can be efficiently simulated by a simulator that sees only the opened messages. SIM-SO notion is the most common one to study [20,22,25,28,29], since it does not require the message distribution to be efficiently conditionally resamplable (cf. [3]). Moreover, previous work showed that SIM-SO-CCA and full-IND-SO-CCA notions are the strongest SO security [2,8,25].

TIGHT REDUCTIONS. When we prove the security of a cryptographic scheme Π, we often construct a reduction to show that breaking the security of Π implies breaking the underlying assumption Γ. For concrete security, we argue that if an adversary $\mathcal{A}$ has advantage ϵ in breaking Π then we have another adversary $\mathcal{B}$ that breaks Γ with advantage $\epsilon' = \epsilon/L$, and the factor L is called the security loss.

A cryptographic scheme is called tightly secure if L is a small constant, assuming that the running time of $\mathcal{A}$ is approximately the same as $\mathcal{B}$ (up to a constant factor). A tight reduction can give quantitatively higher guarantees than a loose one. From a more practical perspective, a tight reduction allows shorter key-length recommendations based on the best known attacks against the underlying assumption. This can potentially yield more efficient schemes. Currently, our community aims to reduce the cost for tight security and construct efficient and tightly secure cryptographic schemes (such as the signature scheme in [12]). Hence, it is more desirable to have an efficient and tightly secure scheme, compared to its non-tight counterparts.

OUR GOAL: COMPACT PKE WITH TIGHT SIM-SO-CCA SECURITY. In this paper, we are interested in efficient and tightly SIM-SO-CCA secure public-key encryption schemes. We aim at schemes with compact ciphertexts and public keys. Here 'compact' means constant-size, and SIM-SO-CCA security provides security against chosen-ciphertext attacks in addition to the SIM-SO security. We discuss the state of the art in approaching this goal as follows:

(ALMOST) TIGHT, YET NON-COMPACT SCHEMES. While there are compact and tightly IND-CCA secure PKE schemes [16,18], known tightly SIM-SO-CCA PKE schemes [27,29] are still non-compact wrt. either ciphertext size or public key size. Moreover, the security reductions in both schemes are not fully tight, but almost tight (in the terminology of [11]), namely, the security loss depends on the message bit-length that is a polynomial of the security parameter. Although almost tightness is already interesting, our goal is to achieve security loss with small constants, and it was unknown even with random oracles.

To provide more details, the scheme of Lyu et al. [29] is a recent PKE scheme with tight SIM-SO-CCA security, and its ciphertexts consist of $O(|m|)$ group elements, where $|m|$ is the bit-length of the message. In a nutshell, their construction is a generic construction that tightly turns a IND-CCA secure key encapsulation mechanism (KEM) to a SIM-SO-CCA secure PKE, and their technique is to encrypt the message "bit-by-bit". Hence, their resulting construction does not preserve the compactness of the underlying KEM in terms of ciphertext overhead. Namely, even if we instantiate it with a compact KEM, it cannot give us a compact PKE with tight SIM-SO-CCA. Furthermore, we note that this bit-wise approach is used in many SIM-SO secure schemes [3,14,28].

While the scheme of Libert et al. [27] has compact ciphertexts, its public keys are not compact. Besides the large public key, their encryption algorithm needs to homomorphically evaluate the evaluation circuit of a PRF over GSW [17] ciphertexts that encrypts a PRF key. Hence, their scheme is very impractical.

COMPACT, YET NON-TIGHT SCHEMES. The work of Heuer et al. [20] is an exception to the bit-wise approach. It is the first work that proves SIM-SO-CCA security of practical PKE schemes, such as DHIES [1], OAEP [5], and Fujisaki-Okamoto (FO) [15], in the random oracle model [4]. All these schemes have compact ciphertexts. However, their security reduction is not tight, due to the guessing strategy in their security proofs. For instance, their proof for the FO transformation lose a factor of $O(\mu \cdot Q_h)$, where μ and Q_h are numbers of challenge ciphertexts and random oracle queries, respectively.

Finally, we stress that, even though there exist compact and (almost) tightly SIM-SO-CPA secure schemes from [3,25], it is not known how to transform them into SIM-SO-CCA by preserving its tightness and compactness. This is the case even in the random oracle model, given the non-tight bounds from the work of Heuer et al. [20].

1.1 Our Contribution

We construct the first compact PKE schemes with tight SIM-SO-CCA security in the random oracle model. More precisely, we propose four PKE schemes following two main ideas. We highlight that our first three schemes achieve tight SIM-SO-CCA security and compact ciphertexts and compact public keys at the same time. Table 1 compares our schemes with other known SO secure PKE schemes under the Diffie-Hellman assumptions.

THREE DIRECT CONSTRUCTIONS. Our first construction, PKE$_{StDH}$, is a direct construction of tightly SIM-SO-CCA secure PKE based on the strong Diffie-Hellman (StDH) assumption [1]. We then use the twinning technique from [10] to remove the decision oracle in the StDH assumption and construct our second tight scheme (called PKE$_{TDH}$) based on the twin DH (TDH) assumption. The TDH assumption is tightly implied by the standard computational DH (CDH)

Table 1. Comparison of our constructions with other SO secure PKE schemes. We ignore schemes that are non-tight and significantly less efficient than ours. $|\mathbb{G}|$ is the bit-length of group $\mathbb{G}$. ℓ is the message bit-length, which is independent of the group size, and it can be any polynomial in the security parameter λ. μ and Q_h are numbers of challenge ciphertexts and random oracle queries, respectively. The SO security losses of DHIES and FO can be found in [20, Theorem 6] and [21, Theorem 6].

Scheme	Security	Ass.	Loss	$|pk|$	$|m|$	$|c| - |m|$	RO?
BHY [3]	IND-SO-CPA	DDH	1	$2|\mathbb{G}|$	$|\mathbb{G}|$	$|\mathbb{G}|$	No
HJR [25]	SIM-SO-CPA	DDH	$\mathbf{O}(\ell)$	$(\ell+1)^2|\mathbb{G}|$	ℓ	$|\mathbb{G}|$	No
LLHG [29]	SIM-SO-CCA	DDH	$\mathbf{O}(\ell)$	$6|\mathbb{G}|$	ℓ	$3\ell|\mathbb{G}|$	No
DHIES proved in [20]	SIM-SO-CCA	StDH	$\mathbf{O}(\mu)$	$|\mathbb{G}|$	ℓ	$|\mathbb{G}|$	Yes
FO proved in [21]	SIM-SO-CCA	DDH	$\mathbf{O}(\mu Q_h)$	$|\mathbb{G}|$	ℓ	$|\mathbb{G}|$	Yes
PKE_{StDH} (Fig. 4)	SIM-SO-CCA	StDH	8	$|\mathbb{G}|$	ℓ	$2|\mathbb{G}|$	Yes
PKE_{TDH} (Fig. 10)	SIM-SO-CCA	CDH	8	$2|\mathbb{G}|$	ℓ	$2|\mathbb{G}|$	Yes
PKE_{DDH} (Fig. 11)	SIM-SO-CCA	DDH	10	$|\mathbb{G}|$	ℓ	$4|\mathbb{G}|$	Yes
FO_1 (in full version [7])	IND-SO-CCA	DDH	2	$2|\mathbb{G}|$	ℓ	$|\mathbb{G}|$	Yes
FO_2 (Fig. 16)	SIM-SO-CCA	DDH	$\mathbf{O}(\ell)$	$(\ell+1)^2|\mathbb{G}|$	ℓ	$|\mathbb{G}|$	Yes

assumption. Hence, this yields the first tightly SIM-SO-CCA secure PKE based on such a standard search assumption.

Both schemes have very short ciphertexts and public keys. Concretely, there are 2 group elements in the ciphertext overhead for PKE_{StDH} and PKE_{TDH}, and 1 element for PKE_{StDH}'s public key and 2 for PKE_{TDH}.

We also show that the decision oracle in the proof of PKE_{StDH} can be removed using the decisional DH assumption. However, the resulting scheme PKE_{DDH} has longer ciphertexts than the previous two, although it is still compact. All these schemes have small-constant security loss and compact ciphertexts and compact public keys.

FOURTH CONSTRUCTION: FUJISAKI-OKAMOTO, REVISITED. Our last contribution is to prove that a lossy encryption [3] can be transformed to a PKE with tight SO security via the well-known Fujisaki-Okamoto (FO) transformation [15]. The transformation preserves the tightness (up to a small constant) and compactness of the underlying lossy encryption.

Roughly speaking, a lossy encryption scheme has normal and lossy keys. Under normal keys, the scheme behave as a normal PKE. But under lossy keys, there exists an opener that can explain a ciphertext to any message by outputting the suitable randomness. An opener is not necessarily efficient. Especially, if the lossy encryption does not have an efficient opener (e.g., the BHY scheme [3]), then we can only show tight IND-SO-CCA security of the FO transformation. However, if the lossy encryption has an efficient opener (e.g., the HJR scheme [25]), then it yields tight SIM-SO-CCA security of the FO transformation.

Our result implies that tight IND-SO-CCA and SIM-SO-CCA security can be achieved from any assumption that has suitable lossy encryption. For a fair comparison, we implement our generic construction with DDH-based lossy encryption schemes from [3,25]. They both have only 1 group element in the cipher-

Table 2. Concrete security and efficiency comparison. All schemes are instantiated with P256, and we consider $\mu = 2^{32}$, $q_H = 2^{32}$, $|m| = 32$ bytes, and the output length of hash is 32 bytes. We consider the concrete security loss in the "Bit Security".

Scheme	Security	Ass.	Bit security	$|pk|$	$|m|$	$|c| - |m|$
BHY [3]	IND-SO-CPA	DDH	128	64	32	32
HJR [25]	SIM-SO-CPA	DDH	120	2113568	32	32
LLHG [29]	SIM-SO-CCA	DDH	120	192	32	24576
DHIES proved in [20]	SIM-SO-CCA	StDH	96	32	32	64
FO proved in [21]	SIM-SO-CCA	CDH	64	32	32	32
PKE$_{StDH}$ (Fig. 4)	SIM-SO-CCA	StDH	125	32	32	96
PKE$_{TDH}$ (Fig. 10)	SIM-SO-CCA	CDH	125	64	32	96
PKE$_{DDH}$ (Fig. 11)	SIM-SO-CCA	DDH	124	32	32	160
FO$_1$ (in full version [7])	IND-SO-CCA	DDH	127	64	32	32
FO$_2$ (Fig. 16)	SIM-SO-CCA	DDH	120	2113568	32	32

text (cf. Table 1). Our proof for the FO transformation is compactness- and tightness-preserving. Hence, for SIM-SO-CCA security, since the HJR scheme has non-compact public keys, it is also the case for our scheme. Similarly, the HJR scheme has only almost tightness, so has ours. We suppose that the size of ciphertexts is more critical than that of public keys, since ciphertexts have to be sent frequently over the internet for each communication, while public keys are stored in a server and can be used for a very long time.

EFFICIENCY COMPARISON. In Table 2 we estimate our concrete efficiency and compare it with other known SO secure schemes. We focus on schemes based the Diffie-Hellman assumptions and ignore those non-tight and significantly less efficient than ours (e.g., [23]). We estimate the efficiency of all schemes using the same NIST P256 curve. According to the corresponding security proofs, we consider the security level achieve by those schemes.

Our schemes significantly reduce the cost for tight SIM-SO-CCA, compared to LLHG. Moreover, our schemes are comparable to the practical PKE schemes, such as FO and DHIES. For instance, our FO$_2$ has the same ciphertext size, but it achieves a higher level of security, thanks to the tight security proof. Both PKE$_{StDH}$ and PKE$_{TDH}$ are comparable to DHIES.

PRACTICAL RELEVANCE. When a RO-based scheme is implemented in practice, one would instantiate the RO with a hash function, such as SHA-3. For SIM-SO-CCA PKE schemes in the ROM (including the previous work of Heuer et al. [20] and ours), we should be more careful and pay extra attention to the impossibility result of Bellare et al. [2]. More precisely, it shows that if a PKE scheme is binding then it cannot be SIM-SO secure. In a nutshell, it uses the binding property to construct an adversary such that there is no simulator can conclude the SIM-SO security. Hence, in the programmable ROM, the work of Heuer et al. and our schemes can all bypass it, since they are not binding according to the definition in [2]. The programmability is crucial for our proofs.

However, if one simply replaces the RO with, for instance, SHA-3, the situation becomes rather complex. For our fourth construction, it is not binding and the security results remain, since it uses lossy encryption and it allows us to generate encryption collisions. This is also the reason why [2] does not apply to lossy encryption schemes. For the scheme of Heuer et al. and our first three direct constructions, they will become binding in this case. Hence, the impossibility result of Bellare et al. applies, and they cannot have SIM-SO-CCA security. But the attack in [2] does not imply an adversary breaking IND-SO security, which means the scheme of Heuer et al. and our first three direct constructions can have IND-SO-CCA security, since SIM-SO-CCA implies IND-SO-CCA. An alternative solution could be finding a suitable programmable hash function in the standard model to instantiate our first three direction constructions. We leave constructing compact and tight SIM-SO-CCA secure PKE in the standard model as an interesting open problem.

1.2   Technical Overview

TECHNICAL GOAL: OPENABILITY AND TIGHTNESS. Selective-opening security is usually difficult to achieve. This is because the simulator $\mathcal{S}$ has to be able to 'open' any challenge ciphertext by producing the corresponding message and randomness. An adversary can verify whether a ciphertext has been correctly opened using the public encryption algorithm. It is not entirely trivial how to provide this openability efficiently. During the security proof, the simulator needs to embed a problem instance into the unopened ciphertexts, since usually it cannot open a ciphertext with a problem instance. Even worse, achieving tightness introduce an additional layer of complexity to the problem, namely, this opening procedure should be done in a tight fashion.

The work of Heuer et al. provides efficient openability by reprogramming the random oracle (RO) and guessing one unopened ciphertext. This unopened ciphertext will be embedded a problem challenge. We recall Heuer et al.'s strategy [20] of proving DHIES as an example to illustrate the aforementioned challenges in achieving tight SIM-SO-CCA security. The work of Heuer et al. is also the starting point of our work.

We consider the DHIES scheme with one-time pad as the symmetric encryption. Let $\mathbb{G} := \langle g \rangle$ be a group with order p, and $\mathsf{pk} := g^x$ be a public key. A ciphertext C of DHIES has the form

$$C := (R := g^r, \mathsf{d} := K \oplus \mathsf{m}, \mathsf{MAC}_k(R, \mathsf{d})),$$

where $(K, k) := H(R, \mathsf{pk}^r)$ and H is modeled as a RO. MAC_k produces a MAC tag using k.

To prove its SIM-SO-CCA security, we use the strong Diffie-Hellman (StDH) assumption which states that given a StDH instance $(X = g^x, Y)$ and oracle access to DHP_X, it is hard to compute Y^x. Here, $\mathrm{DHP}_X(\hat{Y}, \hat{Z})$ outputs the Boolean value of $\hat{Z} = \hat{Y}^x$. The reduction for SIM-SO-CCA security of DHIES firstly define $\mathsf{pk} := X$ and guesses the i^*-th ciphertext will not be opened ($i^* \xleftarrow{\$} [\mu]$). Then Y is embeded into C_{i^*} by $R_{i^*} := Y$. By using the DHP_X oracle and the RO patching technique [20], the reduction simulates the whole security game without knowing the secret x. We can prove that the adversary cannot get any information about $(K_{i^*}, k_{i^*}) = H(Y, Y^x)$ unless it computes Y^x, which breaks the StDH assumption. Thus, d_{i^*} is uniformly random and independent of R_{i^*}.

Unfortunately, since the above strategy needs to guess i^*, it requires a loss of μ, and the resulting security is non-tight and depends on the number of challenge ciphertext. One may consider using the random self-reducibility of StDH and embedding a randomized instance into challenge ciphertext C_i as $R_i := Y \cdot g^{s_i}$ where $s_i \xleftarrow{\$} \mathbb{Z}_p$ (for all $i \in [\mu]$). However, after doing so, one cannot open any ciphertext, since the discrete logarithm of Y is unknown. This is why the guessing approach is required.

OUR SOLUTION I: DHIES WITH DOUBLE RANDOMNESS. Our first solution is a direct improvement on the DHIES scheme by doubling the randomness R in the ciphertext. We only give some rough idea here and refer Sect. 3 for more details.

More precisely, we modify the generation of ciphertexts in DHIES: Instead of sampling a single r, we firstly choose a random bit $b \xleftarrow{\$} \{0, 1\}$, and then we choose $r_b \xleftarrow{\$} \mathbb{Z}_p$ and $R_{1-b} \xleftarrow{\$} \mathbb{G}$ (without knowing R_{1-b}'s discrete logarithm). Our modified DHIES scheme has ciphertexts with form:

$$C = (R_0, R_1, \mathsf{d} = K \oplus \mathsf{m}, h(k, R_0, R_1, \mathsf{d})),$$

where $(K, k) := H(b, R_0, R_1, \mathsf{pk}^{r_b})$, H is a RO, and h is a collision-resistant hash function. We note that sampling a random group element without knowing its discrete logarithm can be done in many widely-used groups like a subgroup of $\mathbb{Z}_q^*$ where q is a safe prime and prime-order elliptic curves.

After the modification, a ciphertext can have two valid randomness, namely, (b, r_b, R_{1-b}) and $(1 - b, r_{1-b}, R_b)$, in the view of an adversary, by carefully programming the RO H. Based on this, our simulator can embed the StDH instances to all challenge ciphertexts and open any ciphertext.

OUR SOLUTION II: LOSSY ENCRYPTION. The idea of having multiple valid randomness can be implemented by a lossy encryption, since under its lossy keys a ciphertext can be explained to different messages. Based on this, we use the lossy encryption as a tool to revise the security proof for the Fujisaki-Okamoto transformation and give a tight proof for its SIM-SO-CCA security. Another view of our second solution is that we transform the lossy-encryption-based SIM-SO-CPA secure PKE to a SIM-SO-CCA secure one, tightly.

OPEN PROBLEMS. We leave constructing (almost) tightly SIM-SO-CCA secure PKE with compact ciphertexts and compact public keys in the standard model as an interesting open problem. Moreover, our direction constructions are based on the Diffie-Hellman assumptions. We will study how to extend them in the post-quantum setting (for instance, with lattices).

2   Preliminaries

Let n be an integer. $[n]$ denotes the set $\{1, ..., n\}$. Let $\mathcal{A}$ be an algorithm. If $\mathcal{A}$ is probabilistic, then $y \xleftarrow{\$} \mathcal{A}(x)$ means that the variable y is assigned to the output of $\mathcal{A}$ on input x. If $\mathcal{A}$ is deterministic, then we write $y := \mathcal{A}(x)$. We write $\mathcal{A}^{\mathcal{O}}$ to indicate that $\mathcal{A}$ has classical access to oracle $\mathcal{O}$. $\mathcal{A} \Rightarrow$ out denotes the event that $\mathcal{A}$ outputs out. Unless we state it explicitly, all our algorithm are probabilistic polynomial-time (PPT). Throughout this paper, λ is the security parameter. The terms such as 'PPT' and 'negligible' are defined wrt λ.

GAMES. We use the code-based games [6] to define and prove security. We implicitly assume that Boolean flags are initialized to false, numerical types are initialized to 0, sets are initialized to $\emptyset$, while strings are initialized to the empty string ϵ. $\Pr[\mathsf{G}^{\mathcal{A}} \Rightarrow 1]$ denotes the probability that the final output $\mathsf{G}^{\mathcal{A}}$ of game G running an adversary $\mathcal{A}$ is 1. Let Ev be an (classical and well-defined) event. We write $\Pr[\mathrm{Ev} : \mathsf{G}]$ to denote the probability that Ev occurs during the game G.

RANDOM ORACLE. We use lazy sampling to simulate random oracles in this paper. Let $\mathcal{X}$ and $\mathcal{Y}$ be two finite sets and $H : \mathcal{X} \to \mathcal{Y}$ be a random oracle in a security game G. During the simulation of G, we use a list H to record all query-respond pairs of H. On query x, the game simulator samples $y \xleftarrow{\$} \mathcal{Y}$, sets $\mathsf{H}[x] := y$ (which means that now $H(x) = y$), and then returns y as the respond. We say x has been queried, or simply $x \in \mathsf{H}$, if and only if $\mathsf{H}[x] = y$ for some $y \in \mathcal{Y}$. For $x \notin \mathsf{H}$, we always have $\mathsf{H}[x] = \bot \notin \mathcal{Y}$.

2.1   Cryptographic Assumptions

Let $\mathbb{G}$ be an cyclic group with a generator g and prime order p. Let $X = g^x$ and $Y = g^y$ for some $x, y \in \mathbb{Z}_p$. The CDH value of X and Y is written as $\mathsf{cdh}(X, Y) = g^{xy}$. Here we suppose that $(\mathbb{G}, g, p)$ is a public parameter.

Definition 1 (Multi-Instance DDH (mDDH)). *We say the* mDDH *problem is hard on* $\mathbb{G}$ *if for any* $\mathcal{A}$, *the* mDDH *advantage of* $\mathcal{A}$ *against* $\mathbb{G}$

$$\mathsf{Adv}_{\mathbb{G}}^{\mathsf{mDDH}}(\mathcal{A}) := \Big| \Pr\big[\mathcal{A}(g_1, (g_0^{r_i}, g_1^{r_i})_{i \in [\mu]}) \Rightarrow 1\big]$$

$$- \Pr\big[\mathcal{A}(g_1, (g_0^{r_i}, g_1^{r_i'})_{i \in [\mu]}) \Rightarrow 1\big] \Big|$$

is negligible, where μ *is the number of challenges,* $g_0 := g$, $g_1 := g_0^{\omega}$ *for some* $\omega \xleftarrow{\$} \mathbb{Z}_p$, *and* $r_i, r_i' \xleftarrow{\$} \mathbb{Z}_p$ *for some* $i \in [\mu]$.

By the random self-reducibility of DDH [13], mDDH assumption is tightly equivalent to DDH assumption (i.e., single-instance version of mDDH).

Definition 2 (Strong Diffie-Hellman (StDH) Problem [1]). *For a fixed* $X \in \mathbb{G}$, *let* DHP$_X$ *be the gap oracle that given* $(Y', Z') \in \mathbb{G}^2$ *outputs whether* $\mathsf{cdh}(X, Y') = Z'$ *or not. We say the* StDH *problem is hard on* $\mathbb{G}$ *if for any* $\mathcal{A}$, *the* StDH *advantage of* $\mathcal{A}$ *against* $\mathbb{G}$, $\mathsf{Adv}_{\mathbb{G}}^{\mathsf{StDH}}(\mathcal{A})$, *is negligible.*

$$\mathsf{Adv}_{\mathbb{G}}^{\mathsf{StDH}}(\mathcal{A}) := \Pr\Big[(X, Y) \xleftarrow{\$} \mathbb{G}^2, \mathcal{A}^{\mathrm{DHP}_X(\cdot, \cdot)}(X, Y) \Rightarrow \mathsf{cdh}(X, Y)\Big]$$

Definition 3 (Twin Diffie-Hellman (TDH) Problem [10]). *For fixed* $X_0, X_1 \in \mathbb{G}$, *let* $2\mathrm{DHP}_{X_0,X_1}$ *be an oracle that on input* $(Y', Z_0', Z_1') \in \mathbb{G}^3$, *determines whether* $\mathsf{cdh}(X_0, Y') = Z_0'$ *and* $\mathsf{cdh}(X_1, Y') = Z_1'$. *We say the* TDH *problem is hard on* $\mathbb{G}$ *if for any* $\mathcal{A}$, *the* TDH *advantage of* $\mathcal{A}$ *against* $\mathbb{G}$

$$\mathsf{Adv}_{\mathbb{G}}^{\mathsf{TDH}}(\mathcal{A}) := \Pr\left[\mathcal{A}^{2\mathrm{DHP}_{X_0,X_1}(\cdot,\cdot,\cdot)}(X_0, X_1, Y) \Rightarrow (\mathsf{cdh}(X_0, Y), \mathsf{cdh}(X_1, Y))\right]$$

is negligible, where $X_0, X_1, Y \xleftarrow{\$} \mathbb{G}$.

Definition 4 (Multi-Instance StDH (mStDH)). *Let* μ *be the number of instance. We say the* mStDH *problem is hard on* $\mathbb{G}$ *if for any* $\mathcal{A}$, *given* $X, Y_1, ..., Y_\mu \xleftarrow{\$} \mathbb{G}$, *the* mStDH *advantage of* $\mathcal{A}$ *against* $\mathbb{G}$, $\mathsf{Adv}_{\mathbb{G}}^{\mathsf{mStDH}}(\mathcal{A})$, *is negligible.*

$$\mathsf{Adv}_{\mathbb{G}}^{\mathsf{mStDH}}(\mathcal{A}) := \Pr\left[\mathcal{A}^{\mathrm{DHP}_X(\cdot,\cdot)}(X, (Y_i)_{i \in [\mu]}) \Rightarrow \mathsf{cdh}(X, Y_i) \text{ for some } i \subset [\mu]\right]$$

Definition 5 (Multi-Instance TDH (mTDH)). *Let* μ *be the number of instance. We say the* mTDH *problem is hard on* $\mathbb{G}$ *if for any* $\mathcal{A}$, *given* $X_0, X_1, Y_1, ...,$ $Y_\mu \xleftarrow{\$} \mathbb{G}$, *the* mTDH *advantage of* $\mathcal{A}$ *against* $\mathbb{G}$, $\mathsf{Adv}_{\mathbb{G}}^{\mathsf{mStDH}}(\mathcal{A})$, *is negligible*

$$\mathsf{Adv}_{\mathbb{G}}^{\mathsf{mStDH}}(\mathcal{A}) := \Pr\Big[\mathcal{A}^{2\mathrm{DHP}_{X_0,X_1}(\cdot,\cdot,\cdot)}(X_0, X_1, (Y_i)_{i \in [\mu]})$$

$$\Rightarrow (\mathsf{cdh}(X_0, Y_i), \mathsf{cdh}(X_1, Y_i)) \text{ for some } i \in [\mu]\Big]$$

The mStDH and mTDH assumptions are tightly implied by the StDH and TDH assumption, respectively. This can be showed naturally by the random self-reducibility of the Diffie-Hellman assumption. We state the lemmas here and leave the proof in our full version paper [7].

Lemma 1 (StDH $\xrightarrow{\text{tight}}$ mStDH). *For any* mStDH *adversary* $\mathcal{A}$, *there exists an* StDH *adversary* $\mathcal{B}$ *such that* $\mathsf{Adv}_{\mathbb{G}}^{\mathsf{mStDH}}(\mathcal{A}) \leq \mathsf{Adv}_{\mathbb{G}}^{\mathsf{StDH}}(\mathcal{B})$.

Lemma 2 (TDH $\xrightarrow{\text{tight}}$ mTDH). *For any* mTDH *adversary* $\mathcal{A}$, *there exists an* TDH *adversary* $\mathcal{B}$ *such that* $\mathsf{Adv}_{\mathbb{G}}^{\mathsf{mTDH}}(\mathcal{A}) \leq \mathsf{Adv}_{\mathbb{G}}^{\mathsf{TDH}}(\mathcal{B})$.

Definition 6 (Collision Resistance). *A hash function* h *has collision resistance if for all adversary* $\mathcal{A}$, *the* CR *advantage of* $\mathcal{A}$ *against* h

$$\mathsf{Adv}_h^{\mathsf{CR}}(\mathcal{A}) := \Pr\left[x \neq x' \wedge h(x) \neq h(x') | (x, x') \xleftarrow{\$} \mathcal{A}(h)\right]$$

is negligible. A hash function family $\mathcal{H}$ *is collision-resistant if for all* $h \xleftarrow{\$} \mathcal{H}$, $\mathsf{Adv}_h^{\mathsf{CR}}(\mathcal{A})$ *is negligible.*

2.2   Public-Key Encryption Scheme

Definition 7 (PKE). *A Public-Key Encryption (PKE) scheme* PKE *consists of three polynomial-time algorithms* (KG, Enc, Dec) *and a message space* $\mathcal{M}$, *a randomness space* $\mathcal{R}$, *and a ciphertext space* $\mathcal{C}$. KG *outputs a public and secret key pair* (pk, sk). *The encryption algorithm* Enc, *on input* pk *and a message* $\mathsf{m} \in \mathcal{M}$, *outputs a ciphertext* $c \in \mathcal{C}$. *We also write* $c := \mathsf{Enc}(\mathsf{pk}, \mathsf{m}; r)$ *to indicate the randomness* $r \in \mathcal{R}$ *explicitly. The decryption algorithm* Dec, *on input* sk *and a ciphertext* c, *outputs a message* $m' \in \mathcal{M}$ *or a rejection symbol* $\perp \notin \mathcal{M}$.

$$
\boxed{
\begin{array}{l}
\textbf{GAME COR}^{\mathcal{A}}_{\textsf{PKE}} \\
\hline
01 \;\; (\textsf{pk}, \textsf{sk}) \leftarrow \textsf{KG} \\
02 \;\; \textsf{m} \xleftarrow{\$} \mathcal{A}^{\mathcal{O}}(\textsf{pk}, \textsf{sk}) \\
03 \;\; c = \textsf{Enc}(\textsf{pk}, \textsf{m}) \\
04 \;\; \textbf{if } \textsf{Dec}(\textsf{sk}, c) = \textsf{m} : \textbf{return } 1 \\
05 \;\; \textbf{return } 0
\end{array}
}
$$

Fig. 1. The COR game for a PKE scheme PKE and $\mathcal{A}$. $\mathcal{A}$ might have access to some oracle $\mathcal{O}$ (e.g., random oracles, decryption oracles). It depends on the specific reduction.

CORRECTNESS OF PKE. Some of our PKE schemes do not have perfect correctness, and the correctness bound of PKE might depend on some computational bound, e.g., the collision bound of hash function. Following [24], we use a game COR to define PKE correctness.

Definition 8 (PKE Correctness). *Let* PKE $:= (\textsf{KG}, \textsf{Enc}, \textsf{Dec})$ *be a PKE scheme with message space* $\mathcal{M}$ *and* $\mathcal{A}$ *be an adversary against* PKE. *The* COR *advantage of* $\mathcal{A}$ *is defined as*

$$
\textsf{Adv}^{\textsf{COR}}_{\textsf{PKE}}(\mathcal{A}) := \Pr\left[\textsf{COR}^{\mathcal{A}}_{\textsf{PKE}} \Rightarrow 1\right],
$$

where the COR *game is defined in Fig. 1. If there exists a constant* δ *such that for all adversary* $\mathcal{A}$, $\textsf{Adv}^{\textsf{COR}}_{\textsf{PKE}}(\mathcal{A}) \leq \delta$, *then we say* PKE *is* $(1 - \delta)$-*correct.*

SELECTIVE OPENING SECURITY. Selective Opening (SO) security preserves confidentiality even if an adversary opens the randomnesses of some ciphertexts. We use simulation-based approach to define SO security as in [20]. We consider two types of SO security definition: Simulation-based SO security against Chosen-Ciphertext Attacks (SIM-SO-CCA, Definition 9) and Indistinguishability-based SO security against Chosen-Ciphertext Attacks (IND-SO-CCA, Definition 10).

Definition 9 (SIM-SO-CCA security). *Let* PKE *be a PKE scheme with message space* $\mathcal{M}$ *and randomness space* $\mathcal{R}$ *and* $\mathcal{A} := (\mathcal{A}_0, \mathcal{A}_1)$ *be an adversary against* PKE. *Let* μ *be the number of challenge ciphertexts Let* Rel *be a relation. We consider two games defined in Fig. 2, where* $\mathcal{A}$ *is run in* REAL-SO-CCA$_{\textsf{PKE}}$ *and a SO simulator* $\mathcal{S} := (\mathcal{S}_0, \mathcal{S}_1)$ *in* IDEAL-SO-CCA$_{\textsf{PKE}}$. $\mathcal{M}_a$ *is a distribution over* $\mathcal{M}$ *chosen by* $\mathcal{A}_0$. *We define the SIM-SO-CCA advantage function*

$$
\textsf{Adv}^{\textsf{SIM-SO-CCA}}_{\textsf{PKE}}(\mathcal{A}, \mathcal{S}, \mu, \textsf{Rel}) := \Big| \Pr\left[\textsf{REAL-SO-CCA}^{\mathcal{A}}_{\textsf{PKE}} \Rightarrow 1\right]
$$
$$
- \Pr\left[\textsf{IDEAL-SO-CCA}^{\mathcal{S}}_{\textsf{PKE}} \Rightarrow 1\right]\Big|,
$$

PKE *is SIM-SO-CCA secure if, for every adversary* $\mathcal{A}$ *and every relation* Rel, *there exists a simulator* $\mathcal{S}$ *such that* $\textsf{Adv}^{\textsf{SIM-SO-CCA}}_{\textsf{PKE}}(\mathcal{A}, \mathcal{S}, \mu, \textsf{Rel})$ *is negligible.*

Definition 10 (IND-SO-CCA security). *Let* PKE *be a PKE scheme with message space* $\mathcal{M}$ *and randomness space* $\mathcal{R}$ *and* $\mathcal{A} := (\mathcal{A}_0, \mathcal{A}_1, \mathcal{A}_2)$ *be an adversary against* PKE. *Let* μ *be the number of challenge ciphertext.*

GAME REAL-SO-CCA$_{\mathsf{PKE}}^{\mathcal{A}}$	GAME IDEAL-SO-CCA$_{\mathsf{PKE}}^{\mathcal{S}}$
01 $(pk, sk) \xleftarrow{\$} \mathsf{KG}$	11 $(\mathcal{M}_a, st) \xleftarrow{\$} \mathcal{S}_0$
02 $(\mathcal{M}_a, st) \xleftarrow{\$} \mathcal{A}_0^{\mathrm{DEC}}(pk)$	12 **for** $i \in [\mu]$:
03 **for** $i \in [\mu]$:	13 $\mathbf{m}[i] := m_i \xleftarrow{\$} \mathcal{M}_a$
04 $\quad \mathbf{m}[i] := m_i \xleftarrow{\$} \mathcal{M}_a$	14 $\quad \mathbf{m}''[i] := |m_i|$
05 $\quad r_i \xleftarrow{\$} \mathcal{R}$	15 $out \xleftarrow{\$} \mathcal{S}_1^{\mathrm{OPEN}}(st, \mathbf{m}'')$
06 $\quad \mathbf{c}[i] := \mathsf{Enc}(pk, m_i; r_i)$	16 **return** $\mathsf{Rel}(\mathcal{M}_a, \mathbf{m}, I, out)$
07 $out \xleftarrow{\$} \mathcal{A}_1^{\mathrm{OPEN, DEC}}(st, \mathbf{c})$	
08 **return** $\mathsf{Rel}(\mathcal{M}_a, \mathbf{m}, I, out)$	OPEN(i) $/\!\!/ i \in [\mu]$
	17 $I := I \cup \{i\}$
DEC(c) $/\!\!/$ for $c \notin \mathbf{c}$	18 **return** (m_i, r_i) $\quad /\!\!/$ REAL-SO-CCA$_{\mathsf{PKE}}$
09 $m := \mathsf{Dec}(sk, c)$	19 **return** m_i $\quad\quad\quad /\!\!/$ IDEAL-SO-CCA$_{\mathsf{PKE}}$
10 **return** m	

Fig. 2. The SO security games for PKE schemes. $\mathcal{S}_1$ only learn the lengths of challenge messages m_i instead of the challenge ciphertexts.

We consider the game defined in Fig. 3. Samp and ReSamp are efficient algorithms output by $\mathcal{A}_0$, where Samp outputs μ messages according to some distribution (determined by $\mathcal{A}_0$) over $\mathcal{M}$, and ReSamp$(I, \mathbf{m}_0)$ resamples $\mathbf{m}_0[i]$ for $i \notin I$ according to the same distribution of Samp and then outputs $\mathbf{m}_1$. For $i \in I$, $\mathbf{m}_0[i] = \mathbf{m}_1[i]$. We define the IND-SO-CCA advantage function

$$\mathsf{Adv}_{\mathsf{PKE}}^{\mathsf{IND\text{-}SO\text{-}CCA}}(\mathcal{A}, \mu) := \left| \Pr\left[\mathsf{IND\text{-}SO\text{-}CCA}_{\mathsf{PKE}, 0}^{\mathcal{A}} \Rightarrow 1 \right] \right.$$
$$\left. - \Pr\left[\mathsf{IND\text{-}SO\text{-}CCA}_{\mathsf{PKE}, 1}^{\mathcal{A}} \Rightarrow 1 \right] \right|.$$

PKE is IND-SO-CCA secure if $\mathsf{Adv}_{\mathsf{PKE}}^{\mathsf{IND\text{-}SO\text{-}CCA}}(\mathcal{A}, \mu)$ is negligible for any $\mathcal{A}$.

3 Direct Constructions

We construct a compact and tightly SIM-SO-CCA PKE, PKE$_{\mathsf{StDH}}$, from the strong Diffie-Hellman assumption. We also weaken this assumption using the twinning technique from [10], and the resulting scheme is only based on the Computational Diffie-Hellman assumption at the cost of being less efficient.

3.1 Construction from the Strong Diffie-Hellman Assumption

Let $\mathbb{G}$ be a group with order p. Let $H : \{0, 1\} \times \mathbb{G}^3 \to \mathcal{M} \times \{0, 1\}^l, h : \{0, 1\}^l \times \mathbb{G}^2 \to \{0, 1\}^\ell$ be hash functions. We construct a compact and tightly SIM-SO-CCA PKE scheme PKE$_{\mathsf{StDH}} = (\mathsf{KG}, \mathsf{Enc}, \mathsf{Dec})$ with message space $\mathcal{M}$ as in Fig. 4. The randomness space of PKE$_{\mathsf{StDH}}$ is the set $\{0, 1\} \times \mathbb{Z}_p \times \mathbb{G}$.

CORRECTNESS. The correctness of PKE$_{\mathsf{StDH}}$ depends on the hash function h. If h is not collision resistant, then there is a decryption error. For instance, a ciphertext

GAME IND-SO-CCA$_{PKE,b}^{\mathcal{A}}$	DEC(c) // for $c \notin \mathbf{c}$
01 $(pk, sk) \xleftarrow{\$} KG$	12 $m := Dec(sk, c)$
02 $(Samp, ReSamp, st_0) \xleftarrow{\$} \mathcal{A}_0(pk)$	13 **return** m
03 $\mathbf{m}_0 \xleftarrow{\$} Samp$	OPEN(i) // $i \in [\mu]$
04 **for** $i \in [\mu]$:	14 $I := I \cup \{i\}$
05 $\quad r_i \xleftarrow{\$} \mathcal{R}$	15 **return** (m_i, r_i)
06 $\quad \mathbf{c}[i] := Enc(pk, \mathbf{m}_0[i]; r_i)$	
07 $st_1 \xleftarrow{\$} \mathcal{A}_1^{\text{OPEN,DEC}}(\mathbf{c}, st_0)$	
08 **for** $i \in [\mu] \backslash I$:	
09 $\quad \mathbf{m}_1[i] := ReSamp(I, \mathbf{m}_0)$	
10 $b' \xleftarrow{\$} \mathcal{A}_1^{\text{DEC}}(st_1, \mathbf{m}_b)$	
11 **return** b'	

Fig. 3. The SO security games for PKE schemes. $\mathcal{S}_1$ only learn the lengths of challenge messages m_i instead of the challenge ciphertexts. For $i \in I, \mathbf{m}_0[i] = \mathbf{m}_1[i]$, and for $i \in [\mu] \backslash I$, $\mathbf{m}_0[i]$ has the same distribution with $\mathbf{m}_1[i]$ but not necessary to be the same.

KG	Enc$(pk, m \in \mathcal{M})$	Dec$(sk, (R_0, R_1, d, \mathcal{T}))$
01 $x \xleftarrow{\$} \mathbb{Z}_p$	06 $b \xleftarrow{\$} \{0, 1\}$	15 $m := \perp$
02 $X := g^x$	07 $r_b \xleftarrow{\$} \mathbb{Z}_p$	16 $Z_0 := R_0^x, Z_1 := R_1^x$
03 $pk := X$	08 $R_b := g^{r_b}$	17 $(K_0, k_0) := H(0, R_0, R_1, Z_0)$
04 $sk := x$	09 $R_{1-b} \xleftarrow{\$} \mathbb{G}$	18 $\mathcal{T}_0 := h(k_0, R_0, R_1, d)$
05 **return** (pk, sk)	10 $Z_b := pk^{r_b}$	19 $(K_1, k_1) := H(1, R_0, R_1, Z_1)$
	11 $(K, k) := H(b, R_0, R_1, Z_b)$	20 $\mathcal{T}_1 := h(k_1, R_0, R_1, d)$
	12 $d := K \oplus m$	21 **if** $\mathcal{T}_0 = \mathcal{T} : m := d \oplus K_0$
	13 $\mathcal{T} := h(k, R_0, R_1, d)$	22 **if** $\mathcal{T}_1 = \mathcal{T} : m := d \oplus K_1$
	14 **return** $(R_0, R_1, d, \mathcal{T})$	23 **return** m

Fig. 4. Our Direct Construction of SIM-SO-CCA secure PKE schemes from the mStDH assumption, $PKE_{StDH} = (KG, Enc, Dec)$

c of m is generated using $b = 1$, which means it uses $\tau_1 = h(k_1, R_0, R_1, d)$ with $(K_1, k_1) := H(1, R_0, R_1, Z_1)$. If there is a collision as $h(k_1, R_0, R_1, d) = h(k_0, R_0, R_1, d)$ and $(K_1, k_1) \neq (K_0, k_0)$, then c will be decrypted incorrectly as $m' := d \oplus K_0 \neq m = d \oplus K_1$. Hence, the correctness error $Adv_{PKE_{StDH}}^{COR}(\mathcal{A})$ is bounded by the collision probability of h. If h is modeled as a random oracle, then $Adv_{PKE_{StDH}}^{COR}(\mathcal{A}) \leq \frac{q_h}{2^\ell}$. In our tight proof, we require collision resistance of a standard hash function, and thus we use the similar requirement here, namely, $Adv_{PKE_{StDH}}^{COR}(\mathcal{A}) \leq Adv_h^{CR}(\mathcal{A})$.

ON SAMPLING OF A GROUP ELEMENT. We require that a group element of $\mathbb{G}$ can be sampled without knowing the corresponding exponent. A concrete example is as follow: Let p be a prime s.t. $q = rp + 1$ is also a prime for some r. Let $\mathbb{G}$ be a subgroup of $\mathbb{Z}_q$ and with order p. Canetti et al. [9, Sect. 4.3.2] showed how to sample a group element from such $\mathbb{G}$ without knowing exponent. Other examples are some widely-used standard elliptic-curve groups, such as NIST P256,

NIST P384, and Curve25519. To generate a random point without knowing the exponent, we can pick a random x-coordinate, compute the point, and then use the cofactor to check whether the point is in its prime subgroup.

Theorem 1. $\mathsf{PKE_{StDH}}$ *in Fig. 4 is SIM-SO-CCA secure (Definition 9) if the mStDH problem is hard on $\mathbb{G}$ and H and h are modeled as random oracles. For any SIM-SO-CCA adversary $\mathcal{A}$ and relation Rel, there exists a simulator $\mathcal{S}$ and an adversary $\mathcal{B}$ such that:*

$$\mathsf{Adv^{SIM\ SO\text{-}CCA}_{PKE_{StDH}}}(\mathcal{A}, \mathcal{S}, \mu, \mathsf{Rel}) \le 8\mathsf{Adv^{mStDH}_{\mathbb{G}}}(\mathcal{B}) + \frac{2n_H^2}{|\mathcal{M}|} + \frac{2(n_H^2 + n_h^2)}{2^l}$$

where q_H and n_{DEC} are the numbers of $\mathcal{A}$'s queries to H and DEC, respectively, and μ is the number of challenge ciphertexts. $n_H = \mu + q_H + 2n_{\mathrm{DEC}}$ and $n_h = \mu + q_h + 2n_{\mathrm{DEC}}$ are the total numbers of queries to H and h, respectively.

By Lemma 1, $\mathsf{PKE_{StDH}}$ in Fig. 4 is SIM-SO-CCA secure under the StDH assumption, and the security reduction is tight.

Corollary 1. $\mathsf{PKE_{StDH}}$ *in Fig. 4 is SIM-SO-CCA secure (Definition 9) if the StDH problem is hard on $\mathbb{G}$ and H and h are modeled as random oracles. For any SIM-SO-CCA adversary $\mathcal{A}$ and relation Rel, there exists a simulator $\mathcal{S}$ and an adversary $\mathcal{B}$ such that:*

$$\mathsf{Adv^{SIM\text{-}SO\text{-}CCA}_{PKE_{StDH}}}(\mathcal{A}, \mathcal{S}, \mu, \mathsf{Rel}) \le 8\mathsf{Adv^{StDH}_{\mathbb{G}}}(\mathcal{B}) + \frac{2n_H^2}{|\mathcal{M}|} + \frac{2(n_H^2 + n_h^2)}{2^l}$$

where q_H and n_{DEC} are the numbers of $\mathcal{A}$'s queries to H and DEC, respectively, and μ is the number of challenge ciphertexts. $n_H = \mu + q_H + 2n_{\mathrm{DEC}}$ and $n_h = \mu + q_h + 2n_{\mathrm{DEC}}$ are the total numbers of queries to H and h, respectively.

Proof (Theorem 1). The theorem is proved by the game sequence in Figs. 5 and 6. In G_0, we use lazy sampling to simulate Random oracles H and h. We assume that from G_0 to G_8, there is no collision among the outputs of random oracle h, the first parts of outputs of H (i.e., K), and the second parts of outputs of H (i.e., k). Let n_H and n_h be the total numbers of queries (including the queries from the game simulator) to H and h, respectively. By collision bounds,

$$\left| \Pr\left[\mathsf{REAL\text{-}SO\text{-}CCA}^{\mathcal{A}}_{PKE_{StDH}} \Rightarrow 1 \right] - \Pr\left[\mathsf{G}^{\mathcal{A}}_0 \Rightarrow 1 \right] \right| \le \frac{n_H^2}{|\mathcal{M}|} + \frac{n_H^2 + n_h^2}{2^l}$$

GAME G_1: We generate $R_{i,1-b_i} := g^{r_{i,1-b_i}}$ by choosing $r_{i,1-b_i} \xleftarrow{\$} \mathbb{Z}_p$, and compute $Z_{i,1-b_i} := X^{r_{i,1-b_i}}$. This modification does not change $\mathcal{A}$'s view since $R_{i,1-b_i}$ is still distributed uniformly at random. Therefore, we have

$$\Pr\left[\mathsf{G}^{\mathcal{A}}_0 \Rightarrow 1 \right] = \Pr\left[\mathsf{G}^{\mathcal{A}}_1 \Rightarrow 1 \right]$$

GAME G_2: We modify DEC oracle. When $\mathcal{A}$ queries DEC on $\mathsf{c} := (R_0, R_1, \mathsf{d}, \mathcal{T})$, if $\mathcal{T}$ is the tag of one of the challenge ciphertexts (i.e., $\mathcal{T} = \mathcal{T}_i$ for some $i \in$

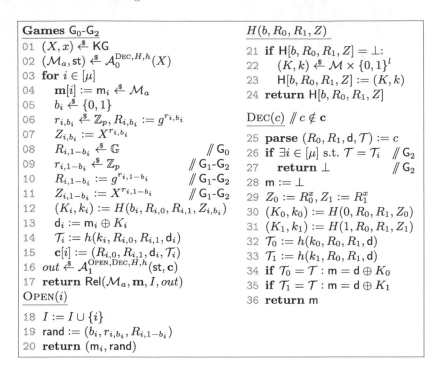

Fig. 5. Games G_0–G_2 for proving Theorem 1. Random oracle h is simulated as usual (i.e., similar to the simulation of H in G_0).

$[\mu]$), then DEC returns $\perp$. By the definition of SIM-SO-CCA security, we have $(R_0, R_1, d, \mathcal{T}) \notin \mathbf{c}$. Thus, if $\mathcal{T} = \mathcal{T}_i$, we have $(R_0, R_1, d) \neq (R_{i,0}, R_{i,1}, d_i)$. From this, we can find a collision for h, since $\mathcal{T}$ must equal to $h(k_0, R_0, R_1, d)$ or $h(k_1, R_0, R_1, d)$. We have assumed there is no collision among the output of h, so we have

$$\Pr\left[G_1^{\mathcal{A}} \Rightarrow 1\right] = \Pr\left[G_2^{\mathcal{A}} \Rightarrow 1\right]$$

GAME G_3: In this game, we simulate DEC by searching for the corresponding keys from the random oracle queries, instead of computing Z_0, Z_1 as in G_2. Intuitively, this does not change the view of $\mathcal{A}$, since a ciphertext is valid if $\mathcal{A}$ has asked the corresponding random oracle queries before. Otherwise, the ciphertext is invalid and the decryption will only output $\perp$.

Concretely, G_3 use the following three lists $H_{\mathsf{val}}, H_{\mathsf{inv}}$, and H_{dec} to keep track of the oracle queries to H, and each of them stores a particular type of oracle queries, namely:

- $(b, R_0, R_1, Z) \in H_{\mathsf{val}}$ if $\mathcal{A}$ has queried H on (b, R_0, R_1, Z) and $Z = R_b^x$. We call this type of hash queries valid.
- $(b, R_0, R_1, Z) \in H_{\mathsf{inv}}$ if $\mathcal{A}$ has queried H on (b, R_0, R_1, Z) and $Z \neq R_b^x$. We call this type of hash queries invalid.
- $(b, R_0, R_1) \in H_{\mathsf{dec}}$ if $\mathcal{A}$ has queried DEC with (R_0, R_1) as parts of a ciphertext. It is clear that $H_{\mathsf{val}} \cap H_{\mathsf{inv}} = \emptyset$.

Games G_3-G_9	$H(b, R_0, R_1, Z)$
01 $(X, x) \xleftarrow{\$} KG$	24 **if** $\exists i \in [\mu]\backslash I$ s.t.
02 $(\mathcal{M}_a, st) \xleftarrow{\$} \mathcal{A}_0^{\mathrm{DEC},H,h}(X)$	$\quad (b, R_0, R_1, Z) = (1 - b_i, R_{i,0}, R_{i,1}, R_{i,1-b_i}^x)$
03 **for** $i \in [\mu]$	$\quad$ **abort** $\qquad\qquad\qquad$ // G_4-G_7
04 $\quad \mathbf{m}[i] := \mathsf{m}_i \xleftarrow{\$} \mathcal{M}_a$	25 **if** $\exists i \in [\mu]\backslash I$ s.t.
05 $\quad b_i \xleftarrow{\$} \{0,1\}$	$\quad (b, R_0, R_1, Z) = (b_i, R_{i,0}, R_{i,1}, R_{i,b_i}^x)$
06 $\quad r_{i,b_i} \xleftarrow{\$} \mathbb{Z}_p, R_{i,b_i} := g^{r_{i,b_i}}$	$\quad$ **abort** $\qquad\qquad\qquad$ // G_5-G_7
07 $\quad Z_{i,b_i} := X^{r_{i,b_i}}$	26 **if** $\exists(K,k)$ s.t. $\mathsf{H_{dec}}[b, R_0, R_1] = (K, k)$
08 $\quad r_{i,1-b_i} \xleftarrow{\$} \mathbb{Z}_p$	$\quad$ **and** $Z = R_b^x$
09 $\quad R_{i,1-b_i} := g^{r_{i,1-b_i}}$	27 $\quad \mathsf{H_{val}}[b, R_0, R_1, Z] := (K, k)$
10 $\quad Z_{i,1-b_i} := X^{r_{i,1-b_i}}$	28 $\quad \mathsf{H_{dec}}[b, R_0, R_1] := \bot$
11 $\quad (K_i, k_i)$	29 **If** $\exists(K,k)$ s.t. $\mathsf{H_{val}}[b, R_0, R_1, Z] = (K, k)$
$\qquad := H(b_i, R_{i,0}, R_{i,1}, Z_{i,b_i})$ // G_3-G_5	$\quad$ **or** $\mathsf{H_{inv}}[b, R_0, R_1, Z] = (K, k)$
12 $\quad (K_i, k_i) \xleftarrow{\$} \mathcal{M} \times \{0,1\}^l$ // G_6-G_9	30 $\quad$ **return** (K, k)
13 $\quad \mathsf{d}_i := \mathsf{m}_i \oplus K_i$	31 **else**
14 $\quad \mathcal{T}_i := h(k_i, R_{i,0}, R_{i,1}, \mathsf{d}_i)$	32 $\quad (K, k) \xleftarrow{\$} \mathcal{M} \times \{0,1\}^l$
15 $\quad \mathbf{c}[i] := (R_{i,0}, R_{i,1}, \mathsf{d}_i, \mathcal{T}_i)$	33 $\quad$ **if** $Z = R_b^x : \mathsf{H_{val}}[b, R_0, R_1, Z] := (K, k)$
16 $out \xleftarrow{\$} \mathcal{A}_1^{\mathrm{OPEN},\mathrm{DEC},H,h}(st, \mathbf{c})$	34 $\quad$ **else** $\mathsf{H_{inv}}[b, R_0, R_1, Z] := (K, k)$
17 **return** $\mathrm{Rel}(\mathcal{M}_a, \mathbf{m}, I, out)$	35 $\quad$ **return** (K, k)
$\mathrm{OPEN}(i)$	$\mathrm{DEC}(c)$ // $c \notin \mathbf{c}$
18 $I := I \cup \{i\}$	36 **parse** $(R_0, R_1, \mathsf{d}, \mathcal{T}) =: c$
19 $\mathsf{H_{val}}[b_i, R_{i,0}, R_{i,1}, Z_{i,b_i}]$	37 **if** $\exists i \in [\mu]$ s.t. $\mathcal{T} = \mathcal{T}_i$: **return** $\bot$ // G_3-G_8
$\quad := (K_i, k_i)$ // G_3,G_6,G_8-G_9	38 $m := \bot$
20 rand $:= (b_i, r_{i,b_i}, R_{i,1-b_i})$ // G_3-G_6,G_8-G_9	39 **for** $b \in \{0,1\}$:
21 $\mathsf{H_{val}}[1 - b_i, R_{i,0}, R_{i,1}, Z_{i,1-b_i}]$	40 $\quad$ **if** $\exists(Z, K, k)$ s.t. $\mathsf{H_{val}}[b, R_0, R_1, Z] = (K, k)$
$\quad := (K_i, k_i)$ // G_7	$\quad$ **or** $\exists(K, k)$ s.t. $\mathsf{H_{dec}}[b, R_0, R_1] = (K, k)$
22 rand $:= (1 - b_i, r_{i,1-b_i}, R_{i,b_i})$ // G_7	41 $\quad (K_b, k_b) := (K, k)$
23 **return** $(\mathsf{m}_i, \text{rand})$	42 $\quad$ **else**
	43 $\quad (K_b, k_b) \xleftarrow{\$} \mathcal{M} \times \{0,1\}^l$
	44 $\quad \mathsf{H_{dec}}[b, R_0, R_1] := (K_b, k_b)$
	45 $\quad \mathcal{T}_b := h(k_b, R_0, R_1, \mathsf{d})$
	46 $\quad$ **if** $\mathcal{T}_b = \mathcal{T} : m = \mathsf{d} \oplus K_b$
	47 **return** m

Fig. 6. Games G_3–G_9 for proving Theorem 1.

Oracles H and DEC in G_3 are simulated in the following ways:

- DEC oracle: On input $(R_0, R_1, \mathsf{d}, \mathcal{T})$, the simulator tries to search (K_b, k_b) ($b \in \{0,1\}$) from $\mathsf{H_{val}}$ (see Items 40 and 41). If it fails, the simulator samples a random key pair (K_b, k_b) and store (b, K_b, k_b) in $\mathsf{H_{dec}}$. Then the simulator decrypts $(R_0, R_1, \mathsf{d}, \mathcal{T})$ as usual.
- H oracle: On input (b, R_0, R_1, Z), the simulator firstly checks if $(b, R_0, R_1) \in \mathsf{H_{dec}}$. If $(b, R_0, R_1) \in \mathsf{H_{dec}}$ and $Z = R_b^x$, then the simulator sets $\mathsf{H_{val}}[b, R_0, R_1, Z] = (K_b, k_b)$ and removes (b, R_0, R_1) from $\mathsf{H_{dec}}$. Then the simulator checks whether (b, R_0, R_1, Z) has been queried, and if so returns the recorded response (see Items 29 and 30). Otherwise, it determines (b, R_0, R_1, Z) should be added to $\mathsf{H_{val}}$ or to $\mathsf{H_{inv}}$ by checking $Z = R_b^x$ (see Items 33 and 34), and samples a fresh (K, k) and records it in $\mathsf{H_{val}}$ or $\mathsf{H_{inv}}$. The output distribution of H in this game is still uniformly random.

Now consider the case that $\mathcal{A}$ queries DEC on $(R_0, R_1, \mathsf{d}, \mathcal{T})$ but $\mathcal{A}$ has not queried H on the corresponding H-query of $(R_0, R_1, \mathsf{d}, \mathcal{T})$. In this case, the simulator cannot extract (K_0, k_0) and (K_1, k_1) from $\mathsf{H_{val}}$. Instead of using x to compute Z_0 and Z_1 as in $\mathsf{G_2}$, the game simulator of $\mathsf{G_3}$ samples fresh key pairs (K_0, k_0) and (K_1, k_1) and adds $(0, R_0, R_1)$ and $(1, R_0, R_1)$ into $\mathsf{H_{dec}}$. Lately, when $\mathcal{A}$ queries H on (b, R_0, R_1, Z) where $Z = R_b^x$, the game simulator "patches" (b, R_0, R_1, Z) into $\mathsf{H_{val}}$, i.e., sets $\mathsf{H_{val}}[b, R_0, R_1, Z] = (K_b, k_b)$, and removes (b, R_0, R_1) from $\mathsf{H_{dec}}$ (see Items 26 to 28).

We note that the use of these three lists is internal but the outputs of H and DEC are the same as in $\mathsf{G_2}$. Thus,

$$\Pr\left[\mathsf{G}_2^{\mathcal{A}} \Rightarrow 1\right] = \Pr\left[\mathsf{G}_3^{\mathcal{A}} \Rightarrow 1\right]$$

GAME $\mathsf{G_4}$: $\mathsf{G_4}$ aborts if $\mathcal{A}$ queries H on $(1 - b_i, R_{i,0}, R_{i,1}, Z_{i,1-b_i})$ with $Z_{i,1-b_i} = R_{i,1-b_i}^x$ and $\mathbf{c}[i]$ is not opened for some $1 \leq i \leq \mu$. We note that this abort condition lead to the CDH value of X and $R_{i,1-b_i}$. Hence, we can bound the probability of this abort event with the multi-challenge strong Diffie-Hellman (mStDH) assumption.

$\mathcal{B}_1^{\mathrm{DHP}_X}(X, Y_1, ..., Y_\mu)$	$H(b, R_0, R_1, Z)$
01 $Z^* := \perp$	16 **if** $\exists i \in [\mu] \backslash I$ s.t.
02 $(\mathcal{M}_a, \mathsf{st}) \xleftarrow{\$} \mathcal{A}_0^{\mathrm{DEC}, H, h}(X)$	$(b, R_0, R_1) = (1 - b_i, R_{i,0}, R_{i,1})$
03 **for** $i \in [\mu]$	**and** $\mathrm{DHP}_X(R_{i,1-b_i}, Z) = 1$
04 $\mathbf{m}[i] := \mathsf{m}_i \xleftarrow{\$} \mathcal{M}_a$	17 $Z^* := Z$ // records the solution
05 $b_i \xleftarrow{\$} \{0, 1\}$	18 Aborts the simulation and returns Z^*
06 $r_{i,b_i} \xleftarrow{\$} \mathbb{Z}_p, R_{i,b_i} := g^{r_{i,b_i}}$	19 **if** $\exists (K, k)$ s.t. $\mathsf{H_{dec}}[b, R_0, R_1] = (K, k)$
07 $Z_{i,b_i} := X^{r_{i,b_i}}$	**and** $\mathrm{DHP}_X(R_b, Z) = 1$
08 $(K_i, k_i) := H(b_i, R_{i,0}, R_{i,1}, Z_{i,b_i})$	20 $\mathsf{H_{val}}[b, R_0, R_1, Z] := (K, k)$
09 $R_{i,1-b_i} := Y_i$	21 $\mathsf{H_{dec}}[b, R_0, R_1] := \perp$
10 $\mathsf{d}_i := \mathsf{m}_i \oplus K_i$	22 **if** $\exists (K, k)$ s.t. $\mathsf{H_{val}}[b, R_0, R_1, Z] = (K, k)$
11 $\mathcal{T}_i := h(k_i, R_{i,0}, R_{i,1}, \mathsf{d}_i)$	**or** $\mathsf{H_{inv}}[b, R_0, R_1, Z] = (K, k)$
12 $\mathbf{c}[i] := (R_{i,0}, R_{i,1}, \mathsf{d}_i, \mathcal{T}_i)$	23 **return** (K, k)
13 $out \xleftarrow{\$} \mathcal{A}_1^{\mathrm{OPEN}, \mathrm{DEC}, H, h}(\mathsf{st}, \mathbf{c})$	24 **else**
14 **if** $Z^* = \perp : Z^* \xleftarrow{\$} \mathbb{G}$	25 $(K, k) \xleftarrow{\$} \mathcal{M} \times \{0, 1\}^l$
15 **return** Z^*	26 **if** $\mathrm{DHP}_X(R_b, Z) = 1$
	27 $\mathsf{H_{val}}[b, R_0, R_1, Z] := (K, k)$
	28 **else** $\mathsf{H_{inv}}[b, R_0, R_1, Z] := (K, k)$
	29 **return** (K, k)

Fig. 7. mStDH adversary $\mathcal{B}_1$ in bounding the difference between $\mathsf{G_3}$ and $\mathsf{G_4}$. The simulation of DEC and h are the same as in $\mathsf{G_4}$ in Fig. 6.

The reduction $\mathcal{B}_1$ against the mStDH assumption is constructed in Fig. 7. On input $(X, Y_1, ..., Y_\mu)$, $\mathcal{B}_1$ sets $R_{i,1-b_i} := Y_i$. It can simulate $\mathsf{G_4}$ without x, since it can use its DHP_X oracle to check whether $Z = \mathsf{cdh}(X, R_{i,1-b_i})$. Therefore,

$$\left| \Pr\left[\mathsf{G}_3^{\mathcal{A}} \Rightarrow 1\right] - \Pr\left[\mathsf{G}_4^{\mathcal{A}} \Rightarrow 1\right] \right| \leq \mathsf{Adv}_{\mathbb{G}}^{\mathsf{mStDH}}(\mathcal{B}_1)$$

GAME G_5: We introduce the abort rule in the H oracle: If $\mathcal{A}$ queries $H(b_i, R_{i,0}, R_{i,1}, R_{i,b_i}^x)$ for some $i \in [\mu]$, then G_5 aborts. Let BAD be this querying event and BAD_j be the event that BAD happens in G_j. The adversary cannot detect this modification unless it triggers BAD_5. We have

$$\left| \Pr\left[G_4^{\mathcal{A}} \Rightarrow 1\right] - \Pr\left[G_5^{\mathcal{A}} \Rightarrow 1\right] \right| \leq \Pr\left[\text{BAD}_5\right]$$

Here we cannot bound $\Pr[\text{BAD}_5]$ using mStDH yet, since if the adversary queries OPEN(i), then the simulator has to returns r_{i,b_i}, where is unknown when constructing reduction from mStDH. We will bound it later. Our strategy is to decouple $\mathbf{c}[i]$ with $H(b_i, R_{i,0}, R_{i,1}, R_{i,b_i}^x)$ and then use the randomness $(1 - b_i, r_{i,1-b_i}, R_{i,b_i})$ to explain $\mathbf{c}[i]$ (and thus we do not need r_{i,b_i} and can construct reduction from mStDH).

GAME G_6: The difference to G_5 is that when generating $\mathbf{c}[i]$, we choose random key pair (K_i, k_i) independent of $H(b_i, R_{i,0}, R_{i,1}, R_{i,b_i}^x)$, and when $\mathcal{A}$ opens $\mathbf{c}[i]$, then we define $H(b_i, R_{i,0}, R_{i,1}, R_{i,b_i}^x)$ as (K_i, k_i).

By abort condition in H, $H(b_i, R_{i,0}, R_{i,1}, R_{i,b_i}^x)$ will not be defined before $\mathbf{c}[i]$ is opened, so this modification does not change $\mathcal{A}$'s view, we have

$$\left| \Pr\left[G_5^{\mathcal{A}} \Rightarrow 1\right] - \Pr\left[G_6^{\mathcal{A}} \Rightarrow 1\right] \right| \leq \Pr\left[\text{BAD}_6\right], \Pr\left[\text{BAD}_5\right] = \Pr\left[\text{BAD}_6\right]$$

GAME G_7: We modify the simulation of OPEN: When $\mathcal{A}$ opens $\mathbf{c}[i]$, we set $H(1 - b_i, R_{i,0}, R_{i,1}, R_{i,1-b_i}^x) := (K_i, k_i)$, but not $H(b_i, R_{i,0}, R_{i,1}, R_{i,b_i}^x)$. Moreover, instead of returning $(b_i, r_{i,b_i}, R_{i,1-b_i})$, we return its complement, $(1 - b_i, r_{i,1-b_i}, R_{i,b_i})$.

We argue that if BAD_7 does not occur, then the view of $\mathcal{A}$ in G_7 is the same as in G_6. This is because G_7 does not abort means that $\mathcal{A}$ has queried neither $H(b_i, R_{i,0}, R_{i,1}, R_{i,b_i}^x)$ for some $i \in [\mu] \backslash I$ nor $H(1 - b_i, R_{i,0}, R_{i,1}, R_{i,1-b_i}^x)$ for some $i \in [\mu] \backslash I$. Hence, $\mathcal{A}$ has no information about these two values, and, as a result, $\mathcal{A}$ cannot tell the change in OPEN. We have

$$\left| \Pr\left[G_6^{\mathcal{A}} \Rightarrow 1\right] - \Pr\left[G_7^{\mathcal{A}} \Rightarrow 1\right] \right| \leq \Pr\left[\text{BAD}_7\right], \Pr\left[\text{BAD}_6\right] = \Pr\left[\text{BAD}_7\right]$$

To conclude our argument, we construct a reduction $\mathcal{B}_2$ against the mStDH assumption to bound $\Pr[\text{BAD}_7]$. $\mathcal{B}_2$ has a similar structure with $\mathcal{B}_1$ in Fig. 7, except that now $\mathcal{B}_2$ embeds Y_i into R_{i,b_i} (by setting $R_{i,b_i} := Y_i$ for all $i \in [\mu]$). The construction of $\mathcal{B}_2$ is shown in Fig. 8.

In $\mathcal{B}_2$'s construction, it does not have r_{i,b_i} and cannot compute $Z_{i,b_i} = R_{i,b_i}^x$. But it is not a problem, since $\mathcal{B}_3$ can program the random oracle H. More precisely, it leaves Z_{i,b_i} as unknown and choose a random pair (K_i, k_i) (cf. Item 10). Now if BAD_7 does not happen then the response of $H(b_i, R_{i,0}, R_{i,1}, Z_{i,b_i})$ is anyway random to $\mathcal{A}$ and it does not change its view. If BAD_7 happens, then $\mathcal{B}_2$ can find out $Z_{i,b_i} = g^{r_{i,b_i} \cdot x}$ by its DHP oracle and extract the solution to the mStDH problem. Thus, we have

$$\Pr\left[\text{BAD}_5\right] = \Pr\left[\text{BAD}_6\right] = \Pr\left[\text{BAD}_7\right] \leq \mathsf{Adv}_{\mathbb{G}}^{\mathsf{mStDH}}(\mathcal{B}_2)$$

$\mathcal{B}_2^{\mathrm{DHP}_X}(X, Y_1, ..., Y_\mu)$

01 $Z^* := \bot$
02 $(\mathcal{M}_a, \mathrm{st}) \xleftarrow{\$} \mathcal{A}_0^{\mathrm{DEC}, H, h}(X)$
03 for $i \in [\mu]$
04      $\mathbf{m}[i] := \mathsf{m}_i \xleftarrow{\$} \mathcal{M}_a$
05      $b_i \xleftarrow{\$} \{0, 1\}$
06      $r_{i,1-b_i} \xleftarrow{\$} \mathbb{Z}_p$
07      $R_{i,1-b_i} := g^{r_{i,1-b_i}}$
08      $Z_{i,1-b_i} := X^{r_{i,1-b_i}}$
09      $R_{i,b_i} := Y_i$
10      $(K_i, k_i) \xleftarrow{\$} K \times \{0, 1\}^l$
11      $\mathsf{d}_i := \mathsf{m}_i \oplus K_i$
12      $\mathcal{T}_i := h(k_i, R_{i,0}, R_{i,1}, \mathsf{d}_i)$
13      $\mathbf{c}[i] := (R_{i,0}, R_{i,1}, \mathsf{d}_i, \mathcal{T}_i)$
14 $out \xleftarrow{\$} \mathcal{A}_1^{\mathrm{OPEN}, \mathrm{DEC}, H, h}(\mathrm{st}, \mathbf{c})$
15 if $Z^* = \bot : Z^* \xleftarrow{\$} \mathbb{G}$
16 return Z^*

$\mathrm{OPEN}(i)$

17 $I := I \cup \{i\}$
18 $H_{\mathsf{val}}[1 - b_i, R_{i,0}, R_{i,1}, Z_{i,1-b_i}]$
        $:= (K_i, k_i)$
19 return $(\mathsf{m}_i, (1 - b_i, r_{i,1-b_i}, R_{i,b_i}))$

$H(b, R_0, R_1, Z)$

20 if $\exists i \in [\mu] \backslash I$ s.t.
        $(b, R_0, R_1, Z) = (1 - b_i, R_{i,0}, R_{i,1}, Z_{i,1-b_i})$
21      $Z^* \xleftarrow{\$} \mathbb{G}$
22   Aborts the simulation and returns Z^*
23 if $\exists i \in [\mu] \backslash I$ s.t.
        $(b, R_0, R_1) = (b_i, R_{i,0}, R_{i,1})$
        and $\mathrm{DHP}_X(R_{i,b_i}, Z) = 1$
24      $Z^* := Z$ // records the solution
25   Aborts the simulation and returns Z^*
26 if $\exists (K, k)$ s.t. $H_{\mathsf{dec}}[b, R_0, R_1] = (K, k)$
        and $\mathrm{DHP}_X(R_b, Z) = 1$
27      $H_{\mathsf{val}}[b, R_0, R_1, Z] := (K, k)$
28      $H_{\mathsf{dec}}[b, R_0, R_1] := \bot$
29 if $\exists (K, k)$ s.t. $H_{\mathsf{val}}[b, R_0, R_1, Z] = (K, k)$
        or $H_{\mathsf{inv}}[b, R_0, R_1, Z] = (K, k)$
30   return (K, k)
31 else
32      $(K, k) \xleftarrow{\$} \mathcal{M} \times \{0, 1\}^l$
33   if $\mathrm{DHP}_X(R_b, Z) = 1$
34          $H_{\mathsf{val}}[b, R_0, R_1, Z] := (K, k)$
35   else $H_{\mathsf{inv}}[b, R_0, R_1, Z] := (K, k)$
36   return (K, k)

Fig. 8. mStDH adversary $\mathcal{B}_2$ in bounding BAD$_7$. It simulates G_7 for $\mathcal{A}$. The simulation of DEC and h are the same as in Fig. 6. If $\mathcal{A}$ queries H on $b_i, R_{i,0}, R_{i,1}, R_{i,1-b_i}$ for some $i \in [\mu] \backslash I$, $\mathcal{B}_2$ aborts the simulation and return a random solution.

Now all challenge ciphertexts are encrypted by random key (K_i, k_i). From G_8 we conclude the proof by undoing the other changes in a reverse order.

GAME G_8: We undo the abort rules in the H oracle, and explain the randomness of $\mathbf{c}[i]$ using $(b_i, r_{i,b_i}, R_{i,1-b_i})$. That is, we withdraw the modifications made in G_7, G_5 and G_4. Since now the computation of (K_i, k_i) is independent of b_i and $1 - b_i$, we can construct reduction from mStDH as we did in G_4 and G_7. Roughly, if we want to embed the challenge into R_{i,b_i}, then we can specify the random bit of $\mathbf{c}[i]$ as $1 - b_i$ and explain the randomness of $\mathbf{c}[i]$ by reprogramming H, and so we do not need the exponent of R_{i,b_i}. We have

$$\left| \Pr[\mathsf{G}_7 \Rightarrow 1] - \Pr[\mathsf{G}_8 \Rightarrow 1] \right| \leq 4\mathsf{Adv}_{\mathbb{G}}^{\mathsf{mStDH}}(\mathcal{B})$$

GAME G_9: We undo the modification made in G_2. We have

$$\Pr[\mathsf{G}_8 \Rightarrow 1] = \Pr[\mathsf{G}_9 \Rightarrow 1]$$

Now we can construct a SIM-SO-CCA simulator $\mathcal{S}$ that simulates G_9 for $\mathcal{A}$ and interacts with the IDEAL-SO-CCA game to conclude the proof. The construction of simulator is shown in Fig. 9.

$\mathcal{S}$ samples d_i uniformly from $\mathcal{M}$ and computes K_i as $\mathsf{d}_i \oplus \mathsf{m}_i$ (when $\mathcal{A}$ opens $\mathbf{c}[i]$), which is equivalent to sampling K_i firstly and then computing $\mathsf{d}_i := K_i \oplus$

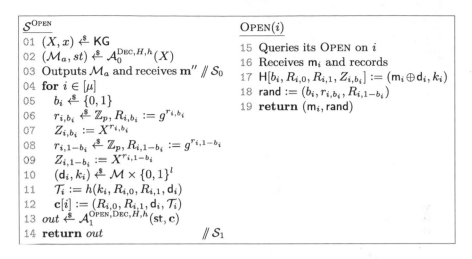

Fig. 9. SIM-SO-CCA simulator $\mathcal{S}$ that simulates G_9 to conclude the proof of Theorem 1. We ignore the simulation of H, h, and DEC which are the same as in G_9 in Fig. 6.

m_i. Therefore, $\mathcal{S}$ perfectly simulates G_9. Note that at the start of the proof we assume that from G_0 to G_8, there is no collision among the outputs of random oracle h, the first parts of outputs of H (i.e., K), and the second parts of outputs of H (i.e., k). Here we need to add back this collision bound. That is,

$$\left| \Pr\left[\mathsf{G}_9^{\mathcal{A}} \Rightarrow 1\right] - \Pr\left[\mathsf{IDEAL\text{-}SO\text{-}CCA}_{\mathsf{PKE_{StDH}}}^{\mathcal{S}} \Rightarrow 1\right] \right| \leq \frac{n_H^2}{|\mathcal{M}|} + \frac{n_H^2 + n_h^2}{2^l}$$

By combining all the probability bounds, we have

$$\left| \Pr\left[\mathsf{REAL\text{-}SO\text{-}CCA}_{\mathsf{PKE_{StDH}}}^{\mathcal{A}} \Rightarrow 1\right] - \Pr\left[\mathsf{IDEAL\text{-}SO\text{-}CCA}_{\mathsf{PKE_{StDH}}}^{\mathcal{S}} \Rightarrow 1\right] \right|$$
$$\leq 8\mathsf{Adv}_{\mathbb{G}}^{\mathsf{mStDH}}(\mathcal{B}) + \frac{2n_H^2}{|\mathcal{M}|} + \frac{2(n_H^2 + n_h^2)}{2^l},$$

as stated in Theorem 1.

3.2 Construction from the Twin Diffie-Hellman Assumption

Using the twinning technique from [10], we can remove the use of StDH assumption in $\mathsf{PKE_{StDH}}$ and have a scheme based on the standard CDH assumption. Let $\mathbb{G}$ be a group with prime order p and generator g. Let $H : \{0,1\} \times \mathbb{G}^3 \to \mathcal{M} \times \{0,1\}^l$, $h : \mathbb{G}^2 \times \{0,1\}^l \to \{0,1\}^\ell$ be hash functions. We propose a PKE scheme $\mathsf{PKE_{TDH}} = (\mathsf{KG}, \mathsf{Enc}, \mathsf{Dec})$ (shown in Fig. 10) based on TDH. The randomness space of $\mathsf{PKE_{TDH}}$ is $\{0,1\} \times \mathbb{Z}_p \times \mathbb{G}$. By [10], the TDH problem is tightly equivalent to the CDH problem. We give the security theorem of $\mathsf{PKE_{TDH}}$ in the full version [7]. The probability bounds are identical to Theorem 1.

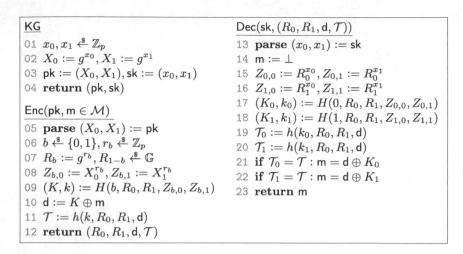

Fig. 10. Our Direction Construction of SIM-SO-CCA secure PKE schemes from the TDH assumption, $\mathsf{PKE}_{\mathsf{TDH}} = (\mathsf{KG}, \mathsf{Enc}, \mathsf{Dec})$

3.3 Direct Construction from the Decisional Diffie-Hellman Assumption

Our third direct construction is based on THE DDH assumption. Let $\mathbb{G}$ be a group with prime order p and two generators g_0 and g_1. Let $H : \{0,1\} \times \mathbb{G}^3 \to \mathcal{M} \times \{0,1\}^l, h : \{0,1\}^l \times \mathbb{G}^2 \to \{0,1\}^\ell$ be hash functions. The PKE scheme $\mathsf{PKE}_{\mathsf{DDH}} = (\mathsf{KG}, \mathsf{Enc}, \mathsf{Dec})$ with message space $\mathcal{M}$ is shown in Fig. 11. The randomness space of $\mathsf{PKE}_{\mathsf{DDH}}$ is the set $\{0,1\} \times \mathbb{Z}_p \times \mathbb{G}^2$.

<u>CORRECTNESS.</u> Similar to $\mathsf{PKE}_{\mathsf{StDH}}$, the correctness of $\mathsf{PKE}_{\mathsf{DDH}}$ depends on the hash function h. The correctness error $\mathsf{Adv}^{\mathsf{COR}}_{\mathsf{PKE}_{\mathsf{StDH}}}(\mathcal{A})$ is bounded by the collision probability of h, namely, $\mathsf{Adv}^{\mathsf{COR}}_{\mathsf{PKE}_{\mathsf{DDH}}}(\mathcal{A}) \leq \mathsf{Adv}^{\mathsf{CR}}_h(\mathcal{A})$.

Theorem 2. $\mathsf{PKE}_{\mathsf{DDH}}$ *in Fig. 11 is SIM-SO-CCA secure (Definition 9) if the mDDH problem is hard on $\mathbb{G}$ and H and h are modeled as random oracles. Concretely, for any SIM-SO-CCA adversary $\mathcal{A}$ and relation* Rel*, there exists a simulator $\mathcal{S}$ and a adversary $\mathcal{B}$ such that:*

$$\mathsf{Adv}^{\mathsf{SIM\text{-}SO\text{-}CCA}}_{\mathsf{PKE}_{\mathsf{DDH}}}(\mathcal{A}, \mathcal{S}, \mu, \mathsf{Rel}) \leq 10\mathsf{Adv}^{\mathsf{mDDH}}_{\mathsf{GGen}}(\mathcal{B}) + \frac{6\mu q_H}{p} + \frac{2n_H^2}{|\mathcal{M}|} + \frac{2(n_H^2 + n_h^2)}{2^l}$$

where q_H and n_{DEC} are the numbers of $\mathcal{A}$'s queries to H and DEC, respectively, and μ is the number of challenge ciphertexts. $n_H = \mu + q_H + 2n_{\mathrm{DEC}}$ and $n_h = \mu + q_H + 2n_{\mathrm{DEC}}$ are the total numbers of queries to H and h, respectively.

$\mathsf{PKE}_{\mathsf{DDH}}$ is based on the DDH-based non-committing KEM in [26], plus the double-randomness technique. The proof of Theorem 2 is similar to Theorem 1. In the reduction, we can embed the DDH challenge into one of $(R_{b,0}, R_{b,1})$ and $(R_{1-b,0}, R_{1-b,1})$, and then claim the ciphertext to another one. Since we always have the secret key (x_0, x_1) in reduction, the decryption oracle can be simulated in a straightforward way. We leave the proof in our full version paper [7].

KG	Dec$(\mathsf{sk}, (R_{0,0}, R_{0,1}, R_{1,0}, R_{1,1}, \mathsf{d}, \mathcal{T}))$
01 $(x_0, x_1) \xleftarrow{\$} \mathbb{Z}_p^2$	14 **parse** $(x_0, x_1) := \mathsf{sk}$
02 $\mathsf{pk} := g_0^{x_0} g_1^{x_1}$	15 $\mathsf{m} := \perp$
03 $\mathsf{sk} := (x_0, x_1)$	16 $Z_0 := R_{0,0}^{x_0} R_{0,1}^{x_1}$
04 **return** $(\mathsf{pk}, \mathsf{sk})$	17 $Z_1 := R_{1,0}^{x_0} R_{1,1}^{x_1}$
	18 $(K_0, k_0) := H(0, R_{0,0}, ..., R_{1,1}, Z_0)$
Enc$(\mathsf{pk}, \mathsf{m} \in \mathcal{M})$	19 $(K_1, k_1) := H(1, R_{0,0}, ..., R_{1,1}, Z_1)$
05 **parse** $(X_0, X_1) := \mathsf{pk}$	20 $\mathcal{T}_0 := h(k_0, R_{0,0}, ..., R_{1,1}, \mathsf{d})$
06 $b \xleftarrow{\$} \{0,1\}, r_b \xleftarrow{\$} \mathbb{Z}_p$	21 $\mathcal{T}_1 := h(k_1, R_{0,0}, ..., R_{1,1}, \mathsf{d})$
07 $R_{b,0} := g_0^{r_b}, R_{b,1} := g_1^{r_b}$	22 **if** $\mathcal{T}_0 = \mathcal{T} : \mathsf{m} = \mathsf{d} \oplus K_0$
08 $R_{1-b,0} \xleftarrow{\$} \mathbb{G}, R_{1-b,1} \xleftarrow{\$} \mathbb{G}$	23 **if** $\mathcal{T}_1 = \mathcal{T} : \mathsf{m} = \mathsf{d} \oplus K_1$
09 $Z_b := \mathsf{pk}^{r_b}$	24 **return** m
10 $(K, k) := H(b, R_{0,0}, ..., R_{1,1}, Z_b)$	
11 $\mathsf{d} := K \oplus \mathsf{m}$	
12 $\mathcal{T} := h(k, R_{0,0}, ..., R_{1,1}, \mathsf{d})$	
13 **return** $(R_{0,0}, ..., R_{1,1}, \mathsf{d}, \mathcal{T})$	

Fig. 11. SIM-SO-CCA secure PKE scheme $\mathsf{PKE}_{\mathsf{DDH}} = (\mathsf{KG}, \mathsf{Enc}, \mathsf{Dec})$

4 Generic Construction: From Lossy Encryption to SO-CCA PKE

In this section, we prove the tight SO security of Fujisaki-Okamoto's (FO) transformation [15] assuming that the underlying PKE is a lossy encryption [3]. More precisely, if the lossy encryption scheme has efficient opener (e.g., the one from [25]), then FO is SIM-SO-CCA-secure. If the lossy encryption does not have efficient opener (e.g., the one from hash proof systems [3,19]), then FO is IND-SO-CCA secure.

We recall the notion of lossy encryption and the FO transformation. Then we prove the tight SO security of FO's transformation in the random oracle model.

Definition 11 (Lossy Encryption [3]). *Let* $\mathsf{wPKE} := (\mathsf{wKG}, \mathsf{wEnc}, \mathsf{wDec})$ *be a PKE scheme with message space* $\mathcal{M}$ *and randomness space* $\mathcal{R}$. wPKE *is lossy if it has the following properties:*

- wPKE *is correct according to Definition 8.*
- *Key indistinguishability: We say* wPKE *has key indistinguishability if there is an algorithm* LKG *such that, for any adversary* $\mathcal{B}$, *the advantage function*

$$\mathsf{Adv}_{\mathsf{wPKE}}^{\mathsf{key-ind}}(\mathcal{B}) := |\Pr[\mathcal{B}(\mathsf{pk}) \Rightarrow 1] - \Pr[\mathcal{B}(\mathsf{pk}') \Rightarrow 1]|$$

is negligible, where $(\mathsf{pk}, \mathsf{sk}) \xleftarrow{\$} \mathsf{wKG}$ *and* $(\mathsf{pk}', \mathsf{td}) \xleftarrow{\$} \mathsf{LKG}$.

- *Lossiness: Let* $(\mathsf{pk}', \mathsf{td}) \xleftarrow{\$} \mathsf{LKG}$ *and* m, m' *be arbitrary messages in* $\mathcal{M}'$, *the statistical distance between* $\mathsf{wEnc}(\mathsf{pk}', \mathsf{m})$ *and* $\mathsf{wEnc}(\mathsf{pk}', \mathsf{m}')$ *is negligible.*

- *Openability: Let* $(\mathsf{pk}', \mathsf{td}) \xleftarrow{\$} \mathsf{LKG}$, m *and* m' *be arbitrary messages, and* r *be arbitrary randomness. For ciphertext* $c := \mathsf{wEnc}(\mathsf{pk}', \mathsf{m}; r)$, *there exists an algorithm* open *such that* $\mathsf{open}(\mathsf{td}, \mathsf{pk}', c, r, \mathsf{m}')$ *outputs* r' *where* $c = \mathsf{wEnc}(\mathsf{pk}', \mathsf{m}'; r')$. *Here* open *can be inefficient.*

We extend the above lossiness definition to a multi-challenge setting. The multi-challenge lossiness is implied by the single-challenge one using hybrid argument. Since it is only a statistical property, the hybrid argument will not affect tightness of the computational advantage.

Definition 12 (Multi-Challenge Lossiness). *Let* $(\mathsf{pk}', \mathsf{td}) \stackrel{\$}{\leftarrow} \mathsf{LKG}$, μ *be the number of challenge, and* $r_1, r_1', ... r_\mu, r_\mu'$ *be arbitrary messages in* $\mathcal{M}'$. *Multi-challenge Lossiness requires that statistical distance between* $\{\mathsf{wEnc}(\mathsf{pk}', r_i)\}_{i \in [\mu]}$ *and* $\{\mathsf{wEnc}(\mathsf{pk}', r_i')\}_{i \in [\mu]}$ *is negligible. We write the distance as* $\varepsilon_{\mathsf{wPKE}}^{\mathsf{m-enc-los}}$.

We require γ-spreadness for our construction.

Definition 13 (γ-Spreadness). *Let* $\mathsf{wPKE} := (\mathsf{wKG}, \mathsf{wEnc}, \mathsf{wDec})$ *be a PKE scheme with message space* $\mathcal{M}$, *randomness space* $\mathcal{R}$, *and ciphertext space* $\mathcal{C}$. *We say* wPKE *is* γ-*spread if for every key pair* $(\mathsf{pk}, \mathsf{sk}) \stackrel{\$}{\leftarrow} \mathsf{wKG}$, *and every message* $\mathsf{m} \in \mathcal{M}$,

$$\max_{\mathsf{c} \in \mathcal{C}} \Pr_{r \stackrel{\$}{\leftarrow} \mathcal{R}} [\mathsf{c} = \mathsf{wEnc}(\mathsf{pk}, \mathsf{m}; r)] \leq 2^{-\gamma}.$$

4.1 Construction

Let $\mathsf{wPKE} := (\mathsf{wKG}, \mathsf{wEnc}, \mathsf{wDec})$ be a lossy encryption scheme with message space $\mathcal{M}'$ and randomness space $\mathcal{R}'$. Let $H : \mathcal{M}' \to \mathcal{M}$ and $G : \mathcal{M}' \times \mathcal{M} \to \mathcal{R}'$ be two hash functions. The FO transformation $\mathsf{FO} := (\mathsf{KG}, \mathsf{Enc}, \mathsf{Dec})$ is defined in Fig. 12. Here we use the one-time pad as the symmetric part to encrypt the message. The randomness space of FO is $\mathcal{R}'$.

KG	Enc(pk, m)	Dec(sk, (e, d))
01 $(\mathsf{pk}, \mathsf{sk}) \stackrel{\$}{\leftarrow} \mathsf{wKG}$	03 $r \leftarrow \mathcal{M}'$	09 $\mathsf{m}' := \perp$
02 **return** $(\mathsf{pk}, \mathsf{sk})$	04 $K := H(r)$	10 $r' := \mathsf{wDec}(\mathsf{sk}, e)$
	05 $\mathsf{d} := K \oplus \mathsf{m}$	11 $R' := G(r', \mathsf{d}), K' := H(r')$
	06 $R := G(r, \mathsf{d})$	12 **if** $e = \mathsf{wEnc}(\mathsf{pk}, r'; R')$
	07 $e := \mathsf{wEnc}(\mathsf{pk}, r; R)$	13 $\mathsf{m}' := \mathsf{d} \oplus K'$
	08 **return** (e, d)	14 **return** m'

Fig. 12. Fujisaki-Okamoto's transformation FO with lossy encryption wPKE.

As shown in [24], if wPKE is $(1 - \delta)$-correct and G is modeled as a random oracle, then FO is $(1 - q_G \delta)$-correct where q_G is the number of queries to G.

Theorems 3 and 4 show the tight SIM-SO-CCA and IND-SO-CCA security of FO, respectively. We only prove Theorem 3 in the main body and leave that of Theorem 4 in our full version paper [7], since both proofs are similar and the SIM-SO-CCA security is more common.

Theorem 3. FO *in Fig. 12 is SIM-SO-CCA secure if* G *and* H *are modeled as random oracles, and* wPKE *is a lossy encryption with efficient openability and* *γ-spreadness. Concretely, for any SIM-SO-CCA adversary $\mathcal{A}$ and relation Rel, there exists a simulator $\mathcal{S}$ and $\mathcal{B}$ such that:*

$$\mathsf{Adv}_{\mathsf{F0}}^{\mathsf{SIM\text{-}SO\text{-}CCA}}(\mathcal{A}, \mathcal{S}, \mu, \mathsf{Rel}) \leq \mathsf{Adv}_{\mathsf{wPKE}}^{\mathsf{key\text{-}ind}}(\mathcal{B}) + 2\varepsilon_{\mathsf{wPKE}}^{\mathsf{m\text{-}enc\text{-}los}}$$

$$+ \frac{\mu n_{\mathrm{DEC}}}{2^{\gamma}} + \frac{2n_H^2}{|\mathcal{M}|} + \frac{2n_G^2}{|\mathcal{R}'|} + \frac{4\mu^2 + 5\mu(q_G + q_H)}{|\mathcal{M}'|},$$

where q_H, q_G, and n_{DEC} are the numbers of $\mathcal{A}$'s queries to G, H, and DEC, respectively, μ is the number of challenge ciphertexts, and $n_G = \mu + n_{\mathrm{DEC}} + q_H$ and $n_H = \mu + n_{\mathrm{DEC}} + q_G$ are the number of queries (including the simulator) to G and H, respectively.

Theorem 4. *F0 in Fig. 12 is IND-SO-CCA secure (Definition 10) if G and H are modeled as random oracles, and wPKE is a lossy encryption and γ-spreadness. Concretely, for any IND-SO-CCA adversary $\mathcal{A}$, there exists $\mathcal{B}$ such that:*

$$\mathsf{Adv}_{\mathsf{F0}}^{\mathsf{IND\text{-}SO\text{-}CCA}}(\mathcal{A}, \mu) \leq 2(\mathsf{Adv}_{\mathsf{wPKE}}^{\mathsf{key\text{-}ind}}(\mathcal{B}) + 3\varepsilon_{\mathsf{wPKE}}^{\mathsf{m\text{-}enc\text{-}los}} + \frac{\mu n_{\mathrm{DEC}}}{2^{\gamma}})$$

$$+ \frac{2n_H^2}{|\mathcal{M}|} + \frac{2n_G^2}{|\mathcal{R}'|} + \frac{6\mu^2 + 5\mu(q_G + q_H)}{|\mathcal{M}'|},$$

where q_H, q_G, and n_{DEC} are the numbers of $\mathcal{A}$'s queries to G, H, and DEC, respectively, μ is the number of challenge ciphertexts, and $n_G = \mu + n_{\mathrm{DEC}} + q_H$ and $n_H = \mu + n_{\mathrm{DEC}} + q_G$ are the number of queries (including the simulator) to G and H, respectively.

4.2   Proof of Theorem 3

We prove it by the game sequence as in Fig. 13. G_0 is the original game except that we use lazy sampling to simulate ROs G and H. We assume that, from G_0 to G_9, there is no collision among r_i's and the outputs of H and G. Let n_G and n_H be the number of queries to G and H, respectively. By the security game in Fig. 13, $n_G = \mu + n_{\mathrm{DEC}} + q_G$ and $n_H = \mu + n_{\mathrm{DEC}} + q_H$. We have

$$\left| \Pr\left[\mathsf{REAL\text{-}SO\text{-}CCA}_{\mathsf{F0}}^{\mathcal{A}} \Rightarrow 1 \right] - \Pr\left[\mathsf{G}_0^{\mathcal{A}} \Rightarrow 1 \right] \right| \leq \frac{n_H^2}{|\mathcal{M}|} + \frac{\mu^2}{|\mathcal{M}'|} + \frac{n_G^2}{|\mathcal{R}'|}$$

GAME G_1: We modify DEC. Instead of using sk to simulate DEC, we use the randomness recorded in G to decrypt given ciphertexts (see Items 40 to 42). This simulation method is exact the same as the one in the original FO transformation [15]. By the argument in [15], if wPKE is γ-spread, then we have

$$\left| \Pr\left[\mathsf{G}_0^{\mathcal{A}} \Rightarrow 1 \right] - \Pr\left[\mathsf{G}_1^{\mathcal{A}} \Rightarrow 1 \right] \right| \leq \frac{\mu \cdot n_{\mathrm{DEC}}}{2^{\gamma}}$$

GAME G_2: We switch the public key to lossy mode by $(\mathsf{pk}', \mathsf{td}) \xleftarrow{\$} \mathsf{LKG}$. Since in this game the decryption oracle are simulated without using sk, we can simulate

Games G_0-G_7		OPEN(i)	
01 $(pk, sk) \stackrel{\$}{\leftarrow} wKG$	// G_0-G_1	27 $G[r_i, d_i] := R_i$	// G_5-G_7
02 $(pk', td) \stackrel{\$}{\leftarrow} LKG$	// G_2-G_7	28 $H[r_i] := K_i$	// G_5-G_7
03 $(pk, sk) := (pk', td)$	// G_2-G_7	29 $R_i' := open(sk, pk, e_i, R_i, r_i')$	// G_7
04 $(\mathcal{M}_a, st) \stackrel{\$}{\leftarrow} \mathcal{A}_0^{H,G}(pk)$		30 $G[r_i', d_i] := R_i'$	// G_7
05 **for** $i \in [\mu]$		31 $H[r_i'] := K_i$	// G_7
06 $\mathbf{m}[i] := m_i \stackrel{\$}{\leftarrow} \mathcal{M}_a$		32 $I := I \cup \{i\}$	
07 $r_i \stackrel{\$}{\leftarrow} \mathcal{M}'$		33 **return** (m_i, r_i)	
08 $r_i' \stackrel{\$}{\leftarrow} \mathcal{M}'$	// G_3-G_7		
09 $K_i := H(r_i)$		DEC(c) // $c \notin \mathbf{c}$	
10 $K_i \stackrel{\$}{\leftarrow} \mathcal{M}$	// G_5	34 **parse** $(e, d) := c$	
11 $d_i := m_i \oplus K_i$		35 $m' := \bot$	
12 $d_i \stackrel{\$}{\leftarrow} \mathcal{M}$	// G_6-G_7	36 $r' := wDec(sk, e)$	// G_0
13 $K_i := d_i \oplus m_i$	// G_6-G_7	37 $R' := G(r', d), K' := H(r')$	// G_0
14 $R_i := G(r_i, d_i)$		38 **if** $e = wEnc(pk, r'; R')$	// G_0
15 $R_i \stackrel{\$}{\leftarrow} \mathcal{R}'$	// G_5-G_7	39 $m' := d \oplus K'$	// G_0
16 $e_i := wEnc(pk, r_i; R_i)$		40 **if** $\exists (r', R')$ s.t. $G[r', d] = R'$	
17 $\mathbf{c}[i] := (e_i, d_i)$		and $e = wPKE(pk, r'; R')$	// G_1-G_7
18 $out \stackrel{\$}{\leftarrow} \mathcal{A}_1^{OPEN,H,G}(st, \mathbf{c})$		41 $K' := H(r')$	// G_1-G_7
19 **return** $Rel(\mathcal{M}_a, \mathbf{m}, I, out)$		42 $m' := d \oplus K'$	// G_1-G_7
		43 **return** m'	
$H(r)$			
20 **if** $\exists i \in [\mu] \backslash I$ s.t. $r = r_i'$	// G_3-G_7	$G(r, d)$	
21 **abort**	// G_3-G_7	44 **if** $\exists i \in [\mu] \backslash I$ s.t. $r = r_i'$	// G_3-G_7
22 **if** $\exists i \in [\mu] \backslash I$ s.t. $r = r_i$	// G_4-G_7	45 **abort**	// G_3-G_7
23 **abort**	// G_4-G_7	46 **if** $\exists i \in [\mu] \backslash I$ s.t. $r = r_i$	// G_4-G_7
24 **if** $H[r] = \bot$		47 **abort**	// G_4-G_7
25 $H[r] := K \stackrel{\$}{\leftarrow} \mathcal{M}$		48 **if** $G[r, d] = \bot$	
26 **return** $H[r]$		49 $G[r, d] := R \stackrel{\$}{\leftarrow} \mathcal{R}'$	
		50 **return** $G[r, d]$	

Fig. 13. Games G_0–G_7 for proving Theorem 3.

G_2 with pk'. By the key indistinguishability of the lossy encryption,

$$\left| \Pr \left[G_1^{\mathcal{A}} \Rightarrow 1 \right] - \Pr \left[G_2^{\mathcal{A}} \Rightarrow 1 \right] \right| \leq \mathsf{Adv}_{wPKE}^{key-ind}(\mathcal{B}_0)$$

GAME G_3: This is a preparation step. We choose some internal randomness r_i' for the opening queries in the next games. We abort G_3 if $\mathcal{A}$ queries either H or G with r_i' before opening $\mathbf{c}[i]$. Since r_i' (for $i \in [\mu]$) are internal and never revealed to $\mathcal{A}$, the probability that $\mathcal{A}$ queries r_i' for some i is $\frac{q_H + q_G}{|\mathcal{M}'|}$. We also require all r_i''s are different. By the union bound and collision bound, we have

$$\left| \Pr \left[G_2^{\mathcal{A}} \Rightarrow 1 \right] - \Pr \left[G_3^{\mathcal{A}} \Rightarrow 1 \right] \right| \leq \frac{\mu \cdot (q_H + q_G)}{|\mathcal{M}'|} + \frac{\mu^2}{|\mathcal{M}'|}$$

GAME G_4: We further modify the abort rules in H and G. If $\mathcal{A}$ queries H or G with r_i and $\mathbf{c}[i]$ is unopened, then G_4 aborts. Let $\mathsf{QueryBad}_j$ be the event that such abort event occurs in G_j, i.e., $\mathcal{A}$ queries H (resp., G) on r_i (resp, (r_i, d_i)) where $\mathbf{c}[i]$ is unopened. Then we have

$$\left| \Pr\left[G_3^{\mathcal{A}} \Rightarrow 1 \right] - \Pr\left[G_4^{\mathcal{A}} \Rightarrow 1 \right] \right| \leq \Pr\left[\mathsf{QueryBad}_4 \right]$$

Here we cannot bound $\Pr\left[\mathsf{QueryBad}_4\right]$ directly yet, since all e_i are correlated to $H(r_i)$ and $G(r_i, \mathsf{d}_i)$. We will bound $\Pr\left[\mathsf{QueryBad}_4\right]$ later. Our strategy for that is to decouple e_i with $G(r_i, \mathsf{d}_i)$ and $H(r_i)$. In the end, $\mathcal{A}$ can query r_i for $i \in [\mu] \backslash I$ (i.e., $\mathbf{c}[i]$ is unopened) with negligible probability.

GAME G_5: We modify the generation of R_i and K_i. In this game, R_i and K_i are chosen uniformly, instead of using H and G. Moreover, upon $\mathrm{OPEN}(i)$, we set $H(r_i) := K_i$ and $G(r_i, \mathsf{d}_i) := R_i$. By the abort rules in G and H, $\mathcal{A}$ can learn neither $H(r_i)$ nor $G(r_i, \mathsf{d}_i)$ before opening $\mathbf{c}[i]$. Thus, we have

$$\Pr\left[G_4^{\mathcal{A}} \Rightarrow 1 \right] = \Pr\left[G_5^{\mathcal{A}} \Rightarrow 1 \right], \ \Pr\left[\mathsf{QueryBad}_4 \right] = \Pr\left[\mathsf{QueryBad}_5 \right]$$

GAME G_6: We further modify the computation of d_i and K_i. In this game, d_i are chosen uniformly at random, and K_i are computed as $K_i := \mathsf{d}_i \oplus \mathsf{m}_i$. In G_5 K_i is distributed uniformly at random. Hence, this modification is conceptual.

$$\Pr\left[G_5^{\mathcal{A}} \Rightarrow 1 \right] = \Pr\left[G_6^{\mathcal{A}} \Rightarrow 1 \right], \ \Pr\left[\mathsf{QueryBad}_5 \right] = \Pr\left[\mathsf{QueryBad}_6 \right]$$

GAME G_7: Upon $\mathrm{OPEN}(i)$, we compute the opened randomness R_i' with respect to r_i' and E_i using the open algorithm (see Item 29), and then set $G(r_i', \mathsf{d}_i) := R_i'$ and $H(r_i') := K_i$. Looking ahead, this modification is necessary for the later modification that $\mathbf{c}[i] = (E_i, \mathsf{d}_i)$ can be claimed to r_i'. $\mathcal{A}$ detects this modification if it queries $H(r_i')$ or $G(r_i', \mathsf{d}_i)$. This modification does not affect the occurring probability of $\mathsf{QueryBad}_7$, since r_i' is perfectly hidden. Therefore,

$$\left| \Pr\left[G_6^{\mathcal{A}} \Rightarrow 1 \right] - \Pr\left[G_7^{\mathcal{A}} \Rightarrow 1 \right] \right| \leq \frac{\mu(q_G + q_H)}{|\mathcal{M}'|}, \ \Pr\left[\mathsf{QueryBad}_6 \right] = \Pr\left[\mathsf{QueryBad}_7 \right]$$

In G_7, we have the following observation: Before $\mathcal{A}$ opens i, R_i are independent of r_i, r_i', K_i, and d_i, so E_i can be viewed as a ciphertext that $E_i := \mathsf{wPKE}(\mathsf{pk}', r_i; R_i)$ where the randomness R_i is sampled independently and uniformly. Therefore, by the *Lossiness* of pk', we can replace $\mathsf{wPKE}(\mathsf{pk}', r_i; R_i)$ as another ciphertext $\mathsf{wPKE}(\mathsf{pk}', r_i''; R_i'')$ where r_i'' and R_i'' are sampled independently and uniformly, and $\mathcal{A}$ cannot distinguish such replacement except with $\varepsilon_{\mathsf{wPKE}}^{\mathsf{m\text{-}enc\text{-}los}}$. We move the description of G_7-G_9 to Fig. 14.

GAME G_8: We modify the generation of ciphertext E_i and simulation of OPEN. In this game, E_i is an encryption of a randomly chosen r_i'' with randomness R_i'' (see Item 14) which are independent of $r_i, r_i', R_i, \mathsf{d}_i$. When $\mathcal{A}$ opens $\mathbf{c}[i] = (E_i, \mathsf{d}_i)$, the game simulator reprograms H and G so that $\mathbf{c}[i]$ can be "explained" by

```
Games G₇-G₉                                    OPEN(i)
```

Games G_7-G_9		OPEN(i)	
01 $(\text{pk}', \text{td}) \xleftarrow{\$} \text{LKG}$		25 $R_i' := \text{open}(\text{sk}, \text{pk}, e_i, R_i, r_i')$	// G_7
02 $(\text{pk}, \text{sk}) := (\text{pk}', \text{td})$		26 $R_i' := \text{open}(\text{sk}, \text{pk}, e_i, R_i'', r_i')$	// G_8-G_9
03 $(\mathcal{M}_a, st) \xleftarrow{\$} \mathcal{A}_0^{\text{DEC},H,G}(\text{pk})$		27 $G[r_i', d_i] := R_i'$	
04 **for** $i \in [\mu]$		28 $H[r_i'] := K_i$	
05 $\quad \mathbf{m}[i] := \mathsf{m}_i \xleftarrow{\$} \mathcal{M}_a$		29 $H[r_i] := K_i$	// G_7-G_8
06 $\quad r_i \xleftarrow{\$} \mathcal{M}'$	// G_7-G_8	30 $G[r_i, d_i] := R_i$	// G_7-G_8
07 $\quad r_i' \xleftarrow{\$} \mathcal{M}'$		31 $I := I \cup \{i\}$	
08 $\quad d_i \xleftarrow{\$} \mathcal{M}$		32 **return** (m_i, r_i)	// G_7
09 $\quad K_i := d_i \oplus \mathsf{m}_i$		33 **return** (m_i, r_i')	// G_8-G_9
10 $\quad R_i \xleftarrow{\$} \mathcal{R}'$			
11 $\quad e_i := \text{wEnc}(\text{pk}, r_i; R_i)$	// G_7	DEC(c) // $c \notin \mathbf{c}$	
12 $\quad r_i'' \xleftarrow{\$} \mathcal{M}'$	// G_8-G_9	34 **parse** $(e, d) := c$	
13 $\quad R_i'' \xleftarrow{\$} \mathcal{R}'$	// G_8-G_9	35 $m' := \bot$	
14 $\quad e_i \xleftarrow{\$} \text{wEnc}(\text{pk}, r_i''; R_i'')$	// G_8-G_9	36 **if** $\exists (r', K')$ s.t. $G[r', d] = R'$	
15 $\quad \mathbf{c}[i] := (e_i, d_i)$		$\quad$ **and** $e = \text{wPKE}(\text{pk}, r'; R')$	
16 $out \xleftarrow{\$} \mathcal{A}_1^{\text{OPEN},\text{DEC},H,G}(st, \mathbf{c})$		37 $\quad K' := H(r')$	
17 **return** $\text{Rel}(\mathcal{M}_a, \mathbf{m}, I, out)$		38 $\quad m' := d \oplus K'$	
		39 **return** m'	
$H(r)$			
18 **if** $\exists i \in [\mu] \backslash I$ s.t. $r = r_i'$	// G_7-G_8	$G(r, d)$	
19 $\quad$ **abort**	// G_7-G_8	40 **if** $\exists i \in [\mu] \backslash I$ s.t. $r = r_i'$	// G_7-G_8
20 **if** $\exists i \in [\mu] \backslash I$ s.t. $r = r_i$	// G_7-G_8	41 $\quad$ **abort**	// G_7-G_8
21 $\quad$ **abort**	// G_7-G_8	42 **if** $\exists i \in [\mu] \backslash I$ s.t. $r = r_i$	// G_7-G_8
22 **if** $H[r] = \bot$		43 $\quad$ **abort**	// G_7-G_8
23 $\quad H[r] := K \xleftarrow{\$} \mathcal{M}$		44 **if** $G[r, d] = \bot : G[r, d] := R \xleftarrow{\$} \mathcal{R}'$	
24 **return** $H[r]$		45 **return** $G[r, d]$	

Fig. 14. Games G_7–G_9 for proving Theorem 3.

message m_i and randomness r_i' (i.e., $\text{Enc}(\text{pk}, \mathsf{m}_i; r_i') = \mathbf{c}[i]$), and returns (m_i, r_i'). By the lossiness of wPKE, the statistical distance between $\{\text{wPKE}(\text{pk}', r_i)\}_{i \in [\mu]}$ with $\{\text{wPKE}(\text{pk}', r_i'')\}_{i \in [\mu]}$ is $\varepsilon_{\text{wPKE}}^{\text{m-enc-los}}$. Hence, we have

$$\left| \Pr\left[G_7^{\mathcal{A}} \Rightarrow 1 \right] - \Pr\left[G_8^{\mathcal{A}} \Rightarrow 1 \right] \right| \leq \varepsilon_{\text{wPKE}}^{\text{m-enc-los}}$$

$$\left| \Pr\left[\text{QueryBad}_7 \right] - \Pr\left[\text{QueryBad}_8 \right] \right| \leq \varepsilon_{\text{wPKE}}^{\text{m-enc-los}}$$

Now $\Pr\left[\text{QueryBad}_8 \right]$ can be bounded. Since r_i and r_i' are chosen uniformly and independent of $\mathbf{c}[i]$ (for $i \in [\mu]$), we have

$$\Pr\left[\text{QueryBad}_8 \right] \leq \frac{\mu(q_G + q_H)}{|\mathcal{M}'|}, \Pr[\text{QueryBad}_7] \leq \varepsilon_{\text{wPKE}}^{\text{m-enc-los}} + \frac{\mu(q_G + q_H)}{|\mathcal{M}'|}$$

Since now r_i' are independent of E_i before opening, and r_i is redundant in the simulation, we withdraw all the abort events defined in H and G, and no longer reprogram $H(r_i)$ and $G(r_i, d_i)$.

GAME G_9: the aborts event defined in H and G are withdraw, and we no longer generate r_i and reprogram $H(r_i)$ and $G(r_i, d_i)$ when $\mathbf{c}[i]$ is opened. Since in

$$
\begin{array}{ll}
\mathcal{S}^{\text{OPEN}'} & \text{OPEN}(i) \\
\hline
\text{01} \ (\mathsf{pk}', \mathsf{td}) \xleftarrow{\$} \mathsf{LKG} & \text{12} \ r_i' \xleftarrow{\$} \mathcal{M}' \\
\text{02} \ (\mathsf{pk}, \mathsf{sk}) := (\mathsf{pk}', \mathsf{td}) & \text{13} \ \text{Queries OPEN}'(i) \\
\text{03} \ (\mathcal{M}_a, st) \xleftarrow{\$} \mathcal{A}_0^{\text{DEC}, H, G}(\mathsf{pk}) & \text{14} \ \text{Receives and records } \mathsf{m}_i \\
\text{04} \ \text{Outputs } \mathcal{M}_a \text{ and receives } \mathbf{m}'' \quad /\!\!/ \, \mathcal{S}_0 & \text{15} \ K_i := \mathsf{d}_i \oplus \mathsf{m}_i \\
\text{05} \ \mathbf{for} \ i \in [\mu] & \text{16} \ R_i' := \mathsf{open}(\mathsf{sk}, \mathsf{pk}, e_i, R_i'', r_i') \\
\text{06} \quad \mathsf{d}_i \xleftarrow{\$} \mathcal{M} & \text{17} \ \mathsf{G}[r_i', \mathsf{d}_i] := R_i' \\
\text{07} \quad r_i'' \xleftarrow{\$} \mathcal{M}', R_i'' \xleftarrow{\$} \mathcal{R}' & \text{18} \ \mathsf{H}[r_i'] := K_i \\
\text{08} \quad e_i \xleftarrow{\$} \mathsf{wEnc}(\mathsf{pk}, r_i'' \| R_i'') & \text{19} \ \mathbf{return} \ (r_i', \mathsf{m}_i) \\
\text{09} \quad \mathbf{c}[i] := (e_i, \mathsf{d}_i) & \\
\text{10} \ out \xleftarrow{\$} \mathcal{A}_1^{\text{OPEN}, \text{DEC}, H, G}(st, \mathbf{c}) & \\
\text{11} \ \mathbf{return} \ out \qquad\qquad\qquad\qquad\quad /\!\!/ \, \mathcal{S}_1 & \\
\end{array}
$$

Fig. 15. SIM-SO-CCA simulator $\mathcal{S}$ that simulates G_9 to conclude the proof of Theorem 3. Here we ignore the details about simulation of Π, G, and DEC which are the same as in Fig. 14.

G_9, for $i \in [\mu]$, r_i are independent of $\mathbf{c}[i]$, and r_i' are independent of $\mathbf{c}[i]$ before opening, the probability that $\mathcal{A}$ can detect this modification is $\frac{2\mu(q_G + q_H)}{|\mathcal{M}'|}$. Note that we have assumed that there is no collision among r_i's. So, we have

$$
\left| \Pr\left[\mathsf{G}_8^{\mathcal{A}} \Rightarrow 1 \right] - \Pr\left[\mathsf{G}_9^{\mathcal{A}} \Rightarrow 1 \right] \right| \leq \frac{2\mu(q_G + q_H)}{|\mathcal{M}'|} + \frac{\mu^2}{|\mathcal{M}'|}
$$

Now we can construct a simulator $\mathcal{S}$ that interacts with the IDEAL-SO-CCA game and simulate G_9 for $\mathcal{A}$. The construction of $\mathcal{S}$ is shown in Fig. 15. The main difference between G_9 and $\mathcal{S}$ is that r_i' is sampled uniformly and K_i is computed when $\mathcal{A}$ queries $\text{OPEN}(i)$, which is conceptual. We have assumed that all r_i''s and all K's are pair-wise distinct, and the outputs of ROs H and G are different. Hence, we have

$$
\left| \Pr\left[\mathsf{G}_9^{\mathcal{A}} \Rightarrow 1 \right] - \Pr\left[\text{IDEAL-SO-CCA}_{\mathsf{F0}}^{\mathcal{S}} \Rightarrow 1 \right] \right| \leq \frac{n_H^2}{|\mathcal{M}|} + \frac{\mu^2}{|\mathcal{M}'|} + \frac{n_G^2}{|\mathcal{R}'|}
$$

Combining all the above difference, we conclude Theorem 3 as

$$
\left| \Pr\left[\text{REAL-SO-CCA}_{\mathsf{F0}}^{\mathcal{A}} \Rightarrow 1 \right] - \Pr\left[\text{IDEAL-SO-CCA}_{\mathsf{F0}}^{\mathcal{S}} \Rightarrow 1 \right] \right|
$$

$$
\leq \mathsf{Adv}_{\mathsf{wPKE}}^{\text{key-ind}}(\mathcal{B}) + 2\varepsilon_{\mathsf{wPKE}}^{\text{m-enc-los}} + \frac{\mu n_{\text{DEC}}}{2^\gamma} + \frac{2n_H^2}{|\mathcal{M}|} + \frac{2n_G^2}{|\mathcal{R}'|} + \frac{4\mu^2 + 5\mu(q_G + q_H)}{|\mathcal{M}'|}
$$

4.3 Instantiations from DDH

We instantiate F0 using the DDH-based lossy encryption from Bellare et al. [3] and Hofheinz et al. [25]. We describe the one with [25] here, since it leads to an (almost) tightly SIM-SO-CCA secure PKE, which is the main focus of this paper. Due to space limitation, we leave the one with [3] in our full version paper [7].

$\mathsf{KG}_2^{\mathsf{fo}}$	$\mathsf{Enc}_2^{\mathsf{fo}}(\mathsf{pk}, \mathsf{m})$	$\mathsf{Dec}_2^{\mathsf{fo}}(\mathsf{sk}, ([\mathbf{R}_0], c), \mathsf{d})$
01 $\mathbf{A}_0 \xleftarrow{\$} \mathbb{Z}_p^{1 \times (N+1)}$	07 $s \leftarrow \{0,1\}^N$	17 $\mathsf{m}' := \perp$
02 $\mathbf{T} \xleftarrow{\$} \mathbb{Z}_p^{N \times 1}$	08 $K := H(s)$	18 $[\mathbf{Z}'] := [\mathbf{T}\mathbf{R}_0]$
03 $\mathbf{A}_1 := \mathbf{T}\mathbf{A}_0 \in \mathbb{Z}_p^{N \times (N+1)}$	09 $\mathsf{d} := K \oplus \mathsf{m}$	19 $c_1 c_2 ... c_N =: c$
04 $\mathsf{pk} := ([\mathbf{A}_0], [\mathbf{A}_1])$	10 $\mathbf{r} := G(s, \mathsf{d}) \in \mathbb{Z}_p^{N+1}$	20 for $i \in [N]$
05 $\mathsf{sk} := \mathbf{T}$	11 $[\mathbf{R}_0] := [\mathbf{A}_0 \mathbf{r}] \in \mathbb{G}$	21 $\quad s'_i := c_i \oplus h([\mathbf{Z}']_i)$
06 return $(\mathsf{pk}, \mathsf{sk})$	12 $[\mathbf{Z}] := [\mathbf{A}_1 \mathbf{r}] \in \mathbb{G}^N$	22 $s' := s'_1 s'_2 ... s'_N$
	13 for $i \in [N]$	23 $K' := H(s'), \mathbf{r}' := G(s', \mathsf{d})$
	14 $\quad c_i := h([\mathbf{Z}]_i) \oplus s_i$	24 if $[\mathbf{R}_0] = [\mathbf{A}_0 \mathbf{r}']$
	15 $c := c_0 c_1 ... c_N$	25 $\quad \mathsf{m}' := \mathsf{d} \oplus K'$
	16 return $(([\mathbf{R}_0], c), \mathsf{d})$	26 return m'

Fig. 16. A DDH-based scheme FO_2 with efficient opener.

AN INSTANTIATION WITH HOFHEINZ ET AL.'S LOSSY ENCRYPTION [25]. We use Hofheinz et al.'s DDH-based lossy encryption to instantiate FO. Following the notation in [25], we use the matrix Diffie-Hellman notation [13] to describe this scheme. Let $\mathbb{G}$ be a group with prime order p and generator g. Let $\mathbf{A} := (a_{i,j})_{(i,j) \in [l] \times [k]}$ be a matrix in $\mathbb{Z}_p^{l \times k}$, then the group representation of $\mathbf{A}$, denoted as $[\mathbf{A}]$, is defined as $(g^{a_{i,j}})_{(i,j) \in [l] \times [k]}$. Given $\mathbf{r}$ and $[\mathbf{A}]$, one can efficiently compute $[\mathbf{A}\mathbf{r}]$ (if their sizes match). We refer [13] for more details .

Let N be a positive integer. Let $H : \{0,1\}^N \to \mathcal{M}$ and $G : \{0,1\}^N \times \mathcal{M} \to \mathbb{Z}_p^{N+1}$ be two hash functions. Let $h : \mathbb{G} \to \{0,1\}$ be a universal hash function. The instantiated PKE scheme FO_2 is shown in Fig. 16. Hofheinz et al.'s DDH-based lossy encryption has efficient opener, and it is $(\log(p))$-spread, thus by Theorem 3, FO_2 has tight SIM-SO-CCA security.

Corollary 2. FO_2 *in Fig. 16 is SIM-SO-CCA secure (Definition 9) if the* DDH *problem is hard on* $\mathbb{G}$. *Concretely, for any SIM-SO-CCA adversary* $\mathcal{A}$ *and relation* Rel, *there exists a simulator* $\mathcal{S}$ *and* $\mathcal{B}$ *such that:*

$$\mathsf{Adv}_{\mathsf{FO}}^{\mathsf{SIM\text{-}SO\text{-}CCA}}(\mathcal{A}, \mathcal{S}, \mu, \mathsf{Rel}) \leq N \cdot \mathsf{Adv}_{\mathbb{G}}^{\mathsf{DDH}}(\mathcal{B}) + \frac{2\mu}{p} + \frac{\mu n_{\mathrm{DEC}}}{p}$$

$$+ \frac{2n_H^2}{|\mathcal{M}|} + \frac{2n_G^2}{p^{N+1}} + \frac{4\mu^2 + 5\mu(q_G + q_H)}{2^N},$$

where q_H, q_G, *and* n_{DEC} *are the numbers of* $\mathcal{A}$'s *queries to* G, H, *and* DEC, *respectively,* μ *is the number of challenge ciphertexts, and* $n_G = \mu + n_{\mathrm{DEC}} + q_H$ *and* $n_H = \mu + n_{\mathrm{DEC}} + q_G$ *are the number of queries (including the simulator) to* G *and* H, *respectively.*

Acknowledgments. This work is supported by the Research Council of Norway under Project No. 324235. We thank the anonymous reviewers from Asiacrypt 2022 for referring us to the work of Bellare et al. [2] and encouraging us to discuss its impacts on previous work in the random oracle model and ours. Moreover, we thank Benedikt Wagner (CISPA, Germany) and one of our reviewers for pointing out a mistake in Game 5 of our previous proof for Theorem 1. Wagner provided very constructive suggestions on it, during his visit at NTNU.

References

1. Abdalla, M., Bellare, M., Rogaway, P.: The oracle Diffie-Hellman assumptions and an analysis of DHIES. In: Naccache, D. (ed.) CT-RSA 2001. LNCS, vol. 2020, pp. 143–158. Springer, Heidelberg (2001). https://doi.org/10.1007/3-540-45353-9_12
2. Bellare, M., Dowsley, R., Waters, B., Yilek, S.: Standard security does not imply security against selective-opening. In: Pointcheval, D., Johansson, T. (eds.) EURO-CRYPT 2012. LNCS, vol. 7237, pp. 645–662. Springer, Heidelberg (2012). https://doi.org/10.1007/978-3-642-29011-4_38
3. Bellare, M., Hofheinz, D., Yilek, S.: Possibility and impossibility results for encryption and commitment secure under selective opening. In: Joux, A. (ed.) EURO-CRYPT 2009. LNCS, vol. 5479, pp. 1–35. Springer, Heidelberg (2009). https://doi.org/10.1007/978-3-642-01001-9_1
4. Bellare, M., Rogaway, P.: Random oracles are practical: a paradigm for designing efficient protocols. In: Denning, D.E., Pyle, R., Ganesan, R., Sandhu, R.S., Ashby, V. (eds.) ACM CCS 1993, pp. 62–73. ACM Press (1993)
5. Bellare, M., Rogaway, P.: Optimal asymmetric encryption. In: De Santis, A. (ed.) EUROCRYPT 1994. LNCS, vol. 950, pp. 92–111. Springer, Heidelberg (1995). https://doi.org/10.1007/BFb0053428
6. Bellare, M., Rogaway, P.: The security of triple encryption and a framework for code-based game-playing proofs. In: Vaudenay, S. (ed.) EUROCRYPT 2006. LNCS, vol. 4004, pp. 409–426. Springer, Heidelberg (2006). https://doi.org/10.1007/11761679_25
7. Bernstein, D.J., Persichetti, E.: Towards KEM unification. Cryptology ePrint Archive (2022)
8. Böhl, F., Hofheinz, D., Kraschewski, D.: On definitions of selective opening security. In: Fischlin, M., Buchmann, J., Manulis, M. (eds.) PKC 2012. LNCS, vol. 7293, pp. 522–539. Springer, Heidelberg (2012). https://doi.org/10.1007/978-3-642-30057-8_31
9. Canetti, R., Fischlin, M.: Universally composable commitments. In: Kilian, J. (ed.) CRYPTO 2001. LNCS, vol. 2139, pp. 19–40. Springer, Heidelberg (2001). https://doi.org/10.1007/3-540-44647-8_2
10. Cash, D., Kiltz, E., Shoup, V.: The twin Diffie-Hellman problem and applications. In: Smart, N. (ed.) EUROCRYPT 2008. LNCS, vol. 4965, pp. 127–145. Springer, Heidelberg (2008). https://doi.org/10.1007/978-3-540-78967-3_8
11. Chen, J., Wee, H.: Fully, (almost) tightly secure IBE and dual system groups. In: Canetti, R., Garay, J.A. (eds.) CRYPTO 2013, Part II. LNCS, vol. 8043, pp. 435–460. Springer, Heidelberg (2013). https://doi.org/10.1007/978-3-642-40084-1_25
12. Diemert, D., Gellert, K., Jager, T., Lyu, L.: More efficient digital signatures with tight multi-user security. In: Garay, J.A. (ed.) PKC 2021, Part II. LNCS, vol. 12711, pp. 1–31. Springer, Cham (2021). https://doi.org/10.1007/978-3-030-75248-4_1
13. Escala, A., Herold, G., Kiltz, E., Ràfols, C., Villar, J.: An algebraic framework for Diffie-Hellman assumptions. In: Canetti, R., Garay, J.A. (eds.) CRYPTO 2013, Part II. LNCS, vol. 8043, pp. 129–147. Springer, Heidelberg (2013). https://doi.org/10.1007/978-3-642-40084-1_8
14. Fehr, S., Hofheinz, D., Kiltz, E., Wee, H.: Encryption schemes secure against chosen-ciphertext selective opening attacks. In: Gilbert, H. (ed.) EUROCRYPT 2010. LNCS, vol. 6110, pp. 381–402. Springer, Heidelberg (2010). https://doi.org/10.1007/978-3-642-13190-5_20

15. Fujisaki, E., Okamoto, T.: Secure integration of asymmetric and symmetric encryption schemes. J. Cryptol. **26**(1), 80–101 (2013)
16. Gay, R., Hofheinz, D., Kohl, L.: Kurosawa-Desmedt meets tight security. In: Katz, J., Shacham, H. (eds.) CRYPTO 2017, Part III. LNCS, vol. 10403, pp. 133–160. Springer, Cham (2017). https://doi.org/10.1007/978-3-319-63697-9_5
17. Gentry, C., Sahai, A., Waters, B.: Homomorphic encryption from learning with errors: conceptually-simpler, asymptotically-faster, attribute-based. In: Canetti, R., Garay, J.A. (eds.) CRYPTO 2013, Part I. LNCS, vol. 8042, pp. 75–92. Springer, Heidelberg (2013). https://doi.org/10.1007/978-3-642-40041-4_5
18. Han, S., Liu, S., Lyu, L., Gu, D.: Tight Leakage-Resilient CCA-Security from Quasi-Adaptive Hash Proof System. In: Boldyreva, A., Micciancio, D. (eds.) CRYPTO 2019, Part II. LNCS, vol. 11693, pp. 417–447. Springer, Cham (2019). https://doi.org/10.1007/978-3-030-26951-7_15
19. Hemenway, B., Libert, B., Ostrovsky, R., Vergnaud, D.: Lossy encryption: constructions from general assumptions and efficient selective opening chosen ciphertext security. In: Lee, D.H., Wang, X. (eds.) ASIACRYPT 2011. LNCS, vol. 7073, pp. 70–88. Springer, Heidelberg (2011). https://doi.org/10.1007/978-3-642-25385-0_4
20. Heuer, F., Jager, T., Kiltz, E., Schäge, S.: On the selective opening security of practical public-key encryption schemes. In: Katz, J. (ed.) PKC 2015. LNCS, vol. 9020, pp. 27–51. Springer, Heidelberg (2015). https://doi.org/10.1007/978-3-662-46447-2_2
21. Heuer, F., Jager, T., Kiltz, E., Schäge, S.: On the selective opening security of practical public-key encryption schemes. Cryptology ePrint Archive, Report 2016/342 (2016). https://eprint.iacr.org/2016/342
22. Heuer, F., Poettering, B.: Selective opening security from simulatable data encapsulation. In: Cheon, J.H., Takagi, T. (eds.) ASIACRYPT 2016, Part II. LNCS, vol. 10032, pp. 248–277. Springer, Heidelberg (2016). https://doi.org/10.1007/978-3-662-53890-6_9
23. Hofheinz, D.: All-but-many lossy trapdoor functions. In: Pointcheval, D., Johansson, T. (eds.) EUROCRYPT 2012. LNCS, vol. 7237, pp. 209–227. Springer, Heidelberg (2012). https://doi.org/10.1007/978-3-642-29011-4_14
24. Hofheinz, D., Hövelmanns, K., Kiltz, E.: A modular analysis of the Fujisaki-Okamoto transformation. In: Kalai, Y., Reyzin, L. (eds.) TCC 2017, Part I. LNCS, vol. 10677, pp. 341–371. Springer, Cham (2017). https://doi.org/10.1007/978-3-319-70500-2_12
25. Hofheinz, D., Jager, T., Rupp, A.: Public-key encryption with simulation-based selective-opening security and compact ciphertexts. In: Hirt, M., Smith, A. (eds.) TCC 2016, Part II. LNCS, vol. 9986, pp. 146–168. Springer, Heidelberg (2016). https://doi.org/10.1007/978-3-662-53644-5_6
26. Jager, T., Kiltz, E., Riepel, D., Schäge, S.: Tightly-secure authenticated key exchange, revisited. In: Canteaut, A., Standaert, F.-X. (eds.) EUROCRYPT 2021, Part I. LNCS, vol. 12696, pp. 117–146. Springer, Cham (2021). https://doi.org/10.1007/978-3-030-77870-5_5
27. Libert, B., Sakzad, A., Stehlé, D., Steinfeld, R.: All-but-many lossy trapdoor functions and selective opening chosen-ciphertext security from LWE. In: Katz, J., Shacham, H. (eds.) CRYPTO 2017, Part III. LNCS, vol. 10403, pp. 332–364. Springer, Cham (2017). https://doi.org/10.1007/978-3-319-63697-9_12

28. Liu, S., Paterson, K.G.: Simulation-Based selective opening CCA security for PKE from key encapsulation mechanisms. In: Katz, J. (ed.) PKC 2015. LNCS, vol. 9020, pp. 3–26. Springer, Heidelberg (2015). https://doi.org/10.1007/978-3-662-46447-2_1

29. Lyu, L., Liu, S., Han, S., Gu, D.: Tightly SIM-SO-CCA secure public key encryption from standard assumptions. In: Abdalla, M., Dahab, R. (eds.) PKC 2018, Part I. LNCS, vol. 10769, pp. 62–92. Springer, Cham (2018). https://doi.org/10.1007/978-3-319-76578-5_3

Identity-Based Matchmaking Encryption from Standard Assumptions

Jie Chen[1(✉)], Yu Li[1], Jinming Wen[2,3], and Jian Weng[2]

[1] Shanghai Key Laboratory of Trustworthy Computing, Software Engineering Institute, East China Normal University, Shanghai 200062, China
s080001@e.ntu.edu.sg, yli@stu.ecnu.edu.cn
[2] College of Information Science and Technology and the College of Cyber Security, Jinan University, Guangzhou 510632, China
jinming.wen@mail.mcgill.ca
[3] State Key Laboratory of Cryptology, P.O. Box 5159, Beijing 100878, China

Abstract. In this work, we propose the first identity-based matchmaking encryption (IB-ME) scheme under the standard assumptions in the standard model. This scheme is proven to be secure under the symmetric external Diffie-Hellman (SXDH) assumption in prime order bilinear pairing groups. In our IB-ME scheme, all parameters have constant number of group elements and are simpler than those of previous constructions. Previous works are either in the random oracle model or based on the q-type assumptions, while ours is built directly in the standard model and based on static assumptions, and does not rely on other crypto tools.

More concretely, our IB-ME scheme is constructed from a variant of two-level anonymous IBE. We observed that this two-level IBE with anonymity and unforgeability satisfies the same functionality of IB-ME, and its security properties cleverly meet the two requirements of IB-ME (Privacy and Authenticity). The privacy property of IB-ME relies on the anonymity of this two-level IBE, while the authenticity property is corresponding to the unforgeability in the 2nd level. This variant of two-level IBE is built from dual pairing vector spaces, and both security reductions rely on dual system encryption.

Keywords: Matchmaking encryption · Identity-based encryption · Standard assumptions · Standard model

1 Introduction

Matchmaking Encryption (ME) is a new form of encryption proposed by Ateniese et al. [3] in Crypto 2019, in which both the sender and the receiver (each with its own attributes) can specify fine-grained access policies the other party must satisfy in order for the message to be revealed. Using ME, a sender with attributes $\sigma \in \{0,1\}^*$ encrypts messages after generating the policy $\mathbb{R}$ of the intended receiver, and a receiver with attributes $\rho \in \{0,1\}^*$ obtains a decryption key $dk_{\mathbb{S}}$ from an authority before decrypting the ciphertext from a sender

S. Agrawal and D. Lin (Eds.): ASIACRYPT 2022, LNCS 13793, pp. 394–422, 2022.
https://doi.org/10.1007/978-3-031-22969-5_14

satisfying the specified policy $\mathbb{S}$. This receiver will correctly decrypt the ciphertext and obtain the message if and only if the sender's attributes σ match the policy $\mathbb{S}$ specified by the receiver, and at the same time the receiver's attributes ρ match the policy $\mathbb{R}$ specified by the sender. The implementation of matchmaking encryption in an identity-based setting is dubbed identity-based matchmaking encryption (IB-ME), where both the sender and the receiver specify a single identity instead of general policies.

Differently from ME, each identity is chosen by the sender or receiver on the fly without talking to the authority in an identity-based setting. Now each identity $x \in \{0,1\}^*$ will represent an access policy $\mathbb{A}$, which means that we use snd and rcv to represent the target policies $\mathbb{S}$ and $\mathbb{R}$ specified by the receiver and the sender, respectively. The sender's identity $\sigma \in \{0,1\}^*$ and its target policies rcv can be embedded in the ciphertext. The receiver with an identity ρ can now additionally specify a target identity snd $\in \{0,1\}^*$ on the fly, and obtain the correct message as long as the sender's identity σ match the receiver's policy snd and vice-versa (i.e., $\rho =$ rcv and $\sigma =$ snd). From this perspective, IB-ME can be considered as a more expressive version and generalization of anonymous identity-based encryption, in which both the sender and the receiver can specify a target communicating entity in a privacy-preserving manner.

Atenicse et al. [3] provide generic frameworks for constructing ME from functional encryption and propose the first IB-ME scheme with provable security under the bilinear Diffie-Hellman (BDH) assumption, but in the random oracle model. They also deploy experiments to prove their construction is practical and created an anonymous bulletin board over a Tor network. Following their work, Francati et al. [16] give the first IB-ME construction satisfying privacy in the plain model (without random oracles), but based on q-ABDHE assumption and non-interactive zero-knowledge (NIZK) proof systems. Meanwhile, they exhibit a generic transform taking as input any private IB-ME and outputting an IB-ME satisfying both enhanced privacy and authenticity. These leave the following problem:

Can we construct IB-ME under the standard assumptions
in the standard model?

1.1 Our Results

In this work, we present the first IB-ME scheme under the standard assumptions in the standard model. This scheme is based on the SXDH assumption in prime order bilinear pairing groups. The construction is direct and does not rely on other cryptographic tools such as non-interactive zero-knowledge proof systems. We summarize existing IB-ME schemes in Table 1 and several salient features of this work from the following two aspects:

- First, we adopt a variant of two-level IBE with anonymity modified from Chen's anonymous IBE and signature scheme [13] to form our construction. This two-level IBE with anonymity and unforgeability satisfies the same functionality of IB-ME, and its security properties cleverly meet the two requirements of IB-ME (privacy and authenticity). The privacy property of IB-ME

Table 1. Comparison with existing IB-ME schemes

Reference	Model	Assumption
AFNV19 [3]	Random Oracle	BDH
FGRV21 [16]	Standard	q-ABDHE + NIZK
Ours	Standard	SXDH

relies on the anonymity of the 1st level IBE, while the authenticity property is corresponding to the unforgeability in the 2nd level. The usage of this variant of two-level anonymous IBE allows our scheme to technically ensure that the identities chosen by the sender and receiver can be checked simultaneously without revealing any information other than whether the match is successful or not.

– Second, this variant of two-level IBE is built from Okamoto and Takashima's dual pairing vector spaces [25] and its security reductions rely on Waters's dual system encryption [32]. During the security proof process, we draw on the idea of delegation functionality in Okamoto and Takashima's hierarchical inner-product encryption [26] and slightly extended the dual system methodology to fit our IB-ME scheme. We rely on an information theoretic argument instead of computational arguments in the final step of the proof. This is the first work to build identity-based matchmaking encryption by combining dual pairing vector spaces and dual system encryption under the standard assumptions.

1.2 Technical Overview

To achieve the above results, we propose a new technique for designing IB-ME schemes, its construction is straightforward and does not rely on other crypto tools. More concretely, we present a variant of two-level IBE with anonymity and unforgeability that satisfies the same functionality of IB-ME. Moreover, its security properties cleverly meet the two requirements of IB-ME (privacy and authenticity). An IB-ME scheme consists of five algorithms, namely Setup that generates the master public key mpk and master secret key msk, SKGen that generates the encryption key ek_σ using the sender's identity σ, RKGen that generates the decryption key dk_ρ using the receiver's identity ρ, Enc that encrypts the message using ek_σ and a target identity rcv, and Dec that decrypts the ciphertext using dk_ρ and a target identity snd. Decryption can be successful if and only if the attributes of the sender and receiver satisfy the target identity respectively, i.e. $\sigma = $ snd $\wedge\ \rho = $ rcv. At the same time, an IB-ME should satisfy two main security properties: privacy and authenticity [3].

Informally, in this variant of two-level IBE, the algorithms RKGen and Enc associated by identities ρ and rcv are the first level, while the second level consists of the algorithms SKGen and Dec associated by identities σ and snd. The privacy property of IB-ME relies on the anonymity of this 2-level IBE, while the

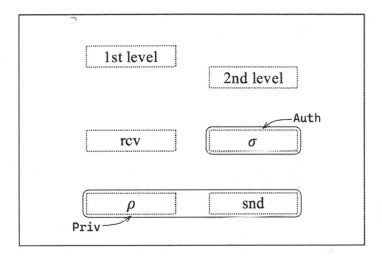

Fig. 1. Security requirements of IB-ME

authenticity property is corresponding to the unforgeability in the 2nd level as all shown in Fig. 1. Decryption can only be done when both levels are matched successfully. Different from IBE, a part of the two keys ek_σ, dk_ρ need to be generated from msk, and each identity-related parameter needs to be generated separately. More concretely, it can be observed that in game $\mathbf{G}_{\Pi,A}^{\text{ib-priv}}(\lambda)$ defining privacy, the adversary outputs two sets of challenge pairs $(m_0, \text{rcv}_0, \sigma_0)$ and $(m_1, \text{rcv}_1, \sigma_1)$ after querying oracles, and then outputs b' for guessing. In game $\mathbf{G}_{\Pi,A}^{\text{ib-auth}}(\lambda)$ defining authenticity, the adversary is actually similar to being unable to forge a ciphertext ct about the message m. The former can be considered as a property of anonymity, while the latter generates an unforgeable signature. Therefore, a two-level anonymous IBE (a signature scheme can be derived from the same IBE) can be used to instantiate this construction and can achieve the security requirements simultaneously.

Our first thought was to use Lewko-Waters composite order IBE scheme [21] for the advantage of its efficiency and shorter parameters, but in view of its difficulty in extending to high-dimensional spaces, we finally decided to choose the prime-order group IBE scheme based on DPVS as the basis of our construction. Meanwhile, since a part of the decryption key dk_ρ needs to be generated from msk, we borrow some ideas from constructing hierarchical inner-product predicate encryption. Specifically, Chen's anonymous IBE [13] and Okamoto and Takashima's HIPE [26] together form the blueprint of this variant of two-level IBE. Thus the message and all identities can be hidden in the high-dimensional basis vectors of the linear subspace. As for security, the privacy property is consistent with proving the full security and anonymity of the 1st level IBE through the dual system encryption methodology. When proving the authenticity property which is similar to the unforgeability of signatures, we make a transformation from IB-ME to IBE and land it on the security of this IBE system. The rest of the proof is similar to that in [13,19].

1.3   Related Work

The idea of using some unique information about the identity of a user as his public encryption key was conceived by Shamir [30] in 1984 and is known as Identity-Based Encryption (IBE). In an identity-based encryption system, a sender who has access to the public parameters can encrypt a message using the target receiver's identity, and the ciphertext can only be decrypted by a receiver who satisfies this identity. We now have constructions of IBE schemes from a large class of assumptions, namely pairings, quadratic residuosity and lattices, starting with the early constructions in the random oracle model [7,15,17], to more recent constructions in the standard model [2,5,6,9,14].

In order to overcome the limitations of partitioning [31], Waters presented a new methodology dubbed dual system encryption [32] for obtaining fully secure IBE and HIBE systems from simple assumptions. It was further developed in several subsequent works [21–24] by Lewko and Waters to enhance the security and the efficiency. Most of these works have used composite order groups as a convenient setting for instantiating the dual system, but with the introduction of dual pairing vector spaces (DPVS) by Okamoto and Takashima [25,26,28], which is a brand new technique based on bilinear pairing groups of prime order, some practical and flexible works emerged. A number of functional encryption schemes [13,20,27,29] that intelligently combine dual system encryption and DPVS have a better performance. Then Lewko et al. successfully explored a general framework [19] based on *pair encoding* [4] summarized by Attrapadung for converting composite order pairing-based cryptosystems into prime order settings and obtained fully secure IBE and HIBE schemes. Chen et al. presented a modular framework [10] based on *predicate encodings* [33] proposed by Wee for the design of efficient adaptively secure attribute-based encryption (ABE) schemes for a large class of predicates under the standard k-Lin assumption in prime-order groups and obtained concrete efficiency improvements for several ABE schemes.

The notion of IB-ME proposed by Ateniese et al. in [3] is a generalization of IBE where the sender and the receiver can both specify a target identity. Following their works, Xu et al. [34] proposed matchmaking attribute-based encryption by extending the IB-ME scheme, and apply it to construct a secure fine-grained bilateral access control data sharing system in cloud-fog computing. They also introduced a new cryptographic tool called lightweight matchmaking encryption [35] and constructed a secure cloud-fog IoT data sharing system with bilateral access control.

Organization: The rest of this paper is organized as follows: Sect. 2 introduces the necessary preliminaries on dual pairing vector spaces and SXDH assumption. We give the definitions of IBE and recall the syntax and security of IB-ME in Sect. 3. We detail our scheme and prove its security in Sect. 4. A brief conclusion and future works are in Sect. 5.

2   Preliminaries

In what follows, we first introduce some notations used in this work. Then we give a few preliminaries related to groups with efficiently computable bilinear maps and define the Symmetric External Diffie-Hellman assumption.

Notation: If S is a finite set, then $r \xleftarrow{R} S$ denotes sampling r uniformly at random from S. If f is an algorithm or function, then $y \leftarrow f(x)$ denotes the output of this algorithm with x as input. $y := x$ denotes that y is defined or substituted by x. Unless otherwise specified, algorithms in this work are randomized and PPT stands for probabilistic polynomial time. We use lowercase letters (e.g., r, s, t) to denote elements in vectors or matrices, bold lowercase letters (e.g., $\mathbf{b_1}, \mathbf{d_2}, \mathbf{f_3^*}$) to denote vectors and bold uppercase letters (e.g., $\mathbf{A}$) to denote matrices. We say a function $\varepsilon(\lambda)$ is *negligible* in λ, if $\varepsilon(\lambda) = o(1/\lambda^c)$ for every $c \in \mathbb{Z}$, and we write $\mathsf{negl}(\lambda)$ to denote a negligible function in λ.

2.1   Dual Pairing Vector Spaces

Our constructions are based on dual pairing vector spaces proposed by Okamoto and Takashima [25,26]. In this work, we concentrate on the asymmetric version [27]. We only briefly describe how to generate random dual orthonormal bases. See [25–27] for a full definition of dual pairing vector spaces.

Definition 1 (Asymmetric bilinear pairing groups). *Asymmetric bilinear pairing groups* $(q, G_1, G_2, G_T, g_1, g_2, e)$ *are a tuple of a prime* q, *cyclic (multiplicative) groups* G_1, G_2 *and* G_T *of order* q, $g_1 \neq 1 \in G_1$, $g_2 \neq 1 \in G_2$, *and a polynomial-time computable nondegenerate bilinear pairing* $e : G_1 \times G_2 \rightarrow G_T$ *i.e.,* $e(g_1^s, g_2^t) = e(g_1, g_2)^{st}$ *and* $e(g_1, g_2) \neq 1$.

In addition to referring to individual elements of G_1 and G_2, we will also consider "vectors" of group elements. For $\mathbf{v} = (v_1, \ldots, v_n) \in \mathbb{Z}_q^n$ and $g_\beta \in G_\beta$, we write $g_\beta^{\mathbf{v}}$ to denote an n-tuple of elements of G_β for $\beta = 1, 2$:

$$g_\beta^{\mathbf{v}} := (g_\beta^{v_1}, \ldots, g_\beta^{v_n}).$$

For any $a \in \mathbb{Z}_q$ and $\mathbf{v}, \mathbf{w} \in \mathbb{Z}_q^n$, we have:

$$g_\beta^{a\mathbf{v}} := (g_\beta^{av_1}, \ldots, g_\beta^{av_n}), \quad g_\beta^{\mathbf{v}+\mathbf{w}} := (g_\beta^{v_1+w_1}, \ldots, g_\beta^{v_n+w_n}).$$

Then we define

$$e(g_1^{\mathbf{v}}, g_2^{\mathbf{w}}) := \prod_{i=1}^{n} e(g_1^{v_i}, g_2^{w_i}) = e(g_1, g_2)^{\mathbf{v}\cdot\mathbf{w}}.$$

Here, the dot product is taken modulo q.

Dual Pairing Vector Spaces. For a fixed (constant) dimension n, we will choose two random bases $\mathbb{B} := (\mathbf{b}_1, \ldots, \mathbf{b}_n)$ and $\mathbb{B}^* := (\mathbf{b}_1^*, \ldots, \mathbf{b}_n^*)$ of $\mathbb{Z}_q^n$, subject to the constraint that they are "dual orthonormal", meaning that

$$\mathbf{b}_j \cdot \mathbf{b}_k^* = 0 \,(\mathsf{mod}\ q)$$

whenever $j \neq k$, and

$$\mathbf{b}_j \cdot \mathbf{b}_j^* = \psi \,(\mathsf{mod}\ q)$$

for all j, where ψ is a random element of $\mathbb{Z}_q$. We denote such algorithm as $\mathsf{Dual}(\mathbb{Z}_q^n)$.

Then for generators $g_1 \in G_1$ and $g_2 \in G_2$, we have

$$e(g_1^{\mathbf{b}_j}, g_2^{\mathbf{b}_k^*}) = 1$$

whenever $j \neq k$, where 1 here denotes the identity element in G_T.

More generally, we can sample multiple tuple of "dual orthonormal" bases. Namely, for fixed (constant) dimension $n_1, \ldots, n_d$, we will choose d tuples of two random bases $\mathbb{B}_i := (\mathbf{b}_{1,i}, \ldots, \mathbf{b}_{n_i,i})$ and $\mathbb{B}_i^* := (\mathbf{b}_{1,i}^*, \ldots, \mathbf{b}_{n_i,i}^*)$ of $\mathbb{Z}_q^{n_i}$, subject to the constraint that they are "dual orthonormal", meaning that

$$\mathbf{b}_{j,i} \cdot \mathbf{b}_{k,i}^* = 0 \,(\mathsf{mod}\ q)$$

whenever $j \neq k$, and

$$\mathbf{b}_{j,i} \cdot \mathbf{b}_{j,i}^* = \psi \,(\mathsf{mod}\ q)$$

for all j, where ψ is a random element of $\mathbb{Z}_q$. We denote such algorithm as $\mathsf{Dual}(\mathbb{Z}_q^{n_1}, \ldots, \mathbb{Z}_q^{n_d})$.

2.2   SXDH Assumptions

Definition 2 (DDH1: Decisional Diffie-Hellman Assumption in G_1).
Given a group generator $\mathcal{G}$, we define the following distribution:

$$\mathbb{G} := (q, G_1, G_2, G_T, g_1, g_2, e) \xleftarrow{R} \mathcal{G},$$
$$a, b, c \xleftarrow{R} \mathbb{Z}_q,$$
$$D := (\mathbb{G}; g_1, g_2, g_1^a, g_1^b).$$

We assume that for any PPT algorithm $\mathcal{A}$(with output in $\{0,1\}$),

$$\mathsf{Adv}_{\mathcal{A}}^{\mathsf{DDH1}}(\lambda) := \left| \Pr[\mathcal{A}(D, g_1^{ab})] - \Pr[\mathcal{A}(D, g_1^{ab+c})] \right|.$$

is negligible in the security parameter λ.

The dual of above assumption is Decisional Diffie-Hellman assumption in G_2 (denoted as DDH2), which is identical to Definition 2 with the roles of G_1 and G_2 reversed. We say that:

Definition 3. *The Symmetric External Diffie-Hellman assumption holds if DDH problems are intractable in both G_1 and G_2.*

2.3   Subspace Assumptions via SXDH

In this subsection, we present subspace assumptions derived from the SXDH assumption. We will rely on these assumptions later to instantiate our encryption schemes. These are analogues of the DLIN-based Subspace assumptions given in [12,19,27].

Definition 4 (DS1: Decisional Subspace Assumption in G_1). *Given a group generator $\mathcal{G}(\cdot)$, define the following distribution:*

$$\mathbb{G} := (q, G_1, G_2, G_T, g_1, g_2, e) \xleftarrow{R} \mathcal{G}(1^\lambda),$$

$$(\mathbb{B}, \mathbb{B}^*) \xleftarrow{R} \mathsf{Dual}(\mathbb{Z}_q^N); \tau_1, \tau_2, \mu_1, \mu_2 \xleftarrow{R} \mathbb{Z}_q,$$

$$U_1 := g_2^{\mu_1 \mathbf{b}_1^* + \mu_2 \mathbf{b}_{K+1}^*}, \dots, U_K := g_2^{\mu_1 \mathbf{b}_K^* + \mu_2 \mathbf{b}_{2K}^*},$$

$$V_1 := g_1^{\tau_1 \mathbf{b}_1}, \dots, V_K := g_1^{\tau_1 \mathbf{b}_K},$$

$$W_1 := g_1^{\tau_1 \mathbf{b}_1 + \tau_2 \mathbf{b}_{K+1}}, \dots, W_K := g_1^{\tau_1 \mathbf{b}_K + \tau_2 \mathbf{b}_{2K}},$$

$$D := (\mathbb{G}; g_2^{\mathbf{b}_1^*}, \dots, g_2^{\mathbf{b}_K^*}, g_2^{\mathbf{b}_{2K+1}^*}, \dots, g_2^{\mathbf{b}_N^*}, g_1^{\mathbf{b}_1}, \dots, g_1^{\mathbf{b}_N}, U_1, \dots, U_K, \mu_2)$$

where K, N are fixed positive integers that satisfy $2K \le N$. We assume that for any PPT algorithm $\mathcal{A}$ (with output in $\{0,1\}$),

$$\mathsf{Adv}_{\mathcal{A}}^{\mathsf{DS1}}(\lambda) := |\Pr[\mathcal{A}(D, V_1, \dots, V_K) = 1] - \Pr[\mathcal{A}(D, W_1, \dots, W_K) = 1]|$$

is negligible in the security parameter λ.

For our construction, we only require the assumption for $K = 4$ and $N = 8$. Furthermore, we do not need to provide μ_2 to the distinguisher. Informally, this means that, given $\tau_1, \tau_2, \mu_1, \mu_2 \xleftarrow{R} \mathbb{Z}_q$ and

$$U_1 = g_2^{\mu_1 \mathbf{b}_1^* + \mu_2 \mathbf{b}_5^*}, U_2 = g_2^{\mu_1 \mathbf{b}_2^* + \mu_2 \mathbf{b}_6^*}, U_3 = g_2^{\mu_1 \mathbf{b}_3^* + \mu_2 \mathbf{b}_7^*}, U_4 = g_2^{\mu_1 \mathbf{b}_4^* + \mu_2 \mathbf{b}_8^*},$$

the distributions (V_1, V_2, V_3, V_4) and (W_1, W_2, W_3, W_4) are computationally indistinguishable, where:

$$V_1 = g_1^{\tau_1 \mathbf{b}_1}, V_2 = g_1^{\tau_1 \mathbf{b}_2}, V_3 = g_1^{\tau_1 \mathbf{b}_3}, V_4 = g_1^{\tau_1 \mathbf{b}_4};$$

$$W_1 = g_1^{\tau_1 \mathbf{b}_1 + \tau_2 \mathbf{b}_5}, W_2 = g_1^{\tau_1 \mathbf{b}_2 + \tau_2 \mathbf{b}_6}, W_3 = g_1^{\tau_1 \mathbf{b}_3 + \tau_2 \mathbf{b}_7}, W_4 = g_1^{\tau_1 \mathbf{b}_4 + \tau_2 \mathbf{b}_8}.$$

Lemma 1. *If the DDH assumption in G_1 holds, then the Subspace assumption in G_1 stated in Definition 4 also holds. More precisely, for any adversary $\mathcal{A}$ against the Subspace assumption in G_1, there exist probabilistic algorithms $\mathcal{B}$ whose running times are essentially the same as that of $\mathcal{A}$, such that*

$$\mathsf{Adv}_{\mathcal{A}}^{\mathsf{DS1}}(\lambda) \le \mathsf{Adv}_{\mathcal{B}}^{\mathsf{DDH1}}(\lambda).$$

Proof. Detailed proofs can be found in [12].

The dual of the Subspace assumption in G_1 is Subspace assumption in G_2 (denoted as DS2), which is identical to Definition 4 with the roles of G_1 and G_2 reversed. Similarly, we can prove that the Subspace assumption holds in G_2 if the DDH assumption in G_2 holds.

2.4 Statistical Indistinguishability Lemma

We require the following lemma for our security proofs, which is derived from [27].

Lemma 2. *For* $p \in \mathbb{Z}_q$, *let* $C_p := \{(\mathbf{x}, \mathbf{v}) | \mathbf{x} \cdot \mathbf{v} = p, \mathbf{0} \neq \mathbf{x}, \mathbf{0} \neq \mathbf{v} \in \mathbb{Z}_q^n\}$. *For all* $(\mathbf{x}, \mathbf{v}) \in C_p$, *for all* $(\mathbf{z}, \mathbf{w}) \in C_p$, *and* $\mathbf{A} \xleftarrow{R} \mathbb{Z}_q^{n \times n}$ (**A** *is invertible with overwhelming probability),*

$$\Pr[\mathbf{x}\mathbf{A}^\top = \mathbf{z} \ \wedge \ \mathbf{v}\mathbf{A}^{-1} = \mathbf{w}] = \frac{1}{\#C_p}.$$

3 Identity-Based Matchmaking Encryption

In what follows, we first recall the definitions of identity-based encryption and signatures. Then we introduce the definition of identity-based matchmaking encryption presented in [3].

3.1 Identity-Based Encryption

In the IBE setting, a functionality $\hat{F}$ is defined over a key space and an index space using sets of identities. The key space $\mathcal{K}$ and index space $\mathcal{I}$ for IBE then corresponds to all identities id. Here

$$\hat{F}(\mathsf{id}, (\mathsf{id}', m)) := \begin{cases} m & \text{if } \mathsf{id}' = \mathsf{id} \\ \bot & \text{otherwise.} \end{cases}$$

An Identity-Based Encryption [7] scheme consists of following four algorithms: Setup, KeyGen, Enc, and Dec.

- Setup$(\lambda) \to (pp, mk)$: The setup algorithm takes in the security parameter λ, and outputs the public parameters pp, and the master key mk.
- KeyGen$(pp, mk, \mathsf{id}) \to sk_{\mathsf{id}}$: The key generation algorithm takes in the public parameters pp, the master key mk, an identity id and produces a secret key sk_{id} for that identity.
- Enc$(pp, \mathsf{id}, m) \to ct_{\mathsf{id}}$: The encryption algorithm takes in the public parameters pp, an identity id, a message m and outputs a ciphertext ct_{id} encrypted under that identity.
- Dec$(pp, sk_{\mathsf{id}}, ct_{\mathsf{id}}) \to m$: The decryption algorithm takes in a secret key sk_{id}, and a ciphertext ct_{id}, and outputs the message m when the ct_{id} is encrypted under the same id.

The security notion of anonymous IBE was formalized by [1], which is defined by the following game, played by a challenger $\mathcal{B}$ and an advertisider $\mathcal{A}$.

- Setup: The challenger $\mathcal{B}$ runs the setup algorithm to generate pp and mk. It gives pp to the adversary $\mathcal{A}$.

- Phase 1: The adversary $\mathcal{A}$ adaptively requests key for identities id, and is provided with corresponding secret key sk_{id}, which the challenger $\mathcal{B}$ generates by running the key generation algorithm.
- Challenge: The adversary $\mathcal{A}$ gives $\mathcal{B}$ two challenge pairs (m_0, id_0^*) and (m_1, id_1^*). The challenge identities must not have been queried in Phase 1. The challenger $\mathcal{B}$ sets $\beta \in \{0,1\}$ randomly, and encrypts m_β under id_β^* by running the encryption algorithm. It send the ciphertext to the adversary $\mathcal{A}$.
- Phase 2: This is the same as Phase 1 with the added restriction a secret key for $\mathsf{id}_0^*, \mathsf{id}_1^*$ cannot be requested.
- Guess: The adversary $\mathcal{A}$ must output a guess β' for β.

The advantage $\mathsf{Adv}_{\mathcal{A}}^{\mathsf{IBE}}(\lambda)$ of an adversary $\mathcal{A}$ is defined to be $\Pr[\beta' = \beta] - 1/2$.

Definition 5. *An Identity-Based Encryption scheme is secure and anonymous if all PPT adversaries achieve at most a negligible advantage in the above security game.*

Remark 1: The security notion of non-anonymous IBE is defined as above with restriction that $\mathsf{id}_0^* = \mathsf{id}_1^*$.

3.2 Signature Schemes

A signature scheme is made up of three algorithms, (KeyGen, Sign, Verify) for generating keys, signing, and verifying signatures, respectively.

- KeyGen(1^λ) : The key generation algorithm takes in the security parameter 1^λ, and outputs the public key pk, and the secret key sk.
- Sign(sk, m) : The signing algorithm takes in the secret key sk and a message m, and produces a signature σ for this message.
- Verify(pk, σ, m) : The verifying algorithm takes in the public key pk and a signature pair (σ, m), and outputs `valid` or `invalid`.

The standard notion of security for a signature scheme is called existential unforgeability under a chosen message attack [18], which is defined using the following game between a challenger $\mathcal{C}$ and an adversary $\mathcal{A}$.

- Setup : The challenger $\mathcal{C}$ runs the key generation algorithm to generate pk and sk. It gives pk to the adversary $\mathcal{A}$.
- Query : The adversary $\mathcal{A}$ adaptively requests for messages $m_1, \ldots, m_\nu \in \{0,1\}^*$, and is provided with corresponding signatures $\sigma_1, \ldots, \sigma_\nu$ by running the sign algorithm Sign.
- Output : Eventually, the adversary $\mathcal{A}$ outputs a pair (σ, m).

The advantage $\mathsf{Adv}_{\mathcal{A}}^{\mathsf{Sig}}(\lambda)$ of an adversary $\mathcal{A}$ is defined to be the probability that $\mathcal{A}$ wins in the above game, namely

(1) m is not any of $m_1, \ldots, m_\nu$;
(2) Verify(pk, σ, m) outputs `valid`.

Definition 6. *A signature scheme is existentially unforgeable under an adaptive chosen message attack if all PPT adversaries achieve at most a negligible advantage in the above security game.*

We assume that for any PPT algorithm $\mathcal{A}$, the probability that $\mathcal{A}$ wins in the above game is negligible in the security parameter 1^λ. We note that the security of the signature scheme can follow from the security of IBE scheme by applying Naor's transform [7,8].

3.3   Syntax of IB-ME

In IB-ME, attributes and policies are treated as binary strings. We denote with rcv and snd the target identities or policies chosed by the sender and the receiver, respectively. We say that a match (resp. mismatch) occurs when $\sigma = $ snd and $\rho = $ rcv (resp. $\sigma \neq$ snd or $\rho \neq$ rcv). The receiver can choose the target identity snd on the fly. More formally, an IB-ME scheme is composed of the following five polynomial-time algorithms:

- Setup(1^λ) $\rightarrow$ (mpk, msk): Upon input the security parameter 1^λ, the randomized setup algorithm outputs the master public key mpk and the master secret key msk.
- SKGen(mpk, msk, σ) $\rightarrow$ ek_σ: Upon input the master secret key msk and the identity σ, the randomized sender-key generator outputs an encryption key ek_σ for σ.
- RKGen(mpk, msk, ρ) $\rightarrow$ dk_ρ: Upon input the master secret key msk and the identity ρ, the randomized receiver-key generator outputs a decryption key dk_ρ for ρ.
- Enc($mpk, ek_\sigma, $ rcv$, m$) $\rightarrow$ ct: Upon input the encryption key ek_σ for identity σ, a target identity rcv and a message $m \in \mathcal{M}$, the randomized encryption algorithm produces a ciphertext ct linked to both σ and rcv.
- Dec($mpk, dk_\rho, $ snd$, ct$) $\rightarrow$ m: Upon input the decryption key dk_ρ for identity ρ, a target identity snd and a ciphertext ct, the deterministic decryption algorithm outputs either a message m or $\bot$.

Correctness. Correctness of IB-ME simply says that in case of a match the receiver obtains the plaintext.

Definition 7 (Correctness of IB-ME). *An IB-ME scheme Π =(Setup, SKGen, RKGen, Enc, Dec) is correct if $\forall \lambda \in \mathbb{N}$, $\forall (mpk, msk)$ output by Setup(1^λ), $\forall m \in \mathcal{M}$, $\forall \sigma, \rho, rcv, snd \in \{0,1\}^*$ such that $\sigma = $ snd and $\rho = $ rcv:*

$$\Pr[\mathsf{Dec}(dk_\rho, \mathsf{snd}, \mathsf{Enc}(ek_\sigma, \mathsf{rcv}, m)) = m] \geq 1 - \mathsf{negl}(\lambda),$$

where $ek_\sigma \xleftarrow{R} $ SKGen(msk, σ) and $dk_\rho \xleftarrow{R} $ RKGen(msk, ρ).

3.4 Security of IB-ME

We now define privacy and authenticity of IB-ME. Recall that privacy captures secrecy of the sender's inputs $(\sigma, \mathsf{rcv}, m)$. This is formalized by asking the adversary to distinguish between $\mathsf{Enc}(ek_{\sigma_0}, \mathsf{rcv}_0, m_0)$ and $\mathsf{Enc}(ek_{\sigma_1}, \mathsf{rcv}_1, m_1)$ where $(m_0, m_1, \sigma_0, \sigma_1, \mathsf{rcv}_1)$ are chosen by the attacker. The definition of authenticity intuitively says that an adversary cannot compute a valid ciphertext under the identity σ, if it does not hold the corresponding encryption key ek_σ produced by the challenger.

$\mathbf{G}^{\mathsf{ib\text{-}priv}}_{\Pi,\mathsf{A}}(\lambda)$	$\mathbf{G}^{\mathsf{ib\text{-}auth}}_{\Pi,\mathsf{A}}(\lambda)$
$(\mathsf{mpk}, \mathsf{msk}) \xleftarrow{R} \mathsf{Setup}(1^\lambda)$	$(\mathsf{mpk}, \mathsf{msk}) \xleftarrow{R} \mathsf{Setup}(1^\lambda)$
$(m_0, m_1, \mathsf{rcv}_0, \mathsf{rcv}_1, \sigma_0, \sigma_1, st) \xleftarrow{R} \mathsf{A}_1^{\mathsf{O}_1, \mathsf{O}_2}(1^\lambda, \mathsf{mpk})$	$(ct, \rho, \mathsf{snd}) \xleftarrow{R} \mathsf{A}^{\mathsf{O}_1, \mathsf{O}_2}(1^\lambda, \mathsf{mpk})$
$b \xleftarrow{R} \{0, 1\}$	$dk_\rho \xleftarrow{R} \mathsf{RKGen}(\mathsf{msk}, \rho)$
$ek_{\sigma_b} \xleftarrow{R} \mathsf{SKGen}(\mathsf{msk}, \sigma_b)$	$m = \mathsf{Dec}(dk_\rho, \mathsf{snd}, ct)$
$ct \xleftarrow{R} \mathsf{Enc}(ek_{\sigma_b}, \mathsf{rcv}_b, m_b)$	If $\forall \sigma \in \mathcal{Q}_{\mathsf{O}_1} : (\sigma \neq \mathsf{snd}) \wedge (m \neq \bot)$
$b' \xleftarrow{R} \mathsf{A}_2^{\mathsf{O}_1, \mathsf{O}_2}(1^\lambda, ct, st)$	$\qquad$ **return** 1
If $(b' = b)$ **return** 1	**Else return** 0
Else return 0	

Fig. 2. Games defining privacy and authenticity security of IB-ME. Oracles $\mathsf{O}_1, \mathsf{O}_2$ are implemented by $\mathsf{SKGen}(\mathsf{msk}, \cdot), \mathsf{RKGen}(\mathsf{msk}, \cdot)$.

Definition 8 (Privacy of IB-ME). *We say that an IB-ME Π satisfies privacy if for all valid PPT adversaries $\mathcal{A} = (\mathsf{A}_1, \mathsf{A}_2)$:*

$$\left| \Pr[\mathbf{G}^{\mathsf{ib\text{-}priv}}_{\Pi,\mathsf{A}}(\lambda) = 1] - \frac{1}{2} \right| \leq \mathsf{negl}(\lambda)$$

where game $\mathbf{G}^{\mathsf{ib\text{-}priv}}_{\Pi,\mathsf{A}}(\lambda)$ is depicted in Fig. 2. Adversary $\mathcal{A}$ is called valid if $\forall \rho \in \mathcal{Q}_{\mathsf{O}_2}$ it satisfies the following invariant:

$$(\textbf{Mismatch condition}) : \rho \neq \mathsf{rcv}_0 \wedge \rho \neq \mathsf{rcv}_1.$$

Definition 9 (Authenticity of IB-ME). *We say that an IB-ME Π satisfies authenticity if for all valid PPT adversaries $\mathcal{A}$:*

$$\Pr[\mathbf{G}^{\mathsf{ib\text{-}auth}}_{\Pi,\mathsf{A}}(\lambda) = 1] \leq \mathsf{negl}(\lambda)$$

where game $\mathbf{G}^{\mathsf{ib\text{-}auth}}_{\Pi,\mathsf{A}}(\lambda)$ is depicted in Fig. 2.

Definition 10 (Secure IB-ME). *We say that an IB-ME Π is secure if it satisfies privacy (Def.8) and authenticity (Def.9).*

4 The Proposed IB-ME Construction

We are now ready to give the concrete construction of our IB-ME scheme.

4.1 Construction

- Setup(1^λ) → (mpk, msk): This algorithm takes in the security parameter 1^λ and generates a bilinear pairing $\mathbb{G} := (q, G_1, G_2, G_T, g_1, g_2, e)$ for sufficiently large prime order q. The algorithm samples random dual orthonormal bases $(\mathbb{D}, \mathbb{D}^*) \xleftarrow{R} \mathsf{Dual}(\mathbb{Z}_q^8)$. Let $\mathbf{d_1}, ..., \mathbf{d_8}$ denote the elements of $\mathbb{D}$ and $\mathbf{d_1^*}, ..., \mathbf{d_8^*}$ denote the elements of $\mathbb{D}^*$. Let $g_T := e(g_1, g_2)^{\mathbf{d_1} \cdot \mathbf{d_1^*}}$. It also picks $\alpha, \eta \xleftarrow{R} \mathbb{Z}_q$ and outputs the master public key as

$$mpk := \{\mathbb{G}; g_T^\alpha, g_T^\eta, g_1^{\mathbf{d_1}}, g_1^{\mathbf{d_2}}\},$$

and the master secret key

$$msk := \{\alpha, \eta, g_1^{\mathbf{d_3}}, g_1^{\mathbf{d_4}}, g_2^{\mathbf{d_1^*}}, g_2^{\mathbf{d_2^*}}, g_2^{\mathbf{d_3^*}}, g_2^{\mathbf{d_4^*}}\}.$$

- SKGen(mpk, msk, σ) → ek_σ: This algorithm picks $r \xleftarrow{R} \mathbb{Z}_q$. The encryption key is computed as

$$ek_\sigma := g_1^{\eta \mathbf{d_3} + r(\sigma \mathbf{d_3} - \mathbf{d_4})}.$$

- RKGen(mpk, msk, ρ) → dk_ρ: This algorithm picks $s, s_1, s_2 \xleftarrow{R} \mathbb{Z}_q$. The decryption key is computed as

$$dk_\rho := \{k_1 = g_2^{\alpha \mathbf{d_1^*} + s_1(\rho \mathbf{d_1^*} - \mathbf{d_2^*}) + s \mathbf{d_3^*}}, k_2 = g_2^{s_2(\rho \mathbf{d_1^*} - \mathbf{d_2^*}) + s \mathbf{d_4^*}}, k_3 = (g_T^\eta)^s\}.$$

- Enc($mpk, ek_\sigma, \mathsf{rcv}, m$) → ct: This algorithm picks $z \xleftarrow{R} \mathbb{Z}_q$ and forms the ciphertext as

$$ct := \{\mathsf{C} = m \cdot (g_T^\alpha)^z, \mathsf{C_0} = ek_\sigma \cdot g_1^{z(\mathbf{d_1} + \mathsf{rcv}\mathbf{d_2})}\}.$$

- Dec($mpk, dk_\rho, \mathsf{snd}, ct$) → m: This algorithm computes the message as

$$m := \frac{\mathsf{C}}{e(\mathsf{C_0}, k_1 \cdot k_2^{\mathsf{snd}}) \cdot k_3^{-1}}.$$

Correctness: Correctness follows when $\mathsf{snd} = \sigma$ and $\mathsf{rcv} = \rho$:

$$e(\mathsf{C_0}, k_1 \cdot k_2^{\mathsf{snd}})$$
$$= e(g_1^{\eta \mathbf{d_3} + r(\sigma \mathbf{d_3} - \mathbf{d_4}) + z(\mathbf{d_1} + \mathsf{rcv}\mathbf{d_2})}, g_2^{\alpha \mathbf{d_1^*} + (s_1 + s_2 \cdot \mathsf{snd})(\rho \mathbf{d_1^*} - \mathbf{d_2^*}) + s(\mathbf{d_3^*} + \mathsf{snd}\mathbf{d_4^*})})$$
$$= e(g_1, g_2)^{\eta s \mathbf{d_3} \cdot \mathbf{d_3^*} + rs(\sigma \mathbf{d_3} \cdot \mathbf{d_3^*} - \mathsf{snd}\mathbf{d_4} \cdot \mathbf{d_4^*}) + \alpha z \mathbf{d_1} \cdot \mathbf{d_1^*} + z(s_1 + s_2 \cdot \mathsf{snd})(\rho \mathbf{d_1} \cdot \mathbf{d_1^*} - \mathsf{rcv}\mathbf{d_2} \cdot \mathbf{d_2^*})}$$
$$= e(g_1, g_2)^{\eta s \mathbf{d_3} \cdot \mathbf{d_3^*} + \alpha z \mathbf{d_1} \cdot \mathbf{d_1^*}} = (g_T)^{\eta s + \alpha z}$$

$$\frac{\mathsf{C}}{e(\mathsf{C_0}, k_1 \cdot k_2^{\mathsf{snd}}) \cdot k_3^{-1}} = \frac{m \cdot (g_T^\alpha)^z}{(g_T)^{\eta s + \alpha z} \cdot g_T^{-\eta s}} = m$$

4.2   Security Analysis

As for security, it can be proved that our proposed IB-ME scheme is secure (Def. 10) according to the Theorem 1 and Theorem 2, namely satisfies privacy (Def. 8) and authenticity (Def. 9) simultaneously.

Theorem 1. *The proposed IB-ME scheme satisfies privacy under the Symmetric External Diffie-Hellman assumption. More precisely, for any PPT adversary $\mathcal{A}$ breaks the privacy property of our IB-ME scheme, there exist probabilistic algorithms $\mathcal{B}_0, \mathcal{B}_{1,1}, \mathcal{B}_{1,2}, \ldots, \mathcal{B}_{\nu,1}, \mathcal{B}_{\nu,2}$ whose running times are essentially the same as that of $\mathcal{A}$, such that*

$$\mathsf{Adv}^{\mathsf{IB\text{-}ME}}_{\mathcal{A}}(\lambda) \leq \mathsf{Adv}^{\mathsf{DDH1}}_{\mathcal{B}_0}(\lambda) + \sum_{\kappa=1}^{\nu}\left(\mathsf{Adv}^{\mathsf{DDH2}}_{\mathcal{B}_{\kappa,1}}(\lambda) + \mathsf{Adv}^{\mathsf{DDH2}}_{\mathcal{B}_{\kappa,2}}(\lambda)\right) + (12\nu + 3)/q$$

where ν is the maximum number of $\mathcal{A}$'s key queries.

Proof Outline: There are many similarities between the proof of our scheme and the anonymous IBE scheme in [12]. We will follow a similar strategy of proving fully secure anonymous IBE, adopting the dual system encryption methodology by Waters [32] to prove that our IB-ME satisfies privacy under the SXDH assumption. The hardest part of the security proof is how to prove the negligible gap between two different forms of dk_ρ, especially when it is composed of three different keys k_1, k_2 and k_3. In order to solve this problem, apart from the concepts of *semi-functional ciphertexts* and *semi-functional keys* in our proof, we introduce the concept of *inter-semi-functional secret key* and provide algorithms that generate them. More precisely, *inter-semi-functional key* means that k_1 is semi-functional and k_2 is normal, while *semi-functional key* means that both k_1 and k_2 are semi-functional, and k_3 always remains the same. We note that these algorithms are only provided in a sequence of security games for the proof, and are not part of the IB-ME scheme. In particular, they do not need to be efficiently computable from the master public key and the master secret key. Meanwhile, another ν games are added into the proof, and for κ from 1 to ν, all the decryption keys will be converted into semi-functional keys step by step according to the sequence of changing normal key to *inter*-semi-functional key first in $\mathsf{Game}_{\kappa,1}$ and then changing it to semi-functional key in $\mathsf{Game}_{\kappa,2}$. In other words, we consider k_1 and k_2 in dk_ρ as two independent keys and generate their semi-functional keys in the **KeyGenSF** algorithm respectively. We first require that the challenger can simulate the two different forms of k_1, and then require it can simulate the two different forms of k_2 with the adversary. Then, we adopt the same procedure as in the security model definition, treating the output of SKGen algorithm as a part of the input to Enc algorithm, from which the corresponding semi-functional ciphertext is generated.

KeyGenSF: The algorithm picks $s, s_1, s_2, \{s_{i,1}\}_{i=5,\ldots,8} \xleftarrow{R} \mathbb{Z}_q$ and forms the inter-semi-functional secret key as

$$\mathsf{dk}_\rho^{(\text{inter-SF})} := \{\ k_1 = g_2^{\alpha \mathbf{d}_1^* + s_1(\rho \mathbf{d}_1^* - \mathbf{d}_2^*) + s\mathbf{d}_3^* + [s_{5,1}\mathbf{d}_5^* + s_{6,1}\mathbf{d}_6^* + s_{7,1}\mathbf{d}_7^*]},$$

$$k_2 = g_2^{s_2(\rho \mathbf{d}_1^* - \mathbf{d}_2^*) + s\mathbf{d}_4^*}, \ k_3 = (g_T^\eta)^s\}; \tag{1}$$

The algorithm picks $s, s_1, s_2, \{s_{i,j}\}_{i=5,\ldots,8;j=1,2} \xleftarrow{R} \mathbb{Z}_q$ and forms the semi-functional secret key as

$$\mathsf{dk}_\rho^{(\text{SF})} := \{\ k_1 = g_2^{\alpha \mathbf{d}_1^* + s_1(\rho \mathbf{d}_1^* - \mathbf{d}_2^*) + s\mathbf{d}_3^* + [s_{5,1}\mathbf{d}_5^* + s_{6,1}\mathbf{d}_6^* + s_{7,1}\mathbf{d}_7^*]},$$

$$k_2 = g_2^{s_2(\rho \mathbf{d}_1^* - \mathbf{d}_2^*) + s\mathbf{d}_4^* + [s_{5,2}\mathbf{d}_5^* + s_{6,2}\mathbf{d}_6^* + s_{8,2}\mathbf{d}_8^*]}, \ k_3 = (g_T^\eta)^s\}. \tag{2}$$

Hereafter we will ignore k_3 since it is always correctly generated.

EncryptSF: The algorithm picks $z, r, r_5, r_6, r_7, r_8 \xleftarrow{R} \mathbb{Z}_q$ and forms a semi-functional ciphertext as

$$ek_\sigma := g_1^{\eta \mathbf{d}_3 + r(\sigma \mathbf{d}_3 - \mathbf{d}_4)}$$

$$\mathsf{CT}_{ek_\sigma,\mathsf{rcv}}^{(\text{SF})} := \{C := m \cdot (g_T^\alpha)^z, C_0 := ek_\sigma \cdot g_1^{z(\mathbf{d}_1 + \mathsf{rcv}\mathbf{d}_2) + [r_5\mathbf{d}_5 + r_6\mathbf{d}_6 + r_7\mathbf{d}_7 + r_8\mathbf{d}_8]}$$

$$= g_1^{\eta \mathbf{d}_3 + r(\sigma \mathbf{d}_3 - \mathbf{d}_4) + z(\mathbf{d}_1 + \mathsf{rcv}\mathbf{d}_2) + [r_5\mathbf{d}_5 + r_6\mathbf{d}_6 + r_7\mathbf{d}_7 + r_8\mathbf{d}_8]}\}. \tag{3}$$

Hereafter we will ignore C since it is always correctly generated. We observe that if one applies the decryption procedure with a (inter) semi-functional key and a normal ciphertext, decryption will succeed because $(\mathbf{d}_5^*, \mathbf{d}_6^*, \mathbf{d}_7^*, \mathbf{d}_8^*)$ are orthogonal to all of the vectors in exponent of C_0, and hence have no effect on decryption. Similarly, decryption of a semi-functional ciphertext by a normal key will also succeed because $(\mathbf{d}_5, \mathbf{d}_6, \mathbf{d}_7, \mathbf{d}_8)$ are orthogonal to all of the vectors in the exponent of the key. When both the ciphertext and key are semi-functional, the result of decryption procedure $e(C_0, k_1 \cdot k_2^{\mathsf{snd}}) \cdot k_3^{-1}$ will have an additional term, namely

$$e(g_1, g_2)^{r_5(s_{5,1} + \mathsf{snd} \cdot s_{5,2})\mathbf{d}_5 \cdot \mathbf{d}_5^* + r_6(s_{6,1} + \mathsf{snd} \cdot s_{6,2})\mathbf{d}_6 \cdot \mathbf{d}_6^* + r_7 s_{7,1}\mathbf{d}_7 \cdot \mathbf{d}_7^* + r_8 \cdot \mathsf{snd} \cdot s_{8,2}\mathbf{d}_8 \cdot \mathbf{d}_8^*}$$

$$= g_T^{(r_5 s_{5,1} + r_6 s_{6,1} + r_7 s_{7,1}) + \mathsf{snd}(r_5 s_{5,2} + r_6 s_{6,2} + r_8 s_{8,2})}.$$

Decryption will then fail unless $r_5 s_{5,1} + r_6 s_{6,1} + r_7 s_{7,1} \equiv 0 \pmod{q}$ and $r_5 s_{5,2} + r_6 s_{6,2} + r_8 s_{8,2} \equiv 0 \pmod{q}$. If this modular equation holds, we say that this key and ciphertext pair is *nominally semi-functional*.

For a probabilistic polynomial-time adversary $\mathcal{A}$ which makes ν key queries $\mathsf{rcv}_1, \ldots, \mathsf{rcv}_\nu$, our proof of security consists of the following sequence of games between $\mathcal{A}$ and a challenger $\mathcal{B}$.

- $\mathsf{Game}_{\mathsf{Real}}$: is the real security game.
- Game_0: is the same as $\mathsf{Game}_{\mathsf{Real}}$ except that the challenge ciphertext is semi-functional.

- $\mathsf{Game}_{\kappa,1}$: for κ from 1 to ν, $\mathsf{Game}_{\kappa,1}$ is the same as Game_0 except that the first κ-1 keys are semi-functional, the κ-th key is inter-semi-functional and the remaining keys are normal.
- $\mathsf{Game}_{\kappa,2}$: for κ from 1 to ν, $\mathsf{Game}_{\kappa,2}$ is the same as Game_0 except that the first κ keys are semi-functional and the remaining keys are normal.
- $\mathsf{Game}_{\mathsf{Final}}$: is the same as $\mathsf{Game}_{\nu,2}$, except that the challenge ciphertext is a semi-functional encryption of a random message in G_T and under two random identities in $\mathbb{Z}_q$. We denote the challenge ciphertext in $\mathsf{Game}_{\mathsf{Final}}$ as $\mathsf{CT}^{(R)}_{ek_{\sigma_R},\mathsf{rcv_R}}$.

We prove following lemmas to show the above games are indistinguishable by following an analogous strategy of [13,19,20]. Our main arguments are computational indistinguishability (guaranteed by the Subspace assumptions, which are implied by the SXDH assumption) and statistical indistinguishability. The advantage gap between $\mathsf{Game}_{\mathsf{Real}}$ and Game_0 is bounded by the advantage of the Subspace assumption in G_1. Additionally, we require a statistical indistinguishability argument to show that the distribution of the challenge ciphertext remains the same from the adversary's view. For κ from 1 to ν, the advantage gaps between $\mathsf{Game}_{\kappa-1,2}$ and $\mathsf{Game}_{\kappa,1}$, and between $\mathsf{Game}_{\kappa,1}$ and $\mathsf{Game}_{\kappa,2}$ are bounded by the advantage of Subspace assumption in G_2. Similarly, we require a statistical indistinguishability argument to show that the distribution of the κ-th semi-functional key remains the same from the adversary's view. Finally, we statistically transform $\mathsf{Game}_{\nu,2}$ to $\mathsf{Game}_{\mathsf{Final}}$ in one step, i.e., we show the joint distributions of

$$\left(\mathsf{mpk}, \mathsf{CT}^{(\mathsf{SF})}_{ek_{\sigma^*_\beta},\mathsf{rcv}^*_\beta}, \left\{\mathsf{dk}^{(\mathsf{SF})}_{\rho_\ell}\right\}_{\ell\in[\nu]}\right) \quad \text{and} \quad \left(\mathsf{mpk}, \mathsf{CT}^{(R)}_{ek_{\sigma_R},\mathsf{rcv_R}}, \left\{\mathsf{dk}^{(\mathsf{SF})}_{\rho_\ell}\right\}_{\ell\in[\nu]}\right)$$

are equivalent for the adversary's view.

We let $\mathsf{Adv}^{\mathsf{Game}_{\mathsf{Real}}}_{\mathcal{A}}$ denote an adversary $\mathcal{A}$'s advantage in the real game.

Lemma 3. *Suppose that there exists an adversary $\mathcal{A}$ where $|\mathsf{Adv}^{\mathsf{Game}_{\mathsf{Real}}}_{\mathcal{A}}(\lambda) - \mathsf{Adv}^{\mathsf{Game}_0}_{\mathcal{A}}(\lambda)| = \epsilon$. Then there exists an algorithm $\mathcal{B}_0$ such that $\mathsf{Adv}^{\mathsf{DS1}}_{\mathcal{B}_0}(\lambda) = \epsilon - 2/q$, with $K = 4$ and $N = 8$.*

Proof. $\mathcal{B}_0$ is given

$$D := \left(\mathbb{G}; g_2^{\mathbf{b}^*_1}, g_2^{\mathbf{b}^*_2}, g_2^{\mathbf{b}^*_3}, g_2^{\mathbf{b}^*_4}, g_1^{\mathbf{b}_1}, \ldots, g_1^{\mathbf{b}_8}, U_1, U_2, U_3, U_4, \mu_2\right)$$

along with (T_1, T_2, T_3, T_4). And in D we have that $U_1 = g_2^{\mu_1\mathbf{b}^*_1+\mu_2\mathbf{b}^*_5}, U_2 = g_2^{\mu_1\mathbf{b}^*_2+\mu_2\mathbf{b}^*_6}, U_3 = g_2^{\mu_1\mathbf{b}^*_3+\mu_2\mathbf{b}^*_7}, U_4 = g_2^{\mu_1\mathbf{b}^*_4+\mu_2\mathbf{b}^*_8}$. We require that $\mathcal{B}_0$ decides whether (T_1, T_2, T_3, T_4) are distributed as

$$(g_1^{\tau_1\mathbf{b}_1}, g_1^{\tau_1\mathbf{b}_2}, g_1^{\tau_1\mathbf{b}_3}, g_1^{\tau_1\mathbf{b}_4}) \quad \text{or} \quad (g_1^{\tau_1\mathbf{b}_1+\tau_2\mathbf{b}_5}, g_1^{\tau_1\mathbf{b}_2+\tau_2\mathbf{b}_6}, g_1^{\tau_1\mathbf{b}_3+\tau_2\mathbf{b}_7}, g_1^{\tau_1\mathbf{b}_4+\tau_2\mathbf{b}_8}).$$

$\mathcal{B}_0$ simulates $\mathsf{Game}_{\mathsf{Real}}$ or Game_0 with $\mathcal{A}$ depending on the distribution of (T_1, T_2, T_3, T_4). To compute the master public key and master secret key, $\mathcal{B}_0$

chooses a random invertible matrix $\mathbf{A} \in \mathbb{Z}_q^{4\times4}$. We then implicitly set dual orthonormal bases $\mathbb{D}, \mathbb{D}^*$ to:

$$\mathbf{d}_1 := \mathbf{b}_1, \ldots, \mathbf{d}_4 := \mathbf{b}_4, \quad (\mathbf{d}_5, \ldots, \mathbf{d}_8) := (\mathbf{b}_5, \ldots, \mathbf{b}_8)\mathbf{A},$$
$$\mathbf{d}_1^* := \mathbf{b}_1^*, \ldots, \mathbf{d}_4^* := \mathbf{b}_4^*, \quad (\mathbf{d}_5^*, \ldots, \mathbf{d}_8^*) := (\mathbf{b}_5^*, \ldots, \mathbf{b}_8^*)(\mathbf{A}^{-1})^\top.$$

We note that $\mathbb{D}, \mathbb{D}^*$ are properly distributed, and reveal no information about $\mathbf{A}$. Moreover, $\mathcal{B}_0$ cannot generate $g_2^{\mathbf{d}_5^*}, g_2^{\mathbf{d}_6^*}, g_2^{\mathbf{d}_7^*}, g_2^{\mathbf{d}_8^*}$, but these will not be needed for creating normal keys. $\mathcal{B}_0$ chooses random value $\alpha, \eta \in \mathbb{Z}_q$ and computes $g_T := e(g_1, g_2)^{\mathbf{d}_1 \cdot \mathbf{d}_1^*}$. It then gives $\mathcal{A}$ the master public key

$$\mathsf{mpk} := \left\{ \mathbb{G}; g_T^\alpha, g_T^\eta, g_1^{\mathbf{d}_1}, g_1^{\mathbf{d}_2} \right\}.$$

The master secret key

$$\mathsf{msk} := \left\{ \alpha, \eta, g_1^{\mathbf{d}_3}, g_1^{\mathbf{d}_4}, g_2^{\mathbf{d}_1^*}, g_2^{\mathbf{d}_2^*}, g_2^{\mathbf{d}_3^*}, g_2^{\mathbf{d}_4^*} \right\}$$

is known to $\mathcal{B}_0$, which allows $\mathcal{B}_0$ to respond to all of $\mathcal{A}$'s key queries by calling the normal key generation algorithm.

$\mathcal{A}$ sends $\mathcal{B}_0$ two pairs $(m_0, \mathsf{rcv}_0^*, \sigma_0^*)$ and $(m_1, \mathsf{rcv}_1^*, \sigma_1^*)$. $\mathcal{B}_0$ chooses a random bit $\beta \in \{0,1\}$ and picks $r' \xleftarrow{R} \mathbb{Z}_q$ and then encrypts m_β under rcv_β^* and $ek_{\sigma_\beta^*}$ as follows:

$$ek_{\sigma_\beta^*} := g_1^{\eta \mathbf{b}_3}(T_3^{\sigma_\beta^*} \cdot T_4^{-1})^{r'},$$

$$\mathsf{C} := m_\beta \cdot \left(e(T_1, g_2^{\mathbf{b}_1^*}) \right)^\alpha = m_\beta \cdot (g_T^\alpha)^z,$$

$$\mathsf{C}_0 := ek_{\sigma_\beta^*} \cdot T_1 \cdot T_2^{\mathsf{rcv}_\beta^*} = g_1^{\eta \mathbf{b}_3}(T_3^{\sigma_\beta^*} \cdot T_4^{-1})^{r'} \cdot T_1 \cdot T_2^{\mathsf{rcv}_\beta^*},$$

where $\mathcal{B}_0$ has implicitly set $r := r'\tau_1$ and $z := \tau_1$. It gives the ciphertext $ct = (\mathsf{C}, \mathsf{C}_0)$ to $\mathcal{A}$.

Now, if (T_1, T_2, T_3, T_4) are equal to $(g_1^{\tau_1 \mathbf{b}_1}, g_1^{\tau_1 \mathbf{b}_2}, g_1^{\tau_1 \mathbf{b}_3}, g_1^{\tau_1 \mathbf{b}_4})$, then this is a properly distributed normal encryption of m_β. In this case, $\mathcal{B}_0$ has properly simulated $\mathsf{Game}_{\mathsf{Real}}$. If (T_1, T_2, T_3, T_4) are equal to $(g_1^{\tau_1 \mathbf{b}_1 + \tau_2 \mathbf{b}_5}, g_1^{\tau_1 \mathbf{b}_2 + \tau_2 \mathbf{b}_6}, g_1^{\tau_1 \mathbf{b}_3 + \tau_2 \mathbf{b}_7}, g_1^{\tau_1 \mathbf{b}_4 + \tau_2 \mathbf{b}_8})$ instead, then the ciphertext element C_0 has an additional term of

$$\tau_2(\mathbf{b}_5 + \mathsf{rcv}_\beta^* \mathbf{b}_6) + r'\tau_2(\sigma_\beta^* \mathbf{b}_7 - \mathbf{b}_8)$$

in its exponent. The coefficients here in the basis $\mathbf{b}_5, \mathbf{b}_6, \mathbf{b}_7, \mathbf{b}_8$ form the vector $\tau_2(1, \mathsf{rcv}_\beta^*, r'\sigma_\beta^*, -r')$. To compute the coefficients in the basis $\mathbf{d}_5, \mathbf{d}_6, \mathbf{d}_7, \mathbf{d}_8$, we multiply the matrix $\mathbf{A}^{-1}$ by the transpose of this vector, obtaining the new vector $\tau_2 \mathbf{A}^{-1}(1, \mathsf{rcv}_\beta^*, r'\sigma_\beta^*, -r')^\top$. Since $\mathbf{A}$ is random (everything else given to $\mathcal{A}$ has been distributed independently of $\mathbf{A}$), these coefficients are uniformly random except with probability $2/q$ (namely, the cases τ_2 defined in Subspace problem is zero, (r_5, r_6, r_7, r_8) defined in Eq. 3 is the zero vector) from Lemma 2. Therefore in this case, $\mathcal{B}_0$ has properly simulated Game_0. This allows $\mathcal{B}_0$ to leverage $\mathcal{A}$'s advantage ϵ between $\mathsf{Game}_{\mathsf{Real}}$ and Game_0 to achieve an advantage $\epsilon - \frac{2}{q}$ against the subspace assumption in G_1, namely $\mathsf{Adv}_{\mathcal{B}_0}^{\mathsf{DS1}}(\lambda) = \epsilon - \frac{2}{q}$. $\square$

Lemma 4. *Suppose that there exists an adversary $\mathcal{A}$ where $|\mathsf{Adv}_{\mathcal{A}}^{\mathsf{Game}_{\kappa-1,2}}(\lambda) - \mathsf{Adv}_{\mathcal{A}}^{\mathsf{Game}_{\kappa,1}}(\lambda)| = \epsilon$. Then there exists an algorithm $\mathcal{B}_{\kappa,1}$ such that $\mathsf{Adv}_{\mathcal{B}_{\kappa,1}}^{\mathsf{DS2}}(\lambda) = \epsilon - 6/q$, with $K = 4$ and $N = 8$.*

Proof. $\mathcal{B}_{\kappa,1}$ is given

$$D := \left(\mathbb{G}; g_1^{\mathbf{b_1}}, g_1^{\mathbf{b_2}}, g_1^{\mathbf{b_3}}, g_1^{\mathbf{b_4}}, g_2^{\mathbf{b_1^*}}, \ldots, g_2^{\mathbf{b_8^*}}, U_1, U_2, U_3, U_4, \mu_2 \right)$$

along with (T_1, T_2, T_3, T_4). And in D we have that $U_1 = g_1^{\mu_1 \mathbf{b_1} + \mu_2 \mathbf{b_5}}, U_2 = g_1^{\mu_1 \mathbf{b_2} + \mu_2 \mathbf{b_6}}, U_3 = g_1^{\mu_1 \mathbf{b_3} + \mu_2 \mathbf{b_7}}, U_4 = g_1^{\mu_1 \mathbf{b_4} + \mu_2 \mathbf{b_8}}$. We require that $\mathcal{B}_{\kappa,1}$ decides whether (T_1, T_2, T_3, T_4) are distributed as

$$(g_2^{\tau_1 \mathbf{b_1^*}}, g_2^{\tau_1 \mathbf{b_2^*}}, g_2^{\tau_1 \mathbf{b_3^*}}, g_2^{\tau_1 \mathbf{b_4^*}}) \quad \text{or} \quad (g_2^{\tau_1 \mathbf{b_1^*} + \tau_2 \mathbf{b_5^*}}, g_2^{\tau_1 \mathbf{b_2^*} + \tau_2 \mathbf{b_6^*}}, g_2^{\tau_1 \mathbf{b_3^*} + \tau_2 \mathbf{b_7^*}}, g_2^{\tau_1 \mathbf{b_4^*} + \tau_2 \mathbf{b_8^*}}).$$

$\mathcal{B}_{\kappa,1}$ simulates $\mathsf{Game}_{\kappa-1,2}$ or $\mathsf{Game}_{\kappa,1}$ with $\mathcal{A}$ depending on the distribution of (T_1, T_2, T_3, T_4). To compute the master public key and master secret key, $\mathcal{B}_{\kappa,1}$ chooses a random invertible matrix $\mathbf{A} \in \mathbb{Z}_q^{4 \times 4}$. We then implicitly set dual orthonormal bases $\mathbb{D}, \mathbb{D}^*$ to:

$$\mathbf{d_1} := \mathbf{b_1}, \ldots, \mathbf{d_4} := \mathbf{b_4}, \quad (\mathbf{d_5}, \ldots, \mathbf{d_8}) := (\mathbf{b_5}, \ldots, \mathbf{b_8})\mathbf{A},$$
$$\mathbf{d_1^*} := \mathbf{b_1^*}, \ldots, \mathbf{d_4^*} := \mathbf{b_4^*}, \quad (\mathbf{d_5^*}, \ldots, \mathbf{d_8^*}) := (\mathbf{b_5^*}, \ldots, \mathbf{b_8^*})(\mathbf{A}^{-1})^{\top}.$$

We note that $\mathbb{D}, \mathbb{D}^*$ are properly distributed, and reveal no information about $\mathbf{A}$. $\mathcal{B}_{\kappa,1}$ chooses random value $\alpha, \eta \in \mathbb{Z}_q$ and compute $g_T := e(g_1, g_2)^{\mathbf{d_1} \cdot \mathbf{d_1^*}}$. It then gives $\mathcal{A}$ the master public key

$$\mathsf{mpk} := \left\{ \mathbb{G}; g_T^{\alpha}, g_T^{\eta}, g_1^{\mathbf{d_1}}, g_1^{\mathbf{d_2}} \right\}.$$

The master secret key

$$\mathsf{msk} := \left\{ \alpha, \eta, g_1^{\mathbf{d_3}}, g_1^{\mathbf{d_4}}, g_2^{\mathbf{d_1^*}}, g_2^{\mathbf{d_2^*}}, g_2^{\mathbf{d_3^*}}, g_2^{\mathbf{d_4^*}} \right\}$$

is known to $\mathcal{B}_{\kappa,1}$, which allows $\mathcal{B}_{\kappa,1}$ to respond to all of $\mathcal{A}$'s key queries by calling the normal key generation algorithm. Since $\mathcal{B}_{\kappa,1}$ also knows $g_2^{\mathbf{d_5^*}}, g_2^{\mathbf{d_6^*}}, g_2^{\mathbf{d_7^*}}, g_2^{\mathbf{d_8^*}}$, it can easily produce (inter) semi-functional keys. To answer the first $\kappa-1$ key queries that $\mathcal{A}$ makes, $\mathcal{B}_{\kappa,1}$ runs the KeyGenSF algorithm to produce semi-functional keys and gives these to $\mathcal{A}$. To answer the κ-th key query for ρ_{κ}, $\mathcal{B}_{\kappa,1}$ picks $s, s_2 \xleftarrow{R} \mathbb{Z}_q$ and responds with:

$$\mathsf{dk}_{\rho_{\kappa}} := \left\{ k_1 = g_2^{\alpha \mathbf{b_1^*} + s \mathbf{b_3^*}} \cdot (T_1^{\rho_{\kappa}} T_2^{-1}), \ k_2 = g_2^{s_2(\rho_{\kappa} \mathbf{b_1^*} - \mathbf{b_2^*}) + s \mathbf{b_4^*}}, \ k_3 = (g_T^{\eta})^s \right\}.$$

Noting that k_2 is a normal key, $\mathcal{B}_{\kappa,1}$ needs to determine whether k_1 is semi-functional or normal key and this implicitly sets $s_1 := \tau_1$. If (T_1, T_2, T_3, T_4) are equal to $(g_2^{\tau_1 \mathbf{b_1^*}}, g_2^{\tau_1 \mathbf{b_2^*}}, g_2^{\tau_1 \mathbf{b_3^*}}, g_2^{\tau_1 \mathbf{b_4^*}})$, then this is a properly distributed normal key.

If (T_1, T_2, T_3, T_4) are equal to $(g_2^{\tau_1 \mathbf{b_1^*} + \tau_2 \mathbf{b_5^*}}, g_2^{\tau_1 \mathbf{b_2^*} + \tau_2 \mathbf{b_6^*}}, g_2^{\tau_1 \mathbf{b_3^*} + \tau_2 \mathbf{b_7^*}}, g_2^{\tau_1 \mathbf{b_4^*} + \tau_2 \mathbf{b_8^*}})$, then this is a inter-semi-functional key, whose exponent vector includes

$$\tau_2(\rho_\kappa \mathbf{b_5}^* - \mathbf{b_6}^* + 0 \cdot \mathbf{b_7}^* + 0 \cdot \mathbf{b_8}^*) \tag{4}$$

as its component in the span of $\mathbf{b_5^*}, \mathbf{b_6^*}, \mathbf{b_7^*}, \mathbf{b_8^*}$. To respond to the remaining key queries, $\mathcal{B}_{\kappa,1}$ simply runs the normal key generation algorithm.

At some point, $\mathcal{A}$ sends $\mathcal{B}_{\kappa,1}$ two pairs $(m_0, \mathsf{rcv}_0^*, \sigma_0^*)$ and $(m_1, \mathsf{rcv}_1^*, \sigma_1^*)$. $\mathcal{B}_{\kappa,1}$ chooses a random bit $\beta \in \{0,1\}$ and picks $r' \xleftarrow{R} \mathbb{Z}_q$ and then encrypts m_β under rcv_β^* and $ek_{\sigma_\beta^*}$ as follows:

$$ek_{\sigma_\beta^*} := g_1^{\eta \mathbf{b_3}}(U_3^{\sigma_\beta^*} \cdot U_4^{-1})^{r'},$$

$$\mathsf{C} := m_\beta \cdot \left(e(U_1, g_2^{\mathbf{b_1^*}})\right)^\alpha = m_\beta \cdot (g_T^\alpha)^z,$$

$$\mathsf{C_0} := ek_{\sigma_\beta^*} \cdot U_1 \cdot U_2^{\mathsf{rcv}_\beta^*} = g_1^{\eta \mathbf{b_3}}(U_3^{\sigma_\beta^*} \cdot U_4^{-1})^{r'} \cdot U_1 \cdot U_2^{\mathsf{rcv}_\beta^*},$$

where $\mathcal{B}_{\kappa,1}$ has implicitly set $r := r' \mu_1$ and $z := \mu_1$. The "semi-functional part" of the exponent vector here is:

$$\mu_2(\mathbf{b_5} + \mathsf{rcv}_\beta^* \mathbf{b_6}) + r' \mu_2(\sigma_\beta^* \mathbf{b_7} - \mathbf{b_8}) \tag{5}$$

We observe that if $\mathsf{rcv}_\beta^* = \rho_\kappa$ (which is not allowed) and the decryption algorithm gives an attribute snd_κ that can correctly decrypt the ciphertext, i.e., $\mathsf{snd}_\kappa = \sigma_\beta^*$, then vectors in Eqs. 4 and 5 would be orthogonal in the decryption algorithm, resulting in a nominally semi-functional ciphertext and key pair. It gives the ciphertext $ct = (\mathsf{C}, \mathsf{C_0})$ to $\mathcal{A}$.

We now argue that since $\mathsf{rcv}_\beta^* \neq \rho_\kappa$, in $\mathcal{A}$'s view the vectors in Eqs. 4 and 5 are distributed as random vectors in the spans of $\mathbf{d_5^*}, \mathbf{d_6^*}, \mathbf{d_7^*}, \mathbf{d_8^*}$ and $\mathbf{d_5}, \mathbf{d_6}, \mathbf{d_7}, \mathbf{d_8}$ respectively. To see this, we take the coefficients of vectors in Eqs. 4 and 5 in terms of the bases $\mathbf{b_5^*}, \mathbf{b_6^*}, \mathbf{b_7^*}, \mathbf{b_8^*}$ and $\mathbf{b_5}, \mathbf{b_6}, \mathbf{b_7}, \mathbf{b_8}$ respectively and translate them into coefficients in terms of the bases $\mathbf{d_5^*}, \mathbf{d_6^*}, \mathbf{d_7^*}, \mathbf{d_8^*}$ and $\mathbf{d_5}, \mathbf{d_6}, \mathbf{d_7}, \mathbf{d_8}$. Using the change of basis matrix $\mathbf{A}$, we obtain the new coefficients (in vector form) as:

$$\tau_2 \mathbf{A}^\top (\rho_\kappa, -1, 0, 0)^\top, \quad \mu_2 \mathbf{A}^{-1}(1, \mathsf{rcv}_\beta^*, r'\sigma_\beta^*, -r')^\top.$$

Since the distribution of everything given to $\mathcal{A}$ except for the κ-th key and the challenge ciphertext is independent of the random matrix $\mathbf{A}$ and $\mathsf{rcv}_\beta^* \neq \rho_\kappa$, we can conclude that these coefficients are uniformly except with probability $4/q$ (namely, the cases τ_2 or μ_2 defined in Subspace problem is zero, $\{s_{i,1}\}_{i=5,\ldots,8}$ or (r_5, r_6, r_7, r_8) defined in Eqs. 1 and 3 is the zero vector) from Lemma 2. Thus, $\mathcal{B}_{\kappa,1}$ has properly simulated $\mathsf{Game}_{\kappa,1}$ in this case.

If (T_1, T_2, T_3, T_4) are equal to $(g_2^{\tau_1 \mathbf{b_1^*}}, g_2^{\tau_1 \mathbf{b_2^*}}, g_2^{\tau_1 \mathbf{b_3^*}}, g_2^{\tau_1 \mathbf{b_4^*}})$, then the coefficients of the vector in Eq. 5 are uniformly except with probability $2/q$ (namely, the cases μ_2 defined in Subspace problem is zero, (r_5, r_6, r_7, r_8) defined in Eq. 3 is the zero vector) from Lemma 2. Thus, $\mathcal{B}_{\kappa,1}$ has properly simulated $\mathsf{Game}_{\kappa-1,2}$ in this case.

In summary, $\mathcal{B}_{\kappa,1}$ has properly simulated either $\mathsf{Game}_{\kappa-1,2}$ or $\mathsf{Game}_{\kappa,1}$ for $\mathcal{A}$, depending on the distribution of (T_1, T_2, T_3, T_4). It can therefore leverage $\mathcal{A}$'s advantage ϵ between these games to obtain an advantage $\epsilon - 6/q$ against the Subspace assumption in G_2, namely $\mathsf{Adv}^{\mathsf{DS2}}_{\mathcal{B}_{\kappa,1}}(\lambda) = \epsilon - 6/q$.   □

Lemma 5. *Suppose that there exists an adversary $\mathcal{A}$ where $|\mathsf{Adv}^{\mathsf{Game}_{\kappa,1}}_{\mathcal{A}}(\lambda) - \mathsf{Adv}^{\mathsf{Game}_{\kappa,2}}_{\mathcal{A}}(\lambda)| = \epsilon$. Then there exists an algorithm $\mathcal{B}_{\kappa,2}$ such that $\mathsf{Adv}^{\mathsf{DS2}}_{\mathcal{B}_{\kappa,2}}(\lambda) = \epsilon - 6/q$, with $K = 4$ and $N = 8$.*

Proof. This proof is very similar to the proof of the previous lemma and $\mathcal{B}_{\kappa,2}$ is given

$$D := \left(\mathbb{G}; g_1^{\mathbf{b_1}}, g_1^{\mathbf{b_2}}, g_1^{\mathbf{b_3}}, g_1^{\mathbf{b_4}}, g_2^{\mathbf{b_1^*}}, \ldots, g_2^{\mathbf{b_8^*}}, U_1, U_2, U_3, U_4, \mu_2 \right)$$

along with (T_1, T_2, T_3, T_4). And in D we have that $U_1 = g_1^{\mu_1 \mathbf{b_1} + \mu_2 \mathbf{b_5}}, U_2 = g_1^{\mu_1 \mathbf{b_2} + \mu_2 \mathbf{b_6}}, U_3 = g_1^{\mu_1 \mathbf{b_3} + \mu_2 \mathbf{b_7}}, U_4 = g_1^{\mu_1 \mathbf{b_4} + \mu_2 \mathbf{b_8}}$. We require that $\mathcal{B}_{\kappa,2}$ decides whether (T_1, T_2, T_3, T_4) are distributed as

$$(g_2^{\tau_1 \mathbf{b_1^*}}, g_2^{\tau_1 \mathbf{b_2^*}}, g_2^{\tau_1 \mathbf{b_3^*}}, g_2^{\tau_1 \mathbf{b_4^*}}) \quad \text{or} \quad (g_2^{\tau_1 \mathbf{b_1^*} + \tau_2 \mathbf{b_5^*}}, g_2^{\tau_1 \mathbf{b_2^*} + \tau_2 \mathbf{b_6^*}}, g_2^{\tau_1 \mathbf{b_3^*} + \tau_2 \mathbf{b_7^*}}, g_2^{\tau_1 \mathbf{b_4^*} + \tau_2 \mathbf{b_8^*}}).$$

$\mathcal{B}_{\kappa,2}$ simulates $\mathsf{Game}_{\kappa,1}$ or $\mathsf{Game}_{\kappa,2}$ with $\mathcal{A}$ depending on the distribution of (T_1, T_2, T_3, T_4). To compute the master public key and master secret key, $\mathcal{B}_{\kappa,2}$ chooses a random invertible matrix $\mathbf{A} \in \mathbb{Z}_q^{4\times4}$. We then implicitly set dual orthonormal bases $\mathbb{D}, \mathbb{D}^*$ to:

$$\mathbf{d_1} := \mathbf{b_1}, \ldots, \mathbf{d_4} := \mathbf{b_4}, \quad (\mathbf{d_5}, \ldots, \mathbf{d_8}) := (\mathbf{b_5}, \ldots, \mathbf{b_8})\mathbf{A},$$

$$\mathbf{d_1^*} := \mathbf{b_1^*}, \ldots, \mathbf{d_4^*} := \mathbf{b_4^*}, \quad (\mathbf{d_5^*}, \ldots, \mathbf{d_8^*}) := (\mathbf{b_5^*}, \ldots, \mathbf{b_8^*})(\mathbf{A}^{-1})^{\top}.$$

We note that $\mathbb{D}, \mathbb{D}^*$ are properly distributed, and reveal no information about $\mathbf{A}$. $\mathcal{B}_{\kappa,2}$ chooses random value $\alpha, \eta \in \mathbb{Z}_q$ and compute $g_T := e(g_1, g_2)^{\mathbf{d_1} \cdot \mathbf{d_1^*}}$. It then gives $\mathcal{A}$ the master public key

$$\mathsf{mpk} := \left\{ \mathbb{G}; g_T^{\alpha}, g_T^{\eta}, g_1^{\mathbf{d_1}}, g_1^{\mathbf{d_2}} \right\}.$$

The master secret key

$$\mathsf{msk} := \left\{ \alpha, \eta, g_1^{\mathbf{d_3}}, g_1^{\mathbf{d_4}}, g_2^{\mathbf{d_1^*}}, g_2^{\mathbf{d_2^*}}, g_2^{\mathbf{d_3^*}}, g_2^{\mathbf{d_4^*}} \right\}$$

is known to $\mathcal{B}_{\kappa,2}$, which allows $\mathcal{B}_{\kappa,2}$ to respond to all of $\mathcal{A}$'s key queries by calling the normal key generation algorithm. Since $\mathcal{B}_{\kappa,2}$ also knows $g_2^{\mathbf{d_5^*}}, g_2^{\mathbf{d_6^*}}, g_2^{\mathbf{d_7^*}}, g_2^{\mathbf{d_8^*}}$, it can easily produce (inter) semi-functional keys. To answer the first $\kappa-1$ key queries that $\mathcal{A}$ makes, $\mathcal{B}_{\kappa,2}$ runs the KeyGenSF algorithm to produce semi-functional keys and gives these to $\mathcal{A}$. To answer the κ-th key query for ρ_κ, $\mathcal{B}_{\kappa,2}$ picks $s, s_1, \{s_{i,1}\}_{i=5,\ldots,8} \xleftarrow{R} \mathbb{Z}_q$ and responds with:

$$\mathsf{dk}_{\rho_\kappa} := \{ k_1 = g_2^{\alpha \mathbf{d_1^*} + s_1(\rho \mathbf{d_1^*} - \mathbf{d_2^*}) + s \mathbf{d_3^*} + [s_{5,1} \mathbf{d_5^*} + s_{6,1} \mathbf{d_6^*} + s_{7,1} \mathbf{d_7^*}]},$$

$$k_2 = g_2^{s \mathbf{b_4^*}} \cdot (T_1^{\rho_\kappa} T_2^{-1}), \quad k_3 = (g_T^{\eta})^s \}.$$

Noting that k_1 is a semi-functional key, $\mathcal{B}_{\kappa,2}$ needs to determine whether k_2 is semi-functional or normal key and this implicitly sets $s_2 := \tau_1$. If (T_1, T_2, T_3, T_4) are equal to $(g_2^{\tau_1 \mathbf{b}_1^*}, g_2^{\tau_1 \mathbf{b}_2^*}, g_2^{\tau_1 \mathbf{b}_3^*}, g_2^{\tau_1 \mathbf{b}_4^*})$, then this is a properly distributed normal key. If (T_1, T_2, T_3, T_4) are equal to $(g_2^{\tau_1 \mathbf{b}_1^* + \tau_2 \mathbf{b}_5^*}, g_2^{\tau_1 \mathbf{b}_2^* + \tau_2 \mathbf{b}_6^*}, g_2^{\tau_1 \mathbf{b}_3^* + \tau_2 \mathbf{b}_7^*}, g_2^{\tau_1 \mathbf{b}_4^* + \tau_2 \mathbf{b}_8^*})$, then this is a semi-functional key, whose exponent vector includes

$$\tau_2(\rho_\kappa \mathbf{b_5}^* - \mathbf{b_6}^* + 0 \cdot \mathbf{b_7}^* + 0 \cdot \mathbf{b_8}^*) \tag{6}$$

as its component in the span of $\mathbf{b}_5^*, \mathbf{b}_6^*, \mathbf{b}_7^*, \mathbf{b}_8^*$. To respond to the remaining key queries, $\mathcal{B}_{\kappa,2}$ simply runs the normal key generation algorithm.

At some point, $\mathcal{A}$ sends $\mathcal{B}_{\kappa,2}$ two pairs $(m_0, \mathsf{rcv}_0^*, \sigma_0^*)$ and $(m_1, \mathsf{rcv}_1^*, \sigma_1^*)$. $\mathcal{B}_{\kappa,2}$ chooses a random bit $\beta \in \{0,1\}$ and picks $r' \xleftarrow{R} \mathbb{Z}_q$ and then encrypts m_β under rcv_β^* and $ek_{\sigma_\beta^*}$ as follows:

$$ek_{\sigma_\beta^*} := g_1^{\eta \mathbf{b_3}}(U_3^{\sigma_\beta^*} \cdot U_4^{-1})^{r'},$$

$$\mathsf{C} := m_\beta \cdot \left(e(U_1, g_2^{\mathbf{b_2^*}})\right)^\alpha = m_\beta \cdot (g_T^\alpha)^z,$$

$$\mathsf{C}_0 := ek_{\sigma_\beta^*} \cdot U_1 \cdot U_2^{\mathsf{rcv}_\beta^*} = g_1^{\eta \mathbf{b_3}}(U_3^{\sigma_\beta^*} \cdot U_4^{-1})^{r'} \cdot U_1 \cdot U_2^{\mathsf{rcv}_\beta^*},$$

where $\mathcal{B}_{\kappa,2}$ has implicitly set $r := r'\mu_1$ and $z := \mu_1$. The "semi-functional part" of the exponent vector here is:

$$\mu_2(\mathbf{b_5} + \mathsf{rcv}_\beta^* \mathbf{b_6}) + r'\mu_2(\sigma_\beta^* \mathbf{b_7} - \mathbf{b_8}) \tag{7}$$

We observe that if $\mathsf{rcv}_\beta^* = \rho_\kappa$ (which is not allowed) and the decryption algorithm gives an attribute snd_κ that can correctly decrypt the ciphertext, i.e., $\mathsf{snd}_\kappa = \sigma_\beta^*$, then vectors in Eqs. 6 and 7 would be orthogonal in the decryption algorithm, resulting in a nominally semi-functional ciphertext and key pair. It gives the ciphertext $ct = (\mathsf{C}, \mathsf{C}_0)$ to $\mathcal{A}$.

We now argue that since $\mathsf{rcv}_\beta^* \neq \rho_\kappa$, in $\mathcal{A}$'s view the vectors in Eqs. 6 and 7 are distributed as random vectors in the spans of $\mathbf{d}_5^*, \mathbf{d}_6^*, \mathbf{d}_7^*, \mathbf{d}_8^*$ and $\mathbf{d}_5, \mathbf{d}_6, \mathbf{d}_7, \mathbf{d}_8$ respectively. To see this, we take the coefficients of vectors in Eqs. 6 and 7 in terms of the bases $\mathbf{b}_5^*, \mathbf{b}_6^*, \mathbf{b}_7^*, \mathbf{b}_8^*$ and $\mathbf{b}_5, \mathbf{b}_6, \mathbf{b}_7, \mathbf{b}_8$ respectively and translate them into coefficients in terms of the bases $\mathbf{d}_5^*, \mathbf{d}_6^*, \mathbf{d}_7^*, \mathbf{d}_8^*$ and $\mathbf{d}_5, \mathbf{d}_6, \mathbf{d}_7, \mathbf{d}_8$. Using the change of basis matrix $\mathbf{A}$, we obtain the new coefficients (in vector form) as:

$$\tau_2 \mathbf{A}^\top (\rho_\kappa, -1, 0, 0)^\top, \quad \mu_2 \mathbf{A}^{-1}(1, \mathsf{rcv}_\beta^*, r'\sigma_\beta^*, -r')^\top.$$

Since the distribution of everything given to $\mathcal{A}$ except for the κ-th key and the challenge ciphertext is independent of the random matrix $\mathbf{A}$ and $\mathsf{rcv}_\beta^* \neq \rho_\kappa$, we can conclude that these coefficients are uniformly except with probability $4/q$ (namely, the cases τ_2 or μ_2 defined in Subspace problem is zero, $\{s_{i,j}\}_{i=5,\ldots,8; j=1,2}$ or (r_5, r_6, r_7, r_8) defined in Eqs. 2 and 3 is the zero vector) from Lemma 2. Thus, $\mathcal{B}_{\kappa,2}$ has properly simulated $\mathsf{Game}_{\kappa,2}$ in this case.

If (T_1, T_2, T_3, T_4) are equal to $(g_2^{\tau_1 \mathbf{b}_1^*}, g_2^{\tau_1 \mathbf{b}_2^*}, g_2^{\tau_1 \mathbf{b}_3^*}, g_2^{\tau_1 \mathbf{b}_4^*})$, then the coefficients of the vector in Eq. 7 are uniformly except with probability $2/q$ (namely, the cases μ_2 defined in Subspace problem is zero, (r_5, r_6, r_7, r_8) defined in Eq. 3 is the zero vector) from Lemma 2. Thus, $\mathcal{B}_{\kappa,2}$ has properly simulated $\mathsf{Game}_{\kappa,1}$ in this case.

In summary, $\mathcal{B}_{\kappa,2}$ has properly simulated either $\mathsf{Game}_{\kappa,1}$ or $\mathsf{Game}_{\kappa,2}$ for $\mathcal{A}$, depending on the distribution of (T_1, T_2, T_3, T_4). It can therefore leverage $\mathcal{A}$'s advantage ϵ between these games to obtain an advantage $\epsilon - 6/q$ against the Subspace assumption in G_2, namely $\mathsf{Adv}_{\mathcal{B}_{\kappa,2}}^{\mathsf{DS2}}(\lambda) = \epsilon - 6/q$. $\qquad\square$

Lemma 6. *For any adversary $\mathcal{A}$, $\mathsf{Adv}_{\mathcal{A}}^{\mathsf{Game_{Final}}}(\lambda) \leq \mathsf{Adv}_{\mathcal{A}}^{\mathsf{Game}_{\nu,2}}(\lambda) + 1/q$.*

Proof. To prove this lemma, we show the joint distributions of

$$\left(\mathsf{mpk}, \mathsf{CT}_{ek_{\sigma_\beta^*}, rcv_\beta^*}^{(\mathsf{SF})}, \left\{ \mathsf{dk}_{\rho_\ell}^{(\mathsf{SF})} \right\}_{\ell \in [\nu]} \right)$$

in $\mathsf{Game}_{\nu,2}$ and that of

$$\left(\mathsf{mpk}, \mathsf{CT}_{ek_{\sigma_R}, rcv_R}^{(\mathsf{R})}, \left\{ \mathsf{dk}_{\rho_\ell}^{(\mathsf{SF})} \right\}_{\ell \in [\nu]} \right)$$

in $\mathsf{Game_{Final}}$ are equivalent for the adversary's view, where $\mathsf{CT}_{ek_{\sigma_R}, rcv_R}^{(\mathsf{R})}$ is a semifunctional encryption of a random message in G_T and under two random identities in $\mathbb{Z}_q$.

For this purpose, we pick $\mathbf{A} := (\xi_{i,j}) \xleftarrow{R} \mathbb{Z}_q^{4 \times 4}$ and define new dual orthonormal bases $\mathbb{F} := (\mathbf{f}_1, \ldots, \mathbf{f}_8)$, and $\mathbb{F}^* := (\mathbf{f}_1^*, \ldots, \mathbf{f}_8^*)$ as follows:

$$
\begin{pmatrix} \mathbf{f}_1 \\ \mathbf{f}_2 \\ \mathbf{f}_3 \\ \mathbf{f}_4 \\ \mathbf{f}_5 \\ \mathbf{f}_6 \\ \mathbf{f}_7 \\ \mathbf{f}_8 \end{pmatrix}
:=
\begin{pmatrix}
1 & 0 & 0 & 0 & 0 & 0 & 0 & 0 \\
0 & 1 & 0 & 0 & 0 & 0 & 0 & 0 \\
0 & 0 & 1 & 0 & 0 & 0 & 0 & 0 \\
0 & 0 & 0 & 1 & 0 & 0 & 0 & 0 \\
\xi_{1,1} & \xi_{1,2} & \xi_{1,3} & \xi_{1,4} & 1 & 0 & 0 & 0 \\
\xi_{2,1} & \xi_{2,2} & \xi_{2,3} & \xi_{2,4} & 0 & 1 & 0 & 0 \\
\xi_{3,1} & \xi_{3,2} & \xi_{3,3} & \xi_{3,4} & 0 & 0 & 1 & 0 \\
\xi_{4,1} & \xi_{4,2} & \xi_{4,3} & \xi_{4,4} & 0 & 0 & 0 & 1
\end{pmatrix}
\begin{pmatrix} \mathbf{d}_1 \\ \mathbf{d}_2 \\ \mathbf{d}_3 \\ \mathbf{d}_4 \\ \mathbf{d}_5 \\ \mathbf{d}_6 \\ \mathbf{d}_7 \\ \mathbf{d}_8 \end{pmatrix},
$$

$$
\begin{pmatrix} \mathbf{f}_1^* \\ \mathbf{f}_2^* \\ \mathbf{f}_3^* \\ \mathbf{f}_4^* \\ \mathbf{f}_5^* \\ \mathbf{f}_6^* \\ \mathbf{f}_7^* \\ \mathbf{f}_8^* \end{pmatrix}
:=
\begin{pmatrix}
1 & 0 & 0 & 0 & -\xi_{1,1} & -\xi_{2,1} & -\xi_{3,1} & -\xi_{4,1} \\
0 & 1 & 0 & 0 & -\xi_{1,2} & -\xi_{2,2} & -\xi_{3,2} & -\xi_{4,2} \\
0 & 0 & 1 & 0 & -\xi_{1,3} & -\xi_{2,3} & -\xi_{3,3} & -\xi_{4,3} \\
0 & 0 & 0 & 1 & -\xi_{1,4} & -\xi_{2,4} & -\xi_{3,4} & -\xi_{4,4} \\
0 & 0 & 0 & 0 & 1 & 0 & 0 & 0 \\
0 & 0 & 0 & 0 & 0 & 1 & 0 & 0 \\
0 & 0 & 0 & 0 & 0 & 0 & 1 & 0 \\
0 & 0 & 0 & 0 & 0 & 0 & 0 & 1
\end{pmatrix}
\begin{pmatrix} \mathbf{d}_1^* \\ \mathbf{d}_2^* \\ \mathbf{d}_3^* \\ \mathbf{d}_4^* \\ \mathbf{d}_5^* \\ \mathbf{d}_6^* \\ \mathbf{d}_7^* \\ \mathbf{d}_8^* \end{pmatrix}.
$$

It is easy to verify that $\mathbb{F}$ and $\mathbb{F}^*$ are also dual orthonormal, and are distributed the same as $\mathbb{D}$ and $\mathbb{D}^*$.

Then the master public key, challenge ciphertext, and queried secret keys, $(\mathsf{mpk}, \mathsf{CT}^{(\mathsf{SF})}_{ek_{\sigma^*_\beta}, \mathsf{rcv}^*_\beta}, \{\mathsf{dk}^{(\mathsf{SF})}_{\rho_\ell}\}_{\ell \in [\nu]})$ in $\mathsf{Game}_{\nu,2}$ are expressed over bases $\mathbb{D}$ and $\mathbb{D}^*$ as

$$\mathsf{mpk} := \left\{ \mathbb{G}; g_T^\alpha, g_T^\eta, g_1^{\mathbf{d_1}}, g_1^{\mathbf{d_2}} \right\}, \quad ek_{\sigma^*_\beta} := g_1^{\eta \mathbf{d_3} + r(\sigma^*_\beta \mathbf{d_3} - \mathbf{d_4})},$$

$$\mathsf{CT}^{(\mathsf{SF})}_{ek_{\sigma^*_\beta}, \mathsf{rcv}^*_\beta} := \{ \mathsf{C} = m \cdot (g_T^\alpha)^z,$$

$$\mathsf{C_0} = ek_{\sigma^*_\beta} \cdot g_1^{z(\mathbf{d_1} + \mathsf{rcv}^*_\beta \mathbf{d_2}) + [r_5 \mathbf{d_5} + r_6 \mathbf{d_6} + r_7 \mathbf{d_7} + r_8 \mathbf{d_8}]}$$

$$= g_1^{\eta \mathbf{d_3} + r(\sigma^*_\beta \mathbf{d_3} - \mathbf{d_4}) + z(\mathbf{d_1} + \mathsf{rcv}^*_\beta \mathbf{d_2}) + [r_5 \mathbf{d_5} + r_6 \mathbf{d_6} + r_7 \mathbf{d_7} + r_8 \mathbf{d_8}]} \},$$

$$\mathsf{dk}^{(\mathsf{SF})}_{\rho_\ell} := \{ k_1 = g_2^{\alpha \mathbf{d_1}^* + s_{1,\ell}(\rho_\ell \mathbf{d_1}^* - \mathbf{d_2}^*) + s_\ell \mathbf{d_3}^* + [s_{5,1,\ell} \mathbf{d_5}^* + s_{6,1,\ell} \mathbf{d_6}^* + s_{7,1,\ell} \mathbf{d_7}^*]},$$

$$k_2 = g_2^{s_{2,\ell}(\rho_\ell \mathbf{d_1}^* - \mathbf{d_2}^*) + s_\ell \mathbf{d_4}^* + [s_{5,2,\ell} \mathbf{d_5}^* + s_{6,2,\ell} \mathbf{d_6}^* + s_{8,2,\ell} \mathbf{d_8}^*]},$$

$$k_3 = (g_T^\eta)^s \}_{\ell \in [\nu]}.$$

Then we can express them over bases $\mathbb{F}$ and $\mathbb{F}^*$ as

$$\mathsf{mpk} := \left\{ \mathbb{G}; g_T^\alpha, g_T^\eta, g_1^{\mathbf{f_1}}, g_1^{\mathbf{f_2}} \right\}, \quad ek_{\sigma^*_\beta} := g_1^{\eta \mathbf{d_3} + r(\sigma^*_\beta \mathbf{d_3} - \mathbf{d_4})},$$

$$\mathsf{CT}^{(\mathsf{R})}_{ek_{\sigma_\mathsf{R}}, \mathsf{rcv}_\mathsf{R}} := \{ \mathsf{C} = m \cdot (g_T^\alpha)^z,$$

$$\mathsf{C_0} = ek_{\sigma^*_\beta} \cdot g_1^{z(\mathbf{d_1} + \mathsf{rcv}^*_\beta \mathbf{d_2}) + [r_5 \mathbf{d_5} + r_6 \mathbf{d_6} + r_7 \mathbf{d_7} + r_8 \mathbf{d_8}]}$$

$$= g_1^{\eta \mathbf{f_3} + (r_3 \mathbf{f_3} + r_4 \mathbf{f_4} + r_1 \mathbf{f_1} + r_2 \mathbf{f_2}) + [r_5 \mathbf{f_5} + r_6 \mathbf{f_6} + r_7 \mathbf{f_7} + r_8 \mathbf{f_8}]} \},$$

$$\mathsf{dk}^{(\mathsf{SF})}_{\rho_\ell} := \{ k_1 = g_2^{\alpha \mathbf{f_1}^* + s_{1,\ell}(\rho_\ell \mathbf{f_1}^* - \mathbf{f_2}^*) + s_\ell \mathbf{f_3}^* + [t_{5,1,\ell} \mathbf{f_5}^* + t_{6,1,\ell} \mathbf{f_6}^* + t_{7,1,\ell} \mathbf{f_7}^* + t_{8,1,\ell} \mathbf{f_8}^*]},$$

$$k_2 = g_2^{s_{2,\ell}(\rho_\ell \mathbf{f_1}^* - \mathbf{f_2}^*) + s_\ell \mathbf{f_4}^* + [t_{5,2,\ell} \mathbf{f_5}^* + t_{6,2,\ell} \mathbf{f_6}^* + t_{7,2,\ell} \mathbf{f_7}^* + t_{8,2,\ell} \mathbf{f_8}^*]},$$

$$k_3 = (g_T^\eta)^s \}_{\ell \in [\nu]}.$$

where

$$r_1 := z - r_5 \xi_{1,1} - r_6 \xi_{2,1} - r_7 \xi_{3,1} - r_8 \xi_{4,1},$$

$$r_2 := z \cdot \mathsf{rcv}^*_\beta - r_5 \xi_{1,2} - r_6 \xi_{2,2} - r_7 \xi_{3,2} - r_8 \xi_{4,2},$$

$$r_3 := r \cdot \sigma^*_\beta - r_5 \xi_{1,3} - r_6 \xi_{2,3} - r_7 \xi_{3,3} - r_8 \xi_{4,3},$$

$$r_4 := -r - r_5 \xi_{1,4} - r_6 \xi_{2,4} - r_7 \xi_{3,4} - r_8 \xi_{4,4};$$

$$\left\{ \begin{array}{l} t_{5,1,\ell} := \alpha \cdot \xi_{1,1} + s_{1,\ell}\rho_\ell \cdot \xi_{1,1} - s_{1,\ell} \cdot \xi_{1,2} + s_\ell \cdot \xi_{1,3} + s_{5,1,\ell} \\ t_{6,1,\ell} := \alpha \cdot \xi_{2,1} + s_{1,\ell}\rho_\ell \cdot \xi_{2,1} - s_{1,\ell} \cdot \xi_{2,2} + s_\ell \cdot \xi_{2,3} + s_{6,1,\ell} \\ t_{7,1,\ell} := \alpha \cdot \xi_{3,1} + s_{1,\ell}\rho_\ell \cdot \xi_{3,1} - s_{1,\ell} \cdot \xi_{3,2} + s_\ell \cdot \xi_{3,3} + s_{7,1,\ell} \\ t_{8,1,\ell} := \alpha \cdot \xi_{4,1} + s_{1,\ell}\rho_\ell \cdot \xi_{4,1} - s_{1,\ell} \cdot \xi_{4,2} + s_\ell \cdot \xi_{4,3} \end{array} \right\}_{\ell \in [\nu]},$$

$$\left\{\begin{aligned}
t_{5,2,\ell} &:= s_{2,\ell}\rho_\ell \cdot \xi_{1,1} - s_{2,\ell} \cdot \xi_{1,2} + s_\ell \cdot \xi_{1,4} + s_{5,2,\ell} \\
t_{6,2,\ell} &:= s_{2,\ell}\rho_\ell \cdot \xi_{2,1} - s_{2,\ell} \cdot \xi_{2,2} + s_\ell \cdot \xi_{2,4} + s_{6,2,\ell} \\
t_{7,2,\ell} &:= s_{2,\ell}\rho_\ell \cdot \xi_{3,1} - s_{2,\ell} \cdot \xi_{3,2} + s_\ell \cdot \xi_{3,4} \\
t_{8,2,\ell} &:= s_{2,\ell}\rho_\ell \cdot \xi_{4,1} - s_{2,\ell} \cdot \xi_{4,2} + s_\ell \cdot \xi_{4,4} + s_{8,2,\ell}
\end{aligned}\right\}_{\ell\in[\nu]},$$

which are all uniformly distributed if (r_5, r_6, r_7, r_8) defined in Eq. 3 is a non-zero vector, since $\left(z, r, \{\xi_{i,j}\}_{i\in[4], j\in[4]}, \{s_{i,j,\ell}\}_{i=5,\ldots,8; j=1,2; \ell\in[\nu]}\right)$ are all uniformly picked from $\mathbb{Z}_q$.

In other words, the coefficients $(z, z \cdot \mathsf{rcv}^*_\beta, r\sigma^*_\beta, -r)$ of $\mathbf{d}_1, \mathbf{d}_2, \mathbf{d}_3, \mathbf{d}_4$ in the C_0 term of the challenge ciphertext is changed to random coefficients $(r_1, r_2, r_3, r_4) \in \mathbb{Z}_q^4$ of $\mathbf{f}_1, \mathbf{f}_2, \mathbf{f}_3, \mathbf{f}_4$, thus the challenge ciphertext can be viewed as a semi-functional encryption of a random message in G_T and under two random identities in $\mathbb{Z}_q$. Moreover, all coefficients $\{t_{i,j,\ell}\}_{i=5,\ldots,8; j=1,2; \ell\in[\nu]}$ of $\mathbf{f}^*_5, \mathbf{f}^*_6, \mathbf{f}^*_7, \mathbf{f}^*_8$ in the $\{\mathsf{dk}^{(\mathsf{SF})}_{\rho_\ell}\}_{\ell\in[\nu]}$ are all uniformly distributed since $\{s_{i,j,\ell}\}_{i=5,\ldots,8; j=1,2; \ell\in[\nu]}$ of $\mathbf{d}^*_5, \mathbf{d}^*_6, \mathbf{d}^*_7, \mathbf{d}^*_8$ are all independent random values. Thus

$$\left(\mathsf{mpk}, \mathsf{CT}^{(\mathsf{SF})}_{ek_{\sigma^*_\beta}, \mathsf{rcv}^*_\beta}, \left\{\mathsf{dk}^{(\mathsf{SF})}_{\rho_\ell}\right\}_{\ell\in[\nu]}\right)$$

expressed over bases $\mathbb{F}$ and $\mathbb{F}^*$ is properly distributed as

$$\left(\mathsf{mpk}, \mathsf{CT}^{(\mathsf{R})}_{ek_{\sigma_\mathsf{R}}, \mathsf{rcv}_\mathsf{R}}, \left\{\mathsf{dk}^{(\mathsf{SF})}_{\rho_\ell}\right\}_{\ell\in[\nu]}\right)$$

in $\mathsf{Game}_{\mathsf{Final}}$.

In the adversary's view, both $(\mathbb{D}, \mathbb{D}^*)$ and $(\mathbb{F}, \mathbb{F}^*)$ are consistent with the same master public key. Therefore, the challenge ciphertext and queried secret keys above can be expressed as keys and ciphertext in two ways, in $\mathsf{Game}_{\nu,2}$ over bases $(\mathbb{D}, \mathbb{D}^*)$ and in $\mathsf{Game}_{\mathsf{Final}}$ over bases $(\mathbb{F}, \mathbb{F}^*)$. Thus, $\mathsf{Game}_{\nu,2}$ and $\mathsf{Game}_{\mathsf{Final}}$ are statistically indistinguishable except with probability $1/q$ (namely, the case (r_5, r_6, r_7, r_8) defined in Eq. 3 is the zero vector). $\qquad\square$

Lemma 7. *For any adversary* $\mathcal{A}$, $\mathsf{Adv}^{\mathsf{Game}_{\mathsf{Final}}}_{\mathcal{A}}(\lambda) = 0$.

Proof. The value of β is independent from the adversary's view in $\mathsf{Game}_{\mathsf{Final}}$. Hence, $\mathsf{Adv}^{\mathsf{Game}_{\mathsf{Final}}}_{\mathcal{A}}(\lambda) = 0$. $\qquad\square$

In $\mathsf{Game}_{\mathsf{Final}}$, the challenge ciphertext is a semi-functional encryption of a random message in G_T and under two random identities in $\mathbb{Z}_q$, independent of the two messages and the challenge identities provided by $\mathcal{A}$. Thus, our IB-ME scheme satisfies the privacy property defined in Def. 8 under the SXDH assumption.

Theorem 2. *The proposed IB-ME scheme satisfies authenticity under the Symmetric External Diffie-Hellman assumption. More precisely, for any PPT adversary* $\mathcal{A}$ *break the authenticity of our IB-ME scheme, its advantage* $\mathsf{Adv}^{\mathsf{Game}_{\mathsf{ib-auth}}}_{\mathcal{A}}(\lambda)$ *is negligible.*

Proof. The authenticity property intuitively says that if an adversary does not hold the corresponding encryption key ek_σ produced by the challenger, it cannot compute a valid ciphertext under the identity σ. Thus, it is corresponding to the unforgeability of signature, and we can directly reduce the authenticity to the security of the IBE system.

Assume that there is a PPT adversary $\mathcal{A}$ which breaks the authenticity property with advantage ϵ, we then employ it to build another PPT algorithm $\mathcal{B}$ to break a fully secure IBE system which consists of the following algorithms:

- IBE.Setup(1^λ) : The same as the Setup algorithm, except that the master public key is $mpk := \{\mathbb{G}; \alpha, g_T^\eta, g_T^\alpha, g_1^{\mathbf{d_1}}, g_1^{\mathbf{d_2}}, g_2^{\mathbf{d_1^*}}, g_2^{\mathbf{d_2^*}}, g_2^{\mathbf{d_3^*}}, g_2^{\mathbf{d_4^*}}\}$, and the master secret key is $msk := \{\eta, g_1^{\mathbf{d_3}}, g_1^{\mathbf{d_4}}\}$.
- IBE.KeyGen(msk, σ) : The same as the SKGen algorithm, and the secret key is $sk_\sigma := g_1^{\eta \mathbf{d_3} + r(\sigma \mathbf{d_3} - \mathbf{d_4})}$.
- IBE.Enc(mpk, σ, m) : Similar to the RKGen algorithm, and the ciphertext is $ct := \{\mathsf{C} = m \cdot (g_T^\eta)^s, \mathsf{C_0} = g_2^{s(\mathbf{d_3^*} + \sigma \mathbf{d_4^*})}\}$.
- IBE.Dec(mpk, sk_σ, ct) : Compute the message as $m := \mathsf{C}/e(\mathsf{C_0}, sk_\sigma)$.

Oracles $\mathsf{O_1}, \mathsf{O_2}$ are implemented by $\mathsf{SKGen}(\mathsf{mpk}, \mathsf{msk}, \cdot)$ and $\mathsf{RKGen}(\mathsf{mpk}, \mathsf{msk}, \cdot)$ and are simulated by $\mathcal{B}$ as follows:

1. $\mathsf{SKGen}(\mathsf{mpk}, \mathsf{msk}, \cdot)$: $\mathcal{A}$ launches a query for identity σ to $\mathsf{O_1}$, then $\mathcal{B}$ transfers this identity σ to the IBE system for generating secret key. It uses the IBE.KeyGen algorithm's output to answer this query and returns the secret key sk_σ to $\mathcal{B}$. Finally, $\mathcal{B}$ uses this secret key sk_σ from IBE as ek_σ to answer $\mathcal{A}$'s query for the encryption key.
2. $\mathsf{RKGen}(\mathsf{mpk}, \mathsf{msk}, \cdot)$: $\mathcal{A}$ launches a query for identity ρ to $\mathsf{O_2}$, then $\mathcal{B}$ transfers this identity ρ to the IBE system. It uses mpk (in IBE) to generate the corresponding keys, randomly picks $s, s_1, s_2 \xleftarrow{R} \mathbb{Z}_q$, computes

$$k_1 = g_2^{\alpha \mathbf{d_1^*} + s_1(\rho \mathbf{d_1^*} - \mathbf{d_2^*}) + s\mathbf{d_3^*}}, k_2 = g_2^{s_2(\rho \mathbf{d_1^*} - \mathbf{d_2^*}) + s\mathbf{d_4^*}}, k_3 = (g_T^\eta)^s,$$

and returns these keys to $\mathcal{B}$. Finally, $\mathcal{B}$ uses $dk_\rho = \{k_1, k_2, k_3\}$ to answer $\mathcal{A}$'s query for the decryption key.

Suoopse that $\mathsf{Adv}_{\mathcal{A}}^{\mathsf{Game_{ib\text{-}auth}}}(\lambda) = \epsilon$, where ϵ is a non-negligible value. Then we can build an algorithm $\mathcal{B}$ whose $\mathsf{Adv}_{\mathcal{B}}^{\mathsf{IBE}}(\lambda) = \epsilon$ as follow:

Upon $\mathcal{A}$ making a query of (σ, ρ), $\mathcal{B}$ generates the encryption key and decryption key to answer this query by simulating $\mathsf{O_1}, \mathsf{O_2}$, and sends (ek_σ, dk_ρ) back to $\mathcal{A}$. Then $\mathcal{A}$ can find another $\sigma^* \neq \mathsf{snd}$ with ϵ probability such that σ^* is also valid for decryption of ct, and sends σ^* to $\mathcal{B}$. Note that the fact snd and σ^* are both valid for $ct_{\sigma, \mathsf{rcv}}$ implies for a ciphertext associated with σ in the underlying IBE, there would be two different secret keys associated with snd and σ^* respectively. The sk_{σ^*} in IBE is identical to the ek_{σ^*} in IB-ME. Therefore, $\mathcal{B}$ can make secret key query for snd, and challenge (m_0, σ_0) and (m_1, σ^*). Then $\mathcal{B}$ can distinguish the challenge ciphertext easily by using the secret key associated with σ^* and break this IBE system.

This means that by this simulation, we have successfully reduced the authenticity of IB-ME to the security of this IBE system. And we have

$$\mathsf{Adv}_{\mathcal{A}}^{\mathsf{Game_{ib-auth}}}(\lambda) \leq \mathsf{Adv}_{\mathcal{B}}^{\mathsf{IBE}}(\lambda).$$

That is to say, if an adversary cannot successfully break the IBE system we constructed, it cannot forge a valid ciphertext in our IB-ME scheme either. We show that this IBE system is fully secure in the full version [11], i.e., the advantage of $\mathcal{B}$ winning the IBE game defined in Def. 5 is negligible. Thus for any PPT adversary $\mathcal{A}$, its advantage of breaking the authenticity property of our IB-ME scheme is negligible. ⊓

Note that we have challenged m and σ at the same time, but in fact, we do not need to challenge identity σ at all. That is to say the security of a trivial IBE is sufficient and anonymity is not required. Because another $\sigma^* \neq \mathsf{snd}$ can be obtained from $\mathcal{A}$ during the proof, and $\mathcal{B}$ sends identity snd and (m_0, m_1) to the challenger, the same result can be obtained.

5   Conclusion

In this paper, we propose the first identity-based matchmaking encryption scheme under the standard assumptions in the standard model. We construct our IB-ME scheme by a variant of two-level anonymous IBE, which is based on Okamoto and Takashima's dual pairing vector spaces, and its security reductions rely on Waters's dual system encryption under the SXDH assumption. Our directly constructed scheme does not rely on other cryptographic tools such as non-interactive zero-knowledge proof systems. Meanwhile, we leave several questions. First, although all parameters in our scheme have constant numbers of group elements, the size should be shorter and the number of pairing for decryption should be reduced to improve efficiency. Second, construct IB-ME schemes that satisfy the enhanced privacy [16] under standard assumptions. Third, practical extensions such as revocability and traceability are further works.

Acknowledgements. We are extremely grateful to the editors and anonymous reviewers for their insightful and constructive comments on this work. Jie Chen is supported by National Natural Science Foundation of China (61972156), NSFC-ISF Joint Scientific Research Program (61961146004), National Key Research and Development Program of China (2018YFA0704701) and Innovation Program of Shanghai Municipal Education Commission (2021-01-07-00-08-E00101). Jinming Wen is partially supported by National Natural Science Foundation of China (No.11871248), Guangdong Major Project of Basic and Applied Basic Research (2019B030302008), and Natural Science Foundation of Guangdong Province of China (2021A515010857, 2022A1515010029). Jian Weng is supported by Major Program of Guangdong Basic and Applied Research Project under Grant No.2019B030302008, National Natural Science Foundation of China under Grant No.61825203, Guangdong Provincial Science and Technology Project under Grant No.2021A0505030033, National Joint Engineering Research Center of Network Security Detection and Protection Technology, and Guangdong Key Laboratory of Data Security and Privacy Preserving.

References

1. Abdalla, M., et al.: Searchable encryption revisited: consistency properties, relation to anonymous IBE, and extensions. J. Cryptol. **21**(3), 350–391 (2008). https://doi.org/10.1007/s00145-007-9006-6
2. Agrawal, S., Boneh, D., Boyen, X.: Efficient lattice (H)IBE in the standard model. In: Gilbert, H. (ed.) EUROCRYPT 2010. LNCS, vol. 6110, pp. 553–572. Springer, Heidelberg (2010). https://doi.org/10.1007/978-3-642-13190-5_28
3. Ateniese, G., Francati, D., Nuñez, D., Venturi, D.: Match me if you can: matchmaking encryption and its applications. In: Boldyreva, A., Micciancio, D. (eds.) CRYPTO 2019. LNCS, vol. 11693, pp. 701–731. Springer, Cham (2019). https://doi.org/10.1007/978-3-030-26951-7_24
4. Attrapadung, N.: Dual system encryption via doubly selective security: framework, fully secure functional encryption for regular languages, and more. In: Nguyen, P.Q., Oswald, E. (eds.) EUROCRYPT 2014. LNCS, vol. 8441, pp. 557–577. Springer, Heidelberg (2014). https://doi.org/10.1007/978-3-642-55220-5_31
5. Boneh, D., Boyen, X.: Efficient selective-ID secure identity-based encryption without random oracles. In: Cachin, C., Camenisch, J.L. (eds.) EUROCRYPT 2004. LNCS, vol. 3027, pp. 223–238. Springer, Heidelberg (2004). https://doi.org/10.1007/978-3-540-24676-3_14
6. Boneh, D., Boyen, X.: Secure identity based encryption without random oracles. In: Franklin, M. (ed.) CRYPTO 2004. LNCS, vol. 3152, pp. 443–459. Springer, Heidelberg (2004). https://doi.org/10.1007/978-3-540-28628-8_27
7. Boneh, D., Franklin, M.: Identity-based encryption from the Weil pairing. In: Kilian, J. (ed.) CRYPTO 2001. LNCS, vol. 2139, pp. 213–229. Springer, Heidelberg (2001). https://doi.org/10.1007/3-540-44647-8_13
8. Boneh, D., Lynn, B., Shacham, H.: Short signatures from the Weil pairing. In: Boyd, C. (ed.) ASIACRYPT 2001. LNCS, vol. 2248, pp. 514–532. Springer, Heidelberg (2001). https://doi.org/10.1007/3-540-45682-1_30
9. Cash, D., Hofheinz, D., Kiltz, E., Peikert, C.: Bonsai trees, or how to delegate a lattice basis. In: Gilbert, H. (ed.) EUROCRYPT 2010. LNCS, vol. 6110, pp. 523–552. Springer, Heidelberg (2010). https://doi.org/10.1007/978-3-642-13190-5_27
10. Chen, J., Gay, R., Wee, H.: Improved Dual System ABE in Prime-Order Groups via Predicate Encodings. In: Oswald, E., Fischlin, M. (eds.) EUROCRYPT 2015. LNCS, vol. 9057, pp. 595–624. Springer, Heidelberg (2015). https://doi.org/10.1007/978-3-662-46803-6_20
11. Chen, J., Li, Y., Wen, J., Weng, J.: Identity-based matchmaking encryption from standard assumptions. Cryptology ePrint Archive, Paper 2022/1246 (2022). https://eprint.iacr.org/2022/12
12. Chen, J., Lim, H.W., Ling, S., Wang, H., Wee, H.: Shorter IBE and signatures via asymmetric pairings. In: Abdalla, M., Lange, T. (eds.) Pairing 2012. LNCS, vol. 7708, pp. 122–140. Springer, Heidelberg (2013). https://doi.org/10.1007/978-3-642-36334-4_8
13. Chen, J., Lim, H.W., Ling, S., Wang, H., Wee, H.: Shorter identity-based encryption via asymmetric pairings. Des. Codes Crypt. **73**(3), 911–947 (2014). https://doi.org/10.1007/s10623-013-9834-3

14. Chen, J., Wee, H.: Fully, (almost) tightly secure IBE and dual system groups. In: Canetti, R., Garay, J.A. (eds.) CRYPTO 2013. LNCS, vol. 8043, pp. 435–460. Springer, Heidelberg (2013). https://doi.org/10.1007/978-3-642-40084-1_25

15. Cocks, C.: An identity based encryption scheme based on quadratic residues. In: Honary, B. (ed.) Cryptography and Coding 2001. LNCS, vol. 2260, pp. 360–363. Springer, Heidelberg (2001). https://doi.org/10.1007/3-540-45325-3_32

16. Francati, D., Guidi, A., Russo, L., Venturi, D.: Identity-Based Matchmaking Encryption Without Random Oracles. In: Adhikari, A., Küsters, R., Preneel, B. (eds.) INDOCRYPT 2021. LNCS, vol. 13143, pp. 415–435. Springer, Cham (2021). https://doi.org/10.1007/978-3-030-92518-5_19

17. Gentry, C., Peikert, C., Vaikuntanathan, V.: Trapdoors for hard lattices and new cryptographic constructions. In: Dwork, C. (ed.) Proceedings of the 40th Annual ACM Symposium on Theory of Computing, Victoria, British Columbia. STOC 2008, Canada, May 17–20, 2008, pp. 197–206. ACM (2008). https://doi.org/10. 1145/1374376.1374407

18. Goldwasser, S., Micali, S., Rivest, R.L.: A digital signature scheme secure against adaptive chosen-message attacks. SIAM J. Comput. **17**(2), 281–308 (1988). https://doi.org/10.1137/0217017

19. Lewko, A.: Tools for simulating features of composite order bilinear groups in the prime order setting. In: Pointcheval, D., Johansson, T. (eds.) EUROCRYPT 2012. LNCS, vol. 7237, pp. 318–335. Springer, Heidelberg (2012). https://doi.org/10. 1007/978-3-642-29011-4_20

20. Lewko, A., Okamoto, T., Sahai, A., Takashima, K., Waters, B.: Fully secure functional encryption: attribute-based encryption and (hierarchical) inner product encryption. In: Gilbert, H. (ed.) EUROCRYPT 2010. LNCS, vol. 6110, pp. 62–91. Springer, Heidelberg (2010). https://doi.org/10.1007/978-3-642-13190-5_4

21. Lewko, A., Waters, B.: New techniques for dual system encryption and fully secure HIBE with short ciphertexts. In: Micciancio, D. (ed.) TCC 2010. LNCS, vol. 5978, pp. 455–479. Springer, Heidelberg (2010). https://doi.org/10.1007/978-3-642-11799-2_27

22. Lewko, A., Waters, B.: Decentralizing attribute-based encryption. In: Paterson, K.G. (ed.) EUROCRYPT 2011. LNCS, vol. 6632, pp. 568–588. Springer, Heidelberg (2011). https://doi.org/10.1007/978-3-642-20465-4_31

23. Lewko, A., Waters, B.: Unbounded HIBE and attribute-based encryption. In: Paterson, K.G. (ed.) EUROCRYPT 2011. LNCS, vol. 6632, pp. 547–567. Springer, Heidelberg (2011). https://doi.org/10.1007/978-3-642-20465-4_30

24. Lewko, A., Waters, B.: New proof methods for attribute-based encryption: achieving full security through selective techniques. In: Safavi-Naini, R., Canetti, R. (eds.) CRYPTO 2012. LNCS, vol. 7417, pp. 180–198. Springer, Heidelberg (2012). https://doi.org/10.1007/978-3-642-32009-5_12

25. Okamoto, T., Takashima, K.: Homomorphic encryption and signatures from vector decomposition. In: Galbraith, S.D., Paterson, K.G. (eds.) Pairing 2008. LNCS, vol. 5209, pp. 57–74. Springer, Heidelberg (2008). https://doi.org/10.1007/978-3-540-85538-5_4

26. Okamoto, T., Takashima, K.: Hierarchical predicate encryption for inner-products. In: Matsui, M. (ed.) ASIACRYPT 2009. LNCS, vol. 5912, pp. 214–231. Springer, Heidelberg (2009). https://doi.org/10.1007/978-3-642-10366-7_13

27. Okamoto, T., Takashima, K.: Fully secure functional encryption with general relations from the decisional linear assumption. In: Rabin, T. (ed.) CRYPTO 2010. LNCS, vol. 6223, pp. 191–208. Springer, Heidelberg (2010). https://doi.org/10. 1007/978-3-642-14623-7_11

28. Okamoto, T., Takashima, K.: Some key techniques on pairing vector spaces. In: Nitaj, A., Pointcheval, D. (eds.) AFRICACRYPT 2011. LNCS, vol. 6737, pp. 380–382. Springer, Heidelberg (2011). https://doi.org/10.1007/978-3-642-21969-6_25

29. Okamoto, T., Takashima, K.: Adaptively attribute-hiding (hierarchical) inner product encryption. In: Pointcheval, D., Johansson, T. (eds.) EUROCRYPT 2012. LNCS, vol. 7237, pp. 591–608. Springer, Heidelberg (2012). https://doi.org/10.1007/978-3-642-29011-4_35

30. Shamir, A.: Identity-based cryptosystems and signature schemes. In: Blakley, G.R., Chaum, D. (eds.) CRYPTO 1984. LNCS, vol. 196, pp. 47–53. Springer, Heidelberg (1985). https://doi.org/10.1007/3-540-39568-7_5

31. Waters, B.: Efficient identity-based encryption without random oracles. In: Cramer, R. (ed.) EUROCRYPT 2005. LNCS, vol. 3494, pp. 114–127. Springer, Heidelberg (2005). https://doi.org/10.1007/11426639_7

32. Waters, B.: Dual system encryption: realizing fully secure IBE and HIBE under simple assumptions. In: Halevi, S. (ed.) CRYPTO 2009. LNCS, vol. 5677, pp. 619–636. Springer, Heidelberg (2009). https://doi.org/10.1007/978-3-642-03356-8_36

33. Wee, H.: Dual system encryption via predicate encodings. In: Lindell, Y. (ed.) TCC 2014. LNCS, vol. 8349, pp. 616–637. Springer, Heidelberg (2014). https://doi.org/10.1007/978-3-642-54242-8_26

34. Xu, S., et al.: Match in my way: fine-grained bilateral access control for secure cloud-fog computing. IEEE Trans. Dependable Secure Comput. **19**(2), 1064–1077 (2022). https://doi.org/10.1109/TDSC.2020.3001557

35. Xu, S., Ning, J., Ma, J., Huang, X., Pang, H., Deng, R.H.: Expressive bilateral access control for internet-of-things in cloud-fog computing. In: Lobo, J., Pietro, R.D., Chowdhury, O., Hu, H. (eds.) SACMAT 2021, Spain, June 16–18, 2021, pp. 143–154. ACM (2021). https://doi.org/10.1145/3450569.3463561

Anonymous Public Key Encryption Under Corruptions

Zhengan Huang[1] ⓘ, Junzuo Lai[2(✉)] ⓘ, Shuai Han[3(✉)] ⓘ, Lin Lyu[4(✉)] ⓘ,
and Jian Weng[2] ⓘ

[1] Peng Cheng Laboratory, Shenzhen, China
[2] College of Information Science and Technology, Jinan University, Guangzhou, China
laijunzuo@gmail.com
[3] School of Electronic Information and Electrical Engineering, Shanghai Jiao Tong University, Shanghai, China
dalen17@sjtu.edu.cn
[4] Bergische Universität Wuppertal, Wuppertal, Germany
lin.lyu@uni-wuppertal.de

Abstract. Anonymity of public key encryption (PKE) requires that, in a multi-user scenario, the PKE ciphertexts do not leak information about which public keys are used to generate them. Corruptions are common threats in the multi-user scenario but anonymity of PKE under corruptions is less studied in the literature. In TCC 2020, Benhamouda et al. first provide a formal characterization for anonymity of PKE under a specific type of corruption. However, no known PKE scheme is proved to meet their characterization.

To the best of our knowledge, all the PKE application scenarios which require anonymity also require confidentiality. However, in the work by Benhamouda et al., different types of corruptions for anonymity and confidentiality are considered, which can cause security pitfalls. What's worse, we are not aware of any PKE scheme which can provide both anonymity and confidentiality under the same types of corruptions.

In this work, we introduce a new security notion for PKE called ANON-RSO$_k$&C security, capturing anonymity under corruptions. We also introduce SIM-RSO$_k$&C security which captures confidentiality under the same types of corruptions. We provide a generic framework of constructing PKE scheme which can achieve the above two security goals simultaneously based on a new primitive called key and message non-committing encryption (KM-NCE). Then we give a general construction of KM-NCE utilizing a variant of hash proof system (HPS) called Key-Openable HPS. We also provide Key-Openable HPS instantiations based on the matrix decisional Diffie-Hellman assumption. Therefore, we can obtain various concrete PKE instantiations achieving the two security goals in the standard model with *compact* ciphertexts. Furthermore, for some PKE instantiation, its security reduction is *tight*.

S. Agrawal and D. Lin (Eds.): ASIACRYPT 2022, LNCS 13793, pp. 423–453, 2022.
https://doi.org/10.1007/978-3-031-22969-5_15

1   Introduction

Anonymity of PKE Under Corruptions. The (single-user) IND-CCA security has been the de facto standard security for public-key encryption (PKE) schemes and is the target security of NIST PKE standardization for the next decades. It provides message confidentiality under CCA attacks. Meanwhile, *anonymity* is another security requirement for PKE and is not provided by the IND-CCA security. Roughly speaking, anonymity of PKE requires that, in a multi-user scenario, the PKE ciphertexts do not leak information about which public keys are used to generate them. The IK-CPA/CCA security given by Bellare et al. [2] is the first formalization of anonymity of PKE.

In such multi-user scenarios, multiple key pairs are generated, potentially correlated plaintexts are encrypted and sent to many receivers. Both the secret keys and the encrypted messages could be leaked due to accidents and/or adversarial attacks, which affects both the confidentiality and the anonymity of the PKE scheme. Researchers capture such threats by formalizing different types of *corruptions* in different multi-user scenarios. Many efforts have been made to establish confidentiality under corruptions and the study to selective-opening attacks are such examples.

However, anonymity of PKE under corruptions is much less studied. To the best of our knowledge, it is not considered until recently by Benhamouda et al. [5] in TCC 2020. They propose anonymity against selective-opening for PKE which is the first (and, to the best of our knowledge, also the only) formal definition of anonymity for PKE under corruptions. We will call this security as ANON-COR (anonymity under corruptions) security in this work. The ANON-COR security defined in [5] is as follows. Given n public keys of n users, an adversary submits t messages of its choice, and then receives t challenge ciphertexts, which are encryptions of the t messages under t distinct random public keys out of the n user public keys. Next, the adversary can adaptively corrupt $Q < n$ users one at a time, obtaining their secret keys. (We will call such kind of corruption as *post-challenge user corruption*.) ANON-COR security requires that no feasible adversary can corrupt more than $\frac{Q}{n} + \epsilon$ (for some constant $\epsilon > 0$) fraction of the ciphertext-encrypting keys with non-negligible probability.

Unfortunately, no known PKE scheme is proved to have ANON-COR security. Actually, Benhamouda et al. [5] only prove that their suggested PKE scheme achieves a simplified version of ANON-COR security (where the adversary is restricted to corrupt some users at once) and conjecture that it also achieves the ANON-COR security. They leave constructing an ANON-COR secure PKE scheme as an interesting problem.

Furthermore, we think the ANON-COR security is restricted in the following sense.

– Non-adaptive. The ANON-COR security is non-adaptive in the sense that the adversary is *not* allowed to obtain any user secret key *before* seeing the challenge ciphertexts. This restricts its application scenario since, in the real-world, some users may be fully controlled by the adversary from the very beginning and the adversary may corrupt other users at any time.

– Single-challenge. The ANON-COR security considers a single-challenge setting where each public key is used *only once* to encrypt a single challenge message. This restriction limits its application scenario since, in practice, each public key is often used multiple times (for example, the application scenario in [5][1]).

Thus, we raise the following research question.

> *Q1: For PKE schemes, can we provide an achievable security formalization which provides anonymity under more adaptive corruptions in the multi-challenge setting?*

Anonymity and Confidentiality Under the Same Types of Corruptions. We are not aware of any application scenario which only requires anonymity but not confidentiality of PKE schemes[2]. To the best of our knowledge, all the PKE application scenarios in the real world which require anonymity also require confidentiality. As an example, Benhamouda et al. [5] consider a blockchain application scenario which requires both of the two security guarantees. However, Benhamouda et al. capture the two security guarantees under *different* types of corruptions. More precisely, as shown in [5, Sect. 2.6], the scheme $\mathcal{E}_1$ requires both anonymity under post-challenge user corruption (ANON-COR security) and confidentiality under *the receiver selective opening (RSO)* corruption.

Although the ANON-COR security is called "anonymous against selective-opening" in [5], we want to note that the post-challenge user corruption considered in ANON-COR security is different from the RSO corruption considered for confidentiality. The RSO corruption [3,14] considers an adversary, after seeing many challenge ciphertexts for different receivers (together with their public keys), is able to open a subset of the challenge ciphertexts (via corrupting a subset of the receivers to obtain their secret keys and received messages). However, the ANON-COR adversary is not able to specify some challenge ciphertexts and open them.

When the two security guarantees (anonymity and confidentiality) are both required, it is more desirable to capture them under the *same* types of corruptions. Taking [5] as an example, where anonymity and confidentiality are both required for the scheme $\mathcal{E}_1$ in [5], it does not make sense for the adversary to attack anonymity *only* using the post-challenge user corruption and attack

---

[1] In the Committee-Selection phase of the evolving-committee proactive secret sharing scheme considered in [5], some users are selected as committee members. Each committee member will encrypt one fresh secret key using its long term public key ($\mathsf{ct} \leftarrow \mathcal{E}_1.\mathsf{Enc}_{\mathsf{pk}}(\mathsf{esk})$). Since the same user may be selected as a committee member multiple times, the user's public key may be used multiple times to encrypt multiple messages.

[2] Actually, it does not make sense to *only* consider the anonymity of some PKE without considering its confidentiality. If confidentiality can be sacrificed, one can trivially achieve anonymity by assigning the identity map as the encryption and decryption algorithm, so that the ciphertext equals the message and is independent of any public key.

confidentiality *only* using the RSO corruption. Actually, there is no anonymity guarantee under the RSO corruption and no confidentiality guarantee under the post-challenge user corruption. This implies that, when the adversary is able to use both post-challenge user corruption and RSO corruption, it is possible that neither anonymity nor confidentiality holds for the PKE scheme. Consequently, when the two security guarantees are required under corruptions, they should be captured under the same types of corruptions.

Unfortunately, we are not aware of any PKE schemes which can provide the two security guarantees under the same types of corruptions. Thus, we raise our second research question.

> *Q2: Can we construct a PKE scheme which provides both anonymity and confidentiality under the same types of corruptions?*

We answer the above two research questions affirmatively in this work.

Our Contributions. In this work:

- We formalize the notion of <u>ANON</u>ymity under <u>R</u>eceiver <u>S</u>elective <u>O</u>pening attacks (in the <u>k</u>-challenge setting), adaptive user <u>C</u>orruptions and <u>C</u>hosen <u>P</u>laintext / <u>C</u>iphertext <u>A</u>ttacks, which we call ANON-RSO$_k$&C-CPA/CCA security for short. To capture confidentiality under the same types of corruptions, we also formalize the notion of SIM-RSO$_k$&C-CPA/CCA security.
- We provide a generic framework of constructing PKE schemes, achieving both ANON-RSO$_k$&C-CCA security and SIM-RSO$_k$&C-CCA security (we denote them as AC-RSO$_k$&C-CCA security for simplicity), based on a new primitive called *key and message non-committing encryption* (KM-NCE).
- We give a general construction of KM-NCE utilizing a variant of hash proof system (HPS) [9] which we call *Key-Openable HPS*.
- Finally, we provide Key-Openable HPS instantiations from the matrix decisional Diffie-Hellman (MDDH) assumption [10].

When plugging the HPS instantiations into the general construction framework, we can obtain an AC-RSO$_k$&C-CCA secure PKE scheme in the standard model which provides anonymity and confidentiality simultaneously under both adaptive user corruptions and RSO corruptions. Moreover, our scheme enjoys the properties that 1) the ciphertext is *compact* (i.e., ciphertext overhead[3] is the size of a constant number of group elements [15], or more generally, is independent of the message length [12]), and 2) the security reduction is *tight*[4]. To the best of our knowledge, our scheme is the first PKE scheme achieving anonymity under adaptive corruptions (which is stronger than the ANON-COR security), thus solving the problem raised by Benhamouda et al. [5] in TCC 2020. Also, our scheme is the first PKE scheme achieving RSO$_k$-CCA security in the standard model with compact ciphertexts and tight security.

---

[3] Ciphertext overhead means the ciphertext bitlength minus plaintext bitlength [15].

[4] Tight reduction means that the security loss of the reduction is independent of the number of users, the number of challenges and the number of queries raised by the adversary.

AC-RSO$_k$ & C security derived from KM-NCE. We take the approach of non-committing encryption [7,8,12] to achieve AC-RSO$_k$&C security. We introduce a new primitive called *key and message non-committing encryption* (KM-NCE), which is some kind of "message & public key-non-committing" encryption. Informally, KM-NCE allows one to generate fake ciphertexts via a fake encryption algorithm, and enables one to open k fake ciphertexts to any k messages under any public key (by showing an appropriate secret key) via an opening algorithm.

We formalize two security properties for KM-NCE. One is a single-user and k-challenge security notion called KMNC$_k$-CPA/CCA security (c.f., Definition 4), and the other is robustness (c.f., Definition 5). Intuitively, KMNC$_k$-CPA/CCA security requires that the real secret key together with k real ciphertexts (encrypting k messages chosen by the adversary) should be computationally indistinguishable from the opened secret key and k fake ciphertexts.

KM-NCE serves as our core technical tool, and we show that KMNC$_k$-CPA/CCA secure and robust KM-NCE implies AC-RSO$_k$&C-CPA/CCA secure PKE. Due to the relative simplicity of KMNC$_k$-CPA/CCA security (single-user, no simulator) in comparison to AC-RSO$_k$&C-CPA/CCA security (multi-user, simulation-based), it is easier and conceptually simpler to construct KM-NCE and prove its security first than constructing AC-RSO$_k$&C-CPA/CCA secure PKE directly.

Generic Construction of KM-NCE. To construct KM-NCE, we propose a new building block called *Key-Openable HPS*, by equipping Hash Proof System (HPS) [9] with a hashing key opening algorithm HOpen$_k$. Informally, given k instances, k hash values and the random coins used to sample them, a projection key (public key of HPS), and a corresponding hashing key (secret key of HPS) as the input, HOpen$_k$ can output another hashing key such that 1) the outputted hashing key corresponds to the same projection key and 2) the given k hash values are exactly hash values of the k instances under the outputted hashing key. We also define some new properties for the key-openable HPS, including *openability$_k$* (c.f., Definition 9) and *universality$_{k+1}$* (c.f., Definition 10). By using key-openable HPS as an essential building block, we present a generic construction of KMNC$_k$-CCA secure KM-NCE.

Instantiations. For concrete instantiations, we provide key-openable HPS instantiations based on the MDDH assumption. Due to the good versatility of the MDDH assumption, we can obtain various concrete instantiations of KM-NCE. Plugging the concrete instantiations into our general framework, we obtain AC-RSO$_k$&C-CCA secure PKE schemes with *compact* ciphertexts in the standard model. For some concrete PKE instantiation, we can even *tightly* prove its AC-RSO$_k$&C-CCA security.

Related Works. The anonymity of PKE is first formalized by Bellare et al. [2] and they call it "key-privacy". Many follow up works continue research in this direction, such as [1,13,21]. Anonymity for PKE under corruptions is firstly considered by Benhamouda et al. [5].

The IND-CCA security in the multi-user setting with adaptive user corruptions except challenge is given in [4,20]. Lee et al. [20] propose the first PKE scheme in the random oracle model with tight IND-CCA security reduction in the multi-user setting with adaptive user corruptions except challenge.

In the research area of receiver selective opening (RSO) corruption for PKE, Bellare et al. [3] point out that IND-CPA security does not imply SIM-RSO-CPA security. Hazay et al. [14] show that RSO security can be achieved from variants of non-committing encryption. Subsequent works [12, 16, 18, 19] consider CCA security in the RSO setting and provide PKE schemes with RSO-CCA security. Yang et al. [23] consider RSO-CCA security in the multi-challenge setting. SIM-RSO-CCA secure PKE schemes with compact ciphertexts are proposed by Hara et al. [12] and Huang et al. [16].

2   Preliminaries

We assume that the security parameter λ is an (implicit) input to all algorithms. For any positive integer n, we use $[n]$ to denote the set $\{1, \cdots, n\}$. For a finite set $\mathcal{S}$, we use $|\mathcal{S}|$ to denote the size of $\mathcal{S}$. For random variables $\mathcal{X}$ and $\mathcal{Y}$ over a finite set $\mathcal{S}$, their statistical distance is $\Delta(\mathcal{X}, \mathcal{Y}) := \frac{1}{2} \sum_{s \in \mathcal{S}} |\Pr[\mathcal{X} = s] - \Pr[\mathcal{Y} = s]|$.

We recall the formal definitions of PKE, collision-resistant hash functions and universal hash functions together with the leftover hash lemma in the full version [17].

3   Anonymity and Confidentiality Under Corruptions

In this section, we firstly introduce the notion of Anonymity under Receiver Selective Opening attacks (in the multi-challenge setting), adaptive user Corruptions and Chosen Plaintext/Ciphertext Attacks, which we call ANON-RSO$_k$&C-CPA/CCA security ($k \in \mathbb{N}$). Then, we introduce the notion of SIM-RSO$_k$&C-CPA/CCA security ($k \in \mathbb{N}$), to capture confidentiality under the same types of corruptions. Finally, we also introduce the notion of AC-RSO$_k$&C-CPA/CCA security, to capture ANON-RSO$_k$&C-CPA/CCA security and SIM-RSO$_k$&C-CPA/CCA security in one notion for convenience.

3.1   Anonymity Under Corruptions

ANON-RSO$_k$& C Security. We formalize a simulation-based anonymity definition under receiver selective opening attacks and adaptive user corruptions, which we call ANON-RSO$_k$&C security ($k \in \mathbb{N}$).

Informally speaking, assume that there are n users, and that a PPT adversary is allowed to (i) adaptively corrupt the users (i.e., obtaining their secret keys) at any time, and (ii) make receiver selective opening queries (i.e., obtaining the corresponding secret keys and the challenge messages) after seeing a challenge ciphertext vector of length $t < n$. ANON-RSO$_k$&C security requires that whatever the adversary (seeing the challenge ciphertext vector) deduces about which public keys are used to generate the challenge ciphertext vector, can also be deduced without seeing any challenge ciphertexts.

Formal definition is as follows.

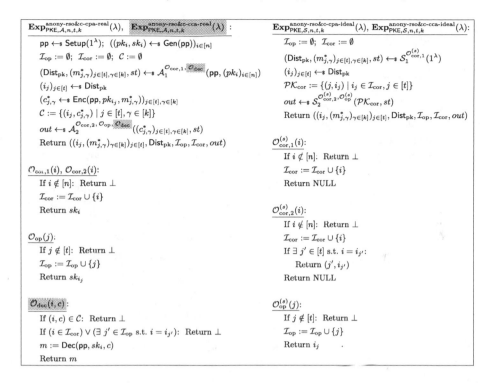

Fig. 1. Experiments for defining ANON-RSO$_k$&C-CPA/CCA security of scheme PKE.

Definition 1. (ANON-RSO$_k$&C-CPA/CCA). *A PKE scheme* PKE = (Setup, Gen, Enc, Dec) *is ANON-RSO$_k$ & C-ATK secure (where ATK $\in$ {CPA, CCA} and $k \in \mathbb{N}$ is a constant), if for any polynomially bounded n, t (where $0 < t \leq n$), and any PPT adversary $\mathcal{A} = (\mathcal{A}_1, \mathcal{A}_2)$, there is a PPT simulator $\mathcal{S} = (\mathcal{S}_1, \mathcal{S}_2)$, such that for any PPT distinguisher $\mathcal{D}$, the advantage* $\mathbf{Adv}_{\mathsf{PKE},\mathcal{A},\mathcal{S},\mathcal{D},n,t,k}^{\mathrm{anon\text{-}rso\&c\text{-}atk}}(\lambda) :=$

$$\left| \Pr[\mathcal{D}(\mathbf{Exp}_{\mathsf{PKE},\mathcal{A},n,t,k}^{\mathrm{anon\text{-}rso\&c\text{-}atk\text{-}real}}(\lambda)) = 1] - \Pr[\mathcal{D}(\mathbf{Exp}_{\mathsf{PKE},\mathcal{S},n,t,k}^{\mathrm{anon\text{-}rso\&c\text{-}atk\text{-}ideal}}(\lambda)) = 1] \right|$$

is negligible, where $\mathbf{Exp}_{\mathsf{PKE},\mathcal{A},n,t,k}^{\mathrm{anon\text{-}rso\&c\text{-}atk\text{-}real}}(\lambda)$ *and* $\mathbf{Exp}_{\mathsf{PKE},\mathcal{S},n,t,k}^{\mathrm{anon\text{-}rso\&c\text{-}atk\text{-}ideal}}(\lambda)$ *are defined in Fig. 1, and atk $\in$ {cpa, cca}. In both of the experiments, we require that for all* Dist$_{\mathrm{pk}}$ *output by $\mathcal{A}_1$ and $\mathcal{S}_1$, it holds that (1)* Dist$_{\mathrm{pk}}$ *is efficiently samplable, and (2) for all* $(i_j)_{j\in[t]} \twoheadleftarrow$ Dist$_{\mathrm{pk}}$, $i_{j_1} \neq i_{j_2}$ *for any distinct $j_1, j_2 \in [t]$.*

Remark 1. Our security notion ANON-RSO$_k$&C-CPA/CCA grants the adversary multiple, adaptive opening queries (i.e., $\mathcal{O}_{\mathrm{op}}$), like [6].

Remark 2. In $\mathbf{Exp}_{\mathsf{PKE},\mathcal{A},n,t,k}^{\mathrm{anon\text{-}rso\&c\text{-}atk\text{-}real}}(\lambda)$ where atk $\in$ {cpa, cca}, there are totally n public keys $(pk_i)_{i\in[n]}$, and only t of them (i.e., $(pk_{i_j})_{j\in[t]}$) are used to generate the challenge ciphertexts $(c_{j,\gamma}^*)_{j\in[t],\gamma\in[k]}$. Note that (i) by querying the opening oracle $\mathcal{O}_{\mathrm{op}}$ on $j \in [t]$ directly, $\mathcal{A}$ can obtain sk_{i_j} corresponding to some specified $(c_{j,\gamma}^*)_{\gamma\in[k]}$; (ii) by querying the corruption oracle $\mathcal{O}_{\mathrm{cor},1}$ or $\mathcal{O}_{\mathrm{cor},2}$, $\mathcal{A}$ can obtain some corresponding secret keys of the n public keys, but cannot ask for the secret key corresponding to some specified $c^*_{j,\gamma}$ since it may not know the value of i_j.

ANON-RSO$_k$&C-CPA $\Rightarrow$ ANON-COR. We show that ANON-RSO$_k$&C-CPA security implies the ANON-COR security [5].

Informally, the experiment for defining ANON-COR security is as follows. At the beginning, the challenger generates n public keys $(pk_i)_{i\in[n]}$, and sends them to an adversary $\mathcal{A}$. After receiving t ($t < n$) messages from $\mathcal{A}$, the challenger randomly samples t distinct public keys from $(pk_i)_{i\in[n]}$, uses them to encrypt the t messages respectively, and sends the t ciphertexts back to $\mathcal{A}$. Then, $\mathcal{A}$ can access to a corruption oracle adaptively, by querying it on any $i \in [n]$ and receiving sk_i as a response. Denote by Q the total number of corruption queries made by $\mathcal{A}$. ANON-COR security requires that for any $\epsilon > 0$ and any $\lambda < t, Q < n(1 - \epsilon)$, no PPT adversary $\mathcal{A}$ can compromise more than $\frac{Q}{n} + \epsilon$ fraction of the ciphertext-encrypting keys with non-negligible probability. Formal definition of ANON-COR security is given in the full version [17].

Note that any ANON-COR adversary can be seen as an ANON-RSO$_k$&C-CPA adversary $\mathcal{A}$ which (i) ignores $(c^*_{j,\gamma})_{2\leq\gamma\leq k}$ for all $j \in [t]$ if $k > 1$, (ii) does not query $\mathcal{O}_{\text{cor},1}$ or $\mathcal{O}_{\text{op}}$, (iii) queries $\mathcal{O}_{\text{cor},2}$ Q times, and (iv) the output distribution Dist_{pk} always samples t distinct indexes $i_1, \cdots, i_t$ uniformly random from $[n]$. The fraction of the ciphertext-encrypting keys that ANON-COR adversary compromises over $(pk_{i_j})_{j\in[t]}$ can be computed directly from experiment $\mathbf{Exp}^{\text{anon-rso\&c-cpa-real}}_{\text{PKE},\mathcal{A},n,t,k}(\lambda)$. ANON-RSO$_k$&C-CPA security guarantees that there is a simulator $\mathcal{S}$ such that $\mathbf{Exp}^{\text{anon-rso\&c-cpa-ideal}}_{\text{PKE},\mathcal{S},n,t,k}(\lambda)$ and $\mathbf{Exp}^{\text{anon-rso\&c-cpa-real}}_{\text{PKE},\mathcal{A},n,t,k}(\lambda)$ are indistinguishable. Note that in $\mathbf{Exp}^{\text{anon-rso\&c-cpa-ideal}}_{\text{PKE},\mathcal{S},n,t,k}(\lambda)$, $\mathcal{S}$ has no information about $(i_j)_{j\in[t]}$ except for the responses obtained via querying $\mathcal{O}^{(s)}_{\text{cor},1}, \mathcal{O}^{(s)}_{\text{cor},2}$. Hence, the fraction of the "ciphertext-encrypting" indexes that $\mathcal{S}$ compromises over $(i_j)_{j\in[t]}$ is nearly $\frac{Q}{n}$. Therefore, the indistinguishability between $\mathbf{Exp}^{\text{anon-rso\&c-cpa-ideal}}_{\text{PKE},\mathcal{S},n,t,k}(\lambda)$ and $\mathbf{Exp}^{\text{anon-rso\&c-cpa-real}}_{\text{PKE},\mathcal{A},n,t,k}(\lambda)$ implies the advantage of the ANON-COR adversary is negligible.

3.2   Confidentiality Under Corruptions

SIM-RSO$_k$& C Security. In order to capture confidentiality under the same corruptions which are considered in ANON-RSO$_k$&C security, we introduce a new security notion, called SIM-RSO$_k$&C security. We stress that SIM-RSO$_k$&C security is similar to SIM-RSO$_k$ security [23], except that the SIM-RSO$_k$&C adversary is allowed to corrupt the receivers *at any time* (i.e., even before seeing the challenge ciphertexts).

Informally, assume that there are n users, and that a PPT adversary is allowed to (i) adaptively corrupt the users (i.e., obtaining their secret keys) at any time, and (ii) make receiver selective opening queries (i.e., obtaining the corresponding secret keys and the challenge messages) after seeing a challenge ciphertext vector of length n. SIM-RSO$_k$&C security requires that whatever the

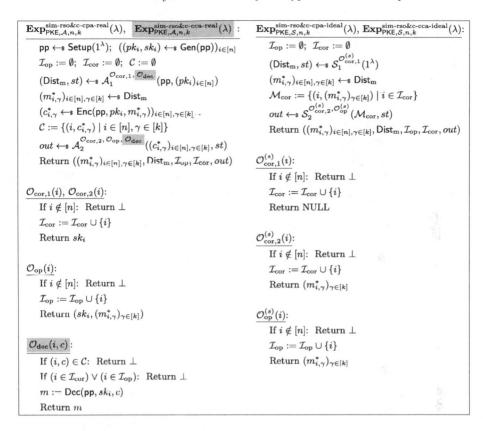

Fig. 2. Experiments for defining SIM-RSO$_k$&C-CPA/CCA security of scheme PKE.

adversary (seeing the challenge ciphertext vector) deduces about the challenge messages, can also be deduced without seeing any challenge ciphertexts.

Formal definition is as follows.

Definition 2. (SIM-RSO$_k$&C-CPA/CCA). *A PKE scheme* PKE = (Setup, Gen, Enc, Dec) *is SIM-RSO$_k$& C-ATK secure (where ATK $\in \{CPA, CCA\}$ and $k \in \mathbb{N}$ is a constant), if for any polynomially bounded $n > 0$, and any PPT adversary $\mathcal{A} = (\mathcal{A}_1, \mathcal{A}_2)$, there is a PPT simulator $\mathcal{S} = (\mathcal{S}_1, \mathcal{S}_2)$, such that for any PPT distinguisher $\mathcal{D}$, the advantage* $\mathbf{Adv}_{\mathsf{PKE},\mathcal{A},\mathcal{S},\mathcal{D},n,k}^{\mathrm{sim\text{-}rso\&c\text{-}atk}}(\lambda) :=$

$$\left| \Pr[\mathcal{D}(\mathbf{Exp}_{\mathsf{PKE},\mathcal{A},n,k}^{\mathrm{sim\text{-}rso\&c\text{-}atk\text{-}real}}(\lambda)) = 1] - \Pr[\mathcal{D}(\mathbf{Exp}_{\mathsf{PKE},\mathcal{S},n,k}^{\mathrm{sim\text{-}rso\&c\text{-}atk\text{-}ideal}}(\lambda)) = 1] \right|$$

is negligible, where $\mathbf{Exp}_{\mathsf{PKE},\mathcal{A},n,k}^{\mathrm{sim\text{-}rso\&c\text{-}atk\text{-}real}}(\lambda)$ *and* $\mathbf{Exp}_{\mathsf{PKE},\mathcal{S},n,k}^{\mathrm{sim\text{-}rso\&c\text{-}atk\text{-}ideal}}(\lambda)$ *are defined in Fig. 2, and atk $\in \{cpa, cca\}$. In both of the experiments, we require that for all $\mathsf{Dist_m}$ output by $\mathcal{A}_1$ and $\mathcal{S}_1$, $\mathsf{Dist_m}$ is efficiently samplable.*

SIM-RSO$_k$&C-ATK $\Rightarrow$ SIM-RSO$_k$-ATK. We claim that SIM-RSO$_k$&C-ATK security (ATK $\in$ {CPA, CCA}) implies simulation-based RSO security in the multi-challenge setting (i.e., SIM-RSO$_k$-ATK security) [23].

Generally, SIM-RSO$_k$-ATK security requires that for any PPT adversary $\mathcal{A}$ in the real experiment of SIM-RSO$_k$-ATK, there is a simulator $\mathcal{S}$, such that the final output of the ideal experiment and that of the real experiment are indistinguishable. Standard SIM-RSO-ATK security [12,14,16] is a special case of SIM-RSO$_k$-ATK security (i.e., $k = 1$). For completeness, formal definition of SIM-RSO$_k$-ATK security is given in the full version [17].

The reason that SIM-RSO$_k$&C-ATK security implies SIM-RSO$_k$-ATK security is as follows. Note that any SIM-RSO$_k$-ATK adversary $\mathcal{A}$ can be seen as a SIM-RSO$_k$&C-ATK adversary which does not query the corruption oracles $\mathcal{O}_{cor,1}, \mathcal{O}_{cor,2}$. SIM-RSO$_k$&C-ATK security guarantees the existence of a simulator $\mathcal{S}'$, such that the final output of the ideal experiment and that of the real experiment are indistinguishable. Hence, for the final output of the ideal experiment $((m_{i,\gamma}^*)_{i \in [n], \gamma \in [k]}, \mathsf{Dist}_m, \mathcal{I}_{op}, \mathcal{I}_{cor}, out)$, it also holds that $\mathcal{I}_{cor} = \emptyset$ (i.e., $\mathcal{S}'$ has never queried $\mathcal{O}_{cor,1}^{(s)}, \mathcal{O}_{cor,2}^{(s)}$). Hence, a SIM-RSO$_k$-ATK simulator $\mathcal{S}$ can be constructed from $\mathcal{S}'$.

3.3 Combining Anonymity and Confidentiality Under Corruptions

We introduce the notion of AC-RSO$_k$&C-CPA/CCA security, to capture ANON-RSO$_k$&C-CPA/CCA security and SIM-RSO$_k$&C-CPA/CCA security in one notion for convenience.

Informally, assume that there are n users, and that a PPT adversary is allowed to (i) adaptively corrupt the users (i.e., obtaining their secret keys) at any time, and (ii) make receiver selective opening queries (i.e., obtaining the corresponding secret keys and the challenge messages) after seeing a challenge ciphertext vector of length $t < n$. AC-RSO$_k$&C security requires that whatever the adversary (seeing the challenge ciphertext vector) deduces about which public keys or messages are used to generate the challenge ciphertext vector, can also be deduced without seeing any challenge ciphertexts.

Formal definition is as follows.

Definition 3. (AC-RSO$_k$&C-CPA/CCA). *A PKE scheme* PKE = (Setup, Gen, Enc, Dec) *is AC-RSO$_k$& C-ATK secure (where ATK $\in$ {CPA, CCA} and $k \in \mathbb{N}$ is a constant), if for any polynomially bounded n, t (where $0 < t \leq n$), and any PPT adversary $\mathcal{A} = (\mathcal{A}_1, \mathcal{A}_2)$, there is a PPT simulator $\mathcal{S} = (\mathcal{S}_1, \mathcal{S}_2)$, such that for any PPT distinguisher $\mathcal{D}$, the advantage* $\mathbf{Adv}_{\mathsf{PKE}, \mathcal{A}, \mathcal{S}, \mathcal{D}, n, t, k}^{\text{ac-rso\&c-atk}}(\lambda) :=$

$$\left| \Pr[\mathcal{D}(\mathbf{Exp}_{\mathsf{PKE}, \mathcal{A}, n, t, k}^{\text{ac-rso\&c-atk-real}}(\lambda)) = 1] - \Pr[\mathcal{D}(\mathbf{Exp}_{\mathsf{PKE}, \mathcal{S}, n, t, k}^{\text{ac-rso\&c-atk-ideal}}(\lambda)) = 1] \right|$$

is negligible, where $\mathbf{Exp}_{\mathsf{PKE}, \mathcal{A}, n, t, k}^{\text{ac-rso\&c-atk-real}}(\lambda)$ *and* $\mathbf{Exp}_{\mathsf{PKE}, \mathcal{S}, n, t, k}^{\text{ac-rso\&c-atk-ideal}}(\lambda)$ *are defined in Fig. 3, and atk $\in$ {cpa, cca}. In both of the experiments, we require that for all* Dist *output by* $\mathcal{A}_1$ *and* $\mathcal{S}_1$, *it holds that (1)* Dist *is efficiently samplable, and (2) for all* $(i_j, (m_{j,\gamma}^*)_{\gamma \in [k]})_{j \in [t]} \leftarrow_\$ $ Dist, $i_{j_1} \neq i_{j_2}$ *for any distinct* $j_1, j_2 \in [t]$.

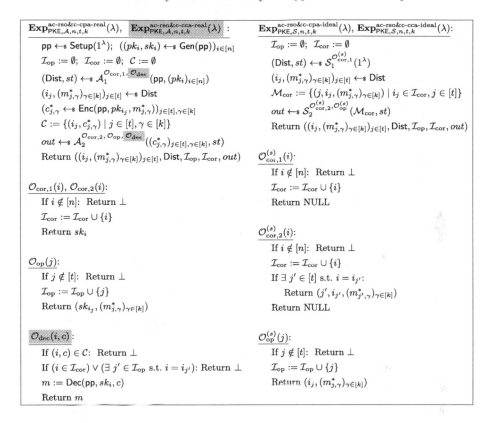

Fig. 3. Experiments for defining AC-RSO$_k$&C-CPA/CCA security of scheme PKE.

Note that, AC-RSO$_k$&C security can be easily simplified to guarantee only ANON-RSO$_k$&C security (when the adversary chooses a distribution Dist that has no entropy in the message part) and can also be simplified to guarantee only SIM-RSO$_k$&C security (by letting $n = t$).

4 AC-RSO$_k$&C Secure PKE from KM-NCE

In this section, we introduce a new primitive called key and message non-committing encryption (KM-NCE), and two security requirements, KMNC$_k$-CPA/CCA and robustness, for it. Then, we show that KMNC$_k$-CPA/CCA secure and robust KM-NCE implies AC-RSO$_k$&C-CPA/CCA secure PKE.

4.1 Key and Message Non-committing Encryption

Now we provide the definition of key and message non-committing encryption (KM-NCE) and security properties for this primitive. Informally, a KM-NCE scheme is a PKE scheme with the property that there is a way to generate fake ciphertexts without any public key, such that any k fake ciphertexts can be later opened to any k messages (by showing an appropriate secret key). This primitive is an extension of receiver non-committing encryption (RNCE) in [8, 12,14]. Generally speaking, the main differences between KM-NCE and RNCE are that (i) KM-NCE is defined in the k-challenge setting, for some constant k, and (ii) the algorithm, generating fake ciphertexts, of KM-NCE does not take any public key as input, while that of RNCE needs the public key.

For $k \in \mathbb{N}$, a key and message non-committing encryption scheme KM-NCE in the k-challenge setting, with a message space $\mathcal{M}$, consists of six PPT algorithms (Setup, Gen, Enc, Dec, Fake, Open$_k$).

- Setup: The setup algorithm, given a security parameter 1^λ, outputs a public parameter pp.
- Gen: The key generation algorithm, given pp, outputs a public key pk, a secret key sk and a trapdoor key tk.
- Enc: The encryption algorithm, given pp, pk and a message $m \in \mathcal{M}$, outputs a ciphertext c.
- Dec: The (deterministic) decryption algorithm, given pp, sk and c, outputs $m \in \mathcal{M} \cup \{\bot\}$.
- Fake: The fake encryption algorithm, given pp, outputs a fake ciphertext c' and a trapdoor td.
- Open$_k$: The opening algorithm, given (pp, tk, pk, sk), k fake ciphertexts $(c'_\gamma)_{\gamma \in [k]}$, k trapdoors $(td_\gamma)_{\gamma \in [k]}$ correponding to $(c'_\gamma)_{\gamma \in [k]}$, and k messages $(m_\gamma)_{\gamma \in [k]}$, outputs a secret key sk'.

For KM-NCE, standard correctness is required. Formally, we require that for any pp generated by Setup, any (pk, sk, tk) generated by Gen(pp) and any $m \in \mathcal{M}$, it holds that Dec(pp, sk, Enc(pp, pk, m)) = m.

Definition 4. (KMNC$_k$-CPA/CCA). *For $k \in \mathbb{N}$, a KM-NCE scheme* KM − NCE = (Setup, Gen, Enc, Dec, Fake, Open$_k$), *in the k-challenge setting, is* KMNC$_k$-ATK *secure (where ATK $\in \{CPA, CCA\}$), if for any PPT adversary* $\mathcal{A} = (\mathcal{A}_1, \mathcal{A}_2, \mathcal{A}_3)$, *the advantage* $\mathbf{Adv}^{kmnc-atk}_{KM-NCE, \mathcal{A}, k}(\lambda) :=$

$$\left| \Pr[\mathbf{Exp}^{kmnc-atk-real}_{KM-NCE, \mathcal{A}, k}(\lambda) = 1] - \Pr[\mathbf{Exp}^{kmnc-atk-sim}_{KM-NCE, \mathcal{A}, k}(\lambda) = 1] \right|$$

is negligible, where experiment $\mathbf{Exp}^{kmnc-atk-real}_{KM-NCE, \mathcal{A}, k}(\lambda)$ *and* $\mathbf{Exp}^{kmnc-atk-sim}_{KM-NCE, \mathcal{A}, k}(\lambda)$ *are defined in Fig. 4, and atk $\in \{cpa, cca\}$.*

We also define a statistical robustness for KM-NCE.

Definition 5 (Robustness). *A KM-NCE scheme* KM-NCE = (Setup, Gen, Enc, Dec, Fake, Open$_k$), *in the k-challenge setting ($k \in \mathbb{N}$), is robust, if the probability* $\epsilon^{rob}_{\text{KM-NCE}}(\lambda) :=$

$$\Pr \left[\begin{array}{c} \text{pp} \leftarrow_\$ \text{Setup}(1^\lambda), (pk, sk, tk) \leftarrow_\$ \text{Gen(pp)}, \\ (c, td) \leftarrow_\$ \text{Fake(pp)} \end{array} : \text{Dec(pp}, sk, c) \neq \bot \right]$$

is negligible.

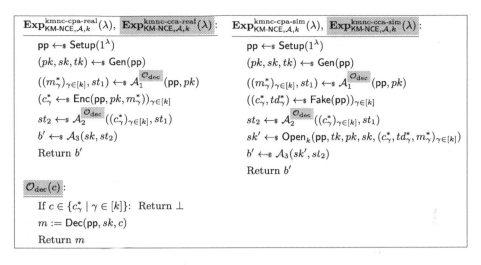

Fig. 4. Experiments for defining KMNC_k-CPA/CCA security of scheme KM-NCE.

4.2 Generic Construction of AC-RSO$_k$&C Secure PKE from KM-NCE

In this section, we show that for $k \in \mathbb{N}$, a KMNC_k-CPA (resp. KMNC_k-CCA) secure and robust KM-NCE scheme implies an AC-RSO$_k$&C-CPA (resp. AC-RSO$_k$&C-CCA) secure PKE scheme. Specifically, we have the following theorem.

Theorem 1. *If a KM-NCE scheme* KM-NCE $=$ (Setup, Gen, Enc, Dec, Fake, Open$_k$), *in the k-challenge setting ($k \in \mathbb{N}$), is KMNC_k-CPA (resp. KMNC_k-CCA) secure and robust, then* PKE $=$ (Setup, Gen, Enc, Dec) *is an AC-RSO$_k$&C-CPA (resp. AC-RSO$_k$&C-CCA) secure PKE scheme.*[5]

Proof of Theorem 1. We just prove that a KMNC_k-CCA secure and robust KM-NCE scheme implies an AC-RSO$_k$&C-CCA secure PKE scheme. The proof for the case of CPA is analogous and much easier, so we omit the details here.

Let n and t be arbitrary polynomials satisfying $0 < t \le n$. Let $\mathcal{A} = (\mathcal{A}_1, \mathcal{A}_2)$ be any PPT adversary attacking PKE $=$ (Setup, Gen, Enc, Dec) in the sense of AC-RSO$_k$&C-CCA, and $\mathcal{D}$ be any PPT distinguisher. Without loss of generality, we assume that $\mathcal{A}$ never repeats an oracle query. Specifically, we assume that if $\mathcal{A}_1$ has queried oracle $\mathcal{O}_{\mathrm{cor},1}$ on some i, then $\mathcal{A}_2$ will not query $\mathcal{O}_{\mathrm{cor},2}$ on i.

We proceed in a series of games.

Game G_{-1}: This is exactly the $\mathbf{Exp}_{\mathsf{PKE},\mathcal{A},n,t,k}^{\mathrm{ac\text{-}rso\&c\text{-}cca\text{-}real}}(\lambda)$ experiment, i.e., $\mathsf{G}_{-1} = \mathbf{Exp}_{\mathsf{PKE},\mathcal{A},n,t,k}^{\mathrm{ac\text{-}rso\&c\text{-}cca\text{-}real}}(\lambda)$.

---

[5] For PKE $=$ (Setup, Gen, Enc, Dec), we require that (i) the public parameter pp generated by Setup can be used for multiple users, and (ii) Gen does not output tk (i.e., the key generation algorithm of PKE firstly invokes the key generation algorithm of KM-NCE to generate (pk, sk, tk), and then outputs (pk, sk), ignoring tk).

More specifically, in G_{-1}, the challenger firstly generates $\mathsf{pp} \leftarrow_\$ \mathsf{Setup}(1^\lambda)$ and $((pk_i, sk_i, tk_i) \leftarrow_\$ \mathsf{Gen}(\mathsf{pp}))_{i \in [n]}$, and sends $(\mathsf{pp}, (pk_i)_{i \in [n]})$ to $\mathcal{A}_1$. The challenger initiates $\mathcal{I}_{\mathrm{op}} := \emptyset$ and $\mathcal{I}_{\mathrm{cor}} := \emptyset$, and keeps track of all $\mathcal{A}$'s issued queries to $\mathcal{O}_{\mathrm{cor},1}, \mathcal{O}_{\mathrm{cor},2}, \mathcal{O}_{\mathrm{op}}$ by maintaining these two sets. Then, the challenger answers $\mathcal{A}_1$'s $\mathcal{O}_{\mathrm{cor},1}, \mathcal{O}_{\mathrm{dec}}$ oracle queries with $(sk_i)_{i \in [n]}$. After receiving Dist, the challenger samples $(i_j, (m_{j,\gamma}^*)_{\gamma \in [k]})_{j \in [t]} \leftarrow \mathsf{Dist}$, computes $(c_{j,\gamma}^* \leftarrow_\$ \mathsf{Enc}(\mathsf{pp}, pk_{i_j}, m_{j,\gamma}^*))_{j \in [t], \gamma \in [k]})$, sets that $\mathcal{C} := \{(i_j, c_{j,\gamma}^*) \mid j \in [t], \gamma \in [k]\}$, and sends $(c_{j,\gamma}^*)_{j \in [t], \gamma \in [k]}$ to $\mathcal{A}_2$. Then, the challenger continues to answer $\mathcal{A}_2$'s $\mathcal{O}_{\mathrm{cor},2}, \mathcal{O}_{\mathrm{op}}, \mathcal{O}_{\mathrm{dec}}$ oracle queries with $(sk_i)_{i \in [n]}$. Finally, when $\mathcal{A}_2$ returns out, the challenger returns $((i_j, (m_{j,\gamma}^*)_{\gamma \in [k]})_{j \in [t]}, \mathsf{Dist}, \mathcal{I}_{\mathrm{op}}, \mathcal{I}_{\mathrm{cor}}, out)$ as its final output.

Game G_0: Game G_0 is the same as G_{-1}, except that two sets $\mathcal{I}_{\mathrm{op\text{-}sk}}$ and $\mathcal{I}_{\mathrm{cor\text{-}sk}}$ are introduced in G_0. Informally, $\mathcal{I}_{\mathrm{op\text{-}sk}}$ is introduced to ensure that if $\mathcal{A}_2$ submits a query $\mathcal{O}_{\mathrm{cor},2}(i)$ such that the secret key corresponding to pk_i has already been given to $\mathcal{A}$ via oracle $\mathcal{O}_{\mathrm{op}}$, then the challenger will directly return the secret key previously given to $\mathcal{A}_2$; $\mathcal{I}_{\mathrm{cor\text{-}sk}}$ is introduced to ensure that if $\mathcal{A}_2$ submits a query $\mathcal{O}_{\mathrm{op}}(j)$ such that the secret key corresponding to pk_{i_j} has already been exposed to $\mathcal{A}$ in a previous corruption query, then the challenger will directly return the secret key previously given to $\mathcal{A}_2$.

Specifically, the differences between G_0 and G_{-1} are as follows. The challenger additionally initiates $\mathcal{I}_{\mathrm{op\text{-}sk}} := \emptyset$ and $\mathcal{I}_{\mathrm{cor\text{-}sk}} := \emptyset$ at the beginning, and answers $\mathcal{A}$'s $\mathcal{O}_{\mathrm{cor},1}, \mathcal{O}_{\mathrm{cor},2}, \mathcal{O}_{\mathrm{op}}$ oracle queries as below:

- on a query $\mathcal{O}_{\mathrm{cor},1}(i)$ where $i \in [n]$, the challenger sets $\mathcal{I}_{\mathrm{cor}} := \mathcal{I}_{\mathrm{cor}} \cup \{i\}$ and $\mathcal{I}_{\mathrm{cor\text{-}sk}} := \mathcal{I}_{\mathrm{cor\text{-}sk}} \cup \{(i, sk_i)\}$, and returns sk_i to $\mathcal{A}_1$;
- on a query $\mathcal{O}_{\mathrm{cor},2}(i)$ where $i \in [n]$, the challenger firstly sets $\mathcal{I}_{\mathrm{cor}} := \mathcal{I}_{\mathrm{cor}} \cup \{i\}$. If there is some $j' \in \mathcal{I}_{\mathrm{op}}$ such that $i_{j'} = i$, then there must be some tuple $(j', i, \overline{sk_i}) \in \mathcal{I}_{\mathrm{op\text{-}sk}}$, and in this case the challenger sets $\mathcal{I}_{\mathrm{cor\text{-}sk}} := \mathcal{I}_{\mathrm{cor\text{-}sk}} \cup \{(i, \overline{sk_i})\}$, and returns $\overline{sk_i}$ to $\mathcal{A}_2$; otherwise, it sets $\mathcal{I}_{\mathrm{cor\text{-}sk}} := \mathcal{I}_{\mathrm{cor\text{-}sk}} \cup \{(i, sk_i)\}$, and returns sk_i to $\mathcal{A}_2$;
- on a query $\mathcal{O}_{\mathrm{op}}(j)$ where $j \in [t]$, the challenger firstly sets $\mathcal{I}_{\mathrm{op}} := \mathcal{I}_{\mathrm{op}} \cup \{j\}$. If $i_j \in \mathcal{I}_{\mathrm{cor}}$, there must be some tuple $(i_j, \overline{sk_{i_j}}) \in \mathcal{I}_{\mathrm{cor\text{-}sk}}$, and in this case the challenger sets $\mathcal{I}_{\mathrm{op\text{-}sk}} := \mathcal{I}_{\mathrm{op\text{-}sk}} \cup \{(j, i_j, \overline{sk_{i_j}})\}$, and returns $\overline{sk_{i_j}}$ to $\mathcal{A}_2$; otherwise, it sets $\mathcal{I}_{\mathrm{op\text{-}sk}} := \mathcal{I}_{\mathrm{op\text{-}sk}} \cup \{(j, i_j, sk_{i_j})\}$, and returns sk_{i_j} to $\mathcal{A}_2$.

Since all the secret keys $(sk_i)_{i \in [n]}$ are generated at the beginning and will not be updated during the proceedings of G_{-1}, the modifications introduced in game G_0 do not change $\mathcal{A}$'s view. Hence, $\Pr[\mathcal{D}(\mathsf{G}_0) = 1] = \Pr[\mathcal{D}(\mathsf{G}_{-1}) = 1]$.

Game $\mathsf{G}_{\widehat{i}}$ $(\widehat{i} \in [n])$: For all $\widehat{i} \in [n]$, $\mathsf{G}_{\widehat{i}}$ is the same as $\mathsf{G}_{\widehat{i}-1}$, except that

(1) when generating the challenge ciphertexts, if there is some $j' \in [t]$ such that $(i_{j'} \notin \mathcal{I}_{\mathrm{cor}}) \wedge (i_{j'} = \widehat{i})$, the challenger generates $(c_{j',\gamma}^*)_{\gamma \in [k]}$ with algorithm Fake instead of Enc, i.e., $((c_{j',\gamma}^*, td_{j',\gamma}^*) \leftarrow_\$ \mathsf{Fake}(\mathsf{pp}))_{\gamma \in [k]}$;

(2) for $\mathcal{A}_2$'s each $\mathcal{O}_{\mathrm{cor},2}$ oracle query i, if there is some $j' \in [t]$ satisfying $(j' \notin \mathcal{I}_{\mathrm{op}}) \wedge (i_{j'} = \widehat{i})$, the challenger returns $sk'_{i_{j'}} \leftarrow_\$ \mathsf{Open}_k(\mathsf{pp}, tk_{i_{j'}}, pk_{i_{j'}}, sk_{i_{j'}}, (c_{j',\gamma}^*, td_{j',\gamma}^*, m_{j',\gamma}^*)_{\gamma \in [k]})$ to $\mathcal{A}_2$; otherwise, it answers this query as in $\mathsf{G}_{\widehat{i}-1}$;

(3) for $\mathcal{A}_2$'s each $\mathcal{O}_{op}$ oracle query j, if the corresponding i_j satisfies $(i_j \notin \mathcal{I}_{cor}) \wedge (i_j = \widehat{i})$, the challenger returns $sk'_{i_j} \leftarrow_\$ \mathsf{Open}_k(pp, tk_{i_j}, pk_{i_j}, sk_{i_j}, (c^*_{j,\gamma}, td^*_{j,\gamma}, m^*_{j,\gamma})_{\gamma \in [k]})$ to $\mathcal{A}_2$; otherwise, it answers this query as in $\mathsf{G}_{\widehat{i}-1}$.

Game $\mathsf{G}_{n+\widehat{i}}$ $(\widehat{i} \in [n])$: For all $\widehat{i} \in [n]$, game $\mathsf{G}_{n+\widehat{i}}$ is the same as $\mathsf{G}_{n+\widehat{i}-1}$, except that for $\mathcal{A}_2$'s each $\mathcal{O}_{dec}$ oracle query (i, c), if $(\exists (i_j, c^*_{j,\gamma}) \in \mathcal{C}$ s.t. $i_j = \widehat{i} \wedge c^*_{j,\gamma} = c) \wedge (i \notin \mathcal{I}_{cor})$, the challenger returns $\bot$ to $\mathcal{A}_2$; otherwise, it answers this query as in game $\mathsf{G}_{n+\widehat{i}-1}$.

We present the following two lemmas whose proofs are given in the full version [17].

Lemma 1. *For each* $\widehat{i} \in [n]$, $|\Pr[\mathcal{D}(\mathsf{G}_{\widehat{i}}) = 1] - \Pr[\mathcal{D}(\mathsf{G}_{\widehat{i}-1}) = 1]| \leq \mathbf{Adv}^{kmnc-cca}_{KM-NCE,\mathcal{B},k}(\lambda)$ *for some PPT adversary* $\mathcal{B}$.

Lemma 2. *For each* $\widehat{i} \in [n]$, $|\Pr[\mathcal{D}(\mathsf{G}_{n+\widehat{i}}) = 1] - \Pr[\mathcal{D}(\mathsf{G}_{n+\widehat{i}-1}) = 1]| \leq t \cdot k \cdot \epsilon^{rob}_{KM-NCE}(\lambda)$.

Note that in game G_{2n}, (i) when generating the challenge ciphertexts, for each $j \in [t]$ such that $i_j \notin \mathcal{I}_{cor}$, the corresponding challenge ciphertexts $(c^*_{j,\gamma})_{\gamma \in [k]}$ are generated with algorithm Fake; (ii) any $\mathcal{O}_{cor,2}$ oracle query $i \in [n]$ such that $i = i_{j'}$ for some $j' \notin \mathcal{I}_{op}$ is answered with algorithm Open_k; (iii) any $\mathcal{O}_{op}$ oracle query $j \in [t]$ such that $i_j \notin \mathcal{I}_{cor}$ is answered with algorithms Open_k; (iv) any $\mathcal{O}_{dec}$ oracle query (i, c) is answered with $\bot$ if there is some $j \in [t]$ and $\gamma \in [k]$ such that $(i_j, c^*_{j,\gamma} = c) \in \mathcal{C}$ and $c^*_{j,\gamma}$ is generated with algorithm Fake. Now, a PPT simulator $\mathcal{S} = (\mathcal{S}_1, \mathcal{S}_2)$ can be constructed, which simulates G_{2n} perfectly for $\mathcal{A}$. Hence, we derive that

$$\mathbf{Exp}^{ac-rso\&c-cca-ideal}_{PKE,\mathcal{S},n,t,k}(\lambda) = \mathsf{G}_{2n}.$$

Due to space limitations, the detailed description of $\mathcal{S}$ will be given in the full version [17].

Therefore, $\mathbf{Adv}^{ac-rso\&c-cca}_{PKE,\mathcal{A},\mathcal{S},\mathcal{D},n,t,k}(\lambda) = |\Pr[\mathcal{D}(\mathsf{G}_{-1}) = 1] - \Pr[\mathcal{D}(\mathsf{G}_{2n}) = 1]|$

$$\leq n \cdot \mathbf{Adv}^{kmnc-cca}_{KM-NCE,\mathcal{B}',k}(\lambda) + n \cdot t \cdot k \cdot \epsilon^{rob}_{KM-NCE}(\lambda) \tag{1}$$

for some PPT adversary $\mathcal{B}'$. This completes the proof of Theorem 1. ∎

5   KM-NCE from Key-Openable Hash Proof System

In this section, we present a generic construction of KM-NCE that is needed in the AC-RSO$_k$&C secure PKE construction in Sect. 4.2. Our main building block is a new variant of Hash Proof System (HPS), called *Key-Openable HPS*. We firstly recall the definition of HPS from [9], and then formalize our new Key-Openable HPS. Next, we show how to construct KM-NCE from Key-Openable HPS. Jumping ahead, we will give concrete instantiations of Key-Openable HPS from the matrix decisional Diffie-Hellman assumption in Sect. 6.

5.1   Recall: Hash Proof System

In this subsection, we recall the formal definition of HPS according to [9]. For applications in constructing KM-NCE, we require that HPS has two parameter generation algorithms, a master parameter generation algorithm MPar and an (ordinary) parameter generation algorithm Par.

Definition 6 (Hash Proof System). *A hash proof system* HPS = (MPar, Par, Pub, Priv) *consists of a tuple of PPT algorithms:*

- *mpar ←$ MPar(1^λ): The master parameter generation algorithm outputs a master public parameter* mpar, *which implicitly defines the universe set* $\mathcal{X}$ *and the hash value space* Π.
  We assume that there are PPT algorithms for sampling $x ←$ $\mathcal{X}$ *uniformly and sampling* $\pi ←$ Π *uniformly. We require* mpar *to be an implicit input of other algorithms.*
- *par ←$ Par(mpar): The (ordinary) parameter generation algorithm takes* mpar *as input, and outputs an (ordinary) public parameter* par, *which implicitly defines* $(\mathcal{L}, \mathcal{SK}, \mathcal{PK}, \Lambda_{(\cdot)}, \alpha)$, *where* $\mathcal{L} \subseteq \mathcal{X}$ *is an NP-language,* $\mathcal{SK}$ *is the hashing key space,* $\mathcal{PK}$ *is the projection key space,* $\Lambda_{(\cdot)} : \mathcal{X} \longrightarrow \Pi$ *is a family of hash functions indexed by a hashing key* sk $\in \mathcal{SK}$, *and* $\alpha : \mathcal{SK} \longrightarrow \mathcal{PK}$ *is the projection function.*
  We assume that $\Lambda_{(\cdot)}$ *and* α *are efficiently computable and there are PPT algorithms for sampling* $x ←$ $\mathcal{L}$ *uniformly together with a witness* w, *and sampling* sk $←$ $\mathcal{SK}$ *uniformly. We require* par *to be an implicit input of other algorithms.*
- $\pi ←$ Pub(pk, x, w): *The public evaluation algorithm outputs the hash value* $\pi = \Lambda_{sk}(x) \in \Pi$ *of* $x \in \mathcal{L}$, *with the help of a projection key* pk $= \alpha(sk)$ *and a witness* w *for* $x \in \mathcal{L}$.
- $\pi ←$ Priv(sk, x): *The private evaluation algorithm outputs the hash value* $\pi = \Lambda_{sk}(x) \in \Pi$ *of* $x \in \mathcal{X}$, *directly using the hashing key* sk.

Perfect correctness *(a.k.a. projectiveness)* of HPS requires that, for all possible mpar ←$ MPar(1^λ) *and* par ←$ Par(mpar), *all hashing keys* sk $\in \mathcal{SK}$ *with* pk := $\alpha(sk)$ *the corresponding projection key, all* $x \in \mathcal{L}$ *with all possible witnesses* w, *it holds that* Pub(pk, x, w) = $\Lambda_{sk}(x)$ = Priv(sk, x).

HPS is associated with a subset membership problem (SMP), which asks whether an element is uniformly chosen from $\mathcal{L}$ or $\mathcal{X}$. SMP can be extended to multi-fold SMP by considering multiple elements.

Definition 7 (Multi-fold SMP). *The multi-fold SMP related to* HPS *is hard, if for any PPT adversary* $\mathcal{A}$ *and any polynomial* Q, *it holds that* $\mathbf{Adv}_{HPS,\mathcal{A}}^{Q\text{-}msmp}(\lambda) := \big| \Pr[\mathcal{A}(mpar, par, \{x_\gamma\}_{\gamma \in [Q]}) = 1] - \Pr[\mathcal{A}(mpar, par, \{x'_\gamma\}_{\gamma \in [Q]})$ = 1]$\big| \leq$ negl(λ), *where* mpar ←$ MPar(1^λ), par ←$ Par(mpar), $x_\gamma ←$ $\mathcal{L}$ *and* $x'_\gamma ←$ $\mathcal{X}$ *for each* $\gamma \in [Q]$.

Tag-based HPS. We recall a tag-based variant of HPS from [9,22], by allowing the hash functions $\Lambda_{(\cdot)}$ to have an additional element called label/tag as input. More precisely, in a tag-based HPS, the public parameter par also implicitly defines a tag space $\mathcal{T}$. Meanwhile, the hash functions $\Lambda_{(\cdot)}$, the public evaluation algorithm Pub and the private evaluation algorithm Priv also take a tag $\tau \in \mathcal{T}$ as input. Accordingly, perfect correctness requires $\mathsf{Pub}(\mathsf{pk}, x, w, \tau) = \Lambda_{\mathsf{sk}}(x, \tau) = \mathsf{Priv}(\mathsf{sk}, x, \tau)$ for all tags $\tau \in \mathcal{T}$.

5.2 Key-Openable HPS

We present the formal definition of our new *Key-Openable HPS*.

Definition 8 (Key-Openable Hash Proof System). *Let $k \in \mathbb{N}$. A key-openable hash proof system* HPS = $(\mathsf{MPar}, \mathsf{Par}, \mathsf{Pub}, \mathsf{Priv}, \mathsf{HOpen}_k)$ *consists of a tuple of PPT algorithms:*

- $(\mathsf{MPar}, \mathsf{Par}, \mathsf{Pub}, \mathsf{Priv})$ *is a hash proof system as per Definition 6. Recall that the master parameter* mpar *output by* $\mathsf{MPar}(1^\lambda)$ *implicitly defines* $(\mathcal{X}, \Pi)$, *and there are PPT algorithms for sampling* $x \leftarrow_\$ \mathcal{X}$ *uniformly and sampling* $\pi \leftarrow_\$ \Pi$ *uniformly. We denote by* $R_\mathcal{X}$ *and* R_Π *the randomness spaces for sampling* $x \leftarrow_\$ \mathcal{X}$ *and* $\pi \leftarrow_\$ \Pi$ *respectively.*
- *In addition to public parameter* par, $\mathsf{Par}(\mathsf{mpar})$ *also outputs a trapdoor information* td, *which will be later used by* HOpen_k.
- $\mathsf{sk}'/\bot \leftarrow_\$ \mathsf{HOpen}_k(\mathsf{td}, \mathsf{pk}, \mathsf{sk}, (x_\gamma, r_{x_\gamma}, \pi_\gamma, r_{\pi_\gamma})_{\gamma \in [k]})$: *The hashing key opening algorithm takes as input the trapdoor* td, *a projection key* $\mathsf{pk} \in \mathcal{PK}$, *a hashing key* $\mathsf{sk} \in \mathcal{SK}$ *satisfying* $\mathsf{pk} = \alpha(\mathsf{sk})$, *and k tuples* $(x_\gamma, r_{x_\gamma}, \pi_\gamma, r_{\pi_\gamma})_{\gamma \in [k]}$ *where* $x_\gamma \in \mathcal{X}$ *with sampling randomness* $r_{x_\gamma} \in R_\mathcal{X}$ *and* $\pi_\gamma \in \Pi$ *with sampling randomness* $r_{\pi_\gamma} \in R_\Pi$ *for each* $\gamma \in [k]$, *and outputs another hashing key* $\mathsf{sk}' \in \mathcal{SK}$ *satisfying* $\mathsf{pk} = \alpha(\mathsf{sk}')$ *and* $\pi_\gamma = \Lambda_{\mathsf{sk}'}(x_\gamma)$ *for each* $\gamma \in [k]$, *or a special symbol* $\bot$ *indicating the failure of opening.*

Tag-based Key-Openable HPS. A key-openable HPS = $(\mathsf{MPar}, \mathsf{Par}, \mathsf{Pub}, \mathsf{Priv}, \mathsf{HOpen}_k)$ is a tag-based key-openable HPS, if $(\mathsf{MPar}, \mathsf{Par}, \mathsf{Pub}, \mathsf{Priv})$ is a tag-based HPS (cf. Sect. 5.1), and HOpen_k also takes a set of tags $(\tau_\gamma)_{\gamma \in [k]}$ as input so that its output sk' satisfies $\mathsf{pk} = \alpha(\mathsf{sk}')$ and $\pi_\gamma = \Lambda_{\mathsf{sk}'}(x_\gamma, \tau_\gamma)$ for each $\gamma \in [k]$.

Below we define a new statistical property for (tag-based) key-openable HPS, called *openability$_k$*. It stipulates the statistical indistinguishability between $(\mathsf{sk}^{(0)}, (\pi_\gamma^{(0)})_{\gamma \in [k]})$ and $(\mathsf{sk}^{(1)}, (\pi_\gamma^{(1)})_{\gamma \in [k]})$, where $\mathsf{sk}^{(0)}$ is a uniformly sampled hashing key, $\pi_\gamma^{(0)} = \Lambda_{\mathsf{sk}_0}(x_\gamma)$ for $x_\gamma \leftarrow_\$ \mathcal{X}$ with randomness r_{x_γ}, $\pi_\gamma^{(1)}$ is uniformly sampled from Π with randomness $r_{\pi_\gamma^{(1)}}$, and $\mathsf{sk}^{(1)}$ is generated by $\mathsf{HOpen}_k(\mathsf{td}, \mathsf{pk}, \mathsf{sk}^{(0)}, (x_\gamma, r_{x_\gamma}, \pi_\gamma^{(1)}, r_{\pi_\gamma^{(1)}})_{\gamma \in [k]})$. Here the subscript k indicates the opening of hashing key w.r.t. k hash values. For tag-based key-openable HPS, the adversary can additionally determine the tags $(\tau_\gamma)_{\gamma \in [k]}$ w.r.t. which the hash values are computed. It is not hard to see that this property implies the usual smoothness property of HPS [9] and also implies that $\mathcal{L}$ is a sparse subset of $\mathcal{X}$.

$$\mathbf{Exp}_{\mathsf{HPS},\mathcal{A}}^{open_k}(\lambda):$$

> $\mathsf{mpar} \leftarrow_\$ \mathsf{MPar}(1^\lambda)$, $(\mathsf{par}, \mathsf{td}) \leftarrow_\$ \mathsf{Par}(\mathsf{mpar})$. $\mathsf{sk} \leftarrow_\$ \mathcal{SK}$, $\mathsf{pk} := \alpha(\mathsf{sk})$.
> For $\gamma \in [k]$, $x_\gamma \leftarrow_\$ \mathcal{X}$ with sampling randomness r_{x_γ}.
> $\boxed{(\tau_\gamma)_{\gamma \in [k]}} \leftarrow_\$ \mathcal{A}(\mathsf{mpar}, \mathsf{par}, \mathsf{td}, \mathsf{pk}, (x_\gamma, r_{x_\gamma})_{\gamma \in [k]})$.
> For $\gamma \in [k]$, $\pi_\gamma^{(0)} := \Lambda_{\mathsf{sk}}(x_\gamma, \boxed{\tau_\gamma})$.
> For $\gamma \in [k]$, $\pi_\gamma^{(1)} \leftarrow_\$ \Pi$ with sampling randomness $r_{\pi_\gamma^{(1)}}$.
> $\mathsf{sk}^{(0)} := \mathsf{sk}$. $\mathsf{sk}^{(1)} \leftarrow_\$ \mathsf{HOpen}_k(\mathsf{td}, \mathsf{pk}, \mathsf{sk}, (x_\gamma, r_{x_\gamma}, \pi_\gamma^{(1)}, r_{\pi_\gamma^{(1)}}, \boxed{\tau_\gamma})_{\gamma \in [k]})$.
> $b \leftarrow_\$ \{0, 1\}$.
> $b' \leftarrow_\$ \mathcal{A}(\mathsf{mpar}, \mathsf{par}, \mathsf{td}, \mathsf{pk}, (x_\gamma, r_{x_\gamma}, \pi_\gamma^{(b)})_{\gamma \in [k]}, \mathsf{sk}^{(b)})$.
> If $b' = b$: Return 1; Else: Return 0.

Fig. 5. Experiment for defining the Openability$_k$ property of (tag-based) key-openable HPS, where the framed parts only appear in the experiment for tag-based HPS.

$$\mathbf{Exp}_{\mathsf{HPS},\mathcal{A}}^{univ_{k+1}}(\lambda):$$

> $\mathsf{mpar} \leftarrow_\$ \mathsf{MPar}(1^\lambda)$, $(\mathsf{par}, \mathsf{td}) \leftarrow_\$ \mathsf{Par}(\mathsf{mpar})$. $\mathsf{sk} \leftarrow_\$ \mathcal{SK}$, $\mathsf{pk} := \alpha(\mathsf{sk})$.
> For $\gamma \in [k]$, $x_\gamma \leftarrow_\$ \mathcal{X}$.
> $(\tau_\gamma)_{\gamma \in [k]} \leftarrow_\$ \mathcal{A}(\mathsf{mpar}, \mathsf{par}, \mathsf{pk}, (x_\gamma)_{\gamma \in [k]})$.
> For $\gamma \in [k]$, $\pi_\gamma := \Lambda_{\mathsf{sk}}(x_\gamma, \tau_\gamma)$.
> $(x, \tau, \pi) \leftarrow_\$ \mathcal{A}(\mathsf{mpar}, \mathsf{par}, \mathsf{pk}, (x_\gamma, \pi_\gamma)_{\gamma \in [k]})$.
> If $(x \in \mathcal{X} \setminus \mathcal{L}) \wedge (\tau \notin \{\tau_\gamma\}_{\gamma \in [k]}) \wedge (\pi = \Lambda_{\mathsf{sk}}(x, \tau))$: Return 1; Else: Return 0.

Fig. 6. Experiment for the Universal$_{k+1}$ property of tag-based key-openable HPS.

Definition 9 (Openability$_k$). *A (tag-based) key-openable HPS is openable$_k$, if for any (unbounded) adversary $\mathcal{A}$, it holds that $\epsilon_{\mathsf{HPS},\mathcal{A}}^{open_k}(\lambda) := | \Pr[\mathbf{Exp}_{\mathsf{HPS},\mathcal{A}}^{open_k}(\lambda) = 1] - 1/2| \leq \mathsf{negl}(\lambda)$, where $\mathbf{Exp}_{\mathsf{HPS},\mathcal{A}}^{open_k}(\lambda)$ is defined in Fig. 5.*

Next we define a statistical property for tag-based HPS, called *universal$_{k+1}$*, which is an extension of the universal$_2$ property proposed in [9].

Definition 10 (Universal$_{k+1}$). *A tag-based key-openable HPS is universal $_{k+1}$, if for any (unbounded) adversary $\mathcal{A}$, it holds that $\epsilon_{\mathsf{HPS},\mathcal{A}}^{univ_{k+1}}(\lambda) := \Pr[\mathbf{Exp}_{\mathsf{HPS},\mathcal{A}}^{univ_{k+1}}(\lambda) = 1] \leq \mathsf{negl}(\lambda)$, where $\mathbf{Exp}_{\mathsf{HPS},\mathcal{A}}^{univ_{k+1}}(\lambda)$ is defined in Fig. 6.*

Finally, we define a statistical property, called *efficient randomness resampling on Π*, which demands that besides the (aforementioned) sampling algorithm of Π which samples uniform element $\pi \in \Pi$ with sampling randomness r_π, there is a randomness resampling algorithm ReSmp_Π that takes as input $\pi \in \Pi$ and outputs a sampling randomness r_π. These two ways of sampling/resampling are statistically indistinguishable.

Definition 11 (Efficient Randomness Resampling on Π). *The hash value space Π of HPS supports efficient randomness resampling, if there*

$pp \leftarrow_s \mathsf{Setup}(1^\lambda)$:

$\quad mpar \leftarrow_s \mathsf{MPar}(1^\lambda).\ \widetilde{mpar} \leftarrow_s \widetilde{\mathsf{MPar}}(1^\lambda).$

$\quad$ // $mpar$ implicitly defines $(\mathcal{X}, \Pi)$.

$\quad$ // $\widetilde{mpar}$ implicitly defines $(\mathcal{X}, \widetilde{\Pi})$.

$\quad H \leftarrow_s \mathcal{H}.$

$\quad$ Return $pp := (mpar, \widetilde{mpar}, H).$

$(pk, sk, tk) \leftarrow_s \mathsf{Gen}(pp)$:

$\quad (par, td) \leftarrow_s \mathsf{Par}(mpar).\ (\widetilde{par}, \widetilde{td}) \leftarrow_s \widetilde{\mathsf{Par}}(\widetilde{mpar}).$

$\quad$ // par implicitly defines $(\mathcal{L}, \mathcal{SK}, \mathcal{PK}, \Lambda_{(\cdot)}, \alpha).$

$\quad$ // $\widetilde{par}$ implicitly defines $(\mathcal{L}, \widetilde{\mathcal{SK}}, \widetilde{\mathcal{PK}}, \widetilde{\Lambda}_{(\cdot)}, \widetilde{\alpha}, \mathcal{T}).$

$\quad sk \leftarrow_s \mathcal{SK},\ pk := \alpha(sk).$

$\quad \widetilde{sk} \leftarrow_s \widetilde{\mathcal{SK}},\ \widetilde{pk} := \widetilde{\alpha}(\widetilde{sk}).$

$\quad$ Return $(pk := (par, \widetilde{par}, pk, \widetilde{pk}), sk := (sk, \widetilde{sk}),$

$\quad\quad tk := (td, \widetilde{td})).$

$c \leftarrow_s \mathsf{Enc}(pp, pk, m \in \Pi)$:

$\quad x \leftarrow_s \mathcal{L}$ with witness w.

$\quad d := \mathsf{Pub}(pk, x, w) + m \in \Pi.$

$\quad \tau := H(x, d) \in \mathcal{T}.$

$\quad \widetilde{\pi} := \widetilde{\mathsf{Pub}}(\widetilde{pk}, x, w, \tau) \in \widetilde{\Pi}.$

$\quad$ Return $c := (x, d, \widetilde{\pi}).$

$m/\bot \leftarrow \mathsf{Dec}(pp, sk, c)$:

$\quad$ Parse $c = (x, d, \widetilde{\pi}).$

$\quad \tau := H(x, d) \in \mathcal{T}.$

$\quad$ If $\widetilde{\pi} \neq \widetilde{\Lambda}_{\widetilde{sk}}(x, \tau)$: Return $\bot$.

$\quad m := d - \Lambda_{sk}(x) \in \Pi.$

$\quad$ Return m.

$(c, td) \leftarrow_s \mathsf{Fake}(pp)$:

$\quad x \leftarrow_s \mathcal{X}$ with sampling randomness r_x.

$\quad d \leftarrow_s \Pi.$

$\quad \widetilde{\pi} \leftarrow_s \widetilde{\Pi}$ with sampling randomness $r_{\widetilde{\pi}}.$

$\quad$ Return $(c := (x, d, \widetilde{\pi}), td := (r_x, r_{\widetilde{\pi}})).$

$sk' \leftarrow_s \mathsf{Open}_k(pp, tk, pk, sk, (c_\gamma, td_\gamma, m_\gamma \in ER_\Pi)_{\gamma \in [k]})$:

$\quad$ Parse $tk = (td, \widetilde{td}),\ c_\gamma = (x_\gamma, d_\gamma, \widetilde{\pi}_\gamma),\ td_\gamma = (r_{x_\gamma}, r_{\widetilde{\pi}_\gamma}).$

$\quad$ For $\gamma \in [k],\ e_\gamma := d_\gamma - m_\gamma \in \Pi.$

$\quad\quad r_{e_\gamma} \leftarrow_s \mathsf{ReSmp}_\Pi(e_\gamma).$

$\quad$ // Note that r_{e_γ} is an samp. rand. for $e_\gamma \in \Pi.$

$\quad sk' \leftarrow_s \mathsf{HOpen}_k(td, pk, sk, (x_\gamma, r_{x_\gamma}, e_\gamma, r_{e_\gamma})_{\gamma \in [k]}).$

$\quad$ For $\gamma \in [k],\ \tau_\gamma := H(x_\gamma, d_\gamma) \in \mathcal{T}.$

$\quad \widetilde{sk}' \leftarrow_s \widetilde{\mathsf{HOpen}}_k(\widetilde{td}, \widetilde{pk}, \widetilde{sk}, (x_\gamma, r_{x_\gamma}, \widetilde{\pi}_\gamma, r_{\widetilde{\pi}_\gamma}, \tau_\gamma)_{\gamma \in [k]}).$

$\quad$ Return $sk' := (sk', \widetilde{sk}').$

Fig. 7. Construct. of $\mathsf{KM\text{-}NCE} = (\mathsf{Setup}, \mathsf{Gen}, \mathsf{Enc}, \mathsf{Dec}, \mathsf{Fake}, \mathsf{Open}_k)$ from HPS, $\widetilde{\mathsf{HPS}}$, $\mathcal{H}$.

exists a PPT algorithm ReSmp_Π, s.t. the statistical distance $\epsilon_{\mathsf{HPS}}^{\Pi\text{-}resmp}(\lambda) := \Delta((\pi, r_\pi), (\pi', r_{\pi'})) \leq \mathsf{negl}(\lambda)$, where $mpar \leftarrow_s \mathsf{MPar}(1^\lambda)$, $\pi \leftarrow_s \Pi$ with sampling randomness r_π, $\pi' \leftarrow_s \Pi$ and $r'_{\pi'} \leftarrow_s \mathsf{ReSmp}_\Pi(\pi')$.

5.3 Generic Construction of KM-NCE from Key-Openable HPS

The building blocks for constructing KM-NCE are as follows.

- Let $\mathsf{HPS} = (\mathsf{MPar}, \mathsf{Par}, \mathsf{Pub}, \mathsf{Priv}, \mathsf{HOpen}_k)$ be a key-openable HPS, whose hash value space Π is an (additive) group and has an efficient randomness resampling algorithm ReSmp_Π.
- Let $\widetilde{\mathsf{HPS}} = (\widetilde{\mathsf{MPar}}, \widetilde{\mathsf{Par}}, \widetilde{\mathsf{Pub}}, \widetilde{\mathsf{Priv}}, \widetilde{\mathsf{HOpen}}_k)$ be a tag-based key-openable HPS, which shares same universe $\mathcal{X}$ and same language $\mathcal{L}$ with HPS.
- Let $\mathcal{H} = \{H : \mathcal{X} \times \Pi \to \mathcal{T}\}$ be a family of collision-resistant hash functions, where Π is the hash value space of HPS and $\mathcal{T}$ is the tag space of $\widetilde{\mathsf{HPS}}$.

We present the generic construction of $\mathsf{KM\text{-}NCE} = (\mathsf{Setup}, \mathsf{Gen}, \mathsf{Enc}, \mathsf{Dec}, \mathsf{Fake}, \mathsf{Open}_k)$ from HPS, $\widetilde{\mathsf{HPS}}$ and $\mathcal{H}$ in Fig. 7. The message space is Π. Note that our generic construction of KM-NCE from key-openable HPS is reminiscent of [11], which constructs PKE scheme from another variant of HPS (the so-called quasi-adaptive HPS).

The perfect correctness of KM-NCE follows from those of HPS and $\widetilde{\mathsf{HPS}}$ directly. Next, we show its $\mathsf{KMNC}_k\text{-}\mathsf{CCA}$ security.

Theorem 2 (KMNC$_k$-CCA security of KM-NCE). *Assume that (1) HPS is openable$_k$, has a hard multi-fold SMP, supports efficient randomness resampling* on Π, (2) $\widetilde{\mathsf{HPS}}$ is universal$_{k+1}$ and openable$_k$, (3) $\mathcal{H}$ is collision-resistant. Then the KM-NCE in Fig. 7 is KMNC$_k$-CCA secure.

Concretely, for any PPT adversary $\mathcal{A}$ against the KMNC$_k$-CCA security of KM-NCE that makes at most Q_d decryption queries, there exist PPT adversaries $\mathcal{B}_1$, $\mathcal{B}_2$ and unbounded adversaries $\mathcal{B}_3$, $\mathcal{B}_4$, $\mathcal{B}_5$, s.t.

$$\mathbf{Adv}_{\mathsf{KM\text{-}NCE},\mathcal{A},k}^{kmnc\text{-}cca}(\lambda) \leq \mathbf{Adv}_{\mathsf{HPS},\mathcal{B}_1}^{k\text{-}msmp}(\lambda) + 2 \cdot \mathbf{Adv}_{\mathcal{H},\mathcal{B}_2}^{cr}(\lambda) + 2Q_d \cdot \epsilon_{\widetilde{\mathsf{HPS}},\mathcal{B}_3}^{univ_{k+1}}(\lambda) \qquad (2)$$
$$+ \ 2\epsilon_{\mathsf{HPS},\mathcal{B}_4}^{open_k}(\lambda) + 2\epsilon_{\mathsf{HPS},\mathcal{B}_5}^{open_k}(\lambda) + 2k \cdot \epsilon_{\mathsf{HPS}}^{\Pi\text{-}resmp}(\lambda).$$

Proof of Theorem 2. We prove the theorem by defining a sequence of games G_0-G_8, with $\mathsf{G}_0 = \mathbf{Exp}_{\mathsf{KM\text{-}NCE},\mathcal{A},k}^{kmnc\text{-}cca\text{-}real}(\lambda)$ and $\mathsf{G}_8 = \mathbf{Exp}_{\mathsf{KM\text{-}NCE},\mathcal{A},k}^{kmnc\text{-}cca\text{-}sim}(\lambda)$, and showing adjacent games indistinguishable. By $\Pr_i[\cdot]$ we denote the probability of a particular event occurring in game G_i.

Game G_0: This is the $\mathbf{Exp}_{\mathsf{KM\text{-}NCE},\mathcal{A},k}^{kmnc\text{-}cca\text{-}real}(\lambda)$ experiment. Thus, $\Pr[\mathsf{G}_0 = 1] = \Pr[\mathbf{Exp}_{\mathsf{KM\text{-}NCE},\mathcal{A},k}^{kmnc\text{-}cca\text{-}real}(\lambda) = 1]$.

In this game, when receiving $(m_\gamma^*)_{\gamma \in [k]}$ from $\mathcal{A}$, the challenger generates c_γ^* using the real encryption algorithm $\mathsf{Enc}(\mathsf{pp}, \mathsf{pk}, m_\gamma^*)$. More precisely, it samples $x_\gamma^* \leftarrow_\$ \mathcal{L}$ with witness w_γ^*, computes $d_\gamma^* := \mathsf{Pub}(\mathsf{pk}, x_\gamma^*, w_\gamma^*) + m_\gamma^*$, $\tau_\gamma^* := H(x_\gamma^*, d_\gamma^*)$, $\widetilde{\pi}_\gamma^* := \widetilde{\mathsf{Pub}}(\widetilde{\mathsf{pk}}, x_\gamma^*, w_\gamma^*, \tau_\gamma^*)$, and sets $c_\gamma^* := (x_\gamma^*, d_\gamma^*, \widetilde{\pi}_\gamma^*)$. It returns $(c_\gamma^*)_{\gamma \in [k]}$ to $\mathcal{A}$. When answering decryption queries $\mathcal{O}_{\text{dec}}(c)$ for $\mathcal{A}$ with $c = (x, d, \widetilde{\pi})$, the challenger computes $\tau := H(x, d)$, and outputs $\perp$ immediately if $c \in \{c_\gamma^*\}_{\gamma \in [k]} \vee \widetilde{\pi} \neq \widetilde{\Lambda}_{\widetilde{\mathsf{sk}}}(x, \tau)$. Otherwise, it computes $m := d - \Lambda_{\mathsf{sk}}(x)$ and returns m to $\mathcal{A}$. In the last step of this game, the challenger sends the real secret key $sk = (\mathsf{sk}, \widetilde{\mathsf{sk}})$ to $\mathcal{A}$.

Game G_1: It is the same as G_0, except that, for each $\gamma \in [k]$, when generating $c_\gamma^* = (x_\gamma^*, d_\gamma^*, \widetilde{\pi}_\gamma^*)$, the challenger computes d_γ^* and $\widetilde{\pi}_\gamma^*$ using $sk = (\mathsf{sk}, \widetilde{\mathsf{sk}})$ instead of using the witness w_γ^* of x_γ^*. Namely, $d_\gamma^* := \Lambda_{\mathsf{sk}}(x_\gamma^*) + m_\gamma^*$ and $\widetilde{\pi}_\gamma^* := \widetilde{\Lambda}_{\widetilde{\mathsf{sk}}}(x_\gamma^*, \tau_\gamma^*)$. By the perfect correctness of HPS and $\widetilde{\mathsf{HPS}}$, this change is conceptual. So $\Pr[\mathsf{G}_1 = 1] = \Pr[\mathsf{G}_0 = 1]$.

Game G_2: It is the same as G_1, except that, for each $\gamma \in [k]$, when generating $c_\gamma^* = (x_\gamma^*, d_\gamma^*, \widetilde{\pi}_\gamma^*)$, the challenger samples $x_\gamma^* \leftarrow_\$ \mathcal{X}$ instead of $x_\gamma^* \leftarrow_\$ \mathcal{L}$. Note that neither the witness w_γ^* of x_γ^* (if $x_\gamma^* \leftarrow_\$ \mathcal{L}$) nor the sampling randomness $r_{x_\gamma^*}$ of x_γ^* (if $x_\gamma^* \leftarrow_\$ \mathcal{X}$) is needed in G_1 and G_2, thus it is straightforward to construct a PPT adversary $\mathcal{B}_1$ against the multi-fold SMP, such that $|\Pr[\mathsf{G}_2 = 1] - \Pr[\mathsf{G}_1 = 1]| \leq \mathbf{Adv}_{\mathsf{HPS},\mathcal{B}_1}^{k\text{-}msmp}(\lambda)$.

Game G_3: It is the same as G_2, except that, when answering decryption queries $\mathcal{O}_{\text{dec}}(c)$ for $\mathcal{A}$ with $c = (x, d, \widetilde{\pi})$, the challenger adds a new rejection rule: it outputs $\perp$ immediately if $\tau \in \{\tau_\gamma^*\}_{\gamma \in [k]}$, where $\tau = H(x, d)$ and $\tau_\gamma^* = H(x_\gamma^*, d_\gamma^*)$ for each $\gamma \in [k]$.

Let Bad denote the event that $\mathcal{A}$ ever queries $\mathcal{O}_{\text{dec}}(c)$ with $c = (x, d, \widetilde{\pi})$, such that $(x, d) \notin \{(x_\gamma^*, d_\gamma^*)\}_{\gamma \in [k]}$ but $\tau \in \{\tau_\gamma^*\}_{\gamma \in [k]}$. We first show that G_2 and G_3 are identical if Bad does not occur, i.e., either $(x, d) = (x_{\gamma_0}^*, d_{\gamma_0}^*)$ for some $\gamma_0 \in [k]$ or $\tau \notin \{\tau_\gamma^*\}_{\gamma \in [k]}$. In the case that $(x, d) = (x_{\gamma_0}^*, d_{\gamma_0}^*)$ for some $\gamma_0 \in [k]$, $\mathcal{O}_{\text{dec}}(c)$

would be rejected both in G_2 and G_3 due to $c = c^*_{\gamma_0} \in \{c^*_\gamma\}_{\gamma \in [k]} \lor \widetilde{\pi} \neq \Lambda_{\widetilde{\mathsf{sk}}}(x, \tau)$. In the case that $\tau \notin \{\tau^*_\gamma\}_{\gamma \in [k]}$, the new rejection rule added in G_3 does not apply, so $\mathcal{O}_{\mathrm{dec}}(c)$ is the same in G_2 and G_3. Overall, G_2 and G_3 are identical when Bad does not occur, thus by the difference lemma, $|\Pr[G_3 = 1] - \Pr[G_2 = 1]| \leq \Pr_3[\mathsf{Bad}]$.

To bound $\Pr_3[\mathsf{Bad}]$, it is straightforward to construct a PPT adversary $\mathcal{B}_2$ against the collision resistance of $\mathcal{H}$, so that $\Pr_3[\mathsf{Bad}] \leq \mathbf{Adv}^{\mathrm{cr}}_{\mathcal{H}, \mathcal{B}_2}(\lambda)$. Consequently, $|\Pr[G_3 = 1] - \Pr[G_2 = 1]| \leq \mathbf{Adv}^{\mathrm{cr}}_{\mathcal{H}, \mathcal{B}_2}(\lambda)$.

Game G_4: It is the same as G_3, except that, when answering decryption queries $\mathcal{O}_{\mathrm{dec}}(c)$ for $\mathcal{A}$ with $c = (x, d, \widetilde{\pi})$, the challenger adds a second new rejection rule: it outputs $\bot$ immediately if $x \in \mathcal{X} \setminus \mathcal{L}$. We note that this new rule may not be PPT checkable, thus the challenger may not be PPT. This does not matter, since the following arguments (before this rule is removed) are statistical.

Let Forge denote the event that $\mathcal{A}$ ever queries $\mathcal{O}_{\mathrm{dec}}(c)$ with $c = (x, d, \widetilde{\pi})$, such that $\tau \notin \{\tau^*_\gamma\}_{\gamma \in [k]}$, $\widetilde{\pi} = \Lambda_{\widetilde{\mathsf{sk}}}(x, \tau)$ but $x \in \mathcal{X} \setminus \mathcal{L}$. Clearly, G_3 and G_4 are identical unless Forge occurs, thus by the difference lemma, $|\Pr[G_4 = 1] - \Pr[G_3 = 1]| \leq \Pr_4[\mathsf{Forge}]$.

To bound $\Pr_4[\mathsf{Forge}]$, we analyze the information about $\widetilde{\mathsf{sk}}$ that $\mathcal{A}$ may obtain in game G_4 before it finishes the $\mathcal{O}_{\mathrm{dec}}$ queries: $\mathcal{A}$ obtains $\mathsf{pk} = \alpha(\widetilde{\mathsf{sk}})$ in pk and obtains $\{\widetilde{\pi}^*_\gamma = \Lambda_{\widetilde{\mathsf{sk}}}(x^*_\gamma, \tau^*_\gamma)\}_{\gamma \in [k]}$ in $\{c^*_\gamma\}_{\gamma \in [k]}$; for $\mathcal{O}_{\mathrm{dec}}$ queries, the challenger will not output m unless $x \in \mathcal{L}$ (due to the new rejection rule added in G_4), thus $\mathcal{O}_{\mathrm{dec}}$ reveals nothing about $\widetilde{\mathsf{sk}}$ beyond $\mathsf{pk} = \alpha(\widetilde{\mathsf{sk}})$.

Then by the universal$_{k+1}$ property of tag-based $\widetilde{\mathsf{HPS}}$, for one $\mathcal{O}_{\mathrm{dec}}(c)$ query made by $\mathcal{A}$, it holds that $\tau \notin \{\tau^*_\gamma\}_{\gamma \in [k]}$, $\widetilde{\pi} = \Lambda_{\widetilde{\mathsf{sk}}}(x, \tau)$ but $x \in \mathcal{X} \setminus \mathcal{L}$ with probability at most $\epsilon^{\mathrm{univ}_{k+1}}_{\widetilde{\mathsf{HPS}}, \mathcal{B}_3}(\lambda)$. By a union bound over at most Q_d number of $\mathcal{O}_{\mathrm{dec}}$ queries, we get that $\Pr_4[\mathsf{Forge}] \leq Q_d \cdot \epsilon^{\mathrm{univ}_{k+1}}_{\widetilde{\mathsf{HPS}}, \mathcal{B}_3}(\lambda)$. Thus, $|\Pr[G_4 = 1] - \Pr[G_3 = 1]| \leq Q_d \cdot \epsilon^{\mathrm{univ}_{k+1}}_{\widetilde{\mathsf{HPS}}, \mathcal{B}_3}(\lambda)$. For completeness, we provide a description of the reduction algorithm $\mathcal{B}_3$ in the full version [17].

Game G_5: It is the same as G_4, except that, for each $\gamma \in [k]$, when generating $c^*_\gamma = (x^*_\gamma, d^*_\gamma, \widetilde{\pi}^*_\gamma)$, the challenger samples $d^*_\gamma \leftarrow_\$ \Pi$ uniformly (instead of $d^*_\gamma := \Lambda_{\mathsf{sk}}(x^*_\gamma) + m^*_\gamma$). Moreover, in the last step of this game, the challenger computes $e^*_\gamma := d^*_\gamma - m^*_\gamma \in \Pi$ and resamples $r_{e^*_\gamma} \leftarrow_\$ \mathsf{ReSmp}_\Pi(e^*_\gamma)$ for each $\gamma \in [k]$, then invokes $\mathsf{sk}' \leftarrow_\$ \mathsf{HOpen}_k(\mathsf{td}, \mathsf{pk}, \mathsf{sk}, (x^*_\gamma, r_{x^*_\gamma}, e^*_\gamma, r_{e^*_\gamma})_{\gamma \in [k]})$, and sends $(\mathsf{sk}', \widetilde{\mathsf{sk}})$ to $\mathcal{A}$. We have the following lemma whose proof is given in the full version [17].

Lemma 3. *There exists an unbounded $\mathcal{B}_4$ against the openable$_k$ property of* HPS, *s.t.* $|\Pr[G_5 = 1] - \Pr[G_4 = 1]| \leq 2 \cdot \epsilon^{\mathrm{open}_k}_{\mathsf{HPS}, \mathcal{B}_4}(\lambda) + 2k \cdot \epsilon^{\Pi\text{-}resmp}_{\mathsf{HPS}}(\lambda)$.

Game G_6: It is the same as G_5, except that, for each $\gamma \in [k]$, when generating $c^*_\gamma = (x^*_\gamma, d^*_\gamma, \widetilde{\pi}^*_\gamma)$, the challenger samples $\widetilde{\pi}^*_\gamma \leftarrow_\$ \widetilde{\Pi}$ uniformly with randomness $r_{\widetilde{\pi}^*_\gamma}$ (instead of $\widetilde{\pi}^*_\gamma := \Lambda_{\widetilde{\mathsf{sk}}}(x^*_\gamma, \tau^*_\gamma)$). Moreover, in the last step of this game, the challenger computes $\widetilde{\mathsf{sk}}' \leftarrow_\$ \widetilde{\mathsf{HOpen}}_k(\widetilde{\mathsf{td}}, \widetilde{\mathsf{pk}}, \widetilde{\mathsf{sk}}, (x^*_\gamma, r_{x^*_\gamma}, \widetilde{\pi}^*_\gamma, r_{\widetilde{\pi}^*_\gamma}, \tau^*_\gamma)_{\gamma \in [k]})$, and sends $(\mathsf{sk}', \widetilde{\mathsf{sk}}')$ to $\mathcal{A}$. We have the following lemma. The proof of this lemma is similar to that of Lemma 3, and is given in the full version [17].

Lemma 4. *There exists an unbounded* $\mathcal{B}_5$ *against the* $openable_k$ *property of tag-based* $\widetilde{\mathsf{HPS}}$, *s.t.* $|\Pr[\mathsf{G}_6 = 1] - \Pr[\mathsf{G}_5 = 1]| \leq 2 \cdot \epsilon_{\widetilde{\mathsf{HPS}},\mathcal{B}_5}^{open_k}(\lambda)$.

Game G_7**:** It is the same as G_6, except that, when answering decryption queries $\mathcal{O}_{\mathrm{dec}}(c)$ for $\mathcal{A}$ with $c = (x, d, \widetilde{\pi})$, the challenger removes the second new rejection rule added in G_4. In other words, it does not check whether $x \in \mathcal{L}$ or $x \in \mathcal{X} \setminus \mathcal{L}$ anymore. We note that the challenger in G_7 is now PPT again.

The change from G_6 to G_7 is symmetric to that from G_3 to G_4. By a similar argument, we get $|\Pr[\mathsf{G}_7 = 1] - \Pr[\mathsf{G}_6 = 1]| \leq Q_d \cdot \epsilon_{\widetilde{\mathsf{HPS}},\mathcal{B}_3}^{univ_{k+1}}(\lambda)$.

Game G_8**:** It is the same as G_7, except that, when answering decryption queries $\mathcal{O}_{\mathrm{dec}}(c)$ for $\mathcal{A}$ with $c = (x, d, \widetilde{\pi})$, the challenger removes the first new rejection rule added in G_3. In other words, it does not check whether $\tau \in \{\tau_\gamma^*\}_{\gamma \in [k]}$ or not anymore.

The change from G_7 to G_8 is symmetric to the change from G_2 to G_3. Similarly, we have that $|\Pr[\mathsf{G}_8 = 1] - \Pr[\mathsf{G}_7 = 1]| \leq \mathbf{Adv}_{\mathcal{H},\mathcal{B}_2}^{\mathrm{cr}}(\lambda)$.

Finally, we note that G_8 is exactly the $\mathbf{Exp}_{\mathsf{KM\text{-}NCE},\mathcal{A},k}^{\mathrm{kmnc\text{-}cca\text{-}sim}}(\lambda)$ experiment.

– For each $\gamma \in [k]$, $c_\gamma^* := (x_\gamma^*, d_\gamma^*, \widetilde{\pi}_\gamma^*)$, where $x_\gamma^* \leftarrow_\$ \mathcal{X}$ with sampling randomness $r_{x_\gamma^*}$, $d_\gamma^* \leftarrow_\$ \Pi$, and $\widetilde{\pi}_\gamma^* \leftarrow_\$ \widetilde{\Pi}$ with randomness $r_{\widetilde{\pi}_\gamma^*}$, the same as the c_γ^* generated by $\mathsf{Fake}(\mathsf{pp})$.
– $\mathcal{O}_{\mathrm{dec}}(c)$ queries are answered by $\mathsf{Dec}(\mathsf{pp}, sk, c)$ when $c \notin \{c_\gamma^*\}_{\gamma \in [k]}$.
– In the last step, $(sk', \widetilde{sk}')$ is generated by first computing $e_\gamma^* := d_\gamma^* - m_\gamma^* \in \Pi$ and resampling $r_{e_\gamma^*} \leftarrow_\$ \mathsf{ReSmp}_\Pi(e_\gamma^*)$ for each $\gamma \in [k]$, then invoking $sk' \leftarrow_\$ \mathsf{HOpen}_k(\mathsf{td}, \mathsf{pk}, sk, (x_\gamma^*, r_{x_\gamma^*}, e_\gamma^*, r_{e_\gamma^*})_{\gamma \in [k]})$ and $\widetilde{sk}' \leftarrow_\$ \widetilde{\mathsf{HOpen}}_k(\widetilde{\mathsf{td}}, \widetilde{\mathsf{pk}}, \widetilde{sk}, (x_\gamma^*, r_{x_\gamma^*}, \widetilde{\pi}_\gamma^*, r_{\widetilde{\pi}_\gamma^*}, \tau_\gamma^*)_{\gamma \in [k]})$ with $\tau_\gamma^* := H(x_\gamma^*, d_\gamma^*)$, the same as $\mathsf{Open}_k(\mathsf{pp}, tk, pk, sk, (c_\gamma^*, r_{c_\gamma^*}, m_\gamma^*)_{\gamma \in [k]})$ where $r_{c_\gamma^*} = (r_{x_\gamma^*}, r_{\widetilde{\pi}_\gamma^*})$.

Thus, $\Pr[\mathsf{G}_8 = 1] = \Pr[\mathbf{Exp}_{\mathsf{KM\text{-}NCE},\mathcal{A},k}^{\mathrm{kmnc\text{-}cca\text{-}sim}}(\lambda) = 1]$.

Taking all things together, we obtain (2), thus Theorem 2 follows. ∎

Finally, we show the robustness.

Theorem 3 (Robustness of KM-NCE). *The proposed* KM-NCE *in Fig. 7 is robust (cf. Definition 5) with* $\epsilon_{\mathsf{KM\text{-}NCE}}^{rob}(\lambda) \leq 1/|\widetilde{\Pi}|$, *where* $\widetilde{\Pi}$ *is the hash value space of* $\widetilde{\mathsf{HPS}}$.

Proof of Theorem 3. For $\mathsf{pp} \leftarrow_\$ \mathsf{Setup}(1^\lambda)$, $(pk, sk, tk) \leftarrow_\$ \mathsf{Gen}(\mathsf{pp})$, $(c, td) \leftarrow_\$ \mathsf{Fake}(\mathsf{pp})$, we analyze the probability $\epsilon_{\mathsf{KM\text{-}NCE}}^{rob}(\lambda) = \Pr[\mathsf{Dec}(\mathsf{pp}, sk, c) \neq \bot]$.

– For $(pk, sk, tk) \leftarrow_\$ \mathsf{Gen}(\mathsf{pp})$, we have $sk = (sk, \widetilde{sk})$ where $\widetilde{sk} \leftarrow_\$ \widetilde{\mathcal{SK}}$.
– For $(c, td) \leftarrow_\$ \mathsf{Fake}(\mathsf{pp})$, we have $c = (x, d, \widetilde{\pi})$ where $x \leftarrow_\$ \mathcal{X}$, $d \leftarrow_\$ \Pi$ and $\widetilde{\pi} \leftarrow_\$ \widetilde{\Pi}$.
– Then in $\mathsf{Dec}(\mathsf{pp}, sk, c)$, it first checks whether or not $\widetilde{\pi} = \widetilde{\Lambda}_{\widetilde{sk}}(x, \tau)$ holds, where $\tau := H(x, d)$, and returns $\bot$ if the check fails.

Since $\widetilde{\pi}$ is uniformly chosen from $\widetilde{\Pi}$ and independent of x, d and $\widetilde{sk}$, so the check $\widetilde{\pi} = \widetilde{\Lambda}_{\widetilde{sk}}(x, \tau)$ passes with probability $1/|\widetilde{\Pi}|$. Overall, we have $\epsilon_{\mathsf{KM\text{-}NCE}}^{rob}(\lambda) = \Pr[\mathsf{Dec}(\mathsf{pp}, sk, c) \neq \bot] \leq \Pr[\widetilde{\pi} = \widetilde{\Lambda}_{\widetilde{sk}}(x, \tau)] = 1/|\widetilde{\Pi}|$. ∎

6 Concrete Instantiations

In this section, we show concrete instantiations of key-openable HPS based on the matrix decisional Diffie-Hellman (MDDH) assumption [10]. As a result, we can obtain concrete instantiations of KM-NCE, which in turn yields AC-RSO$_k$&C-CCA secure PKE schemes with compact ciphertexts. For certain instantiation, the resulting PKE can even achieve tight AC-RSO$_k$&C-CCA security.

6.1 Recall: Matrix Distribution

We recall the definition of matrix distribution defined in [10].

In this section, we use bold uppercase letters to represent matrices and bold lowercase letters to represent (column) vectors. Let GGen be a PPT algorithm that on input 1^λ returns $\mathcal{G} = (\mathbb{G}, q, P)$, a description of an (additive) cyclic group $\mathbb{G}$ with a generator P of order q which is a λ-bit prime. For $a \in \mathbb{Z}_q$, define $[a] := aP \in \mathbb{G}$ as the *implicit representation* of a in $\mathbb{G}$. More generally, for a matrix $\mathbf{A} = (a_{ij}) \in \mathbb{Z}_q^{n \times m}$, we define $[\mathbf{A}]$ as the implicit representation of $\mathbf{A}$ in $\mathbb{G}$, i.e., $[\mathbf{A}] := (a_{ij}P) \in \mathbb{G}^{n \times m}$. Note that from $[a] \in \mathbb{G}$ it is generally hard to compute the value a (discrete logarithm problem is hard in $\mathbb{G}$). Obviously, given $[a], [b] \in \mathbb{G}$ and a scalar $x \in \mathbb{Z}$, one can efficiently compute $[ax] \in \mathbb{G}$ and $[a+b] \in \mathbb{G}$. Similarly, for $\mathbf{A} \in \mathbb{Z}_q^{m \times n}, \mathbf{B} \in \mathbb{Z}_q^{n \times t}, \mathbf{AB} \in \mathbb{Z}_q^{m \times t}$, given $[\mathbf{A}], \mathbf{B}$ one can efficiently compute $[\mathbf{A}]\mathbf{B} := [\mathbf{AB}] \in \mathbb{G}^{m \times t}$ and given $\mathbf{A}, [\mathbf{B}]$, one can efficiently compute $\mathbf{A}[\mathbf{B}] := [\mathbf{AB}] \in \mathbb{G}^{m \times t}$.

Definition 12 (Matrix Distribution). *Let $d, k \in \mathbb{N}^+$. $\mathcal{D}_{d+k,d}$ is called a matrix distribution if it outputs matrices in $\mathbb{Z}_q^{(d+k) \times d}$ of full rank d in polynomial time.*

As in [10], let $\mathcal{U}_{d+k,d}$ be the uniform distribution over $\mathbb{Z}_q^{(d+k) \times d}$. Without loss of generality, for $\mathbf{A} \leftarrow_s \mathcal{D}_{d+k,d}$, we assume that $\overline{\mathbf{A}}$ (the upper square submatrix of $\mathbf{A}$) is invertible.

Definition 13 (The $\mathcal{D}_{d+k,d}$-Matrix Decision Diffie-Hellman Assumption, $\mathcal{D}_{d+k,d}$-MDDH). *Let $\mathcal{D}_{d+k,d}$ be a matrix distribution. The $\mathcal{D}_{d+k,d}$-Matrix Decision Diffie-Hellman ($\mathcal{D}_{d+k,d}$-MDDH) Assumption holds relative to GGen if for each PPT adversary $\mathcal{A}$, the advantage*

$$\mathbf{Adv}_{\mathcal{D}_{d+k,d},\mathsf{GGen},\mathcal{A}}^{\mathrm{mddh}}(\lambda) := |\Pr[\mathcal{A}(\mathcal{G}, [\mathbf{A}], [\mathbf{Aw}]) = 1] - \Pr[\mathcal{A}(\mathcal{G}, [\mathbf{A}], [\mathbf{u}]) = 1]|$$

is negligible, where the probability is taken over $\mathcal{G} \leftarrow_s \mathsf{GGen}(1^\lambda), \mathbf{A} \leftarrow_s \mathcal{D}_{d+k,d}, \mathbf{w} \leftarrow_s \mathbb{Z}_q^d$ and $\mathbf{u} \leftarrow_s \mathbb{Z}_q^{d+k}$.

As shown in [10], $\mathcal{D}_{d+k,d}$-MDDH assumption is a generalization of a large range of assumptions. By setting the matrix distribution $\mathcal{D}_{\ell,k}$ to specific distributions, $\mathcal{D}_{d+k,d}$-MDDH assumption can capture DDH assumption, k-Linear assumption, k-Cascade assumption and many other assumptions.

The MDDH assumption can be generalized into a multi-instance version. We recall the Q-fold MDDH assumption as defined in [10].

Definition 14 (Q-fold $\mathcal{D}_{d+k,d}$-Matrix Decision Diffie-Hellman Assumption). *Let Q be a positive integer and $\mathcal{D}_{d+k,d}$ be a matrix distribution. The Q-fold $\mathcal{D}_{d+k,d}$-Matrix Decision Diffie-Hellman Assumption holds relative to* GGen *if for each PPT adversary $\mathcal{A}$, the advantage*

$$\mathbf{Adv}^{Q\text{-mddh}}_{\mathcal{D}_{d+k,d},\mathsf{GGen},\mathcal{A}}(\lambda) := |\Pr[\mathcal{A}(\mathcal{G},[\mathbf{A}],[\mathbf{AW}]) = 1] - \Pr[\mathcal{A}(\mathcal{G},[\mathbf{A}],[\mathbf{U}]) = 1]|$$

is negligible, where the probability is taken over $\mathcal{G} \leftarrow_{\$} \mathsf{GGen}(1^\lambda)$, $\mathbf{A} \leftarrow_{\$} \mathcal{D}_{d+k,d}$, $\mathbf{W} \leftarrow_{\$} \mathbb{Z}_q^{d \times Q}$ and $\mathbf{U} \leftarrow_{\$} \mathbb{Z}_q^{(d+k) \times Q}$.

6.2 Openable$_k$ HPS Instantiation

In this subsection, we provide a key-openable HPS instantiation with openable$_k$ and efficient randomness resampling properties based on the MDDH assumption. This HPS can be seen as a generalization of the DDH-based HPS in [9]. Inspired by the technique in [12,15], we are able to make the hash value space of our HPS to be compact and efficient randomness resamplable. Meanwhile, this does not affect the openability of our HPS.

More precisely, fixing some group generation algorithm GGen, some positive integers d, k, some matrix distribution $\mathcal{D}_{d+k,d}$ and some polynomial $l = l(\lambda)$ (which can be set as the desired message length of the PKE scheme), consider HPS = (MPar, Par, Pub, Priv, HOpen$_k$) in the following.

– MPar(1^λ). The master parameter generation algorithm runs $\mathcal{G} = (\mathbb{G}, q, P) \leftarrow_{\$} \mathsf{GGen}(1^\lambda)$. Let $\mathcal{H}_\mathsf{u} = \{\mathsf{H}_\mathsf{u} : \mathbb{G} \to \{0,1\}\}$ be a family of universal hash functions based on group $\mathbb{G}$. The algorithm selects $\mathsf{H}_\mathsf{u} \leftarrow_{\$} \mathcal{H}_\mathsf{u}$ and returns mpar $:= (\mathcal{G}, d, k, l, \mathcal{D}_{d+k,d}, \mathsf{H}_\mathsf{u})$ which implicitly defines the instance space $\mathcal{X} := \mathbb{G}^{d+k}$ with randomness space $R_\mathcal{X} := \mathbb{Z}_q^{d+k}$ and the hash value space $\Pi := \{0,1\}^l$ with randomness space $R_\Pi := \mathbb{Z}_q^l$. Given mpar, one can efficiently sample a uniform element x from $\mathcal{X}$ by selecting $r_x = \mathbf{x} \leftarrow_{\$} R_\mathcal{X}$ and setting $x := [r_x] = [\mathbf{x}]$. For simplicity, we define an efficiently computable function $\mathsf{H}_{\mathsf{u},l} : \mathbb{G}^l \to \{0,1\}^l$ where $\mathsf{H}_{\mathsf{u},l}([\mathbf{a}]) := (\mathsf{H}_\mathsf{u}([a_1]), \cdots, \mathsf{H}_\mathsf{u}([a_l]))$ for all $[\mathbf{a}] = [a_1, \cdots, a_l] \in \mathbb{G}^l$. Then, one can also efficiently sample a uniform element π from Π by selecting $r_\pi = \boldsymbol{\pi} \leftarrow_{\$} R_\Pi$ and setting $\pi := \mathsf{H}_{\mathsf{u},l}([\boldsymbol{\pi}]) \in \Pi$.[6]
– Par(mpar). The (ordinary) parameter generation algorithm selects matrix $\mathbf{A} \in \mathbb{Z}_q^{(d+k) \times d} \leftarrow_{\$} \mathcal{D}_{d+k,d}$, then it returns par $:= [\mathbf{A}]$ and td $:= \mathbf{A}$.
  The public parameter par (together with mpar) implicitly defines the language as $\mathcal{L} := [\mathsf{span}(\mathbf{A})] = \{[\mathbf{Aw}] \mid \mathbf{w} \in \mathbb{Z}_q^d\}$. The hashing key space $\mathcal{SK} := \mathbb{Z}_q^{(d+k) \times l}$ and the projection key space $\mathcal{PK} := \mathbb{G}^{d \times l}$. The projection function α maps sk $= \mathbf{S} \in \mathcal{SK}$ to pk $= [\mathbf{P}] \in \mathcal{PK}$ where $[\mathbf{P}] = [\mathbf{A}^\top]\mathbf{S}$ and α is efficiently computable given par and sk. For sk $= \mathbf{S} \in \mathcal{SK}$, the hash function $\Lambda_\mathsf{sk}(\cdot)$ maps

---

[6] Actually, π is only statistical close to uniform. According to the leftover hash lemma together with the union bound, the statistically distance between π and uniform distribution over Π is bounded by $\frac{1}{2}\sqrt{\frac{2}{q}}$, which is exponentially small for polynomially bounded l. Therefore, we omit this statistical distance here.

ReSmp$_\Pi$($\mathbf{b} = (b_1, \cdots, b_l) \in \{0,1\}^l$):	OnebitReSmp(H$_u$, $b_i \in \{0,1\}$):
//Implicit input: H$_u \in$ mpar	For $j \in \{1, \cdots, \lambda\}$:
For $i \in \{1, \cdots, l\}$:	$r_j \leftarrow_\$ \mathbb{Z}_q$
$r_i \leftarrow_\$$ OnebitReSmp(H$_u$, b_i)	If H$_u$($[r_j]$) $= b_i$: Return r_j
Return $\mathbf{r} := (r_1, \cdots, r_l)$	Return $\bot$

Fig. 8. Randomness resample algorithm ReSmp$_\Pi$ for hash value space $\Pi = \{0,1\}^l$ of the hash proof system HPS. The algorithm OnebitReSmp will return $\bot$ and terminate after λ iterations, which makes it a polynomial-time algorithm.

an element $x = [\mathbf{x}] \in \mathcal{X}$ to H$_{u,l}(\mathbf{S}^\top[\mathbf{x}]) \in \Pi$ and it is efficiently computable given sk and x.

Given par, one can efficiently sample a uniform element x from language $\mathcal{L}$ together with a witness w by choosing $w = \mathbf{w} \leftarrow_\$ \mathbb{Z}_q^d$ and computing $x = [\mathbf{x}] = [\mathbf{A}]\mathbf{w}$.

- Pub(pk, x, w). Given public key pk $= [\mathbf{P}] \in \mathcal{PK}$, an instance $x = [\mathbf{x}] = [\mathbf{Aw}] \in \mathcal{L}$, and its witness $w = \mathbf{w}$, the public evaluation algorithm outputs $\pi = $ H$_{u,l}([\mathbf{P}^\top]\mathbf{w}) \in \Pi$.

- Priv(sk, x). Given secret key sk $= \mathbf{S} \in \mathcal{SK}$ and $x = [\mathbf{x}] \in \mathcal{X}$, the private evaluation algorithm outputs $\pi = $ H$_{u,l}(\mathbf{S}^\top[\mathbf{x}]) \in \Pi$.

- HOpen$_k$(td, pk, sk, $(x_\gamma, r_{x_\gamma}, \pi_\gamma, r_{\pi_\gamma})_{\gamma \in \{1, \cdots, k\}}$). Given td $= \mathbf{A}$, pk $= [\mathbf{P}]$, sk $= \mathbf{S}$, $x_\gamma = [\mathbf{x}_\gamma]$, $r_{x_\gamma} = \mathbf{x}_\gamma$, $\pi_\gamma = $ H$_{u,l}([\pi_\gamma])$ and $r_{\pi_\gamma} = \pi_\gamma \in \mathbb{Z}_q^l$ for all $\gamma \in \{1, \cdots, k\}$, the open algorithm computes sk$' = \mathbf{S}' \in \mathbb{Z}_q^{(d+k) \times l}$ by solving the following system of linear equations,

$$\mathbf{S}'^\top (\mathbf{A} \mid \mathbf{x}_1 \mid \cdots \mid \mathbf{x}_k) = (\mathbf{S}^\top \mathbf{A} \mid \pi_1 \mid \cdots \mid \pi_k) \bmod q. \tag{3}$$

Note that, given td $= \mathbf{A}$ and the randomnesses $(r_{x_\gamma} = \mathbf{x}_\gamma)_{\gamma \in \{1, \cdots, k\}}$, one can easily compute the square matrix $\mathbf{M} = (\mathbf{A} \mid \mathbf{x}_1 \mid \cdots \mid \mathbf{x}_k) \in \mathbb{Z}_q^{(d+k) \times (d+k)}$. If $\mathbf{M}$ is invertible, one can easily compute and output $\mathbf{S}'^\top = (\mathbf{S}^\top \mathbf{A} \mid \pi_1 \mid \cdots \mid \pi_k) \cdot \mathbf{M}^{-1} \bmod q$. If $\mathbf{M}$ is not invertible, algorithm HOpen$_k$ outputs $\bot$.

Note that the hash value space $\Pi = \{0,1\}^l$ is an additive group with group operation $\oplus$ (string xor). We define its randomness resample algorithm ReSmp$_\Pi$ in Fig. 8.

Theorem 4. *The above instantiation HPS (1) is a key-openable HPS; (2) has a hard multi-fold SMP under the multi-fold $\mathcal{D}_{d+k,d}$-MDDH assumption (i.e., for any PPT adversary $\mathcal{A}$, there exists a PPT adversary $\mathcal{B}$ such that* $\mathbf{Adv}_{\mathsf{HPS},\mathcal{A}}^{k\text{-msmp}}(\lambda) \leq \mathbf{Adv}_{\mathcal{D}_{d+k,d},\mathsf{GGen},\mathcal{B}}^{k\text{-mddh}}(\lambda)$*); (3) is openable$_k$ and (4) supports efficient randomness resampling on Π with algorithm ReSmp$_\Pi$.*

We put the proof of Theorem 4 in the full version [17].

6.3 Openable$_k$ and Universal$_{k+1}$ Tag-Based HPS Instantiation

In this subsection, we provide a tag-based key-openable HPS instantiation with both openable$_k$ and universal$_{k+1}$ properties based on the MDDH assumption.

This tag-based HPS can be seen as a generalization of the tag-based HPS from the DDH assumption in [9]. More precisely, fixing some group generation algorithm GGen, some positive integers d, k and some matrix distribution $\mathcal{D}_{d+k,d}$, consider instantiation $\widetilde{\mathsf{HPS}} = (\widetilde{\mathsf{MPar}}, \widetilde{\mathsf{Par}}, \widetilde{\mathsf{Pub}}, \widetilde{\mathsf{Priv}}, \widetilde{\mathsf{HOpen}}_k)$ in the following.

– $\widetilde{\mathsf{MPar}}(1^\lambda)$. The master parameter generation algorithm runs $\mathcal{G} = (\mathbb{G}, q, P) \leftarrow_{\$} \mathsf{GGen}(1^\lambda)$ and returns $\widetilde{\mathsf{mpar}} := (\mathcal{G}, d, k, \mathcal{D}_{d+k,d})$ which implicitly defines the instance space $\mathcal{X} := \mathbb{G}^{d+k}$ with randomness space $R_\mathcal{X} := \mathbb{Z}_q^{d+k}$ and the hash value space $\widetilde{\Pi} := \mathbb{G}$ with randomness space $R_{\widetilde{\pi}} := \mathbb{Z}_q$.[7] Given mpar, one can efficiently sample a uniform element x from $\mathcal{X}$ by selecting $r_x = \mathbf{x} \leftarrow_{\$} R_\mathcal{X}$ and set $x = [r_x] = [\mathbf{x}]$. One can also efficiently sample a uniform element $\widetilde{\pi}$ from $\widetilde{\Pi}$ by selecting $r_{\widetilde{\pi}} \leftarrow_{\$} R_{\widetilde{\Pi}}$ and set $\widetilde{\pi} = [r_{\widetilde{\pi}}]$.

– $\widetilde{\mathsf{Par}}(\widetilde{\mathsf{mpar}})$. The (ordinary) parameter generation algorithm selects matrix $\mathbf{A} \in \mathbb{Z}_q^{(d+k)\times d} \leftarrow_{\$} \mathcal{D}_{d+k,d}$, then it returns $\widetilde{\mathsf{par}} := [\mathbf{A}]$ and $\widetilde{\mathsf{td}} := \mathbf{A}$.

The public parameter $\widetilde{\mathsf{par}}$ (together with $\widetilde{\mathsf{mpar}}$) implicitly defines the language as $\mathcal{L} := [\mathrm{span}(\mathbf{A})] = \{[\mathbf{Aw}] \mid \mathbf{w} \in \mathbb{Z}_q^d\}$.[8] The hashing key space $\widetilde{\mathcal{SK}} := \mathbb{Z}_q^{2d+2k}$ and the projection key space $\widetilde{\mathcal{PK}} := \mathbb{G}^{2d}$. The projection function $\widetilde{\alpha}$ maps $\widetilde{\mathsf{sk}} = \mathbf{s} = \begin{pmatrix} \mathbf{s}_1 \\ \mathbf{s}_2 \end{pmatrix} \in \widetilde{\mathcal{SK}}$ (where $\mathbf{s}_1, \mathbf{s}_2 \in \mathbb{Z}_q^{d+k}$) to $\widetilde{\mathsf{pk}} = [\mathbf{p}] = \begin{bmatrix} \mathbf{p}_1 \\ \mathbf{p}_2 \end{bmatrix} = \begin{bmatrix} \mathbf{A}^\top \\ & \mathbf{A}^\top \end{bmatrix} \mathbf{s} \in \widetilde{\mathcal{PK}}$ (where $[\mathbf{p}_i] = [\mathbf{A}^\top \mathbf{s}_i] \in \mathbb{G}^d$ for $i \in \{1,2\}$) and $\widetilde{\alpha}$ is efficiently computable given $\widetilde{\mathsf{par}}$ and $\widetilde{\mathsf{sk}}$. The tag space is $\mathcal{T} := \mathbb{Z}_q$. For $\widetilde{\mathsf{sk}} = \mathbf{s} \in \widetilde{\mathcal{SK}}$, the hash function $\widetilde{\Lambda}_{\widetilde{\mathsf{sk}}}(\cdot, \cdot)$ maps an element $x = [\mathbf{x}] \in \mathcal{X}$ together with a tag $\tau \in \mathcal{T}$ to $\widetilde{\pi} = \mathbf{s}^\top \begin{bmatrix} \mathbf{x} \\ \tau\mathbf{x} \end{bmatrix} = [\mathbf{s}_1^\top \mathbf{x} + \tau \mathbf{s}_2^\top \mathbf{x}] \in \widetilde{\Pi}$ and it is efficiently computable given $\widetilde{\mathsf{sk}}, x$ and τ.

Given $\widetilde{\mathsf{par}}$, one can efficiently sample a uniform element x from language $\mathcal{L}$ together with a witness w by choosing $w = \mathbf{w} \leftarrow_{\$} \mathbb{Z}_q^d$ and computing $x = [\mathbf{x}] = [\mathbf{Aw}]$.

– $\widetilde{\mathsf{Pub}}(\widetilde{\mathsf{pk}}, x, w, \tau)$. Given public key $\widetilde{\mathsf{pk}} = [\mathbf{p}]$, witness $w = \mathbf{w}$ of instance $x = [\mathbf{Aw}]$ and tag τ, the public evaluation algorithm outputs the hash value $\widetilde{\pi} = [\mathbf{p}^\top] \begin{pmatrix} \mathbf{w} \\ \tau\mathbf{w} \end{pmatrix}$.

– $\widetilde{\mathsf{Priv}}(\widetilde{\mathsf{sk}}, x, \tau)$. Given secret key $\widetilde{\mathsf{sk}} = \mathbf{s}$, $x = [\mathbf{x}]$ and tag τ, the private evaluation algorithm outputs $\widetilde{\pi} = \mathbf{s}^\top \begin{bmatrix} \mathbf{x} \\ \tau\mathbf{x} \end{bmatrix}$.

– $\widetilde{\mathsf{HOpen}}_k(\widetilde{\mathsf{td}}, \widetilde{\mathsf{pk}}, \widetilde{\mathsf{sk}}, (x_\gamma, r_{x_\gamma}, \widetilde{\pi}_\gamma, r_{\widetilde{\pi}_\gamma}, \tau_\gamma)_{\gamma \in \{1, \cdots, k\}})$. Given trapdoor $\widetilde{\mathsf{td}} = \mathbf{A}$, public key $\widetilde{\mathsf{pk}} = [\mathbf{p}]$, secret key $\widetilde{\mathsf{sk}} = \mathbf{s}$, instance $x_\gamma = [\mathbf{x}_\gamma]$ with randomness $r_{x_\gamma} = \mathbf{x}_\gamma$, hash value $\widetilde{\pi}_\gamma = [r_{\widetilde{\pi}_\gamma}]$ with randomness $r_{\widetilde{\pi}_\gamma}$ and tag τ_γ for all

---

[7] To get an instantiation $\widetilde{\mathsf{HPS}}$ which satisfies the conditions of Theorem 2, $\widetilde{\mathsf{HPS}}$ needs to share the same universe set $\mathcal{X}$ with HPS. In that way, we can set $(\mathcal{G}, d, k, \mathcal{D}_{d+k,d})$ in $\widetilde{\mathsf{mpar}}$ to be exactly the same with the ones in mpar.

[8] Similarly, we set $\widetilde{\mathsf{par}} := \mathsf{par}$ and $\widetilde{\mathsf{td}} := \mathsf{td}$ to make sure $\widetilde{\mathsf{HPS}}$ shares the same language $\mathcal{L}$ with HPS.

$\gamma \in \{1, \cdots, k\}$, the open algorithm computes $\widetilde{\mathsf{sk}}' = \mathbf{s}' \in \mathbb{Z}_q^{2d+2k}$ by solving the following system of linear equations.

$$\mathbf{s'}^{\top}\mathbf{E} = (\mathbf{s}_1^{\top}\mathbf{A},\ \mathbf{s}_2^{\top}\mathbf{A},\ r_{\widetilde{\pi}_1}, \cdots,\ r_{\widetilde{\pi}_k}) \bmod q,\ \ \mathbf{E} = \begin{pmatrix} \mathbf{A} & & \mathbf{x}_1 & \cdots & \mathbf{x}_k \\ & \mathbf{A} & \tau_1\mathbf{x}_1 & \cdots & \tau_k\mathbf{x}_k \end{pmatrix}. \quad (4)$$

Matrix $\mathbf{E}$ has $2d + 2k$ rows and $2d + k$ columns.

- If matrix $(\mathbf{A} \mid \mathbf{x}_1 \mid \cdots \mid \mathbf{x}_k)$ has full column rank $d + k$, then matrix $\mathbf{E}$ has full column rank $2d + k$ and there are q^k possible solutions for $\mathbf{s}'$ to make Eq. (4) hold. Algorithm $\widetilde{\mathsf{HOpen}}_k$ selects and outputs a uniformly random solution.
- Otherwise, algorithm $\widetilde{\mathsf{HOpen}}_k$ outputs $\perp$.

Note that given $\widetilde{\mathsf{td}} = \mathbf{A}$, tags $(\tau_\gamma)_{\gamma \in \{1, \cdots, k\}}$ and the randomnesses $(r_{x_\gamma} = \mathbf{x}_\gamma)_{\gamma \in \{1, \cdots, k\}}$, one can easily compute the matrix $\mathbf{E}$. The right hand side of Eq. (4) is also efficiently computable given $\widetilde{\mathsf{sk}} = \begin{pmatrix} \mathbf{s}_1 \\ \mathbf{s}_2 \end{pmatrix}$ and randomnesses $(r_{\widetilde{\pi}_\gamma})_{\gamma \in \{1, \cdots, k\}}$.

Theorem 5. *The above instantiation $\widetilde{\mathsf{HPS}}$ (1) is a tag-based key-openable HPS; (2) is universal$_{k+1}$ and (3) is openable$_k$.*

We put the proof of Theorem 5 in the full version [17].

6.4   Concrete AC-RSO$_k$&C-CCA Secure PKE Instantiation

We instantiate our PKE scheme by plugging the instantiations, HPS in Sect. 6.2 and $\widetilde{\mathsf{HPS}}$ in Sect. 6.3, into the generic KM-NCE construction in Fig. 7. By Theorem 1, we immediately get a PKE instantiation that can achieve AC-RSO$_k$&C-CCA security in the standard model with *compact* ciphertexts. If we set the matrix distribution $\mathcal{D}_{d+k,d}$ (i.e., the matrix distribution used to sample matrix $\mathbf{A}$ by the key generation algorithm Gen) to be uniform matrix distribution $\mathcal{U}_{d+k,d}$, the resulting PKE can achieve *tight* AC-RSO$_k$&C-CCA security.

Fixing some group generation algorithm GGen, some positive integers d, k, some matrix distribution $\mathcal{D}_{d+k,d}$ and some polynomial $l = l(\lambda)$, the instantiation PKE = (Setup, Gen, Enc, Dec) with message space $\{0, 1\}^l$ is shown in Fig. 9. This scheme can be viewed as a generalization of the DDH-based scheme in [12, Fig. 3] and both schemes are variants of the Cramer-Shoup encryption scheme [9].

We can see that, for the PKE scheme in Fig. 9, the ciphertext length is $(d + k + 1) \times |\mathbb{G}| + l$ for messages of length l and the ciphertext overhead is the size of a constant number of group elements (since d and k are both fixed constants), which is also independent of the message length. This suggests that the PKE instantiation in Fig. 9 has compact ciphertexts [12,15].

We note that our PKE can achieve *tight* AC-RSO$_k$&C-CCA security for certain instantiation. Taking a closer look at the AC-RSO$_k$&C-CCA security of our

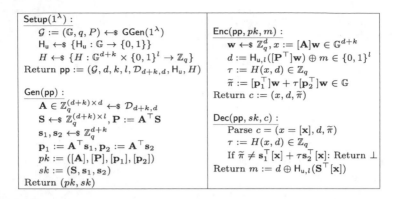

Fig. 9. Concrete AC-RSO$_k$&C-CCA secure PKE instantiation.

MDDH-based PKE instantiation, we obtain the following inequality by combining Eq. (1) in Theorem 1, Eq. (2) in Theorem 2 and Theorem 4 together.

$$\mathbf{Adv}_{\mathsf{PKE},\mathcal{A},\mathcal{S},\mathcal{D},n,t,k}^{\mathrm{ac\text{-}rso\&c\text{-}cca}}(\lambda) \leq n \cdot \mathbf{Adv}_{\mathsf{KM\text{-}NCE},\mathcal{B}',k}^{\mathrm{kmnc\text{-}cca}}(\lambda) + n \cdot t \cdot k \cdot \epsilon_{\mathsf{KM\text{-}NCE}}^{\mathrm{rob}}(\lambda)$$

$$\leq n \cdot \mathbf{Adv}_{\mathcal{D}_{d+k,d},\mathsf{GGen},\mathcal{B}_1}^{k\text{-}\mathrm{mddh}}(\lambda) + 2n \cdot \mathbf{Adv}_{\mathcal{H},\mathcal{B}_2}^{\mathrm{cr}}(\lambda) + 2Q_d n \cdot \epsilon_{\widetilde{\mathsf{HPS}},\mathcal{B}_3}^{\mathrm{univ}_{k+1}}(\lambda) + 2n \cdot \epsilon_{\mathsf{HPS},\mathcal{B}_4}^{\mathrm{open}_k}(\lambda)$$

$$+ 2n \cdot \epsilon_{\widetilde{\mathsf{HPS}},\mathcal{B}_5}^{\mathrm{open}_k}(\lambda) + 2kn \cdot \epsilon_{\mathsf{HPS}}^{\Pi\text{-}\mathrm{resmp}}(\lambda) + n \cdot t \cdot k \cdot \epsilon_{\mathsf{KM\text{-}NCE}}^{\mathrm{rob}}(\lambda). \tag{5}$$

The $2Q_d n \cdot \epsilon_{\widetilde{\mathsf{HPS}},\mathcal{B}_3}^{\mathrm{univ}_{k+1}}(\lambda) + 2n \cdot \epsilon_{\mathsf{HPS},\mathcal{B}_4}^{\mathrm{open}_k}(\lambda) + 2n \cdot \epsilon_{\widetilde{\mathsf{HPS}},\mathcal{B}_5}^{\mathrm{open}_k}(\lambda) + 2kn \cdot \epsilon_{\mathsf{HPS}}^{\Pi\text{-}\mathrm{resmp}}(\lambda) + n \cdot t \cdot k \cdot \epsilon_{\mathsf{KM\text{-}NCE}}^{\mathrm{rob}}(\lambda)$ part in Eq. (5) does not affect tightness of the reduction since it is statistically small. Only reductions to computational properties matter to tightness of the reduction, i.e., the term $n \cdot \mathbf{Adv}_{\mathcal{D}_{d+k,d},\mathsf{GGen},\mathcal{B}_1}^{k\text{-}\mathrm{mddh}}(\lambda) + 2n \cdot \mathbf{Adv}_{\mathcal{H},\mathcal{B}_2}^{\mathrm{cr}}(\lambda)$. This security loss n and $2n$ are introduced because 1) in the proof of Theorem 1 (KMNC$_k$-CCA + robustness $\Rightarrow$ AC-RSO$_k$&C-CCA), we handle one user at a time with n game transitions (cf. Lemma 1), and in each transition, a term $\mathbf{Adv}_{\mathsf{KM\text{-}NCE},\mathcal{B}_1',k}^{\mathrm{kmnc\text{-}cca}}(\lambda)$ is incurred; 2) according to Theorem 2, the term $\mathbf{Adv}_{\mathsf{KM\text{-}NCE},\mathcal{B}_1',k}^{\mathrm{kmnc\text{-}cca}}(\lambda)$ contains $\mathbf{Adv}_{\mathsf{HPS},\mathcal{B}_1'}^{k\text{-}\mathrm{msmp}}(\lambda) + 2 \cdot \mathbf{Adv}_{\mathcal{H},\mathcal{B}_2}^{\mathrm{cr}}(\lambda)$; and 3) according to Theorem 4, $\mathbf{Adv}_{\mathsf{HPS},\mathcal{B}_1''}^{k\text{-}\mathrm{msmp}}(\lambda) \leq \mathbf{Adv}_{\mathcal{D}_{d+k,d},\mathsf{GGen},\mathcal{B}_1}^{k\text{-}\mathrm{mddh}}(\lambda)$.

Alternatively, if we set the matrix distribution to be uniform matrix distribution (i.e., $\mathcal{D}_{d+k,d} := \mathcal{U}_{d+k,d}$), we can avoid such security loss by integrating the proofs of Theorem 1, Theorem 2 and Theorem 4. We can handle the n reductions to the k-fold $\mathcal{U}_{d+k,d}$-MDDH assumption (i.e., $n \cdot \mathbf{Adv}_{\mathcal{U}_{d+k,d},\mathsf{GGen},\mathcal{B}_1}^{k\text{-}\mathrm{mddh}}(\lambda)$) and the $2n$ reductions to the collision-resistance of $\mathcal{H}$ (i.e., $2n \cdot \mathbf{Adv}_{\mathcal{H},\mathcal{B}_2}^{\mathrm{cr}}(\lambda)$) *for all n users at one time* (while keeping the reductions to other statistical properties unchanged, namely one user at a time). Specifically,

– we can change all the kn ciphertexts (of all n users) at one time, corresponding to the game transition G_1 to G_2 in the proof of Theorem 2, and the indistinguishability can be reduced to the $\mathcal{U}_{d+k,d}$-MDDH assumption using Lemma 5 in below;

– we can handle collisions of all users at one time, corresponding to the game transitions G_2 to G_3 and G_7 to G_8 in the proof of Theorem 2.

With this strategy, we obtain a tight reduction with $\mathbf{Adv}^{\mathrm{mddh}}_{\mathcal{U}_{d+k,d},\mathsf{GGen},\mathcal{B}_1}(\lambda) + 2 \cdot \mathbf{Adv}^{\mathrm{cr}}_{\mathcal{H},\mathcal{B}_2}(\lambda)$, instead of $n \cdot \mathbf{Adv}^{k\text{-mddh}}_{\mathcal{U}_{d+k,d},\mathsf{GGen},\mathcal{B}_1}(\lambda) + 2n \cdot \mathbf{Adv}^{\mathrm{cr}}_{\mathcal{H},\mathcal{B}_2}(\lambda)$, to the computational properties. Thus, the PKE scheme enjoys tight security reduction.

Lemma 5. *For any adversary* $\mathcal{A}$*, any positive integer* d, k, n*, any matrix distribution* $\mathcal{D}_{d+k,d}$ *and any group generation algorithm* GGen*, we define the advantage* $\mathbf{Adv}^{(n,k)\text{-mddh}}_{\mathcal{D}_{d+k,d},\mathsf{GGen},\mathcal{A}}(\lambda) := |\Pr[\mathcal{A}(\mathcal{G}, ([\mathbf{A}_i], [\mathbf{X}_i])^n_{i=1}) = 1] - \Pr[\mathcal{A}(\mathcal{G}, ([\mathbf{A}_i], [\mathbf{X}'_i])^n_{i=1}) = 1]|$ *where* $\mathcal{G} \leftarrow_{\$} \mathsf{GGen}(1^\lambda)$*,* $\mathbf{A}_i \leftarrow_{\$} \mathcal{D}_{d+k,d}$*,* $\mathbf{W}_i \leftarrow_{\$} \mathbb{Z}^{d \times k}_q$*,* $\mathbf{X}_i := \mathbf{A}_i \mathbf{W}_i$ *and* $\mathbf{X}'_i \leftarrow_{\$} \mathbb{Z}^{(d+k) \times k}_q$ *for all* $i \in \{1, \cdots, n\}$*. Then, for any PPT adversary* $\mathcal{A}$ *and uniform matrix distribution* $\mathcal{U}_{d+k,d}$*, there exists a PPT adversary* $\mathcal{B}$ *such that*

$$\mathbf{Adv}^{(n,k)\text{-mddh}}_{\mathcal{U}_{d+k,d},\mathsf{GGen},\mathcal{A}}(\lambda) \leq \mathbf{Adv}^{\mathrm{mddh}}_{\mathcal{U}_{d+k,d},\mathsf{GGen},\mathcal{B}}(\lambda) + \frac{k+1}{q-1}.$$

We put the proof of Lemma 5 in the full version [17].

Acknowledgment. We appreciate the anonymous reviewers for their valuable comments. This work was supported by National Natural Science Foundation of China (Grant Nos. 61922036, U2001205, 62002223, 61825203), Major Program of Guangdong Basic and Applied Research Project (Grant No. 2019B030302008), National Joint Engineering Research Center of Network Security Detection and Protection Technology, Guangdong Key Laboratory of Data Security and Privacy Preserving, Guangdong Provincial Science and Technology Project (Grant No. 2021A0505030033), Shanghai Sailing Program (20YF1421100), Young Elite Scientists Sponsorship Program by China Association for Science and Technology (YESS20200185), and the European Research Council (ERC) under the European Union's Horizon 2020 research and innovation programme (Grant agreement 802823).

References

1. Abdalla, M., Bellare, M., Neven, G.: Robust encryption. In: Micciancio, D. (ed.) TCC 2010. LNCS, vol. 5978, pp. 480–497. Springer, Heidelberg (2010). https://doi.org/10.1007/978-3-642-11799-2_28

2. Bellare, M., Boldyreva, A., Desai, A., Pointcheval, D.: Key-privacy in public-key encryption. In: Boyd, C. (ed.) ASIACRYPT 2001. LNCS, vol. 2248, pp. 566–582. Springer, Heidelberg (2001). https://doi.org/10.1007/3-540-45682-1_33

3. Bellare, M., Dowsley, R., Waters, B., Yilek, S.: Standard security does not imply security against selective-opening. In: Pointcheval, D., Johansson, T. (eds.) EUROCRYPT 2012. LNCS, vol. 7237, pp. 645–662. Springer, Heidelberg (2012). https://doi.org/10.1007/978-3-642-29011-4_38

4. Bellare, M., Stepanovs, I.: Security under message-derived keys: Signcryption in iMessage. In: Canteaut, A., Ishai, Y. (eds.) EUROCRYPT 2020. LNCS, vol. 12107, pp. 507–537. Springer, Cham (2020). https://doi.org/10.1007/978-3-030-45727-3_17

5. Benhamouda, F., et al.: Can a public blockchain keep a secret? In: Pass, R., Pietrzak, K. (eds.) TCC 2020. LNCS, vol. 12550, pp. 260–290. Springer, Cham (2020). https://doi.org/10.1007/978-3-030-64375-1_10

6. Böhl, F., Hofheinz, D., Kraschewski, D.: On definitions of selective opening security. In: Fischlin, M., Buchmann, J., Manulis, M. (eds.) PKC 2012. LNCS, vol. 7293, pp. 522–539. Springer, Heidelberg (2012). https://doi.org/10.1007/978-3-642-30057-8_31

7. Canetti, R., Feige, U., Goldreich, O., Naor, M.: Adaptively secure multi-party computation. In: STOC, pp. 639–648 (1996)

8. Canetti, R., Halevi, S., Katz, J.: Adaptively-secure, non-interactive public-key encryption. In: Kilian, J. (ed.) TCC 2005. LNCS, vol. 3378, pp. 150–168. Springer, Heidelberg (2005). https://doi.org/10.1007/978-3-540-30576-7_9

9. Cramer, R., Shoup, V.: Universal hash proofs and a paradigm for adaptive chosen ciphertext secure public-key encryption. In: Knudsen, L.R. (ed.) EUROCRYPT 2002. LNCS, vol. 2332, pp. 45–64. Springer, Heidelberg (2002). https://doi.org/10.1007/3-540-46035-7_4

10. Escala, A., Herold, G., Kiltz, E., Ràfols, C., Villar, J.: An algebraic framework for Diffie–Hellman assumptions. In: Canetti, R., Garay, J.A. (eds.) CRYPTO 2013. LNCS, vol. 8043, pp. 129–147. Springer, Heidelberg (2013). https://doi.org/10.1007/978-3-642-40084-1_8

11. Han, S., Liu, S., Lyu, L., Gu, D.: Tight leakage-resilient CCA-security from quasi-adaptive hash proof system. In: Boldyreva, A., Micciancio, D. (eds.) CRYPTO 2019. LNCS, vol. 11693, pp. 417–447. Springer, Cham (2019). https://doi.org/10.1007/978-3-030-26951-7_15

12. Hara, K., Kitagawa, F., Matsuda, T., Hanaoka, G., Tanaka, K.: Simulation-based receiver selective opening CCA secure PKE from standard computational assumptions. In: Catalano, D., De Prisco, R. (eds.) SCN 2018. LNCS, vol. 11035, pp. 140–159. Springer, Cham (2018). https://doi.org/10.1007/978-3-319-98113-0_8

13. Hayashi, R., Tanaka, K.: The sampling twice technique for the RSA-based cryptosystems with anonymity. In: Vaudenay, S. (ed.) PKC 2005. LNCS, vol. 3386, pp. 216–233. Springer, Heidelberg (2005). https://doi.org/10.1007/978-3-540-30580-4_15

14. Hazay, C., Patra, A., Warinschi, B.: Selective opening security for receivers. In: Iwata, T., Cheon, J.H. (eds.) ASIACRYPT 2015. LNCS, vol. 9452, pp. 443–469. Springer, Heidelberg (2015). https://doi.org/10.1007/978-3-662-48797-6_19

15. Hofheinz, D., Jager, T., Rupp, A.: Public-key encryption with simulation-based selective-opening security and compact ciphertexts. In: Hirt, M., Smith, A. (eds.) TCC 2016. LNCS, vol. 9986, pp. 146–168. Springer, Heidelberg (2016). https://doi.org/10.1007/978-3-662-53644-5_6

16. Huang, Z., Lai, J., Chen, W., Au, M.H., Peng, Z., Li, J.: Simulation-based selective opening security for receivers under chosen-ciphertext attacks. Des. Codes Cryptogr. 87(6), 1345–1371 (2018). https://doi.org/10.1007/s10623-018-0530-1

17. Huang, Z., Lai, J., Han, S., Lyu, L., Weng, J.: Anonymous public key encryption under corruptions. Cryptology ePrint Archive (2022)

18. Jia, D., Lu, X., Li, B.: Receiver selective opening security from indistinguishability obfuscation. In: Dunkelman, O., Sanadhya, S.K. (eds.) INDOCRYPT 2016. LNCS, vol. 10095, pp. 393–410. Springer, Cham (2016). https://doi.org/10.1007/978-3-319-49890-4_22

19. Jia, D., Lu, X., Li, B.: Constructions secure against receiver selective opening and chosen ciphertext attacks. In: Handschuh, H. (ed.) CT-RSA 2017. LNCS, vol.

10159, pp. 417–431. Springer, Cham (2017). https://doi.org/10.1007/978-3-319-52153-4_24

20. Lee, Y., Lee, D.H., Park, J.H.: Tightly CCA-secure encryption scheme in a multi-user setting with corruptions. Des. Codes Cryptogr. **88**(11), 2433–2452 (2020). https://doi.org/10.1007/s10623-020-00794-z

21. Mohassel, P.: A closer look at anonymity and robustness in encryption schemes. In: Abe, M. (ed.) ASIACRYPT 2010. LNCS, vol. 6477, pp. 501–518. Springer, Heidelberg (2010). https://doi.org/10.1007/978-3-642-17373-8_29

22. Qin, B., Liu, S., Chen, K.: Efficient chosen-ciphertext secure public-key encryption scheme with high leakage-resilience. IET Inf. Secur. **9**(1), 32–42 (2015)

23. Yang, R., Lai, J., Huang, Z., Au, M.H., Xu, Q., Susilo, W.: Possibility and impossibility results for receiver selective opening secure PKE in the multi-challenge setting. In: Moriai, S., Wang, H. (eds.) ASIACRYPT 2020. LNCS, vol. 12491, pp. 191–220. Springer, Cham (2020). https://doi.org/10.1007/978-3-030-64837-4_7

Memory-Tight Multi-challenge Security of Public-Key Encryption

Joseph Jaeger$^{(\boxtimes)}$ and Akshaya Kumar

School of Cybersecurity and Privacy Georgia Institute of Technology,
Atlanta, GA, USA
{josephjaeger,akshayakumar}@gatech.edu

Abstract. We give the first examples of public-key encryption schemes which can be proven to achieve multi-challenge, multi-user CCA security via reductions that are tight in time, advantage, *and* memory. Our constructions are obtained by applying the KEM-DEM paradigm to variants of Hashed ElGamal and the Fujisaki-Okamoto transformation that are augmented by adding uniformly random strings to their ciphertexts.

The reductions carefully combine recent proof techniques introduced by Bhattacharyya'20 and Ghoshal-Ghosal-Jaeger-Tessaro'22. Our proofs for the augmented ECIES version of Hashed-ElGamal make use of a new computational Diffie-Hellman assumption wherein the adversary is given access to a pairing to a random group, which we believe may be of independent interest.

Keywords: Public-key cryptography · Provably security · Memory-tightness

1 Introduction

Secure deployment of cryptography requires concrete analysis of schemes to understand how the success probabilities of attackers grow with the amount of resources they employ to attack a system. The use of reduction-based cryptography enables such analysis by using an attacker with running time t and success probability ϵ to construct a related adversary with running time t' and success probability ϵ' against a computational problem whose security is better understood. A gold standard for concrete security reductions are tight reductions for which $t' \approx t$ and $\epsilon' \approx \epsilon$. We refer to such a reduction as *TA-tight* (time-advantage-tight) to distinguish it from other notions of tightness.

Auerbach, Cash, Fersch, and Kiltz [3] argued that the memory usage of an attacker can be crucial in determining its likelihood of success. This kicked of a line of works [7,9–12,17–19,24,28,30,31] on *memory-aware* cryptography which accounts for the memory usage of attackers in security analyses. Auerbach, et al. focused in particular on incorporating memory considerations into the study of reductions. We refer to a reduction as TAM-tight if it is TA-tight and additionally $s \approx s'$ where these variables, respectively, denote the amount of memory used by the original adversary and the reduction adversary.

S. Agrawal and D. Lin (Eds.): ASIACRYPT 2022, LNCS 13793, pp. 454–484, 2022.
https://doi.org/10.1007/978-3-031-22969-5_16

In this work, we construct the first public-key encryption schemes with TAM-tight proofs of multi-challenge (and multi-user) chosen-ciphertext attack (CCA) security. Our schemes are based on variants of the Hashed ElGamal and Fujisaki-Okamoto transformation key encapsulation mechanisms. These variants augment ciphertexts with random strings that are included in hash function calls.

MULTI-CHALLENGE SETTING. As mentioned, our focus in this work is on multi-challenge and multi-user security. This is simply motivated by the fact that encryption schemes get deployed across many different users each of whom will encrypt many messages, so it is important to understand how the security of a scheme degrades as the number of encryptions increase. In particular, the goal of tight proofs is to show that security does not meaningfully degrade. Multiple papers [16, 22, 25] have looked at this in the non-memory-aware setting, providing schemes with TA-tight proofs of security. However, extending any of these proofs to the memory-aware setting is quite difficult.

Prior works on memory-tight CCA secure encryption have identified a primary difficulty in the multi-challenge setting which lies in how the decryption oracle handles challenge ciphertexts. Simply decrypting a challenge ciphertext would lead to trivial attacks against any scheme, so instead the decryption oracle has to recognize these ciphertexts and respond to them in a special manner.[1] This makes writing memory-tight security proofs difficult because the reduction adversary must emulate this differing behavior on decryption queries for challenge or non-challenge ciphertexts, but it is unclear how to go about identifying which are challenge ciphertexts other than remembering and checking against all ciphertexts that were previously returned to encryption queries. In the single-challenge setting is a non-issue, because storing the single challenge ciphertext requires minimal memory.

MEMORY-TIGHTNESS OF HASHED ELGAMAL. In recent years, several papers have discussed the challenge of providing memory-tight security proofs for Hashed ElGamal. Auerbach, et al. [3] gave it at as an example of a proof they considered the memory complexity of, but were unable to improve. Follow-up work by Bhattacharyya [7] and Ghoshal and Tessaro [19] analyzed this further, giving what might seem at first to be contradictory results. Bhattacharyya gave a memory-tight proof for Hashed ElGamal in the single-challenge setting while Ghoshal and Tessaro proved a lower-bound establishing that a memory-tight proof for Hashed ElGamal was not possible.

Resolving this contradiction requires more precisely understanding each result. The lower bound applies specifically for reductions to Strong Computational Diffie-Hellman (CDH) security [2] which are "black-box" in several ways, including that they do not depend on the particular group used. Ghoshal and Tessaro note that Bhattacharyya's result (for single-challenge security) avoids the lower bound by not being black-box in this manner; it depends on the group having an

---

[1] An alternate definitional style would disallow the adversary from querying challenge ciphertexts to its decryption oracle, but prior the works argue this is an inappropriate restriction in the memory-bounded setting [17, 18].

Scheme/Transform	Assumption	Result
Hashed ElGamal (aECIES)	Pair CDH	$CCA KEM
Cramer-Shoup ElGamal (aCS)	Strong CDH	$CCA KEM
Twin ElGamal (aTWIN)	CDH	$CCA KEM
T	$CPA PKE	OW-PCA PKE
aV	OW-PCA PKE	OW-PCVA PKE
aU$^\perp$	OW-PCVA PKE	CCA KEM
KEM/DEM (KD)	$CCA KEM/SKE	$CCA PKE

Fig. 1. TAM-tight reductions we provide. Transformations T, aV, and $\mathsf{aU}^\perp$ are FO transforms discussed in Sect. 5. Results are multi-user, multi-challenge security.

efficient pairing. However, for efficiency it is preferable to implement schemes using elliptical curves for which efficient pairing are believed not to exist.

Our result for Hashed ElGamal is black-box in the sense of Ghoshal and Tessaro. We avoid the lower-bound without requiring an efficient pairing by introducing and using an assumption (Pair CDH) which is stronger than Strong CDH, but is reasonable to assume holds in typical groups based on elliptic curves.[2] We discuss this assumption in more detail momentarily. Indeed, Ghoshal and Tessaro say in their paper [19, Sec. 3.1, p.42], "it appears much harder to extend our result to different types of oracles than [the Strong CDH oracle], as our proof is tailored at this oracle." Our new security notion gives an example of such an oracle to which their result cannot be extended.

1.1 Our Results

We summarize our results in Fig. 1. Omitted proofs and results are provided in the full version of this paper [23].

HASHED ELGAMAL. Our first results consider the security of Hashed ElGamal. Following Bhattacharyya, we actually consider two variants which we refer to as the ECIES [1] and the Cramer-Shoup [8] variants. The negative results of Ghoshal and Tessaro apply only to the ECIES variant. In both, the decryption key is a value $x \in \mathbb{Z}_p^*$ and the encapsulation key is $X = \mathbf{g}^x$. Here $\mathbf{g}$ is a generator of a group of prime order p. For encapsulation, one samples a fresh $y \leftarrow_\$ \mathbb{Z}_p^*$ and returns $\mathbf{g}^y$ as the ciphertext. For ECIES the derived key is $H(X^y)$, while for Cramer-Shoup it is $H(\mathbf{g}^y, X^y)$. Our main results concern "augmented" versions of both of these schemes where the ciphertexts are instead $(a, \mathbf{g}^y)$ where a is a uniformly random bitstring used as an additional input to the hash function.

---

[2] Technically, the lower bound also does not apply because we are considering an augmented scheme which differs from the one analyzed by Ghoshal and Tessaro. However, the augmentation is not important for this comparison, because in the single-challenge setting where the lower-bound was proven, Pair CDH TAM-tightly implies security of the non-augmented scheme as well.

To understand these results, let us discuss the high level idea of proving security for ECIES. A standard, single-challenge proof would work from the Strong CDH assumption in the random oracle model. In Strong CDH an adversary is given $X = \mathbf{g}^x$, $Y = \mathbf{g}^y$, and an oracle O which on input B, C tells whether $B^x = C$. Its goal is to return X^y. The only way to distinguish $H(X^y)$ from random is to query H on input X^y. So a reduction adversary will give X as the encryption key, Y as the challenge ciphertext, simulate random oracle and decapsulation queries, and checks if any of the random oracle queries are X^y in which case it returns that. The oracle O is used for checking whether random oracle queries are X^y (for a random oracle query Z, one queries $O(Y, Z)$ to check) and for maintaining consistency between random oracle and decapsulation queries for non-challenge ciphertexts. A decapsulation query for Y and a random oracle query for Y, Z should return the same result if $Y^x = Z$. The reduction can maintain this consistency by remembering all of the queries made to both oracles and then using O to check for this consistency. This is neither time- nor memory-tight.

Bhattacharyya was able to make this TAM-tight by introducing a new technique for this consistency aspect. They simulate the random oracle $H(C)$ by $h(e(\mathbf{g}, C))$ where h is a random function and e is a pairing. Then the output of a non-challenge decapsulation query B can be simulated as $h(e(X, B))$. In our proof we use a similar technique, but replace the requirement for a pairing-friendly group by using a new variant of CDH we will discuss momentarily.

The first step in making the proof work in the multi-challenge setting is to use Diffie-Hellman rerandomization techniques so we can have multiple Diffie-Hellman challenges. We let the u-th user's public key be X^{x_u} and the i-th ciphertext by Y^{y_i}. For memory-tightness, we pick x_u and y_i using a (pseudo-)random function.[3] Then if the adversary makes a random oracle query $H(C)$ where $C = X^{x_u \cdot y \cdot y_u}$, we have $C^{1/(x_u \cdot y_u)} = X^y$. A challenge here is to know which u and i to use for such a random oracle query. A reduction could check each choice of u, i, but this would lose time-tightness. At the same time, for a decapsulation query B we must be able to identify if B was a prior challenge query. Storing all prior challenge queries loses memory-tightness.

Both of these issues are solved by our addition of an auxiliary string a to each ciphertext and hash query. The idea here is based on memory-tightness techniques of GGJT [17], in that we are going to hide the pertinent information we need in a. Rather than sampling a at random, if the i-th challenge query is made to user u then our reduction adversary picks a to be the "encryption" of (u, i). Now on future random oracle and decapsulation queries we can recover u, i by "decrypting" a. This allows us to properly simulate the view of the adversary.

PAIR CDH SECURITY. As we have been mentioning, we avoid the need for groups with pairing in our result for ECIES by making use of a new computational assumption. This assumption, we refer to as Pair CDH security, extends CDH security by giving the adversary access to an oracle which, on input A and B

---

[3] In the body, this is separated out as a proof that single-challenge CDH tightly implies multi-challenge CDH.

(with discrete logarithms a and b) computes a and b then returns a random function applied to $a \cdot b$. This acts, in essence, as a pairing from the group under consideration to a randomly chosen group. Our use of this in our security proof for ECIES takes advantage of the fact that (i) the pairing is only needed for the proof, not in the construction itself and (ii) the proof does not require the ability to efficiently perform group operations with the output of the pairing. We think this notion may be of further interest if other proofs can be found where better tightness can be achieved using a pairing only in the reduction.

To justify our new assumption we analyze how it compares to existing assumptions. Pair CDH security is implied by CDH security if the group under consideration has an efficient pairing. This holds because we can emulate the random pairing by first applying the efficient pairing and then applying a random function (which may be pseudorandomly instantiated for efficiency). In turn, Pair CDH implies the Gap CDH assumption because a pairing can be used to check whether given group elements form a Diffie-Hellman triple.

These results do not justify the use of Pair CDH security for typical groups based on elliptic curves which do not have pairings. For this, we turn to non-standard models (i.e. algebraic or generic group models [13, 26, 29]). In these models, we are able to show that CDH and Pair CDH are equivalent because learning anything from the oracle requires the ability to find non-trivial collisions in the pairing. The ability to find such collisions can in turn be used to solve the discrete logarithm problem.

FUJISAKI-OKAMOTO TRANSFORMATION. The other KEMs we consider are those derived from the Fujisaki-Okamoto Transformation which starts with a CPA secure public key encryption scheme and applies several random oracle based transformations to construct a CCA secure KEM. Hofheinz, Hövelmanns, and Kiltz [21] gave a nice modular approach for proving the security of several variants of this transformation. Bhattacharyya showed how to make these proofs memory-tight in the single-challenge setting (in some cases requiring one additional intermediate transformation). We extend these to the multi-challenge setting. For the final step of the transform, we need to consider an augmented transform of the existing scheme in which random strings are added to each ciphertext and incorporated into the hash queries. As before, our reduction samples these string as the encryption of the pertinent information it would need to identify challenge ciphertexts and respond to them appropriately.

For these results, we require that the starting CPA scheme have good multi-challenge security. This is a significantly weaker starter point than multi-challenge CCA security because it avoids the issue of having to be able to identify challenge ciphertexts for the decryption oracle.

LIFTING TO PUBLIC-KEY ENCRYPTION. The approaches described above are for key encapsulation mechanisms. This raises the question of whether these tight reductions can be applied to public-key encryption via the KEM-DEM paradigm. It uses a KEM to generate a new symmetric key for a data encapsulation mechanism to encrypt the actual message with. This was previously looked at by GGJT [17], who gave a TAM-tight proof of security. However, because of their particular motivations, the proof assumed the KEM was constructed from a public-key encryption scheme. We show that (with some modifications) the proof works with generic KEMs as well.

2   Preliminaries

2.1   Notation

We recall basic notion and security definitions we will use in our paper.

PSEUDOCODE. For our proofs, we use the code based framework of [6]. If $\mathcal{A}$ is an algorithm, then $x \leftarrow \mathcal{A}^O(x_1, x_2, ...; r)$ denotes running $\mathcal{A}$ on inputs $x_1, x_2, ...$ with coins r and having access to the set of oracles O to produce output x. We use the notation $\leftarrow_\$$ instead of $\leftarrow$ when not explicitly specifying the coins r. If S is a set, $|S|$ denotes its size and $x \leftarrow_\$ S$ denotes sampling x uniformly from S. We use the symbol $\perp$ to indicate rejection. When not specified, tables are initialized empty and integers are initialized to 0.

Security notions are defined with games such as the one in Fig. 3. The probability that the game G outputs true is $\Pr[\mathsf{G}]$. We sometimes use a sequence of "hybrid" games in one figure for our proofs. We use comments of the form $//\mathsf{G}_{[i,j)}$ to indicate that a line of code is included in games G_k for $i \leqslant k < j$. To identify changes made to the kth hybrid, one looks for lines of code commented as $//\mathsf{G}_{[i,k)}$ for code that is no longer included in the kth hybrid and $//\mathsf{G}_{[k,j)}$ for code that is new to the kth hybrid.

COMPLEXITY MEASURES. Following ACFK [3], we measure the local complexities of algorithms and do not include the complexity of oracles that they interact with. We focus on the worst case runtime $\mathbf{Time}(\mathcal{A})$ and memory used for local computation $\mathbf{Mem}(\mathcal{A})$ of any algorithm $\mathcal{A}$.

FUNCTIONS AND IDEAL MODELS. We define $\mathsf{Fcs}(D, R)$ (resp. $\mathsf{Inj}(D, R)$) to be the set of all functions (resp. injections) mapping from D to R. For $f \in \mathsf{Inj}(D, R)$, we define f^{-1} to be its inverse (with $f^{-1}(y) = \perp$ if y has no preimage). If D_t and R_t are sets for each $t \in T$, then we define $\mathsf{Fcs}(T, D, R)$ (resp. $\mathsf{Inj}(T, D, R)$) to be the set of functions f so that $f(t, \cdot) \in \mathsf{Fcs}(D_t, R_t)$ (resp. $f(t, \cdot) \in \mathsf{Inj}(D_t, R_t)$). We let $f_t(\cdot) = f(t, \cdot)$.

For $f \in \mathsf{Inj}(D, R)$ we let $f^\pm$ denote the function defined by $f^\pm(+, x) = f(x)$ and $f^\pm(-, x) = f^{-1}(x)$. We often write $f(x)$ or $f^{-1}(x)$ in place of $f^\pm(+, x)$ or $f^\pm(-, x)$. We let $\mathsf{Inj}^\pm(D, R) = \{f^\pm : f \in \mathsf{Inj}(D, R)\}$ and extend this to define $\mathsf{Inj}^\pm(T, D, R)$ analogously.

Ideal models (e.g. the random oracle or ideal cipher model) are captured by having a scheme S specify a set of functions S.IM. Then, at the beginning of a security game for S, a random $\mathcal{H} \in$ S.IM is sampled. The adversary and some algorithms of the scheme S are then given oracle access to $\mathcal{H}$. The standard model is captured by S.IM being a singleton set containing the identity function.

If F and G are sets of functions, then we define $(\mathsf{F}, \mathsf{G}) = \mathsf{F} \times \mathsf{G} = \{ f \times g : f \in \mathsf{F}, g \in \mathsf{G} \}$. Here, $f \times g$ is the function defined by $f \times g(0, x) = f(x)$

and $f \times g(1, x) = g(x)$. In the code of an algorithm expecting oracle access to $f \times g \in \mathsf{F} \times \mathsf{G}$, we write $f(x)$ or $g(x)$ with the natural meaning. We extend this notation to more than two sets of functions as well.

SWITCHING LEMMA. Our proofs make use of the indistinguishability of random functions and injections, as captured by the following standard result.

Lemma 1 (Switching Lemma). *Fix T, D, R and $N = \min_{t \in T} |R_t|$. For any adversary $\mathcal{A}$ making at most q queries, it holds that $|\Pr[\mathcal{A}^f \Rightarrow 1] - \Pr[\mathcal{A}^g \Rightarrow 1]| \leq 0.5 \cdot q^2 / N$, where the probability is taken over the randomness of $\mathcal{A}$, sampling $f \leftarrow_\$ \mathsf{Fcs}(T, D, R)$, and sampling $g \leftarrow_\$ \mathsf{Inj}(T, D, R)$.*

2.2   Memory-Tightness Background

$\mathcal{F}$-ORACLE ADVERSARIES. We adopt GGJT's [17] oracle adversary formulation for our proofs in the memory-aware setting, i.e., we allow reductions to access uniformly random functions or invertible random injections. Our reductions are of the form shown below for some set of functions $\mathcal{F}$ and algorithm $\mathcal{B}$. We call such an adversary $\mathcal{A}$ an $\mathcal{F}$-oracle adversary.

$$\frac{\text{Adversary } \mathcal{A}^{\mathrm{O}}(\text{in})}{\begin{array}{l} f \leftarrow_\$ \mathcal{F} \\ \text{out} \leftarrow_\$ \mathcal{B}^{\mathrm{O}, f}(\text{in}) \\ \text{Return out} \end{array}}$$

The complexity of adversary $\mathcal{A}$ would include the (large) complexity of sampling, storing, and computing f. However, as proposed in [17], we present theorems in terms of the *reduced complexity* of an oracle aided adversary which is defined as $\mathbf{Time}^*(\mathcal{A}) = \mathbf{Time}(\mathcal{B})$ and $\mathbf{Mem}^*(\mathcal{A}) = \mathbf{Mem}(\mathcal{B})$.

We refer readers to Lemma 2 of [17] which bounds how much an adversary may be aided by a random object by replacing it with a pseudorandom version of the object. Pseudorandom injections can typically be instantiated by appropriately chosen encryption schemes.

There is a small issue when pseudorandomly instantiating a random function if the game $\mathcal{A}$ plays is inefficient. This is the case for some of our reduction adversaries playing CDH variants wherein they have access to some inefficient oracle based on the group. Then the pseudorandomness reduction adversary from [17] will be inefficient because it simulates the game that $\mathcal{A}$ is playing. However, we can simply use pseudorandom schemes believed to be secure even against adversaries with access to the inefficient oracle. This seems reasonable as we can choose a pseudorandom scheme which seems unrelated to the group.

MESSAGE ENCODING TECHNIQUES. The message encoding technique proposed by GGJT in [17] programs randomness that a reduction provides to an adversary in a special way that stores retrievable state information. This is achieved by generating randomness as the output of random injections. The reduction may then invert randomness generated thusly to retrieve state information. For

PKE Syntax	KEM Syntax	SKE Syntax
$(ek, dk) \leftarrow_\$ \mathsf{PKE.K}$	$(ek, dk) \leftarrow_\$ \mathsf{KEM.K}$	$K \leftarrow_\$ \mathsf{SKE.K}$
$c \leftarrow_\$ \mathsf{PKE.E}^{\mathcal{H}}(ek, m)$	$(c, K) \leftarrow_\$ \mathsf{KEM.E}^{\mathcal{H}}(ek)$	$c \leftarrow_\$ \mathsf{SKE.E}^{\mathcal{H}}(K, m)$
$m \leftarrow \mathsf{PKE.D}^{\mathcal{H}}(dk, c)$	$K \leftarrow \mathsf{KEM.D}^{\mathcal{H}}(dk, c)$	$m \leftarrow \mathsf{SKE.D}^{\mathcal{H}}(K, c)$

Fig. 2. Syntax of a public key encryption scheme PKE, key encapsulation mechanism KEM, and symmetric key encryption scheme SKE. The ideal model oracle is $\mathcal{H}$.

example, consider a key encapsulation mechanism that outputs ciphertexts of the form (a, c) where a is uniformly random. Then a reduction can simulate challenge ciphertexts by setting $a = f(i)$ where f is a random injection and i is some pertinent information the reduction would want to know if the adversary later makes oracle queries for the same ciphertext. Then the reduction can recover this information during future queries as $i \leftarrow f^{-1}(a)$.

MAP-THEN-RANDOM-FUNCTION. We describe the main proof technique of Bhattacharyya [7], namely "map-then-rf".[4] This technique allows the reduction to use the composition of an injection and a random function to replace a random function. This relies on the simple fact that if $h \in \mathsf{Inj}(D, S)$, then sampling f according to $f \leftarrow_\$ \mathsf{Fcs}(D, R)$ or $g \leftarrow_\$ \mathsf{Fcs}(S, R); f \leftarrow g \circ h$ are equivalent, meaning, if g is a random function, and h is any injection, then $f \leftarrow g \circ h$ is a random function. This allows a reduction to compute the output $f(x)$ given $h(x)$, even if it does not know x.

2.3 Public Key Encryption

SYNTAX. A public key encryption scheme, PKE, specifies three algorithms - the key generation algorithm (PKE.K) that returns a pair of keys (ek, dk) where ek is the encryption key and dk is the corresponding decryption key, the encryption algorithm (PKE.E) that takes the encryption key ek and a message m and returns ciphertext c, and the decryption algorithm PKE.D that takes the decryption key dk and a ciphertext c and returns message m (or the special symbol $\perp$ to indicate rejection). The syntax of these algorithms is given in Fig. 2.

Perfect correctness requires $\mathsf{PKE.D}^{\mathcal{H}}(dk, c) = m$ for all $(ek, dk) \in [\mathsf{PKE.K}]$, all m, all $\mathcal{H} \in \mathsf{PKE.IM}$, and all $c \in [\mathsf{PKE.E}^{\mathcal{H}}(ek, m)]$. The weaker notion of δ-correctness requires that for all (not necessarily efficient) $\mathcal{D}$,

$$\Pr[\mathsf{PKE.D}^{\mathcal{H}}(dk, \mathsf{PKE.E}^{\mathcal{H}}(ek, m)) \neq m \; : \; m \leftarrow_\$ \mathcal{D}^{\mathcal{H}}(ek, dk)] \leqslant \delta(q)$$

where q upper bounds the number of $\mathcal{H}$ queries $\mathcal{D}$ makes. The probability is over $(ek, dk) \leftarrow_\$ \mathsf{PKE.K}$, $\mathcal{H} \leftarrow_\$ \mathsf{PKE.IM}$, and the coins of $\mathcal{D}$ and PKE.E. When not stated otherwise, schemes are assumed to be perfectly correct.

---

[4] Bhattacharyya actually uses "map-then-prf", as they were not using the oracle adversary formulation.

Game $G_{PKE,b}^{mu\text{-}\$cca}(\mathcal{A})$	$ENC_b(u, m)$	$DEC(u, c)$
$\mathcal{H} \leftarrow_\$ PKE.IM$	$c_1 \leftarrow_\$ PKE.E^{\mathcal{H}}(ek_u, m)$	If $M[u, c] \neq \bot$
$(ek_{(\cdot)}, dk_{(\cdot)}) \leftarrow_\$ PKE.K$	$c_0 \leftarrow_\$ PKE.\mathcal{C}(ek_u, \|m\|)$	Return $M[u, c]$
$b' \leftarrow_\$ \mathcal{A}^{NEW, ENC_b, DEC, \mathcal{H}}$	$M[u, c_b] \leftarrow m$	$m \leftarrow PKE.D^{\mathcal{H}}(dk, c)$
Return $b' = 1$	Return c_b	Return m
$NEW(u)$		
Return ek_u		

Fig. 3. Game defining mu-$cca security of PKE.

We define the encryption keyspace as $PKE.Ek = \{ek : (ek, dk) \in [PKE.K]\}$ and assume that for each $ek \in PKE.Ek$ and allowed message length n, there exists a set $PKE.\mathcal{C}(ek, n)$ such that $PKE.E^{\mathcal{H}}(ek, m) \in PKE.\mathcal{C}(ek, \|m\|)$ always holds. We assume this set is disjoint for distinct message lengths and let $PKE.\mathcal{C}^{-1}(ek, c)$ return n such that $c \in PKE.\mathcal{C}(ek, n)$. We let $PKE.\mathcal{R}$ denote the set from which $PKE.E$ draws its randomness. Sometimes we assume that all messages to be encrypted are drawn from a set $PKE.\mathcal{M}$ of equal length messages and then let $PKE.\mathcal{C}$ simply denote the set of all possible ciphertexts.

INDISTINGUISHABLE FROM RANDOM SECURITY. We consider indistinguishable from random, chosen ciphertext attack ($CCA) security as captured by Fig. 3. The definition multi-user and multi-challenge (allowing multiple challenges per user). It requires ciphertexts output by the encryption scheme be indistinguishable from random, even when given access to a decryption oracle. In this game, the adversary obtains the encryption key ek_u for user u by querying $NEW(u)$. It makes an encryption query $ENC(u, m)$ to receive a challenge encryption of m by u and a decryption query $DEC(u, c)$ to have u decrypt c. The adversary needs to distinguish between the real world ($b = 1$) in which a query to $ENC(u, m)$ returns a real encryption of m and the ideal world ($b = 0$) in which the same query returns a uniformly random element of $PKE.\mathcal{C}(ek_u, \|m\|)$.

Table entry $M[u, c]$ stores the message encrypted in user u's challenge ciphertext c. If the adversary queries DEC with a challenge ciphertext it returns $M[u, c]$ rather than performing the decryption. Prior works on memory-aware cryptography [17, 19] considered other ways a decryption oracle might respond to challenge ciphertexts and argued that this is the "correct" convention. The advantage of an adversary $\mathcal{A}$ is0 $Adv_{PKE}^{mu\text{-}\$cca}(\mathcal{A}) = Pr[G_{PKE,1}^{mu\text{-}\$cca}(\mathcal{A})] - Pr[G_{PKE,0}^{mu\text{-}\$cca}(\mathcal{A})]$. In this and future definition we let $\mathcal{U}$ denote the set of allowed user identifiers u.

The general framework of capturing multi-user security by allowing the attacker to access separate instances of oracles for each user with shared secret bit across them is originally due to Bellare, Boldyreva, and Micali [5] who provided a definition for IND-CPA secure public-key encryption.

ONE-WAYNESS SECURITY. Following the one-wayness security definitions in [21], we define variants of one-wayness security of PKE schemes in the multi-user, multi-challenge setting in Fig. 4. We define three variants - One-Wayness under

Game $\mathsf{G}_{\mathsf{PKE}}^{\mathsf{mu\text{-}ow\text{-}}w}(\mathcal{A})$	$\text{CHAL}(u, i)$	$\text{PCO}(u, m, c)$
$\mathcal{H} \leftarrow_\$ \mathsf{PKE.IM}$	If $C[u, i] \neq \bot$	$m' \leftarrow \mathsf{PKE.D}^{\mathcal{H}}(dk_u, c)$
$(ek_{(\cdot)}, dk_{(\cdot)}) \leftarrow_\$ \mathsf{PKE.K}$	$\quad$ Return $C[u, i]$	Return $m = m'$
$\mathsf{O} \leftarrow \bot \; //w = \mathsf{cpa}$	$m \leftarrow_\$ \mathsf{PKE.M}$	
$\mathsf{O} \leftarrow \text{PCO} \; //w = \mathsf{pca}$	$c \leftarrow_\$ \mathsf{PKE.E}^{\mathcal{H}}(ek_u, m)$	$\text{CVO}(u, c)$
$\mathsf{O} \leftarrow (\text{PCO}, \text{CVO}) \; //w = \mathsf{pcva}$	$C[u, i] \leftarrow c$	$m' \leftarrow \mathsf{PKE.D}^{\mathcal{H}}(dk_u, c)$
$(m', u, i) \leftarrow_\$ \mathcal{A}^{\text{NEW},\text{CHAL},\mathsf{O},\mathcal{H}}$	Return c	Return $(m' \in \mathsf{PKE.M})$
Return $\text{PCO}(u, m', \text{CHAL}(u, i))$	$\text{NEW}(u)$	
	Return ek_u	

Fig. 4. Game defining mu-ow-w security of PKE for $w \in \{\mathsf{cpa}, \mathsf{pca}, \mathsf{pcva}\}$.

Chosen Plaintext Attacks (OW-CPA), One-Wayness under Plaintext Checking Attacks (OW-PCA) and One-Wayness under Plaintext and Validity Checking Attacks (OW-PCVA). The difference between each variant $w \in \{\mathsf{cpa}, \mathsf{pca}, \mathsf{pcva}\}$ is in the auxilliary oracle(s) O that the adversary is given access to.

In each variant, the adversary is tasked with finding the decryption of a challenge ciphertext which encrypt a message randomly sampled from PKE.$\mathcal{M}$. In the game $\mathsf{G}_{\mathsf{PKE}}^{\mathsf{mu\text{-}ow\text{-}cpa}}$, the adversary does not have access to any auxilliary oracle as indicated by $\mathsf{O} \leftarrow \bot$. In the game $\mathsf{G}_{\mathsf{PKE}}^{\mathsf{mu\text{-}ow\text{-}pca}}$, the adversary has access to the Plaintext Checking Oracle PCO which takes as input a valid message-ciphertext pair, and returns true if the message is a valid decryption of the ciphertext and false otherwise. The adversary has access to both oracles, PCO and CVO, in $\mathsf{G}_{\mathsf{PKE}}^{\mathsf{mu\text{-}ow\text{-}va}}$ where CVO takes as input a ciphertext, and returns true if the ciphertext decrypts to a valid message. For each variant, we define $\mathsf{Adv}_{\mathsf{PKE}}^{\mathsf{mu\text{-}ow\text{-}}w}(\mathcal{A}) = \Pr[\mathsf{G}_{\mathsf{PKE}}^{\mathsf{mu\text{-}ow\text{-}}w}]$. Note that an adversary may re-query $\text{CHAL}(u, i)$ to get back the same ciphertext. This makes it hard to prove one-wayness, but easier write proofs starting from one-wayness. We sometime assume challenge identifiers, i, are drawn from a fixed set $\mathcal{I}$.

2.4 Key Encapsulation Mechanisms

SYNTAX. A key encapsulation mechanism, KEM, consists of three algorithms - the key generation algorithm (KEM.K) that returns a pair of keys (ek, dk) where ek is the encapsulation key and dk is the corresponding decapsulation key, the encapsulation algorithm (KEM.E) that takes the encapsulation key ek and returns a ciphertext-key pair (c, K) where $K \in \mathsf{KEM.K}$ and the decapsulation algorithm KEM.D that takes the decapsulation key dk and a ciphertext c and returns a key K (or $\bot$ to indicate rejection). The syntax of these algorithms is shown in Fig. 2. Perfect correctness requires that $\mathsf{KEM.D}^{\mathcal{H}}(dk, c) = K$ for all $(ek, dk) \in [\mathsf{KEM.K}]$, all $\mathcal{H} \in \mathsf{KEM.IM}$, and all $(c, K) \in [\mathsf{KEM.E}^{\mathcal{H}}(ek)]$.

We define encryption keyspace $\mathsf{KEM.Ek} = \{ek : (ek, dk) \in [\mathsf{KEM.K}]\}$. For $ek \in \mathsf{KEM.Ek}$ we let $\mathsf{KEM.C}(ek)$ denote the ciphertext set $\{c : (c, K) \in [\mathsf{KEM.E}(ek)]\}$ and define $|\mathsf{KEM.C}| = \min_{ek \in \mathsf{KEM.Ek}} |\mathsf{KEM.C}(ek)|$. We let $\mathsf{KEM.R}$ denote the set

Game $G_{\text{KEM},b}^{\text{mu-\$cca}}(\mathcal{A})$	$\text{ENCAP}_b(u)$	$\text{DECAP}(u,c)$
$\mathcal{H} \leftarrow\!\!\text{\$ KEM.IM}$	$(c_1, K_1) \leftarrow\!\!\text{\$ KEM.E}^{\mathcal{H}}(ek_u)$	If $T[u,c] \neq \bot$:
$(ek_{(\cdot)}, dk_{(\cdot)}) \leftarrow\!\!\text{\$ KEM.K}$	$c_0 \leftarrow\!\!\text{\$ KEM.}\mathcal{C}(ek_u)$	Return $T[u,c]$
$b' \leftarrow\!\!\text{\$ } \mathcal{A}^{\text{NEW},\text{ENCAP}_b,\text{DECAP},\mathcal{H}}$	$K_0 \leftarrow\!\!\text{\$ KEM.}\mathcal{K}$	$K \leftarrow \text{KEM.D}^{\mathcal{H}}(dk_u, c)$
Return $b' = 1$	$T[u, c_b] \leftarrow K_b$	Return K
$\text{NEW}(u)$	Return (c_b, K_b)	
Return ek_u		

Fig. 5. Game defining mu-$cca security of KEM.

from which KEM.K draws it randomness. We say that KEM is ϵ-*uniform* if for all $ek \in \text{KEM.Ek}$, $\mathcal{H} \in \text{KEM.IM}$, and (not necessarily efficient) $\mathcal{D}$ it holds that

$$\Pr[\mathcal{D}(c) = 1 : c \leftarrow\!\!\text{\$ KEM.}\mathcal{C}(ek)] - \Pr[\mathcal{D}(c) = 1 : (c, \cdot) \leftarrow\!\!\text{\$ KEM.E}^{\mathcal{H}}(ek)] \leqslant \epsilon.$$

INDISTINGUISHABLE FROM RANDOM SECURITY. Our notion of $CCA security for KEMs is presented in Fig. 5, which requires that keys and ciphertexts output by the scheme be indistinguishable from random. The adversary is given a user instantiation oracle NEW, encapsulation oracle ENCAP, and a decapsulation oracle DECAP. Its goal is to distinguish between the real world ($b = 1$) where ENCAP returns true outputs from KEM.E and the ideal world ($b = 0$) where it returns a pair (c, K) chosen uniformly at random from $\text{KEM.}\mathcal{C}(ek) \times \text{KEM.}\mathcal{K}$.

The table T stores the keys corresponding to challenge ciphertexts output by the encapsulation oracle. The decapsulation oracle uses T to respond to challenge queries. The advantage of an adversary $\mathcal{A}$ is defined as $\text{Adv}_{\text{KEM}}^{\text{mu-\$cca}}(\mathcal{A}) = \Pr[G_{\text{KEM},1}^{\text{mu-\$cca}}(\mathcal{A})] - \Pr[G_{\text{KEM},0}^{\text{mu-\$cca}}(\mathcal{A})]$. We also define CCA security (via $\text{Adv}_{\text{KEM}}^{\text{mu-cca}}$ and $G_{\text{KEM},b}^{\text{mu-cca}}$) analogously to $CCA security, except in the ENCAP oracle c_0 is set to equal c_1 rather than being sampled at random.

2.5 Symmetric Key Encryption

SYNTAX. A symmetric key encryption scheme, SKE, consists of three algorithms - the key generation algorithm (SKE.K) that returns a key K, the encryption algorithm (SKE.E) that takes the key K and a message m and returns ciphertext c, and the decryption algorithm SKE.D that takes the key K and a ciphertext c and returns message m (or $\bot$ to indicate rejection). The syntax of these algorithms is given in Fig. 2. Perfect correctness requires that $\text{SKE.D}^{\mathcal{H}}(K, c) = m$ for $K \in [\text{SKE.K}]$, all m, all $\mathcal{H} \in \text{SKE.IM}$, and all $c \in [\text{SKE.E}^{\mathcal{H}}(K, m)]$. We define the ciphertext, message, and expansion lengths of SKE by $\text{SKE.cl}(|m|) = |\text{SKE.E}^{\mathcal{H}}(K, m)|$ (requiring this to hold for all $\mathcal{H}, K, m$), $\text{SKE.ml}(\text{SKE.cl}(l)) = l$, and $\text{SKE.xl} = \min_l \text{SKE.cl}(l) - l$ respectively.

Game $G_{\mathsf{SKE},b}^{\mathsf{mu\text{-}\$cca}}(\mathcal{A})$	$\mathrm{ENC}_b(u,m)$	$\mathrm{DEC}(u,c)$
$\mathcal{H} \leftarrow\!\!\text{\$ } \mathsf{SKE.IM}$	$c_1 \leftarrow\!\!\text{\$ } \mathsf{SKE.E}^{\mathcal{H}}(K_u, m)$	If $M[u,c] \neq \bot$:
$K_{(\cdot)} \leftarrow\!\!\text{\$ } \mathsf{SKE.K}$	$c_0 \leftarrow\!\!\text{\$ } \{0,1\}^{\mathsf{SKE.cl}(|m|)}$	Return $M[u,c]$
$b' \leftarrow\!\!\text{\$ } \mathcal{A}^{\mathrm{ENC}_b, \mathrm{DEC}, \mathcal{H}}$	$M[u, c_b] \leftarrow m$	$m \leftarrow \mathsf{SKE.D}^{\mathcal{H}}(K_u, c)$
Return $b' = 1$	Return c_b	Return m

Fig. 6. Game defining mu-$cca security of SKE.

INDISTINGUISHABLE FROM RANDOM CCA SECURITY. Our notion of $CCA security for SKE schemes is captured by Fig. 6, which requires that ciphertexts output by the encryption scheme be indistinguishable from ciphertexts chosen at random. In this game, the adversary is given access to an encryption oracle ENC and a decryption oracle DEC. The adversary needs to distinguish between the real world ($b = 1$), where ENC returns an encryption of m under K_u and the ideal world ($b = 0$) where the output of ENC is sampled uniformly at random. The advantage of an adversary $\mathcal{A}$ is defined as $\mathsf{Adv}_{\mathsf{SKE}}^{\mathsf{mu\text{-}\$cca}}(\mathcal{A}) = \Pr[\mathsf{G}_{\mathsf{SKE},1}^{\mathsf{mu\text{-}\$cca}}(\mathcal{A})] - \Pr[\mathsf{G}_{\mathsf{SKE},0}^{\mathsf{mu\text{-}\$cca}}(\mathcal{A})]$. We will only need "one-time" security in which the adversary only makes one encryption query per user.

3   Diffie-Hellman Definitions

In this section, we introduce the Computational Diffie-Hellman (CDH) assumptions we need for our later proofs. The first is a multi-user, multi-challenge variant of Strong CDH (which we need for one of our coming KEM proofs). We verify this is TAM-tightly implied by single-challenge variants. The second is a new definition we introduce, Pair CDH, which gives the adversary oracle access to a pairing from the group under consideration to a random group. We provide several results to understand the plausibility of Pair CDH security. We show that it always implies Gap CDH security and is {AM,TM}-tightly equivalent to CDH in algebraic/generic group models [13,26,29] or if the group has a pairing.

3.1   Group Syntax

A prime order group $\mathbb{G}$ is a tuple $(\mathbf{g}, p, \circ)$ where g is a group generator of prime order p under the group operation $\circ$. In our definitions we will treat the group as a priori fixed. We typically omit writing the group operation $\circ$ explicitly and instead write group operations using multiplicative notation. We let $\langle \mathbf{g} \rangle = \{ \mathbf{g}^a : a \in \mathbb{N} \}$. The discrete log(arithm) of an element $X \in \langle \mathbf{g} \rangle$ is the value $\mathrm{dlog}(X) \in \mathbb{Z}_p$ such that $\mathbf{g}^{\mathrm{dlog}(X)} = X$. We let $1_{\mathbb{G}} = \mathbf{g}^0$ denote the identity element. A pairing from $\mathbb{G} = (\mathbf{g}, p, \circ)$ to $\mathbb{G}_2 = (\mathbf{g}_2, p_2, \circ_2)$ is a map $e : \langle \mathbf{g} \rangle \times \langle \mathbf{g} \rangle \rightarrow \langle \mathbf{g}_2 \rangle$ satisfying $e(\mathbf{g}^x, \mathbf{g}^y) = \mathbf{g}_2^{xy}$. We let $\mathbf{Time}(\mathbb{G})$ and $\mathbf{Mem}(\mathbb{G})$ denote the time and memory complexity of computing exponentiations or multiplications in $\langle \mathbf{g} \rangle$.

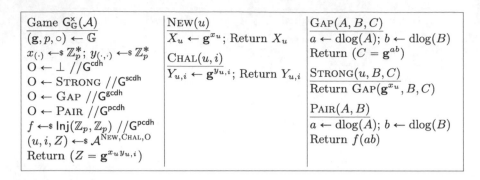

Fig. 7. Security games capturing several variants of the computational Diffie-Hellman problem, namely, CDH, Gap CDH, Strong CDH, and Pair CDH. The last of these is a new notion we introduce which gives the attacker access to a pairing from $\mathbb{G}$ to a random group.

3.2 Computational Diffie-Hellman Variants

In this paper we will make use of several variants of the Computational Diffie-Hellman assumption. These security notions are defined by the game shown in Fig. 7. In each, the adversary is given access to a $\mathbf{g}^x$ and $\mathbf{g}^y$ with the goal of producing $\mathbf{g}^{xy}$. For our later security proofs, it was useful to write "multi-user" and "multi-challenge" version of these games. Thus rather than giving the adversary a single $\mathbf{g}^x$, we give it access to an oracle NEW which on input a string u (which we think of as identifying a user) returns a fresh $\mathbf{g}^{x_u}$. Similarly, the adversary is given access to an oracle CHAL which on inputs string u and i (which we think of as identifying a challenge) returns a fresh $\mathbf{g}^{y_{u,i}}$. For the memory-tightness of future proofs, it is important that the attacker can repeat queries, obtaining the same result as before. The goal of the attacker is to return $\mathbf{g}^{x_u y_{u,i}}$ for any choice of u and i.

The different variants of CDH are captured by the games differing in what (if any) auxiliary oracle O the adversary is given. The standard notion of CDH security is captured by the game G^{cdh} in which the adversary is not given any auxiliary oracle, as expressed by the code $O \leftarrow \bot$. Gap CDH security [27] is captured by G^{gcdh} in which the adversary's oracle GAP takes as input a tuple (A, B, C) and outputs a boolean indicating whether this is a valid Diffie-Hellman tuple (i.e. $C = \mathbf{g}^{\mathrm{dlog}(A)\,\mathrm{dlog}(B)}$). The Strong CDH game G^{scdh} [2] is a weakened version of Gap CDH in which the oracle only allows tuples of the form $(\mathbf{g}^{x_u}, B, C)$.

The final variant is a new security notion we introduce called Pair CDH. In this game G^{pcdh}, the adversary is given access to the oracle PAIR which on input (A, B) returns $f(\mathbf{g}^{ab})$ where a, b are the discrete logs of A, B and f is a random injection. This oracle can be thought of being a pairing to a random group $\mathbb{G}_2 = (\mathbf{g}_2, p, \circ_2)$ where $\mathbf{g}_2 = \mathrm{PAIR}(\mathbf{g}, \mathbf{g})$ and $h \circ_2 h' = f(f^{-1}(h) \circ f^{-1}(h'))$. Note that $\mathcal{A}$ is not able to efficiently compute the operation $\circ_2$.

Adversary $\mathcal{B}_x^{\mathrm{NEW,CHAL,O}}$	$\mathrm{SimNew}(u)$
$(\mathbf{g}, p, \circ) \leftarrow \mathbb{G}$	$x_u' \leftarrow g(u)$
$g \leftarrow_\$ \mathsf{Fcs}(\mathcal{U}, \mathbb{Z}_p^*); \ h \leftarrow_\$ \mathsf{Fcs}(\mathcal{U} \times \mathcal{I}, \mathbb{Z}_p^*)$	Return $X^{x_u'}$
$X \leftarrow \mathrm{NEW}(1); \ Y \leftarrow \mathrm{CHAL}(1, 1)$	
If x = scdh then $\mathrm{SimO} \leftarrow \mathrm{SimStrong}$	$\mathrm{SimChal}(u, i)$
Else $\mathrm{SimO} \leftarrow \mathrm{O}$	$y_{u,i}' \leftarrow h(u, i)$
$(u, i, Z) \leftarrow_\$ \mathcal{A}^{\mathrm{SimNew,SimChal,SimO}}$	Return $Y^{y_{u,i}'}$
$x_u' \leftarrow g(u); \ y_{u,i}' \leftarrow h(u, i)$	
Return $(1, 1, Z^{1/(x_u' y_{u,i}')})$	$\mathrm{SimStrong}(u, B, C)$
	$x_u' \leftarrow g(u)$
	Return $\mathrm{O}(1, B, C^{1/x_u'})$

Fig. 8. Adversary used for Lemma 2.

For $x \in \{\mathsf{cdh}, \mathsf{scdh}, \mathsf{gcdh}, \mathsf{pcdh}\}$ we define $\mathsf{Adv}_{\mathbb{G}}^x(\mathcal{A}) = \Pr[\mathsf{G}_{\mathbb{G}}^x(\mathcal{A})]$. We sometimes need to restrict user identifiers, u, to be from some fixed set $\mathcal{U}$ and challenge identifiers, i, to be a from a fixed set $\mathcal{I}$.

MULTI-CHALLENGE SECURITY. Standard proofs use Diffie-Hellman rerandomization techniques to show that single-challenge security TA-tightly implies multi-challenge security for most variants of Diffie-Hellman-based security notions. The following lemma extends this to TAM-tightness for the notions considered in this paper. The proof is an extension of standard Diffie-Hellman rerandomization techniques that picks the values used for rerandomization as the output of a random function, rather than picking them randomly and storing them.

Lemma 2 (Single-challenge $\Rightarrow$ multi-challenge). *Let $\mathbb{G}$ be a group and $x \in \{\mathsf{cdh}, \mathsf{scdh}, \mathsf{gcdh}, \mathsf{pcdh}\}$. Let $\mathcal{A}$ be an adversary for $\mathsf{G}_{\mathbb{G}}^x$ with $(q_{\mathrm{NEW}}, q_{\mathrm{CHAL}}, q_{\mathrm{O}}) = \mathbf{Query}(\mathcal{A})$. Then we can construct a $(\mathsf{Fcs}(\mathcal{U}, \mathbb{Z}_p^*), \mathsf{Fcs}(\mathcal{U} \times \mathcal{I}, \mathbb{Z}_p^*))$-oracle adversary $\mathcal{B}_x$ (given in the proof) such that*

$$\mathsf{Adv}_{\mathbb{G}}^x(\mathcal{A}) = \mathsf{Adv}_{\mathbb{G}}^x(\mathcal{B}_x)$$
$$\mathbf{Query}(\mathcal{B}_x) = (1, 1, q_{\mathrm{O}})$$
$$\mathbf{Time}^*(\mathcal{B}_x) = O(\mathbf{Time}(\mathcal{A}) + (q_{\mathrm{NEW}} + q_{\mathrm{CHAL}} + q_{\mathrm{O}} + 1)\mathbf{Time}(\mathbb{G}))$$
$$\mathbf{Mem}^*(\mathcal{B}_x) = O(\mathbf{Mem}(\mathcal{A}) + 2\mathbf{Mem}(\mathbb{G})).$$

Proof of Lemma 2). Consider the adversary $\mathcal{B}_x$ shown in Fig. 8. It makes a single query $\mathrm{NEW}(1)$ to obtain a group element X and a single query $\mathrm{CHAL}(1, 1)$ to obtain a group element Y. Then it runs $\mathcal{A}$. Let $x = \mathrm{dlog}(X)$ and $y = \mathrm{dlog}(Y)$.

It responds to $\mathrm{SimNew}(u)$ queries with $X^{x_u'}$ for x_u' the output of its random function. Letting $x_u = x x_u'$, note that $X^{x_u'} = \mathbf{g}^{x_u}$ and x_u is uniformly random because x_u' is. So this oracle has the correct distribution. It responds to $\mathrm{SimChal}$ queries with $Y^{y_{u,i}'}$ for $y_{u,i}'$ output by its random function. Letting $y_{u,i} = y y_{u,i}'$, note that $Y^{y_{u,i}'} = \mathbf{g}^{y_{u,i}}$ and that $y_{u,i}$ is uniformly random because $y_{u,i}'$ is.

For CDH, Gap CDH, or Pair CDH security $\mathcal{B}_\mathsf{x}$ gives $\mathcal{A}$ direct access to O. For Strong CDH security ($\mathsf{x} = \mathsf{scdh}$), $\mathcal{B}_\mathsf{x}$ simulates the oracle by replacing a query (u, B, C) with a query $(1, B, C^{1/x'_u})$ which has the same behavior.

When $\mathcal{A}$ finally halts and outputs (u, i, Z) the adversary $\mathcal{B}_\mathsf{x}$ halts and outputs $(1, 1, Z^{1/(x'_u y'_{u,i})})$. We claim that $\mathcal{B}_\mathsf{x}$ wins whenever $\mathcal{A}$ would. To see this, note that if $\mathcal{A}$ wins then $Z = \mathbf{g}^{x_u y_{u,i}} = \mathbf{g}^{(x x'_u)(y y'_{u,i})}$ and so $Z^{1/(x'_u y'_{u,i})} = \mathbf{g}^{xy}$.   □

3.3   Studying Pair CDH

Pair CDH is a new computational assumption that we've introduced for this work. In this section we provide a few results to give a sense of its difficulty. Namely, we note that Pair CDH security implies Gap CDH security and that it is equivalent to CDH security in certain settings (if $\mathbb{G}$ has an efficient pairing to some $\mathbb{G}_2$, in the algebraic model, and in the generic group model). Given Lemma 2, we focus on the case that the adversary makes only single query each to NEW and CHAL. For brevity, we sketch the relationships here.

PAIR CDH $\Rightarrow$ GAP CDH. To see that Pair CDH security implies Gap CDH security we need only note that $\mathrm{GAP}(A, B, C) = \mathsf{true}$ if and only if $\mathrm{PAIR}(A, B) = \mathrm{PAIR}(\mathbf{g}, C)$. Hence a Pair CDH adversary can efficiently simulate the view of a Gap CDH adversary.

CDH + PAIRING $\Rightarrow$ PAIR CDH. We claim that CDH security implies Pair CDH security if $\mathbb{G}$ has an efficient pairing e to some group $\mathbb{G}_2$ with generator $\mathbf{g}_2$. We can achieve TAM-tightness in this implication by using a $\mathsf{Inj}(\langle \mathbf{g}_2 \rangle, \mathbb{Z}_p)$-oracle adversary. Letting f' denote the random injection, our CDH adversary can simulate PAIR by responding to queries for (A, B) with $f'(e(A, B))$. If $a = \mathrm{dlog}(A)$ and $b = \mathrm{dlog}(B)$, then $f'(e(A, B)) = f'(e(\mathbf{g}, \mathbf{g})^{ab})$. Note that $F(\cdot) = e(\mathbf{g}, \mathbf{g})^{(\cdot)}$ is an injection and so $f(\cdot) = f'(e(\mathbf{g}, \mathbf{g})^{(\cdot)})$ is a random injection. Hence this perfectly emulates PAIR.

CDH + AGM/GGM $\Rightarrow$ PAIR CDH. We claimed that CDH security implies Pair CDH security in the algebraic group model. More precisely, we {AM,TM}-tightly show that CDH security implies Pair CDH security using a $\mathsf{Fcs}(\mathbb{Z}_p^6, \mathbb{Z}_p)$-oracle adversary. We show a way to imperfectly simulate PAIR for algebraic adversaries such that distinguishing this from the real oracle requires the ability to solve the discrete log problem (given $\mathbf{g}^c$ for a random c, find c). Noting that CDH security implies discrete log security gives our claim.

Let X and Y denote the challenge group elements and let $x = \mathrm{dlog}(X)$ and $y = \mathrm{dlog}(Y)$. An algebraic adversary, when making an oracle query (A, B) to PAIR is required to additionally provide "explanations" (a_1, a_2, a_3) and (b_1, b_2, b_3) such that $A = g^{a_1} X^{a_2} Y^{a_3}$ and $B = g^{b_1} X^{b_2} Y^{b_3}$. Then the true PAIR would respond with $f((a_1 + a_2 x + a_3 y) \cdot (b_1 + b_2 x + b_3 y))$. Our CDH adversary will think of this input to f as a degree-two polynomial $P_{A,B}(\mathbf{x}, \mathbf{y}) \in \mathbb{Z}_p[\mathbf{x}, \mathbf{y}]$ whose coefficients it can compute given the explanations for A and B. Letting $(c_1, c_2, \ldots, c_6)$ denote these coefficients and $f' \in \mathsf{Fcs}(\mathbb{Z}_p^6, \mathbb{Z}_p)$, we simulate the output of PAIR as $f'(c_1, c_2, \ldots, c_6)$. Distinguishing this from the true oracle requires finding (A, B) and (A', B') such that $P_{A,B} \neq P_{A',B'}$ (as polynomials), but $P_{A,B}(x, y) = P_{A',B'}(x, y)$. Using analysis techniques from [4], we can use the ability to find such "colliding" polynomials to solve the discrete log problem. We provide details of this analysis in the full version [23].

To achieve TM-tightness, the discrete log reduction picks two of the PAIR oracle queries at random and assumes that they give colliding polynomials. To achieve AM-tightness, we can check every pair of queries for collisions using the memory-tight rewinding technique of Auerbach, et al [3]. Namely, each time we reach a new PAIR oracle query while running the Pair CDH adversary, we pause and run an extra copy of that adversary from the start using the same coins. While running this extra copy, each time it makes a PAIR oracle query we check if this gives a colliding polynomial with the query we paused at in the first adversary. Ignoring memory tightness, $\mathcal{A}$ could simply remember all of the PAIR oracle queries and check them at the end of execution, but then it is not clear how to achieve better time efficiency than checking each pair of queries.

When working in a generic group model [26,29] we can use the same line of reasoning and then information theoretically bound the probability that an adversary finds colliding polynomials by $O(q^2/p)$ where q is the number of queries the Pair CDH adversary makes.

4   Hashed ElGamal KEMs

In this section we present the first example of KEMs with TAM-tight proofs in the multi-challenge setting. The KEMs we consider are variants of the ECIES and Cramer-Shoup Hashed ElGamal KEMs. These variants augment the existing schemes by adding random strings to the ciphertexts and random oracle queries. Our reductions make use of these strings to store pertinent information that will be needed to answer later oracle queries.

4.1   Augmented ECIES

AUGMENTED VERSION. We start with the ECIES [1] variant of Hashed ElGamal. Our augmented version of ECIES includes a random string a in the ciphertext. The augmented ECIES key encapsulation mechanism aECIES[$\mathbb{G}, \mathcal{K}, l$] is parameterized by a group $\mathbb{G} = (\mathbf{g}, p, \circ)$, key space $\mathcal{K}$, and length of the random string, l. The parameters $\mathbb{G}, \mathcal{K}$, and l are fixed for an instantiation of ECIES, so we use aECIES and aECIES[$\mathbb{G}, \mathcal{K}, l$] interchangeably. We define the scheme as follows with aECIES.$\mathcal{K} = \mathcal{K}$ and aECIES.IM = Fcs($\{0, 1\}^l \times \mathbb{G}, \mathcal{K}$).

aECIES.K	aECIES.E$^{\mathcal{H}}(ek)$	aECIES.D$^{\mathcal{H}}(dk, (a, Y))$
$x \leftarrow_\$ \mathbb{Z}_p^*$	$a \leftarrow_\$ \{0,1\}^l$	$Z \leftarrow Y^{dk}$
$ek \leftarrow \mathbf{g}^x$	$y \leftarrow_\$ \mathbb{Z}_p^*$	$K \leftarrow \mathcal{H}(a, Z)$
$dk \leftarrow x$	$Y \leftarrow \mathbf{g}^y$	Return K
Return (ek, dk)	$Z \leftarrow ek^y$	
	$K \leftarrow \mathcal{H}(a, Z)$	
	Return $((a, Y), K)$	

OVERVIEW OF EXISTING TECHNIQUES AND ASSOCIATED CHALLENGES. Bhattacharyya [7] studied ECIES in the memory-aware setting. They pointed out the technique of simulating random oracles with PRFs introduced in [3] cannot be used for this family of KEMs, as in general, decapsulation queries cannot be simulated by the reduction. For example, if a PRF F is used to compute hashes as $F(k, Z)$ instead of the random oracle H, for a decapsulation query Y the reduction would need to return $F(k, Y^{dk})$ which it cannot compute.[5]

Bhattacharyya used the map-then-prf technique as a workaround for groups with pairings. In this technique, the input Z to the random oracle is first operated on by a bilinear map $e(g, Z)$, and then by the PRF F. Hence, the query $H(Z)$ is simulated as $F(k, e(\mathbf{g}, Z))$ and a decapsulation query for Y can be simulated as $F(k, e(ek, Y))$ for all non-challenge ciphertexts. The reduction remembers the challenge ciphertext and returns appropriately when it is queried to DECAP.

This does not scale to the multi-user, multi-challenge setting since it requires that the reduction remembers all the challenge ciphertexts, incurring a memory overhead. Our solution for augmented ECIES combines Ghoshal et al.'s message encoding technique [17] with the map-then-rf technique. We encode the identifying information of challenge ciphertexts in a using a random injection so that this information can be recovered when an appropriate oracle query is made. To avoid the need for an efficiently computable pairing we make use of our new Pair CDH assumption. Our result is captured in Theorem 1.

Theorem 1 (Pair CDH $\Rightarrow$ \$CCA). *Let* aECIES $=$ aECIES$[\mathbb{G}, \mathcal{K}, l]$ *where* $\mathbb{G} = (\mathbf{g}, p, \circ)$ *is a prime order group. Define* $D_1 = \{0, 1\}^l \times \mathbb{G}$ *and* $D_2 = \mathcal{U} \times [q_{\text{ENCAP}}]$. *Let* $\mathcal{A}$ *be an adversary with* $\mathbf{Query}(\mathcal{A}) = (q_{\text{NEW}}, q_{\text{ENCAP}}, q_{\text{DECAP}}, q_H)$ *and assume* $2^l > |\mathcal{U}| \cdot q_{\text{ENCAP}}$. *Then Fig. 12 gives a* $(\mathsf{Fcs}(D_1, \mathcal{K}), \mathsf{Inj}(D_2, \{0, 1\}^l))$-*oracle adversary* $\mathcal{B}$ *such that*

$$\mathsf{Adv}^{\text{mu-\$cca}}_{\text{aECIES}}(\mathcal{A}) \leqslant 2\mathsf{Adv}^{\text{pcdh}}_{\mathbb{G}}(\mathcal{B}) + \frac{q^2_{\text{ENCAP}}}{2^l} + \frac{2q_{\text{ENCAP}}(2q_{\text{DECAP}} + |\mathcal{U}| \cdot q_H)}{2^l(p - 1)}$$

$\mathbf{Query}(\mathcal{B}) = ((q_{\text{NEW}} + q_{\text{ENCAP}}), (q_{\text{ENCAP}} + q_{\text{DECAP}} + q_H), (q_{\text{ENCAP}} + q_{\text{DECAP}} + 2q_H))$
$\mathbf{Time}^*(\mathcal{B}) = O(\mathbf{Time}(\mathcal{A}))$ *and* $\mathbf{Mem}^*(\mathcal{B}) = O(\mathbf{Mem}(\mathcal{A}))$.

CHOOSING THE AUXILIARY STRING LENGTH. When instantiating this scheme one must choose the parameter l which determines the length of a. Larger l incurs a communication cost, while too small of a l can harm the concrete security results. We can expect the $q^2_{\text{ENCAP}}/2^l$ term to dominate the information theoretic part of the bound. With cautious choice of $l = 256$, the size of the ciphertext is not too significantly increased, but even with $q_{\text{ENCAP}} = 2^{64}$ encapsulations (an intentional overestimate of what seems likely) we get $q^2_{\text{ENCAP}}/2^l = 2^{-128}$.

Several of coming theorems use of an auxiliary string a of length l. Similar reasoning applies and $l = 256$ seems like a sufficient choice for all of them.

---

[5] We discuss the use of PRFs to match prior work, but in the oracle adversary framework, we use random function oracles instead.

INTUITION. For each ENCAP query, our Pair CDH adversary programs the random string a as the output of a random injection applied to user identity u and counter i and simulates the random oracle $\mathcal{H}(a, Z)$ as $\mathcal{H}(a, \text{PAIR}(\mathbf{g}, Z))$. This allows us to simulate decapsulations because $\text{PAIR}(\mathbf{g}^x, \mathbf{g}^y) = \text{PAIR}(\mathbf{g}, \mathbf{g}^{xy})$.

Our adversary simulates j-th challenge ciphertexts $Y_{u.i}$ as $\text{CHAL}(u, i)$. To determine if a decapsulation query (u, a, Y) is for a challenge ciphertext, the reduction first inverts a to obtain (v, j). If $v = u$, it re-queries $\text{CHAL}(u, j)$ obtain the corresponding ciphertext $Y_{u,j}$. If $Y = Y_{u,j}$, the reduction assumes this was a challenge ciphertext. Finally, when the adversary $\mathcal{A}$ queries the oracle H with (u, Z) such that $a^{-1} = (u, j)$ and $\text{PAIR}(\mathbf{g}, Z) = \text{PAIR}(\text{NEW}(u), \text{CHAL}(u, j))$, the reduction outputs Z and wins the Pair CDH game.

Proof (of Theorem 1). We use a sequence of hybrids H_0^1 through H_1^1, H_0^2 through H_2^2, and H_0^3 through H_3^3 presented in Figs. 9, 10, and 11 where we establish the following claims that upper bound the advantage of adversary $\mathcal{A}$.

1. $\mathsf{Adv}_{\mathsf{aECIES}}^{\mathsf{mu\text{-}\$cca}}(\mathcal{A}) = 2\Pr[\mathsf{H}_0^1] - 1$
2. $\Pr[\mathsf{H}_0^1] \leqslant \Pr[\mathsf{H}_1^1] + \frac{q_{\text{ENCAP}}^2}{2 \cdot 2^l}$
3. $\Pr[\mathsf{H}_1^1] = \Pr[\mathsf{H}_0^2] = \Pr[\mathsf{H}_1^2]$
4. $\Pr[\mathsf{H}_1^2] = \Pr[\mathsf{H}_2^2] = \Pr[\mathsf{H}_0^3]$
5. $\Pr[\mathsf{H}_0^3] = \Pr[\mathsf{H}_1^3]$
6. $\Pr[\mathsf{H}_1^3] \leqslant \Pr[\mathsf{H}_2^3] + \frac{q_{\text{ENCAP}}(2q_{\text{DECAP}} + |\mathcal{U}| \cdot q_{\mathsf{H}})}{2^l(p-1)}$
7. $\Pr[\mathsf{H}_2^3] \leqslant \Pr[\mathsf{H}_3^3] + \Pr[\mathsf{bad}]$
8. $\Pr[\mathsf{H}_3^3] \leqslant \frac{1}{2}$
9. $\Pr[\mathsf{bad}] \leqslant \mathsf{Adv}_{\mathbb{G}}^{\mathsf{pcdh}}(\mathcal{B})$

TRANSITION TO H_0^1. We claim that the view of $\mathcal{A}$ in H_0^1 is identical to its view in $\mathsf{G}_{\mathsf{aECIES},b}^{\mathsf{mu\text{-}\$cca}}$ (Fig. 5) if b is chosen uniformly. In the latter, ENCAP_1 returns a ciphertext-key pair (c_1, K_1) such that c is the encapsulation of K_1, and ENCAP_0 returns a ciphertext-key pair (c_0, K_0) where c_0 and K_0 are uniformly random elements of the ciphertext space and key space respectively. The same holds for the ENCAP oracle in H_0^1. The table T in $\mathsf{G}_{\mathsf{aECIES},b}^{\mathsf{mu\text{-}\$cca}}$ is indexed by (u, c) and stores the key that was returned by the ENCAP_b oracle for ciphertext c. The table T in H_0^1 behaves analogously. The DECAP oracle in $\mathsf{G}_{\mathsf{aECIES},b}^{\mathsf{mu\text{-}\$cca}}$ returns key K_b that was output by the ENCAP_b oracle when queried on a challenge ciphertext, and returns honest decapsulations otherwise. The same is true for H_0^1. Note that H_0^1's final output is whether $b' = b$, so standard conditional probability calculations give that $\mathsf{Adv}_{\mathsf{aECIES}}^{\mathsf{mu\text{-}\$cca}}(\mathcal{A}) = 2\Pr[\mathsf{H}_0^1] - 1$.

TRANSITION H_0^1 TO H_1^1. In H_1^1, we make the following changes.

1. In the ENCAP oracle we switch from
   (a) sampling the values $a_{(\cdot,\cdot)}$ uniformly to assigning them as the output of a random injection g evaluated on (u, i).
   (b) sampling the values $y_{(\cdot,\cdot)}$ uniformly to assigning them the output of a random function h evaluated on (u, i). Switching to the random function h does not change the view of the adversary because the ordered pair (u, i) never repeats.
2. In the DECAP oracle we switch the If condition from checking whether $T[u, a, Y] \neq \perp$ to evaluating the boolean $u = v \wedge j \in I[u] \wedge Y = \mathbf{g}^{h(u,i)}$. These conditions are equivalent since $T[u, a, Y] \neq \perp$ iff $a = g(u, j)$ and $Y = \mathbf{g}^{h(u,j)}$ for some $j \in I[u]$.

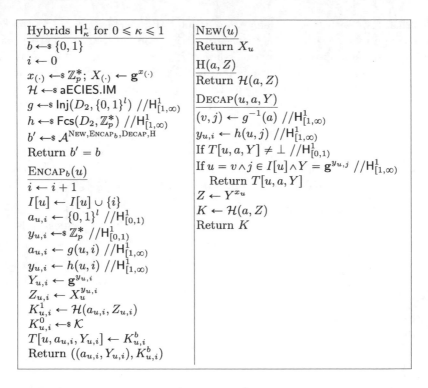

Fig. 9. First set of hybrids H_0^1 through H_1^1 used for proof of Theorem 1.

The only change in the adversary's view comes from 1(a). The switching lemma (Lemma 1) gives us $\Pr[\mathsf{H}_0^1] \leqslant \Pr[\mathsf{H}_1^1] + \frac{q_{\mathrm{ENCAP}}^2}{2 \cdot 2^l}$.

TRANSITION H_1^1 TO H_0^2. The transition to hybrid H_0^2 is shown in Fig. 10. We have highlighted the ways in which H_0^2 differs from H_1^1. Our changes are the following.

1. In the ENCAP oracle, the table $\tilde{T}$ has been added to record the responses to the $\mathcal{H}$ queries made within ENCAP.
2. In the H oracle, an If block is added to check if the input tuple (a, Z) was previously queried to $\mathcal{H}$ from within the ENCAP oracle. In essence, the If block returns $\mathcal{H}(a, Z)$ when queried on a challenge ciphertext. The H oracle would behave the same way without the If block, as in H_1^1.

Hence, $\Pr[\mathsf{H}_1^1] = \Pr[\mathsf{H}_0^2]$.

TRANSITION H_0^2 TO H_1^2. In game H_1^2, we introduce the pairing oracle PAIR. It is only used by oracles within the game; the adversary does not have direct access to it. Note that $\mathrm{PAIR}(\mathbf{g}, Z) = \mathrm{PAIR}(\mathbf{g}^{x_u}, Y)$ iff $Z = Y^{x_u}$.

In the H oracle, we switch the condition from checking $\tilde{T}[a, Z] \neq \bot$ to evaluating the boolean $j \in I[u] \wedge \mathrm{PAIR}(\mathbf{g}, Z) = \mathrm{PAIR}(X_u, Y)$ where $(u, j) = g^{-1}(a), Y = \mathbf{g}^{h(u,j)}$. These two conditions are equivalent. Note that $\tilde{T}[a, Z] \neq \bot$ means it was filled in an encapsulation query. Suppose this was the j-th query and was to u.

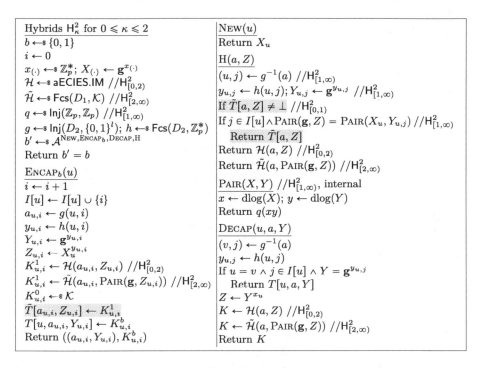

Hybrids H_κ^2 for $0 \leqslant \kappa \leqslant 2$	$\text{NEW}(u)$
$b \leftarrow_\$ \{0,1\}$	Return X_u
$i \leftarrow 0$	$\text{H}(a,Z)$
$x_{(\cdot)} \leftarrow_\$ \mathbb{Z}_p^*;\ X_{(\cdot)} \leftarrow \mathbf{g}^{x_{(\cdot)}}$	$(u,j) \leftarrow g^{-1}(a)\ //\mathsf{H}_{[1,\infty)}^2$
$\mathcal{H} \leftarrow_\$ \text{aECIES.IM}\ //\mathsf{H}_{[0,2)}^2$	$y_{u,j} \leftarrow h(u,j); Y_{u,j} \leftarrow \mathbf{g}^{y_{u,j}}\ //\mathsf{H}_{[1,\infty)}^2$
$\tilde{\mathcal{H}} \leftarrow_\$ \text{Fcs}(D_1,\mathcal{K})\ //\mathsf{H}_{[2,\infty)}^2$	If $\tilde{T}[a,Z] \neq \bot\ //\mathsf{H}_{[0,1)}^2$
$q \leftarrow_\$ \text{Inj}(\mathbb{Z}_p,\mathbb{Z}_p)\ //\mathsf{H}_{[1,\infty)}^2$	If $j \in I[u] \wedge \text{PAIR}(\mathbf{g},Z) = \text{PAIR}(X_u,Y_{u,j})\ //\mathsf{H}_{[1,\infty)}^2$
$g \leftarrow_\$ \text{Inj}(D_2,\{0,1\}^l);\ h \leftarrow_\$ \text{Fcs}(D_2,\mathbb{Z}_p^*)$	$\quad$ Return $\tilde{T}[a,Z]$
$b' \leftarrow_\$ \mathcal{A}^{\text{NEW},\text{ENCAP}_b,\text{DECAP},\text{H}}$	Return $\mathcal{H}(a,Z)\ //\mathsf{H}_{[0,2)}^2$
Return $b' = b$	Return $\tilde{\mathcal{H}}(a,\text{PAIR}(\mathbf{g},Z))\ //\mathsf{H}_{[2,\infty)}^2$
$\text{ENCAP}_b(u)$	$\underline{\text{PAIR}(X,Y)}\ //\mathsf{H}_{[1,\infty)}^2,\ \text{internal}$
$i \leftarrow i+1$	$x \leftarrow \text{dlog}(X);\ y \leftarrow \text{dlog}(Y)$
$I[u] \leftarrow I[u] \cup \{i\}$	Return $q(xy)$
$a_{u,i} \leftarrow g(u,i)$	$\text{DECAP}(u,a,Y)$
$y_{u,i} \leftarrow h(u,i)$	$(v,j) \leftarrow g^{-1}(a)$
$Y_{u,i} \leftarrow \mathbf{g}^{y_{u,i}}$	$y_{u,j} \leftarrow h(u,j)$
$Z_{u,i} \leftarrow X_u^{y_{u,i}}$	If $u = v \wedge j \in I[u] \wedge Y = \mathbf{g}^{y_{u,j}}$
$K_{u,i}^1 \leftarrow \mathcal{H}(a_{u,i}, Z_{u,i})\ //\mathsf{H}_{[0,2)}^2$	$\quad$ Return $T[u,a,Y]$
$K_{u,i}^1 \leftarrow \tilde{\mathcal{H}}(a_{u,i}, \text{PAIR}(\mathbf{g}, Z_{u,i}))\ //\mathsf{H}_{[2,\infty)}^2$	$Z \leftarrow Y^{x_u}$
$K_{u,i}^0 \leftarrow_\$ \mathcal{K}$	$K \leftarrow \mathcal{H}(a,Z)\ //\mathsf{H}_{[0,2)}^2$
$\tilde{T}[a_{u,i}, Z_{u,i}] \leftarrow K_{u,i}^1$	$K \leftarrow \tilde{\mathcal{H}}(a, \text{PAIR}(\mathbf{g},Z))\ //\mathsf{H}_{[2,\infty)}^2$
$T[u, a_{u,i}, Y_{u,i}] \leftarrow K_{u,i}^b$	Return K
Return $((a_{u,i}, Y_{u,i}), K_{u,i}^b)$	

Fig. 10. Second set of hybrids H_0^2 through H_2^2 used for proof of Theorem 1. Grey highlighting is used to show the difference between H_1^1 and H_0^2. Note that PAIR is used internally by other oracles and is not directly accessible to $\mathcal{A}$.

Then $a = g(u,j)$ must hold, j would have been added to $I[u]$, and $Z = X_u^{h(u,j)}$ (so $\text{PAIR}(\mathbf{g}, Z) = \text{PAIR}(X_u, Y)$). Thus, $\Pr[H_0^2] = \Pr[H_1^2]$.

TRANSITION H_1^2 TO H_2^2(map-then-rf). In H_2^2, the random function $\mathcal{H}$ is replaced by a random function $\tilde{\mathcal{H}}$ from $\text{Fcs}(D_1, \mathcal{K})$ where $D_1 = \{0,1\}^l \times \mathbb{Z}_p$. We replaced the function $\mathcal{H}$ as $\mathcal{H}(a,Z) = \tilde{\mathcal{H}}(a, \text{PAIR}(\mathbf{g},Z))$. Then $\mathcal{H}$ is a random function if $\tilde{\mathcal{H}}$ is, $\text{PAIR}(g,.)$ is an injection. Hence, $\Pr[H_1^2] = \Pr[H_2^2]$.

TRANSITION H_2^2 TO H_0^3. Game H_0^3 is shown in Fig. 11. We have highlighted the ways in which H_0^3 differs from H_2^2. In H_0^3, the key $K_{(\cdot,\cdot)}^0$ is assigned the output of a random function $\mathcal{E}$ from $\text{Fcs}(D_1, \mathcal{K})$, instead of sampling it at random. This does not change the adversary's view as a never repeats. Hence, $\Pr[H_2^2] = \Pr[H_0^3]$.

TRANSITION H_0^3 TO H_1^3. In H_1^3, the table T is no longer used. The change is in the DECAP oracle where the If condition evaluates whether the bit b is 0. Note that the DECAP oracle always returns K^b. When $b = 0$, it returns $\mathcal{E}(a, \text{PAIR}(\mathbf{g}, Z))$

$$
\begin{array}{l|l}
\hline
\text{Hybrids } \mathsf{H}^3_\kappa \text{ for } 0 \leqslant \kappa \leqslant 3 & \text{NEW}(u) \\
\hline
b \leftarrow\!\!{\scriptstyle\$}\; \{0,1\} & \text{Return } X_u \\
i \leftarrow 0 & \\
x_{(\cdot)} \leftarrow\!\!{\scriptstyle\$}\; \mathbb{Z}_p^*;\; X_{(\cdot)} \leftarrow \mathbf{g}^{x_{(\cdot)}} & \text{H}(a, Z) \\
\tilde{\mathcal{H}} \leftarrow\!\!{\scriptstyle\$}\; \mathsf{Fcs}(D_1, \mathcal{K}) & (u,j) \leftarrow g^{-1}(a) \\
\mathcal{E} \leftarrow\!\!{\scriptstyle\$}\; \mathsf{Fcs}(D_1, \mathcal{K}) & y_{u,j} \leftarrow h(u,j); Y_{u,j} \leftarrow \mathbf{g}^{y_{u,j}} \\
q \leftarrow\!\!{\scriptstyle\$}\; \mathsf{Inj}(\mathbb{Z}_p, \mathbb{Z}_p) & \text{If } j \in I[u] \wedge \text{PAIR}(\mathbf{g}, Z) = \text{PAIR}(X_u, Y_{u,j}) \;//\mathsf{H}^3_{[0,2)} \\
g \leftarrow\!\!{\scriptstyle\$}\; \mathsf{Inj}(D_2, \{0,1\}^l); & \text{If } \text{PAIR}(\mathbf{g}, Z) = \text{PAIR}(X_u, Y_{u,j}) \;//\mathsf{H}^3_{[2,\infty)} \\
h \leftarrow\!\!{\scriptstyle\$}\; \mathsf{Fcs}(D_2, \mathbb{Z}_p^*) & \qquad \text{Return } \tilde{T}[a, Z] \;//\mathsf{H}^3_{[0,3)} \\
\text{bad} \leftarrow \text{false} \;//\mathsf{H}^3_{[3,\infty)} & \qquad \text{bad} \leftarrow \text{true} \;//\mathsf{H}^3_{[3,\infty)} \\
b' \leftarrow\!\!{\scriptstyle\$}\; \mathcal{A}^{\text{NEW},\text{ENCAP}_b,\text{DECAP},\text{H}} & \qquad \text{ABORT} \;//\mathsf{H}^3_{[3,\infty)} \\
\text{Return } b' = b & \text{Return } \tilde{\mathcal{H}}(a, \text{PAIR}(\mathbf{g}, Z)) \\
\hline
\text{ENCAP}_b(u) & \text{PAIR}(X, Y) \;//\text{Internal} \\
\hline
i \leftarrow i + 1 & x \leftarrow \text{dlog}(X); \; y \leftarrow \text{dlog}(Y) \\
I[u] \leftarrow I[u] \cup \{i\} \;//\mathsf{H}^3_{[0,2)} & \text{Return } q(xy) \\
a_{u,i} \leftarrow g(u,i) & \\
y_{u,i} \leftarrow h(u,i) & \text{DECAP}(u, a, Y) \\
Y_{u,i} \leftarrow \mathbf{g}^{y_{u,i}} & (v,j) \leftarrow g^{-1}(a) \\
Z_{u,i} \leftarrow X_u^{y_{u,i}} & y_{u,j} \leftarrow h(u,j) \\
K_{u,i}^1 \leftarrow \tilde{\mathcal{H}}(a_{u,i}, \text{PAIR}(\mathbf{g}, Z_{u,i})) & Z \leftarrow Y^{x_u} \\
K_{u,i}^0 \leftarrow \mathcal{E}(a_{u,i}, \text{PAIR}(\mathbf{g}, Z_{u,i})) & \text{If } u = v \wedge j \in I[u] \wedge Y = \mathbf{g}^{y_{u,j}} \;//\mathsf{H}^3_{[0,2)} \\
\tilde{T}[a_{u,i}, Z_{u,i}] \leftarrow K_{u,i}^1 \;//\mathsf{H}^3_{[0,3)} & \text{If } u = v \wedge j \in I[u] \wedge Y = \mathbf{g}^{y_{u,j}} \wedge b = 0 \;//\mathsf{H}^3_{[1,2)} \\
T[u, a_{u,i}, Y_{u,i}] \leftarrow K_{u,i}^b \;//\mathsf{H}^3_{[0,1)} & \text{If } u = v \wedge Y = \mathbf{g}^{y_{u,j}} \wedge b = 0 \;//\mathsf{H}^3_{[2,\infty)} \\
\text{Return } ((a_{u,i} Y_{u,i}), K_{u,i}^b) & \qquad \text{Return } T[u, a, Y] \;//\mathsf{H}^3_{[0,1)} \\
 & \qquad \text{Return } \mathcal{E}(a, \text{PAIR}(\mathbf{g}, Z)) \;//\mathsf{H}^3_{[1,\infty)} \\
 & K \leftarrow \tilde{\mathcal{H}}(a, \text{PAIR}(\mathbf{g}, Z)) \\
 & \text{Return } K \\
\hline
\end{array}
$$

Fig. 11. Third set of hybrids H^3_0 through H^3_3 used for proof of Theorem 1.

which is what was used to compute K^0 in ENCAP. The If condition evaluates to false under two cases

1. (a,Y) is a challenge ciphertext and $b = 1$
2. (a,Y) is not a challenge ciphertext.

In both these cases, the DECAP oracle returns $\tilde{\mathcal{H}}(a, \text{PAIR}(\mathbf{g}, Z))$ which is the same as K^1. Therefore, this modification does not change the view of the adversary and $\Pr[\mathsf{H}^3_0] = \Pr[\mathsf{H}^3_1]$.

TRANSITION H^3_1 TO H^3_2. In H^3_2 we change the If statements in the H and DECAP oracles. In both places, we remove the check $j \in I[u]$. H^3_1 and H^3_2 differ only under the following events:

1. The adversary makes a DECAP query for (u, a, Y) such that $(u,j) = g^{-1}(a)$ and $Y = \mathbf{g}^{h(u,j)}$ but $j \notin I[u]$.

2. The adversary makes a H query for (a, Z) such that $\textsc{Pair}(\mathbf{g}, Z) = \textsc{Pair}(X_u, Y)$ and $Y = \mathbf{g}^{h(u,j)}$ (where $(u, j) = g^{-1}(a)$) but $j \notin I[u]$.

To cause either of these events, the adversary must "guess" a point a in the image of g other than the at most $q_{\textsc{Encap}}$ such points for which it was given the corresponding ciphertexts by $\textsc{Encap}$. We analyze the probability of this event in H_1^3 where the behaviors of $\textsc{Decap}$ and H can only depend on the values of $h(u, \cdot)$ and $g(u, \cdot)$ for inputs in $I[u]$. Hence, $\mathcal{A}$ only learns about $h(u, \cdot)$ and $g(u, \cdot)$ through its queries to $\textsc{Encap}$. At some fixed point in time, let n denote the total number of $\textsc{Encap}$ queries that $\mathcal{A}$ has made so far and n_u denote the number of $\textsc{Encap}$ queries it has made to user u.

First consider a query $\textsc{Decap}(u, a, Y)$. For this to differ between the games it must hold that a is in the image of $g(u, \cdot)$. There are $q_{\textsc{Encap}}$ total such values, of which the adversary has already seen n_u from $\textsc{Encap}$. This is out of the 2^l values in the codomain, of which the adversary has already seen n from $\textsc{Encap}$. Thus there is a $(q_{\textsc{Encap}} - n_u)/(2^l - n) \leqslant (q_{\textsc{Encap}} - n_u + n)/2^l \leqslant 2q_{\textsc{Encap}}/2^l$ chance that the adversary picks such an a. The adversary must additionally have guessed the correct $\mathbf{g}^{h(u,j)}$, which it has an at most $1/(p-1)$ chance of having done.

Now consider a query $\mathsf{H}(a, Z)$. For this to differ between the games it must hold that a is in the image of $g(\cdot, \cdot)$. There are $|\mathcal{U}| \cdot q_{\textsc{Encap}}$ total such values, of which the adversary has already seen n from $\textsc{Encap}$. This is out of the 2^l values in the codomain, of which the adversary has already seen n from $\textsc{Encap}$. Thus there is a $(|\mathcal{U}| \cdot q_{\textsc{Encap}} - n)/(2^l - n) \leqslant |\mathcal{U}| \cdot q_{\textsc{Encap}}/2^l$ chance that the adversary picks such an a. The adversary must additionally have guessed the correct $X_u^{h(u,j)}$, which it has an at most $1/(p-1)$ chance of having done.

Applying a union bound over all $\textsc{Decap}$ and H queries gives the claimed bound $\Pr[\mathsf{H}_1^3] \leqslant \Pr[\mathsf{H}_2^3] + (2q_{\textsc{Encap}}q_{\textsc{Decap}} + |\mathcal{U}| \cdot q_{\textsc{Encap}}q_{\mathsf{H}})/(2^l \cdot (p-1))$.

TRANSITION H_2^3 TO H_3^3. In H_3^3, the table $\tilde{T}$ is removed. The change is in the H oracle, wherein if the adversary queries the H oracle with a challenge ciphertext, the H oracle aborts. Using the fundamental lemma of game playing, this probability is bounded as $\Pr[\mathsf{H}_2^3] \leqslant \Pr[\mathsf{H}_3^3] + \Pr[\mathsf{bad}]$.

In H_3^3, the adversary is unable to compute $\mathsf{H}(a, Z)$ for challenge ciphertexts using the H oracle or the $\textsc{Decap}$ oracle. The $\textsc{Decap}$ oracle returns K_b on all challenge ciphertexts (which is the same value that was returned by the $\textsc{Encap}$ oracle) and the H oracle aborts on challenge ciphertexts. Therefore, the adversary's view in H_3^3 is independent of the bit b. Hence, $\Pr[\mathsf{H}_3^3] \leqslant 1/2$.

To bound $\Pr[\mathsf{bad}]$, we construct an adversary $\mathcal{B}$ given in Fig. 12 against the Pair CDH security of $\mathbb{G}$. We claim that $\mathcal{B}$ perfectly simulates H_3^3 for $\mathcal{A}$. Note that the challenge ciphertext is computed using $\mathcal{B}$'s challenge oracle, and the If condition in the H has been replaced with a call to $\mathcal{B}$'s $\textsc{Pair}$ oracle. Whenever the flag bad is set, $\mathcal{B}$ outputs the corresponding (u, j, Z) and wins the Pair CDH game. Therefore, $\Pr[\mathsf{bad}] \leqslant \mathsf{Adv}_{\mathbb{G}}^{\mathsf{pcdh}}(\mathcal{B})$. $\square$

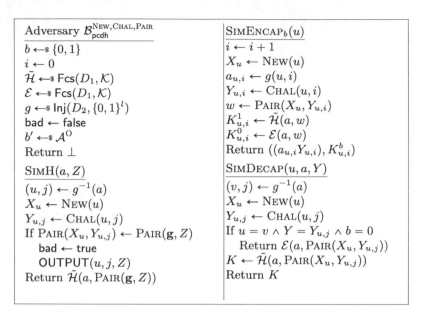

Fig. 12. Adversary $\mathcal{B}$ for Theorem 1; $O = \{\text{New}, \text{SimEncap}_b, \text{SimDecap}, \text{SimH}\}$.

4.2 Augmented Cramer-Shoup KEM

AUGMENTED VERSION. In this section we present a memory-tight reduction for an augmented version of the Cramer-Shoup KEM [8]. The augmented Cramer-Shoup key encapsulation mechanism $\text{aCS}[\mathbb{G}, \mathcal{K}, l]$ is parameterized by a group $\mathbb{G} = (\mathbf{g}, p, \circ)$, key space $\mathcal{K}$, and length of the random string, l. The parameters $\mathbb{G}, \mathcal{K}$, and l are constants for any instantiation of the Cramer-Shoup KEM and hence, we override notation and use aCS and $\text{aCS}[\mathbb{G}, \mathcal{K}, l]$ interchangably. We let $\text{aCS.IM} = \text{Fcs}(\{0,1\}^l \times \mathbb{G}^2, \mathcal{K})$ and define the scheme as follows.

aCS.K	$\text{aCS.E}^{\mathcal{H}}(ek)$	$\text{aCS.D}^{\mathcal{H}}(dk, (a, Y))$
$x \leftarrow_\$ \mathbb{Z}_p^*$	$a \leftarrow_\$ \{0,1\}^l$	$Z \leftarrow Y^{dk}$
$ek \leftarrow \mathbf{g}^x$	$y \leftarrow_\$ \mathbb{Z}_p^*$	$K \leftarrow \mathcal{H}(a, Y, Z)$
$dk \leftarrow x$	$Y \leftarrow \mathbf{g}^y$	Return K
Return (ek, dk)	$Z \leftarrow ek^y$	
	$K \leftarrow \mathcal{H}(a, Y, Z)$	
	Return $((a, Y), K)$	

OVERVIEW OF EXISTING TECHNIQUES AND ASSOCIATED CHALLENGES. A traditional security reduction for the Cramer-Shoup KEM from the Strong CDH problem in the single-user, single-challenge setting would use $l = 0$ and the lazy sampling technique to simulate $\mathcal{H}$ as a random oracle. The reduction would maintain a table T to store $\mathcal{H}$ queries and corresponding responses. When the adversary makes a decapsulation query on Y, the reduction would check the table to see if an entry $T[Y, Z]$ exists such that $\text{GAP}(X, Y, Z) = \text{true}$ where $X = \mathbf{g}^x$ is the public key. If the entry exists, it would return the corresponding value, and if it does not exist, the reduction would sample a new uniformly random element K from the key set $\mathcal{K}$. The reduction would then store $T[Y, \_] \leftarrow K$ and return K. The second entry would be filed in the table T when the adversary makes a hash query for (Y, Z) such that $\text{GAP}(X, Y, Z) = \text{true}$. The reduction wins the Strong CDH game if it outputs a Z such that $Z = \mathbf{g}^{xy}$, which it does by waiting for the Cramer-Shoup adversary to query its hash oracle on inputs (Y, Z) such that $\text{GAP}(X, Y, Z) = \text{true}$.

Like with ECIES, the random oracle simulation using PRF technique cannot be used here as it is not possible for the reduction to simulate decapsulation queries using the PRF. Bhattacharya avoided this issue using the map-then-prf technique, defining $\mathcal{H}(Y, Z)$ so that when $Z = Y^{dk}$, $\mathcal{H}(Y, Z)$ is computable from ek, Y. This allows properly responding to decapsulation queries when the reduction only has access to Y and cannot compute $Z = Y^{dk}$.

This proof breaks in the multi-challenge setting because it is not clear how to identify and respond to challenge ciphertexts without simply storing them all. Once again, augmentation with a random string a allows encoding the information needed to respond to queries appropriately. Our result is captured by the following theorem. The proof of Theorem 2 is given in the full version [23].

Theorem 2 (Strong CDH $\Rightarrow$ \$CCA). *Let* $\mathbb{G} = (\mathbf{g}, p, \circ)$ *be a group of prime order* p. *Let* $\mathcal{K}$ *and* l *be fixed. Define* $\text{aCS} = \text{aCS}[\mathbb{G}, \mathcal{K}, l]$.

Let $D_1 = \{0, 1\}^l \times \mathbb{G} \times \mathbb{G} \cup \{\star\}$ *and* $D_2 = \mathcal{U} \times [q_{\text{ENCAP}}]$. *Let* $\mathcal{A}$ *be a* mu-\$cca *adversary with* $\mathbf{Query}(\mathcal{A}) = (q_{\text{NEW}}, q_{\text{ENCAP}}, q_{\text{DECAP}}, q_{\text{H}})$ *and assume* $2^l > 2q_{\text{ENCAP}}$. *We construct a* $(\text{Fcs}(D_1, \mathcal{K}) \times \text{Inj}(D_2, \{0, 1\}^l))$*-oracle adversary* $\mathcal{B}$ *such that*

$$\text{Adv}_{\text{aCS}}^{\text{mu-\$cca}}(\mathcal{A}) \leqslant 2\text{Adv}_{\mathbb{G}}^{\text{scdh}}(\mathcal{B}) + \frac{q_{\text{ENCAP}}^2}{2^l} + \frac{2q_{\text{ENCAP}}(2q_{\text{DECAP}} + |\mathcal{U}| \cdot q_{\text{H}})}{2^l(p-1)}$$

$\mathbf{Query}(\mathcal{B}) = (q_{\text{NEW}}, (q_{\text{ENCAP}} + q_{\text{DECAP}} + q_{\text{H}}), q_{\text{H}})$

$\mathbf{Time}^*(\mathcal{B}) = O(\mathbf{Time}(\mathcal{A}))$ *and* $\mathbf{Mem}^*(\mathcal{B}) = O(\mathbf{Mem}(\mathcal{A}))$.

INTUITION. For this proof, we program the random string a as the output of a random injection g applied to a user identity u and a counter i and simulate the random oracle H using the random function $\tilde{\mathcal{H}}(\alpha(a, Y, Z))$ where α is an injection from $\{0, 1\}^l \times \mathbb{G} \times \mathbb{G} \cup \{\star\}$ to $\{0, 1\}^l \times \{0, 1\} \times \mathbb{G}^2$ that can be computed with knowing Z when $Z = Y^{ek}$.

Our reduction adversary plays the Strong CDH game where it gets its challenge ciphertexts Y from the oracle $\text{CHAL}(u, i)$. To determine if a decapsulation query (u, a, Y) is for a challenge ciphertext, the reduction first inverts a to obtain (v, j). If $u = v$, it queries $\text{CHAL}(u, j)$ to obtain the ciphertext $Y_{u,j}$. Then it can simply check if $Y = Y_{u,j}$. Finally, when the adversary $\mathcal{A}$ queries the oracle H with (a, Y, Z) such that $g^{-1}(a) = (u, j)$ and $\text{STRONG}(u, Y, Z) = \text{true}$ and $Y = \text{CHAL}(u, j)$, the reduction outputs (u, j, Z) and wins the Strong CDH game.

5   Fujisaki-Okamoto Transformation

The Fujisaki-Okamoto [14,15] transformations use a random oracle to construct an IND-CCA secure KEM from a weakly (IND-CPA) secure PKE scheme. Bhattacharyya presented memory-tight reductions for the modules analyzed in [21] in the single-user, single-challenge setting [7]. In our work, we use the message encoding technique along with map-then-rf technique to prove memory-tight reductions for one version of the Fujisaki-Okamoto transformations in the multi-user, multi-challenge setting. In the following subsections, we present the definitions and memory-tight reductions for the transformations T, aV, and $\mathsf{aU}^\perp$.

5.1   Transformation T [IND-CPA → OW-PCA]

The transformation T constructs a deterministic OW-PCA secure public key encryption scheme $\mathsf{TKE} = \mathsf{T}[\mathsf{PKE}]$ from an IND-CPA secure public key encryption scheme PKE. We define TKE as follows with $\mathsf{TKE.IM} = \mathsf{Fcs}(\mathsf{PKE}.\mathcal{M}, \mathsf{PKE}.\mathcal{R}) \times \mathsf{PKE.IM}$ and $\mathsf{TKE}.\mathcal{M} = \mathsf{PKE}.\mathcal{M}$.

$\mathsf{TKE.E}^{\mathcal{H} \times \mathcal{H}'}(ek, m)$	$\mathsf{TKE.D}^{\mathcal{H} \times \mathcal{H}'}((ek, dk), c)$
$c \leftarrow \mathsf{PKE.E}^{\mathcal{H}'}(ek, m; \mathcal{H}(m))$	$m' \leftarrow \mathsf{PKE.D}^{\mathcal{H}'}(dk, c)$
Return c	If $m' = \perp$ or $\mathsf{PKE.E}^{\mathcal{H}'}(ek, m'; \mathcal{H}(m')) \neq c$
	$\quad$ Return $\perp$
	Return m'

For key generation, $\mathsf{TKE.K}$ samples $(ek, dk) \leftarrow_\$ \mathsf{PKE.K}$ and outputs $(ek, (ek, dk))$. Note that TKE is *tidy*, meaning that if $m = \mathsf{TKE.D}^{\mathcal{H} \times \mathcal{H}'}((ek, dk), c)$, then $c = \mathsf{TKE.E}^{\mathcal{H} \times \mathcal{H}'}(ek, m)$.

We present a memory-tight reduction for T in the multi-user, multi-challenge setting using the randomness programming technique. Our result is captured in Theorem 3, whose proof is given in the full version [23].

Theorem 3 (IND-CPA ⇒ OW-PCA). *Let* $\mathsf{TKE} = \mathsf{T}[\mathsf{PKE}]$. *If* PKE *is* δ-*correct, then* TKE *is* δ'-*correct for* $\delta'(q) = (q+1)\delta(q)$. *Let* $\mathcal{A}$ *be an adversary against* TKE *with* $\mathbf{Query}(\mathcal{A}) = (q_{\mathrm{NEW}}, q_{\mathrm{CHAL}}, q_{\mathrm{PCO}}, q_{\mathrm{H}})$. *Assume* PKE's *algorithms make at most* q_{PKE} *oracle queries and define* $q^* = q_{\mathrm{H}} + q_{\mathrm{CHAL}}(q_{\mathsf{PKE}} + 1) + q_{\mathrm{PCO}}(2q_{\mathsf{PKE}} + 1) + q_{\mathsf{PKE}}$. *We construct an* $(\mathsf{Fcs}(\mathsf{PKE}.\mathcal{M}, \mathsf{PKE}.\mathcal{R}), \mathsf{Inj}^\pm(\mathcal{U} \times \mathcal{I}, \mathsf{PKE}.\mathcal{M}))$-*oracle adversary* $\mathcal{B}$ *such that*

$$\mathbf{Adv}_{\mathsf{TKE}}^{\mathrm{mu\text{-}ow\text{-}pca}}(\mathcal{A}) \leqslant \mathbf{Adv}_{\mathsf{PKE}}^{\mathrm{mu\text{-}cpa}}(\mathcal{B}) + |\mathcal{U}| \cdot \delta'(q^*) + \frac{0.5 q_{\mathrm{CHAL}}^2 + |\mathcal{U}||\mathcal{I}|(q_{\mathrm{H}} + q_{\mathrm{PCO}} + 1)}{|\mathsf{PKE}.\mathcal{M}|}$$

$\mathbf{Query}(\mathcal{B}) = (q_{\mathrm{NEW}}, q_{\mathrm{CHAL}})$
$\mathbf{Time}^*(\mathcal{B}) = O(\mathbf{Time}(\mathcal{A}))$
$\mathbf{Mem}^*(\mathcal{B}) = O(\mathbf{Mem}(\mathcal{A}))$.

The notion of CPA security we require is interesting in that the adversary's encryption queries are of the form (u, i, m) where i is a "challenge identifier" and if it exactly repeats a query (u, i, m) it is given back the ciphertext from the earlier query.

INTUITION. For this proof, we program the random messages m to be the output of a random injection g applied to user identifier u and counter i. Our reduction adversary simulates the CHAL oracle for $\mathcal{A}$ by its own encryption oracle on $g(u, i)$. If any message that $\mathcal{A}$ queries to its random oracle or outputs at the end of execution is in the image of g, then out reduction assumes it is in the real world and outputs 1. In the ideal world, the view of $\mathcal{A}$ is independent of g so we can information theoretically bound the probability it finds such a message.

5.2 Transformation aV [OW-PCA → OW-PCVA]

The augmented transformation aV constructs a deterministic OW-PCVA secure public key encryption scheme VKE = aV[TKE] from a deterministic OW-PCA secure scheme TKE. The unaugmented V transformation was given (with a single-user, single-challenge memory-tight reduction) in [7]. Our augmentation adds a random string to the encryption key which is included with every hash function query. We define VKE as follows with VKE.IM = $\mathsf{Fcs}(\{0,1\}^l \times \mathsf{TKE}.\mathcal{M}, \{0,1\}^\gamma) \times$ TKE.IM and VKE.$\mathcal{M}$ = TKE.$\mathcal{M}$, where l and γ are fixed.

VKE.E$^{\mathcal{H} \times \mathcal{H}'}((a, ek), m)$	VKE.D$^{\mathcal{H} \times \mathcal{H}'}((a, ek, dk), c)$
$c_1 \leftarrow$ TKE.E$^{\mathcal{H}'}(ek, m)$	$(c_1, c_2) \leftarrow c$
$c_2 \leftarrow \mathcal{H}(a, m)$	$m' \leftarrow$ TKE.D$^{\mathcal{H}'}(dk, c_1)$
$c \leftarrow (c_1, c_2)$	If $m' = \perp$ or $\mathcal{H}(a, m') \neq c_2$ or TKE.E$^{\mathcal{H}'}(ek, m') \neq c_1$
Return c	Return $\perp$
	Return m'

For key generation, VKE.K samples $a \leftarrow_\$ \{0,1\}^l$ and $(ek, dk) \leftarrow_\$$ TKE.K, then returns $((a, ek), (a, ek, dk))$. Note that aV is tidy and if T is tidy and δ'-correct, then aV is δ'-correct.

We present a memory-tight reduction for aV in the multi-user, multi-challenge setting using the randomness programming technique. Our result is captured in Theorem 4 whose proof is given in the full version of the paper [23].

Theorem 4 (OW-PCA $\Rightarrow$ OW-PCVA). *Let* VKE = aV[TKE] *and suppose* TKE *is δ'-correct. Let $\mathcal{A}$ be an adversary against* VKE *with* **Query**$(\mathcal{A})$ = $(q_U, q_{CHAL}, q_{PCO}, q_{CVO}, q_H)$. *Assume* TKE*'s algorithms make at most q_{TKE} oracle queries and define $q^* = q_{TKE}(q_H + q_{CHAL} + 2q_{PCO} + 2q_{CVO})$. We construct an $(\mathsf{Fcs}(\{0,1\}^l \times \mathsf{TKE}.\mathcal{M}, \{0,1\}^\gamma), \mathsf{Fcs}(\mathsf{TKE}.\mathcal{C}, \{0,1\}^\gamma), \mathsf{Inj}^{\pm}(\mathcal{U}, \{0,1\}^l))$-oracle adversary $\mathcal{B}$ against* TKE *such that*

$$\mathsf{Adv}_{\mathsf{VKE}}^{\mathsf{mu\text{-}ow\text{-}va}}(\mathcal{A}) \leqslant \mathsf{Adv}_{\mathsf{TKE}}^{\mathsf{mu\text{-}ow\text{-}pca}}(\mathcal{B}) + 4|\mathcal{U}| \cdot \delta'(q^*) + 0.5|\mathcal{U}|^2/2^l + q_{CVO}/2^\gamma$$

$$\mathbf{Query}(\mathcal{B}) = (q_{NEW}, q_{CHAL}, q_{PCO}, q_H \cdot q_{TKE})$$

$$\mathbf{Time}^*(\mathcal{B}) = O(\mathbf{Time}(\mathcal{A})) \text{ and } \mathbf{Mem}^*(\mathcal{B}) = O(\mathbf{Mem}(\mathcal{A})).$$

5.3   Augmented Transformation aU$^\perp$ [OW-PCVA → \$IND-CCA]

The transformation aU$^\perp$ constructs an IND-CCA secure key encapsulation mechanism aUEM = aU$^\perp$[VKE] from a deterministic, OW-PCVA secure public key encryption scheme VKE. We define aUEM as follows where aUEM.K = VKE.K and aUEM.IM = Fcs($\{0,1\}^l \times$ VKE.$\mathcal{M} \times$ VKE.$\mathcal{C}, \mathcal{K}) \times$ VKE.IM ($\mathcal{K}$ is an arbitrary set used as the key set of aUEM).

aUEM.E$^{\mathcal{H} \times \mathcal{H}'}(ek)$	aUEM.D$^{\mathcal{H} \times \mathcal{H}'}(dk, (a, c))$
$m \leftarrow\!\!{}_\$ $ VKE.$\mathcal{M}$	$m \leftarrow$ VKE.D$^{\mathcal{H}'}(dk, c)$
$a \leftarrow\!\!{}_\$ \{0,1\}^l$	If $m = \perp$
$c \leftarrow$ VKE.E$^{\mathcal{H}'}(ek, m)$	Return $\perp$
$K \leftarrow \mathcal{H}(a, m, c)$	$K \leftarrow \mathcal{H}(a, m, c)$
Return $((a, c), K)$	Return K

The following theorem gives our security result.

Theorem 5 (OW-PCVA ⇒CCA). *Let* aUEM = aU$^\perp$[VKE] *where* VKE *is tidy and* δ'*-correct. Let* $\mathcal{A}$ *be an adversary against* aUEM *with* **Query**$(\mathcal{A})$ = $(q_{\text{NEW}}, q_{\text{ENCAP}}, q_{\text{DECAP}}, q_{\text{H}})$. *Assume* VKE*'s algorithms make at most* q_{VKE} *oracle queries and define* $q^* = q_{\text{VKE}}(2q_{\text{H}} + q_{\text{ENCAP}} + 2q_{\text{DECAP}})$. *Let* $D_1 = \{0,1\}^l \times$ VKE.$\mathcal{M} \cup \{\star\} \times$ VKE.$\mathcal{C}$ *and* $D_2 = \mathcal{U} \times \mathcal{I}$,. *We construct an* $(\text{Inj}^\pm(D_2, \{0,1\}^l), \text{Fcs}(D_1, \mathcal{K}), \text{Fcs}(D_1, \mathcal{K}))$*-oracle adversary* $\mathcal{B}$ *such that*

$$\text{Adv}_{\text{aUEM}}^{\text{mu-cca}}(\mathcal{A}) \leqslant 2\text{Adv}_{\text{VKE}}^{\text{mu-ow-va}}(\mathcal{B}) + 2|\mathcal{U}| \cdot \delta'(q^*) + 6|\mathcal{U}| \cdot |\mathcal{I}|/2^l$$

$$\text{Query}(\mathcal{B}) = (q_{\text{NEW}}, (q_{\text{ENCAP}} + q_{\text{DECAP}} + q_{\text{H}}), q_{\text{H}}, q_{\text{DECAP}})$$

$$\text{Time}^*(\mathcal{B}) = O(\text{Time}(\mathcal{A})) \text{ and } \text{Mem}^*(\mathcal{B}) = O(\text{Mem}(\mathcal{A})).$$

INTUITION. This proof is very similar to the proof of Theorem 2. Once again, we program the random string a to be the output of the injective function $g(u, i)$, and the message m to be the output of the random function $h(u, i)$. We use the map-then-rf technique to simulate the oracle H using the random function $\tilde{\mathcal{H}}(\alpha(a, m, c))$ where α is an injective function.

The reduction adversary gets challenge ciphertexts c from the CHAL oracle. To simulate the H oracle it uses the PCO oracle and to simulate the DECAP oracle, it uses its CVO oracle. When the aUEM adversary queries the H oracle with a tuple (a, m, c) such that PCO$(u, m, c) = 1$ and $c = \text{CHAL}(u, j)$ where $(u, j) = g^{-1}(a)$, the reduction outputs (u, j, m).

6   Memory-Tight Reduction for PKE Schemes

In this section, we provide a modified version of the TAM-tight security proof from [17] to show the security of the KEM/DEM construction of public key encryption. Thus, combining one of the KEMs studied in the rest of the paper with an appropriate symmetric encryption scheme gives a PKE scheme with a TAM-tight reduction in the multi-user, multi-challenge setting.

KEM/DEM SCHEME. Let SKE be a symmetric key encryption scheme and KEM be a key encapsulation mechanism. Then the KEM/DEM encryption scheme KD = KD[KEM, SKE] is defined as follows, with KD.IM = KEM.IM × SKE.IM.

KD[KEM, SKE].K	KD[KEM, SKE].$\mathsf{E}^{\mathcal{H} \times \mathcal{H}'}(ek, m)$	KD[KEM, SKE].$\mathsf{D}^{\mathcal{H} \times \mathcal{H}'}(dk, c)$
$(ek, dk) \leftarrow\!\!{\scriptstyle\$}\ \mathsf{KEM.K}$	$(c^k, K) \leftarrow\!\!{\scriptstyle\$}\ \mathsf{KEM.E}^{\mathcal{H}}(ek)$	$(c^k, c^d) \leftarrow c$
Return (ek, dk)	$c^d \leftarrow\!\!{\scriptstyle\$}\ \mathsf{SKE.E}^{\mathcal{H}'}(K, m)$	$K \leftarrow \mathsf{KEM.D}^{\mathcal{H}}(dk, c^k)$
	Return (c^k, c^d)	If $K = \bot$ then return $\bot$
		Return $\mathsf{SKE.D}^{\mathcal{H}'}(K, c^d)$

The following theorem TAM-tightly proves the security of KD.

Theorem 6. *Let* SKE *be a symmetric key encryption scheme,* KEM *be a ϵ-uniform key encapsulation mechanism, and* KD = KD[SKE, KEM]. *Let* $T' = \mathcal{U} \times \bigcup_{ek \in \mathsf{KEM.Ek}} \mathsf{KEM}.\mathcal{C}(ek) \times \mathbb{N}$, *Let* $T = \mathcal{U} \times \mathsf{KEM.Ek}$, $D_{(u,ek)} = [q_{\mathrm{ENC}}]$ *and* $R_{(u,ek)} = \mathsf{KEM}.\mathcal{C}(ek)$. $D'_{(u,c^k,l)} = \{0,1\}^l$, *and* $R'_{(u,c^k,l)} = \{0,1\}^{\mathsf{SKE.cl}(l)}$. *Let* $\mathcal{A}$ *be a* mu-\$cca *adversary against* KD *with* **Query**$(\mathcal{A}) = (q_{\mathrm{NEW}}, q_{\mathrm{ENC}}, q_{\mathrm{DEC}}, q_{\mathrm{H}})$. *Then we can construct a* $(\mathsf{SKE.IM}, \mathsf{Inj}^{\pm}(T', D', R'))$-*oracle adversary* $\mathcal{B}_{\mathsf{KEM}}$ *and* $(\mathsf{KEM.IM}, \mathsf{Fcs}(\mathcal{U}, \mathsf{KEM}.\mathcal{R}), \mathsf{Inj}^{\pm}(T, D, R))$-*oracle adversary against* $\mathcal{B}_{\mathsf{SKE}}$ *such that*

$$\mathsf{Adv}_{\mathsf{KD}}^{\mathsf{mu\text{-}\$cca}}(\mathcal{A}) \leq 2\mathsf{Adv}_{\mathsf{KEM}}^{\mathsf{mu\text{-}\$cca}}(\mathcal{B}_{\mathsf{KEM}}) + \mathsf{Adv}_{\mathsf{SKE}}^{\mathsf{mu\text{-}\$cca}}(\mathcal{B}_{\mathsf{SKE}}) + q_{\mathrm{ENC}} \cdot \epsilon$$

$$+ \frac{1.5q_{\mathrm{ENC}}^2 + 2q_{\mathrm{ENC}}q_{\mathrm{DEC}}}{|\mathsf{KEM}.\mathcal{C}|} + \frac{q_{\mathrm{ENC}}^2 + 2q_{\mathrm{DEC}}}{2^{\mathsf{SKE.xl}}}.$$

The complexities of $\mathcal{B}_{\mathsf{KEM}}$ *and* $\mathcal{B}_{\mathsf{SKE}}$ *basically match those of* $\mathcal{A}$. *Moreover,* $\mathcal{B}_{\mathsf{SKE}}$ *makes at most one encryption query per user.*

INSTANTIATING KD. This result proves the multi-challenge, multi-user security of KD, but requires appropriate choices of KEM and SKE. Naturally, one could choose any of the KEMs studied earlier in this work for the first component.[6] For the symmetric encryption we need a scheme which achieves single-challenge, multi-user security. We are not aware of any TAM-tight multi-user analysis of symmetric encryption scheme, so one instead needs to pick a scheme whose multi-user, single challenge security is sufficiently strong against memory-unbounded adversaries. One reasonable option could be GCM with a random nonce. In the ideal cipher model, Hoang, Tessaro, and Thiruvengadam [20] showed a strong bound for this setting which is essentially independent of the number of users.

References

1. Abdalla, M., Bellare, M., Rogaway, P.: DHIES: an encryption scheme based on the Diffie-Hellman problem. Contributions to IEEE P1363a, September 1998

---

[6] For using the Fujisaki-Okamoto Transformation, which we proved CCA secure, note that $|\mathsf{Adv}_{\mathsf{KEM}}^{\mathsf{mu\text{-}\$cca}}(\mathcal{B}_{\mathsf{KEM}}) - \mathsf{Adv}_{\mathsf{KEM}}^{\mathsf{mu\text{-}cca}}(\mathcal{B}_{\mathsf{KEM}})| \leq q_{\mathrm{ENC}} \cdot \epsilon.$

2. Abdalla, M., Bellare, M., Rogaway, P.: The oracle Diffie-Hellman assumptions and an analysis of DHIES. In: Naccache, D. (ed.) CT-RSA 2001. LNCS, vol. 2020, pp. 143–158. Springer, Heidelberg (2001). https://doi.org/10.1007/3-540-45353-9_12

3. Auerbach, B., Cash, D., Fersch, M., Kiltz, E.: Memory-tight reductions. In: Katz, J., Shacham, H. (eds.) CRYPTO 2017. LNCS, vol. 10401, pp. 101–132. Springer, Cham (2017). https://doi.org/10.1007/978-3-319-63688-7_4

4. Bauer, B., Fuchsbauer, G., Loss, J.: A classification of computational assumptions in the algebraic group model. In: Micciancio, D., Ristenpart, T. (eds.) CRYPTO 2020. LNCS, vol. 12171, pp. 121–151. Springer, Cham (2020). https://doi.org/10.1007/978-3-030-56880-1_5

5. Bellare, M., Boldyreva, A., Micali, S.: Public-key encryption in a multi-user setting: security proofs and improvements. In: Preneel, B. (ed.) EUROCRYPT 2000. LNCS, vol. 1807, pp. 259–274. Springer, Heidelberg (2000). https://doi.org/10.1007/3-540-45539-6_18

6. Bellare, M., Rogaway, P.: The security of triple encryption and a framework for code-based game-playing proofs. In: Vaudenay, S. (ed.) EUROCRYPT 2006. LNCS, vol. 4004, pp. 409–426. Springer, Heidelberg (2006). https://doi.org/10.1007/11761679_25

7. Bhattacharyya, R.: Memory-tight reductions for practical key encapsulation mechanisms. In: Kiayias, A., Kohlweiss, M., Wallden, P., Zikas, V. (eds.) PKC 2020. LNCS, vol. 12110, pp. 249–278. Springer, Cham (2020). https://doi.org/10.1007/978-3-030-45374-9_9

8. Cramer, R., Shoup, V.: Design and analysis of practical public-key encryption schemes secure against adaptive chosen ciphertext attack. Cryptology ePrint Archive (2001). https://eprint.iacr.org/2001/108

9. Dai, W., Tessaro, S., Zhang, X.: Super-linear time-memory trade-offs for symmetric encryption. In: Pass, R., Pietrzak, K. (eds.) TCC 2020. LNCS, vol. 12552, pp. 335–365. Springer, Cham (2020). https://doi.org/10.1007/978-3-030-64381-2_12

10. Diemert, D., Gellert, K., Jager, T., Lyu, L.: Digital signatures with memory-tight security in the multi-challenge setting. In: Tibouchi, M., Wang, H. (eds.) ASIACRYPT 2021. LNCS, vol. 13093, pp. 403–433. Springer, Cham (2021). https://doi.org/10.1007/978-3-030-92068-5_14

11. Dinur, I.: On the streaming indistinguishability of a random permutation and a random function. In: Canteaut, A., Ishai, Y. (eds.) EUROCRYPT 2020. LNCS, vol. 12106, pp. 433–460. Springer, Cham (2020). https://doi.org/10.1007/978-3-030-45724-2_15

12. Dinur, I.: Tight time-space lower bounds for finding multiple collision pairs and their applications. In: Canteaut, A., Ishai, Y. (eds.) EUROCRYPT 2020. LNCS, vol. 12105, pp. 405–434. Springer, Cham (2020). https://doi.org/10.1007/978-3-030-45721-1_15

13. Fuchsbauer, G., Kiltz, E., Loss, J.: The algebraic group model and its applications. In: Shacham, H., Boldyreva, A. (eds.) CRYPTO 2018. LNCS, vol. 10992, pp. 33–62. Springer, Cham (2018). https://doi.org/10.1007/978-3-319-96881-0_2

14. Fujisaki, E., Okamoto, T.: Secure integration of asymmetric and symmetric encryption schemes. In: Wiener, M. (ed.) CRYPTO 1999. LNCS, vol. 1666, pp. 537–554. Springer, Heidelberg (1999). https://doi.org/10.1007/3-540-48405-1_34

15. Fujisaki, E., Okamoto, T.: Secure integration of asymmetric and symmetric encryption schemes. J. Cryptol. 26(1), 80–101 (2011). https://doi.org/10.1007/s00145-011-9114-1

16. Gay, R., Hofheinz, D., Kiltz, E., Wee, H.: Tightly CCA-secure encryption without pairings. In: Fischlin, M., Coron, J.-S. (eds.) EUROCRYPT 2016. LNCS, vol. 9665, pp. 1–27. Springer, Heidelberg (2016). https://doi.org/10.1007/978-3-662-49890-3_1

17. Ghoshal, A., Ghosal, R., Jaeger, J., Tessaro, S.: Hiding in plain sight: memory-tight proofs via randomness programming. In: Dunkelman, O., Dziembowski, S. (eds.) EUROCRYPT 2022. LNCS, vol. 13276, pp. 706–735. Springer, Cham (2022). https://doi.org/10.1007/978-3-031-07085-3_24. https://eprint.iacr.org/2021/1409

18. Ghoshal, A., Jaeger, J., Tessaro, S.: The memory-tightness of authenticated encryption. In: Micciancio, D., Ristenpart, T. (eds.) CRYPTO 2020. LNCS, vol. 12170, pp. 127–156. Springer, Cham (2020). https://doi.org/10.1007/978-3-030-56784-2_5

19. Ghoshal, A., Tessaro, S.: On the memory-tightness of hashed ElGamal. In: Canteaut, A., Ishai, Y. (eds.) EUROCRYPT 2020. LNCS, vol. 12106, pp. 33–62. Springer, Cham (2020). https://doi.org/10.1007/978-3-030-45724-2_2

20. Hoang, V.T., Tessaro, S., Thiruvengadam, A.: The multi-user security of GCM, revisited: tight bounds for nonce randomization. In: Lie, D., Mannan, M., Backes, M., Wang, X. (eds.) ACM CCS 2018, pp. 1429–1440. ACM Press, October 2018

21. Hofheinz, D., Hövelmanns, K., Kiltz, E.: A modular analysis of the Fujisaki-Okamoto transformation. In: Kalai, Y., Reyzin, L. (eds.) TCC 2017. LNCS, vol. 10677, pp. 341–371. Springer, Cham (2017). https://doi.org/10.1007/978-3-319-70500-2_12

22. Hofheinz, D., Jager, T.: Tightly secure signatures and public-key encryption. In: Safavi-Naini, R., Canetti, R. (eds.) CRYPTO 2012. LNCS, vol. 7417, pp. 590–607. Springer, Heidelberg (2012). https://doi.org/10.1007/s00145-020-09356-x

23. Jaeger, J., Kumar, A.: Memory-tight multi-challenge security of public-key encryption. Cryptology ePrint Archive (2022). https://eprint.iacr.org/2022/

24. Jaeger, J., Tessaro, S.: Tight time-memory trade-offs for symmetric encryption. In: Ishai, Y., Rijmen, V. (eds.) EUROCRYPT 2019. LNCS, vol. 11476, pp. 467–497. Springer, Cham (2019). https://doi.org/10.1007/978-3-030-17653-2_16

25. Libert, B., Joye, M., Yung, M., Peters, T.: Concise multi-challenge CCA-secure encryption and signatures with almost tight security. In: Sarkar, P., Iwata, T. (eds.) ASIACRYPT 2014. LNCS, vol. 8874, pp. 1–21. Springer, Heidelberg (2014). https://doi.org/10.1007/978-3-662-45608-8_1

26. Maurer, U.: Abstract models of computation in cryptography. In: Smart, N.P. (ed.) Cryptography and Coding 2005. LNCS, vol. 3796, pp. 1–12. Springer, Heidelberg (2005). https://doi.org/10.1007/11586821_1

27. Okamoto, T., Pointcheval, D.: The gap-problems: a new class of problems for the security of cryptographic schemes. In: Kim, K. (ed.) PKC 2001. LNCS, vol. 1992, pp. 104–118. Springer, Heidelberg (2001). https://doi.org/10.1007/3-540-44586-2_8

28. Shahaf, I., Ordentlich, O., Segev, G.: An information-theoretic proof of the streaming switching lemma for symmetric encryption. In 2020 IEEE International Symposium on Information Theory (ISIT), pp. 858–863 (2020)

29. Shoup, V.: Lower bounds for discrete logarithms and related problems. In: Fumy, W. (ed.) EUROCRYPT'97. LNCS, vol. 1233, pp. 256–266. Springer, Heidelberg (1997). https://doi.org/10.1007/3-540-69053-0_18

30. Tessaro, S., Thiruvengadam, A.: Provable time-memory trade-offs: symmetric cryptography against memory-bounded adversaries. In: Beimel, A., Dziembowski, S. (eds.) TCC 2018. LNCS, vol. 11239, pp. 3–32. Springer, Cham (2018). https://doi.org/10.1007/978-3-030-03807-6_1

31. Wang, Y., Matsuda, T., Hanaoka, G., Tanaka, K.: Memory lower bounds of reductions revisited. In: Nielsen, J.B., Rijmen, V. (eds.) EUROCRYPT 2018. LNCS, vol. 10820, pp. 61–90. Springer, Cham (2018). https://doi.org/10.1007/978-3-319-78381-9_3

Zero Knowledge

Short-lived Zero-Knowledge Proofs and Signatures

Arasu Arun[1]([envelope]), Joseph Bonneau[1,2], and Jeremy Clark[3]

[1] New York University, New York, NY, USA
arasu@nyu.edu
[2] University of Melbourne, Melbourne, VIC, Australia
[3] Concordia University, Montreal, QC, Canada

Abstract. We introduce the short-lived proof, a non-interactive proof of knowledge with a novel feature: after a specified period of time, the proof is no longer convincing. This time-delayed loss of soundness happens "naturally" without further involvement from the prover or any third party. We propose definitions for short-lived proofs as well as the special case of short-lived signatures. We show several practical constructions built using verifiable delay functions (VDFs). The key idea in our approach is to allow any party to forge any proof by executing a large sequential computation. Some constructions achieve a stronger property called reusable forgeability in which one sequential computation allows forging an arbitrary number of proofs of different statements. We also introduces two novel types of VDFs, re-randomizable VDFs and zero-knowledge VDFs, which may be of independent interest. Our constructions for short-lived Σ-protocols and signatures are practically efficient for provers and verifiers, adding a few hundred bytes of overhead and tens to hundreds of milliseconds of proving/verification time.

Keywords: Zero-knowledge proofs · Signatures · VDFs · Time-based crypto

1 Introduction

A digital signature is forever. Or at least, until the underlying signature scheme is broken or the signing key is breached. This is often much more than what is required for real world applications: a signature might need to only provide authenticity for a few seconds to conduct an authenticated key exchange or verify the provenance of an email. At best, the long-lived authentication provided by standard signatures is often unnecessary. In certain cases, however, it may have significant undesirable consequences.

An illustrative example is the DKIM protocol [53] used by modern SMTP servers to sign outgoing email on behalf of the entire domain (e.g., example.com) with a single key. DKIM is primarily intended to prevent email spoofing [27]. As such, these signatures only need a lifetime of minutes for recipient SMTP servers to verify and potentially filter email. However DKIM signatures do not expire

© International Association for Cryptologic Research 2022
S. Agrawal and D. Lin (Eds.): ASIACRYPT 2022, LNCS 13793, pp. 487–516, 2022.
https://doi.org/10.1007/978-3-031-22969-5_17

and instead provide long-lasting evidence of authenticity for old email messages, such as ones leaked through illicit data breaches [72]. As a result, cryptographers have suggested that DKIM servers should periodically rotate keys and reveal old private keys to provide deniability for old DKIM signatures [45].

Our Approach. A short-lived proof convinces the verifier of the following: either a claimed statement x is true or someone expended at least t steps of sequential work to forge the proof. The proof incorporates a random beacon value (e.g., the day's newspaper headline) to ensure it was not created before a specific time T_0. If a verifier observes the proof within Δ units of time after T_0, she will believe it is a valid proof if $\Delta < t$ because it would be impossible to have forged the proof within that time period. Once $\Delta \geq t$, the proof is no longer convincing as it may have been constructed through the forgery process.

Our constructions build on recent advances in time-based cryptography, specifically *verifiable delay functions* (VDFs) [16,64,77]. Under the hood, the sequential computation required for forging a proof or signature in all of our schemes is equivalent to evaluating a VDF on a random input.

Cryptographic Deniability. The idea that signatures should not be permanently verifiable is a special case of *cryptographic deniability*. This is often weaker than intuition suggests. Informally, deniability for signatures means there is no *additional* cryptographic proof that Alice sent a particular message. There may still be circumstantial proof such as logs or testimony, but these would exist whether or not cryptography was used at all.

A simple approach to deniability, as suggested for DKIM, is to publish secret information after running a protocol which enables any party to forge transcripts. Other approaches include deniable key exchange protocols (e.g., OTR messaging [21]) or designated verifier proofs/signatures [49] which limit verifiability to a specified set of parties. By contrast, short-lived proofs are non-interactive and publicly verifiable yet become deniable after a specified period of time without any further action by the prover. For signatures, the signer can maintain a long-lived key even as messages signed with it expire.

The fact that short-lived signatures provide deniability without the sender needing to interact with the recipient (or even know the receivers' public key) makes them uniquely qualified for achieving deniability in several practical scenarios, as we discuss in Sect. 10. An example is sending email with a single signature to a large, potentially unknown group of recipients. To our knowledge, ours is the first primitive to enable this.

To summarize our contributions, as outlined in Table 1:

- We define short-lived zero-knowledge proofs (§4) and short-lived signatures (§8). *Reusable forgeability* captures the useful property of a single slow computation enabling efficient proof forgery for any statement (§4.1).
- We propose a short-lived proof construction (with reusable forgeability) from any generic non-interactive zero-knowledge proof scheme and any VDF (§5).

Table 1. A comparison of our constructions. The symbol ◑ denotes schemes with a time-space tradeoff in the delay parameter t (see §9).

Scheme	Section	Reusable forgeability	No pre-computation	VDF required	Proof/Sign type
		short-lived proofs			
Generic ZK	§5	●	●	any [16]	Generic SNARK
Σ-Precomp	§6.2	○	○	any [16]	Σ-protocol
Σ-rrVDF	§6.3	○	◑	re-randomizable (§A)	Σ-protocol
Σ-zkVDF	§7	◑	●	zero-knowledge (§7.1)	Σ-protocol
		short-lived signatures			
Generic ZK	§8	●	●	any [16]	Generic SNARK
Σsign ∨ zkVDF	§8	◑	●	zero-knowledge (§7.1, §C)	Σ signatures
Sign-Trapdoor	§8.1	○	●	trapdoor ([64,77])	RSA
Sign-Watermark	§8.2	◑	●	watermarkable ([77], §B)	RSA

- We propose a short-lived proof construction (§6.2) for any Σ-protocol (§2.2) and any VDF with a Σ-protocol for verification. Our basic scheme requires precomputation per-proof, which we can eliminate by introducing the concept of re-randomizable VDFs .
- We introduce the notion of zero-knowledge VDFs (§7) and use it to build a short-lived proof with reusable forgeability for Σ-protocols.
- We show that our general Σ-construction can be instantiated with Σ-signatures such as Schnorr (§8). We further introduce highly efficient short-lived signature schemes (§8.1) from trapdoor VDFs [77] and watermarkable VDFs which are as compact as a single VDF proof and offer reusable forgeability.

2   Preliminaries

2.1   Zero-knowledge Proofs and Arguments

We start with a basic background on zero-knowledge proofs, while referring the reader to [74] for more comprehensive introduction. Zero-knowledge proofs concern a relation $R \subset \mathcal{X} \times \mathcal{W}$ — a set of pairs (x, w) where x is called the statement or instance and w is called the witness. For example, for the relation of Diffie-Hellman tuples, we might write: $R_{\mathrm{DH3}} = \{(x = (g, g_1, g_2, g_3), w = (a, b)) | g_1 = g^a \wedge g_2 = g^b \wedge g_3 = g^{ab}\}$. The set of all values x such that there exists a witness w for which $(x, w) \in R$ is the language L_R. It has been shown that all

NP-languages have a zero-knowledge proof system [43]. A non-interactive zero knowledge proof system Π_R for R is a trio of algorithms (we consider R to be hard-coded into all three):

– $\Pi.\mathsf{Setup}(\lambda) \to pp$
– $\Pi.\mathsf{Prove}(pp, x, w) \to \pi$
– $\Pi.\mathsf{Verify}(pp, x, \pi) \to \mathsf{Accept/Reject}$

A proof system is *complete* if, for all $(x, w) \in R$, given a proof $\pi \leftarrow \mathsf{Prove}(pp, x, w)$, the verification algorithm $\mathsf{Verify}(pp, x, \pi)$ outputs Accept. A proof system is *sound* if an unbounded malicious prover (who does not know w) cannot produce an acceptable proof with probability greater than $2^{-\kappa}$ for knowledge error κ. The weaker notion of computational soundness holds for polynomial-time malicious provers; for simplicity we refer to such *argument systems* as proofs.

A *proof of knowledge* has an additional property roughly stating that an adversary must "know" a witness w to compute a proof for $(x, w) \in R$. Knowledge soundness for proofs-of-knowledge is formalized by defining an algorithm $\mathcal{A}$ which outputs an accepted proof π and demonstrating an efficient algorithm $\mathsf{Extract}$ which can interact with $\mathcal{A}$ and output a witness w such that $(x, w) \in R$. Depending on the proof construction, the extractor may need to rewind $\mathcal{A}$ (a rewinding extractor) or inspect $\mathcal{A}$'s internal state (a non-black box extractor).

A proof of knowledge is *zero knowledge* if the proof π reveals nothing about the witness w. Formally, this is established by demonstrating an efficient algorithm $\mathsf{Simulate}$ which, given any statement $x \in L_R$ can output simulated proofs $\tilde{\pi}$ indistinguishable from real proofs such that $\mathsf{Verify}(pp, x, \tilde{\pi}) \to \mathsf{Accept}$. $\mathsf{Simulate}$ may require additional power, such as the ability to *program* the random oracle to give specified responses.

If the system produces *succinct* (e.g., constant or poly-logarithmic sized) arguments, it is a SNARK (or zk-SNARK) for R, of which there are now many known constructions [13,42,46].

2.2   Sigma Protocols

Our constructions in Sects. 6 and 7 target a special class of interactive zero-knowledge proof systems called Σ-protocols [70]. A Σ-protocol [70] is a three-move interactive protocol between a prover P and verifier V:

1. P runs $\Sigma.\mathsf{Commit}(x) \to a$ and sends a to V.
2. V runs $\Sigma.\mathsf{Challenge}() \to c$ and sends c to P.
3. P runs $\Sigma.\mathsf{Respond}(x, w, a, c) \to z$ and sends z to V.
4. V accepts if $\Sigma.\mathsf{Verify}(x, a, c, z) \to \mathsf{Accept}$.

We write Σ_R to denote a Σ-protocol for relation R. A Σ-protocol has *special soundness* if there exists an algorithm $\Sigma.\mathsf{Extract}(x, a, c, c', z, z')$ which outputs a witness w for x given any two accepting transcripts of the form (x, a, c, z) and (x, a, c', z') with $c \neq c'$. A Σ-protocol is honest verifier zero-knowledge if it has an efficient algorithm $\Sigma.\mathsf{Simulate}(x) \rightarrow (\tilde{a}, \tilde{c}, \tilde{z})$ such that $\Sigma.\mathsf{Verify}(x, \tilde{a}, \tilde{c}, \tilde{z}) \rightarrow$ Accept and the distribution of $(x, \tilde{a}, \tilde{c}, \tilde{z})$ is indistinguishable from transcripts of a genuine interaction between a verifier and prover knowing a witness w.

Every Σ-protocol can be transformed into a non-interactive, fully secure (i.e., no honesty assumption on the verifier) zero-knowledge proof in the random oracle model using the Fiat-Shamir heuristic [39], in which the challenge is generated as $c = \mathcal{O}(x, a)$ where $\mathcal{O}$ is the random oracle. $\Sigma.\mathsf{Extract}$ and $\Sigma.\mathsf{Simulate}$ make use of rewinding the other party and programming the random oracle.

2.3   Disjunction of Σ-protocols

The set of relations with Σ-protocols is closed under conjunction and disjunction [32]. The classic protocol for disjunction of Σ-protocols, which we denote Σ-OR, is due to Cramer et al. [32].[1] Let Σ_{R_1} and Σ_{R_2} be Σ-protocols for relations R_1 and R_2 respectively. Assume the prover wants to prove the disjunction of the statement $x = (x_1, x_2)$ and knows a witness w_1 showing that $(x_1, w_1) \in R_1$ (knowing w_2 showing that $(x_2, w_2) \in R_2$ is a symmetric case). The proof is constructed as follows:

Protocol $\Sigma_{R_1 \vee R_2}(x_1, w_1, x_2, -)$:

1. P runs $\Sigma_{R_2}.\mathsf{Simulate}(x_2) \rightarrow (\tilde{a}_2, \tilde{c}_2, \tilde{z}_2)$
2. P runs $\Sigma_{R_1}.\mathsf{Commit}(x_1) \rightarrow a_1$
3. P send $(a_1, \tilde{a}_2)$ to V
4. V sends $c = \Sigma.\mathsf{Challenge}()$ to P
5. P sets $c_1 = c \oplus \tilde{c}_2$
6. P runs $\Sigma_{R_1}.\mathsf{Respond}(x_1, w_1, a_1, c_1) \rightarrow z_1$
7. P sends $(c_1, \tilde{c}_2, z_1, z_2)$ to V
8. V accepts if $c = c_1 \oplus \tilde{c}_2$ and both:
   - $\Sigma_{R_1}.\mathsf{Verify}(x_1, a_1, c_1, z_1) \rightarrow$ Accept
   - $\Sigma_{R_2}.\mathsf{Verify}(x_2, \tilde{a}_2, \tilde{c}_2, \tilde{z}_2) \rightarrow$ Accept

2.4   Beacons

A beacon [65] is a continual source of unpredictable public randomness. A beacon's output at time T_i should be uniformly random and unpredictable as of time T_{i-1}. We assume that beacon values are drawn uniformly randomly from a space $|\mathcal{B}| \geq 2^\lambda$ for the security parameter λ. All our protocols assume an input beacon value denoted b. In practice, NIST operates a centralized beacon which publishes 512 random bits every minute [4]. The drand project operates a public beacon publishing 256 bits every 30 s [2] using a multi-party randomness protocol [73]. Other potential beacons include web server challenges [48], stock market prices [26], and blockchain data [20].

---

[1] Ciampi et al. [25] later introduced a different Σ-OR protocol with certain advantages over the Cramer et al. construction.

2.5   Verifiable Delay Functions

Verifiable delay functions (VDFs) are defined by a trio of algorithms:[2]

- VDF.Setup$(\lambda, t) \rightarrow pp$
- VDF.Eval$(pp, b) \rightarrow (y, \pi)$
- VDF.Verify$(pp, b, y, \pi) \rightarrow$ Accept/Reject

Boneh et al. formalized VDFs and offer formal security definitions [16]. Informally, VDFs satisfy three important properties: (1) Verifiability, meaning that the verification algorithm is efficient (at most polylogarithmic in t and λ) and always accepts when given genuine output from Eval. (2) VDF evaluation must be a function, meaning that Eval is a deterministic algorithm and it is computationally infeasible to find two pairs $(b, y), (b, y')$ with $y \neq y'$ that Eval will accept. And (3) VDFs must impose a computational delay. Roughly speaking, computing a VDF successfully with non-negligible probability over a uniformly distributed challenge b should be impossible without executing t sequential steps. Throughout this paper, we will refer to this property as "t-Sequentiality" to emphasize the time delay parameter. All of our constructions reduce forging a proof to evaluating a VDF on a random input. Formally:

Property 1 (Sequentiality of VDFs (from [16])). For functions $\sigma(t), p(t)$, the VDF is (p, σ)-Sequential if for all randomized algorithms $\mathcal{A}_0$ which run in total time $O(poly(t, \lambda))$, and $\mathcal{A}_1$ which run in parallel time $\sigma(t)$ on at most $p(t)$ processors, the probability of winning the following game is negligible:

1. $pp \leftarrow$ Setup(λ, t)
2. $\alpha \leftarrow \mathcal{A}_0(\lambda, pp, t)$ // advice string
3. Challenger samples VDF input b
4. $y_A \leftarrow \mathcal{A}_1(\alpha, pp, b)$
5. Adversary wins if $y_A = y$ where $y, \pi \leftarrow$ Eval(pp, b)

VDFs constructions have been proposed from generic succinct proofs [16], repeated squaring in groups of unknown order [64,77], permutation polynomials [16], isogenies [37], and homomorphic encryption [50]. Earlier work proposed "weak" VDFs based on computing square roots mod p [35,55]. Proof-of-sequential work (PoSW) [28,56] is a similar primitive that does not require the evaluation to have a unique mapping. Modern VDF constructions are in fact the most efficient known PoSW constructions; for simplicity we present all our constructions using the notation and terminology of VDFs.

An important limitation of all VDF constructions is that they can only guarantee a certain number of steps of sequential computation are required. The real-world or "wall-clock" time needed to execute this computation varies based on the speed of available computing platforms. To manage this limitation, conventional wisdom suggests using a VDF with a relatively simple evaluation function for which highly optimized hardware is available to honest parties, limiting the speedup available to attackers. For this reason, repeated-squaring based VDFs are considered the most practical candidates today.

---

[2] The VDF challenge is traditionally denoted x. We use b to avoid confusion with x as the statement of a zero-knowledge proof.

2.6 VDFs from Repeated Squaring

We focus in particular on VDF constructions which utilize repeated squaring in a group of unknown order, as these have useful algebraic properties for building short-lived proofs and signatures. VDF evaluation is simply $y = \mathsf{VDF}.\mathsf{Eval}(b) = b^{2^t}$. Wesolowski [77] and Pietrzak [64] introduced two distinct approaches for efficiently proving that $y = b^{2^t}$ in a group of unknown order:

Wesolowski Proofs: First, the Prover provides $\tilde{y}$, claiming $\tilde{y} - b^{2^t}$. The verifier provides a random prime ℓ as a challenge. Both parties compute, via long division, the unique values q, r such that $2^t = q\ell + r$ and $0 \le r < \ell$. Finally, the prover outputs a proof $\pi = b^q$. The verifier accepts if and only if $\tilde{y} = q^\ell b^r$.

Pietrzak Proofs: As before, the Prover provides $\tilde{y}$, claiming $\tilde{y} = b^{2^t}$. The prover then provides a value v and asserts that $v = b^{2^{t/2}}$. The verifier chooses a random challenge r and they both compute $\tilde{y}' = \tilde{y} \cdot v^r, b' = v \cdot b^r$. The verifier could manually verify that $\tilde{y}' = (b')^{(2^{t/2})}$ by computing $\frac{t}{2}$ squarings, half as many as the original problem of verifying that $\tilde{y} = b^{(2^t)}$. Alternately, the prover can recursively prove that $\tilde{y}' = (b')^{(2^{t/2})}$. Typically, this is done for d rounds, each halving the size of the exponent, until the verifier manually checks the remaining exponent of size $2^{t/2^d}$.

Boneh et al. [17] provide a detailed comparison of the two proof constructions. Wesolowski proofs are shorter (two group elements instead of $O(\log t)$) but more difficult to compute and rely on slightly stronger assumptions. In this work we observe a new property of both constructions, re-randomizability (§6.3), and introduce a new zero-knowledge variant of Wesolowski proofs (§7).

3 Related Work

Jakobsson et al. introduced the idea of using disjunctive statements to provide deniability [49]. Given a statement x to be proven to Bob in zero knowledge, the proof is transformed into the statement: *{ either x or I know Bob's private key}*. A proof of this compound statement, which is called a *designated verifier proof*, is only convincing to Bob. If Bob is confident that nobody else knows his private key and that he did not compute the proof, then he knows the second clause is false and therefore x is true. Anybody else is unsure if x is true or if Bob *forged* the proof by satisfying the second clause. Another approach to constructing signatures with the designated-verifier property is *chameleon signatures* [52], which use a standard hash-and-sign construction but with a chameleon hash function whose trapdoor is known by the intended verifier.

Many of our constructions[3] follow a similar disjunctive template, with the essential statement being *{ either x or someone solved a VDF of difficulty t on a*

---

[3] An initial version of this work appeared as a Masters thesis [30].

beacon value derived from b which was unknown before time T }. VDFs requires t sequential steps (which approximate elapsed time).

Several other works have used disjunctive proofs to provide different notions of deniability. Baldimtsi et al. showed the constructions and applications for *proofs-of-work-or-knowledge* (PoWorKs) of the form {*either x or someone solved a proof-of-work puzzle*} where the puzzle requires w units of parallelizable computation [8]. Time-traveling simulators, introduced by Goyal et al. [44], provide a similar deniability notion in which a proof is convincing until a blockchain grows to a certain length. Specter et al. proposed {*either x or someone has seen value v released at time T*} for a v to be published at a future time T [72]. This proposal is closest to our own work, as we discuss further in Sect. 10.2.

Similar time-based deniability notions for signatures specifically have been considered by several authors (we believe ours is the first to expand to general proofs). Ferradi et al. [38] in 2015 presented a protocol for what they call *fading signatures* using the RSW time-lock puzzle and a trusted authority to pre-compute some solutions using the trapdoor. Their notion is weaker in that verification is slow, requiring t sequential steps. In hindsight, with the benefit of modern VDFs the slow verification time of their approach could be fixed.

The connection between VDFs and time-based deniability was made by Wesolowski who presented an interactive identification scheme that becomes deniable after the passage of time [77, §8]. Wesolowski also described a time-limited signature protocol which improved on the Ferradi et al. construction in an unpublished 2016 manuscript [76]. Our Sign-Trapdoor construction (§8.1) improves on this protocol by making it transferable, non-interactive, and a true signature (rather than an authentication protocol).

Contemporaneous to our work, Beck et al. [10] propose a construction for what they call *time-deniable signatures*, which utilize time-lock puzzles. Colburn [30] described a folklore construction (called Folk+) in which a time-lock puzzle encapsulating the signing key is published along with a signature. Green et al.'s construction works similarly, except to enable continuous use of the key the time-lock puzzle encapsulates a *restricted signing oracle* which can only sign messages with a timestamp before a chosen expiration date. This construction appears inherently limited to signatures. It also utilizes completely different cryptographic techniques and as a result the reported signing time is over 4 s per message, orders of magnitude slower than our signature constructions.

Other time-based cryptographic primitives have been proposed including encryption, commitments, and signatures [15,66,75]. In the context of the cited literature, a *timed signature* [15] is a commitment to a signature that has been shared and can later be revealed. However if the committer aborts before revealing, the recipient can perform sequential work to uncover the signature. Dodis and Yum introduced a similar idea of *time-capsule signatures* [33] which become valid after a certain period of time when a time-server releases some information. We are essentially solving the inverse problem: instead of a signature being hidden for time Δt and then becomes unforgeable, a short-lived signature is unforgeable for Δt and then becomes deniable.

Definition 1 (Short-Lived Proofs) *Let λ be a security parameter. Let L_R be a language in* NP *and R be a relation such that $(x, w) \in R$ if and only if w is a witness showing $x \in L_R$. Let $\mathcal{B}$ be a space of beacon values where $|\mathcal{B}| \geq 2^\lambda$. A short-lived proof system Π_R^t with time delay $t \in \mathbb{Z}$ is a quartet of randomized algorithms* (Setup, Prove, Forge, Verify)*:*

- Setup$(L, \lambda, t) \to pp$ *produces a set of public parameters pp*
- Prove$(pp, x, w, b) \to \pi$ *produces a proof π if $(x, w) \in R$*
- Forge$(pp, x, b) \to \pi$ *produces a proof π for any x*
- Verify$(pp, x, \pi, b) \to$ Accept/Reject

Π_R^t *must satisfy the following properties:*

- **Completeness:** *For all $(x, w) \in R$ and $b \in \mathcal{B}$, $\pi \leftarrow$ Prove(pp, x, w, b) runs in time less than t and* Verify(pp, x, π, b) *outputs* Accept.
- *t-**Forgeability:*** *For all x, $b \in \mathcal{B}$, $\pi \leftarrow$ Forge(pp, x, b) runs in time $(1 + \epsilon)t$ for some positive constant ϵ and* Verify(pp, x, π, b) *outputs* Accept.
- *t-**Soundness:*** *For all x and for any pair of adversary algorithms $\mathcal{A}_0$ (precomputation) which runs in total time $O(poly(t, \lambda))$ and $\mathcal{A}_1$ (online) which runs in parallel time $\sigma(t)$ with at most $p(t)$ parallel processors, if*

$$
Pr \begin{bmatrix} \alpha \leftarrow \mathcal{A}_0(pp, x); \\ b \xleftarrow{\$} \mathcal{B}; \\ \pi \leftarrow \mathcal{A}_1(pp, x, b, \alpha); \\ \text{Verify}(pp, x, \pi, b) = Accept \end{bmatrix} > \text{neg}(\lambda)
$$

then there exists an algorithm Extract *with rewinding access to $\mathcal{A}_1$ such that with probability $1 - \text{neg}(\lambda)$ the algorithm* Extract(pp, x, b) *outputs a witness w such that $(x, w) \in R$. The probability is over the choice of b and the random coins used by each algorithm.*
- **Indistinguishability:** *For all $(x, w) \in R, b \in \mathcal{B}$ the distributions* {Prove(pp, x, w, b)} *and* {Forge(pp, x, b)} *(taken over the random coins used by each algorithm) are computationally (resp. statistically) indistinguishable.*

4   Definitions

Definition 1 provides our main definition of short-lived proofs. The public parameters *pp* potentially encapsulate both setup needed for an underlying proof system and setup needed for an underlying VDF. Either or both may represent a *trusted setup* if they require a secret parameter that can be used to break security assumptions if not destroyed. The Setup algorithm is also given both a description of the language L and delay parameter t. Some underlying proof systems may require setup specific to L (others may offer *universal setup*) and some underlying VDFs require hard-coding the delay parameter t. In the remainder of the paper, we will generally omit pp to keep notation simpler.

The critical security property, t-Soundness,[4] closely follows the security definition used for VDFs [16]. In our case, the (potentially long-running) pre-processing algorithm $\mathcal{A}_0$ receives not only the public parameters of a VDF function but also the statement x on which the adversary wishes to forge a proof. Once the random beacon value b is known, the attacker's clock starts running and the online algorithm $\mathcal{A}_1$ must attempt to forge a proof in fewer than $\sigma(t)$ time steps (which in all of our constructions reduces to the intractability of solving an underlying VDF in fewer than $\sigma(t)$ time steps).

Note that short-lived proof schemes are inherently zero-knowledge as the Forge algorithm serves as a simulator which produces valid proofs in polynomial-time without knowing a witness. Receiving a proof in time less than t breaks deniability as it must have been produced by Prove, but does not help the verifier break zero-knowledge as the same transcript could still be produced by Forge at a later time. Thus, t-Forgeability and Indistinguishability implies zero-knowledge and we do not to define a separate zero-knowledge property or an additional simulator[5].

4.1   Reusable Forgeability

A basic short-lived proof scheme allows the Forge algorithm time to perform a unique slow computation for each pair (x, b). In practice, this means that forging multiple proofs can be expensive, weakening the deniability as it becomes less plausible that somebody paid the cost of forging. To this end, some short-lived proof schemes may offer a stronger *reusable forgeability* property in which performing one slow computation for a beacon value b enables efficiently forging a proof for *any* statement x without performing a full additional slow computation. Even better, some schemes might allow forging a proof of any statement for any beacon value from a set $B = \{b_1, \ldots b_k\}$ after just one slow computation. We call this property k-reusable forgeability (with basic reusable forgeability being the special case of $k = 1$). In practice, the set B can comprise all prior beacon values, enhancing deniability as one slow computation at any point in the future could enable forgery of all prior proofs.

Definition 2 (k-Reusable Forgeability). *A k-reusably forgeable short-lived proof system Π_R^t is a short-lived proof with two additional functions:*

---

[4] Our notion of t-soundness corresponds to *knowledge soundness*, we denote it as t-soundness for conciseness.

[5] Computational and statistical indistinguishability imply computational and statistical zero-knowledge, respectively. Thus, it might be possible to design a simulator that achieves statistical zero-knowledge while Forge only achieves computational indistinguishability.

- GenAdvice$(pp, B) \rightarrow \alpha$ *takes a set* B *of size* $|B| \leq k$ *and produces (in time* $(1 + \epsilon)t$) *an advice string* α
- FastForge$(pp, x, b, \alpha) \rightarrow \pi$ *produces a proof* π *for any* x

These new functions satisfy the following properties, in addition to all properties of a general short-lived proof system:

- **Reusable Forgeability:** *For all* x *and for all* $B \subseteq \mathcal{B}$ *and* $b \in B$, *given advice string* $\alpha \leftarrow$ GenAdvice(pp, B) *the algorithm* FastForge$(pp, x, b, \alpha) \rightarrow \pi$ *runs in parallel time less than* t *and* Verify(pp, x, π, b) *outputs* Accept.
- **Indistinguishability II:** *For all* $(x, w) \in R$, $B \subseteq \mathcal{B}$ *and* $b \in B$, *given advice string* $\alpha \leftarrow$ GenAdvice(pp, b) *the distributions* $\{$Forge$(pp, x, b)\}$ *and* $\{$FastForge$(pp, x, b, \alpha)\}$ *(taken over the random coins used by each algorithm) are computationally (resp. statistically) indistinguishable.*

Our generic protocol (§5) offers 1-reusable forgeability immediately and extends easily to offer $k-$reusable forgeability for arbitrary k (with proving overhead logarithmic in k). Obtaining 1-reusable forgeability is also possible (though not as easy) for our Σ-constructions.

5   Short-lived Proofs from Generic Zero-Knowledge

Given any VDF scheme and a non-interactive zero-knowledge proof system Π for all languages in NP, we can produce a short-lived proof for any relation R for an NP language L_R. We do this by taking the disjunction ($\vee$) of R with the VDF relation R_{VDF}:

$$R_{\text{VDF}} = \{(x = b, w = (y, \pi)) \mid \text{VDF.Verify}(b, y, \pi)\} \tag{1}$$

The language $L_{R_{\text{VDF}}}$ is in NP because VDF verification must run in polynomial time. Therefore, the disjunction $L_{R \vee R_{\text{VDF}}}$ is in NP and the proof protocol $\Pi_{R \vee R_{\text{VDF}}}$ is a short-lived proof for R:

Theorem 1 (SLP from Generic Zero-Knowlege). *Let* R *be a relation for a language in* NP, Π *be a zero-knowledge argument of knowledge system for languages in* NP *and* VDF *be a verifiable delay function with delay parameter* t. *Then* $\Pi_R^t = \Pi_{R \vee R_{\text{VDF}}}$ *is a short-lived proof protocol with reusable forgeability for* R *with time delay* t

Proof. The required properties follow directly from definitions of the underlying primitives. Completeness of Π_R^t is due to the completeness of Π_R as a prover with a witness can satisfy relation R and ignore the VDF branch. t-Forgeability follows from the correctness property of the underlying VDF, ensuring that Forge can produce convincing forgeries in $(1 + \epsilon)t$ steps by running VDF.Eval(b) and using the output (y, π) to satisfy the VDF branch of the disjunction. The Indistinguishability property followsmmediately from the zero-knowledge property of Π, preventing the adversary from knowing which half of the disjunction was satisfied and meaning an efficient simulator exists as required.

The t-Soundness property relies on the t-Sequentiality (Property 1) of the VDF. The restrictions on algorithms $\mathcal{A}_0, \mathcal{A}_1$ in the t-Soundness definition are identical to those in the t-Sequentiality definition, meaning such algorithms will not be able to solve the VDF with non-negligible probability. This means that any adversary able to produce proofs must know a witness w for x, which the extractor for Π_R can then efficiently extract.

Finally, to show that this scheme offers reusable forgeability, note that the exact same VDF computation VDF.Eval(b) required is independent of the statement x. Thus, it can be computed once and reused across proofs for that beacon.

$\square$

6   Short-lived Proofs from Σ-protocols

While our generic construction offers reusable forgeability and works for all NP-languages, generic zero-knowledge proof systems have practical drawbacks including complexity, high prover costs (§9) and trusted setup in some constructions. We would like to construct short-lived variants for Σ-protocols, an important class of efficient zero-knowledge proofs. They are also natural to consider given that Wesolowski proofs [77] are Σ-protocols and Pietrzak proofs are a multi-round generalization.

6.1   Non-solution: Σ-OR Proofs

A first, insecure attempt at a short-lived proof for a relation R with a Σ-protocol Σ_R is to simply combine Σ_R with the verification protocol Σ_{VDF} for some VDF scheme, for example using the classic Σ-OR construction outlined in Sect. 2.3.

Unfortunately, this generic composition does not yield a short-lived proof system because proofs are distinguishable from forgeries. Standard VDF proofs reveal the unique[6] value $y = $ VDF.Eval(b) to the verifier as part of the proof statement. This means that the algorithm VDF.Prove, which simulates the VDF half of a Σ-OR composition, must provide a fake value $y' \neq y$ as part of the proof whereas the Forge algorithm will simulate the R half of the proof and provide the genuine $y = $ VDF.Eval(b). Our definition of Indistinguishability does not preclude the adversary from running for t steps, meaning they can simply compute the genuine value y themselves by running VDF.Eval(b) and then determine if a proof was constructed using Prove or Forge.

6.2   Short-lived Sigma Proofs from Precomputed VDFs

To ensure indistinguishability, we introduce the construction Σ-Precomp (Protocol 1) which works for any Σ-protocol by modifying the *input* to the VDF instead of the challenge. Assume the prover has precomputed a VDF on a random input value b^*. Just as in the Cramer et al. construction, given a challenge

---

[6] Proofs of sequential work do not have unique solutions, unlike VDFs, meaning they might be used directly in a Σ-OR composition.

Σ-Precomp

Setup
input: relation R, parameters λ, t
output: public parameters pp

 1. $pp \leftarrow$ VDF.Setup(λ, t)

Precompute
input: pp
output: $(b^*, y^*, \pi^*_{\mathrm{VDF}})$

 1. Sample $b^* \xleftarrow{\$} \mathcal{B}$
 2. $(y^*, \pi^*_{\mathrm{VDF}}) \leftarrow$ VDF.Eval(b^*)

Verify
input: pp, x, b, proof $(a, c_1, z, y, c_2, \pi_{\mathrm{VDF}})$
output: Accept/Reject

 1. Obtain $c = \mathcal{O}(x \parallel b \parallel a)$.
 2. Accept if $c = c_1 \oplus c_2$ and both:
    - Σ_R.Verify$(x, a, c_1, z) \rightarrow$ Accept
    - VDF.Verify$(b \oplus c_2, y, \pi_{\mathrm{V}}) \rightarrow$ Accept

Forge
input: pp, statement x, beacon value b
output: proof $(\tilde{a}, \tilde{c}_1, \tilde{z}, y, \pi_{\mathrm{VDF}}, c_2)$

 1. $(\tilde{a}, \tilde{c}_1, \tilde{z}) \leftarrow \Sigma_R$.Simulate$(x)$
 2. Obtain $c = \mathcal{O}(x \parallel b \parallel \tilde{a})$
 3. Set $c_2 = c \oplus c_1$
 4. $(y, \pi_{\mathrm{VDF}}) \leftarrow$ VDF.Eval$(b \oplus c_2)$.

Prove
pre-computed: $(b^*, y^*, \pi^*_{\mathrm{VDF}})$
input: pp, x, w, b
output: proof $(a, c_1, z, y^*, \pi^*_{\mathrm{VDF}}, c_2)$

 1. $a \leftarrow \Sigma_R$.Commit(x)
 2. Challenge $c = \mathcal{O}(x \parallel b \parallel a)$
 3. Set $c_2 = b^* \oplus b$ and $c_1 = c \oplus c_2$
 4. $z = \Sigma_R$.Respond(x, w, a, c_1).

Protocol 1: Short-lived proofs using precomputed VDFs given a relation R with Σ-protocol Σ_R and a VDF scheme VDF

c, the prover chooses two values c_1, c_2 such that $c_1 \oplus c_2 = c$ and c_1 is the challenge used with Σ_R. Instead of using c_2 as the challenge for the VDF proof, it is used to modify the VDF input, evaluating on the point $b \oplus c_2$. The intuition is that the genuine prover can choose c_2 freely and thus set $c_2 = b \oplus b^*$, mapping the VDF input to a value b^* for which it has already precomputed the solution. However, a forger who is simulating Σ_R cannot choose c_1 freely and thus c_2 is an unpredictable random value, which requires the forger to solve a VDF on a random point $(b \oplus c_2)$.

Theorem 2 (Proof for Σ-Precomp). *Let R be a relation for a language in NP with Σ-protocol Σ_R and VDF be a verifiable delay function with delay parameter t. Protocol 1 is a short-lived proof scheme for relation R in the random oracle model.*

Proof. Completeness and t-forgeability of Σ-Precomp follow directly from the completeness of Σ_R and correctness of the VDF.

Indistinguishability: For an input x, witness w and beacon b, let (a, c_1, z, y, π, c_2) and $(\tilde{a}, \tilde{c}_1, \tilde{z}, \tilde{y}, \tilde{\pi}, \tilde{c}_2)$ be the outputs of Prove$(x, b; w)$ and Forge(x, b) and let $c, \tilde{c}$ be the respective challenges.

First, we note that $c_2, \tilde{c}_2$ are uniformly distributed in both algorithms. In Prove, $c_2 = b \oplus b^*$ where b is the beacon output produced only after b^* is sampled (as the honest prover must have pre-computed the VDF output on b^* first). In Force, $\tilde{c}_2 = \tilde{c} \oplus \tilde{c}_1$ where $\tilde{c}$ is generated after $\tilde{c}_1$. This means the VDF proofs are generated on uniformly random inputs $(c_2 \oplus b)$ and $(\tilde{c}_2 \oplus b)$, respectively, making the two VDF sub-transcripts (y, π, c_2) and $(\tilde{y}, \tilde{\pi}, \tilde{c}_2)$ indistinguishable.

Next, we note that $c_1, \tilde{c}_1$ are both uniformly distributed, as well. In Prove, $c_1 = c \oplus c_2$ where c is a random oracle output independent of c_2. By the security of $\Sigma_R.\mathsf{Simulate}$, the value $\tilde{c}_1$ is generated in Forge must be indistinguishable from a randomly generated challenge. Again by the security of $\Sigma_R.\mathsf{Simulate}$, the sub-transcripts (a, c_1, z) and $(\tilde{a}, \tilde{c}_1, \tilde{z})$ are indistinguishable.

Finally, all pairs of sub-challenges (c_1, c_2) are equally likely to be generated by both Prove and Forge. This comes from the fact that each algorithm generates one sub-challenge that is uniformly distributed (c_2 for Prove and c_1 for Forge) and then creates the other challenge using c, which is a random oracle output.

t-Soundness: We define an extractor E which, given a pair of algorithm ($\mathcal{A}_0$, $\mathcal{A}_1$) which output accepting proofs, either (1) extracts a witness w from $\mathcal{A}_1$ for statement x and relation R or (2) computes a VDF output on a random input in fewer than t steps, violating the t-Sequentiality (Property 1) of the underlying VDF.

E first runs $\mathcal{A}_0$ and $\mathcal{A}_1$ to obtain an accepting transcript $(a, c, c_1, c_2, z, y, \pi)$. E then receives a random VDF input b_{chal} from a challenger for the VDF t-sequentiality game. Next, the extractor rewinds $\mathcal{A}_1$ to obtain a new transcript $(a, c', c_1', c_2', z', y', \pi')$ while programming the random oracle to fix $c' = b_{\mathsf{chal}} \oplus b \oplus c_1$. As $c \neq c'$, we have the following two cases:

Case 1: If $c_1 \neq c_1'$, then from the special soundness of Σ_R, a witness for x can be extracted by calling $\Sigma_R.\mathsf{Extract}(x, a, c_1, c_1', z, z')$.

Case 2: If $c_1 = c_1'$, the two VDF proofs are on inputs $d = b \oplus c_2 = b \oplus c \oplus c_1$ and $d' = b \oplus c_2' = b \oplus c' \oplus c_1' = b \oplus c' \oplus c_1$. As the extractor programmed the random oracle to ensure $c' = b_{\mathsf{chal}} \oplus b \oplus c_1$, we have $d' = b_{\mathsf{chal}}$. As both transcripts are accepting, $\mathsf{VDF.Verify}(b_{\mathsf{chal}}, y', \pi') = \mathsf{Accept}$. As $\mathcal{A}_1$ runs in fewer than t steps, E requires fewer than t steps to produce y' as it only rewound and reran the adversary *after* after obtaining b_{chal}. Thus, E can output y' and win the t-Sequentiality game for the underlying VDF.

Assuming the underlying VDF is t-Sequential, Case 2 cannot happen except with non-negligible probability. Therefore E correctly outputs a witness for x (Case 1) with overwhelming probability.

Size Overhead: Transforming a normal proof into a short-lived one using Σ-Precomp adds the VDF output, the VDF proof and the sub-challenge used in the VDF. In the case of Wesolowski's VDF [77], the output and proof are both a group element each and the challenge is λ (security parameter) bits long. For 2048-bit RSA groups with $\lambda = 128$, the total size overhead comes to 528 bytes.

The primary drawback of Σ-Precomp is that each precomputed VDF must only be used once. If the same VDF challenge b^* is visible in two proofs, an adversary can conclude (with overwhelming probability) that both proofs were generated by Prove, breaking Indistinguishability. Additionally it does not offer reusable forgeability as a new VDF evaluation on a random point is required for every run of Forge.

6.3   Optimization with Re-randomizable VDFs

The biggest drawback of this construction is that it requires precomputation before every call to Prove. However, the prover simply needs a fresh, random VDF input/output pair and not a solution on any specific point. We can greatly improve the practicality of this scheme if it is possible to quickly generate VDF solutions (and proofs) on random points. We introduce the notion of a *re-randomizable* VDF that has this property: given a VDF solution (b, y, π) and possibly some auxiliary data α, an efficient algorithm VDF.Randomize$(b, y, \pi, \alpha) \rightarrow (b', y', \pi')$ outputs a randomly distributed solution.

Now, each time Prove is called, instead of precomputing a VDF solution (step 1 of Prove in Protocol 1), a new VDF solution on a random point is produced by calling VDF.Randomize. Indistinguishability of proofs and forgeries reduces to the indistinguishability of random VDF solutions and those generated by re-randomizing a known solution. We propose a definition for re-randomizable VDFs in Appendix A of the full version [7], capturing the necessary indistinguishability property.

For VDFs based on repeated squaring, a random exponent r is chosen and the input/output pair (b, y) is mapped to $(b' - b^r, y' = y^r)$, maintaining the relationship that $y' = (b')^{2^t}$. Unfortunately this homomorphism does not apply to proofs: given $(y, \pi) \leftarrow$ VDF.Eval(b) and a randomized solution (b^r, y^r) we cannot obtain a correct proof by simply computing π^r. However, for repeated-squaring VDFs we can compute a proof for (b^r, y^r) in fewer than t steps using the same advice string α used to compute π when y was originally computed. Wesolowski [77] describes such an advice string of length $O(\sqrt{t})$ that allows a prover to compute a proof in $O(t/\log t)$ steps. This algorithm may still be to slow to re-randomize VDF proofs in reasonable time using commodity hardware. By contrast, Pietrzak proofs can be re-randomized in just $O(\sqrt{t})$ steps using an advice string of length $O(\sqrt{t})$. We provide the details of this re-randomization algorithm (originally suggested by Boneh et al. [17]) in Appendix A.1 of the full version [7].

This homomorphism was observed by Wesolowski, who warned that it was a potential security weakness to be prevented by hashing to a random group element as part of VDF computation [77, Remark 3]; here we use it in a constructive way. It has similarly been used by Thyagarajan et al. to build verifiable timed signatures [75] and by Malavolta and Thygarajan to construct additively homomorphic and fully homomorphic time-lock puzzles [57].

7   Short-lived Proofs from zkVDFs

The previous Σ-based constructions did not provide reusable forgeability (Definition 2). The fundamental problem is that (unlike our generic approach in

Sect. 5), they require Forge to solve the VDF on a new random value b^* derived from b for each forgery, rather than a solution on b itself which could be used for multiple forgeries. We cannot include a standard VDF proof for b in short-lived proofs because all known VDF proof schemes reveal the VDF output $y = \mathsf{Eval}(b)$ which would clearly distinguish proofs from forgeries.

To avoid this distinguishability problem, we propose using a novel *zero-knowledge VDF* (zkVDF) which proves knowledge of the output without revealing it. Of course, since VDF verification is (by definition) an NP statement, it is possible to construct a zkVDF from any VDF using a generic zero-knowledge proof system to prove knowledge of VDF solutions. Our construction in Sect. 5 essentially does this (embedded within a disjunction). Later in this section, we will present a more efficient construction based on Wesolowski proofs [77].

Given a Σ-protocol for R_{zkVDF} and any relation R for which we have a Σ-protocol Σ_R, we can use the standard Σ-OR construction to create a disjunction protocol $\Sigma_{R \vee R_{\mathsf{zkVDF}}}$ which we call Σ-zkVDF. To obtain reusable forgeability, we set the VDF input to be the beacon value b. Thus, $\mathsf{VDF.Eval}(b)$ need be performed only once to generate advice to quickly forge others proofs with b.

Theorem 3 (SLP from zkVDF and Σ-OR). *Let R be a relation for a language in NP, Σ_R be a zero-knowledge Σ-protocol for R and $\Sigma_{R_{zkVDF}}$ be a Σ-protocol for a zkVDF with delay parameter t. Letting x and w be the statement and witness for relation R and b be the beacon, the following is a short-lived proof protocol with reusable forgeability for R with time delay t:*

Σ-zkVDF

- $\mathsf{Prove}(x, w, b)$*: perform $\Sigma_{R \vee R_{zkVDF}}$ by simulating $\Sigma_{R_{zkVDF}}$ with input b and running Σ_R with statement x and witness w*
- $\mathsf{Forge}(x, b)$*: perform $\Sigma_{R \vee R_{zkVDF}}$ by simulating Σ_R for statement x and running $\Sigma_{R_{zkVDF}}$ with input b*
- $\mathsf{Verify}(x, b, \pi)$*: verify $\Sigma_{R \vee R_{zkVDF}}$ with statement x for Σ_R and input b for $\Sigma_{R_{zkVDF}}$*

The proof is identical to that of Theorem 1. Owing to the security of Σ-OR compositions, the verifier cannot tell if the proof was generated by honestly computing the Σ_R branch (requiring knowledge of the witness) or the $\Sigma_{R_{\mathsf{zkVDF}}}$ branch (requiring the VDF solution on beacon b). Appendix D of the full version [7] contains the proof.

7.1   zkVDF Construction

In this section we present a Σ-based zkVDF construction built off of Wesolowski proofs [77]. To do so, we introduce a new zero-knowledge Σ-protocol for proof of knowledge of a power (Protocol 2) using an idea similar to that introduced by Boneh et al. [18, §3.2] for proof of knowledge of discrete log in a group of unknown order. Our zero-knowledge proof that $y = g^u$ sends a blinded value $y' = y \cdot h^v = g^u h^v$ (for a random v) instead of y itself. A proof of the following theorem is provided in Appendix C of the full version [7].

---

zk-PoKP

Parameters: security parameter λ, group of unknown order $\mathbb{G} \leftarrow GGen(\lambda)$, $h \xleftarrow{\$} \mathbb{G}$, $B \geq 2^{2\lambda}|\mathbb{G}|$; random oracle HashToPrime which outputs from the set $\mathsf{Primes}(\lambda)$ of the first 2^λ prime numbers

Prove
input: $g \in \mathbb{G}$, $u \in \mathbb{Z}$, witness $y \in \mathbb{G}$ such that $y = g^u$
output: proof $\pi = \langle a, Q, r_2 \rangle$

1. Sample $v \xleftarrow{\$} [-B, B]$
2. Compute $a = \mathsf{Commit}(y, v) = y \cdot h^v$
3. Compute $\ell = \mathsf{Challenge}(a) = \mathsf{HashToPrime}(a)$
4. Let $u = q_1\ell + r_1$, $v = q_2\ell + r_2$ such that $0 \leq r_1, r_2 \leq \ell$
5. Compute $Q = g^{q_1} h^{q_2}$
6. $\mathsf{Respond}(\ell) = Q, r_2$

Simulate
inputs: $g \in \mathbb{G}$, $u \in \mathbb{Z}$, simulated challenge $\tilde{\ell}$
output: simulated proof $\tilde{\pi} = \langle \tilde{a}, \tilde{Q}, \tilde{r_2} \rangle$

1. Sample $\tilde{q}_1, \tilde{v} \xleftarrow{\$} [-B, B]$
2. Let $\tilde{u} = \tilde{q}_1\tilde{\ell} + \tilde{r}_1$ and $\tilde{v} = \tilde{q}_2\tilde{\ell} + \tilde{r}_2$ such that $0 \leq \tilde{r}_1, \tilde{r}_2 \leq \tilde{\ell}$
3. Compute $\tilde{Q} = g^{\tilde{q}_1} h^{\tilde{q}_2}$
4. Compute $\tilde{a} = \tilde{Q}^{\tilde{\ell}} g^{\tilde{r}_1} h^{\tilde{r}_2}$

Verify
input: $g \in \mathbb{G}$, $u \in \mathbb{Z}$, proof $\pi = \langle a, Q, r_2 \rangle$
output: Accept/Reject

1. Compute $\ell = \mathsf{HashToPrime}(a)$
2. Let $u = q_1\ell + r_1$ such that $0 \leq r_1 \leq \ell$
3. Check that $a \overset{?}{=} Q^\ell g^{r_1} h^{r_2}$

Protocol 2: Σ-protocol for proof-of-knowledge of a power in a group of unknown order.

Theorem 4 (Zero-Knowledge Proof of Knowledge of Power). *Protocol 2 is an honest-verifier zero-knowledge argument of knowledge for the relation $R_{\mathsf{PoKP}} = \{((g, u); y) : g^u = y\}$.*

7.2   Efficiency of zk-PoKP and Σ-zkVDF

The zk-PoKP Simulate algorithm of Protocol 2 is efficient and takes time $O(\lambda \log |\mathbb{G}| + \mathsf{polylog}(t))$. The most significant cost is computing five group exponentiations with small exponents, each involving $O(\log B) = O(\lambda \log |\mathbb{G}|)$ steps. This makes the Σ-zkVDF prover efficient as it runs the zk-PoKP simulation algorithm.

The Forge algorithm for Σ-zkVDF must execute the Prove algorithm of zk-PoKP. This naively takes time $O(t)$, as it involves computing a large power Q^ℓ. For multiple forgeries, this can be improved significantly using a precomputed advice string, identical to that used for re-randomizable VDFs (Sect. 6.3). With an advice string of size $O(\sqrt{t})$, the Prove algorithm requires only $O(t/\log t)$ steps. Unlike the case for general re-randomizable VDFs, our zk-PoKP construction is inherently based off of Wesolowski proofs and cannot utilize the more efficient advice string approach used for re-randomizing Pietrzak proofs. Designing a Pietrzak-style zk-PoKP is an interesting open problem.

Proof Size: The zkVDF proof contains two group elements (a, Q) and the remainder r_2. When using 2048-bit RSA groups and $\lambda = 128$, the total size comes to 529 bytes: 512 bytes for the two group elements and 17 bytes ($\lambda \lg \lambda$) for the value r_2. The Σ-zkVDF construction additionally includes one sub-challenge (the other is implicit), which adds an extra λ bits (16 bytes), making the total overhead for transforming a normal proof into a short-lived one just 545 bytes. The algorithm Σ-zkVDF.Prove requires running the zk-PoKP simulator. The significant operations are raising a group element to a power of size B twice, where $B \approx 2^{2\lambda}|\mathbb{G}|$, and then raising two elements to a power of up to λ twice. In Sect. 9, we evaluate the cost of these operations for 2048-bit RSA groups.

8   Short-lived Signatures

A key special case of zero-knowledge proofs is digital signatures. We define a short-lived signature scheme as follows:

Definition 3 (Short-Lived Signatures). *Let λ be a security parameter and $\mathcal{B}$ be a space of beacon values where $|\mathcal{B}| \geq 2^\lambda$. A short-lived signature scheme with time delay t is a tuple of algorithms:*

- Setup$(\lambda, t) \to pp$
- KeyGen$(pp) \to (pk, sk)$
- Sign$(pp, sk, m, b) \to \sigma$ *takes a message m and beacon b and outputs (in time less than t) a signature σ.*
- Forge$(pp, m, b) \to \sigma$ *takes a message m and beacon b and outputs (in time less than $(1 + \epsilon)t$) a signature σ.*
- Verify$(pp, pk, m, b, \sigma) \to$ Accept/Reject

  The following properties are satisfied:

- **Correctness:**   *For all m, $b \in \mathcal{B}$, if $\sigma \leftarrow$ Sign(pp, sk, m, b), then* Verify$(pp, pk, m, b, \sigma) \to$ Accept.
- **Existential Unforgeability:** *For all pairs of adversary algorithms $\mathcal{A}_0$ (precomputation) which runs in total time $O(poly(t, \lambda))$ and $\mathcal{A}_1$ (online) which runs in parallel time $\sigma(t)$ with at most $p(t)$ processors, the probability that $(\mathcal{A}_0, \mathcal{A}_1)$ win the following game is negligible:*

1. *Challenger C runs $pp \leftarrow \mathsf{Setup}(\lambda, t)$ and $(pk, sk) \leftarrow \mathsf{KeyGen}(pp)$. C sends pp, pk to $(\mathcal{A}_0, \mathcal{A}_1)$.*
2. *The adversary runs $A_0(pp, pk) \leftrightarrow C$ interactively with the challenger, adaptively sending chosen message/beacon queries (m_i, b_i) to the challenger and receiving $\sigma_i \leftarrow \mathsf{Sign}(pp, sk, b_i, m_i)$ in response.*
3. *A_0 outputs an advice string α.*
4. *C samples a random beacon value $b \xleftarrow{\$} \mathcal{B}$ and sends it to the adversary.*
5. *The adversary runs $A_1(pp, pk, \alpha, b) \leftrightarrow C$ interactively with the challenger, adaptively sending chosen message/beacon queries (m_i, b_i) to the challenger and receiving $\sigma_i \leftarrow \mathsf{Sign}(pp, sk, b_i, m_i)$ in response.*
6. *A_1 outputs a claimed forgery (m_*, b, σ_*) and wins if $(m_*, b) \neq (m_i, b_i)$ for all i and $\mathsf{Verify}(pp, pk, m_*, b, \sigma_*) = Accept$.*

- **Indistinguishability:** *For all $m, b \in \mathcal{B}$, given a random $(pk, sk) \leftarrow \mathsf{KeyGen}(pp)$ the distributions $\{\mathsf{Sign}(pp, sk, m, b)\}$ and $\{\mathsf{Forge}(pp, m, b)\}$ (taken over the random coins used by each algorithm and randomly generated private key) are computationally (resp. statistically) indistinguishable.*

This definition closely follows our definition of short-lived proofs and standard security properties for signatures. We present a game-based definition for short-lived signature unforgeability, in contrast with our probabilistic soundness definition for short-lived proofs, to more closely match standard unforgeability definitions for signature schemes. The primary distinction is that the second adversary A_1 is required to run in fewer than t steps (otherwise it could simply run the provided Forge algorithm).

Note that while our Indistinguishability definition compares distributions of output, some signature schemes are deterministic(e.g., RSA [68], BLS [19]). In this case, it is necessary that Sign and Forge produce the same exact signature with overwhelming probability.

We observe that our generic constructions in Sect. 5 can be used to transform any signature scheme into short-lived signature scheme by implementing a zero-knowledge proof for knowledge of a signature. Furthermore, our Σ-based constructions in Sect. 6 can also be used for Σ-based signature schemes such as Schnorr [70] or DSA [1,51].

8.1   Construction from Trapdoor VDFs

We present a short-lived signature construction from *trapdoor VDFs* [77] in Protocol 3. Trapdoor VDFs require a trusted setup which yields a secret evaluation key (the trapdoor) enabling efficient evaluation. Normally, this trapdoor represents a security risk if not destroyed. However, we observe that in the case of short-lived signatures, the trapdoor can serve as a signing key. Repeated-squaring VDFs in RSA groups are trapdoor VDFs: the public parameters include an RSA modulus N and the trapdoor is the factors p, q such that $N = p \cdot q$. With the trapdoor, raising an element to any large exponent z (e.g. $z = 2^t$) is efficient, as z can be reduced modulo $\varphi(N) = (p - 1)(q - 1)$ into an equivalent exponent of size less than N. Note that this trapdoor is equivalent to the private key used for traditional RSA signatures.

Sign-Trapdoor

KeyGen
input: λ, delay parameter t
output: key pair (pk, sk)

1. Generate keys
   $(pk, sk) \leftarrow$ tdVDF.Setup(λ, t)

Sign
input: message m, beacon value b
output: signature σ

1. $x =$ Hash$(m \parallel b)$
2. $\sigma = (y, \pi) \leftarrow$
   tdVDF.TrapdoorEval(sk, x)

Forge
input: message m, beacon value b
output: signature σ

1. $x =$ Hash$(m \parallel b)$
2. Compute with delay:
   $\sigma = (y, \pi) \leftarrow$ tdVDF.Eval(x)

Verify
input: message m, beacon b, $\sigma = (y, \pi)$
output: Accept/Reject

1. $x =$ Hash$(m \parallel b)$
2. Check that tdVDF.Verify(pk, x, y, π, t)

Protocol 3: Short-Lived Signatures from a trapdoor VDF scheme

Theorem 5 (Short-Lived Signatures from Trapdoor VDFs). *Assuming that* Hash *is a random oracle and* tdVDF *is a trapdoor VDF, Protocol 3 is a short-lived signature scheme.*

Proof. The correctness of this scheme comes from the correctness of the underlying trapdoor VDF. Indistinguishability is trivial as signing and forgery produce the exact same VDF output, given that VDFs are deterministic.

Existential unforgeability comes from the definition of a trapdoor VDF and modeling Hash as a random oracle. Since the challenger chooses b randomly during the existential forgery game after the precomputation of $\mathcal{A}_0$, the value $x_* =$ Hash$(m_* \| b)$ will be randomly distributed for any message m_*. Thus, the online algorithm $\mathcal{A}_1$ must evaluate the VDF on a random input in fewer than t steps. The adversary's ability to query for signatures on chosen pairs (m_i, b_i) is new from the traditional VDF security model. However since each such pair leads to a VDF evaluation by the challenger on $x_i =$ Hash$(m_i \| b_i)$, the adversary can only learn VDF evaluations on a polynomial number of random inputs. This ability could be simulated by $\mathcal{A}_0$ precomputing the VDF on a polynomial number of random inputs and passing the results as part of the advice string α. Thus, any pair $(\mathcal{A}_0, \mathcal{A}_1)$ which make queries could be converted into an equivalent pair $(\mathcal{A}_0', \mathcal{A}_1')$ which make no queries but rely on $\mathcal{A}_0'$ to precompute random VDF solutions instead. Winning the signature forgery game with no querying capability is then equivalent to evaluating the VDF on a random input. The VDF security definition states that no suitably bounded algorithms $(\mathcal{A}_0', \mathcal{A}_1')$ can do so with non-negligible probability. □

Table 2. Additional costs to transform a standard proof/signature into a short-lived proof/signature. λ is the security parameter, $\langle \mathbb{G} \rangle$ denotes the size of a group element, $\exp_{\mathbb{G}}(e)$ is the cost of raising a group element to a power of size e, and t is the VDF delay parameter. For concrete evaluations, $\lambda = 128$ and $\mathbb{G}$ is a 2048-bit RSA group. The Generic zk-SNARK method was implemented using Groth16 [47]. All evaluations were performed on a 2.3 GHz 8-Core Intel Core i9 laptop with 16 GB memory.

Protocol		Proof Size Overhead		Proving Time Overhead	
zk-SNARKs	(§5)	0 for Groth16		$\sim$60 s	
Σ Precomp	(§6.2)	$2\langle\mathbb{G}\rangle + \lambda$	528 bytes	$O(T)$ precomputation	
Σ-rrVDF	(§6.3)	$2\langle\mathbb{G}\rangle + \lambda$	528 bytes	$O(T/k)$ precomputation	
Σ-zkVDF	(§7)	$2\langle\mathbb{G}\rangle + 2\lambda$	545 bytes	$2\exp_{\mathbb{G}}(2^{2\lambda}|\mathbb{G}|) + 2\exp_{\mathbb{G}}(2^\lambda)$	120 ms
Sign-Trapdoor	(§8.1)	$\langle\mathbb{G}\rangle + \lambda$	272 bytes	$\exp_{\mathbb{G}}(2^\lambda)$	10 ms
Sign-Watermark	(§8.2)	$\langle\mathbb{G}\rangle + \lambda$	272 bytes	$\exp_{\mathbb{G}}(2^\lambda)$	10 ms

8.2   Construction from Watermarkable VDFs

The construction in Protocol 3 does not offer reusable forgeability, as the Forge evaluates a VDF on a message-dependent value $x = \mathsf{Hash}(m \parallel b)$. We construct an efficient signature scheme (Protocol 5) with reusable forgeability using *watermarkable VDFs* which embed a prover-chosen *watermark* (μ) during proof generation. Watermarkable VDFs were presented informally by Wesolowski [77, §7.2]; we propose a definition capturing the essential security property of *watermark unforgeability* in Appendix B of the full version [7]. The key idea to construct a watermarkable VDF is to embed the watermark into the Fiat-Shamir challenge, computing $\ell \leftarrow \mathsf{HashToPrime}(y \parallel \mu)$ instead of $\ell \leftarrow \mathsf{HashToPrime}(y)$.

Short-lived Signatures from Watermarkable VDFs To build a short-lived signature scheme using a watermarkable VDF, we use the beacon value as the input to the VDF and the message as the watermark. This enables reusable forgeability, as once a forger has computed $y = \mathsf{VDF.Eval}(b)$ for a specific beacon value, along with its associated advice string α, they can sign a new message by computing a new proof using the same advice string. This is equivalent to proof re-randomization, which can be done in significantly fewer than t steps as discussed in Sect. 6.3. We note that reusability is more limited for this signature scheme as the precomputation is specific to an individual user's public key. A single VDF evaluation enables efficient forgery of any statement by a given signer, but will not work between different signers.

9   Implementation and Performance Evaluation

9.1   zk-SNARK Construction

We implement the generic ZK algorithm using zk-SNARKs, which produce succinct non-interactive proofs for large computations. With zk-SNARKs, the statement is represented in a format similar to algebraic circuits and prover efficiency depends on the size of the circuit in gates. Given a base circuit for relation R and an efficient circuit representation of VDF.Verify, it is straightforward to compile a circuit that is the disjunction of the two as outlined in Sect. 5. We implemented a VDF circuit using Wesolowski VDF proofs [77] using a 2048-bit RSA modulus and the "bellman-bignat" library [62].

The total size of the VDF verification circuit is just over 5 million gates. The large size is due to the costly "hash-to-prime" involved in Wesolowski verification. We composed this circuit with an elliptic curve signature verification circuit (acting as the base relation R) of size under 1000 gates. All proofs were generated using the Groth16 construction [47] which produces proofs of constant size around 300 bytes and with verification time under 10 ms. As proofs are constant size and the verification cost is minimal, there is no added overhead on verifiers for short-lived proofs. However, proof generation incurs a significant added cost of around 60 s.

9.2   Σ-based Constructions

Table 2 compares the performance of our algorithms. Our Σ-constructions, which require only a few exponentiations in a group of unknown order, are significantly more efficient than the zk-SNARK method. We evaluated them using Wesolowski proofs in a 2048-bit RSA group, which is conjectured to provide close to $\lambda = 128$ bits of security [9]. We denote the cost of raising a group element to an exponent of magnitude $2^{2\lambda}|\mathbb{G}|$ as $\exp_{\mathbb{G}}(2^{2\lambda}|\mathbb{G}|)$ (this is the value of B in Protocol 4). A single exponentiation takes around 40 ms. The costliest of the Σ-constructions is Σ-zkVDF which takes two $\exp_{\mathbb{G}}(2^{2\lambda}|\mathbb{G}|)$ operations and three $\exp_{\mathbb{G}}(2^{\lambda})$ operations (<10 ms each), leading to a total overhead of under 0.12 s. Section 7.2 contains more details on the size overhead of the Σ-zkVDF construction. Our signature constructions add one or two more group elements to the size of the base proof/signature. With 2048-bit RSA groups, each element is of size $\langle \mathbb{G} \rangle = 256$ bytes.

9.3   Re-randomization Improvements

Three of our constructions (based on re-randomizable VDFs, zero-knowledge VDFs, watermarkable VDFs) can utilize a precomputed advice string α to speed up computation. For Σ-rrVDF, α is used to speed up the Prove algorithm; for Σ-zkVDF and Sign-Watermark, α speeds up FastForge.

In Appendix A.1 of the full version [7], we outline such an advice string of size $O(\sqrt{t})$ for Pietrzak proofs which enables proof computation in $O(\sqrt{t})$ steps. This advice string is applicable to Σ-rrVDF and Sign-Watermark protocols, for which Pietrzak proofs can be used. Table 3 highlights the practical performance of this approach for different delay parameters. We provide numbers for 2048-bit and 1024-bit RSA groups and assume that hardware implementations (e.g.

Table 3. The time taken for re-randomizing a Pietrzak proof on commodity hardware for RSA groups with delay parameters and specialized hardware speed assumptions. The lengths of the advice strings range from 1 MB to 16 MB. The lengths of the proofs range from 2.6 KB to 8.0 KB. The proof verification time is under 50 ms and 150 ms for 1024-bit and 2048-bit RSA groups, respectively.

			Re-Rand Time	
Hardware Speed	Delay	$\log t$	RSA-2048	RSA-1024
2^{25} ops/s	1 min	31	28 s	8 s
	15 min	35	110 s	35 s
2^{30} ops/s	1 min	36	145 s	58 s
	15 min	40	720 s	240 s

FPGAs) for RSA arithmetic can perform up to 2^{25} and 2^{30} operations per second.[7] The improvements achieved by our advice string enable Pietrzak proofs to be rerandomized in minutes with commodity hardware, whereas Wesolowski proofs would still take days with comparably sized advice strings ([77], Section 4.1).

10   Applications

10.1   Deniable Messaging and Email

Deniable messaging protocols aim to ensure that a purported transcript of a secure communication session between Alice and Bob, along with copies of all cryptographic keys used, does not provide convincing evidence of what Alice and Bob actually communicated or (in some cases) that they communicated at all. Generally, a secure messaging (chat) protocol is run between two participants, identified by public keys bound somehow to their real-world identities. When both participants are online, they can use a key agreement protocol to establish an ephemeral shared MAC key for message integrity. Even if the long-term keys are compromised, the transcript could be forged by either party [69], as popularized by Off-the-Record messaging (OTR) [21]. Deliberate publication of the MAC key after the session can extend forgeability to anyone.

Deniability can be extended to offline recipients in a store-and-forward system through non-interactive designated verifier signatures [49] or ring signatures formed between a sender and a recipient [21,69]—however both require prior knowledge of each recipient's public key. Email is a particularly challenging environment, as in addition to being asynchronous and unidirectional the sender cannot assume knowledge of the recipient's public key. OTR's authors believed email is too difficult for an OTR-like protocol [21].

---

[7] Öztürk [78] reported a speed for $\sim 2^{24}$ squarings/second in 2019 in an optimized FPGA implementation used to break the RSA LCS-35 timelock puzzle [67].

Our work suggests a different (and complementary) approach to deniability: a sender with an identifiable public key can provide a short-lived signature on their messages. Recipients within the validity period of the signature can validate the message's authenticity, while the message becomes indistinguishable from a forgery after a period of time and therefore deniable by the original sender. Short-lived messages do not require any knowledge about the recipient, interaction, or follow-up steps, making them very versatile. They can be broadcast asynchronously to a group of unidentified recipients with a single communication, and even forwarded with no additional cryptographic effort, making them suitable for email as well as messaging protocols.

10.2   Deniable Domain Authentication

A specific case of deniable authentication arises with the DomainKeys Identified Mail (DKIM) standard for email. Originally proposed to address email forgeries and spam, DKIM requires that the sender's mail server sign every outbound email with a domain-bound key. For example, all email originating from the mail server for `example.com` would be signed with a key bound to the DNS record of `example.com`, however (unlike the use case above) the signature will not distinguish between mail from `alice@example.com` and `bob@example.com`. By 2015, DKIM headers were present in 83% of all inbound mail to Gmail [34].

Over the past two decades, email dumps—the public release of private email messages from breached servers—have received extensive news coverage [27]. DKIM signatures increase the value of email dumps by certifying their authenticity. The call to periodically release past DKIM private signing keys was popularized by Matthew Green [45]. DKIM signatures do not require validity beyond the network latency of reaching a recipient's mail server.

Specter et al. proposed KeyForge and TimeForge [72] to replace DKIM with *Forward Forgeable Signatures (FFS)* that become *non-attributable* after a specified time (e.g., 15 min). Both KeyForge and TimeForge require future action to ensure deniability: respectively, a secret value released by the signer, or a future signed update to a beacon-like service called a *publicly verifiable timekeeper*. If the time-keeper's private key is lost then all signatures become permanently attributable. Alternately, if the time-keeper is silently compromised then signatures are immediately forgeable. Short-lived signatures can fulfill the same role as a drop-in replacement for DKIM, while requiring no follow-up action by *anyone* and hence deniability is guaranteed at the time of signing. Both TimeForge and short-live signatures expand the current length of a 2048-bit RSA DKIM signature. Our trapdoor RSA-based short-lived signature adds a single group element (200% expansion) while TimeForge signatures expand by 329% [72].

TimeForge also has advantages: no costly VDFs need to be evaluated (or threatened) to provide deniability and the timing of deniability is precise, whereas for short-lived signatures the deniability time period depends on how fast VDFs can be evaluated. Our approaches are complementary: a signature could be both short-lived and forgeable after the release of information as in TimeForge, attaining the advantages of both.

10.3   Receipt-Free Voting

Numerous cryptographic voting protocols involve encoding a voter's selection with an additively homomorphic encryption scheme. A voter wants *ballot casting assurance* [14] that a posted ciphertext decrypts to her choice, however she should not be able to transfer this assurance to anyone else. As the literature moves toward a more realistic view of voters as humans casting ballots at polling places, vote casting needs to be accessible and bare-handed (i.e., no assumption of an additional device at casting time). The dominant approach (exemplified in Helios [5], STAR-Vote [11], and Microsoft's ElectionGuard [3]) is the *Benaloh challenge* [14]: (1) a voter asks for an encryption of a candidate s_i, (2) the voting machine commits (e.g., on paper) a ciphertext c, (3) the voter chooses to audit the ballot or cast it, and (4) if auditing, the voting machine produces the plaintext and randomness (s_i, r) such that $c = \mathsf{Enc}(s_i, r)$; and the voter restarts at (1). Later, aided by a computer, the voter validates all transcripts. This protocol has two drawbacks. If a voter asks for an encryption of candidate Bob, receives one for candidate Alice instead, audits it and receives a proof for Alice, the voter is convinced the machine is malicious (she knows she asked for Bob) but the transcript will not convince a third party that the machine misbehaved. The second drawback is that auditing is probabilistic (and a low audit frequency is observable by the machine itself).

Alternatives to the Benaloh challenge mitigate these drawbacks but add complexity for the voter. A collection of techniques [49,59,60] use quite different protocols to produce a similar outcome: the voter leaves with a receipt that contains the ciphertext c and n proofs that $c = \mathsf{Enc}(s_i)$ for each of the n candidates. One proof is real and the rest are forgeries, but the transcript does not reveal which one is real. These protocols vary, but at a high level, either the machine prepares the forgeries and the voter releases a value (e.g., a challenge) for construction of the real proof; or the machine prepares the real proof, and the voter releases a value (e.g., a trapdoor or private key) for the forgeries.

A short-lived proof can be used in this second paradigm to eliminate all the pre-constructed values (i.e., challenges, keys, trapdoors) the voter must bring into the polling place, replacing them with a simple clock. The voter experience is as follows: (1) a voter selects candidate s_i, (2) the voting machine commits (e.g., on paper) the time T, the name of s_i (in plaintext), a ciphertext c, and a short-lived (e.g., 60 s) proof that $c = \mathsf{Enc}(s_i)$, (3) the voter checks that T and s_i are correct, (4) after two minutes, the machine (possibly a different machine at a different station within the polling place) produces $n-1$ forged receipts for each leftover s_i with the same c and same (and now outdated) T.

Commonly used encryption schemes for voting, like exponential Elgamal and Paillier, have efficient Σ-protocol proofs of plaintext values and be adapted to use any of our Σ-protocol-based short-lived proof constructions. A time parameter of 60 s provides reasonable assurance if the beacon and voter's clock are synchronized to within one second, while not delaying the time to vote substantially. Voters could choose to shred their initial receipt or wait for a set of forged receipts. Attention is required to mitigate side-channel information (like forensics of the paper) to infer the order in which the proofs are printed.

11   Concluding Remarks

We observe that the existence of the Forge algorithm for short-lived proofs circumvents Pass' observation [63] that non-interactive zero-knowledge proofs in the random oracle model are not deniable. Normally, because the simulator requires programmable access to the random oracle, verifiers cannot simulate proofs and hence possession of a proof demonstrates interaction with a genuine prover. In our case, the Forge algorithm does not require programmable random oracle access, only the ability to compute a slow function. Short-lived proofs therefore can offer deniability as they can be forged with no special ability except time.

In practice, this is an important limitation if the time taken to compute a proof is known and is insufficient to produce a forgery. For example, if a short-lived proof π is convincingly timestamped at time T_1 and the beacon value b used to compute π was not known until time T_0, with $T_1 - T_0 < t$, then π could not have been computed via Forge. Thus, deniability for short-lived proofs relies on the assumption that it is not feasible to convincingly timestamp all data. For example, in the case of deniable DKIM signatures, it must be the case that signed emails are not routinely timestamped en masse. We note that other solutions to this problem, including key expiry/rollover or the KeyForge/TimeForge schemes of Specter et al. [72], have the exact same limitation.

Recall that we offer constructions for proofs and signatures based on Σ-protocols, where deniability is added by combining the original statement with a VDF-related statement in a disjunction. As noted in Sect. 3, designated verifier proofs, proofs of work-or-knowledge, and KeyForge/TimeForge also add deniability via disjunction with a second statement. It is straightforward to combine these approaches. For example, a statement could be proven in zero knowledge to be true only if the proof is received by a specific recipient before a signed timekeeper statement is released or a VDF could have been computed. In this way, a proof can gain deniability in the absence of any trusted third party action in the future (as with short-lived proofs) while also gaining deniability without requiring anybody to solve a VDF if the third party acts faithfully.

We conclude with several open problems arising from our work:

- Short lived proofs require *someone* to evaluate a VDF. It might be possible to piggyback off of an existing party computing VDFs, such as a computational time-stamping service [54], some other party computing a long-running continuous VDF [36] or a decentralized protocol using chained VDFs [29, 40].
- Our Σ-zkVDF construction (§7) only provides 1-reusable forgeability. It might be possible to extend this to k-reusable forgeability using an RSA-style accumulator to combine past beacon values (while keeping the accumulator value secret to avoid undermining deniability).

- Our zk-PoKP construction in Protocol 2 is based on Wesolowski proofs. Constructing a zk-PoKP algorithm based on Pietrzak proofs would allow a Σ-zkVDF forger to leverage the more efficient re-randomization algorithm for Pietrzak proofs, enabling significantly faster forging times.
- While our watermarkable VDF signature construction (§8.2) offers reusable forgeability, it requires a VDF computation per-signer (as each signer uses unique public parameters). An ideal scheme might use similar accumulator techniques to forge proofs from a set of signers after just one VDF evaluation.
- Another, potentially much more efficient, approach to achieve generic short-lived proofs is to take an existing generic proof scheme which relies on a Σ-protocol and replace that component with a short-lived equivalent (e.g. our Σ-zkVDF construction). While this would not retain the k-reusability of our generic approach in §5, it would avoid the cost of verifying a VDF within the proof system itself. This approach potentially applies to many popular zero-knowledge proof systems, including Bulletproofs [22], Marlin [24], PLONK [41], Sonic [58], Spartan [71], Supersonic [23], or STARKs [12].
- Another approach to obtain short-lived SNARKs is through composition techniques that construct zero-knowledge proofs of disjunctions of SNARKs with Σ-protocols (zkVDFs in our case) as proposed by Agrawal et al. [6]. This would avoid the costly encoding of VDFs as arithmetic circuits (which require millions of gates) leading to interesting tradeoffs between shorter proving times and larger proof sizes.

Acknowledgments. We give a special acknowledgement to Michael Colburn who initially pursued the idea of short-lived signatures and proofs in his MASc thesis [30] supervised by Jeremy Clark. The constructions in this paper are new, and any ideas or text from [30] incorporated into this paper were originated by Clark. We thank Benjamin Wesolowski, Justin Thaler, Riad Wahby, Benedikt Bünz, Ben Fisch, Dan Boneh, Matthew Green, Michael Specter, and Michael Walfish and for helpful feedback and discussion. We thank Alex Ozdemir for help with implementing our SNARK-based construction.

Arasu Arun and Joseph Bonneau were supported by DARPA under Agreement No. HR00112020022. Any opinions, findings and conclusions or recommendations expressed in this material are those of the authors and do not necessarily reflect the views of the United States Government or DARPA.

Jeremy Clark acknowledges support for this research project from (i) the National Sciences and Engineering Research Council (NSERC), Raymond Chabot Grant Thornton, and Catallaxy Industrial Research Chair in Blockchain Technologies, and (ii) NSERC through a Discovery Grant.

References

1. The digital signature standard: communications of the ACM **35**(7), 36–40 (1992)
2. Drand Randomness Beacon. drand.love (2021)
3. ElectionGuard. https://github.com/microsoft/electionguard (2021)
4. NIST Randomness Beacon Version 2.0. https://beacon.nist.gov/home (2021)
5. Adida, B.: Helios: Web-based Open-audit Voting. In: USENIX Security (2008)

6. Agrawal, S., Ganesh, C., Mohassel, P.: Non-interactive zero-knowledge proofs for composite statements. In: CRYPTO (2018)
7. Arun, A., Bonneau, J., Clark, J.: Short-lived zero-knowledge proofs and signatures. Cryptology ePrint Archive, Paper 2022/190 (2022)
8. Baldimtsi, F., Kiayias, A., Zacharias, T., Zhang, B.: Indistinguishable Proofs of Work or Knowledge. In: Eurocrypt (2016)
9. Barker, E., Dang, Q.: Recommendation for Key Management. NIST Special Publication 800–857 (2015)
10. Beck, G., Choudhuri, A.R., Green, M., Jain, A., Tiwari, P.R.: Time-deniable signatures. Cryptology ePrint Archive, Paper 2022/1018 (2022)
11. Bell, S., et al.: STAR-Vote: a secure, transparent, auditable, and reliable voting system. In: JETS (2013)
12. Ben-Sasson, E., Bentov, I., Horesh, Y., Riabzev, M.: Scalable, transparent, and post-quantum secure computational integrity. Cryptology ePrint Archive, Report 2018/46 (2018)
13. Ben-Sasson, E., Chiesa, A., Genkin, D., Tromer, E., Virza, M.: SNARKs for C: verifying program executions succinctly and in zero knowledge. In: CRYPTO (2013)
14. Benaloh, J.: Ballot casting assurance via voter-initiated poll station auditing. In: EVT (2007)
15. Boneh, D., Naor, M.: Timed commitments. In: CRYPTO (2000)
16. Boneh, D., Bonneau, J., Bünz, B., Fisch, B.: Verifiable delay functions. In: CRYPTO (2018)
17. Boneh, D., Bünz, B., Fisch, B.: A survey of two verifiable delay functions. Cryptology ePrint Archive, Report 2018/712 (2018)
18. Boneh, D., Bünz, B., Fisch, B.: Batching techniques for accumulators with applications to IOPs and stateless blockchains. In: CRYPTO (2019)
19. Boneh, D., Lynn, B., Shacham, H.: Short signatures from the Weil pairing. In: Eurocrypt (2001)
20. Bonneau, J., Clark, J., Goldfeder, S.: On Bitcoin as a public randomness source. Cryptology ePrint Archive, Report 2015/1015 (2015)
21. Borisov, N., Goldberg, I., Brewer, E.: Off-the-record communication, or, why not to use PGP. In: ACM WPES (2004)
22. Bünz, B., Bootle, J., Boneh, D., Poelstra, A., Wuille, P., Maxwell, G.: Bulletproofs: short proofs for confidential transactions and more. In: IEEE Security & Privacy (2018)
23. Bünz, B., Fisch, B., Szepieniec, A.: Transparent SNARKs from DARK compilers. In: CRYPTO (2020)
24. Chiesa, A., Hu, Y., Maller, M., Mishra, P., Vesely, N., Ward, N.: Marlin: preprocessing zkSNARKs with universal and updatable SRS. In: CRYPTO (2020)
25. Ciampi, M., Persiano, G., Scafuro, A., Siniscalchi, L., Visconti, I.: Improved OR Composition of Sigma-Protocols. In: TCC (2016)
26. Clark, J., Hengartner, U.: On the use of financial data as a random beacon. In: EVT/WOTE (2010)
27. Clark, J., van Oorschot, P.C., Ruoti, S., Seamons, K., Zappala, D.: SoK: securing email: a stakeholder-based analysis. In: Financial Cryptography (2021)
28. Cohen, B., Pietrzak, K.: Simple Proofs of Sequential Work. In: CRYPTO (2018)
29. Cohen, B., Pietrzak, K.: The Chia network blockchain (2019)
30. Colburn, M.: Short-lived signatures. Master's thesis, Concordia University (2018)
31. Couteau, G., Klooß, M., Lin, H., Reichle, M.: Efficient range proofs with transparent setup from bounded integer commitments. In: Eurocrypt (2021)

32. Cramer, R., Damgård, I., Schoenmakers, B.: Proofs of partial knowledge and simplified design of witness hiding protocols. In: CRYPTO (1994)
33. Dodis, Y., Yum, D.H.: Time capsule signature. In: Financial Cryptography (2005)
34. Durumeric, Z., et al.: Neither snow nor rain nor MITM: an empirical analysis of email delivery security. In: ACM CCS (2015)
35. Dwork, C., Naor, M.: Pricing via Processing or Combatting Junk Mail. In: CRYPTO (1992)
36. Ephraim, N., Freitag, C., Komargodski, I., Pass, R.: Continuous verifiable delay functions. In: CRYPTO (2020)
37. Feo, L.D., Masson, S., Petit, C., Sanso, A.: Verifiable Delay Functions from Supersingular Isogenies and Pairings. Cryptology ePrint Archive, Report 2019/166 (2019)
38. Ferradi, H., Géraud, R., Naccache, D.: slow motion zero knowledge identifying with colliding commitments. In: ICISC (2015)
39. Fiat, A., Shamir, A.: How to prove yourself: practical solutions to identification and signature problems. In: Odlyzko, A.M. (ed.) CRYPTO 1986. LNCS, vol. 263, pp. 186–194. Springer, Heidelberg (1987). https://doi.org/10.1007/3-540-47721-7_12
40. Foundation, E.: Ethereum 2.0 Beacon Chain (2020)
41. Gabizon, A., Williamson, Z.J., Ciobotaru, O.: PLONK: permutations over lagrange-bases for oecumenical noninteractive arguments of knowledge. Cryptology ePrint Archive, Report 2019/953 (2019)
42. Gennaro, R., Gentry, C., Parno, B., Raykova, M.: Quadratic span programs and succinct NIZKs without PCPs. In: CRYPTO (2013)
43. Goldreich, O., Micali, S., Wigderson, A.: Proofs that yield nothing but their validity or all languages in NP have zero-knowledge proof systems. JACM 38(3) (1991)
44. Goyal, V., Raizes, J., Soni, P.: Time-traveling simulators using blockchains and their applications. Cryptology ePrint Archive, Paper 2022/035 (2022)
45. Green, M.: Ok Google: please publish your DKIM secret keys (November 2020)
46. Groth, J.: Short pairing-based non-interactive zero-knowledge arguments. In: Eurocrypt (2010)
47. Groth, J.: On the size of pairing-based non-interactive arguments. In: CRYPTO (2016)
48. Halderman, J.A., Waters, B.: Harvesting verifiable challenges from oblivious online sources. In: CCS (2007)
49. Jakobsson, M., Sako, K., Impagliazzo, R.: Designated verifier proofs and their applications. In: Eurocrypt (1996)
50. Jaques, S., Montgomery, H., Roy, A.: Time-release cryptography from minimal circuit assumptions. Cryptology ePrint Archive, Report 2020/755 (2020)
51. Johnson, D., Menezes, A., Vanstone, S.: The elliptic curve digital signature algorithm (ECDSA). Int. J. Inf. Secur. 1(1) (2001)
52. Krawczyk, H., Rabin, T.: Chameleon signatures. In: NDSS (2000)
53. Kucherawy, M., Crocker, D., Hansen, T.: DomainKeys identified mail (DKIM) signatures. RFC 6376 (2011)
54. Landerreche, E., Stevens, M., Schaffner, C.: Non-interactive cryptographic timestamping based on verifiable delay functions. In: Financial Cryptography (2020)
55. Lenstra, A.K., Wesolowski, B.: Trustworthy public randomness with sloth, unicorn, and trx. Int. J. Appl. Crypto. 3(4), 330–343 (2017)
56. Mahmoody, M., Moran, T., Vadhan, S.: Publicly verifiable proofs of sequential work. In: ITCS (2013)
57. Malavolta, G., Thyagarajan, S.A.K.: homomorphic time-lock puzzles and applications. In: CRYPTO (2019)

58. Maller, M., Bowe, S., Kohlweiss, M., Meiklejohn, S.: Sonic: Zero-knowledge SNARKs from linear-size universal and updatable structured reference strings. In: ACM CCS (2019)
59. Moran, T., Naor, M.: Receipt-free universally-verifiable voting with everlasting privacy. In: CRYPTO (2006)
60. Neff, C.A.: Practical high certainty intent verification for encrypted votes. Tech. rep, VoteHere Whitepaper (2004)
61. Okamoto, T., Uchiyama, S.: A new public-key cryptosystem as secure as factoring. In: Eurocrypt (1998)
62. Ozdemir, A., Wahby, R., Whitehat, B., Boneh, D.: Scaling verifiable computation using efficient set accumulators. In: USENIX Security (2020)
63. Pass, R.: On deniability in the common reference string and random oracle model. In: CRYPTO (2003)
64. Pietrzak, K.: Simple verifiable delay functions. In: ITCS (2018)
65. Rabin, M.: Transaction protection by Beacons. J. Comput. Syst. Sci. 27(2) (1983)
66. Rivest, R.L., Shamir, A., Wagner, D.A.: Time-lock puzzles and timed-release crypto. Tech. Rep. TR-684, MIT (1996)
67. Rivest, R.L.: Description of the LCS35 time capsule crypto-puzzle. https://people.csail.mit.edu/rivest/lcs35-puzzle-description.txt (1999)
68. Rivest, R.L., Shamir, A., Adleman, L.: A method for obtaining digital signatures and public-key cryptosystems. Commun. ACM 21(2) (1978)
69. Rivest, R.L., Shamir, A., Tauman, Y.: How to leak a secret. In: Asiacrypt (2001)
70. Schnorr, C.P.: Efficient signature generation by smart cards. J. Crypto. 4(3), 161–174 (1991)
71. Setty, S.: Spartan: efficient and general-purpose zkSNARKs without trusted setup. In: CRYPTO (2020)
72. Specter, M.A., Park, S., Green, M.: KeyForge: non-attributable email from forward-forgeable signatures. In: USENIX Security (2021)
73. Syta, E., et al.: Scalable bias-resistant distributed randomness. In: IEEE Security & Privacy (2017)
74. Thaler, J.: Proofs, arguments, and zero-knowledge (2021)
75. Thyagarajan, S.A.K., Bhat, A., Malavolta, G., Döttling, N., Kate, A., Schröder, D.: Verifiable timed signatures made practical. In: ACM CCS (2020)
76. Wesolowski, B.: A proof of time or knowledge. https://hal.archives-ouvertes.fr/hal-03380471
77. Wesolowski, B.: Efficient verifiable delay functions. In: Eurocrypt (2019)
78. Öztürk, E.: Modular multiplication algorithm suitable for low-latency circuit implementations. Cryptology ePrint Archive, Paper 2019/826 (2019)

Non-interactive Zero-Knowledge Proofs to Multiple Verifiers

Kang Yang[1(✉)] and Xiao Wang[2]

[1] State Key Laboratory of Cryptology, Beijing, China
yangk@sklc.org
[2] Northwestern University, Evanston, USA
wangxiao@cs.northwestern.edu

Abstract. In this paper, we study zero-knowledge (ZK) proofs for circuit satisfiability that can prove to n verifiers at a time efficiently. The proofs are secure against the collusion of a prover and a subset of t verifiers. We refer to such ZK proofs as multi-verifier zero-knowledge (MVZK) proofs and focus on the case that a majority of verifiers are honest (i.e., $t < n/2$). We construct efficient MVZK protocols in the random oracle model where the prover sends one message to each verifier, while the verifiers only exchange one round of messages. When the threshold of corrupted verifiers $t < n/2$, the prover sends $1/2 + o(1)$ field elements per multiplication gate to every verifier; when $t < n(1/2 - \epsilon)$ for some constant $0 < \epsilon < 1/2$, we can further reduce the communication to $O(1/n)$ field elements per multiplication gate per verifier. Our MVZK protocols demonstrate particularly high scalability: the proofs are streamable and only require a memory proportional to what is needed to evaluate the circuit in the clear.

1 Introduction

Zero-knowledge (ZK) proofs allow a prover $\mathcal{P}$, who knows a witness w, to convince a verifier $\mathcal{V}$ that $C(w) = 0$ for a circuit C, in the way that $\mathcal{V}$ learns nothing beyond the validity of the statement. One important type of ZK proofs is non-interactive ZK (NIZK), where the prover just needs to send one message to a verifier. This is particularly useful as the prover's message (i.e., the proof) can be reused to convince multiple verifiers. The efficiency of NIZK proofs has been significantly improved in recent years, based on different frameworks (e.g., [3,10–13,15,17,18,34,35,42,44,48,49,54] and references therein). Another important type of ZK proofs is designated-verifier ZK (DVZK), where an interactive protocol needs to be executed between the prover and the verifier. Compared to NIZK, DVZK protocols can often achieve a higher efficiency to prove to one verifier and scale to a very large circuit with a small memory. For example, recent DVZK proof systems [6,8,27,28,50,51] can prove tens of millions of gates per second with very limited bandwidth. However, such an advantage diminishes when the number of verifiers increases: DVZK protocols require the prover to execute the

© International Association for Cryptologic Research 2022
S. Agrawal and D. Lin (Eds.): ASIACRYPT 2022, LNCS 13793, pp. 517–546, 2022.
https://doi.org/10.1007/978-3-031-22969-5_18

protocol with every verifier, while an NIZK proof enables all verifiers to verify the proof concurrently after the prover generates and publishes the proof.

In this work, we explore the middle ground between NIZK and DVZK: we study the efficiency of ZK proofs when a prover wants to prove to multiple verifiers (i.e., multi-verifier ZK, MVZK in short). This setting was first studied by Abe, Cramer and Fehr [2]. Specifically, we consider that a prover $\mathcal{P}$ needs to convince n verifiers $\mathcal{V}_1, \ldots, \mathcal{V}_n$, and the adversary can potentially corrupt a subset of t verifiers *and* optionally the prover. More specifically, we focus on the honest-majority setting, meaning that $t < n/2$ verifiers could be corrupted and can *collude* with the prover. Such an MVZK protocol is closely connected to DVZK in which the prover can only prove to a designated set of verifiers who are known ahead of the protocol execution. However, due to the fact that there is a majority of honest verifiers, it turns out the MVZK protocol can achieve some surprising features, e.g., being *non-interactive* between the prover and the verifiers in the information-theoretic setting.

Because of the involvement of multiple verifiers, there are two types of communications: 1) between the prover and verifiers and 2) between different verifiers. We say that the protocol is a *non-interactive* multi-verifier ZK (NIMVZK) proof if the prover only sends one message to each verifier. We say that the protocol is a *strong* NIMVZK proof if it is an NIMVZK and that there is only one round of communication between verifiers. We allow the verifiers to communicate for one round because without any communication between the verifiers, constructing NIMVZKs appears as difficult as constructing NIZKs.

In the MVZK setting, the known protocol [2] or those that can be implicitly constructed from the known techniques [14,16] either are *not* concretely efficient or *only* prove some specific circuits instead of generic circuits. Furthermore, none of the prior work considers how to stream the MVZK proofs, which is a crucial property to prove large-scale circuits.

1.1   Our Contribution

In this paper, we propose streamable NIMVZK protocols on generic circuits with both theoretical insights and practical implication. The protocols work in the honest-majority setting, meaning that the number t of corrupted verifiers is less than $n/2$, where n is the total number of verifiers. Compared to NIZKs, our NIMVZK protocols are much cheaper in terms of computational cost and use significantly less memory. Compared to DVZK, our protocols have three advantages: 1) the computation is still cheaper; 2) we can achieve the strongly non-interactive property; and 3) the communication is lower, especially when the number of verifiers is large. Specifically, our results are summarized as follows:

1. We present an information-theoretic NIMVZK protocol, where the prover sends $1 + o(1)$ field elements per multiplication gate to every verifier in one message (thus non-interactive), and the verifiers interact in both communication and rounds *logarithmic* to the circuit size. We consider the protocol as a stepping stone to introduce the following two main NIMVZK protocols.

2. Assuming a random oracle (and thus in the computational setting), we construct a strong NIMVZK proof based on Shamir secret sharing, where the verifiers only need to communicate for one round. The prover needs to send $1/2 + o(1)$ field elements per multiplication gate to each verifier and the communication cost between verifiers is still logarithmic.

   The challenge is that the message sent by the prover consists of the shares of every verifier that are private information and thus cannot be revealed. This makes the verifiers have no way to compute the public message that can be used as the input of a random oracle in the Fiat-Shamir transform. We proposed an efficient approach to allow Fiat-Shamir to work across multiple verifiers, i.e., enabling the prover to generate a *small* public message that can be *securely* used in the Fiat-Shamir transform, even if a minority of verifiers collude with the malicious prover.

3. When the corruption threshold is smaller (i.e., $t < n(1/2 - \epsilon)$ for some constant $0 < \epsilon < 1/2$), we use packed secret sharing (PSS) [30] to construct a strong NIMVZK protocol for proving a single generic circuit, which further reduces the communication complexity to $O(1/n)$ field elements per multiplication gate per verifier, while the communication complexity between verifiers is logarithmic to the circuit size. If applying the state-of-art secure multiparty computation (MPC) protocol [39] based on PSS to design an interactive MVZK protocol, the resulting protocol can achieve the same communication complexity. However, the constant in the O notation is significantly larger than our protocol.

   For a single generic circuit, PSS has been used in MPC protocols [24,31,32,39] in the honest-majority setting, but the overhead is often high due to the constraint how to pack the wire values to realize secure evaluation of the circuit. In the ZK setting, our strong NIMVZK protocol can remove the constraint, and achieve optimality for packing wire values and significantly better efficiency for checking correctness of packed sharings. For example, the state-of-the-art PSS-based MPC protocol [39] incurs a total communication cost of $O(n^5 k^2)$ to check the consistency between packed input sharings and output sharings, where $k > \epsilon n + 1/2$ is the number of secrets packed in a single sharing. When being improved in the ZK setting, the total communication cost of our protocol can be reduced to $O(n^2 k^2)$. Furthermore, we develop a non-interactive verification technique for checking correctness of PSS-based multiplication tuples, while the approach used in MPC [39] requires logarithmic rounds.

In summary, we designed concretely efficient MVZK protocols, which provide the attractive properties of NIZK (non-interactivity) and DVZK (memory efficiency and prover-computation efficiency). Although it is not applicable to all settings (due to the assumption of honest-majority verifiers), when it is applicable, the performance improvements to existing protocols are huge.

Streamable Property of Our NIMVZK. Although the communication complexity between the prover and verifiers is linear to the circuit size, all our protocols as described above are *streamable*, meaning that the prover can generate and send the proof on-the-fly and no party needs to store the whole proof during the protocol execution. As a result, the memory consumption of our protocols is proportional to what is needed to evaluate the statement in the clear. Furthermore, we make our strong NIMVZK proofs streamable in the way that the rounds among verifiers keep *unchanged* (i.e., only one round between verifiers is needed for proving multiple batches of gates).

Asymmetric Property of Our Strong NIMVZK. One surprising feature of our two strong NIMVZK protocols in the computational setting is the *asymmetry* among verifiers. Specifically, among all verifiers, a subset of t verifiers only has a *sublinear* communication complexity: each verifier only receives $O(n + \log |C|)$ field elements from the prover, and only needs to send $O(n)$ field elements to other verifiers, where $|C|$ is the number of multiplication gates in a circuit C. It makes the protocols particularly suitable for the applications where the verifiers are a mix of powerful servers and lower-resource mobile devices.

1.2 Applications

Non-interactive MVZK proofs have the following applications:

1. **Drop-in replacement to NIZK and DVZK.** NIMVZK can be used in normal ZK applications as long as the identifies of the verifiers are known ahead of time and satisfy the security requirement (e.g., a majority of verifiers are honest for our NIMVZK protocols). For example, as described in [21], a ZK proof could potentially be used by Apple for auditing their Child Sexual Abuse Material detection protocol. Our NIMVZK protocol can be used when such an auditing needs to be performed to multiple agencies efficiently.
2. **Honest-majority MPC with input predicate check.** In some computational tasks, it is desired to execute the MPC protocol among multiple parties only if the input of every party is valid, where the validity is defined by some predicate. Although generic MPC can realize this functionality, using our NIMVZK protocols could further reduce the overhead of proving the predicate. As our protocols are based on Shamir sharings, it can be seamlessly integrated with MPC protocols also based on Shamir sharings.
3. **Private aggregation systems.** Systems like Prio [23] use a set of servers to collect and aggregate users' data. To prevent mistakes and attacks, users need to prove to the servers that their data is valid, which was done via *secret-shared non-interactive proof* in Prio. However, the protocol assumes the prover not to collude with any verifier for soundness. Our protocol could be a more efficient alternative and is sound even when a user colludes with a minority of servers. On the other hand, for zero-knowledge, Prio can tolerate all-but-one corrupted servers, while our protocols need to assume an honest majority of servers.

1.3 Related Work

The concept of multi-verifier ZK proofs was first discussed by Burmester and Desmedt [19], where they focus on how to save broadcasts. Lepinski, Micali and shelat [45] proposed a fair ZK proof, which ensures that even malicious verifiers who collude with the prover can learn nothing beyond the validity of the statement if the honest verifiers accept the proof. More recently, Baldimtsi et al. [5] proposed a crowd verifiable zero-knowledge proof, where the focus is to transform a Sigma protocol to their setting. All of the above works focus on extending the ZK functionality rather than the concrete efficiency of ZK proofs.

Abe, Cramer and Fehr [2] first studied the MVZK setting, and proposed a strong NIMVZK protocol for circuit satisfiability if at most $t < n/3$ verifiers are corrupted. Their protocol builds on the technique [1], and adopts the Pedersen commitment and verifiable secret sharing. Due to the usage of public-key operations for every non-linear gate, their protocol is *not* concretely efficient. In addition, the ZK proof by Groth and Ostrovsky [43] could be transformed into a strong NIMVZK proof in the corruption threshold of $t < n/2$, but their proof requires public-key operations per gate and thus is *not* concretely efficient. Compared to the NIMVZK proofs [2,43], our proofs do not require any public-key operation and are concretely efficient.

Although Boneh et al. [14] did *not* explicitly consider the MVZK setting, the ZK proofs proposed by them work in this setting. However, these protocols are only efficient applicable for circuits that can be represented by low-degree polynomials (instead of generic circuits). Recently, Boyle et al. [16] shown how to use the ZK proof [14] to design honest-majority MPC protocols with malicious security. However, they only considered how to prove correctness of degree-2 relations, and did *not* involve MVZK proofs on generic circuits yet. In addition, Boyle et al. [16] proposed an approach based on Fiat-Shamir to make the ZK proof on inner-product tuples non-interactive, where the difference between the secret and randomness needs to be sent. One can *generalize* their approach into our MVZK framework, and make the resulting MVZK proof strongly non-interactive. However, their approach requires $3\times$ larger communication than ours. Both works [14,16] did *not* consider how to make the ZK proofs *streamable*, which is addressed by our work.

One can also use maliciously secure MPC protocols in the honest-majority setting (e.g., [16,22,32,38–41,46,47]) to directly obtain interactive MVZK proofs. However, both of communication and computation costs will be significantly larger than our NIMVZK protocols. While NIZK can be transformed into NIMVZK directly, these NIZK proofs with performance similar to our proofs (e.g., recent succinct non-interactive proofs [13,18,48,53,54]) require memory linear to the circuit size, which could lead to a huge memory consumption for circuits with billions of gates. Our NIMVZK protocols are streamable, and the memory consumption of these protocols is proportional to what is needed to evaluate the circuit in the clear (meaning that these protocols only need a small memory cost for proving very large circuits). Compared to MVZK that is constructed from the recent VOLE-based DVZK proofs [6,8,27,28,50,51] by executing the protocol with every verifier, our strong NIMVZK proofs reduce round complexity from $O(1)$ to only one round between the prover and verifiers, and significantly improve efficiency.

Very recently, two works by Applebaum et al. [4] and Baum et al. [7] also presented MVZK protocols in the setting that a majority of verifiers are honest. Applebaum et al. [4] focuses on a theoretical perspective, and gave two strong NIMVZK protocols based on "Minicrypt"-type assumptions in the plain model. Baum et al. [7] adopted an approach similar to ours, and aim to construct concretely efficient MVZK protocols. In particular, they proposed two NIMVZK protocols that allow to identify the cheating verifiers (and thus have stronger security than ours); however, their protocols only tolerate a smaller number of corrupted verifiers (either $t < n/3$ or $t < n/4$). Neither of the works [4,7] adopted packed secret sharings and achieve the communication complexity of $O(1/n)$ per multiplication gate per verifier.

2   Preliminaries

We discuss some important preliminaries here and provide more preliminaries (e.g., security model) in the full version [52].

Notation. We use λ and ρ to denote the computational and statistical security parameters, respectively. We use $x \leftarrow S$ to denote that sampling x uniformly at random from a finite set S. For $a, b \in \mathbb{Z}$ with $a \leq b$, we write $[a, b] = \{a, \ldots, b\}$. We will use bold lower-case letters like $\boldsymbol{x}$ for column vectors, and denote by x_i the i-th component of $\boldsymbol{x}$ with x_1 the first entry. For two vectors $\boldsymbol{x}, \boldsymbol{y}$ of dimension m, $\boldsymbol{x} \odot \boldsymbol{y}$ denotes the inner product of $\boldsymbol{x}$ and $\boldsymbol{y}$ (i.e., $\boldsymbol{x} \odot \boldsymbol{y} = \sum_{i \in [1,m]} x_i \cdot y_i$). Sometimes, when the dimension of vectors $\boldsymbol{x}, \boldsymbol{y}$ is 1 (i.e., $\boldsymbol{x} = x$ and $\boldsymbol{y} = y$), we abuse the notation $\boldsymbol{x} \odot \boldsymbol{y}$ to denote the multiplication $x \cdot y$ for the sake of simplicity. We use $\log_k$ to denote the logarithm in base k, and denote by $\log$ the logarithm notation $\log_2$ for simplicity. For a finite field $\mathbb{F}$, we use $\mathbb{K}$ to denote a degree-r extension field of $\mathbb{F}$. In particular, we fix some monic, irreducible polynomial $f(X)$ of degree r and write $\mathbb{K} \cong \mathbb{F}[X]/f(X)$. Every field element $w \in \mathbb{K}$ can be denoted uniquely as $w = \sum_{h \in [1,r]} w_h \cdot X^{h-1}$ with $w_h \in \mathbb{F}$ for all $h \in [1, r]$. When we write arithmetic expressions involving both elements of $\mathbb{F}$ and elements of $\mathbb{K}$, it is understood that field elements in $\mathbb{F}$ are viewed as the polynomials lying in $\mathbb{K}$ that have only constant terms. For a circuit C, we use $|C|$ to denote the number of multiplication gates.

Zero-Knowledge Functionality. Our ZK functionality for proving circuit satisfiability against multiple verifiers is shown in Fig. 1. Let n be the total number of verifiers. We consider the MVZK protocols in the honest-majority setting, i.e., the adversary allows to corrupt at most $t < n/2$ verifiers. The adversary is also allowed to corrupt the prover. When the prover is honest, functionality $\mathcal{F}_{\mathsf{mvzk}}$ defined in Fig. 1 captures *zero-knowledge*, meaning that t malicious verifiers cannot learn any information on the witness. When the prover is malicious, $\mathcal{F}_{\mathsf{mvzk}}$

Functionality $\mathcal{F}_{\mathsf{mvzk}}$

This functionality runs with a prover $\mathcal{P}$ and n verifiers $\mathcal{V}_1, \ldots, \mathcal{V}_n$. Let $\mathcal{H}$ denote the set of honest verifiers. This functionality operates as follows:

1. Upon receiving $(\mathsf{prove}, C, \boldsymbol{w})$ from $\mathcal{P}$ and (verify, C) from $\mathcal{V}_i$ for all $i \in [1, n]$ where C is a circuit, set $b := \mathsf{true}$ if $C(\boldsymbol{w}) = 0$ and $b := \mathsf{false}$ otherwise.
2. Send b to the adversary. For each $i \in \mathcal{H}$, wait for an input from the adversary, and then do the following:
   - If it is $\mathsf{continue}_i$, send b to the verifier $\mathcal{V}_i$.
   - If it is abort_i, send abort to the verifier $\mathcal{V}_i$.

Fig. 1. Zero-knowledge functionality for honest-majority verifiers.

captures *soundness*, i.e., the malicious prover cannot make the honest verifiers accept if $C(\boldsymbol{w}) \neq 0$, even though it *colludes* with t malicious verifiers.

We can consider MVZK protocols as special MPC protocols. Thus, we adopt the notion of *security with abort* in the MPC setting to define $\mathcal{F}_{\mathsf{mvzk}}$ and other functionalities defined in the subsequent sections, where the corrupted verifiers may receive output while the honest verifiers do not. Our definition does not guarantee *unanimous abort*, meaning that some honest verifiers may receive output while other honest verifiers abort. Nevertheless, it is easy to tune our protocols to satisfy the security notion of *unanimous abort*, by having the verifiers broadcast whether they will abort or not at the end of the protocol execution [36].

Communication Model. The default communication between the prover and verifiers is private channel, unless otherwise specified. We assume that all verifiers are connected via authenticated channels. In the computational setting, the prover sometimes needs to communicate with all verifiers over a broadcast channel. Since we allow abort, the broadcast channel can be established using a standard echo-broadcast protocol [36], where the communication overhead can be improved to be constant small using a collision-resistant hash function. In our strong NIMVZK protocols, the verifiers need to exchange the shares in one round at the end of protocol execution. In parallel with the communication of shares, every verifier can send the hash output of the messages broadcast by the prover to all other verifiers. Therefore, although the echo-broadcast protocol is used in our MVZK proofs, we can still achieve strongly non-interactive.

Linear Secret Sharing Scheme. In our NIMVZK protocols, we will extensively use *linear secret sharing schemes* (LSSSs) with a threshold t. A t-out-of-n LSSS enables a secret x to be shared among n parties, such that no subset of t parties can learn any information on x, while any subset of $t + 1$ parties can reconstruct the secret. To align with the description of our NIMVZK protocols, we let the prover $\mathcal{P}$ play the role of the dealer and let every verifier $\mathcal{V}_i$ obtain the shares. We require that LSSS supports the following procedures:

- $[x] \leftarrow \mathsf{Share}(x)$: In this procedure, a dealer $\mathcal{P}$ shares a secret x among the parties $\mathcal{V}_1, \ldots, \mathcal{V}_n$, such that $\mathcal{V}_i$ gets a share x^i for $i \in [1, n]$. The sharing of x output by this procedure is denoted by $[x]$.
- $x \leftarrow \mathsf{Open}([x])$: Given a sharing $[x]$, this procedure is executed by parties $\mathcal{V}_1, \ldots, \mathcal{V}_n$. At the end of the execution, if $[x]$ is not valid, then all honest parties abort; otherwise, every party will output x.
- *Linear combination*: Given the public coefficients $c_0, c_1, \ldots, c_\ell$ and secret sharings $[x_1], \ldots, [x_\ell]$, $\mathcal{V}_1, \ldots, \mathcal{V}_n$ can *locally* compute $[y] = \sum_{i=1}^{\ell} c_i \cdot [x_i] + c_0$, such that $y = \sum_{i=1}^{\ell} c_i \cdot x_i + c_0$ holds.

We describe two LSSS instantiations shown in the full version [52], where one is Shamir secret sharing and the other is packed secret sharing (a generalization of Shamir secret sharing). For Shamir secret sharing, for a vector $\boldsymbol{x} = (x_1, \ldots, x_m)$, we will use $[\boldsymbol{x}]$ to denote $([x_1], \ldots, [x_m])$. For packed secret sharing, for a vector $\boldsymbol{x} \in \mathbb{F}^k$, we will use $[\boldsymbol{x}]$ to denote a single packed sharing that stores k secrets of $\boldsymbol{x}$. We assume that the shares of any t parties are uniformly random, which is satisfied by the two instantiations.

3 Technical Overview

We describe the ideas in our NIMVZK protocols and how we come up with these constructions in this section. We leave the full details and their proofs of security in later sections.

3.1 Information-Theoretic Non-interactive MVZK

We introduce our non-interactive MVZK proofs starting from an *information-theoretic* NIMVZK protocol that is a warm-up to describe the techniques in our strong NIMVZK protocols.

Our Approach for NIMVZK. Our NIMVZK proofs follow the "commit-and-prove" paradigm, where secrets are committed using Shamir sharings and the security of commitments is guaranteed in the honest-majority setting. At a high level, our information-theoretic protocol (as well as other two protocols discussed later) have the following steps.

1. For the output z of each circuit-input gate or multiplication gate, the prover runs $\mathsf{Share}(z)$ to distribute the shares of $[z]$ to all verifiers. Since LSSS is used, the addition gates can be locally computed by the verifiers.
2. For a circuit with N multiplication gates, we have N multiplication triples $([x_i], [y_i], [z_i])$ over a field $\mathbb{F}$ that the verifiers need to check. All parties jointly sample a uniform element $\chi \in \mathbb{K}$, and then compute the inner-product tuple:

$$[\boldsymbol{x}] := \left([x_1], \ldots, \chi^{N-1} \cdot [x_N]\right), \ [\boldsymbol{y}] := ([y_1], \ldots, [y_N]), \ [z] := \sum_{i=1}^{N} \chi^{i-1} \cdot [z_i].$$

   If there exists one *incorrect* multiplication triple, then the inner-product tuple defined as above is also *incorrect*, except with probability $\frac{N-1}{|\mathbb{K}|}$.

3. The verifiers check correctness of the inner-product tuple $([\boldsymbol{x}], [\boldsymbol{y}], [z])$ with logarithmic communication.

In the information-theoretic setting, the verifiers can call a coin-tossing functionality (shown in the full version [52]) to sample the coefficient χ, but χ is *not* available to the prover while keeping non-interactive between the prover and verifiers. The distributed ZK proofs by Boneh et al. [14] could check correctness of an inner-product tuple, but it only works when the prover knows the secrets. To use their protocol directly, we would need the verifiers to send χ to the prover, and then the round complexity between the prover and verifiers will be at least 3 rounds. Our task is to design a non-interactive protocol that verifies correctness of an inner-product tuple, where the secrets are shared among verifiers and *unknown to the prover*. We adapt the checking approach by Goyal et al. [40,41] (building upon the technique [14]) from the MPC setting to the MVZK setting, and construct a verification protocol to check correctness of inner-product tuples. In particular, our verification protocol makes the prover generate the random sharings and random multiplication triples (instead of letting the verifiers run the DN multiplication protocol [25] that is done in [40,41]), which is sufficient for MVZK as zero-knowledge only needs to hold for an honest prover.

3.2   Distributing Fiat-Shamir for Strong Non-interactive MVZK

With the above preparation, we now discuss how to construct a strong NIMVZK proof, where the verifiers communicate for only one round. This is a highly non-trivial task, as it is even unclear how to sample a random coefficient $\chi \in \mathbb{K}$ as needed in step 2.. Since every verifier can only send one message to other verifiers, using a secure coin-tossing protocol is not possible. The other randomness source that we can use is random oracle (i.e., adopting the Fiat-Shamir heuristic). However, only the shares are sent by the prover where the shares need to be kept secret, and thus the verifiers has no way to compute a public message that can be used as the input of a random oracle. This was in fact attempted in the distributed ZK proof [14] as well, but their non-interactive solution does not allow the prover to collude with any verifier.

Let's first review how Boneh et al. [14] use Fiat-Shamir in the case that all verifiers do not collude with the prover. Suppose that the prover $\mathcal{P}$ sends a message Msg_i along with a randomness r_i to a verifier $\mathcal{V}_i$ for $i \in [1, n]$, where Msg_i and r_i need to be kept secret. Every verifier $\mathcal{V}_i$ can send $\nu_i := \mathsf{H}(\mathsf{Msg}_i, r_i)$ to other verifiers where H is a random oracle, and then generates a random challenge $\chi := \bigoplus_{i \in [1,n]} \nu_i$, when ignoring some details for simplicity. Prover $\mathcal{P}$ can also compute the challenge χ as it knows all messages and randomness. When the prover is corrupted (and thus we are concerning soundness), all verifiers are assumed to be honest and thus can exchange the correct values $\{\nu_i\}_{i \in [1,n]}$, so that the verifiers can compute a random challenge χ to execute the protocol. However, when $\mathcal{P}$ colludes with a verifier $\mathcal{V}_{i^*}$, this method does not work anymore: $\mathcal{P}$ can cheat when the challenge is some value $\chi* \neq \bigoplus_{i \in [1,n]} \mathsf{H}(\mathsf{Msg}_i, r_i)$; after receiving the values of other verifiers, $\mathcal{V}_{i^*}$ can compute $\nu_{i^*} = \mathsf{H}(\mathsf{Msg}_{i^*}, r_{i^*})$ and the correct challenge χ, and then send $\nu'_{i*} = \nu_{i*} \oplus \chi \oplus \chi*$ to every other verifier such that the invalid proof can still go through.

Because of the round-complexity requirement on the verifier side, we cannot let the verifiers to sample χ. So it appears that in order to get a strong NIMVZK, we must find an approach to enable the prover to generate public messages via some sort of Fiat-Shamir transformation in the distributed setting so that: 1) the protocol tolerates the collusion of the prover and a minority of verifiers, and 2) does not require the verifiers to interact more than one round. Let H, H' be two random oracles with exponentially large ranges. Our technique to support Fiat-Shamir is presented as follows.

1. Suppose that the prover $\mathcal{P}$ sends (Msg_i, r_i) to every verifier $\mathcal{V}_i$ over a private channel, where Msg_i consists of the shares held by $\mathcal{V}_i$ for our protocols.
2. Now, $\mathcal{P}$ also broadcasts commitments $com_i := H(\mathsf{Msg}_i, r_i)$ for all $i \in [1, n]$ to all verifiers, where the broadcast does not increase the rounds between verifiers that has been explained in Sect. 2.
3. Every verifier $\mathcal{V}_i$ checks that $com_i = H(\mathsf{Msg}_i, r_i)$. As we assume that $t < n/2$, we can guarantee that a majority of commitments in $com_1, \ldots, com_n$ are computed correctly.
4. The verifiers can generate a random challenge $\chi := H'(com_1, \ldots, com_n)$, as they now know the public messages $com_1, \ldots, com_n$. Then, the verifiers use χ to transform the verification of N multiplication triples into that of an inner-product tuple as described in Sect. 3.1.

If H has a 2λ-bit output length and thus is collision-resistant, then it would make com_i binding and the proof can easily go through, where all the commitments $\{com_i\}$ held by $n - t$ honest verifiers will uniquely define the secrets on all wires. However, we make a key observation that it is sufficient to prove security, if the challenge χ is guaranteed to be defined after the secrets on all wires have been determined (i.e., χ is independent of these secrets). Therefore, it is unnecessary to require the collision resistance for H, but rather we only need H to be second preimage-resistant, which allows to achieve better efficiency, e.g., using the construction [26]. In particular, if χ has been defined and known by the malicious prover, then it must make a query $(com_1, \ldots, com_n)$ to random oracle H'. Then, the malicious prover cheats to find a pair (Msg'_i, r'_i) associated with χ, and then sends it to some honest verifier $\mathcal{V}_i$. The cheat will not be detected only if $com_i = H(\mathsf{Msg}'_i, r'_i)$, which is equivalent to find either a preimage or a second preimage of com_i. The Fiat-Shamir approach as described above only introduces a small communication overhead, i.e., $O(n^2\lambda)$ bits in total between the prover and all verifiers that is independent of the circuit size.

Through the above approach, the verifiers can generate a random challenge non-interactively, and then use it to convert the verification of multiplication triples into that of an inner-product tuple. We can simplify the verification technique (shown in Sect. 3.3) by viewing Shamir secret sharing as a special case of packed secret sharing, and then use it to verify the inner-product tuple in one round between verifiers. The resulting strong NIMVZK protocol is streamable while keeping the round complexity between verifiers unchanged (see below).

3.3   More Efficient Strong NIMVZK from Packed Secret Sharing

The above discussion shows a strong NIMVZK protocol where a prover sends one message to each verifier and the verifiers communicate only one round. It is secure against the adversary corrupting up to a minority of verifiers (i.e., $t < n/2$) and the prover. However, the downside is the communication of $1/2 + o(1)$ field elements per multiplication gate per verifier, and a majority of the proof is used to transmit the shares of wire values. We now discuss the strong NIMVZK protocol that reduces the communication cost to $O(1/n)$ field elements per multiplication gate per verifier, when the threshold of corrupted verifiers $t < n(1/2 - \epsilon)$ for any $0 < \epsilon < 1/2$. This protocol adopts packed secret sharing (PSS) [30] as the underlying LSSS, where each sharing packs $k = O(n)$ secrets.

Using packed secret sharing efficiently for a single generic circuit is a huge challenge, because the layout of the circuit could be complicated for packing k gates, and it is not possible to move around any individual wire when using PSS. In fact, because of this, prior MPC works [9,37] using PSS focus on SIMD operations (i.e., repeated circuits). For a single generic circuit, the state-of-the-art PSS-based MPC protocol [39] requires to evaluate the circuit layer-by-layer that needs the rounds linear to the circuit depth, and splits each output wire into different output wires that each can be used only once. Fortunately, we observe that even a single generic circuit can be packed optimally in the context of zero-knowledge, and can remove the constraints in MPC. Particularly, the prover can prove a circuit in a streamable way without the constraint of proving the circuit layer-by-layer, as the prover knows all the wire values.

Consistency Check of Wire Values. In our NIMVZK protocol, if the out degree of a gate is greater than 1, we allow an output wire to appear multiple times (instead of splitting the output wire into multiple output wires), which enables us to obtain better communication. In this case, we need to use the consistency check to ensure that the same wire is assigned with the same value. Specifically, for each input packed sharing $[y]$, if the j-th secret y_j comes from the i-th secret x_i stored in an output packed sharing $[x]$, then we need to check $x_i = y_j$. This corresponds to the wire that carries the value $x_i = y_j$. Following the work [39], we refer to $([x], [y], i, j)$ as a wire tuple. For the consistency check of wire tuples, we reduce the total communication complexity from $O(n^5 k^2)$ in MPC [39] to $O(n^2 k^2)$ for our strong NIMVZK protocol. For each $i, j \in [1, k]$, let $([x_1], [y_1], i, j), \ldots, ([x_m], [y_m], i, j)$ be the wire tuples with the same indices i, j. We use the random-linear-combination approach to check the consistency. Specifically, the prover $\mathcal{P}$ samples two random vectors x_0, y_0 such that $x_{0,i} = y_{0,j}$, and then distributes the shares of $[x_0]$ and $[y_0]$ to all verifiers. To support Fiat-Shamir, we need $\mathcal{P}$ to generate these shares in two steps: 1) distributing the shares of two random sharings $[r]$ and $[s]$; and 2) broadcasts the differences $u = x_0 + r$ and $v = y_0 + s$ to all verifiers. Then the verifiers can locally compute $[x_0] := u - [r]$ and $[y_0] := v - [s]$. $\mathcal{P}$ and all verifiers can generate a random challenge $\alpha = \mathsf{H}'(\chi, u, v, i, j)$, where χ is another random challenge related to the secrets $\{(x_h, y_h)\}_{h \in [1,m]}$. Then, the verifiers can now check correctness of the following wire tuple:

$$[\boldsymbol{x}] := \sum_{h=1}^{m} \alpha^h \cdot [\boldsymbol{x}_h] + [\boldsymbol{x}_0], \ [\boldsymbol{y}] := \sum_{h=1}^{m} \alpha^h \cdot [\boldsymbol{y}_h] + [\boldsymbol{y}_0].$$

This check can be done by letting the verifiers open $([\boldsymbol{x}], [\boldsymbol{y}])$ and check $x_i = y_j$. When streaming the strong NIMVZK protocol, the verification of wire tuples do *not* increase the rounds between verifiers (see Sect. 6.1 for details).

Verification of PSS-Based Inner-Product Tuples. Once we enable Fiat-Shamir as shown in Sect. 3.2, we also get another benefit that now the challenge χ is also known to the prover $\mathcal{P}$. Thus, we can non-interactively transform the verification of PSS-based multiplication tuples into that of a packed inner-product tuple. We present a *non-interactive* technique to verify the correctness of a packed inner-product tuple, which is inspired by prior work [14,16,39–41]. We also adapt the technique by Baum et al. [8] from the DVZK setting to the MVZK setting in order to further improve computational efficiency. Our verification approach has lower round complexity than that used in PSS-based MPC [39] (one round vs logarithm rounds). At a high level, our protocol for verifying correctness of a packed inner-product tuple works as follows:

1. Suppose that all verifiers hold the shares of a dimension-M packed inner-product tuple $(([\boldsymbol{x}_1], \ldots, [\boldsymbol{x}_M]), ([\boldsymbol{y}_1], \ldots, [\boldsymbol{y}_M]), [\boldsymbol{z}])$, where $\{\boldsymbol{x}_i, \boldsymbol{y}_i\}_{i \in [1,M]}$ and $\boldsymbol{z}$ are secret vectors in $\mathbb{K}^k$. Prover $\mathcal{P}$ knows all the secret vectors, and wants to prove $\boldsymbol{z} = \sum_{i \in [1,M]} \boldsymbol{x}_i * \boldsymbol{y}_i$ where $*$ denotes the component-wise product.

2. The verifiers *recursively* reduce the dimension of $(([\boldsymbol{x}_1], \ldots, [\boldsymbol{x}_M]), ([\boldsymbol{y}_1], \ldots, [\boldsymbol{y}_M]), [\boldsymbol{z}])$ to 2. This is performed by splitting a packed inner-product tuple $(([\boldsymbol{x}_1], \ldots, [\boldsymbol{x}_m]), ([\boldsymbol{y}_1], \ldots, [\boldsymbol{y}_m]), [\boldsymbol{z}])$ into two inner-product tuples $(([\boldsymbol{a}_{1,1}], \ldots, [\boldsymbol{a}_{1,\ell}]), ([\boldsymbol{b}_{1,1}], \ldots, [\boldsymbol{b}_{1,\ell}]), [c_1])$ and $(([\boldsymbol{a}_{2,1}], \ldots, [\boldsymbol{a}_{2,\ell}]), ([\boldsymbol{b}_{2,1}], \ldots, [\boldsymbol{b}_{2,\ell}]), [c_2])$, where $\ell = m/2$ and $[\boldsymbol{z}] = [c_1] + [c_2]$. Then, we use a protocol to compress the two packed inner-product tuples into one inner-product tuple.

3. To realize the above splitting step, $\mathcal{P}$ can directly distribute the shares of $[c_1] = [\sum_{h \in [1,\ell]} \boldsymbol{a}_{1,h} * \boldsymbol{b}_{1,h}]$ to all verifiers. However, this does not support Fiat-Shamir, as no public message is available. Instead, we let $\mathcal{P}$ distribute the shares of a random packed sharing $[\boldsymbol{r}]$, and then broadcast a public message $\boldsymbol{u} = c_1 + \boldsymbol{r}$ to all verifiers, who can locally compute $[c_1] := \boldsymbol{u} - [\boldsymbol{r}]$. Then, the verifiers can locally compute $[c_2] := [\boldsymbol{z}] - [c_1]$.

4. We adopt the polynomial approach to compress two packed inner-product tuples $(([\boldsymbol{a}_{1,1}], \ldots, [\boldsymbol{a}_{1,\ell}]), ([\boldsymbol{b}_{1,1}], \ldots, [\boldsymbol{b}_{1,\ell}]), [c_1])$ and $(([\boldsymbol{a}_{2,1}], \ldots, [\boldsymbol{a}_{2,\ell}]), ([\boldsymbol{b}_{2,1}], \ldots, [\boldsymbol{b}_{2,\ell}]), [c_2])$ into a single tuple $(([\boldsymbol{x}_1], \ldots, [\boldsymbol{x}_\ell]), ([\boldsymbol{y}_1], \ldots, [\boldsymbol{y}_\ell]), [\boldsymbol{z}])$, which has been used in prior work such as [39]. Differently, we will use the Fiat-Shamir transform to realize the non-interactive compression. Specifically, the parties compute the sharings of polynomials $[\boldsymbol{f}_j(\cdot)], [\boldsymbol{g}_j(\cdot)]$ for $j \in [1, \ell]$ and $[\boldsymbol{h}(\cdot)]$, such that $\boldsymbol{f}_j(i) = \boldsymbol{a}_{i,j}, \boldsymbol{g}_j(i) = \boldsymbol{b}_{i,j}$ and $\boldsymbol{h}(i) = c_i$ for $i \in [1, 2]$. Then $\mathcal{P}$ needs to convince the verifiers that $\boldsymbol{h}(X) = \sum_{j \in [1,\ell]} \boldsymbol{f}_j(X) * \boldsymbol{g}_j(X)$, which can be realized by proving $\boldsymbol{h}(\alpha) = \sum_{j \in [1,\ell]} \boldsymbol{f}_j(\alpha) * \boldsymbol{g}_j(\alpha)$ for a random challenge α. $\mathcal{P}$ and all verifiers can generate α by computing $\mathsf{H}'(\gamma, msg)$ where γ is the challenge used in the previous iteration and msg is the public message sent in the

current iteration. Now, the parties can define $([\boldsymbol{x}_j] = [\boldsymbol{f}_j(\alpha)], [\boldsymbol{y}_j] = [\boldsymbol{g}_j(\alpha)])$ for $j \in [1, \ell]$ and $[\boldsymbol{z}] = [\boldsymbol{h}(\alpha)]$, and execute the next iteration.

5. Let $(([\boldsymbol{x}_1], [\boldsymbol{x}_2]), ([\boldsymbol{y}_1], [\boldsymbol{y}_2]), [\boldsymbol{z}])$ be the packed inner-product tuple after the dimension reduction was completed. We can adapt the randomization technique [14,39] to check the correctness of this tuple. In the same way, we can split it into two multiplication tuples $([\boldsymbol{x}_1], [\boldsymbol{y}_1], [\boldsymbol{z}_1])$ and $([\boldsymbol{x}_2], [\boldsymbol{y}_2], [\boldsymbol{z}_2])$ with $[\boldsymbol{z}] = [\boldsymbol{z}_1] + [\boldsymbol{z}_2]$. The prover $\mathcal{P}$ can distribute the shares of a random multiplication tuple $([\boldsymbol{x}_0], [\boldsymbol{y}_0], [\boldsymbol{z}_0])$ with $\boldsymbol{z}_0 = \boldsymbol{x}_0 * \boldsymbol{y}_0$ to all verifiers in the way compatible with Fiat-Shamir. Then, $\mathcal{P}$ and the verifiers can compress $\{([\boldsymbol{x}_i], [\boldsymbol{y}_i], [\boldsymbol{z}_i])\}_{i \in [0,2]}$ into $([\boldsymbol{x}], [\boldsymbol{y}], [\boldsymbol{z}])$. All verifiers can run the Open procedure to obtain $(\boldsymbol{x}, \boldsymbol{y}, \boldsymbol{z})$ and check that $\boldsymbol{z} = \boldsymbol{x} * \boldsymbol{y}$.

Streaming Strong NIMVZK with the Same Round Complexity. We can use the strong NIMVZK protocol to prove a very large circuit in a streamable way, such that between the prover and verifiers are non-interactive for proving a batch of $N = k \cdot M$ multiplication gates each time, and the verifiers *still* communicate only one round for proving the whole circuit. For a batch of $N = k \cdot M$ multiplication gates, the parties can transform M PSS-based multiplication tuples into a packed inner-product tuple with dimension M, and then compress it into a packed inner-product tuple denoted by $\mathsf{IPtuple}_1$ with dimension $M/2^c$ for some integer $c \geq 1$. For another batch of multiplication gates, the parties can generate another packed inner-product tuple $\mathsf{IPtuple}_2$ with dimension $M/2^c$ in the same way. Then, the prover and verifiers can compress $\mathsf{IPtuple}_1$ and $\mathsf{IPtuple}_2$ into a packed inner-product tuple $\mathsf{IPtuple}_3$ with the same dimension $M/2^c$, where the challenge α for this compression is computed with random oracle H' and two challenges to obtain $\mathsf{IPtuple}_1$ and $\mathsf{IPtuple}_2$. After the whole circuit has been evaluated, the verifiers can check correctness of the final packed inner-product tuple (with dimension $M/2^c$) stored in memory by communicating only one round. As a result, all parties only need memory linear to what is needed to evaluate the statement in the clear.

4   Information-Theoretic NIMVZK Proof

We present a non-interactive multi-verifier zero-knowledge (NIMVZK) protocol with information-theoretic security in the $(\mathcal{F}_{\mathsf{coin}}, \mathcal{F}_{\mathsf{verifyprod}})$-hybrid model, assuming an honest majority of verifiers, where $\mathcal{F}_{\mathsf{coin}}$ is a coin-tossing functionality shown in the full version [52]. Functionality $\mathcal{F}_{\mathsf{verifyprod}}$ allows to verify the correctness of an inner-product tuple secretly shared among verifiers. It is possible to instantiate $\mathcal{F}_{\mathsf{verifyprod}}$ using prior work on fully linear PCP (or IOP) [14], but we can improve its communication (or rounds) by adapting the technique by Goyal et al. [40,41] in the MPC setting to the MVZK setting.

Functionality $\mathcal{F}_{\text{verifyprod}}$

This functionality runs with a prover $\mathcal{P}$ and n verifiers $\mathcal{V}_1, \ldots, \mathcal{V}_n$. Let $([\boldsymbol{x}], [\boldsymbol{y}], [z])$ be an inner-product tuple defined over a field $\mathbb{K}$, where the dimension of vectors $\boldsymbol{x}, \boldsymbol{y}$ is N. This functionality operates as follows:

1. Upon receiving the shares of $[\boldsymbol{x}], [\boldsymbol{y}], [z]$ from honest verifiers, run as follows:
   - Reconstruct the secret vectors $\boldsymbol{x}, \boldsymbol{y} \in \mathbb{F}^N$ and secret $z \in \mathbb{F}$ from these shares.
   - Compute the shares of $[\boldsymbol{x}], [\boldsymbol{y}], [z]$ held by corrupted verifiers, and send them to the adversary.
   - If $\mathcal{P}$ is corrupted, send $(\boldsymbol{x}, \boldsymbol{y}, z)$ to the adversary.
2. If $z \neq \boldsymbol{x} \odot \boldsymbol{y}$, then set $b := \text{abort}$, otherwise set $b := \text{accept}$. Send b to the adversary. For $i \in \mathcal{H}$, wait for an input from the adversary, and do the following:
   - If it is continue_i, send b to $\mathcal{V}_i$.
   - If it is abort_i, send abort to $\mathcal{V}_i$.

Fig. 2. Zero-knowledge verification functionality for an inner-product tuple.

4.1   From General Adversaries to Maximal Adversaries for MVZK

Before we describe the NIMVZK protocol, we prove an important lemma that can be used to simplify the proofs of the MVZK protocols in this paper and the future works. Informally, this lemma states that if an MVZK protocol is secure against *exactly* t malicious verifiers, then the protocol is also secure against *at most* t malicious verifiers. The proof of this lemma is based on that of a similar lemma for honest-majority MPC by Genkin et al. [33]. This lemma allows us to only consider the maximum adversaries who corrupt exactly t verifiers, and thus simplifies the security proofs of MVZK protocols. One caveat is that the proof of this lemma needs to specially deal with the case that the honest verifiers will receive output as well as the possible random-oracle queries (e.g., the Fiat-Shamir transform [29] is used).

Lemma 1. *Let Π be an MVZK protocol proving the satisfiability of a circuit C for $n \geq 2t+1$ verifiers. Then, if protocol Π securely realizes $\mathcal{F}_{\text{mvzk}}$ in the presence of any malicious adversary corrupting exactly t verifiers, then Π securely realizes $\mathcal{F}_{\text{mvzk}}$ against any malicious adversary corrupting at most t verifiers.*

The proof of the above lemma is given in the full version [52]. The above lemma can be applied to not only our information-theoretic NIMVZK protocol but also the strong NIMVZK proofs in the computational setting that will be described in Sect. 5 and Sect. 6.

4.2   Our Information-Theoretic NIMVZK Protocol

In Fig. 3, we describe the detailed NIMVZK protocol with information theoretic security in the $(\mathcal{F}_{\text{coin}}, \mathcal{F}_{\text{verifyprod}})$-hybrid model. For each circuit-input gate or

---

Protocol $\Pi_{\mathsf{nimvzk}}^{\mathsf{it}}$

Inputs: A prover $\mathcal{P}$ holds a witness $\boldsymbol{w} \in \mathbb{F}^m$. $\mathcal{P}$ and all verifiers $\mathcal{V}_1, \ldots, \mathcal{V}_n$ hold an arithmetic circuit C over a field $\mathbb{F}$ with $|\mathbb{F}| > n$. Let N denote the number of multiplication gates in the circuit. $\mathcal{P}$ will convince the verifiers that $C(\boldsymbol{w}) = 0$.

Circuit evaluation: In a predetermined topological order, $\mathcal{P}$ and all verifiers evaluate the circuit as follows:

- For each circuit-input wire with input value $w \in \mathbb{F}$, $\mathcal{P}$ (acting as the dealer) runs $[w] \leftarrow \mathsf{Share}(w)$ to distribute the shares to all verifiers.
- For each addition gate with input sharings $[x]$ and $[y]$, all verifiers locally compute $[z] := [x] + [y]$, and $\mathcal{P}$ computes $z := x + y \in \mathbb{F}$.
- For each multiplication gate with input values $x, y \in \mathbb{F}$, $\mathcal{P}$ computes $z := x \cdot y \in \mathbb{F}$, and then executes $[z] \leftarrow \mathsf{Share}(z)$ which distributes the shares to all verifiers.

Verification of multiplication gates: Let $([x_i], [y_i], [z_i])$ be the sharings on the i-th multiplication gate for $i \in [1, N]$. All verifiers and $\mathcal{P}$ execute as follows:

1. The verifiers call the coin-tossing functionality $\mathcal{F}_{\mathsf{coin}}$ to generate a random element $\chi \in \mathbb{K}$, and then set the following inner-product tuple:

$$[\boldsymbol{x}] := \left([x_1], \ldots, \chi^{N-1} \cdot [x_N]\right), \ [\boldsymbol{y}] := ([y_1], \ldots, [y_N]), \ [z] := \sum_{i \in [1,N]} \chi^{i-1} \cdot [z_i].$$

2. The verifiers and $\mathcal{P}$ call functionality $\mathcal{F}_{\mathsf{verifyprod}}$ on $([\boldsymbol{x}], [\boldsymbol{y}], [z])$ to check that $z = \boldsymbol{x} \odot \boldsymbol{y}$. If the verifiers receive **abort** from $\mathcal{F}_{\mathsf{verifyprod}}$, then they abort.

Verification of circuit output: Let $[\eta]$ be the input sharing associated with the single circuit-output gate. All verifiers execute $\eta \leftarrow \mathsf{Open}([\eta])$, and abort if outputting **abort** in the **Open** procedure. If $\eta = 0$, then the verifiers output **true**, otherwise they output **false**.

Fig. 3. Information-theoretic NIMVZK protocol in the $(\mathcal{F}_{\mathsf{coin}}, \mathcal{F}_{\mathsf{verifyprod}})$-hybrid model.

multiplication gate, the prover directly shares the output value to all verifiers. The verifiers can locally compute the shares on the output wires of addition gates. Then, all verifiers check the correctness of all multiplication gates by transforming multiplication triples into an inner-product tuple and then calling functionality $\mathcal{F}_{\mathsf{verifyprod}}$. In parallel, the verifiers also check correctness of the single circuit-output gate via running the **Open** procedure.

In Fig. 2, we give the precise definition of functionality $\mathcal{F}_{\mathsf{verifyprod}}$. In particular, if the prover is honest, the adversary can only obtain the shares of corrupted verifiers from this functionality, which does not reveal any information on the secrets. In other words, this functionality naturally captures zero-knowledge. If the prover is corrupted, this functionality reveals all secrets to the adversary, as the secrets have been known anyway by the adversary. We can view deciding the correctness of an inner-product tuple as a statement which is shared among n verifiers. Functionality $\mathcal{F}_{\mathsf{verifyprod}}$ guarantees that the malicious prover cannot make any honest verifier accept a false statement, and thus captures soundness. In the full version [52], we present an efficient protocol to securely realize $\mathcal{F}_{\mathsf{verifyprod}}$, where the communication and round complexities are $O((n + \tau) \log_\tau |C|)$ field elements per verifier and $\log_\tau |C| + 3$ rounds between verifiers respectively, where $\tau \geq 2$ is a parameter.

Theorem 1. *Protocol $\Pi_{\mathsf{nimvzk}}^{\mathsf{it}}$ shown in Fig. 3 securely realizes functionality $\mathcal{F}_{\mathsf{mvzk}}$ with information-theoretic security and soundness error $\frac{N-1}{|\mathbb{K}|}$ in the $(\mathcal{F}_{\mathsf{coin}}, \mathcal{F}_{\mathsf{verifyprod}})$-hybrid model in the presence of a malicious adversary corrupting up to a prover and t verifiers.*

The proof of this theorem can be found in the full version [52].

5   Strong NIMVZK Proof in the Honest-Majority Setting

In this section, we present a strong NIMVZK proof based on the Fiat-Shamir transform, where a minority of verifiers are allowed to be corrupted and collude with the prover. Our strong NIMVZK protocol adopts a non-interactive commitment based on random oracle to non-interactively transform the verification of multiplication triples into the verification of an inner-product tuple. This protocol still works in the $\mathcal{F}_{\mathsf{verifyprod}}$-hybrid model, where functionality $\mathcal{F}_{\mathsf{verifyprod}}$ can now be non-interactively realized using the Fiat-Shamir transform.

In Fig. 4, we describe the strong NIMVZK protocol $\Pi_{\mathsf{snimvzk}}^{\mathsf{fs}}$ in the $\mathcal{F}_{\mathsf{verifyprod}}$-hybrid model, where the shares are computed over a field $\mathbb{F}$ and the verification of multiplication gates is performed over an extension field $\mathbb{K}$ with $|\mathbb{K}| \geq 2^\lambda$. The strong NIMVZK protocol is the same as the protocol shown in Fig. 3, except for the verification of multiplication gates. In the strong NIMVZK protocol, the verification of multiplication gates is executed non-interactively using a non-interactive commitment based on a random oracle H_1, where a commitment com on a message x is defined as $\mathsf{H}_1(x, r)$ for a randomness $r \in \{0, 1\}^\lambda$. However, we do *not* require that the commitment is binding. Instead, we only need the commitment to be hard to find a pair (x', r') such that $\mathsf{H}_1(x', r') = \mathsf{H}_1(x, r)$ and $x' \neq x$, after $\mathsf{H}_1(x, r)$ has been defined. This has been explained in Sect. 3.2 (see the proof of Theorem 2 for details). The random challenge $\chi \in \mathbb{K}$ is now generated using another random oracle H_2 and the public commitments, instead of calling $\mathcal{F}_{\mathsf{coin}}$. In this case, the prover can compute the secrets $(\boldsymbol{x}, \boldsymbol{y}, z)$ underlying the inner-product tuple using the public coefficient χ and the secret wire values. At first glance, the secrets $(\boldsymbol{x}, \boldsymbol{y}, z)$ seem to be useless for the protocol execution of $\Pi_{\mathsf{snimvzk}}^{\mathsf{fs}}$. Nevertheless, the prover can use $(\boldsymbol{x}, \boldsymbol{y}, z)$ to compute all the secrets involved in the protocol that securely realizes functionality $\mathcal{F}_{\mathsf{verifyprod}}$. In this case, we can securely compute $\mathcal{F}_{\mathsf{verifyprod}}$ in a strongly non-interactive way by making the prover distribute the shares of all secrets non-interactively and all verifiers interact only one round for Open.

Protocol $\Pi_{\text{snimvzk}}^{\text{fs}}$

Inputs: A prover $\mathcal{P}$ holds a witness $\boldsymbol{w} \in \mathbb{F}^m$. $\mathcal{P}$ and the verifiers $\mathcal{V}_1, \ldots, \mathcal{V}_n$ hold an arithmetic circuit C over a field $\mathbb{F}$ with $|\mathbb{F}| > n$ such that $C(\boldsymbol{w}) = 0$. Let N denote the number of multiplication gates in the circuit. Let $\mathsf{H}_1 : \{0,1\}^* \to \{0,1\}^\lambda$ and $\mathsf{H}_2 : \{0,1\}^* \to \mathbb{K}$ be two random oracles.

Circuit evaluation: $\mathcal{P}$ and all verifiers evaluate the circuit in the same way as described in Figure 3.

Verification of multiplication gates: For all m circuit-input wires, the verifiers $\mathcal{V}_1, \ldots, \mathcal{V}_n$ hold the shares of $[w_1], \ldots, [w_m]$, and $\mathcal{P}$ has the witness $\boldsymbol{w} = (w_1, \ldots, w_m)$. For all N multiplication gates, $\mathcal{P}$ and the verifiers respectively hold the secrets and the shares of multiplication triples $([x_1], [y_1], [z_1]), \ldots, ([x_N], [y_N], [z_N])$. For $j \in [1, m]$, let $w_j^1, \ldots, w_j^n$ be the shares of $[w_j]$ held by all verifiers, which are also known by $\mathcal{P}$. For $j \in [1, N]$, let $z_j^1, \ldots, z_j^n$ be the shares of $[z_j]$, which are also obtained by $\mathcal{P}$. All verifiers $\mathcal{V}_1, \ldots, \mathcal{V}_n$ and $\mathcal{P}$ execute as follows:

1. For each $i \in [1, n]$, $\mathcal{P}$ samples $r_i \leftarrow \{0,1\}^\lambda$ and computes

$$com_i := \mathsf{H}_1(w_1^i, \ldots, w_m^i, z_1^i, \ldots, z_N^i, r_i).$$

   Then, $\mathcal{P}$ broadcasts $(com_1, \ldots, com_n)$ to all verifiers, and also sends r_i to every verifier $\mathcal{V}_i$ over a private channel.

2. Every verifier $\mathcal{V}_i$ checks that $com_i = \mathsf{H}_1(w_1^i, \ldots, w_m^i, z_1^i, \ldots, z_N^i, r_i)$, and aborts if the check fails.

3. $\mathcal{P}$ and all verifiers compute $\chi := \mathsf{H}_2(com_1, \ldots, com_n) \in \mathbb{K}$.

4. $\mathcal{P}$ and all verifiers respectively compute the secrets and the shares of the following inner-product tuple:

$$[\boldsymbol{x}] := ([x_1], \ldots, \chi^{N-1} \cdot [x_N]), \quad [\boldsymbol{y}] := ([y_1], \ldots, [y_N]), \quad [z] := \sum_{i \in [1,N]} \chi^{i-1} \cdot [z_i].$$

5. The verifiers and $\mathcal{P}$ call functionality $\mathcal{F}_{\text{verifyprod}}$ on $([\boldsymbol{x}], [\boldsymbol{y}], [z])$ to check that $z = \boldsymbol{x} \odot \boldsymbol{y}$. If the verifiers receive **abort** from $\mathcal{F}_{\text{verifyprod}}$, then they abort.

Verification of circuit output: All verifiers check the correctness of a single circuit-output gate in the same way as shown in Figure 3.

Fig. 4. Strong non-interactive MVZK protocol in the $\mathcal{F}_{\text{verifyprod}}$-hybrid model and random oracle model.

Theorem 2. *Let H_1 and H_2 be two random oracles. Protocol $\Pi_{\text{snimvzk}}^{\text{fs}}$ shown in Fig. 4 securely realizes functionality $\mathcal{F}_{\text{mvzk}}$ with soundness error at most $\frac{Q_1 n + (Q_2 + 1)N}{2^\lambda}$ in the $\mathcal{F}_{\text{verifyprod}}$-hybrid model in the presence of a malicious adversary corrupting up to a prover and t verifiers, where Q_1 and Q_2 are the number of queries to random oracles H_1 and H_2 respectively.*

The proof of the above theorem is given in the full version [52].

Optimizations. For Shamir secret sharing , $\mathcal{P}$ can send a random $\mathsf{seed}_i \in \{0,1\}^\lambda$ to $\mathcal{V}_i$ for each $i \in [1,t]$, who computes all its shares with seed_i and a pseudo-random generator (PRG). This reduces the communication by a half. Furthermore, for each $i \in [1,t]$, $\mathcal{P}$ can send $com_i = \mathsf{H}_1(\mathsf{seed}_i, r_i)$ to $\mathcal{V}_i$, who checks the correctness of com_i using seed_i and r_i. This will reduce the computational cost of generating and verifying t commitments. Using the optimization, the communication among verifiers is *asymmetry*. In particular, among all verifiers, t verifiers $\mathcal{V}_1, \dots, \mathcal{V}_t$ only has a *sublinear* communication complexity. That is, each verifier only receives $O(n + \log|C|)$ field elements from the prover, and only needs to send $O(n)$ field elements to other verifiers. It makes our strong NIMVZK protocol particularly suitable for the applications where $\mathcal{V}_1, \dots, \mathcal{V}_t$ are lower-resource mobile devices and the other verifiers are powerful servers.

Strong NIMVZK Proof for Inner-Product Tuples. Boneh et al. [14] introduced a powerful tool, called distributed zero-knowledge (DZK) proof (a.k.a., ZK proof on a distributed or secret-shared statement), to prove the inner-product statements (and other useful statements). We can use their DZK proof with logarithmic communication to securely realize functionality $\mathcal{F}_{\mathsf{verifyprod}}$ shown in Fig. 2. When applying the Fiat-Shamir transform [16] into their DZK proof, the prover non-interactively sends a proof to all verifiers, and the verifiers execute one-round communication to verify correctness of an inner-product tuple. Note that the proof on the inner-product statement can be sent in parallel with our proof on circuit satisfiability shown in Fig. 4. Therefore, using the DZK proof to instantiate $\mathcal{F}_{\mathsf{verifyprod}}$, our MVZK protocol is strongly non-interactive. While Boneh et al. [14] originally instantiated the DZK proof with replicated secret sharing, Boyle et al. [16] shown that their DZK proof also works for verifiable Shamir secret sharing meaning that a consistency check is needed to guarantee either all verifiers hold a consistent sharing of the secret or honest verifiers abort.

We can simplify the technique by Boneh et al. [14] by avoiding the use of verifiable secret sharing, and slightly optimize the communication from $4.5 \log|C| + 5n$ field elements to $3 \log|C| + 3n$ field elements. We can also improve the hash computation cost for Fiat-Shamir. The improved approach has been described in Sect. 3 by considering Shamir secret sharing as a special case of packed secret sharing. The detailed protocol to strongly non-interactively realize $\mathcal{F}_{\mathsf{verifyprod}}$ can be directly obtained by simplifying the PSS-based protocol $\Pi_{\mathsf{verifyprod}}^{\mathsf{pss}}$ shown in Fig. 8 of Sect. 6.2 via setting the number of packed secrets $k = 1$.

6   Strong NIMVZK Proof with Lower Communication

Based on packed secret sharing (PSS), we present a strong NIMVZK proof with communication complexity $O(|C|/n)$ per verifier, when the threshold of corrupted verifiers $t < n(1/2 - \epsilon)$ for any $0 < \epsilon < 1/2$. Our strong NIMVZK protocol is highly efficient for proving satisfiability of a *single generic* circuit. In the ZK setting, we use PSS optimally. In particular, we eliminate the constraints in the state-of-the-art PSS-based MPC protocol [39] including: 1) evaluating a circuit layer-by-layer, 2) interactively permuting the secrets in a single packed sharing,

---

Functionality $\mathcal{F}_{\text{verifyprod}}^{\text{pss}}$

The packed inner-product tuple over a field $\mathbb{K}$ is denoted by $([\boldsymbol{x}_1], \ldots, [\boldsymbol{x}_\ell])$, $([\boldsymbol{y}_1], \ldots, [\boldsymbol{y}_\ell])$ and $[\boldsymbol{z}]$. This functionality runs with a prover $\mathcal{P}$ and n verifiers $\mathcal{V}_1, \ldots, \mathcal{V}_n$, and operates as follows:

1. Upon receiving the shares of $\{[\boldsymbol{x}_i], [\boldsymbol{y}_i]\}_{i \in [1, \ell]}$ and $[\boldsymbol{z}]$ from all honest verifiers, execute the following:
   - Reconstruct the secrets $(\boldsymbol{x}_1, \ldots, \boldsymbol{x}_\ell)$, $(\boldsymbol{y}_1, \ldots, \boldsymbol{y}_\ell)$ and $\boldsymbol{z}$ from the shares of honest verifiers in $\mathcal{H}_H$.
   - Compute the shares of corrupted verifiers on $([\boldsymbol{x}_1], \ldots, [\boldsymbol{x}_\ell])$, $([\boldsymbol{y}_1], \ldots, [\boldsymbol{y}_\ell])$ and $[\boldsymbol{z}]$, and then send these shares to the adversary.
   - If $\mathcal{P}$ is corrupted, send the shares of $(\{[\boldsymbol{x}_i], [\boldsymbol{y}_i]\}_{i \in [1, \ell]}, [\boldsymbol{z}])$ held by honest verifiers in $\mathcal{H}$ and the secrets $(\{\boldsymbol{x}_i, \boldsymbol{y}_i\}_{i \in [1, \ell]}, \boldsymbol{z})$ to the adversary.

2. If $\boldsymbol{z} \neq \sum_{h \in [1, \ell]} \boldsymbol{x}_h * \boldsymbol{y}_h$ where $*$ denotes the component-wise product, then set $b := \mathsf{abort}$, otherwise set $b := \mathsf{accept}$. Send b to the adversary. For $i \in \mathcal{H}$, wait for an input from the adversary, and do the following:
   - If it is $\mathsf{continue}_i$, send b to $\mathcal{V}_i$.
   - If it is abort_i, send abort to $\mathcal{V}_i$.

Fig. 5. ZK verification functionality for packed inner-product tuples.

3) interactively collecting the secrets from different packed sharings and 4) splitting an output wire into multiple output wires, where all these constraints will make the rounds and communication cost significantly larger than our protocol.

Firstly, we discuss how to transform a general circuit C into another circuit C' with the same output and $|C'| = |C| + O(k)$, such that 1) the number of circuit-input wires, addition gates and multiplication gates is the multiple of k; 2) there are at least k circuit-output wires; 3) the gates with the same type are divided into groups of k. This is done by adding "dummy" wires and gates, and is described in the full version [52]. Then, we present the detailed strong NIMVZK protocol in the $\mathcal{F}_{\text{verifyprod}}^{\text{pss}}$-hybrid model, where $\mathcal{F}_{\text{verifyprod}}^{\text{pss}}$ verifies the correctness of a packed inner-product tuple. Next, we present a strong non-interactive MVZK protocol to securely realize functionality $\mathcal{F}_{\text{verifyprod}}^{\text{pss}}$.

6.1 Strong NIMVZK Based on Packed Secret Sharing

Before showing the detailed strong NIMVZK protocol, we give the definition of functionality $\mathcal{F}_{\text{verifyprod}}^{\text{pss}}$.

Functionality for Verifying Packed Inner-Product Tuples. Let $\mathcal{H}_H \subset \mathcal{H}$ be a fixed $(d+1)$-sized subset of honest verifiers and $\mathcal{H}_C = \mathcal{H} \backslash \mathcal{H}_H$, where recall that $\mathcal{H}$ is the set of all $d + k$ honest verifiers and d is the degree of polynomials for PSS. For a degree-d packed sharing $[\boldsymbol{x}]$, we use $[\boldsymbol{x}]_\mathcal{H}$ to denote the whole sharing that is reconstructed from the shares of honest verifiers in $\mathcal{H}_H$. Given the shares of honest verifiers in $\mathcal{H}$ as input, we can reconstruct the *whole* sharing $[\boldsymbol{x}]$ as follows:

1. Use the $d+1$ shares of honest verifiers in $\mathcal{H}_H$ to reconstruct the whole sharing $[\boldsymbol{x}]_{\mathcal{H}}$. Define the shares of corrupted verifiers on $[\boldsymbol{x}]$ as that on $[\boldsymbol{x}]_{\mathcal{H}}$. Following prior MPC work [39], we always assume that the corrupted verifiers in $\mathcal{C}$ hold the correct shares that they should hold, while they may use incorrect shares during the protocol execution, where $\mathcal{C}$ is the set of corrupted verifiers.
2. Define the secrets of $[\boldsymbol{x}]$ to be that of $[\boldsymbol{x}]_{\mathcal{H}}$.
3. Define the shares of $[\boldsymbol{x}]$ held by honest verifiers in $\mathcal{H}$ as the shares input by the verifiers directly.

The zero-knowledge verification functionality for packed inner-product tuples is shown in Fig. 5. This functionality takes as input a packed inner-product tuple and then checks correctness of the tuple, where each sharing packs k secrets. This functionality sends the shares of corrupted verifiers for each packed sharing to the adversary, where the shares are computed by the above approach based on the shares of honest verifiers in $\mathcal{H}_H$. If the prover is corrupted, this functionality also sends the shares of all honest verifiers and the secrets in all packed sharings to the adversary, as these shares and secrets have been known by the adversary.

PSS-Based Strong NIMVZK Protocol from the Fiat-Shamir Transform. Our PSS-based strong non-interactive MVZK protocol in the $\mathcal{F}_{\text{verifyprod}}^{\text{pss}}$-hybrid model is described in Figs. 6 and 7, where the circuit is defined over a field $\mathbb{F}$ and the verification is performed over an extension field $\mathbb{K}$ with $|\mathbb{K}| \geq 2^{\lambda}$.

The prover and all verifiers first transform the circuit C into an *equivalent* circuit C', which satisfies the requirements of packed secret sharings. For an input vector $\boldsymbol{w} \in \mathbb{F}^k$, if the j-th secret of $\boldsymbol{w}$ for $j \in [1, k]$ corresponds to a dummy circuit-input wire, the secret is set as 0. If $\boldsymbol{w}$ corresponds to k dummy circuit-input wires, then $\boldsymbol{w} = 0^k$ and $[\boldsymbol{w}]$ can be *locally* generated by all verifiers without any communication (as shown in the full version [52]).

Using the non-interactive commitment based on a random oracle, we adopt a similar approach as described in the previous section to transform the check of multiplication tuples into the check of a packed inner-product tuple. Then, by calling functionality $\mathcal{F}_{\text{verifyprod}}^{\text{pss}}$, the verifiers can check correctness of the packed inner-product tuple. Note that the prover can compute the secrets of the packed inner-product tuple, which will be useful for designing a strongly non-interactive protocol to securely realize $\mathcal{F}_{\text{verifyprod}}^{\text{pss}}$ as shown in Sect. 6.2.

During the protocol execution, we need to check the consistency of some secrets stored in two different packed sharings. We perform the consistency check using the random-linear-combination approach based on the Fiat-Shamir transform, which is inspired by the recent checking approach by Goyal et al. [39] for information-theoretic MPC. In particular, the prover will generate a packed input sharing $[\boldsymbol{y}]$ on k multiplication gates, addition gates or circuit-output gates, such that the j-th secret y_j of $[\boldsymbol{y}]$ comes from the i-th secret x_i of a packed output sharing $[\boldsymbol{x}]$ of k circuit-input gates, multiplication gates or addition gates. We need to check that $x_i = y_j$ to guarantee the consistency of y_j. This corresponds to the wire which carries the value $x_i = y_j$ in the circuit. We refer to a tuple $([\boldsymbol{x}], [\boldsymbol{y}], i, j)$ as a *wire tuple* following prior work [39]. We perform the

---

Protocol $\Pi^{\text{pss}}_{\text{snimvzk}}$

Inputs: A prover $\mathcal{P}$ holds a witness $\bar{w}$. $\mathcal{P}$ and n verifiers $\mathcal{V}_1, \ldots, \mathcal{V}_n$ hold a circuit C over a field $\mathbb{F}$. Let k denote the number of secrets that are packed in a single sharing. Let m be the number of packed sharings on circuit-input wires, where each packs k circuit-input gates. Let $M = \lceil |C|/k \rceil$ denote the number of multiplication tuples where each packs k multiplication gates. Let $N = O(|C|/k)$ represent the number of wire tuples in the form of $([\boldsymbol{x}], [\boldsymbol{y}], i, j)$ with $x_i = y_j$. Let $\mathsf{H}_1 : \{0, 1\}^* \to \{0, 1\}^\lambda$ and $\mathsf{H}_2 : \{0, 1\}^* \to \mathbb{K}$ be two random oracles.

Preprocess circuit: All parties run the $\mathsf{PrepCircuit}$ procedure shown in the full version [52] to preprocess the circuit C, and obtain an equivalent circuit C' such that $C'(\bar{w}) = C(\bar{w})$ for any input $\bar{w}$.

Circuit evaluation: In a predetermined topological order, $\mathcal{P}$ and all verifiers evaluate the circuit C' as follows:

- Let $\boldsymbol{w}_1, \ldots, \boldsymbol{w}_m \in \mathbb{F}^k$ be the secret vectors where each associates with k circuit-input wires. For each group of k circuit-input wires with an input vector $\boldsymbol{w} \in \{\boldsymbol{w}_1, \ldots, \boldsymbol{w}_m\}$, $\mathcal{P}$ runs $[\boldsymbol{w}] \leftarrow \mathsf{Share}(\boldsymbol{w})$ to distribute the shares to all verifiers.
- For each group of k addition gates with input sharings $[\boldsymbol{x}]$ and $[\boldsymbol{y}]$, all verifiers *locally* compute $[\boldsymbol{z}] := [\boldsymbol{x}] + [\boldsymbol{y}]$, and $\mathcal{P}$ computes $\boldsymbol{z} := \boldsymbol{x} + \boldsymbol{y} \in \mathbb{F}^k$.
- For each group of k multiplication gates with input sharings $[\boldsymbol{x}]$ and $[\boldsymbol{y}]$, $\mathcal{P}$ computes $\boldsymbol{z} := \boldsymbol{x} * \boldsymbol{y} \in \mathbb{F}^k$ where $*$ denotes the component-wise product, and executes $[\boldsymbol{z}] \leftarrow \mathsf{Share}(\boldsymbol{z})$ which distributes the shares to all verifiers.
- For each group of k input wires of multiplication, addition, or circuit-output gates, such that the corresponding input vector $\boldsymbol{y} \in \mathbb{F}^k$ has not been stored in any single packed input sharing, $\mathcal{P}$ runs $[\boldsymbol{y}] \leftarrow \mathsf{Share}(\boldsymbol{y})$, which distributes the shares to all verifiers.

Locally prepare packed sharings. All verifiers locally do the following:

- For each input sharing $[\boldsymbol{y}]$ on a group of k multiplication, addition, or circuit-output gates, for each position $j \in [1, k]$, suppose that y_j comes from the i-th secret of $[\boldsymbol{x}]$ that is an output sharing on a group of k circuit-input, multiplication, or addition gates. If $[\boldsymbol{y}] \neq [\boldsymbol{x}]$, then set a wire tuple as $([\boldsymbol{x}], [\boldsymbol{y}], i, j)$.
- Denote these wire tuples by $([\boldsymbol{x}_1], [\boldsymbol{y}_1], i_1, j_1), \ldots, ([\boldsymbol{x}_N], [\boldsymbol{y}_N], i_N, j_N)$.
- Remove the repetitive packed sharings in $[\boldsymbol{y}_1], \ldots, [\boldsymbol{y}_N]$, and then denote the resulting sharings as $[\boldsymbol{y}'_1], \ldots, [\boldsymbol{y}'_\ell]$, where ℓ denotes the number of different packed sharings in $[\boldsymbol{y}_1], \ldots, [\boldsymbol{y}_N]$.

All verifiers hold the shares of the following packed sharings:

- The shares of $[\boldsymbol{w}_1], \ldots, [\boldsymbol{w}_m]$ on all circuit-input wires, which are denoted by $(\hat{w}_i^1, \ldots, \hat{w}_i^n)$ for $i \in [1, m]$.
- Let $([\boldsymbol{x}_i], [\boldsymbol{y}_i], [\boldsymbol{z}_i])$ be the i-th multiplication tuple packing the secrets of k multiplication gates for $i \in [1, M]$. The shares of $\{([\boldsymbol{x}_i], [\boldsymbol{y}_i], [\boldsymbol{z}_i])\}_{i \in [1, M]}$, where $\hat{z}_i^1, \ldots, \hat{z}_i^n$ denote the shares of $[\boldsymbol{z}_i]$ for $i \in [1, M]$.
- The shares of $\{[\boldsymbol{y}'_i]\}_{i \in [1, \ell]}$, which are denoted by $\hat{y}_i^1, \ldots, \hat{y}_i^n$ for $i \in [1, \ell]$.

$\mathcal{P}$ holds the whole sharings $\{[\boldsymbol{w}_i]\}_{i \in [1, m]}$, $\{([\boldsymbol{x}_i], [\boldsymbol{y}_i], [\boldsymbol{z}_i])\}_{i \in [1, M]}$ and $\{[\boldsymbol{y}'_i]\}_{i \in [1, \ell]}$.

Fig. 6. PSS-based strong NIMVZK in the $\mathcal{F}^{\text{pss}}_{\text{verifyprod}}$-hybrid model and random oracle model.

Protocol $\Pi^{\mathsf{pss}}_{\mathsf{snimvzk}}$, continued

Procedure for Fiat-Shamir: All verifiers $\mathcal{V}_1, \ldots, \mathcal{V}_n$ and $\mathcal{P}$ execute as follows:

1. For $i \in [1, n]$, $\mathcal{P}$ samples $r_i \leftarrow \{0,1\}^\lambda$ and computes

$$com_i := \mathsf{H}_1(\hat{w}_1^i, \ldots, \hat{w}_m^i, \hat{z}_1^i, \ldots, \hat{z}_M^i, \hat{y}_1^i, \ldots, \hat{y}_\ell^i, r_i).$$

   Then, $\mathcal{P}$ broadcasts $(com_1, \ldots, com_n)$ to all verifiers, and sends r_i to every verifier $\mathcal{V}_i$ over a private channel.
2. Each verifier $\mathcal{V}_i$ checks that $com_i = \mathsf{H}_1(\hat{w}_1^i, \ldots, \hat{w}_m^i, \hat{z}_1^i, \ldots, \hat{z}_M^i, \hat{y}_1^i, \ldots, \hat{y}_\ell^i, r_i)$, and aborts if the check fails.
3. $\mathcal{P}$ and all verifiers compute $\chi := \mathsf{H}_2(com_1, \ldots, com_n) \in \mathbb{K}$.

Verification of multiplication tuples: $\mathcal{P}$ and all verifiers execute as follows:

1. $\mathcal{P}$ and all verifiers respectively compute the secrets and the shares of $[\tilde{\boldsymbol{x}}_i] = \chi^{i-1} \cdot [\boldsymbol{x}_i]$, $[\tilde{\boldsymbol{y}}_i] = [\boldsymbol{y}_i]$ for $i \in [1, M]$ and $[\tilde{z}] := \sum_{i \in [1,M]} \chi^{i-1} \cdot [z_i]$.
2. The verifiers and $\mathcal{P}$ call functionality $\mathcal{F}^{\mathsf{pss}}_{\mathsf{verifyprod}}$ on packed inner-product tuple $(([\tilde{\boldsymbol{x}}_1], \ldots, [\tilde{\boldsymbol{x}}_M]), ([\tilde{\boldsymbol{y}}_1], \ldots, [\tilde{\boldsymbol{y}}_M]), [\tilde{z}])$ to check that $\tilde{z} = \sum_{h \in [1,M]} \tilde{\boldsymbol{x}}_h * \tilde{\boldsymbol{y}}_h$. If the verifiers receive **abort** from $\mathcal{F}^{\mathsf{pss}}_{\mathsf{verifyprod}}$, then they abort.

Verification of consistency of wire tuples: For all $i, j \in [1, k]$, $\mathcal{P}$ and all verifiers initiate an empty list $L(i, j)$. Then, from $h = 1$ to N, they insert $([\boldsymbol{x}_h], [\boldsymbol{y}_h], i_h, j_h)$ into the list $L(i_h, j_h)$. For each $i, j \in [1, k]$, $\mathcal{P}$ and all verifiers check the consistency of the wire tuples in $L(i, j)$ as follows:

1. Let N' be the size of $L(i, j)$. Let $([\boldsymbol{a}_1], [\boldsymbol{b}_1], i, j), \ldots, ([\boldsymbol{a}_{N'}], [\boldsymbol{b}_{N'}], i, j)$ denote the wire tuples in $L(i, j)$.
2. $\mathcal{P}$ samples $\boldsymbol{a}_0, \boldsymbol{b}_0 \leftarrow \mathbb{K}^k$ with $a_{0,i} = b_{0,j}$, and also picks $\boldsymbol{r}, \boldsymbol{r}' \leftarrow \mathbb{K}^k$. Then $\mathcal{P}$ and all verifiers execute the following:
   (a) $\mathcal{P}$ runs $[\boldsymbol{r}] \leftarrow \mathsf{Share}(\boldsymbol{r})$ and $[\boldsymbol{r}'] \leftarrow \mathsf{Share}(\boldsymbol{r}')$, which distributes the shares to all verifiers.
   (b) $\mathcal{P}$ computes $\boldsymbol{u} := \boldsymbol{a}_0 + \boldsymbol{r}$ and $\boldsymbol{u}' := \boldsymbol{b}_0 + \boldsymbol{r}'$, and then broadcasts $(\boldsymbol{u}, \boldsymbol{u}')$ to all verifiers.
   (c) The verifiers compute $[\boldsymbol{a}_0] := \boldsymbol{u} - [\boldsymbol{r}]$ and $[\boldsymbol{b}_0] := \boldsymbol{u}' - [\boldsymbol{r}']$.
3. All verifiers compute $\alpha := \mathsf{H}_2(\chi, \boldsymbol{u}, \boldsymbol{u}', i, j) \in \mathbb{K}$.
4. All verifiers compute $[\boldsymbol{a}] := \sum_{h=0}^{N'} \alpha^h \cdot [\boldsymbol{a}_h]$ and $[\boldsymbol{b}] := \sum_{h=0}^{N'} \alpha^h \cdot [\boldsymbol{b}_h]$.
5. All verifiers run $\boldsymbol{a} \leftarrow \mathsf{Open}([\boldsymbol{a}])$ and $\boldsymbol{b} \leftarrow \mathsf{Open}([\boldsymbol{b}])$. If **abort** is output for the Open procedure, the verifiers abort. Otherwise, they check that $a_i = b_j$, and abort if the check fails.

Verification of circuit output: Let $[\boldsymbol{z}]$ be the sharing on the group of k circuit-output wires including the single actual circuit-output wire. All verifiers execute $\boldsymbol{z} \leftarrow \mathsf{Open}([\boldsymbol{z}])$, and abort if receiving **abort** from the Open procedure. If $\boldsymbol{z} = \boldsymbol{0}^k$, then the verifiers output **true**, otherwise they output **false**.

Fig. 7. PSS-based strong NIMVZK in the $\mathcal{F}^{\mathsf{pss}}_{\mathsf{verifyprod}}$-hybrid model and random oracle model, continued.

consistency check of wire tuples in the *total* communication complexity $O(nk^2)$ elements between the prover and verifiers and $O(n^2k^2)$ elements among verifiers.

Theorem 3. *Let* H_1 *and* H_2 *be two random oracles. Protocol* $\Pi^{\mathsf{pss}}_{\mathsf{snimvzk}}$ *shown in Figs. 6 and 7 securely realizes functionality* $\mathcal{F}_{\mathsf{mvzk}}$ *with soundness error at most* $\frac{Q_1 n + (Q_2+1)(M+N)}{2^\lambda}$ *in the* $\mathcal{F}^{\mathsf{pss}}_{\mathsf{verifyprod}}$*-hybrid model in the presence of a malicious adversary corrupting up to a prover and* $t = d - k + 1$ *verifiers, where degree-d packed sharings are used in protocol* $\Pi^{\mathsf{pss}}_{\mathsf{snimvzk}}$*, each sharing packs* k *secrets, and* Q_1 *and* Q_2 *are the number of queries to* H_1 *and* H_2 *respectively.*

The proof of Theorem 3 can be found in the full version [52].

Protocol $\Pi^{\mathsf{pss}}_{\mathsf{snimvzk}}$ shown in Figs. 6 and 7 are *streamable*, i.e., the circuit can be proved on-the-fly. The prover $\mathcal{P}$ can prove a batch of addition and multiplication gates each time, and stores the secrets that will be used as the input wire values in the next batches of gates. In Sect. 3.3, we give an approach overview on how to stream our protocol $\Pi^{\mathsf{pss}}_{\mathsf{snimvzk}}$ with the *same* round among verifiers. In the full version [52], we provide more details.

6.2 Strong NIMVZK Proof for Packed Inner-Product Tuples

Below, we present a strongly non-interactive MVZK protocol with *logarithmic* communication complexity to verify packed inner-product tuples, which is inspired by the technique by Goyal et al. [39] for MPC that is in turn built on the techniques [14,40,41]. While the MPC protocol [39] requires logarithmic rounds to check correctness of packed inner-product tuples, our strong NIMVZK protocol needs only one round between verifiers. Furthermore, our protocol reduces the communication overhead for verification by making the prover generate the random sharings and messages associated with secrets, compared to the verification of packed inner-product tuples directly using the MPC protocol [39].

Our PSS-based strong NIMVZK protocol $\Pi^{\mathsf{pss}}_{\mathsf{verifyprod}}$ for verifying a packed inner-product tuple is described in Fig. 8. This protocol invokes two sub-protocols $\Pi^{\mathsf{pss}}_{\mathsf{inner\text{-}prod}}$ and $\Pi^{\mathsf{pss}}_{\mathsf{compress}}$ that are described in Figs. 9 and 10 respectively, where $\Pi^{\mathsf{pss}}_{\mathsf{inner\text{-}prod}}$ is used to generate the inner product of two vectors and $\Pi^{\mathsf{pss}}_{\mathsf{compress}}$ is used to compress two packed inner-product tuples into a single tuple. In the dimension-reduction and randomization phases of this protocol, we adapt the approach by Baum et al. [8] used in the DVZK setting to non-interactively generate a challenge α used in sub-protocol $\Pi^{\mathsf{pss}}_{\mathsf{compress}}$ based on the Fiat-Shamir transform. Based on the *round-by-round soundness* [8,20], we can prove that the soundness error of our protocol is negligible (see Theorem 4). In the dimension-reduction phase of $\Pi^{\mathsf{pss}}_{\mathsf{verifyprod}}$, we always assume that the dimension m of the packed inner-product tuple is the multiple of 2 for each iteration. If not, we can pad the dummy zero sharing $[\mathbf{0}]$ into the packed inner-product tuple to satisfy the requirement, where $[\mathbf{0}]$ can be locally computed by all verifiers.

In the protocol $\Pi^{\mathsf{pss}}_{\mathsf{verifyprod}}$ shown in Fig. 8, we assume that $\mathcal{P}$ and all verifiers input a public challenge χ, which is determined after the secrets packed in the input inner-product tuple $((\lbrack\boldsymbol{x}_1\rbrack, \ldots, \lbrack\boldsymbol{x}_M\rbrack), (\lbrack\boldsymbol{y}_1\rbrack, \ldots, \lbrack\boldsymbol{y}_M\rbrack), \lbrack\boldsymbol{z}\rbrack)$ have been

---

Protocol $\Pi_{\text{verifyprod}}^{\text{pss}}$

Inputs: Prover $\mathcal{P}$ and all verifiers $\mathcal{V}_1, \ldots, \mathcal{V}_n$ respectively hold the secrets and shares of a packed inner-product tuple $(([x_1], \ldots, [x_M]), ([y_1], \ldots, [y_M]), [z])$. $\mathcal{P}$ and all verifiers also hold a public value χ, which is determined after these secrets have been defined. The verifiers will check that $z = \sum_{h \in [1,M]} x_h * y_h \in \mathbb{K}^k$. Let $H_2 : \{0,1\}^* \to \mathbb{K}$ be a random oracle where $|\mathbb{K}| \geq 2^\lambda$.

- **Dimension-reduction:** Let m denote the dimension of the packed inner-product tuple in the current iteration, where m is initialized as M. Let γ be the public value in the current iteration, which is initialized as χ.
  While $m > 2$, $\mathcal{P}$ and all verifiers do the following:
  1. Let $\ell = m/2$. $\mathcal{P}$ and all verifiers define $([a_{1,1}], \ldots, [a_{1,\ell}]) := ([x_1], \ldots, [x_\ell])$ and $([a_{2,1}], \ldots, [a_{2,\ell}]) = ([x_{\ell+1}], \ldots, [x_m])$, where $\mathcal{P}$ holds all the secrets and the verifiers hold the shares. Similarly, they define $([b_{1,1}], \ldots, [b_{1,\ell}]) := ([y_1], \ldots, [y_\ell])$ and $([b_{2,1}], \ldots, [b_{2,\ell}]) = ([y_{\ell+1}], \ldots, [y_m])$.
  2. $\mathcal{P}$ and all verifiers execute the sub-protocol $\Pi_{\text{inner-prod}}^{\text{pss}}$ (shown in Figure 9) on $([a_{1,1}], \ldots, [a_{1,\ell}])$ and $([b_{1,1}], \ldots, [b_{1,\ell}])$ to compute the whole sharing $[c_1] = [\sum_{h \in [1,\ell]} a_{1,h} * b_{1,h}]$, where the secrets and the shares are output to $\mathcal{P}$ and the verifiers respectively. $\mathcal{P}$ and all verifiers also obtain $u \in \mathbb{K}^k$.
  3. $\mathcal{P}$ and all verifiers compute $[c_2] := [z] - [c_1]$, where $\mathcal{P}$ obtains c_2 and the verifiers get the shares of $[c_2]$.
  4. $\mathcal{P}$ and all verifiers execute sub-protocol $\Pi_{\text{compress}}^{\text{pss}}$ on $([a_{i,1}], \ldots, [a_{i,\ell}])$, $([b_{i,1}], \ldots, [b_{i,\ell}])$ and $[c_i]$ for $i \in [1,2]$ and (γ, u), where $\Pi_{\text{compress}}^{\text{pss}}$ is described in Figure 10.
  5. $\mathcal{P}$ and all verifiers update γ as the public element α output by $\Pi_{\text{compress}}^{\text{pss}}$, and also set $m := m/2$. Then, $\mathcal{P}$ and the verifiers use the whole output sharings $(([a_1], \ldots, [a_\ell]), ([b_1], \ldots, [b_\ell]), [c])$ from $\Pi_{\text{compress}}^{\text{pss}}$ to update $(([x_1], \ldots, [x_m]), ([y_1], \ldots, [y_m]), [z])$.

- **Randomization:** Let $\gamma \in \mathbb{K}$ along with the inner-product tuple $(([x_1], [x_2]), ([y_1], [y_2]), [z])$ be the final output from the previous phase. $\mathcal{P}$ and all verifiers execute the following procedure:
  1. $\mathcal{P}$ samples $x_0, y_0 \leftarrow \mathbb{K}^k$, and then runs $[x_0] \leftarrow \mathsf{Share}(x_0)$ and $[y_0] \leftarrow \mathsf{Share}(y_0)$, which distribute the shares to all verifiers.
  2. For $i \in [0,1]$, $\mathcal{P}$ and all verifiers execute the sub-protocol $\Pi_{\text{inner-prod}}^{\text{pss}}$ on $([x_i], [y_i])$ to compute the whole sharing $[z_i] = [x_i * y_i]$. Additionally, $\mathcal{P}$ and the verifiers also obtain $u_0, u_1 \in \mathbb{K}^k$.
  3. $\mathcal{P}$ and all verifiers compute $[z_2] := [z] - [z_1]$, where $\mathcal{P}$ obtains z_2 and the verifiers get the shares of $[z_2]$.
  4. $\mathcal{P}$ and all verifiers execute sub-protocol $\Pi_{\text{compress}}^{\text{pss}}$ on $\{([x_i], [y_i], [z_i])\}_{i \in [0,2]}$ and (γ, u_0, u_1). Then, all verifiers obtain the output $([a], [b], [c])$.
  5. All verifiers run $v \leftarrow \mathsf{Open}([v])$ for each $v \in \{a, b, c\}$. If abort is received during the Open procedure, then the verifiers abort. Then, the verifiers check that $c = a * b$. If the check fails, the verifiers output abort. Otherwise, they output accept.

Fig. 8. One-round ZK verification protocol for packed inner-product tuples.

Protocol $\Pi_{\text{inner-prod}}^{\text{pss}}$

Inputs: Prover $\mathcal{P}$ and verifiers $\mathcal{V}_1, \ldots, \mathcal{V}_n$ respectively hold the secrets and the shares of packed sharings $([\boldsymbol{x}_1], \ldots, [\boldsymbol{x}_\ell])$ and $([\boldsymbol{y}_1], \ldots, [\boldsymbol{y}_\ell])$ over a field $\mathbb{K}$.

Protocol execution: $\mathcal{P}$ and all verifiers execute as follows:

1. $\mathcal{P}$ samples $\boldsymbol{r} \leftarrow \mathbb{K}^k$, and then runs $[\boldsymbol{r}] \leftarrow \mathsf{Share}(\boldsymbol{r})$ that distributes the shares to all verifiers.
2. $\mathcal{P}$ computes $\boldsymbol{z} := \sum_{h \in [1,\ell]} \boldsymbol{x}_h * \boldsymbol{y}_h \in \mathbb{K}^k$, and then broadcasts $\boldsymbol{u} := \boldsymbol{z} + \boldsymbol{r} \in \mathbb{K}^k$ to all verifiers.
3. The verifiers compute $[\boldsymbol{z}] := \boldsymbol{u} - [\boldsymbol{r}]$. Then, $\mathcal{P}$ and the verifiers respectively output the secrets and shares of $[\boldsymbol{z}]$, and also output $\boldsymbol{u} \in \mathbb{K}^k$.

Fig. 9. Non-interactive inner-product protocol for packed sharings secure up to additive errors.

Protocol $\Pi_{\text{compress}}^{\text{pss}}$

Inputs: Let m be the number of packed inner-product tuples, and ℓ be the dimension of each packed inner-product tuple. For $i \in [1, m]$, prover $\mathcal{P}$ and all verifiers $\mathcal{V}_1, \ldots, \mathcal{V}_n$ respectively hold the secrets and shares of packed inner-product tuple $((([\boldsymbol{x}_{i,1}], \ldots, [\boldsymbol{x}_{i,\ell}]), ([\boldsymbol{y}_{i,1}], \ldots, [\boldsymbol{y}_{i,\ell}]), [\boldsymbol{z}_i])$. $\mathcal{P}$ and all verifiers also input $\gamma \in \mathbb{K}$ and $\boldsymbol{v}_1, \ldots, \boldsymbol{v}_d \in \mathbb{K}^k$. Let $\mathsf{H}_2 : \{0,1\}^* \to \mathbb{K}$ be a random oracle.

Protocol execution: $\mathcal{P}$ and all verifiers execute as follows:

1. For each $j \in [1, \ell]$, $\mathcal{P}$ computes vectors of degree-$(m-1)$ polynomials $\boldsymbol{f}_j(\cdot)$ and $\boldsymbol{g}_j(\cdot)$, such that $\boldsymbol{f}_j(i) = \boldsymbol{x}_{i,j}$ and $\boldsymbol{g}_j(i) = \boldsymbol{y}_{i,j}$ for all $i \in [1, m]$.
2. For $j \in [1, \ell]$, all verifiers locally compute $[\boldsymbol{f}_j(\cdot)]$ and $[\boldsymbol{g}_j(\cdot)]$ using their shares of $\{[\boldsymbol{x}_{i,j}]\}_{i \in [1,m]}$ and $\{[\boldsymbol{y}_{i,j}]\}_{i \in [1,m]}$ respectively.
3. For $i \in [m+1, 2m-1]$, $\mathcal{P}$ and all verifiers respectively compute the secrets and shares of packed sharings $[\boldsymbol{f}_j(i)]$ and $[\boldsymbol{g}_j(i)]$ for $j \in [1, \ell]$. Then, for $i \in [m+1, 2m-1]$, they execute the sub-protocol $\Pi_{\text{inner-prod}}^{\text{pss}}$ (shown in Figure 9) on $([\boldsymbol{f}_1(i)], \ldots, [\boldsymbol{f}_\ell(i)])$ and $([\boldsymbol{g}_1(i)], \ldots, [\boldsymbol{g}_\ell(i)])$ to compute the whole sharing $[\boldsymbol{z}_i] = [\sum_{j \in [1,\ell]} \boldsymbol{f}_j(i) * \boldsymbol{g}_j(i)]$, where $\mathcal{P}$ obtains $\boldsymbol{z}_i$ and the verifiers get the shares of $[\boldsymbol{z}_i]$. Besides, $\mathcal{P}$ and the verifiers obtain $\boldsymbol{u}_{m+1}, \ldots, \boldsymbol{u}_{2m-1} \in \mathbb{K}^k$.
4. $\mathcal{P}$ and all verifiers locally compute the whole sharing $[\boldsymbol{h}(\cdot)]$ from $[\boldsymbol{z}_1], \ldots, [\boldsymbol{z}_{2m-1}]$, such that $\boldsymbol{h}(\cdot)$ is a vector of degree-$2(m-1)$ polynomial and $\boldsymbol{h}(i) = \boldsymbol{z}_i$ for $i \in [1, 2m-1]$.
5. $\mathcal{P}$ and all verifiers compute $\alpha := \mathsf{H}_2(\gamma, \boldsymbol{v}_1, \ldots, \boldsymbol{v}_d, \boldsymbol{u}_{m+1}, \ldots, \boldsymbol{u}_{2m-1}) \in \mathbb{K}$. If $\alpha \in [1, m]$, the verifiers abort.
6. $\mathcal{P}$ and all verifiers output the secrets and shares of $([\boldsymbol{f}_1(\alpha)], \ldots, [\boldsymbol{f}_\ell(\alpha)])$, $([\boldsymbol{g}_1(\alpha)], \ldots, [\boldsymbol{g}_\ell(\alpha)])$ and $[\boldsymbol{h}(\alpha)]$ respectively, and also output the element α.

Fig. 10. Protocol for compressing packed inner-product tuples.

---

defined. In particular, χ can be defined as $\mathsf{H}_2(com_1, \ldots, com_n)$ as shown in Fig. 7. When using $\Pi_{\text{verifyprod}}^{\text{pss}}$ to realize functionality $\mathcal{F}_{\text{verifyprod}}^{\text{pss}}$, $\chi \in \mathbb{K}$ can be generated by $\mathcal{P}$ and all verifiers in the main NIMVZK protocol shown in Figs. 6

and 7. For the sake of simplicity, we ignore the case that the adversary (who corrupts $\mathcal{P}$) did not make a query to obtain χ but made a query to get a challenge $\alpha = \mathsf{H}_2(\chi, \cdots)$ used in protocol $\Pi^{\mathsf{pss}}_{\mathsf{verifyprod}}$, which occurs with probability at most $\frac{1}{|\mathbb{K}|} \leq \frac{1}{2^\lambda}$. When the adversary makes a query $(com_1, \ldots, com_n)$ to random oracle H_2 and obtains χ, the challenge χ is determined after the secrets stored in the input packed tuple have been defined, except with probability at most $\frac{Q_1 n}{2^\lambda}$ following the proof of Theorem 3.

Sub-protocol for Computing Inner Product. Sub-protocol $\Pi^{\mathsf{pss}}_{\mathsf{inner\text{-}prod}}$ shown in Fig. 9 is used to compute the inner product of two vectors $([\boldsymbol{x}_1], \ldots, [\boldsymbol{x}_\ell])$ and $([\boldsymbol{y}_1], \ldots, [\boldsymbol{y}_\ell])$. The prover now knows all the challenges due to the use of the Fiat-Shamir transform, and thus holds all the secrets involved in the verification procedure. Thus, the prover can directly distribute the shares of $[\boldsymbol{z}]$ with $\boldsymbol{z} = \sum_{h \in [1, \ell]} \boldsymbol{x}_h * \boldsymbol{y}_h$ to all verifiers. To support Fiat-Shamir, the verifiers need to know public messages instead of secret shares. Therefore, we first let the prover generate a random packed sharing $[\boldsymbol{r}]$, and then make it broadcast the *public* difference $\boldsymbol{u} = \boldsymbol{z} + \boldsymbol{r}$ to all verifiers. The prover and verifiers also need to output the message $\boldsymbol{u}$, which will be used in the Fiat-Shamir transform of the main verification protocol. When the prover is malicious, it can introduce an additive error to the sharing $[\boldsymbol{z}]$ output by the verifiers, which is harmless when integrating $\Pi^{\mathsf{pss}}_{\mathsf{inner\text{-}prod}}$ into the main protocol $\Pi^{\mathsf{pss}}_{\mathsf{verifyprod}}$.

Sub-protocol for Compression. Sub-protocol $\Pi^{\mathsf{pss}}_{\mathsf{compress}}$ shown in Fig. 10 is used to compress m packed inner-product tuples into a single packed inner-product tuple. In particular, this protocol invokes the sub-protocol $\Pi^{\mathsf{pss}}_{\mathsf{inner\text{-}prod}}$ instead of calling an inner-product functionality, which seems necessary to support Fiat-Shamir, where the messages related to the secrets need to be used as the input of a random oracle H_2. For every protocol execution of $\Pi^{\mathsf{pss}}_{\mathsf{compress}}$, the prover and all verifiers generate a random challenge $\alpha \in \mathbb{K}$ using the Fiat-Shamir transform. To realize *non-interactively* recursive compression in the main verification protocol, the prover and verifiers also input the public challenge γ from the previous iteration and the public messages produced in the current iteration.

Theorem 4. *Let* $\mathsf{H}_2 : \{0,1\}^* \to \mathbb{K}$ *be a random oracle. Protocol* $\Pi^{\mathsf{pss}}_{\mathsf{verifyprod}}$ *shown in Fig. 8 securely realizes functionality* $\mathcal{F}^{\mathsf{pss}}_{\mathsf{verifyprod}}$ *with soundness error at most* $\frac{4\lceil \log M \rceil + 5Q_2}{2^\lambda - 3}$ *in the presence of a malicious adversary corrupting up to the prover and exactly t verifiers, where Q_2 is the number of queries to random oracle H_2.*

The proof of Theorem 4 is given in the full version [52].

Acknowledgements. Work of Kang Yang is supported by the National Natural Science Foundation of China (Grant Nos. 62102037, 61932019). Work of Xiao Wang is supported in part by DARPA under Contract No. HR001120C0087, NSF award #2016240, and research awards from Meta, Google and PlatON Network. The views, opinions, and/or findings expressed are those of the author(s) and should not be interpreted as representing the official views or policies of the Department of Defense or the U.S. Government.

References

1. Abe, M.: Robust distributed multiplication without interaction. In: Advances in Cryptology-Crypto 1999. LNCS, vol. 1666, pp. 130–147. Springer (1999). https://doi.org/10.1007/3-540-48405-1_9
2. Abe, M., Cramer, R., Fehr, S.: Non-interactive distributed-verifier proofs and proving relations among commitments. In: Advances in Cryptology-Asiacrypt 2002. LNCS, pp. 206–223. Springer (2002). https://doi.org/10.1007/3-540-36178-2_13
3. Ames, S., Hazay, C., Ishai, Y., Venkitasubramaniam, M.: Ligero: lightweight sublinear arguments without a trusted sctup. In: ACM Conference on Computer and Communications Security (CCS) 2017, pp. 2087–2104. ACM Press (2017). https://doi.org/10.1145/3133956.3134104
4. Applebaum, B., Kachlon, E., Patra, A.: Verifiable relation sharing and multi-verifier zero-knowledge in two rounds: trading NIZKs with honest majority. Cryptology ePrint Archive, Paper 2022/167 (2022). https://eprint.iacr.org/2022/167
5. Baldimtsi, F., Kiayias, A., Zacharias, T., Zhang, B.: Crowd verifiable zero-knowledge and end-to-end verifiable multiparty computation. In: Advances in Cryptology-Asiacrypt 2020, Part III. LNCS, pp. 717–748. Springer (2020). https://doi.org/10.1007/978-3-030-64840-4_24
6. Baum, C., Braun, L., Munch-Hansen, A., Razet, B., Scholl, P.: Appenzeller to brie: efficient zero-knowledge proofs for mixed-mode arithmetic and Z2k. In: ACM Conference on Computer and Communications Security (CCS) 2021, pp. 192–211. ACM Press (2021). https://doi.org/10.1145/3460120.3484812
7. Baum, C., Jadoul, R., Orsini, E., Scholl, P., Smart, N.P.: Feta: efficient threshold designated-verifier zero-knowledge proofs. Cryptology ePrint Archive, Paper 2022/082 (2022). https://eprint.iacr.org/2022/082
8. Baum, C., Malozemoff, A.J., Rosen, M.B., Scholl, P.: Mac'n'Cheese: zero-knowledge proofs for Boolean and arithmetic circuits with nested disjunctions. In: Malkin, T., Peikert, C. (eds.) CRYPTO 2021. LNCS, vol. 12828, pp. 92–122. Springer, Cham (2021). https://doi.org/10.1007/978-3-030-84259-8_4
9. Beck, G., Goel, A., Jain, A., Kaptchuk, G.: Order-C secure multiparty computation for highly repetitive circuits. In: Advances in Cryptology-Eurocrypt 2021, Part II. LNCS, pp. 663–693. Springer (2021). https://doi.org/10.1007/978-3-030-77886-6_23
10. Ben-Sasson, E., Bentov, I., Horesh, Y., Riabzev, M.: Scalable zero knowledge with no trusted setup. In: Advances in Cryptology-Crypto 2019, Part III. LNCS, vol. 11694, pp. 701–732. Springer (2019). https://doi.org/10.1007/978-3-030-26954-8_23
11. Ben-Sasson, E., Chiesa, A., Riabzev, M., Spooner, N., Virza, M., Ward, N.P.: Aurora: transparent succinct arguments for R1CS. In: Advances in Cryptology-Eurocrypt 2019, Part I. LNCS, vol. 11476, pp. 103–128. Springer (2019). https://doi.org/10.1007/978-3-030-17653-2_4
12. Ben-Sasson, E., Chiesa, A., Spooner, N.: Interactive oracle proofs. In: 9th Theory of Cryptography Conference–TCC 2016, pp. 31–60. LNCS, Springer (2016). https://doi.org/10.1007/978-3-662-53644-5_2
13. Bhadauria, R., Fang, Z., Hazay, C., Venkitasubramaniam, M., Xie, T., Zhang, Y.: Ligero++: a new optimized sublinear IOP. In: ACM Conference on Computer and Communications Security (CCS) 2020, pp. 2025–2038. ACM Press (2020). https://doi.org/10.1145/3372297.3417893

14. Boneh, D., Boyle, E., Corrigan-Gibbs, H., Gilboa, N., Ishai, Y.: Zero-knowledge proofs on secret-shared data via fully linear PCPs. In: Advances in Cryptology-Crypto 2019, Part III. LNCS, vol. 11694, pp. 67–97. Springer (2019). https://doi. org/10.1007/978-3-030-26954-8_3

15. Bootle, J., Cerulli, A., Chaidos, P., Groth, J., Petit, C.: Efficient zero-knowledge arguments for arithmetic circuits in the discrete log setting. In: Advances in Cryptology-Eurocrypt 2016, Part II. LNCS, vol. 9666, pp. 327–357. Springer (2016). https://doi.org/10.1007/978-3-662-49896-5_12

16. Boyle, E., Gilboa, N., Ishai, Y., Nof, A.: Efficient fully secure computation via distributed zero-knowledge proofs. In: Advances in Cryptology-Asiacrypt 2020, Part III. LNCS, pp. 244–276. Springer (2020). https://doi.org/10.1007/978-3-030-64840-4_9

17. Bünz, B., Bootle, J., Boneh, D., Poelstra, A., Wuille, P., Maxwell, G.: Bulletproofs: short proofs for confidential transactions and more. In: IEEE Symposium Security and Privacy 2018, pp. 315–334. IEEE (2018). https://doi.org/10.1109/SP.2018.00020

18. Bünz, B., Fisch, B., Szepieniec, A.: Transparent SNARKs from DARK compilers. In: Advances in Cryptology-Eurocrypt 2020, Part I. LNCS, vol. 12105, pp. 677–706. Springer (2020). https://doi.org/10.1007/978-3-030-45721-1_24

19. Burmester, M., Desmedt, Y.: Broadcast interactive proofs (extended abstract). In: Advances in Cryptology-Eurocrypt 1991. LNCS, pp. 81–95. Springer (1991). https://doi.org/10.1007/3-540-46416-6_7

20. Canetti, R., et al.: Fiat-Shamir: from practice to theory. In: 51th Annual ACM Symposium on Theory of Computing (STOC), pp. 1082–1090. ACM Press (2019). https://doi.org/10.1145/3313276.3316380

21. Canetti, R., Kaptchuk, G.: The Broken Promise of Apple's Announced Forbidden-photo Reporting System - And How To Fix It. https://www.bu.edu/riscs/2021/08/10/apple-csam/ (2021)

22. Chida, K., et al.: Fast large-scale honest-majority MPC for malicious adversaries. In: Advances in Cryptology-Crypto 2018, Part III. LNCS, vol. 10993, pp. 34–64. Springer (2018). https://doi.org/10.1007/978-3-319-96878-0_2

23. Corrigan-Gibbs, H., Boneh, D.: Prio: private, robust, and scalable computation of aggregate statistics. In: 14th USENIX Symposium on Networked Systems Design and Implementation (NSDI 17), pp. 259–282. USENIX Association, March 2017

24. Damgård, I., Ishai, Y., Krøigaard, M.: Perfectly secure multiparty computation and the computational overhead of cryptography. In: Advances in Cryptology-Eurocrypt 2010. LNCS, pp. 445–465. Springer (2010). https://doi.org/10.1007/978-3-642-13190-5_23

25. Damgård, I., Nielsen, J.B.: Scalable and unconditionally secure multiparty computation. In: Advances in Cryptology-Crypto 2007. LNCS, vol. 4622, pp. 572–590. Springer (2007). https://doi.org/10.1007/978-3-540-74143-5_32

26. Damgård, I., Nielsen, J.B., Nielsen, M., Ranellucci, S.: The TinyTable Protocol for 2-party secure computation, or: gate-scrambling revisited. In: Advances in Cryptology-Crypto 2017, Part I. LNCS, vol. 10401, pp. 167–187. Springer (2017). https://doi.org/10.1007/978-3-319-63688-7_6

27. Dittmer, S., Ishai, Y., Lu, S., Ostrovsky, R.: Improving line-point zero knowledge: two multiplications for the price of one. In: Proceedings of the 2022 ACM SIGSAC Conference on Computer and Communications Security. ACM Press (2022)

28. Dittmer, S., Ishai, Y., Ostrovsky, R.: Line-point zero knowledge and its applications. In: 2nd Conference on Information-Theoretic Cryptography (2021)

29. Fiat, A., Shamir, A.: How to prove yourself: practical solutions to identification and signature problems. In: Advances in Cryptology-Crypto 1986. LNCS, pp. 186–194. Springer (1987). https://doi.org/10.1007/3-540-47721-7_12
30. Franklin, M.K., Yung, M.: Communication complexity of secure computation (extended abstract). In: 24th Annual ACM Symposium on Theory of Computing (STOC), pp. 699–710. ACM Press (1992). https://doi.org/10.1145/129712.129780
31. Garay, J.A., Ishai, Y., Ostrovsky, R., Zikas, V.: The price of low communication in secure multi-party computation. In: Advances in Cryptology-Crypto 2017, Part I. LNCS, vol. 10401, pp. 420–446. Springer (2017). https://doi.org/10.1007/978-3-319-63688-7_14
32. Genkin, D., Ishai, Y., Polychroniadou, A.: Efficient multi-party computation: from passive to active security via secure SIMD circuits. In: Advances in Cryptology-Crypto 2015, Part II. LNCS, vol. 9216, pp. 721–741. Springer (2015). https://doi.org/10.1007/978-3-662-48000-7_35
33. Genkin, D., Ishai, Y., Prabhakaran, M., Sahai, A., Tromer, E.: Circuits resilient to additive attacks with applications to secure computation. In: 46th Annual ACM Symposium on Theory of Computing (STOC), pp. 495–504. ACM Press (2014). https://doi.org/10.1145/2591796.2591861
34. Gennaro, R., Gentry, C., Parno, B., Raykova, M.: Quadratic span programs and succinct NIZKs without PCPs. In: Advances in Cryptology-Eurocrypt 2013. LNCS, pp. 626–645. Springer (2013). https://doi.org/10.1007/978-3-642-38348-9_37
35. Goldwasser, S., Kalai, Y.T., Rothblum, G.N.: Delegating computation: interactive proofs for muggles. In: 40th Annual ACM Symposium on Theory of Computing (STOC), pp. 113–122. ACM Press (2008). https://doi.org/10.1145/1374376.1374396
36. Goldwasser, S., Lindell, Y.: Secure multi-party computation without agreement. J. Cryptol. 18(3), 247–287 (2005). https://doi.org/10.1007/s00145-005-0319-z
37. Gordon, S.D., Starin, D., Yerukhimovich, A.: The more the merrier: reducing the cost of large scale MPC. In: Advances in Cryptology-Eurocrypt 2021, Part II. LNCS, pp. 694–723. Springer (2021). https://doi.org/10.1007/978-3-030-77886-6_24
38. Goyal, V., Li, H., Ostrovsky, R., Polychroniadou, A., Song, Y.: ATLAS: efficient and scalable MPC in the honest majority setting. In: Advances in Cryptology-Crypto 2021, Part II. LNCS, pp. 244–274. Springer (2021). https://doi.org/10.1007/978-3-030-84245-1_9
39. Goyal, V., Polychroniadou, A., Song, Y.: Unconditional communication-efficient MPC via Hall's marriage theorem. In: Advances in Cryptology-Crypto 2021, Part II. LNCS, pp. 275–304. Springer (2021). https://doi.org/10.1007/978-3-030-84245-1_10
40. Goyal, V., Song, Y.: Malicious Security Comes Free in Honest-Majority MPC. Cryptology ePrint Archive, Report 2020/134 (2020). https://eprint.iacr.org/2020/134
41. Goyal, V., Song, Y., Zhu, C.: Guaranteed output delivery comes free in honest majority MPC. In: Advances in Cryptology-Crypto 2020, Part II. LNCS, pp. 618–646. Springer (2020). https://doi.org/10.1007/978-3-030-56880-1_22
42. Groth, J.: Short pairing-based non-interactive zero-knowledge arguments. In: Advances in Cryptology-Asiacrypt 2010. LNCS, pp. 321–340. Springer (2010). https://doi.org/10.1007/978-3-642-17373-8_19
43. Groth, J., Ostrovsky, R.: Cryptography in the multi-string model. In: Advances in Cryptology-Crypto 2007. LNCS, vol. 4622, pp. 323–341. Springer (2007). https://doi.org/10.1007/978-3-540-74143-5_18

44. Ishai, Y., Kushilevitz, E., Ostrovsky, R., Sahai, A.: Zero-knowledge from secure multiparty computation. In: 39th Annual ACM Symposium on Theory of Computing (STOC), pp. 21–30. ACM Press (2007). https://doi.org/10.1145/1250790.1250794

45. Lepinski, M., Micali, S., Shelat, A.: Fair-zero knowledge. In: Theory of Cryptography Conference–TCC 2005. LNCS, vol. 3378, pp. 245–263. Springer (2005). https://doi.org/10.1007/978-3-540-30576-7_14

46. Lindell, Y., Nof, A.: A framework for constructing fast MPC over arithmetic circuits with malicious adversaries and an honest-majority. In: ACM Conference on Computer and Communications Security (CCS) 2017, pp. 259–276. ACM Press (2017). https://doi.org/10.1145/3133956.3133999

47. Nordholt, P.S., Veeningen, M.: Minimising communication in honest-majority MPC by batchwise multiplication verification. In: International Conference on Applied Cryptography and Network Security (ACNS). LNCS, pp. 321–339. Springer (2018). https://doi.org/10.1007/978-3-319-93387-0_17

48. Setty, S.: Spartan: efficient and general-purpose zkSNARKs without trusted setup. In: Advances in Cryptology-Crypto 2020, Part III. LNCS, pp. 704–737. Springer (2020). https://doi.org/10.1007/978-3-030-56877-1_25

49. Wahby, R.S., Tzialla, I., shelat, a., Thaler, J., Walfish, M.: Doubly-efficient zkSNARKs without trusted setup. In: IEEE Symposium Security and Privacy 2018, pp. 926–943. IEEE (2018). https://doi.org/10.1109/SP.2018.00060

50. Weng, C., Yang, K., Katz, J., Wang, X.: Wolverine: fast, scalable, and communication-efficient zero-knowledge proofs for boolean and arithmetic circuits. In: IEEE Symposium Security and Privacy 2021, pp. 1074–1091. IEEE (2021). https://doi.org/10.1109/SP40001.2021.00056

51. Yang, K., Sarkar, P., Weng, C., Wang, X.: QuickSilver: efficient and affordable zero-knowledge proofs for circuits and polynomials over any field. In: ACM Conference on Computer and Communications Security (CCS) 2021, pp. 2986–3001. ACM Press (2021). https://doi.org/10.1145/3460120.3484556

52. Yang, K., Wang, X.: Non-Interactive Zero-Knowledge Proofs to Multiple Verifiers. Cryptology ePrint Archive, Paper 2022/063 (2022). https://eprint.iacr.org/2022/063

53. Zhang, J., et al.: Doubly efficient interactive proofs for general arithmetic circuits with linear prover time. In: ACM Conference on Computer and Communications Security (CCS) 2021, pp. 159–177. ACM Press (2021). https://doi.org/10.1145/3460120.3484767

54. Zhang, J., Xie, T., Zhang, Y., Song, D.: Transparent polynomial delegation and its applications to zero knowledge proof. In: IEEE Symposium Security and Privacy 2020, pp. 859–876. IEEE (2020). https://doi.org/10.1109/SP40000.2020.00052

Rotatable Zero Knowledge Sets
Post Compromise Secure Auditable Dictionaries with Application to Key Transparency

Brian Chen[1]($\boxtimes$), Yevgeniy Dodis[2], Esha Ghosh[3], Eli Goldin[2], Balachandar Kesavan[1], Antonio Marcedone[1], and Merry Ember Mou[1]

[1] Zoom Video Communications, San Jose, USA
{brian.chen,surya.heronhaye,antonio.marcedone,merry.mou}@zoom.us
[2] New York University, New York, USA
dodis@cs.nyu.edu, eg3293@nyu.edu
[3] Microsoft Research, Redmond, USA
esha.ghosh@microsoft.com

Abstract. *Key Transparency* (KT) systems allow end-to-end encrypted service providers (messaging, calls, etc.) to maintain an auditable directory of their users' public keys, producing proofs that all participants have a consistent view of those keys, and allowing each user to check updates to their own keys. KT has lately received a lot of attention, in particular its privacy preserving variants, which also ensure that users and auditors do not learn anything beyond what is necessary to use the service and keep the service provider accountable.

Abstractly, the problem of building such systems reduces to constructing so-called append-only Zero-Knowledge Sets (aZKS). Unfortunately, existing aZKS (and KT) solutions do not allow to adequately restore the privacy guarantees after a server compromise, a form of Post-Compromise Security (PCS), while maintaining the auditability properties. In this work we address this concern through the formalization of an extension of aZKS called *Rotatable ZKS* (RZKS). In addition to providing PCS, our notion of RZKS has several other attractive features, such as a stronger (extractable) soundness notion, and the ability for a communication party with out-of-date data to efficiently "catch up" to the current epoch while ensuring that the server did not erase any of the past data.

Of independent interest, we also introduce a new primitive called a *Rotatable Verifiable Random Function* (VRF), and show how to build RZKS in a modular fashion from a rotatable VRF, ordered accumulator, and append-only vector commitment schemes.

Keywords: Key Transparency · Zero-knowledge sets · Verifiable random functions · Post-compromise security

1 Introduction

End-to-end encrypted communication systems (E2EE), including encrypted chat services (such as WhatsApp [45], Signal [38], Keybase [21], iMessage [2]) and

S. Agrawal and D. Lin (Eds.): ASIACRYPT 2022, LNCS 13793, pp. 547–580, 2022.
https://doi.org/10.1007/978-3-031-22969-5_19

encrypted calls (Zoom [6], Webex [44], Teams [32]), are becoming increasingly common in today's world. E2EE systems require each user to publish a public key, and use the corresponding secret key along with their communication partners' public keys to compute a shared secret which can be used to secure the communication. To enable this, service providers (such as Apple, Zoom, Meta, Microsoft, etc.) need to maintain a directory that maps each user to their public keys, a Public Key Infrastructure (PKI) analogous to the one in place to secure the web. The end-to-end guarantees depend on the authenticity of these public keys, as otherwise a malicious service provider (or one who is hacked or compelled to act maliciously) can replace an honest user's identity public key with another public key whose secret key is known to the provider, and thus implement a meddler-in-the-middle (MitM) attack without the communicating users ever noticing.

KEY TRANSPARENCY. To mitigate this issue, many E2EE communication systems provide users with "security codes", i.e. digests of the communication partners' identity public keys rendered as lists of digits or words, or QR codes. To detect potential meddler-in-the-middle attacks, the communicating users are expected to manually check these codes, either by reading them aloud (in calls), scanning them with their phone apps, or otherwise sharing them out-of-band. It is well understood that this has severe usability challenges [3,18,19,43]. *Key Transparency* (KT)[1] systems augment these checks with a fully automated solution that improves both usability and security.

KT systems enable service providers to maintain an auditable directory that maps each user's identifier (such as a username, phone number or email address) to their identity public keys (analogously to how Certificate Transparency [26] allows to monitor PKI certificates). Providers compute and advertise a short (signed) "commitment" com to the whole directory, and update it (creating a new *epoch*) whenever users join the directory or update their keys. When users query a particular *label* label (a key in the map, such as a username), they get the corresponding *value* val (i.e. public key) and a *proof* π that this (label, val) pair is consistent with com.[2] Clients are then encouraged to periodically monitor the directory to make sure their own identifier maps to the correct keys, thus detecting any attempt to MitM their communications.

Assuming cryptographic soundness of such proofs, to ensure that all clients receive the same answer when they query for the same label, it is enough to ensure they all have the same commitment com. To achieve this, KT relies on clients gossiping the commitment [29], or on public and untamperable ledgers such as blockchains [22]. While the implementation of such *gossiping schemes* is not part of the design (and definition) of KT, and they have seen little practical

---

[1] KT is known under various names in the literature, such as *auditable registries, verifiable key directories, auditable directories* etc. For the purpose of this manuscript, we will stick to using KT.

[2] Additionally, if no (label, val) pair exists for a given label, the proof π becomes an *absence* proof for this label.

deployment[3] [14, 28], improvements in this respect seem feasible, and even the potential for users to independently check might deter the server from misbehaving.

AUDITING.   Although the basic functionality already goes a long way towards holding the server accountable for providing incorrect keys to users, clients would incur a high burden if they had to check the server's consistency at every epoch, especially clients whose keys do not change often as the directory evolves. To mitigate that, most KT systems provide additional *auditing functionality*, where more resourceful parties (called *auditors*) can continuously check that certain properties of the directory are maintained across updates (such as the fact that old keys are never erased, and newer ones are simply appended). Technically, when updating the old commitment com to directory D with a newer commitment com' to D', the server can issue a certain proof π_S asserting that $D \subseteq D'$ (and, ideally, revealing nothing more beyond $|D' \setminus D|$). While any user can be an auditor, in practice it is envisioned that relatively few external auditors would continuously monitor the server in this way, and most clients would rely on that assurance. This also justifies relatively large update proofs (with size proportional to $|D' \setminus D|$). Such KT systems are called *auditable*. In addition to keeping the server honest, auditable KTs might ease the need of clients to check their keys at every epoch, if trusted auditors exist. For example, if a client checked earlier that their keys were correct w.r.t. some (audited) old value com', and later got the current value of com from a trusted auditor, they can be sure their keys are still correct w.r.t. com, thus eliminating the need to ask the server to prove this fact again.

To the best of our knowledge, Keybase [25] is the first deployment of an auditable public key directory; they published the first KT digest on April 2014 [24]. Keybase was created as a more user-friendly and secure replacement for PGP, so their KT favors full transparency and auditability over privacy guarantees. For example, Keybase publicly advertises [23] how many devices each user has the Keybase client app installed on, and how often their keys change (i.e., the app is reinstalled). While this is an acceptable tradeoff for many, this privacy leakage can also be a concern, as surfaced in [27], which studied the privacy concerns of using Keybase for US journalists and lawyers. There could be other important business reasons for requiring privacy as well. A business might not want to use a KT system if doing so means revealing to the world how much churn the company has. If the KT system is used to authenticate group membership as well, revealing which groups a user is part of could leak the organizational structure of the business and facilitate social engineering attacks. In fact, Google and Zoom advocate for adding privacy to KT systems [6, 17]. In addition to being privacy-conscious (which is a good practice anyway), these industry leaders are also concerned about current and future laws and regulations, such as GDPR. Indeed, once a major system is in play, it is extremely hard to change it when a new privacy law/regulation comes into effect. For example, creating a publicly

---

[3] While Keybase posts its KT digests to a blockchain, official Keybase clients do not check them.

visible and immutable trail of a user's encryption key changes in a Key Transparency directory would likely cause a GDPR violation. Similarly, if a user asks the provider to delete their account and all traces, doing so would be very hard without privacy built-in.

PRIVACY-PRESERVING KT.   Motivated by these and other considerations, new KT schemes were developed with privacy. Broadly speaking, privacy can be divided in two categories: *content-privacy* and *metadata-privacy*. Content-privacy hides public keys and usernames from unauthorized parties (e.g., auditors and other users who wouldn't otherwise be able to query for those usernames). KT systems supporting content privacy include [20,28,40–42]. Metadata-privacy also hides information such as when each user first registered in the KT, when and how often their keys change, correlations between multiple updates, etc. on top of content-privacy. We denote metadata-hiding KT schemes as *privacy-preserving KT* (ppKT) [7,17,29]. In ppKT, both external auditors and users should learn as little as possible beyond the data they are actively querying. For example, KT commitments and proofs for a certain user identifier should not reveal information about other users' keys and how often they are changing. Similarly, auditors should enforce that no data is ever deleted from the directory, while learning as little as the total number of keys being updated.

Unlike KT systems without any privacy, in which the key directory data structure can be built entirely on symmetric key primitives like Merkle Trees [10], practical KT systems (with either content-privacy or metadata-privacy) achieve privacy through asymmetric primitives such as Verifiable Random Functions (VRFs) [31].[4] Ignoring some important details, given a (label, val) pair, the server holding the VRF's secret key will use a pseudorandom label $y = \mathsf{VRF}(\mathsf{label})$ in place of the original label. Then: (a) pseudorandomness of y ensures that no information about the original label is leaked; (b) verifiability of y ensures that it can be convincingly opened to the original label; and (c) uniqueness of $y = \mathsf{VRF}(\mathsf{label})$ ensures that each label can be used only once.[5]

KEY ROTATION AND POST COMPROMISE SECURITY.   With the growing popularity and user-base of E2EE communication systems, ppKT is very close to real-world, large-scale deployments [6,13,17]. However, as with any real world system, a ppKT system will likely get compromised at some point, so there should be a robust plan to recover from such a compromise, should it happen. One subtle observation in this regard is that current ppKT systems all require the server to maintain a secret key sk (e.g., the secret key to the VRF, as explained above), in addition to simply storing the users' data. Thus, recovering from such compromise necessitates updating the secret/public keys of the server, which is called key rotation. In addition, even if no evidence of actual compromise is ever found, periodically rotating secret keys is considered an industry best

---

[4] Informally, a VRF [31] is similar to a standard pseudorandom function (PRF), except the secret key owner is also committed to the entire function in advance, and can selectively open some of its outputs in a verifiable manner..

[5] Property (c) is why VRF is needed, and regular commitments to label do not work.

practice and sometimes mandated by regulations [35,36]. For ppKT, rotating the key would ideally ensure that compromise of the server would only violate the privacy of past records (which is unavoidable, as the server stores this data anyway), but not of future records.[6] In other words, the primary goal of key rotation is to achieve what is known as *post compromise security* (PCS): the privacy of ppKT systems should be seamlessly restored in case of (possibly silent) key compromise. This is the main question we address in this work:

How easy is it to add PCS to a ppKT, while maintaining high efficiency?

A NAIVE SOLUTION. To see why this question is non-trivial, let us look at a naive attempt to add key rotation to any existing ppKT, such as SEEMless [7]. The first idea is to simply pick a fresh key pair (sk_t, pk_t) for a ppKT with every rotation number t, and basically view the final database D as a disjoint union of t smaller databases $D_1, \ldots, D_t$, where D_i corresponds to the key pair (sk_i, pk_i). On the surface, this seems to maintain the efficiency of the base ppKT, since the server can figure out which "mini-database" D_i contains a given record (id, pk_{id}) and provide a proof only for this value of i. Unfortunately, this does not work, as the server also needs to provide $(i-1)$ absence proofs that id does not belong to any of the previous databases $D_1, \ldots, D_{i-1}$. Otherwise, the server could insert (id, pk_{id}) in database i, (id, pk'_{id}) in database i', and provide different clients with different answers to the same query, even if a good base ppKT is used. And in the case of id not belonging to the entire database D, the server must provide t such absence proofs. Given that clients might need to lookup many identifiers at once and that providers will have to handle a large volume of queries simultaneously, this multiplicative slowdown is unacceptable for practical use.

A better approach—and indeed the approach we take in this work—is to transfer the entire database D when switching from sk_{t-1} to sk_t, thereby initializing $D_t = D_{t-1}$, and then growing D_t when new data items are appended. This ensures that the efficiency of key lookup, the most frequent and important operation in ppKT, is indeed inherited from the base ppKT to which we are adding PCS, because it is always done w.r.t. the latest public key pk_t. Of course, now the server also needs to prove that it honestly initialized $D_t = D_{t-1}$, so that the users or auditors performing this (potentially expensive, but *rare*) check are convinced that no data was added, removed, or modified. Moreover, this check should be done in a privacy-preserving way, so that auditors learn as little as possible about the database $D = D_{t-1} = D_t$ at this moment, beyond the fact that it was correctly "copied" during key rotation.

Unfortunately, none of the existing ppKT systems appear friendly to such *(key) rotation proofs*, while generic zero-knowledge proofs would be prohibitively

---

[6] The effect of compromise on authenticity/auditability is rather minimal anyway, as the key used to sign the commitments would typically be authenticated using the web PKI, and thus can be revoked upon compromise using existing techniques. Moreover, learning the secret server state doesn't help break the binding of the commitment to the entire set of current records in the directory.

inefficient given large database sizes in typical ppKT systems. As our main technical contribution, we overcome this difficulty by designing a specialized, but still highly efficient, ppKT system which supports efficient key rotation, and hence provides PCS against (possibly silent) server compromises.

1.1   Our Contributions

Before our concrete solution, we list our contributions from the modeling and definitions perspective.

MODELING AND DEFINITIONS. First, much like earlier works on ppKT [7,17,29] we abstract the primitive that we need. Our primitive, which we term *Rotatable Zero Knowledge Set* (RZKS), is a natural extension of the so-called append-only zero-knowledge set (aZKS) from [7].

At a high level, aZKS is a primitive where a prover can incrementally commit to a dictionary D, and later prove (in zero-knowledge) a statement of the form that a certain (label, val) pair belongs to the dictionary, or that a certain label does not belong to the dictionary (for any val). Moreover, there is at most one val for any label, and this val cannot be modified once it is assigned. To model the incremental nature of aZKS, the prover can also prove the "append-only" property to the auditors, such that two commitments com and com' correspond to two dictionaries D and D', where D is a subset of D', in almost[7] zero-knowledge.

Our RZKS notion extends aZKS in several ways. First, and most importantly from the perspective of PCS, we allow a new algorithm for key rotation. Syntax-wise, it is the same as the append algorithm of aZKS: given a (possibly empty) set S of fresh {(label, val)}-pairs to be appended to the current database, we update the commitment com to D to a new commitment com' to $D' = D \cup S$, and output a proof π_R that this operation was done "consistently". However, unlike the regular append operation given by proof π_S, the proof size and time for the rotation operation is allowed to be proportional to the entire database D', as opposed to the number of appended elements $|S|$. What we gain though is the PCS property: unlike with the regular append, compromising the server's state (including D) before the rotation does not help the attacker learn any new information about newly appended elements S, or any elements appended in the future (including those by the standard append operation). (As a bonus, it also wipes out the minimal leakage of regular append mentioned in Footnote 7.)

Second, and of independent interest, we extend the aZKS functionality to support what we call *extension proofs*. Such proofs allow a party to verify that a given newer commitment $\text{com}_{t'}$ commits to a given older commitment com_t (and, therefore, also implies that both $\text{com}_{t'}$ and com_t commit to the same sequence $\text{com}_1, \text{com}_2, \ldots, \text{com}_{t-1}$), for any $t' > t$ (as opposed to the append-only proofs in aZKS only supporting $t' = t + 1$). Here, t and t' are the total number of appends and rotations that were performed to produce the dictionary corresponding to

---

[7] According to a well-defined leakage profile. For [7], the only such leakage reveals if a label known to be missing in D is later inserted in D', which seems acceptable for the main application to KT.

each commitment. These extension proofs are *extremely* efficient (only logarithmic in the number of epochs t'), as they can be instantiated using Merkle Tree append-only proofs [7,41].

Note that, by themselves, extension proofs do not prove that the database has evolved consistently (for example, it is possible that $D_t \not\subseteq D_{t'}$). However, auditors still check that each successive epoch correctly performs append or rotation operations. As a result, extension proofs allow users to confirm that the commitments they receive are authentic and represent a consistently evolving database by occasionally verifying commitments with a trusted auditor, instead of verifying every commitment they receive, which would be a frequent and expensive operation. Concretely, suppose a user receives a series of commitments $com_{t_1}, \ldots, com_{t_n}$ from the server, possibly over a long period of time, along with extension proofs from each com_{t_i} to the next $com_{t_{i+1}}$ (which the user verifies). Then, by verifying just com_{t_n} with an auditor, blockchain, or other gossiping mechanism, the user can guarantee the consistency of every previous com_{t_i} they received with those other sources' views. Furthermore, if the user trusts that the source they are verifying against has also verified that the database evolved consistently at each epoch, they can infer that each $D_{t_i} \subseteq D_{t_{i+1}}$. This also allows auditors and other clients to only gossip about the latest commitment com' and forget any previous commitments. If an older commitment com_{old} is ever needed, the server can always provide com_{old} and the extension proof from com_{old} to com'.

Third, each query also explicitly indicates the epoch at which the queried pair was added to the RZKS directory, which can be verified without any increase in the proof size (obtaining this information in an aZKS would require multiple proofs). We believe that this information can be helpful in practical applications, as older records/keys are often considered more trustworthy than newer ones (the owner has had more time to react to a compromise), and quickly comparing the age of two records can be helpful for more complex applications of RZKS beyond standard KT[8]. Moreover, while previous ppKT do not allow to determine this efficiently, they do not hide this information either.

Finally, our notion of RZKS strengthens the soundness definition of aZKS presented in [7]. Namely, the latter mandates that an adversary cannot produce valid proofs of conflicting statements (for example, proving that the same key maps to different values, possibly in different epochs). Instead, we notice that the soundness of the SEEMless construction of [7] is proven in the Random Oracle model anyway, where we can achieve much stronger forms of soundness. Indeed, our RZKS notion demands a very strong form of *extractability-based* soundness. Roughly, we require the existence of an extractor, which, given any malicious commitment com produced by the attacker (and its random oracle queries), can extract the entire database D for which the attacker can later produce verifying

---

[8] For example, Keybase uses its KT dictionary to also store other statements signed by a user's device, such as when a user wants to add another user to a group: knowing that the statement was signed before the key that signed it is revoked/rotated is important for the security of the system.

membership proofs. We believe this stronger property makes it easier to reason about the security of applications of RZKS.

RZKSCONSTRUCTION. Finally, we show how to build an efficient RZKS system. Our starting point is the aZKS construction from SEEMless [7]. SEEMless uses— in a black-box way—a *verifiable random function* (VRF) [31] and cryptographic hash function to build their aZKS, and recommends the specific DDH-based VRF from the upcoming VRF standard [16].

Recall, a VRF allows the secret key owner (e.g., server) to compactly commit to an entire random-looking function f, but in a way that allows them to convincingly open individual function outputs $f(x)$, without compromising the randomness of yet unopened outputs $f(x')$ for $x' \neq x$. In the aZKS construction of [7], when appending a $(x = \text{label}, v = \text{val})$ pair to D, the server uses the VRF output $f(x)$ to decide where to put a commitment to v in some Merkle Tree T that it builds. If this place is occupied, the server knows that D already contains some v' associated with label x, and can reject the request. Otherwise, it inserts some commitment to v into the Merkle Tree T, and uses the new root of T as the modified commitment value com' to $D' = D \cup \{(x, v)\}$. Intuitively, the use of VRF ensures privacy, as it hides information about the labels that would otherwise be leaked by Merkle proofs of "neighboring" labels. On the other hand, VRF uniqueness and verifiability properties ensure that the server cannot cheat.

One can now consider how to extend the scheme above to support key rotation, provided that the underlying VRF can support what we call *VRF rotation proofs*. Intuitively, a RZKS rotation proof will switch the VRF key from f_1 to f_2, rebuild the Merkle Tree T_1 into T_2 using the same commitments to each of the values, and openly reveal the one-to-one correspondence between leaves of T_1 and T_2 associated with all keys x present in the original database D before the rotation. However, recall that the value x itself should be hidden from auditors verifying consistency of key rotation, which leads to the following problem we solve in this work. We need to design a VRF with a fast zero-knowledge proof showing that two VRF outputs y_1 and y_2 under two independent keys f_1 and f_2 correspond to the same secret input x: $y_1 = f_1(x)$ and $y_2 = f_2(x)$. We call this novel type of VRFs *rotatable*. We discuss them next, and defer the rest of the details of our final RZKS construction to Sect. 5.2, simply highlighting here its modularity: it is built from any rotatable VRF, commitments and other generic building blocks instantiable from Merkle Trees.[9]

ROTATABLE VRFs. Unfortunately, supporting an efficient ZK proof mentioned above is not sufficient for the type of rotatable VRFs we need for RZKS. To achieve PCS for RZKS, our VRF also needs to satisfy a novel type of "non-committing property": upon compromise, the attacker learns of a compact secret key sk for the VRF, which suddenly explains a lot of VRF outputs $\{y\}$ that the attacker saw prior to the corruption (but did not know the corresponding inputs $\{x\}$). More concretely, we use a simulation-based rotatable VRF defini-

---

[9] Namely, so called ordered accumulators, and append-only vector commitment schemes. See Sect. 5.1.

tion, extending the earlier "simulatable VRF" notion of [9] to handle rotations. Under this notion, the simulator must in particular "win" in the following game (which is the most challenging part of our definition explaining the heart of the problem). The simulator must commit to a VRF public key pk, get a bunch of *random* strings $\{y\}$ as various VRF outputs of *unknown* inputs $\{x\}$, answer random oracle queries from the attacker, then learn the hidden set $\{x\}$, and finally produce a secret key sk that correctly explains that $f(x) = y$ for all matching (x, y) pairs of the corresponding sets $\{x\}$ and $\{y\}$.

This problem seems to relate to the area of non-committing encryption (NCE) [5,34], where one compact secret is supposed to open many previously committed ciphertexts in a certain way. As with non-committing encryption [34], building standard model "non-committing" VRF is impossible, as one short secret key sk cannot "explain away" arbitrarily many random looking outputs y. On the other hand, given that several NCE schemes exist in the random oracle model (e.g., [4]), one might hope that the same simple ideas[10] will work in our VRF setting as well. Unfortunately, this does not appear to be the case, due to the inherently algebraic structure of VRF proofs. To understand this inherent tension, let us consider the concrete efficient VRF (from the VRF standard [16]) recommended by the authors of SEEMless. In this VRF, the secret key sk for the VRF is a random exponent α, the public key $pk = g^\alpha$ (for public generator g), and the VRF value $y = f(x) = F'(F(pk, x)^\alpha)$, where F is a random oracle and F' is an "extractor" meant to map a random group element to a random bit-string. (The proof π that $y - f(x)$ is the value $z = F(pk, x)^\alpha$ and the standard Fiat-Shamir variant of the Σ-protocol for the DDH tuple $(g, pk, F(pk, x), z)$ [12].)

When rotating the key pair (α, g^α) to a fresh key pair (β, g^β), first we need to ensure that there exists an efficient ZK proof showing that two random values y and y' satisfy the relation $y = F'(F(g^\alpha, x)^\alpha)$ and $y' = F'(F(g^\beta, x)^\beta)$. As the first obstacle, this seems hard due the outside extractor F'. Fortunately, this problem is trivially solved by getting rid of the "outer extractor" F', and thinking of the VRF as outputting a group element (rather than bit-string) $y = F(pk, x)^\alpha$. Indeed, the standard VRF proof in [16] shows that the above VRF is already secure. The next problem comes from the fact that the old VRF f and the new VRF f' have different public keys g^α and g^β hashed inside the "inner random oracle" F. Once again, it turns out that the VRF proof just needs some domain separation, and goes through if we redefine the output $y = F(salt, x)^\alpha$, where $salt$ is some unpredictable value which does not need to change with any rotation.[11]

This already gives us the ability to construct (at least "syntactically") the required ZK proof of rotation when moving from $sk = \alpha$ to $sk' = \beta$. Indeed, for any unknown x, if $y = F(x)^\alpha$ and $y' = F(x)^\beta$, the server can simply prove that

---

[10] Namely, to a posteriori program random oracle in a manner depending on the strings y, on appropriate inputs involving the secret key sk.

[11] For simplicity of exposition, we omit $salt$ from our description, but recommend that each application uses a fresh salt.

the tuple $(g^\alpha, g^\beta, y, y')$ is a proper DDH tuple (using witness $w = \beta/\alpha$).[12] As we said, though, we also need to provide the PCS property mentioned above. And this appears hopeless at first glance. Indeed, the public key $pk = g^\alpha$ commits to α information-theoretically. Moreover, when programming the random oracle query $F(x)$, the simulator does not know yet which random output y corresponds to x. Hence, the simulator has no chance to correctly program $F(x) = y^{1/\alpha}$. For regular ("non-rotatable") VRFs, we would try to fake the Fiat-Shamir proofs for correctness. In fact, this would extend to rotatable VRFs without the PCS property (i.e., without corruptions of α); but, of course, this is not very interesting from the application perspective. In the case of corruptions, however, the simulator is committed to the secret key α, and will be caught cheating with certainty.

OUR SOLUTION: GGM ANALYSIS. Interestingly, the difficulty of completing the simulation-based PCS proof for our tweaked construction $y = F(x)^\alpha$ does not seem to translate to an explicit attack on the resulting rotatable VRF. Rather, we cannot build a sufficiently adaptive simulator to prevent the type of attack in the previous paragraph. So we ask the question if the construction might be actually be secure, despite the natural proof breaking down. Somewhat surprisingly, we give supporting evidence that this is the indeed the case, by providing such a security analysis in the *generic group model* (GGM) of Shoup [37].[13]

Recall that in Shoup's GGM, all group elements have random bit-string representations, and the group operation $\star$ also has a random multiplication table $\star(a, b)$ (subject to associativity of multiplication). As such, most security assumptions in standard groups (e.g., DDH) will hold in the GGM unconditionally. But now the simulator can commit to the public key $pk = g^\alpha$ *without committing to* α *information-theoretically*. Intuitively, since the attacker does not know value α before the compromise, and has a bounded number of multiplication queries to explore, the simulator can simply choose a random value of α as the secret key, and will have enough freedom to "mess" with the multiplication table $\star(a, b)$ to simultaneously satisfy many equations of the form $y_i = F(x_i)^\alpha$ (as well as $pk = g^\alpha$). However, the formal proof of this statement is rather subtle, and forms one of the main technical novelties of this work. For example, the group laws mandate certain relationships that the attacker can always satisfy, so the simulator has to be extremely careful not to "overplay its hand" and program the multiplication table too aggressively. We present the full simulation proof in Sect. 4.4, and hope that our GGM proof technique will find applications for analyzing other "non-committing" algebraic primitives.

INTERPRETATION OF OUR RESULT. On a philosophical point, we suggest that the value of our GGM security proof should be understood in light of the fact

---

[12] Our final ZK proof will aggregate many such individual input rotation proofs into one compact proof.

[13] We stress that we only use GGM for the ZK property of our construction. Our stronger extractability-based soundness is still proven in the random oracle model, and does not require the GGM.

that ROM-based proofs seem to be inherently stuck, at least for the natural rotatable VRF that we consider. Aside from the obvious consideration that we focused on finding a practical solution to a natural problem for which we could not find an explicit attack, we note that the requirements in our definition of rotatable VRFs are quite strong. Basically, the simulator has to answer all ideal queries without knowing any of the input/output behavior of the VRF, and then must produce a single secret key consistent with not only these ideal queries, but also all fake proofs (including rotation). From this perspective, we feel that it is quite surprising that we managed to overcome these difficulties at all, even relying on the GGM. The GGM proof can also be considered a sanity check that our scheme is likely to be secure under weaker models/assumptions, provided one correspondingly weakens our extremely demanding simulation security definition.

More generally, while the ROM model is obviously preferred to the GGM, practitioners do not mind relying on the GGM, provided it solves an interesting problem. Indeed, we can point to several examples of interesting primitives where standard analyses appear to be stuck, and the GGM provided meaningful answers to these questions. Most notably, Signal leverages in production a protocol which can only be proven secure in the GGM model to achieve group privacy [11,39]. Other important examples include optimal structure-preserving signature schemes [1] and state-restoration soundness analysis of Bulletproofs [15].

2   Notation and Preliminaries

We use square brackets $[a_1, a_2, \ldots, a_n]$ to denote ordered lists of objects, and curly brackets $\{a, b, c, \ldots\}$ for sets. We represent maps $D = \{(a, b), (c, d), \ldots\}$ as sets of label-value pairs. If D is a set of pairs, we denote with $D_{(\cdot)} = \{a, c, \ldots\}$ the set of the first components of each pair (the domain of the corresponding map), and with $D^{(\cdot)} = \{b, d, \ldots\}$ the set of the second components (the range of the map). When clear from context, we slightly abuse notation and write $a \in D$ (instead of $a \in D^{(\cdot)}$) if there is a pair $(a, \cdot)$ in the set, and (when unique) we denote the corresponding value with $D[a]$. Similarly, we use $C[i]$ to denote the i-th element of list C (1-indexed), and $last(C)$ to denote its last element.

We denote with λ the security parameter. Given two security games I and R, each parameterized by an algorithm $\mathcal{A}$ (the adversary), we define the advantage of $\mathcal{A}$ in distinguishing the two experiments as $\left|\Pr[I^{\mathcal{A}} = 1] - \Pr[R^{\mathcal{A}} = 1]\right|$. In each figure defining a security experiment, we denote with $\mathcal{A}^{\mathcal{O}\cdots}(a_1, \ldots, a_n)$ an execution of algorithm $\mathcal{A}$ on input $a_1, \ldots, a_n$ with access to all the oracles defined in that figure.

We use the following conventions to describe algorithms. When a hash function takes more than one input (or a pair), we assume that there is a well defined way to serialize and deserialize such a tuple into a bitstring. Given a boolean b, we use **ensure** b as shorthand for "if not b, **return** 0". We use "**parse** a as $(a_1, \ldots, a_n)$" to denote that an algorithm tries to unpack a tuple of objects, and if the tuple does not have the appropriate length the algorithm returns a dummy output/error. In a security game, we use "**assert** b" to denote that if b is false, the experiment is immediately terminated with a special return value $\bot$; during an oracle call, we use "**require** b" to indicate that if b is false the oracle call by the adversary is interrupted without output, and any effects on the state of this call are reverted.

In the full version of the paper, we recall the Diffie-Hellman assumption, and briefly discuss the Random Oracle Model and Generic Group Model assumptions that our work depends on.

3   Rotatable Zero Knowledge Set

In this section, we formally define Rotatable Zero Knowledge Sets (RZKS). The primitive Zero-Knowledge Set was introduced in [8,30] and extended to append-only ZKS (aZKS) in [7]. We extend the notion of aZKS from SEEMless to add new properties as well as strengthen the soundness guarantees in our new primitive: RZKS.

Definition 1. A Rotatable Zero Knowledge Set (RZKS) consists of a tuple of algorithms $\mathcal{Z} = (\mathcal{Z}.\mathsf{GenPP}, \mathcal{Z}.\mathsf{Init}, \mathcal{Z}.\mathsf{Update}, \mathcal{Z}.\mathsf{PCSUpdate}, \mathcal{Z}.\mathsf{VerifyUpd}, \mathcal{Z}.\mathsf{Query}, \mathcal{Z}.\mathsf{Verify}, \mathcal{Z}.\mathsf{ProveExt}, \mathcal{Z}.\mathsf{VerExt})$ defined as follows:

▷ $\mathsf{pp} \leftarrow \mathcal{Z}.\mathsf{GenPP}(1^\lambda)$: This algorithm takes the security parameter and produces public parameter pp for the scheme. All other algorithms take these pp as input implicitly, even when not explicitly specified.

▷ $(\mathsf{com}, \mathsf{st}) \leftarrow \mathcal{Z}.\mathsf{Init}(\mathsf{pp})$: This algorithm takes as input the public parameters, and produces a commitment com to an empty datastore $\mathsf{D}_0 = \{\}$ and an initial server/prover state st. A datastore D will be a collection of $(\mathsf{label}_i, \mathsf{val}_i, t)$ tuples, where t is an integer indicating that the tuple has been added to the datastore as part of the t-th Update or $\mathsf{PCSUpdate}$ operation (we call this an epoch). Labels will be unique across the datastore (it can be thought of as a map). Each server state st will contain a datastore and a digest, which we will refer to as $\mathbf{D}(\mathsf{st})$ and $\mathbf{com}(\mathsf{st})$. Similarly, each commitment will include the epoch $\mathbf{t}(\mathsf{com})$ of the datastore to which it is referring. (Alternatively, these can be thought of as deterministic functions which are part of the scheme.)

▷ $(\mathsf{com}', \mathsf{st}', \pi_S) \leftarrow \mathcal{Z}.\mathsf{Update}(\mathsf{pp}, \mathsf{st}, S)$, $(\mathsf{com}', \mathsf{st}', \pi_S) \leftarrow \mathcal{Z}.\mathsf{PCSUpdate}(\mathsf{pp}, \mathsf{st}, S)$: Both algorithms take in the public parameters, the current state of the prover st, and a list $S = \{(\mathsf{label}_1, \mathsf{val}_1), (\mathsf{label}_2, \mathsf{val}_2), \ldots, (\mathsf{label}_n, \mathsf{val}_n)\}$ of new (label, value) pairs to insert (the labels must be unique and not already part of $\mathbf{D}(\mathsf{st})$). The algorithm outputs an updated commitment to the datastore, an updated internal state st', and a proof π_S that the update has been done correctly. Intuitively, com' is a commitment to the updated datastore $\mathbf{D}(\mathsf{st}')$ at epoch $\mathbf{t}(\mathsf{st}') = \mathbf{t}(\mathsf{st}) + 1$, which extends $\mathbf{D}(\mathsf{st})$ by also mapping each label_i in S to the pair $(\mathsf{val}_i, \mathbf{t}(\mathsf{st}'))$. As we will see, Update and $\mathsf{PCSUpdate}$ have different tradeoffs between their efficiency and the privacy guarantees they offer.

▷ $0/1 \leftarrow \mathcal{Z}.\mathsf{VerifyUpd}(\mathsf{pp}, \mathsf{com}_{t-1}, \mathsf{com}_t, \pi_S)$: This deterministic algorithm takes in two commitments to the datastore output at successive invocations of Update, and verifies the above proof.

▷ $(\pi, \mathsf{val}, t) \leftarrow \mathcal{Z}.\mathsf{Query}(\mathsf{pp}, \mathsf{st}, u, \mathsf{label})$: This algorithm takes as input a state st, an epoch $u \leq \mathbf{t}(\mathsf{st})$, and a label. If a tuple $(\mathsf{label}, \mathsf{val}, t) \in \mathbf{D}(\mathsf{st})$ and $t \leq u$, it returns val, t and a proof π. Else, it returns $\mathsf{val} = \perp$, $t = \perp$ and a non-membership proof π. In both cases, proofs are meant to be verified against the commitment com_u output during the u-th update.

▷ $1/0 \leftarrow \mathcal{Z}.\mathsf{Verify}(\mathsf{pp}, \mathsf{com}, \mathsf{label}, \mathsf{val}, t, \pi)$: This deterministic algorithm takes a $(\mathsf{label}, \mathsf{val}, t)$ tuple, and verifies the proof π with respect to the commitment com. If $\mathsf{val} = \perp$ and $t = \perp$, this is considered a proof that label is not part of the data structure at epoch $\mathbf{t}(\mathsf{com})$.

▷ $\pi_E \leftarrow \mathcal{Z}.\mathsf{ProveExt}(\mathsf{pp}, \mathsf{st}, t_0, t_1)$: This algorithm takes the state of the prover and two epochs t_0, t_1, and returns a proof π_E that the datastore after the t_1-th update is an extension of the datastore after the t_0-th update. Proofs are meant to be verified against the commitments com_{t_0} and com_{t_1} output by Update during the t_0-th and t_1-th update.

▷ $1/0 \leftarrow \mathcal{Z}.\mathsf{VerExt}(\mathsf{pp}, \mathsf{com}_{t_0}, \mathsf{com}_{t_1}, \pi_E)$: This deterministic algorithm takes two datastore commitments and a proof (generated by ProveExt) and verifies it.

We require a RZKS to satisfy the following security properties:

Completeness. We will say that an RZKS satisfies completeness if for all PPT adversaries $\mathcal{A}$, the probability that the game described in Fig. 1 outputs 0 is negligible in λ.

Intuitively, all updates and queries should behave as expected by their descriptions in the definition. Furthermore, all proofs produced by various updating or querying algorithms should verify when properly queried to the corresponding verification algorithms. More formally, an adversary only breaks completeness if it is able to construct a sequence of queries such that one of the assertions in Fig. 1 fails. For example, the assertion $\mathbf{D}(\mathsf{st}') = \mathbf{D}(\mathsf{st}) \cup \{(\mathsf{label}_i, \mathsf{val}_i, t + 1)\}_{i \in [j]}$ in Update(S) will only fail if the elements added in S are not correctly added to the state of the datastore. Similarly, in Query(label, u) we assert that P.Verify(com_u, label, val', t', π) succeeds, where (val', t', π) are those produced by the corresponding call to P.Query.

Soundness. We will say that an RZKS satisfies soundness if there exists an extractor Extract such that for all PPT adversaries $\mathcal{A}$, the advantage of $\mathcal{A}$ in distinguishing the two experiments described in Fig. 2 is negligible in λ. Note that all the algorithms executed in the experiment get implicit access to the Ideal oracle, as they might need to make, e.g., random oracle calls.

The extractor Extract is required to provide various functionalities based on its first input:

P-Completeness($\mathcal{A}$):

$pp' \leftarrow$ P.GenPP(1^λ)
$(com', st') \leftarrow$ P.Init(pp')
assert $com(st') = com'$ and $t(com') = 0$ and $D(st') = \{\}$
$com_0 \leftarrow com', st \leftarrow st', t \leftarrow 0, pp \leftarrow pp'$
$\mathcal{A}^{\mathcal{O}\cdots}(pp, com_0)$
return 1

Oracles Update(S) and PCSUpdate(S):

parse S as $(label_1, val_1), \ldots, (label_j, val_j)$
require $label_1, \ldots, label_j$ are distinct and do not already appear in $D(st)$
$(com', st', \pi) \leftarrow$ P.Update(st, S) // resp. P.PCSUpdate(st, S)
assert $com(st') = com', t(com') = t + 1$ and $D(st') = D(st) \cup \{(label_i, val_i, t+1)\}_{i \in [j]}$
assert $y \leftarrow$ P.VerifyUpd(com_t, com', π); $y = 1$
$com_{t+1} \leftarrow com', st \leftarrow st', t \leftarrow t + 1$

Oracle Query($label, u$):

require $0 \leq u \leq t$
$(\pi, val', t') \leftarrow$ P.Query($st, u, label$)
If $label \in D(st)$, $(val_D, u_D) \leftarrow D(st)[label]$ and $u_D \leq u$:
    assert $(val', t') = (val_D, u_D)$
Else
    assert $(val', t') = (\bot, \bot)$
assert $y \leftarrow$ P.Verify($com_u, label, val', t', \pi$); $y = 1$

Oracle ProveExt(t_0, t_1): // RZKS only

require $0 \leq t_0 \leq t_1 \leq t$
$\pi_E \leftarrow$ P.ProveExt(st, t_0, t_1)
assert $y \leftarrow$ P.VerExt($com_{t_0}, com_{t_1}, \pi_E$); $y = 1$

Oracle ProveAll(t'): // OA only

$\pi \leftarrow$ P.ProveAll(st, t')
assert $y \leftarrow$ P.VerAll($com_{t'}, D(st)_{\leq t'}, \pi$); $y = 1$

Fig. 1. Completeness for RZKS and OA (an Ordered Accumulator, defined in Sect. 5.1) primitives (denoted with P). Some of the oracles are only applicable to one primitive. In this experiment, the adversary can read all the game's state and the oracle's intermediate variables, such as $com_i \forall i, st, y$. The experiment returns 1 unless one of the assertions is triggered. These checks enforce that the data structure is updated consistently, that the outputs of query reflect the state of the data structure, and that honestly generated proofs pass verification as intended.

- $pp', st \leftarrow$ Extract(Init): Samples public parameters indistinguishable from honestly generated public parameters such that extraction will be possible. Also generates an initial state.
- $D_{com} \leftarrow$ Extract(Extr, st, com): Takes in the internal state and a commitment to the datastore. Outputs the set of $(label, val, i)$ committed to.
- $C_{com} \leftarrow$ Extract(ExtrC, st, com): Takes in the internal state and a commitment to the datastore. Outputs the set of previous commitments, indexed by epoch.

- *out*, st ← Extract(Ideal, st, *in*): Simulates the behavior of some ideal functionality (for example a random oracle or generic group). Takes in any input and
  produces an output indistinguishable from the output the ideal functionality
  would have on that input.

One small subtlety of the definition here is that we do not allow the extractor
to update its state outside of Ideal calls. The only advantage that the extractor
gets over an honest party is its control over the ideal functionality. This allows
for easier composition, since a larger primitive utilizing RZKS will not need to
simulate extractor state.

An adversary breaks soundness if it either distinguishes answers to Ideal
queries in the real game from those produced by the extractor, or if it causes some
assertion to be false in the ideal game. Each assertion in the ideal game captures
some way in which the extractor could be caught in an inconsistent state. For
example, let us consider the assertion D[com][label] $=$ (val*, i^*) in CheckVerD.
This will be false if the adversary can provide a proof that (label, val*, i^*) is in
the datastore with digest com, but the extractor expects this datastore to either
not contain label or to contain (label, val, i) for some different (val, i).

Our soundness definition strengthens the traditional one by providing
extractability. aZKS soundness already guarantees that a (malicious) prover is
unable to produce two verifying proofs for two different values for the same label
with respect to an aZKS commitment it has already produced. However, that
definition does not guarantee that the malicious prover knew the entire collection
of (label, value) pairs at the time it produced the commitment. Extractability
requires that by mandating that the entire datastore can be extracted from the
commitment, except with negligible probability.

We also explicitly guarantee consistency among the RZKS commitments produced over epochs. Informally, consistency guarantees that each commitment to
an epoch also binds the server to all previous commitments (i.e. these can be
extracted from the former). In particular, when the client swaps a commitment
com^a with a more recent one com^b by verifying an extension proof, and then
checks with an auditor that com^b is legitimate, the client can be sure that any
auditor who checked all consecutive audit proofs up to com^b must also have
checked the same com^a for epoch a. This is modeled in the security game by the
assertions in the ExtractC, CheckVerUpdC, and CheckVerExt oracles.

Zero Knowledge. We will say that an RZKS is zero knowledge for leakage function
$L = (L_{\text{Update}}, L_{\text{PCSUpdate}}, L_{\text{Query}}, L_{\text{ProveExt}}, L_{\text{LeakState}})$ if there exists a simulator $\mathcal{S}$
such that every PPT malicious client algorithm $\mathcal{A}$ has negligible advantage in
distinguishing the two experiments of Fig. 3.

The stateful simulator $\mathcal{S}$ is required to provide various functionalities:

- com', pp' ← $\mathcal{S}$(Init): Samples public parameters and an initial commitment
  indistinguishable from honest public parameters such that it will be possible
  to simulate proofs.

P-Sound-IDEAL($\mathcal{A}$):

pp', st $\leftarrow$ Extract(Init)
D $\leftarrow$ {}, $C \leftarrow$ [], pp $\leftarrow$ pp'
$b \leftarrow \mathcal{A}^{Ideal(\cdot),\cdots}$(pp)
return b

Oracle ExtractD(com):

$D_{\mathsf{com}} \leftarrow$ Extract(Extr, st, com)
If com $\in$ D **assert** D[com] $= D_{\mathsf{com}}$
D[com] $\leftarrow D_{\mathsf{com}}$
assert $\forall$ (label, val, i) $\in$ D[com] $: 0 < i \le \mathbf{t}(\mathsf{com})$

Oracle ExtractC(com): // RZKS only

$C_{\mathsf{com}} \leftarrow$ Extract(ExtrC, st, com)
If com $\in$ C **assert** $C[\mathsf{com}] = C_{\mathsf{com}}$
$C[\mathsf{com}] \leftarrow C_{\mathsf{com}}$
assert $|C[\mathsf{com}]| = \mathbf{t}(\mathsf{com})$ and
    $last(C[\mathsf{com}]) = \mathsf{com}$

Oracle CheckVerD(com, label, val*, i^*, π):

require P.Verify(pp, com, label, val*, i^*, π) $= 1$
    and com $\in$ D
If val* $= \perp$ or $i^* = \perp$:
    assert label $\notin$ D[com] $\wedge$ val* $= i^* = \perp$
Else **assert** D[com][label] $= ($val*$, i^*)$

Oracle CheckVerUpdD(com^a, com^b, π):

require P.VerifyUpd(pp, com^a, com^b, π) $= 1$ and
    com^a, com$^b \in$ D
assert D[com^a] $\subseteq$ D[com^b], and
    $\mathbf{t}(\mathsf{com}^b) = \mathbf{t}(\mathsf{com}^a) + 1$, and
    $\forall$(label, val, t) $\in$ D[com^b] \ D[com^a] :
        $t = \mathbf{t}(\mathsf{com}^b)$, and
    $(\mathbf{t}(\mathsf{com}^a) \ne 0$ or D[com^a] $= \{\})$

Oracle CheckVerUpdC(com^a, com^b, π): // RZKS only

require P.VerifyUpd(pp, com^a, com^b, π) $= 1$ and
    com^a, com$^b \in$ C
assert $\mathbf{t}(\mathsf{com}^b) = \mathbf{t}(\mathsf{com}^a) + 1$, and
    $\forall j \le \mathbf{t}(\mathsf{com}^a) : C[\mathsf{com}^a][j] = C[\mathsf{com}^b][j]$

Oracle CheckVerExt(com^a, com^b, π): // RZKS only

require P.VerExt(pp, com^a, com^b, π) $= 1$ and
    com^a, com$^b \in$ C
assert $\forall j \le \mathbf{t}(\mathsf{com}^a) : C[\mathsf{com}^a][j] = C[\mathsf{com}^b][j]$

Oracle CheckVerAll(com, S, π): // OA only

require P.VerAll(pp, com, S, π) $= 1$ and com $\in$ D
assert D[com] $= S$

Oracle Ideal(in):

out, st $\leftarrow$ Extract(Ideal, st, in)
return out

Fig. 2. Soundness for RZKS and OA (both denoted by P). In the ideal world, the map D stores, for each commitment com, the datastore that the Extract algorithm output for that commitment. In addition the map C stores, for each commitment, the (ordered) list of commitments to previous epochs. When the adversary provides proofs, we require that the proofs are consistent with such data structures. In the real world (not pictured), the public parameters are generated as pp $\leftarrow$ P.GenPP(1^λ), and all the oracles do nothing and return no output, except for the Ideal oracle, which implements the ideal objects (such as random oracles) that we abstract to prove security of the primitives (and that are controlled by the extractor in the ideal world). In both cases, P's algorithms implicitly get access to the Ideal oracle as needed.

- (com', π) $\leftarrow \mathcal{S}(($PCS$)$Update, l): Takes in some leakage l about an Update (or, analogously, PCSUpdate) query on input S, i.e. in the experiment $l \leftarrow L_{\mathsf{Update}}(S)$ (or $l \leftarrow L_{\mathsf{PCSUpdate}}(S)$). Outputs a commitment com' indistinguishable from a commitment to the previous datastore with the elements of S appended. Furthermore simulates a proof π that the update was done correctly.
- (π, val', t') $\leftarrow \mathcal{S}(\mathsf{Query}, l)$: Takes in leakage $l \leftarrow L_{\mathsf{Query}}(u, \mathsf{label})$ about the entry indexed by (u, label) in the datastore. Outputs val', t' which would have been returned by an honest query. Also simulates a proof π that D[label] $= ($val', $t')$, or an absence proof if label $\notin$ D.

- $\pi \leftarrow \mathcal{S}(\mathsf{ProveExt}, l)$: Takes in partial information $l \leftarrow L_{\mathsf{ProveExt}}(t_0, t_1)$ from a ProveExt query the between epochs t_0 and t_1. Outputs an extension proof that the commitment provided at epoch t_1 binds to the one at epoch t_0.
- $\mathsf{st} \leftarrow \mathcal{S}(\mathsf{Leak}, l)$: Takes in partial information $l \leftarrow L_{\mathsf{LeakState}}()$ about the datastore and outputs a simulated state consistent with the information given.
- $out \leftarrow \mathcal{S}(\mathsf{Ideal}, in)$: Simulates the behavior of some ideal functionality. Takes in any input and produces an output indistinguishable from the output the ideal functionality would have on that input.

Note that the particular leakage given will be construction specific, but should be designed to be as minimal as possible. Our choice of leakage will be described in detail in Sect. 5.3. In the experiment, the only information the simulator has access to is the output of the leakage function, as well as the queries made to the Ideal oracle. The simulator's ability to control the ideal oracle is crucial for security proofs to go through.

Informally, zero knowledge here means that the proofs generated by any sequence of honest calls to RZKS algorithms can be simulated given access to minimal information about the queries made. The adversary breaks zero knowledge if it is able to generate a sequence of queries such that it can distinguish the output of the simulator from honestly generated outputs and proofs. For example, if the simulator is unable to simulate query proofs, then an adversary could succeed by calling the Update({label, val}) oracle for some (label, val), then the $(\pi, \mathsf{val}, 1) \leftarrow$ Query(label, 1) oracle, and running RZKS.Verify on π. Since the simulator can't simulate query proofs, π generated in the ideal world will not verify and so will be distinguished from π generated in the real world.

Post-compromise security is modelled by allowing for LeakState calls, which reveal the state in its entirety. When the adversary queries this oracle, the simulator is required to output a state that appears consistent with whatever proofs it has revealed before. Healing from compromise is modelled by having a dedicated leakage function for PCSUpdate (different from Update). Note that since all the leakage functions share state, calling LeakState or PCSUpdate might affect the leakage of other future queries.

3.1 Application to Key Transparency

Recall that in an aZKS, the value associated with each label cannot be updated: the prover can only add new (label, value) pairs to the directory. In SEEMless [7], the server uses aZKS to commit to its public key directory by setting the label to (userID || version number) and value to the public key of the user corresponding to that ID. Every update to the underlying public key directory becomes a new label addition to the aZKS. The server collects a batch of these additions and periodically updates the directory, creating a new epoch and publishing a new aZKS commitment. Clients must hold on to all previous commitments until they have double-checked them with the auditors (to ensure that the server is not violating the append-only property and that every client is seeing the same commitments). If clients want to retain the ability to hold the server accountable

RZKS-ZK-REAL($\mathcal{A}$):

$pp' \leftarrow \mathcal{Z}.\mathsf{GenPP}(1^\lambda)$
$(\mathsf{com}', pp', st') \leftarrow \mathcal{Z}.\mathsf{Init}(pp')$
$st \leftarrow st', t \leftarrow 0, pp \leftarrow pp'$
$b \leftarrow \mathcal{A}^{\mathsf{Update}(\cdot),\cdots}(\mathsf{com}', pp)$
return b

Update(S): // analogous for PCSUpdate

parse S as $(\mathsf{label}_1, \mathsf{val}_1), \ldots, (\mathsf{label}_j, \mathsf{val}_j)$
require $\mathsf{label}_1, \ldots, \mathsf{label}_j$ are distinct and do not already appear in $\mathbf{D}(st)$
$(\mathsf{com}', st', \pi) \leftarrow \mathcal{Z}.\mathsf{Update}(st, S)$
$st \leftarrow st', t \leftarrow t + 1$
return (com', π)

Query(label, u):

require $0 \le u \le t$
$(\pi, \mathsf{val}', t') \leftarrow \mathcal{Z}.\mathsf{Query}(pp, st, u, \mathsf{label})$
return (π, val', t')

ProveExt(t_0, t_1):

require $0 \le t_0 \le t_1 \le t$
$\pi \leftarrow \mathcal{Z}.\mathsf{ProveExt}(pp, st, t_0, t_1)$
return π

LeakState():

return st

Ideal(in):

return $\mathsf{Ideal}(in)$

RZKS-ZK-IDEAL($\mathcal{A}$):

$\mathsf{com}', pp' \leftarrow \mathcal{S}(\mathtt{Init})$
$t \leftarrow 0$
$b \leftarrow \mathcal{A}^{\mathsf{Update}(\cdot),\cdots}(\mathsf{com}', pp')$
return b

Update(S): // analogous for PCSUpdate

parse S as $(\mathsf{label}_1, \mathsf{val}_1), \ldots, (\mathsf{label}_j, \mathsf{val}_j)$
require $\mathsf{label}_1, \ldots, \mathsf{label}_j$ are distinct and do not already appear in any of the $S_1, \ldots, S_t$
$(\mathsf{com}', \pi) \leftarrow \mathcal{S}(\mathtt{Update}, L_{\mathsf{Update}}(S))$
$t \leftarrow t + 1, S_t \leftarrow S$
return (com', π)

Query(label, u):

require $0 \le u \le t$
$(\pi, \mathsf{val}', t') \leftarrow \mathcal{S}(\mathtt{Query}, L_{\mathsf{Query}}(u, \mathsf{label}))$
return (π, val', t')

ProveExt(t_0, t_1):

require $0 \le t_0 \le t_1 \le t$
$\pi \leftarrow \mathcal{S}(\mathtt{ProveExt}, L_{\mathsf{ProveExt}}(t_0, t_1))$
return π

LeakState():

return $\mathcal{S}(\mathtt{Leak}, L_{\mathsf{LeakState}}())$

Ideal(in):

return $\mathcal{S}(\mathtt{Ideal}, in)$

Fig. 3. Zero Knowledge (with leakage) security experiments for RZKS. $\mathcal{S}$ is a stateful algorithm (whose state we omit to simplify the notation). The leakage functions $L_{\mathsf{Update}}, L_{\mathsf{Query}}, \ldots$ also share state among each other.

even if auditors are temporarily offline, or if they wish to do the audit themselves in the future, they need to hold on to all the commitments indefinitely, which is inefficient. To solve this problem, SEEMless suggests building a hashchain over all the aZKS commitments, so that the client only needs to remember the tail. This is an improvement, but to skip between two distant commitments, the client has to download all the epochs in between; moreover, the security guarantees deriving from this are not formalized. In contrast, we propose a more efficient solution and formalize its security: we add the ProveExt and VerExt algorithms, which allow the server to directly prove that any given datastore commitment stems from another.

Thus, our advantage over SEEMless lies both in the fact that we give the ability to heal from server state compromise and that we allow the client to only keep the very latest commitment, and to efficiently update to the next one without losing the ability to hold the server accountable later.

4   Rotatable Verifiable Random Functions

In this section, we introduce the notion of a Rotatable Verifiable Random Function, a key component of our RZKS construction. Verifiable Random Functions (VRFs), introduced in [33], are the asymmetric analogues of Pseudorandom Functions: the secret key is necessary to compute the (random-looking) function on any input, as well as a proof that the computation was performed correctly, which can be checked against the corresponding public key. We extend VRFs by adding "rotation" algorithms, which generate a new VRF key pair alongside zero-knowledge proofs that outputs of the new and old VRF on the same (hidden) input are associated. In addition, our rotatable VRFs also satisfy stricter soundness properties.

Definition 2. A Rotatable Verifiable Random Function is a tuple of algorithms VRF = (GenPP, KeyGen, Query, Verify, Rotate, VerRotate) defined as follows:

▷ pp ← VRF.GenPP(1^λ): This algorithm takes the security parameter and produces public parameter pp for the scheme. All other algorithms take these pp as input, even when not explicitly specified.

▷ (sk, pk) ← VRF.KeyGen(pp): The key generation algorithm takes in the global pp and outputs the public key pk and secret key sk.

▷ (y, π) ← VRF.Query(pp, sk, x): The query algorithm takes in pp, the secret key sk and input x, and outputs the evaluation y of the VRF defined by sk on input x, as well as a proof π. We denote with VRF.Eval(sk, x) the first output y of the Query algorithm (i.e. Eval does not return a proof).

▷ $1/0$ ← VRF.Verify(pp, pk, x, y, π): This deterministic function verifies the proof π that y is the output of the VRF defined by pk on input x.

▷ sk', pk', π ← VRF.Rotate(pp, sk, X): Given a secret key[14] and a list of inputs X, this algorithm outputs an updated secret key, an updated public key, and a proof π that the set of VRF output pairs $P = \{(\text{VRF.Eval}(sk, x),$ VRF.Eval(sk', x))$\}_{x \in X}$ satisfies the relationship that each pair corresponds to the same input x (without leaking information about X beyond its size).

▷ $0/1$ ← VRF.VerRotate(pp, pk, pk', P, π): Given two public keys pk, pk' and list of P pairs (y, y'), this deterministic algorithm checks the proof π that each pair consists of the output of the VRFs identified by pk, pk' on the same input x.

For correctness, we require that for all $\lambda, n \in \mathbb{N}$, all sets of inputs $X_1, \ldots, X_n$, and all inputs x:

$$\Pr[\text{pp} \leftarrow \text{VRF.GenPP}(1^\lambda); \ sk_0, pk_0 \leftarrow \text{VRF.KeyGen(pp)};$$
$$sk_i, pk_i, \pi_i \leftarrow \text{VRF.Rotate}(sk_{i-1}, X_i) \text{ for } i = 1, \ldots, n;$$
$$y, \pi \leftarrow \text{VRF.Query}(sk_n, x) : \text{VRF.Verify}(pk_n, x, y, \pi) = 1] = 1.$$

---

[14] Given that the old key sk and new key are independent from one another, we could have equivalently defined Rotate as taking any two secret keys as input.

Moreover, for all $\lambda, n > 0$ and all sets of inputs $X_1, \ldots, X_n$:

$\Pr[\mathsf{pp} \leftarrow \mathsf{VRF.GenPP}(1^\lambda);\ sk_0, pk_0 \leftarrow \mathsf{VRF.KeyGen}(\mathsf{pp});$

$\quad sk_i, pk_i, \pi_i \leftarrow \mathsf{VRF.Rotate}(sk_{i-1}, X_i)$ for $i = 1, \ldots, n : \mathsf{VRF.VerRotate}(pk_n,$

$\quad \{(\mathsf{VRF.Eval}(sk_{n-1}, x), \mathsf{VRF.Eval}(sk_n, x))\}_{x \in X_n}, \pi_n) = 1] = 1.$

4.1   Rotatable VRF Security

Informally, VRFs satisfy two properties. *Uniqueness* mandates that for any public key and input x, there is only one y which can be proven to be output by the function on input x. *Pseudorandomness* guarantees that, for an honestly generated key pair sk, pk and given oracle access to the query oracle on arbitrary inputs, it is hard to distinguish the output of the function on any other (not yet queried) input from a uniformly random value.

We augment the uniqueness and pseudorandomness requirements into soundness and zero-knowledge respectively.

Soundness (Strengthened Uniqueness). We will say that a VRF satisfies soundness if there exists an Extractor such that for all PPT adversaries $\mathcal{A}$, the advantage of $\mathcal{A}$ in distinguishing the experiments of Fig. 4 is negligible.

The extractor Extract is required to provide three functionalities based on its first input:

- $\mathsf{pp}, \mathsf{st} \leftarrow \mathsf{Extract}(\texttt{Init})$: Samples public parameters indistinguishable from honestly generated public parameters such that extraction will be possible. Also generates an initial state.
- $x \leftarrow \mathsf{Extract}(\texttt{Extr}, \mathsf{st}, pk, y)$: Takes in an adversarially chosen public key pk and output y of the function. Outputs the only input x for which the adversary can produce an accepting proof.
- $out, \mathsf{st} \leftarrow \mathsf{Extract}(\texttt{Ideal}, \mathsf{st}, in)$: Simulates the behavior of some ideal functionality (for example a random oracle or generic group). Takes in any input and produces an output indistinguishable from the output the ideal functionality would have on that input.

As with RZKS, we do not allow the extractor to update its state outside `Ideal` calls.

In the ideal experiment, the table T keeps track of the outputs of the extractor. An assertion is triggered (and the adversary can trivially win) if the extractor gives different answers to the same query over time, if the same answer is returned for multiple inputs under the same public key, or if the adversary produces an accepting proof for an input different than what the extractor had predicted (these requirements together capture uniqueness). Moreover, the game also enforces that proofs of rotation are consistent with the extractor's output and the equality condition is respected. In the real experiment, assertions are never triggered, so indistinguishability ensures that public parameters, as well as the answers to ideal queries, give the adversary the same view.

Zero Knowledge (Strengthened Pseudorandomness). We will say that a VRF satisfies zero-knowledge if there exists a simulator such that for all PPT adversaries $\mathcal{A}$, the advantage of $\mathcal{A}$ in distinguishing the experiments of Fig. 5 is negligible.

The stateful simulator $\mathcal{S}$ is required to provide various functionalities:

- $\mathsf{pp}, pk_0 \leftarrow \mathcal{S}(\mathtt{Init})$: Samples public parameters and an initial public key such that it will be possible to simulate proofs.
- $y \leftarrow \mathcal{S}(\mathtt{Corrupted\text{-}Eval}, i, x)$: Takes in a corrupted generation i and input x, and outputs the evaluation of the VRF on i and x. If this is called, the adversary has already obtained the corresponding secret key for generation i, so the simulator is forced to output a value consistent with what the adversary could compute itself.
- $\pi \leftarrow \mathcal{S}(\mathtt{Explain}, i, \mathsf{label}, y)$: Takes in a generation, input, and output. Outputs a simulated proof that the output of the oracle $\mathsf{Eval}(i, x) = y$.
- $pk_{i_{\mathsf{cur}}}, \pi_R \leftarrow \mathcal{S}(\mathtt{Rotate}, P)$: Takes in a set P of pairs (y, y'). Samples a new public key $pk_{i_{\mathsf{cur}}}$ and outputs a simulated proof that for each $(y, y') \in P$ there exists an x such that $\mathsf{Eval}(i_{\mathsf{cur}-1}, x) = y$ and $\mathsf{Eval}(i_{\mathsf{cur}}, x) = y'$.
- $sk_{i_{\mathsf{crpt}}+1}, \ldots, sk_{i_{\mathsf{cur}}} \leftarrow \mathcal{S}(\mathtt{Corrupt}, D)$: Takes in all queries made to Eval. Outputs a collection of secret keys consistent with output of all oracle queries made so far.
- $out \leftarrow \mathcal{S}(\mathtt{Ideal}, in)$: Simulates the behavior of some ideal functionality. Takes in any input and produces an output indistinguishable from the output the ideal functionality would have on that input.

We combine pseudorandomness with a zero knowledge requirement by requiring that in each generation a simulator can sample public parameters such that it can simulate proofs that the VRF is consistent with a new truly random function. Furthermore, the simulator must be able to simulate rotation proofs that the outputs of two random functions stem from the same input. We model post compromise security by requiring that the simulator also be able to sample secret keys consistent with all previous queries. Since it is impossible to sample a secret key consistent with all future queries for a truly random function, after corruption we give the simulator the ability to control the function associated with that epoch. Note that the major difficulty in demonstrating zero knowledge is that the simulator must simulate queries to the ideal oracle without knowing what inputs are asked of the truly random function.

We remark that our definition of zero knowledge is heavily inspired by the notion of a simulatable VRF, introduced in [9]. Simulatable VRFs require that there exists a simulator that can sample simulated public parameters such that for any public key pk, input x in the domain, and y in the range of the VRF, it is possible to simulate a proof π that y is the output of the function on input x (i.e. $\mathsf{Verify}(\mathsf{pp}, pk, x, y, \pi) = 1$). The simulated parameters, outputs and proofs should be indistinguishable from honestly generated ones. Our definition of zero knowledge extends this notion by accounting for rotation proofs and corruptions. Our soundness notion is also stronger as we require extractability.

```
VRF-Sound-IDEAL(𝒜):

T ← [ ]; pp, st ← Extract(Init)
b ← 𝒜^{𝒪···}(pp)
return b


Oracle Extract(pk, y):

x ← Extract(Extr, st, pk, y)
If (pk, y) ∈ T assert T[pk, y] = x
assert x = ⊥ ∨ ∀y′ ≠ y : T[pk, y′] ≠ x
T[pk, y] ← x


Oracle CheckExtraction(pk, y, x, π):

require VRF.Verify(pk, x, y, π) = 1 ∧ (pk, y) ∈ T
assert T[pk, y] = x


Oracle CheckVerRotate(pk₁, pk₂, P, π):

require VRF.VerRotate(pk₁, pk₂, P, π) = 1 ∧
    ∀(u₁, u₂) ∈ P : (pk₁, u₁) ∈ T ∧ (pk₂, u₂) ∈ T
assert ∀(u₁, u₂) ∈ P : T[pk₁, u₁] = T[pk₂, u₂]


Oracle Ideal(in):

out, st ← Extract(Ideal, st, in)
return out
```

Fig. 4. Soundness for VRF. In the real world (not pictured), the public parameters are generated as $\mathsf{pp} \leftarrow \mathsf{VRF.GenPP}(1^\lambda)$, and the oracles do not do anything, except for the Ideal one which implements the necessary ideal objects according to their specification.

4.2 Rotatable VRF Construction

Our rotatable Verifiable Random Function VRF = (GenPP, KeyGen, Query, Verify, Rotate, VerRotate) is instantiated in Fig. 6. In summary, let G be a group of (exponential) prime order p with generator g, and let $F(x)$ be a hash function that maps arbitrary-length bitstrings onto G. Then for a given input x, secret key $sk \in \mathbb{Z}_p^*$, and public key $pk = g^{sk}$, the VRF output is $y = F(x)^{sk}$. To prove this, Query simply produces a Fiat-Shamir zero-knowledge proof that $(g, F(x), pk = g^{sk}, y = F(x)^{sk})$ is a DDH tuple.

Given secret key $sk = \alpha_0 \cdot \cdots \cdot \alpha_i$ and public key $g^{\alpha_0 \cdots \alpha_i}$, Rotate samples α_{i+1} from $\mathbb{Z}_p^*$. It then sets $sk' = \alpha_0 \cdot \cdots \cdot \alpha_{i+1}$ and stores $pk' = pk^{\alpha_{i+1}} = g^{\alpha_0 \cdots \alpha_{i+1}} = g^{sk'}$. Then, it outputs a "batch" Fiat-Shamir zero-knowledge proof that (pk, y, pk', y') is a DDH tuple, where y and y' are random linear combinations of VRF.Eval(sk, x) and VRF.Eval(sk', x) for $x \in X$, respectively. In Fig. 6, the coefficients for the random linear combination are derived as a_u.

4.3 Rotatable VRF Soundness Proof

Soundness of extraction stems directly from soundness of the underlying Fiat-Shamir proof that $(g, F(x), pk = g^{sk}, y = F(x)^{sk})$ is a DDH tuple. To show

VRF-ZK-REAL($\mathcal{A}$):

pp $\leftarrow$ VRF.GenPP(1^λ)
$sk_0, pk_0 \leftarrow$ VRF.KeyGen(pp)
$i_{cur} \leftarrow 0, i_{crpt} \leftarrow -1$
$b \leftarrow \mathcal{A}^{\mathcal{O}\cdots}$(pp, pk_0)
return b

Oracle Eval(i, x):

require $0 \leq i \leq i_{cur}$
$y \leftarrow$ VRF.Eval(sk_i, x)
return y

Oracle Prove(i, x):

require $0 \leq i \leq i_{cur}$
$y, \pi \leftarrow$ VRF.Query(sk_i, x)
return y, π

Oracle Rotate(X):

$sk_{i_{cur}+1}, pk_{i_{cur}+1}, \pi_R \leftarrow$ VRF.Rotate($sk_{i_{cur}}, X$)
$i_{cur} \leftarrow i_{cur} + 1$
return $pk_{i_{cur}}, \pi_R$

Oracle Corrupt():

return $sk_{i_{crpt}+1}, \ldots, sk_{i_{cur}}$
$i_{crpt} \leftarrow i_{cur}$

Oracle Ideal(in):

return Ideal(in)

VRF-ZK-IDEAL($\mathcal{A}$):

pp, $pk_0 \leftarrow \mathcal{S}$(Init)
$i_{cur} \leftarrow 0, i_{crpt} \leftarrow -1, D = \{0 : \{\}\}$
$b \leftarrow \mathcal{A}^{\mathcal{O}\cdots}$(pp, pk_0)
return b

Oracle Eval(i, x):

require $0 \leq i \leq i_{cur}$
If $i \leq i_{crpt}$:
    return $\mathcal{S}$(Corrupted-Eval, i, x)
If $x \notin D[i]$:
    $y \overset{\$}{\leftarrow} Y; D[i][x] \leftarrow y$
return $D[i][x]$

Oracle Prove(i, x):

require $0 \leq i \leq i_{cur}$
$\pi \leftarrow \mathcal{S}$(Explain, $i, x,$ Eval(i, x))
return Eval(i, x), π

Oracle Rotate(X):

$i_{cur} \leftarrow i_{cur} + 1$
$P \leftarrow \{($Eval($i_{cur} - 1, x$), Eval(i_{cur}, x)) $\mid x \in X\}$
$pk_{i_{cur}}, \pi_R \leftarrow \mathcal{S}$(Rotate, P)
return $pk_{i_{cur}}, \pi_R$

Oracle Corrupt():

$sk_{i_{crpt}+1}, \ldots, sk_{i_{cur}} \leftarrow \mathcal{S}$(Corrupt, D)
return $sk_{i_{crpt}+1}, \ldots, sk_{i_{cur}}$
$i_{crpt} \leftarrow i_{cur}$

Oracle Ideal(in):

return $\mathcal{S}$(Ideal, in)

Fig. 5. Zero Knowledge experiments for the Rotatable VRF.

soundness of rotation, again we use the fact that the underlying Fiat-Shamir proof that (pk, y, pk', y') is a DDH tuple is sound. The only subtlety is to show that batching the rotation proofs in the manner we do works. That is, we need to show that if (y, y') are a random linear combination of $\{($VRF.Eval(sk, x), VRF.Eval(sk', x))$\}_{x \in X}$, then if $y' = y^\alpha$, with all but negligible probability we also have VRF.Eval(sk', x) = VRF.Eval(sk, x)$^\alpha$ for all $x \in X$.

Taking the contrapositive, we just need to show that if there is any (y_0, y_0') in $\{($VRF.Eval(sk, x), VRF.Eval(sk', x))$\}_{x \in X}$ such that $y_0' \neq y_0^\alpha$, then the probability that a random linear combination (y, y') satisfies $y' = y^\alpha$ must be negligible. Note that if there exists a pair (y_1, y_1') in $\{($VRF.Eval(sk, x), VRF.Eval(sk', x))$\}_{x \in X}$ such that (y_0, y_0') and (y_1, y_1') are linearly independent as elements of $G \times G$, then (y, y') will be uniformly random and so will satisfy $y' = y^\alpha$ with only negligible probability. But if there is no such pair, then $(y, y') = (y_0^c, y_0'^c)$ for some c, so $y' = y^\alpha$ with probability 1. A detailed formal proof of the following theorem is deferred to the full version of this paper.

▷ pp ← VRF.GenPP(1^λ):
 – p ← prime exponential in λ
 – G ← group of order p
 – g ← generator of G
 – Sample hash function $F : \{0,1\}^* \to G$
 – Sample hash function $F' : \{0,1\}^* \to \mathbb{Z}_p$
 – **return** pp ← (p, G, g, F, F')

▷ (sk, pk) ← VRF.KeyGen(pp):
 – **parse** pp as (p, G, g, F, F')
 – $\alpha_0 \stackrel{\$}{\leftarrow} \mathbb{Z}_p^*$
 – $sk \leftarrow \alpha_0$
 – $pk \leftarrow g^{\alpha_0}$
 – **return** (sk, pk)

▷ (y, π) ← VRF.Query(pp, sk, x):
 – **parse** pp as (p, G, g, F, F')
 – $y \leftarrow F(x)^{sk}$
 – $r \stackrel{\$}{\leftarrow} \mathbb{Z}_p$
 – $c \leftarrow F'(g, F(x), g^{sk}, F(x)^{sk}, g^r, F(x)^r)$
 – $z \leftarrow r - c \cdot sk$
 – $\pi \leftarrow (g^r, F(x)^r, z)$
 – **return** (y, π)

▷ 1/0 ← VRF.Verify(pp, pk, x, y, π):
 – **parse** pp as (p, G, g, F, F')
 – **ensure** $pk \neq g^0$
 – **parse** π as (h_1, h_2, z)
 – $c \leftarrow F'(g, F(x), pk, y, h_1, h_2)$
 – **ensure** $h_1 = g^z \cdot pk^c$
 – **ensure** $h_2 = F(x)^z \cdot y^c$
 – **return** 1

▷ sk', pk', π ← VRF.Rotate(pp, sk, X):
 – **parse** pp as (p, G, g, F, F')
 – $\alpha \stackrel{\$}{\leftarrow} \mathbb{Z}_p^*$
 – $sk' \leftarrow sk \cdot \alpha$
 – $pk' \leftarrow g^{sk'}$
 – $P \leftarrow \{(\text{VRF.Eval}(sk, x),$
      $\text{VRF.Eval}(sk', x))\}_{x \in X}$
 – For each $(u, u') \in P$:
  • $a_u \leftarrow F'(u, u', pk, pk', P)$
 – $y \leftarrow \prod_{(u,u') \in P} u^{a_u}$
 – $y' \leftarrow \prod_{(u,u') \in P} (u')^{a_u}$
 – $r \stackrel{\$}{\leftarrow} \mathbb{Z}_p$
 – $c \leftarrow F'(pk, y, pk', y', pk^r, y^r)$
 – $z \leftarrow r - c\alpha$
 – $\pi \leftarrow (pk^r, y^r, z)$
 – **return** (sk', pk', π)

▷ 0/1 ← VRF.VerRotate(pp, pk, pk', P, π):
 – **parse** pp as (p, G, g, F, F')
 – **ensure** $pk, pk' \neq g^0$
 – For each $(u, u') \in P$:
  • $a_u \leftarrow F'(u, u', pk, pk', P)$
 – $y \leftarrow \prod_{(u,u') \in P} u^{a_u}$
 – $y' \leftarrow \prod_{(u,u') \in P} (u')^{a_u}$
 – **parse** π as (h_1, h_2, z)
 – $c \leftarrow F'(pk, y, pk', y', h_1, h_2)$
 – **ensure** $h_1 = pk^z \cdot (pk')^c$
 – **ensure** $h_2 = y^z \cdot (y')^c$
 – **return** 1

Fig. 6. Our Rotatable VRF construction.

Theorem 1. *If F and F' are modeled as random oracles, and if the DDH assumption holds, then there exists a simulator* Extract *such that for any efficient adversary $\mathcal{A}$,*

$$|\Pr[\text{VRF-Sound-REAL}(\mathcal{A}) \to 1] - \Pr[\text{VRF-Sound-IDEAL}(\mathcal{A}) \to 1]| \leq negl(\lambda).$$

4.4   Rotatable VRF Zero Knowledge Proof

Since our construction generates zero-knowledge proofs in Prove and Rotate, one would hope that simulating these proofs would be enough to prove zero-knowledge of the construction. In fact, if there were no Corrupt oracle, then simply programming the random oracle F' would be enough to simulate these proofs and achieve zero-knowledge. However, once an adversary has called Corrupt and obtained some secret key sk_i, it can then easily distinguish previously outputted Eval(i, x) from the true VRF output $F(x)^{sk_i}$ by simply calculating $F(x)^{sk_i}$ itself and comparing the two.

This intuition extends to arbitrary simulation strategies. Consider for example an adversary who asks for $F(x)$ and $F(x')$, $y \leftarrow \mathsf{Eval}(i, x)$, $y' \leftarrow \mathsf{Eval}(i, x')$ in this order, for two distinct x, x' and some i. At the time of the F queries, the uniformly random outputs of the VRF have not yet been sampled, and so the simulator's output cannot depend on them. Once the adversary calls the Corrupt oracle, the simulator can produce a value for sk_i only if $\log_{F(x)}(y) = \log_{F(x')}(y')$, which only happens with negligible probability. While this specific problem could be solved by adding an additional hash at the end of the VRF computation, i.e. defining $\mathsf{VRF.Eval}(sk, x) = H(F(x)^{sk})$ as in [16] and treating H as a programmable random oracle, similar issues arise when considering our efficient rotation proofs, which would force the game to reveal preimages for the hash before the corruption happens (and so before the simulator knows what algebraic relations should exists between the outputs of F and the group elements revealed in rotation proofs).

To solve this problem, we need to treat G as a generic group. This allows the simulator to hold off on sampling sk_i until Corrupt is called. Until this point, the simulator will treat pk_i as an arbitrary group element, but it will keep track of all algebraic relationships between unknown arbitrary group elements. Then, when Corrupt is called, the simulator will have access to a list of all group elements h such that the adversary expects $h = g^{f(sk_i)}$ for some function f of sk_i. At this point, the simulator will choose sk_i uniformly at random, and can program the generic group such that $g^{f(sk_i)} = h$ for all such f. A detailed formal proof of the following theorem is deferred to the full version of this paper.

Theorem 2. *If the group G is modeled as a generic group, and F, F' are modeled as random oracles, then there exists a simulator S such that for any efficient $\mathcal{A}$,*

$$|\Pr[\mathsf{VRF\text{-}ZK\text{-}REAL}(\mathcal{A}) \rightarrow 1] - \Pr[\mathsf{VRF\text{-}ZK\text{-}IDEAL}(\mathcal{A}) \rightarrow 1]| \leq negl(\lambda).$$

5 RZKS-Construction

5.1 Relevant Primitives

In order to construct RZKS, we rely on a number of building blocks aside from Rotatable VRFs. Security definitions and constructions are included in the full version of the paper, but we include the syntax and a short description here for ease of reference.

Simulatable Commitments. A commitment is a scheme which allows a prover to publish a commitment to any given value such that the prover may later publish a proof that the commitment was indeed generated from the initial value. Furthermore, the simulatability requirement states that the commitment reveals no information about the committed value. A full definition and construction is included in the full version of the paper.

Definition 3 (Simulatable Commitments). A Simulatable Commitment Scheme çonsists of 3 algorithms (C.Init, C.Commit, C.Verify) defined as follows:

▷ pp ← C.GenPP(1^λ): On input the security parameter, GenPP outputs public parameters pp.

▷ com, aux ← C.Commit(pp, m): Using the global parameters pp, the (randomized) commit algorithm produces commitment com to message m, and decommitment information aux.

▷ 1/0 ← C.Verify(pp, com, m, aux): This deterministic algorithm checks whether com is a valid commitment to message m, given the decommitment aux.

Ordered Accumulator (OA). An ordered accumulator is a scheme which allows a prover to commit to a sequence of label/value pairs. Furthermore, an ordered accumulator allows the prover to verifiably append label/value pairs to a previously committed sequence to generate a new commitment. The prover can later provide proofs that a given label/value pair is in the committed sequence or that a given label is not included in the committed sequence. A construction is given in the full version of the paper. Completeness and soundness are defined analogously to RZKS in Figs. 1 and 2 respectively.

Definition 4. An Ordered Accumulator is a tuple of algorithms OA = (GenPP, Init, Update, VerifyUpd, Query, Verify, ProveAll, VerAll) defined as follows:

▷ GenPP, Init, Update, VerifyUpd, Query, Verify are defined analogously to the RZKS in Definition 1.

▷ π ← OA.ProveAll(pp, st, u): This algorithm outputs π which can be verified against the commitment com_u output by the u-th call to Update. It proves the set of label value pairs included in the datastore up to epoch u.

▷ 1/0 ← OA.VerAll(pp, com_u, P, π): This deterministic algorithm takes a digest com_u, a set P of (label, val, t) pairs, and a proof. It checks that P is the set of all pairs that com_u commits to.

Append-Only Vector Commitments (AVC). An append-only vector commitment can be used to commit to a list of values, extend the list without recomputing the commitment from scratch, prove what the value is at a specific position in the list, and prove that two commitments have been obtained by extending the same list.

We briefly discuss the syntax of this primitive here, and defer the security definitions and construction to the full version of the paper.

Definition 5. An Append-only Vector Commitment is a tuple of algorithms AVC = (GenPP, Init, Update, ProveExt, VerExt, Query, Verify) defined as follows:

▷ pp ← AVC.GenPP(1^λ): This algorithm takes the security parameter and produces public parameter pp for the scheme. All other algorithms take these pp as input, even when not explicitly specified.

▷ (com, st) ← AVC.Init(pp): This algorithm produces an initial commitment com to an empty list $D_0 = \{\}$, and an initial server/prover state st. Each server state st will contain a list and a digest, which we will refer to as $\mathbf{D}$(st) and com(st). Similarly, each commitment will include an integer t(com) (also called an epoch for consistency with other primitives) representing the size of the list it commits to. (Alternatively, these can be thought of as deterministic functions which are part of the scheme.)

▷ (com′, st′, π_S) ← AVC.Update(pp, st, val): This algorithm takes in the current state of the prover st, and a value val. The algorithm outputs an updated commitment to the datastore, an updated internal state st′, and proof π (to be verified with VerExt) that the update has been done correctly. Intuitively, com′ is a commitment to the list $\mathbf{D}(st′) = \mathbf{D}(st)\|$val of size t(com′) = t(com(st)) + 1.

▷ π ← AVC.ProveExt(pp, st, $t′, t$): Given the prover's state st and two integers, the algorithm produces a proof that the list committed to by com$_t$ (output at the t-th invocation of Update) extends the one committed to by com$_{t′}$.

▷ 0/1 ← AVC.VerExt(pp, com′, com, π): This deterministic algorithm takes in two digests and proves that the list committed to by com extends the one committed to by com′. The proofs can be produced by either Update or ProveExt.

▷ (π, val) ← AVC.Query(pp, st, $u, t′$): This algorithm takes as input a state st and epochs u and $l′$ such that $u \leq t′ \leq t$(st). It returns val = $\mathbf{D}(st)[u]$ and a membership proof π to be verified against the commitment com$_{t′}$ output by Update during the $t′$-th update.

▷ 0/1 ← AVC.Verify(pp, com, u, val, π): This deterministic algorithm checks the proof π (produced by Query) that val is the is the u-th element of the list committed by com.

5.2 RZKS Construction

We describe our RZKS construction in Fig. 7. The RZKS commits to a set of (label, val) pairs by storing in an ordered accumulator (tlbl, tval) pairs, where a given tlbl is the VRF output[15] for a given label, and a given tval is the commitment to a given val. Elements are added to the OA in batches, where the i-th update to the OA produces a digest at the i-th epoch. At each epoch, the OA digest and VRF public key are stored in the corresponding index of the AVC. The resulting AVC digest is returned as the RZKS digest.

Updating the RZKS produces an append-only proof, which contains the append-only proofs for the underlying OA and AVC. To verify the presence of a (label, val), inclusion/exclusion proofs include the VRF proof, commitment opening, OA digest, a proof that the label/value pair is consistent with that digest, and a proof that the digest is at the expected index of the vector that the AVC digest commits to.

The AVC data structure allows the RZKS to support the ProveExt and VerExt algorithms, in which the server proves that a recent RZKS digest commits to an older one that the verifier currently holds (therefore, the client can forget the old digest without losing the ability to hold the server accountable later).

---

[15] The Rotatable VRF presented in this work outputs group elements, while the ordered accumulator takes as input bit-strings, so we implicitly assume that these group elements have a unique bit-string representation.

The RZKS construction is similar to the append-only ZKS described in SEEMless [7], but i) each leaf also contains the epoch number at which such leaf was inserted, and ii) it uses a rotatable VRF instead of a standard one. To perform a rotation, the prover rotates the VRF key and builds a brand new ordered accumulator using the same commitments as the old one, but uses the new VRF outputs as labels. The audit proof for such a rotation involves the VRF rotation proof for all the pre-existing labels, plus an append-only proof for any new labels that were added.

Finally, we summarize the state that the RZKS maintains (note that some values in the state are redundant for the sake of readability). It maintains D, a map of all the (label, val) pairs in the RZKS, and epno, the latest epoch number. It also contains st_{OA} and st_{AVC}, the underlying state of the OA and AVC, respectively. It stores com_{epno}, which is the latest value stored at the epno-th position in the AVC (recall that it contains the latest OA digest and VRF public key). And, the RZKS state stores K_{VRF}, a map of the VRF keypair for each VRF keypair generation; G, a map of the corresponding VRF generation for each epoch number; and g, the latest VRF keypair generation number.

5.3 RZKS Protocol Security

Theorem 3. *The scheme described in Fig. 7 satisfies completeness according to definition 1.*

This is easy to see by inspection.

Theorem 4. *Let OA be an Ordered Accumulator, C be a Commitment scheme, VRF be a VRF, and AVC be an Append-only Vector Commitment, all satisfying their respective definitions of soundness w.r.t. their own idealized objects. Then the RZKS construction of Fig. 7 satisfies soundness, w.r.t. the set of all such idealized objects.*

Proof Sketch: To prove soundness, we define an RZKS extractor that trivially combines those for the underlying building blocks. It extracts a dictionary from an RZKS digest by feeding the output of each extractor as input to the next, and answers Ideal oracle queries for a primitive's ideal object by running the appropriate extractor. Given this extractor, we make a hybrid argument: we first need to add extra assertions to the ideal RZKS game enforcing that the individual components of an RZKS proof match the output of the corresponding extractors (indistinguishability can be proven based on the soundness of those primitives). This prevents an adversary from submitting proofs for the same tuples that the combined extractor outputs, but that disagree with the internal extractors. After that, we can start removing the individual extractors and honestly implementing the corresponding ideal objects (relying a second time on the same soundness properties of the underlying primitives) to get to the real game. The full proof is in the full version of the paper.

Last, we prove that the RZKS construction satisfies zero-knowledge with leakage. The leakage function provides the simulator, for Update queries, with

▷ $pp \leftarrow$ RZKS.GenPP(1^λ):
- $pp_{VRF} \leftarrow$ VRF.GenPP(1^λ)
- $pp_{OA} \leftarrow$ OA.GenPP(1^λ)
- $pp_C \leftarrow$ C.GenPP(1^λ)
- $pp_{AVC} \leftarrow$ AVC.GenPP(1^λ)
- **return** pp $\leftarrow$ ($pp_{VRF}, pp_{OA}, pp_C, pp_{AVC}$)

▷ (com, st) $\leftarrow$ RZKS.Init(pp):
- **parse** pp as ($pp_{VRF}, pp_{OA}, pp_C, pp_{AVC}$)
- epno $\leftarrow 0, g \leftarrow 0, K_{VRF} \leftarrow \{\}, D \leftarrow \{\}$,
  $st_{OA} \leftarrow \{\}, G \leftarrow \{\}$
- $sk_0, pk_0 \leftarrow$ VRF.KeyGen(pp_{VRF});
  $K_{VRF}[g] \leftarrow (sk_0, pk_0), G[epno] \leftarrow g$
- $(st', com_{OA}^0) \leftarrow$ OA.Init(pp_{OA});
  $com_{INT}^0 \leftarrow (com_{OA}^0, pk_0); st_{OA}[g] \leftarrow st'$
- $(st_{AVC}, \_) \leftarrow$ AVC.Init(pp_{AVC});
- $com^1, st_{AVC}, \pi^0 \leftarrow$ AVC.Update(st_{AVC}, com_{INT}^0)
- st $\leftarrow (K_{VRF}, D, com^1, epno, g, G, st_{OA}, st_{AVC})$
- **return** com^1, st

▷ (com, st', π) $\leftarrow$ RZKS.Update(st, S_{update}):
  (com, st', π) $\leftarrow$ RZKS.PCSUpdate(st, S_{update}):
  // bullets with □ only apply to PCSUpdate
- **parse** st as
  $(K_{VRF}, D, com, epno, g, G, st_{OA}, st_{AVC})$;
  set epno $\leftarrow$ epno $+ 1$
- **parse** S_{update} as
  $(label_1, val_1), \ldots, (label_n, val_n)$
- **ensure** $label_1, \ldots, label_n$ are distinct and $\notin D$
- □ $L \leftarrow \{label \mid (label, (\ldots)) \in D\}$
- □ $sk_{g+1}, pk_{g+1}, \pi_{VRF} \leftarrow$ VRF.Rotate($K_{VRF}[g], L$)
- □ $K_{VRF}[g+1] \leftarrow (sk_{g+1}, pk_{g+1})$
- □ $g \leftarrow g + 1$
- □ For $g' \in \{g, g-1\}$: $\{tlbl_j^{g'}\}_{j \in L} \leftarrow$
  $\{$VRF.Eval($K_{VRF}[g'].sk, j$)$\}_{j \in L}$
- □ $\pi_{OA}^{g-1} \leftarrow$ OA.ProveAll($st_{OA}[g-1]$, epno -1)
- □ $st', \_ \leftarrow$ OA.Init(pp_{OA}); For $i \in [epno-1]$:
  - $S_{OA} \leftarrow \{(tlbl_j^g, tval_j) \mid (j, (\cdot, i, tval_j, \cdot)) \in D\}$
  - $st', com_{OA}', \_ \leftarrow$ OA.Update(st', S_{OA})
- □ $st_{OA}[g] \leftarrow st'; com_{OA}^{epno-1} \leftarrow com_{OA}$
- □ $\pi_{OA}^g \leftarrow$ OA.ProveAll($st_{OA}[g]$, epno -1)
- $S_{OA} \leftarrow \{\}$; For each $(label_i, val_i) \in S_{update}$:
  - $tlbl_i \leftarrow$ VRF.Eval($K_{VRF}[g].sk, label_i$)
  - $tval_i, aux_i \leftarrow$ C.Commit(val_i)
  - $S_{OA} \leftarrow S_{OA} \cup \{(tlbl_i, tval_i)\}$
  - $D \leftarrow D \cup \{(label_i, (val, epno, tval_i, aux_i))\}$
- $st_{OA}[g], com_{OA}^{epno}, \pi_{OA} \leftarrow$ OA.Update($st_{OA}[g], S_{OA}$);
  $com_{INT}^{epno} \leftarrow (com_{OA}^{epno}, K_{VRF}[g].pk); G[epno] \leftarrow g$;
  $\pi' \leftarrow \pi_{OA}$
- com, $st_{AVC}, \pi_{AVC} \leftarrow$ AVC.Update($st_{AVC}, com_{INT}^{epno}$)
- $\_, \pi_{AVC}^{epno} \leftarrow$ AVC.Query($st_{AVC}, \mathbf{t}(com), \mathbf{t}(com)$)
- $com_{INT}^{epno-1}, \pi_{AVC}^{epno-1} \leftarrow$
  AVC.Query($st_{AVC}, \mathbf{t}(com) - 1, \mathbf{t}(com) - 1$)
- □ $\pi' \leftarrow (\pi_{OA}, \pi_{OA}^{g-1}, \pi_{OA}^g, \pi_{VRF}, com_{OA}^{epno-1},$
  $\{(tlbl_j^{g-1}, tlbl_j^g, tval_j, epno_j)\}_{j \in L})$
- $\pi \leftarrow (\pi', \pi_{AVC}, com_{INT}^{epno}, com_{INT}^{epno-1},$
  $\pi_{AVC}^{epno}, \pi_{AVC}^{epno-1})$
- st $\leftarrow (K_{VRF}, D, com, epno, g, G, st_{OA}, st_{AVC})$
- **return** (com, st, π)

▷ 0/1 $\leftarrow$ RZKS.VerifyUpd($com^{t_0}, com^{t_1}, \pi$):
- **parse** π as
  $(\pi', \pi_{AVC}, com_{INT}^{t_1}, com_{INT}^{t_0}, \pi_{AVC}^{t_1}, \pi_{AVC}^{t_0})$
- **parse** $com_{INT}^{t_0}$ as $(com_{OA}^{t_0}, pk_{t_0})$
- **parse** $com_{INT}^{t_1}$ as $(com_{OA}^{t_1}, pk_{t_1})$
- **ensure** OA.$\mathbf{t}(com_{OA}^{t_0}) + 2 =$ AVC.$\mathbf{t}(com^{t_0}) +$
  $1 =$ AVC.$\mathbf{t}(com^{t_1}) =$ OA.$\mathbf{t}(com_{OA}^{t_1}) + 1$
- **ensure** AVC.VerExt($com^{t_0}, com^{t_1}, \pi_{AVC}) = 1$
- For $t \in \{t_0, t_1\}$: **ensure** AVC.Verify(com^t,
  AVC.$\mathbf{t}(com^t), com_{INT}^t, \pi_{AVC}^t) = 1$
- If $pk_{t_0} = pk_{t_1}$:
  - **parse** π' as π_{OA}; set $com_{OA}' \leftarrow com_{OA}^{t_0}$
  Else:
  - **parse** π' as $(\pi_{OA}, \pi_{OA}^{g-1}, \pi_{OA}^g, \pi_{VRF}, com_{OA}',$
    $\{(tlbl_j^{g-1}, tlbl_j^g, tval_j, epno_j)\}_{j \in L})$
  - **ensure** VRF.VerRotate(pk_{t_0}, pk_{t_1},
    $\{(tlbl_j^{g-1}, tlbl_j^g)\}_{j \in L}, \pi_{VRF}) = 1$
  - **ensure** OA.VerAll($com_{OA}^{t_0}, \{(tlbl_j^{g-1}, tval_j,$
    $epno_j)\}_{j \in L}, \pi_{OA}^{g-1}) = 1$
  - **ensure** OA.VerAll($com_{OA}', \{(tlbl_j^g, tval_j,$
    $epno_j)\}_{j \in L}, \pi_{OA}^g) = 1$
- **ensure**
  OA.VerifyUpd($com_{OA}', com_{OA}^{t_1}, \pi_{OA}) = 1$
- **return** 1

▷ $(\pi, val, t) \leftarrow$ RZKS.Query(st, u, label):
- **parse** st as
  $(K_{VRF}, D, com, epno, g, G, st_{OA}, st_{AVC})$
- **ensure** $u \leq$ epno
- $(tlbl, \pi_{VRF}) \leftarrow$
  VRF.Query($K_{VRF}[G[u]].sk$, label)
- If label $\in D$ and $D[label].epno_{label} \leq u$:
  - $(val, epno_{label}, tval, aux) \leftarrow D[label]$
  Else:
  - $(val, epno_{label}, tval, aux) \leftarrow (\bot, \bot, \bot, \bot)$
- $\pi_{OA}, \_ \leftarrow$ OA.Query($st_{OA}[G[u]], u, tlbl$)
- $\pi_{AVC}, com_{INT} \leftarrow$ AVC.Query(st_{AVC}, u, u)
- $\pi \leftarrow (\pi_{AVC}, \pi_{OA}, \pi_{VRF}, tlbl, tval, aux, com_{INT})$
- **return** $\pi, val, epno_{label}$

▷ 0/1 $\leftarrow$ RZKS.Verify(com, label, val, t, π):
- **parse** π as
  $(\pi_{AVC}, \pi_{OA}, \pi_{VRF}, tlbl, tval, aux, com_{INT})$
- **parse** com_{INT} as (com_{OA}, pk)
- **ensure** VRF.Verify(pk, label, $tlbl, \pi_{VRF}) = 1$
- **ensure** AVC.$\mathbf{t}(com) =$ OA.$\mathbf{t}(com_{OA}) + 1$
- If $t = \bot \lor val = \bot \lor tval = \bot$:
  Then **ensure** $val = tval = t = \bot$
  Else **ensure** C.Verify($val, tval, aux) = 1$
- **ensure** OA.Verify($com_{OA}, tlbl, tval, t, \pi_{OA}) = 1$
- **ensure** AVC.Verify(com, AVC.$\mathbf{t}(com), com_{INT}$,
  $\pi_{AVC}) = 1$
- **return** 1

▷ $(\pi, val, t) \leftarrow$ RZKS.ProveExt(st, t_0, t_1):
- **parse** st as
  $(K_{VRF}, D, com, epno, g, G, st_{OA}, st_{AVC})$
- **return** AVC.ProveExt(st_{AVC}, t_0, t_1)

▷ 0/1 $\leftarrow$ RZKS.VerExt($com^{t_0}, com^{t_1}, \pi$):
- **return** AVC.VerExt($com^{t_0}, com^{t_1}, \pi$)

Fig. 7. Our RZKS construction. We implicitly assume that the public parameters output by GenPP are input to all other algorithms, parsed into their components and input to the VRF, OA and C, AVC algorithms as appropriate (as shown in Init). Moreover, since the OA commitment to the empty datastore ends up as the first element of the AVC, in this construction we define RZKS.$\mathbf{t}(com)$ as AVC.$\mathbf{t}(com) - 1$.

the number of elements that are being added to the data structure, as well as the labels (but not values) of any added pair that the adversary has queried since the last PCSUpdate (and was given absence proofs for). PCSUpdate queries only include the number of added pairs. When the adversary calls the Query oracle, the simulator is given the queried label, as well as the epoch it was added at (if the label is in the RZKS) and value (if it was added no later than the queried epoch). On LeakState queries, the simulator is given the full contents of the data structure, and subsequent Update queries until the next PCSUpdate also reveal all the added labels (but not the values). Finally, ProveExt queries just reveal the queried epochs. A formal definition follows:

Leakage **L.**

- The shared state consists of a set of labels X, a datastore D, a counter t for the current epoch (initialized to 0), a counter g for the current generation (i.e. the number of PCSUpdate operations performed, also starting at 0), a map G that matches each epoch to the respective generation, and a boolean *leaked* (initially false).
- $L_{\mathsf{Query}}(\mathsf{label}, u)$: If $\exists(\mathsf{label}, \mathsf{val}, t') \in \mathsf{D}$ such that $t' \leq u$, the function returns $(\mathsf{label}, \mathsf{val}, t', u)$. If $\exists(\mathsf{label}, \mathsf{val}, t') \in \mathsf{D}$ such that $G[t'] = G[u]$, the function returns $(\mathsf{label}, \bot, t', u)$. Otherwise, it returns $(\mathsf{label}, \bot, \bot, u)$ and, if $G[u] = g$, adds label to X.
- $L_{\mathsf{Update}}(S)$: Parse $S = \{(\mathsf{label}_i, \mathsf{val}_i)\}$. If S contains any duplicate label, or any label which appears in D, this function returns $\bot$. Else, it increments t, sets $G[t] \leftarrow g$, and adds the pairs from S to the datastore D at epoch t. If *leaked*, it returns the labels in S. Else, it returns $|S|$ and the set of labels from S which are also in X.
- $L_{\mathsf{PCSUpdate}}(S)$: Parse $S = \{(\mathsf{label}_i, \mathsf{val}_i)\}$. If S contains any duplicate label, or any label which appears in D, this function returns $\bot$. Else, it increments t, adds the pairs from S to the datastore D at epoch t, and updates $X \leftarrow \{\}$, *leaked* $\leftarrow$ *false*, and $g \leftarrow g + 1$, $G[t] \leftarrow g$. It returns $|S|$.
- $L_{\mathsf{LeakState}}(S)$: Set *leaked* $\leftarrow$ *true*. **return** D.
- $L_{\mathsf{ProveExt}}(t_0, t_1)$: **return** (t_0, t_1).

Theorem 5. *Let* VRF *be a rotatable VRF in some idealized model,* C *be a simulatable commitments scheme in some idealized model, and* AVC *be any Append-only Vector Commitment. Then, our* $\mathcal{Z}$ *construction satisfies zero-knowledge with leakage* **L** *as above in the idealized models used by the underlying protocols.*

Proof Sketch: The proof is structured as a hybrid argument. Starting from the real game, one can first substitute Commitments and (Rotatable) VRF outputs and proofs with random strings or those produced by the respective simulators, and then notice that, at this point, the information provided by the leakage function **L** is sufficient to produce these simulated values without relying on the full input to the oracle calls. For example, when an Update oracle query happens (for a non compromised key), the simulator receives the number of pairs that the adversary wants to add to the RZKS, and can itself generate enough random strings (to use as VRF outputs) and simulated commitments to add to the OA, and then adds the new OA commitment to the AVC. Upon corruption or queries, the simulator learns the actual values corresponding to these queries, and can simulate commitment openings and VRF proofs accordingly, and provide honestly generated OA and AVC proofs. A full proof is deferred to the full version of the paper.

5.4 Instantiation and Complexity

If we allow each building block to be instantiated as constructed in the full version of the paper, we can define an instantiation for the entire scheme. We then calculate the efficiency of each building block, which then gives us the efficiency of the entire scheme.

We calculate an upper bound on the number of hash computations and group exponentiations under the constructions of our building blocks and RZKS as follows: we define n to be the size of the stored datastore, and s to be the size of the update query (when relevant). Let ℓ be the number of bits needed to represent a group element. We assume without loss of generality that ℓ is also the number of bits needed to represent a group exponent. Let ℓ' be the number of bits to represent a label. Note that when some algorithm ignores the proof output from another, we skip the proof calculation. The full list of complexities is displayed in Fig. 8.

	Hashes	Exponentiations	Proof length
VRF.Query	2	4	3ℓ
VRF.Verify	2	4	-
VRF.Rotate	$3s+1$	$4s+3$	3ℓ
VRF.VerRotate	$s+1$	$2s+4$	-
C.Commit	1	-	2λ
C.Verify	1	-	-
OA.Update	$2(s+\lceil \log n\rceil)$	-	$s\ell' + 2(s+1)\lceil \log(n+s)\rceil\lambda$
OA.VerifyUpd	$4(s+1)\lceil \log(n+s)\rceil + 2s$	-	-
OA.Query	$2n-1$	-	$4\lceil \log(n+1)\rceil\lambda$
OA.Verify	$4\lceil \log(n+1)\rceil + 2$	-	-
OA.ProveAll	-	-	-
OA.VerAll	$2n$	-	-
AVC.Update	$2(1+\lceil \log n\rceil)$	-	$4\lambda\lceil \log n\rceil$
AVC.ProveExt	$2\lceil \log n\rceil$	-	$4\lambda\lceil \log n\rceil$
AVC.VerExt	$6\lceil \log n\rceil$	-	-
AVC.Query	$2\lceil \log n\rceil$	-	$4\lambda\lceil \log n\rceil$
AVC.Verify	$4\lceil \log n\rceil + 1$	-	-
RZKS.Update	$4s + 6\lceil \log n\rceil + 2$	s	$s\ell' + (2(s+1)\lceil \log(n+s)\rceil + 12\lceil \log n\rceil)\lambda$
RZKS.PCSUpdate	$6s + 8\lceil \log n\rceil + 3n + 3$	$s+4n+3$	$3\ell + s\ell' + (2(s+1)\lceil \log(n+s)\rceil + 12\lceil \log n\rceil)\lambda$
RZKS.VerifyUpd	$6\lceil \log(n+s)\rceil + 8\lceil \log n\rceil + 4n + 2s + 3$	$2s+4$	-
RZKS.Query	$2\lceil \log n\rceil + 2n + 1$	4	$3\ell + 8\lambda\lceil \log(n+1)\rceil$
RZKS.Verify	$8\lceil \log(n+1)\rceil + 6$	4	-
RZKS.ProveExt	$2\lceil \log n\rceil$	-	-
RZKS.VerExt	$6\lceil \log n\rceil$	-	-

Fig. 8. The complexity of our various constructions.

Acknowledgements. At the commencement of the work leading to this paper, the authors had discussions with Melissa Chase (of Microsoft), and Julia Len (an intern at Zoom). The authors are appreciative of their contributions.

References

1. Abe, M., Groth, J., Haralambiev, K., Ohkubo, M.: Optimal structure-preserving signatures in asymmetric bilinear groups. In: Rogaway, P. (ed.) CRYPTO 2011. LNCS, vol. 6841, pp. 649–666. Springer, Heidelberg (2011). https://doi.org/10.1007/978-3-642-22792-9_37
2. apple.com. Apple privacy. https://www.apple.com/privacy/features. Accessed 03 Aug 2022
3. Assal, H., Hurtado, S., Imran, A., Chiasson, S.: What's the deal with privacy apps? A comprehensive exploration of user perception and usability. In: Proceedings of the 14th International Conference on Mobile and Ubiquitous Multimedia, MUM 2015, pp. 25–36. Association for Computing Machinery, New York (2015)
4. Black, J., Rogaway, P., Shrimpton, T.: Black-box analysis of the block-cipher-based hash-function constructions from PGV. Cryptology ePrint Archive, Report 2002/066 (2002). https://eprint.iacr.org/2002/066
5. Black, J., Rogaway, P., Shrimpton, T.: Encryption-scheme security in the presence of key-dependent messages. In: Nyberg, K., Heys, H. (eds.) SAC 2002. LNCS, vol. 2595, pp. 62–75. Springer, Heidelberg (2003). https://doi.org/10.1007/3-540-36492-7_6
6. Blum, J., et al.: E2e encryption for zoom meetings. In: White paper (2021). https://github.com/zoom/zoom-e2e-whitepaper/blob/master/zoom_e2e.pdf
7. Chase, M., Deshpande, A., Ghosh, E., Malvai, H.: H.: SEEMless: secure end-to-end encrypted messaging with less trust. In: Cavallaro, L., Kinder, J., Wang, X.F., Katz, J. (eds.) ACM CCS 2019, pp. 1639–1656. ACM Press, November 2019
8. Chase, M., Healy, A., Lysyanskaya, A., Malkin, T., Reyzin, L.: Mercurial commitments with applications to zero-knowledge sets. In: Cramer, R. (ed.) EUROCRYPT 2005. LNCS, vol. 3494, pp. 422–439. Springer, Heidelberg (2005). https://doi.org/10.1007/11426639_25
9. Chase, M., Lysyanskaya, A.: Simulatable VRFs with applications to multi-theorem NIZK. In: Menezes, A. (ed.) CRYPTO 2007. LNCS, vol. 4622, pp. 303–322. Springer, Heidelberg (2007). https://doi.org/10.1007/978-3-540-74143-5_17
10. Chase, M., Meiklejohn, S.: Transparency overlays and applications. In: Weippl, E.R., Katzenbeisser, S., Kruegel, C., Myers, A.C., Halevi, S. (eds.) ACM CCS 2016, pp. 168–179. ACM Press, October 2016
11. Chase, M., Perrin, T., Zaverucha, G.: The signal private group system and anonymous credentials supporting efficient verifiable encryption. In: ACM CCS 2020, pp. 1445–1459. ACM Press (2020)
12. Chaum, D., Pedersen, T.P.: Wallet databases with observers. In: Brickell, E.F. (ed.) CRYPTO 1992. LNCS, vol. 740, pp. 89–105. Springer, Heidelberg (1993). https://doi.org/10.1007/3-540-48071-4_7
13. Novi Financial. Auditable key directory (2021). https://github.com/novifinancial/akd/. Accessed 26 May 2022

14. Gasser, O., Hof, B., Helm, M., Korczynski, M., Holz, R., Carle, G.: In log we trust: revealing poor security practices with certificate transparency logs and internet measurements. In: Beverly, R., Smaragdakis, G., Feldmann, A. (eds.) PAM 2018. LNCS, vol. 10771, pp. 173–185. Springer, Cham (2018). https://doi.org/10.1007/978-3-319-76481-8_13

15. Ghoshal, A., Tessaro, S.: Tight state-restoration soundness in the algebraic group model. In: Malkin, T., Peikert, C. (eds.) CRYPTO 2021. LNCS, vol. 12827, pp. 64–93. Springer, Cham (2021). https://doi.org/10.1007/978-3-030-84252-9_3

16. Goldberg, S., Reyzin, L., Papadopoulos, D., Včelák, J.: Verifiable random functions (VRFs). Internet Draft draft-irtf-cfrg-vrf-12, Internet Engineering Task Force, May 2022. Work in Progress

17. Google: Key transparency overview. https://github.com/google/keytransparency/blob/master/docs/overview.md. Accessed 31 Aug 2022

18. Herzberg, A., Leibowitz, H.: Can Johnny finally encrypt? Evaluating E2E-encryption in popular IM applications. In: Proceedings of the 6th Workshop on Socio-technical Aspects in Security and Trust, STAST 2016, pp. 17–28. Association for Computing Machinery, New York (2016)

19. Herzberg, A., Leibowitz, H., Seamons, K., Vaziripour, E., Justin, W., Zappala, D.: Secure messaging authentication ceremonies are broken. IEEE Secur. Privacy **19**(2), 29–37 (2021)

20. Hu, Y., Hooshmand, K., Kalidhindi, H., Yang, S.J., Popa, R.A.: Merkle2: a low-latency transparency log system. In: 2021 IEEE Symposium on Security and Privacy (SP), pp. 285–303 (2021)

21. Keybase.io: Keybase chat. https://book.keybase.io/docs/chat. Accessed 03 Aug 2022

22. keybase.io: Keybase is now writing to the stellar blockchain. https://book.keybase.io/docs/server/stellar. Accessed 29 July 2022

23. Keybase.io: Meet your sigchain (and everyone else's). https://book.keybase.io/docs/server#meet-your-sigchain-and-everyone-elses. Accessed 29 July 2022

24. keybase.io: Keybase first commitment (2014). https://keybase.io/_/api/1.0/merkle/root.json?seqno=1. Accessed 26 May 2022

25. Keybase.io: Keybase is not softer than tofu (2019). https://keybase.io/blog/chat-apps-softer-than-tofu. Accessed 05 May 2019

26. Laurie, B., Langley, A., Kasper, E., Messeri, E., Stradling, R.: Certificate Transparency Version 2.0. RFC 9162, December 2021

27. Lerner, A., Zeng, E., Roesner, F.: Confidante: usable encrypted email: a case study with lawyers and journalists. In: 2017 IEEE European Symposium on Security and Privacy, EuroS&P 2017, Paris, France, 26–28 April 2017, pp. 385–400. IEEE (2017)

28. Meiklejohn, S., et al.: Think global, act local: gossip and client audits in verifiable data structures (2020)

29. Melara, M.S., Blankstein, A., Bonneau, J., Felten, E.W., Freedman, M.J.: Coniks: bringing key transparency to end users. In: Usenix Security, pp. 383–398 (2015)

30. Micali, S., Rabin, M., Kilian, J.: Zero-knowledge sets. In: Proceedings of the 44th Annual IEEE Symposium on Foundations of Computer Science, FOCS 2003, p. 80. IEEE Computer Society, USA (2003)

31. Micali, S., Rabin, M.O., Vadhan, S.P.: Verifiable random functions. In: 40th FOCS, pp. 120–130. IEEE Computer Society Press, October 1999

32. microsoft.com: Teams end-to-end encryption (2022). https://docs.microsoft.com/en-us/microsoftteams/teams-end-to-end-encryption. Accessed 26 May 2022

33. Muthukrishnan, S., Rajaraman, R., Shaheen, A., Gehrke, J.: Online scheduling to minimize average stretch. In: 40th FOCS, pp. 433–442. IEEE Computer Society Press, October 1999

34. Nielsen, J.B.: Separating random oracle proofs from complexity theoretic proofs: the non-committing encryption case. In: Yung, M. (ed.) CRYPTO 2002. LNCS, vol. 2442, pp. 111–126. Springer, Heidelberg (2002). https://doi.org/10.1007/3-540-45708-9_8

35. Elaine Barker (NIST): Nist sp 800-57 part 1 rev. 5 recommendation for key management: Part 1 - general (2022). https://nvlpubs.nist.gov/nistpubs/SpecialPublications/NIST.SP.800-57pt1r5.pdf. Accessed 10 Aug 2022

36. LLC. PCI Security Standards Council: Payment card industry data security standard: Requirements and testing procedures, v4.0 (2022). https://listings.pcisecuritystandards.org/documents/PCI-DSS-v4_0.pdf. Accessed 10 Aug 2022

37. Shoup, V.: Lower bounds for discrete logarithms and related problems. In: Fumy, W. (ed.) EUROCRYPT 1997. LNCS, vol. 1233, pp. 256–266. Springer, Heidelberg (1997). https://doi.org/10.1007/3-540-69053-0_18

38. signal.org: Technical information (2016). https://www.signal.org/docs. Accessed 03 Aug 2022

39. signal.org: Technology preview: signal private group system (2019). https://signal.org/blog/signal-private-group-system/. Accessed 22 Aug 2022

40. Tomescu, A., Bhupatiraju, V., Papadopoulos, D., Papamanthou, C., Triandopoulos, N., Devadas, S.: Transparency logs via append-only authenticated dictionaries. In: Cavallaro, L., Kinder, J., Wang, X.F., Katz, J. (eds.) ACM CCS 2019, pp. 1299–1316. ACM Press, November 2019

41. Tyagi, N., Fisch, B., Bonneau, J., Tessaro, S.: Client-auditable verifiable registries. Cryptology ePrint Archive, Paper 2021/627 (2021). https://eprint.iacr.org/2021/627

42. Tzialla, I., Kothapalli, A., Parno, B., Setty, S.: Transparency dictionaries with succinct proofs of correct operation. Cryptology ePrint Archive, Paper 2021/1263 (2021). https://eprint.iacr.org/2021/1263

43. Vaziripour, E., et al.: Is that you, Alice? A usability study of the authentication ceremony of secure messaging applications. : Thirteenth Symposium on Usable Privacy and Security (SOUPS 2017), pp. 29–47. USENIX Association, Santa Clara, July 2017

44. webex.com: Webex end-to-end encryption (2022). https://help.webex.com/en-us/article/WBX44739/What-Does-End-to-End-Encryption-Do?. Accessed 26 May 2022

45. whatsapp.com: Whatsapp encryption overview. In: White paper (2021). Accessed 03 Aug 2022

Quantum Algorithms

Nostradamus Goes Quantum

Barbara Jiabao Benedikt$^{(\boxtimes)}$, Marc Fischlin🆔, and Moritz Huppert

Cryptoplexity, Technische Universität Darmstadt, Darmstadt, Germany
{barbara_jiabao.benedikt,marc.fischlin}@tu-darmstadt.de,
moritz.huppert@proton.me

Abstract. In the Nostradamus attack, introduced by Kelsey and Kohno (Eurocrypt 2006), the adversary has to commit to a hash value y of an iterated hash function H such that, when later given a message prefix P, the adversary is able to find a suitable "suffix explanation" S with $H(P \| S) = y$. Kelsey and Kohno show a herding attack with $2^{2n/3}$ evaluations of the compression function of H (with n bits output and state), locating the attack between preimage attacks and collision search in terms of complexity. Here we investigate the security of Nostradamus attacks for quantum adversaries. We present a quantum herding algorithm for the Nostradamus problem making approximately $\sqrt[3]{n} \cdot 2^{3n/7}$ compression function evaluations, significantly improving over the classical bound. We also prove that quantum herding attacks cannot do better than $2^{3n/7}$ evaluations for random compression functions, showing that our algorithm is (essentially) optimal. We also discuss a slightly less tight bound of roughly $2^{3n/7-s}$ for general Nostradamus attacks against random compression functions, where s is the maximal block length of the adversarially chosen suffix S.

Keywords: Hash function · Herding attack · Lower bound · Nostradamus · Quantum · Grover

1 Introduction

Hash functions serve as a versatile tool in cryptography, thus coming with several security requirements like collision resistance, preimage resistance, or second preimage resistance. In 2006 Kelsey and Kohno [26] introduced a new kind of attack and security property for iterated hash functions H, based on a compression function h with state and output size n. The attack requires the adversary to first commit to a hash value y_{trgt} and later, after given a message prefix P, to find a message suffix S such that $H(P \| S) = y_{\text{trgt}}$. The attack is known under the technical term *chosen-target forced-prefix* (CTFP) preimage attack, but is often referred to by the more picturesque title *Nostradamus attack*. The latter is via the connection to forecasting scenarios: The hash value can be seen as a commitment to a allegedly correct prediction of some event P in the future, which the attacker aims to attest by finding a suitable suffix S.

© International Association for Cryptologic Research 2022
S. Agrawal and D. Lin (Eds.): ASIACRYPT 2022, LNCS 13793, pp. 583–613, 2022.
https://doi.org/10.1007/978-3-031-22969-5_20

1.1 Herding Attacks

The so-called herding attack of Kelsey and Kohno [26] is a Nostradamus attack with roughly $\mathcal{O}(2^{2n/3})$ evaluations of the compression function h.[1] This is still far from the birthday bound for collision resistance, but it is clearly better than a preimage search with $\mathcal{O}(2^n)$ evaluations. The herding attack can be divided into two phases:

Offline phase: In the offline phase the adversary first determines the target hash value y_{trgt} and builds a diamond structure, which is a hash tree of height k. The tree connects 2^k distinct leaves to the root value y_{trgt} via iterating h on different message blocks. Kelsey and Kohno discuss that the overall effort to build such a tree is $\mathcal{O}(2^{(n+k)/2})$. Blackburn et al. [9] later pointed out a flaw in the analysis and gave a bound of $\mathcal{O}(\sqrt{k} \cdot 2^{(n+k)/2})$.

Online phase: In the online phase the adversary is then presented the prefix P. It searches for a linking message part m_{link} to one of the leaves in the diamond structure, such that m_{link} and the message blocks on the tree path m_{path} yield the suffix $S = m_{\mathrm{link}} \| m_{\mathrm{path}}$. Kelsey and Kohno discuss that this step requires $\mathcal{O}(2^{n-k})$ evaluations of h.

Choosing $k = \frac{n}{3}$ then yields an overall effort of $\mathcal{O}(\sqrt{n} \cdot 2^{2n/3})$ of both phases together.

We note that there are variations of the above fundamental attack, also discussed in [26]. One is to use expandable messages [17,27] to accommodate variable-length suffixes when the message length is included in the padding. Such expandable messages can be used in combination with elongated diamond structures. These elongated structures significantly increase the suffix length, by a term 2^r for parameter r, but reduce the effort to roughly $2^{2n/3-2r/3}$. We do not look into such variations here, since our attacks already work well with basic diamond structures.

1.2 Quantum Herding Attack

By a result of Brassard et al. [13] it is known that quantum computers facilitate the search for collisions in hash functions, reducing the effort from $\mathcal{O}(2^{n/2})$ in the classical setting to $\mathcal{O}(2^{n/3})$ in the quantum case. The algorithm itself is based on Grover's quantum search algorithm [23]. The question we address in this work here is if quantum search or collision finding helps in improving herding attacks. Can we expect the same speed-up of a factor 2/3 in the exponent as in the collision case?

As a very fundamental result we first argue that quantum collision search gives an easy attack with $\mathcal{O}(2^{n/2})$ evaluations of h, without the need to construct a diamond structure. Namely, pick an arbitrary target value y_{trgt} and, once receiving the prefix P, use Grover's search algorithm to find the linking message block $m_{\mathrm{link}} \in \{0,1\}^B$ of $B \gg n$ bits. If we assume that the hash function is

---

[1] Unless stated otherwise, all bounds refer to expected numbers of evaluations.

approximately regular, then there are roughly $t = 2^{B-n}$ such message blocks mapping to the target value. But then Grover's algorithm requires $\mathcal{O}(\sqrt{2^B/t}) = \mathcal{O}(2^{n/2})$ quantum evaluations of h to find m_{link}. This already improves over the classical bound (and requires no storage for the diamond structure).

The next, more elaborate attempt is to replace the collision search to create the diamond structure in the attack of Kelsey and Kohno [26] by a quantum algorithm. The improvement may, however, be less expedient than envisioned at first, because the original diamond structure generation throws many values in parallel and then "sieves" for a sufficient number of simultaneous collisions. For the quantum case we proceed step by step. Nonetheless, we show that with this approach we indeed achieve an improvement factor of 2/3 in the exponent compared to the classical attack, requiring $\mathcal{O}(2^{4n/9})$ compression function evaluation.

Our main result is an enhanced version of the quantum attack with a diamond structure. We show that we can actually build a diamond structure more efficiently if we wisely re-use some of the previous evaluations when searching for collisions. Optimizing the parameters we achieve an attack with $\mathcal{O}(\sqrt[3]{k} \cdot 2^{3n/7})$ evaluations of h. The bounds for the attacks are displayed in Fig. 1.

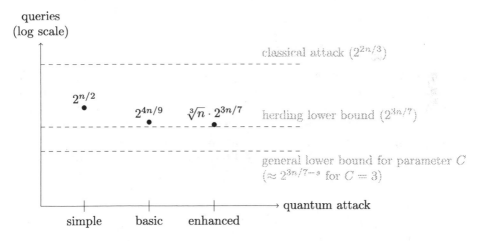

Fig. 1. Upper and lower bounds on expected number of compression function evaluations for quantum attacks (neglecting constants). The simple attack is the straightforward quantum attack, the basic attack uses a diamond structure formed by basic quantum collision search, and the enhanced attack optimizes generation of diamond structure. The parameter s denotes the maximal number of message blocks in the adversarial suffix S, n is the hash function's output size, and the parameter C denotes the number of multicollisions for the lower bound.

We have implemented our attacks in IBM's Qiskit software development kit.[2] The classical simulation of such quantum algorithms in Qiskit, however, restricts the number of available qubits. Therefore, we were only able to run our

---

[2] Available via https://git.rwth-aachen.de/marc.fischlin/quantum-nostradamus.

algorithms against a toy hash function with very limited block and output length $B = n = 8$, based on similar attempts in [21,33]. For this hash function our experiments confirm that the enhanced attack is superior to the basic attack in terms of actual run time, mainly in the offline phase, with almost equal statistics in the online phase. Still, due to the restricted choice of n these results must be taken with prudence. This is the more true as the simple attack with Grover's algorithm outperforms both algorithms for $n = 8$, presumably because it does not require the additional overhead for the herding step.

Our attacks are primarily designed for iterated hash functions of the Merkle-Damgård type [15,30], such as SHA2 [16]. The quantum herding attack is in principle also applicable to sponge-based hash functions [8], as we discuss in in the full version. In this case, our simple quantum attack also yields a bound of $\mathcal{O}(2^{n/2})$ for n-bit outputs. For the basic and enhanced attack the capacity c of the sponge becomes the relevant parameter for building the diamond structure with the hash collisions on the intermediate values, yielding the overall bounds $\mathcal{O}(2^{4c/9})$ resp. $\mathcal{O}(\sqrt[3]{c} \cdot 2^{3c/7})$. For SHA3 [20], however, we have $c = 2n$, such that the latter bounds are inferior to the one of the simple attack. For extendable output functions like SHAKE [20] the choice of the best attack depends on the relationship of n and c.

1.3 Quantum Lower Bounds

Can we go below the bound of $2^{3n/7}$ evaluations for our enhanced attack? We argue that for herding attacks this is impossible, at least generically. For this we use a lower bound of Liu and Zhandry [29] for the quantum query complexity of finding C-collisions in random functions, i.e., C distinct values all mapping to the same function value. The bound states that one needs at least $\Omega(2^{n(1-\frac{1}{2^C-1})/2})$ queries to find such collisions. We argue below that a successful Nostradamus attack essentially allows to find such C-collisions such that the bound transfers to our scenario accordingly.

We first give a general lower bound for Nostradamus attacks for random function h, independently of how the adversary operates. The idea is as follows. Recall that the Nostradamus attacker first commits to the target hash value y_{trgt}, and in the second phase computes the suffix S for the given value P. We can hence repeat the second phase multiple times with different prefixes $P_1, P_2, \ldots$ to generate suffixes $S_1, S_2, \ldots$, such that all inputs $P_i \| S_i$ map to the same target hash values y_{trgt}. If we run the adversary sufficiently often, roughly $(C-1)^s$ times where s is the maximal number of blocks in the suffix S, then at some point we derive a C-collision. It follows that the Nostradamus attack must make at least $\Omega(2^{n(1-\frac{1}{2^C-1})/2})$ queries, divided by $(C-1)^s$. For $C = 3$ the bound simplifies to approximately $2^{3n/7-s}$.

We next argue that the factor $(C-1)^s$ can be avoided for herding attacks using a diamond structure. Specifically, we consider $C = 3$ and use a single solution S of the adversary together with the a specially crafted diamond structure to form a 3-collision. Since we do not need repetitions the extra factor $(C-1)^s$

disappears. We hence turn a Nostradamus attack with diamond structures into a 3-collision finder, with the same number of compression function evaluations (up to constants). It follows from the bound of Liu and Zhandry [29] for $C = 3$ that at least $\Omega 2^{3n/7}$ quantum oracle queries are necessary for a random compression function. This shows that our attack is essentially optimal.

1.4 Related Work

Several other works have further refined the herding attack in the classical setting. Andreeva and Mennink [3] generalized the chosen-target forced-prefix attack and discuss the case that the part P may appear in the middle of the adversary's final output, $S_1 \| P \| S_2$. This covers for example attacks on zipper hash functions and similar constructs [2]. Kortelainen and Kortelainen [28] show how to remove the extra factor $\sqrt{k}$ when building the diamond structure, using a sophisticated construction. Weizman et al. [35] subsequently improve the constant in the $\mathcal{O}$-notation for generating the diamond. We do not pursue such generalizations here.

Aiming at general improvements of quantum attacks for arbitrary hash functions, Chailloux et al. [14] discuss different memory-time trade-offs for finding collisions and a preimage in a list of values. Using variations of Grover's algorithm they show collision attacks with $\widetilde{\mathcal{O}}\left(2^{2n/5}\right)$ evaluations, and multi-target preimage attacks with $\widetilde{\mathcal{O}}\left(2^{3n/7}\right)$ evaluations. Both algorithms only use $\mathcal{O}(n)$ quantum memory, and can be parallelized. Parallelization of multi-target preimage search, also in realistic communication models, has also been considered in [4]. They show that, with realistic communication models, they can find a preimage in a list of t values with p processors with $\mathcal{O}(\sqrt{2^n/pt^{1/2}})$ evaluation steps (where the bound improves from $t^{1/2}$ to t for models with free communication). We do not aim to optimize memory usage for our algorithms, which already store the diamond structure, but focus on the number of hash evaluations here.

Dedicated quantum attacks against specific hash function, improving over the generic bounds, have gained more attention recently. The work of Hosoyamada and Sasaki [24] discusses collision-finding attacks against AES-MMO and Whirlpool. Refined collision and preimage attacks on AES-like hash functions have been presented subsequently by Dong et al. [18,19], Florez Gutierrez et al. [22], as well as Ni et al. [31]. Hosoyamada and Sasaki [25] devised dedicated quantum collision attacks against reduced versions of SHA-256 and SHA-512. Wang et al. [34] present preimage attacks on 4-round versions of Keccak. While neither of these works considers the Nostradamus attack, the results and methods may be also useful to devise improved Nostradamus attacks against specific hash functions. In this work here, however, we are interested in the complexity of generic attacks.

2 Preliminaries

2.1 Hash Functions

Analogously to [26] we consider hash functions $H : \{0,1\}^* \to \{0,1\}^n$ based on the Merkle-Damgård construction with a compression function $h : \{0,1\}^B \times$

$\{0,1\}^n \rightarrow \{0,1\}^n$, where B is the size of the message blocks and n the size of the final hash value and the intermediate hash states. We implicitly assume a (public) initialization vector IV, and for some input $m = m_1 m_2 \ldots m_\ell$, $m_i \in \{0,1\}^B$, aligned to block length B, we define the iterated compression function as

$$h^*(m) := y_\ell, \text{ where } y_0 = IV, y_i = h(m_i, y_{i-1}) \text{for } i = 1, 2 \ldots, \ell.$$

For non-aligned inputs m we assume that the hash function uses a form of suffix-padding function pad which only depends on the input length, such that $m \, \| \, \mathsf{pad}(|m|) \in (\{0,1\}^B)^*$. For example, for Merkle-Damgård hash functions like SHA2 one appends $10^d \, \| \, \langle |m| \rangle$ for a sufficient number d of 0-bits, where $\langle i \rangle$ is a fixed-length binary encoding of the integer i. We assume that the padding extends the message by at most one block.

We assume additionally that the compression function for any fixed intermediate value $y \in \{0,1\}^n$, given as $h_y(\cdot) := h(\cdot, y)$, is surjective and sufficiently close to regular. More specifically, we assume that this function is β-*balanced*, i.e., $|h_y^{-1}(h_y(m))| \geq \beta \cdot 2^{B-n}$ holds for all $m \in \{0,1\}^B$. We note that collisions would be easier to find for the compression function if the preimage sets are significantly skewed [5]. Indeed, Bellare and Kohno [5] also define a more fine grained balance notion for hash functions resp. compression functions. If $h_y(\cdot)$ is β-balanced according to our notion, then it has a balance factor of at least $1 - 2\log_{2^n} \beta$ according to their notion. The simpler balance notion here suffices to give precise performance guarantees when using Grover's quantum algorithm:

Definition 1. *A compression function $h : \{0,1\}^B \times \{0,1\}^n \rightarrow \{0,1\}^n$ for $B \geq n$ of an iterated hash function H is called β-balanced (where $\beta \in (0,1]$) if for any $y \in \{0,1\}^n$ and any $z \in \{0,1\}^n$ the number of preimages satisfies $|h_y^{-1}(z)| \geq \beta \cdot 2^{B-n}$ for the function $h_y(\cdot) = h(\cdot, y)$.*

We observe that, by definition, $\beta > 0$. This means that a β-balanced compression function satisfies $|h_y^{-1}(z)| \geq \beta \cdot 2^{B-n} > 0$ for any y, z. In other words, any image z has a preimage under $h_y(\cdot)$. This in particular means that the function $h_y(\cdot)$ is surjective for any y. Let us stress that we only need the β-balance property for the formal analysis. Our attacks may still succeed if the function is less balanced. This also complies with our implementation results where our example hash function is not even surjective.

We briefly discuss that, if we assume h to be random, then it is β-balanced for $\beta = \frac{1}{2}$ with overwhelming probability for $B \gg n$. For this note that for any fixed image z, the probability that z has less than $\beta \cdot 2^{B-n}$ of the expected number 2^{B-n} of preimages, is at most $\exp(-2^{B-n}/8)$ by the Chernoff bound, and thus double exponentially small. Hence, the probability that there exists any z among the 2^n images violating the bound is still double exponentially small. This means that for a random h we can almost surely assume that each value $z \in \{0,1\}^n$ is hit by at least $\beta \cdot 2^{B-n}$ preimages.

2.2   Quantum Collision Finding

In the following chapter we will see that the Nostradamus attack is based on finding collisions of the compression function h. Therefore we will introduce a generalization of Grover's algorithm as $\mathsf{Grover}(F, y_0)$ and the specific quantum model, which is required for the efficient collision finding algorithm from [13]:

Theorem 2. (Grover, [12]). *Let $F : X \to Y$ be a function, $N := |X|$ be the cardinality of the domain, $y_0 \in Y$ be fixed and*

$$t = |F^{-1}(y_0)| = |\{x \in X | F(x) = y_0\}|$$

be the cardinality of the preimage of y_0 under F. If $t \geq 1$, then the algorithm $\mathsf{Grover}(F, y_0)$ outputs an $x \in X$ with $F(x) = y_0$ after $\mathcal{O}\left(\sqrt{N/t}\right)$ expected evaluations of F.

We consider an adversary with access to a (local) quantum computer and assume that the compression function h is quantum accessible, i.e., the adversary can implement h efficiently on its quantum computer. This allows the adversary to query this h-oracle with arbitrary superposition of the inputs, akin to the quantum random oracle model [10]. This enables us to also use the algorithm $\mathsf{Grover}(F, y_0)$ in cases where the function F depends on the compression function h.

Note that Boyer et al. [12] discuss that the above theorem even holds if the number t of solutions is not known in advance. In our attacks against hash functions we will later take advantage of the fact the compression function h is β-balanced (where β is usually assumed to be constant). Hence, using Grover's algorithm for searching a preimage for some value z among the $N = 2^B$ many inputs to h, of which at least $t \geq \beta \cdot 2^{B-n}$ solutions map to z, takes an expected number $\mathcal{O}\left(\sqrt{2^B/\beta 2^{B-n}}\right) = \mathcal{O}\left(\beta^{-1/2} \cdot 2^{n/2}\right)$ of function evaluations. For constant β this equals $\mathcal{O}\left(2^{n/2}\right)$. We also point out that in most of our attacks we apply Grover's algorithm multiple times, in sequential order, such that the expected number of total evaluations of F is given by the sum of the expected evaluations of the individual calls, by the linearity of expectations.

3   Classical Nostradamus Attack

In this section we review the idea of Kelsey and Kohno [26] for the Nostradamus attack.

Attack. The Nostradamus attack lets the adversary first commit to a target hash value y_{trgt}. Then it receives a prefix P where we assume for simplicity that P is aligned to block length and that the number ℓ_P of blocks of P is known in advance. If this was not the case the adversary could simply guess the actual length of P and append 0's if necessary. The task of the adversary is now to find a suffix S such that $\mathsf{H}(P \| S) = y_{\text{trgt}}$.

We have to specify how the prefix P is chosen in the attack. Kelsey and Kohno [26] assume that the prefix is chosen randomly from a specified set. To avoid assumptions about the distribution of P we ask the adversary to succeed for any given value P. The latter matches the fact that known attacks indeed achieve this. We specify this by quantifying over all possible prefixes P, demanding the adversary to win in all cases. We also require the adversary to succeed with probability 1 (over the internal randomness). Since we allow the adversary to run in expected time, we can always enforce this by iterating till success, compensating for smaller success probabilities by larger run times:

Definition 3 (Successful Nostradamus Attack). *A successful Nostradamus attack $\mathcal{A}$ for an iterated hash function H based on compression function $h : \{0,1\}^B \times \{0,1\}^n \to \{0,1\}^n$ consists of two algorithms $(\mathcal{A}^{off}, \mathcal{A}^{onl})$ such that, for any $\ell_P \in \mathbb{N}$, any $P \in \{0,1\}^{\ell_P \cdot B}$, the following experiment $\mathbf{Exp}_{\mathsf{H},\mathcal{A},P}^{Nostr}(\lambda)$ always returns 1:*

$$\underline{\mathbf{Exp}_{\mathsf{H},\mathcal{A},P}^{Nostr}(\lambda)\text{:}}$$

$2\text{:}\quad (st, y_{trgt}) \leftarrow\!\!\$\; \mathcal{A}^{off}(1^\lambda, \ell_P) \quad /\!\!/ \textit{offline}$

$3\text{:}\quad S \leftarrow \mathcal{A}^{onl}(st, P) \quad /\!\!/ \textit{online}$

$4\text{:}\quad \textbf{return}\; [\mathsf{H}(P \,\|\, S) = y_{trgt}]$

Note that we do not make any stipulations on the run time of adversary $\mathcal{A}$. In this sense there is always the trivial attack which executes an exhaustive search. We are interested in more efficient attacks, of course. The parameter ℓ_P usually enters the run time of the adversary, but only mildly compared to the search for the initial state and for the suffix.

Offline Phase. The Nostradamus attack of Kelsey and Kohno [26] consists of an offline and an online phase. In the offline phase the adversary creates a (binary) tree structure (V, E), also referred to as a diamond structure. Each node $v \in V \subseteq \{0,1\}^n$ in the tree represents a hash state and each (directed and labeled) edge $e = (y_0, m, y_1) \in E \subseteq V \times \{0,1\}^B \times V$ represents a transition from one hash state to the next one via the message label, $h(m, y_0) = y_1$.[3]

The tree consists of 2^k *distinct* leaves where k is chosen appropriately. The algorithm of Kelsey and Kohno starts by sampling the leaf nodes, building up the tree level by level, by trying message blocks for each node in order to find collisions. A level of the tree is a set of nodes with identical (edge) distance to the root node. These levels and its nodes get numbered by the distance, e.g., level k contains the leaf nodes $y_{k,1}, y_{k,2}, \ldots$ and level 0 the root node $y_{0,1}$. In general, node $y_{i,j}$ is the jth node of level i. See Fig. 2.

---

[3] Note that we simply identify a node with its hash value label. Formally, to make nodes with identical hash values distinct, we add a position in the tree to each node (given by the level and its order within the nodes of the same level) but usually omit mentioning the position value.

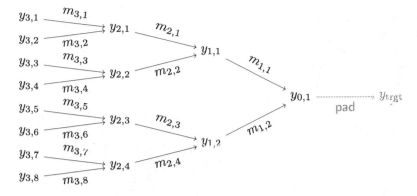

Fig. 2. Tree structure of height 3 in offline phase of classical Nostradamus attack, where $y_{\text{trgt}} = h(\text{pad}(B(\ell_P + 4)), y_{0,1})$.

The original analysis in [26] argued that roughly $2^{\frac{n+k}{2}}$ compression function evaluations are necessary to form such a tree of height k. This, however, ignored that one node value $y_{i,j}$ may collide with multiple other values, such that we would not get a full binary tree. Luckily, in [9] it is shown that we can capture this by an extra factor $\sqrt{k}$. That is, in [9] it is proven that $\sqrt{k} \cdot \ln 2 \cdot 2^{\frac{n-k}{2}}$ evaluations of the function h suffice per node, and $\Theta(\sqrt{k} \cdot 2^{\frac{n+k}{2}})$ in total for building the complete diamond structure. Afterwards the hash value $y_{\text{trgt}} = h(\text{pad}(B \cdot (\ell_P + k + 1)), y_{0,1})$ can be computed and submitted, where $y_{0,1}$ is the root of the tree structure.

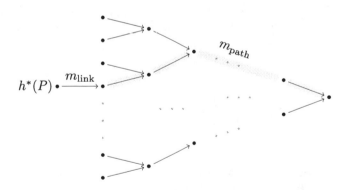

Fig. 3. Linking message m_{link} for given prefix P.

Online Phase. In the online phase the adversary uses the tree structure to construct the suffix S. For this the adversary searches for a linking message block $m_{\text{link}} \in \{0,1\}^B$ such that $h(m_{\text{link}}, h^*(P))$ hashes to a leaf in the tree (see Fig. 3). If such a message block m_{link} has been found, then the suffix S is given by $S := m_{\text{link}} \| m_{\text{path}}$, where m_{path} denotes the (concatenation of the) message labels on the path from the leaf to the root of the tree. See Fig. 3. Note that by construction $P \| S$ consists of ℓ_P blocks in P, one linking block m_{link}, and the k blocks in m_{path}, such that the message is of bit length $B \cdot (\ell_P + k + 1)$. Together with the appended padding the hash function thus maps $P \| S$ to the pre-selected target value y_{trgt}.

Since there are 2^k randomly and independently sampled leaves in the tree one can expect to find such a linking message m_{link} with 2^{n-k} evaluations of h. Moreover [9] proves that the height of the tree can be optimally chosen with $k \approx \frac{n}{3}$, so that the entire attack needs $\mathcal{O}(\sqrt{n} \cdot 2^{\frac{2n}{3}})$ evaluations of h.

4   Quantum-Based Nostradamus Attacks

We start by giving a straightforward quantum version of the Nostradamus attack, exploiting that one can find collisions for hash functions in about $2^{n/2}$ quantum steps. This already beats the classical bound of roughly $2^{2n/3}$ but we show afterwards that one can go lower with more advanced strategies. The first advanced attack follows the same structure as the classical herding attack but reduces the work load to approximately $2^{4n/9}$ by using Grover's algorithm. The second advanced algorithm then optimizes the step to build the diamond structure, resulting in a total number of roughly $2^{3n/7}$ compression function evaluations.

4.1   Simple Quantum Attack

We first describe a very simple quantum Nostradamus attack $\mathcal{A}_{\text{simple}}$. The algorithm receives as input the block length ℓ_P of the unknown prefix, picks a random value y (to which $P \| S$ will map under h^*) and applies the final iteration for the padding. In the online phase it simply runs Grover's algorithm to find a linking message block m_{link}:

$\mathcal{A}^{\text{off}}_{\text{simple}}(1^\lambda, \ell_P)$: // offline	$\mathcal{A}^{\text{onl}}_{\text{simple}}(y, P)$: // online
1 : $y \xleftarrow{\$} \{0,1\}^n$	1 : $p \leftarrow h^*(P)$
2 : $y_{\text{trgt}} \leftarrow h(\text{pad}(B \cdot (\ell_P + 1)), y)$	2 : $S \leftarrow \text{Grover}(h_p, y)$ $/\!/ h_p(\cdot) = h(\cdot, p)$
3 : **return** (y, y_{trgt})	3 : **return** S

Proposition 4. *Let* H *be a hash function with a β-balanced compression function* $h : \{0,1\}^B \times \{0,1\}^n \to \{0,1\}^n$. *Algorithm* $\mathcal{A}_{simple}$ *mounts a successful Nostradamus attack against* h *with* $\mathcal{O}(\beta^{-1/2} \cdot 2^{n/2} + \ell_P + 1)$ *expected evaluations of* h *for any prefix* P *of block length* ℓ_P.

Note that for constant β (e.g., recall that $\beta = \frac{1}{2}$ works with overwhelming probability for random function h) we obtain approximately the bound $\mathcal{O}(2^{n/2})$.

Proof. Correctness follows directly by construction and the fact that the compression function h is surjective according to the balance property: For the chosen value y there exists at least one preimage S such that

$$y = h_p(S) = h(S, p) = h^*(P \| S),$$

and $y_{\text{trgt}} = h(\text{pad}(|P \| S|), y)$ by construction. As for performance, Theorem 2 implies that the search for S needs $\mathcal{O}\left(\sqrt{N/t}\right)$ evaluations in expectation where

$$N = |\{0, 1\}^B| = 2^B \quad \text{and} \quad t = |h_p^{-1}(y)| = |\{m \in \{0, 1\}^B | h_p(m) = y\}|.$$

Since the function h_p is β-balanced it follows that $t \geq \beta \cdot 2^{B-n}$, yielding an overall effort of $\mathcal{O}\left(\beta^{-\frac{1}{2}} \cdot 2^{n/2}\right)$ evaluations for Grover's search. The computation of y_{trgt} in the offline phase and the initial computation of p in the online phase need at most $\ell_P + 1$ additional evaluations of h. □

4.2 Basic Quantum Attack with Diamond Structure

This section shows that the construction of a diamond structure is rewarding and leads to a more efficient quantum attack. We first describe a basic version of this attack and optimize it in the next section. We present the algorithm by dividing into several sub algorithms, following the structure of the classical attack: In the offline phase we use algorithm Diamond$_{\text{basic}}$ to build the diamond structure. The algorithm itself uses a collision finder Claw as a subroutine. In the online phase we once more use the Link algorithm to find the linking message m_{link}. Finally we put all algorithms together to derive our adversary.

We first describe the claw finding algorithm Claw, which is in fact the well-known BHT algorithm [13] adapted to our setting. The algorithm takes as input a parameter ℓ and two functions h_y and $h_{y'}$ and returns m, m' such that $h_y(m) = h_{y'}(m')$. It does so by first sampling a random list of 2^ℓ input messages m_i for h_y, and uses Grover's algorithm to find the matching value m' for $h_{y'}$.

$\text{Claw}_\ell(h_y, h_{y'})$:

1: $m_1, \ldots, m_{2^\ell} \leftarrow_\$ \{0, 1\}^B$ such that $h_y(m_i) \neq h_y(m_j)$ for all $i \neq j$

2: $m' \leftarrow \text{Grover}(F, 1)$ // F as in Proposition 5

3: $i \leftarrow \{1, \ldots, 2^\ell\}$ such that $h_y(m_i) = h_{y'}(m')$

4: **return** (m_i, m')

Note that for a well-balanced compression function and for $\ell \leq n/3$, as we need below, the message sampling in the first step can be implemented by picking all the m_i's randomly, at once. A hash collision among the at most $2^\ell \leq 2^{n/3}$ hash values happens with sufficiently small probability, and in the rare case of a collision we repeat the process. On average we do not need to make more than two iterations for avoiding collisions. In the theorem below, and also in our implementation, we nonetheless consider the general case of a β-balanced compression function, in which case we re-sample each m_i individually if it collides with a previous choice.

Proposition 5. ([13]). *Let* H *be a hash function with β-balanced compression function* $h : \{0,1\}^B \times \{0,1\}^n \to \{0,1\}^n$. *For* $\ell \in \mathbb{N}$, *a set* $m_1, \ldots, m_{2^\ell} \in \{0,1\}^B$ *of distinct messages, and values* $y, y' \in \{0,1\}^n$. *Let* $F : \{0,1\}^B \to \{0,1\}$ *be the function defined by*

$$F(m) = 1 \quad :\Longleftrightarrow \quad \exists i \in \{1, \ldots, 2^\ell\} : h_y(m_i) = h_{y'}(m).$$

Then algorithm $\mathsf{Claw}_\ell(h_y, h_{y'})$ *outputs messages* $m, m' \in \{0,1\}^B$ *with* $h_y(m) = h_{y'}(m')$ *and needs* $\mathcal{O}\left(\beta^{-1} \cdot 2^\ell + \beta^{-\frac{1}{2}} \cdot 2^{\frac{n-\ell}{2}}\right)$ *expected evaluations of* h *for* $\ell < n$. *In particular,* $\mathsf{Claw}_{n/3}$ *needs* $\mathcal{O}\left(\beta^{-1} \cdot 2^{\frac{n}{3}}\right)$ *expected evaluations of* h.

Note that for $\beta \leq 1$ we always have $\beta^{-1/2} = \mathcal{O}\left(\beta^{-1}\right)$ and from now on bound the factor $\beta^{-1/2}$ by the inverse of β.

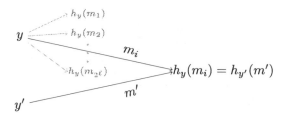

Fig. 4. Illustration of algorithm $\mathsf{Claw}_\ell(h_y, h_{y'})$.

Proof. Let us first discuss how we generate the 2^ℓ message blocks m_i in the first step for general compression functions. For this we iterate through $i = 1, 2, 3, \ldots$ and for each i we repeatedly pick $m_i \leftarrow_\$ \{0,1\}^B$ randomly, until $h_y(m_i)$ is different from all the previous hash values. For the i-th step there are $2^n - (i-1)$ hash values unoccupied, and each image has at least $\beta \cdot 2^{B-n}$ preimages. Hence, we pick such a good preimage with probability at least

$$2^{-B} \cdot (2^n - i + 1) \cdot \beta \cdot 2^{B-n} \geq (2^n - 2^\ell) \cdot \beta \cdot 2^{-n} = (1 - 2^{\ell-n}) \cdot \beta \geq \tfrac{1}{2} \cdot \beta$$

for $\ell < n$. On average, we thus only need to sample $\mathcal{O}(\beta^{-1})$ times for each of the 2^ℓ message blocks.

The proof now follows straightforwardly from Theorem 2, noting that the input size of function F equals $N = 2^B$, and since all 2^ℓ hash values are distinct and each hash value has at least $\beta \cdot 2^{B-n}$ preimages by the balance property of h, there are at least $t \geq \beta \cdot 2^\ell \cdot 2^{B-n}$ possible solutions.   □

Given the claw finding algorithm we present our basic algorithm for deriving the diamond structure. The algorithm takes as input a parameter k determining the height of the tree (V, E) which it outputs. The algorithm uses $\mathsf{Claw}_{n/3}$ to determine collisions for neighbored values:

Diamond$_{\mathrm{basic}}(k)$:

1 :	$(V, E) \leftarrow \emptyset$
2 :	$y_{k,1}, y_{k,2}, \ldots, y_{k,2^k} \leftarrow\$ \{0,1\}^n$ pairwise different $\quad /\!/$ leaf nodes
3 :	**for** $s = k, \ldots, 1$ **do** $\quad /\!/$ for constructing level $s - 1$
4 :	$\quad V \leftarrow V \cup \{y_{s,1}, \ldots, y_{s,2^s}\}$
5 :	$\quad$ **for** $i = 1, \ldots, 2^{s-1}$ **do**
6 :	$\quad\quad (m, m') \leftarrow \mathsf{Claw}_{n/3}(h_{y_{s,2i-1}}, h_{y_{s,2i}})$
7 :	$\quad\quad E \leftarrow E \cup \{(y_{s,2i-1}, m, h_{y_{s,2i-1}}(m)), (y_{s,2i}, m', h_{y_{s,2i-1}}(m))\}$
8 :	$\quad\quad y_{s-1,i} \leftarrow h_{y_{s,2i-1}}(m) \quad /\!/$ for next iteration
9 :	$\quad$ **endfor**
10 :	**endfor**
11 :	**return** (V, E)

Proposition 6. *Let* H *be a hash function with* β-*balanced compression function* $h : \{0,1\}^B \times \{0,1\}^n \to \{0,1\}^n$. *Algorithm* Diamond$_{\mathrm{basic}}(k)$ *outputs a diamond structure of height* $k \le n$ *after* $\mathcal{O}(\beta^{-1} \cdot 2^{\frac{n}{3}+k})$ *expected evaluations of* h.

Once more, for constant β the bound simplifies to $\mathcal{O}(2^{\frac{n}{3}+k})$. We note that the sampling of the 2^k values $y_{k,i}$'s in Line 2 must yield pairwise distinct strings. Since we later choose $k = n/9$ a collision among the random values happens with negligible probability only. Alternatively, we may either pick arbitrary distinct values, e.g., by incrementing a counter value, or sample each $y_{k,i}$ as often till we have a fresh value. In our implementation we use the latter. We remark that this sampling step in either case does not account for the number of hash evaluations.

Proof. The output graph is indeed a diamond structure of height k, because on the one hand the connecting rule is fulfilled since $h_{y_{s,2i-1}}(m) = h_{y_{s,2i}}(m')$ for any s, i and the edges are directed from higher to lower levels since the algorithm starts with level k. On the other hand the graph has an underlying binary tree structure with 2^k leaf nodes. The algorithm constructs, for a fixed value s, the next level $s-1$ by connecting the nodes of level s in a pairwise manner. Therefore the next level contains 2^{s-1} nodes. Within the last iteration, where $s = 1$, a level with $2^{1-1} = 1$ nodes is constructed, which is the root node of the resulting graph.

As for performance, we mainly need to consider the repetitions of Line 6, the executions of algorithm Claw. Individually each search on average needs $\mathcal{O}(\beta^{-1} \cdot 2^{\frac{n}{3}})$ evaluations of h according to Proposition 5. Thus, for building the entire structure,

$$\sum_{s=1}^{k} \sum_{i=1}^{2^{s-1}} \mathcal{O}(\beta^{-1} \cdot 2^{\frac{n}{3}}) = \mathcal{O}\left(\beta^{-1} \cdot 2^{\frac{n}{3}} \cdot \sum_{s=1}^{k} 2^{s-1}\right)$$

$$= \mathcal{O}(\beta^{-1} \cdot 2^{\frac{n}{3}} \cdot (2^k - 1)) = \mathcal{O}(\beta^{-1} \cdot 2^{\frac{n}{3}+k})$$

expected evaluations are required in total. $\qquad\square$

We next describe our algorithm Link finding the linking message m_{link}. The algorithm takes as input the 2^k leaves of the diamond structure and the prefix P, and finds a message block m_{link} that connects P to one of the leaves. This is done via Grover's algorithm by defining a suitable function F identifying such links:

$\text{Link}(P, y_{k,1}, \dots, y_{k,2^k})$:

1 : $p \leftarrow h^*(P)$

2 : $m_{\text{link}} \leftarrow \text{Grover}(F, 1)$ $\mathbin{/\!\!/}$ F as in Proposition 7

3 :   let $y_{k,i}$ be a leaf with $h(m_{\text{link}}, p) = y_{k,i}$

4 :   **return** $(m_{\text{link}}, y_{k,i})$

Proposition 7. *Let* H *be a hash function with β-balanced compression function* $h : \{0,1\}^B \times \{0,1\}^n \to \{0,1\}^n$. *Let* $P \in \{0,1\}^{\ell_P}$, $p \leftarrow h^*(P)$ *and* (V, E) *be a diamond structure of height k with (distinct) leaves $(y_{k,1}, \dots, y_{k,2^k})$. Define the function* $F : \{0,1\}^B \to \{0,1\}$ *as*

$$F(m) = 1 \;:\Longleftrightarrow\; \exists i \in \{1, \dots, 2^k\} : h(m, p) = y_{k,i}.$$

Then algorithm $\text{Link}(k, y_{k,1}, \dots, y_{k,2^k})$ *outputs a linking message block m_{link} and the connecting leaf node $y_{k,i}$ with $h(m_{\text{link}}, p) = y_{k,i}$, requiring* $\mathcal{O}\!\left(\beta^{-\frac{1}{2}} \cdot 2^{\frac{n-k}{2}} + \ell_P\right)$ *expected evaluations of h in total.*

Proof. The correctness follows directly by the construction of the function F. As for performance, Theorem 2 implies that the search for m_{link} needs $\mathcal{O}\!\left(\sqrt{N/t}\right)$ evaluations where

$$N = 2^B \quad \text{and} \quad t = |\bigcup_{i=1}^{2^k} h_p^{-1}(y_{k,i})| \geq \beta \cdot 2^{k+B-n},$$

since the function h_p is β-balanced, and the leaves are distinct in a diamond structure. The first step to compute p needs at most ℓ_P evaluations of h. $\qquad\square$

At this point, $\text{Diamond}_{\text{basic}}$ and Link can be composed to form our basic quantum attack $\mathcal{A}_{\text{basic}}$ (for parameter k), using a diamond structure of height k. We presume that $\text{Path}((V, E), y)$ is the algorithm, which concatenates the message labels on the edges from leaf y to the root of the tree:

$\mathcal{A}_{\text{basic}}^{k,\text{off}}(1^\lambda, \ell_P)$:

1 : $(V, E) \leftarrow\!\!\$\; \text{Diamond}_{\text{basic}}(k)$

2 :   let $y_{0,1} \in V$ be the root of (V, E)

3 : $y_{\text{trgt}} \leftarrow h(\text{pad}(B \cdot (\ell_P + k + 1)), y_{0,1})$

4 :   **return** $((V, E), y_{\text{trgt}})$

$\mathcal{A}_{\text{basic}}^{k,\text{onl}}((V, E), P)$:

1 :   leaves $y_{k,1}, \dots, y_{k,2^k}$ of (V, E)

2 : $(m_{\text{link}}, y) \leftarrow \text{Link}(P, y_{k,1}, \dots, y_{k,2^k})$

3 : $m_{\text{path}} \leftarrow \text{Path}((V, E), y)$

4 : $S \leftarrow m_{\text{link}} \| m_{\text{path}}$

5 :   **return** S

Theorem 8. (Basic Quantum Attack with Diamond Structure). *Let* H *be a hash function with β-balanced compression function* $h : \{0,1\}^B \times \{0,1\}^n \to \{0,1\}^n$. *Let* $k \in \mathbb{N}$. *Then adversary* $\mathcal{A}_{basic}^k$ *mounts a successful Nostradamus attack and needs*

$$\mathcal{O}\left(\beta^{-1} \cdot 2^{\frac{n}{3}+k} + \beta^{-\frac{1}{2}} \cdot 2^{\frac{n-k}{2}} + \ell_P + 1\right)$$

expected evaluations of h. *In particular, for* $k = \frac{n}{9}$ *the adversary* $\mathcal{A}_{basic}^{n/9}$ *in total needs* $\mathcal{O}\left(\beta^{-1} \cdot 2^{\frac{4n}{9}} + \ell_P + 1\right)$ *expected evaluations of* h.

4.3 Attack with Enhanced Diamond Structure Generation

We next present an advanced attack, essentially reaching the lower bound of $\Omega(2^{3n/7})$ for herding attacks discussed in Sect. 5.2. The algorithm still applies the same general strategy as in the previous section, but uses an enhanced algorithm to create the diamond structure. The idea there was basically to connect two nodes in the tree via algorithm Claw for parameter $n/3$, resulting in $2^{n/3}$ compression function evaluations for each connection. We now discuss how we can speed up this process by re-using data across the various connection steps.

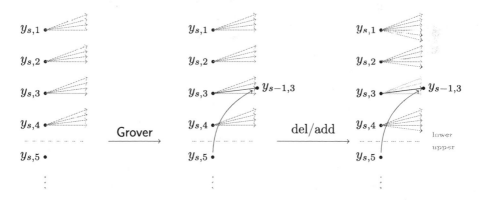

Fig. 5. Iteration of algorithm Diamond$_{\text{enhanced}}$. Left: At the beginning of each iteration the nodes in the lower half have 2^ℓ potential successors, roughly distributed equally over all remaining nodes (here 16 successors, assigned evenly to the 4 nodes $y_{s,1}, \ldots, y_{s,4}$). The next node from the upper half (here $y_{s,5}$) shall be connected to a node in the lower part. Middle: Grover's algorithm finds a match to one of the 2^ℓ successors (called $y_{s-1,3}$ here) of a node from the lower part (here $y_{s,3}$) such that we can connect the two nodes. Right: We remove all potential successor pointers from the connected node from the lower half and add, step by step, the same number of new pointers to the remaining nodes (here displayed as dotted arrows).

Consider a level s of the tree for which we try to connect the 2^s nodes $y_{s,1}, \ldots, y_{s,2^s}$ in a pairwise manner. We split the 2^s nodes into a lower and an upper half of 2^{s-1} nodes each. For the lower half we will compute a list Y of 2^ℓ hash evaluations $h(m_j, y_{s,i})$, equally spread out over the 2^{s-1} values. Here ℓ will be an appropriate parameter to be determined later. Then we use Grover's search to connect the first value $y_{s,2^{s-1}+1}$ of the upper half to some of these 2^ℓ values via some message block m'. Once we have found such a connection we are going to remove the partner node from the lower half and all of its $2^\ell/2^{s-1}$ entries in Y. We add this amount of new values, again equally spread out over the remaining $2^{s-1} - 1$ values paired up, to fill the list Y up to 2^ℓ elements again. This idea is displayed in Fig. 5. Then we are going to connect the second node from the upper half to an entry in the updated list Y, as before. We continue till we have all 2^s values and then proceed with the next level $s - 1$ till we have eventually built the entire tree.

We present the enhanced algorithm for creating the diamond structure next. The algorithm takes as input the parameter k (for the tree height) and outputs the tree (V, E). It internally uses the parameter ℓ (for the list size) in dependence of the tree level s it currently considers. The algorithm uses Grover's algorithm as a subroutine for a function $F_{y,Y}(m)$ with parameters y and Y, which checks if $h_y(m)$ is in the list Y. Put differently, Grover's algorithm returns such a matching message block m.

$\text{Diamond}_{\text{enhanced}}(k)$:

---

1 : $\quad (V, E) \leftarrow \emptyset$

2 : $\quad y_{k,1},\ y_{k,2},\ \ldots, y_{k,2^k} \leftarrow\!\$ \ \{0,1\}^n$ pairwise different $\quad$ // leaf nodes

3 : $\quad$ **for** $s = k, \ldots, 1$ **do** $\quad$ // for constructing level $s - 1$

4 : $\qquad \ell \leftarrow \lceil (n + 2s - 2\log_2 s)/3 \rceil$

5 : $\qquad V \leftarrow V \cup \{y_{s,1}, \ldots, y_{s,2^s}\}$

6 : $\qquad L \leftarrow \{1, \ldots, 2^{s-1}\} \quad$ // list of unprocessed nodes in lower half

7 : $\qquad Y \leftarrow \emptyset \quad$ // storing precomputed values

8 : $\qquad \gamma \leftarrow 1 \quad$ // state counter where to add next elements to Y

9 : $\qquad$ **foreach** $y \in \{y_{s,2^{s-1}+1}, \ldots, y_{s,2^s}\}$ **do** $\quad$ // process each node in upper half

10 : $\qquad\quad$ **while** $|Y| < 2^\ell$ **do** $\quad$ // Add elements to list of precomputed values?

11 : $\qquad\qquad$ **while** $\gamma \notin L$ **do** $\gamma \leftarrow (\gamma \bmod 2^{s-1}) + 1 \quad$ // choose (circularly) next node

12 : $\qquad\qquad m \leftarrow\!\$ \ \{0,1\}^B$ such that $(*, *, h(m, y_{s,\gamma})) \notin Y \quad$ // unique in Y

13 : $\qquad\qquad Y \leftarrow Y \cup \{(m,\ \gamma,\ h(m, y_{s,\gamma})\} $

14 : $\qquad\qquad \gamma \leftarrow \gamma + 1$

15 : $\qquad\quad$ **endwhile** $\qquad m \leftarrow \text{Grover}(F_{y,Y}, 1) \quad$ // F as in Theorem 9

16 : $\qquad\quad$ search for $(m', i, y') \in Y$ with $y' = h_y(m) \quad$ // unique due to Line 12

17 : $\qquad\quad E \leftarrow E \cup \{(y_{s,i},\ m',\ y'), (y,\ m,\ y')\} \quad$ // connect nodes

18 : $\qquad\quad y_{s-1,i} \leftarrow y' \quad$ // for next iteration, noting that $i \leq 2^{s-1}$

19 : $\qquad\quad$ delete any $(*, i, *)$ from $Y \quad$ // only values for other nodes remain in Y

20 : $\qquad\quad$ delete i from $L \quad$ // $y_{s,i}$ has been processed

21 : $\qquad$ **endforeach**

22 : $\quad$ **endfor**

23 : $\quad$ **return** (V, E)

Note that we need to find a message $\bar{m}$ in Line 12 for which the hash value has not been assigned yet. Recall that ℓ is in the order of $(n+2s)/3 \leq (n+2k)/3$. We will later set $k = n/7$ such that $\ell \leq 3n/7$. Hence, assuming that the hash value of m is uniformly distributed, the probability of having collisions is at most $2^{2\ell} \cdot 2^{-n} \leq 2^{-n/7}$ and thus negligible. Below, for general β-balanced compression function, we follow once more the idea to sample-till-success to find m, as also done in the algorithm Claw.

Theorem 9. *Let* H *be a hash function with β-balanced compression function* $h : \{0,1\}^B \times \{0,1\}^n \to \{0,1\}^n$. *For a list Y and a fixed value y let the function* $F : \{0,1\}^n \to \{0,1\}$ *be defined as*

$$F(m) = 1 \quad :\Longleftrightarrow \quad \exists y' \in Y : h_y(m) = y'.$$

Then algorithm Diamond$_{\text{enhanced}}(k)$ *outputs a diamond structure (V, E) of height k with*

$$\mathcal{O}\left(\beta^{-1} \cdot 2^{\frac{n+2k+\log_2(k)}{3}}\right)$$

expected evaluations of h *for $k < n$.*

Proof. Correctness follows by construction. Starting with the random leaves the algorithm connects one node from the lower part with one node from the upper part in each iteration, yielding a binary tree. The exact matching of values in the lower and upper half also implies that the search for the next γ-value always terminates (because L has at least one value if there is some y in the upper half left).

We next look at the performance of the algorithm. Each application of the Grover algorithm (in Line 15) needs $\mathcal{O}\left(\sqrt{N/t}\right)$ evaluations of h, where

$$N = |\{0,1\}^B| = 2^B, \qquad t \geq \beta \cdot 2^\ell \cdot 2^{B-n}$$

by construction, since the set Y contains exactly 2^ℓ pairwise different elements from $\{0,1\}^n$ in each iteration and the function h is β-balanced. This means that the algorithm requires $\beta^{-\frac{1}{2}} \cdot 2^{\frac{n-\ell}{2}}$ evaluations of the compression function on the average. Note that we run Grover on level s for 2^s times, yielding an overall number of $\beta^{-\frac{1}{2}} \cdot 2^{\frac{n-\ell+2s}{2}}$ expected evaluations of h for this level.

Next we consider the number of hash evaluations for filling up the list Y (**while**-loop in Line 10) when iterating through all values y in the upper half (**foreach**-loop in Line 9). For this we start by looking at the search for a message block with a fresh hash value in Line 12. As in the Claw algorithm we can sample a message block $m \leftarrow\!\!\$\ \{0,1\}^B$ till we found one whose hash value is not in Y. Since we have at most $2^\ell - 1$ many values in Y at this point, where $\ell \leq (n+2s)/3 \leq (n+2k)/3 < n$ for $k < n$, we can conclude as in the analysis of Proposition 5 that each search requires $\mathcal{O}(\beta^{-1})$ attempts on the average.

For the overall analysis of the loops we first discuss a simplified version, neglecting rounding of fractions. Since we start with Y being empty for the first value y, we need 2^ℓ many values to fill the list in the first iteration. Then, at the end of the iteration, we remove at most $2^\ell/2^{s-1}$ entries in Y for the identified element $y_{s,i}$ in the lower part. This implies that in the next step we need to make $2^\ell/2^{s-1}$ hash evaluations to fill the list Y again up to 2^ℓ elements. In the next step, however, we have only $2^{s-1} - 1$ elements left, and thus remove at most $2^\ell/(2^{s-1} - 1)$ elements for the subsequent iteration. This continues for $2^{s-1} - 2$, $2^{s-1} - 3$, ..., until all the 2^{s-1} iterations for values in the upper half are completed. For the final step we need to add $2^\ell/(2^{s-1} - (2^{s-1} - 2)) = 2^\ell/2$ elements.

We can write the total number of elements to be sampled for the resurrection of the complete list Y thus as:

$$2^\ell + 2^\ell \cdot \sum_{j=0}^{2^{s-1}-2} \frac{1}{2^{s-1} - j} = 2^\ell \cdot \sum_{j=1}^{2^{s-1}} \frac{1}{j} = \mathcal{O}\left(2^\ell \cdot \ln 2^s\right) = \mathcal{O}\left(s \cdot 2^\ell\right),$$

using $\sum_{i=1}^q \frac{1}{j} \le \ln q + c$ for the harmonic series. On average, we need $\mathcal{O}\left(\beta^{-1}\right)$ h-evaluations for each element.

Note that, so far, we have not taken into account that we may not be able to spread out all 2^ℓ elements in Y evenly on the remaining entries in L. By construction, however, the difference of assigned elements from Y can differ by at most 1. We can incorporate this into our analysis by using an extra factor 2 for the number of elements which need to be added to Y in each iteration, resulting in the same asymptotic bound. Similarly, since we round up ℓ in the algorithm, it can grow by an additive term 1 at most, which is also "swallowed" by a constant in the asymptotic notation.

In total, for level s of the tree we thus need

$$\mathcal{O}\left(\beta^{-1} \cdot s \cdot 2^\ell + \beta^{-\frac{1}{2}} \cdot 2^{\frac{n-\ell+2s}{2}}\right) = \mathcal{O}\left(\beta^{-1} \cdot 2^{\ell+\log_2 s} + \beta^{-\frac{1}{2}} \cdot 2^{\frac{n-\ell+2s}{2}}\right)$$

evaluations on the average. By our choice of ℓ (in dependence of s) the two terms become equal (except for the β factor), such that together with $s \le k$ the bound simplifies to

$$\mathcal{O}\left(\beta^{-1} \cdot 2^{\frac{n+2s+\log_2 s}{3}}\right) = \mathcal{O}\left(\beta^{-1} \cdot 2^{\frac{n+2s+\log_2 k}{3}}\right).$$

Summing over all k stages we thus get a total number of

$$\sum_{s=1}^k \mathcal{O}\left(\beta^{-1} \cdot 2^{\frac{n+2s+\log_2 k}{3}}\right) = \mathcal{O}\left(\beta^{-1} \cdot 2^{\frac{n+\log_2 k}{3}} \cdot \sum_{s=1}^k 2^{\frac{2s}{3}}\right)$$

$$= \mathcal{O}\left(\beta^{-1} \cdot 2^{\frac{n+\log_2 k}{3}} \cdot 2^{\frac{2k}{3}}\right)$$

$$= \mathcal{O}\left(\beta^{-1} \cdot 2^{\frac{n+2k+\log_2 k}{3}}\right)$$

compression function evaluations on the average.   □

With this new algorithm our adversary is now a straightforward adaption of the basic one, with the enhanced diamond structure generation replacing the basic one:

$\mathcal{A}_{\text{enhanced}}^{k,\text{off}}(1^{\lambda}, \ell_P)$:

1 : $(V, E) \leftarrow\$ \text{Diamond}_{\text{enhanced}}(k)$

2 : let $y_{0,1} \in V$ be the root of (V, E)

3 : $y_{\text{trgt}} \leftarrow h(\text{pad}(B \cdot (\ell_P + k + 1)), y_{0,1})$

4 : **return** $((V, E), y_{\text{trgt}})$

$\mathcal{A}_{\text{enhanced}}^{k,\text{onl}}((V, E), P)$:

1 : leaves $y_{k,1}, \ldots, y_{k,2^k}$ of (V, E)

2 : $(m_{\text{link}}, y) \leftarrow \text{Link}(P, y_{k,1}, \ldots, y_{k,2^k})$

3 : $m_{\text{path}} \leftarrow \text{Path}((V, E), y)$

4 : $S \leftarrow m_{\text{link}} \| m_{\text{path}}$

5 : **return** S

Theorem 10. *Let* H *be a hash function with β-balanced compression function* $h : \{0,1\}^B \times \{0,1\}^n \rightarrow \{0,1\}^n$. *Let* $k \in \mathbb{N}$. *Then adversary* $\mathcal{A}_{\text{enhanced}}^k$ *mounts a successful Nostradamus attack and needs*

$$\mathcal{O}\Big(\beta^{-1} \cdot 2^{\frac{n+2k+\log_2(k)}{3}} + \beta^{-\frac{1}{2}} \cdot 2^{\frac{n-k}{2}} + \ell_P + 1\Big)$$

evaluations of h. *In particular, for* $k = \frac{n}{7}$ *we get a total number of*

$$\mathcal{O}\Big(\beta^{-1} \cdot \sqrt[3]{n} \cdot 2^{3n/7} + \ell_P + 1\Big)$$

evaluations of h *on the average.*

We note that, ignoring the factor $\sqrt[3]{k}$ and the term ℓ_P, and assuming β to be constant, this upper bound matches our lower bound for herding attacks shown in the next section. This means, in order to significantly improve the attack, a different strategy than building a diamond structure must be used. Even then, the general lower bound still applies.

5 Quantum Lower Bound for Nostradamus Attacks

In this section we show a lower bound on the number of hash queries for mounting a quantum Nostradamus attack, assuming that the compression function h behaves like a random function. Our result is based on a query lower bound for finding C-collisions for a random function f, i.e., distinct $x_1, \ldots, x_C$ such that $f(x_1) = f(x_2) = \cdots = f(x_C)$. Liu and Zhandry [29] gave such a lower bound:

Theorem 11. ([29]). *Given a random function $f : X \rightarrow Y$ any quantum algorithm needs to make at least $\Omega\left(|Y|^{\frac{1}{2} \cdot \left(1 - \frac{1}{2^C - 1}\right)}\right)$ quantum queries to oracle f to find a C-collision with constant probability.*

For example, for a threefold collision $C = 3$ one needs at least $|Y|^{3/7}$ many queries. Liu and Zhandry [29] also give a matching upper bound (but which is irrelevant in our setting here).

5.1   General Lower Bound for Nostradamus Attacks

Recall that the general structure of an adversary $\mathcal{A}$ in the Nostradamus attack consists of two stages, on offline stage in which the adversary outputs the target hash value y_{trgt} and some state information st, and then the online-stage adversary receives the prefix P and the state, and outputs the suffix S to match the target hash value. The idea for the lower bound is now to repeatedly run the second-stage adversary to create multiple collisions for the same target hash value y_{trgt}, but for varying prefixes P_i, hoping to collect a C-collision. The total number of queries we make is $q_{off} + R \cdot q_{onl}$, where q_{off} is the number of hash evaluations of $\mathcal{A}$ in the offline phase, R is the number of repetitions, and q_{onl} the number of queries of $\mathcal{A}$ in the online phase. If we are able to show that we get a C-collision with this approach, then it follows that this total number of queries must exceed the lower bound in [29], also indicating a lower bound for $\mathcal{A}$'s queries in relation to R.

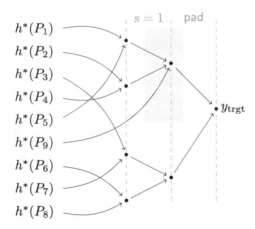

Fig. 6. Finding a C-collision by repeating online phase with different values $P_1, \ldots, P_R$. In the example $C = 3$ and $s = 1$. When running $\mathcal{A}^{onl}$ with inputs $P_1, \ldots, P_{R-1}$ in this order, we may fill up a $(C-1)$-ary tree with root y_{trgt} from the answers. After $R = (C-1)^{s+2} + 1 = 9$ repetitions we lastly get a C-collision somewhere in the tree (gray box).

Let us explain the process for the concrete example $C = 3$ in Fig. 6. Assume that we have already obtained the target hash value y_{trgt} from the offline adversary $\mathcal{A}^{off}$. Now we run the online part $\mathcal{A}^{onl}$ several times, presenting the adversary different prefixes $P_1, P_2, \ldots$ in the repetitions. Each time the adversary will give us a suffix path to reach y_{trgt}. Some of these paths may collide before reaching y_{trgt}, e.g., all the suffixes for $P_1, \ldots, P_8$ in Fig. 6 have length $s = 1$ and collide earlier or latest in y_{trgt}. The worst case for us occurs if these paths do not yield a C-collision yet. This can only happen if they form a full $(C-1)$-ary tree with root y_{trgt}, with all $(C-1)^{s+2}$ intermediate values

$h^*(P_1), h^*(P_2), \ldots, h^*(P_{R-1})$ already forming $(C-1)$-collisions in one of the $(C-1)^{s+1}$ leaves, and the adversary connecting these leaves optimally with the suffix block and the padding to yield y_{trgt}. But then, if we make the R-th repetition for P_R, where $R = (C-1)^{s+2} + 1$, this link must connect to a node which now forms a C-collision. In the example in Fig. 6 this happens for $R = 9$.

We next state the above idea formally in the following theorem. Recall that we assume that the adversary succeeds with probability 1 in the Nostradamus attack for any prefix P. We discuss afterwards relaxations for random prefixes and algorithms with lower success probabilities (over the choice of the compression function).

Theorem 12. *Let* H *be a hash function with compression function* $h : \{0,1\}^B \times \{0,1\}^n \to \{0,1\}^n$, *and assume that* h *is a random function. Let* $\mathcal{A}$ *be a quantum Nostradamus attacker, making at most* q_{off} *quantum queries to oracle* h *in the offline phase, and at most* q_{onl} *quantum queries to oracle* h *in the online phase. Assume further that* $\mathcal{A}$ *outputs a suffix* S *of at most* s *blocks. Then for any integer* $C \geq 3$ *such that* $(C-1)^{s+2} < 2^B$ *we have*

$$q_{\mathit{off}} + ((C-1)^{s+2} + 1) \cdot q_{\mathit{onl}} = \Omega\left(2^{\left(1 - \frac{1}{2^C-1}\right)n/2}\right).$$

Let us interpret this bound for some concrete cases. For sake of simplicity let us assume that the number of offline and online queries are roughly equal (as is the case in our herding attack). We can thus ignore q_{off} for now, and we also simplify the bound further to:

$$(C-1)^s \cdot q_{\mathrm{onl}} = \Omega\left(2^{\left(1 - \frac{1}{2^C-1}\right)n/2}\right).$$

If we fix $C = 3$, for example, then this bound becomes $q_{\mathrm{onl}} = \Omega\left(2^{3n/7-s}\right)$.

It seems as if we can push the lower bound on the right hand side as close to $2^{n/2}$ as desired, by increasing C, in contradiction to our basic attack with $2^{4n/9}$ evaluations and the advanced one with $2^{3n/7}$ evaluations. However, recall that the basic herding attack chooses a tree of height roughly $s = k = n/9$ such that the pre-factor $(C-1)^s$ is at least $2^{n/9}$, yielding an overall product of $2^{5n/9}$ on the left hand side. The same holds for the enhanced attack, where $s = k = n/7$ such that the factor becomes $2^{n/7}$ and hence lifts the value from $2^{3n/7}$ to $2^{4n/7}$.

A remarkable conclusion from the theorem's bound is that, choosing short suffixes for attacks, e.g., of constant block size s, one cannot go significantly below the bound $2^{n/2}$. This holds unless the attack exploits knowledge of the compression function h and does not treat it as a black-box random function. But this square-root bound is already achieved with our simple quantum attack— where indeed the suffix consists of the single linking message block m_{link}. In other words, sophisticated attacks need to rely on long suffixes.

Proof of Theorem 12. Consider an arbitrary Nostradamus adversary $\mathcal{A}$ with the above restrictions. We construct a c-collision finder $\mathcal{C}$ against h as follows.

Algorithm $\mathcal{C}$ runs $\mathcal{A}^{\mathsf{off}}$ to derive (st, y_{trgt}). Then it runs $\mathcal{A}^{\mathsf{onl}}$ for $R = (C-1)^{s+2}+1$ times for the same state but different values $P_1, \ldots, P_R \in \{0,1\}^B$. Note that the requirement on C guarantees that there are so many distinct input blocks. These executions create R suffixes $S_1, \ldots, S_R$ such that for all i the hash evaluations yield the target value, $\mathsf{H}(P_i \| S_i) = y_{\mathsf{trgt}}$. Recall that the final block may contain the padding information such that the suffix part actually consists of up to $s+1$ blocks. Below we simply consider this to be part of the suffix S_i and think of the hash function using no padding.

Assume now that, in the last iteration of h of the R hash evaluations, we have C distinct inputs (m_j, y_j) with $h(m_j, y_j) = y_{\mathsf{trgt}}$. Then we have found the C-collision for h and can stop here. If not, there are at most $C-1$ distinct pairs (m_j, y_j) resulting in y_{trgt}. But then there must exist one value y_j among those pairs which $\lceil R/(C-1) \rceil$ of the total number of hash evaluations reach in the second-to-last iteration. Note that $\lceil R/(C-1) \rceil = (C-1)^{s+1}+1$ such that we can recursively apply the argument, losing a factor $(C-1)$ in the remaining set of collision with each unsuccessful iteration. After at most $s+1$ iterations we have then either found our C-collision, or have a value y which at least $(C-1)+1 = C$ distinct input blocks $P_{j_1}, P_{j_2}, \ldots$ reach, i.e., $h(IV, P_{j_q}) = y$ for $q = 1, 2, \ldots, C$. Fortunately, this would then constitute our sought-after C-collision.   □

We finally discuss how to attenuate the assumption about the adversary always winning in the Nostradamus attack. As remarked earlier, if the adversary's success probability was only over its internal randomness, then we could easily account for this by repeating the adversary. However, now the probability space also comprises the choice of the random compression function (which is not chosen freshly if we would rerun the attack). Hence, assume that the adversary $\mathcal{A}$ has a success probability of ϵ (over the choice of the random function h and its internal coin tosses, like measurements of quantum states). One may think of ϵ to be some small constant, although the approach below also works with other values for ϵ.

We first use the splitting lemma [32] to conclude that our once sampled offline-stage output (st, y_{trgt}) is often good enough to make the online stage succeed with probability $\epsilon/2$. The probability of obtaining such a good offline-stage output is at least $\epsilon/2$ itself. Condition on this being the case. Next recall that the target value y_{trgt} is from now on fixed and that $\mathcal{A}$ succeeds with probability $\epsilon/2$ if we give it a prefix P. Hence, to collect $R = (C-1)^{s+2}+1$ such successful samples as in the proof, we need on average $2R/\epsilon$ attempts now. In other words, we have an expected number of $q_{\mathsf{off}} + \frac{2R}{\epsilon}((C-1)^{s+2} + 1) \cdot q_{\mathsf{onl}}$ queries in order to succeed with probability $\epsilon/2$ (with which the offline-phase output is good).

We can even go one step further and take the prefix to be a random element from $\{0,1\}^B$ (as it was assumed in the original work by Kelsey and Kohno [26]). Let us assume it is uniform in $\{0,1\}^B$. Since we expect no collisions to occur before reaching $2^{B/2}$ samples—and B is usually significantly larger than $\log R$—we can assume that all sampled values P_i are distinct. This is sufficient for the argument in the proof.

5.2   Improved Lower Bound for Herding Attacks

In this section we discuss that we get a better lower bound for quantum herding attacks. For this we assume that $\mathcal{A}^{\mathrm{off}}$ creates a diamond tree structure of height k and that $\mathcal{A}^{\mathrm{onl}}$ then tries to find a linking message part m_{link} to connect the prefix P to the tree. We assume here for simplicity of presentation that P is aligned to block length and that m_{link} is a single block. As before, we show how to build a 3-collision finder $\mathcal{C}$ for $C = 3$ from such a herding adversary $\mathcal{A}$. Our approach, however, is more direct and omits the large number of repetitions, thus yielding a better bound.

Our collision finder $\mathcal{C}$ will run the two phases of the herding adversary, $\mathcal{A}^{\mathrm{off}}$ and $\mathcal{A}^{\mathrm{onl}}$, but for different height parameters. For $\mathcal{A}^{\mathrm{off}}$ it will pretend that the height of the tree should be $k + 1$, but then prune the last layer of the tree and give $\mathcal{A}^{\mathrm{onl}}$ the pruned tree of height k. The advantage of this approach is that we already have a 2-collision on the k-th level via the tree for height $k + 1$ from $\mathcal{A}^{\mathrm{off}}$. Together with the new link to level k which we receive from $\mathcal{A}^{\mathrm{onl}}$, we immediately have a 3-collision (see Fig. 7).

More formally, $\mathcal{C}$ first lets $\mathcal{A}^{\mathrm{off}}$ create a diamond structure, but uses the parameter $k + 1$. By this we get a tree of height $k + 1$ from $\mathcal{A}^{\mathrm{off}}$. Then we cut level $k+1$ to get a tree of height k as required. Note that here the node values $y_{k,i}$ are, strictly speaking, not picked randomly. But they are the result of applying the random function h to a uniformly chosen input $(y_{k+1,i}, m_{k+1,i})$, such that we assume for simplicity that the herding attacks also works for such generated values.

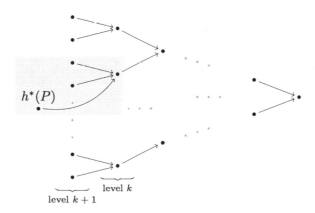

Fig. 7. Finding a 3-collision (gray box) using a diamond structure for height k for $\mathcal{A}$, generated by truncating a $k + 1$-height structure.

Algorithm $\mathcal{C}$ next runs $\mathcal{A}^{\mathrm{onl}}$ for the pruned tree of height k, and an arbitrary prefix P. This adversary eventually outputs a message part m_{link} that connects the iterated prefix value $h^*(P)$ to a tree node $y_{k,i}$ at *level k* (see Fig. 7). Algorithm $\mathcal{C}$ outputs the two children $y_{k+1,2i-1}$ and $y_{k+1,2i}$ with their corresponding labels

$m_{k+1,2i-1}$ and $m_{k+1,2i}$ pointing to $y_{k,i}$, together with $(h^*(P), m_{\text{link}})$ as the 3-collision for h.

For the analysis of the success probability of $\mathcal{C}$ we would have to make another assumption about the number of preimages under h. But since on average each image has 2^B preimages, we simply neglect the probability of $(h^*(P), m_{\text{link}})$ being equal to $(y_{k+1,2i-1}, m_{k+1,2i-1})$ or $(y_{k+1,2i}, m_{k+1,2i})$. In this case $\mathcal{C}$ succeeds whenever $\mathcal{A}$ wins. For the run time of $\mathcal{C}$ note that creating the diamond structure for $k + 1$ instead of k is, asymptotically, equally expensive. Hence, except for constants, our algorithm $\mathcal{C}$ obtains a 3-collision with the same number of oracle queries to h as $\mathcal{A}$. It follows that the total number of quantum queries of $\mathcal{C}$—and thus of $\mathcal{A}$—to random oracle h is at least $\Omega(2^{3n/7})$.

6   Implementation Results

In this section, we empirically evaluate the algorithms from Sects. 4.1, 4.2, and 4.3. For this purpose we have implemented the quantum algorithms in IBM's Qiskit.[4] The open-source software development kit Qiskit makes it possible to design quantum circuits in Python and to simulate their execution on a classical computer. Potentially, these algorithms can also be later run on quantum computing devices.

6.1   Algorithms

The local simulation of the quantum algorithm severely restricts the number of simultaneously accessible qubits. Hence, instead of using output-truncated versions of SHA2 or SHA3 with large internal states, we use an iterative toy hash function with adjustable block size B and output n as the attack target for the algorithms.

We build this toy hash function in a similar way as the open-source project Qibo [21], which is used to evaluate generic quantum attacks by Ramos-Calderer et al. [33] and utilizes a ChaCha-permutation [6] as sponge function. In particular, we use this permutation $f : \{0,1\}^{B+n} \to \{0,1\}^{B+n}$ as the basis for our hash function and truncate it (to the last n bits) to derive our compression function

$$h(m, y) := \text{trunc}(f(m\|y))$$

for the Merkle-Damgård construction.

The quantum circuit corresponding to the permutation f is shown in (the right hand side of) Fig. 8. We use parameters $B = 8$ and $n = 8$ for the compression function throughout the evaluation. We note that empirically evaluations show that the compression function h_y is not surjective for all y, such that it not β-balanced for any $\beta > 0$. For measuring the run times we ignore such failures. However, as we discuss in more detail later, our advanced attacks still work well, showing that the theoretical results are on the conservative side.

---

[4] https://qiskit.org/.

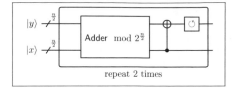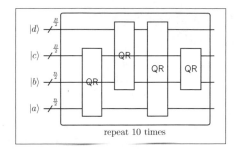

Fig. 8. The description of the quantum circuit for the permutation that is based on the ChaCha permutation. The left part presents the description of the sub algorithm Quarter Round, which is denoted as QR. A quarter round consists of an adder, CNOT-gates, and qubit shuffling. The latter is denoted by the ↻ symbol. The right part describes the permutation f. In the 10 repetitions of the core function, four QR executions are applied to the input registers. We note that a line which is drawn through a gate is not an input to that specific gate but rather routed through.

Recall that our algorithms apply Grover's algorithm for different functions F for finding claws in a list of values. To implement these functions F from Proposition 5 and Theorem 9 in a quantum circuit, we hardcode the list into the circuit, using one n-bit Toffoli gate for each of the 2^ℓ distinct bit strings $h_y(m_i)$ in the list resp. the 2^k distinct leaves $y_{k,i}$. More precisely, given a bit string that either represents a hash evaluation $h_y(m_i)$ or a leaf $y_{k,i}$, an n-bit Toffoli gate is constructed such that the corresponding classical logical operator exclusively maps this bit string to 1. These Toffoli gates are chained together to form the existential quantification in the proposition. The resulting quantum circuit is illustrated in Fig. 9.

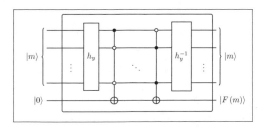

Fig. 9. Description of the quantum circuits that implements the functions F from from Proposition 5 and Theorem 9. The string y in $h_y(\cdot) = h(\cdot, y)$ corresponds either to y' or to p, depending on the context. Each n-bit Toffoli gate represents one (unique) string to which the input is compared to, such that the exclusive-or on the final qubit is only set at most once. Applying h_y^{-1} at the end restores the original content of m.

We note that implementing Grover's algorithm for our functions F, also allowing to recover the matching message m, as well as the other steps of the algorithms are straightforward to implement. We thus omit their description here for sake of brevity.

6.2   Experiments

The evaluation was carried out with commodity hardware, namely, a 2.6 GHz 6-core Intel Core i7 processor, an AMD Radeon Pro 5300M 4 GB graphics card, and a 16 GB 2667 MHz MHz DDR4 RAM. The evaluation process consists of running all three algorithms (simple, basic, and enhanced). We measure the offline and online run times, as well as the number of sampling operations and function calls. The latter includes the number of calls to the quantum compression function and the number of quantum calls to the functions from Proposition 5 and Theorem 9. The results of the evaluation are shown in Fig. 10.

Simple (Section 4.1)					
k	Offline [s]	Online [s]	F_{claw}	F_{link}	Samples
—	0	147.93	—	—	—

Basic (Section 4.2)					
k	Offl. [s]	Onl. [s]	F_{claw}	F_{link}	Sampl.
1	89.55	232.47	3	8	10
2	260.73	142.52	9	5	28
3	626.13	90.95	21	3	68
4	1363.76	62.13	45	2	138
5	2824.66	34.17	93	1	290

Enhanced (Section 4.3)					
k	Offl. [s]	Onl. [s]	F_{claw}	F_{link}	Sampl.
1	62.73	235.34	2	8	18
2	190.2	145.49	6	5	50
3	429.95	90.27	14	3	84
4	927.26	63.83	30	2	138
5	1459.14	31.69	46	1	285

Fig. 10. Evaluation results for the algorithms from Sects. 4.1, 4.2, and 4.3 and our compression function $h(m, y)$. The parameter k denotes the height of the tree of the diamond. *Online* and *Offline* denote the online and offline run time, respectively, in seconds. F_{claw} and F_{link} denote the quantum function calls to the functions F from Proposition 5 and 7, respectively. *Samples* denotes the number of sampling operations for messages and leaves for building the diamond structure during an attack. Note that the simple attack fails in some cases for our specific hash function in which case we do not measure its run time.

Note that for $k = 1$, i.e., trees consisting only of one level, the sum of calls (i.e., $F_{claw} + F_{link}$) to the functions F from Proposition 5 and 7 is minimal. This is true for the attack with the basic and the enhanced diamond structure, confirming the optimal choice of $k = \lceil n/9 \rceil$ for $n = 8$ according to Theorems 8 and 9. Furthermore, for $k = 1$ the sum of function calls for the basic and enhanced online attacks (i.e., F_{link}) already is less or equal to the 16 quantum function calls in the simple quantum attack based on Grover's algorithm. Nevertheless, the run time of the attacks with a diamond structure is larger compared to the run time of the simple attack. We expect this to be related to the fact that the quantum circuit of Grover's algorithm for the functions of Proposition 5 and 7 is much more complex than when only applied to the compression function in the simple attack. More precisely, with our design of the circuit in Fig. 9, Qiskit simulates the compression function twice and an exponential number of n-bit Toffoli gates, compared to only one simulated compression function call in the simple quantum attack, such that the advantage of using a diamond structure does not pay off for the small value of n yet.

Initially, the enhanced diamond construction requires more samples than the basic diamond construction, while it also exhibits a smaller sum of calls (i.e., $F_{claw} + F_{link}$) to the functions F. For the construction of larger diamonds, such as $k = 4$, the sample count of the enhanced attack is surpassed by the basic diamond construction. This is due to the fact that the enhanced attacker re-uses data across the connection steps of a single layer in the diamond, while the basic attacker resamples all data for each quantum collision finding. Thus, our experiments confirm that the advanced attacker requires less sampled data to construct a large diamond.

Instead of using the optimal run time, an attacker may also deploy a trade-off and, for a slight decrease in online run time, accept an increase in the offline run time. In this case, the attacker creates a larger diamond for $k > 1$ and thereby achieves an improvement for the online run time. Our experiments confirm this expected behavior, with F_{link} dropping and F_{claw} increasing.

We previously noted that $h(m, y)$ is not even surjective and not β-balanced. Since the simple attack from Sect. 4.1 picks a random value $y \leftarrow_\$ \{0, 1\}^n$ this results in the attack failing for values y without preimage. In this case, Grover's algorithm cannot find a corresponding preimage. However, the more advanced attacks from Sects. 4.2 and 4.3 always succeeded. The reason is that the trees are built in forward direction, such that each value y has at least two preimages.

We note that all the observations are specific to the parameters $n = B = 8$. According to our theoretical results the asymptotic complexity of the simple quantum attack increases faster in n than it does for the diamond attacks. As a consequence, for higher values of n a larger difference in online run time should occur even for smaller k. Unfortunately, the simulation of higher n in Qiskit is expensive and, in some cases, technically infeasible due to the large sizes of quantum circuits. Thus, the empirical demonstration remains open.

7   Conclusion

Our results show that fundamental quantum algorithms for finding collisions can be used to speed up the classical Nostradamus attack. Our algorithms have been designed and analyzed in an "idealized" quantum model, but our implementation of the toy example indicates that they can be run on in principle on a quantum computer. Turning these attacks into real quantum programs may still entail a lot of engineering aspects which can significantly influence the run time, e.g., [1, 7, 11]. It remains an interesting open question how fast our attacks can be made on real quantum computers.

As mentioned in the related work section, other works like [4, 14] have aimed to give time-memory trade-offs, especially in order to reduce quantum memory, and to parallelize the search for collisions and preimages. We have not inves-tigates such trade-offs for Nostradamus attacks, especially in light of the large quantum memory requirements, inherited from the BHT collision search algorithm [13]. Nonetheless, we expect similar techniques as in [4,14] to be applicable here as well.

Acknowledgments. We thank the anonymous reviewers for valuable comments.

This research work has been funded by the German Federal Ministry of Education and Research and the Hessian Ministry of Higher Education, Research, Science and the Arts within their joint support of the National Research Center for Applied Cybersecurity ATHENE.

References

1. Amy, M., Matteo, O.D., Gheorghiu, V., Mosca, M., Parent, A., Schanck, J.M.: Estimating the cost of generic quantum pre-image attacks on SHA-2 and SHA-3. In: Avanzi, R., Heys, H.M. (eds.) Selected Areas in Cryptography - SAC 2016–23rd International Conference, St. John's, NL, Canada, 10–12 August 2016, Revised Selected Papers. Lecture Notes in Computer Science, vol. 10532, pp. 317–337. Springer (2016). https://doi.org/10.1007/978-3-319-69453-5_18

2. Andreeva, E., Bouillaguet, C., Dunkelman, O., Kelsey, J.: Herding, second preimage and trojan message attacks beyond Merkle-damgård. In: Jr., M.J.J., Rijmen, V., Safavi-Naini, R. (eds.) Selected Areas in Cryptography, 16th Annual International Workshop, SAC 2009, Calgary, Alberta, Canada, 13–14 August 2009, Revised Selected Papers. Lecture Notes in Computer Science, vol. 5867, pp. 393–414. Springer (2009). https://doi.org/10.1007/978-3-642-05445-7_25

3. Andreeva, E., Mennink, B.: Provable chosen-target-forced-midfix preimage resistance. In: Miri, A., Vaudenay, S. (eds.) Selected Areas in Cryptography - 18th International Workshop, SAC 2011, Toronto, ON, Canada, 11–12 August 2011, Revised Selected Papers. Lecture Notes in Computer Science, vol. 7118, pp. 37–54. Springer (2011). https://doi.org/10.1007/978-3-642-28496-0_3

4. Banegas, G., Bernstein, D.J.: Low-communication parallel quantum multi-target preimage search. In: Adams, C., Camenisch, J. (eds.) Selected Areas in Cryptography - SAC 2017–24th International Conference, Ottawa, ON, Canada, 16–18 August 2017, Revised Selected Papers. Lecture Notes in Computer Science, vol. 10719, pp. 325–335. Springer (2017). https://doi.org/10.1007/978-3-319-72565-9_16

5. Bellare, M., Kohno, T.: Hash function balance and its impact on birthday attacks. In: Cachin, C., Camenisch, J. (eds.) Advances in Cryptology - EUROCRYPT 2004. Lecture Notes in Computer Science, vol. 3027, pp. 401–418. Springer, Heidelberg, Germany, Interlaken, Switzerland, 2–6 May 2004. https://doi.org/10.1007/978-3-540-24676-3_24

6. Bernstein, D.: ChaCha, a variant of Salsa20 (2008). https://cr.yp.to/chacha/chacha-20080128.pdf

7. Bernstein, D.J.: Cost analysis of hash collisions : will quantum computers make SHARCS obsolete? In: SHARCS 2009 Workshop Record (Proceedings 4th Workshop on Special-purpose Hardware for Attacking Cryptograhic Systems, Lausanne, Switserland, 9–10 September 2009), pp. 105–116 (2009)

8. Bertoni, G., Daemen, J., Peeters, M., Van Assche, G.: Sponge functions. Ecrypt Hash Workshop (2007)

9. Blackburn, S.R., Stinson, D.R., Upadhyay, J.: On the complexity of the herding attack and some related attacks on hash functions. Des. Codes Cryptogr. **64**(1–2), 171–193 (2012)

10. Boneh, D., Dagdelen, Ö., Fischlin, M., Lehmann, A., Schaffner, C., Zhandry, M.: Random oracles in a quantum world. In: Lee, D.H., Wang, X. (eds.) Advances in Cryptology - ASIACRYPT 2011–17th International Conference on the Theory and Application of Cryptology and Information Security, Seoul, South Korea, 4–8 December 2011. Proceedings. Lecture Notes in Computer Science, vol. 7073, pp. 41–69. Springer (2011). https://doi.org/10.1007/978-3-642-25385-0_3

11. Bonnetain, X., Jaques, S.: Quantum period finding against symmetric primitives in practice. IACR Trans. Cryptogr. Hardw. Embed. Syst. **2022**(1), 1–27 (2022)

12. Boyer, M., Brassard, G., Høyer, P., Tapp, A.: Tight bounds on quantum searching. Fortschritte der Physik **46**(4–5), 493–505 (1998)

13. Brassard, G., Høyer, P., Tapp, A.: Quantum cryptanalysis of hash and claw-free functions. In: Lucchesi, C.L., Moura, A.V. (eds.) LATIN 1998: Theoretical Informatics, Third Latin American Symposium, Campinas, Brazil, 20–24 April 1998, Proceedings. Lecture Notes in Computer Science, vol. 1380, pp. 163–169. Springer (1998). https://doi.org/10.1007/BFb0054319

14. Chailloux, A., Naya-Plasencia, M., Schrottenloher, A.: An efficient quantum collision search algorithm and implications on symmetric cryptography. In: Takagi, T., Peyrin, T. (eds.) Advances in Cryptology - ASIACRYPT 2017–23rd International Conference on the Theory and Applications of Cryptology and Information Security, Hong Kong, China, 3–7 December 2017, Proceedings, Part II. Lecture Notes in Computer Science, vol. 10625, pp. 211–240. Springer (2017). https://doi.org/10. 1007/978-3-319-70697-9_8

15. Damgård, I.: A design principle for hash functions. In: Brassard, G. (ed.) Advances in Cryptology - CRYPTO'89. Lecture Notes in Computer Science, vol. 435, pp. 416–427. Springer, Heidelberg, Germany, Santa Barbara, CA, USA, 20–24 August 1990. https://doi.org/10.1007/0-387-34805-0_39

16. Dang, Q.: Secure hash standard. Federal Inf. Process. Stds. (NIST FIPS), National Institute of Standards and Technology, Gaithersburg, MD (2015–08-04 2015)

17. Dean, R.D.: Formal Aspects of Mobile Code Security. Ph.D. thesis, Computer Science Department, Princeton University (1999)

18. Dong, X., Sun, S., Shi, D., Gao, F., Wang, X., Hu, L.: Quantum collision attacks on AES-like hashing with low quantum random access memories. In: Moriai, S., Wang, H. (eds.) Advances in Cryptology - ASIACRYPT 2020, Part II. Lecture Notes in Computer Science, vol. 12492, pp. 727–757. Springer, Heidelberg, Germany, Daejeon, South Korea, 7–11 December 2020. https://doi.org/10.1007/978-3-030-64834-3_25

19. Dong, X., Zhang, Z., Sun, S., Wei, C., Wang, X., Hu, L.: Automatic classical and quantum rebound attacks on aes-like hashing by exploiting related-key differentials. In: Tibouchi, M., Wang, H. (eds.) Advances in Cryptology - ASIACRYPT 2021–27th International Conference on the Theory and Application of Cryptology and Information Security, Singapore, 6–10 December 2021, Proceedings, Part I. Lecture Notes in Computer Science, vol. 13090, pp. 241–271. Springer (2021). https://doi.org/10.1007/978-3-030-92062-3_9

20. Dworkin, M.: SHA-3 standard: permutation-based hash and extendable-output functions. Federal Inf. Process. Stds. (NIST FIPS), National Institute of Standards and Technology, Gaithersburg, MD (2015–08-04 2015)

21. Efthymiou, S., et al.: Qibo: An open-source full stack API for quantum simulation and quantum hardware control (2022). https://github.com/qiboteam/qibo

22. Flórez-Gutiérrez, A., Leurent, G., Naya-Plasencia, M., Perrin, L., Schrottenloher, A., Sibleyras, F.: New results on Gimli: full-permutation distinguishers and
improved collisions. In: Moriai, S., Wang, H. (eds.) Advances in Cryptology - ASI-
ACRYPT 2020, Part I. Lecture Notes in Computer Science, vol. 12491, pp. 33–
63. Springer, Heidelberg, Germany, Daejeon, South Korea, 7–11 December 2020.
https://doi.org/10.1007/978-3-030-64837-4_2

23. Grover, L.K.: A fast quantum mechanical algorithm for database search. In: Miller,
G.L. (ed.) Proceedings of the Twenty-Eighth Annual ACM Symposium on the
Theory of Computing, Philadelphia, Pennsylvania, USA, 22–24 May 1996, pp.
212–219. ACM (1996)

24. Hosoyamada, A., Sasaki, Y.: Finding hash collisions with quantum computers by
using differential trails with smaller probability than birthday bound. In: Canteaut, A., Ishai, Y. (eds.) Advances in Cryptology - EUROCRYPT 2020, Part II.
Lecture Notes in Computer Science, vol. 12106, pp. 249–279. Springer, Heidelberg,
Germany, Zagreb, Croatia, 10–14 May 2020. https://doi.org/10.1007/978-3-030-
45724-2_9

25. Hosoyamada, A., Sasaki, Y.: Quantum collision attacks on reduced SHA-256 and
SHA-512. In: Malkin, T., Peikert, C. (eds.) Advances in Cryptology - CRYPTO
2021, Part I. Lecture Notes in Computer Science, vol. 12825, pp. 616–646. Springer,
Heidelberg, Germany, Virtual Event, 16–20 August 2021. https://doi.org/10.1007/
978-3-030-84242-0_22

26. Kelsey, J., Kohno, T.: Herding hash functions and the nostradamus attack. In:
Vaudenay, S. (ed.) Advances in Cryptology - EUROCRYPT 2006, 25th Annual
International Conference on the Theory and Applications of Cryptographic Techniques, St. Petersburg, Russia, 28 May–1 June 2006, Proceedings. Lecture Notes
in Computer Science, vol. 4004, pp. 183–200. Springer (2006). https://doi.org/10.
1007/11761679_12

27. Kelsey, J., Schneier, B.: Second preimages on n-bit hash functions for much less
than 2^n work. In: Cramer, R. (ed.) Advances in Cryptology - EUROCRYPT 2005,
24th Annual International Conference on the Theory and Applications of Cryptographic Techniques, Aarhus, Denmark, 22–26 May 2005, Proceedings. Lecture
Notes in Computer Science, vol. 3494, pp. 474–490. Springer (2005). https://doi.
org/10.1007/11426639_28

28. Kortelainen, T., Kortelainen, J.: On diamond structures and trojan message
attacks. In: Sako, K., Sarkar, P. (eds.) Advances in Cryptology - ASIACRYPT
2013–19th International Conference on the Theory and Application of Cryptology
and Information Security, Bengaluru, India, 1–5 December 2013, Proceedings, Part
II. Lecture Notes in Computer Science, vol. 8270, pp. 524–539. Springer (2013).
https://doi.org/10.1007/978-3-642-42045-0_27

29. Liu, Q., Zhandry, M.: On finding quantum multi-collisions. In: Ishai, Y., Rijmen,
V. (eds.) EUROCRYPT 2019. LNCS, vol. 11478, pp. 189–218. Springer, Cham
(2019). https://doi.org/10.1007/978-3-030-17659-4_7

30. Merkle, R.C.: A certified digital signature. In: Brassard, G. (ed.) Advances in
Cryptology - CRYPTO'89. Lecture Notes in Computer Science, vol. 435, pp. 218–
238. Springer, Heidelberg, Germany, Santa Barbara, CA, USA, 20–24 August 1990

31. Ni, B., Dong, X., Jia, K., You, Q.: (quantum) collision attacks on reduced simpira
v2. IACR Trans. Symmetric Cryptol. 2021(2), 222–248 (2021)

32. Pointcheval, D., Stern, J.: Security arguments for digital signatures and blind signatures. J. Cryptol. 13(3), 361–396 (2000)

33. Ramos-Calderer, S., Bellini, E., Latorre, J.I., Manzano, M., Mateu, V.: Quantum search for scaled hash function preimages. Quantum Inf. Process. **20**(5), 1–28 (2021). https://doi.org/10.1007/s11128-021-03118-9

34. Wang, R., Li, X., Gao, J., Li, H., Wang, B.: Quantum rotational cryptanalysis for preimage recovery of round-reduced keccak. IACR Cryptol. ePrint Arch, p. 13 (2022). https://eprint.iacr.org/2022/013

35. Weizman, A., Dunkelman, O., Haber, S.: Efficient construction of diamond structures. In: Patra, A., Smart, N.P. (eds.) Progress in Cryptology - INDOCRYPT 2017–18th International Conference on Cryptology in India, Chennai, India, 10–13 December 2017, Proceedings. Lecture Notes in Computer Science, vol. 10698, pp. 166–185. Springer (2017). https://doi.org/10.1007/978-3-319-71667-1_9

Synthesizing Quantum Circuits of AES with Lower T-depth and Less Qubits

Zhenyu Huang[1,2] and Siwei Sun[3,4(✉)]

[1] SKLOIS, Institute of Information Engineering, Chinese Academy of Sciences,
Beijing, China
huangzhenyu@iie.ac.cn
[2] School of Cyber Security, University of Chinese Academy of Sciences,
Beijing, China
[3] School of Cryptology, University of Chinese Academy of Sciences, Beijing, China
sunsiwei@ucas.ac.cn
[4] State Key Laboratory of Cryptology, P.O. Box 5159, Beijing 100878, China

Abstract. The significant progress in the development of quantum computers has made the study of cryptanalysis based on quantum computing an active topic. To accurately estimate the resources required to carry out quantum attacks, the involved quantum algorithms have to be synthesized into quantum circuits with basic quantum gates. In this work, we present several generic synthesis and optimization techniques for circuits implementing the quantum oracles of iterative symmetric-key ciphers that are commonly employed in quantum attacks based on Grover and Simon's algorithms. Firstly, a general structure for implementing the round functions of block ciphers in-place is proposed. Then, we present some novel techniques for synthesizing efficient quantum circuits of linear and non-linear cryptographic building blocks. We apply these techniques to AES and systematically investigate the strategies for depth-width trade-offs. Along the way, we derive a quantum circuit for the AES S-box with *provably* minimal T-depth based on some new observations on its classical circuit. As a result, the T-depth and width (number of qubits) required for implementing the quantum circuits of AES are significantly reduced. Compared with the circuit proposed in EURO-CRYPT 2020, the T-depth is reduced from 60 to 40 without increasing the width or 30 with a slight increase in width. These circuits are fully implemented in Microsoft Q# and the source code is publicly available. Compared with the circuit proposed in ASIACRYPT 2020, the width of one of our circuits is reduced from 512 to 371, and the Toffoli-depth is reduced from 2016 to 1558 at the same time. Actually, we can reduce the width to 270 at the cost of increased depth. Moreover, a full spectrum of depth-width trade-offs is provided, setting new records for the synthesis and optimization of quantum circuits of AES.

Keywords: Quantum circuit · T-depth · Grover's algorithm · AES

ⓒ International Association for Cryptologic Research 2022
S. Agrawal and D. Lin (Eds.): ASIACRYPT 2022, LNCS 13793, pp. 614–644, 2022.
https://doi.org/10.1007/978-3-031-22969-5_21

1 Introduction

The rapid and fruitful development of the theory and practice on building computing machines that exploit quantum mechanical phenomena has made the research on algorithms running on quantum computers a topic with potential practical consequences. This especially attracts substantial attention from the cryptographic community, where the security of many primitives relies on the computational hardness of solving certain number theoretical or combinatorial problems. Shor's algorithm [Sho99] is probably the most influential research in this aspect. It will compromise the security of many widely deployed public-key cryptosystems (including RSA, DSA, and ECC) if large-scale quantum computers are ever built.

For symmetric-key ciphers, a trivial application of Grover's algorithm [Gro96] results in a quadratic speedup of the exhaustive search attack. If the attackers have access to the keyed quantum oracle, it is shown that many symmetric-key schemes can be broken with Simon's period-finding algorithm [KLLN16a, BLNS21]. Since the practical relevance of querying online keyed quantum oracles is questionable, some subsequent work investigates techniques limited to classical queries and offline quantum computations [HS18a, BHN+19, CNS17]. Also, quantum attacks derived from dedicated cryptanalytic techniques are extensively studied [BNS19b, BNS19a, HS18b, KLLN16b, NS, HSa]. To concretely estimate the complexities in the standard quantum circuit model [NC16], the quantum circuits for these attacks have to be constructed based on some basic quantum gates. Our community is especially interested in constructing efficient quantum circuits for cryptographic primitives fulfilling specific input-output behaviors since such circuits typically work as sub-circuits of the quantum attacks. The National Institute of Standards and Technology (NIST) used the complexity of the quantum circuit for AES with a bound of depth called MAXDEPTH as a baseline to categorize the post-quantum public-key schemes into different security levels in the call for proposals to the standardization of post-quantum cryptography [NIS16]. Note that when the quantum circuits are applied in exhaustive key search, we can use parallelization by dividing the search space, which naturally decreases the depth but increases the number of quantum gates and qubits, and in fact, to perform exhaustive key search attacks on AES on a quantum computer with Grover's algorithm under NIST's MAXDEPTH bound, parallelization is required for the majority of values of MAXDEPTH. For parallelization, we have the following observation.

Observation 1. *Let $\mathfrak{C}$ be the quantum circuit implementing the Grover oracle and $\mathbb{V}$ be the search space. If we divide $\mathbb{V}$ into k^2 equal parts and execute k^2 parallel Grover searches, the number of iterations of $\mathfrak{C}$ required in each search will decrease by a factor of k. The number of qubits required in each search will not change, while the number of gates used in each search will decrease by a factor of k. Therefore, the number of qubits required in all searches will increase by a factor of k^2, and the number of gates used in all searches will increase by a factor of k.*

Let $\mathfrak{C}'$ be a new quantum circuit whose depth is reduced to $depth(\mathfrak{C})/k$, and the numbers of qubits and gates are respectively increased by a factor less than k^2 and a factor less than k. Then, Observation 1 implies that using $\mathfrak{C}'$ in a parallel approach is better than applying $\mathfrak{C}$. This is the very reason that [JNRV20] claims: Grover's algorithm does not parallelize well, meaning that minimizing depth rather than width is crucial to make the most out of the available depth. Finally, NIST states that the MAXDEPTH restriction is motivated by the difficulty of running extremely long serial computations. Plausible values for MAXDEPTH range from 2^{40} logical gates through 2^{64} logical gates, to no more than 2^{96} logical gates.

Related Work. To perform quantum attacks based on Grover's and Simon's algorithm, one has to implement the actual device that executes the attack, whose cost is evaluated in the quantum circuit model. Especially, in [HB20], an economic model which can be used to determine whether a quantum key-recovery attack is profitable was developed. A critical parameter of this model is the cost of a concrete quantum encryption circuit. From an attacker's perspective, it is important to reduce the cost. From a designer's perspective, it is important to have an accurate understanding of the cost to evaluate the security margin and to guide future designs. In particular, the classical and quantum implementations of AES receive most attention from our community.

The first quantum circuit of AES [GLRS16] was proposed by Grassl et al., where the so-called zig-zag structure was introduced to reduce the width (the number of qubits required) of the resulting quantum circuits. The width was further reduced in a follow-up work [ASAM18]. In [LPS19], Langenberg et al. presented improved quantum circuits for the S-box and key expansion of AES, leading to significantly improved AES circuits. At ASIACRYPT 2020, by tweaking the zig-zag structure, together with new quantum circuits for the AES S-box and its inverse constructed based on improved classical circuits Zou et al. significantly improved the width of the quantum circuit of AES [ZWS+20].

While the primary goal of the above works is to reduce the width, the cryptographic community is more concerned with the depth, since in NIST's ongoing post-quantum standardization effort, different security categories are defined according to the quantum resources needed to attack AES with a depth bound. At EUROCRYPT 2020, Jaques et al. proposed several techniques to improve the depth, and presented the currently known sallowest quantum circuit for AES [JNRV20]. Besides, we also see works considering the implementation of quantum circuits for other primitives (e.g., SHA-2 and SHA3 [AMG+16], LowMC [JNRV20], and ECC [BBvHL21]).

Note that this line of research is not only interested by the cryptographic community, but also contributes to a much broader subject known as synthesis and optimization of quantum circuits. As realistic quantum computers will likely require some fault tolerance schemes where the amount of error correction is proportional to the resources used, the effect of quantum circuit optimizations becomes even more profound than its classical counterpart. In fact, the industrial community has already invested huge resources to develop the tool chain for synthesis and optimization of quantum circuits [Mic, IBM].

Our Contributions. We propose an *in-place* quantum circuit for the (invertible) round-function of a block cipher or other iterative designs. With this type of in-place structure, the circuits implementing the round functions can be connected together to form the whole design without using additional ancilla qubits. In addition, we present a generic method for constructing such *in-place* circuits with *out-of-place* sub-circuits. Then, a systematic comparison of this structure with previous designs (the pipeline structure and the zig-zag structure) is made with respect to the depth and width (the number of qubits) requirements.

We then consider how to implement the building blocks of a symmetric-key cipher efficiently. Specifically, a SAT-based technique for synthesizing small linear components is presented, which can output the CNOT network with the minimal gate count. For nonlinear components, a systematic method for constructing a circuit mapping $|x\rangle|a\rangle$ to $|x\rangle|a \oplus f(x)\rangle$ based on a circuit mapping $|x\rangle|0\rangle$ to $|x\rangle|f(x)\rangle$ meeting certain clearly defined conditions with only additional CNOT gates is given. Based on this method, we present circuits for AES S-box and it inverse with both T-depth and width improved compared to the one given in [ZWS+20].

To further reduce the T-depth, we formulate a technique for converting an AND-depth-t classical circuit into a T-depth-t quantum circuit. We note that this is not a trivial conversion due to the peculiarities of a quantum circuit, and the natural order of the classical circuit has to be rearranged to achieve this goal. Based on this method with a new observation on the classical circuit for the AES S-box, we obtain two circuits for the AES S-box with T-depth-4 and T-depth-3, respectively, both of which have lower T-depth than the one presented at EUROCRYPT 2020 [JNRV20]. Since the degree of the algebraic normal form of the AES S-box is 7, with less than 3 stages of multiplications, one cannot generate polynomials with degree 7, which implies that our implementation reaches the theoretical lower bound.

By applying the method presented in this paper, we significantly improve the efficiency of the quantum circuits for AES. Compared with the circuit proposed in EUROCRYPT 2020, the T-depth is reduced from 60 to 40 without increasing the width and the T-gate count, or 30 with a slight increase in width and T-gate count. These circuits are fully implemented in Microsoft Q# and the source code is publicly available. Compared with the circuit proposed in ASIACRYPT 2020, the width of one of our circuits is reduced from 512 to 371, the number of Toffoli gates is reduced from 19788 to 17888, and the Toffoli-depth is reduced from 2016 to 1558 at the same time. Actually, we can reduce the width to 270 at the cost of increased depth. Moreover, by varying the local and global circuit structures, a full spectrum of depth-width trade-offs are provided and illustrated in Fig. 19 (Sect. 8), setting new records for the synthesis and optimization of quantum circuits of AES.

2   Preliminaries on Synthesizing Quantum Circuits

The states of an n-qubit quantum system can be described by the unit vectors in $\mathbb{C}^{2^n}$. A quantum state is typically written as $|u\rangle$, and in this paper we denote it by $|u\rangle_n$ to emphasize that this state has n qubits. When n is clear from the context, we abbreviate $|0\cdots0\rangle_n$ as $|0\rangle$.

A quantum algorithm manipulates the state of an n-qubit system through a series of unitary transformations and measurements, where a unitary transformation is a linear map U over $\mathbb{C}^{2^n}$ with $UU^\dagger = U^\dagger U = I$. Any unitary transformation can be constructed with a finite set of single-qubit and two-qubit unitary transformations through composition and tensor product. In the standard quantum circuit model [NC16], we call these simple single-qubit and two-qubit unitary transformations quantum gates. In particular, we consider how to synthesize a quantum circuit with the commonly used universal fault-tolerant gate set Clifford $+$ T, which contains the Clifford gates:

$$H = \frac{1}{\sqrt{2}}\begin{pmatrix} 1 & 1 \\ 1 & -1 \end{pmatrix}, \quad S = \begin{pmatrix} 1 & 0 \\ 0 & i \end{pmatrix}, \quad \text{CNOT} = \begin{pmatrix} 1 & 0 & 0 & 0 \\ 0 & 1 & 0 & 0 \\ 0 & 0 & 0 & 1 \\ 0 & 0 & 1 & 0 \end{pmatrix},$$

and the non-Clifford gate $T = \begin{pmatrix} 1 & 0 \\ 0 & e^{i\pi/4} \end{pmatrix}$.

We also frequently employ the Pauli-X gate $X = HS^2H = \begin{pmatrix} 0 & 1 \\ 1 & 0 \end{pmatrix}$ and the Toffoli gate

$$\text{ToF} = \begin{pmatrix} 1 & 0 & 0 & 0 & 0 & 0 & 0 & 0 \\ 0 & 1 & 0 & 0 & 0 & 0 & 0 & 0 \\ 0 & 0 & 1 & 0 & 0 & 0 & 0 & 0 \\ 0 & 0 & 0 & 1 & 0 & 0 & 0 & 0 \\ 0 & 0 & 0 & 0 & 1 & 0 & 0 & 0 \\ 0 & 0 & 0 & 0 & 0 & 1 & 0 & 0 \\ 0 & 0 & 0 & 0 & 0 & 0 & 0 & 1 \\ 0 & 0 & 0 & 0 & 0 & 0 & 1 & 0 \end{pmatrix}.$$

In this work, we are mainly concerned with the quantum circuits that can compute a classical vectorial Boolean function when the input is in the computational basis. Since the Toffoli gate can be used to simulate a universal gate set for classical computation, all these circuits can be constructed by using only Toffoli gates with additional qubits (potentially set to appropriate values). For example, the multiplication operation $a\cdot b$ can be directly implemented by the Toffoli gate $|a\rangle|b\rangle|c\rangle \rightarrow |a\rangle|b\rangle|c \oplus a\cdot b\rangle$. There is another quantum circuit for implementing the functionality of a classical AND gate, and we call it a quantum AND gate. This gate together with its adjoint is illustrated in Fig. 1.

Optimization Goals. The complexity of a quantum circuit can be measured in terms of its width (number of qubits), gate count, and depth. The cryptographic

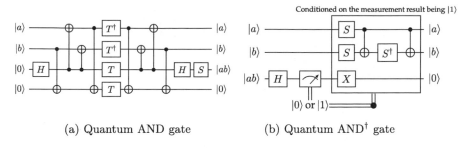

(a) Quantum AND gate   (b) Quantum AND† gate

Fig. 1. The quantum AND gate together with its adjoint

community is mainly concerned with the depth metric, and in particular the T-depth is the most interested parameter of a quantum circuit. The reason can be summarized as follows.

Firstly, as indicated in [JNRV20] and our introduction, favoring lower depth at the cost of a slightly larger width in the circuit leads to costs that are smaller in several metrics than for the circuits presented in [GLRS16, ASAM18, LPS19]. Secondly, the time for fault tolerant quantum computation is proportional to one round of measurement per layer of T-gates, and so the runtime is dominated by T-depth, rather than gate count, circuit depth, or even measurement depth [Fow12]. This is why in the adjoint of the quantum AND gate (see Fig. 1) we avoid using T-gates at the cost of quantum measurement.

The T-depth is defined as the minimum number of stages of parallel applications of T-gates in a circuit, where parallel T-gates are allowed when they are acting on different qubits. Note that we can implement the Toffoli gate with several different circuits based on the Clifford $+T$ gates with different T-depth, and these circuits with T-depth 1, 2, and 3 can be found in [Sel12, AMMR13].

3   The Round-in-Place Structure for Iterative Primitives

We start by reviewing the two main structures used in previous work on the quantum circuits for AES, including the pipeline structure [JNRV20] and the zig-zag structure [GLRS16] illustrated in Fig. 2 and Fig. 3, respectively. In the Figures we can see that each pair of neighbouring sub-circuits (represented as small rectangles) are not perfectly aligned horizontally, but forming a stepladder pattern. This interconnection pattern is due to the *out-of-place* nature of the circuit $\mathcal{R}_i$, which implements the i-th round function of AES, mapping $|k_i\rangle|x\rangle|0\rangle$ to $|k_i\rangle|x\rangle|O(\mathcal{R}_s)\rangle$, where $|k_i\rangle$ is the round key, $|x\rangle$ is the input state, and $|O(\mathcal{R}_i)\rangle$ is the output state of the round function. Here by *out-of-place* we mean that the output $|O(\mathcal{R}_i)\rangle$ of the round function is not carried in the qubits that encode the input $|x\rangle$.

It is easy to see that, since the round transformation is implemented by the out-of-place circuit $\mathcal{R}_s$, the pipeline structure needs lots of qubits to preserve the input states of all rounds, and the zig-zag structure is designed to reduce

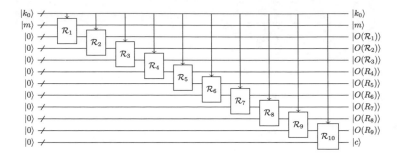

Fig. 2. The pipeline structure, where k_0 is the initial key, m is the plaintext, c is the ciphertext. For the sake of simplicity, the ancilla qubits and the key expansion process are omitted, and $|k_0\rangle$ is used as the round key in each round.

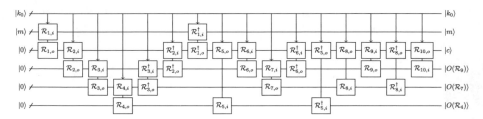

Fig. 3. The zig-zag structure, where $\mathcal{R}_s^\dagger$ is used to erase the redundant input states of $\mathcal{R}_{s+1}$.

the cost of qubits by using the reverse circuit for the last round to erase these inputs. In [JNRV20], the pipeline structure was used for designing low-depth circuits of AES, and in [GLRS16, ASAM18, LPS19], the zig-zag structure was used for designing low-width circuits of AES.

3.1 The Round-in-Place Structure

For iterative designs with invertible round functions, an *in-place* implementation (with some ancillae) of the round function maps $|k_s\rangle|x\rangle|0\rangle$ to $|k_s\rangle|O(\mathcal{R}_s)\rangle|0\rangle$. Such in-place implementations of the round functions can be connected naturally to form the compositions of the round functions without the need of additional qubits. However, directly designing an in-place circuit with low T-depth for the round transformation involved in typical ciphers is difficult. In contrast, a compact out-of-place circuit for a round transformation can be efficiently derived from a compact classical circuit by implementing additions by CNOT gates and multiplications by Toffoli gates.

A natural idea is to construct an in-place circuit based on out-of-place sub-circuits, and this construction is depicted in Fig. 4. This structure is a general form of the circuits used in [AMG+16], and later we will show that previous work neglect important things in implementing $U_{R^{-1}}$.

We write the classical round transformation of a symmetric cipher into the following form $\texttt{Round} : (x, k) \to (T(x, k), k)$. Let $\texttt{Round}^{-1} : (z, k) \to (T'(z, k), k)$ be the inverse of the round transformation, where $T'(T(x, k), k) = x$. Suppose we have an out-of-place circuit U_R for $\texttt{Round}$ and an out-of-place circuit $U_{R^{-1}}$ for $\texttt{Round}^{-1}$. Actually, by implementing the classical Boolean operations with Toffoli gates we can always construct such out-of-place circuits. Then the circuit in Fig. 4 in-place implements $\texttt{Round}$ with some ancilla qubits.

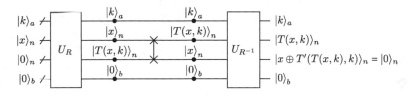

Fig. 4. In-place implementation of an invertible round transformation based on out-of-place sub-circuits

Figure 4 provides an efficient way to implement an invertible round transformation in-place. We call this kind of circuit an *out-of-place based* (abbreviated as OP-based) in-place circuit. Based on this circuit, we can implement an iterative cipher with the structure shown in Fig. 5. We call this structure the OP-based round-in-place structure, abbreviated as the *round-in-place* structure.

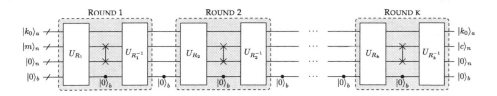

Fig. 5. The OP-based round-in-place structure

Remark 1. In Fig. 4, the functionality of $U_{R^{-1}}$ is to compute T', and then XOR T' into the third register. Moreover, this functionality should work when the state on the third register is $|x\rangle$, which is related with the state in the second register. Therefore,

- If we use a $U_{R^{-1}}$ which only works for $|0\rangle$, the output will be wrong.
- If we use a $U_{R^{-1}}$ which works for any $|y\rangle$, the output will be correct. However, since the relationship between $|x\rangle$ and $|T(x, k)\rangle$ is not sufficiently used, such $U_{R^{-1}}$ will cost more quantum resources.

In [AMG+16], an in-place circuit of the χ function of SHA3 was presented with a structure which is similar with our OP-based in-place circuit. However, their implementation of χ^{-1} is a straightforward implementation of the classical circuit from KECCAK tools, which is equivalent to a $U_{R^{-1}}$ that only works for $|0\rangle$, leading to incorrect output. In [ZWS+20], the implementation is equivalent to a $U_{R^{-1}}$ that works for any $|y\rangle$, which needs more quantum resources.

Remark 2. Note that $U_{R^{-1}}$ and $U_R^\dagger$ are different. $U_R^\dagger$ is the circuit for the adjoint of the unitary transformation U_R. It can be obtained by placing the adjoints of the gates in U_R in the reverse order, and thus $U_R^\dagger$ and U_R cost the same quantum resources in most times. In comparison, $U_{R^{-1}}$ is the circuit for the reverse transformation R^{-1}. Hence, the costs of U_R and $U_{R^{-1}}$ are different in general. However, in our context, R is for encryption, while R^{-1} is for decryption. For most symmetric ciphers, the complexities of the encryption and decryption are similar. Hence the quantum resources for implementing U_R and $U_{R^{-1}}$ are almost the same.

3.2   A Depth-Width Comparison of Different Structures

Given an iterative block cipher whose block size is n-bit, a rough estimation of the width of the circuits with the pipeline, zig-zag, and round-in-place structures, when they use the same out-of-place circuit U_R as their main component, can be easily obtained from Figs. 2, 3, 4 and 5. We suppose U_R requires n qubits for the input data block, n qubits for the output data block, and αn for the round key and ancillae. Note that in Fig. 2 and Fig. 3, we omit the possible ancilla qubits used in $\mathcal{R}_s$. Then, the widths of the three structures for implementing all r-round operations are presented in Table 1.

Table 1. The widths of different structures, where t is the minimal number such that $\sum_{i=1}^{t} i > r$.

Pipeline	Zig-zag	Round-in-place
$(r + \alpha + 1)n$	$(t + 1 + \alpha)n \approx (\sqrt{2r} + \alpha)n$	$(2 + \alpha)n$

To estimate the depths of these structures, we need to consider two different scenarios. In the first scenario, we build circuits for the Grover oracles used in exhaustive key search attacks. In the second scenario, we construct circuits for the encryption oracles used in [KLLN16a].

Circuits for Grover Oracles. First, we consider the Grover oracle: $|y\rangle|q\rangle \rightarrow |y\rangle|q \oplus f(y)\rangle$, where $f(y)$ is a Boolean function that outputs one bit 1 or 0. When given some pairs of plaintext and ciphertext, by constructing a Grover oracle with the key $|k\rangle$ as the input, one can use Grover's algorithm to search the correct key. For simplicity, we consider the case of using one pair of plaintext and ciphertext (m, c_0). In this case, the circuits of the Grover oracle based on different structures are shown in Fig. 6. In this figure, the out-of-place sub-circuit

E_{OP} denotes the encryption circuit generated by the pipeline or zig-zag structure, while the in-place sub-circuit E_{IP} denotes the encryption circuit generated by the round-in-place structure. Besides the ciphertext $|c\rangle$, E_{OP} outputs the redundant state $|r\rangle$ corresponding to those $O(R_s)$ in Fig. 2. Since the plaintext m is fixed, $|m\rangle$ is a computational basis state, which can be viewed as ancilla qubits in this circuit. The sub-circuit "COM" compares $|c\rangle$ with the provided ciphertext c_0, if they are equal, then flips the target qubit $|q\rangle$. Apparently, in these two circuits, the depth of the oracle is roughly two times of the depth of the encryption circuit (E_{OP} or E_{IP}).

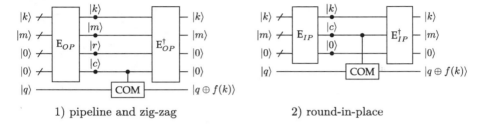

1) pipeline and zig-zag   2) round-in-place

Fig. 6. The Grover oracle based on different structure

Circuits for Encryption Oracles. Now we consider the encryption oracle defined in [KLLN16a]: $|m\rangle|0\rangle \rightarrow |m\rangle|E(m)\rangle$, where m is the plaintext, and $E(m)$ is the corresponding ciphertext. For this oracle, if its input is a superposition $\sum_m |m\rangle|0\rangle$, then its output will be a superposition $\sum_m |m\rangle|E(m)\rangle$. Figure 7 shows how to construct quantum cryptographic oracles based on different structures. Here $|c\rangle = |E(m)\rangle$ is the ciphertext. Note that, in this oracle, we do not need to store $|k\rangle$ by qubits, since we can pre-compute all the round keys via classical computation, and add them on the input of each round by Pauli-X gates. For the pipeline and the zig-zag structures, since we need uncomputation to erase the redundant state $|r\rangle$, the depth of the oracle is twice of that of the encryption process. However, for the round-in-place structure, since we do not generate $|r\rangle$, we do not need the uncomputation process.

By summarizing the above discussion, we have the following results. Given a symmetric cipher with r rounds, suppose the depths (or T-depths) of U_R and $U_{R^{-1}}$ in Fig. 4 are both d. This is reasonable according to Remark 2. If we ignore the cost of the compare process in the Grover oracle and the copy process in the quantum cryptographic oracle, then the depths (or T-depths) and the DW-costs (the product of depth and width) of the oracles based on these three structures are as shown in Table 2.

From the results in Table 1 and Table 2, we have the following observations.

Observation 2. *The round-in-place structure has the smallest width in any cases. For the Grover oracle, the pipeline structure has the lowest depth. When $r \leq 3 + \alpha$, the DW-cost of the pipeline structure is lowest, and when $r > 3 + \alpha$,*

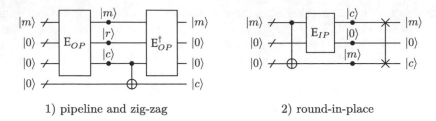

1) pipeline and zig-zag   2) round-in-place

Fig. 7. The encryption oracle based on different structure

Table 2. The depths and DW-costs of the oracles based on different structures

Metric	Type	Pipeline	Zig-zag	Round-in-place
Depth	Grover	$2r \cdot d$	$2 \cdot \sum_{i=1}^{t}(2(t-i)+1) \approx 4r \cdot d$	$4r \cdot d$
	Encrypt	$2r \cdot d$	$\approx 4r \cdot d$	$2r \cdot d$
DW-cost	Grover	$2r(r+1+\alpha)nd$	$2r(\sqrt{2r}+\alpha)nd$	$4r(2+\alpha)nd$
	Encrypt	$2r(r+1+\alpha)nd$	$2r(\sqrt{2r}+\alpha)nd$	$2r(2+\alpha)nd$

the DW-cost of the round-in-place structure is lowest. For the quantum crypto-graphic oracle, the pipeline structure and the round-in-place structure have the same depth, and the DW-cost of the round-in-place structure is always the lowest.

4   Synthesizing Optimal CNOT Circuits with SAT

It is well known that an invertible linear transformation over $\mathbb{F}_2^n$ can be implemented in-place with n qubits by the CNOT gates [PMH08]. Given an invertible transformation represented as a binary matrix, the PLU decomposition technique [GLRS16, JNRV20, ZWS+20] and the heuristic algorithm proposed at FSE 2020 [XZL+20] are typically employed to produce a compact CNOT circuit implementing the linear transformation. However, these methods are far from being optimal. In the following, we present a SAT-based method to generate the most compact CNOT circuit for invertible linear transformations over $\mathbb{F}_2^n$. Due to the difficulty of solving large scale SAT models in practice, the SAT-based technique only works when n is small.

The idea of our algorithm is to convert the problem of finding a circuit with k CNOT gates into the problem of solving a system of Boolean equations. Similar ideas were used in [FS10, Sto16, MSM18], where different classical and quantum circuit synthesis problems were considered. A brief introduction of our method is given in the following, and a detailed description can be found in the full version of this paper [HSb].

Given a positive integer k and a linear transformation which can be expressed as n linear forms $L_i(x_1, x_2, \ldots, x_n)$ in x_i, we generate a model with the following sets of variables: $B = (b_{ij})_{k \times n}$, $C = (c_{ij})_{k \times n}$, $F = (f_{ij})_{n \times n}$, and $\Psi = \{\psi_{i,j,s}\}_{k \times n \times n}$. These variables are of different semantics. For B and C,

$b_{ij_1} = c_{i,j_2} = 1$ means the i-th gate is a CNOT gate with control wire j_1 and target wire j_2. For F, $f_{ij} = 1$ implies that the final output of the j-th wire is L_i. For Ψ, $\psi_{i,j,k} = 1$ indicates that after the i-th gate, the coefficient of x_k in the Boolean expression for the j-th wire is 1. We generate the equations of these variables to encode the gates and their outputs. It can be shown that the obtained system of equations has a solution (which can be tested with SAT solvers) if and only if there is a CNOT-circuit with k CNOT gates. By incrementally increasing k, we can identify the minimal k such that the corresponding equations can be satisfied simultaneously. According to our experiments, linear transformations with less than 9 variables can be solved in a reasonable time. Hence, we employ this method to identify the optimal CNOT sub-circuits in our implementations of the AES S-box presented in Sect. 5.

5   In-Place Circuits for Nonlinear Components

It is well known that for any $\mathbb{F}_2^n \to \mathbb{F}_2^n$ permutation, there is an in-place quantum circuit implementing it using only Pauli-X, CNOT, and Toffoli gates with at most one ancilla qubit [SPMH03]. To obtain this in-place implementation with minimal width, one needs to solve a corresponding permutation factorization problem, which is computationally difficult for large permutations like the AES S-box. Moreover, the complexity in terms of the gate count and depth of the circuits produced by this method is typically far from being satisfactory. For example, in [GLRS16], the authors estimated that the in-place circuit of the AES S-box with 9 qubits would cost about 9695 T gates. In contrast, with the method provided in this section, we can construct an in-place circuit that requires only 22 qubits and 728 T gates.

In what follows, we consider special quantum circuits (named as $\mathfrak{C}^0$- and $\mathfrak{C}^*$-circuits) implementing a vectorial Boolean function, based on which in-place quantum circuits for nonlinear transformations with different shapes can be constructed.

5.1 $\mathfrak{C}^0$- And $\mathfrak{C}^*$-Circuits for a Vectorial Boolean Function

Given a vectorial Boolean function $f\colon \mathbb{F}_2^a \to \mathbb{F}_2^b$, a $\mathfrak{C}^0$-circuit for f is a quantum circuit mapping $|x\rangle_a |0\rangle_b |0\rangle_c$ to $|x\rangle_a |f(x)\rangle_b |0\rangle_c$ for any $x \in \mathbb{F}_2^a$, and a $\mathfrak{C}^*$-circuit for f is a quantum circuit mapping $|x\rangle_a |y\rangle_b |0\rangle_c$ to $|x\rangle_a |y \oplus f(x)\rangle_b |0\rangle_c$ for any $(x, y) \in \mathbb{F}_2^{a+b}$. Obviously, a $\mathfrak{C}^*$-circuit for f is always a $\mathfrak{C}^0$-circuit for f. Moreover, building a $\mathfrak{C}^0$-circuit is much easier than building a $\mathfrak{C}^*$-circuit since a $\mathfrak{C}^*$-circuit has more restrictions on its input-output behavior. For example, the circuits for the AES S-box proposed in [GLRS16, ASAM18, LPS19] are $\mathfrak{C}^0$-circuits, but not $\mathfrak{C}^*$-circuits.

Next, we present a generic method that can convert a $\mathfrak{C}^0$-circuit for f with some clearly defined properties (called *simplex* $\mathfrak{C}^0$-circuits) into a $\mathfrak{C}^*$-circuit for f efficiently. In particular, the obtained $\mathfrak{C}^*$-circuit do not increase the T-depth of the corresponding $\mathfrak{C}^0$-circuit.

Simplex $\mathcal{C}^0$-Circuits. A $\mathcal{C}^0$-circuit for f is *simplex* if it maps $|x\rangle_a|y\rangle_b|0\rangle_c$ to

$$|x\rangle_a|A(y) \oplus f(x)\rangle_b|0\rangle_c,$$

where $A : \mathbb{F}_2^b \to \mathbb{F}_2^b$ is an invertible linear function. We now consider the gate-level structures of (simplex) $\mathcal{C}^0$-circuits, which gives some intuitive ideas on how to construct efficient simplex $\mathcal{C}^0$-circuits and the sufficient condition for a $\mathcal{C}^0$-circuit to be simplex.

Suppose we have a quantum circuit built with Pauli-X, CNOT and Toffoli gates. Let $|x_1, x_2, \cdots, x_a\rangle|y_1, y_2, \cdots, y_b\rangle|0\rangle_c$ and $|t_1, \cdots, t_a\rangle|z_1, \cdots, z_b\rangle|0\rangle_c$ be the input and output of the circuit with x_i, y_i and z_i in $\mathbb{F}_2$. For $j \in \{1, 2, \cdots, b\}$, we have

$$z_j(x, y) = \sum_{u,v} a_{u,v}^j x^u y^v + \sum_u b_u^j x^u + \sum_u d_u^j y^u + T^j(x) + L^j(y) + c^j,$$

where $a_{u,v}^j$, b_u^j, d_u^j, and $c^j \in \mathbb{F}_2$, $u \in \mathbb{F}_2^a$, $v \in \mathbb{F}_2^b$, T^j and L^j are linear functions, and x^u is the monomial $\prod_{u_i=1} x_i$. If this circuit is a $\mathcal{C}^0$-circuit of the vectorial function $f(x) = (f_1(x), \cdots, f_b(x))$, then

$$z_j(x, 0) = \sum_u b_u^j x^u + T^j(x) + c^j = f_j(x),$$

which implies that $z_j(x, y) = f_j(x) + \sum_{u,v} a_{u,v}^j x^u y^v + \sum_s d_u^j y^u + L^j(y)$. Thus, if the quantum gates applied to the input qubits $|x_1, x_2, \cdots, x_a\rangle|y_1, y_2, \cdots, y_b\rangle|0\rangle_c$ do not produce any nontrivial $x^u y^v$ and y^u terms whose degree are greater or equal to 2, the $\mathcal{C}^0$-circuit must be simplex.

Now we analyze which operations may generate these $x^u y^v$ and y^u terms. We denote the set of the b output wires in this circuit as $\mathcal{W}$, and the set of other $a + c$ wires as $\mathcal{V}$. Then, the algebraic expressions of the initial states on the wires in $\mathcal{W}$ are y. All operations that operate on at least one wire in $\mathcal{W}$ can be classified into the following 7 types of operations illustrated in Fig. 8.

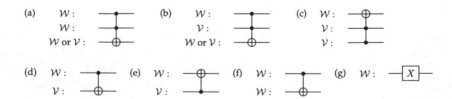

Fig. 8. Operations that operate on at least one wire in $\mathcal{W}$

Denote $\sum_i a_{u,v}^j x^u y^v + \sum_s d_u^j y^u + L^j(y)$ by $h_j(x, y)$, then $z_j(x, y) = h_j(x, y) + f_j(x)$. From the above analysis, to design a $\mathcal{C}^0$-circuit, since there is no constraint on $h_j(x, y)$, we do not need to care about the generation of $x^u y^v$ and y^u. This means the above 7 types of operations are all permitted. The designer only needs to focus on efficiently constructing $f(x)$. From the view of algebraic expression, the qubits on the wires in $\mathcal{W}$ can be seen as newly defined variables. Additions and multiplications about these variables can be used to generate $f(x)$.

To design a simplex $\mathfrak{C}^0$-circuit, we should guarantee that $h_j(x, y) = L^j(y)$, which means that $x^u y^v$ and y^u should not appear. We have the following observation:

Observation 3. *If some $x^u y^v$ was generated, it is hard to eliminate this $x^u y^v$ in the following steps, unless we repeat the same Toffoli gate which generates it.*

This means generating $x^u y^v$ will likely increase the cost of the circuits. Hence a natural criterion for designing a compact $\mathfrak{C}^*$-circuit is to avoid generating $x^u y^v$.

Under this criterion, operations (a) and (b) should obviously be avoided. For operation (d), it can only be applied when we use the qubit on the target wire as a dirty qubit for some CNOT gates. Note that operation (e), (f), and (d) under this constrain can be described together as the following operation.

(h): Apply s CNOT gates, $s \geq 1$, which map $|u\rangle_a |w\rangle_b |v\rangle_c$ to $|u\rangle_a |L(u, v, w)\rangle_b |v\rangle_c$ for any u, v, w, where L is a linear function w.r.t. u, v, w.

Thus, our criterion for designing a compact $\mathfrak{C}^*$-circuit is: *only operations (c), (h), (g) can be applied on the output wires.* Note that without applying operation (a), y^{γ_s} will not be generated either. Therefore, under this criterion, if we construct U_f, which is a $\mathfrak{C}^0$-circuit of f, then the output of U_f is

$$|x\rangle |h_1(x, y) + f_1(x), \dots, h_b(x, y) + f_b(x)\rangle |0\rangle,$$

where $h_k(x, y) = L_k(y)$ is a linear function with respect to y. Let $A(y) = (L_1(y), L_2(y), \dots, L_b(y))$, then the output can be denoted by $|x\rangle |A(y) \oplus f(x)\rangle |0\rangle$, which implies this is a simplex $\mathfrak{C}^0$-circuit.

Converting Simplex $\mathfrak{C}^0$-circuits into $\mathfrak{C}^*$-Circuits. Let U_f be a $\mathfrak{C}^0$-circuit of $f(x)$. We can construct a $\mathfrak{C}^*$-circuit of $f(x)$ as shown in Fig. 9. Here $U_{A^{-1}}$ is an in-place sub-circuit, which implements $A^{-1}(y)$. $|y\rangle$ will be converted to $|A^{-1}(y)\rangle$ after passing $U_{A^{-1}}$, and the output of this modified circuit will be $|x\rangle |A(A^{-1}(y)) \oplus f(x)\rangle |0\rangle = |x\rangle |y \oplus f(x)\rangle |0\rangle$, which means this is a $\mathfrak{C}^*$-circuit of $f(x)$.

It is easy to see that the $\mathfrak{C}^*$-circuit constructed by the above method has the same width as the $\mathfrak{C}^0$-circuit U_f. Moreover, the numbers and the depths of Toffoli gates (or T gates) are the same for these two circuits. From this construction, we can see that the $\mathfrak{C}^*$-circuit and the simplex $\mathfrak{C}^0$-circuit are almost the same. Hence, to efficiently construct a $\mathfrak{C}^*$-circuit, we should also follow the above criterion of designing a simplex $\mathfrak{C}^0$-circuit, and by this way we will always obtain a simplex $\mathfrak{C}^0$-circuit. Then, the process of designing a $\mathfrak{C}^*$-circuit of $f(x)$ can be summarized as following steps:

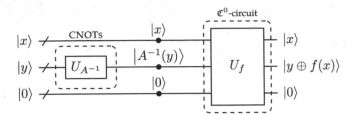

Fig. 9. A $\mathfrak{C}^*$-circuit based on a simplex $\mathfrak{C}^0$-circuit

1) We design U_f, a $\mathfrak{C}^0$-circuit of $f(x)$, in which only operations (c), (g), and (h) can be applied on the wires in $\mathcal{W}$. Then U_f will be a simplex $\mathfrak{C}^0$-circuit.
2) We determine A. Note that, in a simplex $\mathfrak{C}^0$-circuit, operations (c), (g) will not generate any term containing y, which means A is determined by (h) operations. Hence, we can obtain A by computing the composition of the linear transformations corresponding to all (h) operations.
3) If A is identity, this is already a $\mathfrak{C}^*$-circuit. Otherwise, we implement A^{-1} by an in-place CNOT circuit $U_{A^{-1}}$, then construct a $\mathfrak{C}^*$-circuit as Fig. 9.

For most S-box implementation problems, the number of the output wires is not bigger than 8, which means, by our SAT-based algorithm, we can make sure $U_{A^{-1}}$ uses the minimal number of CNOT gates.

5.2 In-Place Implementations of Nonlinear Transformations of Different Shapes with $\mathfrak{C}^0$- And $\mathfrak{C}^*$-Circuits

We show how to implement nonlinear transformations with typical shapes encountered in practice with $\mathfrak{C}^0$- and $\mathfrak{C}^*$-Circuits. In most symmetric ciphers, a nonlinear component can correspond to one of the classical invertible nonlinear transformations presented in Fig. 10.

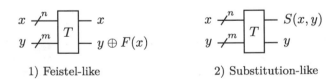

1) Feistel-like   2) Substitution-like

Fig. 10. Two kinds of classical invertible nonlinear transformations

Feistel-Like Transformations. First, we consider Feistel-like classical invertible nonlinear transformations of the form

$$\Psi : (x, y) \mapsto (x, y \oplus F(x)).$$

The quantum circuit of this type of nonlinear transformation can be realized by a $\mathfrak{C}^*$-circuit of F, which in turn can be derived from a simplex $\mathfrak{C}^0$-circuit of F. We note that a $\mathfrak{C}^*$-circuit of F, mapping $|x\rangle|y\rangle|0\rangle$ to

$$|x\rangle|y \oplus F(x)\rangle|0\rangle = |\Psi(x,y)\rangle|0\rangle,$$

is an *out-of-place* implementation of F but an *in-place* implementation of Ψ. Feistel-like structures are frequently seen in Feistel ciphers, NFSR-based designs, and key-schedule algorithms of block ciphers.

For example, SubByte and the following XOR operation in the AES key schedule can be seen as a Feistel-like transformation. Therefore to in-place implement this transformation, we can construct a simplex $\mathfrak{C}^0$-circuit of the AES S-box, then extend it to a $\mathfrak{C}^*$-circuit. In the previous works about AES, a lot of proposed quantum circuits of the AES S-box are simplex $\mathfrak{C}^0$-circuits [ASAM18, LPS19, ZWS+20], since in these circuits only operations (c),(h),(g) are applied. Hence, by the method proposed in Sect. 5.1, we can easily extend them to the $\mathfrak{C}^*$-circuits. For example, based on the simplex $\mathfrak{C}^0$-circuit of the AES S-box proposed in [ZWS+20], we construct a compact $\mathfrak{C}^*$-circuit[1]. In this $\mathfrak{C}^*$-circuit, the in-place circuit implementing A^{-1} with minimum number of CNOT gates is achieved by our SAT-based algorithm. This circuit costs 10 CNOT gates, and in the full version of this paper [HSb], we present the matrix corresponding to A^{-1} and the specific form of this circuit.

In Table 3, we compare the quantum resources used in our $\mathfrak{C}^*$-circuit and those used in the $\mathfrak{C}^*$-circuit proposed in [ZWS+20].

Table 3. Quantum resources for implementing the S-box of AES

	#ancilla	Toffoli-depth	#Toffoli	#CNOT	#Pauli-X	Source
$\mathfrak{C}^0$-S-box	6	41	52	326	4	[ZWS+20]
$\mathfrak{C}^*$-S-box	7	60	68	352	4	[ZWS+20]
	6	41	52	336	4	This paper

Substitution-Like Transformations. Next, we consider classical invertible substitution-like transformations of the form

$$\Phi : (x,y) \rightarrow (S(x,y), y).$$

It is easy to see that the description of such nonlinear transformation is the same as that of the round transformation discussed in Sect. 3, and thus we can implement such nonlinear transformation by the OP-based in-place circuit in Fig. 11.

---

[1] The C code for checking the correctness of this $\mathfrak{C}^*$-circuit is available at https://github.com/hzy-cas/AES-quantum-circuit.

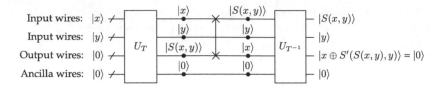

Fig. 11. An OP-based in-place circuit for a substitution-like nonlinear transformation, where S' is a function satisfying $S'(S(x,y),y) = x$ for any y.

We now consider how to implement the building blocks of the circuit depicted in Fig. 11. For U_T, it can be implemented as a $\mathfrak{C}^0$-circuit of S. For $U_{T^{-1}}$, a circuit which mapping $|S(x)\rangle|y\rangle|x\rangle|0\rangle$ to $|S(x)\rangle|y\rangle|x \oplus S'(S(x),y)\rangle|0\rangle$, one may attempt to implement this part as a $\mathfrak{C}^*$-circuit of S'. We now show that this is an overshoot.

For $U_{T^{-1}}$, if we set $z = S(x)$, then $S'(z,y) = x$, and $U_{T^{-1}}$ maps $|z\rangle|y\rangle|S'(z,y)\rangle|0\rangle$ to $|z\rangle|y\rangle|0\rangle|0\rangle$. Suppose U_{T_0} is a circuit that maps $|z\rangle|y\rangle|0\rangle|0\rangle$ to $|z\rangle|y\rangle|S'(z,y)\rangle|0\rangle$, then obviously, $U_{T_0}^\dagger$, the reverse circuit of U_{T_0}, is equivalent to $U_{T^{-1}}$. Therefore, to implement a substitution-like transformation, we only need to design a $\mathfrak{C}^0$-circuit of S, and a $\mathfrak{C}^0$-circuit of S', whose reverse circuit is used.

6   A Method for Constructing Low T-Depth Circuits

As discussed in Sect. 2, the T-depth is the most concerned parameter. In our context, T gates only appear in the Toffoli gates, the quantum AND gates and their adjoints, which are employed to implement the quantum correspondences of the multiplications in the classical computation. In the following, we first show that there always exists a quantum circuit with T-depth equal to the AND-depth of the corresponding classical circuit. Therefore, we can first construct a classical circuit with low AND-depth, and then convert it into a low T-depth quantum circuit.

6.1   Classical AND-depths v.s. Quantum T-depths

The AND-depth of a classical circuit (a.k.a. the multiplicative depth) constructed with AND, XOR, and NOT gates is the largest number of AND gates on any path from a primary input to a primary output. For example, the AND-depth of the classical circuit shown in Fig. 12 is 1.

The readers may think that it is trivial to build a quantum version of a given classical circuit such that the T-depth of the quantum circuit is equal to the AND-depth of the classical circuit by just properly replacing the classical AND gates with Toffoli gates or quantum AND gates, all of which have T-depth 1 implementations. However, for quantum circuits, a direct copy as the "b" signal in Fig. 12 is not allowed and a qubit cannot be used in different quantum gates simultaneously. Therefore, a quantum circuit obtained from a classical one by the simple replacement strategy mentioned above maintaining the natural order of operations may result in increased T-depth. For example, the quantum circuits with different Toffoli-depth depicted in (2) and (3) of Fig. 12 both implement the functionality of the classical circuit given in (1) of Fig. 12. Next, we show that the AND-depth of a classical circuit set a lower bound for the T-depth of its quantum counterpart, and this lower bound is achievable.

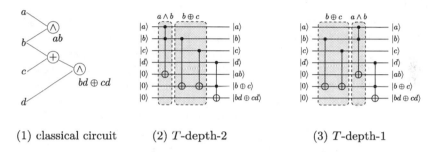

| (1) classical circuit | (2) T-depth-2 | (3) T-depth-1 |

Fig. 12. Quantum implementations of a classical circuit with multiplicative depth 1

Theorem 1. *Given a classical circuit with AND-depth s, the T-depth of the quantum circuit implementing all the nodes of the classical circuit is not smaller than s. Moreover, with sufficiently many ancillae, we can construct a quantum circuit implementing all the nodes of the classical circuit with T-depth s.*

Theorem 1 can be proved by a constructive method with some notations and terminologies introduced below. Here we illustrate this method with the classical circuit given by Example 1. A full proof of Theorem 1 can be found in the full version [HSb], which also provides a generic method to convert an AND-depth t classical circuit into a T-depth t quantum circuit.

Example 1. $M_4 = M_1 \cdot M_2, \quad M_5 = M_2 \cdot M_3, \quad M_6 = M_4 \oplus M_3, \quad M_7 = M_5 \oplus M_1,$
$M_8 = M_7 \cdot M_2, \quad M_9 = M_7 \cdot M_3, \quad M_{10} = M_8 \oplus M_6, \quad M_{11} = M_{10} \oplus M_9, \quad M_{12} = M_7 \cdot M_6,$
$M_{13} = M_{11} \cdot M_3$

The AND-depth of the circuit given by Example 1 is 3. Before we present the method for building the corresponding T-depth-3 quantum circuit, we define the AND-depth for each intermediate node (or signal) appearing in the circuit. We call a variable an AND-variable if it represents the output of an AND gate. In Example 1, M_4, M_5, M_8, M_9, M_{12}, and M_{13} are AND-variables.

Let M_i and M_j be two AND-variables. M_j is said to be an AND-successor of M_i if $M_j = M_i \cdot M_k$ for some k, or $M_j = M_u \cdot M_v$ for some u and v such that M_u is generated from M_i by some XOR operations, which is denoted by $M_i \rightarrow M_j$. Also, we call M_i is an AND-predecessor of M_j. In our Example 1, we have $M_5 \rightarrow M_8$ and $M_8 \rightarrow M_{13}$, forming a directed path $M_5 \rightarrow M_8 \rightarrow M_{13}$. By generating all such paths, a directed acyclic graph is obtained with the nodes representing the AND-variables. Then, the AND-depth of an AND-variable M, denoted as $d_\wedge(M)$ is defined as k, if M is the k-th node in the longest path containing M. Apparently, the AND-depth of a classical circuit is equal to the maximum AND-depth of its AND-variables. It is easy to see that, for an AND-variable M, if M has no AND-predecessor, then $d_\wedge(M) = 1$, otherwise $d_\wedge(M) = 1 + \max_{v \in \mathrm{Pre}(M)} d_\wedge(v)$, where $\mathrm{Pre}(M)$ denotes the set of all predecessors of M. In Example 1, $d_\wedge(M_4) = d_\wedge(M_5) = 1, d_\wedge(M_8) = d_\wedge(M_9) = d_\wedge(M_{12}) = 2$, and $d_\wedge(M_{13}) = 3$.

Now, we are ready to describe our method for building the quantum circuit. Note that since we aim at reducing the T-depth, in our constructions, we always use the quantum AND gate with T-depth 1 and its adjoint with T-depth 0 depicted in Fig. 1 whenever possible, while in the figures illustrating the quantum circuits, we use Toffoli gates due to the compactness of its visualization. The circuit generated by our method has the following features, Firstly, for AND-variables with the same AND-depth, a layer of quantum AND gates which generate these AND-variables are applied in parallel. Secondly, before the quantum AND layer, all necessary input are generated with a CNOT network. In particular, when a variable is used as inputs of different AND gates of the subsequent AND layer, we can copy it into an ancilla qubit with the application of a CNOT gate, and clean the effect of the CNOT gate after the quantum AND layer.

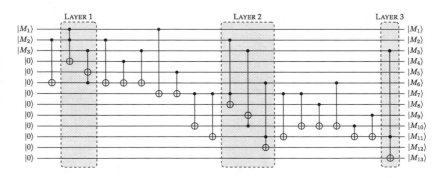

Fig. 13. An AND-depth-3 implementation for the classical circuit in Example 1

Figure 13 present a quantum circuit corresponding to Example 1. In this circuit, we have three layers of quantum AND gates, within each layer the gates are applied in parallel. In Layer 1, we generate M_4 and M_5. In Layer 2, we generate M_8, M_9, and M_{12}. In Layer 3, we generate M_{13}. The variables required by Layer 1 are M_1, M_2, and M_3. Since $M_4 = M_1 \cdot M_2$, and $M_5 = M_2 \cdot M_3$, M_2 is needed in two different AND gates. Therefore, before Layer 1, we copy $|M_2\rangle$ to another qubit by a CNOT gate. This is an idle qubit, which will be used to store $|M_6\rangle$. We clean this qubit after Layer 1. The variables required by Layer 2 are M_2, M_3,

M_6, and M_7. Hence, we have to generate M_6 and M_7. We do this by applying 4 CNOT gates before Layer 2. Similarly, M_7 is required for computing M_8, M_9, and M_{12}. We copy it into two idle qubits by 2 CNOT gates before Layer 2, which are cleaned after Layer 2. The variables required by Layer 3 are M_2, and M_{11}. Thus, before Layer 3, we apply 4 CNOT gates to generate M_{11}. This leads to a quantum circuit computing all the nodes in Example 1 with T-depth 3.

6.2   A Trick for Reducing the AND-depth of Classical Circuits

According to the discussion of the previous section, low AND-depth classical circuits imply low T-depth quantum circuits. In this section, we show how to reduce the AND-depth of a classical circuit without changing the functionalities of its primary outputs based on a simple observation. Let $M_4 = M_1 \cdot M_2$ and $M = M_4 \cdot M_3$ with $d_\wedge(M_1) = 2, d_\wedge(M_2) = 1$, and $d_\wedge(M_3) = 1$. Then $d_\wedge(M_4) = 3$, and $d_\wedge(M) = 4$. In addition, We can deduce that $M = (M_1 \cdot M_2) \cdot M_3$. Obviously, we also have $M = M_1 \cdot (M_2 \cdot M_3)$. Therefore, if we first compute $M_4 = M_2 \cdot M_3$, and then $M = M_1 \cdot M_4$. The AND-depth of M is reduced from 4 to 3.

We now show how this idea works for a more complicated case based on Example 1, where M_1, M_2, M_3 are primary inputs and M_{12}, M_{13} are primary outputs. For M_{13}, we have

$$M_{13} = (M_{10} \oplus M_9)M_3 = (M_8 \oplus M_6 \oplus M_9)M_3 = M_7(M_2M_3) \oplus M_6M_3 \oplus M_7M_3.$$

We modify the circuit by using the following steps to generate M_{13}: $N_1 = M_2 \cdot M_3$, $N_2 = M_6 \cdot M_3$, $N_3 = M_7 \cdot M_3$, $N_4 = M_7 \cdot N_1$, $N_5 = N_4 \oplus N_2$, $M_{13} = N_5 \oplus N_3$. M_{13} is not an AND-variable anymore. It is easy to check that $d_\wedge(N_1) = 1$, $d_\wedge(N_2) = 2$, $d_\wedge(N_3) = 2$, and $d_\wedge(N_4) = 2$. Therefore, the AND-depth of this new circuit is 2. The modified circuit is given by Example 2, where M_{12} and M_{13} are the primary outputs.

Example 2. $M_4 = M_1 \cdot M_2$, $M_5 = M_2 \cdot M_3$, $M_6 = M_4 \oplus M_3$, $M_7 = M_5 \oplus M_1$, $M_8' = M_6 \cdot M_3$, $M_9' = M_7 \cdot M_3$, $M_{10}' = M_7 \cdot M_5$, $M_{11}' = M_{10}' \oplus M_8'$, $M_{12} = M_7 \cdot M_6$, $M_{13} = M_{11}' \oplus M_9'$.

More generally, given a classical circuit, we can try to reduce its AND-depth as follows. Firstly, for a M' which is not an AND-variable, we extend the definition of $d_\wedge(M')$, by setting $d_\wedge(M')$ to be $\max_i\{d_\wedge(M_i)\}$, where M_i is an AND-variables and there is a path from M_i to M' in the classical circuit.

For an AND-variable M with $d_\wedge(M) = d \geq 3$, we have $M = M_1M_2$ for some M_1 with $d_\wedge(M_1) = d - 1$. If $d_\wedge(M_2) \leq d - 3$, we further decompose M_1 to variables with lower AND-depth. That is we write M_1 as $\sum_{i,j} M_i'M_j' + \sum_k M_k'$, where $d_\wedge(M_i') \leq d - 2, d_\wedge(M_j') \leq d - 2, d_\wedge(M_k') \leq d - 2$, for any i, j, k. Then, we have

$$M = \sum_{i,j} M_i'(M_j'M_2) + \sum_k M_k'M_2.$$

For any M_i' with $d_\wedge(M_i') = d - 2$, if the corresponding M_j' always satisfies $d_\wedge(M_j^1) \leq d - 3$, then by constructing some new operations which generate those $M_j'M_2$ first, we can reduce the AND-depth of M from d to $d - 1$.

6.3 *T*-depth-4 and *T*-depth-3 Quantum Circuits for the AES S-box

Firstly, based on the classical circuit proposed in [BP12] with AND-depth 4 (see the full version [HSb]), we can build a quantum circuit for the AES S-box with *T*-depth 4 by employing the method given in Sect. 6.1. In comparison, the *T*-depth of the quantum circuit presented by Jaques et al. [JNRV20] based on the same classical circuit is 6, and its width is the same as ours.

Furthermore, based on the trick given in Sect. 6.2, we transform the classical circuit proposed in [BP12] into a circuit with AND-depth 3 (shown in the full version [HSb]), based on which a *T*-depth-3 quantum circuit for the AES S-box can be constructed. Note that, the algebraic degree of the AES S-box is 7. With two layers of multiplications, we can only obtain polynomials with degree 4. This means that we need at least three layers of multiplications to generate the output of the AES S-box. Therefore, if we implement the AES S-box by firstly designing a classical reversible circuit, and then decomposing each gate into the Clifford+*T* gates, the minimum *T*-depth is 3, which is achieved by our circuit. A comparison of our circuits and the one presented in [JNRV20] is presented in Table 4. In this table, the gate counts and depths are obtained by summing the corresponding values for the forward circuit which computes the S-box output and the uncomputation circuit. The Q# code for our *T*-depth-4 and *T*-depth-3 quantum circuits are available at https://github.com/hzy-cas/AES-quantum-circuit.

Table 4. Quantum resources for different AES S-box circuits

#CNOT	#1qClifford	#T	# Measure	*T*-depth	Full depth	Width	Source
664	205	136	34	6	**117**	136	[JNRV20]
718	208	136	34	4	**109**	136	*T*-depth-4
1395	467	312	78	3	**113**	218	*T*-depth-3

Remark 3. The widths presented in Table 4 are not obtained by the Q# resource estimator. As mentioned in the latest ePrint version of [JNRV20] and https://github.com/microsoft/qsharp-runtime/issues/192. There was a bug in Q# that produces conflict width and depth estimations, and this issue was solved in the latest version of Q#. However, when Q# tries to optimize the *T*-depth, the width obtained is not optimal (https://github.com/microsoft/qsharp-runtime/pull/446). Therefore, we manually estimate the widths to obtain more accurate figures. The specific estimation process can be found in the full version of this paper [HSb].

7   Efficient Quantum Circuits for AES

To implement an iterative block cipher, we proceed as follows. Firstly, we choose the pipeline structure or the round-in-place structure according to our optimization objective (low depth or low width). Then, implement the linear layers with Xiang et al.'s method, the PLU decomposition method, or the SAT-based technique presented in this paper. For the nonlinear components, construct $\mathfrak{C}^0$-circuits and then convert them to $\mathfrak{C}^*$ circuits. Finally, plug these sub-circuits into the round-level structure. We will show case this procedure with AES-128, and all the techniques can be easily extended to AES-192, AES-256, and other iterative block ciphers.

7.1 Low-Width Quantum Circuits for AES

First of all, we choose the round-in-place structure according to observation 2 of Sect. 3. Then, we show how to implement the building blocks of AES in-place. The ShiftRow and RotByte operations can be easily implemented by rewiring. For the MixColumns operation, regarded as a 32×32 binary matrix, we employ the in-place circuit from [XZL+20], which requires 92 CNOT gates. In the following, we consider the implementations of the S-boxes, for which different circuits are used in different situations.

S-boxes in the Key Schedule Data Path. Since the S-box is immediately followed by an XOR operation in the key schedule (a Feistel-like transformation), we only need a $\mathfrak{C}^*$-circuit of the S-box. In our implementation, we used the $\mathfrak{C}^*$-circuit introduced in Sect. 5.2.

For the sake of simplicity, we call a $\mathfrak{C}^*$-circuit (or $\mathfrak{C}^0$-circuit) of the AES S-box a $\mathfrak{C}^*$ (or $\mathfrak{C}^0$) S-box. Figure 14 illustrates the structure of the in-place circuit for the AES key schedule. In this figure, SubByte represents the sub-circuit for the parallel application of four $\mathfrak{C}^*$ S-boxes. Note that while the implementations of the S-boxes are improved in this paper, the high-level structure is attributed to [JNRV20].

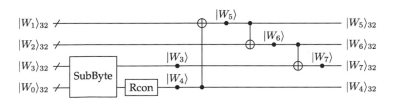

Fig. 14. An in-place circuit for generating the first round key

S-boxes in the Encryption Data Path. In the encryption process of AES, ByteSub can be regarded as a substitution-like transformation defined in Sect. 4, and thus we can implement it with the OP-based in-place circuit. This means we need to construct a $\mathfrak{C}^0$ S-box and a $\mathfrak{C}^0$ S-box^{-1}. Here, we use the $\mathfrak{C}^0$ S-box proposed in [ZWS+20], based on which a $\mathfrak{C}^0$ S-box^{-1} can be constructed as follows.

Suppose $x \in \mathbb{F}_2^8$ is the input of the S-Box, then y, the output of the S-box, is equal to $LS_0(x) + c$, where L is a linear function and $S_0(x)$ is the inverse of x in $\mathbb{F}_2^8$. Hence, we have $x = S_0^{-1}L^{-1}(y + c) = S_0L^{-1}(y + c) = L^{-1}(LS_0)L^{-1}(y + c)$.

Suppose U_0 is the circuit that implements $|x\rangle|0\rangle|0\rangle \to |x\rangle|LS_0(x)\rangle|0\rangle$. Obviously, it can be generated from a $\mathfrak{C}^0$ S-box by deleting the last 4 Pauli-X gates. Then, it is easy to check that the circuit in Fig. 15 is a $\mathfrak{C}^0$-circuit of S-box^{-1}.

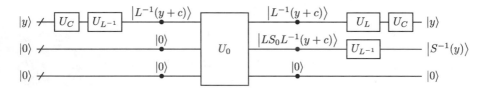

Fig. 15. The circuit for implementing the S-box^{-1} of AES

In Fig. 15, U_C is the circuit consisting of 4 Pauli-X gates, and it implements constant addition of c. U_L and $U_{L^{-1}}$ are the circuits consisting of CNOT gates, and they implement the linear transformation L and L^{-1} respectively. According to our SAT-based method, L can be implemented by 14 CNOT gates. Consequently, we can implement a $\mathfrak{C}^0$-circuit of S-Box^{-1} with 6 ancilla qubits, 52 Toffoli gates, 41 Toffoli depth, 368 CNOT gates, and 8 NOT gates[2]. In the full version [HSb], we present the specific form of the matrix corresponding to L, and the quantum circuit that implements L with minimal number of CNOT gates.

Based on the above circuits, we can in-place implement `ByteSub` in each round by 16 OP-based in-place circuits of the S-box. We suppose these 16 in-place circuits are implemented in parallel, then our implementation of `ByteSub` has the following two phases:

Phase 1: Implement 16 $\mathfrak{C}^0$ S-box, denoted by `ByteSub`$_1$;
Phase 2: Implement 16 reverse circuits of $\mathfrak{C}^0$ S-box^{-1}, denoted by `ByteSub`$^{-1}$.

Note that in the key schedule process of each round, we need to apply 4 $\mathfrak{C}^*$ S-boxes. Obviously, by applying 2 of them in **Phase 1**, and another 2 of them in **Phase 2**, we can reduce the DW-cost of the whole circuit. Under this strategy, our implementation of the i-th round of AES can be illustrated by Fig. 16.

In this figure, k_{i-1} denotes the round key in the $(i-1)$-th round, and c_{i-1} denotes the output state of the $(i-1)$-th round. The last step of each round is `AddRoundKey`, which can be implemented by applying 128 CNOT gates in parallel, and denoted by a CNOT gate in Fig. 16. For the final round, we do not need the sub-circuit `Mixcol`. Round 0, which performs a bitwise XOR of k_0 to the plaintext, can be implemented by applying 128 CNOT gates in parallel.

---

[2] The C code for checking the correctness of this $\mathfrak{C}^0$-circuits of S-box^{-1} is available at https://github.com/hzy-cas/AES-quantum-circuit.

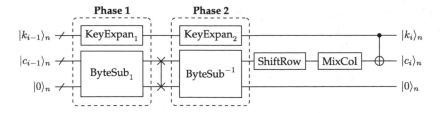

Fig. 16. The in-place implementation of the i-th round of AES-128

Note that the ancilla qubits used in these S-box circuits are ignored in this figure. From Table 3, we know that the number of ancilla qubits used in our out-of-place ($\mathfrak{C}^0$ or $\mathfrak{C}^*$) S-box circuit is 6, and the Toffoli-depth of this circuit is 41. We implement 18 out-of-place S-box circuits simultaneously in a phase, and the third register uses $8 \times 16 = 128$ qubits to store the outputs of the 16 out-of-place S-box circuits in ByteSub$_1$, hence the width of this one round circuit is $18 \times 6 + 3 \times 128 = 492$, and the Toffoli-depth is $41 \times 2 = 82$.

Apparently, for this one round circuit, we can make a tradeoff between width and depth by reducing the number of S-box circuits applied in parallel. In Fig. 16, the Toffoli-depth for sequentially implementing one $\mathfrak{C}^*$ S-box in KeyExpan$_1$ and one $\mathfrak{C}^*$ S-box in KeyExpan$_2$ is the same as the Toffoli-depth of an OP-based in-place S-box circuit. Hence, we see two sequential $\mathfrak{C}^*$ S-box as a whole circuit, and in the following call such circuit and the OP-based in-place S-box circuit, *double-depth S-box circuits*. In this case, the process of **Phase 1** and **Phase 2** in Fig. 16 implements 18 double-depth S-box circuits in parallel. Now suppose we implement p double-depth S-box circuits in parallel, where $p|18$.

- If $p = 9$, the Toffoli-depth is $82 \times 2 = 164$. In the 9 double-depth S-box circuits applied in the same layer, one of them is in KeyExpan and eight of them are in ByteSub, hence the width is $128 \times 2 + 6 + (8 + 6) \times 8 = 374$.
- If $18/p \geq 3$, the Toffoli-depth is $82 \times 18/p = 1476/p$. the widest part of such one round circuit is a phase in which all p double-depth S-box circuits are in ByteSub, and the width is $2 \times 128 + (8 + 6)p = 256 + 14p$.

Table 5 present the numbers of different quantum gates used in each component and one round. Obviously, these numbers are irrelevant to p.

Table 5. Quantum resources for implementing different components of AES

	KeyExpan	MixCol	AddRoundKey	ByteSub$_1$	ByteSub^{-1}	One round
#Toffoli	208	0	0	832	832	1872
#CNOT	1440	368	128	5216	6096	13248
# Pauli-X	48	0	0	64	128	240

Implement the Grover Oracle. If we want to construct a Grover oracle to search k_0, since the plaintext m is fixed and Round 0 is adding k_0 on m, we can apply Pauli-X gates on some specific ones of the wires carrying $|k_0\rangle$ to obtain $|k_0 \oplus m\rangle$, then when $|k_0\rangle$ is needed later, apply Pauli-X gates on these wires again to convert $|k_0 \oplus m\rangle$ back to $|k_0\rangle$. Therefore, we can use the circuit in Fig. 17 to implement Round 0 and Round 1 together.

Compare to Fig. 16, in this circuit, we do not need to implement $\texttt{ByteSub}^{-1}$, hence we can save $16 \times 52 = 832$ Toffoli gates. Note that, the a qubits in the third register, which will be used in the following rounds, is idle in these two rounds. Hence if $p \geq 9$, we have $a > 96$, then the 16 S-box circuits in $\texttt{ByteSub}_1$ can be implemented in parallel (need 96 ancilla qubits), Similarly, $\texttt{KeyExpan}$, which contains 4 S-box circuits, can be implemented in parallel. However, $\texttt{ByteSub}_1$ and $\texttt{KeyExpan}$ can not be implemented simultaneously. Therefore, if $p \geq 9$, the width and Toffoli-depth of these two rounds are 384 and 82 respectively. Moreover, these two rounds use $256+64+48 = 368$ Pauli-X gates, $5126+1440+368+128 = 7152$ CNOT gates and $208 + 832 = 1040$ Toffoli gates.

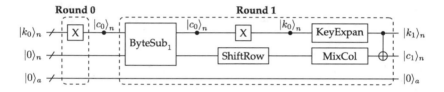

Fig. 17. The implementation of the round 0 and round 1 of AES

Then, by combining the circuits in Fig. 17 and Fig. 16, we can implement encryption circuit of the AES Grover oracle. In Table 6, we present the quantum resources needed for this circuit with $p = 18$ and $p = 9$, and compare our results with the results presented in [ZWS+20].

Table 6. Quantum resources for implementing AES-128

Width	Toffoli-Depth	#Toffoli	#CNOT	#Pauli-X	source
512	2016	19788	128517	4528	[ZWS+20]
492	820	17888	126016	2528	$p = 18$
374	1558	17888	126016	2528	$p = 9$

7.2 Low-Depth Quantum Circuits for AES

To reduce the depth, we should use the pipeline structure. First, we consider the nonlinear components. Since the round transformation is implemented out-of-place in the pipeline structure, for implementing $\texttt{ByteSub}$, we only need a low

T-depth $\mathfrak{C}^0$ S-box. For the AES key schedule, since it is the Feistel-like non-linear transformation, as shown in Sect. 7.1, it can be implemented in-place with the circuit in Fig. 14. It is easy to see that compared to other out-of-place implementations of the AES key schedule, the depth of such in-place implementation is also lower, since no extra operations are needed. As a consequence, we also in-place implement the AES key schedule in our shallower circuit of AES based on a $\mathfrak{C}^*$ S-box.

In Sect. 6.3, we presented two circuits of the AES S-box, which have T-depth 4 and T-depth 3 respectively. In these two circuits, the output wires are only be used as target wires. Therefore, these circuits are both $\mathfrak{C}^*$-circuits, and can be used in ByteSub and the key schedule.

For MixColoumns, we use the in-place circuit in Sect. 7.1, which has the optimal width, and the lowest CNOT-count until now. The Q# resource estimator shows that the depth of this in-place circuit is 30. In our resource estimation model, the CNOT-depth metric is less important than other metrics. For these reasons, we use this in-place circuit to implement MixColoumns.

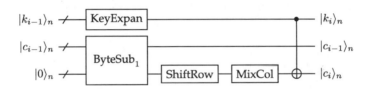

Fig. 18. The out-of-place implementation of the i-th round of AES-128

Figure 18 presents our implementation of the i-th round. In KeyExpan and ByteSub$_1$, 20 S-box circuits are applied in parallel. The round 0 is implemented by applying 128 CNOT gates in parallel, which maps $|k_0\rangle|m\rangle$ to $|k_0\rangle|c_0\rangle$. Note that, we don't use the circuit in Fig. 17 to implement the round 0 and round 1 together. The reason is in Fig. 17, KeyExpan and ByteSub$_1$ cannot be applied in parallel, hence the T-depth and full depth will be higher.

We implemented our AES circuits by Q# based on the code proposed in [JNRV20], and our code of Mixcolumn and the S-box (https://github.com/hzy-cas/AES-quantum-circuit). Table 7 shows the quantum resources of our circuits based on different S-box implementations. Same as in [JNRV20], the results presented here are the quantum resources required for implementing the forward circuit, which outputs the ciphertext, and the reverse circuit, which is used for uncomputation. As in Table 4, except the width[3], other values are obtain from Q# resource estimator. We can see that, similarly as the results of the S-box circuits, the T-depths and the full depths of our circuits are all lower than those in [JNRV20].

---

[3] In the full version [HSb], we show how to obtain these values of widths.

Table 7. Quantum resources for implementing AES and AES†

#CNOT	#1qClifford	#T	#M	T-depth	Full depth	width	source
291150	83116	54400	13600	**120**	**2827**	3936	[JNRV20]
298720	83295	54400	13600	**80**	**2198**	3936	with T-depth-4 S-box
570785	189026	124800	31200	**60**	**2312**	5576	with T-depth-3 S-box

8 General Width and *T*-depth Trade-offs

By combing different S-box circuits with different structures, and adjusting the number of S-box circuits applied in parallel, we can have a spectrum of trade-offs between width and T-depth.

Since we consider the T-depth, the Toffoli gates in the out-of-place S-box circuits used in Sect. 7.1 should be decomposed into Clifford+T gates. Note that, we cannot replace these Toffoli gates with quantum AND gates, since the output wires of the multiplication operations are not initialized to $|0\rangle$. Here, we use the T-depth-1 Toffoli gate proposed in [Sel12], where 4 ancilla qubits are required. In these S-box circuits, we apply at most two Toffoli gates in parallel, hence we need 8 extra ancilla qubits. In all, we have a Clifford+T implementation of the S-box (or the S-box^{-1}) with $8 + 6 = 14$ ancilla qubits and T-depth 41. We name these S-box circuits as Circuit 0. Moreover, to use the T-depth-4 (or the T-depth-3) S-box circuit in the round-in-place structure, we need to construct a $\mathfrak{C}^0$-circuit of the S-box^{-1}. Obviously, we can construct such circuit with T-depth-4 (or T-depth-3), since the nonlinear parts in the classical circuits of the S-box and S-box^{-1} are the same. We name these T-depth-4 circuits as Circuit 1, and these T-depth-3 circuits as Circuit 2.

We obtain the trade-off curve shown in Fig. 19 by applying Circuit 0, Circuit 1, and Circuit 2 in different structures. In this figure, Strategy 1, 2, 3, respectively correspond to the use of Circuit 0, Circuit 1, Circuit 3 in the round-in-place structure. Strategy 4, 5, 6 respectively correspond to the use of Circuit 0, Circuit 1, Circuit 3 in the pipeline structure. Different points on a curve are obtained by applying different number of S-box circuits in parallel. We also list the results of previous works [ZWS+20, JNRV20, GLRS16, LPS19] in this figure. For [ZWS+20], since the T-depth is not presented, here we decompose the Toffoli gate by the same Clifford+T gates as in Circuit 0, hence slightly increase the width. For the point corresponding to [JNRV20], the width is fixed as we mentioned. The detailed process for estimating these T-depths and widths is presented in the full version of this paper [HSb].

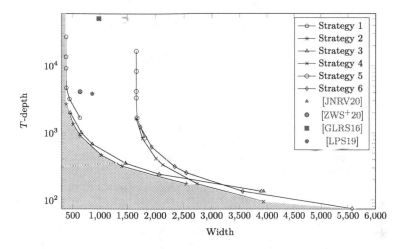

Fig. 19. The width and T-depth for implementing the Grover oracle of AES-128

9 Conclusion and Discussion

We propose the round-in-place structure for the quantum circuits of iterative ciphers, and manage to find a generic way to efficiently realize this structure in practice. We give guidelines in how to synthesize quantum circuits with specific optimization objectives based on a detailed analysis of the pipeline, zig-zag, and round-in-place structures. Moreover, new techniques for implementing the quantum circuits for linear and non-linear building blocks are presented. In particular, based on a new observation on the classical circuit of the AES S-box, we obtain a quantum circuit for the AES S-box with T-depth 3, reaching its theoretical minimum. Based on these techniques and results, we produce significantly improved quantum circuits for AES with respect to both depth and width. Finally, we conjecture that without optimizing across the natural hierarchical boundaries formed by the round functions of AES, the T-depth of the quantum circuit cannot be further improved.

Acknowledgments. This work is supported by the National Key Research and Development Program of China (2022YFB2700014), the National Natural Science Foundation of China (Grants No. 61977060, 62032014), the Fundamental Research Funds for the Central Universities.

References

[AMG+16] Amy, M., Di Matteo, O., Gheorghiu, V., Mosca, M., Parent, A., Schanck, J.: Estimating the cost of generic quantum pre-image attacks on SHA-2 and SHA-3. In: Avanzi, R., Heys, H. (eds.) SAC 2016. LNCS, vol. 10532, pp. 317–337. Springer, Cham (2017). https://doi.org/10.1007/978-3-319-69453-5_18

[AMMR13] Amy, M., Maslov, D., Mosca, M., Roetteler, M.: A meet-in-the-middle algorithm for fast synthesis of depth-optimal quantum circuits. IEEE Trans. Comput. Aided Des. Integr. Circuits Syst. **32**(6), 818–830 (2013)

[ASAM18] Almazrooie, M., Samsudin, A., Abdullah, R., Mutter, K.N.: Quantum reversible circuit of AES-128. Quantum Inf. Process. **17**(5), 1–30 (2018). https://doi.org/10.1007/s11128-018-1864-3

[BBvHL21] Banegas, G., Bernstein, D.J., van Hoof, I., Lange, T.: Concrete quantum cryptanalysis of binary elliptic curves. IACR Trans. Cryptogr. Hardw. Embed. Syst. **2021**(1), 451–472 (2021)

[BHN+19] Bonnetain, X., Hosoyamada, A., Naya-Plasencia, M., Sasaki, Yu., Schrottenloher, A.: Quantum attacks without superposition queries: the offline Simon's algorithm. In: Galbraith, S.D., Moriai, S. (eds.) ASIACRYPT 2019. LNCS, vol. 11921, pp. 552–583. Springer, Cham (2019). https://doi.org/10.1007/978-3-030-34578-5_20

[BLNS21] Bonnetain, X., Leurent, G., Naya-Plasencia, M., Schrottenloher, A.: Quantum linearization attacks. In: Tibouchi, M., Wang, H. (eds.) ASIACRYPT 2021. LNCS, vol. 13090, pp. 422–452. Springer, Cham (2021). https://doi.org/10.1007/978-3-030-92062-3_15

[BNS19a] Bonnetain, X., Naya-Plasencia, M., Schrottenloher, A.: On quantum slide attacks. In: Paterson, K.G., Stebila, D. (eds.) SAC 2019. LNCS, vol. 11959, pp. 492–519. Springer, Cham (2020). https://doi.org/10.1007/978-3-030-38471-5_20

[BNS19b] Bonnetain, X., Naya-Plasencia, M., Schrottenloher, A.: Quantum security analysis of AES. IACR Trans. Symmetric Cryptol. **2019**(2), 55–93 (2019)

[BP12] Boyar, J., Peralta, R.: A Small Depth-16 Circuit for the AES S-Box. In: Gritzalis, D., Furnell, S., Theoharidou, M. (eds.) SEC 2012. IAICT, vol. 376, pp. 287–298. Springer, Heidelberg (2012). https://doi.org/10.1007/978-3-642-30436-1_24

[CNS17] Chailloux, A., Naya-Plasencia, M., Schrottenloher, A.: An efficient quantum collision search algorithm and implications on symmetric cryptography. In: Takagi, T., Peyrin, T. (eds.) ASIACRYPT 2017. LNCS, vol. 10625, pp. 211–240. Springer, Cham (2017). https://doi.org/10.1007/978-3-319-70697-9_8

[Fow12] Fowler, A.G.: Time-optimal quantum computation. arXiv preprint arXiv:1210.4626 (2012)

[FS10] Fuhs, C., Schneider-Kamp, P.: Synthesizing shortest linear straight-line programs over GF(2) using SAT. In: Strichman, O., Szeider, S. (eds.) SAT 2010. LNCS, vol. 6175, pp. 71–84. Springer, Heidelberg (2010). https://doi.org/10.1007/978-3-642-14186-7_8

[GLRS16] Grassl, M., Langenberg, B., Roetteler, M., Steinwandt, R.: Applying Grover's algorithm to AES: quantum resource estimates. In: Takagi, T. (ed.) PQCrypto 2016. LNCS, vol. 9606, pp. 29–43. Springer, Cham (2016). https://doi.org/10.1007/978-3-319-29360-8_3

[Gro96] Grover, L.K.: A fast quantum mechanical algorithm for database search. In: Miller, G.L. (ed.) Proceedings of the Twenty-Eighth Annual ACM Symposium on the Theory of Computing, pp. 212–219. ACM (1996)

[HB20] Harsha, B., Blocki, J.: An economic model for quantum key-recovery attacks against ideal ciphers. In: 20th Annual Workshop on the Economics of Information Security, Brussels, 14–15 December 2020

[HSa]   Hosoyamada, A., Sasaki, Yu.: Finding hash collisions with quantum computers by using differential trails with smaller probability than birthday bound. In: Canteaut, A., Ishai, Y. (eds.) EUROCRYPT 2020. LNCS, vol. 12106, pp. 249–279. Springer, Cham (2020). https://doi.org/10.1007/978-3-030-45724-2_9

[HSb]   Huang, Z., Sun, S.: Synthesizing quantum circuits of AES with lower t-depth and less qubits. https://eprint.iacr.org/2022/620

[HS18a] Hosoyamada, A., Sasaki, Y.: Cryptanalysis against symmetric-key schemes with online classical queries and offline quantum computations. In: CT-RSA 2018, Proceedings, pp. 198–218 (2018)

[HS18b] Hosoyamada, A., Sasaki, Yu.: Quantum Demiric-Selçuk meet-in-the-middle attacks: applications to 6-round generic feistel constructions. In: Catalano, D., De Prisco, R. (eds.) SCN 2018. LNCS, vol. 11035, pp. 386–403. Springer, Cham (2018). https://doi.org/10.1007/978-3-319-98113-0_21

[IBM]   IBM QiskitL Open-source quantum development. https://qiskit.org/

[JNRV20] Jaques, S., Naehrig, M., Roetteler, M., Virdia, F.: Implementing Grover oracles for quantum key search on AES and LowMC. In: Canteaut, A., Ishai, Y. (eds.) EUROCRYPT 2020. LNCS, vol. 12106, pp. 280–310. Springer, Cham (2020). https://doi.org/10.1007/978-3-030-45724-2_10

[KLLN16a] Kaplan, M., Leurent, G., Leverrier, A., Naya-Plasencia, M.: Breaking symmetric cryptosystems using quantum period finding. In: Robshaw, M., Katz, J. (eds.) CRYPTO 2016. LNCS, vol. 9815, pp. 207–237. Springer, Heidelberg (2016). https://doi.org/10.1007/978-3-662-53008-5_8

[KLLN16b] Kaplan, M., Leurent, G., Leverrier, A., Naya-Plasencia, M.: Quantum differential and linear cryptanalysis. IACR Trans. Symmetric Cryptol. **2016**(1), 71–94 (2016)

[LPS19] Langenberg, B., Pham, H., Steinwandt, R.: Reducing the cost of implementing AES as a quantum circuit. IACR Cryptology ePrint Archive, p. 854 (2019)

[Mic]   Microsoftt Q#. Quantum development. https://devblogs.microsoft.com/qsharp/

[MSM18] Meuli, G., Soeken, M., De Micheli, G.: Sat-based CNOT, T quantum circuit synthesis. In: Reversible Computation, RC 2018, Leicester, UK, pp. 175–188 (2018)

[NC16] Nielsen, M.A., Chuang, I.L.: Quantum Computation and Quantum Information. Cambridge University Press, Cambridge (2016)

[NIS16] NIST: Submission requirements and evaluation criteria for the post-quantum cryptography standardization process (2016). https://csrc.nist.gov/projects/post-quantum-cryptography

[NS]   Naya-Plasencia, M., Schrottenloher, A.: Optimal merging in quantum k-xor and k-sum algorithms. In: Canteaut, A., Ishai, Y. (eds.) EUROCRYPT 2020. LNCS, vol. 12106, pp. 311–340. Springer, Cham (2020). https://doi.org/10.1007/978-3-030-45724-2_11

[PMH08] Patel, K.N., Markov, I.L., Hayes, J.P.: Optimal synthesis of linear reversible circuits. Quantum Inf. Comput. **8**(3), 282–294 (2008)

[Sel12] Selinger, P.: Quantum circuits of t-depth one. CoRR, abs/1210.0974 (2012)

[Sho99] Shor, P.W.: Polynomial-time algorithms for prime factorization and discrete logarithms on a quantum computer. SIAM Rev. **41**(2), 303–332 (1999)

[SPMH03] Shende, V.V., Prasad, A.K., Markov, I.L., Hayes, J.P.: Synthesis of reversible logic circuits. IEEE Trans. Comput. Aided Des. Integr. Circuits Syst. **22**(6), 710–722 (2003)

[Sto16] Stoffelen, K.: Optimizing S-box implementations for several criteria using SAT solvers. In: Peyrin, T. (ed.) FSE 2016. LNCS, vol. 9783, pp. 140–160. Springer, Heidelberg (2016). https://doi.org/10.1007/978-3-662-52993-5_8

[XZL+20] Xiang, Z., Zeng, X., Lin, D., Bao, Z., Zhang, S.: Optimizing implementations of linear layers. IACR Trans. Symmetric Cryptol. **2020**(2), 120–145 (2020)

[ZWS+20] Zou, J., Wei, Z., Sun, S., Liu, X., Wu, W.: Quantum circuit implementations of AES with fewer qubits. In: Moriai, S., Wang, H. (eds.) ASIACRYPT 2020. LNCS, vol. 12492, pp. 697–726. Springer, Cham (2020). https://doi.org/10.1007/978-3-030-64834-3_24

Exploring SAT for Cryptanalysis: (Quantum) Collision Attacks Against 6-Round SHA-3

Jian Guo[1]($\boxtimes$) (ID), Guozhen Liu[1] (ID), Ling Song[2] (ID), and Yi Tu[1] (ID)

[1] School of Physical and Mathematical Sciences, Nanyang Technological University, Singapore, Singapore
{guojian,guozhen.liu}@ntu.edu.sg, tuyi0002@e.ntu.edu.sg
[2] College of Cyber Security, Jinan University, Guangzhou, China

Abstract. In this work, we focus on collision attacks against instances of SHA-3 hash family in both classical and quantum settings. Since the 5-round collision attacks on SHA3-256 and other variants proposed by Guo *et al.* at JoC 2020, no other essential progress has been published. With a thorough investigation, we identify that the challenges of extending such collision attacks on SHA-3 to more rounds lie in the inefficiency of differential trail search. To overcome this obstacle, we develop a SAT-based automatic search toolkit. The tool is used in multiple intermediate steps of the collision attacks and exhibits surprisingly high efficiency in differential trail search and other optimization problems encountered in the process. As a result, we present the first 6-round classical collision attack on SHAKE128 with time complexity $2^{123.5}$, which also forms a quantum collision attack with quantum time $2^{67.25}/\sqrt{S}$, and the first 6-round quantum collision attack on SHA3-224 and SHA3-256 with quantum time $2^{97.75}/\sqrt{S}$ and $2^{104.25}/\sqrt{S}$, where S represents the hardware resources of the quantum computer. The fact that classical collision attacks do not apply to 6-round SHA3-224 and SHA3-256 shows the higher coverage of quantum collision attacks, which is consistent with that on SHA-2 observed by Hosoyamada and Sasaki at CRYPTO 2021.

Keywords: SHA-3 · SAT-based automatic search tool · Collision attacks · Quantum cryptanalysis

1 Introduction

The KECCAK hash function [BDPVA13], designed by Bertoni *et al.* in 2008, was standardized as the Secure Hash Algorithm-3 (SHA-3) [Dwo15] in 2015 by the National Institute of Standards and Technology (NIST) of the U.S. The SHA-3 family has four instances with fixed digest lengths, namely, SHA3-224, SHA3-256, SHA3-384 and SHA3-512, and two eXtendable-Output Functions (XOFs) SHAKE128 and SHAKE256. Being one of the most important cryptographic hash functions, SHA-3 (KECCAK) has received intensive security analysis. The most relevant security criteria for cryptographic hash functions include

More details are available in the full version of this paper: https://ia.cr/2022/184.

S. Agrawal and D. Lin (Eds.): ASIACRYPT 2022, LNCS 13793, pp. 645–674, 2022.
https://doi.org/10.1007/978-3-031-22969-5_22

preimage resistance and collision resistance. Preimage attacks of SHA-3 were investigated in [NRM11, MPS13, GLS16, LSLW17, LS19, Raj19, LHY21, HLY21]. The best-known practical attacks reach 3 rounds of SHAKE128 and SHA3-224 [GLS16, LS19][1] while the best-known theoretical ones can reach 4 rounds of all its instances [MPS13, Raj19, HLY21]. With marginal time complexity gains over bruteforce, theoretical preimage attacks cover up to 7/8/9 rounds for KECCAK-224/256/512, respectively [CKMS14, Ber10, MPS13].

More relevant to this research are the collision attacks on SHA-3 (KECCAK) with reduced number of rounds. In [DDS12, DDS14], Dinur et al. presented practical collision attacks on 4 rounds of KECCAK-224 and KECCAK-256. The actual collisions were found by combining a 3-round differential trail and a 1-round connector (which connects the differential trail to valid initial values). The same authors also presented practical collision attacks on 3-round KEC-CAK-384/KECCAK-512, and theoretical collision attacks on 5/4-round KECCAK-256/KECCAK-384 using internal differentials [DDS13]. Following the framework proposed by Dinur et al. in [DDS12], Qiao et al. introduced 2-round connectors by prepending a fully linearized round to the 1-round connectors and obtained actual collisions for 5-round SHAKE128 [QSLG17]. Further, these connectors were improved in [SLG17, GLL+20] to consume fewer degrees of freedom by using partial linearization. Consequently, 3-round connectors became possible and practical collision attacks on 5-round SHA3-224 and SHA3-256 were obtained.

Collision Attack in Quantum Settings. In the previous works, collision attacks of SHA-3 were studied only in classical settings. Recently, quantum collision attacks are attracting more attention and showing unexpected efficiencies.

The generic security margin of collision attacks in quantum settings has been investigated with the recent progress in post-quantum security of cryptographic schemes and primitives. Several quantum collision algorithms [BHT98, CNPS17] were introduced to provide security bounds for generic hash functions. However, the quantum collision attack against concrete hash functions was not published until 2020 [HS20]. In this work, Hosoyamada and Sasaki demonstrated that differential trails of low probability that couldn't be utilized in classical collision attacks were exploited to mount quantum collision attacks of more rounds. Later, the authors extended their quantum collision search algorithms to other hash functions and proposed the first quantum collision attacks on SHA-2 at CRYPTO 2021 [HS21]. Additionally, results of quantum rebound attacks on AES hashing modes [DSS+20] and quantum multi-collision distinguishers [BGLP] on dedicated hash functions were also presented.

Challenges. There are two major challenges in mounting quantum collision attacks on SHA-3. The first is to search for differential trails that are more suitable for quantum collision attacks, *i.e.*, trails that cover as many rounds as possible with the bound on the probability relaxed to 2^{-n}. As a consequence, the search space expands drastically which calls for more advanced and efficient

---

[1] The preimage attack on 3-round KECCAK-256 in [LHY21] has a time complexity 2^{65}, but no concrete preimage is given.

searching techniques. The second challenge lies in connecting the differential trail with the initial state. When differential trails with lower probability are used, more conditions are imposed on the internal state which should be satisfied by the connector. Thus, to avoid being the bottle neck of the whole attack, connectors must be constructed in more efficient way than before.

SAT-Based Cryptanalysis. Great attention from the cryptography community has been paid on automatic tools for linear and differential trail search. Normally, mathematical problems such as Boolean Satisfiability Problem (SAT), Mixed Integer Linear Programming (MILP), Satisfiability Modulo Theories (SMT), and other related methods are employed to construct such automatic tools. Since the performance of automatic search is determined by the power of the corresponding mathematical solvers, the efficiency is not particularly satisfactory when cryptographic ciphers with large state sizes are analyzed. Practically, most of the previous related works focus on lightweight ciphers where the automatic tools showed incredible strength.

The SAT problem decides whether a set of constraints could be satisfied by giving valid assignments to variables. In the research line of SAT-based cryptanalysis, Mouha and Preneel searched differential trails of ARX ciphers with SAT method in [MP13]. Based on SAT, Sun *et al.* [SWW18] put forward an automatic search method for ciphers with Sboxes to obtain differential trails of more accurate as well as high probability. In [SWW21], Sun *et al.* proposed a new encoding method to convert the Matsui's bounding conditions into Boolean formulas, which could reduce clauses and speed up the SAT solving phase. Besides, Morawiecki and Srebrny presented preimage attack on 3-round KECCAK hash functions by developing a SAT toolkit [MS13].

Our Contributions. Inspired by Hosoyamada and Sasaki's findings from [HS20, HS21] that collision attacks in quantum settings can take advantage of differential trails of low probability, we develop an automatic trail search toolkit based on SAT and propose advanced collision attacks on SHA-3 in both classical and quantum settings. The results of our work and the comparison with previous works are listed in Table 1. Main contributions are summarized in the following.

1. **The SAT-based automatic trail search toolkit** To facilitate differential trail search of the underlying permutation KECCAK-f of SHA-3, an SAT-based automatic search toolkit is developed. The toolkit is not only simple to implement but also provides more flexibility and better efficiency in generating various differential trails compared to dedicated trail search strategies in [DVA12, MDA17, LQT19]. It's interesting to note that for cryptographic primitives of large state size like KECCAK-f, automatic tools such as the MILP-based ones are unlikely to provide advantage in trail search. That's why specialized search techniques were proposed for SHA-3. Surprisingly, the SAT-based automatic toolkit fills the vacancy and shows excellent performance in trail search of the large-state KECCAK-f.

Table 1. Summary of collision attacks against the SHA-3 family

Target	Type	Rounds	Time complexity	Reference
SHA3-224	Classical	5	Practical	[GLL+20]
	Quantum	**6**	$2^{97.75}/\sqrt{S}$	Sect. 4.4
SHA3-256	Classical	5	Practical	[GLL+20]
	Quantum	**6**	$2^{104.25}/\sqrt{S}$	Sect. 4.3
SHA3-384	Classical	4	2^{147}	[DDS13]
SHA3-512	Classical	3	Practical	
SHAKE128	Classical	5	Practical	[GLL+20]
	Classical	**6**	$2^{123.5}$	Sect. 4.2
	Quantum	**6**	$2^{67.25}/\sqrt{S}$	
SHAKE256	-	-	-	-

2. **Advanced collision attack algorithms for** SHA-3 Augmented with the SAT-based automatic tool, the collision attack methods used in [DDS12, DDS14, QSLG17, SLG17, GLL+20] are improved in multiple ways. Collision attacks proposed in those works primarily consist of two phases, i.e., a phase of differential trail search that ensures collision on the digest bits, also referred to as the *colliding trail* search phase in our work, and a second phase of constructing "connectors" that generates message pairs satisfying the constraints imposed by the padding rule and initial value of SHA-3 and the input difference of the colliding trail at the same time. Both phases are considerably improved utilizing our automatic tool.

   – Colliding trail search algorithms that generate colliding trails of *any rounds*, *any digest length*, and *high probability* are presented. In other words, search space of colliding trails is covered efficiently which has been impossible in previous works.
   – Improved connector construction algorithms are proposed. Differential trails of the connectors (which are called *connecting differential trails* in the rest of the paper) can not only be directly generated but also produce sufficient degrees of freedom which has been the bottleneck in extending the collision attacks to more rounds.

3. **The first 6-round collision attacks on** SHA-3 With the novel automatic tool and the improved algorithms, we finally extend the 5-round collision attacks on SHA-3 instances to 6-round. In detail, 6-round classical collision attacks on SHAKE128 with complexity $2^{123.5}$, 6-round quantum collision attacks on SHA3-224 and SHA3-256 with complexity $2^{97.75}/\sqrt{S}$ and $2^{104.25}/\sqrt{S}$ respectively, are mounted. To the best of our knowledge, this is the first time that quantum collision attacks are mounted on SHA-3 and one more round is covered compared with previous results in classical setting.

Organization. The rest of the paper is organized as follows. In Sect. 2, an overview of the SAT-aided collision attacks on SHA-3 instances is provided. In

Sect. 3, specifications of SHA-3 hash functions and implementations of the SAT-based automatic search toolkit are presented. Section 4 exhibits the first 6-round collision attacks on SHA-3 in both classical and quantum settings. Section 5 concludes the paper. Details of differential trails and message pairs are given in the supplementary material.

2   Overview of SAT-Based Collision Attacks Against SHA-3

In this section, limitations of previous collision attacks are discussed. Subsequently, the SAT-based automatic trail search toolkit that can be conveniently applied to all kinds of cryptanalytic scenarios are introduced. Basic ideas used to extend previous collision attacks by one round in both classical and quantum settings are also presented.

2.1   Limitations of Previous Collision Attacks

As depicted in Fig. 1, the collision attacks on SHA-3 and KECCAK instances take a 3-stage analytic framework, $i.e.$,

- at stage 1, prepare n_{r_2}-round colliding trails of high probability that ensure d-bit digest collision. ΔS_I and ΔS_O stand for the input and output difference of colliding trails.
- at stage 2, construct n_{r_1}-round connectors that promise a subspace of message pairs which meet both the message difference $\Delta \overline{M}$ imposed by the sponge construction[2] and the input difference ΔS_I of the colliding trails.
- at the last stage, exhaustively enumerate the messages pairs generated with the connectors until one message pair that collides in digest bits is found.

A continuous series of investigations [DDS12, DDS14, QSLG17, SLG17, GLL+20] have been conducted on collision attacks against SHA-3. Both the colliding trail search phase and the connector construction phase have been intensively inspected. At first glance, it seems that there is no room for further improvements. Actually, no essential progress has ever been published since the last work [SLG17] presented five years ago. The lack of new results can be explained from two aspects, $i.e.$, the constrained and low-efficiency colliding trail search algorithms, and the quick consumption of degrees of freedom from the connectors by (full) linearization.

2.1.1   Difficulty in Generating Colliding Trails of More Rounds

Due to the huge state size of KECCAK-f, trail search of any kind, be it the general truncated differential trail or the colliding trail, is a difficult task. In previous collision attacks, the strategy to search colliding trails is quite simple, $i.e.$,

---

[2] In this attack model, collision messages of 1-block are generated. The constraints imposed by the sponge construction include (1) the c-bit capacity, $i.e.$, c continuous "0" bits, and (2) 2-bit padding "11" which is concatenated with a "01" or "1111" string at the tail of the message block.

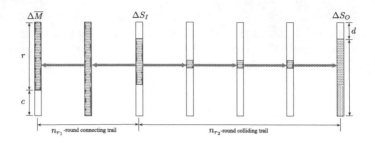

Fig. 1. Overview of $(n_{r_1}+n_{r_2})$-round collision attack on SHA-3

1. General 3-round differential trails obtained from dedicated search algorithms, *e.g.*, [DVA12, MDA17, LQT19], are extended forward by one round and exhaustively searched for possible d-bit collision.
2. When sufficient 3-round trails with digest collision are collected, extend them backward by one round to determine satisfactory output differences for connectors, which at the same time are the input differences ΔS_I of the n_{r_2}-round colliding trails.

There are two problems regarding to this colliding trail search strategy. On one hand, the exhaustive colliding trail search, especially the backward extension, drain computing resources significantly. In practice, sophisticated implementation techniques and even GPU resources [SLG17] are introduced to speed up the colliding trail search. However, without dramatical increase in computing power, it's unlikely that the search efficiency can be improved further. On the other hand, the colliding trails are limited by the results of general truncated differential trail search. For example, the 5-round practical collision attack on SHA3-256 [GLL+20] is not possible until new results on general 3-round trails [LQT19] are published. Particularly, even with ultimate computing power, better colliding trails won't be possible unless results of general trail search are updated. Then it comes to the common trail search problem again which is a challenging task.

2.1.2   Quick Consumption of Degrees of Freedom in Connector Construction

The connector construction is comprised of two parts. In the first part, as depicted in Fig. 1, connecting trails whose input difference (*i.e.*, $\Delta \overline{M}$) and output difference (*i.e.*, ΔS_I) are partially or fully fixed are constructed. In the second part, data structures that output a subspace of message pairs following the connecting trail are generated. Essentially, as long as the connecting trail is determined, systems of equations (*i.e.*, the data structures) on messages are listed in which the *degree of freedom* (DF for short) are quickly consumed[3]. As

---

[3] The practical algorithms are much more complex. We just describe in this abstract way to express basic ideas.

conditions of both $\Delta \overline{M}$ and ΔS_I are strict, the sophisticated Target Difference Algorithms (TDA for short) are devised to determine the exact connecting trails. When we try to extend the connector by one more round, $\Delta \overline{M}$ and ΔS_I are so heavy that connecting trails are hard to generate. Even if the TDA generates connecting trails, data structures become impossible to construct as almost all DF is consumed to meet conditions of the heavy connecting trail. Therefore, developing new connecting trail search methods to generate lighter connecting trails would be a feasible way to save DF and possibly allow to extend collision attacks for more rounds.

Summary. Limitations of collision attacks lie in inefficiency of differential trail search, more specifically, the lack of effective search techniques for trails of special requirements.

2.2   SAT-based Automatic Trail Search Toolkit

Automatic search has long been introduced to evaluate robustness of symmetric primitives. However, it's not the case of KECCAK-f permutation. Indeed, the initial attempts with MILP method failed to generate good trails due to the large KECCAK-f state. Researchers have to develop dedicated techniques to investigate the propagation properties of KECCAK-f. On the other hand, automatic search based on other mathematical problems, such as SAT and SMT, is not properly studied. In this work, SAT-based automatic search shows productivity in generating trails involved in collision attacks on SHA-3.

2.2.1   SAT-based Colliding Trail Search

With the SAT-based toolkit, differential trails that (1) satisfy the d-bit digest collision, (2) cover more rounds, (3) follow any specific differential pattern, and (4) meet any probability constraint can be effectively generated. The search space is expanded to the extent that efficiency of automatic search tool outperforms dedicated search strategy. Moreover, as the new method does not rely on truncated differential trails, colliding trail search will not be limited by progress of such general trails any more. Cryptanalysts are also free from devising and implementing sophisticated trail search algorithms. We emphasize that colliding trails of low probability, *e.g.*, with complexity near or even beyond the birthday bound, are easily generated. Such trails are utilized to mount collision attacks in quantum settings.

2.2.2   SAT-based Connecting Trail Search

Similar to the case of colliding trail search, the SAT-based connecting trail search is effortlessly implemented. Good connecting trails that (1) follow the fixed input and output differences of the connectors and (2) provide adequate DF for messages are generated. The idea of finding connecting trails with SAT gives insights to the constrained-input constrained-output (CICO) problem [BPVA+11] of sponge constructions. As the input and output differences of connecting trails are partially or fully fixed, this is generally a difficult problem. Except for the sophisticated approach used in [DDS12, DDS14, QSLG17, SLG17, GLL+20], there is no other progress on constructing connecting trails. The SAT-based connecting trail search method presents the first general solution for the problem of bypassing the constraints imposed by the sponge construction in collision attacks on SHA-3.

2.3   Improved (Quantum) Collision Attacks on SHA-3

With the SAT-based automatic tool, collision attacks on SHA-3 instances that cover one more round are mounted in both quantum and classical settings.

2.3.1   6-Round Collision Attacks on SHAKE128

With the SAT-based tool, 4-round colliding trails of 256-bit digest collision are generated. Although one round is extended compared to trails used in previous works, the 4-round colliding trails are of low probability. To mount valid collision attacks, one round of the colliding trails is merged into the connectors, *i.e.*, the 6-round collision attacks consist of a 3-round connecting trail and a 3-round colliding trail. Due to the low probability, the 3-round connectors can only be partially constructed, *i.e.*, only a fraction of the third round conditions are treated while the other constraints are left for the brute force stage. Ultimately, a theoretical 6-round collision attack on SHAKE128 are mounted with complexity $2^{123.5}$ which is slightly better than the generic attack.

2.3.2   6-Round Quantum Collision Attacks on SHA3-224 and SHA3-256

The identical 4-round colliding trail is used to mount 6-round collision attacks on SHA3-224 and SHA3-256. Constrained by the great amount of DF consumed, it becomes impossible to construct even partial 3-round connectors for these instances. Therefore, for SHA3-224 and SHA3-256, only 2-round connectors are feasible. 6-round collision attacks on SHA3-224 and SHA3-256 cannot be mounted in classical setting as complexity of the 4-round colliding trail exceeds the birthday bound. Fortunately, colliding trails of low complexity can be employed to mount quantum collision attacks. In a nutshell, 6-round quantum collision attacks on SHA3-224 and SHA3-256 with complexity $2^{97.75}/\sqrt{S}$ and $2^{104.25}/\sqrt{S}$ are presented.

3   SHA-3 and SAT-based Automatic Search Toolkit

In this section, we describe notations used in the collision attacks and specifications of the SHA-3 family hash functions. Afterwards, the SAT-based automatic search toolkit developed for KECCAK-f permutation is presented.

3.1   Notations

Most of the notations to be used in this paper are listed below.

c	Capacity of a sponge function
r	Rate of a sponge function
d	Length of the digest in bits
p	Number of fixed bits in the initial state due to padding
n_r	Number of rounds
KECCAK-f	The underlying permutation of SHA-3 hahs functions
$\theta, \rho, \pi, \chi, \iota$	The five operations of the round function of KECCAK-f. A subscript i denotes the operation at the i-th round, e.g., χ_i denotes the χ layer at the i-th round where $i = 0, 1, 2, \cdots$
λ	Composition of θ, ρ, π and its inverse denoted by λ^{-1}
RC_i	Round constant of the i-th round, where $i = 0, 1, 2, \cdots$
$R^i(\cdot)$	KECCAK-f permutation reduced to the first i rounds
$S(\cdot)$	5-bit Sbox operating on each row of KECCAK-f state
$\delta_{in}, \delta_{out}$	5-bit input and output differences of an Sbox
DDT	Differential distribution table, and $\text{DDT}(\delta_{in}, \delta_{out}) = |\{x : S(x) + S(x + \delta_{in}) = \delta_{out}\}|$, where $|\cdot|$ denotes the size of a set
α_i	Input difference of the i-th round, where $i = 0, 1, 2, \cdots$
β_i	Input difference of χ in the i-th round, where $i = 0, 1, 2, \cdots$
w_i	Propagation weight (weight for short) of the i-th round
$w(\beta_i)$	Weight of β_i, where β_i is the input difference of χ
$w^{rev}(\alpha_i)$	Minimal reverse weight of α_i
DF	Degree of freedom of the solution space of connectors
$\overline{M}$	Padded message of M. Note that $\overline{M}$ is of one block in our attacks
$M_1 \| M_2$	Concatenation of strings M_1 and M_2
x_i	Bit value vector before λ of each round, where $i = 0, 1, 2, \cdots$
y_i	Bit value vector before χ of each round, where $i = 0, 1, 2, \cdots$
E_{y_i}	System of equations on y_i of each round, where $i = 0, 1, 2, \cdots$

3.2   Description of SHA-3 Family

The SHA-3 family [Dwo15] consists of a subset of KECCAK [BDPVA13] hash functions that are built upon the sponge construction [BDPVA07, GJMG11] with an internal permutation called KECCAK-f.

3.2.1   Specification of KECCAK-f Permutation

The underlying permutation KECCAK-f takes a large state size of 1600 bits and there are 24 iterative rounds in total. Each round of KECCAK-f is comprised of five operations, namely, the four linear operations denoted by θ, ρ, π and ι, and one solely nonlinear operation denoted by χ. The 1600-bit state is organized as a 3-dimensional array of bits, i.e., $5 \times 5 \times 64$, denoted with $A[5][5][64]$. Each of the state bits indexed by the coordinate (i, j, k) in the state array is denoted by $A[i][j][k]$ where $0 \le i, j < 5$, and $0 \le k < 64$. The 5 step mappings of the KECCAK-f round are specified with the following transformations.

$$\theta: \quad A[i][j][k] \leftarrow A[i][j][k] \oplus \sum_{j'=0}^{4} A[i-1][j'][k] \oplus \sum_{j'=0}^{4} A[i+1][j'][k-1].$$

ρ: $A[i][j] \leftarrow A[i][j] \lll T(i,j)$, where $T(i,j)$s are constants.

π: $A[j][2i+3j] \leftarrow A[i][j]$.

χ: $A[i][j][k] \leftarrow A[i][j][k] \oplus (A[i+1][j][k] \oplus 1) \cdot A[i+2][j][k]$.

ι: $A[0][0] \leftarrow A[0][0] \oplus RC_{i_r}$, where RC_{i_r} is the i_r-th round constant.

The multiplication used in χ operation is in $\mathsf{GF}(2)$. As ι won't affect differences, we ignore it in the rest of the paper unless otherwise stated.

3.2.2   Instances of SHA-3 Family

According to the bit length of digest, SHA-3 contains 6 instances, *i.e.*, the four variants SHA3-224/256/384/512 that have a fixed hash length (where the numbers 224/256/384/512 stand for the hash size) and the two variants SHAKE128 and SHAKE256 of extendable outputs. A multirate padding rule 10*1 is defined for all SHA-3 instances. For the four standardized instances SHA3-224/256/384/512, a 2-bit string "01" is concatenated to the message before padded while the capacity is specified as $c = 2 \times d$. In regards to the two extendable variants, a 4-bit string "1111" is concatenated to the messages, and the capacity is 256 and 512 bits for SHAKE128 and SHAKE256 respectively. The digest size d of SHAKE128 and SHAKE256 can vary, and therefore the collision resistance level is given by $\min(d/2, 128)$ and $\min(d/2, 256)$ correspondingly.

3.3   SAT Implementation

In the following, the SAT solver, the descriptions of the KECCAK-f permutation and its differential propagation, and the objective functions are illustrated.

CryptoMiniSAT. We choose CryptoMiniSAT as the underlying solver to implement our automatic toolkit. Since proposed in [SNC09], the conflict-driven clause-learning(CLDL) SAT solver has been improved greatly [SNC10, Soo14, Soo16, SBH+19, SDG+20, SSK+20]. Enhanced with a sequence of advanced search strategies such as Gauss-Jordan elimination and target phases [QUE19], CryptoMiniSAT shows outstanding performances among other SAT solvers. Except for high performance, CryptoMiniSAT also provides a neat interface for XOR expressions. In fact, most well-performed SAT solvers only understand constraints in *conjunctive normal form* (CNF for short) and users must consider the complicated problem of describing cryptographic primitives with CNFs. By contrast, CryptoMiniSAT allows attackers concentrate on attacks while providing high performance and simple implementation.

To implement SAT-based automatic trail search method, two kinds of constraints are fed into CryptoMiniSAT, namely, conditions imposed by (1) differential propagation over round functions (or in other words the description of round functions), and (2) objective functions such as the number of active Sboxes and the propagation probability.

Round Function. As depicted in the following model,

$$\alpha_r \xrightarrow{\theta} c_r \xrightarrow{\pi \circ \rho} \beta_r \xrightarrow{\chi} \alpha_{r+1}$$

two state differences, *i.e.*, α_r (the input difference of the r-th round) and β_r (the input difference of the χ operation of the r-th round) are introduced to the SAT implementation for a single r-th round. The 1600-bit difference α_r is represented by 1600 variables, *i.e.*, variable of each bit (whose coordinate is $\alpha_r[i][j][k]$ where $0 \le i, j < 5$ and $0 \le k < 64$) is indexed with $(320 \times j + 64 \times i + k)_{\alpha_r}$. This way we establish the mapping relationship between the 1600 variables and the corresponding state difference.

Recall that ρ and π are simply bit permutations. Differential propagations over the two linear operations are described through mapping the indexes of variables. For example, assuming that an active bit $c_r[i][j][k]$ is transformed to $\beta_r[i'][j'][k']$ through $\pi \circ \rho$, then the index mapping of the two variables is $(320 \times j + 64 \times i + k)_{c_r} \xrightarrow{\pi \circ \rho} (320 \times (2 \times i + 3 \times j) + 64 \times j + (k - T(i,j))\%64)_{\beta_r}$. These operations are described with plain index transformation and no additional SAT computation is required.

By definition, θ operation updates each bit through XORing itself to two columns. Accordingly, θ is described with XOR clauses that could be directly understood by CryptoMiniSAT. That is, the XOR sums of 320 columns (denoted by $\alpha[i][k]$) are described with 320 variables each of which is indexed by $64 \times i + k$. As a result, the mapping of variable indexes induced by θ operation is captured with $(320 \times j + 64 \times i + k)_{c_r} = (320 \times j + 64 \times i + k)_{\alpha_r} \oplus (64 \times (i-1) \mid k)_{ColumnSum} \oplus (64 \times (i+1) + (k-1))_{ColumnSum}$. Here, the subscript $ColumnSum$ indicates the variables of column sums.

Practically, the three linear operations (*i.e.*, θ, ρ and π) are treated as a whole. The total index mapping of variables is described with $(320 \times (2 \times i + 3 \times j) + 64 \times j + (k - T(i,j))\%64)_{\beta_r} = (320 \times j + 64 \times i + k)_{\alpha_r} \oplus (64 \times (i-1) + k)_{ColumnSum} \oplus (64 \times (i+1) + (k-1))_{ColumnSum}$.

In regard to the only nonlinear operation χ which is generally considered as 5-bit Sbox, both the *difference distribution table* (DDT for short) and the operation itself are interpreted with truth tables. Specifically,

- The DDT is described with listing a truth table of 11 variables, including 10 variables that represent input and output difference and 1 variable marking compatibility of DDT entries. When fed into Logical Friday (refer to http://sontrack.com), 46 CNFs are generated to describe the DDT. Differential propagation over χ, *i.e.*, relationship between the input difference β_r and output difference α_{r+1}, is then depicted with simply writing CNFs of each Sbox.
- Similarly, variables that correspond to the input and output values of χ are connected with CNFs generated from χ truth table. Empirically, 11 variables are needed to construct truth tables and 29 CNFs are produced.

In summary, $1600 \times 2 + 320 = 3520$ variables are used to describe one round of KECCAK-f permutation in the SAT-based implementation. The relationship among variables are specified with methods illustrated above. Identical round description that is of different variable sets is implemented for each round. Multiple rounds are described by connecting each round, i.e., (1) the input variables of each round are the output variables of its previous round and (2) the output variables of each round are the input variables of its next round.

Objective Function. In the context of 6-round collision attacks on SHA-3, the number of active Sboxes and the propagation weight (*weight* for short)[4] are the two mainly considered objectives. To describe the objectives, constraints on integers (i.e., number of active Sboxes and weights) are transformed to CNFs. The sequential encoding method [Sin05] is employed to describe addition over integers, e.g., $\sum_{i=0}^{n-1} x_i \leq w$ where $w \geq 1$. In this process, $(n \times (w+1) - w)$ auxiliary variables are introduced. More specifically,

- *Constraint on the Number of Active Sbox.* To describe the number of active Sboxes of each χ, 320 variables are introduced to indicate whether an Sbox is active or not. The sum of all the variables needs to satisfy a threshold weight (say w), e.g., $\sum_{i=0}^{319} x_i \leq w$. Accordingly, $(320 \times (w+1) - w)$ extra variables are introduced to transform the constraint on the number of active Sboxes to CNFs.
- *Constraint on the Propagation Weight.* The DDT entries take 4 possible values (i.e., 2, 4, 8, and 32), and the corresponding propagation weights belong to $\{0, 2, 3, 4\}$. As shown in Eq. 1, four auxiliary variables denoted by (p_0, p_1, p_2, p_3) are introduced to represent the weight of each Sbox, meaning that $(320 \times 4 \times (w+1) - w)$ extra variables are added to describe constraints on the weight of a whole state. Likewise, the weight constraint which is obtained through summing up all the variables is then transformed to CNFs.

$$(p_0, p_1, p_2, p_3) = \begin{cases} (1,1,1,1), & \text{DDT}(\delta_{in}, \delta_{out}) = 2; \\ (0,1,1,1), & \text{DDT}(\delta_{in}, \delta_{out}) = 4; \\ (0,0,1,1), & \text{DDT}(\delta_{in}, \delta_{out}) = 8; \\ (0,0,0,0), & \text{DDT}(\delta_{in}, \delta_{out}) = 32. \end{cases} \tag{1}$$

3.4   SAT-based Automatic Search Toolkit

In this section, we explain how to implement various trail search algorithms based on the SAT implementation. Let's first review some definitions and concepts introduced in [DVA12, BPVA+11]. The 6-round attack model presented in Sect. 4.1.3 is placed here in advance to better explain definitions.

Probabilistic Property of χ. As the algebraic degree of χ is 2, its DDT shows some interesting properties. For a given input difference, all its compatible output differences share equal propagation probability. Correspondingly, for a given β_i, all its compatible α_{i+1} take the same probability or weight. For a

---

[4] The *propagation weight* is defined as the opposite of the binary logarithm of the propagation probability. For example, if the propagation probability of a differential trail is 2^{-32}, the corresponding weight is 32.

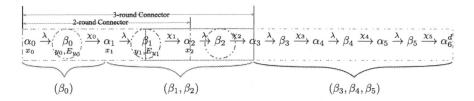

Fig. 2. The 6-round collision attack model

given output difference, as the degree of χ^{-1} is 3, there exist one or several compatible input differences that hold a better probability than the other input differences. Likewise, for a given α_i, there exist some compatible β_{i-1} that have the best differential probability, which is also called the *minimum reverse weight* (and generally denoted by $w^{rev}(\alpha_i)$).

Trail core. As depicted in Fig. 2, a general 4-round differential trail consists of input and output differences of all four rounds, *i.e.*, $(\alpha_2, \alpha_3, \alpha_4, \alpha_5, \alpha_6)$. Recall that as λ is a linear transformation, α_i propagates to β_i deterministically. The 4-round differential trail is also denoted by $(\beta_2, \beta_3, \beta_4, \beta_5)$. Comparatively, the 4-round ***trail core*** is composed of three differences, i.e., $(\beta_3, \beta_4, \beta_5)$, taking advantage of the property that the minimal reverse weight of α_3 can be directly computed to evaluate the family of 4-round trails that have $(\beta_3, \beta_4, \beta_5)$ as their tail.

In the Fig. 2 model, $(\beta_3, \beta_4, \beta_5)$ represents the colliding trail

3.4.1   SAT-based Colliding Trail Search

To set up the colliding trail search model, description of differential trail $(\alpha_3, \beta_3, \alpha_4, \beta_4, \alpha_5, \beta_5, \alpha_6^d)$ needs to be added into the SAT model. Differential propagation over the round functions is implemented in the way introduced in last section. At this stage, only constraints that are exclusively imposed by the colliding trails are introduced. Aligned with the requirements for constructing colliding trails in [GLL+20], the SAT-based search method is implemented from two aspects, *i.e.*, the digest collision and the connector construction.

From the perspective of collision search, we don't have to check α_6 for d-bit collision (denoted by α_6^d). Rather, extra constraints on β_5 that ensure α_6^d collision are considered. Take colliding trail search of SHAKE128 as an example, to guarantee the first 4 lanes of α_6 to be 0, the input difference to the first 64 Sboxes of β_5 must belong to the set {00000, 00001, 00101, 10101, 00011, 01011, 00111, 10111, 01111, 11111}. The candidate input differences listed above form a space which is represented by CNFs. Through adding the corresponding CNFs on variables of β_5 to the system, constraints on digest collision is implemented.

On the other hand, to maximally facilitate the connector, the *minimum reverse weight* of α_3 (denoted by $w^{rev}(\alpha_3)$) and propagation weight $w(\beta_3) + w(\beta_4) + w(\beta_5^d)$ of the colliding trail are taken into consideration. Altogether, the objective function of $w^{rev}(\alpha_3) + w(\beta_3) + w(\beta_4) + w(\beta_5^d)$ is described with CNFs and added to the system. In some situations, the constraints on weight are replaced by the constraints on the number of active Sboxes, *i.e.*, $AS(\alpha_3) + AS(\alpha_4) + AS(\beta_4) + AS(\beta_5^d)$ which results in $(320 \times 3 \times (w+1) - w) + (64 \times (w+1) - w)$ auxiliary variables included to the SAT system.

With this implementation, 3-round colliding trails are not only generated more efficiently but also of better probability. In contrast, the best 3-round colliding trail used in previous collision attack on SHA3-256 is of probability 2^{-43}. It's worth noticing that 4-round colliding trails which could be utilized to mount collision attacks of 6 rounds is generated for the first time. Table 2 gives comparison of search efficiency. It demonstrates that the new SAT-based trail search is superior to earlier strategies in both efficiency and effectiveness.

Table 2. Comparison of the SAT-based tools with other dedicated approaches

Type	Permutation	Rounds	Weight	Time	Reference
Colliding trail	$Keccak\text{-}f[1600]$	3	43	Several weeks[1]	[GLL+20]
		3	32	$2\,s^{[2]}$	Sect. 3.4.1
		4	141	$5\,mins^{[2]}$	Sect. 3.4.1
General trail	$Keccak\text{-}f[1600]$	4	134	-	[MDA17]
			133	47.76 h	Sect. 3.4.3
	$Keccak\text{-}f[800]$	4	104	-	[MDA17]
			95	28.42 h	Sect. 3.4.3

[a] There are two stages, *i.e.*, the forward extension executed with one CPU core and the backward extension deployed with three NVIDIA GeForce GTX970 GPUs.
[b] The SAT-based implementation is deployed with one 3.6 GHz Intel Core i9.

3.4.2   SAT-based Connecting Trail Search
In accordance with the considerations for constructing connecting trails that promise valid connectors, the trail search of $(\alpha_0,\beta_0,\alpha_1,\beta_1,\alpha_2,\beta_2)$ is specified with two phases.

Phase 1. In the first phase, (β_1, β_2) are to be determined for given α_3. First, description of the differential trail $(\beta_1,\alpha_2,\beta_2,\alpha_3)$ are added to the SAT system. Afterward, constraints on propagation weight of β_1 and β_2 are established, *i.e.*, CNFs of a minimal $w(\beta_1) + w(\beta_2)$ are listed. By now, $6400 + 320$ variables are used to describe the connecting trail where 6400 variables are introduced for the 2-round propagation and 320 variables correspond to conditions of the summed weight. And we also restrict weight of each round, namely, $w(\beta_1) \leq w_1$ and $w(\beta_2) \leq w_2$ which results in an extra $(1280 \times (w_1 + 1) - w_1) + (1280 \times (w_2 + 1) - w_2)$ variables. The objective function of weight is described with the method illustrated in the last section. Overall, this model needs $6400 + 320 + (1280 \times (w_1 + 1) - w_1) + (1280 \times (w_2 + 1) - w_2)$ variables.

Phase 2. The input difference of χ_0 of the first round is determined in this phase with the SAT-based implementation. Given the output difference α_1, variables that represent a pair of messages (x_0^1, x_0^2) and the input difference β_0 are introduced to describe the half round propagation. Precisely, constraints on bit positions of capacity and padding are depicted by fixing the corresponding variables to be 0 or some settled value. Constraints on $w(\beta_0)$, the weight of β_0, are also covered to make sure that the degree of freedom will be maximally produced for connectors. Simply put, CNFs for objective function of a minimal $w(\beta_0)$ are added to the SAT model. With the SAT-based implementation, connecting trails that yield much greater DF are generated.

3.4.3 SAT-based Truncated Trail Search

Except for the special trail search scenarios, SAT-based solution also performs well in general truncated differential trail search. As can be seen from the experimental results, SAT-based implementation handles 3-round KECCAK-f permutation quickly. It turns out that 3-round trail cores generated with the SAT-based automatic trail search method are consistent with results from previous works [DVA12, MDA17, LQT19].

We take 4-round differential trail search as an example to explain the SAT-based trail search implementation. The 4-round trail is modelled with

$$\beta_2 \xrightarrow{\chi} \alpha_3 \xrightarrow{\lambda} \beta_3 \xrightarrow{\chi} \alpha_4 \xrightarrow{\lambda} \beta_4 \xrightarrow{\chi} \alpha_5 \xrightarrow{\lambda} \beta_5 \xrightarrow{\chi} \alpha_6.$$

First, CNF description of the differential trail $(\alpha_3, \beta_3, \alpha_4, \beta_4, \alpha_5, \beta_5)$ is added to the SAT system. As 6 differences are involved, $10560 = 1600 \times 6 + 3 \times 320$ variables are required to describe the difference propagation. Similar to the colliding trail search implementation, constraint on the sum of weight $w = w^{rev}(\alpha_3) + w(\beta_3) + w(\beta_4) + w(\beta_5)$ where $w \leq 133$ is also added to the SAT system. Another $685947 = (1280 \times 4 \times (133 + 1) - 133)$ auxiliary variables are included in the process of transforming the objective function to CNFs. In total, there are 696507 variables in this SAT-based 4-round differential trail search implementation.

With respect to search efficiency, although it displays unexpectedly well performance in 3-round trail search, it cannot traverse the search space of 4-round trails efficiently. A tight lower bound on propagation weight for 4-round differential trails is unfortunately not settled in this paper. However, two better 4-round trails of weight 133 which is the lowest known weight so far are generated. Table 8 in supplementary material B shows the two trails (refer to the full version [GLST22]).

The SAT-based differential trail search implementation is further extended to other KECCAK permutations [BPVA+11] such as KECCAK-f[800]. Analogous to KECCAK-f (which is also denoted by KECCAK-f[1600]), similar round functions are iterated for multiple rounds in KECCAK-f[800] only that its state size is of 800 bits. Table 9 in supplementary material B shows a good trail that improves the lower bound of 4-round trails for KECCAK-f[800]. Table 2 gives an overview of the advantage of the automatic search compared to previous works.

Summary. By picking up different compositions of constraints on the number of active Sboxes and weight or even considering a single state not in the whole, we obtain variant SAT models with different efficiency. The SAT-based automatic search toolkit helps us understand the differential propagation property of KECCAK-f in a distinct viewpoint. It also demonstrates that automatic solvers perform efficiently on cryptographic primitives with large state size.

4   Collision Attacks Against SHA-3 Instances in Classical and Quantum Settings

In this section, a classical 6-round collision attack on SHAKE128, and two 6-round quantum collision attacks on SHA3-224/SHA3-256 are mounted.

4.1   Basic Attack Strategy

Aided by the SAT-based automatic search toolkit, we propose advanced collision attacks on SHA-3 instances based on the analytic framework described in Sect. 2. The enhanced collision attack is comprised of three phases, *i.e.*,

- Phase 1, generate n_{r_2}-round colliding trails of d-bit digest with the SAT-based tool.
- Phase 2, generate n_{r_1}-round connecting trails that link the conditions of sponge construction and the input difference of the colliding trail with the SAT-based tool.
- Phase 3, construct connectors that generate a subspace of messages which follow the n_{r_1}-round connecting trails.

The brute force phase where collision messages are generated will not be included as only theoretical collision attacks are presented in this work.

4.1.1   Generating Colliding Trails

Based on the SAT implementation techniques elaborated in Sect. 3, we add the implementation of colliding trail search algorithms to the toolkit. Except that the d-bit collision must be satisfied, the propagation weight of the 4-round colliding trail core must also be small enough to promise a possible 6-round collision attack. Eventually, several 4-round colliding trail cores are generated. We select the best one to mount collision attacks. Without considering the connector, weight of the 4-round colliding trail is 141 (*i.e.*, $89 + 24 + 20 + 8 = 141$). The propagation weight of the 4-round trail core is shown in Fig. 3 while the exact differences are listed in Trail No.1 (shown in Table 5) of supplementary material B.

$$\alpha_2 \xrightarrow{\lambda} \beta_2 \underset{\underset{w^{rev}=89}{}}{\xrightarrow{\chi_2}} \alpha_3 \xrightarrow{\lambda} \beta_3 \underset{\underset{w=24}{}}{\xrightarrow{\chi_3}} \alpha_4 \xrightarrow{\lambda} \beta_4 \underset{\underset{w=20}{}}{\xrightarrow{\chi_4}} \alpha_5 \xrightarrow{\lambda} \beta_5 \underset{\underset{w^d=8}{}}{\xrightarrow{\chi_5}} \alpha_6^d$$

Fig. 3. The 4-round colliding trail model. The 4-round trail is purposely placed at the last 4 rounds of a 6-round differential trail to be consistent with the collision attack model. In the last round, only d-bit collision is concerned and denoted by α_6^d.

4.1.2   Generating Connecting Trails

As shown in Fig. 3, even the minimal weight (*i.e.*, ≥ 141) of 4-round colliding trails exceeds the birthday bound (*e.g.*, 128 for SHAKE128 and SHA3-256). It's impractical to randomly select a 4-round colliding trail and generate the corresponding 2-round connecting trail. We develop a two-step approach to determine the connecting trails. The input difference of the 4-round colliding trail core is generated in combination with the differences of connecting trails. Let's explain the idea with the 6-round collision attack model shown in Fig. 4.

– In the first step, the input difference (*i.e.*, β_2) of the 4-round colliding trail core ($\beta_3, \beta_4, \beta_5$) is determined together with the input difference (*i.e.*, β_1) of the second round of the connecting trails. Practically, the 2-round differential trails (β_1, β_2) that are not only compatible with α_3, but also of minimal weight are generated with the SAT-based tool.
– In the second step, the lightest β_0 (in terms of weight) that are compatible with α_1 and meet the restrictions on α_0 imposed by the sponge construction are generated with the SAT-based tool.

To demonstrate the strength of the SAT-based method, we compare experimental results on SHA3-256 with previous work. In previous results, when the first round of the connector is processed, the DF remained is estimated to be around 124 (for more illustration refer to Sect. 5.2 of [GLL+20]). In comparison, the new connecting trails provide a DF up to $330 \sim 430$ which is surprisingly superior. This accords with the number of active Sboxes of β_0. Almost all of the 320 Sboxes of β_0 are active (*e.g.*, the number of nonactive Sboxes is around 10) with the previous *target difference algorithm*, while with our SAT-based strategy there are around $40 \sim 50$ nonactive Sboxes in β_0. Without the extra gain of DF, it's impossible to extend the attack by one round.

Remark 1. The three undetermined differences β_0, β_1, and β_2 cannot be generated all at once. On one hand, even if ($\beta_0, \beta_1, \beta_2$) are determined in one step, the distribution of weights (*i.e.*, $w(\beta_0)$, $w(\beta_1)$, and $w(\beta_2)$) is random. In our experiments, such ($\beta_0, \beta_1, \beta_2$) cannot sustain a good connector in general. On the other hand, the SAT-based toolkit cannot support searching such trails efficiently.

4.1.3   Constructing Connectors

The connecting trails, combined with the colliding trails, constitute the full 6-round differential trail with which the connectors that generate a subspace of messages that follow the connecting trails can be constructed. Considering that weight of the 4-round colliding trail exceeds the birthday bound, to mount a valid attack, we transfer the first round of the colliding trail to the connector. In detail, the 6-round collision attack on SHAKE128 consists of a 3-round connector and a 3-round colliding trail (refer to Fig. 4). As for SHA3-224 and SHA3-256, 6-round quantum collision attacks that consist of a 2-round connector and a 4-round colliding trail are mounted (refer to Fig. 4). We highlight that the connecting trails cannot provide enough DF to satisfy all the constraints in connectors even for theoretical attacks. Therefore, merely a fraction of constraints of the last round of 2/3-round connectors are picked up to be processed.

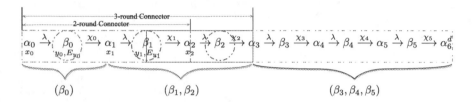

Fig. 4. The 6-round collision attack model

2-Round Connectors. We improve the algebraic-aided method adopted by previous works [DDS12, DDS14, QSLG17, SLG17, GLL+20] to construct connectors that generate message pairs following partially the output difference of the connectors.

Principally, the systems of linear equations on messages are listed and solved. The linear equations correspond to the conditions of sponge functions and differences of the connecting trail. The 2-round connector model exhibited in Fig. 5 illustrates how the system of linear equations is established.

1. First, linear equations of the $(c + p)$-bit conditions imposed by the sponge construction are listed, where c and p correspond to the *capacity* and *padding* bits respectively. Take the case of SHA3-256 as an example, the capacity is $c = 256 \times 2 = 512$ bits, and the padding rule is 10*1. To provide as many DF as possible, we set the padding as fixed "11" string. Also the 2-bit string "01" is concatenated to the tail of the message block. In total, a 4-bit fixed string (*i.e.*, "0111") is considered as the p-bit condition.

   Linear equations on the $(c + p)$-bit conditions are directly listed on the input messages x_0. As y_0 and x_0 are linked with the linear transformation λ, the linear equations on x_0 are easily transferred to equations on y_0. In the case of 2-round connectors, the systems of linear equations on y_0 are listed and denoted by E_{y_0}.

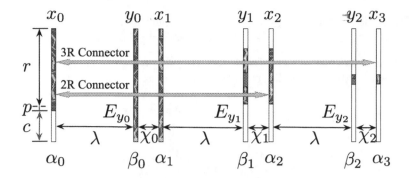

Fig. 5. The 2-round and 3-round connectors (Color figure online)

2. Next, linear equations on y_0 that meet conditions imposed by first round differential (β_0, α_1) are added to E_{y_0}. Message pairs constructed from the solutions of the current E_{y_0} system must follow the (β_0, α_1) differential. Details on how the equations can be listed are illustrated with Property 1 of the supplementary material A.

3. To list equations of conditions imposed by the second round differential (β_1, α_2), the first round must be bypassed. Linearization and partial linearization techniques on χ operation proposed in [QSLG17,SLG17] are borrowed directly to ensure that the y_1 bits can be expressed by the linear combinations of involved y_0 bits. Consequently, E_{y_1}, the system of linear equations on y_1 for (β_1, α_2), is transferred to a group of linear equations on y_0.

   To this end, extra equations on y_0 that allows the involved y_1 bits linear with respect to the χ operation must be added to E_{y_0}. Practically, as there is a whole round between y_1 and y_0, the x_1 bits that are involved to the corresponding y_1 bits according to λ operation are linearized. The principal property exploited to linearize x_1 bits is briefly summarized in Property 2 of the supplementary material A.

   The DF left after the last two steps cannot sustain solving all the β_1 active Sboxes. A greedy algorithm that sorts the active Sboxes of β_1 by the number of unlinearized x_1 bits is utilized to choose the β_1 Sboxes to be treated[5].

   To sum up, linear equations on y_0 that linearize the involved x_1 bits of partially chosen β_1 Sboxes are added to E_{y_0} in this step.

4. At last, the system of equations on y_1 (*i.e.*, E_{y_1}) of the partially treated β_1 Sboxes is transferred to linear equations on y_0 with the linearization equations generated in the last step, and added to the system E_{y_0}.

The Algorithm 1 shown in supplementary material A provides a concise description on construction of the 2-round connector. When a consistent system of linear equations on y_0 (*i.e.*, E_{y_0}) is successfully generated, the alleged 2-round connector is constructed. The solution space of E_{y_0} is composed of a subspace of messages, *i.e.*, y_0. A pair of messages (y_0^1, y_0^2) generated through XOR-ing y_0^1

---

[5] The other β_1 Sboxes that are not treated are indicated with red block in Fig. 5.

with β_0, while y_0^1 is a random solution of E_{y_0}, follows (1) the input difference α_0 and (2) a fraction of the output difference α_2 of the 2-round connector.

3-round Connector. In constructing 3-round connector, χ_0 of the first round is fully linearized, making the first round a linear layer. As a result, the 3-round connector can be viewed as a 2-round connector. We adopt the model shown in Fig. 5 to explain how the system of linear equations of the 3-round connector is constructed.

1. First, list linear equations on y_0 for (1) the $(c + p)$-bit conditions and (2) the constraints imposed by the first round (β_0, α_1) differential. The system of linear equations is denoted by E_{y_0}.
2. Next, fully linearize the χ_0 layer of the first round and transfer the equations on y_0 to equations on y_1. Namely, additional equations on y_0 that corresponds to linearizing each active and non-active Sbox of (β_0, α_1) differential are added to the current E_{y_0}. Expressions of the linearized χ_0 are utilized to convert the system of linear equations on y_0 (*i.e.*, E_{y_0}) to the system of linear equations on y_1 (*i.e.*, E_{y_1}).
3. List linear equations on y_1 for constraints imposed by the second round (β_1, α_2) differential. Add those equations to the present system of equations E_{y_1}.
4. With the same greedy algorithm utilized in 2-round connector construction, select a fraction of conditions of β_2 to solve and linearize the related x_2 bits. Add the equations on y_1 that linearize the involved x_2 bits of the partially treated (β_2, α_3) differential to the current E_{y_1} system.
5. List equations on y_2 for conditions imposed by the partially solved (β_2, α_3) differential of the last round of the 3-round connector. Convert the system of linear equations on y_2 to equations on y_1 based on the linearization of involved x_2 bit in the last step. Add the y_1 equations generated at this step to the whole E_{y_1} system.

When all equations are listed and organized in the system of equations on y_1 (*i.e.*, E_{y_1}), the 3-round connector is successfully constructed. A subspace of message pairs generated from the solution space of E_{y_1} satisfy that (1) the input conditions imposed by sponge constructions are met and (2) the output difference of the 3-round connector is partially met as expected. The Algorithm 2 in supplementary material A illustrates construction of the 3-round connector.

4.2   Collision Attack Against 6-Round `SHAKE128`

Following the basic attack strategy, a collision attack on 6-round `SHAKE128` is mounted. The model in Fig. 6 gives basic details of the attack.

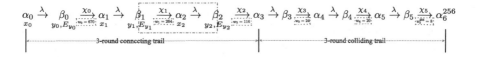

Fig. 6. The 6-round collision attack model for `SHAKE128`

As discussed in Sect. 4.1.2, the minimal weight of the best 4-round colliding trail core exceeds the birthday bound. To make the collision attack feasible, the first round of the 4-round colliding trail is transferred to the connector. Hence, the 6-round collision attack consists of a 3-round connector and a 3-round colliding trail. Propagation weight of each round is identified in Fig. 6. The 4-round colliding trail core is specified in Table 5 of supplementary material B, more specifically, the $(\beta_3, \beta_4, \beta_5)$ differences of Trail No.1. The probability of the 3-round colliding trail is 2^{-52} (where $2^{-52} = 2^{-24} \cdot 2^{-20} \cdot 2^{-8}$). The two-step SAT-based connecting trail search method described in Sect. 4.1.2 is applied to first determine (β_1, β_2) differences and fix β_0 difference subsequently. The connecting trail is listed in Table 7, *i.e.*, Trail No.3 in supplementary material B.

Now that the whole 6-round differential trail is determined, the 3-round connector can be constructed with the method illustrated in Sect. 4.1.3. The third round of the 3-round connector is partially solved, *e.g.*, in our experiment, 36 out of the 116 constraints of (β_2, α_3) are solved. The DF of the 3-round connector is 27^6. Alternatively, the 3-round connector generates a subspace of 2^{27} messages that satisfy the 36 conditions of the input difference α_3 of the colliding trail. A pair of solution messages are given in Table 10 of supplementary material B.

The unsolved conditions of (β_2, α_3) are treated together with the colliding trail through exhaustive search. In the brute force phase, message pairs generated from connectors are verified for whether satisfying α_3 or not. If not, simply abandon the current pair and try another one. Otherwise, further check the 256-bit digests of the pair until a collision is encountered.

Remark 2. Apart from the current work that exemplifies the collision resistance of a typical 128-bit security level, inner collisions [GJMG11] could also be analyzed with the same idea. As indicated in [GLL+20] (an inner collision of a 160-bit Keccak Challenge), the inner collision attack that constructs collision on capacity bits yields collisions of any digest length.

Complexity. The overall complexity includes complexity of both the connector construction phase and the exhaustive search phase.

---

[6] Indeed, the size of solution space is not always 2^{27} (or DF=27). This is an average number calculated from our experiments repeated on $2^{14.3}$ connectors.

- *In the exhaustive search phase*, the time complexity is 2^{132} 6-round SHAKE128 computations (where $2^{132} = 2^{116-36} \cdot 2^{52}$). However, taking advantage of the early-abort technique, the search process is sped up by iteratively filtering out half of the message pairs at each step. The cost of computing each additional bit constraint on β_2 equals to $\frac{11}{1600} \cdot \frac{1}{6} = 2^{-9.8}$ 6-round SHAKE128 computation as 11 bits of α_2 states are involved. When checking all the 2^{132} message pairs with one bit constraint, only half of the pairs satisfy the restriction while the other half are discarded, *i.e.*, the so-called *early-abort*. For the remaining message pairs, another bit constraint will be checked and filter out half of those message pairs. This iterative process continues on the surviving message pairs until all the bit constraints on β_2 are checked. $1/2$ of the messages stop by first bit constraint, $1/4$ by the second bit constraint, $1/8$ by the third bit etc. Hence the time complexity would be $2^{132} \cdot 2^{-9.8} \cdot (1 \cdot 1/2 + 2 \cdot 1/4 + 3 \cdot 1/8 + \cdots)$ $= 2^{123.2}$ 6-round SHAKE128 computations.

- *In the connector construction phase*, the time complexity corresponds to the time used to construct 2^{105} (*i.e.*, $2^{132}/2^{27} = 2^{105}$) connectors. Let's first discuss the equivalent conversion of implementation efficiency between connector construction and 6-round SHAKE128. The computation cost of 6-round SHAKE128 is $6 \cdot (\underbrace{(4 \cdot 320 + 2 \cdot 1600)}_{\theta} + \underbrace{3 \cdot 1600}_{\chi} + \underbrace{64}_{\iota}) = 56064$ bitwise operations. Further, solving systems of linear equations dominates the time of connector construction[7]. The time complexity of Gauss-Jordan elimination for system of boolean equations is $\mathcal{O}(m^2 n)$ bitwise operations [HJ12], where m is the number of equations and n is the number of variables. In the worst case, there are 1600 non-redundant equations in the final system, i.e., $m = 1600$. The complexity would be no greater than $1600^3 = 4.096 \times 10^9$ bitwise operations. Consequently, time cost of constructing a connector equals to $4.096 \times 10^9 / 56064 = 2^{16.2}$ 6-round SHAKE128. The time complexity in connector construction is equivalent to $2^{105} \cdot 2^{16.2} = 2^{121.2}$ 6-round SHAKE128 computations.

In total, time complexity of the classical collision attack is $2^{123.2} + 2^{121.2} = 2^{123.5}$ 6-round SHAKE128 computations. Complexity of quantum collision attack[8] is $2^{67.25}/\sqrt{S}$.

Table 3 gives an overview of the time complexity tradeoff between brute force search phase and connector construction phase according to the number of constraints on β_2 solved. The more the constraints are solved, the smaller the DF of connectors is, the better the brute force complexity is and the worse the connector complexity is.

---

[7] Refer to Remark 3 for more discussion on the cost of connectors.

[8] Complexity analysis of quantum collision attack will be illustrated in Sect. 4.3.

Table 3. Summary of complexity corresponding to the number of constraints solved

#constraints	DF of connector	Data complexity	Connector complexity	Brute force complexity	Total complexity
35	28	133	121.2	124.2	124.4
36	**27**	**132**	**121.2**	**123.2**	**123.5**
37	23	131	124.2	122.2	124.5
38	22	130	124.2	121.2	124.3
39	20	129	125.2	120.2	125.2
40	15	128	129.2	119.2	129.2
41	13	127	130.2	118.2	130.2
42	10	126	132.2	117.2	132.2
43	7	125	134.2	116.2	134.2
44	4	124	136.2	115.2	136.2
45	1	123	138.2	114.2	138.2

Remark 3. Experiments on $2^{14.3}$ connectors show that solving systems of equations dominates the time of connector construction. In particular,

- when fully linearizing the first round, due to the large DF, almost all Sboxes are successfully linearized in the first try and very occasionally it needs extra tries;
- when partially linearizing the second round where no more than 40 constraints are treated, about $1/3$ tests succeed with the first or a second try for each Sbox while around $2/3$ tests collapse and we should start the partial linearizing process again. But as this process consumes 0.01s on average (compared with 0.8s used to construct the whole connector) it won't affect the complexity analysis.

Overall, neglected time is consumed in listing equations which is consistent with the observations from [GLL+20].

Remark 4. Experimental results outlined in Table 4 conforms to the theoretical complexity analysis of the connector construction phase. The average execution time of each connector construction (denoted by T_c) is 0.8s. In our C++ implementation, around 2^{20} 6-round SHAKE128 are computed in each second. The time of connector construction equals to $2^{105} \cdot 2^{19.67} = 2^{124.67}$ SHAKE128 computations which validates the attack.

4.3 Quantum Collision Attack Against 6-Round SHA3-256

The colliding trail used in 6-round collision attacks on SHAKE128 is also used in attacks on SHA3-256 and SHA3-224. As shown in Fig. 7, the 6-round collision attack on SHA3-256 consists of a 2-round connector and a 4-round colliding

Table 4. Summary of the results of collision attacks on 6-Round SHA-3 instances

Target	Type	Trail core	T_c	DF	Complexity	Solution
SHAKE128	Classical	No. 3	0.8s	27	$2^{123.5}$	Table 10
	Quantum				$2^{67.25}/\sqrt{S}$	
SHA3-256	Quantum	No. 1	3s	5	$2^{104.25}/\sqrt{S}$	Table 11
SHA3-224	Quantum	No. 2	3s	22	$2^{97.75}/\sqrt{S}$	Table 12

Fig. 7. The 6-round collision attack model for SHA3-256

trail. Note that, the (β_1, β_2) used in the attack on SHAKE128 is also applied here. The entire 6-round differential trail is given in Table 5, *i.e.*, Trail No.1 in supplementary material B. The 2-round connector solves 226 out of the total 264 conditions imposed by (β_1, α_2). The solution space of the 2-round connector ensures a subspace of message pairs that follow partial α_2 difference as expected. In our experiment, the 2-round connector is constructed in 3 s on average. The DF of the connector is 5. Or to put it differently, the size of the solution space is 2^5. Example of a pair of messages that follow the connector is given in Table 11 of supplementary material B.

The unsolved conditions (*i.e.*, 38 left) of (β_1, α_2) are treated together with the colliding trail whose weight is $116 + 24 + 20 + 8 = 168$. In classical settings, the time complexity of the brute force phase is $2^{38} \cdot 2^{168} = 2^{206}$ 6-round SHA3-256 computations with which a valid collision attack cannot be conducted. However, such differential trails of low probability can be exploited in quantum settings.

Quantum Collision Attack. As stated in [HS21], no existing quantum collision attack on a random function could outperform classical attack based on parallel rho method [VOW94] in terms of time-space tradeoff. We follow their way and consider a quantum collision attack valid if its time complexity is less than $2^{n/2}/s$, where n denotes the digest length, and S is the hardware size required for the attack (or in other words, S is the maximum size of quantum computers and classical computers). Note that instead of designing concrete quantum circuits matching the theoretical bound of time-space tradeoff, the authors of [HS21] assume such quantum circuits exist already and concentrate on complexity evaluation of the quantum attacks. We adopt the same strategy in [HS21] to mount the 6-round quantum collision attack on SHA3-256.

Suppose there exists a quantum circuit $\mathcal{C}_1$ for the connector construction of depth T_c and width S_c. That is, the quantum circuit constructs a connector in time T_c with S_c qubits. Similarly, suppose there exists another quantum circuit $\mathcal{C}_2$ of depth T_s and width S_s for the one-block SHA-3 variants, *i.e.*, the quantum implementation of the 6-round targets (in this case SHA3-256). The idea that converts the classical attacks to the quantum collision attacks is described as follows.

1. Prepare message pairs (M, M') with the quantum circuit $\mathcal{C}_1$.
2. For each (M, M') pair, compute the digests with quantum circuit $\mathcal{C}_2$, and check whether they are identical.
3. Repeat the above two steps until a collision is found.

Complexity. Considering the solution space of the 2-round connector (which is 2^5), 2^{201} connectors are needed in theory. There are simply two kinds of operations in the quantum implementation of connectors, namely, listing the system of boolean equations and solving it with Gaussian-Jordan elimination, both of which are linear operations. Compared with T_s of the nonlinear SHA-3 variants (or more specific the χ operation), the depth T_c of $\mathcal{C}_1$ where only linear operations are involved is negligible [AMG+16]. Hence, time complexity of quantum collision attack is dominated by the time complexity of the exhaustive search phase.

Suppose we have a quantum computer of size S, taking parallelization into account, the time complexity of Grover search [Gro96] in the exhaustive search phase is

$$T_A \cdot (\pi/4) \cdot \sqrt{S_A/(p \cdot S)},$$

where p is the probability of finding a collision in the classical setting, and T_A (resp. S_A) is the depth (resp. width) of the quantum collision attack. The depth (resp. width) of the quantum circuits of the SHA-3 variants (*i.e.*, $\mathcal{C}_2$) are defined as the unit depth (resp. width), meaning that $T_s = 1$ and $S_s = 1$. Specifically, as the state size and the digest size are $2 \times 1600 + 256 = 3456$ bits, we regard at least 3456 qubits are required in circuit $\mathcal{C}_2$. The overall depth and width are evaluated with the following analysis.

– *Depth* (T_A). As T_c is negligible, $T_A = T_s = 1$.
– *Width* (S_A). In the quantum circuits of connectors (*i.e.*, $\mathcal{C}_1$), the quantum states include (1) the auxiliary m qubits (as there are 264 conditions, $m = 264$) that mark whether a condition will be treated or not in the partial linearizing step and (2) the $k \times 1601$ qubits that store the k boolean equations ($k \leq 1600$) of the system of linear equations. The overall $S_A = S_c + S_s = (m + k \times 1601 + 3456)/3456 \leq (264 + 1600 \times 1601 + 3456)/3456 = 742^9$.

Therefore, the total time complexity of the quantum collision attack on 6-round SHA3-256 is

$$1 \cdot (\pi/4) \cdot \sqrt{(742 \times 2^{206})/S} = 2^{104.25}/\sqrt{S}.$$

---

[9] More auxiliary qubits may be required for intermediate variables (*e.g.*, in greedy algorithm and Gaussian-Jordan elimination) in $\mathcal{C}_1$. Those variables are of the state size multiplied by a constant. As the worst case of Gaussian-Jordan elimination is considered and $\mathcal{C}_2$ also contains intermediate variables, this evaluation is reasonable.

Comparing to the generic attack cost under the time-space metric which is $2^{128}/S$, our quantum collision attack is valid as long as $S \leq 2^{47.5}$.

Remark 5. In the quantum search, we should prepare 2^{206} messages which brings to the concern that whether it's possible to construct so many connectors. This concern could be answered through introducing multi-blocks. The first block (which is identical for the two messages) provides distinct capacity bits at each time which are used to construct different connectors of the same connecting trails. We can try as many as 2^{512} first blocks which are sufficient for the attack.

4.4   Quantum Collision Attack Against 6-Round SHA3-224

As shown in Fig. 8, the 6-round trail of SHA3-224 (which is listed in Table 6 of supplementary material B) is comprised of the same colliding trail used in attacks on SHAKE128 and SHA3-256 and a 2-round connecting trail searched with the SAT-based tool. In our experiment, the 2-round connectors are averagely constructed in 3 s. The size of the solution space is 2^{22}. Example of a pair of messages that follow the connector is given in Table 12 of supplementary material B. The 2-round connector solves 240 out of the 268 conditions imposed by the (β_1, α_2) differential. Therefore, the classical complexity of the brute-force phase is $2^{28+113+24+20+8} = 2^{193}$ 6-round SHA3-224 computations. Similar to the attack on SHA3-256, we mount 6-round quantum collision attack on SHA3-224. Likewise, we adopt the strategy utilized in [HS21]. Suppose we have a quantum computer of size S, the complexity of our attack is

$$1 \cdot (\pi/4) \cdot \sqrt{(((268 + 1600 \times 1601 + 3424)/3424) \times 2^{193})/S} = 2^{97.75}/\sqrt{S}.$$

under the time-space metric $2^{112}/S$, and the quantum collision attack is faster than the generic attack when $S \leq 2^{28.5}$.

Fig. 8. The 6-round collision attack model for SHA3-224

5   Conclusion

We investigate the previous collision attacks on SHA-3, identify the limitations of ideas, methods, and techniques employed in those attacks, and summarize directions that can be improved to mount collision attacks on SHA-3 that cover more rounds. Briefly, if the colliding trails that cover more rounds and connecting trails that promise more degree of freedom in constructing connectors are generated, the collision attacks are most likely to be improved. The major challenge lies in the fact that differential trails of KECCAK-f permutation are difficult to search as the large state size results in a search space that is too enormous to be covered effectively. Luckily, we observe that the automatic search tool, *i.e.*, the SAT solver performs extraordinarily well in modeling the differential propagation of KECCAK-f. In this work, a powerful SAT-based automatic search toolkit is proposed to overcome the clarified challenges. We demonstrate that the SAT-based trail search methods are applicable to all kind of analytic scenarios where trails are involved. With the SAT-based toolkit, advanced collision attacks on SHA-3 instances are presented. Totally, a 6-round collision attack on SHAKE128 of complexity $2^{123.5}$, a 6-round quantum collision attack on SHA3-256 of complexity $2^{104.25}/\sqrt{s}$, and a 6-round quantum collision attack on SHA3-224 of complexity $2^{97.75}/\sqrt{s}$ are proposed. It's not only that the 6-round classical and quantum collision attacks are introduced for the first time but also shows that quantum collision attack is able to cover more rounds or targets than classical collision attacks.

Acknowledgements. This research is partially supported by Nanyang Technological University in Singapore under Start-up Grant 04INS000397C230, and Ministry of Education in Singapore under Grants RG91/20 and MOE2019-T2-1-060. Ling Song is supported by the National Natural Science Foundation of China (Grants 62022036, 62132008).

References

[AMG+16] Amy, M., Di Matteo, O., Gheorghiu, V., Mosca, M., Parent, A., Schanck, J.: Estimating the cost of generic quantum pre-image attacks on SHA-2 and SHA-3. In: Avanzi, R., Heys, H. (eds.) SAC 2016. LNCS, vol. 10532, pp. 317–337. Springer, Cham (2017). https://doi.org/10.1007/978-3-319-69453-5_18

[BDPVA07] Bertoni, G., Daemen, J., Peeters, M., Van Assche, G.: Sponge functions. In: ECRYPT Hash Workshop, vol. 2007. Citeseer (2007)

[BDPVA13] Bertoni, G., Daemen, J., Peeters, M., Van Assche, G.: Keccak. In: Johansson, T., Nguyen, P.Q. (eds.) EUROCRYPT 2013. LNCS, vol. 7881, pp. 313–314. Springer, Heidelberg (2013). https://doi.org/10.1007/978-3-642-38348-9_19

[Ber10] Bernstein, D.J.: Second preimages for 6 (7?(8??)) rounds of keccak. NIST mailing list (2010)

[BGLP] Bao, Z., Guo, J., Li, S., Pham, P.: Quantum multi-collision distinguishers (2020)

[BHT98] Brassard, G., HØyer, P., Tapp, A.: Quantum cryptanalysis of hash and claw-free functions. In: Lucchesi, C.L., Moura, A.V. (eds.) LATIN 1998. LNCS, vol. 1380, pp. 163–169. Springer, Heidelberg (1998). https://doi.org/10.1007/BFb0054319

[BPVA+11] Bertoni, G., Peeters, M., Van Assche, G., et al. The keccak reference (2011)

[CKMS14] Chang, D., Kumar, A., Morawiecki, P., Sanadhya, S.K.: 1st and 2nd Preimage Attacks on 7, 8 and 9 Rounds of Keccak-224,256,384,512. SHA-3 workshop, August 2014

672   J. Guo et al.

[CNPS17]   Chailloux, A., Naya-Plasencia, M., Schrottenloher, A.: An efficient quantum collision search algorithm and implications on symmetric cryptography. In: Takagi, T., Peyrin, T. (eds.) ASIACRYPT 2017. LNCS, vol. 10625, pp. 211–240. Springer, Cham (2017). https://doi.org/10.1007/978-3-319-70697-9_8

[DDS12]   Dinur, I., Dunkelman, O., Shamir, A.: New attacks on keccak-224 and keccak-256. In: Canteaut, A. (ed.) FSE 2012. LNCS, vol. 7549, pp. 442–461. Springer, Heidelberg (2012). https://doi.org/10.1007/978-3-642-34047-5_25

[DDS13]   Dinur, I., Dunkelman, O., Shamir, A.: Collision attacks on up to 5 rounds of SHA-3 using generalized internal differentials. In: Moriai, S. (ed.) FSE 2013. LNCS, vol. 8424, pp. 219–240. Springer, Heidelberg (2014). https://doi.org/10.1007/978-3-662-43933-3_12

[DDS14]   Dinur, I.: Dunkelman, orr, shamir, adi: improved practical attacks on round-reduced keccak. J. Cryptol. 27(2), 183–209 (2014)

[DSS+20]   Dong, X., Sun, S., Shi, D., Gao, F., Wang, X., Hu, L.: Quantum aHashing with Low Quantum Random Access Memories. In: Moriai, S., Wang, H. (eds.) ASIACRYPT 2020. LNCS, vol. 12492, pp. 727–757. Springer, Cham (2020). https://doi.org/10.1007/978-3-030-64834-3_25

[DVA12]   Daemen, Joan, Van Assche, Gilles: Differential propagation analysis of Keccak. In: Canteaut, Anne (ed.) FSE 2012. LNCS, vol. 7549, pp. 422–441. Springer, Heidelberg (2012). https://doi.org/10.1007/978-3-642-34047-5_24

[Dwo15]   Dworkin, M.J.: SHA-3 standard: Permutation-based hash and extendable-output functions (2015)

[GJMG11]   Guido, B., Joan, D., Michaël, P., Gilles, V.A.: Cryptographic sponge functions (2011)

[GLL+20]   Guo, J.: Liao, Guohong, Liu, Guozhen, Liu, Meicheng, Qiao, Kexin, Song, Ling: Practical collision attacks against round-reduced sha-3. J. Cryptol. 33(1), 228–270 (2020)

[GLS16]   Guo, J., Liu, M., Song, L.: Linear structures: applications to cryptanalysis of round-reduced KECCAK. In: Cheon, J.H., Takagi, T. (eds.) ASIACRYPT 2016. LNCS, vol. 10031, pp. 249–274. Springer, Heidelberg (2016). https://doi.org/10.1007/978-3-662-53887-6_9

[GLST22]   Guo, J., Liu, G., Song, L., Tu, Y.: Exploring SAT for cryptanalysis: (Quantum) collision attacks against 6-Round SHA-3 (Full Version) (2022). https://eprint.iacr.org/2022/184

[Gro96]   Grover, L.K.: A fast quantum mechanical algorithm for database search. In: Proceedings of the Twenty-Eighth Annual ACM Symposium on Theory of Computing, pp. 212–219 (1996)

[HJ12]   Han, C.-S., Jiang, J.-H.R.: When Boolean satisfiability meets gaussian elimination in a simplex way. In: Madhusudan, P., Seshia, S.A. (eds.) CAV 2012. LNCS, vol. 7358, pp. 410–426. Springer, Heidelberg (2012). https://doi.org/10.1007/978-3-642-31424-7_31

[HLY21]   He, L., Lin, X., Hongbo, Yu.: Improved preimage attacks on 4-round keccak-224/256. IACR Trans. Symmetric Cryptol. 2021(1), 217–238 (2021)

[HS20]   Hosoyamada, A., Sasaki, Y.: Finding hash collisions with quantum computers by using differential trails with smaller probability than birthday bound. In: Advances Cryptology-EUROCRYPT, vol. 249, p. 12106 (2020)

[HS21] Hosoyamada, A., Sasaki, Y.: Quantum collision attacks on reduced sha-256 and sha-512. IACR Cryptol. ePrint Arch. **292** (2021)

[LHY21] Lin, X., He, L., Hongbo, Y.: Improved preimage attacks on 3-round KECCAK-224/256. IACR Trans. Symmetric Cryptol.**2021**(3), 84–101 (2021)

[LQT19] Liu, G., Qiu, W., Tu, T.: New techniques for searching differential trails in keccak. IACR Trans. Symmet. Cryptol. **2019,** 407–437 (2019)

[LS19] Ting Li and Yao Sun. Preimage attacks on round-reduced KECCAK-224/256 via an allocating approach. In Yuval Ishai and Vincent Rijmen, editors, Advances in Cryptology - EUROCRYPT 2019-38th Annual International Conference on the Theory and Applications of Cryptographic Techniques, Darmstadt, Germany, May 19–23, 2019, Proceedings, Part III, volume 11478 of LNCS, pages 556–584. Springer, 2019

[LSLW17] Li, T.: Sun, Yao, Liao, Maodong, Wang, Dingkang: Preimage attacks on the round-reduced KECCAK with cross-linear structures. IACR Trans. Symmetric Cryptol. **2017**(4), 39–57 (2017)

[MDA17] Mella, S., Daemen, J.J.C., Van Assche, G.: New techniques for trail bounds and application to differential trails in Keccak . IACR Trans. Symmet. Cryptol. **2017**(1), 329–357 (2017)

[MP13] Mouha, N., Preneel, B.: Towards finding optimal differential characteristics for ARX: application to salsa20. Cryptology ePrint Archive, Report 2013/328 (2013). https://eprint.iacr.org/2013/328

[MPS13] Morawiecki, P., Pieprzyk, J., Srebrny, M.: Rotational cryptanalysis of round-reduced KECCAK. In: Moriai, S. (ed.) FSE 2013. LNCS, vol. 8424, pp. 241–262. Springer, Heidelberg (2014). https://doi.org/10.1007/978-3-662-43933-3_13

[MS13] Morawiecki, P.: Srebrny, Marian: a sat-based preimage analysis of reduced Keccak hash functions. Inf. Process. Lett. **113**(10–11), 392–397 (2013)

[NRM11] Naya-Plasencia, M., Röck, A., Meier, W.: Practical analysis of reduced-round KECCAK. In: Bernstein, D.J., Chatterjee, S. (eds.) INDOCRYPT 2011. LNCS, vol. 7107, pp. 236–254. Springer, Heidelberg (2011). https://doi.org/10.1007/978-3-642-25578-6_18

[QSLG17] Qiao, K., Song, L., Liu, M., Guo, J.: New collision attacks on round-reduced keccak. In: Coron, J.-S., Nielsen, J.B. (eds.) EUROCRYPT 2017. LNCS, vol. 10212, pp. 216–243. Springer, Cham (2017). https://doi.org/10.1007/978-3-319-56617-7_8

[QUE19] SEPARATE DECISION QUEUE. Cadical at the sat race 2019. SAT RACE 2019, p. 8 (2019)

[Raj19] Rajasree, M.S.: Cryptanalysis of round-reduced KECCAK using non-linear structures. In: Hao, F., Ruj, S., Sen Gupta, S. (eds.) INDOCRYPT 2019. LNCS, vol. 11898, pp. 175–192. Springer, Cham (2019). https://doi.org/10.1007/978-3-030-35423-7_9

[SBH+19] Soos, M., Biere, A., Heule, M., Järvisalo, M., Suda, M.: Cryptominisat 5.6 with yalsat at the sat race 2019. In: Proceedings of SAT Race, pp. 14–15 (2019)

[SDG+20] Soos, M., Devriendt, J., Gocht, S.,. Shaw, A., Meel, K.S.: CryptoMiniSat with CCAnr at the sat competition 2020. In: SAT Competition , p. 27 (2020)

[Sin05] Sinz, C.: Towards an optimal CNF encoding of Boolean cardinality constraints. In: van Beek, P. (ed.) CP 2005. LNCS, vol. 3709, pp. 827–831. Springer, Heidelberg (2005). https://doi.org/10.1007/11564751_73

[SLG17] Song, L., Liao, G., Guo, J.: Non-full Sbox linearization: applications to collision attacks on round-reduced KECCAK. In: Katz, J., Shacham, H. (eds.) CRYPTO 2017. LNCS, vol. 10402, pp. 428–451. Springer, Cham (2017). https://doi.org/10.1007/978-3-319-63715-0_15

[SNC09] Soos, M., Nohl, K., Castelluccia, C.: Extending SAT solvers to cryptographic problems. In: Kullmann, O. (ed.) SAT 2009. LNCS, vol. 5584, pp. 244–257. Springer, Heidelberg (2009). https://doi.org/10.1007/978-3-642-02777-2_24

[SNC10] Soos, M., Nohl, K., Castelluccia, K.: Cryptominisat, SAT Race solver descriptions (2010)

[Soo14] Soos, M.: Cryptominisat v4. SAT Competition, p. 23 (2014)

[Soo16] Soos, M.: The CryptoMiniSat 5 set of solvers at sat competition 2016. In: Proceedings of SAT Competition, p. 28 (2016)

[SSK+20] Soos, M., Selman, B., Kautz, H., Devriendt, J., Gocht, S.: CryptoMiniSat with Walksat at the SAT competition 2020. In: SAT Competition 2020, pp. 29 (2020)

[SWW18] Sun, L., Wang, W., Wang. M.: More accurate differential properties of led64 and midori64. IACR Trans. Symmet. Cryptol. **2018**, 93–123 (2018)

[SWW21] Sun, L., Wang, W., Wang, W.: Accelerating the search of differential and linear characteristics with the sat method. IACR Trans. Symmet. Cryptol. **2021**, 269–315 (2021)

[VOW94] Van Oorschot, P.C., Wiener, M.J.: Parallel collision search with application to hash functions and discrete logarithms. In: Proceedings of the 2nd ACM Conference on Computer and Communications Security, pp. 210–218 (1994)

Lattice Cryptanalysis

Log-𝒮-unit Lattices Using Explicit Stickelberger Generators to Solve Approx Ideal-SVP

Olivier Bernard[1,2(✉)] ⓘ, Andrea Lesavourey[1] ⓘ, Tuong-Huy Nguyen[1,3],
and Adeline Roux-Langlois[1] ⓘ

[1] Univ Rennes, CNRS, IRISA, Rennes, France
olivier.bernard@normalesup.org,
{andrea.lesavourey,tuong-huy.nguyen,adeline.roux-langlois}@irisa.fr
[2] Thales, Gennevilliers, France
[3] DGA Maîtrise de l'Information, Bruz, France

Abstract. In 2020, Bernard and Roux-Langlois introduced the Twisted-PHS algorithm to solve Approx-Svp for ideal lattices on any number field, based on the PHS algorithm by Pellet-Mary, Hanrot and Stehlé. They performed experiments for prime conductors cyclotomic fields of degrees at most 70, one of the main bottlenecks being the computation of a log-𝒮-unit lattice which requires subexponential time.

Our main contribution is to extend these experiments to cyclotomic fields of degree up to 210 for most conductors m. Building upon new results from Bernard and Kučera on the Stickelberger ideal, we use explicit generators to construct full-rank log-𝒮-unit sublattices fulfilling the role of approximating the full Twisted-PHS lattice. In our best approximate regime, our results show that the Twisted-PHS algorithm outperforms, over our experimental range, the CDW algorithm by Cramer, Ducas and Wesolowski, and sometimes beats its asymptotic volumetric lower bound.

Additionally, we use these explicit Stickelberger generators to remove almost all quantum steps in the CDW algorithm, under the mild restriction that the plus part of the class number verifies $h_m^+ \leq O(\sqrt{m})$.

Keywords: Ideal lattices · Approx-SVP · Stickelberger ideal · S-unit attacks · Twisted-PHS algorithm

1 Introduction

The ongoing NIST Post-Quantum Cryptography competition illustrates the importance of the *Learning With Errors* (Lwe) problem as an intermediate building block for a wide variety of cryptographic schemes. Most of these cryptographic schemes rely on a structured version of the Lwe problem allowing for much more satisfactory performance, compared to schemes based on the unstructured Lwe problem. The first structured variant of Lwe, later known

ⓒ International Association for Cryptologic Research 2022
S. Agrawal and D. Lin (Eds.): ASIACRYPT 2022, LNCS 13793, pp. 677–708, 2022.
https://doi.org/10.1007/978-3-031-22969-5_23

as the Ring-LWE problem, is shown to be at least as hard as the *Approximate Shortest Vector Problem* on ideal lattices (Approx-id-Svp) using quantum worst-case to average-case reductions [SSTX09,LPR10]. One important matter is to determine whether using this structured version of LWE could lower the hardness hypothesis of the scheme. Notably, an efficient solver for Approx-id-SVP would render the worst-case to average-case reduction to Ring-LWE meaningless as a security argument. Note however that even in this case, this would not directly imply an efficient solver for the Ring-LWE problem.

In the case of arbitrary lattices, Approx-Svp is a well-studied hard problem. It consists in finding relatively short vectors of a given lattice, within an approximation factor of the shortest vector. The best theoretical trade-off between runtime and approximation factor is known as Schnorr's hierarchy [Sch87]: one can reach, for any $\alpha \in (0,1)$, an approximation factor $2^{\widetilde{O}(n^{\alpha})}$ in time $2^{\widetilde{O}(n^{1-\alpha})}$. The closest known practical algorithm to this trade-off is the BKZ algorithm [SE94], a generalization of the well-known LLL algorithm [LLL82]. In the particular case of ideal lattices, i.e., lattices that correspond to ideals of the ring of integers $\mathcal{O}_K$ of a number field K, one could hope that the best reduction algorithms would remain those associated with arbitrary lattices. However, this simplifying assumption seems questionable, since the underlying number-theoretic structure is precisely what makes Ring-LWE a more efficient building block. Thus, going beyond the BKZ algorithm and estimating the hardness of Approx-id-Svp using algebraic ideas has gathered more attention, starting by works from [EHKS14,CGS14,BS16,CDPR16]. Earlier works aimed at the more restricted case of Approx-id-Svp for principal ideals. A strategy for this case is devised as a two parts algorithm. The first part requires solving the Principal Ideal Problem (PIP), i.e., finding any generator of the ideal; the second part aims at finding the shortest one, by solving a Closest Vector Problem (CVP) in the so-called *log-unit lattice*. This shortest generator is expected to solve Approx-SVP for a sufficiently small approximation factor. Ultimately, for the particular case of cyclotomic fields of prime power conductors, [CDPR16] proved that Approx-id-SVP on principal ideals is solvable in quantum polynomial time, but only reaching an approximation factor $2^{\widetilde{O}(\sqrt{n})}$.

Subsequent works in a more general case can be divided in two different paths. The first one [CDW17,CDW21] aimed at extending the results from [CDPR16] to arbitrary ideal lattices over any cyclotomic fields, while still reaching in quantum polynomial time an approximation factor $2^{\widetilde{O}(\sqrt{n})}$. One of their contributions is to reduce the arbitrary ideal case to the principal ideal case by solving the *Close Principal Multiple Problem* (CPMP): given an ideal $\mathfrak{b}$, one computes an ideal $\mathfrak{c}$ of small algebraic norm s.t. $\mathfrak{bc}$ is a principal ideal. In order to ensure that $\mathfrak{c}$ has a small norm, a new key technical ingredient, specific to cyclotomic fields, was the use of the Stickelberger lattice, which has good geometric properties. Then, the results from [CDPR16] are applied to $\mathfrak{bc}$ to obtain a candidate short element of $\mathfrak{b}$, using the fact that $\mathfrak{c}$ has a small norm. The concrete consequences of this method were experimented in [DPW19], under different regimes which mainly differ upon which CVP solver is used. The first regime (called "Naive") uses Babai's Nearest Plane algorithm, whereas the second regime uses a heuristic CVP algorithm relatively to *ad hoc* pseudo-norms. From these experiments, the asymptotic performance of those decoding algorithms was estimated, which led to simulated approximation factors reached by the CDW algorithm. Finally, given experimentally verified constants, a volumetric lower bound was derived for the approximation factors that could be reached in the best scenario. According to this lower bound, the CDW algorithm is expected to beat the BKZ_{300} algorithm for cyclotomic fields of degrees at least larger than 7000. Since NIST submissions based on structured lattices rely on cyclotomic fields of degree at most 1024, this could be perceived as somewhat reassuring.

The second path is explored in [PHS19, BR20]. Those works, applying to arbitrary number fields, replace the two reductions steps from [CDW21] with a single CVP instance, so as to find a principal multiple ideal which is not only of small algebraic norm, but is also generated by a small element. A key ingredient achieving this is to use a generalization of the units of $\mathcal{O}_K$, called $\mathcal{S}$-units; this formalism was an underlying feature of [PHS19] and was later made explicit in [BR20]. The PHS algorithm splits into a preprocessing phase and a query phase. The preprocessing phase consists in preparing the decoding of a particular lattice depending only on the number field K, *via* the computation of a hint following Laarhoven's CVP with preprocessing algorithm [Laa16], which takes exponential time. Then, any Approx-id-SVP instance in K can be interpreted as an Approx-CVP instance in this lattice, efficiently solved thanks to the hint. Up to the preprocessing, the query phase yields new time/quality trade-offs: as in [CDW21] for cyclotomic fields, it reaches approximation factor $2^{\tilde{O}(\sqrt{n})}$ in quantum polynomial time; however, the PHS algorithm also allows for better trade-offs than Schnorr's hierarchy, from polynomial to $2^{\tilde{O}(\sqrt{n})}$ approximation factors. On the downside, the computation of the lattice itself takes classically subexponential time, which is a serious obstacle for studying their geometry and obtaining concrete asymptotic estimations as was done in [DPW19] for the CDW algorithm.

Then, [BR20] introduced Twisted-PHS, a "Twisted" version of the PHS algorithm whose main difference lies in a fundamental modification of the underlying lattice, thanks to a natural normalization coming from the *Product Formula*. The problem of finding a short vector is expected to be better encoded within this new lattice, ultimately leading to smaller outputs. Even though the proven trade-offs between runtime and approximation factor remain the same for the Twisted-PHS algorithm as for the PHS algorithm, very significant improvements have been experimentally illustrated in [BR20, Fig. 5.3], showing much better approximation factors compared to the PHS algorithm for number fields of degree up to 60, where Laarhoven's CVP algorithm is replaced in practice by Babai's Nearest Plane algorithm [Bab86]. These were to our knowledge the first experimental evidence of the geometric peculiarity of normalized log-$\mathcal{S}$-unit lattices and of the practical potential of this type of attack. In this practical version, experiments are solely limited by the classical complexity of computing the lattice.

Unfortunately, the attained dimensions, up to 60, are not sufficient to assess the practical limits of the Twisted-PHS algorithm: its heuristic analysis [BR20] could give only a loose upper bound, or miss unexpected performance in practical dimensions due to its asymptotic nature, even in the cryptographical range.

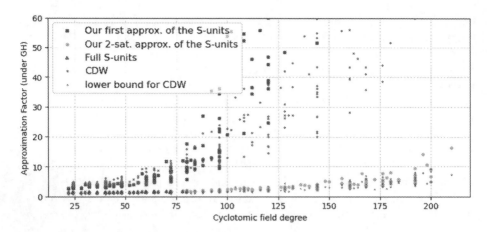

Fig. 1. Average of approximation factors achieved by our implementation of Twisted-PHS, using log-$\mathcal{S}$-unit sublattices in cyclotomic fields over random simulated instances, compared to those achieved by CDW [DPW19], assuming the Gaussian Heuristic throughout all instances.

Our Contributions. We develop theoretical and practical improvements regarding algorithms for solving Approx-id-Svp, in both lines of work following the CDW algorithm and the Twisted-PHS algorithm. Even though the hardness of the Approx-id-Svp does not concretely impact the security of cryptographic schemes, it is important to get a better understanding of both approaches, which are the only ones successfully exploiting the structure of a lattice.

Our core ingredient is the introduction of a full-rank family of independent $\mathcal{S}$-units, whose algebraic properties are proven in Sect. 3. In Sect. 4, we use this family to remove most quantum steps of the CDW algorithm, leaving only one step during a preprocessing phase done once for any given field, and one step for each query.

In Sect. 5, this family allows us to achieve experiments on algorithms in the (Twisted-)PHS family, for most cyclotomic fields of dimension up to 210. By comparison, previous experiments [DPW19, BR20] only considered cyclotomic fields of conductors $m = p > 2$ prime and $m = 2^e > 2$. Our work comes with an improved implementation of the initial Twisted-PHS algorithm, allowing us to extend the experiments conducted in [BR20] up to dimension 80 and for all cyclotomic fields. It also includes different regimes of approximation for this algorithm, using sublattices of the log-$\mathcal{S}$-unit lattice obtained thanks to our new construction beyond dimension 80 up to 210. These regimes yield concrete upper bounds for the approximation factors that could be reached by the full Twisted-PHS algorithm up to dimension 210, as illustrated in Fig. 1:

1. The depicted approximation factors were estimated using the *Gaussian Heuristic*, matching the *exact ones* obtained by [BR20] without this hypothesis.

2. Our best approximate regime yields approximation factors that are comparable (sometimes even smaller) to the asymptotical volumetric lower bound regime of the CDW algorithm.

In [DPW19], it was already noted that the PHS approach should outperform the lower bound, but at the cost of computing Laarhoven's hint in exponential time. Our work show that for medium dimensions, where asymptotical results should start to be meaningful, the Twisted-PHS algorithm is at least comparable to the CDW lower bound, though without this exponential hint precomputation.

As suggested in [BR20], and illustrated in small dimensions, the Twisted-PHS algorithm performance may be explained by the peculiar geometric nature of the log-$\mathcal{S}$-unit lattice. In our work, this is confirmed by the computations of several geometrical parameters on the basis obtained by our implementation, across all considered cyclotomic fields, sublattices and factor bases. This specificity, observed in a wide variety of regimes and even in medium dimensions, suggest a deeper explanation, a possibility recently explored by Bernstein and Lange [BL21]. We provide a full implementation of all our experiments at https://github.com/ob3rnard/Tw-Sti.

Technical Overview. In [BR20], the log-$\mathcal{S}$-unit lattice needed for the preprocessing phase was built using generic number theory tools. Our main idea is to shortcut this generic computation by considering a maximal family $\mathfrak{F}$ of independent $\mathcal{S}$-units, where $\mathcal{S}$ verifies some conditions (detailed in Sect. 3), leading to sublattices of the log-$\mathcal{S}$-unit lattice. The family $\mathfrak{F}$ is composed of three parts:

1. Circular units, also known as cyclotomic units, e.g. in [Was97, §8];
2. Generators coming from the explicit proof of Stickelberger's theorem proof;
3. Real $\mathcal{S}$-units coming from the maximal real subfield K_m^+ of K_m, where K_m is the cyclotomic field of conductor m.

The first two parts are classically easy to compute. In particular, the effectiveness of the second part comes from two recent results of [BK21]: the knowledge of an explicit *short* $\mathbb{Z}$*-basis* of the Stickelberger ideal for *any* conductor [BK21, Th. 3.6], and the effective computations of generators corresponding to these short relations using Jacobi sums [BK21, §5]. On the contrary, the last part still relies on generic number theory tools which are classically costly, but are now performed in a number field of half degree, which propels us to degree 210.

As an important theoretical contribution, we prove in Theorem 3.11 that $\mathfrak{F}$ is indeed a full-rank family of multiplicatively independent $\mathcal{S}$-units, by computing explicitly its (finite) index in the full $\mathcal{S}$-unit group. This can be seen as a generalization of the strategy of [CDW17, Def. 2] to obtain a full-rank lattice of class relations, restricted to the relative class group. In particular, our result proves the experimentally conjectured value [DPW19, Rem. 3] of the index of their family.

Finally, the index of $\mathfrak{F}$ contains a large power of 2 that can be removed using classical 2-saturation techniques of Sect. 3.5, leading to a family $\mathfrak{F}_{\mathsf{sat}}$. We then use the explicit knowledge of these special $\mathcal{S}$-units in two different situations.

Theoretical Improvements of the CDW Algorithm. In Sect. 4, we remove almost all quantum steps of the CDW algorithm while still guaranteeing its approximation factor [CDW21, Th. 5.1], at the small price of restricting to cyclotomic fields s.t. $h_m^+ \leq O(\sqrt{m})$ ([BLNR21, Hyp. B.1]), where h^+ denotes the plus part of the class number (defined in Sect. 2.2), whereas [CDW21, Ass. 2] uses $h_m^+ \leq \mathrm{poly}(m)$.

For that purpose, we state an equivalent rewriting of [CDW21, Alg. 7], making explicit some hidden steps useful for subsequent modifications. Then, the explicit Stickelberger generators and real $\mathcal{S}$-units are used to remove the last call to the quantum PIP solver. Finally, considering the module of *all* real class group relations allows us to remove the quantum random walk mapping any ideal of K_m into the relative class group. This last part uses our Theorem 3.11 and needs [BLNR21, Hyp. B.1] to obtain the same bound on the approximation factor.

Only two quantum steps remain: the first is performed once to compute real $\mathcal{S}$-units in K_m^+, of degree only half, the second is solving the ClDL for each query.

Experimenting the Twisted-PHS Algorithm in Medium Dimensions. We apply Twisted-PHS [BR20] on our full-rank sublattices of the log-$\mathcal{S}$-unit lattice, yielding *approximated* regimes of the Twisted-PHS algorithm. Up to degree 210, for most conductors, the newly implemented algorithm is used to compute the sublattices associated with $\mathfrak{F}$ and $\mathfrak{F}_{\mathsf{sat}}$, for varying subsets $\mathcal{S}$ according to the number of Galois orbits of totally split primes used. In particular, we explicitly compute the Stickelberger generators and real generators of $\mathfrak{F}$ and effectively perform the 2-saturation of $\mathfrak{F}$ to get $\mathfrak{F}_{\mathsf{sat}}$. Up to degree 80, the whole log-$\mathcal{S}$-unit lattice is also computed, corresponding to a fundamental system $\mathfrak{F}_{\mathsf{su}}$ of $\mathcal{S}$-units. This last computation of $\mathfrak{F}_{\mathsf{su}}$ remains unfeasible at higher dimensions. We evaluate the geometry of all these lattices with standard indicators described in Sect. 2.5: the root-Hermite factor δ_0, the orthogonality defect δ and the logarithm of the Gram-Schmidt norms. We consistently observe the same phenomena already pointed out in [BR20, §5.1 and 5.2], that indicate close to orthogonal lattices.

Next, since computing ClDL solutions for random ideals quickly becomes intractable, we simulate this step by sampling random outputs similarly to what was done in [DPW19, Hyp. 8]. Given those targets and the preprocessed lattices associated with $\mathfrak{F}$, $\mathfrak{F}_{\mathsf{sat}}$ and $\mathfrak{F}_{\mathsf{su}}$, we evaluate the approximation factors reached by these different regimes, by assuming the Gaussian Heuristic. These two assumptions, i.e., using simulated targets and the Gaussian Heuristic, are validated by the fact that up to degree 80, where it is feasible to compute the full $\mathcal{S}$-unit group generated by $\mathfrak{F}_{\mathsf{su}}$, our approximation factors match the *exact* approximation factors obtained in [BR20, Fig. 1.1], where those heuristics were not used. Finally, we compare our results to the approximation factors obtained by the CDW algorithm [CDW21] in the "Naive" regime of [DPW19], under the same working assumptions as above. We observe that in our best approximate regime, using $\mathfrak{F}_{\mathsf{sat}}$, our estimated approximation factors are close, and sometimes smaller, than the theoretical lower bound derived in [DPW19]. This suggests that the crossover with BKZ_{300} could be lower than expected for the Twisted-PHS algorithm.

Relations to Other Works Related to $\mathcal{S}$-units. Some recent mathematical results regarding the Stickelberger lattice were established in [BK21]. The authors described, for *any* conductor, an easily computable *short basis* for this lattice, and how to explicitly compute the associated principal ideal generators through Jacobi sums. In our work, this result is brought into fruition to solve Approx-id-SVP. The completion of this short basis into a full-rank lattice of class relations, the effective computation of the explicit generators and the 2-saturation of these elements, yielded the different approximated regimes of Twisted-PHS and allowed us to remove many quantum steps from the CDW algorithm.

In a talk on August 2021 at SIAM Conference,[1] Bernstein announced a joint work with Eisenträger, Rubin, Silverberg and van Vredendaal, by illustrating the construction of small $\mathcal{S}$-units using Jacobi sums that lead to an "$\mathcal{S}$-unit attack" in the power-of-2 conductor case up to degree 64, assuming $h_{2^e}^+ = 1$. The talk also announced a paper that has yet to appear. In this light, we are not able to compare our use of explicit Stickelberger generators to their work. However, this talk does neither mention a short basis of the Stickelberger lattice, which is at the heart of our work, nor lift all obstructions to apply it to *any* conductor.

In December 2021, a "filtered-$\mathcal{S}$-unit software" was released by Bernstein, treating the prime $p \leq 43$ conductor case, on a webpage[2] describing the "simplest $\mathcal{S}$-unit attack" using a technique described in [BL21]. This work is not related to our construction. Finally, the authors of [BL21] argued that "spherical models" should not be applied to log-$\mathcal{S}$-unit lattices, which may have particular geometric properties. This phenomenon was experimentally observed already in [BR20], and is confirmed by all of our experiments in medium dimensions.

2 Preliminaries

Notations. For any $i, j \in \mathbb{Z}$ with $i \leq j$, the set of all integers between i and j is denoted by $[\![i, j]\!]$. For any $x \in \mathbb{Q}$, let $\{x\}$ denote its fractional part, i.e., such that $0 \leq \{x\} < 1$ and $x - \{x\} \in \mathbb{Z}$. A vector is represented by a bold letter $\mathbf{v}$, and for any $p \in \mathbb{N}^* \cup \{\infty\}$, its ℓ_p-norm is written $\|\mathbf{v}\|_p$. The n-dimensional vector with all 1's is denoted by $\mathbf{1}_n$. All matrices are given using *row* vectors.

2.1 Cyclotomic Fields

We denote the cyclotomic field of conductor m, $m \not\equiv 2 \mod 4$, by $K_m = \mathbb{Q}[\zeta_m]$, where ζ_m is a primitive m-th root of unity. It has degree $n = \varphi(m)$, its maximal order is $\mathcal{O}_{K_m} = \mathbb{Z}[\zeta_m]$ ([Was97, Th. 2.6]), and its discriminant is given precisely by $\Delta_{K_m} = (-1)^{\varphi(m)/2} \frac{m^{\varphi(m)}}{\prod_{p|m} p^{\varphi(m)/(p-1)}}$ ([Was97, Pr. 2.7]), which is of order n^n.

In this paper, we consider *any* conductor $m > 1$ of the general prime factorization $m = p_1^{e_1} p_2^{e_2} \cdots p_t^{e_t}$, $m \not\equiv 2 \mod 4$, and let $q_i = p_i^{e_i}$ for all $i \in [\![1, t]\!]$.

---

[1] The slides are available at https://cr.yp.to/talks.html#2021.08.20.

[2] This is hosted by https://s-unit.attacks.cr.yp.to/filtered.html.

In particular, m has exactly t distinct prime divisors. Let G_m denote the Galois group of K_m, which can be made explicit by ([Was97, Th. 2.5]):

$$G_m = \left\{\sigma_s : \zeta_m \longmapsto \zeta_m^s;\ 0 < s < m, (s, m) = 1\right\} \simeq \left(\mathbb{Z}/m\mathbb{Z}\right)^{\times}.$$

In particular, we denote by $\sigma_s \in G_m$ the automorphism sending any m-th root of unity to its s-th power. For convenience, the automorphism induced by complex conjugation is written $\tau = \sigma_{-1}$. The algebraic norm of $\alpha \in K_m$ is defined by $\mathcal{N}(\alpha) = \prod_{\sigma \in G_m} \sigma(\alpha)$, hence the absolute norm element in the integral group ring $\mathbb{Z}[G_m]$ is $N_m = \sum_{\sigma \in G_m} \sigma$.

Maximal Real Subfield. The maximal real subfield of K_m, written K_m^+, is the fixed subfield of K_m under complex conjugation, i.e., $K_m^+ := K_m^{\langle \tau \rangle} = \mathbb{Q}(\zeta_m + \zeta_m^{-1})$. Its maximal order is $\mathcal{O}_{K_m^+} = \mathbb{Z}[\zeta_m + \zeta_m^{-1}]$ (see e.g. [Was97, Pr. 2.16]).

By Galois theory, since $\langle \tau \rangle$ is a normal subgroup of G_m, the maximal real subfield of K_m is a Galois extension of $\mathbb{Q}$ with Galois group $G_m^+ := \text{Gal}(K_m^+/\mathbb{Q})$ isomorphic to $G_m/\langle \tau \rangle$. We identify G_m^+ with the following system of representatives modulo τ restricted to K_m^+: $G_m^+ = \left\{\sigma_s|_{K_m^+};\ 0 < s < \frac{m}{2}, (s, m) = 1\right\}$. Technically, each $\sigma_s|_{K_m^+} \in G_m^+$ extends in G_m to either σ_s or $\tau\sigma_s = \sigma_{-s}$. For simplicity, we always choose to lift $\sigma_s|_{K_m^+} \in G_m^+$ to $\sigma_s \in G_m$ and drop the restriction to K_m^+ which should be clear from the context. This slight abuse of notation appears to be very practical. For example, the corestriction $\text{Cor}_{K_m/K_m^+}(\sigma_s|_{K_m^+})$, defined as the sum of all elements of G_m that restricts to $\sigma_s|_{K_m^+}$, namely $\sigma_s + \tau\sigma_s$, is written using the much simpler expression $(1 + \tau) \cdot \sigma_s$.

2.2   Real and Relative Class Groups

Fractional ideals of K_m form a multiplicative group $\mathcal{I}_m$ containing the normal subgroup $\mathcal{P}_m := \{\langle \alpha \rangle;\ \alpha \in K_m\}$ of principal ideals. The quotient group $\mathcal{I}_m/\mathcal{P}_m$ is called the *class group* of K_m and denoted by Cl_m. It is finite and its cardinal h_m is the *class number* of K_m. For any $\mathfrak{b} \in \mathcal{I}_m$, the class of $\mathfrak{b}$ in Cl_m is written $[\mathfrak{b}]$.

The integral group ring $\mathbb{Z}[G_m]$ acts naturally on $\mathcal{I}_m$; more precisely, for any element $\alpha = \sum_{\sigma \in G_m} a_\sigma \sigma \in \mathbb{Z}[G_m]$, and any $\mathfrak{b} \in \mathcal{I}_m$, $\mathfrak{b}^\alpha := \prod_{\sigma \in G_m} \sigma(\mathfrak{b})^{a_\sigma}$. The class group and class number of the maximal real subfield K_m^+ are denoted respectively by Cl_m^+ and h_m^+. The relative norm map $\mathcal{N}_{K_m/K_m^+}$ induces a homomorphism from Cl_m to Cl_m^+, whose kernel is the so-called *relative class group*, written Cl_m^- and of cardinal the *relative class number* h_m^-. Hence, by construction, for any $\mathfrak{b}$ s.t. $[\mathfrak{b}] \in \text{Cl}_m^-$, $\mathfrak{b}^{1+\tau} \cap K_m^+$ is principal. One important specificity of cyclotomic fields is that the real class group Cl_m^+ embeds into Cl_m *via* the natural inclusion map, which to each ideal class $[\mathfrak{b}] \in \text{Cl}_m^+$ associates the ideal class $[\mathfrak{b} \cdot \mathcal{O}_{K_m}] \in \text{Cl}_m$ [Was97, Th. 4.14]. Concretely, it implies that $h_m = h_m^+ \cdot h_m^-$ is the product of the plus part and the relative part of the class number.

Plus Part and Relative Part of the Class Number. Generally, not much is known about the class number of a number field, and the analytic class number formula [Neu99, Cor. 5.11(ii)] allows obtaining a rough upper bound $h_m \leq \widetilde{O}(\sqrt{|\Delta_{K_m}|})$.

In the case of cyclotomic fields though, the structure of the relative class group is better understood. Using analytic means, the relative class number has the following explicit expression [Was97, Th. 4.17]: $h_m^- = Qw \cdot \prod_{\chi \text{odd}} \left(-\frac{1}{2} B_{1,\chi} \right)$, where $w = 2m$ if m is odd and $w = m$ if m is even, $Q = 1$ if m is a prime power and $Q = 2$ otherwise, and $B_{1,\chi}$ is defined by $\frac{1}{f} \sum_{a=1}^{f} a \cdot \chi(a)$ for any odd primitive character χ modulo m of conductor f dividing m. Computing this value is in practice very efficient, using adequate representations of Dirichlet characters.

The hard part of cyclotomic class numbers computations is to obtain the plus part h_m^+, and relatively few of them are known. We use the values from [Was97, Tab. §4], [Mil14, Th. 1.1 and 1.2] and [BFHP21, Tab. 1], consistently assuming the *Generalized Riemann Hypothesis* (GRH). We also provide 58 additional values of h_m^+ in [BLNR21, Tab. 2.1] for completeness.

The fact that the plus part of the class number seems much smaller than the relative part is striking. Weber's conjecture claims that $h_{2^e}^+ = 1$ for any $e > 1$, and Buhler, Pomerance and Robertson [BPR04] argue, based on Cohen-Lenstra heuristics, that for all but finitely many pairs (p, e), where p is a prime and e is a positive integer, $h_{p^{e+1}}^+ = h_{p^e}^+$. For prime power conductors, this conjecture claims that the plus part is asymptotically constant. These conjectures are backed up by Schoof's extensive calculations [Sch03] in the prime conductor case, and by the above explicit values. In particular, under GRH, Miller proved Weber's conjecture up to $m = 512$, and we note that according to Schoof's table, $h_m^+ \leq \sqrt{m}$ holds for more than 96.6% of all prime conductors $m = p < 10000$.

Prime Ideal Classes Generators. When picking a set of prime ideals in the algorithms of this paper, an important feature is that they generate the class group. In general, even assuming GRH, only a large bound on the norm of generators is known, indeed Bach proved [Bac90, Th. 4] that $\mathcal{N}(\mathfrak{L}_{\max}) \leq 12 \ln^2 |\Delta_{K_m}|$, where $\mathfrak{L}_{\max}$ is the biggest ideal inside a generating set of Cl_m of minimum norm. In practice though, this bound seems very pessimistic [BDF08, §6].

On the other hand, as prime ideals belong to Cl_m^- only with probability roughly $1/h_m^+$, searching for generators of the *subgroup* Cl_m^- mechanically increases the provable upper bound on generators. More precisely, writing as $\mathfrak{L}_{\max}^-$ the biggest ideal of a generating set of Cl_m^-, Wesolowski proved [Wes18, Rem. 2] that $\mathcal{N}(\mathfrak{L}_{\max}^-) \leq \left(2.71 h_m^+ \cdot \ln |\Delta_{K_m}| + 4.13 \right)^2$.

Finally, we use the notation $h_{m,(\mathfrak{L}_1,\ldots,\mathfrak{L}_k)}$ to denote the cardinal of the subgroup of Cl_m generated by the k classes $[\mathfrak{L}_i]$, i.e., the determinant of the kernel of $\mathfrak{f}_{\mathfrak{L}_1,\ldots,\mathfrak{L}_k} : (e_1, \ldots, e_k) \in \mathbb{Z}^k \longmapsto \prod_{1 \leq i \leq k} [\mathfrak{L}_i]^{e_i} \in \mathrm{Cl}_m$.

2.3   Logarithmic $\mathcal{S}$-embeddings

We introduce log-$\mathcal{S}$-unit lattices and discuss proper normalization by the Product Formula that was at the heart of the practical improvements of [BR20] compared to [PHS19].

Places of the cyclotomic field K_m are usually split into two parts: the set $\mathcal{S}_\infty$ of *infinite* places can be identified with the (complex) embeddings of K_m into $\mathbb{C}$, up to conjugation; the set $\mathcal{S}_0$ of *finite* places is specified by the infinite set of prime ideals of K_m, each prime ideal $\mathfrak{p}$ inducing an embedding of K_m into its $\mathfrak{p}$-adic completion $K_{m,\mathfrak{p}}$. Hence, any place $v \in \mathcal{S}_\infty \cup \mathcal{S}_0$ induces an absolute value $|\cdot|_v$ on K_m, and Ostrowski's theorem for number fields [Nar04, Th. 3.3] shows that all possible absolute values on K_m are obtained in this way. Concretely, for $\alpha \in K_m$: $\forall \sigma \in \mathcal{S}_\infty, |\alpha|_\sigma = |\sigma(\alpha)|$ and $\forall \mathfrak{p} \in \mathcal{S}_0, |\alpha|_\mathfrak{p} = p^{-v_\mathfrak{p}(\alpha)}$, where $v_\mathfrak{p}(\cdot)$ is the valuation of α at $\mathfrak{p}$ and $\langle p \rangle = \mathfrak{p} \cap \mathbb{Z}$. A remarkable fact is that all these absolute values are tied by the *Product Formula* [Nar04, Th. 3.5]:

$$\forall \alpha \in K_m, \qquad \prod_{v \in \mathcal{S}_\infty \cup \mathcal{S}_0} |\alpha|_v^{[K_{m,v}:\mathbb{Q}_v]} = 1. \qquad (2.1)$$

The $\mathcal{S}_\infty$-part of this product is $|\mathcal{N}(\alpha)|$, as for $\sigma \in \mathcal{S}_\infty$, $K_{m,\sigma} = \mathbb{C}$ and $\mathbb{Q}_\sigma = \mathbb{R}$, so that $[K_{m,\sigma} : \mathbb{Q}_\sigma] = 2$. Similarly, for $\mathfrak{p} \in \mathcal{S}_0$, we have $|\alpha|_\mathfrak{p}^{[K_{m,\mathfrak{p}}:\mathbb{Q}_p]} = \mathcal{N}(\mathfrak{p})^{-v_\mathfrak{p}(\alpha)}$.

$\mathcal{S}$-unit Group Structure. Fix a finite set $\mathcal{S}$ of places; in this paper we shall consider that $\mathcal{S}$ *always* contains $\mathcal{S}_\infty$. The so-called $\mathcal{S}$-unit group of K_m, denoted by $\mathcal{O}_{K_m,\mathcal{S}}^\times$, is the multiplicative subgroup of K_m generated by all elements whose valuations are non zero only at the finite places of $\mathcal{S}$. Formally:

$$\mathcal{O}_{K_m,\mathcal{S}}^\times = \left\{ \alpha \in K_m; \ \langle \alpha \rangle = \prod_{\mathfrak{p} \in \mathcal{S} \cap \mathcal{S}_0} \mathfrak{p}^{v_\mathfrak{p}(\alpha)} \right\}.$$

Theorem 2.1 (Dirichlet-Chevalley-Hasse [Nar04, Th. III.3.12, Cor. 1]**).** *The $\mathcal{S}$-unit group is the direct product of the group of roots of unity $\mu(\mathcal{O}_{K_m}^\times)$ and a free abelian group with $|\mathcal{S}| - 1$ generators. There exists a fundamental system of $\mathcal{S}$-units $\varepsilon_1, \ldots, \varepsilon_{|\mathcal{S}|-1}$ such that any $\varepsilon \in \mathcal{O}_{K_m,\mathcal{S}}^\times$ is uniquely written as: $\varepsilon = \mu \cdot \prod_{i=1}^{|\mathcal{S}|-1} \varepsilon_i^{k_i}$, where $\mu \in \langle \pm\zeta_m \rangle$ is a root of unity and $k_i \in \mathbb{Z}$.*

Log-$\mathcal{S}$-unit Lattice. A fundamental ingredient of the proof of this theorem is to build an embedding of $\mathcal{O}_{K_m,\mathcal{S}}^\times$ into the real space of dimension $|\mathcal{S}|$, whose kernel is $\mu(\mathcal{O}_{K_m}^\times)$ and whose image is a lattice of dimension $(|\mathcal{S}|-1)$. This embedding is called the *logarithmic $\mathcal{S}$-embedding*, and its image is called the log-$\mathcal{S}$-unit lattice.

Several equivalent definitions of this logarithmic $\mathcal{S}$-embedding are acceptable for the proof. However, for cryptanalytic purposes, experimental evidence [BR20] suggests that it is crucial to use a properly normalized embedding for the decodability of the log-$\mathcal{S}$-unit lattice. Thus, we define [Nar04, §3, p.98]:

$$\mathrm{Log}_\mathcal{S}\, \alpha = \left([K_{m,v} : \mathbb{Q}_v] \cdot \ln|\alpha|_v \right)_{v \in \mathcal{S}} = \left(\{2\ln|\sigma(\alpha)|\}_{\sigma \in \mathcal{S}_\infty}, \{-v_\mathfrak{p}(\alpha) \ln \mathcal{N}(\mathfrak{p})\}_{\mathfrak{p} \in \mathcal{S} \cap \mathcal{S}_0} \right).$$

By definition of $\mathcal{O}_{K_m,\mathcal{S}}^\times$, $\mathbb{R} \otimes \mathrm{Log}_\mathcal{S}\, \mathcal{O}_{K_m,\mathcal{S}}^\times$ is included in the hyperplane orthogonal to $\mathbf{1}_{|\mathcal{S}|}$. Showing that its dimension is at least $|\mathcal{S}| - 1$ is more involved.

A basis of the log-$\mathcal{S}$-unit lattice is given by the images $\mathrm{Log}_\mathcal{S}\, \varepsilon_i$ of the fundamental system of $\mathcal{S}$-units of Theorem 2.1, as in [BR20, Eq. (2.7)]. Actually, we shall use later that for any maximal set of independent $\mathcal{S}$-units, their images under any logarithmic $\mathcal{S}$-embedding form a full rank sublattice of the corresponding log-$\mathcal{S}$-unit lattice. Its volume is given by [BR20, Pr. 2.2 and Eq. (2.8)].

As mentioned in [PHS19, BDPW20, BR20], a convenient trick in the context of the cryptanalysis of id-SVP is to consider an *expanded* version of the logarithmic $\mathcal{S}$-embedding, halving and repeating twice $\mathcal{S}_\infty$-coordinates, namely:

$$\overline{\mathrm{Log}}_{\mathcal{S}}\,\alpha = \Big(\{\ln|\sigma(\alpha)|, \ln|\sigma(\alpha)|\}_{\sigma \in \mathcal{S}_\infty}\,, \{[K_{m,\mathfrak{p}} : \mathbb{Q}_p] \cdot \ln|\alpha|_{\mathfrak{p}}\}_{\mathfrak{p} \in \mathcal{S} \cap \mathcal{S}_0} \Big).$$

In particular, this reduces the volume of the log-$\mathcal{S}$-unit lattice, as shown by [BR20, Pr. 2.3]. In practice though, we did not observe any fundamental difference between the approximation factors obtained using $\mathrm{Log}_{\mathcal{S}}$ or $\overline{\mathrm{Log}}_{\mathcal{S}}$.

2.4   Hard Problems in Number Theory

One of the most difficult classical steps of the Approx-id-SVP algorithms proposed in [CDW17, PHS19, BR20, CDW21] is to find a solution to the ClDL defined as:

Problem 2.2 (Class Group Discrete Logarithm (ClDL)). Given a basis of prime ideals $\{\mathfrak{L}_1, \ldots, \mathfrak{L}_k\}$, and a challenge ideal $\mathfrak{b}$, find $\alpha \in K_m$ and integers $e_1, \ldots, e_k$ such that $\langle\alpha\rangle = \mathfrak{b} \cdot \prod_i \mathfrak{L}_i^{e_i}$, if this decomposition exists.

In this definition, we also ask for an *explicit* element α of the field, contrary to the definition of, e.g., [CDW17, Pr. 2]. Nevertheless, we note that in both quantum and classical worlds, the standard way to solve this problem boils down to computing $\mathcal{S}$-units, for $\mathcal{S}$ containing $\mathfrak{b}$ and the $\mathfrak{L}_i$'s, so that this explicit element is a byproduct of the resolution. Furthermore, put in this form it encompasses the well-known *Principal Ideal Problem* (PIP), using an empty set of ideals.

The *Shortest Generator Problem* (SGP) asks, from a generator α of a principal ideal, for the shortest generator α' such that $\langle\alpha\rangle = \langle\alpha'\rangle$. Similarly, we define:

Problem 2.3 (Shortest Class Group Discrete Logarithm (S-ClDL)). Given a solution $\langle\alpha\rangle = \mathfrak{b} \cdot \prod_i \mathfrak{L}_i^{e_i}$ to the ClDL problem, find $w_1, \ldots, w_k \in \mathbb{Z}_{\geq 0}$ and $\alpha' \in K_m$ such that $\langle\alpha'\rangle = \mathfrak{b} \cdot \prod_i \mathfrak{L}_i^{w_i}$ and α' is the smallest possible one.

The condition for the w_i's to be positive is crucial. Note that all recent algorithms for Approx-id-SVP that are not bound to principal ideals eventually output an approximate solution of the S-ClDL [CDW21, PHS19, BR20]. If the set of prime ideals is sufficiently large compared to $\mathfrak{b}$, then S-ClDL is exactly id-SVP.

We also mention the Close Principal Multiple (CPM) problem which, given an ideal $\mathfrak{b}$, asks to find $\mathfrak{c}$ such that $\mathfrak{b}\mathfrak{c}$ is principal and $\mathcal{N}(\mathfrak{c})$ is small. This specific problem is used in [CDW21], and the authors prove that under GRH and using a factor base containing all prime ideals of norm up to $m^{4+o(1)}$, there exists a solution $\mathfrak{c}$ with $\mathcal{N}(\mathfrak{c}) \leq \exp(\widetilde{O}(m^{1+o(1)}))$ [CDW21, §1.3.4].

Complexities. As shown in [BS16], class groups, unit groups, class group discrete logarithms and principal ideal generator computations can be reduced to $\mathcal{S}$-unit groups computations for appropriate sets of places $\mathcal{S}$. Denote by $\mathrm{T}_{\mathcal{S}}(K_m)$ the running time of the computation of the $\mathcal{S}$-unit group in K_m. Under GRH, in a quantum setting, $\mathrm{T}_{\mathcal{S}}(K_m) = \mathrm{poly}\big(\ln|\Delta_{K_m}|, |\mathcal{S}|, \max_{\mathfrak{p}\in\mathcal{S}} \ln \mathcal{N}(\mathfrak{p})\big)$ by [EHKS14, BS16]. In a classical setting, $\mathrm{T}_{\mathcal{S}}(K_m) = \mathrm{poly}\big(|\mathcal{S}|, \max_{\mathfrak{p}\in\mathcal{S}} \ln \mathcal{N}(\mathfrak{p})\big) \cdot \exp \widetilde{O}\big(\ln^{2/3}(|\Delta_K|)\big)$ is mainly subexponential in the degree of the cyclotomic field K_m [BF14, PHS19]. The exponent can be lowered to $1/2$ when m is a prime power [BEF+17].

2.5   Lattices

Let L be a Euclidean lattice of full rank n. The first minimum $\lambda_1(L)$ of L is defined as the ℓ_2-norm of the smallest vector $\mathbf{v} \in L^*$, and the ℓ_2-distance from $\mathbf{t}$ to L, for any $\mathbf{t}$ in the span $L \otimes \mathbb{R}$ of L, is defined by $\mathrm{dist}_2(L, \mathbf{t}) = \min_{\mathbf{v}\in L}\|\mathbf{t} - \mathbf{v}\|_2$.

The *Approximate Shortest Vector Problem* (Approx-Svp) is, given a lattice L and an approximation factor af, to find $\mathbf{v} \in L$ such that $\|\mathbf{v}\|_2 \leq \mathrm{af} \cdot \lambda_1(L)$. Similarly, the *Approximate Closest Vector Problem* (Approx-Cvp) asks, given a lattice L, an approximation factor af and a target $\mathbf{t}$ in the span $L \otimes \mathbb{R}$ of L, for a vector $\mathbf{v} \in L$ such that $\|\mathbf{t} - \mathbf{v}\|_2 \leq \mathrm{af} \cdot \mathrm{dist}_2(L, \mathbf{t})$. A practical Approx-Cvp oracle is given by Babai's Nearest Plane algorithm [Bab86].

Bounding Approximation Factors. An ideal lattice of K_m is the full-rank image under the Minkowski embedding in $\mathbb{R}^{\varphi(m)}$ of a fractional ideal $\mathfrak{b}$ of K_m. Unlike generic lattices, a lower bound of the first minimum is implied by the arithmetic-geometric mean inequality, using that for any $b \in \mathfrak{b}$, $\mathcal{N}(\mathfrak{b})$ divides $|\mathcal{N}(b)|$. Thus:

$$\sqrt{n} \cdot \mathcal{N}(\mathfrak{b})^{1/n} \leq \lambda_1(\mathfrak{b}) \leq \sqrt{n} \cdot \mathcal{N}(\mathfrak{b})^{1/n}\sqrt{|\Delta_{K_m}|}^{1/n}, \qquad (2.2)$$

where $n = \varphi(m) = \deg K_m$ and the right inequality is Minkowski's inequality. Actually, applying the Gaussian Heuristic to ideal lattices would give that on average, $\lambda_1(\mathfrak{b}) \approx \sqrt{\frac{n}{2\pi e}} \cdot \mathrm{Vol}^{1/n}(\mathfrak{b})$, where $\mathrm{Vol}(\mathfrak{b}) = \mathcal{N}(\mathfrak{b})\sqrt{|\Delta_{K_m}|}$. This hypothesis is commonly used for the analysis of cryptosystems based on structured lattices, and we note that the *exact* approximation factors reached by the Twisted-PHS algorithm in [BR20] match this heuristic.

For any $\mathbf{x} \in \mathfrak{b}$, let $\mathrm{af}(\mathbf{x}) = \|\mathbf{x}\|_2/\lambda_1(\mathfrak{b})$ denote the approximation factor reached by $\mathbf{x}$ for the Svp in the ideal lattice $\mathfrak{b}$. In general, $\lambda_1(\mathfrak{b})$ is not known, but Eq. (2.2) imply the bounds $\mathrm{af}_{\mathsf{inf}}(\mathbf{x}) \leq \mathrm{af}(\mathbf{x}) \approx \mathrm{af}_{\mathsf{gh}}(\mathbf{x}) \leq \mathrm{af}_{\mathsf{sup}}(\mathbf{x})$, where:

$$\mathrm{af}_{\mathsf{inf}}(\mathbf{x}) := \frac{\|\mathbf{x}\|_2}{\sqrt{n} \cdot \mathrm{Vol}^{1/n}(\mathfrak{b})}, \qquad \mathrm{af}_{\mathsf{sup}}(\mathbf{x}) := \frac{\|\mathbf{x}\|_2}{\sqrt{n} \cdot \mathcal{N}(\mathfrak{b})^{1/n}},$$

$$\mathrm{af}_{\mathsf{gh}}(\mathbf{x}) := \sqrt{2\pi e} \cdot \mathrm{af}_{\mathsf{inf}}(\mathbf{x}). \qquad (2.3)$$

Quality of a Lattice Basis. Several indicators have been used in the literature to attempt to measure the quality of a lattice basis $B = (\mathbf{b}_1, \ldots, \mathbf{b}_n)$ relatively to the Svp or the Cvp. We will focus on the following three standard quantities:

1. the root-Hermite Factor $\delta_0(B)$, defined by $\delta_0^n(B) = \|\mathbf{b}_1\|_2/\mathrm{Vol}^{1/n} B$, is commonly used to compare lattice reduction algorithms like LLL [LLL82] or BKZ [CN11]. On average, LLL reaches $\delta_0 \approx 1.022$ [GN08] whereas BKZ with block-size $b \geq 50$ heuristically yields $\delta_0 \approx \big(\frac{b}{2\pi e}(\pi b)^{1/b}\big)^{1/(2b-2)}$ [Che13].

2. the (normalized) orthogonality defect $\delta(B)$, given by $\delta^n(B) = \prod_i \left(\frac{\|\mathbf{b}_i\|_2}{\mathrm{Vol}^{1/n} B} \right)$ [MG02, Def. 7.5] involves all vectors of the basis. By Minkowski's second theorem, its smallest possible value is upper bounded by $\sqrt{1 + \frac{n}{4}}$.

3. the logarithms of the norms of *Gram-Schmidt Orthogonalization* (GSO) vectors $\mathbf{b}_i^\star$ give also valuable information. For example, a rapid decrease in the sequence $\ln \|\mathbf{b}_i^\star\|_2$ at $i \geq 2$ indicates that $\mathbf{b}_i$ is rather not orthogonal to the previously generated subspace $\langle \mathbf{b}_1, \ldots, \mathbf{b}_{i-1} \rangle$.

3   An Explicit Full-Rank Family of Independent $\mathcal{S}$-units

In this section, we exhibit a full rank family of *independent* $\mathcal{S}$-units, where the finite places $\mathcal{S}$ correspond to a collection of full Galois orbits of split prime ideals. As mentioned in introduction, this family is composed of three parts:

1. Circular units are recalled in Sect. 3.1 using the material from [Kuč92, Th. 6.1];
2. Stickelberger generators are in Sect. 3.2, sticking to the exposition of [BK21];
3. Real $\mathcal{S}^+$-units (apart from real units), where $\mathcal{S}^+$ is the set $\mathcal{S} \cap K_m^+$ of places of $\mathcal{S}$ restricted to K_m^+, are in Sect. 3.3.

Considering real $\mathcal{S}^+$-units and proving in Sect. 3.4 the multiplicative index of our family in the full $\mathcal{S}$-unit group constitute our main theoretical contributions. Finally, the saturation process used to mitigate this index is described in Sect. 3.5.

Remark 3.1 Recall that m has prime factorization $m = q_1 q_2 \cdots q_t \not\equiv 2 \mod 4$, where $q_i = p_i^{c_i} > 2$ for $i \in [\![1, t]\!]$. In this section, we will use subsets M_m^+ and M_m' of $[\![1, m]\!]$ that are useful to describe resp. a fundamental family of circular units and a short $\mathbb{Z}$-basis of the Stickelberger ideal of K_m. Their precise definitions from resp. [Kuč92, p.293] and [BK21, Eq. (11)] can be found in [BLNR21, §A.1].

3.1   Circular Units

Circular units are sometimes called *cyclotomic units* in the literature, as in [Was97, §8]. We prefer to use the historical terminology from algebraic number theory, e.g. Sinnott [Sin78, §4] and Kučera [Kuč92, §2], in order to avoid any confusion with the whole unit group $\mathcal{O}_{K_m}^\times$ of the m-th cyclotomic field.

Definition 3.2 (Circular units [Was97, §8.1]). *Let V_m be the multiplicative subgroup of $K_m^\times$ generated by $\{1 - \zeta_m^a; \ 1 \leq a \leq m\}$. The group of circular units is the intersection $C_m := V_m \cap \mathcal{O}_{K_m}^\times$.*

Note that V_m contains the torsion of K_m, since $-\zeta_m = (1 - \zeta_m) / (1 - \zeta_m^{-1})$. The circular units form a subgroup of $\mathcal{O}_{K_m}^\times$ of finite index, more precisely:

Proposition 3.3 ([Sin78, Th. p.107]). *The index of C_m in $\mathcal{O}_{K_m}^\times$ is finite:*

$$[\mathcal{O}_{K_m}^\times : C_m] = 2^b \cdot h_m^+, \qquad \text{with } b = \begin{cases} 0 & \text{if } t = 1, \\ 2^{t-2} + 1 - t & \text{otherwise,} \end{cases}$$

where t is the number of distinct prime divisors of m.

Hence, circular units provide a very large subgroup of $\mathcal{O}_{K_m}^\times$: indeed, the real part of the class number is expected to be small (Sect. 2.2), and the other factor *generically* grows linearly in m (see [HW38, Th. 430 and 431] for a precise statement).

An explicit system of fundamental circular units for any m has been given in [GK89] and independently in [Kuč92, Th. 6.1]. More precisely, for $0 < a < m$, define the following special circular units, where $m_i = m/p_i^{e_i}$ [Kuč92, p.176]:

$$v_a = \begin{cases} 1 - \zeta_m^a & \text{if } \forall i \in [\![1, t]\!], m_i \nmid a, \\ \dfrac{1 - \zeta_m^a}{1 - \zeta_m^{m_i}} & \text{otherwise, for the unique } m_i \mid a. \end{cases} \tag{3.1}$$

Theorem 3.4 ([Kuč92, Th. 6.1]). *Recall the definition of $M_m^+ \subsetneq [\![1, m]\!]$ can be found in [BLNR21, §A.1]. The set $\{v_a; \ a \in M_m^+\}$ is a system of fundamental circular units of K_m: for any circular unit $\eta \in C_m$, there exist a uniquely determined map $k : M_m^+ \to \mathbb{Z}$, and a root of unity $\mu \in \langle \pm \zeta_m \rangle$ s.t. $\eta = \mu \cdot \prod_{a \in M_m^+} v_a^{k(a)}$.*

A crucial point for the cryptanalysis of id-Svp in [CDW21] is that the logarithmic embedding of these elements is short. Namely, explicitly writing the constants that appear in the proof of [CDW21, Lem. 3.5], we have, for any $0 < a < m$, that $\|\mathrm{Log}_{\mathcal{S}_\infty}(1 - \zeta_m^a)\|_2 \leq 1.32 \cdot \sqrt{m}$.

3.2   Stickelberger Generators

In this section, we use [BK21, Th. 3.1] to describe a short *basis* of the so-called Stickelberger ideal, viewed as a $\mathbb{Z}$-module. These Stickelberger short relations correspond to principal ideals whose generators are surprisingly easy to compute using Jacobi sums as in [BK21, §6]. Following Sinnott [Sin80], for all $a \in \mathbb{Z}$, let:

$$\theta_m(a) = \sum_{s \in (\mathbb{Z}/m\mathbb{Z})^\times} \left\{ -\frac{as}{m} \right\} \cdot \sigma_s^{-1} \in \mathbb{Q}[G_m], \tag{3.2}$$

and let N_m be the absolute norm element $N_m = \sum_{\sigma \in G_m} \sigma$.

Definition 3.5 (Stickelberger ideal [Sin80, p.189]**).** *Let $\mathcal{S}_m'$ be the $\mathbb{Z}$-module of $\mathbb{Q}[G_m]$ generated by $\{\theta_m(a); \ 0 < a < m\} \cup \{\frac{1}{2} N_m\}$. The Stickelberger ideal of K_m is the intersection $\mathcal{S}_m = \mathcal{S}_m' \cap \mathbb{Z}[G_m]$.*

As in [CDW21], we shall refer to the *Stickelberger lattice* when $\mathcal{S}_m$ is viewed as a $\mathbb{Z}$-module. Note that in some references, like in [Was97, §6.2], the Stickelberger ideal is defined as the smaller ideal $\mathbb{Z}[G_m] \cap \theta_m(-1)\mathbb{Z}[G_m]$, which coincides with Def. 3.5 if and only if m is a prime power [Kuč86, Pr. 4.3].

Theorem 3.6 (Stickelberger's theorem [Sin80, Th. 3.1]**).** *The Stickelberger ideal $\mathcal{S}_m$ of K_m annihilates the class group of K_m. Hence, for any ideal $\mathfrak{b}$ of K_m and any $\alpha = \sum_{\sigma \in G_m} a_\sigma \sigma \in \mathcal{S}_m$, the ideal $\mathfrak{b}^\alpha = \prod_{\sigma \in G_m} \sigma(\mathfrak{b})^{a_\sigma}$ is principal.*

An outstanding point is that the proof of this important result is completely explicit, i.e., for any $\alpha \in \mathcal{S}_m$, and any fractional ideal $\mathfrak{b}$ of K_m, an explicit $\gamma \in K_m$ s.t. $\langle \gamma \rangle = \mathfrak{b}^\alpha$ is constructed. It appears that when α is a short element of $\mathcal{S}_m$, this explicit generator is very efficiently computable.

A Short Basis of the Stickelberger Lattice. An element of the integral group ring $\mathbb{Z}[G_m]$ is called *short* if it is of the form $\sum_{\sigma \in G_m} a_\sigma \sigma \in \mathbb{Z}[G_m]$, where $a_\sigma \in \{0, 1\}$ for all $\sigma \in G_m$. Short elements of $\mathcal{S}_m$ have been identified in [Sch08, Th. 9.3(i) and Ex. 9.3] in the prime conductor case, and the proof has been adapted to any conductor in [CDW21, Lem. 4.4] to prove the shortness of the following generating set of $\mathcal{S}_m$:

$$W = \{w_a; \ a \in [\![2, m]\!]\}, \quad \text{with } w_a = \theta_m(1) + \theta_m(a-1) - \theta_m(a). \tag{3.3}$$

Note that using $\theta_m(a) + \theta_m(-a) = N_m$ when $m \nmid a$, we obtain $w_a = w_{m-a+1}$ whenever $1 < a < m$, and that $w_m = N_m$ using also $\theta_m(m) = 0$. Hence, W is the set $\{w_a; \ 2 \leq a \leq \lceil \frac{m}{2} \rceil\} \cup \{N_m\}$.

We emphasize that only knowing a generating set of short elements as in [CDW21] is not necessarily sufficient. Though it would be possible to build a basis from this generating set to solve the CVP like in [CDW21, Cor. 2.2], without any geometric loss using e.g. [MG02, Lem. 7.1], we observed that the slight euclidean norm growth of the obtained basis vectors translates into a dramatic increase of the size of the (possibly rational) coefficients of the corresponding generators, in a way that significantly hinders subsequent computations. In particular, in order to climb dimensions as far as possible and best approach log-$\mathcal{S}$-unit lattices using the saturation process described in Sect. 3.5, it is crucial to constrain both the number of elements we use and their size, i.e., to use a *basis* of the Stickelberger lattice containing only *short* elements. In [BK21], a very large family of short elements [BK21, Pr. 3.1] encompassing $W \setminus \{N_m\}$ is made explicit:

Proposition 3.7 ([BK21, Pr. 3.1]**).** *Let a, $b \in \mathbb{Z}$ satisfying $m \nmid a$, $m \nmid b$ and $m \nmid (a+b)$. Then $\alpha = \theta_m(a) + \theta_m(b) - \theta_m(a+b)$ is a short element of $\mathcal{S}_m$. Moreover, $(1 + \tau) \cdot \alpha = N_m$, so exactly one half of the coefficients of α are zeros.*

Then, from this family, a short basis is computationally easy to extract:

Theorem 3.8 ([BK21, Th. 3.6]**).** *Recall $M'_m \subsetneq [\![1, m]\!]$ is defined in [BLNR21, §A.1]. There exists an efficiently computable map $\alpha_m(\cdot)$ from $[\![1, m]\!]$ to the family of short elements of $\mathcal{S}_m$ described in Pr. 3.7, s.t. $\{\alpha_m(c); \ c \in M'_m\} \cup \{N_m\}$ is a $\mathbb{Z}$-basis of the Stickelberger lattice $\mathcal{S}_m$ of K_m having only short elements.*

The explicit definition of $\alpha_m(\cdot)$ is given in [BK21, §3.2], and included for completeness in [BLNR21, §A.2]. We stress that when m is a prime, this basis coincides with the one given by [Sch08, Th. 9.3(i)] and with the set W in Eq. (3.3).

Effective Stickelberger Generators Using Jacobi Sums. As previously mentioned, the proof of Theorem 3.6 is explicit, i.e., for any $\alpha \in \mathcal{S}_m$ and any fractional ideal $\mathfrak{b}$ of K_m, it builds an explicit $\gamma \in K_m$ s.t. $\langle \gamma \rangle = \mathfrak{b}^\alpha$ [Was97, §6.2], [Sin80, §3.1]. Moreover, when α is a short basis element from Theorem 3.8, it turns out that γ has a simple expression using Jacobi sums [BK21, §5].

We briefly treat the split case here. Let $\ell \in \mathbb{Z}$ be a prime s.t. $\ell \equiv 1 \mod m$, and let $\mathfrak{L}$ be any fixed (split) prime ideal of K_m above ℓ. Let a, b be such as in Proposition 3.7, then for $\alpha = \theta_m(a) + \theta_m(b) - \theta_m(a+b)$, we have that $\mathfrak{L}^\alpha$ is a principal ideal generated by the following Jacobi sum [BK21, Pr. 5.1]:

$$\mathcal{J}_{\mathfrak{L}}(a,b) = - \sum_{u \in \mathcal{O}_{K_m}/\mathfrak{L}} \chi_{\mathfrak{L}}^a(u)\chi_{\mathfrak{L}}^b(1-u) \quad \in K_m, \tag{3.4}$$

where $\chi_{\mathfrak{L}}(u) \in \langle \zeta_m \rangle$ verifies $\chi_{\mathfrak{L}}(u) \equiv u^{(\ell-1)/m} \mod \mathfrak{L}$, for any $u \in \left(\mathcal{O}_{K_m}/\mathfrak{L} \right)^\times$, and $\chi_{\mathfrak{L}}(0) = 0$. When $\alpha = \alpha_m(c)$ for $c \in M'_m$, we shall write $\gamma_{\mathfrak{L},c}^-$ for the generator of $\mathfrak{L}^{\alpha_m(c)}$. Using a discrete logarithm table for elements of $(\mathcal{O}_{K_m}/\mathfrak{L})^\times$, the computation, for a fixed prime $\mathfrak{L}$, of all Jacobi sums corresponding to the short basis $\{\alpha_m(c); \ c \in M'_m\}$ is very fast.

3.3   Real $\mathcal{S}^+$-units

A consequence of Theorem 3.8, since $|M'_m| = \frac{\varphi(m)}{2}$, is that the Stickelberger lattice only has rank $\frac{\varphi(m)}{2} + 1$ in $\mathbb{Z}[G_m]$; in particular, it is not full rank, hence cannot be directly used as a lattice of class relations. In previous works, obtaining a full rank lattice in $\mathbb{Z}[G_m]$ from $\mathcal{S}_m$ was done by projecting into $(1-\tau)\mathcal{S}_m$ [CDW21, §4.3], or by the adjunction of $(1+\tau)\mathbb{Z}[G_m]$ [CDW17, Def. 2]. Both can be used as a lattice of class relations for the *relative* class group Cl_m^-. In particular, the so-called *augmented* Stickelberger lattice $\mathcal{S}_m + (1+\tau)\mathbb{Z}[G_m]$ annihilates the relative class group and has full rank in $\mathbb{Z}[G_m]$, as shown in [CDW17, Lem. 2].

We generalize this result by considering the module of all real class group relations between relative norm ideals of ideals from the entire class group Cl_m. In Sect. 3.4, we shall prove that the Stickelberger lattice augmented with these real class group relations yields a lattice of class relations for the *whole* class group. Note that, as opposed to other modules like $(1-\tau)\mathcal{S}_m$ or $\mathcal{S}_m + (1+\tau)\mathbb{Z}[G_m]$, real class group relations actually depend on the underlying prime ideals.

On one hand, this affects negatively the shortness of the obtained relation vectors: putting those in Hermite Normal Form, we shall see later that each relation, viewed as a vector of integer valuations, has ℓ_2-norm at most h_m^+. On the other hand, removing the constraint to belong to the relative class group brings a significant practical and theoretical gap: first, it allows choosing prime ideals of smallest possible norms, which as shown in [BR20, §3.3] or [CDW21, Th. 4.8] lowers in practice the obtained approximation factor; second, whereas prime ideals of norm at most Bach's bound are sufficient to generate the entire class group, prime generators for the *relative* class group are only proven to be of norm bounded by the *larger* bound $(2.71 \cdot h_m^+ \cdot \ln \Delta_{K_m} + 4.13)^2$ from [Wes18].

Lifting Real Class Group Relations. Let $\ell_1, \ldots, \ell_d$ be distinct prime integers satisfying $\ell_i \equiv 1 \mod m$, so that ℓ_i splits in K_m, for all i in $[\![1, d]\!]$. For each i, fix a prime ideal $\mathfrak{L}_i \mid \ell_i$ in K_m of norm ℓ_i, and let $\mathfrak{l}_i = \mathcal{N}_{K_m/K_m^+}(\mathfrak{L}_i) = \mathfrak{L}_i^{1+\tau} \cap K_m^+$ be the relative norm ideal of $\mathfrak{L}_i$. Since $\mathfrak{L}_i$ is a split prime ideal of K_m dividing ℓ_i, the ideal $\mathfrak{l}_i$ is a split prime ideal of K_m^+ of norm ℓ_i, and by Kummer-Dedekind's theorem we have $\mathfrak{l}_i \cdot \mathcal{O}_{K_m} = \mathfrak{L}_i^{1+\tau}$. This justifies the slight abuse of notation of writing $\mathfrak{l}_i^\sigma = \mathfrak{L}_i^{(1+\tau)\sigma} \cap K_m^+$, for any $\sigma \in G_m$.

We are interested in the real class group relations between all prime ideals in the G_m^+-orbits of the $\mathfrak{l}_i$, i.e., between the following prime ideals of K_m^+:

$$\{\mathfrak{l}_i^{\sigma_s};\ i \in [\![1, d]\!], 0 < s < \tfrac{m}{2}, (s, m) = 1\}. \tag{3.5}$$

The important point is, any class relation in K_m^+ between ideals from Eq. (3.5) translates to a class relation in K_m using repeatedly $\mathfrak{l}_i^\sigma \cdot \mathcal{O}_{K_m} = \mathfrak{L}_i^{(1+\tau)\sigma}$. More precisely, let $(r_1, \ldots, r_d) \in \mathbb{Z}[G_m^+]^d$ represent a real class relation in K_m^+ between ideals $\{\mathfrak{l}_i^{\sigma_s}\}$ of Eq. (3.5), i.e., there exists $\gamma_r^+ \in K_m^+$ s.t. $\gamma_r^+ \cdot \mathcal{O}_{K_m^+} = \prod_{i=1}^d \mathfrak{l}_i^{r_i}$. Then, this relation lifts naturally to a class relation $((1+\tau) \cdot r_1, \ldots, (1+\tau) \cdot r_d)$ in K_m between prime ideals in the G_m-orbits $\{\mathfrak{L}_i^\sigma;\ i \in [\![1, d]\!], \sigma \in G_m\}$ as:

$$\gamma_r^+ \cdot \mathcal{O}_{K_m} = \prod_{i=1}^d \mathfrak{L}_i^{(1+\tau)r_i}. \tag{3.6}$$

Let $C_{\mathfrak{l}_1,\ldots,\mathfrak{l}_d}^+$ denote the lattice of class relations between elements of all G_m^+-orbits of $\{\mathfrak{l}_i;\ i \in [\![1, d]\!]\}$. Concretely, it is the kernel of the following map:

$$\mathfrak{f}_{\mathfrak{l}_1,\ldots,\mathfrak{l}_d} : (r_{i,s})_{\substack{1 \le i \le d, \\ 0 < s < m/2, (s,m)=1}} \in \mathbb{Z}^{d \cdot \frac{\varphi(m)}{2}} \longmapsto \prod_{i,s} [\mathfrak{l}_i^{\sigma_s}]^{r_{i,s}} \in \mathrm{Cl}_m^+. \tag{3.7}$$

Using the canonical isomorphism of $\mathbb{Z}$-modules $\mathbb{Z}^{d \cdot \frac{\varphi(m)}{2}} \simeq_\mathbb{Z} \mathbb{Z}[G_m^+]^d$, the lattice of class relations $C_{\mathfrak{l}_1,\ldots,\mathfrak{l}_d}^+$ may be viewed as a $\mathbb{Z}$-submodule of $\mathbb{Z}[G_m^+]^d$. Lifting all these relations back to K_m as in Eq. (3.6), we therefore obtain the submodule $(1+\tau) \cdot C_{\mathfrak{l}_1,\ldots,\mathfrak{l}_d}^+ \subseteq (1+\tau)\mathbb{Z}[G_m]^d$, that we shall call the lattice of *real class relations* between the G_m-orbits of $\{\mathfrak{L}_i;\ i \in [\![1, d]\!]\}$.

Remark 3.9. When $h_m^+ = 1$, $C_{\mathfrak{l}_1,\ldots,\mathfrak{l}_d}^+$ is isomorphic to d copies of the integral group ring $\mathbb{Z}[G_m^+]$ and the lattice of real class relations is simply $(1+\tau)\mathbb{Z}[G_m]^d$.

Euclidean Norm of Real Class Relations. We now identify a real class group relation from $C_{\mathfrak{l}_1,\ldots,\mathfrak{l}_d}^+$ to a vector in $\mathbb{Z}^{d \cdot \frac{\varphi(m)}{2}}$. In other words, we consider only the valuations of these relations on the G_m^+-orbits of the prime ideals $\mathfrak{l}_1, \ldots, \mathfrak{l}_d$. Furthermore, $C_{\mathfrak{l}_1,\ldots,\mathfrak{l}_d}^+$ is put in Hermite Normal Form, conveniently for the proof of the following proposition, provided in the full version of this paper [BLNR21], but better bounds might easily be obtained using e.g. the LLL algorithm.

Proposition 3.10. *Suppose the lattice $C_{\mathfrak{l}_1,\ldots,\mathfrak{l}_d}^+$ of real class relations is in HNF. Then, for all $\mathbf{w} \in C_{\mathfrak{l}_1,\ldots,\mathfrak{l}_d}^+ \subseteq \mathbb{Z}[G_m^+]^d$, we have $\|\mathbf{w}\|_2 \le \|\mathbf{w}\|_1 \le h_m^+$.*

This means that $(1+\tau)\cdot C^+_{l_1,\ldots,l_d}$ can be used in the CDW algorithm instead of $(1+\tau)\mathbb{Z}[G_m]^d$, as we will see in Sect. 4, while still reaching the same asymptotic approximation factor under the same assumption on the Galois-module structure of Cl_m [CDW21, Ass. 1], as long as $h^+_m \leq O(\sqrt{m})$. This slightly more restrictive hypothesis (see the discussion in Sect. 2.2) will be more than compensated by the fact that it removes the need for the l_i's to be principal, which has a significant impact in practice on the algebraic norm of the chosen ideals, and thus on the final approximation factor reached in [CDW21, Alg. 6].

Explicit Real Generators. For each relation $r = (r_1,\ldots,r_d) \in C^+_{l_1,\ldots,l_d}$, we compute an explicit $\gamma^+_r \in K^+_m \subsetneq K_m$ that verifies Eq. (3.6). Together with the unit group $\mathcal{O}^\times_{K^+_m}$ of K^+_m, they form a fundamental system of $\mathcal{S}^+$-units, where the finite places of $\mathcal{S}^+$ are the G^+_m-orbits of the relative norm ideals l_i.

In the next section, we shall see that adding the explicit Stickelberger generators of Sect. 3.2 to these real generators yields a maximal set of independent $\mathcal{S}$-units in the degree $\varphi(m)$ cyclotomic field K_m, at the much smaller cost of computing a fundamental system of real $\mathcal{S}^+$-units in K^+_m of degree only $\frac{\varphi(m)}{2}$.

In practice, though this remains the main bottleneck of our experimental setting, it allows us to push effectively our experiments up to degree $\varphi(m) = 210$, whereas the full $\mathcal{S}$-unit group computations of [BR20] were bound to $\varphi(m) = 70$.

3.4   A $\mathcal{S}$-unit Subgroup of Finite Index

As in Sect. 3.3, let $\ell_1,\ldots,\ell_d$ be prime integers satisfying $\ell_i \equiv 1 \mod m$; for each i, fix a (split) prime ideal $\mathfrak{L}_i \mid \ell_i$ in K_m and let $l_i = \mathfrak{L}_i \cap K^+_m$. Let $\mathcal{S}$ be a set of places containing, apart from the infinite places of K_m, all G_m-orbits of the $\mathfrak{L}_i$'s. Combining the results of Sect. 3.1, Sect. 3.2 and Sect. 3.3, we get the following family of $\mathcal{S}$-units:

$$\mathfrak{F} = \{v_a;\ a \in M^+_m\} \cup \{\gamma^-_{\mathfrak{L}_i,b};\ i \in [\![1,d]\!], b \in M'_m\} \cup \{\gamma^+_r;\ r \in C^+_{l_1,\ldots,l_d}\} \quad (3.8)$$

where the first set is the set of *circular units* given by Theorem 3.4, the second is the set of explicit *Stickelberger generators* stated at the end of Sect. 3.2 and the last one is the set of *real generators* as in Eq. (3.6).

This family has $(\varphi(m)/2 - 1) + d\cdot\varphi(m)$ elements, which matches precisely the multiplicative rank of the full $\mathcal{S}$-unit group modulo torsion $\mathcal{O}^\times_{K_m,\mathcal{S}}/\mu(\mathcal{O}^\times_{K_m})$.[3] In this section, we prove that these $\mathcal{S}$-units are indeed independent and we compute the index of the subgroup of $\mathcal{O}^\times_{K_m,\mathcal{S}}$ generated by those elements.

Theorem 3.11. *Let $h_{m,(\mathfrak{L}_1,\ldots,\mathfrak{L}_d)}$ (resp. $h^+_{m,(l_1,\ldots,l_d)}$) be the cardinal of the subgroup of Cl_m (resp. Cl^+_m) generated by the G_m-orbits of $\mathfrak{L}_1,\ldots,\mathfrak{L}_d$ (resp. the G^+_m-orbits of $l_1,\ldots,l_d$). The family $\mathfrak{F}$ given in Eq. (3.8) is a maximal set of independent $\mathcal{S}$-units. The subgroup generated by $\mathfrak{F}$ in $\mathcal{O}^\times_{K_m,\mathcal{S}}/\mu(\mathcal{O}^\times_{K_m})$ has index:*

$$\left(\frac{h_m \cdot h^+_{m,(l_1,\ldots,l_d)}}{h_{m,(\mathfrak{L}_1,\ldots,\mathfrak{L}_d)}}\right)\cdot 2^b \cdot \left(h^-_m\right)^{d-1} \cdot \left(2^{\frac{\varphi(m)}{2}-1}\cdot 2^a\right)^d,$$

---

[3] Note that for our purpose, the torsion units play no role and can thus be put aside.

where $a = b = 0$ if m is a prime power, and $a = 2^{t-2} - 1$, $b = 2^{t-2} + 1 - t$ when m has t distinct prime divisors.

When the G_m-orbits of the $\mathfrak{L}_i$'s generate Cl_m, the first term in this index equals h_m^+. As we shall see in Sect. 3.5, the powers of 2 can be killed by saturation techniques, so the problem comes from the $(h_m^-)^{d-1}$ part, which has generically *huge* prime factors. Intuitively, this is because the Stickelberger relations miss all class group relations that exist between two (or more) distinct G_m-orbits.

First, we show that the lattice obtained by adding one copy of the Stickelberger ideal per G_m-orbit, to the lattice $(1 + \tau) \cdot C_{\mathfrak{l}_1,\ldots,\mathfrak{l}_d}^{\prime+}$ of real class relations, yields a full-rank submodule of $\mathbb{Z}[G_m]^d$. Hence, we have obtained a full-rank lattice of class relations for the union of all G_m-orbits above $\ell_1, \ldots, \ell_d$.

We begin by restricting our attention to the case $d = 1$. We need the following lemma, which extends and proves an observation already made in [DPW19, Rem. 3] in the prime conductor case (see [BLNR21, §3.4] for the full proofs):

Lemma 3.12. *The index of $\mathcal{S}_m + (1 + \tau) \cdot \mathbb{Z}[G_m^+]$ in $\mathbb{Z}[G_m]$ is finite:*

$$[\mathbb{Z}[G_m] : \mathcal{S}_m + (1 + \tau) \cdot \mathbb{Z}[G_m^+]] = 2^{\varphi(m)/2 - 1} \cdot 2^a \cdot h_m^-,$$

where $a = 0$ if $t = 1$ and $a = 2^{t-2} - 1$ else, where m has t prime divisors.

When $h_m^+ = 1$, the lattice of real class relations is always $(1 + \tau) \cdot \mathbb{Z}[G_m^+]$, and Lemma 3.12 gives the whole story. In the general case $h_m^+ \neq 1$, we deduce:

Lemma 3.13. *Let ℓ be a prime integer that splits in K_m, let $\mathfrak{L} \mid \ell$ in K_m and let $\mathfrak{l} = \mathfrak{L}^{1+\tau} \cap K_m^+$. Let $h_{m,(\mathfrak{l})}^+$ be the cardinal of the subgroup of Cl_m^+ generated by the G_m^+-orbit of $\mathfrak{l}$ in K_m^+. The $\mathbb{Z}$-module generated by $\mathcal{S}_m$ and the lattice $(1+\tau) \cdot C_{\mathfrak{l}}^+$ of real class relations of the G_m-orbit of $\mathfrak{L}$, has finite index in $\mathbb{Z}[G_m]$:*

$$[\mathbb{Z}[G_m] : \mathcal{S}_m + (1 + \tau) \cdot C_{\mathfrak{l}}^+] = 2^{\varphi(m)/2 - 1} \cdot 2^a \cdot h_m^- \cdot h_{m,(\mathfrak{l})}^+,$$

where $a = 0$ if $t = 1$ and $a = 2^{t-2} - 1$ else, where m has t prime divisors.

Finally, for the case where there are $d \geq 1$ orbits, a reasoning very similar to the proofs of Lemma 3.12 and 3.13 leads to:

Proposition 3.14. *Let $h_{m,(\mathfrak{l}_1,\ldots,\mathfrak{l}_d)}^+$ be the cardinal of the subgroup of Cl_m^+ generated by all G_m^+-orbits of $\mathfrak{l}_1,\ldots,\mathfrak{l}_d$. Then, the $\mathbb{Z}$-module generated by the lattice $(1+\tau) \cdot C_{\mathfrak{l}_1,\ldots,\mathfrak{l}_d}^+ \subseteq (1+\tau) \cdot \mathbb{Z}[G_m^+]^d$ of real class relations between the G_m-orbits of the $\mathfrak{L}_i$'s, and the diagonal block matrix of d copies of $(\mathcal{S}_m \setminus N_m \mathbb{Z})$, verifies:*

$$[\mathbb{Z}[G_m]^d : \mathcal{S}_m^d + (1 + \tau) \cdot C_{\mathfrak{l}_1,\ldots,\mathfrak{l}_d}^+] = \left(2^{\varphi(m)/2 - 1} \cdot 2^a \cdot h_m^-\right)^d \cdot h_{m,(\mathfrak{l}_1,\ldots,\mathfrak{l}_d)}^+.$$

Proof of Theorem 3.11. The independence comes from Proposition 3.14 and the trivial fact that circular units are independent from Stickelberger and real generators. The index of the subgroup generated by $\mathfrak{F}$ in $\mathcal{O}_{K_m,\mathcal{S}}^\times / \mu(\mathcal{O}_{K_m}^\times)$ is given by:

$$[\mathcal{O}_{K_m}^\times : C_m] \cdot \frac{[\mathbb{Z}[G_m]^d : \mathcal{S}_m^d + (1 + \tau) \cdot C_{\mathfrak{l}_1,\ldots,\mathfrak{l}_d}^+]}{|\det(\ker \mathfrak{f}_\mathcal{S})|},$$

where $\ker \mathfrak{f}_{\mathcal{S}}$ is the lattice of all class group relations between finite places of $\mathcal{S}$. The first term is given by Proposition 3.3 and the numerator of the second term by Proposition 3.14. By definition of $\mathcal{O}_{K_m,\mathcal{S}}^{\times}$, the denominator is precisely $h_{m,(\mathfrak{L}_1,\dots,\mathfrak{L}_d)}$. Rearranging terms adequately yields the result.   □

3.5   Saturation

Saturation is a standard tool of computational algebraic number theory that has been used in various contexts like unit and class group computations, and can be traced back at least to [PZ89, §5.7]. This procedure is described in more detail in [BLNR21, §3.5], and we refer to e.g. [BFHP21, §4.3] for a formal exposition.

Intuitively, the e-saturation procedure applied to $\mathfrak{F}$ consists in detecting e-th powers in the subgroup generated by $\mathfrak{F}$, including their e-th roots in the set and rebuilding a basis of multiplicatively independent elements. At the end, the index of the new basis is no longer divisible by e. Remark that the output size does not depend on e, but only on the number and size of the elements of $\mathfrak{F}$.

As the index given by Theorem 3.11 is divisible by a large power of 2, it is therefore natural to 2-saturate $\mathfrak{F}$ in order to mitigate its exponential growth, obtaining the 2-saturated family $\mathfrak{F}_{\mathsf{sat}}$. Note however that the relative class number h_m^- in the index of Theorem 3.11 hides *huge* prime factors that at first glance render this strategy hopeless in general to obtain the full $\mathcal{S}$-unit group from $\mathfrak{F}$.

4   Removing Quantum Steps from the CDW Algorithm

The full material for this section is given in [BLNR21, §B], we summarize the main points here. The CDW algorithm for solving Approx-SVP was introduced in [CDW17] for cyclotomic fields of prime power conductors, using short relations of the Stickelberger lattice as a keystone. [CDW21] extended it to all conductors.

In this section, we show how to use the results of Sect. 3.2, Sect. 3.3 and Sect. 3.4 to remove most quantum steps of [CDW21]. More precisely, we first propose an equivalent rewriting of [CDW21, Alg. 7] that enlightens some hidden steps that reveal useful for subsequent modifications. Then, we plug in the explicit generators of Sect. 3.2 ([BK21]) and Eq. (3.6) for relative class group orbits, to remove the last call to the quantum PIP solver. Finally, by considering the module of *all* real class group relations, using Proposition 3.14 and Theorem 3.11, we remove the need of a random walk mapping any ideal of K_m into Cl_m^-, at the (small) additional price of restricting to cyclotomic fields such that $h_m^+ \leq O(\sqrt{m})$.

An Equivalent Rewriting of CDW [BLNR21, §B.2]. Omitting details, the CDW algorithm works as follows, for any challenge ideal $\mathfrak{a}$ of K_m [CDW21, Alg. 7]:

1. Random walk to Cl_m^-: find $\mathfrak{b}$ such that $[\mathfrak{a}\mathfrak{b}] \in \mathrm{Cl}_m^-$.
2. Solve the ClDL of $\mathfrak{a}\mathfrak{b}$ on G_m-orbits of the prime ideals $\mathfrak{L}_1,\dots,\mathfrak{L}_d$ of Cl_m^-. This gives a vector[4] $\epsilon = (\epsilon_1,\dots,\epsilon_d) \in \mathbb{Z}[G_m]^d$ such that $\mathfrak{a}\mathfrak{b} \cdot \prod_i \mathfrak{L}_i^{\epsilon_i}$ is principal.

---

[4] In the CDW algorithm, the explicit generator given by the ClDL solver is discarded.

3. Solve the CPMP by projecting each ϵ_i in $\pi(\mathcal{S}_m) = (1 - \tau)\mathcal{S}_m$, find a close vector $v_i = y_i \cdot \pi(\mathcal{S}_m)$ and lift v_i to get some η_i s.t. $\pi(\eta_i) = v_i$, $\|\epsilon - \eta\|_1$ is small *with positive coordinates*, and $\mathfrak{a}\mathfrak{b} \cdot \prod_i \mathcal{L}_i^{\epsilon_i - \eta_i}$ is principal.
4. Apply the PIP algorithm of [BS16] to get a generator of this principal ideal.
5. Reduce the obtained generator by circular units like in [CDPR16].

This eventually outputs $h \in \mathfrak{a}$ of length $\|h\|_2 \leq \exp\big(\widetilde{O}(\sqrt{m})\big) \cdot \mathcal{N}(\mathfrak{a})^{1/\varphi(m)}$.

We focus on the lift procedure of Step 3. In [CDW21], $v \in \pi(\mathcal{S}_m)$ is lifted to $\eta \in \mathcal{S}_m$ with non-negative coordinates by setting $(\eta_\sigma, \eta_{\tau\sigma}) = (v_\sigma, 0)$ if $v_\sigma \geq 0$ and $(0, -v_\sigma)$ otherwise, for all $\sigma \in G_m^+$. This works because $[\mathfrak{c}]^{-1} = [\mathfrak{c}^\tau]$ for any $\mathfrak{c} \in \mathrm{Cl}_m^-$, but hides which exact product of relative norm ideals is involved. We propose a totally equivalent lift procedure: from $v = y \cdot \pi(\mathcal{S}_m)$, consider the preimage $\tilde{\eta} = y \cdot \mathcal{S}_m$. Define η by removing $\min\{\tilde{\eta}_\sigma, \tilde{\eta}_{\tau\sigma}\}$ to each $\tilde{\eta}_\sigma$ coordinate. Now, it is obvious that η is a combination y of relations in $\mathcal{S}_m$, and of relative norm relations given by the min part. Details are in [BLNR21, Alg. B.6].

Using Explicit Stickelberger Generators [BLNR21, §B.3]. Each element w_a of the generating set W of $\mathcal{S}_m$ corresponds to a generator $\mathcal{J}_{\mathcal{L}}(1, a - 1)$ (see Sect. 3.2). Similarly, each relative norm ideal writes $\langle \gamma_s^+ \rangle = \mathcal{L}^{(1 \mid \tau)\sigma_s}$ (see Sect. 3.3). Hence, from an (explicit) ClDL solution $\langle \alpha \rangle = \mathfrak{a}\mathfrak{b} \cdot \mathcal{L}^\epsilon$, and given a CPMP solution, explicitly written as above as $\eta = y \cdot W + u \cdot (1 + \tau) \cdot \mathbb{Z}[G_m^+]$, we have that a generator of $\mathfrak{a}\mathfrak{b} \cdot \mathcal{L}^{\epsilon - \eta}$ is directly given by $\alpha / \big(\prod_a \mathcal{J}_{\mathcal{L}}(1, a - 1)^{y_a} \prod_s (\gamma_s^+)^{u_s}\big)$. This allows us to remove the quantum PIP in dimension n in step 4 (for each query). In exchange, we need to compute (only once) all real generators for relative norm relations, which can be done in dimension $\varphi(m)/2$ by [BS16, Alg. 2].

Avoiding the Random Walk [BLNR21, §B.4]. Finally, note that several quantum steps are performed (for each query) in the random walk that maps ideals to Cl_m^-. Using the results of Sect. 3.3, we replace the module $(1+\tau) \cdot \mathbb{Z}[G_m]^d$ by the module of all real class group relations. Asymptotically, we prove in [BLNR21, Pr. B.7] that this does not change the bound on the approximation factor obtained in [CDW21, Th. 5.1], under the same assumption on the Galois-module structure of Cl_m [CDW21, Ass. 1], as long as we restrict to fields K_m with $h_m^+ \leq O(\sqrt{m})$. This additional tiny assumption is largely compensated by the fact that only two quantum steps remain: one is performed only once in dimension $\varphi(m)/2$ to compute real class group relations and generators, and the second is solving the ClDL for each query (see [BLNR21, Tab. B.1]).

5   Computing Log-$\mathcal{S}$-unit Sublattices in Higher Dimension

Our main goal is to simulate the Twisted-PHS algorithm for high degree cyclotomic fields. To this end, we compute full-rank sublattices of the full log-$\mathcal{S}$-unit lattice using the knowledge of the maximal set $\mathfrak{F}$ of independent $\mathcal{S}$-units defined by Eq. (3.8) and its 2-saturated counterpart $\mathfrak{F}_{\mathrm{sat}}$ from Sect. 3.5. These sets are lifted from a complete set of real $\mathcal{S}^+$-units (see Sect. 3.3), hence are obtained at the classically subexponential cost of working in the half degree maximal real subfield. We note that by Theorem 3.11, the index of these families grows rapidly as the number of orbits increases, hence these approximated modes give an upper bound on the approximation factors that can be expected when using Twisted-PHS.

The Twisted-PHS algorithm is briefly recalled in Sect. 5.1, and our experimental setting is detailed in Sect. 5.2. Then, we analyse in Sect. 5.3 the geometric characteristics of our log-$\mathcal{S}$-unit sublattices and the obtained approximation factors in Sect. 5.4.

5.1   The Twisted-PHS Algorithm

The Twisted-PHS algorithm [BR20] was introduced as an improvement of the PHS algorithm [PHS19]. Both aim at solving Approx-id-SVP in any number field and have the same theoretically proven bounds for running time and reached approximation factors. However, the explicit $\mathcal{S}$-units formalism in [BR20] leads to a proper normalization of the used log-$\mathcal{S}$-embedding, weighting coordinates according to finite places norms. This turned out to give experimentally significant improvements on the lattices' decodability and on reached approximation factors.

Both algorithms are split in a *preprocessing phase*, performed only once for a fixed number field, and a *query phase*, for each challenge ideal. More precisely:

1. The preprocessing phase consists in choosing a set of finite places $\mathcal{S}$ generating the class group, computing the corresponding log-$\mathcal{S}$-unit lattice for an appropriate log-$\mathcal{S}$-embedding, and preparing the lattice for subsequent Approx-CVP requests using the Laarhoven's algorithm from [Laa16];
2. For each challenge ideal $\mathfrak{b}$, the query phase consists in first solving the CLDL relatively to $\mathcal{S}$, obtaining $\langle \alpha \rangle = \mathfrak{b} \cdot \prod_{\mathcal{L} \in \mathcal{S}} \mathcal{L}^{v_{\mathcal{L}}}$. Then, this element is projected onto the span of the above log-$\mathcal{S}$-unit lattice, and a close vector of this lattice gives a $\mathcal{S}$-unit s s.t. α/s is hopefully small. Here, guaranteeing that $\alpha/s \in \mathfrak{b}$ is achieved by applying a drift parameterized by some β on the target.

In the Twisted-PHS case, since the obtained lattice, after proper normalization, appears to have exceptionally good geometric characteristics, it was proposed to replace Laarhoven's algorithm by a lazy BKZ reduction in the preprocessing phase and Babai's Nearest Plane algorithm in the query phase [BR20, Alg. 4.2 and 4.3]. We will consider only this practical version in our experiments.

In details, for a number field K, the log-$\mathcal{S}$-unit lattice used in the Twisted-PHS algorithm is defined as $\varphi_{\mathsf{tw}}(\mathcal{O}_{K,\mathcal{S}}^{\times})$, where φ_{tw} is the log-$\mathcal{S}$-embedding given by $f_H \circ \overline{\mathrm{Log}}_{\mathcal{S}}$ [BR20, Eq. (4.1)], for an isometry f_H from the span H of $\overline{\mathrm{Log}}_{\mathcal{S}}$ to $\mathbb{R}^k$, where k equals the multiplicative rank of $\mathcal{O}_{K,\mathcal{S}}^{\times}$ modulo torsion.

Among the consequences of the proper normalization induced by $\overline{\mathrm{Log}}_{\mathcal{S}}$, the authors showed how to optimally choose a set of finite places that generate the class group [BR20, Alg. 4.1]. Namely, taking ideals of increasing prime norms in the set $\mathcal{S}$, they noticed that the density of the associated (twisted) log-$\mathcal{S}$-unit lattice $\varphi_{\mathsf{tw}}(\mathcal{O}_{K,\mathcal{S}}^{\times})$ increases up to an optimal value before decreasing.

Finally, a tricky aspect of the resolution resides in guaranteeing that the output solution is indeed an element of the challenge ideal, i.e., that $v_{\mathfrak{L}}(\alpha/s) \geq 0$ for all $\mathfrak{L} \in \mathcal{S} \cap \mathcal{S}_0$. In [BR20], this is done by applying a drift vector in the span of the log-$\mathcal{S}$-unit lattice, parameterized by some β whose optimal value is searched using a dichotomic strategy in the query phase. Concretely [BR20, Eq. (4.7)]:

$$\mathbf{t} = f_H\left(\left\{\ln|\alpha|_\sigma - \frac{k\beta + \ln\mathcal{N}(\mathfrak{b}) - \sum_{\mathfrak{L}\in\mathcal{S}}\ln\mathcal{N}(\mathfrak{L})}{[K:\mathbb{Q}]}\right\}_\sigma, \left\{\ln|\alpha|_{\mathfrak{L}}^{[K_{\mathfrak{L}}:\mathbb{Q}_\ell]} + \beta - \ln\mathcal{N}(\mathfrak{L})\right\}_{\mathfrak{L}\in\mathcal{S}}\right).$$

5.2 Experimental Settings

Computing the full group of $\mathcal{S}$-units in a classical way is rapidly intractable, even in the case of cyclotomic fields; therefore, experiments performed in [BR20] on Twisted-PHS were bound to $\varphi(m) \leq 70$. We apply the Twisted-PHS algorithm using our full-rank sublattices of the whole log-$\mathcal{S}$-unit lattice induced by the independent family $\mathfrak{F}$ of Eq. (3.8), its 2-saturated counterpart $\mathfrak{F}_{\mathsf{sat}}$ (Sect. 3.5) and, when feasible, a fundamental system $\mathfrak{F}_{\mathsf{su}}$ for the full $\mathcal{S}$-unit group. Approximated modes with $\mathfrak{F}$ or $\mathfrak{F}_{\mathsf{sat}}$ give a glimpse on how Twisted-PHS scales in higher dimensions, where asymptotic phenomena like the growth of h_m start to express.

Source Code and Hardware Description. All experiments have been implemented using SAGEMATH v9.0 [Sag20], except for the full $\mathcal{S}$-unit groups computations for which we used MAGMA [BCP97], which appears much faster for this particular task and also offers an indispensable product ("Raw") representation. Moreover, fplll [FpL16] was used to perform all lattice reduction algorithms. The entire source code is provided on https://github.com/ob3rnard/Tw-Sti.

Most of the computations were performed in less than two weeks on a server with 72 Intel® Xeon® E5-2695v4 @2.1 GHz with 768 GB of RAM, using 2 TB of storage for the precomputations. Real class group computations were performed on a single Intel® Core™ i7-8650U @3.2 GHz CPU using 10 GB of RAM.

Table 1. List of ignored conductors (†: failure to compute Cl_m^+ within a day).

m	$\varphi(m)$	h_m^+	m	$\varphi(m)$	h_m^+	m	$\varphi(m)$	h_m^+	m	$\varphi(m)$	h_m^+	m	$\varphi(m)$	h_m^+	m	$\varphi(m)$	h_m^+
136	64	2	408	128	2	205	160	2	356	176	†	520	192	4	265	208	†
212	104	5	268	132	†	328	160	†	376	184	†	840	192	†	424	208	†
145	112	2	284	140	†	440	160	5	191	190	11	303	200	†	636	208	†
183	120	4	292	144	†	163	162	4	221	192	†	404	200	†			
248	120	4	504	144	4	332	164	†	388	192	†	309	204	†			
272	128	2	316	156	†	344	168	†	476	192	†	412	204	†			

Targeted Cyclotomic Fields. We consider cyclotomic fields of *any* conductor m s.t. $20 < \varphi(m) \leq 210$ with known real class number $h_m^+ = 1$, including those from [BLNR21, Tab. 2.1]. The restriction to $h_m^+ = 1$ is only due to technical interface obstructions, i.e., we are not aware of how to access the non-trivial real class group relations internally computed by SAGEMATH. Additionally, for some of the conductors, we were not able to obtain the real class group within a day. Thus, we are left with 210 distinct cyclotomics fields, and Table 1 lists all ignored conductors.

Finite Places Choice. The optimal set of places computed by [BR20, Alg. 4.1] yields a number $d_{\max}$ of split G_m-orbits of smallest norms maximizing the density of the corresponding full log-$\mathcal{S}$-unit lattice. However, the index of our log-$\mathcal{S}$-unit *sublattices*, given by Theorem 3.11, grows too quickly, roughly in $(h_m^-)^{d-1}$, so that their density always decreases as soon as $d > 1$. This remark motivates us to compute all log-$\mathcal{S}$-unit sublattices for $d = 1$ to $d_{\max}$ first split G_m-orbits.

Full Rank log-$\mathcal{S}$-unit *Sublattices.* The first maximal set of independent $\mathcal{S}$-units that we consider is $\mathfrak{F}$ from Eq. (3.8). The 2-saturation process of Sect. 3.5 mitigates the huge index of $\mathfrak{F}$, yielding family $\mathfrak{F}_{\mathsf{sat}}$. A fundamental system $\mathfrak{F}_{\mathsf{su}}$ of the full $\mathcal{S}$-unit group $\mathcal{O}_{K_m,\mathcal{S}}^\times$ (modulo torsion) is also used whenever it is computable in reasonable time, i.e., up to $\varphi(m) < 80$. As noted in Sect. 2.3, their images under any log-$\mathcal{S}$-embedding φ form full-rank sublattices resp. L_{urs}, L_{sat}, L_{su}, generated by resp. $\varphi(\mathfrak{F})$, $\varphi(\mathfrak{F}_{\mathsf{sat}})$, $\varphi(\mathfrak{F}_{\mathsf{su}})$, of the corresponding full log-$\mathcal{S}$-unit lattice $\varphi(\mathcal{O}_{K_m,\mathcal{S}}^\times)$.

We consider several choices of the log-$\mathcal{S}$-embedding φ. Namely, we tried to evaluate the advantage of using the expanded $\overline{\mathrm{Log}}_\mathcal{S}$ (exp) over $\mathrm{Log}_\mathcal{S}$, labelled tw (as twisted by $[\mathbb{C} : \mathbb{R}] = 2$). We also considered versions with (iso) or without (noiso) the isometry f_H of [BR20, Eq. (4.2)]. This yields four choices for φ, e.g. tag noiso/tw is $\varphi = \mathrm{Log}_\mathcal{S}$ and iso/exp gives the original $\varphi_{\mathsf{tw}} = f_H \circ \overline{\mathrm{Log}}_\mathcal{S}$.

Compact Product Representation. In order to avoid the exponential growth of algebraic integers viewed in $\mathbb{Z}[x]/\langle\Phi_m(x)\rangle$, we use a compact product representation, so that any element α in $\mathfrak{F}$ (resp. $\mathfrak{F}_{\mathsf{sat}}$ or $\mathfrak{F}_{\mathsf{su}}$) is written on a set $g_1, \ldots, g_N$ of N small elements as $\alpha = \prod_{i=1}^N g_i^{e_i}$. Hence, besides the g_i's, each α is stored as a vector $e \in \mathbb{Z}^N$, and for any choice of φ, we have $\varphi(\alpha) = \sum_{i=1}^N e_i \cdot \varphi(g_i)$. This allows us to compute φ without the coefficient explosion encountered in [BR20, §5], which unlocks the full log-$\mathcal{S}$-unit lattices computations beyond degree 60.

Table 2. Geometric characteristics of L_{urs}, L_{sat} and L_{su} for $\mathbb{Q}(\zeta_{152})$ and $\mathbb{Q}(\zeta_{211})$ with log-$\mathcal{S}$-embedding φ_{tw} (of type iso/exp). For *all* bases, the root-Hermite factor verifies $|\delta_0 - 1| < 10^{-3}$.

m	d	set	k	$\mathrm{Vol}^{1/k}$	δ			$\max_{1 \leq i \leq k} \|\mathbf{b}_i\|_2$		
					raw	LLL	bkz$_{40}$	raw	LLL	bkz$_{40}$
152	1	urs	107	8.691	2.016	1.570	1.551	45.007	38.466	38.202
		sat	107	6.928	4.398	1.787	1.822	752.306	23.280	21.720
		su	107	6.928	28.396	1.805	1.828	3163.723	21.953	21.446
	2	urs	179	9.683	2.157	1.623	1.590	48.754	41.313	41.404
		sat	179	7.384	7.670	1.885	1.896	6273.562	23.280	22.772
		su	179	6.816	65.355	2.226	2.322	3427.134	23.221	24.741
211	1	urs	314	14.325	2.672	2.291	2.257	96.068	97.930	96.569
		sat	314	11.386	9.998	2.581	2.562	9742.552	59.387	59.578
	5	urs	1154	18.232	3.118	2.542	2.497	118.124	119.160	115.888
		sat	1154	13.341	19.443	2.918	2.901	32067.612	71.428	72.752
	7	urs	1574	18.976	3.161	2.557	2.512	120.838	121.129	119.020
		sat	1574	13.771	26.841	2.927	2.910	530646.708	71.428	72.752

Lattice Reductions. For each of the constructed log-$\mathcal{S}$-unit sublattices, i.e. for each number of orbits $d \in [\![1, d_{\max}]\!]$, for each family of independent $\mathcal{S}$-units $\mathfrak{F}$, $\mathfrak{F}_{\mathsf{sat}}$ and (when feasible) $\mathfrak{F}_{\mathsf{su}}$, and for each choice of log-$\mathcal{S}$-embedding, we compare several levels of reduction: no reduction ("raw"), LLL-reduction and BKZ$_{40}$-reduction.

5.3 Geometry of the Lattices

For all described choices of log-$\mathcal{S}$-unit sublattices, we first evaluate several geometrical parameters (see Sect. 2.5): reduced volume $V^{1/k}$, root-Hermite factor δ_0, orthogonality defect δ. We only give here a few examples giving a glimpse of what happens in general, and additional data can be found in [BLNR21, §C.1].

Table 2 contains data for cyclotomic fields $\mathbb{Q}(\zeta_{152})$ and $\mathbb{Q}(\zeta_{211})$ of degrees resp. 72 and 210. All values correspond to the iso/exp log-$\mathcal{S}$-embedding, i.e., $\varphi = \varphi_{\mathsf{tw}}$. Indeed, as illustrated by [BLNR21, Tab. C.2], we experimentally note that using (no)iso/exp seems geometrically slightly better than using (no)iso/tw. Notice how small is the normalized orthogonality defect after only LLL reduction, unambiguously below the tight Minkowski bound $\sqrt{1 + \frac{k}{4}}$.

We then look at the logarithm of the Gram-Schmidt norms, for every described choice of log-$\mathcal{S}$-unit sublattices. Figure 2 plots the Gram-Schmidt log norms before and after BKZ reduction of the lattices L_{sat}, using the original iso/exp log-$\mathcal{S}$-embedding φ_{tw}. As in [BR20, Fig. B.1–10], for each field the two curves are almost superposed, which is consistent with the observations on the orthogonality defect. We also checked the impact of the log-$\mathcal{S}$-embedding choice

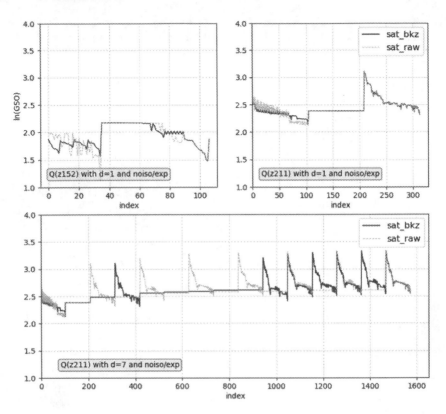

Fig. 2. L_{sat} lattices for $\mathbb{Q}(\zeta_{152})$ and $\mathbb{Q}(\zeta_{211})$: Gram-Schmidt log norms before and after reduction by BKZ_{40}.

among all four options on the Gram-Schmidt logarithm norms of the *unreduced* basis $\varphi(\mathfrak{F}_{sat})$. As expected, the isometry f_H has no influence on the Gram-Schmidt norms. On the other hand, using $\text{Log}_{\mathcal{S}}$ or $\overline{\text{Log}}_{\mathcal{S}}$ seems to alter only the first norms, and in a very small way. This can be seen in [BLNR21, Tab. C.4]. Again, increasing the number of orbits does not influence these behaviours.

We stress that these very peculiar geometric characteristics – shape of the logarithm of the norms of the Gram-Schmidt basis, ease of reduction, very small orthogonality defect (after LLL) – already observed in [BR20, §5.1-2], are consistently viewed across all conductors, degrees, log-$\mathcal{S}$-unit sublattices and number of orbits. To give a concrete idea of e.g. the striking ease of reduction of these log-$\mathcal{S}$-unit sublattices, we report that for $m = 211$, BKZ_{40} terminates in around 7 min (resp. 30 min) on the log-$\mathcal{S}$-unit sublattice of dimension $k = 1154$ (resp. 1574) corresponding to $d = 5$ (resp. $d_{\max} = 7$), which is unusually fast.

This very broad phenomenon suggests that the explanation is possibly deep, an observation that has been recently developed by Bernstein and Lange [BL21].

5.4   Evaluation of the Approximation Factor

In [BR20], evaluating in practice the approximation factors reached by Twisted-PHS is done by choosing random split ideals of prime norm, solving the ClDL for these challenges and comparing the length of the obtained algebraic integer with the length of the exact shortest element. As the degrees of the fields grow, solving the ClDL and exact id-SVP becomes rapidly intractable. Hence, we resort to simulating random outputs of the ClDL, similarly to [DPW19, Hyp. 8], and estimate the obtained approximation factors with inequalities from Eq. (2.3).

Simulation of ClDL Solutions. To simulate targets that heuristically correspond to explicit generators α output by the ClDL, we assume that for each ideal $\mathfrak{L}_i \in \mathcal{S}$, the vector $\left(v_{\mathfrak{L}_i^\sigma}(\alpha)\right)_{\sigma \in G_m}$ of $\mathbb{Z}^{\frac{\varphi(m)}{2}}$ is uniform modulo the lattice of class relations, and that after projection along the $\mathbf{1}$-axis, $\left(\ln|\sigma(\alpha)|\right)_\sigma$ is uniform modulo the log-unit lattice. These hypotheses have already been used in [DPW19, Hyp. 8] or [BR20, H. 4.8], and are backed up by theoretical results in [BDPW20, Th. 3.3].

Drawing random elements modulo a lattice of rank k is done by following a Gaussian distribution of sufficiently large deviation. Concretely, we first choose a random split prime p in $[\![2^{97}, 2^{103}]\!]$. Then, for each $\mathfrak{L} \in \mathcal{S} \cap \mathcal{S}_0$, we pick random valuations $v_{\mathfrak{L}}(\alpha)$ modulo the lattice of class relations of rank $|\mathcal{S} \cap \mathcal{S}_0|$ and random elements $(u_\sigma)_{\sigma \in G_m^+} \in \mathbb{R}^{\varphi(m)/2}$ in the span of the log-unit lattice of rank $\frac{\varphi(m)}{2} - 1$.

Finally, we simulate $(\ln|\sigma(\alpha)|)_\sigma$ by adding $\frac{\ln p + \sum_{\mathfrak{L} \in \mathcal{S}} v_{\mathfrak{L}} \ln \mathcal{N}(\mathfrak{L})}{\varphi(m)}$ to each coordinate u_σ, so that their sum is $\frac{\ln|\mathcal{N}(\alpha)|}{2}$. For each field, we thereby generate 100 random targets on which to test Twisted-PHS on all lattice versions.

Reconstruction of a Solution. For each simulated ClDL generator α, given as a random vector $(\{\ln|\sigma(\alpha)|\}_{\sigma \in G_m^+}, \{v_{\mathfrak{L}}(\alpha)\}_{\mathfrak{L} \in \mathcal{S} \cap \mathcal{S}_0})$, it is easy to compute $\varphi(\alpha)$ for any log-$\mathcal{S}$-embedding φ and to derive a target as in [BR20, Eq. (4.7)], including a drift parameterized by some β. Then, considering e.g. $L_{\mathsf{sat}} = \varphi(\mathfrak{F}_{\mathsf{sat}})$, given by the BKZ_{40}-*reduced* basis $U_{\mathsf{bkz}} \cdot \varphi(\mathfrak{F}_{\mathsf{sat}})$, we find a close vector $v = (y \cdot U_{\mathsf{bkz}}) \cdot \varphi(\mathfrak{F}_{\mathsf{sat}})$ to this target using Babai's Nearest Plane algorithm, and from y, U_{bkz} and $\mathfrak{F}_{\mathsf{sat}}$ we easily recover, in compact representation, $s \in \mathcal{O}_{K_m, \mathcal{S}}^\times$ s.t. $v = \varphi(s)$ and also α/s.

The purpose of the drift parameter β is to guarantee $v_{\mathfrak{L}}(\alpha/s) \geq 0$ on all finite places. As mentioned in [BR20], the length of α/s is extremely sensitive to the value of β, so that they searched for an optimal value by dichotomy. However, this positiveness property actually does not seem to be monotonic in β, and in practice, using the same β on each finite place coordinate is too coarse when the dimension grows, resulting in unnecessarily large approximation factors. We instead obtained best results using random drifts in ℓ_∞-norm balls of radius 1 centered on the $\mathbf{1}$ axis. A first sampling of $O(\varphi(m))$ random points $\beta \cdot \mathbf{1} + \mathcal{B}_\infty(1)$ for a wide range of random β's allows us to select a β_0 around which we found the best $\|\alpha/s\|_2$ with all $v_{\mathfrak{L}}(\alpha/s)$ being positive. Then we sample $O(\varphi(m))$ uniform random points in the neighbourhood of β_0, namely in $[0.9\beta_0, 1.1\beta_0] \cdot \mathbf{1} + \mathcal{B}_\infty(1)$, and output the overall optimal $\|\alpha/s\|_2$ having all $v_{\mathfrak{L}}(\alpha/s) \geq 0$.

Estimator of the Approximation Factor. Since we do not have access to the shortest element of a challenge ideal, we cannot compute an exact approximation

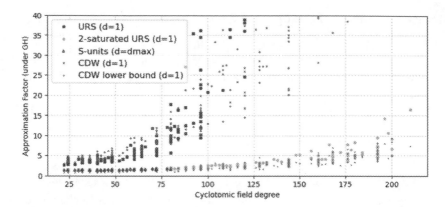

Fig. 3. Approximation factors, with Gaussian Heuristic, reached by Tw-PHS for cyclotomic fields of degree up to 210, on lattices L_{urs}, L_{sat} and L_{su}.

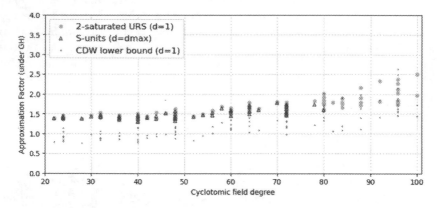

Fig. 4. Approximation factors, with Gaussian Heuristic, reached by Tw-PHS for cyclotomic fields of degree up to 100, on lattices L_{sat} and L_{su}.

factor as in [BR20]. Instead, we estimate the retrieved approximation factor using the inequalities implied by Eq. (2.3). We focus on the Gaussian Heuristic, which gives consistent results with the exact approximation factors found in [BR20], in small dimensions. For each cyclotomic field, the plotted points are the means, over the 100 simulated random targets, of the minimal approximation factors obtained using options iso/noiso and exp/tw. For each family $\mathfrak{F}$, $\mathfrak{F}_{\mathsf{sat}}$ and $\mathfrak{F}_{\mathsf{su}}$, we chose to keep only the factor base that gives the best result. This systematically translated into using $d = 1$ G_m-orbit for $\mathfrak{F}$ and $\mathfrak{F}_{\mathsf{sat}}$, whereas we had to use $d = d_{\max}$ for $\mathfrak{F}_{\mathsf{su}}$, as predicted by the Twisted-PHS algorithm.

Figure 3 shows the approximation factor $\mathsf{af}_{\mathsf{gh}}$ obtained for all lattices L_{urs}, L_{sat} and L_{su} (when feasible) after BKZ_{40} reduction. Figure 4 is a zoom of Fig. 3 that focuses on L_{sat} and L_{su} on small dimensions. First, we remark that using $\mathfrak{F}$ from Eq. (3.8), the retrieved approximation factors are increasing rapidly. Using the 2-saturated family $\mathfrak{F}_{\mathsf{sat}}$ yields much better results, and looking closely at

Fig. 4 shows that using a basis $\mathfrak{F}_{\mathsf{su}}$ of the full $\mathcal{S}$-unit group, when feasible, even improves the picture if $d_{\max} > 1$, in which case L_{su} is denser than L_{sat}. For L_{su}, we stress that we obtain estimated approximation factors very similar to the exact ones observed in [BR20].

More generally, we observe a very strong correlation between the density of our lattices and the obtained approximation factors – the denser, the better. As an important related remark, the variance seen for $\mathsf{af}_{\mathsf{gh}}$ in Fig. 3 for distinct fields of same degree follows the variations of the norm of the first split prime, thus of the reduced volume of the considered log-$\mathcal{S}$-unit sublattice. We expect this variance to be smoothed through conductors for the full log-$\mathcal{S}$-unit lattice.

Furthermore, considering $m = 211$, the $\mathfrak{F}$ family gives $\mathrm{Vol}^{1/314} L_{\mathsf{urs}} \approx 14.325$ and an estimated $\mathsf{af}_{\mathsf{gh}} \approx 13170$, for $\mathfrak{F}_{\mathsf{sat}}$ we get $\mathrm{Vol}^{1/314} L_{\mathsf{sat}} \approx 11.386$ and a much smaller estimated $\mathsf{af}_{\mathsf{gh}} \approx 16.4$, whereas the optimal number of orbits predicted by the Twisted-PHS factor base choice algorithm [BR20, Alg. 4.1] is $d_{\max} = 7$, which yields a full log-$\mathcal{S}$-unit lattice of reduced volume only $\mathrm{Vol}^{1/1574} L_{\mathsf{su}} \approx 9.635$.

Comparison to the CDW Algorithm. Using the same experimental setting, we compute the approximation factors obtained using the CDW algorithm as implemented in [DPW19] ("Naive version") with additional BKZ$_{40}$ lattice reductions, as well as the experimentally derived *volumetric lower bound* from [DPW19, Eq. (5) and Tab. 1]. Those values are also represented in Figs. 3 and 4.

We note that our experimental results using the $\mathfrak{F}_{\mathsf{sat}}$ family are comparable to this volumetric lower bound. Moreover, for some fields, e.g. in dimensions 96, 160, 168, 200, this lower bound is defeated by the (approximated version of the) Twisted-PHS algorithm. Note that this does not invalidate the lower bound itself, which is stated for the two-phase CDW algorithm, but indicates the power of combining both steps in only one lattice as in the Twisted-PHS algorithm.

Acknowledgements. The first author is deeply indebted to Radan Kučera for the proof of Lemma 3.12, and for thorough and invaluable discussions about the Stickelberger ideal. Andrea Lesavourey is funded by the Direction Générale de l'Armement (Pôle de Recherche CYBER), with the support of Région Bretagne. This work is supported by the European Union PROMETHEUS project (Horizon 2020 Research and Innovation Program, grant 780701) and by the PEPR quantique France 2030 programme (ANR-22-PETQ-0008).

References

[Bab86]   Babai, L.: On Lovász' lattice reduction and the nearest lattice point problem. Combinatorica **6**(1), 1–13 (1986)

[Bac90]   Bach, É.: Explicit bounds for primality testing and related problems. Math. Comput. **55**(191), 355–380 (1990)

[BCP97]   Bosma, W., Cannon, J., Playoust, C.: The Magma algebra system. I. The user language. J. Symbol. Comput. **24**(3–4), 235–265 (1997). Computational Algebra and Number Theory (London, 1993)

[BDF08]   Belabas, K., Diaz, F., Diaz, E.: Friedman: small generators of the ideal class group. Math. Comput. **77**(262), 1185–1197 (2008)

[BDPW20] de Boer, K., Ducas, L., Pellet-Mary, A., Wesolowski, B.: Random self-reducibility of ideal-SVP via Arakelov random walks. In: Micciancio, D., Ristenpart, T. (eds.) CRYPTO 2020. LNCS, vol. 12171, pp. 243–273. Springer, Cham (2020). https://doi.org/10.1007/978-3-030-56880-1_9

[BEF+17] Biasse, J.-F., Espitau, T., Fouque, P.-A., Gélin, A., Kirchner, P.: Computing generator in cyclotomic integer rings. In: Coron, J.-S., Nielsen, J.B. (eds.) EUROCRYPT 2017. LNCS, vol. 10210, pp. 60–88. Springer, Cham (2017). https://doi.org/10.1007/978-3-319-56620-7_3

[BF14] Biasse, J., Fieker, C.: Subexponential class group and unit group computation in large degree number fields. LMS J. Comput. Math. **17**(A), 385–403 (2014)

[BFHP21] Biasse, J., Fieker, C., Hofmann, T., Page, A.: Norm relations and computational problems in number fields. arXiv:2002.12332v3 [math.NT] (2021)

[BK21] Bernard, O., Kučera, R., A short basis of the Stickelberger ideal of a cyclotomic field. arXiv:2109.13329 [math.NT] (2021)

[BL21] Bernstein, D.J., Lange, T.: Non-randomness of s-unit lattices. Cryptology ePrint Archive, Report 2021/1428 (2021)

[BLNR21] Bernard, O., Lesavourey, A., Nguyen, T., Roux-Langlois, A.: Log-$\mathcal{S}$-unit lattices using explicit Stickelberger generators to solve Approx Ideal-SVP (full version). Cryptology ePrint Archive, Report 2021/1384 (2021)

[BPR04] Buhler, J., Pomerance, C., Robertson, L.: Heuristics for class numbers of prime-power real cyclotomic fields. Fields Inst. Commun. **41**, 149–157 (2004)

[BR20] Bernard, O., Roux-Langlois, A.: Twisted-PHS: using the product formula to solve approx-SVP in ideal lattices. In: Moriai, S., Wang, H. (eds.) ASIACRYPT 2020. LNCS, vol. 12492, pp. 349–380. Springer, Cham (2020). https://doi.org/10.1007/978-3-030-64834-3_12

[BS16] Biasse, J.-F., Song, F.: Efficient quantum algorithms for computing class groups and solving the principal ideal problem in arbitrary degree number fields. In: SODA, pp. 893–902. SIAM (2016)

[CDPR16] Cramer, R., Ducas, L., Peikert, C., Regev, O.: Recovering short generators of principal ideals in cyclotomic rings. In: Fischlin, M., Coron, J.-S. (eds.) EUROCRYPT 2016. LNCS, vol. 9666, pp. 559–585. Springer, Heidelberg (2016). https://doi.org/10.1007/978-3-662-49896-5_20

[CDW17] Cramer, R., Ducas, L., Wesolowski, B.: Short stickelberger class relations and application to ideal-SVP. In: Coron, J.-S., Nielsen, J.B. (eds.) EUROCRYPT 2017. LNCS, vol. 10210, pp. 324–348. Springer, Cham (2017). https://doi.org/10.1007/978-3-319-56620-7_12

[CDW21] Cramer, R., Ducas, L., Wesolowski, B.: Mildly short vectors in cyclotomic ideal lattices in quantum polynomial time. J. ACM **68**(2) (2021)

[CGS14] Campbell, P., Groves, M., Shepherd, D.: Soliloquy: a cautionary tale (2014). http://docbox.etsi.org/Workshop/2014/201410_CRYPTO/S07_Systems_and_Attacks/S07_Groves_Annex.pdf

[Che13] Chen, Y.: Réduction de réseau et sécurité concrète du chiffrement complètement homomorphe. Ph.D. thesis, Paris 7 (2013)

[CN11] Chen, Y., Nguyen, P.Q.: BKZ 2.0: better lattice security estimates. In: Lee, D.H., Wang, X. (eds.) ASIACRYPT 2011. LNCS, vol. 7073, pp. 1–20. Springer, Heidelberg (2011). https://doi.org/10.1007/978-3-642-25385-0_1

[DPW19] Ducas, L., Plançon, M., Wesolowski, B.: On the shortness of vectors to be found by the ideal-SVP quantum algorithm. In: Boldyreva, A., Micciancio,

D. (eds.) CRYPTO 2019. LNCS, vol. 11692, pp. 322–351. Springer, Cham (2019). https://doi.org/10.1007/978-3-030-26948-7_12

[EHKS14] Eisenträger, K., Hallgren, S., Kitaev, A.Y., Song, F.: A quantum algorithm for computing the unit group of an arbitrary degree number field. In: STOC, pp. 293–302. ACM (2014)

[FpL16] FpLLL development team: FpLLL, a lattice reduction library (2016). https://github.com/fplll/fplll

[GK89] Gold, R., Kim, J.: Bases for cyclotomic units. Compos. Math. **71**(1), 13–27 (1989)

[GN08] Gama, N., Nguyen, P.Q.: Predicting lattice reduction. In: Smart, N. (ed.) EUROCRYPT 2008. LNCS, vol. 4965, pp. 31–51. Springer, Heidelberg (2008). https://doi.org/10.1007/978-3-540-78967-3_3

[HW38] Hardy, G.H., Wright, E.M.: An Introduction to the Theory of Numbers, 4th edn. Oxford University Press, Oxford (1938)

[Kuč86] Kučera, R.: On a certain subideal of the Stickelberger ideal of a cyclotomic field. Archivum Mathematicum **22**(1), 7–19 (1986)

[Kuč92] Kučera, R.: On bases of the Stickelberger ideal and of the group of circular units of a cyclotomic field. J. Number Theory **40**(3), 284–316 (1992)

[Laa16] Laarhoven, T.: Sieving for closest lattice vectors (with preprocessing). In: Avanzi, R., Heys, H. (eds.) SAC 2016. LNCS, vol. 10532, pp. 523–542. Springer, Cham (2017). https://doi.org/10.1007/978-3-319-69453-5_28

[LLL82] Lenstra, A.K., Lenstra, H.W., Lovász, L.: Factoring polynomials with rational coefficients. Math. Ann. **261**, 515–534 (1982)

[LPR10] Lyubashevsky, V., Peikert, C., Regev, O.: On ideal lattices and learning with errors over rings. In: Gilbert, H. (ed.) EUROCRYPT 2010. LNCS, vol. 6110, pp. 1–23. Springer, Heidelberg (2010). https://doi.org/10.1007/978-3-642-13190-5_1

[MG02] Micciancio, D., Goldwasser, S.: Complexity of Lattice Problems. Kluwer International Series in Engineering and Computer Science, vol. 671. Springer, New York (2002). https://doi.org/10.1007/978-1-4615-0897-7

[Mil14] Miller, J.C.: Class numbers of real cyclotomic fields of composite conductor. LMS J. Comput. Math. **17**, 404–417 (2014)

[Nar04] Narkiewicz, W.: Elementary and Analytic Theory of Algebraic Numbers. Springer Monographs in Mathematics, 3rd edn. Springer, Heidelberg (2004). https://doi.org/10.1007/978-3-662-07001-7

[Neu99] Neukirch, J.: Algebraic Number Theory. Grundlehren des mathematischen Wissenschaften, vol. 322. Springer, Cham (1999). https://doi.org/10.1007/978-3-319-07545-7

[PHS19] Pellet-Mary, A., Hanrot, G., Stehlé, D.: Approx-SVP in ideal lattices with pre-processing. In: Ishai, Y., Rijmen, V. (eds.) EUROCRYPT 2019. LNCS, vol. 11477, pp. 685–716. Springer, Cham (2019). https://doi.org/10.1007/978-3-030-17656-3_24

[PZ89] Pohst, M., Zassenhaus, H.: Algorithmic Algebraic Number Theory. Encyclopedia of Mathematics and its Applications. Cambridge University Press, Cambridge (1989)

[Sag20] Sage Developers: SageMath, the Sage Mathematics Software System (Version 9.0), (2020). https://www.sagemath.org

[Sch87] Schnorr, C.: A hierarchy of polynomial time lattice basis reduction algorithms. Theor. Comput. Sci. **53**, 201–224 (1987)

[Sch03] Schoof, R.: Class numbers of real cyclotomic fields of prime conductor. Math. Comput. **72**(242), 913–937 (2003)

[Sch08] Schoof, R.: Catalan's Conjecture. Universitext, Springer, London (2008). https://doi.org/10.1007/978-1-84800-185-5

[SE94] Schnorr, C., Euchner, M.: Lattice basis reduction: improved practical algorithms and solving subset sum problems. Math. Program. **66**, 181–199 (1994)

[Sin78] Sinnott, W.: On the Stickelberger ideal and the circular units of a cyclotomic field. Ann. Math. **108**(1), 107–134 (1978)

[Sin80] Sinnott, W.: On the Stickelberger ideal and the circular units of an abelian field. Invent. Math. **62**, 181–234 (1980)

[SSTX09] Stehlé, D., Steinfeld, R., Tanaka, K., Xagawa, K.: Efficient public key encryption based on ideal lattices. In: Matsui, M. (ed.) ASIACRYPT 2009. LNCS, vol. 5912, pp. 617–635. Springer, Heidelberg (2009). https://doi.org/10.1007/978-3-642-10366-7_36

[Was97] Washington, L.C.: Introduction to Cyclotomic Fields. Graduate Texts in Mathematics, 2 edn, vol. 83. Springer, New York (1997)

[Wes18] Wesolowski, B.: Generating subgroups of ray class groups with small prime ideals. In: ANTS-XIII. The Open Book Series, vol. 2, pp. 461–478, Mathematical Sciences Publisher (2018)

On Module Unique-SVP and NTRU

Joël Felderhoff[1,2](✉), Alice Pellet-Mary[3], and Damien Stehlé[2,4]

[1] Inria Lyon, Lyon, France
[2] ENS de Lyon, Lyon, France
`joel.felderhoff@ens-lyon.fr`
[3] Univ. Bordeaux, CNRS, Inria, Bordeaux INP, IMB, Talence, France
[4] Institut Universitaire de France, Paris, France

Abstract. The NTRU problem can be viewed as an instance of finding a short non-zero vector in a lattice, under the promise that it contains an exceptionally short vector. Further, the lattice under scope has the structure of a rank-2 module over the ring of integers of a number field. Let us refer to this problem as the module unique Shortest Vector Problem, or mod-uSVP for short. We exhibit two reductions that together provide evidence the NTRU problem is not just a particular case of mod-uSVP, but representative of it from a computational perspective.

First, we reduce worst-case mod-uSVP to worst-case NTRU. For this, we rely on an oracle for id-SVP, the problem of finding short non-zero vectors in ideal lattices. Using the worst-case id-SVP to worst-case NTRU reduction from Pellet-Mary and Stehlé [ASIACRYPT'21], this shows that worst-case NTRU is equivalent to worst-case mod-uSVP.

Second, we give a random self-reduction for mod-uSVP. We put forward a distribution D^{uSVP} over mod-uSVP instances such that solving mod-uSVP with a non-negligible probability for samples from D^{uSVP} allows to solve mod-uSVP in the worst-case. With the first result, this gives a reduction from worst-case mod-uSVP to an average-case version of NTRU where the NTRU instance distribution is inherited from D^{uSVP}. This worst-case to average-case reduction requires an oracle for id-SVP.

1 Introduction

Let K be a number field, $\mathcal{O}_K$ its ring of integers and $\|\cdot\|$ the ℓ_2-norm in the complex embedding vector space. A notable example is $K = \mathbb{Q}[x]/(x^d+1)$ with d a power of 2: in this case, we have $\mathcal{O}_K = \mathbb{Z}[X]/\Phi(X)$ and $\|a\| = (d\sum_i |a_i|^2)^{1/2}$ for all $a = \sum_{0 \le i < d} a_i x^i \in K$. In the (search) NTRU problem, one is given $h \in R_q := \mathcal{O}_K/q\mathcal{O}_K$ with the promise that there exists a pair $(f,g) \in \mathcal{O}_K^2$ such that $gh = f \mod q\mathcal{O}_K$ and $\|f\|, \|g\|$ are significantly smaller than $\sqrt{q}$ (by a factor γ called the gap of the NTRU instance, see Definition 2.15 for a formal definition). The goal is to find a short multiple of the pair (f,g). An efficient algorithm for the NTRU problem for appropriate parameters would lead to a cryptanalysis of the seminal NTRU encryption scheme [HPS98], a variant of which appears among the finalists of the NIST post-quantum cryptography standardization process [CDH+20].

© International Association for Cryptologic Research 2022
S. Agrawal and D. Lin (Eds.): ASIACRYPT 2022, LNCS 13793, pp. 709–740, 2022.
https://doi.org/10.1007/978-3-031-22969-5_24

It was noticed very early that the NTRU problem can be interpreted in terms of Euclidean lattices [HPS98, CS97]. Indeed, the set $L_h := \{(a,b)^T \in K^2 : bh = a \bmod q\mathcal{O}_K\}$ forms a $(2d)$-dimensional lattice, when viewing $\mathcal{O}_K$ as a d-dimensional lattice via the embedding map (or, more elementarily for the running example, using the polynomial expressions). The lattice is described by h, from which a basis can be computed. This lattice has two peculiar properties. First, it contains an unusually short non-zero vector (f,g). Indeed, for most h's, we have $\det L_h = \Delta_K \cdot q^d$, where Δ_K refers to the field discriminant; our running example satisfies $\Delta_K = d^d$. As a result, one would expect the shortest non-zero vectors to have ℓ_2-norm around $q^{1/2}$, up to limited factors depending on Δ_K and d; but $(f,g)^T$ is much shorter, by assumption. However, this is not quite an instance of the unique Shortest Vector Problem (uSVP), as L_h does not contain just one exceptionally short non-zero vector (up to sign), but d linearly independent short vectors: in our running example, the $(x^i \cdot f, x^i \cdot g)^T$'s for $i \in [d]$ are linearly independent and belong to L_h and; in the general case, a short $\mathbb{Z}$-basis of $\mathcal{O}_K$ can be used in place of the x^i's. This leads us to the second peculiarity of the L_h lattice: as it is invariant under multiplication by elements of $\mathcal{O}_K$, it is a rank-2 $\mathcal{O}_K$-module. We hence have a rank-2 $\mathcal{O}_K$-module with the promise that it contains an unusually short non-zero vector, i.e., an unusually dense rank-1 submodule. We call mod-uSVP the problem of finding a short non-zero vector in rank-2 module containing an unusually short vector. In this introduction, we call gap of the mod-uSVP instance the ratio between the root determinant of the lattice (which predicts what would be expected for the euclidean norm of the shortest vector) and the actual euclidean norm of a shortest non-zero vector (see Definition 2.12 for a formal definition).

Search NTRU and mod-uSVP actually come with two flavors. The most natural one, described above, asks to recover a short vector of the corresponding rank-2 module. This is the variant we implicitly consider in this introduction when we discuss NTRU and mod-uSVP. As mentioned above, the NTRU and mod-uSVP lattices not only contain an unexpectedly short vector, but also an unexpectedly dense rank-1 sublattice. The second variant, which we refer to as NTRU$_{mod}$ or mod-uSVP$_{mod}$, asks to recover a basis of this dense submodule.

As seen above, the NTRU problem can be viewed as a special case of a lattice problem. It is however unclear if its instances are representative instances of some standard lattice problem, or, more precisely, if they are computationally equivalent to general instances of such a problem. In [Pei16, Section 4.4.4], Peikert sketched a reduction from a decision version of the NTRU problem to the Ring Learning With Errors (RLWE) problem [SSTX09, LPR10]; this reduction can be adapted to the search NTRU problem we consider here. Note that under some parameter constraints, RLWE is computationally equivalent to the Shortest Independent Vectors Problem for rank-2 modules [LS15, AD17] (mod-SIVP), which consists in finding $2d$ linearly independent vectors whose longest one is not much longer than optimal. Oppositely, in a recent work, Pellet-Mary and Stehlé [PS21] exhibited a reduction from the Shortest Vector Problem for lattices corresponding to ideals of $\mathcal{O}_K$ (id-SVP) to NTRU. Enhanced by the id-

SVP self-reducibility from [dBDPW20], this leads to a reduction from worst-case id-SVP to an average-case version of the NTRU problem.

Overall, we see that NTRU sits between id-SVP and mod-SIVP. Interestingly, id-SVP admits algorithms that outperform generic lattice reduction algorithms [LLL82, Sch87] for some parameter ranges [CDW21, PHS19]. As such a phenomenon is unknown in the case for mod-SIVP, there is potentially quite some room between id-SVP and mod-SIVP. With this state of affairs, it is unclear which of these problems captures the true hardness of NTRU, or if NTRU lies somewhere strictly in between.

Contributions. We give evidence that the NTRU problem is not just a particular case of mod-uSVP, but actually representative of it. More precisely, we show that worst-case NTRU is computationally equivalent to worst-case mod-uSVP, and that worst-case and an appropriately defined average-case mod-uSVP are also computationally equivalent, provided we have an oracle for id-SVP in both cases (and up to reduction losses). Together, these results imply that worst-case mod-uSVP reduces to average-case NTRU, provided we have an oracle for id-SVP. Combining this result with the reduction from worst-case id-SVP to worst-case NTRU from [PS21], this also implies that worst-case NTRU is computationally equivalent to worst-case mod-uSVP, without an id-SVP oracle.

Our first result is a collection of four reductions from the four variants of mod-uSVP (average case vs worst-case and vector vs module) to the corresponding four variants of NTRU, relying on an approximate id-SVP oracle. We give below a simplified version of one of these reductions, in the special case of power-of-two cyclotomic fields. More details and the other reductions can be found in Theorem 4.1.

Theorem 1.1 (Simplified version of Theorem 4.1). *Let K be a power-of-two cyclotomic field of degree d. Let $\gamma_{\mathrm{SVP}}, \gamma^+, \gamma_{\mathrm{NTRU}} > 1$. For all $q \geq 2^d \cdot \mathrm{poly}(\gamma^+)$ and $\gamma^- \geq \mathrm{poly}(d) \cdot \gamma_{\mathrm{NTRU}} \cdot \sqrt{\gamma_{\mathrm{HSVP}}}$, (worst-case) mod-uSVP$_{\mathrm{mod}}$ with gap in $[\gamma^-, \gamma^+]$ reduces in polynomial time to (worst-case) NTRU$_{\mathrm{mod}}$ with modulus q and gap $\geq \gamma_{\mathrm{NTRU}}$ and (worst-case) id-SVP with approximation factor γ_{SVP}.*

More concretely, when starting from a mod-uSVP instance for which the shortest non-zero vectors are $\approx \gamma$ times smaller than the root determinant, the reduction produces an NTRU instance satisfying $\sqrt{q}/(\|f\| + \|g\|) \approx \gamma^{O(1)}$, up to factors depending on field invariants. This transformation can be used to derive a reduction from average-case mod-uSVP to average-case NTRU (where the NTRU distribution is induced by the mod-uSVP distribution) and a reduction from worst-case mod-uSVP to worst-case NTRU (and similarly for the variants searching a dense rank-1 submodule). To achieve this transformation, an id-SVP oracle is required to find non-zero vectors in ideals within a factor $\gamma^{O(1)}$ from optimal. Note that for cyclotomic fields, the algorithm from [CDW21] allows to implement the oracle in quantum polynomial time when $\gamma \approx 2^{\sqrt{d}}$. Note also that [PS21] showed a reduction from worst-case id-SVP to worst-case NTRU, which is compatible with the reduction from worst-case mod-uSVP to worst-case

NTRU (relying on an id-SVP oracle). Combining both, we then obtain a reduction from worst-case mod-uSVP to worst-case search NTRU which *does not* rely on an id-SVP oracle. A drawback of the reduction is that it results in an NTRU modulus q of the order of $\approx 2^d$, even for small gap parameters γ. The modulus can be decreased by allowing the reduction to be more costly. Using lattice reduction algorithms [Sch87], one can reach $q \approx \gamma^{O(1)} \cdot \beta^{O(d/\beta)}$ if allowing for a reduction that runs in time polynomial in d, 2^β, $\log \Delta_K$ and $\zeta_K(2)$ (where ζ_K refers to the Dedekind zeta function). The quantities $\log \Delta_K$ and $\zeta_K(2)$ depend on the number field, and may not be polynomially bounded in the field degree d. In our running example, we have $\log \Delta_K = O(d)$ and $\zeta_K(2) = O(1)$ (see [SS13]).

Second, we exhibit a random self-reducibility property for mod-uSVP$_{\mathrm{mod}}$. More explicitly, we give a reduction from worst-case mod-uSVP$_{\mathrm{mod}}$ for rank-2 modules to an average-case version of itself, whose instances can be sampled from efficiently. The reduction preserves the gap parameter γ, up to factors depending on field invariants, and runs in time polynomial in $\log \Delta_K$.

Theorem 1.2 (Simplified version of Theorem 6.1, under ERH). *Let K be a power-of-two cyclotomic field of degree d. For any gap $\mathrm{poly}(d) < \gamma \le 2^{O(d)}$, there exists an efficiently samplable distribution D_γ^{uSVP} over* uSVP *instances with gap $\ge \gamma$ such that worst-case* mod-uSVP$_{\mathrm{mod}}$ *with gap $\ge \gamma' = \gamma \cdot \mathrm{poly}(d)$ reduces in polynomial time to average-case* mod-uSVP$_{\mathrm{mod}}$ *for instance distribution D_γ^{uSVP}.*

Combined with the first reduction, the above allows to map a worst-case instance of mod-uSVP$_{\mathrm{mod}}$ to an average-case instance of NTRU$_{\mathrm{mod}}$, where the NTRU$_{\mathrm{mod}}$ instance distribution is inherited from the average-case mod-uSVP distribution. This reduction relies on an id-SVP oracle. Since mod-uSVP$_{\mathrm{mod}}$ and mod-uSVP are computationally equivalent (up to polynomial losses) when we have an id-SVP oracle, this also provides a reduction from worst-case uSVP to average-case NTRU. Contrary to the reduction from worst-case uSVP to worst-case NTRU, we cannot use the result of [PS21] to get rid of the id-SVP oracle. This is because the average-case distribution of NTRU instances that is produced by our reduction may not be compatible with the one used in [PS21].

We summarize the known reductions between variants of mod-uSVP and NTRU in Fig. 1. Note that the reductions may not be composable due to incompatible parameter restrictions or instance distributions.

Technical Overview. The NTRU problem is a restriction of mod-uSVP modules with a basis of a specific shape. In general, a rank-2 module M is represented by a pseudo-basis, i.e., two vectors $(\mathbf{b}_1, \mathbf{b}_2)$ in K^2 and two ideals I_1, I_2 of $\mathcal{O}_K$ such that $M = \mathbf{b}_1 I_1 + \mathbf{b}_2 I_2$. When the two ideals I_1 and I_2 are both equal to $\mathcal{O}_K$, the pseudo-basis is a basis, and the module is said to be free (note that a free module is a module that has at least one basis, but not all of its pseudo-bases will satisfy $I_1 = I_2 = \mathcal{O}_K$). In the NTRU problem, the instance is a basis $(\mathbf{b}_1, \mathbf{b}_2)$ of a free module contained in $\mathcal{O}_K^2$, with $\mathbf{b}_1 = (1, h)^T$ for some $h \in \mathcal{O}_K$ and $\mathbf{b}_2 = (0, q)^T$ for some integer q which is a parameter of the NTRU problem. Hence, the only degree of freedom in this basis comes from the choice of h. The NTRU problem then asks to solve mod-uSVP in this very specific module.

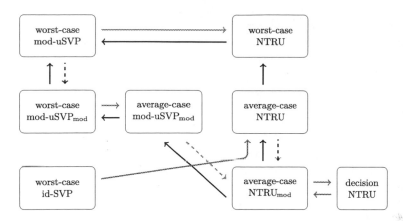

Fig. 1. Known reductions between NTRU and mod-uSVP variants. Dashed arrows require an id-SVP oracle. Blue arrows are proven in [PS21] and red arrows are proven in this article. The black arrows are folklore. (Color figure online)

In the reduction from mod-uSVP to NTRU, we start with an arbitrary pseudo-basis of an arbitrary module M, and transform it into an NTRU basis. We then call the NTRU solver on this NTRU instance and lift the solution back to the original mod-uSVP module. In order to meaningfully lift a short vector (or a dense rank-1 submodule) back, we require our transformation to preserve the geometry of the rank-2 module M as much as possible. Our transformation proceeds in four main steps.

First of all, we transform the input module $M \subset K^2$ into an integral module whose volume is bounded from below and above by quantities depending only on the parameters of the reduction (NTRU modules are in $\mathcal{O}_K^2$ and have volume q^d). This is done by scaling M to the desired volume, and then rounding it to an integral module with a very close geometry. This rounding is performed by sampling two quasi-orthogonal vectors in the dual of M, and multiplying M on the left by the matrix whose rows are these two vectors. Multiplication on the left corresponds to a distortion of the ambient space, but since the two vectors are quasi orthogonal, this does not change the geometry too much. Also, as the row vectors of the sampled matrix belong to the dual of M, the resulting module is integral.

Our second step aims at obtaining the triangular shape of the NTRU basis. To do so, we compute the Hermite Normal Form of the pseudo-basis. With some probability, the two coefficients on the first row of the pseudo-basis will be coprime, leading to an HNF basis with a 1 as a top-left coefficient, exactly what we need for an NTRU instance. This is where $\zeta_K(2)$ comes into play, as it closely relates to the probability that two random elements of $\mathcal{O}_K$ are coprime.

At this point, our pseudo-basis still has coefficient ideals. We remove them with an id-SVP solver: we compute short x_1 and x_2 in the ideals I_1 and I_2, respectively, and then replace the pseudo-basis $((\mathbf{b}_1, \mathbf{b}_2), (I_1, I_2))$ by the basis $(x_1 \mathbf{b}_1, x_2 \mathbf{b}_2)$. This step has the effect of slightly sparsifying the module, i.e., it leads to a rank-2 submodule whose determinant is not much larger. If our gap is sufficiently large compared to the approximation factor of the id-SVP solver, our sparsified module will still contain an unexpectedly short non-zero vector.

We now have a basis of a free module with vectors of the form $(1, h')^T$ and $(0, b)^T$, with h' and b in $\mathcal{O}_K$. Our last step consists in replacing b by the NTRU parameter q. This is done by multiplying the second coordinates of both our basis vectors by q/b. If $q/b \approx 1$ (which we can ensure thanks to the id-SVP solver), then this does not change the geometry of the module too much.

To conclude, the transformation we have described allows us to transform any module of rank-2 with an unexpectedly short vector into an NTRU module with roughly the same geometry. The transformation is reversible, hence, we can lift any short vector or dense module found in the NTRU module back to the original rank-2 module. Since this transformation is a Karp reduction, it can be used to reduce average-case variants of mod-uSVP to average-case variants of NTRU where the NTRU distribution is inherited from the one on the uSVP instances.

For the random self-reducibility of mod-uSVP$_{\mathrm{mod}}$, we start with an arbitrary rank-2 module M and want to randomize it so that the distribution of the output module M' does not depend on M. Once again, we design the transformation so that it preserves the geometry of the module, to be able to meaningfully lift any dense rank-1 submodule of M' back to a dense rank-1 submodule of M. For this reduction, we assume that all our worst-case modules live in $K_{\mathbb{R}}^2 = (K \otimes_{\mathbb{Q}} \mathbb{R})^2$ and have fixed volume (which we can always achieve by scaling the module). We also assume that the ℓ_2-norm of their shortest non-zero vectors is exactly $1/\gamma < 1$. This restriction to modules with a known gap can be waived, by guessing the gap and sparsifying the module (see Sect. 6).

Let us explain the main ideas behind the randomization in the simpler case of $K = \mathbb{Q}$. We have a lattice $M \subset \mathbb{R}^2$ with volume 1 and shortest non-zero vector $\mathbf{s}$ with $\|\mathbf{s}\| = 1/\gamma$. Up to rotation of the ambient space, we can assume that $\mathbf{s} = (1/\gamma, 0)^T$. Let us take $\mathbf{t} \in \mathbb{R}^2$ such that $(\mathbf{s}, \mathbf{t})$ forms a basis of M. Since the volume of M is 1, we know that $\mathbf{t} = (t_0, \gamma)^T$ for some $t_0 \in \mathbb{R}$. Up to the rotation of the ambient space, the quantity t_0 is the only degree of freedom. Note also that the lattice only depends on $t_0 \bmod 1/\gamma$. Let $\pi_{\mathbf{s}}(\mathbf{t})$ denote the quantity t_0, i.e., the norm of the orthogonal projection of $\mathbf{t}$ onto span($\mathbf{s}$). This discussion shows that the lattice M is uniquely determined by the span of its shortest non-zero vector and the quantity $\gamma \cdot \pi_{\mathbf{s}}(\mathbf{t}) \bmod 1$. Hence, to "hide" the lattice M, it suffices to "hide" these two quantities. Note that we use the vectors $\mathbf{s}$ and $\mathbf{t}$ for our reasoning, but we usually do not have access to them: we randomize our module by performing only operations that can be done on any of the bases of M (for $K_{\mathbb{R}}^2$ instead of $\mathbb{R}^2$, we expect that finding the analogue of $(\mathbf{s}, \mathbf{t})$ is difficult).

In order to hide the span of $\mathbf{s}$, one can apply a uniform orthonormal transformation to the ambient space. To hide the quantity $\gamma \cdot \pi_{\mathbf{s}}(\mathbf{t}) \bmod 1$, we "blur" the ambient space, by applying to it a transformation that is close to orthogonal, but not fully so. By appropriately choosing the transformation, one can obliviously transform the quantity $\gamma \cdot \pi_{\mathbf{s}}(\mathbf{t})$ into $x \cdot \gamma \cdot \pi_{\mathbf{s}}(\mathbf{t}) + y$, where x and y are some random variables. Recall that this quantity only matters modulo 1. Hence, if the standard deviation of y is sufficiently large compared to 1, then $y \bmod 1$ will be uniformly distributed and will hide the original value of $\pi_{\mathbf{s}}(\mathbf{t})$. The existence of a gap ensures that a close-to-orthogonal transformation suffices for this purpose.

This intuition over $\mathbb{R}^2$ explains one component of our randomization procedure, which we call the geometric randomization (see Sect. 5.2). Another important part of our randomization, which we call the coefficient randomization (Sect. 5.1), focuses on the coefficient ideals of the pseudo-basis (which are just $\mathbb{Z}$ for lattices). The transformation described above will have the effect of randomizing the vectors $\mathbf{b}_1$ and $\mathbf{b}_2$ of a pseudo-basis of our module M, but will have no impact on the coefficients ideals I_1 and I_2.

In order to hide those ideals, the first step is to multiply the module M by some uniformly distributed ideal I, using [dBDPW20]. Our new coefficient ideals $I \cdot I_1$ and $I \cdot I_2$ will then be uniformly distributed too. This is however not sufficient to fully hide the ideals, since the quotient $(I \cdot I_1)/(I \cdot I_2)$ is constant. In order to hide this last quantity, or decouple the ideals, we sparsify the module with respect to some prime ideal $\mathfrak{p}$: concretely, we take a uniformly random rank-2 submodule of M among those of index $\mathfrak{p}$.[1] This process generalizes lattice sparsification as introduced in [Kho06]. Lattice sparsification is a classic tool to remove one (or several) annoying vectors in a lattice. Here, the purpose is different: it has the effect of obliviously multiplying I_1 by $\mathfrak{p}$ while leaving I_2 unchanged (with probability close to 1). By [dBDPW20], the uniform distribution over bounded-norm prime ideals is close to the uniform distribution over norm-1 ideals (after renormalization of their norm), in the sense that little remains to be done to obtain the latter distribution. As a result, this sparsification enables us to (almost) randomize both I_1 and I_2, independently of one another. The gap to perfect randomization is handled by carefully studying the distribution resulting from the geometric and coefficient randomization (Sect. 5.3).

Summing up, our randomization consists in two main steps: a distortion of the ambient space, which randomizes the vectors $(\mathbf{b}_1, \mathbf{b}_2)$ and a sparsification, which hides the coefficient ideals I_1 and I_2 (together with the multiplication of the module by a random ideal I). Interestingly, we note that these two operations are similar (though adapted to rank-2 modules) to the ones that were used in [dBDPW20] to randomize ideal lattices.

The transformation described above allows us to transform an arbitrary module M of $K_{\mathbb{R}}^2$ into a random module M' of $K_{\mathbb{R}}^2$ whose distribution is independent of the input module. One last subtlety to handle in order to have a full worst-case to average-case reduction is to compute a canonical representation of the module M'. Indeed, the pseudo-basis of the properly distributed module M' that we have at the end of the randomization procedure might leak information about the input module M. Unfortunately, one cannot compute HNF bases in $K_{\mathbb{R}}^2$ (the HNF gives a canonical representation of rational lattices). In order to obtain a

---

[1] For two rank-2 modules $M' \subseteq M$ with pseudo-bases $((\mathbf{b}'_1, I'_1), (\mathbf{b}'_2, I'_2))$ and $((\mathbf{b}_1, I_1), (\mathbf{b}_2, I_2))$ respectively, we say that M' has index $\mathfrak{p}$ in M if $\det_K(\mathbf{b}'_1, \mathbf{b}'_2) \cdot I'_1 I'_2 = \mathfrak{p} \cdot \det_K(\mathbf{b}_1, \mathbf{b}_2) \cdot I_1 I_2$.

canonical representation of M', we then round it to a close module in $\mathcal{O}_K^2$ for which we will be able to compute an HNF pseudo-basis. The rounding procedure is the same as the one described in the reduction from uSVP to NTRU, and the distribution of the output pseudo-basis only depends on the input module and not on the specific pseudo-basis that is provided to represent it.

Discussion. A question arising from our reduction concerns the possibility to sample an NTRU instance from the distribution obtained at the end of the reduction, together with a short secret vector of the corresponding NTRU module. The difficulty stems from the fact that the output NTRU distribution we obtain after the reduction is not easy to describe, except as "the distribution obtained by running the reduction". The same difficulty also appeared in [PS21], where it was tackled by running the reduction to sample from the average-case NTRU distribution (and keeping in mind some quantities generated during the reduction in order to create a short vector of the output NTRU module). In our case, we face two additional difficulties when trying to apply the same strategy. First, we note that even sampling from the NTRU distribution, without asking for a short vector of the corresponding module, does not seem straightfoward. Since our mod-uSVP to NTRU reduction requires an id-SVP solver and takes subexponential time if one wants to reach small NTRU modulus q, it does not provide an efficient sampling algorithm for our final NTRU distribution. Secondly, our reduction allows us to lift a short vector from the NTRU module back to the uSVP module, but it is not so clear whether the converse is also possible (i.e., starting with a known vector of the uSVP module and obtaining a short vector of the final NTRU module). This is because of the sparsification step: when we sparsify a lattice, we can lift a vector from the sparser lattice back to the denser lattice (this is actually the same vector), but the converse seems more difficult.

Another question we leave open is about the compatibility of our reduction with those from [PS21]. Our worst-case mod-uSVP$_{mod}$ to average-case NTRU$_{mod}$ reduction produces a new distribution over NTRU instances. It is unclear whether this distribution can be used in the search to decision reduction from [PS21]. It is also unclear how it compares to the one produced by the worst-case id-SVP to average-case NTRU reduction from [PS21].

It should be noted that the regime where NTRU is provably secure (see [SS13]) is completely distinct from the regime required by our reductions. Indeed, the regime of [SS13] requires that f and g are slightly larger than $\sqrt{q}$, whereas our reduction requires f and g to be significantly smaller than $\sqrt{q}$. In other words, we are in a regime where NTRU is a uSVP instance (and we are trying to show that in this regime, it is representative of all uSVP instances), whereas [SS13] works in a regime where an NTRU instance is statistically close to uniform; in particular, in that regime, the underlying lattice is not a uSVP instance. The regime of the overstretch-NTRU attacks (including [KF17]) is also distinct from ours, but in the opposite direction. In these attacks, it is assumed that $\|f\|$ and $\|g\|$ are poly(d) and q grows; whereas in our case, we have $\|f\|$ and $\|g\|$ of the form $\sqrt{q}/\text{poly}(d)$. Said differently, in those attacks, the short vector is short in absolute terms, whereas in our case it is short relative to what it would be for a random lattice of the same volume. We leave as an open problem to check whether these two regimes can be made to intersect.

2   Preliminaries

We use standard Landau notations, with underlying constants that are absolute (e.g., they do not depend on the specific choice of number field). We consider column vectors (unless they are explicitly transposed). Vectors and matrices are respectively denoted in bold lowercase and uppercase fonts. For a vector $\mathbf{x} \in \mathbb{C}^k$, we let $\|\mathbf{x}\|$ denote its Hermitian norm.

We let $\mathcal{D}(c, s)$ refer to the normal distribution over $\mathbb{R}$ of center c and standard deviation $s > 0$. For X a set that is finite or has finite Lebesgue measure, we let $\mathcal{U}(X)$ denote the uniform distribution over X. For two distributions D_1, D_2 with compatible supports, we let $\mathrm{SD}(D_1, D_2) = \int |D_1(t) - D_2(t)| \, \mathrm{d}t / 2$ refer to their statistical distance. For D_1, D_2 with $\mathrm{Supp}(D_1) \subseteq \mathrm{Supp}(D_2)$, we let $\mathrm{RD}(D_1 \parallel D_2) = \int D_1(t)^2 / D_2(t) \, \mathrm{d}t$ refer to their Rényi divergence of order 2. The probability preservation property states that for any event E, the inequality $D_2(E) \geq D_1(E)^2 / \mathrm{RD}(D_1 \parallel D_2)$ holds.

For a lattice L, we let $D_{L,s,\mathbf{c}}$ denote the Gaussian distribution of support L, standard deviation parameter s and center parameter $\mathbf{c} \in \mathrm{span} L$. We will use the following lemma, to sample discrete (tail-cut) Gaussian distributions. This lemma is adapted from [GPV08, Theorem 4.1]. A proof of this precise formulation can be found in [PS21, Lemma 2.2].

Lemma 2.1. *There exists a polynomial time algorithm that takes as input a basis* $\mathbf{B} = (\mathbf{b}_1, \ldots, \mathbf{b}_n)$ *of an* n-*dimensional lattice* L, *a parameter* $s \geq \sqrt{n} \cdot \max_i \|\mathbf{b}_i\|$ *and a center* $\mathbf{c} \in \mathrm{span} L$ *and outputs a sample from a distribution* $\hat{D}_{\mathbf{B},s,\mathbf{c}}$ *such that*

- $\mathrm{SD}(D_{L,s,\mathbf{c}}, \hat{D}_{\mathbf{B},s,\mathbf{c}}) \leq 2^{-\Omega(n)}$;
- *for all* $\mathbf{v} \leftarrow \hat{D}_{\mathbf{B},s,\mathbf{c}}$, *it holds that* $\|\mathbf{v} - \mathbf{c}\| \leq \sqrt{n} \cdot s$.

Some results are obtained under the Extended Riemann Hypothesis (ERH).

2.1   Number Fields

Let K be a number field of degree $d \geq 2$ and ring of integers $\mathcal{O}_K$. Let $K_{\mathbb{R}} = K \otimes_{\mathbb{Q}} \mathbb{R}$. We identify any element of K with its canonical embedding vector $\sigma : x \mapsto (\sigma_1(x), \cdots, \sigma_d(x))^T \in \mathbb{C}^d$. This leads to an identification of $K_{\mathbb{R}}$ with $\{\mathbf{y} \in \mathbb{C}^d : \forall i \in [r_1], y_i \in \mathbb{R} \text{ and } \forall i \in [r_2], \overline{y_{r_1+r_2+i}} = y_{r_1+i}\}$, where r_1 and r_2 respectively denote the number of real and pairs of complex embeddings. Note that the set $K_{\mathbb{R}}$ is a real vector subspace of dimension d embedded (via σ) in $\mathbb{C}^d$ and that $\sigma(\mathcal{O}_K)$ is a full rank lattice in $K_{\mathbb{R}}$. The (absolute) discriminant Δ_K is defined as $\Delta_K = |\det(\sigma(\mathcal{O}_K))^2|$. We have $d = O(\log \Delta_K)$, for Δ_K growing to infinity.

For $x \in K_{\mathbb{R}}$, we define $\overline{x} \in K_{\mathbb{R}}$ as the element obtained by componentwise complex conjugation of the canonical embedding vector of x. We extend this notation to vectors and matrices over $K_{\mathbb{R}}$, and let $\mathbf{x}^{\dagger}$ denote $\overline{\mathbf{x}}^T$ for any $\mathbf{x} \in K_{\mathbb{R}}^n$. We define $\overline{K}$ and $\overline{\mathcal{O}_K}$ as the subsets of $K_{\mathbb{R}}$ obtained by applying complex conjugation to elements of K and $\mathcal{O}_K$, respectively. For $\mathbf{x}, \mathbf{y} \in K_{\mathbb{R}}^n$, we define $\langle \mathbf{x}, \mathbf{y} \rangle_{K_{\mathbb{R}}} = \mathbf{x}^{\dagger} \cdot \mathbf{y} \in K_{\mathbb{R}}$ and $\|\mathbf{x}\| = \|\sigma(\langle \mathbf{x}, \mathbf{x} \rangle_{K_{\mathbb{R}}})\|^{1/2}$. The (absolute value of the) algebraic norm of $x \in K_{\mathbb{R}}$ is defined as $\mathcal{N}(x) = \prod_i |\sigma_i(x)|$. The algebraic norm of $\mathbf{x} \in K_{\mathbb{R}}^n$ is defined as $\mathcal{N}(\mathbf{x}) = \mathcal{N}(\langle \mathbf{x}, \mathbf{x} \rangle_{K_{\mathbb{R}}})^{1/2}$.

We define $K_{\mathbb{R}}^+$ as the subset of $K_{\mathbb{R}}$ corresponding to having all y_i's being positive real numbers. For $x \in K_{\mathbb{R}}^+$, we define $x^{1/2}$ as the element of $K_{\mathbb{R}}^+$ obtained by taking the square-roots of the embeddings.

We let $\mathcal{O}_K^{\times} = \{x \in \mathcal{O}_K : \mathcal{N}(x) = 1\}$ denote the set of units of $\mathcal{O}_K$ and $\mathrm{Log}\mathcal{O}_K^{\times} = \{(\log |\sigma_i(x)|)_i : x \in \mathcal{O}_K^{\times}\} \subset \mathbb{R}^d$ denote the log-unit lattice. Note that $\mathrm{span}_{\mathbb{R}}(\mathrm{Log}\mathcal{O}_K^{\times}) = E := \{\mathbf{y} \in \mathbb{R}^d : \sum y_i = 0 \wedge \forall i \in [r_2], y_{r_1+r_2+i} = y_{r_1+i}\}$, by Dirichlet's unit theorem. For $\zeta \in E$, we define $\exp(\zeta)$ as the element of $K_{\mathbb{R}}^+$ whose i-th embedding is $\exp(\zeta_i)$, for all i.

In this work, we assume that we know a LLL-reduced [LLL82] $\mathbb{Z}$-basis $(r_i)_{i \leq d}$ of $\mathcal{O}_K$. We define $\delta_K = \max_i \|r_i\|_{\infty}$. We have $1 \leq \delta_K \leq \Delta_K^{O(1)}$: the left inequality follows from the fact that $\|r\|_{\infty} \geq 1$ for all $r \in \mathcal{O}_K \setminus \{0\}$, whereas the right inequality derives from Minkowski's second theorem and the LLL-reducedness of the r_i's. In the case of cyclotomic number fields, taking the power basis gives $\delta_K = 1$. For $x = \sum_i x_i r_i \in K_{\mathbb{R}}$, we define $\lfloor x \rceil = \sum_i \lfloor x_i \rceil r_i$. We will use the notation $\{x\} = x - \lfloor x \rceil$. We have $\|\{x\}\|_{\infty} \leq d \cdot \delta_K$, and hence $\|\{x\}\| \leq d^{3/2} \cdot \delta_K$.

We will consider the following distributions over $K_{\mathbb{R}}$. Note that for $r \in K_{\mathbb{R}}^+$, the distribution of $r \cdot x$ for $x \sim \mathcal{D}_{K_{\mathbb{R}}}(c, \mathbf{s})$ is $\mathcal{D}_{K_{\mathbb{R}}}(r \cdot c, (\sigma_i(r) \cdot \mathbf{s}_i)_i)$.

Definition 2.2. *Let $\mathbf{s} \in \mathbb{R}_{>0}^{r_1+r_2}$. We define the normal distribution $\mathcal{D}_{K_{\mathbb{R}}}(c, \mathbf{s})$ of center $c \in K_{\mathbb{R}}$ and standard deviation vector $\mathbf{s}$ as the distribution obtained by independently sampling real numbers $(y)_{i \in [d]}$ with*

$$\begin{cases} y_j \sim \mathcal{D}(0, s_j) & \text{for } j \in [r_1] \\ y_{r_1+j}, y_{r_1+r_2+j} \sim \mathcal{D}(0, s_{r_1+j}) & \text{for } j \in [r_2] \end{cases}$$

and then returning $c + y$ where $y \in K_{\mathbb{R}}$ is such that $\sigma_j(y) = y_j$ for $j \in [r_1]$ and $\sigma_{r_1+j}(y) = y_{r_1+j} + iy_{r_1+j}$ for $j \in [r_2]$.

We define $\chi_{K_{\mathbb{R}}}$ as the distribution of $(\langle \mathbf{x}, \mathbf{x} \rangle_{K_{\mathbb{R}}})^{1/2}$ for $\mathbf{x} \in K_{\mathbb{R}}^2$ sampled according to $\mathcal{D}_{K_{\mathbb{R}}}(0, 1)^2$.

For a matrix $\mathbf{B} \in K_{\mathbb{R}}^{n \times n}$, we define $\det(\mathbf{B}) = \mathcal{N}(\det_{K_{\mathbb{R}}}(\mathbf{B}))$. We say that $\mathbf{B}$ is orthogonal if $\mathbf{B}^{\dagger} \cdot \mathbf{B} = \mathbf{I}$, which implies that $\det(\mathbf{B}) = 1$. We let $\mathcal{O}_n(K_{\mathbb{R}})$ denote the set of orthogonal matrices. If a matrix $\mathbf{B} \in K_{\mathbb{R}}^{n \times n}$ has $K_{\mathbb{R}}$-linearly independent columns (i.e., no non-trivial linear combination is zero), then it admits a QR-factorization $\mathbf{B} = \mathbf{Q}\mathbf{R}$ with $\mathbf{Q} \in \mathcal{O}_n(K_{\mathbb{R}})$ and $\mathbf{R} \in K_{\mathbb{R}}^{n \times n}$ upper triangular with diagonal elements in $K_{\mathbb{R}}^+$ (see, e.g., [LPSW19, Section 2.3]).

2.2 Ideals

A fractional ideal (resp. oriented replete ideal) is a subset of K of the form $x \cdot I$ for some $x \in K^{\times}$ (resp. $x \in K_{\mathbb{R}}^{\times}$) and $I \subseteq \mathcal{O}_K$ an integral ideal. Unless specified otherwise, by default, an ideal will refer to an oriented replete ideal. For I ideal of K, we define the ideal $\overline{I} = \{\overline{x} : x \in I\}$ of $\overline{K}$. Using the canonical embedding, any non-zero ideal is identified to a d-dimensional lattice, called ideal lattice. The algebraic norm of an integral ideal I is $\mathcal{N}(I) := |\mathcal{O}_K/I|$ if it is non-zero and zero otherwise. This is extended to oriented replete ideals xI with $x \in K_{\mathbb{R}}^{\times}$ and I an integral ideal by setting $\mathcal{N}(xI) = \mathcal{N}(x) \cdot \mathcal{N}(I)$.

For I_1 and I_2 integral, the product ideal $I_1 I_2$ is the ideal spanned by all $x_1 \cdot x_2$ with $x_1 \in I_1$ and $x_2 \in I_2$. An integral ideal I is said prime if it cannot be written as $I = I_1 \cdot I_2$ with I_1, I_2 integral and both distinct from $\mathcal{O}_K$. For any $B \geq 0$, we let $\pi_K(B)$ denote the number of prime ideals with algebraic norm $\leq B$. Under the ERH, there exists an absolute constant c such that for any $B \geq (\log \Delta_K)^c$, we have $\pi_K(B) \in (B/\log B) \cdot [0.9, 1.1]$ (see [BS96, Theorem 8.7.4]). If $x_1 I_1$ and $x_2 I_2$ are two ideals with I_1 and I_2 integral, we define their product as $(x_1 I_1) \cdot (x_2 I_2) = (x_1 x_2)(I_1 I_2)$. The inverse of an ideal I is $I^{-1} = \{x \in K_{\mathbb{R}}^{\times} : xI \subseteq \mathcal{O}_K\}$.

We will use algorithms from [dBDPW20] to sample among different classes of ideals.

Lemma 2.3 (Adapted from [dBDPW20, Lemma 2.2], ERH). *There exists an algorithm $\mathcal{A}$ and an absolute constant c such that for any $B \geq (\log \Delta_K)^c$, algorithm $\mathcal{A}$ on input B runs in time $\mathrm{poly}(\log B, d)$ and returns a prime ideal uniformly among prime ideals of norm $\leq B$.*

We will also rely on Algorithm 2.1, which is adapted from [dBDPW20, Theorem 3.3], to sample (essentially) uniformly in the set $\mathcal{I}_1$ of norm-1 ideals, in time polynomial in $\log B$. Note that [dBDPW20] considers norm-1 ideals xI with I integral and all $\sigma_i(x)$'s being positive integers. This discrepancy is handled by introducing u at Step 3. The standard deviation in Step 2 and tailcut may seem arbitrary at first sight: these choices simplify the analysis of the module randomization (in Sect. 5.3). A proof of the following lemma is given in the full version of this work.

Algorithm 2.1. Ideal-Sample$_B$

1: Sample $\mathfrak{p}$ uniformly among prime ideals of norms $\leq B$, using Lemma 2.3;
2: Sample $\zeta \in E$ from the centered normal law with standard deviation $d^{-3/2}$, conditioned on $\|\zeta\| \leq 1/d$;
3: Sample u uniform in $\{x \in K_{\mathbb{R}}, \forall i \in [d] : \|\sigma_i(x)\| = 1\}$;
4: Return $u \cdot \exp(\zeta) \cdot \mathfrak{p}/\mathcal{N}^{1/d}(\mathfrak{p})$.

Lemma 2.4 (Adapted from [dBDPW20, Theorem 3.3], ERH). *There exists an absolute constant c such that for any $B \geq (d^d \Delta_k)^c$, Ideal-Sample$_B$ runs in time polynomial in $\log B$ and its output distribution is within $2^{-\Omega(d)}$ statistical distance from $\mathcal{U}(\mathcal{I}_1)$.*

2.3 Modules

A module is a subset of some $K_{\mathbb{R}}^m$ of the form $M = \sum_{i \leq k} \mathbf{b}_i I_i$ where the I_i's are non-zero ideals and the $\mathbf{b}_i$'s are $K_{\mathbb{R}}$-linearly independent. This is written compactly as $M = \mathbf{B} \cdot \mathbb{I}$ (where $\mathbf{B}$ is the matrix whose columns are the $\mathbf{b}_i$ and $\mathbb{I} = (I_1, \ldots, I_k)$). The tuple $((I_1, \mathbf{b}_1), \ldots, (I_k, \mathbf{b}_k))$ is called a pseudo-basis of M and is written compactly as $(\mathbf{B}, \mathbb{I})$. The integer k is the rank of M. We define $\mathcal{N}(M) = \det(\mathbf{B}) \cdot \prod_{i \leq k} \mathcal{N}(I_i)$. Note that for $d = m = 1$, this matches the norm of an ideal. Using the canonical embedding, any rank-k module is identified to a (kd)-dimensional lattice, called module lattice. In particular, we define $\det(M)$ as the determinant of the module lattice. Note that $\det(M) = \mathcal{N}(M) \cdot \Delta_K^{k/2}$. The module successive minima $\lambda_i(M)$ for $i \in [kd]$ are defined similarly. We will also be interested in the module norm-minimum $\lambda_1^{\mathcal{N}}(M) = \inf\{\mathcal{N}(N) : N \text{ rank-1 submodule of } M\}$. A rank-1 submodule of M is said densest if it reaches $\lambda_1^{\mathcal{N}}(M)$.

The dual of a module M is defined as $M^{\vee} = \{\mathbf{b}^{\vee} \in \mathrm{span}_{K_{\mathbb{R}}}(M) : \forall \mathbf{b} \in M, \langle \mathbf{b}^{\vee}, \mathbf{b} \rangle_{K_{\mathbb{R}}} \in \mathcal{O}_K\}$: note that $M^{\vee}$ is an $\overline{\mathcal{O}_K}$-module, $\sigma(M^{\vee})$ is the dual lattice of $\sigma(M)$ and $(\mathbf{B} \cdot \mathbb{I})^{\vee} = (\mathbf{B}^{-\dagger} \cdot \mathbb{J})$, where $J_i = (\overline{I_i})^{-1}$ for all $i \leq k$.

For any full-rank module $M \subseteq K^m$, there exists a pseudo-basis $(\mathbf{B}, \mathbb{I})$ such that $\mathbf{B} \in K^{m \times m}$ is lower-triangular with ones on the diagonal. It is called a Hermite Normal Form of M and can be computed in polynomial time from any finite set of pairs $\{(I_i, \mathbf{b}_i)\}_i$ such that $M = \sum_i \mathbf{b}_i I_i$ and the $\mathbf{b}_i$'s are not necessarily independent [BP91, Coh96, BFH17].

Definition 2.5. *Let M be a module. A submodule $N \subseteq M$ is said to be primitive if it satisfies any of the three equivalent conditions:*

- *the module N is maximal for the inclusion in the set of submodules of M of rank at most $\mathrm{rank}(N)$;*
- *there is a module N' with $M = N + N'$ and $\mathrm{rank}(M) = \mathrm{rank}(N) + \mathrm{rank}(N')$;*
- *we have $N = M \cap \mathrm{span}_K(N)$.*

In particular, any densest rank-1 submodule of M is primitive.

A proof that the three conditions are equivalent is provided in the full version of this work. The last statement follows from Condition 1.

The latter lemma allows us to conclude that the module norm-minimum is reached (see the full version of this work for a proof).

Lemma 2.6. *For any module M, there exists a rank-1 submodule N of M such that $\mathcal{N}(N) = \lambda_1^{\mathcal{N}}(M)$.*

The following result provides a lower bound on the probability that a rank-1 module $\mathbf{v} \cdot \mathcal{O}_K$ is primitive in a rank-k module M, when $\mathbf{v} \in M$ is sampled from a sufficiently wide Gaussian distribution. Taking $M = \mathcal{O}_K^k$, this provides in particular a lower bound on the probability that k elements sampled independently of a Gaussian distribution in $\mathcal{O}_K$ are relatively coprime. This result generalizes [SS13, Lemma 4.4], which proved the result for $k = 2$ and $M = \mathcal{O}_K^2$

(with a proof inspired from [Sit10]). The proof for the general case with rank-k modules is very similar to the special case $M = \mathcal{O}_K^2$, hence we leave it to the full version. In this work, we will only use Lemma 2.7 for modules of rank-2, however, for the sake of re-usability, we state and prove it for modules of arbitrary ranks.

Lemma 2.7. *There exists an absolute polynomial P such that the following holds. For any $\delta \geq 0$, degree-d number field K, integer $k \geq 2$, rank-k module $M \subset K_{\mathbb{R}}^k$, if $\mathbf{c} \in \mathrm{span}_{K_{\mathbb{R}}}(M)$ and $\varsigma > 0$ are such that $\|\mathbf{c}\| \leq \delta \cdot \varsigma$ and $\varsigma \geq \lambda_{kd}(M) \cdot P(\Delta_K^{1/d}, k, d, \delta, \lambda_{kd}(M)/\lambda_1(M))$, then it holds that*

$$\Pr_{\mathbf{v} \leftarrow D_{M,\varsigma,\mathbf{c}}} \left(\mathbf{v} \cdot \mathcal{O}_K \text{ is primitive in } M\right) \geq \frac{1}{4\zeta_K(k)},$$

where $\zeta_K(\cdot)$ is the Dedekind zeta function of K and the λ_i's refer to the minima of the lattice $\sigma(M)$.

2.4   Rank-2 Modules with a Gap

In this work, we are interested in rank-2 modules that contain an unexpectedly dense rank-1 submodule, i.e., in modules M with $\lambda_1^{\mathcal{N}}(M)$ significantly smaller than $\sqrt{\mathcal{N}(M)}$. We define the gap of M by

$$\gamma(M) = \left(\frac{\mathcal{N}(M)^{\frac{1}{2}}}{\lambda_1^{\mathcal{N}}(M)}\right)^{\frac{1}{d}}.$$

The following lemma shows that if the gap is sufficiently large, then the densest rank-1 submodule is unique. A proof may be found in the full version of this work.

Lemma 2.8. *Let M be a rank-2 module with gap $\gamma > 0$ and N a densest rank-1 submodule of M. If N' is a rank-1 submodule of M with $\mathcal{N}(N') < \gamma^d \sqrt{\mathcal{N}(M)}$, then $N' \subseteq N$.*

In particular, for $\gamma > 1$, the densest rank-1 submodule is unique and any vector $\mathbf{b} \in M$ with $\|\mathbf{b}\| < \gamma \cdot \mathcal{N}(M)^{1/(2d)}$ belongs to it.

In the following, when a rank-2 module M has a gap larger than 1, we will implicitly use Lemma 2.8 when referring to the densest rank-1 submodule of M. Most rank-2 modules we will consider will have gap larger than 1.

This can be used to show that we can use the QR-factorization to precisely describe rank-2 modules (see the full version for a proof).

Lemma 2.9. *Let M be a rank-2 module with gap $\gamma > 0$. Then M can be written as*

$$\frac{\mathcal{N}^{\frac{1}{2d}}(M)}{\gamma} \cdot \mathbf{Q} \cdot \left(\begin{bmatrix} 1 \\ 0 \end{bmatrix} \cdot J_1 + \begin{bmatrix} r \\ 1 \end{bmatrix} \cdot \gamma^2 \cdot J_2\right),$$

where $\mathbf{Q} \in \mathcal{O}_2(K_{\mathbb{R}})$, $r \in K_{\mathbb{R}}$, J_1 and J_2 are norm-1 ideals. We call this a QR-standard-form for M.

We note that there are multiple QR-standard forms for any module M, as units of $\mathbb{C}$ can be transferred from the ideal coefficients to the matrix $\mathbf{Q}$. In the following section, we will be interested in modules with specific distributions expressed in terms of QR-standard forms. It will then be convenient to define a module by a (well-distributed) QR-standard form. Note that the modules we define this way have norm 1.

Definition 2.10. *For any* $\mathbf{Q} \in \mathcal{O}_2(K_{\mathbb{R}}), \gamma > 0, r \in K_{\mathbb{R}}$ *and norm-1 ideals* J_1, J_2, *we define*

$$\mathtt{QRSF\text{-}2\text{-}Mod}(\mathbf{Q}, \gamma, J_1, J_2, r) = \frac{1}{\gamma} \cdot \mathbf{Q} \cdot \left(\begin{bmatrix} 1 \\ 0 \end{bmatrix} \cdot J_1 + \begin{bmatrix} r \\ 1 \end{bmatrix} \cdot \gamma^2 \cdot J_2 \right).$$

We will use the following result on the first and last minimum of the dual of a rank-2 module with a gap. The proof is provided in the full version of this paper.

Lemma 2.11. *Let* M *be a rank-2 module in* $K_{\mathbb{R}}^2$ *with gap* $\gamma(M) \geq 1$. *Then*

$$\lambda_{2d}(M^{\vee}) \leq 2\sqrt{d} \cdot \gamma(M) \cdot \mathcal{N}(M)^{-\frac{1}{2d}}$$

$$\lambda_1(M^{\vee})^{-1} \leq 2d \cdot \gamma(M) \cdot \mathcal{N}(M)^{1/(2d)} \cdot \delta_K \cdot \Delta_K^{\frac{1}{2d}}.$$

2.5 Algorithmic Problems

In this section, we define different variants of the unique-SVP problem for rank-2 modules, as well as variants of the NTRU problem. The definitions of the different NTRU problems differ slightly from the ones defined in [PS21]: this is to emphasize the resemblance between uSVP and NTRU. The difference between the NTRU definitions in this work and the ones in [PS21] are sufficiently minor that they can be reduced to one another without difficulty, and we hence opted to keep the same names.

Definition 2.12 (γ-uSVP **instance**). *Let* $\gamma > 0$. *A* γ-uSVP *instance consists in a pseudo-basis* $(\mathbf{B}, \mathbb{I})$ *of a rank-2 module* $M \subset K^2$ *such that* M *contains a non-zero vector* $\mathbf{s}$ *with* $\|\mathbf{s}\| \leq \gamma^{-1} \cdot \mathcal{N}(M)^{1/(2d)}$.

Note that any module M associated to a γ-uSVP instance contains the rank-1 submodule $s\mathcal{O}_K$ whose norm is $\leq \sqrt{\mathcal{N}(M)}/\gamma^d$. By Lemma 2.8, this implies that if $\gamma > 1$, then the module M has a unique densest rank-1 submodule.

Definition 2.13 (($\mathcal{D}, \gamma, \gamma'$)-uSVP$_{\mathrm{vec}}$ **and** (γ, γ')-wc-uSVP$_{\mathrm{vec}}$). *Let* $\gamma \geq \gamma' > 0$ *and* $\mathcal{D}$ *a distribution over* γ-uSVP *instances. The* ($\mathcal{D}, \gamma, \gamma'$) *average-case unique SVP problem for rank-2 modules* (($\mathcal{D}, \gamma, \gamma'$)-uSVP$_{\mathrm{vec}}$ *for short) asks, given as input a pseudo-basis of some rank-2 module* M *sampled from* $\mathcal{D}$, *to compute a vector* $\mathbf{s} \in M \setminus \{\mathbf{0}\}$ *such that* $\|\mathbf{s}\| \leq \mathcal{N}(M)^{1/(2d)}/\gamma'$. *The advantage of an algorithm* $\mathcal{A}$ *against the* ($\mathcal{D}, \gamma, \gamma'$)-uSVP$_{\mathrm{vec}}$ *problem is defined as*

$$\mathrm{Adv}(\mathcal{A}) = \Pr_{(\mathbf{B}, \mathbb{I}) \leftarrow \mathcal{D}} \left(\mathcal{A}((\mathbf{B}, \mathbb{I})) = \mathbf{s} \text{ with } \left| \begin{matrix} \mathbf{s} \in M \setminus \{\mathbf{0}\} \\ \|\mathbf{s}\| \leq \mathcal{N}(M)^{1/(2d)}/\gamma' \end{matrix} \right. \right),$$

where the probability is also taken over the internal randomness of $\mathcal{A}$.

The worst-case variant $((\gamma, \gamma')$-wc-uSVP$_{\text{vec}})$ asks to solve this problem for any γ-uSVP instance $(\mathbf{B}, \mathbb{I})$.

Definition 2.14 $((\mathcal{D}, \gamma)$-uSVP$_{\text{mod}}$ **and** γ-wc-uSVP$_{\text{mod}})$. *Let $\gamma > 0$ and $\mathcal{D}$ a distribution over γ-uSVP instances. The $(\mathcal{D}, \gamma)$ unique SVP problem for rank-2 modules $((\mathcal{D}, \gamma)$-uSVP$_{\text{mod}}$ for short) asks, given as input a γ-uSVP module M sampled from $\mathcal{D}$, to recover a densest rank-1 submodule $N \subset M$. The advantage of an algorithm $\mathcal{A}$ against the $(\mathcal{D}, \gamma)$-uSVP$_{\text{mod}}$ problem is defined as*

$$\text{Adv}(\mathcal{A}) = \Pr_{(\mathbf{B}, \mathbb{I}) \leftarrow \mathcal{D}} \left(\mathcal{A}((\mathbf{B}, \mathbb{I})) = N \text{ with } \begin{vmatrix} N \subset M \text{ with } \text{rk}(N) = 1 \\ \mathcal{N}(N) = \lambda_1^{\mathcal{N}}(M) \end{vmatrix} \right),$$

where the probability is also taken over the internal randomness of $\mathcal{A}$.

The worst-case variant $(\gamma$-wc-uSVP$_{\text{mod}})$ asks to solve this problem for any γ-uSVP instance $(\mathbf{B}, \mathbb{I})$.

We can now define the NTRU problems, as special cases of the uSVP variants above.

Definition 2.15 (NTRU instance). *Let $q \geq 2$ be an integer, and $\gamma > 0$ a real number. A (γ, q)-NTRU instance is a γ-uSVP instance whose pseudo-basis is required to be of the form $((\mathbf{b}_1, \mathcal{O}_K), (\mathbf{b}_2, \mathcal{O}_K))$ with $\mathbf{b}_1 = (1, h)^T$ for some $h \in \mathcal{O}_K$ and $\mathbf{b}_2 = (0, q)^T$.*

Comparison with [PS21]. In [PS21], an NTRU instance consists in the single element $h \in R_q$, whereas we consider it as a basis of a rank-2 module in this work. Both formalisms are equivalent, since one can reconstruct the basis of the rank-2 module from h (and also q, which is a parameter of the problem). A second difference comes from the fact that [PS21] requires the short vector $\mathbf{s} = (s_1, s_2)^T$ to satisfy $\|s_1\|, \|s_2\| \leq \sqrt{q}/\gamma$, whereas we require that $\|\mathbf{s}\| \leq \sqrt{q}/\gamma$. This means that a (γ, q)-NTRU instance for us is a (γ, q)-NTRU instance for [PS21], but the converse does not hold: a (γ, q)-NTRU instance for [PS21] is only guaranteed to be a $(\sqrt{2} \cdot \gamma, q)$-NTRU instance for us.

Definition 2.16 (NTRU problems). *Let $q \geq 2$, $\gamma \geq \gamma' > 0$ and $\mathcal{D}$ a distribution over (γ, q)-NTRU instances. The $(\mathcal{D}, \gamma, \gamma', q)$-NTRU$_{\text{vec}}$ problem, (γ, γ', q)-wc-NTRU$_{\text{vec}}$ problem, $(\mathcal{D}, \gamma, q)$-NTRU$_{\text{mod}}$ problem and (γ, q)-wc-NTRU$_{\text{mod}}$ problem are the restrictions of the uSVP problems to (γ, q)-NTRU instances.*

From the definitions of the NTRU and uSVP problems, one can see that the average case NTRU$_{\text{vec}}$ and NTRU$_{\text{mod}}$ problems reduce to the worst-case uSVP$_{\text{vec}}$ and uSVP$_{\text{mod}}$ problems. In the next sections, we will show that the converse also holds, provided we have an oracle solving ideal-SVP.

Finally, we also recall the definition of the Hermite shortest vector problem in ideal lattices (id-HSVP).

Definition 2.17 (γ-id-HSVP). *Let* $\gamma \geq \sqrt{d} \cdot \Delta_K^{1/(2d)}$. *Given as input a fractional ideal* $I \subset K$, *the* γ-id-HSVP *problem asks to find an element* $x \in I \setminus \{0\}$ *such that* $\|x\| \leq \gamma \cdot \mathcal{N}(I)^{1/d}$.

By Minkowski's theorem, this problem is well-defined for any $\gamma \geq \sqrt{d} \cdot \Delta_K^{1/(2d)}$.

3   New Tools on Module Lattices

In this section, we present new tools to manipulate module lattices. For the sake of re-usability, we describe them for modules of arbitrary ranks, but we will use them only in rank 2 in the reductions of the present work. The missing proofs of this section are available in the full version of this paper.

3.1   Module Sparsification

An essential ingredient in the module randomization of Sect. 5 is sparsification. In this subsection, we extend to modules the definition and some properties of sparsification over lattices [Kho06].

Definition 3.1. *Let* M *a module,* $\mathfrak{p}$ *a prime ideal,* $\overline{\mathbf{b}^\vee} \in (M^\vee/\mathfrak{p}M^\vee) \setminus \{\mathbf{0}\}$ *and* $\mathbf{b}^\vee$ *a lift of* $\overline{\mathbf{b}^\vee}$ *in* $M^\vee$. *The sparsification of* M *by* $(\overline{\mathbf{b}^\vee}, \mathfrak{p})$ *is the submodule*

$$M' = \left\{ \mathbf{m} \in M, \ \langle \mathbf{b}^\vee, \mathbf{m} \rangle_{K_\mathbb{R}} \in \mathfrak{p} \right\}.$$

The submodule M' *does not depend on the choice of the vector* $\mathbf{b}^\vee$ *lifting* $\overline{\mathbf{b}^\vee}$.

Note that $M \subseteq M' \subseteq \mathfrak{p}M$, implying that M' has the same rank as M. As showed by the following two lemmas, sparsification increases the module norm by a factor $\mathcal{N}(\mathfrak{p})$ and an arbitrary rank-1 submodule of M is not contained in M' (except with probability $\leq 1/\mathcal{N}(\mathfrak{p})$).

Lemma 3.2. *Let* M *a module,* $\mathfrak{p}$ *a prime ideal and* $\overline{\mathbf{b}^\vee} \in (M^\vee/\mathfrak{p}M^\vee) \setminus \{\mathbf{0}\}$. *Let* M' *be the sparsification of* M *by* $(\overline{\mathbf{b}^\vee}, \mathfrak{p})$. *Then* $\mathcal{N}(M') = \mathcal{N}(\mathfrak{p}) \cdot \mathcal{N}(M)$.

Lemma 3.3. *Let* M *a rank-k module,* $\mathfrak{p}$ *a prime ideal and* $\mathbf{b}I$ *a primitive rank-1 submodule of* M. *Let* $\overline{\mathbf{b}^\vee}$ *be uniformly distributed in* $(M^\vee/\mathfrak{p}M^\vee) \setminus \{\mathbf{0}\}$ *and* M' *be the sparsification of* M *by* $(\overline{\mathbf{b}^\vee}, \mathfrak{p})$. *Then* $\mathbf{b}\mathfrak{p}I \subseteq M'$ *and, except with probability* $1/\mathcal{N}(\mathfrak{p}) - 1/\mathcal{N}(\mathfrak{p})^k$, *we have* $\mathbf{b}I \not\subseteq M'$.

The following lemma states that a module sparsification can be efficiently computed. The algorithm generalizes the one for lattice sparsification, detailed, e.g., in [BSW16].

Lemma 3.4. *There exists a polynomial-time algorithm taking as inputs an arbitrary pseudo-basis of* $M \subset K_\mathbb{R}^k$, *a prime ideal* $\mathfrak{p}$ *and* $\overline{\mathbf{b}^\vee} \in (M^\vee/\mathfrak{p}M^\vee) \setminus \{\mathbf{0}\}$ *and computing a pseudo-basis of the sparsification of* M *by* $(\overline{\mathbf{b}^\vee}, \mathfrak{p})$.

3.2   Module Rounding

In this section, we describe the `DualRound` algorithm that rounds a rank-k module contained in $K_\mathbb{R}^k$ into a module contained in $\mathcal{O}_K^k$ (with a close geometry), in a way that does not depend on how the module in $K_\mathbb{R}^k$ was represented. We do that by sampling almost orthogonal vectors in the dual lattice, in a similar fashion to what was done in [dBDPW20] in the context of ideal lattices. We believe that this technique of rounding via the dual might have other applications, especially in situations where one would like to have the analogue of an HNF basis for lattices with real coefficients.

`DualRound` is parameterized by a standard deviation parameter $\varsigma > 0$, a BKZ block-size $\beta \in \{2, \ldots, kd\}$ and an error bound $\varepsilon > 0$. It starts by computing a short $\mathbb{Z}$-basis of $\mathbf{C}^\vee$, by using a provable variant of the BKZ algorithm [Sch87, HPS11, GN08, ALNS20]. This offers different runtime-quality trade-offs. It then uses the discrete Gaussian sampler from Lemma 2.1 with orthogonal center parameters $\mathbf{t}_i$.

Algorithm 3.1. Algorithm `DualRound`$_{\varsigma, \beta, \varepsilon}$

Input: A pseudo-basis $(\mathbf{B}, \mathbb{I})$ of a rank-k module $M \subset K_\mathbb{R}^k$.
1: Compute a $\mathbb{Z}$-basis of $M^\vee$;
2: Run BKZ with block-size β on it to obtain a new $\mathbb{Z}$-basis $\mathbf{C}^\vee$ of $M^\vee$;
3: Set $R = \varepsilon^{-1}\sqrt{kd}\varsigma$;
4: For $i \in [k]$, set $\mathbf{t}_i = R \cdot \mathbf{e}_i$, where $\mathbf{e}_i$ is the i-th canonical unit vector of $K_\mathbb{R}^k$;
5: For $i \in [k]$, sample $\mathbf{y}_i \leftarrow \hat{D}_{\mathbf{C}^\vee, \varsigma, \mathbf{t}_i}$;
6: Return $\mathbf{Y} = (\mathbf{y}_1 | \ldots | \mathbf{y}_k)^\dagger$.

Lemma 3.5. *Let* $(\mathbf{B}, \mathbb{I})$ *be a pseudo-basis of a rank-k module* $M \subset K_\mathbb{R}^k$. *Let* $\beta \in \{2, \cdots, kd\}$, $\varepsilon > 0$, *and* ς *be such that* $\varsigma \geq (kd)^{kd/\beta + 3/2} \cdot \lambda_{kd}(M^\vee)$. *Algorithm* `DualRound` *runs in time polynomial in* 2^β, $\log(\varsigma/\varepsilon)$ *and the bitsize of its input. Further, on input* $(\mathbf{B}, \mathbb{I})$, `DualRound`$_{\varsigma, \beta, \varepsilon}$ *outputs a matrix* $\mathbf{Y} \in \mathrm{M}_k(K_\mathbb{R})$ *such that*

- $(\mathbf{Y} \cdot \mathbf{B}) \cdot \mathbb{I}$ *is contained in* $\mathcal{O}_K^k$;
- $\mathbf{Y} = R \cdot \mathbf{I}_k + \mathbf{E}$ *for* $R = \varepsilon^{-1}\sqrt{kd}\varsigma > 0$ *and* $\|e_{ij}\| \leq \varepsilon R$ *for all* $i, j \in [k]$.

Moreover, if $(\mathbf{B}', \mathbb{I}')$ *is another pseudo-basis of* M *and if* $\mathbf{Y}'$ *is the output of* `DualRound` *given this pseudo-basis as input, then*

$$\mathrm{SD}(\mathbf{Y}, \mathbf{Y}') \leq 2^{-\Omega(kd)}.$$

Note that Lemma 3.5 does not necessarily ensure that the matrix $\mathbf{Y}$ is invertible, hence the module $\mathbf{Y} \cdot \mathbf{B} \cdot \mathbb{I}$ might not be of rank k. However, by choosing ε sufficiently small and using the second condition on $\mathbf{Y}$, one can make sure that $\mathbf{Y}$ is indeed invertible. This is the purpose of Lemma 3.6.

Lemma 3.6. *Let* $\mathbf{Y} \in K_\mathbb{R}^{k \times k}$ *be such that* $\mathbf{Y} = R \cdot \mathbf{I}_k + \mathbf{E}$ *for some* $R > 0$ *and* $\|e_{ij}\| \leq \varepsilon \cdot R$ *for all* $i, j \in [k]$. *Assume that* $\varepsilon \leq 1/(2k)$. *Then* $\mathbf{Y}$ *is invertible and we have* $\mathbf{Y}^{-1} = R^{-1} \cdot \mathbf{I}_k + \mathbf{E}'$, *with* $\|e'_{ij}\| \leq (k+1) \cdot \varepsilon \cdot R^{-1}$ *for all* $i, j \in [k]$. *Further, it holds that* $\det(\mathbf{Y}) \in [(1 + (k+1)(k+2)\varepsilon)^{-d/2}, (1 + 3\varepsilon)^{d/2}] \cdot R^{kd}$.

4   From uSVP to NTRU

In this section, we prove the following result

Theorem 4.1. *Let K be a number field of degree d with $\zeta_K(2) = 2^{o(d)}$ and let $\gamma^+ > 0$ (recall that $\zeta_K(\cdot)$ denotes the Dedekind zeta function of K). There exists $q_0 = \mathrm{poly}(\Delta_K^{1/d}, d, \delta_K, \gamma^+) \in \mathbb{R}_{\geq 0}$ and an algorithm* uSVP-to-NTRU *such that the following holds. For any $q \geq q_0$, $\gamma_{\mathrm{NTRU}} \geq \gamma'_{\mathrm{NTRU}} > 1$, $\gamma_{\mathrm{HSVP}} \geq \sqrt{d}\Delta_K^{1/(2d)}$, let*

$$\gamma_{\mathrm{uSVP}} = \gamma_{\mathrm{NTRU}} \cdot \sqrt{\gamma_{\mathrm{HSVP}}} \cdot 16\sqrt{2} \cdot d^{3/2} \cdot \delta_K$$

$$\gamma'_{\mathrm{uSVP}} = \frac{\gamma'_{\mathrm{NTRU}}}{\gamma_{\mathrm{HSVP}}^{3/2} \cdot 4 \cdot d^{9/2} \cdot \delta_K^2}.$$

For any distribution $\mathcal{D}_{\mathrm{uSVP}}$ over γ_{uSVP}-uSVP instances with gap $\leq \gamma^+$, let $\mathcal{D}_{\mathrm{NTRU}}$ be the distribution uSVP-to-NTRU$(\mathcal{D}_{\mathrm{uSVP}}, q, \gamma_{\mathrm{HSVP}})$. *We have four reductions*

- *from $(\mathcal{D}_{\mathrm{uSVP}}, \gamma_{\mathrm{uSVP}})$-uSVP$_{\mathrm{mod}}$ to $(\mathcal{D}_{\mathrm{NTRU}}, \gamma_{\mathrm{NTRU}}, q)$-NTRU$_{\mathrm{mod}}$;*
- *from γ_{uSVP}-wc-uSVP$_{\mathrm{mod}}$ restricted to modules with gap $\leq \gamma^+$ to $(\gamma_{\mathrm{NTRU}}, q)$-wc-NTRU$_{\mathrm{mod}}$;*
- *from $(\mathcal{D}_{\mathrm{uSVP}}, \gamma_{\mathrm{uSVP}}, \gamma'_{\mathrm{uSVP}})$-uSVP$_{\mathrm{vec}}$ to $(\mathcal{D}_{\mathrm{NTRU}}, \gamma_{\mathrm{NTRU}}, \gamma'_{\mathrm{NTRU}}, q)$-NTRU$_{\mathrm{vec}}$;*
- *from $(\gamma_{\mathrm{uSVP}}, \gamma'_{\mathrm{uSVP}})$-wc-uSVP$_{\mathrm{vec}}$ restricted to modules with gap $\leq \gamma^+$ to $(\gamma_{\mathrm{NTRU}}, \gamma'_{\mathrm{NTRU}}, q)$-wc-NTRU$_{\mathrm{vec}}$.*

Given access to an oracle solving γ_{HSVP}-id-HSVP, the four reductions run in time polynomial in their input size, in $\exp(\frac{d\log(d)}{\log(2q/q_0)})$ and in $\zeta_K(2)$.

The outline of the reduction is given in Fig. 2. Note that the quantity $\zeta_K(2)$ may be exponential in d for some number fields (which may impact on the run-time of the reduction, or even on the applicability of the reduction since we require $\zeta_K(2) = 2^{o(d)}$). In the case of power-of-two cyclotomic fields, it was proven in [SS13, Lemma 4.2] that $\zeta_K(2) = O(1)$. The missing proofs of this section are available in the full version of this work.

4.1   Pre-conditioning the uSVP Instance

In this section, we use algorithm DualRound to pre-process the input module and control its volume. In order to have the Hermite Normal Form of our integral module look like an NTRU instance, we slightly modify the geometry of our input module to make it have what we call the coprime property (see Definition 4.2). Hence, we describe an algorithm, called PreCond (available in the full version of this paper), which combines all this and transform any uSVP instance (with a lower bounded gap) into a new uSVP instance with roughly the same geometry and with all the properties we will require in Sect. 4.2.

Definition 4.2 (Coprime property). *We say that a rank-2 module $M \subseteq \mathcal{O}_K^2$ has the* coprime property *if it holds that*

$$\{x \in \mathcal{O}_K \mid \exists y \in \mathcal{O}_K, \ (x, y)^T \in M\} = \mathcal{O}_K.$$

In other words, the module M has the coprime property if the ideal spanned by the first coordinate of all the vectors of M is equal to $\mathcal{O}_K$.

We note that having the coprime property is not very constraining. In fact, any module can be applied a small distorsion in order to ensure the coprime property. This is formalized in Lemma 4.3 below.

Lemma 4.3. *Let $(\mathbf{B}, \mathbb{I})$ be a pseudo-basis of a rank-2 module $M \subset K^2$ with gap $\gamma(M) \geq 1$. There exists some $V_0 > 0$ with $V_0^{1/(2d)} = \mathrm{poly}(\Delta_K^{1/d}, d, \delta_K, \gamma(M))$ and an algorithm* `PreCond` *such that the following holds. Let $\beta \in \{2, \cdots, 2d\}$ and $V > 0$ be such that $V^{1/(2d)} \geq (2d)^{2d/\beta} \cdot V_0^{1/(2d)}$. Then, on input $(\mathbf{B}, \mathbb{I})$, V and β, algorithm* `PreCond` *outputs a matrix $\mathbf{Y} \in \mathrm{GL}_2(K)$ such that*

- *if $(\mathbf{B}, \mathbb{I})$ is a γ_{uSVP}-uSVP instance, then $(\mathbf{YB}, \mathbb{I})$ is a γ'_{uSVP}-uSVP instance for $\gamma'_{\mathrm{uSVP}} = \gamma_{\mathrm{uSVP}}/(2\sqrt{2})$;*
- *the rank-2 module $M' := \mathbf{YB} \cdot \mathbb{I}$ is contained in $\mathcal{O}_K^2$;*
- *$\mathcal{N}(M') \in [1/2^d, 2^d] \cdot V$;*
- *M' has the coprime property;*
- *$\mathbf{Y} = R \cdot \mathbf{I}_2 + \mathbf{E}$ for some $R = V^{1/(2d)} \cdot \mathcal{N}(M)^{-1/(2d)} > 0$ and $\|e_{ij}\| \leq R/5$ for all $1 \leq i, j \leq 2$.*

Assume that $\zeta_K(2) \leq 2^{o(d)}$. Then Algorithm `PreCond` *runs in expected time polynomial in its input bitsize, in 2^β and in $\zeta_K(2)$.*

4.2 Transforming a uSVP Instance into an NTRU Instance

As the NTRU modules are free, the second step of our reduction finds a free module containing our uSVP instance and transforms it into an NTRU instance. For this purpose, we use the `BalanceIdeal` algorithm (available in the full version of this work) that takes as input any fractional ideal I and uses a γ_{HSVP}-id-HSVP oracle to output a balanced element x such that $\langle x \rangle$ contains I but is not much larger than it.

Lemma 4.4. *There exists an algorithm* `BalanceIdeal` *that takes as input a fractional ideal $I \subset K$ and a parameter $\gamma_{\mathrm{HSVP}} \geq \sqrt{d} \cdot \Delta_K^{1/(2d)}$, and outputs an element $x \in K$ such that $I \subseteq \langle x \rangle$ and $|\sigma_i(x)| \in [1 - 1/d, 1 + 1/d] \cdot \sigma^{-1}$ for all $i \leq d$, where $\sigma = \gamma_{\mathrm{HSVP}} \cdot d^2 \cdot \delta_K \cdot \mathcal{N}(I)^{-1/d}$.*

Moreover, given access to a γ_{HSVP}-id-HSVP oracle, it runs in polynomial time and makes one call to the γ_{HSVP}-id-HSVP oracle.

We can now describe our algorithm transforming a uSVP instance into an NTRU instance: Algorithm 4.1. The operations done by this algorithm are summarised in Fig. 2 and proven in Lemma 4.6.

Algorithm 4.1. Algorithm `Conditioned-to-NTRU`

Input: A pseudo-basis $\mathbf{B}_1 \cdot \mathbb{I}$ of a rank-2 module in $\mathcal{O}_K^2$ and some parameters q and γ_{HSVP}

Output: A basis $\mathbf{B}_4$ of a free rank-2 module and some auxiliary information `aux`

1: Compute the HNF pseudo-basis $\mathbf{B}_2 \cdot \mathbb{J}$ of the rank-2 module spanned by $\mathbf{B}_1 \cdot \mathbb{I}$

   Let $a = \mathbf{B}_2[1,0] \not\approx \mathbb{B}_2 = \begin{pmatrix} 1 & 0 \\ a & 1 \end{pmatrix}$

2: Sample $b \leftarrow \mathtt{BalanceIdeal}(J_2, \gamma_{\mathrm{HSVP}})$

3: Compute $h = \lfloor a \cdot q/b \rceil$

4: **Return** $\mathbf{B}_4 = \begin{pmatrix} 1 & 0 \\ h & q \end{pmatrix}$ and $\mathtt{aux} = (a, b, J_1, J_2)$

Lemma 4.5. *Let* $\gamma_{\mathrm{HSVP}} \geq \sqrt{d}\Delta_K^{1/(2d)}$, $q \in \mathbb{Z}_{>0}$ *and* $(\mathbf{B}, \mathbb{I})$ *be a pseudo-basis of a rank-2 module* $M \subseteq \mathcal{O}_K^2$. *Assume that we have access to a* γ_{HSVP}-id-HSVP *oracle. On input* $\gamma_{\mathrm{HSVP}}, q$ *and* $(\mathbf{B}, \mathbb{I})$, *algorithm* `Conditioned-to-NTRU` *runs in polynomial time in the bitsize of its input and makes one call to the* γ_{HSVP}-id-HSVP *oracle.*

Lemma 4.6. *Let* $\gamma_{\mathrm{HSVP}} \geq \sqrt{d} \cdot \Delta_K^{1/(2d)}$, $\gamma_{\mathrm{NTRU}} > 1$ *and* $q \in \mathbb{Z}_{>0}$ *be some parameters. Define*

$$V = \gamma_{\mathrm{HSVP}}^d \cdot q^d \cdot d^d$$

$$\text{and } \gamma_{\mathrm{uSVP}} = \gamma_{\mathrm{NTRU}} \cdot \sqrt{\gamma_{\mathrm{HSVP}}} \cdot 8 \cdot d^{3/2} \cdot \delta_K.$$

Let $(\mathbf{B}, \mathbb{I})$ *be any* γ_{uSVP}-uSVP *instance in* $\mathcal{O}_K^2$, *with the coprime property and with norm in* $[1/2^{2d} \cdot V, 2^{2d} \cdot V]$. *Then on input* $(\mathbf{B}, \mathbb{I}), \gamma_{\mathrm{HSVP}}, q$, *the algorithm* `Conditioned-to-NTRU` *outputs* $(\mathbf{B}_4, \boldsymbol{aux})$ *such that* $\mathbf{B}_4$ *is a* $(\gamma_{\mathrm{NTRU}}, q)$-NTRU *instance.*

The `aux` information output by algorithm `Conditioned-to-NTRU` will be used to lift any short vector/dense submodule from the NTRU instance back to the uSVP instance. The proofs of Lemmas 4.5 and 4.6 are available in the full version of this work.

4.3 Lifting Back Short Vectors and Dense Submodules

In this section, we prove that using the auxiliary information `aux` produced by Algorithm `Conditioned-to-NTRU`, one can lift a short vector or a densest submodule from the output NTRU instance back to the input uSVP instance. The proofs of Lemmas 4.7 and 4.8 are available in the full version of this work.

Lemma 4.7. *There exists an algorithm* `LiftMod` *such that the following holds. Let* $q, \gamma_{\mathrm{HSVP}}$ *and* $(\mathbf{B}, \mathbb{I})$ *be as in Lemma 4.6. Let* M_1 *denote the rank-2 module generated by* $(\mathbf{B}, \mathbb{I})$, $[\mathbf{C}, (a, b, J_1, J_2)] \leftarrow$ `Conditioned-to-NTRU`$((\mathbf{B}, \mathbb{I}), q, \gamma_{\mathrm{HSVP}})$ *and let* M_4 *denote the rank-2 free module generated by* $\mathbf{C}$.

Let $(\mathbf{v}, J)$ *be a pseudo-basis of the densest rank-1 submodule of* M_4. *Then, on input* $a, b, (\mathbf{C}, \mathcal{O}_K^2)$ *and* $(\mathbf{v}, J)$, *algorithm* `LiftMod` *outputs* $\mathbf{w} \in K$ *such that* $\mathrm{span}_K(\mathbf{w}) \cap M_1$ *is the densest rank-1 submodule of* M_1.

Module	Pseudo-basis	Short vector
M_1	$\begin{bmatrix} \begin{pmatrix} I_1 & I_2 \\ b_{11} & b_{12} \\ b_{21} & b_{22} \end{pmatrix} \end{bmatrix}$	$\mathbf{s}_1 = \begin{pmatrix} u \\ v \end{pmatrix}$

$\Big\downarrow$ Step 1 — HNF

Module	Pseudo-basis	Short vector
$M_2 = M_1$	$\begin{bmatrix} \begin{pmatrix} J_1 & J_2 \\ 1 & 0 \\ a & 1 \end{pmatrix} \end{bmatrix}$	$\mathbf{s}_2 = \mathbf{s}_1$

$\Big\downarrow$ Step 2 — Principalization

Module	Pseudo-basis	Short vector
$M_3 \supseteq M_2$	$\begin{bmatrix} \begin{pmatrix} \mathcal{O}_K & \mathcal{O}_K \\ 1 & 0 \\ a & b \end{pmatrix} \end{bmatrix}$	$\mathbf{s}_3 = \mathbf{s}_2$

$\Big\downarrow$ Step 3 — distorsion + rounding

Module	Pseudo-basis	Short vector
M_4	$\begin{bmatrix} \begin{pmatrix} \mathcal{O}_K & \mathcal{O}_K \\ 1 & 0 \\ \lfloor a \cdot q/b \rceil & q \end{pmatrix} \end{bmatrix}$	$\mathbf{s}_4 = \begin{pmatrix} u \\ v \cdot q/b - u \cdot \{a \cdot q/b\} \end{pmatrix}$

Fig. 2. Outline of algorithm `Conditioned-to-NTRU`.

Moreover, algorithm `LiftMod` *runs in polynomial time.*

Lemma 4.8. *There exists an algorithm* `LiftVec` *such that the following holds. Let $q, \gamma_{\mathrm{HSVP}}$ and $(\mathbf{B}, \mathbb{I})$ be as in Lemma 4.6. Let M_1 denote the rank-2 module generated by $(\mathbf{B}, \mathbb{I})$, $[\mathbf{C}, \boldsymbol{aux}] \leftarrow$ `Conditioned-to-NTRU`$((\mathbf{B}, \mathbb{I}), q, \gamma_{\mathrm{HSVP}})$ and let M_4 denote the rank-2 free module generated by $\mathbf{C}$.*

Let $\mathbf{s} \in M_4$. Then, on input $\boldsymbol{aux}, \gamma_{\mathrm{HSVP}}, (\mathbf{C}, \mathcal{O}_K^2)$ and $\mathbf{s}$, algorithm `LiftVec` *outputs a vector $\mathbf{t} \in M$ such that $\|\mathbf{t}\| \leq \|\mathbf{s}\| \cdot 68 \cdot \gamma_{\mathrm{HSVP}}^2 \cdot d^4 \cdot \delta_K^2$.*

If given access to a γ_{HSVP}-id-HSVP oracle, algorithm `LiftVec` *runs in polynomial time and makes 1 call to the oracle.*

Combining all the results of this section, one can prove Theorem 4.1.

5   Randomization of Rank-2 Modules with Gaps

A rank-2 module with a gap can, by Lemma 2.9 and the fact that densest submodules are primitive, be written as $M = \mathbf{u} \cdot J_1 + \mathbf{v} \cdot J_2$ where $\mathbf{u} \cdot J_1$ is a densest rank-1 submodule of M. Informally, the goal of this section is to randomize $\mathbf{u}, \mathbf{v}, J_1, J_2$ without changing the gap too much. The missing proofs of this section are available in the full version of this work.

We first describe the average-case distribution we are considering. Note that the gap parameter γ' is itself a random variable.

Definition 5.1. *Let $\gamma > 0$ and $B > 2$. We define the distribution $D_{B,\gamma}^{\mathrm{module}}$ over rank-2 and norm-1 modules by:*

$$D_{B,\gamma}^{\mathrm{module}} = \mathtt{QRSF\text{-}2\text{-}Mod}(\mathbf{Q}, \gamma', I_1, I_2, r),$$

where

- *the matrix $\mathbf{Q}$ is uniform in $\mathcal{O}_2(K_{\mathbb{R}})$;*
- *the gap parameter γ' is set as $\gamma' = \gamma \cdot \mathcal{N}(c/a)^{1/(2d)}/B^{1/(2d)}$ with $(a,c) \in K_{\mathbb{R}}^2$ distributed as $\chi_{K_{\mathbb{R}}} \times \mathcal{D}(0,1)$ conditioned on the event that for all $i \in [d]$ we have $|\sigma_i(a \cdot c)| \geq 1/d$;*
- *the ideals I_1, I_2 are uniform in $\mathcal{I}_1$ (the set of norm-1 ideals);*
- *the element r is uniform in $K_{\mathbb{R}}$ mod $\gamma'^{-2} \cdot I_1 I_2^{-1}$.*

We now state the main theorem of this section, which can be viewed as a worst-case to average-case reduction for rank-2 modules with a gap.

Theorem 5.2 (ERH). *For all $B \geq (d^d \Delta_k)^{\Omega(1)}$ and $\gamma \geq B^{1/(2d)}$ there exists a procedure $\mathtt{Randomize}_B$ that runs in time polynomial in $\log B$ and the bitsize of its input, and such that on input a pseudo-basis $(\mathbf{B}, \mathbb{I})$ of a rank-2 and norm-1 module M of gap γ outputs a pair $((\mathbf{B}', \mathbb{I}'), \boldsymbol{aux})$ such that*

- *the pseudo-basis $(\mathbf{B}', \mathbb{I}')$ spans a rank-2 and norm-1 module M';*
- *any event that holds for $D_{B,\gamma}^{\mathrm{module}}$ with probability $\varepsilon \geq 2^{-o(d)}$ also holds for M' with probability $\Omega(\varepsilon^4)$ over the internal randomness of $\mathtt{Randomize}_B$.*

Further, there exists a deterministic algorithm $\mathtt{Recover}$ that runs in time polynomial in the bitsize of its input such that for M' as above, if U' is a densest rank-1 submodule of M', then $\mathtt{Recover}(U', \boldsymbol{aux})$ returns the densest rank-1 submodule of M, with probability $1 - 2^{-\Omega(d)}$ over the randomness of $\mathtt{Randomize}_B$.

We note that the theorem does not state that the output distribution of $\mathtt{Randomize}_B$ is $D_{B,\gamma}^{\mathrm{module}}$, but only that they are close in the sense that any event that holds with sufficient probability for $D_{B,\gamma}^{\mathrm{module}}$ also holds for the output distribution of $\mathtt{Randomize}_B$ with a polynomially related probability.

$\mathtt{Randomize}_B$ is described in Algorithm 5.6. It consists of two main steps: a coefficient randomization (described in Sect. 5.1), whose purpose is to randomize the ideal coefficients; and a geometric randomization (described in Sect. 5.2), whose purpose is to randomize the pseudo-basis matrix. Section 5.3 compares the distribution that would ideally be returned by the composition of the coefficient and geometric randomizations, with the distribution of the pseudo-basis in Definition 5.1. Finally, we complete the proof of Theorem 5.2 in Sect. 5.4.

5.1   Coefficient Randomization

In the coefficient randomization step, our aim is to randomize the ideal coefficients of a good pseudo-basis (i.e., whose first pair corresponds to the densest rank-1 submodule), given an arbitrary pseudo-basis of a rank-2 module. One may multiply the whole pseudo-basis by a random ideal, but this only randomizes the pair of ideal coefficients. More precisely, this leaves the ratio of the ideal coefficients unchanged. To decouple the ideal coefficients, we use module sparsification, as described in Sect. 3. This first step towards coefficient randomization is formally described in Algorithm 5.1. Steps 1 and 3 are respectively performed using Lemmas 2.3 and 3.4.

Algorithm 5.1. Partial Coefficient Randomization: $\mathtt{Partial\text{-}CR}_B$

Input: A pseudo-basis of a rank-2 module M.

1: Sample $\mathfrak{p}$ uniformly among prime ideals of norms $\leq B$;
2: Sample $\overline{\mathbf{b}^\vee}$ uniformly in $(M^\vee/\mathfrak{p}M^\vee) \setminus \{\mathbf{0}\}$;
3: Return a pseudo-basis of the sparsification of M by $(\overline{\mathbf{b}^\vee}, \mathfrak{p})$ along with $\mathfrak{p}$.

Theorem 5.3 (ERH). *Let $B \geq (\log \Delta_K)^{\Omega(1)}$. The runtime of $\mathtt{Partial\text{-}CR}_B$ is polynomial in $\log B$ and the bitsize of its input. Let $(\mathbf{B}, \mathbb{I})$ be a pseudo-basis of a rank-2 module M, and let $(J_1, \mathbf{u}), (J_2, \mathbf{v})$ be an arbitrary pseudo-basis of M. Let M' be the rank-2 module spanned by the pseudo-basis output by $\mathtt{Partial\text{-}CR}_B$ when given $(\mathbf{B}, \mathbb{I})$ as input, let $\overline{\mathbf{b}^\vee}$ be the element of $(M^\vee/\mathfrak{p}M^\vee) \setminus \{\mathbf{0}\}$ sampled in $\mathtt{Partial\text{-}CR}_B$ and let $\mathbf{b}^\vee$ be a lift of $\overline{\mathbf{b}^\vee}$ in $M^\vee$.*
Then, with probability $1 - (1/B)^{\Omega(1)}$, we have $\langle \mathbf{b}^\vee, \mathbf{u} \rangle_{K_\mathbb{R}} \notin \mathfrak{p}J_1^{-1}$. In that case, there exists $x \in J_1 J_2^{-1}$ such that

$$M' = \mathbf{u} \cdot \mathfrak{p}J_1 + (\mathbf{v} + x\mathbf{u}) \cdot J_2.$$

Assume further that $\gamma(M) \geq B^{1/(2d)}$ and that $\mathbf{u} \cdot J_1$ is the densest rank-1 submodule of M. Then, still when $\langle \mathbf{b}^\vee, \mathbf{u} \rangle_{K_\mathbb{R}} \notin \mathfrak{p}J_1^{-1}$, we have that $\gamma(M') = \gamma(M)/\mathcal{N}(\mathfrak{p})^{1/(2d)} > 1$ and $\mathbf{u} \cdot \mathfrak{p}J_1$ is the densest rank-1 submodule of M'.

The result follows from Lemmas 5.4 and 5.5, whose proofs are postponed to the full version of this work.

Lemma 5.4. *Borrowing the notations of Theorem 5.3, we have*

$$\mathbf{u} \cdot \mathfrak{p}J_1 \subset M' \quad and \quad \mathbf{u} \cdot J_1 \not\subset M',$$

with probability $1 - (1/B)^{\Omega(1)}$ over the choices of $\mathfrak{p}$ and $\overline{\mathbf{b}^\vee}$.

Lemma 5.5. *Borrowing the notations of Theorem 5.3 and assuming that $\mathbf{u} \cdot J_1 \not\subset M'$, there exists $x \in J_1 J_2^{-1}$ such that $(\mathbf{v} + x\mathbf{u}) \cdot J_2 \subset M'$.*

We now describe the coefficient randomization. Ideally, we would have access to a pseudo-basis $((J_1, \mathbf{u}), (J_2, \mathbf{v}))$ of the module M under scope, for which the densest rank-1 submodule is $\mathbf{u} \cdot J_1$. We would multiply J_1 by a random ideal and J_2 by another random ideal. Unfortunately, given only access to an arbitrary pseudo-basis $((I_1, \mathbf{b}_1), (I_2, \mathbf{b}_2))$ of M, this seems difficult to achieve obliviously. Instead, we use algorithm $\mathtt{Ideal\text{-}Sample}$ (Algorithm 2.1) to obtain a uniform norm-1 ideal I, and multiply M by it. This will obliviously multiply J_1 and J_2 by I. As this distribution is invariant by ideal multiplication, the ideal $J_2 I / \mathcal{N}(J_2)^{1/d}$ will be uniform among norm-1 ideals. It remains to obliviously randomize the first ideal independently of the second one. For this purpose, we use `Partial-CR` (Algorithm 5.1), which has the effect of obliviously multiplying the first ideal with a random prime ideal $\mathfrak{p}$ while leaving the second one unchanged (with overwhelming probability). Note that multiplying by a random prime ideal is the main component of the ideal randomization algorithm `Ideal-Sample`. In a sense, this "almost" randomizes J_1.

Algorithm 5.2 describes the process on the input basis $((I_1, \mathbf{b}_1), (I_2, \mathbf{b}_2))$. The corresponding randomization performed on the hidden pseudo-basis $((J_1, \mathbf{u}), (J_2, \mathbf{v}))$ is described in Algorithm 5.3. Note that there is no need for Algorithm 5.3 to be efficient as its sole purpose is to describe the behavior of Algorithm 5.2 on the hidden pseudo-basis.

In Theorem 5.6, we show that the resulting distributions on the output modules are statistically close, and describe the evolution of the densest rank-1 submodule.

Algorithm 5.2. Real Coefficient Randomization: $\text{Real-CR}_{B,B'}$

Input: A pseudo-basis $((I_1, \mathbf{b}_1), (I_2, \mathbf{b}_2))$ of a module $M \subset K_{\mathbb{R}}^2$.
1: Let $((I_1', \mathbf{b}_1'), (I_2', \mathbf{b}_2')), \mathfrak{p}$ be the output of Partial-CR_B on input $((I_1, \mathbf{b}_1), (I_2, \mathbf{b}_2))$;
2: Sample $\mathfrak{q}$ using $\text{Ideal-Sample}_{B'}$;
3: Let $\mathbf{b}_i'' = \mathbf{b}_i' / \mathcal{N}(\mathfrak{p})^{1/(2d)}$ for $i \in [2]$;
4: Return $((\mathfrak{q}I_1', \mathbf{b}_1''), (\mathfrak{q}I_2', \mathbf{b}_2'')), \mathfrak{p}, \mathfrak{q}$.

Algorithm 5.3. Ideal Coefficient Randomization: Ideal-CR_B

Input: $\mathbf{Q} \in \mathcal{O}_2(K_{\mathbb{R}}), \gamma > 1, J_1, J_2$ ideals of norm 1, $r \in K_{\mathbb{R}}$;
1: Let $M = \text{QRSF-2-Mod}(\mathbf{Q}, \gamma, J_1, J_2, r)$;
2: Let $\mathbf{u} = 1/\gamma \cdot \mathbf{Q} \cdot (1, 0)^T$ and $\mathbf{v} = \gamma \cdot \mathbf{Q} \cdot (r, 1)^T$;
3: Sample $\mathfrak{p}$ uniformly among prime ideals of norms $\leq B$;
4: Sample $\mathbf{b}^\vee$ in $M^\vee$, uniform in $M^\vee / \mathfrak{p} M^\vee$ conditioned on $\langle \mathbf{b}^\vee, \mathbf{u} \rangle_{K_{\mathbb{R}}} \notin \mathfrak{p} J_1^{-1}$;
5: Find $x \in J_1 J_2^{-1}$ such that $\langle \mathbf{b}^\vee, \mathbf{v} + x \cdot \mathbf{u} \rangle_{K_{\mathbb{R}}} \in \mathfrak{p} J_2^{-1}$;
6: Sample J uniformly among norm-1 ideals;
7: Return $(\mathbf{Q}, \gamma/\mathcal{N}(\mathfrak{p})^{1/(2d)}, J_1 J_2^{-1} J \mathfrak{p}/\mathcal{N}^{1/d}(\mathfrak{p}), J, r + x)$.

Theorem 5.6 (ERH). *Assume that $B' \geq (d^d \Delta_K)^{\Omega(1)}$ and $B \geq (\log \Delta_K)^{\Omega(1)}$. The runtime of $\text{Real-CR}_{B,B'}$ is polynomial in $\log(BB')$ and the bitsize of its input.*

Let $M = \frac{1}{\gamma} \cdot \mathbf{Q} \cdot \left(\begin{bmatrix} 1 \\ 0 \end{bmatrix} \cdot J_1 + \begin{bmatrix} r \\ 1 \end{bmatrix} \cdot \gamma^2 \cdot J_2 \right) \subset K_{\mathbb{R}}^2$ a module with norm 1, in QR-standard form. Then the distribution of the module output by $\text{Real-CR}_{B,B'}$ on input an arbitrary pseudo-basis of M is within statistical distance $(1/B)^{\Omega(1)} + 2^{-d}$ of $\text{QRSF-2-Mod}(\text{Ideal-CR}_B(\mathbf{Q}, \gamma, J_1, J_2, r))$.

Assume further that $\gamma \geq B^{1/(2d)}$ and let U denote the densest rank-1 sub-module of M. Let $(M', \mathfrak{p}, \mathfrak{q})$ be the output of `Real-CR`$_{B,B'}$ *on input M. Then, with probability $1 - (1/B)^{\Omega(1)}$, we have that $\gamma(M') = \gamma(M)/\mathcal{N}(\mathfrak{p})^{1/(2d)} > 1$ and the densest rank-1 submodule of M' is*

$$\mathcal{N}(\mathfrak{p})^{\frac{1}{2d}} \cdot U \cdot \mathfrak{q} \frac{\mathfrak{p}}{\mathcal{N}^{1/d}(\mathfrak{p})}.$$

5.2   Geometric Randomization

In the geometric module randomization, we will use a distribution D_{distort} over $K_{\mathbb{R}}^{2 \times 2}$ whose purpose is to distort the geometric relationship between the densest rank-1 submodule and the complementing rank-1 submodule of the rank-2 module under scope. We define D_{distort} as $\mathcal{D}_{K_{\mathbb{R}}}(0,1)^{2 \times 2}$ conditioned on the event that $\|\det(\sigma_i(\mathbf{D}))\| > 1/d$ holds for all $i \in [d]$.

The following lemmas describe useful properties of the distribution D_{distort}.

Lemma 5.7. *The following properties hold.*

- *The distribution D_{distort} can be sampled from in time polynomial in d.*
- *The distribution D_{distort} is invariant by multiplication on the left and the right by matrices in $\mathcal{O}_2(K_{\mathbb{R}})$.*

Lemma 5.8. *Let D be the distribution over $K_{\mathbb{R}}^{2 \times 2}$ of*

$$\mathbf{Q} \cdot \begin{pmatrix} a & b \\ 0 & c \end{pmatrix}$$

where $\mathbf{Q} \leftarrow \mathcal{U}(\mathcal{O}_2(K_{\mathbb{R}}))$, $a \leftarrow \chi_{K_{\mathbb{R}}}$ and $b, c \leftarrow \mathcal{D}_{K_{\mathbb{R}}}(0,1)$, conditioned on the event that for all $i \in [d]$ we have $|\sigma_i(a \cdot c)| \geq 1/d$. Then $D = D_{\text{distort}}$.

Let $((J_1, \mathbf{u}), (J_2, \mathbf{v}))$ be a pseudo-basis of a rank-2 module M. Assume that $\mathbf{u} \cdot J_1$ is the densest rank-1 submodule, but that we have access to this pseudo-basis only indirectly, via an arbitrary pseudo-basis of M. Write

$$(\mathbf{u}|\mathbf{v}) = \mathbf{Q} \cdot \begin{pmatrix} 1 & r \\ 0 & 1 \end{pmatrix},$$

for some $r \in K_{\mathbb{R}}$. The purpose of the geometric randomization is to map r to some r' that is uniform modulo $J_1 J_2^{-1}$, while at the same time not distorting the module M too much, so that the randomized M still has a gap and its rank-1 densest submodule is related to $\mathbf{u} \cdot J_1$. For this purpose, we multiply M on the left by a matrix sampled from D_{distort}. For the analysis, it is convenient to take it Gaussian, and to avoid a potentially large distortion, we avoid matrix samples with small determinant. This corresponds to algorithm `Real-GR` (Algorithm 5.4). The effect on the hidden pseudo-basis $((J_1, \mathbf{u}), (J_2, \mathbf{v}))$ is described in algorithm `Ideal-GR` (Algorithm 5.5). In Theorem 5.9, we show that the resulting module distributions are identical, and describe the evolution of the densest rank-1 sublattice.

Algorithm 5.4. Real Geometric Randomization: `Real-GR`

Input: A pseudo-basis $((I_1, \mathbf{b}_1), (I_2, \mathbf{b}_2))$ of a norm-1 module $M \subset K_\mathbb{R}^2$.
1: Sample $\mathbf{D} \leftarrow D_\text{distort}$ (using Lemma 5.7);
2: $(\mathbf{b}_1' | \mathbf{b}_2') \leftarrow \det(\mathbf{D})^{-1/(2d)} \cdot \mathbf{D} \cdot (\mathbf{b}_1 | \mathbf{b}_2)$;
3: Return $((I_1, \mathbf{b}_1'), (I_2, \mathbf{b}_2')), \mathbf{D}$.

Algorithm 5.5. Ideal Geometric Randomization: `Ideal-GR`

Input: $\mathbf{Q} \in \mathcal{O}_2(K_\mathbb{R}), \gamma > 1, J_1, J_2$ ideals of norm 1, $r \in K_\mathbb{R}$;
1: Sample $a \leftarrow \chi_{K_\mathbb{R}}$ and $c \leftarrow \mathcal{D}(0,1)$ conditioned on the event that for all $i \in [d]$ we have $|\sigma_i(a \cdot c)| \geq 1/d$;
2: Sample $b \leftarrow \mathcal{D}(0,1)$;
3: Sample $\mathbf{Q}' \leftarrow \mathcal{U}(\mathcal{O}_2(K_\mathbb{R}))$;
4: Set $J_1' = a/\mathcal{N}^{1/d}(a) \cdot J_1$ and $J_2' = c/\mathcal{N}^{1/d}(c) \cdot J_2$;
5: Set $\gamma' = \gamma \cdot \mathcal{N}(c/a)^{1/(2d)}$;
6: Set $r' = (b + ar)/c$;
7: Return $(\mathbf{Q}', \gamma', J_1', J_2', r')$.

Theorem 5.9. *Algorithm* `Real-GR` *runs in polynomial time. Let* $M = \frac{1}{\gamma} \cdot \mathbf{Q} \cdot \left(\begin{bmatrix} 1 \\ 0 \end{bmatrix} \cdot J_1 + \begin{bmatrix} r \\ 1 \end{bmatrix} \cdot \gamma^2 \cdot J_2 \right) \subset K_\mathbb{R}^2$ *a module with norm 1, in QR-standard-form. Let* M' *be the module spanned by the output of* `Real-GR` *on input an arbitrary pseudo-basis of* M. *Then the distribution of* M' *is identical to the distribution* `QRSF-2-Mod(Ideal-GR(`$\mathbf{Q}, \gamma, J_1, J_2, r$`))`.

Further, if $\gamma > d$ *and* U *is the densest rank-1 submodule of* M, *then, with probability* $1 - 2^{-\Omega(d)}$, *we have* $\gamma(M') > 1$ *and the densest rank-1 submodule of* M' *is* $\det(\mathbf{D})^{-1/(2d)} \cdot \mathbf{D} \cdot U$, *where* $\mathbf{D}$ *is the Gaussian matrix sampled during the execution of* `Real-GR`.

5.3   On the `Ideal-GR` o `Ideal-CR` Distribution

We define a few probability distributions over the inputs of `QRSF-2-Mod`, which we will use to show that the operations performed on the available arbitrary pseudo-basis randomize the rank-2 module, so that the input module is "forgotten" in the output module distribution while at the same time controlling the evolution of the densest rank-1 submodule.

Definition 5.10. *Let* $B \geq 2$ *and* $\gamma > 0$. *We consider the following random variables, which are assumed independent (unless stated otherwise).*

- $\mathbf{Q}$ *uniform in* $\mathcal{O}_2(K_\mathbb{R})$;
- $b \in K_\mathbb{R}$ *distributed as* $\mathcal{D}_{K_\mathbb{R}}(0,1)$;
- $(a, c) \in K_\mathbb{R}^2$ *distributed as* $\chi_{K_\mathbb{R}} \times \mathcal{D}_{K_\mathbb{R}}(0,1)$ *conditioned on the event that for all* $i \in [d]$ *we have* $|\sigma_i(a \cdot c)| \geq 1/d$; *we define* $\gamma' = \gamma \cdot \mathcal{N}(c/a)^{1/(2d)}/B^{1/(2d)}$;
- $\mathfrak{p}$ *uniform among prime ideals of norms* $\leq B$;
- I_1, I_2, J *uniform in* $\mathcal{I}_1$ *(the set of norm-1 ideals)*;

- $\zeta \in E$ sampled from the centered normal law of standard deviation $d^{-3/2}$, conditioned on $\|\zeta\| \leq 1/d$;
- u uniform in $\{x \in K_{\mathbb{R}}, \forall i \in [d] : \|\sigma_i(x)\| = 1\}$;
- r' uniform in $K_{\mathbb{R}} \bmod \gamma'^{-2} \cdot I_1 I_2^{-1}$.

Let $J_1, J_2 \in \mathcal{I}_1$ and $r \in K_{\mathbb{R}}$ arbitrary. Let x be as in Step 5 of $\mathtt{Ideal\text{-}CR}_B$, when given $(\mathbf{Q}, \gamma, J_1, J_2, r)$ as input and with the variable $\mathfrak{p}$ of $\mathtt{Ideal\text{-}CR}_B$ being the random variable above. In order to simplify the notations, we define the random variable:

$$I(J_1, J_2) = \mathcal{N}^{\frac{1}{d}}\left(\frac{c}{a}\right) \cdot \frac{au}{c \exp(\zeta)} \cdot J_1 J_2^{-1} J \frac{\mathfrak{p}}{\mathcal{N}^{1/d}(\mathfrak{p})} \in \mathcal{I}_1.$$

Let $r''(J_1, J_2)$ be uniformly distributed in $K_{\mathbb{R}} \bmod \gamma'^{-2} \cdot I(J_1, J_2) \cdot J^{-1}$.

We define the following distributions of the form $(\tilde{\mathbf{Q}}, \tilde{\gamma}, \tilde{I}_1, \tilde{I}_2, \tilde{r})$, where the random variables $\tilde{r}$ is defined modulo $\tilde{\gamma}^{-2} \cdot \tilde{I}_1 \cdot \tilde{I}_2^{-1}$:

$$D_{B,\gamma}^{\mathrm{rand}} : \left(\mathbf{Q},\ \gamma \frac{\mathcal{N}\left(\frac{c}{a}\right)^{\frac{1}{2d}}}{\mathcal{N}(\mathfrak{p})^{\frac{1}{2d}}},\ \frac{a}{\mathcal{N}^{1/d}(a)} J_1 J_2^{-1} J \frac{\mathfrak{p}}{\mathcal{N}^{1/d}(\mathfrak{p})},\ \frac{c}{\mathcal{N}^{1/d}(c)} \cdot J,\ \frac{b + a(r + x)}{c}\right),$$

$$D_{B,\gamma}^{(1)} : \left(\mathbf{Q},\ \gamma \frac{\mathcal{N}\left(\frac{c}{a}\right)^{\frac{1}{2d}}}{\mathcal{N}(\mathfrak{p})^{\frac{1}{2d}}},\ \mathcal{N}^{\frac{1}{d}}\left(\frac{c}{a}\right) \cdot \frac{au}{c} \cdot J_1 J_2^{-1} J \frac{\mathfrak{p}}{\mathcal{N}^{1/d}(\mathfrak{p})},\ J,\ u\frac{b + a(r + x)}{c}\right),$$

$$D_{B,\gamma}^{(2)} : \left(\mathbf{Q},\ \gamma \cdot \frac{\mathcal{N}\left(\frac{c}{a}\right)^{\frac{1}{2d}}}{\mathcal{N}(\mathfrak{p})^{\frac{1}{2d}}},\ I(J_1, J_2),\ J,\ u\frac{b + a(r + x)}{c \exp(\zeta)}\right),$$

$$D_{B,\gamma}^{(3)} : \left(\mathbf{Q},\ \gamma',\ I(J_1, J_2),\ J,\ \frac{B^{\frac{1}{d}}}{\mathcal{N}^{1/d}(\mathfrak{p})} \cdot u\frac{b + a(r + x)}{c \exp(\zeta)}\right),$$

$$D_{B,\gamma}^{(4)} : (\mathbf{Q},\ \gamma',\ I(J_1, J_2),\ J,\ r''(J_1, J_2)),$$

$$D_{B,\gamma}^{\mathrm{target}} : (\mathbf{Q},\ \gamma',\ I_1,\ I_2,\ r').$$

Note that $D_{B,\gamma}^{\mathrm{rand}}$ is the distribution obtained by composing $\mathtt{Ideal\text{-}CR}_B$ (Algorithm 5.3) and $\mathtt{Ideal\text{-}GR}$ (Algorithm 5.5), on an input of the form $(\mathbf{Q}_0, \gamma, J_1, J_2, r)$ with (γ, J_1, J_2, r) as above and $\mathbf{Q}_0 \in \mathcal{O}_2(K_{\mathbb{R}})$ arbitrary. These algorithms significantly randomize the QR-standard form, but it still depends on (J_1, J_2, r). On the other hand, the distribution $D_{B,\gamma}^{\mathrm{target}}$ is independent of (J_1, J_2, r). Our goal is to show that these two distributions are similar, in the sense that any event that holds with some probability $\varepsilon \geq 2^{-o(d)}$ for one holds with probability $\varepsilon^{O(1)}$ for the other one.

For this purpose, we consider the intermediate (hybrid) distributions of Definition 5.10. To help the reader, we use two colours in the definition of the successive distributions. The entries of the tuples that are in red are those that change compared to the previous distribution. The variables with blue background are those that depend on (J_1, J_2, r). The relations between the distributions of Definition 5.10 are pictorially summarized in Fig. 3. The lemmas formally stating these relations and their proofs are provided in the full version of this paper. Some of the relations require $B \geq (d^d \Delta_K)^{\Omega(1)}$ or $\gamma \geq d^{1/4} \cdot \Delta_K^{1/(2d)}$.

$$D_{B,\gamma}^{\text{rand}} = D_{B,\gamma}^{(1)} \xrightarrow{\text{RD}_2 = O(1)} D_{B,\gamma}^{(2)} \xrightarrow{\text{RD}_2 = O(1)} D_{B,\gamma}^{(3)} \xleftarrow{\text{SD} = 2^{-\Omega(d)}} D_{B,\gamma}^{(4)} \xleftarrow{\text{SD} = 2^{-\Omega(d)}} D_{B,\gamma}^{\text{target}}$$

Fig. 3. The relations between the distributions of Definition 5.10, proved in the full version of this paper. Here $D \xrightarrow{\text{RD}_2 = O(1)} D'$ means $\text{RD}(D' \parallel D) = O(1)$ and $D \xrightarrow{\text{SD} = 2^{-\Omega(d)}} D'$ means $\text{SD}(D, D') = 2^{-\Omega(d)}$.

5.4 Full Module Randomization

The full randomization algorithm $\texttt{Randomize}_B$ (Algorithm 5.6) is the composition of algorithms $\texttt{Real-CR}$ and $\texttt{Real-GR}$.

Algorithm 5.6. (Real) Full Randomization: $\texttt{Randomize}_B$

Input: A pseudo-basis $(\mathbf{B}, \mathbb{I})$ of a norm-1 module $M \subset K_{\mathbb{R}}^2$.
1: Apply $\texttt{Real-CR}_{B,(d^d \Delta_K)^{\Omega(1)}}$ to $(\mathbf{B}, \mathbb{I})$ and let $((\mathbf{B}^\circ, \mathbb{I}^\circ), \mathfrak{p}, \mathfrak{q})$ be the output;
2: Apply $\texttt{Real-GR}$ to $(\mathbf{B}^\circ, \mathbb{I}^\circ)$ and let $((\mathbf{B}', \mathbb{I}'), \mathbf{D})$ be the output;
3: Return $((\mathbf{B}', \mathbb{I}'), \texttt{aux})$ with $\texttt{aux} = (\mathfrak{p}, \mathfrak{q}, \mathbf{D})$.

Let $((\mathbf{B}', \mathbb{I}'), \texttt{aux})$ be an output of $\texttt{Randomize}_B$, and U' be a rank-1 submodule of the module spanned by $(\mathbf{B}', \mathbb{I}')$. We define:

$$\texttt{Recover}(U', \texttt{aux} = (\mathfrak{p}, \mathfrak{q}, D)) = (\mathcal{N}(\mathfrak{p}) \cdot \det(\mathbf{D}))^{\frac{1}{2d}} \cdot \mathbf{D}^{-1} \cdot U' \cdot \mathfrak{q}^{-1} \mathfrak{p}^{-1}.$$

With these choices of algorithms $\texttt{Randomize}_B$ and $\texttt{Recover}$, we can finally prove Theorem 5.2. For this purpose, we show that the module distribution that is output from the randomization algorithm (on an arbitrary input) and the distribution $D_{B,\gamma}^{\text{module}}$ from Definition 5.1 are close in the mixed "SD plus RD" sense of Fig. 3. The full proof is available in the full version of this work.

6 Random Self-reducibility of Module uSVP

The main result of this section is the following worst-case to average-case reduction for uSVP_{mod}.

Theorem 6.1 (ERH). *There exist $\gamma_0 = (d \Delta_K^{1/d})^{O(1)}$ and a family of distributions $(D_\gamma^{\text{uSVP}})_{\gamma \geq \gamma_0}$ such that the following properties hold for any $\gamma \geq \gamma_0$:*

- *if $\gamma \leq (2^d \Delta_K^{1/d})^{O(1)}$, then D_γ^{uSVP} can be sampled from in time polynomial in $\log \Delta_K$;*
- *with probability $1 - 2^{-\Omega(d)}$, a sample from D_γ^{uSVP} is a pseudo-basis of a rank-2 module $M \subseteq \mathcal{O}_K^2$ with gap $\gamma(M) \geq \gamma \cdot \sqrt{d} \Delta_K^{1/(2d)}$; in particular, these are γ-uSVP instances;*
- *there exists a Karp reduction from γ'-wc-uSVP_{mod} to $(D_\gamma^{\text{uSVP}}, \gamma)$-$\text{uSVP}_{\text{mod}}$, with $\gamma' = \gamma \cdot (d \cdot \Delta_K^{1/d})^{O(1)}$; the reduction runs in time polynomial in $\log \Delta_K$ and the input bitsize.*

Note that the restriction on γ for the first condition is very mild, as in this parameter range, uSVP_{mod} can be solved in polynomial time using the LLL algorithm [LLL82]. We now proceed in two steps. We first define and study the distribution D^{uSVP}, and then prove Theorem 6.1.

6.1 A Distribution over uSVP Instances

Let $\gamma > 1$. The distribution D^{uSVP}_γ is defined as follows:

- sample a module from $D^{\text{module}}_{B,\gamma'}$ along with a pseudo-basis $(\mathbf{B}, \mathbb{I})$, with $B = (d^d \Delta_K)^{O(1)}$ and $\gamma' = 2\gamma \cdot \sqrt{d} \Delta_K^{1/(2d)} \cdot \sqrt{d} B^{1/d}$ (see Definition 5.1) and using Ideal-Sample to sample from $\mathcal{I}_1$;
- call $\text{DualRound}_{\varsigma,\beta,\varepsilon}(\mathbf{B}, \mathbb{I})$ with $\varsigma = (2^d \Delta_K^{1/d})^{O(1)}$, $\beta = 2$ and $\varepsilon = 1/(2d)^{3/2}$, and let $\mathbf{Y}$ denote the output;
- return $\text{HNF}(\mathbf{Y} \cdot \mathbf{B}, \mathbb{I})$.

The first two statements of Theorem 6.1 are implied by the following lemmas, whose proofs can be found in the full version of this work.

Lemma 6.2. *A sample M from $D^{\text{module}}_{B,\gamma'}$ has gap $\gamma(M) \geq \gamma'/(\sqrt{d} B^{1/d})$, with probability $1 - 2^{-\Omega(d)}$.*

Using the latter result and Lemma 2.11, we obtain that the assumptions of Lemma 3.5 are satisfied. This implies that the above sampling algorithm runs in time polynomial in $\log \Delta_K$. By Lemmas 3.5 and 3.6, the output is a pseudo-basis of a rank-2 module in $\mathcal{O}_K^2$.

Lemma 6.3. *Let $\gamma > 2$. Let $(\mathbf{B}, \mathbb{I})$ be a pseudo-basis of a rank-2 module M with gap γ. Let $\mathbf{Y}$ denote the output of $\text{DualRound}_{\varsigma,\beta,\varepsilon}(\mathbf{B}, \mathbb{I})$ with $\varsigma = \gamma \cdot (2d)^{2d+3}$, $\beta = 2$ and $\varepsilon = 1/(2d)^{3/2}$. Then the module spanned by $(\mathbf{Y} \cdot \mathbf{B}, \mathbb{I})$ has gap $\geq \gamma/2$.*

The definition of D^{uSVP}_γ and Lemmas 6.2 and 6.3 implies that the modules whose pseudo-basis are sampled from D^{uSVP}_γ have gap $\geq \gamma \cdot \sqrt{d} \Delta_K^{1/(2d)}$, and hence are γ-uSVP instances with overwhelming probability.

6.2 Reducing Worst-Case Instances to D^{uSVP} Instances

We first introduce intermediate problems, that will allow us to split the reduction into several steps.

Definition 6.4. *Let $\gamma > 1$. A γ-uSVP$^{\mathcal{N}}$ instance consists in a pseudo-basis $(\mathbf{B}, \mathbb{I})$ of a rank-2 module $M \subset K^2$ such that $\gamma(M) \geq \gamma$.*

Let $\mathcal{D}$ a distribution over γ-uSVP$^{\mathcal{N}}$ instances. The (D, γ)-uSVP$^{\mathcal{N}}_{\text{mod}}$ problem asks, given as input a sample $(\mathbf{B}, \mathbb{I})$ from $\mathcal{D}$, to recover a densest rank-1 submodule of the module spanned by $(\mathbf{B}, \mathbb{I})$.

The worst-case variant γ-wc-uSVP$^{\mathcal{N}}_{\text{mod}}$ asks to solve this problem for any γ-uSVP$^{\mathcal{N}}$ instance.

The $\gamma^{\approx}$-wc-uSVP$^{\mathcal{N}}_{\text{mod}}$ variant is the restriction of γ-wc-uSVP$^{\mathcal{N}}_{\text{mod}}$ to the γ-uSVP$^{\mathcal{N}}$ instances whose spanned modules M satisfy $\gamma(M) \in [\gamma, \gamma \cdot (1 + 1/d)]$.

Note that worst-case wc-uSVP$_{\text{mod}}$ reduces to wc-uSVP$^{\mathcal{N}}_{\text{mod}}$ as the existence of a short non-zero vector implies the one of a dense rank-1 module. Similarly, uSVP$^{\mathcal{N}}_{\text{mod}}$ reduces to uSVP$_{\text{mod}}$ with a loss of a $(\sqrt{d}\Delta_K^{1/d})$ factor in the parameters, thanks to Minkoswki's theorem. To prove the third statement of Theorem 6.1, it hence suffices to reduce wc-uSVP$^{\mathcal{N}}_{\text{mod}}$ to uSVP$^{\mathcal{N}}_{\text{mod}}$ for distribution D_γ^{uSVP}. The result follows from Lemmas 6.5 and 6.7.

The first lemma states that to solve γ-wc-uSVP$^{\mathcal{N}}_{\text{mod}}$ (in which the gap is only bounded from below), then it suffices to solve $\gamma^{\approx}$-wc-uSVP$^{\mathcal{N}}_{\text{mod}}$ (in which the gap is almost known). It relies on sparsification.

Lemma 6.5 (ERH). *Let $\gamma, \gamma' > 1$ satisfying $\gamma' \geq 2\log(\Delta_K)^{O(1/d)} \cdot \gamma$. Then γ'-wc-uSVP$^{\mathcal{N}}_{\text{mod}}$ reduces to $\gamma^{\approx}$-wc-uSVP$^{\mathcal{N}}_{\text{mod}}$. The reduction runs in time polynomial in $(\log \Delta_K)^{O(1)}$ and its input bitsize and succeeds with probability $\Omega(1/(d^2 + \log \Delta_K))$.*

Using the Rényi divergence, it is possible to relate the success probability of an algorithm towards solving uSVP$^{\mathcal{N}}_{\text{mod}}$ for samples from D_γ^{uSVP} with the same probability for $D_{\gamma'}^{\text{uSVP}}$, when γ and γ' are sufficiently close.

Lemma 6.6. *Let $\gamma, \gamma', \gamma'' > 1$ with $\gamma' \in \gamma \cdot [1, 1+1/d]$ and $\gamma'' = \gamma/(d\Delta_K^{1/d})^{O(1)}$. Then any algorithm that solves $(D_\gamma^{\text{uSVP}}, \gamma'')$-uSVP$^{\mathcal{N}}_{\text{mod}}$ with probability ε also solves $(D_{\gamma'}^{\text{uSVP}}, \gamma'')$-uSVP$^{\mathcal{N}}_{\text{mod}}$ with probability $\Omega(\varepsilon^2)$.*

Equipped with the latter result, we are now able to state the worst-case to average case component of the reduction.

Lemma 6.7 (ERH). *Let $\gamma, \gamma', \gamma'' > 1$ with $\gamma' = \gamma \cdot (d\Delta_K^{1/d})^{O(1)}$ and $\gamma'' = \gamma/(d\Delta_K^{1/d})^{O(1)}$. Then there is a reduction from $\gamma^{\approx}$-wc-uSVP$^{\mathcal{N}}_{\text{mod}}$ to $(D_{\gamma'}^{\text{uSVP}}, \gamma'')$-uSVP$^{\mathcal{N}}_{\text{mod}}$. The reduction runs in time polynomial in $\log \Delta_K$ and the input bitsize, and if the $(D_{\gamma'}^{\text{uSVP}}, \gamma'')$-uSVP$^{\mathcal{N}}_{\text{mod}}$ oracle succeeds with probability $\varepsilon \geq 2^{-o(d)}$, then the reduction succeeds with probability $\varepsilon^{O(1)}$.*

Acknowledgments. The authors thank Koen de Boer, Guillaume Hanrot and Aurel Page for insightful discussions. Joël Felderhoff is funded by the Direction Générale de l'Armement (Pôle de Recherche CYBER). The authors were supported by the CHARM ANR-NSF grant (ANR-21-CE94-0003) and by the PEPR quantique France 2030 programme (ANR-22-PETQ-0008). The last author was supported in part by the European Union Horizon 2020 Research and Innovation Program Grant 780701.

References

[AD17] Albrecht, M.R., Deo, A.: Large modulus ring-LWE $\geq$ module-LWE. In: Takagi, T., Peyrin, T. (eds.) ASIACRYPT 2017. LNCS, vol. 10624, pp. 267–296. Springer, Cham (2017). https://doi.org/10.1007/978-3-319-70694-8_10

[ALNS20] Aggarwal, D., Li, J., Nguyen, P.Q., Stephens-Davidowitz, N.: Slide reduction, revisited—filling the gaps in SVP approximation. In: Micciancio, D., Ristenpart, T. (eds.) CRYPTO 2020. LNCS, vol. 12171, pp. 274–295. Springer, Cham (2020). https://doi.org/10.1007/978-3-030-56880-1_10

[BFH17] Biasse, J.-F., Fieker, C., Hofmann, T.: On the computation of the HNF of a module over the ring of integers of a number field. J. Symb. Comput. (2017)

[BP91] Bosma, W., Pohst, M.: Computations with finitely generated modules over Dedekind domains. In: ISSAC (1991)

[BS96] Bach, E., Shallit, J.O.: Algorithmic Number Theory: Efficient Algorithms (1996)

[BSW16] Bai, S., Stehlé, D., Wen, W.: Improved reduction from the bounded distance decoding problem to the unique shortest vector problem in lattices. In: ICALP (2016)

[CDH+20] Chen, C., et al.: NTRU: a submission to the NIST post-quantum standardization effort (2020). https://www.ntru.org/

[CDW21] Cramer, R., Ducas, L., Wesolowski, B.: Mildly short vectors in cyclotomic ideal lattices in quantum polynomial time. J. ACM (2021)

[Coh96] Cohen, H.: Hermite and Smith normal form algorithms over Dedekind domains. Math. Comput. (1996)

[CS97] Coppersmith, D., Shamir, A.: Lattice attacks on NTRU. In: Fumy, W. (ed.) EUROCRYPT 1997. LNCS, vol. 1233, pp. 52–61. Springer, Heidelberg (1997). https://doi.org/10.1007/3-540-69053-0_5

[dBDPW20] de Boer, K., Ducas, L., Pellet-Mary, A., Wesolowski, B.: Random self-reducibility of ideal-SVP via Arakelov random walks. In: Micciancio, D., Ristenpart, T. (eds.) CRYPTO 2020. LNCS, vol. 12171, pp. 243–273. Springer, Cham (2020). https://doi.org/10.1007/978-3-030-56880-1_9

[GN08] Gama, N., Nguyen, P.Q.: Finding short lattice vectors within Mordell's inequality. In: STOC (2008)

[GPV08] Gentry, C., Peikert, C., Vaikuntanathan, V.: Trapdoors for hard lattices and new cryptographic constructions. In: STOC (2008)

[HPS98] Hoffstein, J., Pipher, J., Silverman, J.H.: NTRU: a ring-based public key cryptosystem. In: Buhler, J.P. (ed.) ANTS 1998. LNCS, vol. 1423, pp. 267–288. Springer, Heidelberg (1998). https://doi.org/10.1007/BFb0054868

[HPS11] Hanrot, G., Pujol, X., Stehlé, D.: Analyzing blockwise lattice algorithms using dynamical systems. In: Rogaway, P. (ed.) CRYPTO 2011. LNCS, vol. 6841, pp. 447–464. Springer, Heidelberg (2011). https://doi.org/10.1007/978-3-642-22792-9_25

[KF17] Kirchner, P., Fouque, P.-A.: Revisiting lattice attacks on overstretched NTRU parameters. In: Coron, J.-S., Nielsen, J.B. (eds.) EUROCRYPT 2017. LNCS, vol. 10210, pp. 3–26. Springer, Cham (2017). https://doi.org/10.1007/978-3-319-56620-7_1

[Kho06] Khot, S.: Hardness of approximating the shortest vector problem in high ℓ_p norms. J. Comput. Syst. Sci. (2006)

[LLL82] Lenstra, A.K., Lenstra Jr., H.W., Lovász, L.: Factoring polynomials with rational coefficients. Math. Ann. (1982)

[LPR10] Lyubashevsky, V., Peikert, C., Regev, O.: On ideal lattices and learning with errors over rings. In: Gilbert, H. (ed.) EUROCRYPT 2010. LNCS, vol. 6110, pp. 1–23. Springer, Heidelberg (2010). https://doi.org/10.1007/978-3-642-13190-5_1

[LPSW19] Lee, C., Pellet-Mary, A., Stehlé, D., Wallet, A.: An LLL algorithm for module lattices. In: Galbraith, S.D., Moriai, S. (eds.) ASIACRYPT 2019. LNCS, vol. 11922, pp. 59–90. Springer, Cham (2019). https://doi.org/10.1007/978-3-030-34621-8_3

[LS15] Langlois, A., Stehlé, D.: Worst-case to average-case reductions for module lattices. Des. Code Cryptogr. (2015)

[Pei16] Peikert. C.: A decade of lattice cryptography. Found. Trends Theor. Comput. Sci. (2016)

[PHS19] Pellet-Mary, A., Hanrot, G., Stehlé, D.: Approx-SVP in ideal lattices with pre-processing. In: Ishai, Y., Rijmen, V. (eds.) EUROCRYPT 2019. LNCS, vol. 11477, pp. 685–716. Springer, Cham (2019). https://doi.org/10.1007/978-3-030-17656-3_24

[PS21] Pellet-Mary, A., Stehlé, D.: On the hardness of the NTRU problem. In: Tibouchi, M., Wang, H. (eds.) ASIACRYPT 2021. LNCS, vol. 13090, pp. 3–35. Springer, Cham (2021). https://doi.org/10.1007/978-3-030-92062-3_1

[Sch87] Schnorr, C.-P.: A hierarchy of polynomial lattice basis reduction algorithms. Theor. Comput. Sci. (1987)

[Sit10] Sittinger, B.D.: The probability that random algebraic integers are relatively r-prime. J. Number Theory (2010)

[SS13] Stehlé, D., Steinfeld, R.: Making NTRUEncrypt and NTRUSign as secure as standard worst-case problems over ideal lattices (2013). https://eprint.iacr.org/2013/004

[SSTX09] Stehlé, D., Steinfeld, R., Tanaka, K., Xagawa, K.: Efficient public key encryption based on ideal lattices. In: Matsui, M. (ed.) ASIACRYPT 2009. LNCS, vol. 5912, pp. 617–635. Springer, Heidelberg (2009). https://doi.org/10.1007/978-3-642-10366-7_36

A Non-heuristic Approach to Time-Space Tradeoffs and Optimizations for BKW

Hanlin Liu[1] and Yu Yu[1,2]([📧])

[1] Shanghai Jiao Tong University, Shanghai 200240, China
hans1024@sjtu.edu.cn, yuyu@yuyu.hk
[2] Shanghai Qi Zhi Institute, 701 Yunjin Road, Shanghai 200232, China

Abstract. Blum, Kalai and Wasserman (JACM 2003) gave the first sub-exponential algorithm to solve the Learning Parity with Noise (LPN) problem. In particular, consider the LPN problem with constant noise and dimension n. The BKW solves it with space complexity $2^{\frac{(1+\epsilon)n}{\log(n)}}$ and time/sample complexity $2^{\frac{(1+\epsilon)n}{\log(n)}} \cdot 2^{\Omega(n^{\frac{1}{1+\epsilon}})}$ for small constant $\epsilon \to 0^{+}$.

We propose a variant of the BKW by tweaking Wagner's generalized birthday problem (Crypto 2002) and adapting the technique to a c-ary tree structure. In summary, our algorithm achieves the following:

1. **(Time-space tradeoff).** We obtain the same time-space tradeoffs for LPN and LWE as those given by Esser et al. (Crypto 2018), but without resorting to any heuristics. For any $2 \leq c \in \mathbb{N}$, our algorithm solves the LPN problem with time complexity $2^{\frac{\log(c)(1+\epsilon)n}{\log(n)}} \cdot 2^{\Omega(n^{\frac{1}{1+\epsilon}})}$ and space complexity $2^{\frac{\log(c)(1+\epsilon)n}{(c-1)\log(n)}}$ for $\epsilon \to 0^{+}$, where one can use Grover's quantum algorithm or Dinur et al.'s dissection technique (Crypto 2012) to further accelerate/optimize the time complexity.

2. **(Time/sample optimization).** A further adjusted variant of our algorithm solves the LPN problem with sample, time and space complexities all kept at $2^{\frac{(1+\epsilon)n}{\log(n)}}$, saving factor $2^{\Omega(n^{\frac{1}{1+\epsilon}})}$ for $\epsilon \to 0^{+}$ in time/sample compared to the original BKW, and the variant of Devadas et al. (TCC 2017).

3. **(Sample reduction).** Our algorithm provides an alternative to Lyubashevsky's BKW variant (RANDOM 2005) for LPN with a restricted amount of samples. In particular, given $Q = n^{1+\epsilon}$ (resp., $Q = 2^{n^{\epsilon}}$) samples for any constant $\epsilon > 0$, our algorithm saves a factor of $2^{\Omega(n)/\log(n)^{1-\kappa}}$ (resp., $2^{\Omega(n^{\kappa})}$) for constant $\kappa \to 1^{-}$ in running time while consuming roughly the same space, compared with Lyubashevsky's algorithm.

In particular, the time/sample optimization benefits from a careful analysis of the error distribution among the correlated candidates, which was not studied by previous rigorous approaches such as the analysis of Minder and Sinclair (J.Cryptology 2012) or Devadas et al. (TCC 2017).

© International Association for Cryptologic Research 2022
S. Agrawal and D. Lin (Eds.): ASIACRYPT 2022, LNCS 13793, pp. 741–770, 2022.
https://doi.org/10.1007/978-3-031-22969-5_25

1   Introduction

1.1   The LPN Problem and the BKW Algorithm

The LPN problem with dimension $n \in \mathbb{N}$ and noise rate $0 < \mu < 1/2$ asks to recover the $\mathbf{s} \xleftarrow{\$} \mathbb{F}_2^n$ given an oracle that for each query responds with $(\mathbf{a}_i, \langle \mathbf{a}_i, \mathbf{s} \rangle + e_i)$ for uniformly random $\mathbf{a}_i \xleftarrow{\$} \mathbb{F}_2^n$ and Bernoulli distributed error e_i, i.e., $\Pr[e_i = 1] = \mu$. Equivalently, LPN can be rephrased in the matrix-vector format, i.e., to recover $\mathbf{s}$ given $(\mathbf{A}, \mathbf{A} \cdot \mathbf{s} + \mathbf{e})$, where $\mathbf{A}$ is a random $Q \times n$ Boolean matrix, $\mathbf{e} \leftarrow \mathcal{B}_\mu^Q$, '$\cdot$' and '$+$' denotes matrix vector multiplication and bitwise addition over $\mathbb{F}_2$. It is worth mentioning that a candidate solution can be verified with high confidence in polynomial time and space for any non-trivial noise rate $\mu \leq 1/2 - 1/\mathsf{poly}(n)$. A straightforward algorithm exhaustively searches for $\mathbf{s}$ (or any n-bit substring of $\mathbf{e}$ whose corresponding submatrix of $\mathbf{A}$ is invertible), which takes exponential time but consumes only polynomial-size space and thus can be applied in extreme space-constrained situations.

Blum, Kalai and Wassermann [6] gave the first sub-exponential algorithm (the BKW algorithm) that solves the LPN problem via an iterative block-wise Gaussian elimination method. Consider the $\mathsf{LPN}_{n,\mu}$ problem with dimension n, and noise rate $\mu = \frac{1-\gamma}{2}$. For block size b, and number of iterations a such that $ab = n$, the algorithm does the following (see Sect. 2.3 for more formal details):

1. Runs for a iterations and reduces the dimension by b bits in each iteration (by XORing LPN sample pairs whose corresponding block sum to zero). This results in samples in the form of $(\mathbf{u}_1, \langle \mathbf{u}_1, \mathbf{s} \rangle + \tilde{e}_j) = (\mathbf{u}_1, \mathbf{s}_1 + \tilde{e}_j)$, where $\mathbf{s}_1$ is the first bit of $\mathbf{s}$, and $\tilde{e}_j$ is the sum of noise from 2^a original LPN samples.
2. Repeats step 1 on fresh new LPN samples for $m \approx (1/\gamma)^{2^{a+1}}$ times, obtaining at least one candidate $(\mathbf{u}_1, \mathbf{s}_1 + \tilde{e}_j)$ each time.
3. Majority votes on the m samples obtained in step 2 and produces a candidate for $\mathbf{s}_1$. Repeats the process for other bits of $\mathbf{s}$ (on previously used samples).

The BKW solves the LPN problem in time T, using space of size M and up to Q samples, and succeeds with the probability P as below

$$T \approx 2^b \cdot (1/\gamma)^{2^{a+1}}, \quad M \approx 2^b, \quad Q \approx 2^b \cdot (1/\gamma)^{2^{a+1}}, \quad P = 1 - \mathsf{negl}(n) ,$$

where throughout the paper "$\approx$" denotes the approximate relation that omits a multiplicative $\mathsf{poly}(n)$ factor. For any constant $0 < \gamma < 1$, we set $a = \frac{\log(n)}{1+\epsilon}$ and $b = \frac{(1+\epsilon)n}{\log(n)}$ such that $T \approx 2^{\frac{(1+\epsilon)n}{\log(n)}} \cdot 2^{\Omega(n^{\frac{1}{1+\epsilon}})}$ and $M \approx 2^{\frac{(1+\epsilon)n}{\log(n)}}$, where constant $\epsilon \to 0^+$. Quite naturally, one may raise the following questions:

1. **(Time-space tradeoff).** *Is it possible to achieve meaningful time-space tradeoffs for BKW to deal with bounded space in practice?*
2. **(Time/sample optimization).** *Is it possible to optimize the time/sample without sacrificing space, in particular, to eliminate the $(1/\gamma)^{2^{a+1}}$ factor?*

3. (**Sample reduction**). *Is it possible to push the sample complexity to a much lower order of magnitude than the time/space complexities?*

Below we first survey related works and progress made in tackling the above problems followed by a summary of our contributions.

1.2   Time-Space Tradeoff for BKW

The huge space consumption of BKW has become an obstacle for a realistic security evaluation of LPN/LWE-based crypto-systems. As discussed in [18], while performing 2^{60} or more steps is considered doable with a reasonable budget, an algorithm consuming a space of size 2^{60} is definitely out of reach in practice. Likewise, in the lattice setting the enumeration method (e.g., the Kannan's algorithm [26] that takes time $2^{O(n\log(n))}$ and space $\mathsf{poly}(n)$) often beats the lattice sieving [16,29–31] with time and space $2^{O(n)}$ in practice, and there is a renewed interest in the time-space trade-offs, e.g., lattice tuple sieving [4,22,23].

Esser et al. [18] discussed time-space tradeoff for BKW, but their algorithm already needs exponential time for space requirement below $2^{n/\log(n)}$. Later, they [17] introduced another variant of the BKW with better support for time-space trade-offs, called the c-sum BKW, where $2 \leq c \in \mathbb{N}$. Initially, it starts with a list of independent and uniformly random vectors $L_0 = (\mathbf{a}_{0,1}, \cdots, \mathbf{a}_{0,N})$, omitting the noisy parity bits for succinctness. It iteratively takes sums of c samples from the previous list L_i and stores those (that zero out the $(i+1)$-th b-bit block) into the next L_{i+1}, until at last it reaches a given target (typically of Hamming weight 1). The rest of the steps (repeating the above process m times, majority voting, etc.) are similar to the original BKW [6]. Note that c is the parameter to tune the tradeoff between space and time. In particular, $\binom{N}{c}$ increases exponentially with c, so with larger c one may use a smaller space at the cost of increasing time.

Nevertheless, the intermediate samples during each iteration of the c-sum BKW are somehow correlated, e.g., $\mathbf{a}_1+\mathbf{a}_2$, $\mathbf{a}_2+\mathbf{a}_3$ and $\mathbf{a}_1+\mathbf{a}_3$ are correlated in that they jointly sum to $\mathbf{0}$ regardless of the values of $\mathbf{a}_1, \mathbf{a}_2, \mathbf{a}_3$. Note that the original BKW [6] resolves the independence issue by using 2^b reference vectors (whose i-th block take all values over $\mathbb{F}_2^b$) in each i-th iteration, and XORing the rest vectors with one of the reference vector (zeroing out the i-th block), which produces independent vectors for the next iteration. In the generalized c-sum setting [17], it is not clear how the independence can be guaranteed to obtain a rigorous analysis of the running time, space consumption and success rate. Esser et al. [17] resorted to the independence heuristic that simply assumes independence among those vectors, and they also provided some empirical evidences that the results (for certain parameter choices) behave close to the analysis under the idealized heuristics. We remark that similar independence heuristics were already used in the optimized analysis of concrete LPN instances (e.g., [7,8,20]).

Under the independence heuristics, Esser et al. [17] obtained various variants of the c-sum BKW, such as the naive c-sum BKW, dissection c-sum BKW, tailored dissection c-sum BKW, and quantum c-sum BKW, as shown in Table 1.

The naive c-sum BKW is the most generic one that admits time-space tradeoffs for arbitrary $2 \leq c \in \mathbb{N}$, the dissection c-sum BKW is the time-optimized version of the naive c-sum BKW for $c \in \{(i^2 + 3i + 4)/2 : 0 \leq i \in \mathbb{N}\}$, the tailored dissection c-BKW is a fine-grained version of the dissection c-sum BKW (by adjusting the value of β, see also a visual illustration in [33, Figure 4]) that relies on additional heuristics, and the quantum c-sum BKW is the quantumly accelerated version of the naive c-sum BKW via the Grover algorithm [9,15,19]. They also applied the c-sum BKW to the LWE problem [36] and got similar results (see Table 5). We refer to Sect. 2.4 for more details about the c-sum BKW algorithm. Looking ahead, we provide unconditional algorithms with essentially the same complexities (see Sect. 3.2 through Sect. 3.7).

Table 1. The time and space complexities of the c-sum BKW [17] (and our c-sum$^+$ BKW) for solving the $\text{LPN}_{n,\mu}$ problem, where $N_c = 2^{\frac{\log(c)}{c-1} \cdot \frac{n}{\log(n)} \cdot (1+\epsilon)}$ and constant $\epsilon > 0$.

	c-sum (c-sum$^+$) BKW	Space	Time	for
Classic	Original BKW [6]	N_2	N_2	$c = 2$
	Naive	N_c	N_c^{c-1}	$c \geq 2$
	Dissection	N_c	$N_c^{c-\sqrt{2c}}$	$c = 4, 7, 11, \cdots$
	Tailored Dissection	N_c^{β}	$N_c^{c-\beta\sqrt{2c}}$	$c = 4, 7, 11, \cdots$ $\beta \in [1, \frac{\sqrt{c}}{\sqrt{c}-1}]$
Quantum	Naive + Grover	N_c	$N_c^{c/2}$	$c \geq 2$

Table 2. A comparison of our time-space tradeoff and the heuristic state of the art [11, 13] for solving the $\text{LPN}_{n,\mu}$ problem, where $N_c = 2^{\frac{\log(c)}{c-1} \cdot \frac{n}{\log(n)} \cdot (1+\epsilon)}$ and constant $\epsilon > 0$.

	Time-space tradeoff		for
Our dissection version	$T = N_c^{c-\sqrt{2c}}$	$M = N_c$	$c = 4, 7, 11, \cdots$
Dinur [13]	$T^{\log(n)/2+1} \cdot M^{\log(n)/2} = N_2^{\log(n)}$		$\sqrt{T} < M < T$
	$T^2 \cdot M^{3\log(n)/2-2} = N_2^{\log(n)}$		$M < \sqrt{T}$
Delaplace et al. [11]	$T = N_2^{(\frac{1}{2}+\frac{1}{c})\log(c)}$	$M = N_2^{\frac{2}{c}\log(c)}$	$c \geq 2$

We give the same end results as [17] while removing its underlying heuristics (see Table 1). We mention that [11,13] further advanced the heuristic-based state-of-the-art with the aid of parallel collision search (PCS). PCS used the similar independence heuristics such as $H_A(\mathbf{y}) = \mathbf{y}^T \cdot \mathbf{A}$ behaves like a random oracle or pseudorandom function, where $\mathbf{A}$ is the public matrix of the LPN problem. Note that the assumption doesn't hold in general even for secret $\mathbf{A}$, e.g., for $\mathbf{y}_1 = \mathbf{y}_2 + \mathbf{y}_3$ we have $H_A(\mathbf{y}_1) = H_A(\mathbf{y}_2) + H_A(\mathbf{y}_3)$. As depicted in Table 2, it is quite challenging to do a comprehensive comparison with [11,13]. For instance, when $\sqrt{T} < M < T$, Dinur [13] achieves roughly $T \cdot M \approx N_2^2$, e.g., $T \approx N_2^{4/3}$

and $M \approx N_2^{2/3}$. In contrast, our dissection version (see Theorem 6) achieves the same $T \approx N_2^{4/3}$ and $M \approx N_2^{2/3}$ by setting $c = 4$. Further, for $M < \sqrt{T}$ our result seems better than [13] (i.e., our $T^2 \cdot M^{3\log(n)/2-2} < N_2^{\log(n)}$), but the comparison is unfair as the explicit result analyzed/stated in [13] only considers $c = 4$ (generalizing to other c may yield different results). Delaplace et al. [11] is less superior to ours for $c < 11$, and it outperforms ours when $M < 2^{0.35n/\log(n)}$ ($c \geq 11$ in our case). Therefore, as far as time-space tradeoff is concerned, we mainly focus on [17], and leave it as future work on how to remove all the heuristics of [11,13].

1.3   Time/Sample Optimization and Sample Reduction for BKW

As discussed in Sect. 1.1, the BKW [6] repeats step 1 for $(1/\gamma)^{2^{a+1}} = 2^{\Omega(n^{\frac{1}{1+\epsilon}})}$ times and thus increases the time and sample complexities by the same factor. In fact, step 1 may have already produced sufficiently many samples $(\mathbf{s}_1 + \tilde{e}_j)$, and intuitively one just needs a majority vote to decode out $\mathbf{s}_1$. However, those noise, say $\tilde{e}_j$ and $\tilde{e}_{j'}$, are both the XOR sums of noise from the LPN samples, and they might not be (even pairwise) independent. Levieil and Fouque [32] used the LF1 technique to replace the majority voting and recover multiple secret bits (instead of a single one) at the same time. However, the BKW variant of [32] employs the LF2 technique when mixing up the vectors and heuristically assumes that the mixed up vectors behave as if they were independent, which is what we want to avoid in this paper. Devadas et al. [12] proposed a (non-heuristic) single-list pair-wise iterative collision search method to optimize the BKW, where they showed that the distribution of solutions is close to a Poisson distribution and applied the Chen-Stein method [3] of the second moment analysis to bound the difference. As a result, their variant solves the LPN problem (with overwhelming probability) in time T, using space of size M and sample complexity Q as below

$$T \approx 2^b \cdot (1/\gamma)^{2^a}, \quad M \approx 2^b \cdot (1/\gamma)^{2^a}, \quad Q \approx 2^b \ ,$$

where their sample complexity gets rid of the $(1/\gamma)^{2^{a+1}}$ factor as desired, time complexity is only mitigated (factor $(1/\gamma)^{2^{a+1}}$ squared to $(1/\gamma)^{2^a}$), and space complexity even deteriorates by factor $(1/\gamma)^{2^a}$ compared to the original BKW [6].

Lyubashevsky [34] studied how to solve the LPN problem with fewer samples. In particular, he used $Q = n^{1+\epsilon}$ (for constant $\epsilon > 0$) LPN samples as a basis to generate as many samples as needed, and feed them to the original BKW [6]. Concretely, let $(\mathbf{A}, \mathbf{t}^\mathsf{T} = (\mathbf{s}^\mathsf{T}\mathbf{A} + \mathbf{x}^\mathsf{T}))$ be the initial LPN samples, where $\mathbf{A}$ is the $n \times Q$ matrix, and vectors with 'T' denote row vectors. A "re-randomized LPN" oracle take as input $(\mathbf{A}, \mathbf{t}^\mathsf{T})$ and responds with $(\mathbf{A}\mathbf{r}_i, \mathbf{t}^\mathsf{T}\mathbf{r}_i = \mathbf{s}^\mathsf{T}\mathbf{A}\mathbf{r}_i + \mathbf{x}^\mathsf{T}\mathbf{r}_i)$ as the i-th re-randomized LPN sample, where every $\mathbf{r}_i$ is drawn from the set of length-Q-weight-w strings uniformly at random. For an appropriate value of w, $(\mathbf{A}, \mathbf{A}\mathbf{r}_i, \mathbf{x}^\mathsf{T}\mathbf{r}_i)$ is statistically close to $(\mathbf{A}, \mathbf{U}_n, \mathbf{x}^\mathsf{T}\mathbf{r}_i)$ by the leftover hash lemma [25] with mildly strong noise $\mathbf{x}^\mathsf{T}\mathbf{r}_i$. In the end, Lyubashevsky's variant of BKW solves the LPN problem (with overwhelming probability) in time T, using space of size M and sample complexity Q as below

$$T \approx 2^b \cdot (4/\gamma)^{2^{a+2} \cdot n/(\epsilon \log(n))}, \quad M \approx 2^b, \quad Q = n^{1+\epsilon} \ .$$

For constant $0 < \gamma < 1$, we set $a = \kappa \cdot \log \log(n)$ and $b = \frac{n}{\kappa \log \log(n)}$ for constant $0 < \kappa < 1$ and thus $T = 2^{\frac{n}{\kappa \log \log(n)}} \cdot 2^{\Omega(n)/\log(n)^{1-\kappa}}$, which is optimized when $\kappa \to 1^-$. Let us mention that Lyubashevsky's technique [34] also implies that LPN with $Q = 2^{n^\epsilon}$ (constant $0 < \epsilon < 1$) samples can be solved in time and space complexity $2^{O(n/\log(n))}$. We refer to Sect. 4.2 for more details.

1.4   Our Contributions

In this paper, we consider a problem that can be seen as a variant of Wagner's generalized birthday problem [38]. We recall the generalized birthday problem that, given $k = 2^a$ independent lists of i.i.d. uniformly random vectors, challenges to find out k vectors, one from each list, summing to a specified target, where the k vectors constitute a solution to the problem. The problem we consider further generalizes and differs to the generalized birthday problem in the following ways.

– (**Generalization**). We consider the case of $k = c^a$ for $2 \leq c \in \mathbb{N}$ and $a \in \mathbb{N}^+$.
– (**Pairwise-independence**). Each list consists of pairwise independent (instead of i.i.d. random) vectors, and all the lists are mutually independent.
– (**Bias analysis**). Our analysis framework extends to the case where each random vector is labelled with a true/false flag (to fully represent the LPN problem). We show that as long as the initial bias (the difference between the number of true and false samples) is bounded, the resulting bias among the solutions will be bounded (with reasonable blowup) as well, a feature not studied by the previous algorithms for the generalized birthday problem.[1]

As visualized in Fig. 1(b), our algorithm, referred to as the c-sum$^+$ BKW, breaks down the above problem on c^a lists into $(c^{a-1} + \cdots + c^0)$ subproblems on c lists, called the c-sum$^+$ problems. Further, we show that as long as the pairwise-independence (for vectors within each list) and mutual independence (among the lists) are satisfied for the c^a lists at the input level, the conditions will be satisfied by the lists at every other level (e.g., $L_{1,1}, L_{1,2}, L_{1,3}$ in Fig. 1(b)). We give analysis of the time, space and success probability without resorting to heuristics, thank to the pairwise-independence condition. Under our unified framework, the three tweaks, i.e., generalization, bias analysis and pairwise-independence, lead to the following advantages respectively.

1. (**Time-space tradeoff**). Our algorithm admits various time-space tradeoffs for solving LPN (shown in Table 1) and LWE (see Table 5), same as those achieved by the c-sum BKW [17], but without relying on any heuristiscs.

---

[1] The original generalized birthday problem omits LPN's noise labels. Even if many solutions are found, the correlations among the accumulated noise do not support majority voting. Therefore, non-heuristic analysis typically repeats the process on fresh new samples for $2^{n^{1-\epsilon}}$ times and thus incurs the same overhead on time/sample.

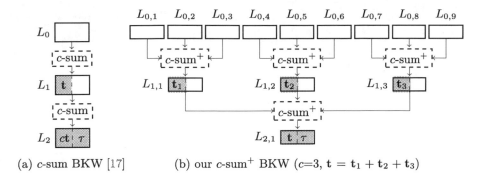

(a) c-sum BKW [17]   (b) our c-sum$^+$ BKW (c=3, $t = t_1 + t_2 + t_3$)

Fig. 1. An illustration of the c-sum BKW [17] and our c-sum$^+$ BKW.

2. **(Time/sample optimization).** We carefully analyze and bound the error distribution of the correlated solutions in step 1 (e.g., $L_{2,1}$ in Fig. 1(b)), and therefore avoid the "repeat-m-times loop" in step 2. This saves a factor of $N_2 = (1/\gamma)^{2^{a+1}} = 2^{\Omega(n^{\frac{1}{1+\epsilon}})}$ for small constant $\epsilon \to 0^+$ in time and sample complexities compared to the original BKW [6]. Our algorithm also enjoys a sub-exponential $\sqrt{N_2}$ advantage in time and space complexities compared to the optimized BKW of Devadas et al. [12]. See Table 3 for more details.

3. **(Sample reduction).** By using pairwise independent samples for the initial lists, we provide an alternative to Lyubashevsky's BKW variant [34] with improved time complexity. In particular, given $Q = n^{1+\epsilon}$ (resp., $Q = 2^{n^\epsilon}$) samples for constant $\epsilon > 0$, our algorithm saves a factor of $2^{\Omega(n)/\log(n)^{1-\kappa}}$ (resp., $2^{\Omega(n^\kappa)}$) for constant $\kappa \to 1^-$ in running time compared with the counterpart in [34]. We refer to Table 4 and Sect. 4.2 for details.

Table 3. The space, time and sample complexities of different variants of the BKW for solving the LPN$_{n,\mu}$ problem with $\mu = (1 - \gamma)/2$, under condition $N_1 \approx N_2$, where $ab = n$, $N_1 = 2^b$ and $N_2 = (1/\gamma)^{2^{a+1}}$ disregarding poly(n) factors.

Algorithm	Space	Time	Sample	Condition
The original BKW [6]	N_1	$N_1 \cdot N_2$	$N_1 \cdot N_2$	$N_1 \approx N_2$
Devadas et al.'s [12]	$N_1 \cdot \sqrt{N_2}$	$N_1 \cdot \sqrt{N_2}$	N_1	$N_1 \approx N_2$
Ours	N_1	N_1	N_1	$N_1 \approx N_2$

It might seem counter-intuitive that our results listed in Table 3 and Table 4 only depend on N_1 but still needs to satisfy $N_1 \approx N_2$ (or similar ones in Table 4) for optimized time complexity. As we will see, the condition $N_1 \geq N_2$ (or alike) is translated from the condition that sufficient amount of samples are needed to ensure the correctness of majority voting (see Theorem 9), and we thus let $N_1 \approx N_2$ for optimized complexity and fair comparison.

Table 4. The space, time and sample complexities of different variants of the BKW for solving the $\mathsf{LPN}_{n,\mu}$ problem with $\mu = (1 - \gamma)/2$, where $ab = n$, $N_1 = 2^b$, $N_2 = (4/\gamma)^{2^{a+2} \cdot n/(\epsilon \log(n))}$ and $N_2' = (4/\gamma)^{2^{a+2} \cdot n^{1-\epsilon}}$ and constant $\epsilon > 0$.

Sample	Algorithm	Space	Time	Condition
$n^{1+\epsilon}$	Lyubashevsky's [34]	N_1	$N_1 \cdot N_2$	$N_1 \approx N_2$
	Ours	N_1	N_1	$(N_1)^{\log \log(n)} \approx N_2$
2^{n^ϵ}	Lyubashevsky's [34]	N_1	$N_1 \cdot N_2'$	$N_1 \approx N_2'$
	Ours	N_1	N_1	$(N_1)^{\log(n)} \approx N_2'$

RELATED WORK. Minder and Sinclair [35] used second-moment analysis, and gave rigorous time/space bounds of Wagner's generalized problem. While it looks promising that the analysis of Minder and Sinclair [35] can be adapted to our further generalized case of $c \geq 2$ and pairwise-independent vectors (within each list), their approach does not support bias analysis. Recall that the original generalized birthday problem omits the noise labels, $e_1, \cdots, e_N$, from the LPN. Therefore, even many k-sum solutions are found, the correlations among the accumulated noise may not support majority voting. Note that even the pairwise independence condition does not hold for the noise, e.g., e_1, e_2, $e_1 + e_2$ are pairwise independent only for uniformly random (not for biased) e_1 and e_2. This is why previous non-heuristic algorithms have to repeat the process on fresh new samples for $N_2 = 2^{n^{1-\epsilon}}$ times and thus incurs the same overhead on time/sample. Recently, Devadas et al. [12] partially mitigated the issue by bounding the voting difference using the Chen-Stein method [3], but their bound is not as good as ours. As shown in Table 3, our result removes this penalty factor $m = N_2$ almost for free (without significantly increasing the time/sample complexity).

2   Preliminary

2.1   Notation

We use $\log(\cdot)$ to denote the binary logarithm. For $a \leq b \in \mathbb{N}$, $[a, b] \overset{\text{def}}{=} \{a, a + 1, \cdots, b\}$ and $[a] := [1, a]$. $|\mathcal{S}|$ is the cardinality of the set $\mathcal{S}$. For any set $\mathcal{S}$ and $0 \leq s \leq |\mathcal{S}|$, $\binom{\mathcal{S}}{s}$ denotes the set of all size-s subsets of $\mathcal{S}$. A list $L = (l_1, \cdots, l_N)$ is an element from set $\mathcal{S}^N$ with length $|L| = N$. We denote the empty list by $\emptyset$.

For $\mathbf{x} \in \mathbb{F}_2^n$ and $b < n$ we denote the last b coordinates of $\mathbf{x}$ by $\mathsf{low}_b(\mathbf{x})$. $\mathbf{u}_i$ denotes the i-th unit vector, and $\mathbf{0}^b$ denotes the zero vector of dimension b. We use ' $:=$ ' to denote deterministic value assignment. $\mathcal{U}_\mathcal{S}$ denotes the uniform distribution over set $\mathcal{S}$. $\mathcal{B}_\mu$ denotes the Bernoulli distribution with parameter μ, i.e., for $x \leftarrow \mathcal{B}_\mu$ we have $\Pr[x = 1] = \mu$ and $\Pr[x = 0] = 1 - \mu$. We use $\mathbf{s} \overset{\$}{\leftarrow} \mathcal{S}$ (resp., $\mathbf{s} \leftarrow S$) to denote sampling $\mathbf{s}$ from set $\mathcal{S}$ uniformly at random (resp., according to distribution S). For $L = (l_1, \cdots, l_N)$ with every l_i uniformly distributed over $\mathbb{F}_2^b$, we say that L consists of pairwise independent elements if for every $1 \leq i < j \leq N$ the corresponding (l_i, l_j) is uniform over $\mathbb{F}_2^{2b}$.

Lemma 1 (Piling-up Lemma). *For $0 < \mu < 1/2$ and random variables e_1, $e_2, \cdots, e_\ell$ that are i.i.d. to $\mathcal{B}_\mu$ we have $\Pr[\bigoplus_{i=1}^{\ell} e_i = 1] = \frac{1}{2}(1 - (1-2\mu)^\ell)$.*

Lemma 2 (Chebyshev's Inequality). *Let X be any random variable (taking real number values) with expectation μ and standard deviation σ (i.e., $Var[X] = \sigma^2 = \mathbb{E}[(X - \mu)^2]$). Then, for any $\delta > 0$ we have $\Pr\left[|X - \mu| \geq \delta\sigma\right] \leq \frac{1}{\delta^2}$.*

Lemma 3. *For pairwise independent real-valued r.v.s $X_1, \cdots, X_m$ it holds that*

$$Var\left[\sum_{i=1}^{m} X_i\right] = \sum_{i=1}^{m} Var\left[X_i\right] .$$

We defer the proof of Lemma 3 to the full version of the paper [33].

2.2   The Learning Parity with Noise Problem

The LPN problem comes with two versions, the decisional LPN and the search LPN, which are polynomially equivalent [2,5,27]. Therefore, we only state the search version for simplicity.

Definition 1 (Learning Parity with Noise). *For $n \in \mathbb{N}$, $\mathbf{s} \in \mathbb{F}_2^n$ and $0 < \mu < 1/2$, denote by* Sample *an oracle that, when queried, picks $\mathbf{a} \xleftarrow{\$} \mathbb{F}_2^n$, $e \leftarrow \mathcal{B}_\mu$ and outputs a sample of the form $(\mathbf{a}, l = \langle \mathbf{a}, \mathbf{s}\rangle + e)$. The $\mathsf{LPN}_{n,\mu}$ problem refers to recovering the random secret[2] $\mathbf{s}$ given access to* Sample. *We call n the dimension, $\mathbf{s}$ the secret, μ the error rate, l the label of $\mathbf{a}$ and e the noise.*

2.3   The Original BKW

The BKW algorithm [6] works in iterations, and during each i-th iteration, it uses 2^b reference vectors (whose i-th block take all values over $\mathbb{F}_2^b$). The rest vectors are added with the corresponding reference vector to zero out the i-th block, which yields new labels with doubled noise (the sum of a reference vector and another) and losing 2^b vectors in each iteration. The procedure repeats for b iterations (i.e., zeros out ab bits) until reaching a unit vector, say $\mathbf{u}_1$, and let the corresponding label be a candidate for $\langle \mathbf{u}_1, \mathbf{s}\rangle = \mathbf{s}_1$. One further repeats the above on new samples and does a majority vote to recover $\mathbf{s}_1$ with overwhelming probability. The procedure to recover other bits of $\mathbf{s}$ is likewise.

Theorem 1 (The BKW algorithm [6]). *For dimension n, block size b and number of blocks a such that $ab \geq n$, there is an algorithm that succeeds (with an overwhelming probability) in solving the $\mathsf{LPN}_{n,\mu}$ problem in time $T \approx 2^b \cdot (1/\gamma)^{2^{a+1}}$ and using space of size $M \approx 2^b$, where the noise rate $\mu = 1/2 - \gamma/2$.*

Concretely, for constant $0 < \epsilon < 1$, we set $a = \frac{\log(n)}{1+\epsilon}$ and $b = \frac{(1+\epsilon)n}{\log(n)}$ such that T and M are both on the order of $2^{\frac{(1+\epsilon)n}{\log(n)} + O(1)n^{\frac{1}{1+\epsilon}}} \approx 2^{\frac{(1+\epsilon+o(1))n}{\log(n)}}$.

---

[2] The distribution of the secret is typically uniform over $\mathbb{F}_2^n$, but it has no effect on the complexity of the BKW-style algorithms and thus is irrelevant in our context.

2.4   The c-sum Problem and c-sum BKW

Given a list of N (typically uniformly random) vectors, the c-sum problem challenges to find out c of them whose (XOR) sum equals a specified target (typically $\mathbf{0}^b$). Esser et al. [17] considered the variant that aims to find sufficiently many (at least N) such solutions. Notice that N is both the number of vectors in the input list and the amount of solutions produced as output. As we will later see, this (together with the independence heuristics) enables the c-sum BKW algorithm [17] to work from one iteration to another without losing samples.

Definition 2 (The c-sum Problem (c-SP) [17]). *Let $b, c, N \in \mathbb{N}$ with $c \geq 2$.*
Let $L \overset{\text{def}}{=} (\mathbf{a}_1, \cdots, \mathbf{a}_N)$ be a list where $\mathbf{a}_i \overset{\$}{\leftarrow} \mathbb{F}_2^b$ independently for all i and let
$\mathbf{t} \in \mathbb{F}_2^b$ be a target. A single-solution of the c-sum problem is a size-c set $\mathcal{L} \in \binom{[N]}{c}$
such that $\bigoplus_{j \in \mathcal{L}} \mathbf{a}_j = \mathbf{t}$. A complete-solution is a set of at least N distinct single-
solutions.

Esser et al. [17] proposed a variant of the BKW, referred to as the c-sum BKW, that admits time-space tradeoffs. This is achieved by generalizing the original BKW [6], which zeroes out one block per iteration by taking the sum of two vectors (i.e., 2-sum), to one that generates new samples that are the sum of c samples from previous iterations for arbitrary $2 \leq c \in \mathbb{N}$. It turns out that the c-sum BKW algorithm significantly reduces the space needed, as $\binom{N}{c}$ blows up exponentially with respect to c, at the cost of increased running time.

Algorithm 1: The c-sum BKW

Input: access to the oracle $\mathsf{LPN}_{n,\mu}$
Output: $\mathbf{s} \in \mathbb{F}_2^n$

1 $a := \frac{\log(n)}{(1+\epsilon_a)\log(c)}$, $b := \frac{n}{a}$, $m := \frac{8(1-\mu)n}{(1-2\mu)^{2c^a}}$, $N := 2^{\frac{b+c\log(c)+1}{c-1}}$;

2 **for** $i \leftarrow 1, \cdots, m$ **do**
3 $\quad$ Get N fresh LPN samples and save them in L;
4 $\quad$ **for** $j \leftarrow 1, \cdots, a-1$ **do**
5 $\quad\quad$ $L \leftarrow c\text{-sum}(L, j, \mathbf{0}^b)$;
6 $\quad$ $L \leftarrow c\text{-sum}(L, a, \mathbf{u}_1)$;
7 $\quad$ **if** $L = \emptyset$ **then**
8 $\quad\quad$ **Return** $\perp$;
9 $\quad$ Pick $(\mathbf{u}_1, b_i)$ uniformly from L;

10 $\mathbf{s}_1 \leftarrow \text{majorityvote}(b_1, \cdots, b_m)$;
11 Determine $\mathbf{s}_2, \cdots, \mathbf{s}_n$ the same way;
12 **Return** $\mathbf{s} = \mathbf{s}_1 \ldots \mathbf{s}_n$;

We recall the c-sum BKW in Algorithm 1. For a block size b and $j \in [a]$, let the coordinates $[n - jb + 1, n - (j - 1)b]$ denote the j-th stripe. The important component of the c-sum BKW algorithm is the c-sum algorithm (see line 5 and

6) that generates some refresh samples whose j-th stripe for $j \in [a-1]$ (resp. the a-th stripe) is zeros (resp. the first unit vector). If the above steps generate some label-$\mathbf{u}_1$ samples, we pick one of these $(\mathbf{u}_1, b_i)$ sample uniformly at random (see line 9). Determining the first bit $\mathbf{s}_1$ with overwhelming probability needs sufficiently many independent label-$\mathbf{u}_1$ samples via the for-loop (see line 2). The process of recovering other bits $\mathbf{s}_i$ is likewise (by reusing the LPN samples).

INDEPENDENCE HEURISTIC [17]. However, the output samples of the c-sum algorithm are somehow correlated and may not feed into the next c-sum algorithm, which requires independent samples for its input (see Definition 2). For instance, the output of a 2-sum algorithm $\mathbf{a}_1+\mathbf{a}_2$, $\mathbf{a}_2+\mathbf{a}_3$ and $\mathbf{a}_1+\mathbf{a}_3$ are correlated in the sense that they sum to $\mathbf{0}$ regardless of the values of $\mathbf{a}_1, \mathbf{a}_2, \mathbf{a}_3$. Esser et al. [17] introduced the independence heuristic that assumes independence among those vectors. Similar independence heuristics were already used in the optimized analysis of concrete LPN instances [7,8,20].

3   The C-Sum$^+$ BKW and Time-Space Tradeoffs

In this section, we introduce the k-Generalized Birthday Problem [38], and breaks it down into many instances of c-sum$^+$ problems, where $k = c^a$. By giving solutions, optimizations, and speedups to the c-sum$^+$ problems, we get many variants of BKW algorithm (referred to as the c-sum$^+$ BKW), which achieve the same complexities (up to polynomial factors) as the counterparts of c-sum BKW by Esser et al. [17] without relying on heuristics.

We consider the k-Generalized Birthday Problem: there are $k = c^a$ lists $L_{0,1}$, ..., L_{0,c^a}, where each $L_{0,i} \overset{\text{def}}{=} (\mathbf{a}_{i,1}, \cdots, \mathbf{a}_{i,N})$ has N vectors, and satisfies

1. **(Intra-list pairwise independence).** Within each list $L_{0,i}$, each $\mathbf{a}_{i,j}$ is uniformly random, and every pair of distinct vectors is pairwise independent, i.e., for all $j \neq k$ $(\mathbf{a}_{i,j}, \mathbf{a}_{i,k})$ is uniformly random.
2. **(Inter-list independence).** $L_{0,1}, \cdots, L_{0,c^a}$, each seen as a random variable, are all mutually independent.

A solution of the problem is to find k vectors, one from each list, that sum to a specified target $\mathbf{t}$, i.e., $(j_1, \cdots, j_k) \in [N]^k$ such that $\bigoplus_{i=1}^{k} \mathbf{a}_{i,j_i} = \mathbf{t}$. The goal of the problem is to find as many (N or more) solutions as possible.

Further, in the extended k-Generalized Birthday Problem, we associate lists $E_{0,1}$, ..., E_{0,c^a} with $L_{0,1}$, ..., L_{0,c^a} respectively, where $E_{0,i} \overset{\text{def}}{=} (\mathbf{e}_{i,1}, \cdots, \mathbf{e}_{i,N}) \in \mathbb{F}_2^N$ is the list of noise labels, and define the noise label of a solution as $\bigoplus_{i=1}^{k} \mathbf{e}_{i,j_i}$ accordingly. In addition to finding out N or more solutions, the extended problem also requires the noise labels of the solutions are biased (i.e., more 0-labels than 1-labels) as long as the noise of each $L_{0,1}$, ..., L_{0,c^a} is sufficiently biased. Since this subsection serves to remove the heuristics of (and gives results fully comparable to) [17], we defer the extended problem and contributions of optimized time/sample optimization to Sect. 4.

3.1   The c-sum$^+$ Problem

Before presenting our c-**sum**$^+$ BKW, we first define the c-sum$^+$ problem below. Unlike the c-sum problem (see Definition 2) that produces c-sums from a single list, the c-sum$^+$ problem takes as input c lists and asks to find c vectors, one from each list, that sum to a given target. Furthermore, we require that the c lists are mutually independent, each consisting of pairwise independent vectors.

Definition 3 (The c-sum$^+$ Problem $(c\text{-}SP^+)$). *Let $b, c, N \in \mathbb{N}$ with $c \geq 2$. Let $L_1, \cdots, L_c$, $L_i \overset{\text{def}}{=} (\mathbf{a}_{i,1}, \cdots, \mathbf{a}_{i,N}) \in \mathbb{F}_2^{b \cdot N}$, satisfy the Intra-list pairwise independence and Inter-list independence conditions (as defined in the k-Generalized Birthday Problem).*

Further, let $\mathbf{t} \in \mathbb{F}_2^b$ be a target. A solution of the c-sum$^+$ problem is a size-c list $K \overset{\text{def}}{=} (k_1, \cdots, k_c) \in [N]^c$ such that $\bigoplus_{i=1}^{c} \mathbf{a}_{i,k_i} = \mathbf{t}$.

In fact, we will need the c-sum$^+$ problem to give at least N solutions (instead of a single one) in order to form another list for the subsequent iterations in our BKW algorithm. As stated in the lemma below, the pairwise independence already ensures the existence of sufficiently many (i.e., N) solutions albeit with less strong error probability, i.e., $2/N$ instead of $2^{-\Omega(N)}$ assumed under the independence heuristic [17]. As we will see, $2/N = \mathsf{negl}(n)$ for a super-polynomial N already suffices.

Lemma 4. *For $N = 2^{\frac{b+1}{c-1}}$, the $c\text{-}SP^+$ problem (as per Definition 3) has at least N and at most $3N$ solutions with the probability more than $1 - 2/N$.*

Proof. For every $K = (k_1, \cdots, k_c) \in [N]^c$ define a 0/1-valued variable X_K that takes value $X_K = 1$ iff $\bigoplus_{i=1}^{c} \mathbf{a}_{i,k_i} = \mathbf{t}$. Thus, $X = \sum_K X_K$ is the number of solutions to the c-sum$^+$ problem, where every $K \in [N]^c$ has expectation $\mathbb{E}[X_K] = 2^{-b}$ and all the X_K are pairwise independent. Therefore,

$$\Pr\left[|X - 2N| > N\right] \leq \Pr\left[|X - \mathbb{E}[X]| > N\right] \leq \frac{Var[X]}{N^2} = \frac{\sum_S Var[X_S]}{N^2} \leq \frac{2}{N} \ ,$$

where the first inequality is due to $N^{c-1} = 2^{b+1}$ and $\mathbb{E}[X] = N^c \cdot 2^{-b} = 2N$, and the second inequality is based on Chebyshev's inequality, the first equality is due to Lemma 3, and the last inequality is due to $Var[X_i] = \mathbb{E}[X_i^2] - \mathbb{E}[X_i]^2 \leq \mathbb{E}[X_i^2] = \mathbb{E}[X_i]$. □

3.2   The c-sum$^+$ BKW

We introduced the c-sum$^+$ problem in Definition 3, and we show in Lemma 4 that it has at least N solutions (except with the probability $2/N$). We defer the concrete algorithms (and optimizations) for finding out the N solutions to a later stage. Instead, we assume a solver for c-sum$^+$ with time $T_{c,N,b}$ and space $M_{c,N,b}$, and then show how our c-sum$^+$ BKW algorithm breaks down the LPN problem into many instances of the c-sum$^+$ problem.

Abstractly speaking, our c-sum$^+$ BKW algorithm employs a c-ary tree of depth a (see Fig. 2 for an illustration of $a = 2$, $c = 3$), where each node represents a list of vectors, and each parent-node list consists of vectors each of which is the sum of c vectors from its c child nodes respectively (one from each child node). Further, we assume that for every parent node list $\left\{ \bigoplus_{i=1}^{c} \mathbf{a}_{i,k_i} \middle| (k_1, \cdots, k_c) \in [N]^c \right\}$ the choices $(k_1, \cdots, k_c)$ of the c-sums are independent of the values of its child lists L_1, ..., L_c, where $L_i = (\mathbf{a}_{i,1}, \cdots, \mathbf{a}_{i,N})$. While this independence assumption may seem contradictory to the c-sum$^+$ problem that seeks solutions satisfying $\bigoplus_{i=1}^{c} \mathbf{a}_{i,k_i} = \mathbf{t}$, we stress that this is due to the simplification of the problem. That is, our c-sum$^+$ BKW algorithm, just like the original BKW [6], zeros out the coordinates in iterations: at the j-the iteration, it finds the linear combinations of the j-th stripes that sum to zero, and produces the same combinations of the $(j + 1)$-th stripes as the resulting list for the next iteration, i.e., $\left\{ \bigoplus_{i=1}^{c} \mathbf{a}_{i,k_i}^{j+1} \middle| (k_1, \cdots, k_c) \in [N]^c, \bigoplus_{i=1}^{c} \mathbf{a}_{i,k_i}^{j} = \mathbf{t} \right\}$, where the choice $(k_1, \cdots, k_c)$ is independent of the set of $(j+1)$-th stripe vectors $\{\mathbf{a}_{i,k}^{j+1} | i \in [c], k \in [N]\}$ to be combined.

Under the above simplified model, we have the following lemma that states that the leaf-level lists satisfy the intra-list pairwise independence and inter-list independence conditions (see Definition 3), then the conditions will preserved and propagated to all the non-leaf list nodes, all the way down to the root.

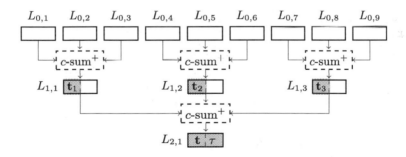

Fig. 2. An illustration of the c-sum$^+$ BKW for $c = 3$, where $\mathbf{t} = \mathbf{t}_1 + \mathbf{t}_2 + \mathbf{t}_3$

Lemma 5 (Pairwise independence preserving). *If the leaf-level lists $L_{0,1}$, ..., L_{0,c^a} are all mutually independent, and each $L_{0,i}$ consists of pairwise independent vectors. Then, for every $1 \leq j \leq a$ it holds that $L_{j,1}$, ..., $L_{j,c^{a-j}}$ are mutually independent, and every $L_{j,i}$ (for $1 \leq i \leq c^{a-j}$) consists of pairwise independent vectors.*

Proof. The proof follows by induction, namely, if the condition holds for level j, then it is also true for level $j + 1$. The mutual independence follows from the tree structure, i.e., if $L_{j,1}$, ..., $L_{j,c^{a-j}}$ are all mutually independent, then so are the next-level parents $L_{j+1,1}$, ..., $L_{j+1,c^{a-j-1}}$ since each parent only depends on its own children nodes. Moreover, if at level j, $L_{j,1}$, ..., $L_{j,c^{a-j}}$ are all mutually independent and every list $L_{j,i}$ (for $1 \leq i \leq c^{a-j}$) consists of pairwise independent vectors, then at level $j + 1$ we need to show that every list $L_{j+1,i'}$ (for $i' \in [c^{a-j-1}]$) consists of pairwise independent vectors as well. Consider any two vectors from $L_{j+1,i'}$ that are distinct c-sums of its child lists, say $\bigoplus_{\ell=1}^{c} \mathbf{a}_{\ell,k_\ell}$ and $\bigoplus_{\ell=1}^{c} \mathbf{a}_{\ell,k'_\ell}$. Then, there exists at least one $\ell \in [c]$ such that $k_\ell \neq k'_\ell$ and $(\mathbf{a}_{\ell,k_\ell}, \mathbf{a}_{\ell,k'_\ell}) \sim \mathcal{U}_{\mathbb{F}_2^b}$ (as they are from the same list at level j which has pairwise independent vectors) and they are independent from other summand vectors in the c-sum (since the lists at level j are all mutually independent). It follows that $(\bigoplus_{\ell=1}^{c} \mathbf{a}_{\ell,k_\ell}, \bigoplus_{\ell=1}^{c} \mathbf{a}_{\ell,k'_\ell})$ is jointly uniform over $\mathbb{F}_2^{2b}$ and thus are pairwise independent. □

We can now reduce the problem of solving LPN to (many instances of) the c-sum$^+$ problem without relying on any heuristics (thanks to the pairwise independence preserving property by Lemma 5). The algorithm is formally described in Algorithm 2. For a block size b and $j \in [a]$, let the coordinates $[n - jb + 1, n - (j-1)b]$ denote the j-th stripe. Our algorithm proceeds level by level. At the 0-th level, the algorithm gets fresh LPN sample to initialize every list $L_{0,k}$ for $k \in [c^a]$ with $|L_{0,k}| = N = 2^{\frac{b+1}{c-1}}$ (see line 1). Then, at each j-th level ($1 \leq j \leq a - 1$) our algorithm invokes c-sum$^+$ that takes as input the lists $L_{j-1,c(k-1)+1}, \cdots L_{j-1,ck}$ at the $(j-1)$-th level, and produces as output list $L_{j,k}$ at the j-th level (see lines $4-6$). The execution on the a-th (root) level is slightly different, i.e., we only need to solve a single instance of the c-sum$^+$ with target $\mathbf{u}_1$ (instead of zero), and produces a single solution (instead of N solutions). In other words, the code at line 10 is somewhat unnecessary in that it first produces N solutions (stored in $L_{a,1}$) but only (randomly) picks one of them, which is another problem we are going to tackle in the next section. Finally, we repeat the above many times on fresh LPN samples, and majority vote to decode out first secret bit. The recovery of other secret bits is likewise (reusing the samples). The c-sum$^+$ algorithm is an important building block of the c-sum$^+$ BKW. We state below their relations in terms of correctness and complexity.

Theorem 2 (The c-sum$^+$BKW).   *The* LPN$_{n,\mu}$ *problem with* $\mu = 1/2 - \gamma/2$ *can be solved in time T and space M with the probability P as below*

$$T \approx T_{c,N,b} \cdot c^a \cdot \left(\frac{1}{\gamma}\right)^{2 \cdot c^a}, \quad M \approx M_{c,N,b} \cdot c^a, \quad P \geq 1 - \frac{1}{N} \cdot c^a \cdot \left(\frac{1}{\gamma}\right)^{2 \cdot c^a} \cdot \mathsf{poly}(n) - \frac{n}{2^n},$$

where $T_{c,N,b}$ and $M_{c,N,b}$ are respectively the time and space complexities of the c-sum$^+$ algorithm that aims for N distinct solutions to the c-sum$^+$ problem with block (target) size b, $ab \geq n$, and $N = 2^{\frac{b+1}{c-1}}$ for $2 \leq c \in \mathbb{N}$.

Notice: for now we omit the sample complexity since $Q \approx T$ under the scenario of unlimited samples, as opposed to the setting considered in Sect. 4.2.

Proof. The c-sum$^+$ algorithm is used to instantiate the c-sum$^+$ subroutine in Algorithm 2. As discussed in Lemma 4, the c-sum$^+$ problem (implicitly defined in the j-th stripe of samples and the target vector $\mathbf{0}^b$ or $\mathbf{u}_i$ for $i \in [b]$ and $j \in [a]$) has at least N distinct solutions with the probability at least $1 - 2/N$.

Algorithm 2: The c-sum$^+$ BKW

Input: access to the oracle LPN$_{n,\mu}$
Output: $\mathbf{s} \in \mathbb{F}_2^n$

1 $a := \frac{\log(n)}{(1+\epsilon_a)\log(c)}$, $b := \frac{n}{a}$, $m := \frac{8(1-\mu)n}{(1-2\mu)^{2c^a}}$, $N := 2^{\frac{b+1}{c-1}}$;

2 **for** $i \leftarrow 1, \cdots, m$ **do**

3   Save fresh LPN samples in $L_{0,1}, \ldots, L_{0,c^a}$, each of size N;

4   **for** $j \leftarrow 1, \cdots, a-1$ **do**

5   **for** $k \leftarrow 1, \cdots, c^{a-j}$ **do**

6              $L_{j,k} \leftarrow c\text{-sum}^+(L_{j-1,c(k-1)+1}, \cdots, L_{j-1,ck}, j, 0^b)$;

7      $L_{a,1} \leftarrow c\text{-sum}^+(L_{a-1,1}, \cdots, L_{a-1,c}, a, \mathbf{u}_1)$;

8   **if** $L_{a,1} = \emptyset$ **then**

9   **Return** $\perp$;

10   Pick $(\mathbf{u}_1, b_i)$ uniformly from $L_{a,1}$;

11 $\mathbf{s}_1 \leftarrow$ majorityvote$(b_1, \cdots, b_m)$;
12 Determine $\mathbf{s}_2, \cdots, \mathbf{s}_n$ the same way;
13 **Return** $\mathbf{s} = \mathbf{s}_1 \ldots \mathbf{s}_n$;

Therefore, the corresponding BKW algorithm aborts with the probability at most $\frac{1}{N} \cdot c^a \cdot (\frac{1}{\gamma})^{2 \cdot c^a} \cdot \mathsf{poly}(n)$ via the union bound.

We now analyze the probability of the event that a single bit of the secret (e.g., $\mathbf{s}_1$) can be recovered correctly. Let the labels $b_1, \cdots, b_m$ (generated in line 10) have the corresponding noise $e_1, \cdots, e_m$, i.e., $b_i = \mathbf{s}_1 \oplus e_i$ for $i \in [m]$. The c-sum$^+$ subroutines are invoked $\frac{m \cdot c^a}{c-1}$ times, and each final resulting vector (that are a sum of c^a initial vectors) bears a noise of rate $\frac{1}{2} - \frac{1}{2}\gamma^{c^a}$ via the Piling-up Lemma (see Lemma 1). Moreover, $e_1, \cdots, e_m$ are all independent. Then, a single secret bit can be recovered with error rate $\frac{1}{2^n}$ (by a Chernoff Bound). Therefore, the probability of recovering secret key is $P \geq 1 - 1/N \cdot c^a \cdot (1/\gamma)^{2 \cdot c^a} \cdot \mathsf{poly}(n) - \frac{n}{2^n}$. Since it runs the c-sum$^+$ subroutine $c^a \cdot (\frac{1}{\gamma})^{2 \cdot c^a} \cdot \mathsf{poly}(n)$ times, we have $M \approx M_{c,N,b} \cdot c^a$ and $T \approx T_{c,N,b} \cdot c^a \cdot (\frac{1}{\gamma})^{2 \cdot c^a}$. $\qquad \square$

Next, we show different variants of the c-sum$^+$ BKW via instantiating the corresponding c-sum$^+$ algorithm.

3.3 Naive c-sum$^+$ BKW Algorithm

Our naive c-sum$^+$ algorithm is showed in [33, Algorithm 3]. Similar to the naive approach [17], it first enumerates all possible $\mathbf{p} = \bigoplus_{j=1}^{c-1} \mathbf{a}_{j,i_j} \in \mathbb{F}_2^b$ for all $\mathbf{a}_{j,i_j} \in L_j$ and $j \in [c-1]$, and checks whether $\mathbf{p} \oplus \mathbf{t}$ appears in the sorted list L_c or not, where the target vector $\mathbf{t} \in \mathbb{F}_2^b$. We obtain Theorem 3 by combining Lemma 6 with Theorem 2.

Lemma 6. *The naive c-sum$^+$ algorithm solves the c-sum$^+$ problem with target length b and list size $N \geq 2^{\frac{b+1}{c-1}}$ ($2 \leq c \in \mathbb{N}$) in time $N^{c-1} \cdot \mathsf{poly}(b,c)$ and space $N \cdot \mathsf{poly}(b,c)$, and it returns N distinct solutions with the probability $1 - 2/N$.*

Proof. Sorting out the list L_c is a one-time effort that takes time $\tilde{O}(N)$, and enumerating all possible combinations of the $c-1$ lists takes time $N^{c-1} \cdot \mathsf{poly}(b,c) \cdot \log(N) = N^{c-1} \cdot \mathsf{poly}(b,c)$ where $O(b\log(N))$ accounts for the time complexity of the binary search for $\mathbf{p} \oplus \mathbf{t}$ in the sorted L_c. The algorithm consumes space of size $N \cdot \mathsf{poly}(b,c)$ since it only stores up to N solutions. □

Theorem 3 (Naive c-sum$^+$ BKW). *The* $\mathsf{LPN}_{n,\mu}$ *problem with* $\mu = 1/2 - \gamma/2$ *can be solved in time* $T \approx N^{c-1} \cdot c^a \cdot (\frac{1}{\gamma})^{2 \cdot c^a}$ *and space* $M \approx N \cdot c^a$ *with the probability* $P \geq 1 - \frac{1}{N} \cdot c^a \cdot (\frac{1}{\gamma})^{2 \cdot c^a} \cdot \mathsf{poly}(n) - \frac{n}{2^n}$, *where* $ab \geq n$, *and* $N = 2^{\frac{b+1}{c-1}}$.

Concretely, for noise rate $\mu = 1/4$, we set $a = \frac{\log(n)}{\log(c)(1+\epsilon)}$ and $b = \frac{\log(c)(1+\epsilon)n}{\log(n)}$ to get $\log(M) = \frac{\log(c)}{c-1} \cdot \frac{n(1+\epsilon)}{\log(n)}$, $\log(T) = \log(c) \cdot \frac{n(1+\epsilon+o(1))}{\log(n)}$ and $P \geq 1 - \mathsf{negl}(n)$.

3.4   Quantum c-sum$^+$ BKW Algorithm

Following the steps in [17], we adopt the Grover's algorithm [19] (see Theorem 4) to quantumly speed up the crucial (and time-consuming) first step in the naive c-sum$^+$ (see [33, Algorithm 4]). To this end, we define

$$f_{\mathbf{t}} : [N]^{c-1} \to \{0,1\}, \quad f_{\mathbf{t}} : (i_1, \cdots, i_{c-1}) \mapsto \begin{cases} 1, & \exists \mathbf{a}_{c,i_c} \in L_c : \sum_{j=1}^c \mathbf{a}_{j,i_j} = \mathbf{t} \\ 0 & \text{otherwise} \end{cases}.$$

Once given $(i_1, \cdots, i_{c-1}) \in f^{-1}(1)$ we can recover all i_c such that $(i_1, \cdots, i_c)$ constitutes a solution to c-sum$^+$ in time $\tilde{O}(\log(|L|))$ from a sorted list L_c. Lemma 7 follows from Theorem 4 and Lemma 4.

Theorem 4 (Grover Algorithm [9,15,19]). *Let* $f : D \to \{0,1\}$ *be a function with non-empty support. Then, Grover outputs with overwhelming probability a uniformly random preimage of 1, making q queries to f, where* $q = \tilde{O}\left(\sqrt{\frac{|D|}{|f^{-1}(1)|}}\right)$.

Lemma 7. *The quantum c-sum$^+$ algorithm solves the c-sum$^+$ problem with target length b and list size $N \geq 2^{\frac{b+1}{c-1}}$ ($2 \leq c \in \mathbb{N}$) in time $N^{\frac{c}{2}} \cdot \mathsf{poly}(b,c)$ and space $N \cdot \mathsf{poly}(b,c)$, and it returns N distinct solutions with the probability $1 - 2/N$.*

Combining Lemma 7 and Theorem 2, we obtain Theorem 5.

Theorem 5 (Quantum c-sum$^+$ BKW). *The* $\mathsf{LPN}_{n,\mu}$ *problem with* $\mu = 1/2 - \gamma/2$ *can be quantumly solved in time* $T \approx N^{\frac{c}{2}} \cdot c^a \cdot (\frac{1}{\gamma})^{2 \cdot c^a}$ *and space* $M \approx N c^a$ *with the probability* $P \geq 1 - \frac{1}{N} c^a (\frac{1}{\gamma})^{2c^a} \mathsf{poly}(n) - \frac{n}{2^n}$, *where* $ab \geq n$, *and* $N = 2^{\frac{b+1}{c-1}}$.

Again, with noise rate $\mu = 1/4$ we set $a = \frac{\log(n)}{\log(c)(1+\epsilon)}$ and $b = \frac{\log(c)(1+\epsilon)n}{\log(n)}$ to get $\log(M) = \frac{\log(c)}{c-1} \cdot \frac{n(1+\epsilon)}{\log(n)}$, $\log(T) = \frac{c \cdot \log(c)}{2(c-1)} \cdot \frac{n(1+\epsilon+o(1))}{\log(n)}$, $P \geq 1 - \mathsf{negl}(n)$, where factor $\frac{c}{2(c-1)}$ represents the quantum speedup over the classic algorithm.

3.5   Dissection c-sum$^+$ BKW Algorithm

Esser et al. [17] borrowed the dissection technique from [14,37] to optimize the running time of their c-sum algorithm, referred to as dissection c-sum. The dissection c-sum perfectly fits into our c-sum$^+$ problem even better with only minor adaptions. Below we briefly introduce the dissection c-sum, and analyze its running time and space consumption in solving the c-sum$^+$ problem. We defer the redundancy to the appendix and reproduced the (slightly adapted) proofs for completeness.

Following [17] we introduce the join operation (see Definition 4) to facilitate the description of the dissection c-sum algorithm. We slightly abuse the notation in Fig. 3 by extending the operation to multiple lists, e.g., $\bowtie_{\tau_3}$ operates on L_8, L_9, L_{10}, L_{11} with target τ_3. This operation can be implemented in a space friendly way without storing the intermediate lists. We simply adapt the naive $(i+1)$-sum$^+$ algorithm on lists $L_{c_{i-1}+1}, \cdots, L_{c_i}$ whose target vector τ_i may not be of full length b, in which case the algorithm returns all the combinations whose lowest $|\tau_i|$-bit sum is τ_i.

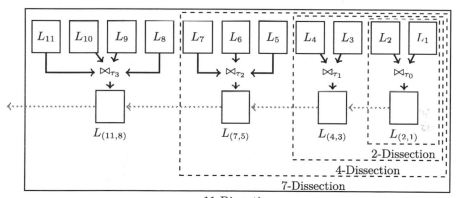

Fig. 3. An illustration of the dissection 11-sum on input lists $L_{11}, \cdots, L_1$ that recursively invokes dissection 7- and 4-sum (in dashed boxes), where $\bowtie_\tau$ is the join operator (as per Definition 4) and implemented by Naive c-sum$^+$ (as per [33, Algorithm 4]), the blank box stores the intermediate results of $\bowtie_{\tau_j}$ operation, combine results from previous invocations on-the-fly, and returns the found match through the red dotted arrows. (Color figure online)

Definition 4 (Join Operator [17]). *Let $d \in \mathbb{N}$ and $L_1, \cdots, L_k \in (\mathbb{F}_2^d)^*$ be lists. The joins of two and multiple lists are respectively defined as*

$$L_1 \bowtie L_2 \stackrel{\text{def}}{=} (\mathbf{a}_1 \oplus \mathbf{a}_2 : \mathbf{a}_1 \in L_1, \mathbf{a}_2 \in L_2),$$

$$L_1 \bowtie L_2 \bowtie \cdots \bowtie L_k \stackrel{\text{def}}{=} \Big(\big((L_1 \bowtie L_2) \bowtie L_3 \big) \cdots \bowtie L_k \Big) .$$

For $\mathbf{t} \in \mathbb{F}_2^{d'}$ with $d' \le d$, the join of L_1 and L_2 on target $\mathbf{t}$ is defined as

$$L_1 \bowtie_{\mathbf{t}} L_2 \stackrel{\text{def}}{=} (\mathbf{a}_1 \oplus \mathbf{a}_2 : \mathbf{a}_1 \in L_1, \mathbf{a}_2 \in L_2 \wedge \text{low}_{|\mathbf{t}|}(\mathbf{a}_1 \oplus \mathbf{a}_2) = \mathbf{t}) .$$

Definition 5 (The Magic Sequence [14]). *Let* $c_{-1} \stackrel{\text{def}}{=} 1$ *and define the magic sequence via the recurrence* $\forall i \in \mathbb{N}^+ \cup \{0\} : c_i \stackrel{\text{def}}{=} c_{i-1} + i + 1$, *which leads to the general formula for the magic sequence:* $\text{magic} \stackrel{\text{def}}{=} \left\{ c_i \stackrel{\text{def}}{=} \left(\frac{1}{2} \cdot (i^2 + 3i + 4) \right) \right\}_{i \in \mathbb{N}^+}$.

The parameter c of the dissection c-sum can no longer be an arbitrary integer but belongs to the "magic sequence" (Definition 5), i.e., $c_i \stackrel{\text{def}}{=} (i^2 + 3i + 4)/2$. Fix a certain i (and c_i), we recall the list size $\forall j \in [c_j] : |L_j| = N = 2^{\frac{b+1}{c_i - 1}}$. For convenience, let $\lambda \stackrel{\text{def}}{=} \frac{b+1}{c_i - 1}$ so that block size $b = (c_i - 1)\lambda - 1$. The algorithm employs the meet-in-the-middle strategy with (intermediate) targets of smaller sizes $\tau_j \in \mathbb{F}_2^{j\lambda}$ (for $j \in [i]$), and $\tau_0 \in \mathbb{F}_2^{\lambda}$ in its iterations.

We now give a high-level recursive description about the Dissection c_i-sum algorithm that aims to find out N solutions to the c_i-sum$^+$ problem for a target $\mathbf{t} \in \mathbb{F}_2^b$, which recursively invokes the dissection c_j-sum algorithm ($j < i$) to get all the combinations whose lowest $j\lambda$-bit sum is τ_j. The base case ($i = 0$, $c_0 = 2$), i.e., the Dissection 2-sum degenerates into the naive 2-sum$^+$ algorithm with a minor exception that the target τ_0 may be not of full length b. We illustrate the general case with a concrete example ($i = 3$, $c_3 = 11$) in Fig. 3. Taking as input lists $L_1, \cdots, L_{c_i}$ and a target $\mathbf{t}$, the algorithm divides the lists into two groups $L_1, \cdots, L_{c_{i-1}}$ and $L_{c_{i-1}+1}, \cdots, L_{c_i}$, where $c_i = c_{i-1} + i + 1$ due to the magic sequence. For each intermediate target $\tau_i \in \mathbb{F}_2^{i\cdot\lambda}$, do the following:

1. Invoke the (adapted) naive $(i + 1)$-sum$^+$ algorithm on lists $L_{c_{i-1}+1}, \cdots, L_{c_i}$ with the target vector τ_i to get all the combinations whose lowest $(i \cdot \lambda)$-bit sum is τ_i. Store all the solutions in list $L_{(c_i, c_{i-1}+1)}$.
2. Invoke the dissection $c_{(i-1)}$-sum algorithm on lists $L_1, \cdots, L_{c_{i-1}}$ with target $\text{low}_{(i-1)\cdot\lambda}(\tau_i) \oplus \text{low}_{(i-1)\cdot\lambda}(\mathbf{t})$. The results are passed to the parent call on-the-fly (see the red dotted line in Fig. 3), and combined with those in $L_{(c_i, c_{i-1}+1)}$, producing only those summing to $\mathbf{t}$ as output.
3. Repeat the above for all possible values of $\tau_i \in \mathbb{F}_2^{i\cdot\lambda}$.

ON SPACE CONSUMPTION. We stress that the above provides only an oversimplified description, and the actual algorithm (see [33, Algorithm 6 & 7]) is slightly more complicated to keep the space consumption within $O(iN)$. First, for each $0 \leq j \leq i$ we use $L_{(c_j, c_{j-1}+1)}$ to store the results of the naive $(j+1)$-sum$^+$ on lists $L_{c_{j-1}+1}, \ldots, L_{c_j}$ (see the $\bowtie_{\tau_j}$ operation and the blank boxes in Fig. 3). Second, every single result from $L_{(2,1)}$ is passed to $L_{(4,3)}$, and so on, all the way to $L_{(c_i, c_{i-1}+1)}$ on-the-fly to form the final output (or be discarded if it fails the checking). In other words, no additional space will be allocated for merging $L_{(2,1)}$ with $L_{(4,3)}$, and then $L_{(7,5)}$, etc., to avoid a blowup in space consumption. Finally, one can observe that the intermediate target size τ_j ($0 \leq j \leq i$) are chosen such that the expected size of $L_{(c_j, c_{j-1}+1)}$ is N. That is, $(j + 1)$-sum$^+$ on $(j + 1)$ lists, each of size $N = 2^{\lambda}$, yields N^{j+1} combinations, each having a chance of $2^{-|\tau_j|}$ to hit target τ_j. Thus, we have $N^{j+1}/2^{|\tau_j|} = N$ (more formally in the full version [33]), and the overall space consumption is $O(iN)$.

The dissection c_i-sum$^+$ [33, Algorithm 6] invokes the interative procedure c_j-Dissect [33, Algorithm 7] for $j \leq i$ to solve the c_i-sum$^+$ problem for $c_i \in$ magic. We already show in Lemma 4 that for any $2 \leq c \in \mathbb{N}$ the problem has at least N solutions (except with the probability $2/N$). Esser et al. [17] showed that the dissection c_i-sum$^+$ does an exhaustive search over all solutions.

Compared with the naive c-sum$^+$ algorithm that also exhausts all solutions, dissection c_i-sum$^+$ enjoys optimized time complexity as stated in [33, Lemma 17]. Esser et al. [17] analyzed the c_i-Dissect [33, Algorithm 7] subroutine (essentially the $\bowtie_{\tau_j}$ operation in Fig. 3) in terms of expected time and space, and we further give their upper bounds in [33, Lemma 14 & 16] to reach a more formal statement in [33, Lemma 17]. Combining [33, Lemma 17] and Theorem 2, we obtain Theorem 6.

Theorem 6 (Dissection c-sum$^+$ BKW Algorithm). *For any $c_i \in$ magic, the* $\mathsf{LPN}_{n,\mu}$ *problem with* $\mu = 1/2 - \gamma/2$ *can be solved in time* $T \approx N^{c_i-1} \cdot c_i^a \cdot (\frac{1}{\gamma})^{2 \cdot c_i^a}$ *and space* $M \approx N \cdot c_i^a$ *with the probability* $P \geq 1 - \frac{1}{N} \cdot c_i^a \cdot (\frac{1}{\gamma})^{2 \cdot c_i^a} \cdot \mathrm{poly}(n) - \frac{n}{2^n}$, *where* $ab \geq n$, *and* $N = 2^{\frac{b+1}{c_i-1}}$.

Concretely, for $\mu = 1/4$, we can set $a = \frac{\log(n)}{\log(c_i)(1+\epsilon)}$ and $b = \frac{\log(c_i)(1+\epsilon)n}{\log(n)}$ so that $\log(M) = \frac{\log(c_i)}{c_i-1} \cdot \frac{n(1+\epsilon)}{\log(n)}$, $\log(T) = (1 - \frac{i}{c_i-1}) \cdot \log(c_i) \cdot \frac{n(1+\epsilon+o(1))}{\log(n)}$, $P \geq 1 - \mathsf{negl}(n)$, where the optimization over the naive c-sum$^+$ BKW is highlighted.

3.6 Tailored Dissection c-sum$^+$ BKW

The dissection c-sum$^+$ trades time for space of smaller size $M_i \approx 2^{\left(\frac{\log(c_i)}{c_i-1}\right)\frac{n(1+\epsilon)}{\log(n)}}$ where $c_i = (i^2 + 3i + 4)/2$. In practice, it may turn out that the size of actual usable space $M \in (M_i, M_{i-1})$, leaving an unused space of size $(M_{i-1} - M)$. To address this issue, Esser et al. [17] introduced the tailored dissection c_i-sum technique to enable more fine-grained time-space tradeoffs. That is, still use $N = 2^{\frac{b+1}{c_i-1}}$, but increase the list size 2^λ from N to $N^\beta \approx M$ ($\beta > 1$) to fully utilize the available space. However, the optimized running time of their algorithm needs not only the independence heuristic but also relies on the tailoring heuristic [17] (see [33, Appendix B]), which postulates that one needs only to go through the first 2^y (for $y = b - c_{i-1} \cdot \lambda + 1$) constraints $\tau_i \in \mathbb{F}_2^{i \cdot \lambda}$ (in the outmost for-loop of [33, Algorithm 7]) to recover at least N^β distinct solutions (with high probability). In a similar vein, we present an unconditional version called tailored dissection c_i-sum$^+$ that aims for the first N^β (instead of all) distinct solutions and halts as soon as $2^\lambda = N^\beta$ solutions are found (see line 9 of [33, Algorithm 7]). Instead of relying on any heuristics, we prove in [33, Lemma 18] unconditionally that the outmost for-loop needs only 2^y iterations for $y = b - c_{i-1}\lambda + 1$. Combining [33, Lemma 19] and Theorem 2, we obtain Theorem 7.

Theorem 7 (Tailored Dissection c-sum$^+$ BKW). *For any $c_i \in$ magic, the* LPN$_{n,\mu}$ *problem with $\mu = 1/2 - \gamma/2$ can be solved in time $T \approx N^{c_i-1+(1-\beta)\cdot i}$.* $c_i^a \cdot (\frac{1}{\gamma})^{2\cdot c_i^a}$ *and space $M \approx N^\beta \cdot c_i^a$ with the probability $P \geq 1 - \frac{1}{N^\beta} \cdot c_i^a \cdot (\frac{1}{\gamma})^{2\cdot c_i^a} \cdot$* poly$(n) - \frac{n}{2^n}$, *where $ab \geq n$, $N = 2^{\frac{b+1}{c_i-1}}$ and $\beta \in [1, \frac{c_i-1}{c_{i-1}}]$.*

Concretely, for $\mu = 1/4$ we can set $a = \frac{\log(n)}{\log(c_i)(1+\epsilon)}$ and $b = \frac{\log(c_i)(1+\epsilon)n}{\log(n)}$ so that $\log(M) = \frac{\beta \cdot \log(c_i)}{c_i-1} \cdot \frac{n(1+\epsilon)}{\log(n)}$, $\log(T) = (1 - \frac{\beta \cdot i}{c_i-1}) \cdot \log(c_i) \cdot \frac{n(1+\epsilon+o(1))}{\log(n)}$ and $P = 1 - \mathsf{negl}(n)$ where the difference to the dissection c-sum$^+$ BKW was highlighted.

3.7 Time-Space Trade-Offs for Solving LWE

Regev [36] introduced the Learning With Errors (LWE) problem, generalizing LPN over arbitrarily large moduli in presence of Gaussian-like noise.

Definition 6 (Learning With Errors). *Let $\mathcal{D}_\sigma$ be a discrete Gaussian distribution with mean zero and variance σ^2. For $n \in \mathbb{N}$, prime $p \in \mathbb{N}$, $\mathbf{s} \in \mathbb{F}_p^n$, denote by* Sample *an oracle that, when queried, samples $\mathbf{a} \xleftarrow{\$} \mathbb{F}_p^n$, $e \leftarrow \mathcal{D}_\sigma$ and outputs a sample of the form $(\mathbf{a}, l) := (\mathbf{a}, \langle \mathbf{a}, \mathbf{s}\rangle + e)$. The* LWE$_{n,\sigma,p}$ *problem refers to recovering the random secret $\mathbf{s}$ given access to* Sample.

Albrecht et al. [1] adapted the BKW algorithm to solve the LWE problem, with subsequent improvement by [10,21,24]. Similarly, the BKW reduces the dimension of LWE by summing up samples and cancelling out the corresponding blocks in iterations. The number of samples needed for the majority vote is $m = \mathrm{e}^{\frac{4\pi^2\sigma^2 2^a}{p^2}}$ after a BKW steps [28]. Herold et al. [24] showed that setting $a = (1 - \epsilon_a)\log(n) + 2\log(p) - 2\log(\sigma)$ for constant $\epsilon_a > 0$ yields $m = \mathrm{e}^{4\pi^2 n^{1-\epsilon_a}}$ and results in time, space and sample complexities $\tilde{O}(p^b \cdot \mathrm{e}^{4\pi^2 n^{1-\epsilon_a}}) = p^{b\cdot(1+\epsilon)} = 2^{\frac{n\cdot\log(p)\cdot(1+\epsilon)}{\log(n)+2\log(p)-2\log(\sigma)}}$.

Following the steps of Esser et al. [17], we also generalize the c-sum$^+$ problem to arbitrary moduli p and employed (slightly tweaked versions of) the aforementioned algorithms to solve the c-sum$^+$ problem with arbitrary moduli p whose elementary operations (e.g., addition, sorting and binary search) are now over $\mathbb{F}_p$. Compared with [24], we adjust a by a factor of $\log(c)$ and set $a = \frac{(1-\epsilon_a)\log(n)+2\log(p)-2\log(\sigma)}{\log(c)}$ for constant $\epsilon_a > 0$. We summarize the results in Table 5, which are essentially the same as that of the c-sum BKW for LWE [17] but without using heuristics.

Table 5. The time and space complexities of the c-sum (c-sum$^+$) BKW algorithms for solving the $\mathsf{LWE}_{n,\sigma,p}$ problem, where $N_c = 2^{\frac{\log(c)}{c-1} \cdot \frac{n \cdot \log(p) \cdot (1+\epsilon)}{\log(n)+2\log(p)-2\log(\sigma)}}$, n is the dimension, and constant $\epsilon > 0$.

	c-sum (c-sum$^+$) BKW	Space	Time	for
Classic	Original BKW [6]	N_2	N_2	$c = 2$
	Naive	N_c	N_c^{c-1}	$c \geq 2$
	Dissection	N_c	$N_c^{c-\sqrt{2c}}$	$c \in$ magic
	Tailored Dissection	N_c^β	$N_c^{c-\beta\sqrt{2c}}$	$c \in$ magic, $\beta \in [1, \frac{\sqrt{c}}{\sqrt{c-1}}]$
Quantum	Naive + Grover	N_c	$N_c^{c/2}$	$c \geq 2$

4   The c-sum$^\#$ BKW with Time/Sample Optimizations

In this section, we consider the extended k-Generalized Birthday Problem (see Sect. 3), and give the full-fledged variant, called c-sum$^\#$ BKW, to optimize the time, space and sample complexities of the original BKW algorithm [6]. Moreover, it further pushes the sample complexity to 2^{n^ϵ} or even $n^{1+\epsilon}$, which also optimize the complexities over Lyubashevsky's BKW variant [34].

4.1   Time, Space, and Sample Optimizations

As shown in Table 6, we compare the results of the original BKW [6], Devadas et al.'s optimized version [12] and our c-sum$^\#$ BKW (for $c = 2$ as in Theorem 8).

Table 6. The space, time and sample complexities of different variants of the BKW algorithms for solving the $\mathsf{LPN}_{n,\mu}$ problem with $\mu = (1-\gamma)/2$, $\gamma \geq 2^{-n^\sigma}$ and constant $0 < \sigma < 1$ under condition $N_1 \approx N_2$, where $ab = n$, $N_1 = 2^b$ and $N_2 = (1/\gamma)^{2^{a+1}}$ disregarding $\mathsf{poly}(n)$ factors for convenience.

Algorithm	Space	Time	Sample	Condition
The original BKW [6]	N_1	$N_1 \cdot N_2$	$N_1 \cdot N_2$	$N_1 \approx N_2$
Devadas et al.'s [12]	$N_1 \cdot \sqrt{N_2}$	$N_1 \cdot \sqrt{N_2}$	N_1	$N_1 \approx N_2$
Our 2-sum$^\#$ BKW	N_1	N_1	N_1	$N_1 \approx N_2$

We know that the last step of the BKW involves balancing the two factors $N_1 = 2^b$ and $N_2 = (1/\gamma)^{2^{a+1}}$ to roughly the same magnitude given $ab = n$. As specified in 2-sum$^\#$ BKW (see Theorem 8 for $c = 2$), it requires essentially the same condition, i.e., $b = 2^{a+1}\log(1/\gamma) + O(\log(n))$. Asymptotically, for constant $0 < \gamma < 1$, we typically set $a = \frac{\log(n)}{1+\epsilon}$ and $b = \frac{(1+\epsilon)n}{\log(n)}$, and thus our algorithm speeds up the running time of the original BKW [6] by a factor of $2^{n^{\frac{1}{1+\epsilon}}}$ while using roughly the same amount of space, where constant ϵ is arbitrarily close to 0 for optimized time complexity. Recently, Devadas et al. [12] optimized the running time of the original BKW from $N_1 \cdot N_2$ to $N_1 \cdot \sqrt{N_2}$ at the cost of increasing the space complexity from N_1 to $N_1 \cdot \sqrt{N_2}$. Thus, the 2-sum$^\#$ BKW enjoys a sub-exponential factor advantage both in time/space complexities compared to [12].

Algorithm 3: The c-sum$^\#$ BKW

Input: access to the oracle $\mathsf{LPN}_{n,\mu}$
Output: $\mathbf{s} \in \mathbb{F}_2^n$
1   $b := \frac{n}{a}$, $N := 2^{\frac{b}{c-1}}$;
2   Save fresh LPN samples in $L_{0,1}, \ldots, L_{0,c^a}$, each of size N;
3   **for** $j \leftarrow 1, \cdots, a-1$ **do**
4   **for** $k \leftarrow 1, \cdots, c^{a-j}$ **do**
5         $L_{j,k} \leftarrow c\text{-sum}^+(L_{j-1,c(k-1)+1}, \cdots, L_{j-1,ck}, j, 0^b)$;

6   $L_{a,1} \leftarrow c\text{-sum}^+(L_{a-1,1}, \cdots, L_{a-1,c}, a, \mathbf{u}_1)$;
7   $\mathbf{s}_1 \leftarrow \text{majorityvote}(b_1, \cdots b_{|L_{a,1}|})$;
8   Determine $\mathbf{s}_2, \cdots, \mathbf{s}_n$ the same way over the same LPN samples;
9   **Return** $\mathbf{s} = \mathbf{s}_1 \ldots \mathbf{s}_n$;

MAJORITY VOTING ON CORRELATED SAMPLES. The c-sum BKW [17] and our c-sum$^+$ BKW (Algorithm 2) pick a single sample from $L_{a,1}$ and repeat the process for $m \approx (1/\gamma)^{2^{a+1}}$ times on fresh LPN samples (see line 2–10 in Algorithm 2). We argue that this step can be removed with a careful adaption, and therefore reduces the time/sample complexities by factor $2^{\Omega(n^{\frac{1}{1+\epsilon}})}$. This is the motivation of introducing the extended k-generalized birthday problem (see Sect. 3). Hopefully, we recover the single bit of secret via a majority voting on the elements in $L_{a,1}$ (line 7 in Algorithm 3). This is non-trivial since the noise bits in $L_{a,1}$ are linear combinations of individual noises of the LPN samples, and thus they are not even pairwise independent[3]. We observe that in order to majority-vote for the correct result it suffices that the resulting noise remains biased-to-zero. For every sample list $L_{j,k}$ we define the corresponding noise-indicator list $E_{j,k}$, whose every i-th element $(-1)^{e_i}$ corresponds to the i-th element of $L_{j,k}$, i.e., $(\mathbf{a}_i, \mathbf{a}_i \cdot \mathbf{s} \oplus e_i)$. $\text{bias}(E_{j,k}) = \sum_{i=1}^{|E_{j,k}|}(-1)^{e_i}$ refers to the difference between the numbers of 0's and 1's in the noise of $L_{j,k}$. Therefore, the majority voting is successful if and only if the final $\text{bias}(E_{a,1}) > 0$.

THE c-SUM$^\#$ BKW. We now describe how to adapt the c-sum$^+$ BKW (Algorithm refalg:cspssumspsbkwspsplus) to avoid the outmost repeat-m-times loop. The c-sum$^+$ BKW is sample-preserving, i.e., it invokes subroutines such as the naive c-sum$^+$ [33, Algorithm 4] that halt as soon as N solutions are found. In

---

[3] Unlike uniformly random vectors, the linear combinations of i.i.d. biased bits are not pairwise independent, e.g., $e_1 + e_2$ and e_2 for $e_1, e_2 \leftarrow \mathcal{B}_\mu$ with $0 < \mu < 0.5$.

contrast, we let the c-sum$^{\#}$ BKW be exhaustive, i.e., the underlying c-sum$^+$ solver (e.g., [33, Algorithm 9]) must output all solutions. We start with the initial leaf-level lists $E_{0,1}, \cdots, E_{0,c^a}$ with $|E_{0,k}| = N$ and sufficiently large bias$(E_{0,k})$ for every $k \in [c^a]$. Then, as shown in Lemma 9, for every $j \in [a]$ and $k \in [c^{a-j}]$ the $|E_{j,k}|$ will be bounded within $N(1 \pm o(1))$ and bias$(E_{j,k})$ stays positive. To achieve this, we set $N = 2^{b/(c-1)}$ (instead of $N = 2^{(b+1)/(c-1)}$). Consider the c-sum$^+$ problem instance whose input noise-indicator lists are $E_{j-1,1}$, $\cdots, E_{j-1,c}$ and output noise-indicator list $E_{j,1}$, whose elements are chosen from $JE_{j,1} \overset{\text{def}}{=} E_{j-1,1} \bowtie \cdots \bowtie E_{j-1,c}$ (all possible c sums). In particular, each element from list $JE_{j,1}$ is included into $E_{j,1}$ iff the corresponding c-sum$^+$ hits the target, which occurs with the probability 2^{-b}. Further, whether an element in $JE_{j,1}$ hits the specified target or not is a pairwise independent event (see Lemma 5). With $|E_{j-1,k}| \approx N$ for every $k \in [c]$, we have that $|E_{j,1}|$ has expected value roughly $N^c/2^b = N$ and thus remains around N by Chebyshev's inequality. We also lower bound the corresponding bias$(E_{j,1})$ for every $j \in [a]$. We state the results in Lemma 9, and prior to that we introduce Lemma 8 as an analogue of the piling-up lemma that characterizes how the bias is amplified through the c-sum$^+$ operations.

Lemma 8. *For* $JE_{j+1} \overset{\text{def}}{=} E_{j,k+1} \bowtie E_{j,k+2} \cdots \bowtie E_{j,k+c}$, *we have* bias$(JE_{j+1}) = \prod_{i=1}^{c}$ bias$(E_{j,k+i})$.

Proof. It follows from the definitions of bias and $\bowtie$ by rearranging the terms:

$$\text{bias}(JE_{j+1}) = \sum_{l_1 \in [n_1], \cdots, l_c \in [n_c]} (-1)^{e_{l_1}^1} \times \cdots \times (-1)^{e_{l_c}^c}$$

$$= \left(\sum_{l_1 \in [n_1]} (-1)^{e_{l_1}^1} \right) \times \cdots \times \left(\sum_{l_c \in [n_c]} (-1)^{e_{l_c}^c} \right) = \prod_{i=1}^{c} \text{bias}(E_{j,k+i}) \ ,$$

where we use shorthand $n_i \overset{\text{def}}{=} |E_{j,k+i}|$ for $1 \leq i \leq c$ for notational convenience. $\qquad \square$

Lemma 9. *For* $N = 2^{\frac{b}{c-1}}$, *any* $2 \leq c \in \mathbb{N}$, $0 < \varepsilon < 1$ *and* $0 < \delta < 1$ *such that* $\delta^{c^a} \sqrt{N} \varepsilon \geq 2^a c^{2a}$, *if the level-0 lists* $E_{0,1}, \ldots, E_{0,c^a}$ *satisfy* $|E_{0,k}| = N$, bias$(E_{0,k}) \geq \delta N$ *for* $1 \leq k \leq c^a$. *Then, at every level* $j \in [a]$, *for every* k-*th list* $E_{j,k}$ $(1 \leq k \leq c^{a-j})$ *we have*

$$\Pr\left[\text{bias}(E_{j,k}) \leq \left(\delta^{c^j} N - \frac{2^j \sqrt{N} c^{2j}}{\varepsilon} \right) \right] \leq c^{4j} \cdot \varepsilon \ ,$$

$$\Pr\left[\left| |E_{j,k}| - N \right| \geq \frac{2^j \sqrt{N} c^{2j}}{\varepsilon} \right] \leq c^{4j} \cdot \varepsilon \ .$$

Proof. The base case $j = 0$ holds by assumption, i.e., bias$(E_{0,k}) \geq \delta N$ and $|E_{0,k}| = N$ for every $1 \leq k \leq c^a$. We prove the rest by induction, i.e., if it holds for level j, then it also true for level $j + 1$. It suffices to consider the first list

$E_{j+1,1}$ on level $j+1$ whose elements are selected from the set of all c-sum$^+$ of the c lists, i.e., $JE_{j+1,1}=E_{j,1} \bowtie \cdots \bowtie E_{j,c}$ with the probability at least $1 - c^{4j+1}\varepsilon$, we have (by the definition of $\bowtie$) $N^c(1 - \frac{2^j c^{2j+1}}{\sqrt{N}\varepsilon}) \leq N^c(1 - \frac{2^j c^{2j}}{\sqrt{N}\varepsilon})^c < |JE_{j+1,1}| < N^c(1 + \frac{2^j c^{2j}}{\sqrt{N}\varepsilon})^c \leq N^c(1 + \frac{2^{j+1} c^{2j+1}}{\sqrt{N}\varepsilon})$, where by [33, Lemma 21] $(1+d)^c \leq 1 + 2cd$ and $(1-d)^c \geq 1 - cd$ for $0 < cd < 1$, $c \geq 2$. Every element from list $JE_{j+1,1}$ has a chance of 2^{-b} to be selected into $E_{j+1,1}$ in a pair-wise independent manner among the elements of $JE_{j+1,1}$ (see Lemma 5). Thus, the above implies (recall $N^{c-1} = 2^b$) $\Pr\left[\left|\mathbb{E}[|E_{j+1,1}|] - N\right| < \frac{2^{j+1}\sqrt{N}c^{2j+1}}{\varepsilon}\right] \geq 1 - c^{4j+1}\varepsilon$. Similar to the proof of Lemma 4 (except for a different value of N), we have

$$
\begin{aligned}
&\Pr\left[\left||E_{j+1,1}| - N\right| \geq \frac{2^{j+1}\sqrt{N}c^{2j+2}}{\varepsilon}\right] \\
&\leq \Pr\left[\left||E_{j+1,1}| - \mathbb{E}[|E_{j+1,1}|]\right| \geq \frac{2^{j+1}\sqrt{N}c^{2j+1}(c-1)}{\varepsilon}\right] \\
&\quad + \Pr\left[\left|\mathbb{E}[|E_{j+1,1}|] - N\right| \geq \frac{2^{j+1}\sqrt{N}c^{2j+1}}{\varepsilon}\right] \\
&\leq \frac{Var[|E_{j+1,1}|]}{N/\varepsilon^2} + c^{4j+1}\cdot\varepsilon \leq \frac{\mathbb{E}[|E_{j+1,k}|]}{N/\varepsilon^2} + c^{4j+1}\cdot\varepsilon \leq c^{4j+3}\cdot\varepsilon \;.
\end{aligned}
\tag{1}
$$

By Lemma 8 the following holds with the probability at least $1 - c^{4j+1}\varepsilon$

$$
\mathsf{bias}(JE_{j+1,1}) > \delta^{c^{j+1}} N^c\left(1 - \frac{2^j c^{2j}}{\delta^{c^j}\sqrt{N}\varepsilon}\right)^c \geq \delta^{c^{j+1}} N^c\left(1 - \frac{2^j c^{2j+1}}{\delta^{c^j}\sqrt{N}\varepsilon}\right) \;,
$$

where the Bernoulli's inequality $(1-d)^c \geq 1 - cd$ is applicable since $c \geq 2$ and $d = \frac{2^j c^{2j}}{\delta^{c^j}\sqrt{N}\varepsilon} < \frac{2^a c^{2a}}{\delta^{c^a}\sqrt{N}\varepsilon} \leq 1$. We recall $\mathsf{bias}(E_{j+1,1}) \overset{def}{=} \sum_{l=1}^{|JE_{j+1,1}|} \mathbf{v}_l \cdot (-1)^{e_l}$, where random variable $\mathbf{v}_l$ is 1 if the corresponding c-sum$^+$ hits the specified target (so that the corresponding $(-1)^{e_l}$ is included in $E_{j+1,1}$) or is 0 otherwise. By Lemma 5 all the $\mathbf{v}_l$'s are pairwise independent, each with expectation 2^{-b}, and therefore $\mathbb{E}[\mathsf{bias}(E_{j+1,1})] = 2^{-b} \cdot \mathsf{bias}(JE_{j+1,1})$. We have $\Pr\left[\mathbb{E}[\mathsf{bias}(E_{j+1,1})] > \delta^{c^{j+1}} N - \frac{2^j\sqrt{N}c^{2j+1}}{\varepsilon}\right] \geq 1 - c^{4j+1}\varepsilon$, and thus

$$
\begin{aligned}
&\Pr\left[\mathsf{bias}(E_{j+1,1}) \leq \delta^{c^{j+1}} N - \frac{2^{j+1}\sqrt{N}c^{2j+2}}{\varepsilon}\right] \\
&\leq \Pr\left[\mathsf{bias}(E_{j+1,1}) - \mathbb{E}[\mathsf{bias}(E_{j+1,1})] \leq \frac{2^j\sqrt{N}c^{2j+1}(2c-1)}{\varepsilon}\right] \\
&\quad + \Pr\left[\mathbb{E}[\mathsf{bias}(E_{j+1,1})] < \delta^{c^{j+1}} N - \frac{2^j\sqrt{N}c^{2j+1}}{\varepsilon}\right] \\
&\leq \frac{Var[\mathsf{bias}(E_{j+1,1})]}{N/\varepsilon^2} + c^{4j+1}\cdot\varepsilon \leq \frac{\mathbb{E}[|E_{j+1,k}|]}{N/\varepsilon^2} + c^{4j+1}\cdot\varepsilon \leq c^{4j+3}\cdot\varepsilon \;,
\end{aligned}
\tag{2}
$$

where the analysis is essential the same as that for bounding $|E_{j+1,1}|$ except that $Var[\mathsf{bias}(E_{j+1,1})] = \sum_{l=1}^{|JE_{j+1,1}|} Var[\mathbf{v}_l \cdot (-1)^{e_l}] \leq \sum_{l=1}^{|JE_{j+1,1}|} \mathbb{E}[\mathbf{v}_l] = \mathbb{E}\left[\sum_{l=1}^{|JE_{j+1,1}|} \mathbf{v}_l\right]$. $\qquad\square$

Now we state the fully optimized algorithm in Theorem 8, and compare the case $c = 2$ (no time-space tradeoff) with the original BKW [6] and the one by Devadas et al. [12] in Table 6.

Theorem 8 (The c-sum$^{\#}$ BKW). *The* $\mathsf{LPN}_{n,\mu}$ *problem with* $\mu = 1/2 - \gamma/2$ *can be solved in time* T *and space* M *with the probability* P *as below*

$$T \approx T_{c,N,b} \cdot c^a, \quad M \approx M_{c,N,b} \cdot c^a, \quad P \geq 1 - 2c^{5a} \cdot n \cdot \varepsilon,$$

where $T_{c,N,b}$ *and* $M_{c,N,b}$ *are respectively the time and space complexities of the* c-sum^{+} *algorithm that aims for all distinct solutions to the* c-sum^{+} *problem with block (target) size* b, $ab \geq n$, $b > n^{0.6}$, $\gamma > 2^{-b/3}$, $b/\big(2(c-1)\big) \geq c^a \log(1/\gamma) + 3a \log(c) + 2 \log(1/\varepsilon) + \mathsf{negl}(n)$ *and* $N = 2^{\frac{b}{c-1}}$ *for* $2 \leq c \in \mathbb{N}$.

Notice: for now we omit the sample complexity since $Q \approx M$ *under the scenario of unlimited samples.*

Proof. Set the δ in Lemma 9 to $\gamma - 2^{-\frac{b}{2}}\sqrt{\log(1/\varepsilon)}$, and we have by Chernoff bound

$$\Pr\left[\mathsf{bias}(E^0_{0,k}) \leq N \cdot \delta\right] \leq \Pr\left[\mathsf{bias}(E^0_{0,k})/N - \gamma \leq (\delta - \gamma)\right] \leq 2^{-2^{-b}\log(1/\varepsilon)N} = \varepsilon,$$

where $N = 2^{\frac{b}{c-1}}$. The condition $\delta^{c^a}\sqrt{N}\varepsilon \geq 2^a c^{2a}$ in Lemma 9 is now

$$\frac{b}{2(c-1)} \geq c^a \log(1/\delta) + a + 2a \log(c) + \log(1/\varepsilon)$$

$$= c^a \log(1/\gamma) + a + 2a \log(c) + \log(1/\varepsilon) + c^a \log\left(1 + \frac{2^{-b/2}O(\sqrt{\log(1/\varepsilon)})}{\gamma}\right)$$

$$\geq c^a \log(1/\gamma) + a + 2a \log(c) + \log(1/\varepsilon) + c^a 2^{-b/6} \cdot O\big(\sqrt{(1/\varepsilon)}\big).$$

By Lemma 9 the size of every list $E_{j,k}$ is at most $N + N^{0.5} \cdot c^{3a}/\varepsilon = O(N)$ with the probability at least $1 - c^{4a} \cdot \varepsilon$, and thus all lists have size $O(N)$ with the probability at least $1 - c^{5a} \cdot n \cdot \varepsilon$. As for the correctness, the bias of the final list $E_{a,1}$ is positive with the probability at least $1 - c^{4a} \cdot \varepsilon$ in order to successfully recover a single bit of the secret. Overall, it recovers the whole secret correctly with the probability more than $1 - c^{4a} \cdot n \cdot \varepsilon$ by the union bound.   □

4.2 Sample Reduction for BKW

Lyubashevsky [34] introduced the "sample amplification" technique to further push the sample complexity to $Q = n^{1+\epsilon}$. Let $(\mathbf{A}, \mathbf{t}^{\mathsf{T}} = (\mathbf{s}^{\mathsf{T}}\mathbf{A} + \mathbf{x}^{\mathsf{T}}))$ be all the LPN samples one can have, where $\mathbf{A}$ is the $n \times Q$ matrix, and vectors with 'T' denote row vectors. A "sample amplification" oracle takes as input $(\mathbf{A}, \mathbf{t}^{\mathsf{T}})$ and responds with $(\mathbf{A}\mathbf{r}_i, \mathbf{t}^{\mathsf{T}}\mathbf{r}_i = \mathbf{s}^{\mathsf{T}}\mathbf{A}\mathbf{r}_i + \mathbf{x}^{\mathsf{T}}\mathbf{r}_i)$ as the i-th re-randomized LPN sample, and generates as many LPN sample as needed, where every $\mathbf{r}_i \xleftarrow{\$} \mathcal{R}_{Q,w}$ is drawn from the set of length-Q-weight-w strings uniformly at random. Finally, invoke the original BKW [6] on the generated samples. In order to make the approach work provably, $(\mathbf{A}, \mathbf{A}\mathbf{r}_i, \mathbf{x}^\mathsf{T}\mathbf{r}_i)$ should be statistically close to $(\mathbf{A}, \mathbf{U}_n, \mathbf{x}^\mathsf{T}\mathbf{r}_i)$ by the leftover hash lemma [25], which requires min-entropy $\mathbf{H}_\infty(\mathbf{r}_i) = \log\binom{Q}{w} > n$. Therefore, Lyubashevsky [34] chose $w = \frac{2n}{\epsilon \log(n)}$ for $Q = n^{1+\epsilon}$.

Our c-sum$^\#$ BKW supports sample amplification in a different and slightly more efficient way. The c-sum$^\#$ BKW (Algorithm 3) initializes the lists $L_{0,1}, \ldots, L_{0,c^a}$, with independent fresh LPN samples. However, the pairwise independence preserving lemma (Lemma 5) only requires each $L_{0,k}$ (for $k \in [2^a]$) has pairwise independent vectors. Our sample amplification simply divides $\mathbf{A}$ into $n \times \frac{Q}{2^a}$ submatrices $\mathbf{A}_1, \cdots, \mathbf{A}_{2^a}$ accordingly, and loads each $L_{0,k}$ with distinct w-linear combinations of the $(\mathbf{A}_k, \mathbf{s}^\mathsf{T}\mathbf{A}_k + \mathbf{x}_k^\mathsf{T})$, i.e.,

$$\forall k \in [2^a] : L_{0,k} := \left((\mathbf{A}_k\mathbf{r}_1, \mathbf{s}^\mathsf{T}\mathbf{A}_k\mathbf{r}_1 + \mathbf{x}_k^\mathsf{T}\mathbf{r}_1), \cdots, (\mathbf{A}_k\mathbf{r}_N, \mathbf{s}^\mathsf{T}\mathbf{A}_k\mathbf{r}_N + \mathbf{x}_k^\mathsf{T}\mathbf{r}_N) \right)$$

where $\mathbf{r}_1, \cdots, \mathbf{r}_N$ are distinct vectors of weight w, and $N = 2^b \leq \binom{Q/2^a}{w}$. So far we essentially override the LPN sample oracle of the c-sum$^\#$ BKW (line 2 of Algorithm 3), which takes time and space 2^{a+b}. The rest of the steps are the same as those in Algorithm 3.

Lemma 10. *For $k = o(m)$ we have $\log\binom{m}{k} = (1 + o(1))k\log\left(\frac{m}{k}\right)$.*

Lemma 11 ([34]). *If a bucket contains m balls, $(\frac{1}{2} + p)m$ of which are colored white, and the rest colored black, and we select k balls at random without replacement, then the probability that we selected an even number of black balls is at least $\frac{1}{2} + \frac{1}{2}\left(\frac{2mp-k+1}{m-k+1}\right)^k$.*

Theorem 9 (The 2-sum$^\#$ BKW with fewer samples). *The $\mathsf{LPN}_{n,\mu}$ problem with $\mu = 1/2 - \gamma/2$ and given up to Q samples can be solved in time T, space M with the probability P as below*

$$T \approx 2^{a+b}, \quad M \approx 2^{a+b}, \quad P \geq 1 - 2^{6a} \cdot n \cdot \varepsilon - 2^a \cdot 2^{-\Omega\left(\frac{Q\gamma^2}{2^a}\right)},$$

where $a, b, w \in \mathbb{N}$ and $0 < \varepsilon < 1$ satisfy $ab = n$, $Q\gamma \geq 2^{a+2}w$, and $\log\binom{Q/2^a}{w} \geq b \geq 2^{a+1}w\log(4/\gamma) + 6a + 2\log(1/\varepsilon)$.

Proof. Let $Q' \stackrel{\text{def}}{=} Q/2^a$, and define $E_{0,k} \stackrel{\text{def}}{=} ((-1)^{\mathbf{x}_k^\mathsf{T}\mathbf{r}_1}, \cdots, (-1)^{\mathbf{x}_k^\mathsf{T}\mathbf{r}_N})$. We have by the Chernoff bound that $\Pr[|\mathbf{x}_k^\mathsf{T}| > (1/2 - \gamma/4)Q'] \leq 2^{-\Omega(Q'\gamma^2)}$. Then, by Lemma 11 with the probability at least $1 - 2^{-\Omega(Q'\gamma^2)}$ and for $\gamma \geq 4w/Q'$

$$\mathsf{bias}(E_{0,k}) \geq N \cdot \left(\frac{2Q'(\gamma/4) - w + 1}{Q' - w + 1}\right)^w \geq N \cdot \left(\frac{\gamma}{2} - \frac{w}{Q'}\right)^w \geq N\left(\frac{\gamma}{4}\right)^w .$$

The condition $\delta^{c^a}\sqrt{N}\varepsilon \geq 2^a c^{2a}$ in Lemma 9 becomes $b \geq 2^{a+1}w\log(4/\gamma) + 6a + 2\log(1/\varepsilon)$, where we set $\delta = (\gamma/4)^w$. The probability argument (and the rest of the proof) is similar Theorem 8 by adding the extra term $2^a \cdot 2^{-\Omega(Q'\gamma^2)}$. $\qquad\square$

Table 7. The space, time and sample complexities of different variants of the BKW algorithms for solving the $\mathsf{LPN}_{n,\mu}$ problem with $\mu = (1 - \gamma)/2$ and sample complexity $Q = n^{1+\epsilon}$, where $ab = n$, $N_1 = 2^b$, $N_2 = (4/\gamma)^{2^{a+2} \cdot n/(\epsilon \log(n))}$ and constant $\epsilon > 0$ disregarding $\mathsf{poly}(n)$ factors for convenience.

Algorithm	Space	Time	Sample	Condition
Lyubashevsky's [34]	N_1	$N_1 \cdot N_2$	$n^{1+\epsilon}$	$N_1 \approx N_2$
Ours	N_1	N_1	$n^{1+\epsilon}$	$(N_1)^{\log \log(n)} \approx N_2$

As shown in Table 7, we compare [34] with our algorithm for solving $\mathsf{LPN}_{n,\mu}$ problem with $Q = n^{1+\epsilon}$, $\mu = 1/2 - \gamma/2$ and $\gamma \geq 2^{-\log(n)^\sigma}$. Lyubashevsky's technique [34] requires $\log \binom{Q}{w} > n$ to satisfy the entropy condition of the leftover hash lemma, and thus picks $w - 2n/(\epsilon \log(n))$, $a = \kappa \cdot \log \log(n)$ and $b = \frac{n}{\kappa \log \log(n)}$ for positive constants σ, κ satisfying $0 < \kappa + \sigma < 1$. Concretely, consider the extreme case $\gamma = 2^{-\log(n)^\sigma}$ whose running time (omitting $\mathsf{poly}(n)$ factors) $T^{n^{1+\epsilon}}_{Lyu05} \approx 2^b \cdot (1/\gamma)^{2^a \cdot n/\log(n)} \leq 2^{\frac{n}{\kappa \log \log(n)}} \cdot 2^{\frac{n}{\log(n)^{1-\sigma-\kappa}}}$.

In contrast, our algorithm uses all the w-linear combinations and do not require them to look jointly independent, and therefore only need $\log \binom{Q'}{w} \geq b$. As a result, for same values $a = \kappa \cdot \log \log(n)$ and $b = \frac{n}{\kappa \log \log(n)}$, we let $w = 2n/(\epsilon \kappa \log(n) \log \log(n))$ for positive constants κ and σ satisfying $\kappa + \sigma < 1$. One can verify that the three inequalities (for $Q\gamma$, $\log \binom{Q/2^a}{w}$, and b) in Theorem 9 are all satisfied with running time and success probability (where $c = 2^{-\log^2 n}$):

$$T^{n^{1+\epsilon}}_{\text{c-sum+bkw}} \approx 2^b = 2^{\frac{n}{\kappa \log \log(n)}}$$

$$P^{n^{1+\epsilon}}_{\text{c-sum+bkw}} \geq 1 - 2^{6a} \cdot n \cdot \varepsilon - 2^a \cdot 2^{-\Omega(\frac{Q\gamma^2}{2^a})} = 1 - \mathsf{negl}(n) \ .$$

That is, for the same parameter choices our algorithm saves a sub-exponential multiplicative factor $2^{\frac{n}{\log(n)^{1-\sigma-\kappa}}}$ over [34] in running time, where constant $1 - \sigma - \kappa$ arbitrarily close to 0 for optimized time complexity. We refer to Table 7 below for a comparison in the general case, which enjoys (for constant $0 < \gamma < 1$) a sub-exponential factor $(4/\gamma)^{2^{a+2} \cdot n/(\epsilon \log(n))}/\mathsf{poly}(n) = 2^{\Omega(n)/\log(n)^{1-\kappa}}$ speedup in running time without consuming (substantially) more space. Note that our N_1 could be even smaller in magnitude than N_2 by using a smaller w and thus produces less stronger noise for majority voting.

Table 8. The space, time and sample complexities of different variants of the BKW algorithms for solving the $\mathsf{LPN}_{n,\mu}$ problem with $\mu = (1 - \gamma)/2$ and sample complexity $Q = 2^{n^\epsilon}$, where $ab = n$, $N_1 = 2^b$, $N_2 = (4/\gamma)^{2^{a+2} \cdot n^{1-\epsilon}}$ and constant $\epsilon > 0$ disregarding $\mathsf{poly}(n)$ factors.

Algorithm	Space	Time	Sample	Condition
Lyubashevsky's [34]	N_1	$N_1 \cdot N_2$	2^{n^ϵ}	$N_1 \approx N_2$
Ours	N_1	N_1	2^{n^ϵ}	$(N_1)^{\log(n)} \approx N_2$

Another interesting setting is $\mathsf{LPN}_{n,\mu}$ with $\mu = 1/2 - \gamma/2$, $\gamma \geq 2^{-n^\sigma}$, and $Q = 2^{n^\epsilon}$ for constant $0 < \epsilon < 1$, for which we can keep the time complexity within $2^{O(n/\log(n))}$ as depicted in Table 8. Lyubashevsky's technique [34] picks $w = 2n^{1-\epsilon}$ (to satisfy $\log\binom{Q}{w} > n$), $a = \kappa \cdot \log(n)$ and $b = \frac{n}{\kappa \log(n)}$ for positive constants σ, κ and ϵ satisfying $\sigma + \kappa < \epsilon$. Concretely, consider the extreme case $\gamma = 2^{-n^\sigma}$ whose running time $T_{Lyu05}^{2^{n^\epsilon}} \approx 2^b \cdot (1/\gamma)^{2^a \cdot n^{1-\epsilon}} \leq 2^{\frac{n}{\kappa \log(n)}} \cdot 2^{n^{1-(\epsilon-\sigma-\kappa)}}$.

In contrast, our algorithm uses the same $a = \kappa \cdot \log(n)$ and $b = \frac{n}{\kappa \log(n)}$ but set $w = 2n^{1-\epsilon}/(\kappa \log(n))$, where positive constants κ, σ and ϵ satisfying $\sigma + \kappa < \epsilon$. This meets all the three conditions (for $Q\gamma$, $\log\binom{Q/2^a}{w}$, and b) in Theorem 9. The resulting running time and success probability (where $\varepsilon = 2^{-\log^2 n}$):

$$T_{\text{c-sum+bkw}}^{2^{n^\epsilon}} \approx 2^b = 2^{\frac{n}{\kappa \log(n)}} \qquad P_{\text{c-sum+bkw}}^{2^{n^\epsilon}} = 1 - \mathsf{negl}(n) \ .$$

That is, for the same parameter choices our algorithm enjoys a sub-exponential factor $2^{n^{1-(\epsilon-\sigma-\kappa)}}$ advantage over [34] in running time, where constant $(\epsilon - \sigma - \kappa)$ is arbitrarily close to 0 for optimized time complexity. We refer to Table 8 below for a comparison in the general case, where for constant $0 < \gamma < 1$ our algorithm saves a sub-exponential factor $(4/\gamma)^{2^{a+2} \cdot n^{1-\epsilon}}/\mathsf{poly}(n) = 2^{O(n^{1-(\epsilon-\kappa)})}$ for arbitrarily small constant $(\epsilon - \kappa)$ with roughly the same space. Note that our N_1 could be even smaller in magnitude than N_2, thanks to the smaller w in use.

Acknowledgements. Yu Yu was supported by the National Key Research and Development Program of China (Grant Nos. 2020YFA0309705 and 2018YFA0704701), the National Natural Science Foundation of China (Grant Nos. 62125204 and 61872236), and the Major Program of Guangdong Basic and Applied Research (Grant No. 2019B030302008). Yu Yu also acknowledges the support from the XPLORER PRIZE and Shanghai Key Laboratory of Privacy-Preserving Computation.

References

1. Albrecht, M.R., Cid, C., Faugère, J., Fitzpatrick, R., Perret, L.: On the complexity of the BKW algorithm on LWE. DCC **74**(2), 325–354 (2015)
2. Applebaum, B., Ishai, Y., Kushilevitz, E.: Cryptography with constant input locality. In: Menezes, A. (ed.) CRYPTO 2007. LNCS, vol. 4622, pp. 92–110. Springer, Heidelberg (2007). https://doi.org/10.1007/978-3-540-74143-5_6
3. Arratia, R., Goldstein, L., Gordon, L.: Two moments suffice for Poisson approximations: the Chen-Stein method. Ann. Probab. 9–25 (1989)
4. Bai, S., Laarhoven, T., Stehlé, D.: Tuple lattice sieving. Cryptology ePrint Archive, Report 2016/713 (2016). https://eprint.iacr.org/2016/713
5. Blum, A., Furst, M., Kearns, M., Lipton, R.J.: Cryptographic primitives based on hard learning problems. In: Stinson, D.R. (ed.) CRYPTO 1993. LNCS, vol. 773, pp. 278–291. Springer, Heidelberg (1994). https://doi.org/10.1007/3-540-48329-2_24
6. Blum, A., Kalai, A., Wasserman, H.: Noise-tolerant learning, the parity problem, and the statistical query model. In: 32nd Annual ACM Symposium on Theory of Computing (STOC), pp. 435–440. ACM Press (2000)
7. Bogos, S., Tramèr, F., Vaudenay, S.: On solving LPN using BKW and variants - implementation and analysis. Cryptogr. Commun. **8**(3), 331–369 (2016)

8. Bogos, S., Vaudenay, S.: Optimization of LPN solving algorithms. In: Cheon, J.H., Takagi, T. (eds.) ASIACRYPT 2016. LNCS, vol. 10031, pp. 703–728. Springer, Heidelberg (2016). https://doi.org/10.1007/978-3-662-53887-6_26

9. Boyer, M., Brassard, G., Høyer, P., Tapp, A.: Tight bounds on quantum searching (p493–505). Fortschritte Der Physik 46(4–5) (2010)

10. Budroni, A., Guo, Q., Johansson, T., Mårtensson, E., Wagner, P.S.: Making the BKW algorithm practical for LWE. In: Bhargavan, K., Oswald, E., Prabhakaran, M. (eds.) INDOCRYPT 2020. LNCS, vol. 12578, pp. 417–439. Springer, Cham (2020). https://doi.org/10.1007/978-3-030-65277-7_19

11. Delaplace, C., Esser, A., May, A.: Improved low-memory subset sum and LPN algorithms via multiple collisions. In: Albrecht, M. (ed.) IMACC 2019. LNCS, vol. 11929, pp. 178–199. Springer, Cham (2019). https://doi.org/10.1007/978-3-030-35199-1_9

12. Devadas, S., Ren, L., Xiao, H.: On iterative collision search for LPN and subset sum. In: Kalai, Y., Reyzin, L. (eds.) TCC 2017. LNCS, vol. 10678, pp. 729–746. Springer, Cham (2017). https://doi.org/10.1007/978-3-319-70503-3_24

13. Dinur, I.: An algorithmic framework for the generalized birthday problem. DCC 87(8), 1897–1926 (2019)

14. Dinur, I., Dunkelman, O., Keller, N., Shamir, A.: Efficient dissection of composite problems, with applications to cryptanalysis, knapsacks, and combinatorial search problems. In: Safavi-Naini, R., Canetti, R. (eds.) CRYPTO 2012. LNCS, vol. 7417, pp. 719–740. Springer, Heidelberg (2012). https://doi.org/10.1007/978-3-642-32009-5_42

15. Dohotaru, C., Høyer, P.: Exact quantum lower bound for Grover's problem. Quantum Inf. Comput. 9(5&6), 533–540 (2009)

16. Ducas, L.: Shortest vector from lattice sieving: a few dimensions for free. In: Nielsen, J.B., Rijmen, V. (eds.) EUROCRYPT 2018. LNCS, vol. 10820, pp. 125–145. Springer, Cham (2018). https://doi.org/10.1007/978-3-319-78381-9_5

17. Esser, A., Heuer, F., Kübler, R., May, A., Sohler, C.: Dissection-BKW. In: Shacham, H., Boldyreva, A. (eds.) CRYPTO 2018. LNCS, vol. 10992, pp. 638–666. Springer, Cham (2018). https://doi.org/10.1007/978-3-319-96881-0_22

18. Esser, A., Kübler, R., May, A.: LPN decoded. In: Katz, J., Shacham, H. (eds.) CRYPTO 2017. LNCS, vol. 10402, pp. 486–514. Springer, Cham (2017). https://doi.org/10.1007/978-3-319-63715-0_17

19. Grover, L.K.: A fast quantum mechanical algorithm for database search. In: 28th Annual ACM Symposium on Theory of Computing (STOC), pp. 212–219. ACM Press (1996)

20. Guo, Q., Johansson, T., Löndahl, C.: Solving LPN using covering codes. J. Cryptol. 33(1), 1–33 (2020)

21. Guo, Q., Johansson, T., Mårtensson, E., Wagner, P.S.: On the asymptotics of solving the LWE problem using coded-BKW with sieving. Cryptology ePrint Archive, Report 2019/009 (2019). https://eprint.iacr.org/2019/009

22. Herold, G., Kirshanova, E.: Improved algorithms for the approximate k-list problem in Euclidean norm. In: Fehr, S. (ed.) PKC 2017. LNCS, vol. 10174, pp. 16–40. Springer, Heidelberg (2017). https://doi.org/10.1007/978-3-662-54365-8_2

23. Herold, G., Kirshanova, E., Laarhoven, T.: Speed-ups and time–memory trade-offs for tuple lattice sieving. In: Abdalla, M., Dahab, R. (eds.) PKC 2018. LNCS, vol. 10769, pp. 407–436. Springer, Cham (2018). https://doi.org/10.1007/978-3-319-76578-5_14

24. Herold, G., Kirshanova, E., May, A.: On the asymptotic complexity of solving LWE. DCC 86(1), 55–83 (2018)

25. Impagliazzo, R., Zuckerman, D.: How to recycle random bits. In: 30th Annual Symposium on Foundations of Computer Science (FOCS), pp. 248–253. IEEE (1989)
26. Kannan, R.: Improved algorithms for integer programming and related lattice problems. In: 15th Annual ACM Symposium on Theory of Computing (STOC), pp. 193–206. ACM Press (1983)
27. Katz, J., Shin, J.S.: Parallel and concurrent security of the HB and HB$^+$ protocols. In: Vaudenay, S. (ed.) EUROCRYPT 2006. LNCS, vol. 4004, pp. 73–87. Springer, Heidelberg (2006). https://doi.org/10.1007/11761679_6
28. Kirchner, P., Fouque, P.-A.: An improved BKW algorithm for LWE with applications to cryptography and lattices. In: Gennaro, R., Robshaw, M. (eds.) CRYPTO 2015. LNCS, vol. 9215, pp. 43–62. Springer, Heidelberg (2015). https://doi.org/10.1007/978-3-662-47989-6_3
29. Laarhoven, T.: Sieving for shortest vectors in lattices using angular locality-sensitive hashing. In: Gennaro, R., Robshaw, M. (eds.) CRYPTO 2015. LNCS, vol. 9215, pp. 3–22. Springer, Heidelberg (2015). https://doi.org/10.1007/978-3-662-47989-6_1
30. Laarhoven, T., Mariano, A.: Progressive lattice sieving. In: Lange, T., Steinwandt, R. (eds.) PQCrypto 2018. LNCS, vol. 10786, pp. 292–311. Springer, Cham (2018). https://doi.org/10.1007/978-3-319-79063-3_14
31. Laarhoven, T., de Weger, B.: Faster sieving for shortest lattice vectors using spherical locality-sensitive hashing. In: Lauter, K., Rodríguez-Henríquez, F. (eds.) LATINCRYPT 2015. LNCS, vol. 9230, pp. 101–118. Springer, Cham (2015). https://doi.org/10.1007/978-3-319-22174-8_6
32. Levieil, É., Fouque, P.-A.: An improved LPN algorithm. In: De Prisco, R., Yung, M. (eds.) SCN 2006. LNCS, vol. 4116, pp. 348–359. Springer, Heidelberg (2006). https://doi.org/10.1007/11832072_24
33. Liu, H., Yu, Y.: A non-heuristic approach to time-space tradeoffs and optimizations for BKW. Cryptology ePrint Archive, Paper 2021/1343 (2021). https://eprint.iacr.org/2021/1343
34. Lyubashevsky, V.: The parity problem in the presence of noise, decoding random linear codes, and the subset sum problem. In: Chekuri, C., Jansen, K., Rolim, J.D.P., Trevisan, L. (eds.) APPROX/RANDOM -2005. LNCS, vol. 3624, pp. 378–389. Springer, Heidelberg (2005). https://doi.org/10.1007/11538462_32
35. Minder, L., Sinclair, A.: The extended k-tree algorithm. J. Cryptol. **25**(2), 349–382 (2012)
36. Regev, O.: On lattices, learning with errors, random linear codes, and cryptography. In: 37th Annual ACM Symposium on Theory of Computing (STOC), pp. 84–93. ACM Press (2005)
37. Schroeppel, R., Shamir, A.: A t=o($2^{n/2}$), s=o($2^{n/4}$) algorithm for certain np-complete problems. SIAM J. Comput. **10**(3), 456–464 (1981)
38. Wagner, D.: A generalized birthday problem. In: Yung, M. (ed.) CRYPTO 2002. LNCS, vol. 2442, pp. 288–304. Springer, Heidelberg (2002). https://doi.org/10.1007/3-540-45708-9_19

Improving Bounds on Elliptic Curve Hidden Number Problem for ECDH Key Exchange

Jun Xu[1,2], Santanu Sarkar[3](✉), Huaxiong Wang[4], and Lei Hu[1,2]

[1] State Key Laboratory of Information Security, Institute of Information Engineering, Chinese Academy of Sciences, Beijing 100093, China
{xujun,hulei}@iie.ac.cn
[2] School of Cyber Security, University of Chinese Academy of Sciences, Beijing 100049, China
[3] Indian Institute of Technology Madras, Sardar Patel Road, Chennai 600036, India
santanu@iitm.ac.in
[4] Division of Mathematical Sciences, School of Physical and Mathematical Sciences, Nanyang Technological University, Singapore, Singapore
hxwang@ntu.edu.sg

Abstract. Elliptic Curve Hidden Number Problem (EC-HNP) was first introduced by Boneh, Halevi and Howgrave-Graham at Asiacrypt 2001. To rigorously assess the bit security of the Diffie–Hellman key exchange with elliptic curves (ECDH), the Diffie–Hellman variant of EC-HNP, regarded as an elliptic curve analogy of the Hidden Number Problem (HNP), was presented at PKC 2017. This variant can also be used for practical cryptanalysis of ECDH key exchange in the situation of side-channel attacks.

In this paper, we revisit the Coppersmith method for solving the involved modular multivariate polynomials in the Diffie–Hellman variant of EC-HNP and demonstrate that, for any given positive integer d, a given sufficiently large prime p, and a fixed elliptic curve over the prime field $\mathbb{F}_p$, if there is an oracle that outputs about $\frac{1}{d+1}$ of the most (least) significant bits of the x-coordinate of the ECDH key, then one can give a heuristic algorithm to compute all the bits within polynomial time in $\log_2 p$. When $d > 1$, the heuristic result $\frac{1}{d+1}$ significantly outperforms both the rigorous bound $\frac{5}{6}$ and heuristic bound $\frac{1}{2}$. Due to the heuristics involved in the Coppersmith method, we do not get the ECDH bit security on a fixed curve. However, we experimentally verify the effectiveness of the heuristics on NIST curves for small dimension lattices.

Keywords: Hidden number problem · Elliptic curve hidden number problem · Modular inversion hidden number problem · Lattice · Coppersmith method

S. Agrawal and D. Lin (Eds.): ASIACRYPT 2022, LNCS 13793, pp. 771–799, 2022.
https://doi.org/10.1007/978-3-031-22969-5_26

1   Introduction

1.1   Background

At CRYPTO 1996, Boneh and Venkatesan [6] first proposed the hidden number problem (HNP) to prove that computing the most significant bits of the Diffie-Hellman (DH) key is as hard as computing the entire key in the DH key exchange for a prime field. It is called the bit security of the DH key exchange. There are a lot of follow-up works, such as [1,7] and [12, Chapter 21.7.1]. HNP has been proven to be an extremely useful tool in many cryptographic areas. One example is its vast use for analysis of DSA and ECDSA in side-channel attacks, such as [15,27]. At USENIX Security 2021, Merget et al. presented the first practical HNP-based attack on the DH key exchange [23]. Albrecht and Heninger presented a new result for solving HNP [2] at Eurocrypt 2021.

The ECDH key exchange is an analog of the DH key exchange, which adopts the group of points on an elliptic curve to enhance performance and security. Roughly speaking, for a given elliptic curve $\mathcal{E}$ over some finite field and a given point $Q \in \mathcal{E}$, two participants with private keys a, b compute $[a]Q, [b]Q$ separately, then send the computed value to each other, and finally, the two participants generate the shared key $[ab]Q$. Naturally, one may want to assess the difficulty of computing partial bits of ECDH key exchange. At ANTS 1998, Boneh [3, Section 5] proposed the open problem: *Does a similar result to the bit security of Diffie-Hellman key exchange* [6] *hold in the group of points of an elliptic curve?* The issue has been raised for 20 years, but few results have been presented because of the complexity associated with the addition formula of points of an elliptic curve. The reason is also presented in the introduction of papers such as [5,16,28,32].

EC-HNP. In [4, Section 5], Boneh, Halevi and Howgrave-Graham presented the elliptic curve hidden number problem (EC-HNP) to study the bit security of ECDH. The authors stated that EC-HNP can be heuristically solved using the idea from Method II for Modular Inversion Hidden Number Problem (MIHNP). Furthermore, they mentioned that the heuristic approach can be converted into a rigorous one in some cases, which corresponds to the following bit security result. Computing $(1 - \epsilon)$ of the most significant bits of the x-coordinate of the ECDH key is as hard as computing the ECDH key itself for a given curve over a prime field, where $\epsilon \approx 0.02$. The detailed proofs were not presented.

Shani [28] demonstrated at PKC 2017 that solving EC-HNP$_x$, which can be viewed as the Diffie-Hellman variant of EC-HNP, is sufficient to demonstrate the bit security of ECDH. The involved strategy is similar to the idea of HNP [6].

Definition 1 (EC-HNP$_x$ [28]). *Fix a prime p, a given elliptic curve $\mathcal{E}$ over $\mathbb{F}_p$, a given point $R \in \mathcal{E}$ and a positive number δ. Let $P \in \mathcal{E}$ be a hidden point. Let $O_{P,R}$ be an oracle that on input m outputs the δ most significant bits of the x-coordinate of $P + [m]R$. That is, $O_{P,R}(m) = \mathrm{MSB}_\delta(x_{P+[m]R})$. The goal is to recover the hidden point P, given query access to the oracle.*

Suppose there is an oracle that outputs some partial information of $[uv]Q$ on input $[u]Q$ and $[v]Q$. For given points Q, $[a]Q$ and $[b]Q$ in the ECDH key exchange, an attacker first selects an integer m, computes $[m]Q$, and then obtains $[a+m]Q$ from $[a]Q + [m]Q = [a+m]Q$. Querying the oracle on input $[a+m]Q$ and $[b]Q$, the attacker can get partial information of $[(a+m)b]Q = [ab]Q + [m][b]Q = P + [m]R$ where $P := [ab]Q$ and $R := [b]Q$. By repeating this process for several m's, the attacker will be able to recover the ECDH key $P = [ab]Q$ if the EC-HNP$_x$ is solved.

In [23, Section 8], Merget et al. mentioned that *this may result in a small timing side-channel information that leaks the MSB of the x-coordinate of the shared point in ECDH. The EC-HNP is related to the HNP and could potentially be applied here.* We contend that the aforementioned attack scenario falls within the scope of EC-HNP$_x$. This attack scenario specifically considers whether the server reuses the same ECDH value $R = [b]Q$ across sessions, where b is the server's static key in TLS-ECDH or a reusable ephemeral key in TLS-ECDHE. A client generates secret a and transmits the value $[a]Q$. Hence, the ECDH key between the server and the client is $P = [ab]Q$. An attacker first chooses some integer m and computes $[a+m]Q$. Then, session's ECDH secret is $[(a+m)b]Q = P + [m]R$. (The above process is very similar to [23, Figure 1]). As a result, if the MSBs of the x-coordinate of $P + [m]R$ are leaked by the small timing side-channel attack for several m, the attacker can obtain the ECDH key P by solving EC-HNP$_x$. EC-HNP$_x$, like HNP, can play an important role in side-channel attacks.

Hardcore Bits. Shani rigorously solved EC-HNP$_x$ and then obtained the following bit security result by combining the underlying idea from Method I for MIHNP [4,21]. For a given curve over a prime field, computing about $\frac{5}{6}$ of the most (least) significant bits of the x-coordinate of the ECDH key is as hard as computing the entire ECDH key. Besides, Shani also analyzed the case of extension fields and generalized the result of Jao, Jetchev and Venkatesan [16].

Papers such as [6,28] demonstrated that DH and ECDH have hardcore bits, which are bits that are difficult to compute as the full shared key.

Heuristic Algorithm. In [32], Xu et al. used the Coppersmith method to solve EC-HNP$_x$, which was inspired by Method II of MIHNP [4,33]. For a fixed curve over a prime field, if there is an oracle that outputs about $\frac{1}{2}$ of the most (least) significant bits of the x-coordinate of the ECDH key, then there is a heuristic algorithm to compute all the bits in polynomial time.

The Coppersmith method is used to calculate small solutions of polynomials. In 1996, Coppersmith proposed rigorous methods for finding the small roots of a modular univariate polynomial and an integer bivariate polynomial [8,9]. In 2006, Jochemsz and May [18] presented heuristic strategies for finding the small roots of modular (and integer) multivariate polynomials. The Coppersmith method is widely used in the security analysis of cryptosystems, the computational complexity analysis of mathematical problems, and the security proof of cryptosystems; see the survey [22] and recent papers, such as [10,24,30,34].

Since the Coppersmith method for modular multivariate polynomials is heuristic, the result in [32] cannot prove that ECDH has hardcore bits. It is important to note that EC-HNP$_x$ is directly related to the actual cryptanalysis of ECDH key exchange for a fixed curve in the work of side-channel attacks [23]. The problem of solving EC-HNP$_x$ is essentially the problem of finding the desired small root of modular multivariate polynomials. The advantage of the Coppersmith method is that it utilizes algebraic structures of polynomials to improve the ability to find small roots. A natural motivation is that one wants to know the best result if the Coppersmith method is used to deal with EC-HNP$_x$.

Related Works. At CRYPTO 2001, Boneh and Shparlinksi [5] showed that if there is an efficient algorithm to predict the least significant bit (LSB) of the ECDH secrets on a non-negligible fraction of a family of curves isomorphic to a curve $\mathcal{E}_0$, then the ECDH key for the curve $\mathcal{E}_0$ can be computed in polynomial time. It does not imply that computing a single LSB of the ECDH key is as hard as computing the entire ECDH key for the same curve $\mathcal{E}_0$. At CRYPTO 2008, Jetchev and Venkatesan [17] utilized isogenies to enlarge the applicability of the method in [5] based on the generalized Riemann hypothesis. However, neither [5] nor [17] provides the hardness of bits for ECDH for a fixed curve. In [5, Section 7], Boneh and Shparlinksi mentioned that they hope their methods will eventually show that a single LSB of ECDH is the hardcore bit for a fixed curve.

1.2   Our Contribution

In this paper, we revisit the Coppersmith method to solve modular multivariate polynomials derived from EC-HNP$_x$ and obtain a new bound.

Result 1. *Let d be any given positive integer. Given a sufficiently large prime $p = 2^{\omega(d^{(2+c)d})}$, and a positive $n = d^{3+c}$ for any constant $c > 0$. For $2n + 1$ given calls to the oracle in EC-HNP$_x$, under Assumption 1 (see Page 8), one can recover the hidden point for EC-HNP$_x$ when the number δ of known MSBs (LSBs) satisfies*

$$\frac{\delta}{\log_2 p} > \frac{1}{d+1} + \varepsilon, \tag{1}$$

where $\varepsilon > 0$ and $\varepsilon = o(\frac{1}{d+1})$. The total time complexity is polynomial in $\log_2 p$ for any constant d.

Corresponding to the ECDH case, we have the following result.

Result 2. *Define d, p as in Result 1. Under Assumption 1, one can compute all the bits in polynomial time for a given elliptic curve $\mathcal{E}$ over the prime field $\mathbb{F}_p$ if there is an oracle that outputs about $\frac{1}{d+1}$ of the most (least) significant bits of the x-coordinate of the ECDH key.*

The bound (1) tends to $\delta/\log_2 p > 0$ as d grows large. It means that the ratio of known MSBs or LSBs number can be infinitesimal. When d becomes large, the modulus $p = 2^{\omega(d^{(2+c)d})}$, the involved lattice dimension $w = \mathcal{O}(n^{d+1})$, and the running time of the algorithm become enormous, with the time complexities of the LLL algorithm and the Gröbner basis computation increasing as $d^{\mathcal{O}(d)}$ and $d^{\mathcal{O}(n)}$, respectively.

The heuristic bound (1) for $d > 1$ is better than the rigorous bound $\delta/\log_2 p > \frac{5}{6}$ [28] and the heuristic result $\delta/\log_2 p > \frac{1}{2}$ [32]. Due to the heuristics of the Coppersmith method, the results in this paper and [32] are not rigorous. It should be noted that the $\frac{1}{2}$ bound on $\delta/\log_2 p$ in [32] is asymptotic. That is, the $\frac{1}{2}$ bound can only be reached when the involved lattice dimension and modulus p tend to infinity (see the analysis of Sect. 1.3). In this work, the smallest dimensions of our lattice to achieve the $\frac{1}{2}$ bound is 2879 for a sufficiently large $p = 2^{\omega(d^{(2+c)d})}$, where $d = 2$. The LLL algorithm terminates within $\mathcal{O}(w^{4+\gamma} b^{1+\gamma})$ bit operations for any $\gamma > 0$ [25], where w is the involved dimension and b is the maximal bit size in the input basis matrix. For our case, $w = 2879$, $w^4 \approx 2^{46}$ and b is bounded by $3d \log_2 p$. Therefore, the LLL algorithm needs a considerable time to get the desired vector. Thus, we do not experimentally show that the $\frac{1}{2}$ barrier is broken.

1.3   Technical Overview

As mentioned before, we revisit the Coppersmith method to find the desired root $(c_0, \tilde{e}_1, \cdots, \tilde{e}_n)$ in n given polynomials

$$\mathcal{F}_j(x_0, y_j) := A_j + B_j x_0 + C_j x_0^2 + D_j y_j + E_j x_0 y_j + x_0^2 y_j$$

derived from EC-HNP$_x$, satisfying $\mathcal{F}_j(e_0, \tilde{e}_j) = 0 \bmod p$ for $1 \le j \le n$, where the value X is the upper bound of $|e_0|, |\tilde{e}_1|, \cdots, |\tilde{e}_n|$, i.e., $|e_0| < X, |\tilde{e}_1| < X, \cdots, |\tilde{e}_n| < X$. Since $X = p/2^\delta$ for EC-HNP$_x$, where p is the modulus and δ is the number of known MSBs (LSBs), we can see that for a fixed p, X and δ are inversely related. To make δ as small as possible, X must be as large as possible.

For any given positive integer d, we construct w multivariate polynomials $G_1(x_0, y_1, \cdots, y_n), \cdots, G_w(x_0, y_1, \cdots, y_n)$ satisfying

$$G_j(e_0, \tilde{e}_1, \cdots, \tilde{e}_n) = 0 \quad \bmod p^d \text{ for all } 1 \le j \le w.$$

Let $\mathcal{L}$ be a Coppersmith lattice, which is spanned by the coefficient vectors of $G_j(x_0 X, y_1 X, \cdots, y_n X)$ for all $1 \le j \le w$, where w and $\det(\mathcal{L})$ are the dimension and determinant of the lattice $\mathcal{L}$, respectively. The basis matrix of $\mathcal{L}$ can be arranged into a triangular matrix.

After the lattice basis reduction, we expect to get $n+1$ multivariate polynomials $Q_1(x_0, y_1, \cdots, y_n), \cdots, Q_{n+1}(x_0, y_1, \cdots, y_n)$ such that $Q_j(e_0, \tilde{e}_1, \cdots, \tilde{e}_n) = 0$ over the integers for all $1 \le j \le n$. Under Assumption 1, we can efficiently recover the desired root $(e_0, \tilde{e}_1, \cdots, \tilde{e}_n)$.

In the Coppersmith method, for a sufficiently large modulus p, the condition for finding the target root $(e_0, \tilde{e}_1, \cdots, \tilde{e}_n)$ can be briefly written as

$$(\det(\mathcal{L}))^{\frac{1}{w}} < p^d. \tag{2}$$

As shown in [28,32], the strategy of solving MIHNP can help to solve EC-HNP$_x$. Inspired by the approach for MIHNP [34], we expect to add enough helpful vectors into the lattice of [32].

In [32], a lattice $\mathcal{L}'$ with triangular basis matrix was constructed. For any given positive integer d, take $n = d^3$. Then we can write $\dim(\mathcal{L}') = (2d + 1)\binom{n}{d}(1+o(1))$, and $\det(\mathcal{L}') = X^{\overline{\alpha}}p^{\overline{\beta}}$, where $\overline{\alpha} = 2d(2d+1)\binom{n}{d}(1+o(1))$ and $\overline{\beta} = 2d\binom{n}{d}(1+o(1))$. For a sufficiently large $p = 2^{\omega(2^n)}$, the Coppersmith condition (2) states: $|\det(\mathcal{L}')|^{\frac{1}{\dim(\mathcal{L}')}} < p^d$, which reduces to $X < p^{\frac{1}{2}-\frac{1}{2d}-\overline{\varepsilon}}$, where $\overline{\varepsilon} > 0$ and $\overline{\varepsilon} = o(\frac{1}{d})$. Plugging $X = p/2^\delta$ into the above relation, we get $\delta/\log_2 p > \frac{1}{2}+\frac{1}{2d}+\overline{\varepsilon}$, which becomes $\delta/\log_2 p > \frac{1}{2}$ when d tends to infinity. It means that, in order to achieve $1/2$ bound, the involved lattice dimension $\dim(\mathcal{L}')$ and the size of modulus p tend to infinity.

In this paper, we first consider $\binom{n}{d+1}$ of helpful polynomials. To be specific, we randomly choose $d+1$ different integers from the set $\{1, \cdots, n\}$. Without loss of generality, let $d+1$ integers be $j_1, \cdots, j_{d+1}$, where $1 \leq j_1 < \cdots < j_{d+1} \leq n$. For any fixed tuple $(j_1, \cdots, j_{d+1})$, we choose a linear combination (with the leading term $y_{j_1} \cdots y_{j_{d+1}}$) of the following polynomials:

$$\sum_{u=1}^{d+1}\sum_{v=0}^{1} K_{u,v} \cdot x_0^v \mathcal{F}_{j_1} \cdots \mathcal{F}_{j_{u-1}} y_{j_u} \mathcal{F}_{j_{u+1}} \cdots \mathcal{F}_{j_{d+1}} \text{ for some } K_{u,v} \in \mathbb{Z}. \quad (3)$$

We then consider the algebraic structure of linear combinations (3) and design a lattice. We construct more compact linear combinations compared to (3) so that all monomials related to x_0^{2d} and x_0^{2d+1} are removed. That is, the monomials $x_0^{i_0} y_1^{i_1} \cdots y_n^{i_n}$ for all $(i_0, i_1, \cdots, i_n) \in \mathcal{I}_3$ are deleted from new linear combinations, where $\mathcal{I}_3 := (\{(i_0, i_1, \cdots, i_n) \mid 2d \leq i_0 \leq 2d + 1, 0 \leq i_1, \cdots, i_n \leq 1, 0 \leq i_1 + \cdots + i_n \leq d\}$. Then we get a lattice with triangular basis matrix. In this case, we can deduce that the upper bound $X < p^{1-\frac{1}{d+1}-\varepsilon}$, where $\varepsilon > 0$ and $\varepsilon = o(\frac{1}{d+1})$. Based on $X = p/2^\delta$, we obtain $\delta/\log_2 p > \frac{1}{d+1} + \varepsilon$, which becomes $\delta/\log_2 p > 0$ when d tends to infinity.

The polynomial construction for the lattice in this work looks similar to that in [32]. However, this does not mean that our lattice construction is ordinary. When it comes to the Coppersmith method, small differences in parameter selection can lead to significant differences in efficiency. While dealing with multivariate Coppersmith method, the core point and technical difficulty is constructing as many helpful polynomials as possible. The rest is a conventional technique.

1.4   Organization

The rest of this paper is organized as follows. In Sect. 2, we review some results on lattice, the Coppersmith method, elliptic curves, the transformation from EC-HNP$_x$ to a class of modular polynomials, and orders of monomials. The existing method is revisited in Sect. 3. In Sect. 4, we use algebraic structure of polynomials to design a lattice. In Sect. 5, we prove that the involved basis matrix is triangular. In Sect. 6, we compare our result with the existing work. We present our experimental results in Sect. 7.

2   Preliminaries

Throughout the paper, p is a prime where $p > 3$.

2.1   Lattice

A lattice $\mathcal{L}$ is a discrete subgroup of $\mathbb{R}^m$. An alternative equivalent definition of an integer lattice can be given using a basis. Let $\mathbf{b_1}, \cdots, \mathbf{b_w}$ be linear independent row vectors in $\mathbb{R}^m$, a lattice $\mathcal{L}$ spanned by them is

$$\mathcal{L} = \left\{ \sum_{i=1}^{w} k_i \mathbf{b_i} \mid k_i \in \mathbb{Z} \right\}.$$

The set $\{\mathbf{b_1}, \cdots, \mathbf{b_w}\}$ is called a basis of $\mathcal{L}$ and the matrix $\mathbf{B} = [\mathbf{b_1}^T, \cdots, \mathbf{b_w}^T]^T$ is the corresponding basis matrix. The dimension and determinant of $\mathcal{L}$ are respectively

$$\dim(\mathcal{L}) = w, \det(\mathcal{L}) = \sqrt{\det(\mathbf{BB}^T)}.$$

When $m = w$, lattice is called full rank. In this paper, the involved lattices are full-rank integer lattices.

The well-known LLL lattice reduction algorithm [20] can produce a reduced basis that has the following property.

Lemma 1 (LLL). *Let $\mathcal{L}$ be a w-dimensional integer lattice. Within polynomial time, the LLL algorithm outputs reduced basis vectors $\mathbf{v}_1, \ldots, \mathbf{v}_w$ that satisfy*

$$\|\mathbf{v}_i\| \leq 2^{\frac{w(w-1)}{4(w+1-i)}} (\det(\mathcal{L}))^{\frac{1}{w+1-i}}, 1 \leq i \leq w.$$

2.2   The Coppersmith Method

We briefly review how to use the Coppersmith method to solve multivariate modular polynomials.

Problem Definition. Let $f_1(x_0, x_1, \cdots, x_n)$, $\cdots$, $f_m(x_0, x_1, \cdots, x_n)$ be original polynomials, which are irreducible multivariate polynomials defined over $\mathbb{Z}$, with a common root $(\tilde{x}_0, \tilde{x}_1, \cdots, \tilde{x}_n)$ modulo a known integer p such that $|\tilde{x}_0| < X_0$, $\cdots$, $|\tilde{x}_n| < X_n$. The goal is to recover the desired root $(\tilde{x}_0, \cdots, \tilde{x}_n)$ in polynomial time. To ensure recovery of the desired root, the size of values $X_0, \cdots, X_n$ must be bound.

Polynomials Collection. One chooses polynomials,

$$g_1(x_0, x_1, \cdots, x_n), \cdots, g_w(x_0, x_1, \cdots, x_n)$$

such that $(\tilde{x}_0, \tilde{x}_1, \cdots, \tilde{x}_n)$ is a common root modulo a power of p. Generally, multiples of lifting polynomials are selected, where a lifting polynomial is defined as the product of some powers of original polynomials and variables. For example,

$$g_j(x_0, x_1, \cdots, x_n) := p^{d-(\beta_1^j+\cdots+\beta_m^j)} x_0^{\alpha_0^j} x_1^{\alpha_1^j} \cdots x_n^{\alpha_n^j} f_1^{\beta_1^j} \cdots f_m^{\beta_m^j},$$

where $j \in \{1, \cdots, w\}$, $d \in \mathbb{Z}^+$, and $\alpha_0^j, \alpha_1^j, \cdots, \alpha_n^j, \beta_1^j, \cdots, \beta_m^j \in \mathbb{Z}^+ \cup \{0\}$ satisfying $0 \leq \beta_1^j + \cdots + \beta_m^j \leq d$. It is not hard to see that $g_j(\tilde{x}_0, \tilde{x}_1, \cdots, \tilde{x}_n) \equiv 0 \bmod p^d$ for every $j \in \{1, \cdots, w\}$. For the Coppersmith method, the most complex step is the selection of polynomials $g_1, \cdots, g_w$ when dealing with multiple original polynomials. The difference between this paper's polynomial selection and the above strategy is that linear combinations of lifting polynomials are considered.

Lattice Construction. Let the vector $\mathbf{b}_j$ $(1 \leq j \leq w)$ be the coefficient vector of the polynomial $g_j(x_0 X_0, x_1 X_1, \ldots, x_n X_n)$ with variables $x_0, x_1, \ldots, x_n$. Then one constructs the lattice $\mathcal{L} = \left\{ \sum_{j=1}^{w} k_j \mathbf{b}_j \mid k_j \in \mathbb{Z} \right\}$.

Reduced Basis. One runs the LLL algorithm and obtains the w reduced basis vectors $\mathbf{v}_1, \ldots, \mathbf{v}_w$, where $\mathbf{v}_j$ is the coefficient vector of the polynomial $h_j(x_0 X_0, x_1 X_1, \ldots, x_n X_n)$ for $j \in \{1, \cdots, w\}$. Note that the LLL algorithm performs linear operations. Hence, $\mathbf{v}_j$ is a linear combination of the vectors $\mathbf{b}_1, \cdots, \mathbf{b}_w$. That is, $h_j(x_0, x_1, \ldots, x_n)$ is a linear combination of $g_1(x_0, x_1, \ldots, x_n), \cdots, g_w(x_0, x_1, \ldots, x_n)$. Then, $h_j(\tilde{x}_0, \tilde{x}_1, \cdots, \tilde{x}_n) = 0 \pmod{p^d}$ for every $j \in [1, \cdots, w]$. In order to get $h_j(\tilde{x}_0, \tilde{x}_1, \cdots, \tilde{x}_n) = 0$ over $\mathbb{Z}$ for some $j \in \{1, \cdots, w\}$, we need the following lemma in this process.

Lemma 2 ([14]). *Let $h(x_0, x_1, \ldots, x_n)$ be an integer polynomial that consists of at most w monomials. Let d be a positive integer and the integers X_i be the upper bound of $|\tilde{x}_i|$ for $i = 0, 1, \cdots, n$. Let $\|h(x_0 X_0, x_1 X_1, \ldots, x_n X_n)\|$ be the Euclidean length of the coefficient vector of $h(x_0 X_0, x_1 X_1, \ldots, x_n X_n)$ with variables $x_0, x_1, \ldots, x_n$. Suppose that*

1. $h(\tilde{x}_0, \tilde{x}_1, \cdots, \tilde{x}_n) = 0 \pmod{p^d}$,
2. $\|h(x_0 X_0, x_1 X_1, \ldots, x_n X_n)\| < \frac{p^d}{\sqrt{w}}$,

then $h(\tilde{x}_0, \tilde{x}_1, \cdots, \tilde{x}_n) = 0$ holds over $\mathbb{Z}$.

To make $h_j(\tilde{x}_0, \tilde{x}_1, \cdots, \tilde{x}_n) = 0$ for all $1 \leq j \leq n + 1$ hold, from Lemma 1 and Lemma 2, we need the Euclidean lengths of the $n + 1$ reduced basis vectors $\mathbf{v}_1, \ldots, \mathbf{v}_{n+1}$ satisfy the condition

$$2^{\frac{w(w-1)}{4(w-n)}} \cdot \left(\det(\mathcal{L}) \right)^{\frac{1}{w-n}} < \frac{p^d}{\sqrt{w}}, \quad w = \dim(\mathcal{L}). \tag{4}$$

Based on Condition (4), one can determine the size of bounds $X_0, \cdots, X_n$.

Desired Root Recovery. We have no assurance that the $n + 1$ obtained polynomials $h_1, \cdots, h_{n+1}$ are algebraically independent. Under Assumption 1, the corresponding equations can be solved using elimination techniques such as the Gröbner basis computation, and then the desired root $(\widetilde{x}_0, \widetilde{x}_1, \cdots, \widetilde{x}_n)$ is recovered. In this paper, we use computer experiments to show that our heuristic approach works.

Assumption 1 ([19]). *Let $h_1, \cdots, h_{n+1} \in \mathbb{Z}[x_0, x_1, \cdots, x_n]$ be the polynomials that are found by the Coppersmith method. Then the ideal generated by the polynomial equations $h_1(x_0, x_1, \cdots, x_n) = 0, \cdots, h_{n+1}(x_0, x_1, \cdots, x_n) = 0$ has dimension zero.*

The involved Assumption 1 is called the zero-dimensional ideal assumption, which is a relaxation of algebraically independent assumption, first appeared in [19]. We consider a zero-dimensional ideal, namely, an ideal I such that the number of common zeros of the polynomials in I is finite in the algebraic closure of the field of coefficients [11]. It seems very difficult to verify whether there are finite number of common zeros or not.

Helpful Polynomials. An important strategy of choosing the above polynomials $g_1(x_0, x_1, \cdots, x_n), \cdots, g_w(x_0, x_1, \cdots, x_n)$ is to choose as many *helpful polynomials* as possible.

Definition 2 ([22,29]). *Define d and $\mathcal{L}$ as above. A vector in the triangular basis matrix, which is the coefficient vector of $g(x_0 X, x_1 X, \cdots, x_n X)$, is called a helpful vector if the absolute value of its diagonal component[1] is greater than 0 and less than p^d. That is, $g(x_0, x_1, \cdots, x_n)$ is called a helpful polynomial[2]. Else, $g(x_0, x_1, \cdots, x_n)$ is called a non-helpful polynomial.*

Next, we show why helpful polynomials can work. We obtain the simplified condition $(\det(\mathcal{L}))^{\frac{1}{w}} < p^d$ by ignoring low-order terms in Condition (4). For a triangular basis matrix, the left side of the simplified condition is regarded as the geometric mean of all diagonals of the basis matrix. A helpful polynomial contributes to the determinant with a factor greater than 0 and less than p^d. The more helpful polynomials in the lattice, the easier the condition for solving modular equations is to be satisfied. It implies that the Coppersmith method becomes more and more effective, and the above bounds X_i become larger and larger. Therefore, one should choose as many helpful polynomials as possible.

---

[1] The diagonal component of the coefficient vector of $g(x_0 X, x_1 X, \cdots, x_n X)$ corresponds to the leading term of $g(x_0, x_1, \cdots, x_n)$. Specifically, the diagonal component is equal to the leading coefficient of $g(x_0 X, x_1 X, \cdots, x_n X)$.

[2] There is a one-to-one correspondence between helpful polynomials and helpful vectors. The coefficient vector of $g(x_0 X, x_1 X, \cdots, x_n X)$ is a helpful vector if and only if $g(x_0, x_1, \cdots, x_n)$ is a helpful polynomial.

2.3 Elliptic Curves

For a prime field $\mathbb{F}_p$, consider an elliptic curve $\mathcal{E}$ over $\mathbb{F}_p$, given in a Weierstrass form $\mathcal{E} : y^2 = x^3 + ax + b$ over $\mathbb{F}_p$ with $a, b \in \mathbb{F}_p$ and $4a^3 + 27b^2 \neq 0$. Let $P = (x_P, y_P) \in \mathbb{F}_p^2$ be a point on the curve $\mathcal{E}$, where x_P (resp. y_P) is called the x-coordinate (resp. y-coordinate) of point P. The set of points on $\mathcal{E}$, together with the point at infinity O, forms an additive abelian group. Hasse's theorem shows that the number of points $\#\mathcal{E}$ on the curve $\mathcal{E}(\mathbb{F}_p)$ satisfies the relation: $|\#\mathcal{E} - p - 1| \leq 2\sqrt{p}$. The additive inverse of point P is $-P = (x_P, -y_P)$. For an integer m, $[m]P$ denotes successive m-time addition of the point P, and $[-m]P = m[-P]$. Given two points $P = (x_P, y_P)$ and $Q = (x_Q, y_Q)$ on $\mathcal{E}$, where $P \neq \pm Q$, consider the addition $P + Q = (x_{P+Q}, y_{P+Q})$. Let $s_{P+Q} = \frac{y_P - y_Q}{x_P - x_Q}$. The x-coordinate and y-coordinate of $P + Q$ are respectively

$$x_{P+Q} = s_{P+Q}^2 - x_P - x_Q, \ y_{P+Q} = s_{P+Q}(x_P - x_{P+Q}) - y_P. \tag{5}$$

2.4 From EC-HNP$_x$ to Modular Polynomials

We present the transformation in [28] from the problem of recovering x_P in EC-HNP$_x$ (see Definition 1), the x-coordinate of the hidden point $P = (x_P, y_P)$, to the problem of finding small solutions of modular polynomials. In brief, our target is to find the desired small root $(e_0, \tilde{e}_i)$ of the following modular polynomial

$$\mathcal{F}_i(x_0, y_i) := A_i + B_i x_0 + C_i x_0^2 + D_i y_i + E_i x_0 y_i + x_0^2 y_i = 0 \ (\mathrm{mod} \ p), \ 1 \leq i \leq n. \tag{6}$$

Here coefficients A_i, B_i, C_i, D_i, E_i are known, and unknown integers $e_0, \tilde{e}_1$, $\cdots, \tilde{e}_n$ are all bounded by the value $X := p/2^\delta$. The specific analysis is as follows.

Eliminating y_P. For a given point R in an elliptic curve $\mathcal{E}$ over $\mathbb{F}_p$, we produce $Q = [m]R = (x_Q, y_Q)$ and $-Q = [-m]R = (x_Q, -y_Q)$, where m is a positive integer. According to $y_P^2 = x_P^3 + ax_P + b$, $y_Q^2 = x_Q^3 + ax_Q + b$ and (5), we obtain

$$
\begin{aligned}
x_{P+Q} + x_{P-Q} &= (s_{P+Q}^2 - x_P - x_Q) + (s_{P-Q}^2 - x_P - x_Q) \\
&= \left(\frac{y_P - y_Q}{x_P - x_Q}\right)^2 + \left(\frac{y_P + y_Q}{x_P - x_Q}\right)^2 - 2x_P - 2x_Q \\
&= 2\left(\frac{y_P^2 + y_Q^2}{(x_P - x_Q)^2} - x_P - x_Q\right) \\
&= 2\left(\frac{x_Q x_P^2 + (a + x_Q^2)x_P + ax_Q + 2b}{(x_P - x_Q)^2}\right).
\end{aligned}
\tag{7}
$$

Constructing Modular Polynomials. Query the oracle $O_{P,R}$ in EC-HNP$_x$ on $2n + 1$ different inputs 0 and $\pm m_i$ for $i = 1, \cdots, n$. Then we obtain $O_{P,R}(0)$ and $O_{P,R}(\pm m_i)$. We write $h_i = O_{P,R}(m_i) = \mathrm{MSB}_\delta(x_{P+Q_i}) = x_{P+Q_i} - e_i$ and $h_i' = O_{P,R}(-m_i) = \mathrm{MSB}_\delta(x_{P-Q_i}) = x_{P-Q_i} - e_i'$, where $|e_i| < p/2^{\delta+1}$ and $|e_i'| < p/2^{\delta+1}$ for all $1 \leq i \leq n$. Let $\tilde{h}_i = h_i + h_i'$ and $\tilde{e}_i = e_i + e_i'$, we have

$\tilde{h}_i + \tilde{e}_i = x_{P+Q_i} + x_{P-Q_i}$, where $|\tilde{e}_i| < p/2^\delta$ for $i = 1, \cdots, n$. According to (7), we get

$$\tilde{h}_i + \tilde{e}_i = 2\left(\frac{x_{Q_i}x_P^2 + (a+x_{Q_i}^2)x_P + ax_{Q_i} + 2b}{(x_P - x_{Q_i})^2}\right), \quad 1 \le i \le n. \tag{8}$$

Moreover, we write $h_0 = O_{P,R}(0) = \mathrm{MSB}_\delta(x_P) = x_P - e_0$, where $|e_0| < p/2^{\delta+1}$. Hence, $\tilde{h}_i + \tilde{e}_i = 2(\frac{x_{Q_i}(h_0+e_0)^2 + (a+x_{Q_i}^2)(h_0+e_0) + ax_{Q_i} + 2b}{(h_0+e_0-x_{Q_i})^2})$. After multiplying by $(h_0 + e_0 - x_{Q_i})^2$, we get $A_i + B_i e_0 + C_i e_0^2 + D_i \tilde{e}_i + E_i e_0 \tilde{e}_i + e_0^2 \tilde{e}_i = 0 \bmod p$, $1 \le i \le n$, where known coefficients A_i, B_i, C_i, D_i, E_i satisfy (in the field $\mathbb{F}_p$)

$$
\begin{aligned}
A_i &= \left(\tilde{h}_i(h_0 - x_{Q_i})^2 - 2h_0^2 x_{Q_i} - 2(a + x_{Q_i}^2)h_0 - 2ax_{Q_i} - 4b\right), \\
B_i &= 2(\tilde{h}_i(h_0 - x_{Q_i}) - 2h_0 x_{Q_i} - a - x_{Q_i}^2), C_i = (\tilde{h}_i - 2x_{Q_i}), \\
D_i &= (h_0 - x_{Q_i})^2, E_i = 2(h_0 - x_{Q_i}).
\end{aligned}
\tag{9}
$$

Therefore, $(e_0, \tilde{e}_i)$ is a small root of the polynomial

$$\mathcal{F}_i(x_0, y_i) = A_i + B_i x_0 + C_i x_0^2 + D_i y_i + E_i x_0 y_i + x_0^2 y_i = 0 \pmod{p},$$

where $1 \le i \le n$ and $e_0, \tilde{e}_1, \cdots, \tilde{e}_n$ are all bounded by $X := p/2^\delta$. Once the desired vector $(e_0, \tilde{e}_1, \cdots, \tilde{e}_n)$ is obtained, x_P can be recovered based on $x_P = e_0 + h_0$. After x_P is recovered, y_P will be extracted due to $y_P^2 = x_P^3 + ax_P + b \bmod p$.

2.5   Order of Monomials

We first recall reverse lexicographic order and graded lexicographic reverse order respectively. For more details, please refer to [31, Section 21.2]. Let $i_0, i_1, \cdots, i_n, i_0', i_1', \cdots, i_n'$ be nonnegative integers.

Reverse Lexicographic Order: $(i_1', \cdots, i_n') \prec_{revlex} (i_1, \cdots, i_n) \Leftrightarrow$ the rightmost nonzero entry in $(i_1' - i_1, \cdots, i_n' - i_n)$ is negative.

Graded Reverse Lexicographic Order: $(i_1', \cdots, i_n') \prec_{grevlex} (i_1, \cdots, i_n) \Leftrightarrow$

$$\sum_{m=1}^n i_m' < \sum_{m=1}^n i_m \text{ or } \left(\sum_{m=1}^n i_m' = \sum_{m=1}^n i_m \text{ and } (i_1', \cdots, i_n') \prec_{revlex} (i_1, \cdots, i_n)\right).$$

In this paper, we utilize the following order of monomials, which is also used in [34].

$$
\begin{aligned}
&x_0^{i_0'} y_1^{i_1'} \cdots y_n^{i_n'} \prec x_0^{i_0} y_1^{i_1} \cdots y_n^{i_n} \Leftrightarrow \\
&(i_1', \cdots, i_n') \prec_{grevlex} (i_1, \cdots, i_n) \text{ or } \left((i_1', \cdots, i_n') = (i_1, \cdots, i_n) \text{ and } i_0' < i_0\right).
\end{aligned}
\tag{10}
$$

It is noteworthy that i_0 and i_0' are treated differently than $i_1, \cdots, i_n$ and $i_1', \cdots, i_n'$ respectively. According to (10), we can determine the leading term of a multivariate polynomial. For example, for $\mathcal{F}_j = A_j + B_j x_0 + C_j x_0^2 + D_j y_j + E_j x_0 y_j + x_0^2 y_j$ for $1 \le j \le n$ in (6), we have

$$1 \prec x_0 \prec x_0^2 \prec y_j \prec x_0 y_j \prec x_0^2 y_j. \tag{11}$$

Hence, the leading monomial of $\mathcal{F}_j$ is $x_0^2 y_j$. Further, the leading coefficient of $\mathcal{F}_j$ is 1, and the leading term of $\mathcal{F}_j$ is $x_0^2 y_j$.

3   Existing Lattice

In this section, we review the lattice in [32] for solving (6). Here we provide a different description of the lattice, closer to the lattice we introduce later. First, we recall the index set

$$\mathcal{I}_{[\text{XHS20}]}(n,d) = \{(i_0, i_1, \cdots, i_n) \mid 0 \leq i_0 \leq 2d,$$
$$0 \leq i_1, \cdots, i_n \leq 1, 0 \leq l \leq d\}, \tag{12}$$

where integers n, d satisfying $1 \leq d \leq n$, and $l := i_1 + \cdots + i_n$ satisfying $0 \leq l \leq d$.

3.1   Lattice $\mathcal{L}_{[\text{XHS20}]}(n,d)$

For any fixed tuple $(i_0, i_1, \cdots, i_n) \in \mathcal{I}_{[\text{XHS20}]}(n,d)$, we construct polynomial $f_{i_0,i_1,\ldots,i_n}(x_0, y_1, \cdots, y_n)$ as follows.

Case a: When $l = 0$ and $0 \leq i_0 \leq 2d$, define

$$f_{i_0,i_1,\cdots,i_n}(x_0, y_1, \cdots, y_n) := x_0^{i_0}.$$

Case b: When $l = 1$ and $0 \leq i_0 \leq 1$, define

$$f_{i_0,i_1,\cdots,i_n}(x_0, y_1, \cdots, y_n) := x_0^{i_0} y_1^{i_1} \cdots y_n^{i_n}.$$

Case c: When $1 \leq l \leq d$ and $2l \leq i_0 \leq 2d$, define

$$f_{i_0,i_1,\cdots,i_n}(x_0, y_1, \cdots, y_n) := x_0^{i_0-2l} \mathcal{F}_1^{i_1} \cdots \mathcal{F}_n^{i_n}.$$

Case d: When $2 \leq l \leq d$ and $0 \leq i_0 \leq 2l - 1$, define

$$f_{i_0,i_1,\cdots,i_n}(x_0, y_1, \cdots, y_n) := \sum_{u=1}^{l} \sum_{v=0}^{1} w_{i_0+1,u+lv} \cdot x_0^v \mathcal{F}_{j_1} \cdots \mathcal{F}_{j_{u-1}} y_{j_u} \mathcal{F}_{j_{u+1}} \cdots \mathcal{F}_{j_l}, \tag{13}$$

where $\mathcal{F}_i(x_0, y_i) = A_i + B_i x_0 + C_i x_0^2 + D_i y_i + E_i x_0 y_i + x_0^2 y_i = 0 \pmod{p}$ for $1 \leq i \leq n$ defined in (6), integers $j_1, \cdots, j_l$ are defined in Lemma 3, and $w_{i_0+1,u+lv}$ is element of the $(i_0 + 1)$-th row and the $(u + lv)$-th column of the matrix $\mathbf{W}_{j_1,\cdots,j_l}$, which is also defined in Lemma 3.

Lemma 3 ([32]). *Let $i_1, \cdots, i_n$ be integers satisfying $0 \leq i_1, \cdots, i_n \leq 1$. Denote $l = i_1 + \cdots + i_n$, where $2 \leq l \leq n$. Let $j_1, \cdots, j_l$ be integers satisfying $1 \leq j_1 < \cdots < j_l \leq n$ and $y_{j_1} \cdots y_{j_l} = y_1^{i_1} \cdots y_n^{i_n}$. Let a $2l \times 2l$ integer matrix $\mathbf{M}_{j_1,\cdots,j_l}$ be the following coefficient matrix:*

$$\begin{pmatrix} \prod_{u \neq 1} (x_0^2 + E_{j_u} x_0 + D_{j_u}) \\ \vdots \\ \prod_{u \neq l} (x_0^2 + E_{j_u} x_0 + D_{j_u}) \\ x_0 \prod_{u \neq 1} (x_0^2 + E_{j_u} x_0 + D_{j_u}) \\ \ddots \\ x_0 \prod_{u \neq l} (x_0^2 + E_{j_u} x_0 + D_{j_u}) \end{pmatrix} = \mathbf{M}_{j_1,\cdots,j_l} \begin{pmatrix} 1 \\ \vdots \\ x_0^{l-1} \\ x_0^l \\ \vdots \\ x_0^{2l-1} \end{pmatrix} \bmod p^{l-1}, \tag{14}$$

where integers D_{j_u} and E_{j_u} are the coefficients in the polynomial $\mathcal{F}_{j_u} = A_{j_u} + B_{j_u}x_0 + C_{j_u}x_0^2 + D_{j_u}y_{j_u} + E_{j_u}x_0y_{j_u} + x_0^2y_{j_u}$ for $1 \le u \le l$. Then the matrix $\mathbf{M}_{j_1,\cdots,j_l}$ is invertible over $\mathbb{Z}_{p^{l-1}}$. Denote $\mathbf{W}_{j_1,\cdots,j_l}$ as its inverse matrix. Hence,

$$\mathbf{W}_{j_1,\cdots,j_l} \cdot \mathbf{M}_{j_1,\cdots,j_l} = I_{2l} \bmod p^{l-1}, \tag{15}$$

where I_{2l} is the $2l \times 2l$ identity matrix.

Lemma 4 ([32]). *Based on the order (10), the monomial $x_0^{i_0}y_1^{i_1}\cdots y_n^{i_n}$ is the leading term of the polynomial $f_{i_0,i_1,\cdots,i_n}(x_0,y_1,\cdots,y_n)$ for $(i_0,i_1,\cdots,i_n) \in \mathcal{I}_{[\mathrm{XHS20}]}(n,d)$. Let*

$$F_{i_0,i_1,\cdots,i_n}(x_0,y_1,\cdots,y_n) := \begin{cases} p^{d+1-l}f_{i_0,i_1,\cdots,i_n} & \text{for } 1 \le l \le d, 0 \le i_0 \le 2l-1, \\ p^{d-l}f_{i_0,i_1,\cdots,i_n} & \text{for } 0 \le l \le d, 2l \le i_0 \le 2d. \end{cases}$$
$$\tag{16}$$

Let $\mathcal{L}_{[\mathrm{XHS20}]}(n,d)$ be the lattice which is spanned by the coefficient vectors of polynomials

$$F_{i_0,i_1,\cdots,i_n}(x_0X,y_1X,\cdots,y_nX) \text{ for all } (i_0,i_1,\cdots,i_n) \in \mathcal{I}_{[\mathrm{XHS20}]}(n,d),$$

where the value X is the upper bound of $|e_0|, |\tilde{e}_1|, \cdots, |\tilde{e}_n|$. The diagonal elements in triangular basis matrix of lattice $\mathcal{L}_{[\mathrm{XHS20}]}(n,d)$ are as follows:

$$\begin{cases} p^{d+1-l}X^{i_0+l} & \text{for } 1 \le l \le d, 0 \le i_0 \le 2l-1, \\ p^{d-l}X^{i_0+l} & \text{for } 0 \le l \le d, 2l \le i_0 \le 2d. \end{cases}$$

According to Lemma 4, the dimension and determinant of $\mathcal{L}_{[\mathrm{XHS20}]}(n,d)$ are respectively

$$\dim(\mathcal{L}_{[\mathrm{XHS20}]}(n,d)) = (2d+1)\sum_{l=0}^{d}\binom{n}{l} \text{ and } \det(\mathcal{L}_{[\mathrm{XHS20}]}(n,d)) = X^{\overline{\alpha}}p^{\overline{\beta}},$$

where

$$\overline{\alpha} = d(2d+1)\sum_{l=0}^{d}\binom{n}{l} + (2d+1)\sum_{l=0}^{d}l\binom{n}{l}, \ \overline{\beta} = d(2d+1)\sum_{l=0}^{d}\binom{n}{l} - (2d-1)\sum_{l=0}^{d}l\binom{n}{l}.$$

For a sufficiently large modulus p, one can use the simplified Coppersmith condition (2), which does not affect the asymptotic bound. Based on (2), we get the condition $\left(\det(\mathcal{L}_{[\mathrm{XHS20}]}(n,d))\right)^{\frac{1}{\overline{w}}} < p^d$, where $\overline{w} = \dim(\mathcal{L}_{[\mathrm{XHS20}]}(n,d))$, which is equivalent to

$$X < p^{\frac{d\overline{w}-\overline{\beta}}{\overline{\alpha}}}. \tag{17}$$

We omit the tedious calculation and give the following results directly. For any $1 \le d \le n$, $\frac{d\overline{w}-\overline{\beta}}{\overline{\alpha}} < \frac{1}{2}$. For any given positive integer d, take $n = d^3$. Then we have $\overline{w} = (2d+1)\binom{n}{d}(1+o(1))$, $\overline{\alpha} = 2d(2d+1)\binom{n}{d}(1+o(1))$ and $\overline{\beta} = 2d\binom{n}{d}(1+o(1))$. For a sufficiently large $p = 2^{\omega(2^n)}$, the condition (17) becomes $X < p^{\frac{1}{2}-\frac{1}{2d}-\overline{\varepsilon}}$, where $\overline{\varepsilon} > 0$ and $\overline{\varepsilon} = o(\frac{1}{d})$. Plugging $X = p/2^\delta$ for EC-HNP$_x$ into the above inequality, we have $\delta/\log_2 p > \frac{1}{2} + \frac{1}{2d} + \overline{\varepsilon}$. When d tends to infinity, this condition reduces to

$$\frac{\delta}{\log_2 p} > \frac{1}{2}. \tag{18}$$

4   New Lattice

In this section, we design a new lattice by mining the algebraic structure.

4.1   Lattice $\mathcal{L}(n, d, t)$

Let $\mathcal{I}(n, d, t)$ be an index set which is equal to $\mathcal{I}(n, d, t) = \mathcal{I}_1 \cup \mathcal{I}_2$, where

$$\mathcal{I}_1 := \{(i_0, i_1, \cdots, i_n) \mid 0 \leq i_0 \leq 2d - 1, 0 \leq i_1, \cdots, i_n \leq 1, 0 \leq l \leq d\},$$
$$\mathcal{I}_2 := \{(i_0, i_1, \cdots, i_n) \mid 0 \leq i_0 \leq t, 0 \leq i_1, \cdots, i_n \leq 1, l = d + 1\}.$$

Here, $1 \leq d < n$, $0 \leq t \leq 2d - 1$ and $l = i_1 + \cdots + i_n$ satisfying $0 \leq l \leq d + 1$.

Remark 1. According to (12), we get that the index set $\mathcal{I}_{[\text{XHS20}]}(n, d)$ equals

$$\{(i_0, i_1, \cdots, i_n) \mid 0 \leq i_0 \leq 2d, 0 \leq i_1, \cdots, i_n \leq 1, 0 \leq l \leq d\}.$$

It is obvious that $\mathcal{I}_1$ is a subset of $\mathcal{I}_{[\text{XHS20}]}(n, d)$, whereas $\mathcal{I}_2$ is not.

Based on $F_{i_0, i_1, \cdots, i_n}(x_0, y_1, \cdots, y_n)$ in Lemma 4, we construct the polynomial $G_{i_0, i_1, \cdots, i_n}(x_0, y_1, \cdots, y_n)$ as follows.

Case A: For any given $(i_0, i_1, \cdots, i_n) \in \mathcal{I}_1$, we define

$$G_{i_0, i_1, \cdots, i_n}(x_0, y_1, \cdots, y_n) = F_{i_0, i_1, \cdots, i_n}(x_0, y_1, \cdots, y_n).$$

Since $F_{i_0, i_1, \cdots, i_n}(e_0, \tilde{e}_1, \cdots, \tilde{e}_n) = 0 \bmod p^d$, we have $G_{i_0, i_1, \cdots, i_n}(e_0, \tilde{e}_1, \cdots, \tilde{e}_n) = 0 \bmod p^d$.

Case B: For any given $(i_0, i_1, \cdots, i_n) \in \mathcal{I}_2$, we define

$$G_{i_0, i_1, \cdots, i_n}(x_0, y_1, \cdots, y_n) = \left(H_{i_0, i_1, \cdots, i_n} + J_{i_0, i_1, \cdots, i_n} + K_{i_0, i_1, \cdots, i_n}\right) \bmod p^d,$$

which is considered to be the corresponding polynomial over $\mathbb{Z}$. Without loss of generality, we let $j_1, \cdots, j_{d+1}$ be integers satisfying $1 \leq j_1 \leq \cdots \leq j_{d+1} \leq n$ and $y_{j_1} y_{j_2} \cdots y_{j_{d+1}} = y_1^{i_1} y_2^{i_2} \cdots y_n^{i_n}$, and

$$H_{i_0, i_1, \cdots, i_n} = \sum_{u=1}^{d+1} \sum_{v=0}^{1} w_{i_0+1, u+v(d+1)} \cdot x_0^v \mathcal{F}_{j_1} \cdots \mathcal{F}_{j_{u-1}} y_{j_u} \mathcal{F}_{j_{u+1}} \cdots \mathcal{F}_{j_{d+1}},$$

$$J_{i_0, i_1, \cdots, i_n} = \sum_{u=1}^{d+1} \sum_{v=0}^{1} w_{i_0+1, u+v(d+1)} \cdot x_0^v \mathcal{F}_{j_1} \cdots \mathcal{F}_{j_{u-1}} C_{j_u} \mathcal{F}_{j_{u+1}} \cdots \mathcal{F}_{j_{d+1}},$$

$$K_{i_0, i_1, \cdots, i_n} = \sum_{u=1}^{d+1} w_{i_0+1, u+(d+1)} \cdot \mathcal{F}_{j_1} \cdots \mathcal{F}_{j_{u-1}} (B_{j_u} - C_{j_u} E_{j_u}) \mathcal{F}_{j_{u+1}} \cdots \mathcal{F}_{j_{d+1}},$$

where the integers B_{j_u}, C_{j_u} and E_{j_u} are the coefficients in the polynomial $\mathcal{F}_{j_u} = A_{j_u} + B_{j_u} x_0 + C_{j_u} x_0^2 + D_{j_u} y_{j_u} + E_{j_u} x_0 y_{j_u} + x_0^2 y_{j_u}$ for $1 \leq u \leq d + 1$, and the integer $w_{i_0+1, m}(1 \leq m \leq 2d + 2)$ is the m-th component of the $(i_0 + 1)$-th row vector in the inverse matrix $\mathbf{W}_{j_1, \cdots, j_{d+1}}$, which is defined in Lemma 3.

For Case B, the desired vector $(e_0, \tilde{e}_1, \cdots, \tilde{e}_n)$ is common root of $H_{i_0, i_1, \cdots, i_n}$, $J_{i_0, i_1, \cdots, i_n}$ and $K_{i_0, i_1, \cdots, i_n}$ modulo p^d. Hence, $G_{i_0, i_1, \cdots, i_n}(e_0, \tilde{e}_1, \cdots, \tilde{e}_n) = 0 \bmod p^d$.

Lemma 5. *Define* $G_{i_0,i_1,\cdots,i_n}(x_0, y_1, \cdots, y_n)$ *and* $\mathcal{I}(n, d, t)$ *as above. Let* $\mathcal{L}(n, d, t)$ *be a lattice spanned by the coefficient vectors of* $G_{i_0,i_1,\cdots,i_n}(x_0 X, y_1 X, \cdots, y_n X)$ *for all* $(i_0, i_1, \cdots, i_n) \in \mathcal{I}(n, d, t)$, *where the value* X *is the upper bound of* $|e_0|, |\tilde{e}_1|, \cdots, |\tilde{e}_n|$. *Then the basis matrix is triangular if the coefficient vectors of* $G_{i_0,i_1,\cdots,i_n}(x_0 X, y_1 X, \cdots, y_n X)$ *are arranged based on the order of the corresponding* $x_0^{i_0} y_1^{i_1} \cdots y_n^{i_n}$ *from low to high. The diagonal elements in the triangular basis matrix of* $\mathcal{L}(n, d, t)$ *are as follows:*

$$\begin{cases} p^{d+1-l} X^{i_0+l} & \text{for } 0 \le l \le d, 0 \le i_0 \le 2l-1, \\ p^{d-l} X^{i_0+l} & \text{for } 0 \le l < d, 2l \le i_0 \le 2d-1, \\ X^{i_0+d+1} & \text{for } l = d+1, 0 \le i_0 \le t. \end{cases} \quad (19)$$

The dimension of $\mathcal{L}(n, d, t)$ is equal to the number of $\mathcal{I}(n, d, t)$. Namely,

$$\dim(\mathcal{L}(n, d, t)) = (t+1) \binom{n}{d+1} + 2d \sum_{l=0}^{d} \binom{n}{l}. \quad (20)$$

The determinant of $\mathcal{L}(n, d, l)$ is equal to

$$\det(\mathcal{L}(n, d, t)) =: X^{\alpha} p^{\beta}, \quad (21)$$

where

$$\alpha = \frac{(2d+t+2)(t+1)}{2} \binom{n}{d+1} + d \sum_{l=0}^{d} (2d-1+2l)\binom{n}{l},$$

$$\beta = 2d^2 \sum_{l=0}^{d} \binom{n}{l} - (2d-2) \sum_{l=0}^{d} l\binom{n}{l}.$$

4.2 Improved Bound

According to the steps in the Coppersmith method in Sect. 2.2, the Coppersmith condition (4) must be satisfied for the polynomials $h_i(x_0, y_1, \ldots, y_n)$ for all $1 \le i \le n+1$, corresponding to the first $n+1$ LLL reduced basis vectors, to contain the desired root $(e_0, \tilde{e}_1, \ldots, \tilde{e}_n)$ over integers. That is,

$$2^{\frac{w(w-1)}{4(w-n)}} \det(\mathcal{L}(n, d, t))^{\frac{1}{w-n}} < \frac{p^d}{\sqrt{w}}, \quad (22)$$

where $w = \dim(\mathcal{L}(n, d, t))$. Once we get the above $n+1$ polynomials h_i's, under Assumption 1, we can compute the wanted root $(e_0, \tilde{e}_1, \ldots, \tilde{e}_n)$ using the Gröbner basis.

Plugging (20) and (21) into (22), we obtain

$$X < \left(2^{-\frac{w(w-1)}{4\alpha}} \cdot w^{-\frac{w-n}{2\alpha}}\right) \cdot p^{S(n,d,t)}, \quad (23)$$

where

$$S(n, d, t) := \frac{d(w-n)-\beta}{\alpha} = \frac{d(t+1)\binom{n}{d+1}+(2d-2)\sum_{l=0}^{d} l\binom{n}{l}-dn}{\frac{(2d+t+2)(t+1)}{2}\binom{n}{d+1}+d\sum_{l=0}^{d}(2d-1+2l)\binom{n}{l}}.$$

For a given sufficiently large $p = 2^{\omega(d^{(2+c)d})}$ for any positive integer d and any constant $c > 0$, the condition (23) can be simplified as

$$X < p^{S(n,d,t)}.$$

By taking integers $t = 0$ and $n = d^{3+c}$, the condition becomes

$$X < p^{1 - \frac{1}{d+1} - \varepsilon}. \tag{24}$$

Here, $\varepsilon = o(\frac{1}{d+1}) = \dfrac{d^2(2d-1)\sum\limits_{l=0}^{d}\binom{n}{l} + 2\sum\limits_{l=0}^{d} l\binom{n}{l} + d(d+1)n}{(d+1)^2\binom{n}{d+1} + d(d+1)(2d-1)\sum\limits_{l=0}^{d}\binom{n}{l} + 2d(d+1)\sum\limits_{l=0}^{d} l\binom{n}{l}} > 0.$

The running time of the LLL algorithm depends on the dimension and the maximal bit size of the input triangular basis matrix. For $t = 0$ and $n = d^{3+c}$, the dimension of $\mathcal{L}(n, d, t)$ is equal to $\binom{n}{d+1} + 2d\sum_{l=0}^{d}\binom{n}{l} = \mathcal{O}(n^{d+1}) = \mathcal{O}(d^{(3+c)d})$, and the bit size of the entries in the triangular basis matrix is bounded by $3d\log_2 p$ from (19). Based on [25], the time complexity of the LLL algorithm is

$$\text{poly}\big(3d\log_2 p, \mathcal{O}(d^{(3+c)d})\big) = \mathcal{O}((\log_2 p)^{\mathcal{O}(1)} d^{\mathcal{O}(d)}) \tag{25}$$

which is polynomial in $\log_2 p$ for any constant d.

The running time of the Gröbner basis computation relies on the degrees and number of variables of input polynomials as well as the size of input polynomials. Based on [13], the time complexity of the Gröbner basis computation for a zero-dimensional system is polynomial in $\max\{S, D^N\} < Nh(eD)^N$, where N is the number of variables, and S is the size of the input polynomials in dense representation, h is the maximal size of the coefficients of the input polynomials, D is arithmetic mean value of the degrees of input polynomials and e is Euler constant. For our lattice $\mathcal{L}(n, d, t)$, when $t = 0$ and $n = d^{3+c}$, the number of variables is $n + 1$, the degree of input polynomials h_i's ($1 \le i \le n+1$) is $3d - 1$ according to (38), and the maximal size h is less than $d\log_2 p$ based on Lemma 2. That is, $N = n + 1$, $D = N(3d - 1)/N = 3d - 1$, and $h < d\log_2 p$. Hence, the time complexity of the Gröbner basis computation is bounded by

$$\text{poly}(Nh(eD)^N) = \mathcal{O}((\log_2 p)^{\mathcal{O}(1)} d^{\mathcal{O}(n)}) \tag{26}$$

which is polynomial in $\log_2 p$ for any constant d. From (25) and (26), the overall complexity is polynomial in $\log_2 p$ for any constant d.

Finally, if any vector $(x_0, \tilde{y}_1, \cdots, \tilde{y}_n) \in \mathbb{Z}^{n+1}$ such that $\mathcal{F}_j(x_0, \tilde{y}_j) = 0 \bmod p$ for all $1 \le j \le n$ in (6), where the upper bound of $|x_0|, |\tilde{y}_1|, \cdots, |\tilde{y}_n|$ satisfies (24), then $(x_0, \tilde{y}_1, \cdots, \tilde{y}_n)$ is also a common root over $\mathbb{Z}$ of the input polynomials $h_1, \cdots, h_{n+1}$ of Gröbner basis computation. The following result shows that the number of these roots is not only limited, but also only one with an overwhelming probability.

Lemma 6. *For a given sufficiently large prime $p = 2^{\omega(d^{(2+c)d})}$ for any positive integer d and any constant $c > 0$, given $n = d^{3+c}$ polynomials $\mathcal{F}_j(x_0, y_j)$ satisfying $\mathcal{F}_j(e_0, \tilde{e}_j) = 0 \bmod p$ for $1 \le j \le n$ in (6), the probability that there is an* integer vector $(e_0', \tilde{e}_1', \cdots, \tilde{e}_n') \neq (e_0, \tilde{e}_1, \cdots, \tilde{e}_n)$, such that $\mathcal{F}_i(e_0', \tilde{e}_i') = 0 \pmod{p}$ for all $1 \leq i \leq n$, where the upper bound of $|e_0'|, |\tilde{e}_1'|, \cdots, |\tilde{e}_n'|$ satisfies (24), does not exceed $\mathcal{O}(\frac{1}{p})$.

According to the above analysis, we get the following result.

Theorem 1. *For a given sufficiently large prime* $p = 2^{\omega(d^{(2+c)d})}$ *for any positive integer* d *and any constant* $c > 0$, *given* $n = d^{3+c}$ *polynomials* $\mathcal{F}_j(x_0, y_j)$ *satisfying* $\mathcal{F}_j(e_0, \tilde{c}_j) - 0 \bmod p$ *for* $1 \leq j \leq n$ *in (6), under Assumption 1, one can compute the desired root* $(e_0, \tilde{e}_1, \cdots, \tilde{e}_n)$, *if the bound* X *of* $|e_0|, |\tilde{e}_1|, \cdots, |\tilde{c}_n|$ *satisfies*

$$X < p^{1 - \frac{1}{d+1} - \varepsilon},$$

where $\varepsilon = o(\frac{1}{d+1}) > 0$. *The overall time complexity is polynomial in* $\log_2 p$ *for any constant* d.

Since $X = p/2^\delta$ for the case of EC-HNP$_x$, we get a new bound for EC-HNP$_x$ from Theorem 1.

Theorem 2. *Define* d, n, p, ε *as in Theorem 1. For* $2n + 1$ *given calls to the oracle* $O_{P,R}(m)$ *in EC-HNP$_x$, under Assumption 1, one can recover the hidden point* P *when the number* δ *of known MSBs satisfies*

$$\frac{\delta}{\log_2 p} > \frac{1}{d+1} + \varepsilon.$$

For the least significant bits (LSBs) case, the problem of solving the corresponding EC-HNP$_x$ can be converted into finding the desired root $(e_0, \tilde{e}_1, \cdots, \tilde{e}_n)$ of the involved polynomials based on [28, Section 6.1]. Note that the forms of these polynomials as well as the size of the desired root are the same as those in (6). Therefore, we obtain the same bound as in the MSBs case.

For the case of ECDH, we get the following result from Theorem 2.

Theorem 3. *Define* d, p *as in Theorem 1. For a given elliptic curve* $\mathcal{E}$ *over the prime field* $\mathbb{F}_p$, *if there is an oracle that outputs about* $\frac{1}{d+1}$ *of the most (least) significant bits of the* x-coordinate of the ECDH key, under Assumption 1, one can compute all the bits in polynomial time.*

5   Proof of Triangular Basis Matrix

First, we present the following relation, which can be utilized to construct triangular basis matrix.

Lemma 7. *Define the matrices* $\mathbf{M}_{j_1, \cdots, j_{d+1}}$ *and* $\mathbf{W}_{j_1, \cdots, j_{d+1}}$ *as in Lemma 3, where* $1 \leq j_1 < \cdots < j_{d+1} \leq n$. *Let* $w_{i_0+1, m}$ *be the entry of the* $(i_0 + 1)$-*th* *row and the m-th column of* $\mathbf{W}_{j_1,\cdots,j_{d+1}}$, *where* $0 \le i_0 \le 2d+1, 1 \le m \le 2d+2$. *Then we have*

$$
\begin{cases}
\sum_{u=1}^{d+1} w_{i_0+1,d+1+u} = 0 \bmod p^d, & \text{for } 0 \le i_0 \le 2d, \\
\sum_{u=1}^{d+1} \left(w_{i_0+1,u} + w_{i_0+1,d+1+u} \sum_{m \ne u} E_{j_m} \right) = 0 \bmod p^d, & \text{for } 0 \le i_0 \le 2d-1,
\end{cases}
\tag{27}
$$

where E_{j_m} *is the coefficient of the polynomial* $\mathcal{F}_{j_m} = A_{j_m} + B_{j_m}x_0 + C_{j_m}x_0^2 + D_{j_m}y_{j_m} + E_{j_m}x_0 y_{j_m} + x_0^2 y_{j_m}$ *for* $1 \le m \le d+1$.

Proof. According to (14), we get that the $(2d+2) \times (2d+2)$ matrix $\mathbf{M}_{j_1,\cdots,j_{d+1}}$ is the following coefficient matrix:

$$
\begin{pmatrix}
\prod_{u \ne 1} (x_0^2 + E_{j_u}x_0 + D_{j_u}) \\
\vdots \\
\prod_{u \ne d+1} (x_0^2 + E_{j_u}x_0 + D_{j_u}) \\
x_0 \prod_{u \ne 1} (x_0^2 + E_{j_u}x_0 + D_{j_u}) \\
\ddots \\
x_0 \prod_{u \ne d+1} (x_0^2 + E_{j_u}x_0 + D_{j_u})
\end{pmatrix}
= \mathbf{M}_{j_1,\cdots,j_{d+1}}
\begin{pmatrix}
1 \\
\vdots \\
x_0^d \\
x_0^{d+1} \\
\vdots \\
x_0^{2d+1}
\end{pmatrix}
\bmod p^d.
\tag{28}
$$

For the sake of discussion, let $\widetilde{F}_{j_m} = \prod_{u \ne m}(x_0^2 + E_{j_u}x_0 + D_{j_u})$ for all $1 \le m \le d+1$. The last column of $\mathbf{M}_{j_1,\cdots,j_{d+1}}$ corresponds to the vector whose elements are respectively the coefficients of x_0^{2d+1} in the following polynomials

$$
\widetilde{F}_{j_1}, \cdots, \widetilde{F}_{j_{d+1}}, x_0 \cdot \widetilde{F}_{j_1}, \cdots, x_0 \cdot \widetilde{F}_{j_{d+1}}.
$$

Note that the coefficient of x_0^{2d+1} in the polynomial $\widetilde{F}_{j_m}$ is 0 for all $1 \le m \le d+1$, and the coefficient of x_0^{2d+1} in the polynomial $x_0\widetilde{F}_{j_m}$ is 1 for all $1 \le m \le d+1$. That is, the last column of $\mathbf{M}_{j_1,\cdots,j_{d+1}}$ is $(0,\cdots,0,1,\cdots,1)^T$, where the number of components 1 is $d+1$. Since $(w_{i_0+1,1},\cdots,w_{i_0+1,2d+2})$ is the (i_0+1)-th row of the inverse matrix $\mathbf{W}_{j_1,\cdots,j_{d+1}}$ modulo p^d, for $0 \le i_0 \le 2d$, we get that

$$
(w_{i_0+1,1},\cdots,w_{i_0+1,2d+2}) \cdot (0,\cdots,0,1,\cdots,1)^T = 0 \bmod p^d,
$$

i.e. $\sum_{u=1}^{d+1} w_{i_0+1,d+1+u} = 0 \bmod p^d$.

The penultimate column of $\mathbf{M}_{j_1,\cdots,j_{d+1}}$ corresponds to the vector whose elements are respectively the coefficients of x_0^{2d} in the following polynomials

$$
\widetilde{F}_{j_1}, \cdots, \widetilde{F}_{j_{d+1}}, x_0 \cdot \widetilde{F}_{j_1}, \cdots, x_0 \cdot \widetilde{F}_{j_{d+1}}.
$$

Note that the coefficient of x_0^{2d} in $\widetilde{F}_{j_m}$ is 1 for all $1 \le m \le d+1$, and the coefficient of x_0^{2d} in $x_0\widetilde{F}_{j_m}$ is $E_{j_1} + \cdots + E_{j_{m-1}} + E_{j_{m+1}} + \cdots + E_{j_{d+1}}$

for $1 \le m \le d+1$. It implies that the penultimate column of $\mathbf{M}_{j_1,\cdots,j_{d+1}}$ is $(1,\cdots,1,\sum_{m\neq 1}E_{j_m},\cdots,\sum_{m\neq d+1}E_{j_m})^T$, where the number of components 1 is $d+1$. Based on $(w_{i_0+1,1},\cdots,w_{i_0+1,2d+2})$ is the (i_0+1)-th row of $\mathbf{W}_{j_1,\cdots,j_{d+1}}$ modulo p^d, for $0 \le i_0 \le 2d-1$, we obtain that

$$(w_{i_0+1,1},\cdots,w_{i_0+1,2d+2}) \cdot (1,\cdots,1,\sum_{m\neq 1}E_{j_m},\cdots,\sum_{m\neq d+1}E_{j_m})^T = 0 \bmod p^d.$$

That is, $\sum_{u=1}^{d+1}\left(w_{i_0+1,u} + w_{i_0+1,d+1+u}\sum_{m\neq u}E_{j_m}\right) = 0 \bmod p^d$.

The above lemma is now used to show the form of $G_{i_0,i_1,\cdots,i_n}(x_0,y_1,\cdots,y_n)$ for $(i_0,i_1,\cdots,i_n) \in \mathcal{I}_2$.

Lemma 8. *Define $G_{i_0,i_1,\cdots,i_n}(x_0,y_1,\cdots,y_n)$ and $\mathcal{I}_1,\mathcal{I}_2$ as in Sect. 4. If the tuple $(i_0,i_1,\cdots,i_n) \in \mathcal{I}_2$, then we have*

$$G_{i_0,i_1,\cdots,i_n} = x_0^{i_0}y_1^{i_1}\cdots y_n^{i_n} + \sum_{(i_0',i_1',\cdots,i_n')\in\mathcal{I}_1} a_{i_0',i_1',\cdots,i_n'}x_0^{i_0'}y_1^{i_1'}\cdots y_n^{i_n'},$$

where $a_{i_0',i_1',\cdots,i_n'} \in \mathbb{Z}$.

Proof. First, we present that the leading term of $G_{i_0,i_1,\cdots,i_n}(x_0,y_1,\cdots,y_n)$ is $x_0^{i_0}y_1^{i_1}\cdots y_n^{i_n}$ for $(i_0,i_1,\cdots,i_n) \in \mathcal{I}_2$. In this case,

$$G_{i_0,i_1,\cdots,i_n} = H_{i_0,i_1,\cdots,i_n} + J_{i_0,i_1,\cdots,i_n} + K_{i_0,i_1,\cdots,i_n}$$

in the sense of modulo p^d. Here,

$$H_{i_0,i_1,\cdots,i_n} = \sum_{u=1}^{d+1}\sum_{v=0}^{1}w_{i_0+1,u+v(d+1)}\cdot x_0^v\mathcal{F}_{j_1}\cdots\mathcal{F}_{j_{u-1}}y_{j_u}\mathcal{F}_{j_{u+1}}\cdots\mathcal{F}_{j_{d+1}},$$

$$J_{i_0,i_1,\cdots,i_n} = \sum_{u=1}^{d+1}\sum_{v=0}^{1}w_{i_0+1,u+v(d+1)}\cdot x_0^v\mathcal{F}_{j_1}\cdots\mathcal{F}_{j_{u-1}}C_{j_u}\mathcal{F}_{j_{u+1}}\cdots\mathcal{F}_{j_{d+1}},$$

$$K_{i_0,i_1,\cdots,i_n} = \sum_{u=1}^{d+1}w_{i_0+1,u}\cdot\mathcal{F}_{j_1}\cdots\mathcal{F}_{j_{u-1}}(B_{j_u}-C_{j_u}E_{j_u})\mathcal{F}_{j_{u+1}}\cdots\mathcal{F}_{j_{d+1}},$$

where integers $1 \le j_1 < \cdots < j_{d+1} \le n$ satisfy $y_{j_1}\cdots y_{j_{d+1}} = y_1^{i_1}\cdots y_n^{i_n}$.

In order to show the case of $H_{i_0,i_1,\cdots,i_n}(x_0,y_1,\cdots,y_n)$, we first consider the following equations:

$$\begin{pmatrix} y_{j_1}\cdot\mathcal{F}_{j_2}\cdots\mathcal{F}_{j_{d+1}} \\ \ddots \\ \mathcal{F}_{j_1}\cdots\mathcal{F}_{j_d}y_{j_{d+1}} \\ x_0\cdot y_{j_1}\mathcal{F}_{j_2}\cdots\mathcal{F}_{j_{d+1}} \\ \ddots \\ x_0\cdot\mathcal{F}_{j_1}\cdots\mathcal{F}_{j_d}y_{j_{d+1}} \end{pmatrix} = \begin{pmatrix} \mathcal{H}_{1,0} \\ \vdots \\ \mathcal{H}_{d+1,0} \\ \mathcal{H}_{1,1} \\ \vdots \\ \mathcal{H}_{d+1,1} \end{pmatrix} + \mathbf{M}_{j_1,\cdots,j_{d+1}}\begin{pmatrix} y_{j_1}\cdot y_{j_2}\cdots y_{j_{d+1}} \\ x_0\cdot y_{j_1}y_{j_2}\cdots y_{j_{d+1}} \\ \vdots \\ x_0^{2d+1}\cdot y_{j_1}y_{j_2}\cdots y_{j_{d+1}} \end{pmatrix} \bmod p^d.$$

$$(29)$$

Here, the matrix $\mathbf{M}_{j_1,\cdots,j_{d+1}}$ is defined in (28), and the polynomial $\mathcal{H}_{u,v}$ ($1 \leq u \leq d+1$, $0 \leq v \leq 1$) is composed of the terms in $x_0^v \mathcal{F}_{j_1} \cdots \mathcal{F}_{j_{u-1}} y_{j_u} \mathcal{F}_{j_{u+1}} \cdots \mathcal{F}_{j_{d+1}}$ except the terms of monomials

$$y_{j_1} \cdots y_{j_{d+1}}, x_0 y_{j_1} \cdots y_{j_{d+1}}, \cdots, x_0^{2d+1} y_{j_1} \cdots y_{j_{d+1}}.$$

It implies that the leading monomial in $\mathcal{H}_{u,v}$ is $x_0^{i_0'} y_{k_1} \cdots y_{k_m}$, where $0 \leq i_0' \leq 2d+1$ and $\{k_1, \cdots, k_m\} \subsetneq \{j_1, \cdots, j_{d+1}\}$. Hence, $m < d+1$. According to the order (10), we get

$$x_0^{i_0'} y_{k_1} \cdots y_{k_m} \prec y_{j_1} \cdots y_{j_{d+1}} \prec x_0 y_{j_1} \cdots y_{j_{d+1}} \prec \cdots \prec x_0^{2d+1} y_{j_1} \cdots y_{j_{d+1}}. \quad (30)$$

Note that $\mathbf{W}_{j_1,\cdots,j_{d+1}}$ is the inverse matrix of $\mathbf{M}_{j_1,\cdots,j_{d+1}}$ modulo p^d. Multiplying the two sides of Eq. (29) by $\mathbf{W}_{j_1,\cdots,j_{d+1}}$ to the left, we get

$$\mathbf{W}_{j_1,\cdots,j_{d+1}} \begin{pmatrix} y_{j_1} \cdot \mathcal{F}_{j_2} \cdots \mathcal{F}_{j_{d+1}} \\ \ddots \\ \mathcal{F}_{j_1} \cdots \mathcal{F}_{j_d} y_{j_{d+1}} \\ x_0 \cdot y_{j_1} \mathcal{F}_{j_2} \cdots \mathcal{F}_{j_{d+1}} \\ \ddots \\ x_0 \cdot \mathcal{F}_{j_1} \cdots \mathcal{F}_{j_d} y_{j_{d+1}} \end{pmatrix} = \mathbf{W}_{j_1,\cdots,j_{d+1}} \begin{pmatrix} \mathcal{H}_{1,0} \\ \vdots \\ \mathcal{H}_{d+1,0} \\ \mathcal{H}_{1,1} \\ \vdots \\ \mathcal{H}_{d+1,1} \end{pmatrix} + \begin{pmatrix} y_{j_1} \cdot y_{j_2} \cdots y_{j_{d+1}} \\ x_0 \cdot y_{j_1} y_{j_2} \cdots y_{j_{d+1}} \\ \vdots \\ x_0^{2d+1} \cdot y_{j_1} y_{j_2} \cdots y_{j_{d+1}} \end{pmatrix}$$
$$(31)$$

(in the sense of modulo p^d). Since $(w_{i_0+1,1}, \cdots, w_{i_0+1,2d+2}$ is the (i_0+1)-th row of $\mathbf{W}_{j_1,\cdots,j_{d+1}}$, where $0 \leq i_0 \leq t$, from (31), we have

$$H_{i_0,i_1,\cdots,i_n} = \sum_{u=1}^{d+1} \sum_{v=0}^{1} w_{i_0+1,u+(d+1)v} \cdot x_0^v \mathcal{F}_{j_1} \cdots \mathcal{F}_{j_{u-1}} y_{j_u} \mathcal{F}_{j_{u+1}} \cdots \mathcal{F}_{j_{d+1}}$$
$$= x_0^{i_0} y_{j_1} y_{j_2} \cdots y_{j_{d+1}} + \sum_{u=1}^{d+1} \sum_{v=0}^{1} w_{i_0+1,u+(d+1)v} \mathcal{H}_{u,v} \bmod p^d. \quad (32)$$

Based on $x_0^{i_0} y_{j_1} y_{j_2} \cdots y_{j_{d+1}} = x_0^{i_0} y_1^{i_1} \cdots y_n^{i_n}$ and (30), we obtain that $x_0^{i_0} y_1^{i_1} \cdots y_n^{i_n}$ is the leading term of $H_{i_0,i_1,\cdots,i_n}$. Moreover, all monomials except $x_0^{i_0} y_1^{i_1} \cdots y_n^{i_n}$ in $H_{i_0,i_1,\cdots,i_n}$ belong to the set

$$\{x_0^{i_0'} y_1^{i_1'} \cdots y_n^{i_n'} \mid 0 \leq i_0' \leq 2d+1, 0 \leq i_1', \cdots, i_n' \leq 1, 0 \leq i_1' + \cdots + i_n' \leq d\}. \quad (33)$$

For the case of $J_{i_0,i_1,\cdots,i_n}$, let $x_0^{r_0} y_{s_1} \cdots y_{s_m}$ be the leading monomial of $J_{i_0,i_1,\cdots,i_n}$, where $0 \leq r_0 \leq 2d+1$ and $\{s_1, \cdots, s_m\} \subsetneq \{j_1, \cdots, j_{d+1}\}$. Thus, $m < d+1$. Based on the order (10), we get $x_0^{r_0} y_{s_1} \cdots y_{s_m} \prec x_0^{i_0} y_{j_1} \cdots y_{j_{d+1}}$. That is, $x_0^{r_0} y_{s_1} \cdots y_{s_m} \prec x_0^{i_0} y_1^{i_1} \cdots y_n^{i_n}$.

Similarly, we can also prove that the order of the leading monomial of $K_{i_0,i_1,\cdots,i_n}$ is less than the order of $x_0^{i_0} y_1^{i_1} \cdots y_n^{i_n}$.

To sum up, we get that $x_0^{i_0} y_1^{i_1} \cdots y_n^{i_n}$ is the leading term of $G_{i_0,i_1,\cdots,i_n}(x_0, y_1, \cdots, y_n)$. In addition, all monomials except the leading monomial $x_0^{i_0} y_1^{i_1} \cdots y_n^{i_n}$ in $G_{i_0,i_1,\cdots,i_n}$ lie in the set (33).

Then, we prove that $G_{i_0,i_1,\cdots,i_n}(x_0, y_1, \cdots, y_n)$ does not contain any term related to x_0^{2d+1} and x_0^{2d}. It means that all monomials except $x_0^{i_0} y_1^{i_1} \cdots y_n^{i_n}$ in

$G_{i_0,i_1,\cdots,i_n}$ lie in $\{x_0^{i'_0} y_1^{i'_1} \cdots y_n^{i'_n} \mid (i'_1,\cdots,i'_n) \in \mathcal{I}_1\}$. That is, we can rewrite $G_{i_0,i_1,\cdots,i_n}$ as

$$x_0^{i_0} y_1^{i_1} \cdots y_n^{i_n} + \sum_{(i'_0,i'_1,\cdots,i'_n) \in \mathcal{I}_1} a_{i'_0,i'_1,\cdots,i'_n} x_0^{i'_0} y_1^{i'_1} \cdots y_n^{i'_n},$$

where $a_{i'_0,i'_1,\cdots,i'_n} \in \mathbb{Z}$, and $\mathcal{I}_1 = \{(i'_0,i'_1,\cdots,i'_n) \mid 0 \le i'_0 \le 2d-1, 0 \le i'_1,\cdots,i'_n \le 1, 0 \le i'_1 + \cdots + i'_n \le d\}$.

For the convenience of subsequent analysis, we rewrite $\mathcal{F}_{j_u} = A_{j_u} + B_{j_u} x_0 + C_{j_u} x_0^2 + D_{j_u} y_{j_u} + E_{j_u} x_0 y_{j_u} + x_0^2 y_{j_u}$ as

$$x_0^2 (y_{j_u} + C_{j_u}) + x_0 (E_{j_u} y_{j_u} + B_{j_u}) + (D_{j_u} y_{j_u} + A_{j_u}), 1 \le u \le d+1.$$

We rewrite $G_{i_0,i_1,\cdots,i_n}$ for $(i_0,i_1,\cdots,i_n) \in \mathcal{I}_2$ as

$$G_{i_0,i_1,\cdots,i_n} = T_1 + T_2 + T_3$$

in the sense of modulo p^d, where

$$T_1 := \sum_{u=1}^{d+1} w_{i_0+1,u+(d+1)} \cdot x_0 \mathcal{F}_{j_1} \cdots \mathcal{F}_{j_{u-1}} (y_{j_u} + C_{j_u}) \mathcal{F}_{j_{u+1}} \cdots \mathcal{F}_{j_{d+1}}$$

$$T_2 := \sum_{u=1}^{d+1} w_{i_0+1,u} \cdot \mathcal{F}_{j_1} \cdots \mathcal{F}_{j_{u-1}} (y_{j_u} + C_{j_u}) \mathcal{F}_{j_{u+1}} \cdots \mathcal{F}_{j_{d+1}}$$

$$T_3 := \sum_{u=1}^{d+1} w_{i_0+1,u+(d+1)} \cdot \mathcal{F}_{j_1} \cdots \mathcal{F}_{j_{u-1}} (D_{j_u} - C_{j_u} E_{j_u}) \mathcal{F}_{j_{u+1}} \cdots \mathcal{F}_{j_{d+1}}.$$

Since $\deg(x_0) = 2$ in $\mathcal{F}_{j_u}$ for $1 \le u \le d+1$, we have that $\deg(x_0) \le 2d+1$ for T_1, and $\deg(x_0) \le 2d$ for T_2 and T_3.

We can deduce that the x_0^{2d+1}-related term in $G_{i_0,i_1,\cdots,i_n}$ only appears in T_1. Specifically, the x_0^{2d+1}-related term is

$$\sum_{u=1}^{d+1} w_{i_0+1,u+(d+1)} \cdot x_0^{2d+1} (y_{j_1} + C_{j_1}) \cdots (y_{j_{d+1}} + C_{j_{d+1}})$$

in sense of modulo p^d. According to (27), we have $\sum_{u=1}^{d+1} w_{i_0+1,u+d+1} = 0 \bmod p^d$, where $0 \le i_0 \le 2d-1$. Therefore, $G_{i_0,i_1,\cdots,i_n}$ does not have any term related to x_0^{2d+1}.

We can deduce that the x_0^{2d}-related term in $G_{i_0,i_1,\cdots,i_n}$ appears in T_1, T_2 and T_3.

For the case of $T_1 = \sum_{u=1}^{d+1} w_{i_0+1,u+(d+1)} \cdot x_0 \mathcal{F}_{j_1} \cdots \mathcal{F}_{j_{u-1}} (y_{j_u} + C_{j_u}) \mathcal{F}_{j_{u+1}} \cdots \mathcal{F}_{j_{d+1}}$, based on $\mathcal{F}_{j_u} = x_0^2(y_{j_u} + C_{j_u}) + x_0(E_{j_u}(y_{j_u} + C_{j_u}) + (B_{j_u} - C_{j_u} E_{j_u})) + (A_{j_u} + D_{j_u} y_{j_u})$ for $1 \le u \le d+1$, the x_0^{2d}-related term of T_1 is

$$\sum_{u=1}^{d+1} w_{i_0+1,u+(d+1)} (\sum_{m \ne u} E_{j_m}) \cdot x_0^{2d} (y_{j_1} + C_{j_1}) \cdots (y_{j_{d+1}} + C_{j_{d+1}})$$
$$+ \sum_{u=1}^{d+1} (\sum_{m \ne u} w_{i_0+1,m+(d+1)}) \cdot x_0^{2d} (B_{j_u} - C_{j_u} E_{j_u}) \prod_{m \ne u} (y_{j_m} + C_{j_m}). \tag{34}$$

For the case of $\mathcal{T}_2 = \sum_{u=1}^{d+1} w_{i_0+1,u} \cdot \mathcal{F}_{j_1} \cdots \mathcal{F}_{j_{u-1}}(y_{j_u} + C_{j_u})\mathcal{F}_{j_{u+1}} \cdots \mathcal{F}_{j_{d+1}}$, the x_0^{2d}-related term of $\mathcal{T}_2$ is

$$\sum_{u=1}^{d+1} w_{i_0+1,u} \cdot x_0^{2d}(y_{j_1} + C_{j_1}) \cdots (y_{j_{d+1}} + C_{j_{d+1}}). \tag{35}$$

For the case of $\mathcal{T}_3 = \sum_{u=1}^{d+1} w_{i_0+1,u+(d+1)} \cdot \mathcal{F}_{j_1} \cdots \mathcal{F}_{j_{u-1}}(B_{j_u} - C_{j_u}E_{j_u})\mathcal{F}_{j_{u+1}} \cdots \mathcal{F}_{j_{d+1}}$, the x_0^{2d}-related term of $\mathcal{T}_3$ is

$$\sum_{u=1}^{d+1} w_{i_0+1,u+(d+1)} \cdot x_0^{2d}(B_{j_u} - C_{j_u}E_{j_u}) \prod_{m \neq u} (y_{j_m} + C_{j_m}). \tag{36}$$

According to (34), (35) and (36), we get that the x_0^{2d}-related term in $G_{i_0,i_1,\cdots,i_n}$ is equal to

$$\begin{aligned}
&\sum_{u=1}^{d+1}(\sum_{u=1}^{d+1} w_{i_0+1,d+1+u}) \cdot x_0^{2d}(B_{j_u} - C_{j_u}E_{j_u}) \prod_{m \neq u} (y_{j_m} + C_{j_m}) \\
&+ \sum_{u=1}^{d+1}(w_{i_0+1,u} + w_{i_0+1,d+1+u} \sum_{m \neq u} E_{j_m}) \cdot x_0^{2d}(y_{j_1} + C_{j_1}) \cdots (y_{j_{d+1}} + C_{j_{d+1}})
\end{aligned} \tag{37}$$

in sense of modulo p^d. According to (27), we have that $\sum_{u=1}^{d+1} w_{i_0+1,d+1+u} = 0 \pmod{p^d}$ and $\sum_{u=1}^{d+1} \left(w_{i_0+1,u} + w_{i_0+1,d+1+u} \sum_{m \neq u} E_{j_m} \right) = 0 \pmod{p^d}$ for $0 \leq i_0 \leq 2d - 1$. Hence, $G_{i_0,i_1,\cdots,i_n}$ does not have any term related to x_0^{2d}, where $(i_0,i_1,\cdots,i_n) \in \mathcal{I}_2$.

Finally, we show that the involved basis matrix of $\mathcal{L}(n,d,t)$ is triangular. That is, we provide proof for Lemma 5.

Proof. First, we present that the leading term of $G_{i_0,i_1,\cdots,i_n}(x_0,y_1,\cdots,y_n)$ is $x_0^{i_0} y_1^{i_1} \cdots y_n^{i_n}$ for $(i_0,i_1,\cdots,i_n) \in \mathcal{I}(n,d,t)$. We respectively consider **Case A** and **Case B**.

For **Case A**, the corresponding $(i_0,i_1,\cdots,i_n) \in \mathcal{I}_1$. We define

$$G_{i_0,i_1,\cdots,i_n}(x_0,y_1,\cdots,y_n) = F_{i_0,i_1,\cdots,i_n}(x_0,y_1,\cdots,y_n)$$

From Lemma 4, and $\mathcal{I}_1 \subset \mathcal{I}_{[\text{XHS20}]}(n,d)$, we obtain that the leading term of $G_{i_0,i_1,\cdots,i_n}(x_0,y_1,\cdots,y_n)$ is as follows:

$$\begin{cases} p^{d+1-l} x_0^{i_0} y_1^{i_1} \cdots y_n^{i_n} & \text{for } 1 \leq l \leq d \text{ and } 0 \leq i_0 \leq 2l - 1, \\ p^{d-l} x_0^{i_0} y_1^{i_1} \cdots y_n^{i_n} & \text{for } 0 \leq l < d \text{ and } 2l \leq i_0 \leq 2d - 1. \end{cases}$$

For **Case B**, the corresponding $(i_0,i_1,\cdots,i_n) \in \mathcal{I}_2$. From Lemma 8, we get that the leading term of $G_{i_0,i_1,\cdots,i_n}(x_0,y_1,\cdots,y_n)$ is $x_0^{i_0} y_1^{i_1} \cdots y_n^{i_n}$, where $l = i_1 + \cdots + i_n = d + 1$ and $0 \leq i_0 \leq t$.

To sum up, the leading term of $G_{i_0,i_1,\cdots,i_n}(x_0,y_1,\cdots,y_n)$ is equal to

$$\begin{cases} p^{d+1-l} x_0^{i_0} y_1^{i_1} \cdots y_n^{i_n} & \text{for } 1 \leq l \leq d \text{ and } 0 \leq i_0 \leq 2l - 1, \\ p^{d-l} x_0^{i_0} y_1^{i_1} \cdots y_n^{i_n} & \text{for } 0 \leq l < d \text{ and } 2l \leq i_0 \leq 2d - 1, \\ x_0^{i_0} y_1^{i_1} \cdots y_n^{i_n} & \text{for } l = d + 1 \text{ and } 0 \leq i_0 \leq t. \end{cases} \tag{38}$$

Next, we prove that the basis matrix of $\mathcal{L}(n,d,t)$ can be arranged into a triangular matrix. Since the basis matrix of $\mathcal{L}(n,d,t)$ is made up of the coefficient vectors of polynomials $G_{i_0,i_1,\cdots,i_n}(x_0X, y_1X,\cdots,y_nX)$ for all $(i_0,i_1,\cdots,i_n) \in \mathcal{I}(n,d,t)$, and there is a one-to-one correspondence between the polynomial $G_{i_0,i_1,\cdots,i_n}(x_0,y_1,\cdots,y_n)$ and the corresponding polynomial $G_{i_0,i_1,\cdots,i_n}(x_0X, y_1X, \cdots, y_nX)$, our goal translates to show that $G_{i_0,i_1,\cdots,i_n}(x_0,y_1,\cdots,y_n)$ for all $(i_0,i_1,\cdots,i_n) \in \mathcal{I}(n,d,t)$ form a triangular matrix.

For the level $l = 0$, the corresponding polynomial $G_{i_0,i_1,\cdots,i_n}(x_0,y_1,\cdots,y_n)$ is equal to $p^d x_0^{i_0}$ for $i_0 = 0,1,\cdots,2d-1$. From the order (10), we have $p^d \prec p^d x_0 \prec \cdots \prec p^d x_0^{2d-1}$. It implies that all $G_{i_0,i_1,\cdots,i_n}(x_0,y_1,\cdots,y_n)$ for $l = 0$ generate a triangular matrix. The remaining proof is inductive. For any fixed tuple $(i_0,i_1,\cdots,i_n) \in \mathcal{I}(n,d,t)$, suppose that all polynomials $G_{i_0',i_1',\cdots,i_n'}(x_0,y_1,\cdots,y_n)$, satisfying $x_0^{i_0'} y_1^{i_1'} \cdots y_n^{i_n'} \prec x_0^{i_0} y_1^{i_1} \cdots y_n^{i_n}$, have produced a triangular matrix as stated in Lemma 5. Then we prove that all polynomials added after the polynomial $G_{i_0,i_1,\cdots,i_n}(x_0,y_1,\cdots,y_n)$ still form a triangular matrix. Based on the above analysis, $x_0^{i_0} y_1^{i_1} \cdots y_n^{i_n}$ is the leading monomial of the polynomial $G_{i_0,i_1,\cdots,i_n}(x_0, y_1,\cdots,y_n)$. Let $x_0^{k_0} y_1^{k_1} \cdots y_n^{k_n}$ be any given monomial of $G_{i_0,i_1,\cdots,i_n}(x_0,y_1,\cdots,y_n)$ other than the leading monomial $x_0^{i_0} y_1^{i_1} \cdots y_n^{i_n}$. Obviously, we have $x_0^{k_0} y_1^{k_1} \cdots y_n^{k_n} \prec x_0^{i_0} y_1^{i_1} \cdots y_n^{i_n}$. Since $x_0^{k_0} y_1^{k_1} \cdots y_n^{k_n}$ is the leading monomial of polynomial $G_{k_0,k_1,\cdots,k_n}(x_0,y_1,\cdots,y_n)$, we get that all monomials except $x_0^{i_0} y_1^{i_1} \cdots y_n^{i_n}$ already appeared in the diagonals of a triangular matrix. Thus, all polynomials after $G_{i_0,i_1,\cdots,i_n}(x_0,y_1,\cdots,y_n)$ is added still produce a triangular matrix. To summarize, the basis matrix of $\mathcal{L}(n,d,t)$ is triangular according to the order of $x_0^{i_0} y_1^{i_1} \cdots y_n^{i_n}$ for all $(i_0,i_1,\cdots,i_n) \in \mathcal{I}(n,d,t)$ from low to high.

The diagonal elements in the triangular basis matrix of $\mathcal{L}(n,d,t)$ are all from the leading coefficients of $G_{i_0,i_1,\cdots,i_n}(x_0X, y_1X,\cdots,y_nX)$ for $(i_0,i_1,\cdots,i_n) \in \mathcal{I}(n,d,t)$. Based on (38), the diagonal elements of triangular basis matrix are as follows:

$$\begin{cases} p^{d+1-l}X^{i_0+l} & \text{for } 1 \le l \le d \text{ and } 0 \le i_0 \le 2l-1, \\ p^{d-l}X^{i_0+l} & \text{for } 0 \le l < d \text{ and } 2l \le i_0 \le 2d-1, \\ X^{i_0+d+1} & \text{for } l = d+1 \text{ and } 0 \le i_0 \le t. \end{cases}$$

6   Comparison with the Existing Work

Figure 1 compares the theoretical upper bound X for the lattice in Sect. 4.1 and that in [32]. We can see that our lattice is significantly better than that in [32]. In Fig. 1, we take the smallest lattice dimension among different n,d,t for the fixed upper bound. For example, to cross the bound 0.45, the minimum lattice is 940 ($n = 13, d = 2, t = 1$) whereas the minimum dimension in [32] is $2^{39.06}$ ($n = 40, d = 13$).

In Table 1, we present a theoretical comparison of the smallest lattice dimension on the fixed percentage $\delta/\log_2 p$ for a sufficiently large $p = 2^{\omega(d^{(2+c)d})}$. The symbol "−" means that even with a huge lattice dimension, the corresponding $\delta/\log_2 p \leq 0.50$ can not be obtained.

From the second row of Table 1, we can see that in order to reach the 0.60 bound of $\delta/\log_2 p$, the smallest dimension of [32] is 394995 ($n = 16, d = 7$), and the smallest dimensions of our lattice is 326 ($n = 24, d = 1, t = 0$). Therefore, our lattice is practical, while the lattice in [32] is not practical.

Based on the fourth row of Table 1, the smallest lattice dimension is 2879 ($n = 23, d = 2, t = 0$) to obtain the 0.50 bound of $\delta/\log_2 p$. The LLL algorithm terminates within $\mathcal{O}(w^{4+\gamma}b^{1+\gamma})$ bit operations for any $\gamma > 0$ [25], where w is the lattice dimension, and b is the maximal bit-size in the input basis matrix. For $w = 2879$, $w^4 \approx 2^{46}$. The bit-size b for our lattice is bounded by $3d\log_2 p$ (see (19) in Lemma 5). Hence, for a sufficiently large p, it takes a considerable amount of time for the LLL algorithm to output the desired short vector.

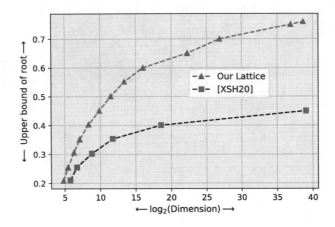

Fig. 1. Comparison of the theoretical upper bound of the root for different dimensions.

Table 1. Comparison of the smallest dimensions for known bit percentages.

$\delta/\log_2 p$	Our		[32]	
	Lattice in Sect. 4.1		Lattice	
	(n, d, t)	Dimension	(n, d)	Dimension
0.65	(15,1,0)	137	(10,4)	3474
0.60	(24,1,0)	326	(16,7)	394995
0.55	(13,2,1)	940	(40,13)	$2^{39.06}$
0.50	(23,2,0)	2879	−	−
0.45	(37,2,0)	10586	−	−
0.40	(71,2,0)	67383	−	−

7   Experiments

We have implemented our experiments in SAGE 9.3 using Linux Ubuntu with Intel® Core™i7-7920HQ CPU 3.67 GHz. We have used the L^2 algorithm [26] for lattice reduction. We tested the algorithm up to lattice dimension 298. In our experiments, the zero-dimensional ideal assumption, i.e. Assumption 1 is always valid. Our experimental results are shown in Table 2. We run 100 experiments for each parameter.

Table 2. Experimental results of Sect. 4.1 on NIST curves. From Equation (23), the required bounds is $X < p^{S(n,d,t)}$ for the lattice $\mathcal{L}(n,d,t)$. Thus the number of known bits should be lower bounded by $(1 - S(n,d,t))\log_2 p$. The column of Theo. represents this value. The column of Exp. gives corresponding experimental values.

Curve	n	d	t	Dim.	Theo.	Exp.	Given	Known MSBs			Known LSBs		
								Suc.	LLL (sec.)	GB (sec.)	Suc.	LLL (sec.)	GB (sec.)
NIST-192				143	132	69%	94%	0.51	0.04	96%	0.53	0.04
NIST-224				167	154	69%	99%	0.51	0.04	95%	0.64	0.05
NIST-256	6	1	1	44	191	176	69%	100%	0.57	0.05	100%	0.65	0.05
NIST-384				286	263	68%	100%	0.74	0.07	100%	0.99	0.07
NIST-521				388	357	69%	100%	1.06	0.08	100%	1.23	0.11
NIST-192				137	125	65%	100%	2.26	0.11	100%	2.41	0.11
NIST-224				160	145	65%	100%	2.44	0.15	100%	2.92	0.12
NIST-256	10	1	0	67	182	165	64%	100%	2.79	0.13	100%	3.13	0.13
NIST-384				272	245	64%	100%	4.06	0.17	100%	4.97	0.19
NIST-521				371	330	63%	100%	6.49	0.23	100%	6.60	0.23
NIST-192				135	129	67%	100%	10.64	0.17	100%	10.08	0.18
NIST-224				157	150	67%	100%	13.86	0.18	100%	13.54	0.21
NIST-256	5	2	1	84	180	172	67%	100%	18.78	0.21	100%	18.92	0.23
NIST-384				269	256	67%	100%	32.69	0.28	100%	31.92	0.36
NIST-521				365	347	67%	100%	38.43	0.34	100%	38.67	0.37
NIST-192				129	120	63%	100%	14.44	0.40	100%	11.90	0.33
NIST-224				150	139	62%	100%	17.17	0.49	100%	14.12	0.39
NIST-256	13	1	0	106	172	159	62%	100%	18.17	0.56	100%	17.09	0.43
NIST-384				257	235	61%	100%	26.69	0.76	100%	27.20	0.58
NIST-521				349	320	61%	100%	41.83	0.92	100%	42.51	0.78
NIST-192				135	130	68%	100%	19.12	0.34	100%	22.64	0.36
NIST-224				158	152	68%	100%	25.70	0.42	100%	26.76	0.41
NIST-256	6	2	0	108	180	174	68%	100%	29.42	0.48	100%	31.77	0.45
NIST-384				270	263	68%	100%	49.65	0.65	100%	52.67	0.59
NIST-521				366	360	69%	100%	78.84	0.82	100%	80.13	0.73
NIST-192				123	116	60%	99%	47.61	1.27	98%	48.77	1.00
NIST-224				144	135	60%	100%	54.27	1.39	100%	55.35	1.12
NIST-256	16	1	0	154	164	155	61%	100%	66.70	1.45	100%	67.10	1.21
NIST-384				246	230	60%	100%	119.05	2.13	100%	118.08	1.79
NIST-521				334	310	60%	100%	164.07	2.73	100%	166.56	2.03
NIST-192				130	126	66%	99%	111.52	1.27	99%	114.83	0.98
NIST-224				152	148	66%	100%	133.61	1.29	100%	138.78	1.17
NIST-256	7	2	0	151	174	168	66%	100%	145.50	1.52	100%	147.39	1.25
NIST-384				260	253	66%	100%	264.65	1.97	100%	262.15	1.65
NIST-521				353	340	65%	100%	357.88	2.53	100%	363.22	2.07

(*continued*)

Table 2. (*continued*)

Curve	n	d	t	Dim.	Theo.	Exp.	Given	Known MSBs			Known LSBs		
								Suc.	LLL (sec.)	GB (sec.)	Suc.	LLL (sec.)	GB (sec.)
NIST-192				135	128	67%	100%	59.41	0.27	100%	64.74	0.22
NIST-224				158	150	67%	100%	64.67	0.29	100%	67.65	0.24
NIST-256	5	3	0	161	180	170	66%	100%	73.62	0.33	100%	71.92	0.27
NIST-384				270	255	66%	100%	120.58	0.43	100%	124.39	0.37
NIST-521				367	345	66%	100%	175.77	0.51	100%	176.14	0.46
NIST-192				134	125	65%	100%	82.25	0.21	100%	84.92	0.20
NIST-224				156	145	65%	100%	88.77	0.27	100%	89.34	0.23
NIST-256	5	3	1	166	178	166	65%	100%	100.87	0.29	100%	104.57	0.25
NIST-384				267	250	65%	100%	144.94	0.41	100%	140.31	0.34
NIST-521				361	339	65%	100%	211.27	0.51	100%	214.37	0.41
NIST-192				132	124	65%	100%	94.37	0.21	99%	98.16	0.20
NIST-224				154	144	64%	95%	106.45	0.22	95%	107.29	0.22
NIST-256	5	3	2	171	176	165	64%	100%	106.31	0.25	100%	103.60	0.24
NIST-384				264	247	64%	100%	175.18	0.34	100%	170.94	0.34
NIST-521				358	335	64%	100%	260.96	0.42	100%	263.96	0.42
NIST-192				118	114	59%	97%	320.58	4.30	95%	313.52	4.19
NIST-224				137	132	59%	94%	444.92	4.78	94%	452.65	4.79
NIST-256	21	1	0	254	157	152	59%	100%	524.03	5.21	100%	544.92	5.22
NIST-384				235	225	59%	100%	864.33	7.11	100%	880.24	6.82
NIST-521				318	301	58%	100%	1272.32	9.37	100%	1280.23	9.50

We always get more than $\frac{w}{2}$ polynomials that satisfy the desired root over $\mathbb{Z}$ after lattice reduction, where w is the dimension of the lattice. Intermediate coefficient swell is a well-known difficulty for computing Gröbner bases over integers. To overcome this problem, we compute Gröbner basis over small prime fields $GF(q)$ such that the product of these primes is larger than the size of unknown values. Then we use the Chinese Remainder Theorem to find the desired root. Using this method, we can find the root after lattice reduction in a few seconds for all parameters. If X is the upper bound of root, we need to consider primes up to N such that $\prod_{\text{prime } q \leq N} q > X$. Since $\prod_{\text{prime } q \leq N} q = e^{\theta(N)}$, we need $e^{\theta(N)} > X$, where $\theta(N) = \sum_{\text{prime } q \leq N} \log q$ is the first Chebyshev function. Since $\theta(N)$ asymptotically approaches to N for large values of N, considering first $\log_e X$ many prime fields will be sufficient for large N for our attack.

After Gröbner basis computation, we get polynomials of the form $x_0 - e_0, y_1 - \tilde{e}_1, y_2 - \tilde{e}_2, \ldots, y_n - \tilde{e}_n$ in $GF(q)$. Let $T = \prod_{q \leq N} q$. Hence using Chinese Remainder Theorem we get $\hat{e}_i \equiv e_i \bmod T$ for $i \in [0, n]$. Thus $e_i = \hat{e}_i$ or $e_i = \hat{e}_i - T$. Hence we can easily collect secrets. We always collect the root for our theoretical values. In fact, experimentally we are able to cross these bounds. In these situations also, success rate is close to 100% in all cases.

One can see from Table 2 that it is possible to find the hidden point P by querying the oracle $2n + 1 = 2 \cdot 21 + 1 = 43$ times for the case of NIST-521 and $(n, d, t) = (21, 1, 0)$. Theoretically, knowing 318 MSBs/LSBs of the x-coordinate of $P + [m]R$ in each query should be sufficient for our attack, where the x-coordinate has 521 bits in total. In practice, we are getting better results. Experimentally, knowledge of 301 bits is sufficient to find the hidden point.

Xu et al. [32] used a dimension 294 lattice to recover the hidden point when the number of exposed bits is 333 (see the last row of [32, Table 1], where $333 \approx 0.64 \cdot 521$). Here using a 254-dimension lattice, we can recover the hidden point when the number of exposed bits is 301.

Acknowledgment. The authors would like to thank anonymous reviewers for their helpful comments and suggestions. Jun Xu and Lei Hu was supported the National Natural Science Foundation of China (Grants 61732021, 62272454). Huaxiong Wang was supported by the National Research Foundation, Singapore under its Strategic Capability Research Centres Funding Initiative and Singapore Ministry of Education under Research Grant MOE2019-T2-2-083.

References

1. Akavia, A.: Solving hidden number problem with one bit oracle and advice. In: Halevi, S. (ed.) CRYPTO 2009. LNCS, vol. 5677, pp. 337–354. Springer, Heidelberg (2009). https://doi.org/10.1007/978-3-642-03356-8_20

2. Albrecht, M.R., Heninger, N.: On bounded distance decoding with predicate: breaking the "Lattice Barrier" for the hidden number problem. In: Canteaut, A., Standaert, F.-X. (eds.) EUROCRYPT 2021. LNCS, vol. 12696, pp. 528–558. Springer, Cham (2021). https://doi.org/10.1007/978-3-030-77870-5_19

3. Boneh, D.: The decision Diffie-Hellman problem. In: Buhler, J.P. (ed.) ANTS 1998. LNCS, vol. 1423, pp. 48–63. Springer, Heidelberg (1998). https://doi.org/10.1007/BFb0054851

4. Boneh, D., Halevi, S., Howgrave-Graham, N.: The modular inversion hidden number problem. In: Boyd, C. (ed.) ASIACRYPT 2001. LNCS, vol. 2248, pp. 36–51. Springer, Heidelberg (2001). https://doi.org/10.1007/3-540-45682-1_3

5. Boneh, D., Shparlinski, I.E.: On the unpredictability of bits of the elliptic curve Diffie-Hellman scheme. In: Kilian, J. (ed.) CRYPTO 2001. LNCS, vol. 2139, pp. 201–212. Springer, Heidelberg (2001). https://doi.org/10.1007/3-540-44647-8_12

6. Boneh, D., Venkatesan, R.: Hardness of computing the most significant bits of secret keys in Diffie-Hellman and related schemes. In: Koblitz, N. (ed.) CRYPTO 1996. LNCS, vol. 1109, pp. 129–142. Springer, Heidelberg (1996). https://doi.org/10.1007/3-540-68697-5_11

7. Boneh, D., Venkatesan, R.: Rounding in lattices and its cryptographic applications. In: Saks, M.E. (ed.) Proceedings of the Eighth Annual ACM-SIAM Symposium on Discrete Algorithms, 5–7 January 1997, New Orleans, Louisiana, USA, pp. 675–681. ACM/SIAM (1997)

8. Coppersmith, D.: Finding a small root of a bivariate integer equation; factoring with high bits known. In: Maurer, U. (ed.) EUROCRYPT 1996. LNCS, vol. 1070, pp. 178–189. Springer, Heidelberg (1996). https://doi.org/10.1007/3-540-68339-9_16

9. Coppersmith, D.: Finding a small root of a univariate modular equation. In: Maurer, U. (ed.) EUROCRYPT 1996. LNCS, vol. 1070, pp. 155–165. Springer, Heidelberg (1996). https://doi.org/10.1007/3-540-68339-9_14

10. Coron, J.-S., Zeitoun, R.: Improved factorization of $N = p^r q^s$. In: Smart, N.P. (ed.) CT-RSA 2018. LNCS, vol. 10808, pp. 65–79. Springer, Cham (2018). https://doi.org/10.1007/978-3-319-76953-0_4

11. Faugère, J.-C., Gianni, P.M., Lazard, D., Mora, T.: Efficient computation of zero-dimensional Gröbner Bases by change of ordering. J. Symb. Comput. **16**(4), 329–344 (1993)

12. Galbraith, S.D.: Mathematics of Public Key Cryptography. Cambridge University Press, Cambridge (2012)

13. Hashemi, A., Lazard, D.: Sharper complexity bounds for zero-dimensional Gröbner bases and polynomial system solving. Int. J. Algebra Comput. **21**(5), 703–713 (2011)

14. Howgrave-Graham, N.: Finding small roots of univariate modular equations revisited. In: Darnell, M. (ed.) Cryptography and Coding 1997. LNCS, vol. 1355, pp. 131–142. Springer, Heidelberg (1997). https://doi.org/10.1007/BFb0024458

15. Jancar, J., Sedlacek, V., Svenda, P., Sýs, M.: Minerva: the curse of ECDSA nonces systematic analysis of lattice attacks on noisy leakage of bit-length of ECDSA nonces. IACR Trans. Cryptogr. Hardw. Embed. Syst. **2020**(4), 281–308 (2020)

16. Jao, D., Jetchev, D., Venkatesan, R.: On the bits of elliptic curve Diffie-Hellman keys. In: Srinathan, K., Rangan, C.P., Yung, M. (eds.) INDOCRYPT 2007. LNCS, vol. 4859, pp. 33–47. Springer, Heidelberg (2007). https://doi.org/10.1007/978-3-540-77026-8_4

17. Jetchev, D., Venkatesan, R.: Bits security of the elliptic curve Diffie–Hellman secret keys. In: Wagner, D. (ed.) CRYPTO 2008. LNCS, vol. 5157, pp. 75–92. Springer, Heidelberg (2008). https://doi.org/10.1007/978-3-540-85174-5_5

18. Jochemsz, E., May, A.: A strategy for finding roots of multivariate polynomials with new applications in attacking RSA variants. In: Lai, X., Chen, K. (eds.) ASIACRYPT 2006. LNCS, vol. 4284, pp. 267–282. Springer, Heidelberg (2006). https://doi.org/10.1007/11935230_18

19. Jochemsz, E., May, A.: A polynomial time attack on rsa with private CRT-exponents smaller than $N^{0.073}$. In: Menezes, A. (ed.) CRYPTO 2007. LNCS, vol. 4622, pp. 395–411. Springer, Heidelberg (2007). https://doi.org/10.1007/978-3-540-74143-5_22

20. Lenstra, A.K., Lenstra, H.W., Lovász, L.: Factoring polynomials with rational coefficients. Mathematische Annalen **261**(4), 515–534 (1982)

21. Ling, S., Shparlinski, I.E., Steinfeld, R., Wang, H.: On the modular inversion hidden number problem. J. Symbol. Comput. **47**(4), 358–367 (2012)

22. May, A.: Using LLL-reduction for solving RSA and factorization problems. In: Nguyen, P., Vallée, B. (eds.) The LLL Algorithm. Information Security and Cryptography, pp. 315–348. Springer, Heidelberg (2010). https://doi.org/10.1007/978-3-642-02295-1_10

23. Merget, R., Brinkmann, M., Aviram, N., Somorovsky, J., Mittmann, J., Schwenk, J.: Raccoon attack: finding and exploiting most-significant-bit-oracles in TLS-DH(E). In: 30th USENIX Security Symposium (USENIX Security 2021). USENIX Association, Vancouver, B.C., August 2021

24. Nemec, M., Sýs, M., Svenda, P., Klinec, D., Matyas, V.: The return of Coppersmith's attack: practical factorization of widely used RSA moduli. In: Proceedings of the 2017 ACM SIGSAC Conference on Computer and Communications Security, CCS 2017, Dallas, TX, USA, 30 October–03 November 2017, pp. 1631–1648 (2017)

25. Neumaier, A., Stehlé, D.: Faster LLL-type reduction of lattice bases. In: Abramov, S.A., Zima, E.V., Gao, X.-S. (eds.) Proceedings of the ACM on International Symposium on Symbolic and Algebraic Computation, ISSAC 2016, Waterloo, ON, Canada, 19–22 July 2016, pp. 373–380. ACM (2016)
26. Nguyen, P.Q., Stehlé, D.: An LLL algorithm with quadratic complexity. SIAM J. Comput. **39**(3), 874–903 (2009)
27. Ryan, K.: Return of the hidden number problem. A widespread and novel key extraction attack on ECDSA and DSA. IACR Trans. Cryptogr. Hardw. Embed. Syst. **2019**(1), 146–168 (2019)
28. Shani, B.: On the bit security of elliptic curve Diffie–Hellman. In: Fehr, S. (ed.) PKC 2017. LNCS, vol. 10174, pp. 361–387. Springer, Heidelberg (2017). https://doi.org/10.1007/978-3-662-54365-8_15
29. Takayasu, A., Kunihiro, N.: Better lattice constructions for solving multivariate linear equations modulo unknown divisors. In: Information Security and Privacy - 18th Australasian Conference, ACISP 2013, Brisbane, Australia, 1–3 July 2013. Proceedings, pp. 118–135 (2013)
30. Takayasu, A., Lu, Y., Peng, L.: Small CRT-exponent RSA revisited. In: Coron, J.-S., Nielsen, J.B. (eds.) EUROCRYPT 2017. LNCS, vol. 10211, pp. 130–159. Springer, Cham (2017). https://doi.org/10.1007/978-3-319-56614-6_5
31. von zur Gathen, J., Gerhard, J.: Modern Computer Algebra, 3rd edn. Cambridge University Press, Cambridge (2013)
32. Jun, X., Lei, H., Sarkar, S.: Cryptanalysis of elliptic curve hidden number problem from PKC 2017. Des. Codes Cryptogr. **88**(2), 341–361 (2020)
33. Jun, X., Sarkar, S., Lei, H., Huang, Z., Peng, L.: Solving a class of modular polynomial equations and its relation to modular inversion hidden number problem and inversive congruential generator. Des. Codes Cryptogr. **86**(9), 1997–2033 (2018)
34. Xu, J., Sarkar, S., Hu, L., Wang, H., Pan, Y.: New results on modular inversion hidden number problem and inversive congruential generator. In: Boldyreva, A., Micciancio, D. (eds.) CRYPTO 2019. LNCS, vol. 11692, pp. 297–321. Springer, Cham (2019). https://doi.org/10.1007/978-3-030-26948-7_11

Author Index

Arun, Arasu   487

Benedikt, Barbara Jiabao   583
Bernard, Olivier   677
Bonneau, Joseph   487

Campanelli, Matteo   151
Chen, Brian   547
Chen, Jie   394
Clark, Jeremy   487
Cosseron, Orel   32
Cui, Jiamin   241

David, Bernardo   151
Deng, Yi   334
Derbez, Patrick   68
Devillez, Henri   273
Dodis, Yevgeniy   547
Dowling, Benjamin   119

Euler, Marie   68

Felderhoff, Joël   709
Fischlin, Marc   583
Fouque, Pierre-Alain   68

Ghosh, Esha   547
Goldin, Eli   547
Grubbs, Paul   181
Gu, Dawu   210
Guo, Jian   645

Han, Shuai   210, 423
Hauck, Eduard   119
Hoffmann, Clément   32
Hu, Kai   241
Hu, Lei   771
Huang, Zhengan   423
Huang, Zhenyu   614
Huppert, Moritz   583

Jaeger, Joseph   454
Jutla, Charanjit   304

Kesavan, Balachandar   547
Khoshakhlagh, Hamidreza   151
Konring, Anders   151
Kumar, Akshaya   454

Lai, Junzuo   423
Len, Julia   181
Lesavourey, Andrea   677
Li, Yu   394
Liu, Guozhen   645
Liu, Hanlin   741
Liu, Shengli   210
Lyu, Lin   423
Lyu, You   210

Marcedone, Antonio   547
Méaux, Pierrick   32
Miracle, Sarah   3
Mou, Merry Ember   547

Nguyen, Phuong Hoa   68
Nguyen, Tuong-Huy   677
Nielsen, Jesper Buus   151

Pan, Jiaxin   363
Patranabis, Sikhar   304
Pellet-Mary, Alice   709
Pereira, Olivier   273
Peters, Thomas   273
Pijnenburg, Jeroen   89
Poettering, Bertram   89

Riepel, Doreen   119
Ristenpart, Thomas   181
Rösler, Paul   119
Roux-Langlois, Adeline   677

Sarkar, Santanu   771
Song, Ling   645
Standaert, François-Xavier   32
Stehlé, Damien   709
Sun, Siwei   614

Tu, Yi   645

Wang, Huaxiong   771
Wang, Meiqin   241
Wang, Xiao   517
Wei, Puwen   241
Wen, Jinming   394
Weng, Jian   394, 423

Xu, Jun   771

Yang, Kang   517
Yilek, Scott   3
Yu, Yu   741

Zeng, Runzhi   363
Zhang, Xinxuan   334

Printed in the United States
by Baker & Taylor Publisher Services